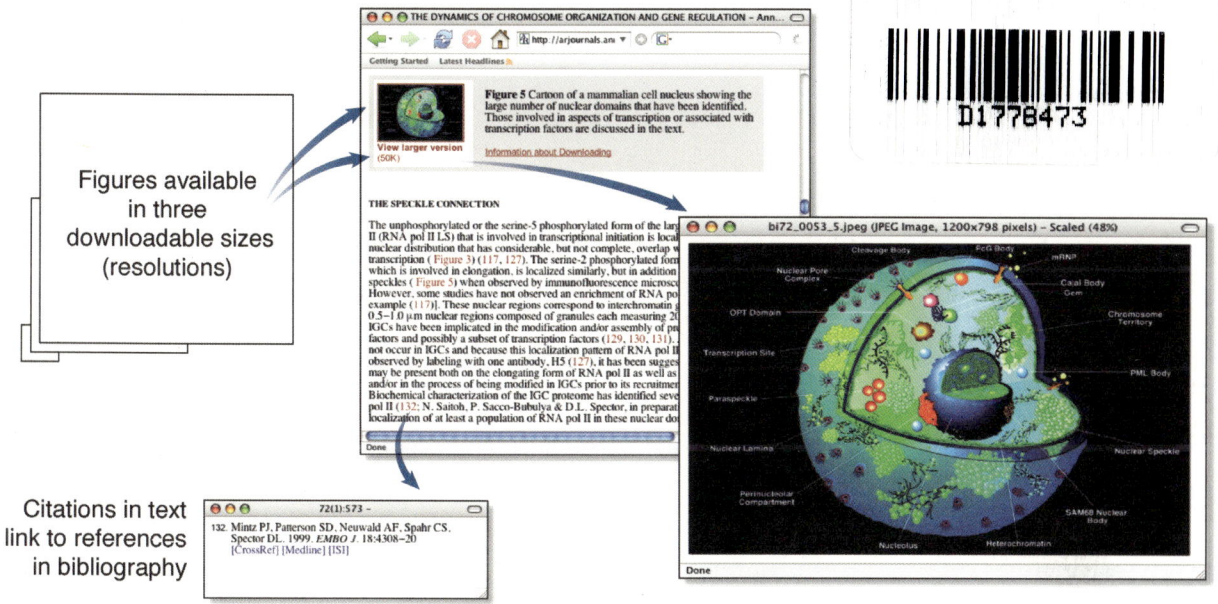

Figures available in three downloadable sizes (resolutions)
THE DYNAMICS OF CHROMOSOME ORGANIZATION AND GENE REGULATION – Ann...
Figure 5 Cartoon of a mammalian cell nucleus showing the large number of nuclear domains that have been identified. Those involved in aspects of transcription or associated with transcription factors are discussed in the text.
View larger version (50K)
Information about Downloading
THE SPECKLE CONNECTION
bi72_0053_5.jpeg (JPEG Image, 1200x798 pixels) - Scaled (48%)
Cleavage Body
PcG Body
mRNP
Nuclear Pore Complex
Cajal Body Gem
OPT Domain
Chromosome Territory
Transcription Site
PML Body
Paraspeckle
Nuclear Lamina
Nuclear Speckle
Perinucleolar Compartment
SAM68 Nuclear Body
Nucleolus
Heterochromatin
Citations in text link to references in bibliography
132. Mintz PJ, Patterson SD, Neuwald AF, Spahr CS, Spector DL. 1999. EMBO J. 18:4308–20 [CrossRef] [Medline] [ISI]

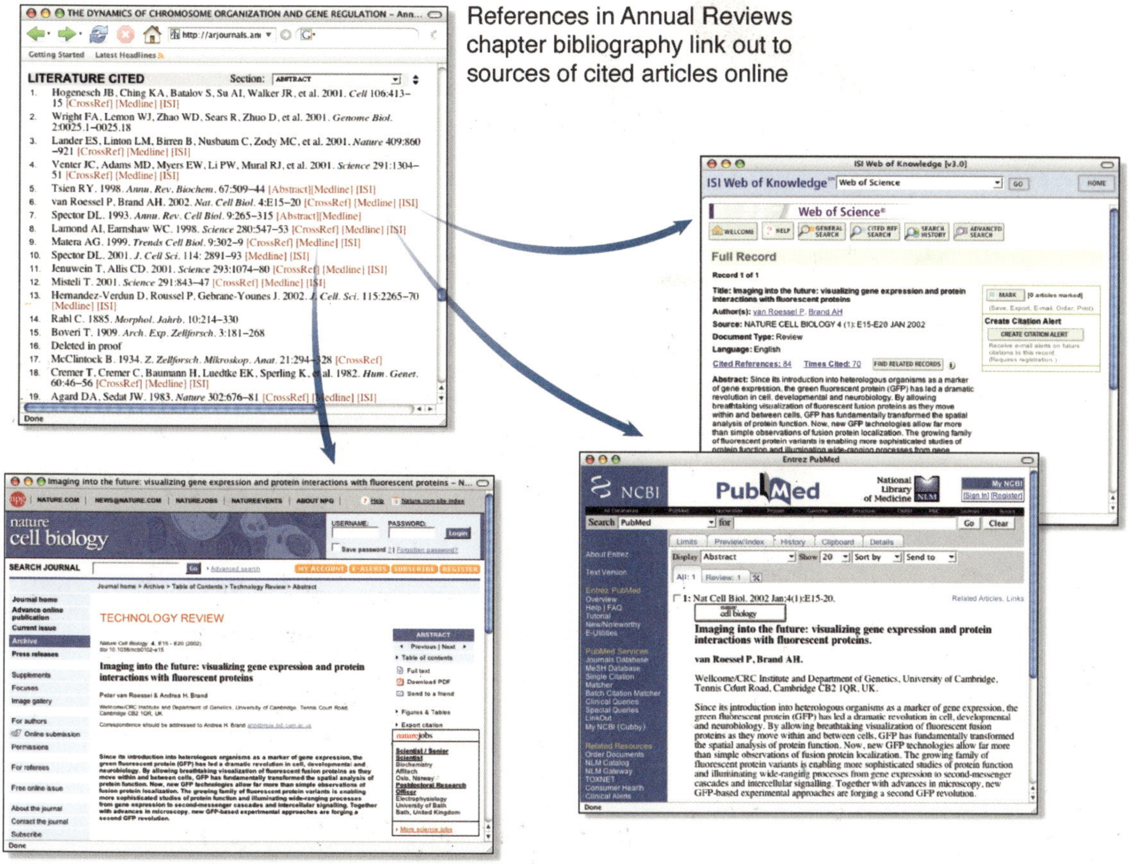

References in Annual Reviews chapter bibliography link out to sources of cited articles online
LITERATURE CITED
ISI Web of Knowledge
Web of Science
Full Record
Imaging into the future: visualizing gene expression and protein interactions with fluorescent proteins
nature cell biology
TECHNOLOGY REVIEW
NCBI
PubMed
Entrez PubMed

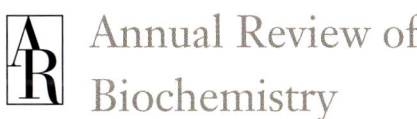
Annual Review of Biochemistry

Editorial Committee (2006)

Tania A. Baker, Massachusetts Institute of Technology
Christopher Dobson, University of Cambridge
Robert H. Fillingame, University of Wisconsin
F. Ulrich Hartl, Max Planck Institut für Biochemie
Laura L. Kiessling, University of Wisconsin
Roger D. Kornberg, Stanford University
Paul Modrich, Duke University
Christian R.H. Raetz, Duke University
James E. Rothman, Columbia University
JoAnne Stubbe, Massachusetts Institute of Technology
Jeremy W. Thorner, University of California, Berkeley

Responsible for the Organization of Volume 75 (Editorial Committee, 2004)

Tania A. Baker
Robert H. Fillingame
Roger D. Kornberg
Rowena G. Matthews
Gregory Petsko
Christian R.H. Raetz
James E. Rothman
JoAnne Stubbe
Jeremy W. Thorner
Gerald R. Crabtree (Guest)
Judith Frydman (Guest)
Julie A. Theriot (Guest)

Production Editor: Jesslyn S. Holombo
Bibliographic Quality Control: Mary A. Glass
Electronic Content Coordinator: Suzanne K. Moses
Illustration Editor: Douglas Beckner
Subject Indexer: Kyra Kitts

Annual Review of Biochemistry

Volume 75, 2006

Roger D. Kornberg, *Editor*
Stanford University School of Medicine

Christian R.H. Raetz, *Associate Editor*
Duke University Medical Center

James E. Rothman, *Associate Editor*
Columbia University College of Physicians and Surgeons

Jeremy W. Thorner, *Associate Editor*
University of California, Berkeley

www.annualreviews.org • science@annualreviews.org • 650-493-4400

Annual Reviews
4139 El Camino Way • P.O. Box 10139 • Palo Alto, California 94303-0139

Annual Reviews
Palo Alto, California, USA

COPYRIGHT © 2006 BY ANNUAL REVIEWS, PALO ALTO, CALIFORNIA, USA. ALL RIGHTS RESERVED. The appearance of the code at the bottom of the first page of an article in this serial indicates the copyright owner's consent that copies of the article may be made for personal or internal use, or for the personal or internal use of specific clients. This consent is given on the condition that the copier pay the stated per-copy fee of $20.00 per article through the Copyright Clearance Center, Inc. (222 Rosewood Drive, Danvers, MA 01923) for copying beyond that permitted by Section 107 or 108 of the U.S. Copyright Law. The per-copy fee of $20.00 per article also applies to the copying, under the stated conditions, of articles published in any *Annual Review* serial before January 1, 1978. Individual readers, and nonprofit libraries acting for them, are permitted to make a single copy of an article without charge for use in research or teaching. This consent does not extend to other kinds of copying, such as copying for general distribution, for advertising or promotional purposes, for creating new collective works, or for resale. For such uses, written permission is required. Write to Permissions Dept., Annual Reviews, 4139 El Camino Way, P.O. Box 10139, Palo Alto, CA 94303-0139 USA.

International Standard Serial Number: 0066-1154
International Standard Book Number: 0-8243-0875-1
Library of Congress Catalog Card Number: 32-25093

All Annual Reviews and publication titles are registered trademarks of Annual Reviews.

⊚ The paper used in this publication meets the minimum requirements of American National Standards for Information Sciences—Permanence of Paper for Printed Library Materials, ANSI Z39.48-1992.

Annual Reviews and the Editors of its publications assume no responsibility for the statements expressed by the contributors to this *Annual Review*.

TYPESET BY TECHBOOKS, FALLS CHURCH, VA
PRINTED AND BOUND BY QUEBECOR WORLD PRINTING, TAUNTON, MA

Contents

Annual Review
of Biochemistry
Volume 75, 2006

Wanderings of a DNA Enzymologist: From DNA Polymerase to Viral
Latency
I. Robert Lehman 1

Signaling Pathways in Skeletal Muscle Remodeling
Rhonda Bassel-Duby and Eric N. Olson 19

Biosynthesis and Assembly of Capsular Polysaccharides in
Escherichia coli
Chris Whitfield 39

Energy Converting NADH:Quinone Oxidoreductase (Complex I)
Ulrich Brandt 69

Tyrphostins and Other Tyrosine Kinase Inhibitors
Alexander Levitzki and Eyal Mishani 93

Break-Induced Replication and Recombinational Telomere Elongation
in Yeast
Michael J. McEachern and James E. Haber 111

LKB1-Dependent Signaling Pathways
Dario R. Alessi, Kei Sakamoto, and Jose R. Bayascas 137

Energy Transduction: Proton Transfer Through the Respiratory
Complexes
Jonathan P. Hosler, Shelagh Ferguson-Miller, and Denise A. Mills ... 165

The Death-Associated Protein Kinases: Structure, Function, and
Beyond
Shani Bialik and Adi Kimchi 189

Mechanisms for Chromosome and Plasmid Segregation
*Santanu Kumar Ghosh, Sujata Hajra, Andrew Paek,
and Makkuni Jayaram* 211

Chromatin Modifications by Methylation and Ubiquitination:
Implications in the Regulation of Gene Expression
Ali Shilatifard 243

v

Structure and Mechanism of the Hsp90 Molecular Chaperone Machinery
Laurence H. Pearl and Chrisostomos Prodromou 271

Biochemistry of Mammalian Peroxisomes Revisited
Ronald J.A. Wanders and Hans R. Waterham 295

Protein Misfolding, Functional Amyloid, and Human Disease
Fabrizio Chiti and Christopher M. Dobson 333

Obesity-Related Derangements in Metabolic Regulation
Deborah M. Muoio and Christopher B. Newgard 367

Cold-Adapted Enzymes
Khawar Sohail Siddiqui and Ricardo Cavicchioli 403

The Biochemistry of Sirtuins
Anthony A. Sauve, Cynthia Wolberger, Vern L. Schramm, and Jef D. Boeke 435

Dynamic Filaments of the Bacterial Cytoskeleton
Katharine A. Michie and Jan Löwe 467

The Structure and Function of Telomerase Reverse Transcriptase
Chantal Autexier and Neal F. Lue 493

Relating Protein Motion to Catalysis
Sharon Hammes-Schiffer and Stephen J. Benkovic 519

Animal Cytokinesis: From Parts List to Mechanisms
Ulrike S. Eggert, Timothy J. Mitchison, and Christine M. Field 543

Mechanisms of Site-Specific Recombination
Nigel D.F. Grindley, Katrine L. Whiteson, and Phoebe A. Rice 567

Axonal Transport and Alzheimer's Disease
Gorazd B. Stokin and Lawrence S.B. Goldstein 607

Asparagine Synthetase Chemotherapy
Nigel G.J. Richards and Michael S. Kilberg 629

Domains, Motifs, and Scaffolds: The Role of Modular Interactions in the Evolution and Wiring of Cell Signaling Circuits
Roby P. Bhattacharyya, Attila Reményi, Brian J. Yeh, and Wendell A. Lim 655

Ribonucleotide Reductases
Pär Nordlund and Peter Reichard 681

Introduction to the Membrane Protein Reviews: The Interplay of Structure, Dynamics, and Environment in Membrane Protein Function
Jonathan N. Sachs and Donald M. Engelman 707

Relations Between Structure and Function of the Mitochondrial
ADP/ATP Carrier
*H. Nury, C. Dahout-Gonzalez, V. Trézéguet, G.J.M. Lauquin,
G. Brandolin, and E. Pebay-Peyroula* 713

G Protein–Coupled Receptor Rhodopsin
Krzysztof Palczewski 743

Transmembrane Traffic in the Cytochrome $b_6 f$ Complex
*William A. Cramer, Huamin Zhang, Jiusheng Yan, Genji Kurisu,
and Janet L. Smith* 769

INDEXES

Subject Index 791

Author Index 825

ERRATA

An online log of corrections to *Annual Review of Biochemistry* chapters (if any, 1977 to the present) may be found at http://biochem.annualreviews.org/errata.shtml

Related Articles

From the *Annual Review of Biomedical Engineering*, Volume 7 (2005)

DNA Mechanics
Craig J. Benham and Steven P. Mielke

From the *Annual Review of Biophysics and Biomolecular Structure*, Volume 35 (2006)

Lessons from Lactose Permease
Lan Guan and H. Ronald Kaback

Evolutionary Relationships and Structural Mechanisms of AAA+ Proteins
Jan P. Erzberger and James M. Berger

Expanding the Genetic Code
Lei Wang, Jianming Xie, and Peter G. Schultz

Radiolytic Protein Footprinting with Mass Spectrometry to Probe the Structure of Macromolecular Complexes
Keiji Takamoto and Mark R. Chance

Ribosome Dynamics: Insights from Atomic Structure Modeling into Cryo-Electron Microscopy Maps
Kakoli Mitra and Joachim Frank

Single-Molecule Analysis of RNA Polymerase Transcription
Lu Bai, Thomas J. Santangelo, and Michelle D. Wang

From the *Annual Review of Cell and Developmental Biology*, Volume 21 (2005)

Regulation of Protein Activities by Phosphoinositide Phosphates
Verena Niggli

Assembly of Variant Histones into Chromatin
Steven Henikoff and Kami Ahmad

From the *Annual Review of Genetics*, Volume 39 (2005)

Complexity in Regulation of Tryptophan Biosynthesis in Bacillus subtilis
Paul Gollnick, Paul Babitzke, Alfred Antson, and Charles Yanofsky

From the *Annual Review of Immunology*, Volume 24 (2006)

How TCRs Bind MHCs, Peptides, and Coreceptors

Markus G. Rudolph, Robyn L. Stanfield, and Ian A. Wilson

  Mechanism and Control of V(D)J Recombination at the Immunoglobulin Heavy Chain Locus
  David Jung, Cosmas Giallourakis, Raul Mostoslavsky, and Frederick W. Alt

From the *Annual Review of Medicine*, Volume 57 (2006)

  Enzyme Replacement for Lysosomal Diseases
  Roscoe O. Brady

From the *Annual Review of Microbiology*, Volume 59 (2005)

  Translational Regulation of GCN4 and the General Amino Acid Control of Yeast
  Alan G. Hinnebusch

  Regulation of Bacterial Gene Expression by Riboswitches
  Wade C. Winkler and Ronald R. Breaker

From the *Annual Review of Neuroscience*, Volume 29 (2006)

  Noncoding RNAs in the Mammalian Central Nervous System
  Xinwei Cao, Gene Yeo, Alysson R. Muotri, Tomoko Kuwabara, and Fred H. Gage

From the *Annual Review of Pharmacology and Toxicology*, Volume 46 (2006)

  Accessory Proteins for G Proteins: Partners in Signaling
  Motohiko Sato, Joe B. Blumer, Violaine Simon, and Stephen M. Lanier

  Regulation of Phospholipase C Isozymes by Ras Superfamily GTPases
  T. Kendall Harden and John Sondek

  Function of Retinoid Nuclear Receptors: Lessons from Genetic and Pharmacological Dissections of the Retinoic Acid Signaling Pathway During Mouse Embryogenesis
  Manuel Mark, Norbert B. Ghyselinck, and Pierre Chambon

From the *Annual Review of Plant Biology*, Volume 57 (2006)

  MicroRNAs and Their Regulatory Roles in Plants
  Matthew W. Jones-Rhoades, David P. Bartel, and Bonnie Bartel

Annual Reviews is a nonprofit scientific publisher established to promote the advancement of the sciences. Beginning in 1932 with the *Annual Review of Biochemistry*, the Company has pursued as its principal function the publication of high-quality, reasonably priced *Annual Review* volumes. The volumes are organized by Editors and Editorial Committees who invite qualified authors to contribute critical articles reviewing significant developments within each major discipline. The Editor-in-Chief invites those interested in serving as future Editorial Committee members to communicate directly with him. Annual Reviews is administered by a Board of Directors, whose members serve without compensation.

2006 Board of Directors, Annual Reviews

Richard N. Zare, *Chairman of Annual Reviews, Marguerite Blake Wilbur Professor of Chemistry, Stanford University*
John I. Brauman, *J.G. Jackson–C.J. Wood Professor of Chemistry, Stanford University*
Peter F. Carpenter, *Founder, Mission and Values Institute, Atherton, California*
Sandra M. Faber, *Professor of Astronomy and Astronomer at Lick Observatory, University of California at Santa Cruz*
Susan T. Fiske, *Professor of Psychology, Princeton University*
Eugene Garfield, *Publisher,* The Scientist
Samuel Gubins, *President and Editor-in-Chief, Annual Reviews*
Steven E. Hyman, *Provost, Harvard University*
Daniel E. Koshland Jr., *Professor of Biochemistry, University of California at Berkeley*
Joshua Lederberg, *University Professor, The Rockefeller University*
Sharon R. Long, *Professor of Biological Sciences, Stanford University*
J. Boyce Nute, *Palo Alto, California*
Michael E. Peskin, *Professor of Theoretical Physics, Stanford Linear Accelerator Center*
Harriet A. Zuckerman, *Vice President, The Andrew W. Mellon Foundation*

Management of Annual Reviews

Samuel Gubins, President and Editor-in-Chief
Richard L. Burke, Director for Production
Paul J. Calvi Jr., Director of Information Technology
Steven J. Castro, Chief Financial Officer and Director of Marketing & Sales
Jeanne M. Kunz, Human Resources Manager and Secretary to the Board

Annual Reviews of

Anthropology
Astronomy and Astrophysics
Biochemistry
Biomedical Engineering
Biophysics and Biomolecular
    Structure
Cell and Developmental Biology
Clinical Psychology
Earth and Planetary Sciences
Ecology, Evolution, and
    Systematics
Entomology
Environment and Resources

Fluid Mechanics
Genetics
Genomics and Human Genetics
Immunology
Law and Social Science
Materials Research
Medicine
Microbiology
Neuroscience
Nuclear and Particle Science
Nutrition
Pathology: Mechanisms of
    Disease

Pharmacology and Toxicology
Physical Chemistry
Physiology
Phytopathology
Plant Biology
Political Science
Psychology
Public Health
Sociology

SPECIAL PUBLICATIONS
Excitement and Fascination of
    Science, Vols. 1, 2, 3, and 4

Wanderings of a DNA Enzymologist: From DNA Polymerase to Viral Latency

I. Robert Lehman

Department of Biochemistry, Stanford University, Stanford, California 94305;
email: blehman@cmgm.stanford.edu

Key Words

DNA ligase, DNA replication, DNA repair, recombination, viruses

Abstract

I am a member of what has been called, perhaps too grandiosely, "The Greatest Generation." I grew up during the Great Depression and served in the U.S. Army during World War II. Because of my military service and the benefits of the GI Bill, I was able to attend college and, later, graduate school. Early in my graduate studies, I became fascinated with enzymes and the biochemical reactions that they catalyze. This fascination has never left me during the 50 years I have been a "DNA enzymologist." I was fortunate to have had as a mentor Arthur Kornberg, one of the great biochemists of the twentieth century, and a splendid group of postdocs and graduate students. I have studied DNA polymerases, DNA nucleases, DNA ligases, and DNA recombinases, enzymes that are critical to our understanding of DNA replication, repair, and recombination. Most recently, I have been studying herpes virus replication and inadvertently wandered into an entirely new area—viral latency.

Contents

INTRODUCTION	2
EARLY YEARS IN BALTIMORE	2
DNA POLYMERASE IN ST. LOUIS	4
FIRST INDEPENDENT RESEARCH— EXONUCLEASE I	8
THE MOVE WEST—STANFORD	9
NUCLEASES	9
BACK TO DNA POLYMERASE	9
DNA LIGASE	10
BACK TO DNA POLYMERASE AGAIN	11
EUKARYOTIC DNA REPLICATION	12
REC A	13
HERPES VIRUS DNA REPLICATION	14
CURRENT WORK ON HSV-1 LATENCY	15
ENVOI	15

INTRODUCTION

I was born in the town of Tauroggen in what was then known as Memel Territory. Memel Territory, which had been part of German East Prussia, was ceded to Lithuania by the Versailles Treaty following World War I. It was, when I was born, a German-speaking enclave in the southwestern part of Lithuania. My father, who had served in the German army during World War I (he had been wounded and was awarded the Iron Cross for valor), was deeply troubled by the anti-Semitism to which he and his family were increasingly subjected by both the ethnic Germans and the native Lithuanians. In 1927, he made the fortunate decision to emigrate to the United States. Most of the members of our extended family chose to remain and perished in the Holocaust that began a decade later. The Great Depression started within two years of our arrival in Baltimore, but because my father managed to find and keep a job, we were spared the worst of that dreadful period.

EARLY YEARS IN BALTIMORE

My early years were uneventful. I was a good, but not outstanding, student. During my last two years of high school, to help support the family, I worked part-time in a large meat market as a butcher's assistant. I became quite a skillful meat cutter and was promised a full-time job after high school graduation, a not inconsequential prospect because the United States was still mired in the Depression, and the possibility of college after high school graduation was very remote. All of this changed with the attack on Pearl Harbor by the Japanese on Sunday, December 7, 1941. I can still recall the radio announcer breaking in on the broadcast of the New York Philharmonic Symphony to report that Pearl Harbor had been attacked. The following day, the entire student body of my high school assembled to hear the radio broadcast of President Franklin D. Roosevelt's speech, "December 7, 1941 a date that will live in infamy." I remember discussing this momentous event after the assembly with my best friend Bobby Schwartz. Bobby and I optimistically concluded that we would probably not be affected by the war, which still seemed very far away, and it would almost certainly be over by the time we graduated. How wrong we were! Bobby became a paratrooper in the Eighty-Second Airborne Division and was killed in Holland in September 1944, just before his twentieth birthday.

I was drafted into the army at age 18, in the summer of 1943, less than three months after graduating from high school. Possessed of few if any of the skills that might have gotten me a safe desk job, I was assigned to the infantry. As a member of the Third Infantry Division, I participated in the invasion of southern France, an invasion that few know about today, and later in the battle for Germany. These were brutal experiences, incomprehensible to anyone who has not faced enemy fire in ground combat. Miraculously,

I survived without any serious wounds. However, because we were not issued watertight boots, with the onset of what turned out to be an unusually cold and snowy winter in the Vosges mountains of eastern France, I developed a gangrenous condition known as "trench foot." After about six weeks in an evacuation hospital and a rehabilitation center I was pronounced "fit for duty" and ordered to return to my unit, which was preparing for the Rhine River crossing and the assault on Germany. Of the approximately 200 men in my company who landed in southern France in August 1944, only about 30 of us remained at war's end, in May of 1945.

In contrast to the Viet Nam War veterans, we were treated as heroes, with very tangible economic benefits, collectively termed "the GI Bill of Rights." These benefits included a year of unemployment benefits ($20 per week for 52 weeks) and free college tuition for periods up to six years, depending upon length of service. Included was a monthly stipend of $50, a more than adequate sum at the time.

The chance to attend college at government expense was a dream come true. Fortunately, I was accepted at Johns Hopkins University in Baltimore, despite a less than sterling high school record. I decided to major in chemistry, a subject that I enjoyed in high school. My ambition was to become an industrial chemist, like an uncle whom I admired.

I was an excellent student and compiled a nearly 4.0 grade-point average. In my senior year, I was very much influenced by a biology course, given by Bill McElroy, which had a strong biochemistry component. I was fascinated by the pathways of carbohydrate, lipid, and energy metabolism that McElroy revealed to us. He was an exciting teacher, and he opened my eyes to the wonders of metabolism. I abandoned my plans to become an organic chemist following graduation from college and applied to the Doctoral Program in the Department of Biochemistry at the Johns Hopkins School of Public Health.

My thesis advisor was Roger Herriott, an eminent protein chemist who had been attracted to the study of bacteriophages. Roger had been at the Rockeller Institute, the forerunner of the Rockefeller University, and was aware of and, more importantly, appreciated the significance of the discovery by Avery, McLeod, and McCarty published in 1944 that DNA was the transforming principle of pneumococcus and indeed represented the chemical nature of the gene. I recall Herriott telling me that the T2 phage with its hexagonal head and long tail was really a hypodermic syringe containing "a bag of transforming principles," which were injected into the bacterial host by the protein coat. This was several years before the famous Hershey-Chase experiment that showed that this was indeed the case.

Because of my interest in metabolism, Herriott suggested that I study the metabolic changes that occurred in the host *Escherichia coli* after infection with an intact T2 phage and the isolated protein coats or "ghosts" of T2 that he had recently found could be produced from the phage by osmotic shock. My doctoral dissertation involved a study of the metabolic changes that occurred in *E. coli* following infection with phage T2 and T2 ghosts. There were indeed major changes in oxygen uptake and CO_2 release that probably reflected changes in glycolysis and the pentose phosphate pathway. There were also changes in phosphate metabolism (1). However, I failed to detect the profound alterations in nucleotide metabolism that were needed for the synthesis of T2 phage DNA and, in particular, for the synthesis hydroxymethylcytosine, which replaced cytosine in T2, T4, and T6 bacteriophages (2).

I remained convinced, after completing my doctorate, that the most excitement in biochemistry was in intermediary and energy metabolism. This was reinforced by a talk given by Irving Lieberman, then a postdoctoral fellow with Arthur Kornberg, at the 1954 Federation Meeting in Atlantic City on the discovery of phosphoribosylpyrophosphate and its role in nucleotide biosynthesis. I still recall my awe and excitement at hearing this remarkable work.

I began my postdoctoral training in the fall of 1954 with Alvin Nason at the McCollum-Pratt Institute at Johns Hopkins. Al was doing interesting work on the role of metal ions, notably molybdenum, in electron transport in *Neurospora* and suggested that I work on a particulate NAD-linked cytochome *c* reductase from rat muscle with the aim of identifying a metal cofactor. Although I approached the project with considerable enthusiasm, I soon decided that it was not for me. Lieberman's 10-minute Federation talk was still very much on my mind, and I went to the library and read every paper Kornberg had published on coenzyme and nucleotide biosynthesis that I could find. I then wrote Arthur who had moved from the National Institutes of Health (NIH) to Washington University in St. Louis, asking if I could join his lab as a postdoctoral fellow. To my great good fortune, he accepted me. We have remained colleagues and close friends ever since.

DNA POLYMERASE IN ST. LOUIS

I arrived in St. Louis in the summer of 1955, a summer notable for having broken a long-standing record for the number of consecutive days in which the temperature reached or exceeded 100° F. Arthur was chairman of the Department of Microbiology, whose faculty, in addition to himself, consisted of Paul Berg, Melvin Cohn, Robert De Mars, David Hogness, Irving Lieberman, and Philip Varney (the sole holdover from the previous department). Later, Dale Kaiser replaced De Mars, who had left to discharge his military obligation with service at the NIH. Shortly thereafter, Jerard Hurwitz, who had been at the NIH, joined the faculty. The department was housed in rather dilapidated quarters in the top floor of the old Washington University Clinic Building. The elevator was reputed to be the oldest still in operation in the city of St. Louis, and riding it could often be an adventure. There was a journal club that met daily at lunch. All the members of the department participated, and we were joined by Martin Kamen, who was in the Department of Radiology, and Stanley Cohen, who was working on nerve growth factor in Rita Levi-Montalcini's lab on the main Washington University campus. The discussions were spirited, critical, often combative, and, occasionally, somewhat intimidating. But they were great fun and helped to keep us up to date in the current literature. Arthur's group, in addition to myself, consisted of Uriel Littauer, a postdoc from Israel; Jose Fernandez, a postdoc from Brazil; Arthur's late wife, Sylvy Kornberg; and Ernie Simms, a technician.

The discovery that the DNA of the T-even phages contained hydroxymethycytosine in place of cytosine seemed terribly important, and I was eager to learn how hydroxymethylcytosine was made. Arthur felt that this was a good project and managed to acquire a small amount of [^{14}C] β-carbon-labeled serine, a presumed donor of the hydroxymethyl group, for my experiments. Within a short time, I found that extracts of T2-infected *E. coli* did incorporate the ^{14}C-label into dCMP, and I began fractionation to purify the enzyme responsible for this activity. At about the same time, Arthur had observed that small amounts of ^{14}C-labeled thymidine, which he had obtained from Morris Friedkin in the Pharmacology Department, were incorporated into an acid-insoluble product by *E. coli* extracts. The counts were low (fewer than 100), but they were made acid soluble upon treatment with pancreatic DNase. I recall that when Arthur told me of this tantalizing finding, which suggested that he had achieved DNA synthesis in vitro. I was tremendously excited and asked if I could put my project on hold and join him. He agreed. Later, we learned that Joel Flaks and Seymour Cohen had discovered the T4 phage dCMP hydroxymethylase (3) and had opened up the whole field of virus-induced enzymes. But I have never regretted my decision.

Several months after this, Maurice Bessman, a new postdoc, arrived, and together with Ernie Simms and Arthur, the four of us began to fractionate the activity responsible for the incorporation of the labeled thymidine into an acid-insoluble, DNase I–sensitive product. Later, we were joined by Julius Adler, another postdoc, and Steve Zimmerman, Arthur's first graduate student.

Earlier that year, Arthur and Ernie Simms had begun work on the purification of an activity in *E. coli* that converted thymidine in the presence of ATP to what they referred to as thymidine-X, later identified as dTMP. The activity was thymidine kinase. Additional products, presumed to be dTDP and dTTP, were also observed. The ability to make ^{32}P-labeled dTMP was a significant step forward because we were not limited by the low radioactivity of the [^{14}C]thymidine that was available, and the [^{32}P]dTMP incorporated into the acid-insoluble product was now in the hundreds and occasionally in the thousands of counts/min. We bet that dTTP was the true substrate for our enzyme rather than dTMP or dTDP, although the latter was a distinct possibility because of the finding a year earlier by Grunberg-Manago and Ochoa that the nucleoside diphosphates rather than the triphosphates were the substrates for their ribonucleotide-polymerizing enzyme from *Azotobacter* (4), which later turned out to be polynucleotide phosphorylase, an enzyme involved in messenger RNA degradation. Once we had [^{32}P]dTMP, we prepared α-[^{32}P]dTTP by incubating our ^{32}P-labeled dTMP with a partially purified nucleoside-diphosphate kinase and ATP and then isolating the dTTP. Our assay mixture now consisted of a crude sonic extract of *E. coli*, α-[^{32}P]dTTP, ATP, Mg^{2+}, and buffer. As in the original experiment, acid-insoluble ^{32}P was measured. With this assay, we began to fractionate the crude extract for dTTP incorporation into DNA.

To begin the fractionation we added streptomycin sulfate to the extract to produce a precipitate that contained the cellular nucleic acids and a nucleic acid-free supernatant. Streptomycin sulfate was used frequently at the time to remove nucleic acids, often a hindrance to protein purification in bacterial extracts. Assay of the nucleic acid-free supernatant (S-fraction) and the nucleic acid-containing precipitate (P-fraction) showed them to be devoid of dTTP incorporation activity. However, when the two fractions were combined, activity was restored. We also observed that prior incubation of the extract or the P-fraction for a few minutes at 37°C increased activity substantially. Clearly more than one enzyme was required for the incorporation of dTTP into an acid-insoluble product. The complexity of the system became even more apparent when we began to fractionate S and P. The P-fraction could be subfractionated into two fractions, one heat-labile and the other heat-stable, both of which (in combination with the S-fraction) were necessary for activity. The S-fraction could be separated into both a heat-labile fraction and a heat-stable fraction that could pass through a dialysis membrane, i.e., was dialyzable. The latter could be further fractionated by Dowex-1 chromatography into three discrete fractions. (Dowex-1, an anion exchange resin used at the time, separated low-molecular-weight acidic compounds.) Thus, incorporation of dTTP into an acid-insoluble product presumably DNA, required (*a*) two heat-labile fractions; (*b*) a heat-stable fraction; (*c*) three heat-stable, dialyzable, chromatographically distinct fractions; and (*d*) ATP. In the absence of any one of these components, the activity was significantly diminished. Clearly, a lot was going on. The heat-labile component in the P-fraction turned out to be the enzyme that catalyzed phosphodiester bond synthesis, and this enzyme we named DNA polymerase. The heat-stable, nondialyzable component in the P-fraction was DNA. The heat-labile, nondialyzable component of the S-fraction was a mixture of deoxynucleotide kinases, which together with nucleoside-diphosphate kinase produced the heat-stable, dialyzable mixture of dCTP, dATP, and dGTP.

We reconciled these complex requirements for the incorporation of ^{32}P-labeled dTTP into an acid-insoluble product as follows. The DNA in the extract and the P-fraction were degraded by endogenous nucleases to the deoxynucleoside monophosphates (dNMPs). (Recall that preincubation of the extract or the P-fraction significantly enhanced activity.) These were converted to the corresponding deoxynucleoside triphosphates (dNTPs) by the kinases in the S-fraction and ATP to generate dCTP, dATP, and dGTP. The heat-labile component of the P-fraction was the DNA polymerase. We speculated that the DNA in the P-fraction served two functions in addition to being the source of the deoxynucleoside monophosphates. First, it protected the miniscule amount of the labeled DNA that was synthesized from degradation by the nucleases in the extract. Second, Arthur had been strongly influenced by the work on glycogen phosphorylase in the Cori laboratory. In the case of glycogen phosphorylase, glycogen served as a "primer" for the addition of glucosyl units from glucose 1-phosphate to extend the glycogen chain. Similarly, he felt that dTMP from dTTP was being added to pre-existing DNA chains.

Once the outlines of the reaction became clear, we set about the task of reconstituting the reaction with purified components. We partially purified each of the deoxynucleotide kinases in the S-fraction, and with these kinases and nucleoside-diphosphate kinase, we synthesized and characterized the four dNTPs (dTTP, dCTP, dGTP, and dATP). This alone was a substantial advance, because with the exception of dTTP, none of the other dNTPs had previously been described.

To prepare the four ^{32}P-labeled dNTPs, we started with ^{32}P-labeled DNA, isolated from ^{32}P-labeled *E. coli*, from which we generated the four ^{32}P-labeled dNMPs (dAMP, dCMP, dTMP, and dGMP) by treatment with pancreatic DNase and snake venom phosphodiesterase. The dNMPs were individually purified and converted enzymatically to the corresponding ^{32}P-labeled dNTPs. Because the procedure usually consumed two to three weeks, the 14-day half-life of ^{32}P necessitated the use of large quantities of ^{32}P (50–100 mCi) in the 100-ml low-phosphate culture medium that we used for the growth of *E. coli*.

Purification of the DNA polymerase was a difficult and demanding task. The enzyme was present in relatively small amounts even in rapidly growing *E. coli* (about 300 molecules/cell). Fortunately, a fermenter, which had been installed in the department, for the large-scale growth of *E. coli* supplied hundreds of grams of log phase *E. coli*. Later 100-pound batches of *E. coli* cell paste were obtained from the Grain Processing Corporation in Muscatine, Iowa. DEAE-cellulose and phosphocellulose, invented by Herbert Sober at the NIH, liberated us from the sole reliance on the ammonium sulfate, alumina Cγ, and acetone, which were the major protein fractionation tools at the time. With the aid of chromatography performed with these ion exchangers, we were able to obtain a several 1000-fold purified but not yet homogeneous preparation of the DNA polymerase. A vexing problem at the time was our inability to remove deoxyribonuclease activity from the enzyme. It was found later that a 3′ to 5′ exonuclease, which serves a vital proofreading function, is a component of *E. coli* DNA polymerase and, indeed, virtually all DNA polymerases, excising incorrectly incorporated nucleotides as replication proceeds.

With the progress in fractionation, the reaction requirements were now considerably simplified. Conversion of α-[^{32}P]dTTP into an acid-insoluble product, i.e., DNA, required only the partially purified DNA polymerase, Mg^{2+}, DNA, dCTP, dGTP, and dATP (5, 6). We further found that all four dNTPs were absolutely required. With omission of any one of the other three dNTPs, incorporation of α-[^{32}P]dTTP fell to background levels (7).

The requirement for all four dNTPs was puzzling. If the DNA that we added was simply serving as a primer, why would all four dNTPs be needed? Was it possible that the DNA polymerase was performing the template-directed replication proposed by Watson and Crick for their double-stranded structure of DNA (8). To test this seemingly wild idea, we used DNAs with A+T/G+C ratios ranging from 0.5 to 1.9 as "primers." The result was stunning. The ratio in the product corresponded closely to that of the added DNA throughout the synthesis and was independent of the relative concentrations of the individual dNTPs. Clearly, the added DNA was serving as a template to direct the polymerase as it synthesized new DNA chains, or as we cautiously put it in our initial publication, "These results suggest that enzymatic synthesis of DNA by the 'polymerase' of *E. coli* represents the replication of a DNA template" (9).

Having referred to the DNA added to their action as the primer, by analogy to glycogen phosphorylase, it was now clear that it also was serving as a template. However, we now know that all DNA polymerases require a primer to initiate a DNA chain. Various priming mechanisms (short RNA chains, proteins) have evolved to make up for this shortcoming in an otherwise magnificent enzyme (10, 11).

By 1958, we had established that a DNA primer, a template, and all four dNTPs were required by our DNA polymerase for the synthesis of DNA. However, we received a rude shock one day when we observed, to our surprise, that with our most highly purified enzyme DNA synthesis could proceed in the apparent absence of DNA. This episode is worth recounting in more detail. As I mentioned earlier, an important aim in purifying the DNA polymerase was to remove the contaminating DNase activity, which destroyed the product that we synthesized. Howard Schachman, the distinguished physical chemist, had come from Berkeley to spend a sabbatical year in the laboratory to analyze the DNA product of the polymerase reaction. Using the tools of the polymer chemist, the ultracentrifuge and the viscometer, Howard quickly demonstrated that the product of the reaction was indeed a large polymer. It then occurred to us that viscometry might be a very sensitive method to assay for nuclease activity in our most purified DNA polymerase preparations. I set up a nuclease reaction that contained calf thymus DNA, Mg^{2+}, and the DNA polymerase. At Howard's suggestion, dCTP, dTTP, and dATP were added to closely mimic the standard synthetic reaction conditions. dGTP (which was the most difficult of the triphosphates to prepare) was omitted to prevent DNA synthesis. The viscosity of the reaction mixture was then measured over an extended period. To my disappointment, the viscosity of the solution fell to that of the reaction buffer within about an hour. Obviously, the DNA was completely degraded; our best polymerase preparation was still contaminated with nuclease(s). This experiment was performed on a Saturday morning at the same time that I was proctoring a microbiology exam for second-year medical students down the hall from the physical chemistry laboratory, which Arthur had set up for Howard. (All of us in the department, including postdoctoral fellows, participated in medical school teaching at the time). After spending some time answering student questions and collecting exams, I returned to the laboratory to discard the reaction mixture and clean the viscometer. Before doing so, I absent mindedly took one last reading. To my amazement the viscosity of the solution had actually increased, and with repeated readings, the viscosity of the solution increased even further but then eventually fell back to that of the buffer. Could we be observing template-independent DNA synthesis? A number of control experiments were hastily performed that ruled out bacterial growth and contamination of the three dNTPs with dGTP. Howard Schachman and Julius Adler quickly found that the increase in viscosity required only the DNA polymerase, Mg^{2+}, dATP, and dTTP and occurred only after a lengthy lag. The product was a copolymer of alternating dAMP and dTMP, d(A-T) (deoxyadenylate-deoxythymidylate) (12). Work several years later by Arthur Kornberg, Gobind Khorana, and coworkers (13) showed that the rules had not been violated. The synthesis of the d(A-T) copolymer was indeed template directed, but the template consisted of trace amounts of DNA present in the DNA polymerase preparation. The polymer was the result of a reiterative mode of DNA replication in which slippage of one stretch of alternating A and T residues within the contaminating DNA generated overlapping ends, and these, when filled in by the polymerase, increased the chain length; ultimately, the high-molecular-weight d(A-T) copolymer was produced in quantity (13).

The DNA polymerase that we discovered is now called DNA polymerase I. In the intervening years, four additional DNA polymerases, DNA polymerase II, DNA polymerase III holoenzyme, and DNA polymerases IV and V, have been identified in *E. coli* and purified (14). The multisubunit DNA polymerase III holoenzyme is actually the enzyme that catalyzes the synthesis of the *E. coli* chromosome (15). DNA polymerase I, by virtue of its intrinsic ribonuclease H activity, together with its DNA polymerase activity plays an essential role in processing the nascent Okazaki fragments produced during the discontinuous replication of the lagging strand at the replication fork to prepare them for ligation (16). DNA polymerases II, IV, and V serve in the repair of DNA (14).

In eukaryotes, the situation is even more complex. Fifteen distinct cellular DNA polymerases have been identified, and the list continues to grow (17). Three of these are devoted to replication of the genome (DNA polymerases α, δ, and ϵ); DNA polymerase γ replicates the mitochondrial genome. The rest are all devoted to the repair of specific lesions in DNA. There are, in addition, virally encoded polymerases that replicate viral DNA genomes and, in the case of the retroviruses, reverse transcribe their RNA genomes into DNA. Despite the number and diversity of DNA polymerases, all of these enzymes show the same requirements that we observed nearly 50 years ago for the polymerase of *E. coli*: a template (DNA or RNA) to guide the polymerase in its base selection, a primer onto which deoxynucleotides are added, the four dNTPs, and Mg^{2+}. There are factors associated with DNA polymerases, e.g., clamps, clamp loaders, and exonucleases, which increase the efficiency and fidelity of DNA replication, but the basic mechanisms of replicating a DNA chain are all the same.

With the crystallization of many DNA polymerases, the determination of their three-dimensional structures, and the application of presteady-state kinetic analyses, much is now known about the detailed chemical mechanism of the polymerase reaction (18). This information has been invaluable in the design of effective chemotherapeutic agents, in particular antiviral drugs.

I view those days in the mid-1950s in the Department of Microbiology on the fourth floor of the old Clinic Building at Washington University to be among the most thrilling and enjoyable of my scientific career. There were new and unexpected findings being made virtually every day, and all of us in our small group shared in the joy and excitement of those discoveries.

In the summer of 1957, I received an offer of an assistant professorship at the McCollum-Pratt Institute at Johns Hopkins from Bill McElroy, the director. I had been a postdoc for two years, and under normal circumstances, it would have been time to move on. Moreover, academic positions were hard to come by. However, it was clear that I was working a gold mine, and I had no intention of leaving. I continued working as a postdoc for an additional year and then began independent research as an instructor in the department.

FIRST INDEPENDENT RESEARCH—EXONUCLEASE I

In looking around for a project to begin independent research, I had been impressed by the importance of specific proteases in the analysis of protein structure and sequence, and I was struck by the absence of comparable enzymes that acted on DNA. Our purification of DNA polymerase from *E. coli* extracts and constant attempts to rid the enzyme of nucleases that degraded the DNA product suggested that *E. coli* might be a good source of such enzymes. The first of these that I purified, exonuclease I, turned out to be extraordinarily useful. It was absolutely specific for single-stranded DNA and was therefore a very effective, in fact, the only reagent at the time that could distinguish single-stranded from double-stranded DNA (19). A notable example of its usefulness was in the early DNA renaturation studies of Paul Doty, Julius Marmur, and Carl Schildkraut (20). After heat or alkaline denaturation of rather heterogeneous preparations of calf thymus or salmon sperm DNA, renaturation yielded duplex DNA with single-stranded tails. These produced rather messy, often uninterpretable, banding patterns in the CsCl-density gradients that were used to measure renaturation. However, upon treatment with exonuclease I, to remove the single-stranded tails, sharp bands appeared, making it clear that true renaturation had occurred (20). Exonuclease I was also instrumental in the discovery by Fiers and Sinsheimer that the single-stranded DNA of phage ΘX174 was circular (21). Although single stranded, it was degraded by exonuclease I only after being nicked by an endonuclease. Another fortunate property of exonuclease I, which I was able to exploit, was its ability, in contrast to all other known nucleases, to degrade glycosylated T-even phage DNA completely to mononucleotides. It had been found some years before that the T-even phage DNAs contained glucose linked to the hydroxymethylcytosine that replaced cytosine. With the use of exonuclease I, my research assistant, Ann Pratt, and I were able to determine the patterns of glycosylation of the hydroxmethylcytosine residues, and we discovered to our amazement that in T4, in which all of the hydroxymethylcytosines are glucosylated, half contained glucose in the α configuration, and in the other half, it was in the β configuration (22). Later, Kornberg and coworkers (23) discovered that T4 actually encodes distinct α and β glucosyl transferases. In phage T6 DNA, the dissacharide gentiobiose was the predominant glucosyl residue. This was all very exciting, and I was convinced that these elaborate patterns of glycosylation must be terribly important. It now appears that one of their functions is to protect the phage DNA from restriction.

THE MOVE WEST—STANFORD

In the summer of 1957, Arthur was offered the Chair of Biochemistry at the Stanford School of Medicine, and he invited the Microbiology faculty, including me, its most junior member, and Robert (Buzz) Baldwin, then at the University of Wisconsin, to join him in forming the new Biochemistry Department at Stanford. I had never been west of St. Louis, but the opportunity was too good to pass up. We arrived at Stanford in the spring of 1959 and quickly set up shop in the newly constructed Stanford University Medical Center.

NUCLEASES

Buoyed by the success with exonuclease, I decided to begin research at Stanford by continuing the search for DNases in *E. coli* and also branched out to other microorganisms. With Stuart Linn, a graduate student, and Ian Kerr, a postdoc, we purified several of these enzymes; many were specific for single-stranded DNA and showed preferential cleavage of certain sequences, but none were truly base or sequence specific (24). We missed the jackpot—the restriction endonucleases. Stuart Linn did, however, in his subsequent postdoctoral work with Werner Arber in Geneva, discover the first restriction endonuclease, EcoB.

BACK TO DNA POLYMERASE

In the mid-1960s, I was invited by Bob Sinsheimer to present a seminar at Caltech. During my visit, I met with Bob Edgar who told me of work in his lab with *amber* and temperature-sensitive mutants of phage T4 that were defective in DNA replication (25). As a consequence of work in the Cohen, Kornberg, and Bessman labs, it was clear that the T phages encoded a variety of enzymes required for the replication of their DNA, including a novel DNA polymerase. Edgar offered to send me a series of DNA negative T4 mutants, in the hope that we would be able to determine which was defective in the phage DNA polymerase. Adrian De Waard, a Dutch postdoc, and Aniko Vessey Paul, my first graduate student, quickly demonstrated that extracts prepared from cells infected by amber mutants in gene 43 were lacking in the T4 DNA polymerase (26). We then went on to show that temperature-sensitive gene 43 mutants produced a temperature-sensitive DNA polymerase, thus demonstrating that gene 43 was indeed the structural gene for the T4 DNA polymerase. This was an important result because it demonstrated for the first time that the DNA polymerase activity that we and others were measuring in vitro is required for DNA synthesis in vivo (27). The role of T4 DNA polymerase in T4 DNA replication in vivo was further reinforced by experiments carried out by Zach Hall, a temporarily transplanted neurobiologist, who showed that the DNA polymerase purified from cells infected with T4 mutator mutants in gene 43 was mutagenic in vitro. These results became all the more important when DeLucia and Cairns several years later, in 1969, found a mutant of E. coli lacking DNA polymerase (i.e., DNA polymerase I) and questioned the role of DNA polymerase in DNA replication (28). More about that later.

DNA LIGASE

While casting about for a new area of research in the late 1960s, I heard Matthew Meselson describe his work with Jean Weigle, which demonstrated quite clearly that genetic recombination occurred by the breakage and rejoining of the recombining DNA molecules and not by a replication mechanism in which portions of the two molecules were copied alternately by a DNA polymerase. What were the enzymes that were able to catalyze the joining of DNA molecules? Baldomero (Toto) Olivera joined the lab as a postdoc after training in DNA physical chemistry with Norman Davidson at Caltech, and I suggested to Toto that we look for an enzyme in E. coli that could promote such a joining reaction. The substrate we devised was a poly (dA) chain of about 1000 nucleotides to which was annealed multiple 100-nucleotide-long poly (dT) segments each labeled with ^{32}P at its 5′ terminus. This, in effect, produced a duplex DNA molecule with nicks spaced at 100 nucleotide intervals. Joining was measured by conversion of the ^{32}P-labeled 5 phosphomonoester to a phosphodiester with E. coli alkaline phosphatase, which could hydrolyze the monoester substrate but not the diester product. Our very first experiment demonstrated a joining activity in our E. coli extracts (29). With this rather simple assay, we began to fractionate these extracts with the aim of purifying the responsible activity.

Early on, we noted that increasing amounts of extract did not produce a corresponding increase in activity. At low levels of extract, there was virtually no joining activity. The reactions were performed in the presence of ATP, and increasing the level of ATP did not eliminate the lag. Something was limiting. On the assumption that the limiting factor was a heat-stable cofactor, we added a small amount of "kochsaft," a boiled extract of E. coli, a classic maneuver that dates back to the days of Warburg and Meyerhof and the resolution of the glycolytic pathway. The lag disappeared. This, of course, provided us with an assay with which to purify the cofactor. It turned out to be NAD, whose pyrophosphate bond was cleaved to produce AMP and nicotinamide mononucleotide, an extraordinary and at that time unprecedented use of a redox coenzyme (30).

As we were purifying the polynucleotide joining activity, we became aware that a similar activity had been observed in four other labs. Martin Gellert at the NIH had also found it in *E. coli* extracts, Charles Richardson and Bernard Weiss at Harvard, Jerard Hurwitz and Malcolm Gefter at Albert Einstein, and Arthur Kornberg and Nicholas Cozzarelli, my next door neighbors at Stanford, had all found it in extracts of T4-infected cells. The phage enzyme used ATP rather than NAD. Each group had its own name for the enzyme (Hurwitz's "sealase" was the most colorful), but we all settled on DNA ligase, the name coined by Richardson and Weiss.

There was an exciting period of intense, but generally friendly, competition among the labs studying the enzyme. My group, Toto Olivera, Zach Hall, Paul Modrich, and Richard Gumport, were able to work out the mechanism of the joining reaction. In the first intermediate, enzyme-AMP, the adenylyl group of NAD is linked by a phosphoamide bond to a lysine in the active site of the enzyme, releasing a nicotinamide mononucleotide. Then the adenylyl group is transferred to the 5′ phosphate group of the DNA, where it is linked by a pyrophosphate bond. In the final step, there is an attack of the 3′ hydroxyl of the DNA on the activated 5′ phosphoryl group to form a phosphodiester bond with the release of AMP. The mechanism for the *E. coli* enzyme also held for the T4 DNA ligase and, in fact, for the mammalian ligases of which there are as many as four now known. The use of NAD as a cofactor seems to be confined to bacteria; the T4 phage and mammalian ligases all use ATP instead of NAD (31).

Subsequent studies of *E. coli* and T4 mutants by several labs, including Tom Broker and Yasahiro and Naoyo Anraku in my lab, showed that DNA ligase is essential in vivo for the joining of Okazaki fragments during DNA replication, the joining of DNA chains during nucleotide and base excision repair of DNA, and the joining of DNA segments following cleavage of the Holiday junction in homologous recombination (32–34). And, of course, DNA ligase became a key reagent in the construction of recombinant DNA molecules. I do not believe that any of us working on the enzyme at that time foresaw the central role it would play in the genetic engineering revolution. That was left to Paul Berg, Peter Lobban, Stan Cohen, and Herb Boyer.

BACK TO DNA POLYMERASE AGAIN

The DeLucia–Cairns mutant announced in *Nature* in 1968 was a bombshell (28). Although ostensibly lacking DNA polymerase activity, the mutant was fully viable; its only defect was an increased sensitivity to ultraviolet irradiation. Now our DNA polymerase was relegated to the lowly role of a "repair enzyme." Some different system was thought by some to be responsible for chromosomal replication, possibly using substrates other than the deoxynucleoside triphosphates (16). Ironically, with the realization that many human cancers result from defects in DNA repair, this function is currently regarded as one of the hottest fields in biology.

Although Cairns was right in believing that our DNA polymerase (Pol I) was not responsible for chromosomal replication in *E. coli*, he was wrong in evaluating its role. With *E. coli* extracts apparently deficient in Pol I, DNA polymerase II and subsequently DNA polymerase III holoenzyme, each present in relatively few copies per cell, could be detected and purified from Pol I mutant extracts. DNA polymerase III was identified as part of a machine, responsible for replication of the *E. coli* chromosome. How did Pol I fit into the picture?

I had been deeply distressed by the Cairns paper and all of the ensuing publicity. I was determined to see if the Cairns mutant was really lacking in Pol I. It was a risky project; one that I felt was best to work on myself. With the help of my technician, Janice Chien, I was able to show that the extracts of the *pol A$^-$* strain, which bore an *amber* mutation, did contain a low level of Pol I (1%–2% of wild type), resulting from readthrough of the nonsense codon. However, the levels of the 5′ → 3′ exonuclease activity associated with Pol I were normal (35). The 5′ → 3′ exonuclease polypeptide was, however, far smaller than the intact enzyme. Pol I contains two domains: one with polymerase activity associated with 3′ → 5′ proofreading exonuclease and a second domain with a 5′ → 3′ exonuclease. Protease treatment of Pol I yields the "large" or "Klenow" fragment with polymerase and 3′ → 5′ exonuclease activities as well as a small fragment with the 5′ → 3′ exonuclease (36). It turned out that the *amber* mutation in the Cairns mutant was positioned in such a way that translation yielded the intact small fragment, explaining our observation that the mutant extracts contained normal levels of 5′ → 3′ exonuclease activity, of a size corresponding to that of the small fragment (35).

Bruce Konrad and I subsequently isolated a temperature-sensitive mutant defective in the 5′ → 3′ exonuclease activity but normal in its polymerase activity. Under standard growth conditions, the 5′ → 3′ exonuclease of Pol I is essential for DNA replication (37). The Cairns mutant was viable because it retained this activity. We now know that the essential role of Pol I in DNA replication is the 5′ → 3′ exonuclease removal of the RNA primers that initiate Okazaki fragment synthesis, followed by filling in the resulting gap by DNA polymerase action prior to their being joined by DNA ligase (37).

Robert Bambara, a postdoc, and Dennis Uyemura, a graduate student, purified Pol I from several of *pol A* mutants with defects in the polymerase domain. Because their 5′ → 3′ exonuclease activity was unchanged, they were fully viable; however, they were abnormally sensitive to UV radiation, presumably as a result of their polymerase defect. In examining various features of the polymerase activity in these mutants, Bambara and Uyemura developed what I believe to be the first quantitative way to assess the processivity of deoxynucleotide polymerization by DNA polymerases (38). High processivity is of course essential for chromosome replication, and we now know that there are complex protein assemblies that interact with DNA polymerases to tether them to the template in order to prevent their dissociation during polymerization.

As another extension of our work on DNA polymerase mutants, Duane Eichler, a postdoc, and then later Per Olaf Nyman, a sabbatical visitor from Sweden, and Bik Kwoon Tye, a postdoc, examined the formation and processing of Okazaki fragments in these mutants. Their joining was clearly retarded to an extent that depended on the severity of the defect. We could support the claims of Reiji and Tuneko Okazaki that their pulse-labeled fragments were intermediates in DNA replication (on the lagging strand) rather than artifacts of DNA strand scissions, a possibility that had not yet been excluded (39).

EUKARYOTIC DNA REPLICATION

At about this time, I began to think about eukaryotic DNA replication. An obvious place to start was with a DNA polymerase. Two nuclear DNA polymerases had been described: DNA polymerase α (Pol α) was believed to be involved in chromosomal replication and DNA polymerase β (Pol β) in DNA repair. The field was highly populated and rather contentious. Pol α seemed to come in various sizes, and there was no agreement about its molecular weight and number of subunits. Its cellular abundance is low, so that barely microgram quantities emerged from the rather elaborate purification procedures that had been devised.

In casting about for an abundant source of Pol α, I decided on *Drosophila melanogaster*. Kriegstein and Hogness (40), in our department, had shown earlier that chromosomal replication in early stage embryos of *D. melanogaster* proceeded at a frenetic pace: the entire *Drosophila* genome was replicated in about 3 min. This was a consequence of the large number of replication forks operating in tandem. It struck me that such embryo extracts must be highly enriched in DNA polymerase (presumably Pol α). Assays of the extracts showed that this was indeed the case.

Our purification was begun by two sabbatical visitors, the late John Boezi, from Michigan State, and the late Geoffrey Banks, from Mill Hill in London (41). The project then passed on to three postdocs: Laurie Kaguni, Guiseppi Villani, and Brian Sauer. They showed that Pol α consisted of four subunits; the largest, at 180 kDA, contained the DNA polymerase activity (42). Ron Conaway, a graduate student, then demonstrated that the most purified preparations of Pol α contained primase activity (primase had not yet been demonstrated in eukaryotes); he and Laurie Kaguni showed that the primase was associated with the two smaller of the four subunits (43). The association of primase with Pol α in *Drosophila* was quickly confirmed by several laboratories, in eukaryotes as diverse as baker's yeast and humans.

REC A

While this work on Pol α was in progress, we had actually begun a completely new line of investigation on genetic recombination in *E. coli*, an interest of mine since the ligase days. Kevin McEntee, who joined my lab as a postdoc, had constructed, as part of his doctoral research at the University of Chicago, a specialized lambda transducing phage, containing the *rec A* gene. The *rec A* gene had been identified by John Clark in the early 1960s as essential for homologous recombination in *E. coli*, but its product had never been isolated, and no one had any idea of what it did. Kevin's postdoctoral fellowship application had involved a study of some of our Pol I mutant enzymes. But when deciding on a specific project, the idea of isolating the *rec A* gene product and determining its function seemed far more attractive. George Weinstock, who arrived in the lab at about the same time, joined forces with McEntee.

The *rec A* gene product had been known to be involved in radiation-induced mutagenesis or error-prone repair as it was known. We, therefore, devised an assay for its purification that looked for nucleotide misincorporation during replication of single-stranded θX174 DNA in vitro (44). After nearly a year of failure, we decided, in desperation, to try something heretical. Induction of Kevin's transducing phage yielded large amounts of the Rec A protein, easily visible at the predicted size of about 40 kDa in a stained polyacrylamide gel of the induced crude extracts. The idea was to simply purify the 40-kDA polypeptide without a functional assay. The danger, of course, was that we would end up with a pure but totally inactive protein. This was a risky strategy particularly with a protein of unknown function. Unconventional then, this has now become almost standard procedure with cloned and overexpressed gene products.

With pure Rec A protein in hand, we could test it for various activities that are known to be associated with enzymes that act on DNA, such as DNA and RNA polymerase, nuclease, and DNA-dependent ATPase. However, before we could perform any of these assays, we heard a presentation by Tomoko Ogawa, of the University of Osaka, at the 1978 Cold Spring Harbor Meeting, in which she showed that a purified preparation of the Rec A protein had ATPase activity dependent on single-stranded DNA; it was a DNA-dependent ATPase. At about the same time, Jeffrey Roberts, Christine Roberts, and Nancy Craig, who had been studying the regulatory role of the *rec A* gene and identified the Rec A protein as a protease, also discovered its DNA-dependent ATPase activity (45). On returning to Stanford, Kevin quickly confirmed their findings with our Rec A protein preparation. He then went on to demonstrate that the DNA-dependent ATPase associated with the Rec A protein purified from a cold-sensitive *rec A* mutant was also cold sensitive. This result was important because it ruled out contamination of Rec A protein preparations with one or more of the many DNA-dependent ATPases that are present in *E. coli* (46). Coincident with these studies, George Weinstock was investigating the fate of the single-stranded DNA (heat-denatured P22 DNA) during the ATP hydrolysis and found that the complementary P22 single strands were being reannealed. McEntee and Weinstock then quickly showed that not only could the Rec A protein promote the ATP-dependent renaturation of complementary single strands, but it could also promote ATP-driven strand exchange between a single strand and a homologous DNA duplex (47). In essence, the Rec A protein could form a Holiday junction, the key intermediate in homologous recombination, thereby explaining the essentialness of the *recA* gene for homologous recombination in *E. coli*. Findings similar to ours were made at about the same time by Charles Radding and by Steve West and Paul Howard-Flanders at Yale. Again, there was an intense, but largely friendly, competition between the various groups.

The Rec A group, which initially consisted of McEntee and Weinstock, quickly grew once we published our findings. Michael Cox, Peter Riddles, Randy Bryant, Zvi Livneh, Douglas Julin, and Daniel Soltis joined the lab in the next five years and contributed to our attempts to understand the mechanism by which the Rec A protein promotes strand exchange, and several, notably Cox and Bryant, have continued to work productively on the Rec A protein after leaving my lab.

The nine-year period that we worked on the Rec A protein was exciting, productive, and ultimately very gratifying. Although, important features of Rec A-promoted strand exchange still remain elusive. For example, a key step in the process, the mechanism by which the Rec A protein searches for and finds the homologous regions between recombining DNA molecules, is still a mystery. Nevertheless, we did go a long way toward solving a fundamental problem in biology. There may also be broader consequences of this work. The human Rec A analogue, the Rad51 protein, seems to play an important role in embryogenesis: embryos of Rad51 "knockout" mice survive for only a few days, and Rad51 has been shown to interact with BRCA1 and BRCA2, the human breast cancer susceptibility genes (48).

During this period, we continued to work on eukaryotic DNA replication. Although our expectation that *Drosophila* embryos would be an abundant source of replication enzymes proved correct, we were unsuccessful, despite the rapid rate of DNA replication in those embryos, in finding extracts that could promote origin-dependent DNA replication. Part of the problem was the uncertainty about a *Drosophila* origin; no origin had been unequivocally identified. At the same time, several labs, notably those of Tom Kelly, Jerry Hurwitz, and Bruce Stillman, had demonstrated SV40 origin-dependent DNA replication in mammalian cell extracts and were well on their way to identifying and resolving the components that were involved. Was there another viral chromosome with a defined origin that we might explore?

HERPES VIRUS DNA REPLICATION

Edward Mocarski, from the microbiology department at Stanford, had told us about his work on the herpes simplex type 1 virus (HSV-1) and its life cycle. It contained not one, but three identifiable origins of replication; but it was rather large and unwieldy (152 kb). Nevertheless, introduction of one of these origins into a 5-kb plasmid permitted its replication in HSV-1-infected cells. Replication generated long concatamers, indicating that it proceeded predominantly by a rolling circle mechanism.

In recent years, we have tried to reconstitute HSV-1 DNA replication with purified enzymes. Unlike SV40, HSV-1 encodes most of the enzymes it needs to replicate its genome, including a DNA polymerase, a single-strand DNA-binding protein, a polymerase processivity factor, a primosome (helicase-primase), and an origin-binding protein (49). The last three were discovered in my lab. Again, I was fortunate to have been able to recruit a splendid group of postdocs and students: Per Elias and Mike O'Donnell started the project and were joined later by James Crute, Robert Bruckner, Mark Dodson, Tatsuya Tsurumi, Rami Skaliter, Sam S-K Lee, Ke-Jung Huang, and Don He, as postdocs, and Tom Hernandez, Rebecca Dutch, and Boris Zemelman, as graduate students. We are enormously aided in our efforts by Ed Mocarski, a seemingly inexhaustible source of information and ideas about herpes viruses. Although we have succeeded in reconstituting rolling circle replication, the initial, origin-dependent phase of replication has eluded us and remains a challenge for the future.

CURRENT WORK ON HSV-1 LATENCY

Very recently as an outgrowth of our work on HSV-1 DNA replication, we have begun an investigation of the molecular basis of viral latency. An important feature of the HSV-1 life cycle is its ability to establish latency in neurons. Following primary infection, HSV-1 gains access to sensory neurons and travels via axons to establish infection in the sensory neurons that innervate the site of infection. Once within the neurons, the virus can either enter a productive (lytic) cycle, resulting in the release of progeny viruses or establish latency (50).

During latency, the viral DNA is present in the neuronal nuclei and is maintained in a nonlinear (circular?) nonintegrated nucleosome-bound state; transcription is restricted to several latency-associated transcripts, whose function is unclear. During neuronal latency, HSV-1 has no apparent impact on the infected individual. However, the latent virus can be reactivated throughout the life of the individual by a variety of stimuli and produce recurrent disease. For example, recurrent HSV-1 infection can cause corneal scarring, leading to the loss of vision. In fact, HSV-1 is one of the most common infectious causes of corneal blindness in the developed countries. Although the molecular mechanisms underlying the establishment of latency and the reactivation from latency have been extensively investigated, they are poorly understood.

In a search for binding partners for the HSV-1 replication initiator, the UL9 protein, by using a yeast two-hybrid screen, we observed that NFB42, a neuron-specific protein, which forms part of the E3 ubiquitin-ligase complex that catalyzes ubiquitination and proteosome-mediated degradation of phosphorylated proteins, formed a tight complex with the UL9 protein. The fact that NFB42 is neuron specific immediately suggested a link between the UL9 protein and HSV-1 neuronal latency. In pursuing this lead, Chi-Yong Eom and I were able to demonstrate that the phosphorylated UL9 protein is polyubiquinated in neurons and undergoes proteosomal degradation. These findings, which are still at an early stage, suggest that in neurons the UL9 protein is phosphorylated by cellular kinase(s), is recognized by the neuron-specific NFB42, and then is degraded via the ubiquitin-proteosome pathway. They further suggest that this degradation leads to the inhibition of initiation of HSV-1 replication and possibly to the establishment of neuronal latency (51). Time and further work will tell if this conjecture is correct. In any event, in our attempts to understand the HSV-1 replication initiator, we unpredictably wandered into uncharted and, for us, exciting new territory.

ENVOI

Looking back at my more than 50 years in biochemical research, I am grateful for having lived through an extraordinary period of biological discovery. When I began graduate studies, there was still a lively debate over whether citrate was an integral component of the Krebs citric acid cycle or only an offshoot. The structure of DNA was a matter of conjecture, and it was not completely accepted that the gene is composed of DNA. Nothing was known about how DNA was replicated. RNA was a rather ill-defined substance, and there were only suggestions that it was somehow required for protein synthesis. Although the pathways of carbohydrate, lipid, and amino acid synthesis were being rapidly elucidated, virtually nothing was known about how they were regulated. Today, we know their regulation in great detail. The structure of DNA and the various RNAs are known, and the human genome has been sequenced. The cloning of genes, their expression, and their repression are exercises in college biology courses. Enormous strides have been made in understanding the workings of the nervous and immune systems. These discoveries and many others have had a profound effect on our understanding and treatment of human disease, and there are certainly many more discoveries to come.

Having grown up in the Great Depression and served in World War II, I consider myself very fortunate to have survived that dreadful period and to have been able to participate, if only in a very minor way, in the extraordinary scientific advances of the past 50 years. I consider myself particularly fortunate in having had an inspirational teacher, Arthur Kornberg, and to have been associated with a talented and hard-working group of graduate students and postdocs—not all of whom I have been able to mention. None of the successful "wanderings" that I have described would have been possible without them.

ACKNOWLEDGMENT

Almost all of the work described in this account has been supported by grants from the National Institutes of Health.

LITERATURE CITED

1. Lehman IR, Herriott RM. 1958. *J. Gen. Physiol.* 41:1067–82
2. Wyatt GR, Cohen SS. 1953. *Biochem. J.* 55:744–82
3. Flaks JG, Cohen SS. 1957. *Biochim. Biophys. Acta* 25:667–72
4. Grunberg-Manago M, Ochoa S. 1955. *J. Am. Chem. Soc.* 77:3165–66
5. Lehman IR, Bessman MJ, Simms ES, Kornberg A. 1958. *J. Biol. Chem.* 233:163–70
6. Bessman MJ, Lehman IR, Simms ES, Kornberg A. 1958. *J. Biol. Chem.* 233:171–77
7. Kornberg A. 1957. In *The Chemical Basis of Heredity*, ed. WD McElroy, B Glass, pp. 579–608. Baltimore, MD: Johns Hopkins Univ. Press
8. Watson JD, Crick FHC. 1953. *Nature* 171:737–39
9. Lehman IR, Zimmerman SB, Adler J, Bessman MJ, Simms ES, Kornberg A. 1958. *Proc. Natl. Acad. Sci. USA* 44:1191–96
10. Chalberg MD, Kelly TJ Jr. 1979. *Proc. Natl. Acad. Sci. USA* 76:655–59
11. Schekman R, Wickner W, Westergaard O, Brutlag D, Greider K, et al. 1972. *Proc. Natl. Acad. Sci. USA* 69:2691–95
12. Schachman HK, Adler N, Radding CM, Lehman IR, Kornberg A. 1960. *J. Biol. Chem.* 235:3242–49
13. Kornberg A, Bertsch L, Jackson JF, Khorana HG. 1964. *Proc. Natl. Acad. Sci. USA* 51:315–23
14. Goodman MF. 2002. *Annu. Rev. Biochem.* 71:17–50
15. Kornberg A. 1988. *J. Biol. Chem.* 263:1–4
16. Lehman IR, Uyemura DG. 1976. *Science* 193:963–69
17. Burgers PMJ, Koonin EV, Bruford E, Blanaco L, Burtis KC, et al. 2001. *J. Biol. Chem.* 276:43487–90

18. Kuchta RD, Mizrahi V, Benkovic PA, Johnson KA, Benkovic S. 1987. *Biochemistry* 26:8410–17
19. Lehman IR. 1960. *J. Biol. Chem.* 235:1479–87
20. Schildkraut CL, Marmur J, Doty P. 1961. *J. Mol. Biol.* 3:595–609
21. Fiers W, Sinsheimer RL. 1962. *J. Mol. Biol.* 5:408–17
22. Lehman IR, Pratt EA. 1960. *J. Biol. Chem.* 235:3254–59
23. Kornberg SR, Zimmerman SB, Kornberg A. 1961. *J. Biol. Chem.* 236:1487–93
24. Lehman IR. 1971. *Annu. Rev. Biochem.* 4:251–70
25. Edgar RS, Denhardt GH, Epstein RH. 1964. *Genetics* 49:635–46
26. DeWaard A, Paul AV, Lehman IR. 1965. *Proc. Natl. Acad. Sci. USA* 54:1241–48
27. Hall ZW, Lehman IR. 1968. *J. Mol. Biol.* 36:321–33
28. De Lucia P, Cairns J. 1969. *Nature* 224:1164–66
29. Olivera BM, Lehman IR. 1967. *Proc. Natl. Acad. Sci. USA* 57:1426–33
30. Olivera BM, Lehman IR. 1967. *Proc. Natl. Acad. Sci. USA* 57:1700–4
31. Lehman IR. 1974. *Science* 186:790–97
32. Anraku N, Lehman IR. 1969. *J. Mol. Biol.* 46:467–79
33. Konrad EB, Modrich P, Lehman IR. 1974. *J. Mol. Biol.* 90:115–26
34. Modrich P, Lehman IR. 1971. *Proc. Natl. Acad. Sci. USA* 68:1002–5
35. Lehman IR, Chen JR. 1973. *J. Biol. Chem.* 248:7502–11
36. Setlow P, Brutlag D, Kornberg A. 1972. *J. Biol. Chem.* 247:224–31
37. Konrad EB, Lehman IR. 1974. *Proc. Natl. Acad. Sci. USA* 71:2048–51
38. Uyemura D, Bambara R, Lehman IR. 1975. *J. Biol. Chem.* 250:8577–84
39. Lehman IR, Tye BK, Nyman PO. 1979. *Cold Spring Harbor Symp. Quant. Biol.* 43:221–30
40. Kriegstein H, Hogness D. 1974. *Proc. Natl. Acad. Sci. USA* 71:135–39
41. Banks GR, Boezi JA, Lehman IR. 1979. *J. Biol. Chem.* 254:9886–92
42. Conaway RC, Lehman IR. 1982. *Proc. Natl. Acad. Sci. USA* 79:2523–27
43. Kaguni LS, Rossignol JM, Conaway RC, Banks GR, Lehman IR. 1983. *J. Biol. Chem.* 258:9037–39
44. Schekman R, Weiner A, Kornberg A. 1974. *Science* 186:987–93
45. Roberts JW, Roberts CW, Craig NL. 1978. *Proc. Natl. Acad. Sci. USA* 75:4714–18
46. Weinstock GM, McEntee K, Lehman IR. 1979. *Proc. Natl. Acad. Sci. USA* 76:126–30
47. McEntee K, Weinstock GM, Lehman IR. 1979. *Proc. Natl. Acad. Sci. USA* 76:2615–19
48. Scully R, Chen J, Plug A, Xiao Y, Weaver D, et al. 1997. *Cell* 88:265–75
49. Lehman IR, Boehmer PE. 1999. *J. Biol. Chem.* 274:28059–62
50. Nahmias AJ, Roizman B. 1973. *N. Engl. J. Med.* 289:781–89
51. Eom C-Y, Lehman IR. 2003. *Proc. Natl. Acad. Sci. USA* 100:9803–7

Signaling Pathways in Skeletal Muscle Remodeling

Rhonda Bassel-Duby and Eric N. Olson

Department of Molecular Biology, University of Texas Southwestern Medical Center, Dallas, Texas 75390-9148; email: rhonda.bassel-duby@utsouthwestern.edu, eric.olson@utsouthwestern.edu

Key Words

myofiber, MEF2, calcineurin, calcium-dependent protein kinase, exercise adaptation, hypertrophy

Abstract

Skeletal muscle is comprised of heterogeneous muscle fibers that differ in their physiological and metabolic parameters. It is this diversity that enables different muscle groups to provide a variety of functional properties. In response to environmental demands, skeletal muscle remodels by activating signaling pathways to reprogram gene expression to sustain muscle performance. Studies have been performed using exercise, electrical stimulation, transgenic animal models, disease states, and microgravity to show genetic alterations and transitions of muscle fibers in response to functional demands. Various components of calcium-dependent signaling pathways and multiple transcription factors, coactivators and corepressors have been shown to be involved in skeletal muscle remodeling. Understanding the mechanisms involved in modulating skeletal muscle phenotypes can potentiate the development of new therapeutic measures to ameliorate muscular diseases.

Contents

INTRODUCTION	20
PROPERTIES OF MYOFIBERS	21
Myofiber Diversity	21
Myofiber Adaptability	22
SIGNALING PATHWAYS IN MYOFIBER REMODELING	22
Myocyte Enhancer Factor-2 and Histone Deacetylases	23
Calcineurin/Nuclear Factor of Activated T Cells	25
Calcium/Calmodulin-Dependent Protein Kinase, Protein Kinase C, and PKCmu/Protein Kinase D	26
Peroxisome Proliferator-Activated Receptor Delta and Peroxisome-Proliferator-Activated Receptor Gamma Coactivator-1 alpha	27
Ras/Mitogen-Activated Protein Kinase	28
Insulin-Like Growth Factor, Akt, and Mammalian Target of Rapamycin	28
CLINICAL SIGNIFICANCE	29
Muscular Dystrophy	29
Type 2 Diabetes Mellitus and Obesity	30
Muscle Atrophy	31
Anabolic Steroids	32
CONCLUSIONS	32

Calcineurin: a heterodimeric protein phosphatase (PP2B) comprised of calmodulin-binding catalytic A and regulatory B subunits

INTRODUCTION

Skeletal muscle is composed of heterogeneous specialized myofibers that enable the body to maintain posture and perform a wide range of movements and motions. It is this diversity of myofibers that enables different muscle groups to fulfill a variety of functions. In addition to its obvious roles in motility, skeletal muscle plays a central role in the control of whole-body metabolism. These seemingly different functions are controlled by signaling pathways that enable muscle fibers to respond to the changing metabolic and functional demands of the organism.

The premise that myofibers remodel and modify their phenotype was demonstrated over 45 years ago when cross-reinnervation studies were shown to alter the contractile properties of myofibers (1). Similarly, skeletal muscle responds to exercise training by remodeling the biochemical, morphological, and physiological states of individual myofibers. The remodeling process provides an adaptive response that serves to maintain a balance between physiological demands for contractile work and the capacity of skeletal muscle to meet those demands. Many of the remodeling responses involve activation of intracellular signaling pathways and consequent genetic reprogramming, resulting in alterations of muscle mass, contractile properties, and metabolic states.

Advances in genetic engineering have allowed the introduction or depletion of factors within the myofiber, facilitating the evaluation of signaling factors during muscle remodeling. In particular, myofiber transformation has successfully been achieved in transgenic mouse models using muscle-specific promoters to drive expression of calcineurin (protein phosphatase 2B) and various calcium-dependent kinases. Activation of specific signaling pathways in myofibers has profound effects, not only on contractile proteins, but also on alterations of metabolic states leading to changes in muscle performance. Because of space limitations, this review does not discuss the pathways controlling muscle development (2) nor the contribution of satellite cells to skeletal muscle regeneration (3) but focuses on the signaling mechanisms that modify myofiber function with emphasis on clinical significance and therapeutic approaches to ameliorate muscle diseases.

PROPERTIES OF MYOFIBERS

The musculature of the body is composed of a variety of muscle groups, such as soleus, extensor digitorum longus, and plantaris. Each muscle group is comprised of heterogeneous myofibers that differ in their biochemical, physiological, and metabolic parameters. The myofiber content is a determinant of muscle heterogeneity in contraction speed and fatigue resistance. A striking feature of the myofiber is the ability to transform and remodel in response to environmental demands.

Myofiber Diversity

Although histologically skeletal muscle appears uniform (**Figure 1a**), it is comprised of myofibers that are heterogeneous with respect to size, metabolism, and contractile function (4). On the basis of specific myosin heavy-chain isoform expression, myofibers are classified into type I, type IIa, type IId/x, and type IIb fibers, with types I and IIa exhibiting oxidative metabolism and types IIx and IIb being primarily glycolytic (5, 6). Type I myofibers, also termed slow-twitch fibers, exert a slow contraction owing to the ATPase activity associated with the type I myosin. Slow-twitch myofibers are rich in mitochondria, have more capillaries surrounding each fiber, exhibit oxidative metabolism, have a low velocity of shortening, and have a high resistance to fatigue. Type II fibers, termed fast-twitch myofibers, exert quick contractions and fatigue rapidly. The slow oxidative fibers are required for maintenance of posture and tasks involving endurance, whereas fast-glycolytic fibers are required for movements involving strength and speed. Different myofiber subtypes are detected during embryonic life (7), and patterning of fiber types within major muscle groups is established postnatally (8).

In addition to the variability seen in myosin heavy-chain gene expression, fiber-type differences are observed with the expression profile of other muscle proteins, such as tropomyosin, myosin light chain, parvalbumin, phospholambin, and sarcoplasmic reticulum calcium ATPase (SERCA). Although there are multiple levels of distinction among myofibers, classically, fiber type is defined on the basis of its myosin heavy-chain isoform expression profile. Fiber type is determined using assays that delineate the differences in ATPase activity that correlate with specific myosin heavy-chain isoforms. The basis of the

Slow-twitch myofibers: muscle fibers that express type I myosin heavy chain

Fast-twitch myofibers: muscle fibers expressing type IIa, type IId/x, and type IIb myosin heavy chain

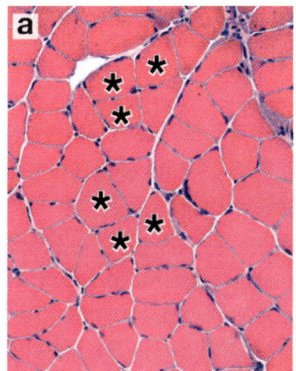 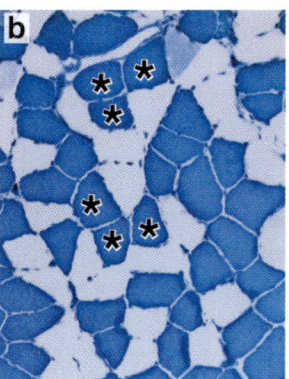 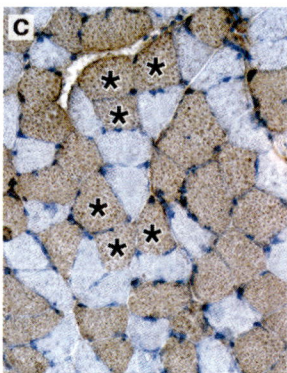

Figure 1
Heterogeneous distribution of skeletal muscle fibers. Fiber-type analysis of serial transverse sections of mouse soleus by (*a*) hematoxylin and eosin stain showed a checkerboard pattern of fibers, (*b*) metachromatic dye-ATPase method showed type I fibers (stained *dark blue*) and type IIA (stained *light blue*), and (*c*) immunohistochemistry using a monoclonal antibody recognized type I myosin heavy chain. The asterisks mark the same type I fibers in each panel.

Motor neuron: a neuron that innervates muscle fibers

Signal transduction: an extracellular signal stimulates a receptor, activating a messenger and changing gene expression and phenotype

reaction is the deposition of insoluble salts of inorganic phosphate cleaved from ATP by myofibrillar ATPase(s) followed by substitution of the phosphates with less soluble chromogenic salts (9) (**Figure 1b**). Immunohistochemistry using monoclonal antibodies that recognize isoform-specific myosin heavy chain is another method used to determine fiber type (**Figure 1c**).

Slow-twitch oxidative myofibers (type I) are involved in sustained, tonic contractile events and maintain intracellular calcium concentrations at relatively high levels (100–300 nM) (10, 11). In contrast, fast-twitch glycolytic myofibers (type IIb) are used for sudden bursts of contraction and are characterized by brief, high-amplitude calcium transients and lower ambient calcium levels (less than 50 nM) (12). These properties of skeletal muscle fibers are dependent on the pattern of motor nerve stimulation, such that tonic motor neuron activity at low frequency (10–20 Hz) promotes the slow fiber phenotype, whereas phasic motor neuron firing at high frequency (100–150 Hz) results in fast fibers.

Myofiber Adaptability

The ability of skeletal muscle to remodel and change phenotypically can be demonstrated by cross-innervation experiments in which slow-twitch muscle (soleus) reinnervated with nerve fibers that normally supply fast-twitch muscle (flexor digitorum longus) results in an increase in contractile speed. Conversely, innervation of fast-twitch muscle with nerve fibers normally found on soleus muscle causes slower contraction (1). These studies established that specific impulse patterns delivered by motor neurons exert a phenotypic influence on the muscles they innervate and that myofibers are capable of remodeling. Further studies using electrical stimulation to modify neural activity delivered to a target muscle corroborate the cross-reinnervation data by showing predicted changes in myosin isoforms (13, 14). Exercise training also induces changes in skeletal muscle by transforming the myofibers to an increased oxidative metabolism and inducing fiber-type transitions from type IIb → type IId/x → type IIa → type I. To everyone's chagrin, upon cessation of exercise training these myosin heavy-chain isoform transitions and metabolic changes are reversed.

Neuronal stimulation reprograms gene expression in the myofiber primarily by using calcium as a second messenger. The input received from motor neurons via acetylcholine receptors generates a depolarization of the membrane, which reaches the sarcolemma transverse (T)-tubular membrane (15). The voltage-operated calcium channel or L-type calcium channel (the dihydropyridine receptor) in the T-tubules interacts with a skeletal muscle-specific sarcoplasmic reticulum calcium-release channel, the ryanodine receptor (RyR1) (16). This physical interaction causes the RyR1 to open and release calcium from the sarcoplasmic reticulum. The changes in intracellular calcium concentrations determine muscle contraction and activate signaling pathways. The process of myofiber transformation is regulated by multiple signaling pathways, many of which converge on each other, culminating in the activation and, perhaps, repression of a myriad of genes involved in remodeling of skeletal muscle.

SIGNALING PATHWAYS IN MYOFIBER REMODELING

Myofibers respond to physiological and pathological signals by transforming and remodeling to adapt to the environmental demands. This adaptation is accomplished through signal transduction by which an extracellular signal interacts with receptors at the cell surface, activating factors in signaling pathways and ultimately remodeling the myofiber by effecting a change in gene expression.

Myocyte Enhancer Factor-2 and Histone Deacetylases

It is well recognized that the myocyte enhancer factor-2 (MEF) transcription factors, in conjunction with multiple myogenic regulatory factors, play a dominant role in muscle formation by activating muscle-specific genes and that the MEF2/histone deacetylase (HDAC) signaling pathway plays an important role in the transformation of myofibers in response to intracellular calcium fluctuations incurred by the effects of external physiological signals (**Figure 2**). MEF2 is a muscle-enriched transcription factor that binds to an A/T-rich DNA sequence in the control regions of numerous muscle-specific genes (17). There are four vertebrate *MEF2* genes, *MEF2A*, *-B*, *-C*, and *-D*, which are expressed in distinct, but overlapping, patterns during embryogenesis and in adult tissues. In the mouse, *Mef2c* gene expression is detected in developing skeletal muscle concomitant with activation of the skeletal muscle differentiation gene program. High levels of MEF2 proteins are clearly detectable in developing muscle lineages during embryogenesis (18). MEF2A protein appears as cells enter the differentiation pathway, and MEF2C is expressed late in the differentiation program. Studies in primary cell cultures of human skeletal muscle cells showed that various stimuli, such as addition of insulin, hydrogen peroxide, osmotic stress, and activation of AMP-activated protein kinase (AMPK) resulted in activation of MEF2D DNA binding (19).

Myocyte enhancer factor-2 (MEF2): a family of transcription factors that activates muscle-specific genes

HDAC: histone deacetylase

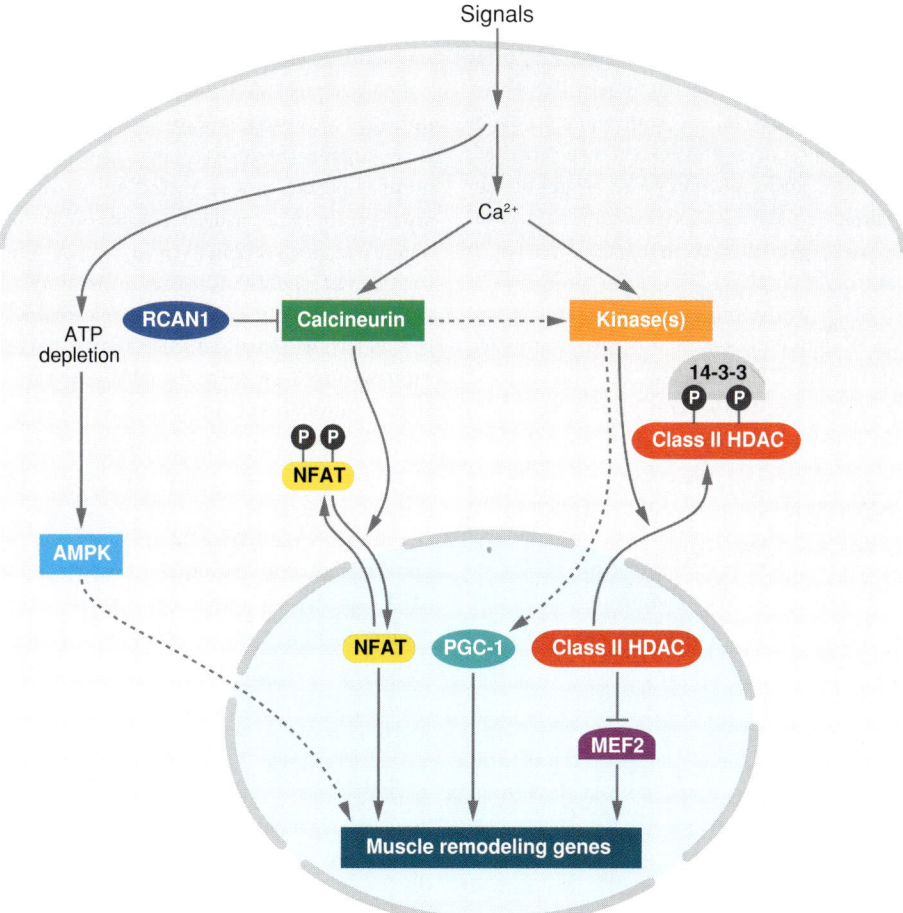

Figure 2

Signaling pathways activate skeletal muscle remodeling genes. In response to physiological demands, intracellular calcium concentration is elevated, activating the calcineurin/nuclear factor of activated T cells (NFAT) and MEF2/HDAC signaling pathways. In response to workload, ATP is depleted activating AMPK.

Figure 3

Exercise stimulates transcriptional activation of MEF2. (*a*) Soleus from sedentary MEF2 indicator mouse was stained with X-gal to detect *lacZ* expression. (*b*) Soleus from MEF2 indicator mouse subjected to three days of voluntary wheel running was stained with X-gal to detect *lacZ* expression.

To monitor the transcriptional activity of MEF2 in vivo, a transgenic MEF2 sensor mouse that harbors a *lacZ* transgene under the control of three tandem copies of the MEF2 consensus DNA-binding site was generated (20). During embryogenesis, these mice express *lacZ* in developing cardiac, skeletal, smooth muscle, and neuronal cells. After birth, transgene expression is downregulated, although MEF2 protein levels are high, suggesting that MEF2 activity is repressed. A series of studies showed that MEF2 activity is controlled through association with class II HDACs, which bind to MEF2 and repress MEF2 activity (21–25). In response to various signals, HDAC kinases are activated and phosphorylate these HDACs, creating a docking site for intracellular chaperone protein 14-3-3 to bind HDAC and mask the nuclear localization sequence as well as induce a conformational change in HDAC that unmasks a nuclear export sequence, causing HDAC to exit from the nucleus and promoting MEF2 activity (26–28). Signal-dependent release of class II HDACs from MEF2 appears to play a role in skeletal muscle differentiation (22). Transitioning myofibers to a slow-twitch phenotype, using 10-Hz electrical stimulation, translocates HDAC4 from the nucleus to the cytoplasm and increases MEF2 activity, further supporting the role of class II HDACs in signaling pathways during skeletal muscle remodeling (29).

To identify factors that stimulate MEF2 activity, MEF2 sensor mice were subjected to voluntary wheel running and electrical stimulation of the sciatic nerve (30). These stimuli have been shown to promote a substantial degree of fiber-type transformation (type IIb to IIa to I) and to upregulate expression of proteins associated with oxidative metabolism, such as myoglobin. Following these exercise regimens, MEF2 is activated, as demonstrated by the expression of *lacZ* in the MEF2 sensor mice (**Figure 3**).

Further studies on MEF2 activation by exercise showed that this response is blocked when cyclosporine A, an inhibitor of the serine/threonine protein phosphatase 2B, calcineurin, is administered. In addition, crossing transgenic mice overexpressing activated calcineurin in skeletal mice with the MEF2 sensor mice showed that MEF2 activity was dramatically activated by calcineurin signaling (30, 31). Furthermore, it was shown that the activation of both the MEF2 and calcineurin pathways promotes expression of muscle-specific genes, including *myoglobin*, *myosin heavy chain*, and *slow troponin I* (32, 33). These findings revealed cross talk between the MEF2/HDAC and calcineurin signaling pathways and delineated a molecular pathway in which calcineurin and MEF2 participate in the adaptive mechanisms by which myofibers acquire specialized contractile and metabolic properties as a function of changing patterns of muscle contraction induced by exercise (**Figure 2**).

Although many MEF2 gene targets are known, signaling pathways downstream of MEF2 are largely unknown. Analysis of the gene expression profile of mice lacking *Mef2c* identified a decrease in expression of a novel MEF2-regulated gene encoding a muscle-specific protein kinase Stk23/Srpk3, belonging to the serine arginine protein kinase (SRPK) family, which phosphorylates serine/arginine repeat-containing proteins (34). The *Srpk3* gene is specifically expressed in the heart and skeletal muscle from embryogenesis to adulthood and is controlled by a muscle-specific enhancer with an essential MEF2-binding site. When the *Srpk3* gene is disrupted in mice, myofibers show an increase in centrally placed nuclei, a characteristic of many myopathies, and disorganized intermyofibrillar network in type II fibers. Overexpression of *Srpk3* in skeletal muscle causes severe myofiber degeneration and early lethality. These findings show that SRPK-mediated signaling plays important roles in muscle growth and homeostasis downstream of muscle-specific transcription regulated by MEF2.

Calcineurin/Nuclear Factor of Activated T Cells

Calcineurin, a heterodimeric protein phosphatase comprised of a calmodulin-binding catalytic A subunit and a calcium-binding regulatory B subunit, is specifically activated by sustained, low-amplitude calcium waves and is a sensor of contractile activity by sensing calcium fluctuations (35, 36). Signaling is initiated by sustained, low-amplitude calcium waves allowing calcium to bind calmodulin, which activates calcineurin via the regulatory subunit (37). Upon activation, calcineurin dephosphorylates nuclear factor of activated T cells (NFAT), resulting in translocation of NFAT from the cytoplasm to the nucleus where it associates with other transcription factors to activate specific sets of calcium-dependent target genes (38) (**Figure 2**). Among the transcription factors that may serve as an important partner for NFAT proteins in myocytes is MEF2 (17, 30, 31).

Overexpression of activated calcineurin in myoblasts modulates myofiber gene expression by activating a subset of genes, which are associated with type I myofibers, such as myoglobin and troponin I slow (32, 39). To examine the effect of calcineurin in vivo, transgenic mice, harboring a muscle creatine kinase promoter driving activated calcineurin, were generated and shown to upregulate endogenous oxidative proteins, such as myoglobin, in a dose-dependent manner and drive fast to slow myofiber transformation (30, 40), supporting the role of the calcineurin/NFAT pathway in myofiber remodeling. Additional evidence of the role of the calcineurin/NFAT pathway in fiber-type specificity is seen by a reduction in oxidative/slow type I fibers in mice lacking calcineurin A isoforms alpha or beta (41). In addition, conditional calcineurin β1 knockout mice, lacking the calcium-binding regulatory subunit specifically in skeletal muscle, display dramatic deficiencies in both myosin heavy-chain type I and IIa protein expression and a decrease in the number of slow fibers (42). These results further demonstrate that calcineurin activity regulates the slow fiber program. Notably, the conventional calcineurin A alpha or beta knockout mice show a reduction in muscle weight; in contrast, no significant weight reduction was seen with the skeletal muscle-specific conditional knockout calcineurin β1 mice. The difference observed in muscle mass between these two calcineurin knockout lines is most likely attributable to the elimination of calcineurin in all cells, including myogenic progenitors and myoblasts, in the conventional calcineurin knockout mice, whereas elimination of calcineurin is restricted to post-differentiated myofibers in the mice lacking calcineurin β1 specifically in skeletal muscle.

Further evidence for a role of calcineurin in maintenance of the slow fiber phenotype is seen in the treatment of rats with cyclosporine A, an inhibitor of calcineurin activity, which results in an induction of glycolytic enzymes and a decrease in slow type I contractile proteins with a transformation toward a fast phenotype (32, 43). These findings are further supported by a study showing that calcineurin inhibitors block upregulation of type I isoforms of myosin in a muscle regenerating system (44) and in a primary skeletal muscle cell culture (45), showing that calcineurin activity is required for induction and maintenance of the slow type I myofiber gene program. Furthermore, studies in mice have shown that NFAT activity is higher in slow compared to fast muscles

NFAT: nuclear factor of activated T cells

Skeletal muscle hypertrophy: an increase in the size of muscle fiber, which increases muscle mass

(46). Introduction of a synthetic peptide inhibitor of calcineurin-mediated NFAT activation into the soleus leads to downregulation of slow myosin heavy-chain expression and an upregulation of myosin heavy-chain type IId/x (46). These results indicate that NFAT activity is required for maintenance of slow myosin heavy-chain gene expression and potentially is involved in the repression of the fast myosin heavy-chain IIx gene. Although the mechanism whereby calcineurin signaling induces the slow fiber gene via NFAT and MEF2 seems clear, it remains unclear how the fast fiber gene program is repressed by such signals.

Overexpression of a protein inhibitor of calcineurin, RCAN1 (previously known as MCIP-1) (47), has been shown to inhibit calcineurin activity in vivo. Stable mouse lines containing a conditional RCAN1 transgene were generated and crossed with a transgenic mouse line containing a skeletal muscle-specific promoter driving Cre recombinase. This strategy results in the expression of RCAN1 in skeletal muscle by using skeletal muscle-specific Cre recombinase to excise a region of DNA and place RCAN1 cDNA in the open reading frame. Using this Cre-ON approach, it was shown that the skeletal muscle of the mice overexpressing RCAN-1 has a decrease in calcineurin activity compared to wild-type mice (47), and most notably, these transgenic mice lack type 1 fibers. These findings show that calcineurin activity is essential for maintaining type I fibers.

In contrast, inconsistencies with the calcineurin/NFAT pathway model were shown by slow fiber-specific expression of a reporter gene (luciferase) controlled by expression of mutated forms of the slow troponin I promoter that lack NFAT- or MEF2-binding sites (48), and another group (49) showed that in vivo injections of a plasmid expressing activated calcineurin did not activate the slow myosin light chain promoter in soleus or extensor digitorum longus muscle. In addition, mice lacking NFATc2 or -c3 exhibit reduced muscle fiber size or number, respectively, but no significant change in the proportions of fiber types (50, 51). However, multiple studies have clearly shown and confirmed that skeletal muscle hypertrophy is not dependent on calcineurin activity (42, 52, 53).

Calsarcins, a family of sarcomeric proteins, have been identified as regulators of calcineurin by interacting with calcineurin and colocalizing with the z-disk protein, alpha-actinin (54). Cell culture experiments demonstrate that calsarcin-1 suppresses calcineurin activity in vitro. Mice deficient in calsarcin-1 showed that in vivo calcineurin activity and signaling are enhanced in striated muscle, indicating that the absence of calsarcin-1 relieves calcineurin inhibition (55). Consistent with the hypothesis that calcineurin activity promotes type I fibers, calsarcin-1 deficient mice show an increase in type I fibers. Through protein interactions, calsarcins serve to tether calcineurin to the sarcomere, placing it in proximity to a unique intracellular calcium pool where it can interact with specific upstream activators and downstream substrates. These findings identify the sarcomere as a site of regulation of the calcineurin/NFAT signaling pathway, via calsarcin-1, and implicate the sarcomere as an active modulator of myofiber remodeling at the level of gene transcription.

Calcium/Calmodulin-Dependent Protein Kinase, Protein Kinase C, and PKCmu/Protein Kinase D

Class II HDACs (HDAC4, HDAC5, HDAC7, and HDAC9) are highly expressed in skeletal muscle and directly bind MEF2, repressing expression of MEF2-dependent genes. It has been shown that binding of class II HDACs to MEF2 is mediated by 18 conserved amino acids in the amino-terminal extensions of class II HDACs, a domain that is lacking in class I HDACs (56). Phosphorylation of class II HDACs results in their export from the nucleus and activation of MEF2-dependent genes (22), leading to muscle remodeling. Because of the critical role of HDAC phosphorylation in regulating myocyte differentiation and remodeling, there has been intense interest in identifying the kinase(s) responsible for class II HDAC nuclear export and inactivation in vivo. In vitro studies have shown that signaling by calcium/calmodulin-dependent protein kinase (CaMK) results in phosphorylation of class II HDACs, promoting shuttling of HDACs from the nucleus to the cytoplasm and activation of MEF2 (22). Further evidence supporting the role of CaMK in skeletal muscle remodeling is seen when addition of a CaMK inhibitor, KN-62, blocks HDAC-green fluorescent protein (GFP) translocation from the nucleus to the cytoplasm in response to slow fiber-type electrical stimulation in isolated myofibers (29). In addition, CaMKII is known to be sensitive to the frequency of calcium oscillations (57) and is activated during hypertrophic growth and endurance adaptations (58). The notion that CaMK is involved in muscle remodeling is supported by ectopically overexpressing CaMKIV in skeletal muscle and observing an increase in type I fibers. However, CaMKIV is not expressed endogenously in skeletal muscle, and mice lacking CaMKIV have normal fiber-type composition with an increase in slow myosin heavy-chain isoform in the soleus muscle (59). Therefore, although exogenous CaMKIV promotes transformation of myofibers to a slow phenotype, it is unlikely that CaMKIV plays a role in physiological skeletal muscle remodeling. Furthermore on the basis of a biochemical assay for the HDAC kinase, there appears to be another HDAC kinase that is induced in response to calcineurin signaling (at least in the heart); this kinase is resistant to CaMK inhibitors and does not bind to calmodulin (60).

To further define the signaling pathways leading to the phosphorylation of class II HDACs, the potential of multiple kinase pathways to stimulate HDAC5 nuclear export was examined and showed that the protein kinase C (PKC) pathway promotes nuclear export of HDAC5 by stimulating phosphorylation of the 14-3-3 docking sites (61). Further studies showed that PKCmu/protein kinase D (PKD) acts as a downstream effector kinase of PKC and stimulates the nuclear export of HDAC5. On the basis of expression of PKD in skeletal muscle, in vitro studies, and transgenic mouse lines (M.S. Kim, R. Bassel-Duby, and E.N. Olson, unpublished data), we speculate that PKD is an important skeletal muscle HDAC kinase.

Exercise studies performed in humans showed that atypical PKC isoforms (aPKCzeta, -lambda, -mu), but not conventional PKC isoforms (cPKCalpha, -beta1, -beta2, and -delta), are activated by exercise in contracting muscle (61a, 61b). These findings are consistent with the transgenic mouse data showing a role for PKDmu/PKD in skeletal muscle remodeling and suggesting a potential role for atypical PKC in the regulation of skeletal muscle function and metabolism during exercise in both mice and humans.

Involvement of skeletal muscle signaling pathways is seen with other members of the PKC family. PKC-theta, a member of the novel PKC subfamily, is the predominant PKC isoform expressed in skeletal muscle (62, 63). In adult skeletal muscle, PKC-theta is expressed primarily in type II glycolytic fibers (64). Studies using lipid emulsion infusion in rats showed that activation of PKC-theta is associated with skeletal muscle insulin resistance (65, 66). Most recently, mice lacking PKC-theta were shown to be protected against fat-induced defects in skeletal muscle insulin signaling (67), indicating that PKC-theta is a crucial component mediating fat-induced insulin resistance in skeletal muscle.

Peroxisome Proliferator-Activated Receptor Delta and Peroxisome-Proliferator-Activated Receptor Gamma Coactivator-1 alpha

Enhanced oxidative capacity and metabolic efficiency of skeletal muscle is seen

PKC: protein kinase C

PKD: PKCmu/protein kinase D

PGC-1α: peroxisome-proliferator-activated receptor-gamma coactivator-1

PPAR: peroxisome proliferator-activated receptor

mTOR: mammalian target of rapamycin

IGF-1: insulin-like growth factor

following exercise training, in part owing to a dramatic increase in mitochondrial content resulting from changes in the expression of genes that increase mitochondrial biogenesis. The transcriptional coactivator peroxisome-proliferator-activated receptor-gamma coactivator-1 (PGC-1α) is considered a master regulator of mitochondrial gene expression and has been shown to activate mitochondrial biogenesis and oxidative metabolism (68–70). PGC-1α, expressed in brown fat and skeletal muscle, is preferentially enriched in type I myofibers. Studies performed in humans and rodents show that endurance exercise induces PGC-1α mRNA and protein expression (71–74). Skeletal muscle-specific overexpression of PGC-1α in transgenic mice resulted in an increase in type I fibers in white vastus and plantaris muscles (75). These transgenic mice also exhibited an increase in proteins involved in metabolic oxidation and, most importantly, displayed an increase in muscle performance and a decrease in muscle fatigue. Using fiber-type-specific promoters, it was shown that PGC-1α activates transcription in cooperation with MEF2 proteins and serves as a target for calcineurin signaling, which has been implicated in slow fiber gene expression. These findings indicate that PGC-1α is a principle factor modulating muscle fiber type and outline a combinatorial effect of activation of multiple signaling pathways evoked during skeletal muscle remodeling.

Peroxisome proliferator-activated receptor (PPAR) delta is a major transcriptional regulator of fat burning in adipose tissue through activation of enzymes associated with long-chain fatty-acid β-oxidation (76) and is the predominant PPAR isoform present in skeletal muscle. PPAR delta was overexpressed in skeletal muscle, resulting in a fiber-type switch to increase the number of oxidative myofibers (77), and the mice with activated PPAR delta showed an increase specifically in type I fibers and the ability to continuously run up to twice the distance of wild-type littermates (78). Because PPARs associate with PGC-1, it is conceivable that exercise induction of PGC-1α may activate PPAR delta and induce myofiber remodeling.

Ras/Mitogen-Activated Protein Kinase

High-intensity exercise (79) and electrostimulation (80) have been shown to activate the Ras/mitogen-activated protein kinase (MAPK) pathway. In vivo studies showed that Ras-dependent pathways affect both fiber size and fiber type (81). Introduction of exogenous MAPK-activating Ras (RasV12S35) into denervated regenerating muscle fibers induced the expression of type I myosin heavy chain but did not affect myofiber size. The Ras/MAPK pathway mediates the switch in a myosin heavy-chain gene induced by slow motor neurons in regenerating muscle. In contrast, activation of the PI3K/protein kinase B (Akt) pathway by Ras induces muscle growth but does not alter fiber-type distribution, corroborating the studies performed with overexpression of activated Akt in skeletal muscle (82).

Insulin-Like Growth Factor, Akt, and Mammalian Target of Rapamycin

As exemplified by the physique of a bodybuilder, skeletal muscle can adapt to workload by changing myofiber size. Studies using a functional overload model of the rat plantaris showed that the Akt/mammalian target of rapamycin (mTOR) signaling pathway is activated during hypertrophy (52), corroborating studies that showed hypertrophy of cultured myoblasts in response to insulin-like growth factor (IGF-1) to be dependent on a PI3K/Akt/mTOR pathway (83) (**Figure 4**). Transgenic mice overexpressing constitutively active Akt, specifically in skeletal muscle, showed an increase in muscle mass owing to an increase in muscle fiber size (84). Direct and indirect targets of Akt (also referred to as protein kinase B) include mTOR and glycogen synthase kinase 3.

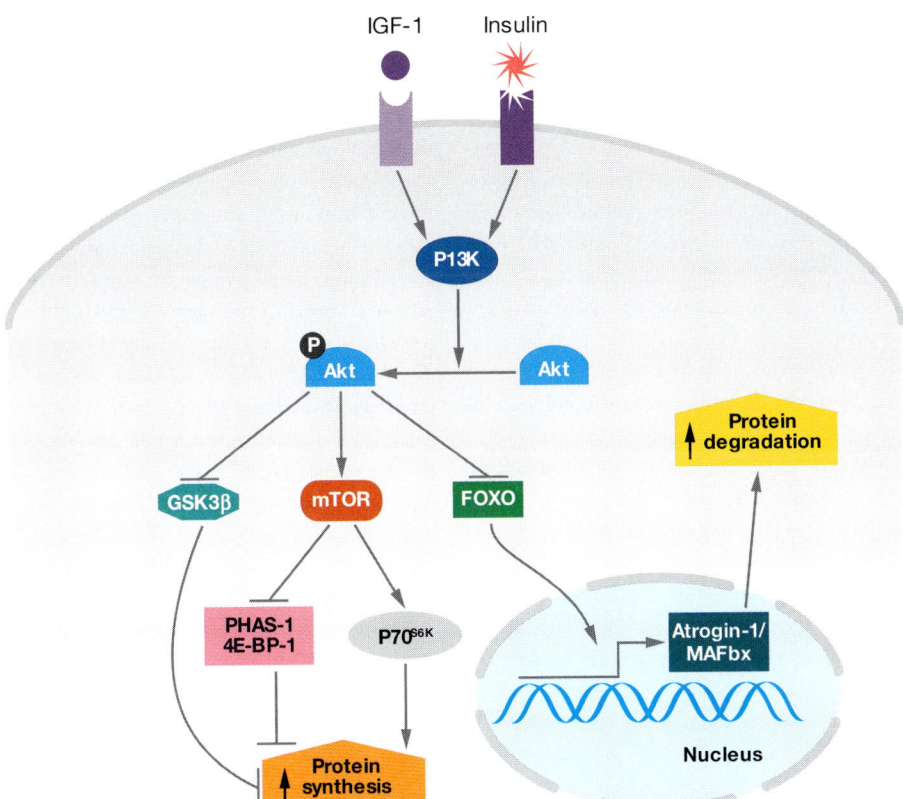

Figure 4
Signaling pathways in hypertrophy and atrophy. In response to IGF, the Akt/mTOR signaling pathway is activated. Phosphorylated Akt phosphorylates FOXO, inhibiting FOXO nuclear entry. Activation of mTOR by Akt promotes protein synthesis and increases muscle mass, resulting in hypertrophy. In disease states, Akt is not activated, and unphosphorylated FOXO enters the nucleus and induces the muscle atrophy F-box (MAFbx)/atrogin-1/ expression gene, promoting muscle atrophy.

mTOR is a kinase, sensitive to rapamycin, whose downstream targets, $p70^{S6K}$ and PHS-1/4E-BP1 increase protein translation initiation and elongation, promoting protein synthesis. Plantaris muscle from rats subjected to muscle overload and treated with rapamycin, an inhibitor of mTOR activity, showed similar activation of Akt in response to increased workload but did not show any change in myofiber size or weight, demonstrating that activation of mTOR is necessary for skeletal muscle hypertrophy.

CLINICAL SIGNIFICANCE

Understanding the signaling pathways that control myofiber remodeling is pertinent to several important human diseases, including inherited myopathies, systemic metabolic diseases, and common cardiovascular disorders. In muscular dystrophy, certain fibers are preferentially affected with degenerative changes; in diabetes, skeletal muscle contributes to exercise intolerance; and in heart failure patients, skeletal muscle atrophy is associated with a subgroup of patients at extremely high risk. The factors in signaling pathways in muscle remodeling may be viable therapeutic targets for the treatment of skeletal muscle disease.

Muscular Dystrophy

Duchenne muscle dystrophy (DMD) is a debilitating, life-threatening X-linked recessive muscular disorder, caused by mutations in the dystrophin gene. A strategy used to alleviate DMD involves upregulation of utrophin, an autosomal homolog of dystrophin. Activation of calcineurin stimulates the expression of utrophin through an NFAT site in the utrophin promoter (85). Moreover,

Skeletal muscle atrophy: a decrease in myofiber size, ultimately generating a decrease in total muscle mass

GLUT4: glucose transporter 4

overexpressing activated calcineurin in skeletal muscle of *mdx* mice, lacking the dystrophin gene, results in an increase in utrophin expression, an increase in oxidative fibers, and a decrease in pathology, suggesting that expression of exogenous calcineurin in skeletal muscles provides substantial beneficial effects on dystrophic muscle fibers (86). In addition, it was observed that in skeletal muscle of DMD patients, fast myofibers are preferentially affected with degenerative changes, whereas slow myofibers are relatively spared (87). Introduction of calcineurin in skeletal muscle not only activates utrophin expression but should also promote the formation of type I fibers, displacing the fast fibers that are more prone to damage. It will be of interest to examine whether the induction of slow type I fibers is sufficient to ameliorate DMD. It is encouraging that overexpressing IGF-1 within skeletal muscles reduces the severity of the dystrophy, demonstrating that modification of the myofiber has beneficial therapeutic effects in DMD (88).

Type 2 Diabetes Mellitus and Obesity

Skeletal muscle accounts for the majority of insulin-stimulated glucose uptake in humans and rodents. The insulin signaling pathway in skeletal muscle is controlled by a series of phosphorylation events linking initial activation of the insulin receptor to downstream substrates and ultimately translocating glucose transporter 4 (GLUT4) to the plasma membrane to bind and uptake glucose. A major contributing factor to the progressive development of type 2 diabetes is reduced insulin-stimulated whole-body glucose disposal, with the greatest defects attributed to skeletal muscle. Impaired insulin signal transduction (89) and defects in GLUT4 trafficking (90) are associated with skeletal muscle insulin resistance in individuals with type 2 diabetes. Fiber-type specific differences are seen in the insulin signal transduction pathway. In human skeletal muscle, insulin-stimulated glucose transport directly correlates with the percentage of slow-twitch muscle fibers, suggesting that a reduced skeletal muscle type I myofiber population may be one component of a multifactorial process involved in the development of insulin resistance (91). In fact, slow-twitch oxidative skeletal muscle has greater insulin binding capacity as well as increased insulin receptor kinase activity and autophosphorylation compared with fast-twitch glycolytic skeletal muscle (92). Furthermore, muscles with a greater percentage of oxidative myofibers have a higher content of GLUT4 (93). Overexpression of activated calcineurin in skeletal muscle of transgenic mice evokes an increase in type I myofibers and leads to improved insulin-stimulated glucose uptake (in association with increased expression of the insulin receptor, Akt, and GLUT4) compared to wild-type littermates (94). Interestingly, such mice are protected against glucose intolerance when maintained on a high-fat diet. These results validate calcineurin as a target to improve insulin signal transduction, enhance GLUT4 to correct glucose transport defects, and improve glucose homeostasis in diabetic individuals.

A non-insulin-dependent pathway regulating glucose transport and GLUT4 translocation to the plasma membrane and T-tubules in skeletal muscle involves AMPK a heterotrimeric protein that senses increases in AMP-to-ATP and creatine-to-phosphocreatine ratios via a mechanism that involves allosteric and phosphorylation modifications (95). AMPK is activated in skeletal muscle in response to exercise, phosphorylating target proteins along diverse metabolic pathways, resulting in an increase of ATP-generating pathways, such as glucose uptake and fatty-acid oxidation (96, 97). Studies using AICAR, a pharmacological activator of AMPK, in addition to transgenic overexpression of dominant-negative mutants of AMPK showed conclusively that AMPK activation increases skeletal muscle glucose transport by translocating GLUT4 to the membrane, comparable to the effect seen with exercise. These findings point to the AMPK pathway as a potential target for therapeutic strategies to restore metabolic balance to type 2 diabetic patients.

Differences in muscle fiber composition may also play a role in determining susceptibility to dietary obesity. Skeletal muscle in obese individuals exhibits reduced oxidative capacity, increased glycolytic capacity, and a decreased percentage of type I fibers (91, 98). PPARs comprise a family of nuclear hormone receptors that mediate the transcriptional effects of fatty acids and fatty-acid metabolites. Transgenic mice overexpressing PPAR delta in skeletal muscle exhibited an increase in oxidative myofibers and a reduction in adipocyte size (99). Treatment of mice with the PPAR delta agonist GW501516 promoted an increase in expression of genes involved in oxidative fibers and mitochondrial biogenesis (78), and transgenic mice overexpressing activated PPAR delta in skeletal muscle kept on a high-fat diet gained 50% less weight than wild-type littermates, implying that expression of PPAR delta in skeletal muscle has a protective role against obesity.

Muscle Atrophy

Muscle atrophy is defined as a decrease in myofiber size, ultimately generating a decrease in total muscle mass, resulting from disuse, disease, or injury. Sarcopenia is an age-related chronic loss of muscle and strength; and cachexia is a form of muscle atrophy associated with muscle disease or damage to the nerve associated with the muscle, commonly leading to severe muscle wasting. Atrophic myofibers have a smaller cross-sectional area than normal myofibers and generate a reduced force. However, they generally do not undergo apoptosis but retain most of the structural features of normal muscle. There is much interest in understanding the signaling pathways that mediate atrophy in order to design therapies to inhibit these pathways and ultimately to alleviate muscle atrophy. Gene expression profiling of muscles harvested from multiple atrophy mouse models identified two genes, *muscle ring finger* (*MuRF*)1 and *muscle atrophy F-box* (*MAFbx*)/*atrogin-1*, to be upregulated in atrophied muscle (100–102), and genetic deletion of these genes partially alleviated muscle atrophy. Both MuRF1 and atrogin-1/MAFbx proteins are E3 ubiquitin ligases responsible for the substrate specificity of ubiquitin conjugation as part of the ATP-dependent ubiquitin-proteosome proteolysis pathway involved in protein breakdown and degradation, which may conceivably result in a decrease of myofiber size. Interestingly, FOXO transcription factors, substrates of Akt, have been shown to induce atrogin-1/MAFbx expression (103, 104), connecting the molecular mediators of atrophy and the IGF-1/PI3K/Akt hypertrophy pathway (**Figure 4**). In the presence of IGF-1, PI3K/Akt is activated and phosphorylates FOXO, preventing it from entering the nucleus to activate atrophy-related genes. Skeletal muscle hypertrophy, following administration of IGF-1, is mediated by an increase in protein synthesis owing to Akt-induced phosphorylation, activation of mTOR, as well as a lack of MAFbx/atrogin-1 expression caused by Akt-induced phosphorylation of FOXO and nuclear exclusion. Muscle disuse leads to a reduction in PI3K/Akt activity and a decrease in FOXO phosphorylation, triggering nuclear import of FOXO and activation of the atrogin-1/MAFbx.

NF-κB, a mediator of cytokine tumor necrosis factor (TNF) alpha during the inflammatory response, is activated during muscular disuse. Myofibers treated with TNF plus interferon-gamma fail to maintain contractile activities and show significant reductions in both MyoD and myosin heavy-chain gene expression, suggesting NF-κB involvement in cachexia by suppression of muscle-specific gene expression (105). Other studies using two separate mouse models, one designed to activate NF-κB and the other to inhibit NF-κB activity selectively in skeletal muscle, demonstrated that activation of the NF-κB pathway is sufficient to induce severe skeletal atrophy, resembling cachexia (106).

MuRF: muscle ring finger

Interestingly, it was shown that activation of NF-κB in muscle promotes proteolysis, as evidenced by elevated MuRF1 transcripts and protein levels, but does not activate cytokine signaling (106). Blocking the NF-κB pathway was shown to ameliorate muscle atrophy, suggesting new drug targets for clinical intervention during cachexia and other skeletal muscle atrophies.

Short periods of myofiber dennervation and muscle disuse provoke muscle atrophy, which in certain cases is reversible, leading to the concept of compensatory mechanisms to sustain myofiber composition following limited episodes of inactivity. Expression of Runx1, a DNA-binding protein, is strongly induced following myofiber denervation (100, 107). Using mice lacking Runx1 specifically in skeletal muscle, it was shown that expression of Runx1 is required to sustain denervated muscles from undergoing autophagy and severe muscle wasting (108).

A novel transcription factor named Mus-TRD1 (muscle TFII-I repeat domain-containing protein 1) was isolated because of its ability to bind the enhancer region of the troponin I slow gene (109). There are 11 mouse MusTRD isoforms, and studies have shown that MusTRD1 can act as a repressor of the troponin I (TnI) slow enhancer. It is hypothesized that modulation of the Mus-TRD isoform content within muscle fibers provides a means of differentially regulating downstream target genes in muscles of different fiber composition.

Anabolic Steroids

A timely issue in muscle remodeling is the use of androgens as anabolic agents to increase skeletal muscle mass and reduce body fat. Testosterone effects on skeletal muscle mass are dose dependent, with administration of supraphysiological doses leading to a substantial increase in muscle size and strength. Androgen receptors reside in muscle cells and most likely mediate the response to androgens. Interestingly, studies determining the effects of testosterone on muscle performance showed that testosterone administration is associated with an increase in leg power and strength but showed no change in muscle fatigability and no change in specific tension, indicating that testosterone-induced gains in muscle strength are reflective of an increase of muscle mass (110). The increase in muscle mass is hypertrophic growth, as it is associated with an increase in myofiber cross-sectional area, observed both in type I and type II myofibers (111) and is not due to an increase in the number of myofibers. No significant transition of myofiber specificity is seen because the relative proportion of type I and type II fibers does not change after administration of testosterone. In addition, no change is observed in the number of fibers per unit of muscle; however, an increase in myonuclear number is apparent and is hypothesized to be attributable to fusion with satellite cells (111). A study showing the long-term effects (about 10 years) of anabolic steroids on high-level power-lifter athletes showed a larger myofiber area with more myonuclei per fiber and more centralized nuclei in athletes using steroids (112). In addition, no differences were seen with regard to fiber-type proportions; however, the type I fibers had 61% larger area, and the type II fibers had a 44% larger area than athletes without steroids.

CONCLUSIONS

We are advised by physicians, family members, and various government agencies to improve our health status by exercising. During exercise, the motor neuron is stimulated, resulting in activation of multiple signaling pathways in the myofibers and remodeling skeletal muscle to adapt to the physiological demand. Great strides have been made in animal models to understand the signaling pathways involved in muscle remodeling. However, whether these signaling pathways are physiologically valid in humans needs to be confirmed, and the identity of additional transcription factors and target genes remains to be determined. Using multiple approaches, it has been elegantly shown that various components of signaling pathways promote fiber-type transitions, and it will be challenging and informative to determine how these pathways intercalate and connect to remodel myofibers. Once further advances are made, we will depend on somatic cell delivery systems to target muscle fibers and provide components that have been shown to reduce the severity of muscular and metabolic diseases. In the future, exercising might mean taking a "pill" to activate skeletal muscle remodeling via signaling pathways. But for now, it is no pain, no gain. Keep on running! Remember, it is better to burn out than fade away.

SUMMARY POINTS

1. Skeletal muscle is comprised of a complex array of heterogeneous muscle fibers that differ in their physiological and metabolic parameters.
2. In response to environmental demands, skeletal muscle remodels and changes phenotypically in order to sustain muscle performance.
3. Skeletal muscle remodels by activating signaling pathways to reprogram gene expression.
4. Changes in calcium-dependent signaling pathways play key roles in regulating muscle growth and metabolism.
5. Genetic and pharmacological modulation of skeletal muscle signaling pathways offer therapeutic opportunities for the treatment of muscle diseases.

FUTURE ISSUES TO BE RESOLVED

1. Many of the defined signaling pathways in skeletal muscle remodeling have been determined using transgenic or knockout mouse models. Confirmation is needed to determine whether these signaling pathways are physiologically valid and are involved in humans.
2. Although many signaling pathways have been identified in remodeling skeletal muscle, it remains unclear how these pathways are initiated by the motor neuron and how the pathways are intercalated.
3. The identity of additional transcription factors and target genes that are involved in skeletal muscle remodeling remains to be determined.
4. Discovery of a skeletal muscle somatic cell delivery system is needed to target muscle fibers and provide components that have been shown to reduce the severity of muscular and metabolic diseases.

ACKNOWLEDGMENTS

We thank Alisha Tizenor and John M. Shelton for assistance with the figures.

LITERATURE CITED

1. Buller AJ, Eccles JC, Eccles RM. 1960. *J. Physiol.* 150:407–39
2. Buckingham M, Bajard L, Chang T, Daubas P, Hadchouel J, et al. 2003. *J. Anat.* 202:59–68

3. Charge SB, Rudnicki MA. 2004. *Physiol. Rev.* 84:209–38
4. Williams RS, Neufer PD. 1996. In *The Handbook of Physiology. Exercise: Regulation and Integration of Multiple Systems*, ed. LB Rowell, JT Shepherd, pp. 1124–50. Bethesda, MD: Am. Physiol. Soc.
5. Pette D, Staron RS. 2000. *Microsc. Res. Tech.* 50:500–9
6. Schiaffino S, Reggiani C. 1996. *Physiol. Rev.* 76:371–423
7. Stockdale FE. 1997. *Cell Struct. Funct.* 22:37–43
8. Garry DJ, Bassel-Duby RS, Richardson JA, Grayson J, Neufer PD, Williams RS. 1996. *Dev. Genet.* 19:146–56
9. Ogilvie RW, Feeback DL. 1990. *Stain Technol.* 65:231–41
10. Chin ER, Allen DG. 1996. *J. Physiol.* 491(Pt. 3):813–24
11. Hennig R, Lomo T. 1985. *Nature* 314:164–66
12. Westerblad H, Allen DG. 1991. *J. Gen. Physiol.* 98:615–35
13. Lomo T, Westgaard RH, Dahl HA. 1974. *Proc. R. Soc. London Ser. B* 187:99–103
14. Salmons S, Vrbova G. 1969. *J. Physiol.* 201:535–49
15. Berchtold MW, Brinkmeier H, Muntener M. 2000. *Physiol. Rev.* 80:1215–65
16. Murayama T, Ogawa Y. 2002. *Trends Cardiovasc. Med.* 12:305–11
17. Black BL, Olson EN. 1998. *Annu. Rev. Cell. Dev. Biol.* 14:167–96
18. Subramanian SV, Nadal-Ginard B. 1996. *Mech. Dev.* 57:103–12
19. Al-Khalili L, Chibalin AV, Yu M, Sjodin B, Nylen C, et al. 2004. *Am. J. Physiol. Cell Physiol.* 286:C1410–16
20. Naya FJ, Wu C, Richardson JA, Overbeek P, Olson EN. 1999. *Development* 126:2045–52
21. Lu J, McKinsey TA, Zhang CL, Olson EN. 2000. *Mol. Cell* 6:233–44
22. McKinsey TA, Zhang CL, Lu J, Olson EN. 2000. *Nature* 408:106–11
23. Miska EA, Karlsson C, Langley E, Nielsen SJ, Pines J, Kouzarides T. 1999. *EMBO J.* 18:5099–107
24. Wang AH, Bertos NR, Vezmar M, Pelletier N, Crosato M, et al. 1999. *Mol. Cell. Biol.* 19:7816–27
25. Lemercier C, Verdel A, Galloo B, Curtet S, Brocard MP, Khochbin S. 2000. *J. Biol. Chem.* 275:15594–99
26. McKinsey TA, Zhang CL, Olson EN. 2000. *Proc. Natl. Acad. Sci. USA* 97:14400–5
27. Kao HY, Verdel A, Tsai CC, Simon C, Juguilon H, Khochbin S. 2001. *J. Biol. Chem.* 276:47496–507
28. Wang AH, Kruhlak MJ, Wu J, Bertos NR, Vezmar M, et al. 2000. *Mol. Cell. Biol.* 20:6904–12
29. Liu Y, Randall WR, Schneider MF. 2005. *J. Cell Biol.* 168:887–97
30. Wu H, Rothermel B, Kanatous S, Rosenberg P, Naya FJ, et al. 2001. *EMBO J.* 20:6414–23
31. Wu H, Naya FJ, McKinsey TA, Mercer B, Shelton JM, et al. 2000. *EMBO J.* 19:1963–73
32. Chin ER, Olson EN, Richardson JA, Yang Q, Humphries C, et al. 1998. *Genes Dev.* 12:2499–509
33. Friday BB, Mitchell PO, Kegley KM, Pavlath GK. 2003. *Differentiation* 71:217–27
34. Nakagawa O, Arnold M, Nakagawa M, Hamada H, Shelton JM, et al. 2005. *Genes Dev.* 19:2066–77
35. Crabtree GR. 1999. *Cell* 96:611–14
36. Dolmetsch RE, Xu K, Lewis RS. 1998. *Nature* 392:933–36
37. Dolmetsch RE, Lewis RS, Goodnow CC, Healy JI. 1997. *Nature* 386:855–58
38. Rao A, Luo C, Hogan PG. 1997. *Annu. Rev. Immunol.* 15:707–47
39. Delling U, Tureckova J, Lim HW, De Windt LJ, Rotwein P, Molkentin JD. 2000. *Mol. Cell. Biol.* 20:6600–11

40. Naya FJ, Mercer B, Shelton J, Richardson JA, Williams RS, Olson EN. 2000. *J. Biol. Chem.* 275:4545–48
41. Parsons SA, Wilkins BJ, Bueno OF, Molkentin JD. 2003. *Mol. Cell. Biol.* 23:4331–43
42. Parsons SA, Millay DP, Wilkins BJ, Bueno OF, Tsika GL, et al. 2004. *J. Biol. Chem.* 279:26192–200
43. Bigard X, Sanchez H, Zoll J, Mateo P, Rousseau V, et al. 2000. *J. Biol. Chem.* 275:19653–60
44. Serrano AL, Murgia M, Pallafacchina G, Calabria E, Coniglio P, et al. 2001. *Proc. Natl. Acad. Sci. USA* 98:13108–13
45. Higginson J, Wackerhage H, Woods N, Schjerling P, Ratkevicius A, et al. 2002. *Pflügers Arch.* 445:437–43
46. McCullagh KJ, Calabria E, Pallafacchina G, Ciciliot S, Serrano AL, et al. 2004. *Proc. Natl. Acad. Sci. USA* 101:10590–95
47. Oh M, Rybkin II, Copeland V, Czubryt MP, Shelton JM, et al. 2005. *Mol. Cell. Biol.* 25:6629–38
48. Calvo S, Venepally P, Cheng J, Buonanno A. 1999. *Mol. Cell. Biol.* 19:515–25
49. Swoap SJ, Hunter RB, Stevenson EJ, Felton HM, Kansagra NV, et al. 2000. *Am. J. Physiol. Cell Physiol.* 279:C915–24
50. Horsley V, Friday BB, Matteson S, Kegley KM, Gephart J, Pavlath GK. 2001. *J. Cell Biol.* 153:329–38
51. Kegley KM, Gephart J, Warren GL, Pavlath GK. 2001. *Dev. Biol.* 232:115–26
52. Bodine SC, Stitt TN, Gonzalez M, Kline WO, Stover GL, et al. 2001. *Nat. Cell Biol.* 3:1014–19
53. Dupont-Versteegden EE, Knox M, Gurley CM, Houle JD, Peterson CA. 2002. *Am. J. Physiol. Cell Physiol.* 282:C1387–95
54. Frey N, Richardson JA, Olson EN. 2000. *Proc. Natl. Acad. Sci. USA* 97:14632–37
55. Frey N, Barrientos T, Shelton JM, Frank D, Rutten H, et al. 2004. *Nat. Med.* 10:1336–43
56. McKinsey TA, Zhang CL, Olson EN. 2001. *Curr. Opin. Genet. Dev.* 11:497–504
57. De Koninck P, Schulman H. 1998. *Science* 279:227–30
58. Chin ER. 2004. *Proc. Nutr. Soc.* 63:279–86
59. Akimoto T, Ribar TJ, Williams RS, Yan Z. 2004. *Am. J. Physiol. Cell Physiol.* 287:C1311–19
60. Zhang CL, McKinsey TA, Chang S, Antos CL, Hill JA, Olsen EN. 2000. *Cell* 110:479–88
61. Vega RB, Harrison BC, Meadows E, Roberts CR, Papst PJ, et al. 2004. *Mol. Cell. Biol.* 24:8374–85
61a. Rose AJ, Michell BJ, Kemp BE, Hargreaves M. 2004. *J. Physiol.* 561:861–70
61b. Perrini S, Henriksson J, Zierath JR, Widegren U. 2004. *Diabetes* 53:21–24
62. Chang JD, Xu Y, Raychowdhury MK, Ware JA. 1993. *J. Biol. Chem.* 268:14208–14
63. Osada S, Mizuno K, Saido TC, Suzuki K, Kuroki T, Ohno S. 1992. *Mol. Cell. Biol.* 12:3930–38
64. Donnelly R, Reed MJ, Azhar S, Reaven GM. 1994. *Endocrinology* 135:2369–74
65. Griffin ME, Marcucci MJ, Cline GW, Bell K, Barucci N, et al. 1999. *Diabetes* 48:1270–74
66. Yu C, Chen Y, Cline GW, Zhang D, Zong H, et al. 2002. *J. Biol. Chem.* 277:50230–36
67. Kim JK, Fillmore JJ, Sunshine MJ, Albrecht B, Higashimori T, et al. 2004. *J. Clin. Investig.* 114:823–27
68. Puigserver P, Wu Z, Park CW, Graves R, Wright M, Spiegelman BM. 1998. *Cell* 92:829–39
69. Wu Z, Puigserver P, Andersson U, Zhang C, Adelmant G, et al. 1999. *Cell* 98:115–24

70. Vega RB, Huss JM, Kelly DP. 2000. *Mol. Cell. Biol.* 20:1868–76
71. Akimoto T, Pohnert SC, Li P, Zhang M, Gumbs C, et al. 2005. *J. Biol. Chem.* 280:19587–93
72. Baar K, Wende AR, Jones TE, Marison M, Nolte LA, et al. 2002. *FASEB J.* 16:1879–86
73. Pilegaard H, Saltin B, Neufer PD. 2003. *J. Physiol.* 546:851–58
74. Terada S, Tabata I. 2004. *Am. J. Physiol. Endocrinol. Metab.* 286: E208–16
75. Lin J, Wu H, Tarr PT, Zhang CY, Wu Z, et al. 2002. *Nature* 418:797–801
76. Wang YX, Lee CH, Tiep S, Yu RT, Ham J, et al. 2003. *Cell* 113:159–70
77. Grimaldi PA. 2003. *Biochem. Soc. Trans.* 31:1130–32
78. Wang YX, Zhang CL, Yu RT, Cho HK, Nelson MC, et al. 2004. *PLoS Biol.* 2:e294
79. Goodyear LJ, Chang PY, Sherwood DJ, Dufresne SD, Moller DE. 1996. *Am. J. Physiol. Endocrinol. Metab.* 271:E403–8
80. Aronson D, Dufresne SD, Goodyear LJ. 1997. *J. Biol. Chem.* 272:25636–40
81. Murgia M, Serrano AL, Calabria E, Pallafacchina G, Lomo T, Schiaffino S. 2000. *Nat. Cell Biol.* 2:142–47
82. Pallafacchina G, Calabria E, Serrano AL, Kalhovde JM, Schiaffino S. 2002. *Proc. Natl. Acad. Sci. USA* 99:9213–18
83. Rommel C, Bodine SC, Clarke BA, Rossman R, Nunez L, et al. 2001. *Nat. Cell Biol.* 3:1009–13
84. Lai KM, Gonzalez M, Poueymirou WT, Kline WO, Na E, et al. 2004. *Mol. Cell. Biol.* 24:9295–304
85. Chakkalakal JV, Stocksley MA, Harrison MA, Angus LM, Deschenes-Furry J, et al. 2003. *Proc. Natl. Acad. Sci. USA* 100:7791–96
86. Chakkalakal JV, Harrison MA, Carbonetto S, Chin E, Michel RN, et al. 2004. *Hum. Mol. Genet.* 13:379–88
87. Webster C, Silberstein L, Hays AP, Blau HM. 1988. *Cell* 52:503–13
88. Shavlakadze T, White J, Hoh JF, Rosenthal N, Grounds MD. 2004. *Mol. Ther.* 10:829–43
89. Krook A, Bjornholm M, Galuska D, Jiang XJ, Fahlman R, et al. 2000. *Diabetes* 49:284–92
90. Ryder JW, Yang J, Galuska D, Rincon J, Bjornholm M, et al. 2000. *Diabetes* 49:647–54
91. Hickey MS, Carey JO, Azevedo JL, Houmard JA, Pories WJ, et al. 1995. *Am. J. Physiol. Endocrinol. Metab.* 268:E453–57
92. Song XM, Ryder JW, Kawano Y, Chibalin AV, Krook A, Zierath JR. 1999. *Am. J. Physiol. Regul. Integr. Comp. Physiol.* 277:R1690–96
93. Henriksen EJ, Bourey RE, Rodnick KJ, Koranyi L, Permutt MA, Holloszy JO. 1990. *Am. J. Physiol. Endocrinol. Metab.* 259:E593–98
94. Ryder JW, Bassel-Duby R, Olson EN, Zierath JR. 2003. *J. Biol. Chem.* 278:44298–304
95. Carling D. 2004. *Trends Biochem. Sci.* 29:18–24
96. Merrill GF, Kurth EJ, Hardie DG, Winder WW. 1997. *Am. J. Physiol. Endocrinol. Metab.* 273:E1107–12
97. Mu J, Brozinick JT Jr, Valladares O, Bucan M, Birnbaum MJ. 2001. *Mol. Cell* 7:1085–94
98. Tanner CJ, Barakat HA, Dohm GL, Pories WJ, MacDonald KG, et al. 2002. *Am. J. Physiol. Endocrinol. Metab.* 282:E1191–96
99. Luquet S, Lopez-Soriano J, Holst D, Fredenrich A, Melki J, et al. 2003. *FASEB J.* 17:2299–301
100. Bodine SC, Latres E, Baumhueter S, Lai VK, Nunez L, et al. 2001. *Science* 294:1704–8
101. Gomes MD, Lecker SH, Jagoe RT, Navon A, Goldberg AL. 2001. *Proc. Natl. Acad. Sci. USA* 98:14440–45
102. Stevenson EJ, Giresi PG, Koncarevic A, Kandarian SC. 2003. *J. Physiol.* 551:33–48

103. Latres E, Amini AR, Amini AA, Griffiths J, Martin FJ, et al. 2005. *J. Biol. Chem.* 280:2737–44
104. Sandri M, Sandri C, Gilbert A, Skurk C, Calabria E, et al. 2004. *Cell* 117:399–412
105. Guttridge DC, Mayo MW, Madrid LV, Wang CY, Baldwin AS Jr. 2000. *Science* 289:2363–66
106. Cai D, Frantz JD, Tawa NE Jr, Melendez PA, Oh BC, et al. 2004. *Cell* 119:285–98
107. Zhu X, Yeadon JE, Burden SJ. 1994. *Mol. Cell. Biol.* 14:8051–57
108. Wang XX, Blagden C, Fan JH, Nowak SJ, Taniuchi I, et al. 2005. *Genes Dev.* 19:1715–22
109. O'Mahoney JV, Guven KL, Lin J, Joya JE, Robinson CS, et al. 1998. *Mol. Cell. Biol.* 18:6641–52
110. Storer TW, Magliano L, Woodhouse L, Lee ML, Dzekov C, et al. 2003. *J. Clin. Endocrinol. Metab.* 88:1478–85
111. Sinha-Hikim I, Artaza J, Woodhouse L, Gonzalez-Cadavid N, Singh AB, et al. 2002. *Am. J. Physiol. Endocrinol. Metab.* 283:E154–64
112. Eriksson A, Kadi F, Malm C, Thornell LE. 2005. *Histochem. Cell Biol.* 124:167–75

RELATED REVIEWS

1. Lee SJ. 2004. *Annu. Rev. Cell Dev. Biol.* 20:61–86
2. Rennie MJ, Wackerhage H, Spangenburg EE, Booth FW. 2004. *Annu. Rev. Physiol.* 66:799–28
3. Pownall ME, Gustafsson MK, Emerson CP Jr. 2002. *Annu. Rev. Cell Dev. Biol.* 18:747–83
4. Clark KA, Mcelhinny AS, Beckerle MC, Gregorio CC. 2002. *Annu. Rev. Cell Dev. Biol.* 18:637–706
5. Geeves MA, Holmes KC. 1999. *Annu. Rev. Biochem.* 68:687–728

Biosynthesis and Assembly of Capsular Polysaccharides in *Escherichia coli*

Chris Whitfield

Department of Molecular and Cellular Biology, University of Guelph, Guelph, Ontario N1G 2W1, Canada; email: cwhitfie@uoguelph.ca

Key Words

cell-surface biogenesis, *trans*-envelope assembly complex, glycosyltransferases, polysaccharide export

Abstract

Capsules are protective structures on the surfaces of many bacteria. The remarkable structural diversity in capsular polysaccharides is illustrated by almost 80 capsular serotypes in *Escherichia coli*. Despite this variation, the range of strategies used for capsule biosynthesis and assembly is limited, and *E. coli* isolates provide critical prototypes for other bacterial species. Related pathways are also used for synthesis and export of other bacterial glycoconjugates and some enzymes/processes have counterparts in eukaryotes. In gram-negative bacteria, it is proposed that biosynthesis and translocation of capsular polysaccharides to the cell surface are temporally and spatially coupled by multiprotein complexes that span the cell envelope. These systems have an impact on both a general understanding of membrane trafficking in bacteria and on bacterial pathogenesis.

Contents

INTRODUCTION 40
STRUCTURES AND SURFACE ASSOCIATION OF *ESCHERICHIA COLI* CAPSULES 41
   Group 1 and 4 Capsules Are Related to LPS O Antigens 42
   Colanic Acid Is Related to Group 1 Capsules 44
   Group 2 and 3 Capsules 44
BIOSYNTHESIS AND ASSEMBLY OF CAPSULES BELONGING TO GROUPS 1 AND 4 45
   Genetic Organization of Group 1 *cps* Loci 45
   The Genetic Determinants for Group 4 Capsules and Wzy-Dependent O Antigens Are Allelic 47
   Regulatory Features Distinguish Expression of Colanic Acid from Group 1 K Antigens 47
   Biosynthesis of Group 1 and 4 Capsular Polysaccharides in a Wzy-Dependent Process 47
   Regulation of K_{LPS} Chain Length by Wzz 49
   Wzc Controls High-Level Polymerization of Group 1 Capsular Polysaccharides and Colanic Acid 50
   Wza-Dependent Translocation of Group 1 Capsular Polysaccharides Across the Outer Membrane 51
   Assembly of Group 4 Capsules Also Requires Wza, Wzb, and Wzc 52
   Wzi Is a Component Unique to Group 1 Capsules 53
BIOSYNTHESIS AND ASSEMBLY OF CAPSULES BELONGING TO GROUPS 2 AND 3 53
   Genetic Organization of Group 2 *kps* Loci 54
   Reorganization of the *kps* Locus in Isolates with Group 3 Capsules 55
   Chain Elongation of Group 2 Capsular Polymers 55
   Initiation Reactions and the Nature of the Endogenous Acceptor ... 56
   ABC Transporter-Dependent Export of Nascent Group 2 Capsular Polysaccharide 57
   Additional Proteins Contribute to a Complex that Couples Biosynthesis to Export 59
   KpsD and KpsE Mediate Translocation of Group 2 and 3 Capsular Polysaccharides to the Cell Surface 61
CONCLUDING REMARKS 62

INTRODUCTION

Cell-surface glycoconjugates play critical roles in interactions between bacteria and their immediate environment(s). Given that *Escherichia coli* isolates cause a range of infections and may have to withstand a transition between an animal host and a soil or water environment, it is perhaps not surprising that the surface architectures of these bacteria are diverse. *E. coli* isolates produce two serotype-specific surface polysaccharides: the lipopolysaccharide (LPS) O antigen and capsular polysaccharide K antigen. Variations in structures of these polysaccharides give rise to ~170 different O antigens and ~80 K antigens (1). Other polymers are not serotype specific. For example, most (if not all) isolates produce a polysaccharide known as enterobacterial common antigen (2), and many produce an extracellular polysaccharide called

K antigen: a major surface antigen used in *E. coli* serotyping, determined by capsular polysaccharide structure

colanic acid (or M antigen) under specific growth conditions (1) (see below). Two additional exopolymers have been identified more recently in *E. coli* because of their roles in cellular aggregation and the formation of biofilms on abiotic surfaces: the (1-4)-β-glucan bacterial cellulose (3) and a regulated (1-6)-β-GlcNAc polymer similar to a product made by staphylococci (4). The extent of their distribution within the species is unknown.

E. coli capsules are surface-enveloping structures comprising high-molecular-weight (capsular) polysaccharides that are firmly attached to the cell (**Figure 1**). They are well-established virulence factors, often acting by protecting the cell from opsonophagocytosis and complement-mediated killing (reviewed in 2, 5). The 80 different capsular serotypes in *E. coli* were originally divided into groups based on serological properties, and later revisions incorporated genetic and biochemical criteria (reviewed in 1, 5). The classification has since been expanded to four groups (6) (**Table 1**). *E. coli* group 1 and 4 capsules share a common assembly system, and this is fundamentally different from the one used for group 2 and 3 capsules.

Biosynthesis and assembly of capsular polysaccharides is a complex process. Activated precursors (nucleotide monophospho and diphospho sugars) in the cytoplasm are assembled into the nascent polysaccharide ($M_r > 100{,}000$) by enzymes associated with the inner membrane. A dedicated translocation pathway moves nascent polymer through the periplasm and across the outer membrane to the cell surface. Emerging evidence points to the existence of *trans*-envelope assembly complexes that coordinate the biosynthesis of polymer with the export and translocation steps in both space and time. This may provide continuity between the cytoplasm and outer surface of the cell at the site of synthesis, reducing the problems associated with crossing a cell envelope comprising two different membranes, a periplasm, and the peptidoglycan layer (**Figure 1**). Despite the diversity in bacterial glycoconjugates, bacteria use a limited repertoire of biosynthesis and assembly strategies, and *E. coli* capsules have proved to be influential prototypes. The purpose of this review is to provide a contemporary overview of the model systems.

STRUCTURES AND SURFACE ASSOCIATION OF *ESCHERICHIA COLI* CAPSULES

The major groups in the early capsule classification systems were distinguished by physical properties, including retention of the masking K antigen after heating cell suspensions (reviewed in 1, 5). The thermostability or thermolability properties of K antigens reflect differences in the means by which the capsule is linked to the cell surface. Despite extensive studies describing the repeat-unit structure of

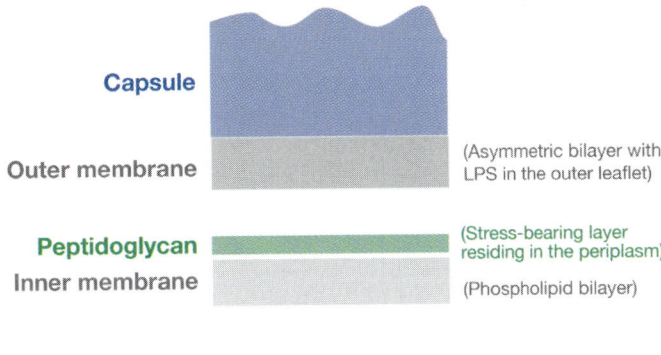
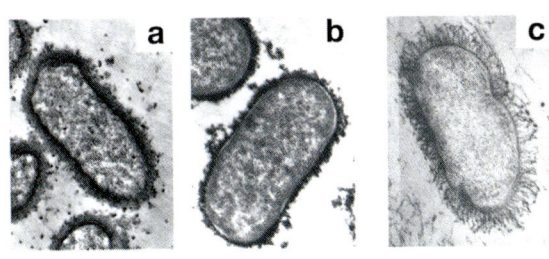

Figure 1

Electron micrographs of encapsulated *E. coli*. The schematic on top shows the organization of the cell envelope in gram-negative bacteria. The micrographs show the results of different procedures used to visualize highly hydrated capsule structures. Panel *a* shows serotype K1 (group 2) with the capsule structure preserved (or stabilized) using antibodies specific for the capsule (132). Panels *b* and *c* show serotype K30 (group 1) labeled with cationized ferritin (133) or after freeze substitution (134). Image in panel *c* courtesy of R. Harris and T.J. Beveridge.

Capsule: the surface layer on many bacteria, usually formed from capsular polysaccharide

Table 1 Classification of *Escherichia coli* capsules incorporating features of the biosynthesis and assembly systems

Characteristic	Group 1	Group 2	Group 3	Group 4
Former K-antigen group	IA	II	I/II or III	IB (O-antigen capsules)
Thermostability of K antigen	Yes	No	No	Yes
Coexpressed with O serogroups	Limited range (O8, O9, O20, O101)	Many	Many	Often O8, O9 but occasionally none
Coexpressed with colanic acid	No	Yes	Yes	Yes
Genetic locus	*cps* near *his*	*kps* near *ser A*	*kps* near *ser A*	Near *his*
Thermoregulated expression	No	Yes	No	No
Elevated levels of CMP-Kdo synthetase	No	Yes	No	No
Terminal lipid moiety	LPS lipid A core in K_{LPS}; unknown for capsular K antigen	α-glycerophosphate	α-glycerophosphate?	LPS lipid A core in K_{LPS}; unknown for capsular K antigen
Polymer chain grows at	Reducing terminus	Nonreducing terminus	Nonreducing terminus?	Reducing terminus
Polymerization system	Wzy dependent	Processive glycosyltransferase activity	Processive glycosyltransferase activity?	Wzy dependent
PST-1 protein	Wzx	None	None	Wzx
ABC transporter	None	KpsMT	KpsMT?	None
MPA-1 protein	Wzc	None	None	Wzc
MPA-2 protein	None	KpsE	KpsE?	None
OMA protein	Wza	KpsD	KpsD?	Wza
Model system(s)	Serotype K30	Serotypes K1, K5	Serotypes K10, K54	Serotypes K40, O111
Similar to capsules in	*Klebsiella, Erwinia*	*Neisseria, Haemophilus*	*Neisseria, Haemophilus*	None known

K antigens, it is perhaps surprising that the precise linkage(s) of capsules to the cell surface is (are) still not fully resolved in all *E. coli* groups.

Group 1 and 4 Capsules Are Related to LPS O Antigens

Group 1 and 4 capsules are found in *E. coli* isolates that cause intestinal infections, including representatives of enteropathogenic (EPEC), enterotoxigenic (ETEC) and enterohemorrhagic (EHEC) *E. coli*. Group 1 capsules are acidic polysaccharides, typically containing uronic acids, and tend to be rather similar in structure (**Figure 2**). Similar (and occasionally identical) capsules are found in *Klebsiella pneumoniae*. Group 4 capsule structures are more diverse and are distinguished from those in group 1 by the presence of acetamido sugars in their repeat-unit structures.

Group 1 and 4 K antigens are expressed on the cell surface in two forms. One form is linked to a LPS lipid A core and is termed K_{LPS} to distinguish it from LPS molecules carrying the serological O antigen in the same isolate. In group 4 capsule producers, K_{LPS} contains long chains of K antigen, whereas

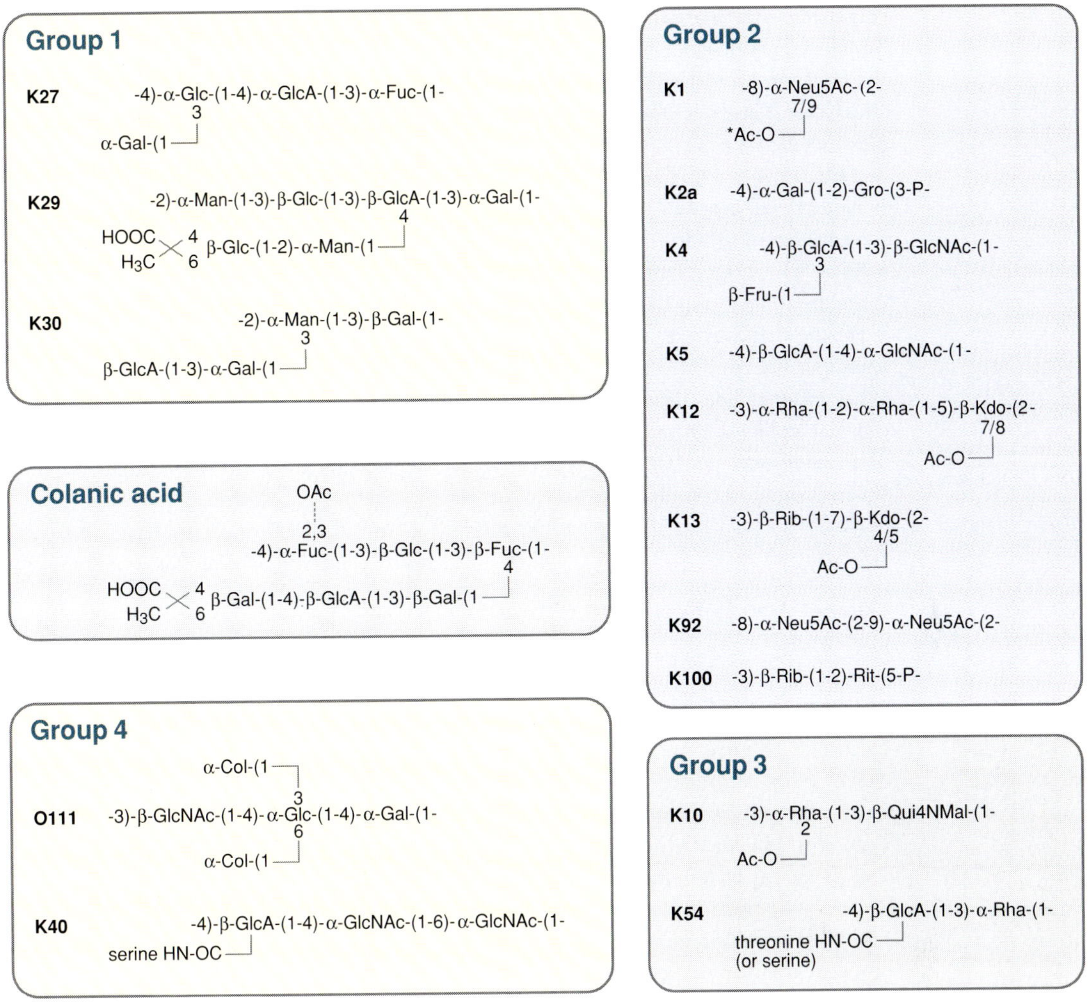

Group 1

K27   -4)-α-Glc-(1-4)-α-GlcA-(1-3)-α-Fuc-(1-
                                           3
        α-Gal-(1⏌

K29   -2)-α-Man-(1-3)-β-Glc-(1-3)-β-GlcA-(1-3)-α-Gal-(1-
                                                       4
     HOOC⟩⟨4 β-Glc-(1-2)-α-Man-(1⏌
     H$_3$C   6

K30   -2)-α-Man-(1-3)-β-Gal-(1-
                                             3
    β-GlcA-(1-3)-α-Gal-(1⏌

Group 2

K1   -8)-α-Neu5Ac-(2-
                    7/9
       *Ac-O⏌

K2a   -4)-α-Gal-(1-2)-Gro-(3-P-

K4   -4)-β-GlcA-(1-3)-β-GlcNAc-(1-
                              3
       β-Fru-(1⏌

K5   -4)-β-GlcA-(1-4)-α-GlcNAc-(1-

K12   -3)-α-Rha-(1-2)-α-Rha-(1-5)-β-Kdo-(2-
                                                   7/8
                           Ac-O⏌

K13   -3)-β-Rib-(1-7)-β-Kdo-(2-
                                   4/5
           Ac-O⏌

K92   -8)-α-Neu5Ac-(2-9)-α-Neu5Ac-(2-

K100   -3)-β-Rib-(1-2)-Rit-(5-P-

Colanic acid

                          OAc
                         ⋮
                         2,3
     -4)-α-Fuc-(1-3)-β-Glc-(1-3)-β-Fuc-(1-
                                               4
   HOOC⟩⟨4 β-Gal-(1-4)-β-GlcA-(1-3)-β-Gal-(1⏌
   H$_3$C   6

Group 4

             α-Col-(1⏋
                        3
O111   -3)-β-GlcNAc-(1-4)-α-Glc-(1-4)-α-Gal-(1-
                                   6
             α-Col-(1⏌

K40   -4)-β-GlcA-(1-4)-α-GlcNAc-(1-6)-α-GlcNAc-(1-
       serine HN-OC⏌

Group 3

K10   -3)-α-Rha-(1-3)-β-Qui4NMal-(1-
                             2
      Ac-O⏌

K54   -4)-β-GlcA-(1-3)-α-Rha-(1-
    threonine HN-OC⏌
    (or serine)

Figure 2

Repeat-unit structures of representative *E. coli* capsules and colanic acid exopolysaccharide. (Qui4NMal, 4-(2-carboxyacetamido)-4,6-dideoxyglucose; Col, colitose, 3,6-dideoxygalactose.) The asterisk in the K1 structure denotes form-variable (on-off) O-acetylation, and the dashed line in the colanic structure represents nonstoichiometric O-acetylation. The structures have been published elsewhere: K27 (135), K29 (136), K30 (137), K1 (138), K2a (139), K4 (140), K5 (141), K12 (142), K13 (143), K92 (144), K100 (145), K10 (146), K54 (147), O111 (148), K40 (149), and colanic acid (150).

group 1 K$_{LPS}$ is limited to a short oligosaccharide containing only one or a few K-antigen repeat units. The capsule evident in electron micrographs (**Figure 1**) is comprised of high-molecular-weight capsular K antigen, and interactions between the polysaccharide chains on the cell surface create a higher-order structure (7). Although the precise linkage between the capsular K antigen and the cell surface has not been established, LPS is not involved. Furthermore, there is no precursor-product relationship between the K$_{LPS}$ and high-molecular-weight capsular forms (8–10); capsule assembly for groups 1 and 4 requires a separate dedicated translocation system that is not used by K$_{LPS}$.

Given the structural similarity between group 4 K$_{LPS}$ and LPS O antigens, it is not surprising that the serology is often confusing. Some isolates produce a group 4 capsule as the only serotype-specific polysaccharide, in which case it is given O-antigen status (examples include O26, O55, O100, O111, O113, and O127) (8, 9, 11). L. Lieve and colleagues first described these as "O-antigen capsules" in their studies of O111 (8). Other group 4 capsules have K-antigen status (e.g., K40) (12) because they (like group 1 K antigens) are found in isolates that coexpress an additional neutral O antigen (i.e., one of O8, O9, O9a, O20, and O101 group). The distribution between the capsular and K_{LPS} forms and the involvement of an additional neutral O antigen in some isolates point to a complex interplay in surface polymers that may be critical in pathogenicity.

Colanic Acid Is Related to Group 1 Capsules

Colanic acid production is widespread in *E. coli* isolates. Its structure resembles group 1 capsules (**Figure 2**), and they are assembled by essentially identical processes. In contrast to the authentic serotype-specific group 1 capsules, a substantial amount of the colanic acid produced by a culture is secreted into the growth medium as an exopolysaccharide. Whether this is due to physico-chemical properties of the polymer itself or a subtle difference in biosynthesis and assembly has not been established (13). The most obvious feature distinguishing colanic acid from group 1 capsules is the absence of colanic acid production in wild-type isolates grown at 37°C on typical lab media. This is due to complex transcriptional regulation of the colanic acid biosynthesis locus. As might be expected, colanic acid has no known role in virulence, and the biological role for colanic acid lies primarily in the lifestyle of *E. coli* outside the host (reviewed in 14). Colanic acid biosynthesis genes are part of an extensive regulon responding to alterations in (or damage to) cell envelope structure, osmotic shock, and growth on surfaces. This exopolysaccharide is essential for the later stages of *E. coli* K-12 biofilm development on abiotic surfaces and is important for withstanding desiccation.

Group 2 and 3 Capsules

Group 2 and 3 capsules are found in *E. coli* isolates that cause extraintestinal infections. The structural features and components of the repeat units of group 2 and 3 capsules vary extensively (**Figure 2**). Some (e.g., K2a and K100) contain phosphate residues in their backbone structures and are reminiscent of gram-positive teichoic acids. Several group 2 capsular polysaccharides resemble vertebrate glycoconjugates. Examples include the K1 antigen [α-(2-8)-linked polysialic acid], K4 (a substituted chondroitin backbone), and K5 (an N-acetylheparosan backbone) (5). These K antigens occur in isolates causing significant extraintestinal infections, and their inability to elicit a strong and protective antibody response limits effective vaccination strategies (reviewed in 5).

In terms of structure and assembly, *E. coli* group 2 and 3 capsules are reminiscent of capsules in *Neisseria meningitidis* and *Haemophilus influenzae*. A unifying structural theme in the related *E. coli* and meningococcal capsules is the presence of diacylglycerophosphate at the reducing terminus of a proportion of the polymer isolated from cultures (15). In the polysialic acid capsules of *E. coli* serotype K92 and meningococcal serotype B, the available evidence indicates that the lipid is linked directly to the reducing terminal Neu5Ac residue via a phosphodiester bridge (15). Unfortunately, an unequivocal resolution of the linkage structure was not possible. The same lipid moiety was later found in several other *E. coli* group 2 and 3 capsules, but in contrast to K92, preliminary evidence was reported for a 3-deoxy-D-*manno*-2-octulosonic acid (Kdo) residue located between the lipid and the reducing terminal sugar of the K12 and K82 polymer chains (16). Again, primary data with a definitive structure for the linkage region have not been published. The proposed Kdo-containing linkage appears throughout the group 2 capsule literature and is supported by additional biochemical data. However, it is puzzling why the linkage would differ in some serotypes, and the structure(s) of the termini need to be revisited because this uncertainty compromises interpretation of data for critical biosynthetic steps (see below). The lipid terminus is thought to anchor the capsule to the cell surface, but only 20% to 50% of the isolated polymer has the phospholipid substitution (17). This may reflect lability of the phosphodiester linkage, but it raises questions about the extent and integrity of surface association. It has been suggested that nonlipidated polymer may be retained at the surface via ionic and other interactions.

BIOSYNTHESIS AND ASSEMBLY OF CAPSULES BELONGING TO GROUPS 1 AND 4

The sequences of genetic loci for group 1 and 4 capsules have identified important features shared with many LPS O-antigen biosynthesis loci. They all map (entirely or in part) to a polymorphic chromosomal region near the *his* (histidine-biosynthesis) operon. The genes contributing to capsule expression in groups 1 and 4 encode several conserved proteins (**Table 2**), reflecting their common biosynthesis pathways, as well as additional proteins that determine the unique repeat-unit structure of each K antigen. Variations in the organization of the genetic loci, and the distribution of critical genes in more than one chromosomal region, distinguish the capsule biosynthesis systems in groups 1 and 4.

Genetic Organization of Group 1 *cps* Loci

The group 1 capsule biosynthesis locus (*cps*) comprises two regions (**Figure 3**) separated by a putative stem-loop transcriptional attenuator (18). The same locus is found in *Klebsiella* sp., presumably as a result of horizontal gene transfer (19). The 5′ part of the locus contains four conserved genes (*wzi*, *wza*, *wzb*, and *wzc*) present in all group 1 *cps* loci. Three of the four gene products (Wza, Wzb, and Wzc) are involved in polymerization control and translocation of the product from the inner membrane to the cell surface. They operate independently of capsule structure. The 3′ region of the locus is serotype specific and encodes enzymes for a Wzy-dependent biosynthesis system. These include enzymes for producing any sugar nucleotide precursors dedicated to capsule synthesis, glycosyltransferases (GTs), and two integral inner membrane proteins (Wzy and Wzx). Although the specific gene products are dictated by the serotype (and repeat-unit structure of the resulting polysaccharide), genes encoding

GT: glycosyltransferase

Wzy: an integral inner membrane protein, required for polymerization of und-PP-linked repeat units in Wzy-dependent synthesis

Wzx: an integral inner membrane protein, required for export of und-PP-linked repeat units in Wzy-dependent synthesis

Table 2 Conserved proteins involved in the biosynthesis of colanic acid and capsules belonging to groups 1 and 4

Protein	Protein family	Location	Function (or putative function)
Wzx	PST-1	Inner membrane (integral)	Transfers nascent undecaprenyl diphosphate-linked repeat units across the inner membrane
Wzy		Inner membrane (integral) with periplasmic catalytic site	Putative polymerase; assembles undecaprenyl diphosphate-linked polymers using lipid-linked repeat units exported by Wzx
Wzc	MPA-1 (PCP-2a)	Inner membrane (integral) with a large periplasmic domain and cytosolic N and C termini	Participates in high-level polymerization of capsular polysaccharide and forms part of a *trans*-envelope capsule translocation complex; Wzc activity is determined by cycling of its phosphorylation state via the cytosolic C-terminal tyrosine autokinase domain.
Wzb	PTP	Cytoplasm	Protein tyrosine phosphatase; dephosphorylates Wzc
Wza	OMA	Outer membrane	Forms a multimeric putative translocation channel and interacts with the periplasmic domain of Wzc

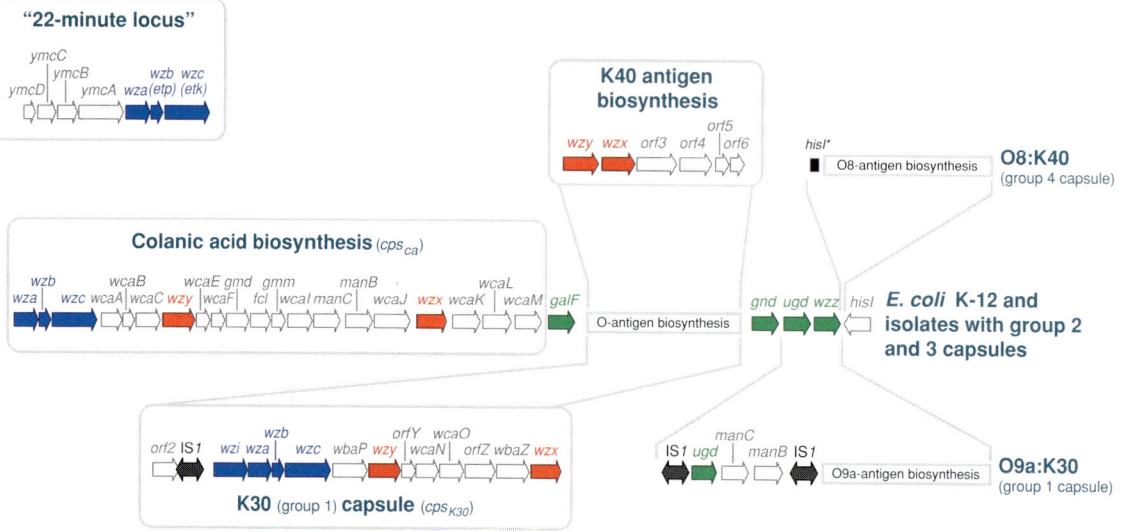

Figure 3

Organization of the genes required for expression of colanic acid and capsules belonging to groups 1 and 4. The central part of the figure shows the conserved chromosomal *his*-linked region in *E. coli* K-12 and isolates with group 2 and 3 capsules. These isolates are able to produce colanic acid. This region has undergone substantial rearrangement in the group 1 prototype (serotype O9a:K30) with insertion/replacements introducing genes for expression of both O and K antigens. Group 1-producing isolates are unable to produce colanic acid, and the locus may have been lost during the chromosomal rearrangements. The genes for group 4 capsule biosynthesis are found in a typical O-antigen biosynthesis locus, and the colanic acid locus is intact in isolates producing group 4 capsules. An additional locus at a region corresponding to 22-minutes on the *E. coli* chromosome is essential for group 4 capsule expression and duplicates some genes in the group 1 *cps* locus. The characteristic Wzx and Wzy genes are highlighted in red, and known genes involved in regulation of high-level polymerization and translocation are in blue. "Housekeeping" genes (*green*) are identified as reference points.

characteristic Wzx and Wzy homologs define the pathway and are always present.

For all intents and purposes, the 3′ region is identical in gene content to loci found in many bacteria with LPS O antigens synthesized via Wzy-dependent pathways (20). Furthermore, the group 1 *cps* locus effectively occupies the same location (between *galF* and *gnd*) on the genome as the LPS O-antigen biosynthesis locus in *E. coli* K-12 and in isolates with group 2, 3, and 4 capsules (**Figure 3**). It is the presence of the *wzi*, *wza*, *wzb*, and *wzc* genes (**Table 2**) that distinguishes the loci for Wzy-dependent group 1 capsules from those for O antigens. These conserved genes form a surface assembly system devoted to capsular K antigen and do not participate in assembly of K_{LPS}. An unlinked copy of the *wza*, *wzb*, and *wzc* genes is found in an operon at a location corresponding to 22 minutes on the *E. coli* K-12 chromosome (**Figure 3**).

Transcription of the *cps* locus is driven from a constitutive promoter upstream of *wzi* (18) and involves an RfaH-dependent antitermination mechanism common to several genetic loci, including many polysaccharide biosynthesis operons (21). RfaH is recruited by an 8-nucleotide *ops* (operon polarity suppressor) sequence located at the 5′ end of the transcript and interacts with RNA polymerase to favor transcript elongation (reviewed in 22). In the absence of RfaH, transcriptional read-through past the stem-loop transcriptional attenuator in *cps* is diminished, and capsule production is significantly reduced (18).

The Genetic Determinants for Group 4 Capsules and Wzy-Dependent O Antigens Are Allelic

Group 4 capsule biosynthesis in serotype K40 (12) (**Figure 3**) and O111 (23) involves a gene locus indistinguishable from those responsible for expression of many different Wzy-dependent O antigens, and genes encoding Wza, Wzb, and Wzc are absent from the *his*-linked locus. Instead, these functions are contributed by close homologs in the "22-minute locus" (**Figure 3**), and all seven genes in this transcriptional unit are required for group 4 capsule assembly in serotype O127 (11). Thus, although the organization of genes required for expression of group 1 and 4 capsules differs, the principal gene products are conserved, and the overall features of the biosynthesis and assembly pathways are predicted to be essentially the same.

Regulatory Features Distinguish Expression of Colanic Acid from Group 1 K Antigens

The locus for colanic acid biosynthesis is located upstream of *galF* in *E. coli* K-12 and in isolates with group 2, 3, and 4 capsules (24) (**Figure 3**). The organization of this locus is, in most respects, identical to the group 1 capsule locus. The 5′ part of the colanic acid locus contains highly conserved homologs of *wza*, *wzb*, and *wzc* (note the absence of *wzi*) and is separated from the biosynthesis region containing *wzx* and *wzy* by a predicted stem-loop transcriptional attenuator. Isolates with group 1 capsules are unable to produce colanic acid (25), and the colanic acid genes may be absent as a result of past genetic rearrangement in these regions. However, colanic acid expression can be induced in isolates with group 2, 3, and 4 capsules (25, 26).

The complex transcriptional regulation of colanic acid production is controlled by the Rcs (regulation of capsule synthesis) proteins. The Rcs system is a complex phosphorelay system that is now known to extend well beyond colanic acid regulation (reviewed in 14). The Rcs transcriptome encodes proteins targeted to the envelope or involved in envelope modifications (such as colanic acid formation). They include cell-envelope proteins induced by shock and osmostress conditions as well as others associated with swarming behavior and biofilm formation. The Rcs system is integrated into other cellular regulatory circuits and may regulate surface remodeling in *E. coli* in response to a change in lifestyle.

Biosynthesis of Group 1 and 4 Capsular Polysaccharides in a Wzy-Dependent Process

Much of our current knowledge of the Wzy-dependent pathway results from studies on LPS O antigens, particularly those of *Salmonella enterica* (reviewed in 20, 27) (**Figure 4**).

Group 1 capsular polysaccharides are assembled on a carrier lipid comprising the C_{55}-polyisoprenoid lipid derivative, undecaprenyl phosphate (und-P). The general features of this reaction series, and the identity and involvement of und-P, were first established in the group 1 capsule representative in *Klebsiella* (28). Initiation of the *E. coli* serotype K30 prototype requires the WbaP enzyme and involves the reversible transfer of Gal-1-P from UDP-galactose to und-P (29). WbaP is a member of a family of polyisoprenyl-phosphate hexose-1-phosphate transferases (27) that initiate O antigen and capsule biosynthesis in many bacteria by transfer of Gal-1-P, or Glc-1-P, to und-P (20, 27, 30). In contrast, group 4 capsule initiation involves transfer of GlcNAc-1-P by WecA, a representative of the polyisoprenyl-phosphate N-acetylhexosamine-1-phosphate transferase family that includes both prokaryotic and eukaryotic proteins (reviewed in 31). The *wecA* gene is encoded by the locus for enterobacterial common antigen biosynthesis, and although it was first characterized in this context, WecA is required for biosynthesis

> **Undecaprenyl phosphate (und-P):** this C_{55} polyisoprenoid lipid derivative serves as a carrier for assembly of surface polysaccharides in *E. coli*

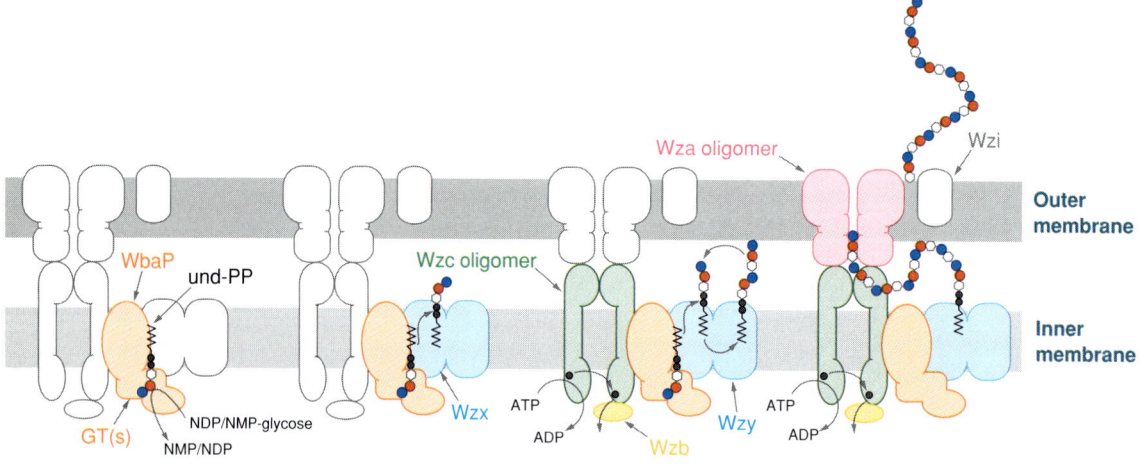

Figure 4

A model for biosynthesis and assembly of group 1 and 4 capsules. Beginning at the left, und-PP-linked repeat units are assembled at the interface between the cytoplasm and the inner membrane. Newly synthesized und-PP-linked repeats are then flipped across the membrane in a process requiring Wzx. This provides the substrates for Wzy-dependent polymerization wherein the polymer grows by transfer of the growing chain to the incoming und-PP-linked repeat unit. Continued polymerization requires transphosphorylation of C-terminal tyrosine residues in the Wzc oligomer and dephosphorylation by the Wzb phosphatase. Polymer is translocated by Wza, which likely acts as a channel. Wzi is unique to group 1 capsules and appears to be involved in modulating surface association.

of many *E. coli* O antigens. WecA and WbaP both contain several transmembrane segments, and this may be related to their need to interact with an obligatory lipid acceptor. The putative catalytic site of the *Salmonella* O antigen WbaP and its homologs is located in the C-terminal cytoplasmic domain (32, 33). Although the assignment of WbaP and WecA as initiating transferases is consistent with all available data, neither enzyme has been studied in purified form. Completion of the und-PP-linked repeat unit is catalyzed by a series of peripheral monofunctional GTs that transfer additional glycoses to the lipid intermediate.

Wzy-dependent polymerization was first described in classic experiments involving the assembly of the *Salmonella* O antigens (reviewed in 20, 27). The donors for the polymerization reaction are und-PP-linked repeat units, and polymer elongation involves transfer of the nascent polymer from its und-PP carrier to the new lipid-linked repeat unit, effectively increasing chain length in a blockwise manner by adding new repeat units at the reducing terminus. The available evidence for the *Salmonella* O antigens suggests the same lipid (und-PP) is used throughout, but this has not been confirmed by definitive structural studies for longer-chain intermediates; such analyses are technically challenging owing to the complexity of the molecules and their low abundance. In serotype K30 (group 1) (29) and K40 (group 4) (12), *wzy* mutants lack capsules and add only a single K-antigen repeat unit to the lipid A core in K_{LPS}. These results provided the first proof that the capsular and K_{LPS} forms of group 1 and 4 K antigens share common repeat-unit donors and polymerization machinery. Wzy is an integral membrane protein containing ~12 transmembrane segments and a large periplasmic loop (34). The catalytic mechanism of Wzy is unknown, and it is important to remember that no Wzy homolog has been purified and studied at a biochemical level to directly confirm the widely assumed "polymerase" activity.

Wzy-dependent polymerization occurs in the periplasm (35), and as a consequence, the assembly pathway is dependent on the export of lipid-linked repeat units across the inner membrane (**Figure 4**). In O-antigen biosynthesis systems, this process involves the *wzx* gene product (36, 37), and the same is likely true in the capsule biosynthesis systems. Wzx homologs are integral membrane proteins with multiple predicted transmembrane segments, and sequence similarities define a family of putative polysaccharide-specific transport (PST) proteins, designated PST-1 for the capsule-assembly homologs and PST-2 for those involved in O-antigen synthesis (38). However, the sequences do not provide any insight into the biochemical activity of Wzx. Although the Wzx homolog from the related enterobacterial common antigen biosynthesis system is required for transmembrane flipping of a water-soluble isoprenyl-PP-GlcNAc derivative in vesicles (39), it is not certain whether Wzx is the only component required in the process. Indeed, the N-terminal transmembrane domain of WbaP influences export of some *Salmonella* O antigens (32). It has been proposed that Wzx proteins may have specificity for the initial sugar in the lipid-linked repeat unit, perhaps via recognition of WbaP or WecA (40). The interaction may create a scaffold required for forming the lipid-linked repeat unit and then for releasing it to the export pathway (reviewed in 27). Undecaprenol has also been implicated as a scaffold for organizing proteins including GTs (41), and the lipid: protein complexes may alter the biophysical properties of the local membrane environment, perhaps aiding the flipping activity (42). The export process is a complex one, and the system may be correspondingly complicated.

Given that the polymer is apparently elongated in an undecaprenol-linked form, there is a requirement for its release into the translocation pathway once an appropriate chain length has been achieved. This process could be a side reaction of either Wzy or WaaL. Wzy must have the capability to release polymer from the lipid carrier during polymerization. The WaaL protein performs essentially the same cleavage step when it transfers nascent glycans to the lipid A core (e.g., in the formation of K_{LPS}) (reviewed in 20, 27). The Wzy and WaaL enzymes share motifs in a periplasmic loop and may share a similar reaction mechanism (43). In either event, the process must be regulated in some manner to ensure the appropriate chain length is produced, and this could be dictated by protein-protein interactions in the context of a larger assembly complex. These processes are currently under investigation.

Regulation of K_{LPS} Chain Length by Wzz

Polymerization of Wzy-dependent LPS O antigens and K_{LPS} is terminated by transfer of the polymer (or oligosaccharide) from the lipid intermediate to lipid A-core acceptor. The reaction is catalyzed by WaaL. The extent of heterogeneity of the O-antigen chain lengths is dictated by the O-antigen chain-length determinant, the Wzz protein (previously designated as Cld or Rol) (reviewed in 20, 44). In the absence of Wzz activity, short unregulated O-antigen chains are formed, rather than a characteristic cluster of modal lengths. The *wzz* gene is located near the *his* locus in *E. coli* K-12 and isolates with group 2 and 3 capsules (**Figure 3**). The difference in K_{LPS} chain lengths in groups 1 and 4 is simply due to the absence of *wzz* in isolates with group 1 capsules (45, 46). In fact, modality and a longer chain length can be imparted on group 1 K_{LPS} by introduction of a heterologous *wzz* gene (45).

Wzz proteins have a characteristic membrane topology and are grouped in a family called polysaccharide copolymerase-1 (PCP-1) (47). These proteins have two transmembrane helices flanking a periplasmic domain. The periplasmic domain is predicted to form coiled-coil structure and is implicated in determining chain-length modality. However the mechanism of action of Wzz proteins is unknown.

Wzz: an inner-membrane PCP-1 protein, which participates in regulating chain length in Wzy-dependent biosynthesis of LPS-linked glycans

PCP: polysaccharide copolymerase protein family

MPA-1 protein (aka PCP-2a, e.g., Wzc): it participates in synthesis and assembly of high-molecular-weight group 1 and 4 capsular polysaccharides

Wzc Controls High-Level Polymerization of Group 1 Capsular Polysaccharides and Colanic Acid

The characterization of Wzz has raised the question of how the capsular K-antigen polymerization is controlled. Interest in the characteristic Wzc protein as a candidate chain-length regulator for group 1 capsular polysaccharides was stimulated by its similarity to Wzz in terms of predicted membrane topology (38, 44). *E. coli* K30 mutants with *wzc* defects are unable to make detectable amounts of capsular K antigen but are still able to polymerize K_{LPS} (29, 48). Wzc proteins are grouped in the inner (cytoplasmic) membrane-periplasmic auxiliary-1 (MPA-1) (38) or PCP-2a (47) families, whose members are involved in a growing number of related capsular and exopolysaccharide systems in gram-positive and gram-negative bacteria. Wzc is distinguished from Wzz by its possession of a C-terminal cytoplasmic domain harboring ATP-binding motifs (Walker A and B) as well as a tyrosine-rich region (7 of the last 17 residues in *E. coli* K30). Work in A.J. Cozzone's laboratory (49) established that Wzc homologs are tyrosine autokinases that phosphorylate at multiple residues at the expense of ATP and are dephosphorylated by cognate phosphatases. In *E. coli*, Wzb is the protein tyrosine phosphatase (PTP family). Wzc phosphorylation involves a transphosphorylation process (48, 50). Isolated Wzc shows heterogeneity in phosphorylation (A. Reid and C. Whitfield, unpublished results), and mutational analyses support a model wherein phosphorylation load on the C-terminal tyrosines (rather than any specific residue) is important for its function in capsule assembly (51). A minimum of four modifiable tyrosine residues are required in the Wzc C terminus for its competence in capsule assembly. The observation that *wzb* and *wzc* mutations both result in an acapsular phenotype led to the hypothesis that Wzc function requires its cycling between phosphorylated and nonphosphorylated forms (48). The Wzc-Wzb proteins are highly conserved in group 1 capsules and colanic acid biosynthesis systems. Although there is evidence suggesting differential effects of phosphorylated Wzc in these systems (52), the homologs do function in the same manner when examined in the context of the group 1 K30 capsule-assembly systems (13).

Wzc proteins are known to oligomerize independent of phosphorylation (51, 53). Using cryo-electron microscopy with single-particle analysis, a structure of the Wzc tetramer has been resolved at 14 Å (54) (**Figure 5**). The structure does not support the proposed role of the C-terminal kinase domain in oligomerization (53), but it is consistent with the participation of the periplasmic domain in oligomerization of Wzz proteins (55). The exact contribution of the putative periplasmic coiled-coil motifs to the Wzc oligomerization is now being assessed.

The precise function of Wzc is still unknown, as is that of Wzz. One possibility is

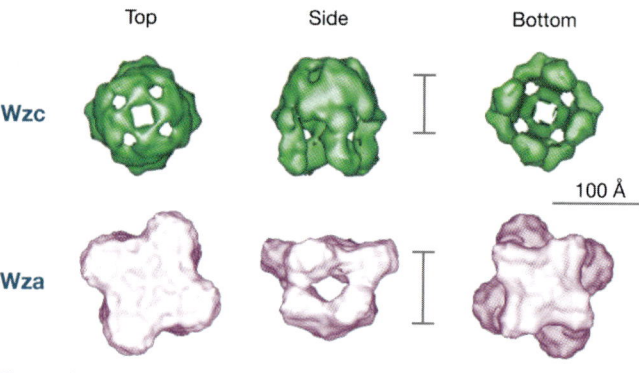

Figure 5
Surface-rendered three-dimensional structures of Wzc and Wza at 14 Å and 15.5 Å, respectively. The structures are derived from single-particle analysis of negatively stained samples in cryo-transmission electron microscopy (cryo-TEM). Wzc forms a tetramer with C4 rotational symmetry (54). The upper domain contains a region of connected density ~20-Å thick around periphery of the structure. Four unconnected roots extend from the upper ring. These contain the N terminus and each can accommodate one C-terminal kinase domain based on molecular modeling. Wza is an octameric structure that adopts a tetrameric symmetry (59). This generates a structure containing two stacked rings with the upper one being slightly larger in diameter. A small elliptical cavity in the center (~40-Å diameter) is enclosed by four symmetrical arms. The cavity is closed at the top and bottom of the structure. The vertical bars indicate reference sizes for the inner and outer membranes measured in cryo-TEM images of frozen hydrated samples (131).

that these proteins really are "copolymerases" and interact with Wzy. However, in preliminary studies, *wzc* mutants also result in dramatic reduction in the initiating Gal-1-P (WbaP) transferase activity (A. Reid and C. Whitfield, unpublished results), and the acapsular phenotype could be due to dramatically reduced flow in the polymer biosynthesis reactions. In this respect, Wzc might interact directly with WbaP to modulate its catalytic activity, or alternatively, Wzc might play a more structural role by coordinating the organization of an efficient multienzyme complex including WbaP. It is tempting to speculate that Wzc and Wzz have similar functions in polymer biosynthesis, but why does Wzc have the additional complexity of the C-terminal autokinase domain? The answer may lie in additional interactions involving Wzc and outer-membrane components in the assembly system (see below).

Wza-Dependent Translocation of Group 1 Capsular Polysaccharides Across the Outer Membrane

Wza is essential for surface assembly of group 1 capsules (56, 57). The Wza protein is a member of a family of outer-membrane auxiliary (OMA) proteins associated with capsule and exopolysaccharide assembly systems (38). The Wza homologs encoded by the group 1 capsule and colanic acid loci are highly conserved and functionally exchangeable (13).

Wza is a lipoprotein that forms sodium dodecyl sulfate-stable multimeric ring-like structures (56) resembling the "secretins" associated with filamentous phage assembly and protein secretion through type II, III and IV systems in a range of gram-negative bacteria (58). Secretins exist as large channels formed by multimeric complexes of 10 or more monomers, and the structure of Wza in two-dimensional electron crystallography revealed an octameric organization (57). A three-dimensional structure with 15.5-Å resolution was obtained from single-particle analysis and showed an arrangement reflecting a tetrameric symmetry in which the protein complex, encloses a central cavity (59) (**Figure 5**). No channel "openings" are apparent at the presumed external and periplasmic faces, and it is conceivable that isolated Wza oligomers adopt a closed state because large permanently open channels would compromise outer-membrane integrity. The open-closed state could be dictated by essential interactions with additional components in the assembly complex (see below). The best-characterized export channel in *E. coli* is TolC, which is involved in drug efflux and type 1 protein secretion (reviewed in 60). TolC is a trimer with an extensive α-helical periplasmic domain that opens the channel via an iris-like conformational transition. OMA proteins are clearly different in overall structure and, presumably, in function. Crystals of Wza oligomers diffracting to 3 Å have been generated (61), and a high-resolution structure will offer critical insight into capsule biogenesis and may be very informative for some protein secretion systems.

The conclusion that Wza provides the secretin for group 1 capsules is complicated by similar acapsular phenotypes and loss of high-molecular-weight polymerization in both *wza* and *wzc* mutants (56). This suggests some interactions between the outer- and inner-membrane components in a tightly coordinated process coupling biosynthesis to translocation. Strains expressing a nonacylated derivative of Wza are acapsular, but unlike *wza*-null mutants, they accumulate periplasmic polymer (57). The nonacylated Wza localizes to the outer membrane but forms unstable oligomers. These may be recognized by the rest of the complex so biosynthesis is maintained but cannot support translocation and effectively uncouple the coordination. Further evidence for a *trans*-envelope complex is provided by the interaction of Wzc and Wza identified in cross-linking experiments (57). Phosphorylation of Wzc does not dictate interaction but could potentially modulate conformation of the complex to open or close the

Outer-membrane auxiliary (OMA) protein: by oligomerization it forms the putative secretin for translocation of polymer

TYPE 4 SECRETION SYSTEMS

Type 4 secretion systems (T4SSs) are involved in DNA transfer by conjugation, DNA uptake, and the transfer of DNA and effector proteins to eukaryotic target cells. In *Agrobacterium* spp, a T4SS is responsible for the delivery of both effector proteins and single-stranded DNA molecules (T-DNA) into plant cells. T4SSs provide an interesting parallel to capsule assembly because both must overcome the challenge of moving a high-molecular-weight anionic polymer across the bacterial cell envelope. In T4SSs, this is achieved by a *trans*-envelope complex comprised of 12 or more different proteins, some present in multiple copies. The complex provides a coordinated conduit from the cytoplasm to the cell exterior. The components include a hetero-oligomeric component associated with the outer membrane that is linked to proteins spanning the periplasm and interacting with those in the inner membrane. The complex also contains at least three ATP-binding proteins required for recruitment of the DNA substrate and its ATP-hydrolysis-dependent transfer to the inner-membrane channel. They may play roles in assembly of a functional complex. The organization of the complex and the protein-protein interactions within it are being dissected by systematic genetic and biochemical approaches, and insight into the DNA translocation pathway is emerging.

translocation channel. The recent isolation of higher-order structures containing Wza and Wzc for cryo-EM structural studies (R.F. Collins, K. Beis, R.C. Ford, J.H. Naismith, and C. Whitfield, unpublished results) will provide a critical next step in resolving the structure and function of the assembly system. There are some interesting parallels in terms of the requirement for ATP-binding proteins in this system and *trans*-envelope type IV secretion systems (T4SS) involved in translocation of DNA and effector proteins (reviewed in 61a). The T4SS seems to involve more components than the group 1 capsule system, but both translocate a hydrophilic polymer via a multiprotein complex that includes an oligomeric outer membrane secretin.

The biochemical and structural evidence for an envelope-spanning multienzyme complex brings into context classic early studies on sites of capsule translocation carried out by M.E. Bayer. Translocation of group 1 (K29) capsule occurs at a limited number of sites on the cell surface that coincide with domains where the inner and outer membranes come into close apposition (62). The interpretation of these "zones of adhesion" has been controversial because of the need for specific preparation techniques for their visualization (reviewed in 63), but there are increasing examples in the literature of cell envelope-spanning multienzyme complexes for export of proteins and drugs. A coordinated capsule assembly complex (as depicted in **Figure 4**) would provide a physical and functional connection between the cell surface and the polymerization machinery in the inner membrane and would overcome the practical problem of transferring high-molecular-weight capsular polymers ($M_r > 100,000$) to the surface in multiple steps.

Assembly of Group 4 Capsules also Requires Wza, Wzb, and Wzc

Analysis of Wza, Wzb, and Wzc functions in *E. coli* K30 (group 1) has been complicated by the additional copies of *wza*, *wzb*, and *wzc* in the 22-minute locus (**Figure 3**). These genes are only functional in certain backgrounds, including EPEC, ETEC, and EHEC isolates (11, 64). The kinase and phosphatase activities of Wzc (also known as Etk) and Wzb (Etp) have been confirmed, and they have been shown to participate with low efficiency in production of K30 capsular polysaccharide and colanic acid (48, 52). The Wza homolog encoded by the 22-minute locus can also function in group 1 capsule assembly (56). Given the overall similarity of group 1 and 4 polymer biosynthesis, it seemed logical that homologs of Wza, Wzb, and Wzc would be involved in group 4 capsule assembly, and this has been confirmed (11). Perhaps more interesting is the finding that the locus contains a single transcriptional unit including the four additional genes (*ymcABCD*) required for group 4 capsule expression. All four are predicted to be exported proteins, and YmcA and YmcC may be lipoproteins. Database searches are not informative as to function, although they do identify hypothetical proteins from polysaccharide systems in other bacteria. Paralogs of *ymcABCD* (*yjbHGFE*) are found elsewhere on the *E. coli* chromosome (11). Given the common theme in assembly of group 1 and 4 capsules, it is interesting that neither set of paralogs is essential for group 1 K30 capsule formation (A.N. Reid and C. Whitfield, unpublished data). It is conceivable that other (unidentified) genes fulfill the same role in group 1 systems; otherwise this may represent a point of divergence between the group 1 and group 4 assembly pathways.

Wzi Is a Component Unique to Group 1 Capsules

The Wzi protein is a heat-modifiable monomeric β-barrel protein that plays a role in the final stages of capsule assembly (65). Wzi is the only component of the group 1 *cps* locus that is not essential for capsule biosynthesis and assembly. It is also not found in the colanic acid and group 4 capsule systems. Mutants lacking Wzi show a significant reduction in surface-associated capsule and a corresponding increase in cell-free polymer. Although the exact mode of linkage of group 1 capsules is still unknown, the mutant phenotype is consistent with Wzi playing a role (direct or indirect) in surface attachment of the capsule. It is striking that *wzi* is confined to those systems wherein the polymer product is tightly associated with the cell surface in a discrete capsular structure.

BIOSYNTHESIS AND ASSEMBLY OF CAPSULES BELONGING TO GROUPS 2 AND 3

The biosynthesis of group 2 and 3 capsules is performed by proteins encoded by *kps* loci located near *serA*. Like the group 1 and 4 capsule loci, group 2 and 3 loci encode several conserved proteins (**Table 3**) that serve as diagnostic markers for their biosynthesis system. The group 2 and 3 loci differ in genetic organization and by regulatory features that

Table 3 Conserved proteins involved in the biosynthesis of groups 2 and 3 capsules

Protein	Protein family or homolog	Location	Function
KpsF	YrbH	Cytoplasm	Homolog of arabinose-5-phosphate epimerase; involved in CMP-Kdo biosynthesis
KpsU	KdsB	Cytoplasm	Homolog of CMP-Kdo synthetase; involved in CMP-Kdo biosynthesis
KpsC		Cytoplasm	Precise function not established but essential for capsule export
KpsS		Cytoplasm	Precise function not established but essential for capsule export
Wzm	ABC-A2 (CPSE) transporter TMD	Inner membrane (integral)	Transmembrane domain component of the ABC transporter; exports nascent polymer across the inner membrane
Wzt	ABC-A2 (CPSE) transporter NBD	Inner membrane (peripheral)	Nucleotide-binding domain component of the ABC transporter; exports nascent polymer across the inner membrane
KpsE	MPA-2	Periplasm (associated with outer face of inner membrane)	Putative membrane-fusion (or adaptor) protein; couples ABC transporter to later translocation steps
KpsD	Putative OMA	Outer membrane	Candidate for capsular polysaccharide translocation channel; requires KpsE for its proper localization

ABC transporter: a transporter superfamily containing representatives that export group 2 and 3 capsular polysaccharides

facilitate the characteristic thermoregulation of group 2 capsules.

Genetic Organization of Group 2 *kps* Loci

In 1981, R.P. Silver and colleagues (66) reported the first cloning and expression of a capsule gene cluster from any bacterium, with their studies on *E. coli* group 2 serotype K1 genetics. The chromosomal loci for group 2 capsule expression (designated *kps*) have a conserved structure comprising three regions (reviewed in 6, 67, 68) (**Figure 6**). The serotype-specific central region (region 2) encodes GTs and any specialized sugar nucleotide synthetases required for capsule biosynthesis. The size and gene content of region 2 is therefore serotype specific and varies according to the complexity of the repeat-unit structure of the polymer formed. The absence of *wzx* and *wzy* reflects the fundamental differences in the biosynthesis mechanism, compared to group 1 and 4 capsules.

Region 2 is flanked by genes whose products are conserved in different serotypes. They act independently of the structure of the capsular polysaccharide and are involved in a range of activities encompassing export and assembly of the capsule on the cell surface. The completed polymer is exported across the inner membrane by an ATP-binding-cassette (ABC) transporter comprised of the region 3 gene products (KpsMT), but there are

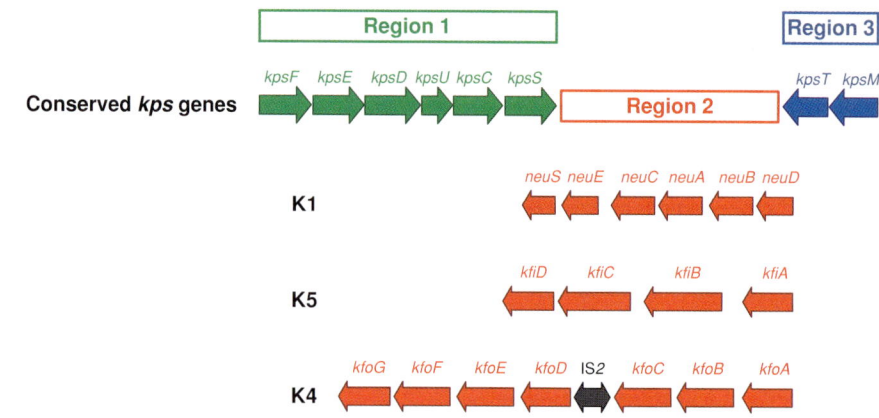

Figure 6

Organization of the genes required for expression of capsules belonging to groups 2 and 3. The group 2 *kps* locus comprises a serotype-specific region 2 flanked by two regions (1 and 3) conserved across group 2 serotypes. Region 3 encodes the ATP-binding-cassette (ABC) transporter, and region 1 gene products are involved in polymer export and translocation. Region 2 genes encode enzymes for polymer biosynthesis, and the complexity of this region corresponds to the repeat-unit structures (**Figure 2**). Group 3 loci contain region 3 genes and some (but not all) region 1 genes, and the loci show extensive evidence of rearrangement.

unanswered questions concerning the functions of some of the six proteins encoded by region 1 (KpsFEDUCS).

Transcription from a σ^{70}-dependent promoter (or promoters) upstream of *kpsF* yields a polycistronic message that covers the complete region 1 (69, 70). The region 3 promoter generates a large transcript that reads through into region 2 (71) and is dependent on RfaH-mediated antitermination (71), as in the transcription of group 1 capsule loci. The region 1 and 3 promoters are transcriptionally silent at 20°C (69, 70, 72). This thermoregulation is a defining feature of group 2 capsules, and although a detailed understanding of the process is not yet available, current information points to a complex and multifactorial system (72).

In at least one case, genes outside *kps* also influence the repeat-unit structure. The K1 antigen is subject to form variation dictated by the presence or absence of O-acetylation (**Figure 2**). The O-acetyltransferase (NeuO) is part of a lysogenic bacteriophage-like element that is unlinked to *kps* (73). Form variation involves the on-off modulation of *neuO* expression via slipped-strand DNA mispairing.

Reorganization of the *kps* Locus in Isolates with Group 3 Capsules

Information for group 3 capsule loci is limited to partial sequences for serotypes K10 and K54 (reviewed in 68). These group 3 clusters have an organization with a central serotype-specific domain flanked by some (but not all) of the characteristic conserved genes from group 2 (*kpsMTEDSC*). However, their positions and relative order differ from group 2, suggesting recombination events have occurred with a locus that is allelic in isolates with group 2 and 3 capsules (74). Direct information for group 3 capsule biosynthesis and assembly is scarce, and the system is largely inferred from conservation of genes encoding critical export and translocation functions and complementation studies exploiting mutants in group 2 systems (74). A striking feature differentiating expression of group 2 and 3 capsules is that the group 3 loci lack the characteristic thermoregulation and are produced at all growth temperatures.

Chain Elongation of Group 2 Capsular Polymers

Considerable progress has been made in understanding individual group 2 capsule GTs by exploiting the ability of these enzymes to extend exogenous polymeric and oligosaccharide acceptors. GTs expressed in the absence of the remaining capsule biosynthesis machinery and in the absence of initiation show in vitro chain extension activity with an appropriate acceptor. As an added incentive, several representative systems contain GTs with biotechnological relevance. For example, the poly-α-2–8-NeuNAc sialyltransferase from *E. coli* K1 has the capacity to generate engineered polysialylgangliosides (75), and the glycosaminoglycan backbones of the K4 and K5 polymers provide templates for chemical and enzymatic modification to generate products of biomedical importance (76).

The polysialyltransferases from serotypes K1 and K92 represent perhaps the best characterized of these GTs. Research over several decades beginning with the work in the laboratories of Roseman and Troy, and extended by Vimr, Vann, and others, has culminated in the identification of NeuS as a processive GT that transfers Neu5Ac residues from CMP-Neu5Ac to the nonreducing terminus of the nascent glycan (reviewed in 77, 78). The poly-α-2,8-sialyltransferase (NeuS) from K1 elongates both exogenous and endogenous acceptors in vitro (79, 80). The same is true of the highly conserved K92 NeuS homolog that generates a polymer with alternating α-2,8/α-2,9 linkages in serotype K92 (81, 82). In a *neuS*-null mutant of *E. coli* K1, the K92 NeuS enzyme forms its cognate serotype-specific product with alternating α-2,8/α-2,9 linkages (80–82). The collective data indicate that NeuS enzymes are the sole determinants of serotype specificity, and the K92 NeuS enzyme has dual linkage specificity. The high degree of conservation in NeuS proteins has allowed the generation of informative chimeras to establish regions responsible for linkage specificity (82), but the mechanism of processive transferases is unknown. In fact, in the absence of a solved structure of a representative with defined processive GT activity, even the mechanism for an enzyme as well studied as cellulose synthase is still contentious (reviewed in 83).

The NeuS enzyme from K1 elongates oligosialyl exogenous acceptors (79, 80), with most evidence pointing to a preference for a tetramer or larger (79). Interestingly, soluble oligomers are typically poor acceptors for processive extension by K1 and K92 NeuS enzymes, but sialylgangliosides and acceptors with a terminal hydrophobic aglycone are efficiently elongated (75, 81, 84). The reason for the enhancing effect of the lipid terminus is unknown.

Polymer elongation at the nonreducing terminus is a conserved feature in group 2 capsules, with comparable processes shown for the K4 (85) and K5 (86, 87) polysaccharides. The processive bifunctional chondroitin polymerase (KfoC), from serotype K4 has both β-GalNAc and β-GlcA GT activities, and in contrast to NeuS, it efficiently elongates soluble tetra- and hexasaccharide acceptors to generate a high-molecular-weight product (85, 88). The extension of the K5 oligosaccharide acceptors (87) requires the action of two separate GT activities, but unlike the K4 system, these are found in separate polypeptides. These two K5 GT activities were initially both assigned to KfiC (89), but subsequent studies identify KfiA as the α-GlcNAc GT, with KfiC being the β-GlcA GT (90). In an interesting confirmation of these GT assignments, sequences similar to both KfiA and KfiC are present in the single bifunctional heparosan synthase from *Pasteurella multocida* serotype D (91).

Form variation in O-acetylation of K1 polysialic acid indicates that this modification is not essential for polymerization. The O-acetyltransferase utilizes acetyl coenzyme A as the donor and can modify larger oligosialyl acceptors [$>$(Neu5Ac)$_{14}$] in vitro (92). The enzyme also modifies polysialic acid in an export-deficient mutant, indicating the reaction occurs in the cytoplasm (73). Along similar lines, chain elongation (and chondroitin formation) by KfoC can occur independent of the addition of side chain fructose residues in K4. Moreover, the observation that fructosylated acceptors cannot serve as exogenous acceptors lends additional support to the conclusion that side chain addition may even occur postpolymerization (85).

The termination of an efficient processive enzyme activity represents an equally interesting unresolved problem. In *E. coli* K1, the majority of the chains fall within a reasonably narrow size range with a maximum chain length of 160–230 residues (93), suggesting an active process in size determination. In some LPS O antigens assembled by processive GTs, chain termination and coupling to an ABC transporter is achieved by enzymes that add novel residues to the nonreducing terminus of the nascent glycan (94). A terminal residue on group 2 capsules would potentially be easily overlooked in structural analyses, but there are no obvious candidates for equivalent terminating enzymes encoded by the *kps* locus. Alternative possibilities for chain termination include loss of affinity of the GT for the polymer beyond a certain chain length, an abortive chain translocation process within the catalytic site similar to that proposed for chain termination in the type 3 polysaccharide synthase from *Streptococcus pneumoniae* (95), or an allosteric effect mediated by other components of the assembly system as proposed elsewhere (78). Regardless, the actual process is expected to have a significant impact on virulence and remains an important area for further study.

Initiation Reactions and the Nature of the Endogenous Acceptor

Although NeuS enzymes are sufficient for elongation, they are unable to initiate polysialic acid formation, and de novo synthesis requires other proteins from the *kps* locus (79–82). This indicates a requirement for an "initiase" enzyme, but identification of the players and characterization of the process are currently limited because the nature of the endogenous acceptor for group 2 capsule biosynthesis is equivocal. An early analysis of K1 biosynthesis led to the identification of und-P-Neu5Ac as a potential intermediate, suggesting a process resembling other bacterial glycoconjugates (96). However, lipid-linked polysialic acid has not been identified, and the existing data does not discriminate between a model wherein the nascent chain grows on und-P and one wherein und-P-Neu5Ac serves as a donor to another acceptor, such as diacylglycerophosphate. Superficially, the latter possibility would resemble the extension of lipomannan chains in the gram-positive bacterium, *Micrococcus luteus*, in which und-P-Man is the donor (97). Attempts to demonstrate the involvement of an equivalent und-P-linked intermediate in K5 biosynthesis using several different approaches have proved unsuccessful (86). The data are consistent with this particular system operating without lipid intermediates, although conclusions based on negative results must be interpreted with caution. For example, it is conceivable that all of the in vitro activity involves rapid extension of preexisting acceptors (similar to the majority of the activity seen in K1) (98). Exclusively long lipid-linked chains would not be efficiently extracted with solvent. Further studies are clearly warranted to definitively resolve the exact role (or not) of und-P in group 2 and 3 capsular polysaccharide biosynthesis.

If the endogenous acceptor is not und-P, what other molecules may fulfill this role? In the K1 system, there have also been reports of the involvement of a 20-kDa polypeptide as an endogenous acceptor (99, 100). However, the precise identity of this protein is unknown, as is the role of the protein: Does it provide an acceptor for the growing glycan, or is it an additional intermediate involved in the translocation of the polymer from cytoplasm to cell surface? NeuE and (77) and KpsT (101) have both been considered as candidates for the 20-kDa protein, and both are important for polymer export (see below).

It is also formally possible that the endogenous acceptor is diacylglycerophosphate (or diacylglycerophosphate-Kdo). Several export-deficient mutants in group 2 systems accumulate intracellular polymer with an added lipid terminus (see below). Furthermore, the K5 polymer synthesized in vitro has a reducing terminal residue whose properties are consistent with Kdo (86). Although not definitive, the data certainly suggest that the Kdo residue is present at the onset of chain elongation. Whether the same is true of the diacylglycerophosphate moiety remains to be established. The full range of potential acceptors for bacterial polysaccharides is currently unknown, but there is precedent for initiation of bacterial polysaccharides on a glycerophospholipid acceptors: The processive streptococcal type 3 capsular polysaccharide synthase uses phosphatidylglycerol as its acceptor in the native streptococcal background and when the synthase is expressed in *E. coli* (102).

Given the overall conservation in the assembly components and processes (to the extent they are understood), completely different endogenous acceptors within representative group 2 systems would be rather surprising. The ambiguity in the existing data serves to underline the need for reinvestigation of the precise structures of the reducing termini of representative capsular polysaccharides (and of polymers accumulating in defined mutants). Until this information is available, the critical chain initiation reaction for group 2 (and 3) capsule biosynthesis cannot be resolved.

ABC Transporter-Dependent Export of Nascent Group 2 Capsular Polysaccharide

ABC transporters (or traffic ATPases) drive the import or export of substrates at the

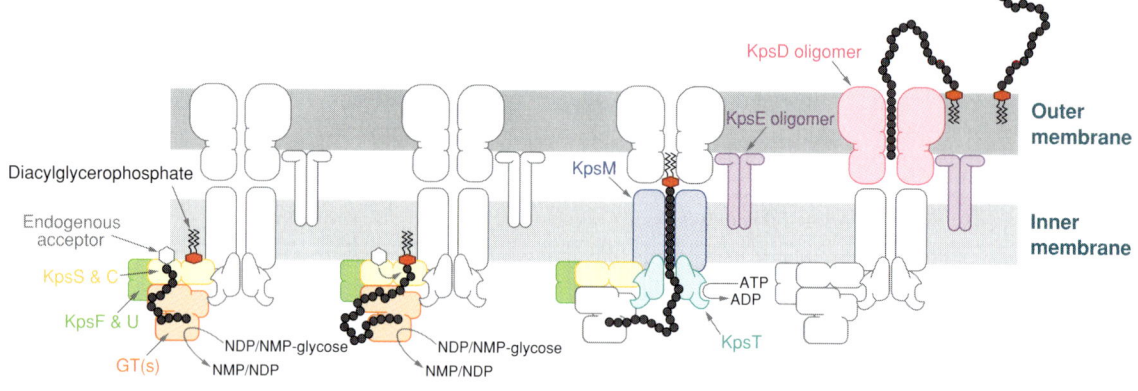

Figure 7

A model for biosynthesis and assembly of group 2 capsules. Beginning at the left, polymer formation is initiated on an unknown endogenous acceptor (*open hexagon*) and is extended by processive GTs, adding residues to the nonreducing terminus of the chain. The final product is linked to diacylglycerophosphate (or diacylglycerophosphate-Kdo), but it is unclear whether this is present at initiation or added during, or after (as shown here), polymerization. The polymer is exported via the ABC transporter (KpsM and KpsT). KpsS and C are essential for this process, and KpsF and U also participate, but the details are unknown. The orientation of the polymeric substrate during export has not been established, and biosynthesis and export may be temporally coupled. Translocation across the periplasm and outer membrane requires KpsE and KpsD, which provide putative membrane-fusion (adaptor) protein and OMA protein functions.

expense of ATP hydrolysis. There are 80 members of the ABC transporter superfamily in *E. coli* (reviewed in 103). They are comprised of two domains, a transmembrane domain (TMD) and a nucleotide-binding domain (NBD) that contains the Walker A and B sequences (indicative of ATP-binding), and additional motifs that characterize the ABC transporter superfamily. The two domains can be present as individual polypeptides or may be fused in larger proteins. KpsM and KpsT were initially identified as the components of an ABC transporter on the basis of sequence data, and their identities were then confirmed by biochemical approaches (reviewed in 104, 105). KpsM is the integral inner-membrane TMD component, with six transmembrane helices, and KpsT is the cytoplasmic NBD. These proteins form an ABC transporter classified in the ABC-A2 (106) or CPSE (107) families. The functional group 2 (and 3) capsule transporter is proposed to consist of two subunits each of KpsM and KpsT.

Mutations in *kpsT* result in accumulation of intracellular polymer located (sometimes in patches) at the periphery of the cytoplasm in proximity to the inner membrane (108–110), which is consistent with an inability to complete the export process. An attractive working model for export has been proposed on the basis of studies of a dominant-negative KpsT mutant defective in ATP hydrolysis and analogies to other ABC transporters (104, 105). In this model, KpsT from *E. coli* K1 associates with polysialic acid in the context of the additional capsule biosynthesis machinery (101) and may introduce the polymer into the export channel. KpsT binds ATP, undergoes a conformational change, and interacts with the TMD. Deinsertion requires ATP hydrolysis, but a mutant unable to perform hydrolysis remains locked into the complex and is nonfunctional (110). In this locked state, the KpsT mutant protein is exposed at the periplasmic face (101). Repetitive cycles of insertion-deinsertion may be necessary to complete the export process. The involvement of the polymeric substrate and other Kps proteins in supporting this process must also be considered.

The nature of the substrate for the ABC transporter has been controversial because of varying reports regarding the lipidation status of the intracellular polymer accumulating in various mutants. In initial studies with the K5 system, the polymer in a region 3 mutant was reported to lack the lipid terminus (108). However, using a different isolation method, the polymer in a K1 *kpsT* mutant was lipidated (70). It has been proposed that the results are very sensitive to the experimental protocols (70). Several other mutants (see below) also accumulate lipid-modified polymer, suggesting that lipidation precedes transport. From the available data, it is apparent that lipidation, by itself, is not sufficient for export via the KpsMT ABC transporter (70). However, the current data does not rule out the modification being essential for export. Another more complex question is the orientation of the exported product. In a coupled biosynthesis and export process, the transporter could potentially export the substrate before elongation is complete, but which end enters the exporter first—the diacylglycerophosphate or the glycan? KpsMT proteins have no specificity for polymer structure, suggesting they recognize a conserved element, assumed to be the lipid modification (as depicted in the model in **Figure 7**), but this is still conjecture. However, recent studies on the MsbA ABC transporter, which exports LPS lipid A core, have resulted in an attractive model whereby the hydrophilic domains of the export substrate are sequestered within the lumen of the channel, and the hydrophobic acyl chains are effectively excluded and are dragged through the surrounding lipid bilayer (110a).

Additional Proteins Contribute to a Complex that Couples Biosynthesis to Export

Several mutations result in the accumulation of intracellular polymer within the cytoplasm. The phenotypes of the *kpsMT* mutations are entirely predictable given the function of the ABC transporter, but those involving *kpsFU*, *kpsSC* and *neuE* are less obvious. These mutants accumulate unusual large electron-transparent domains containing aggregates of polymer within the cytoplasm (70, 111); these are often referred to as "lacunae." The distribution of lacunae within the cell appears to vary in the different mutants, perhaps reflecting defects at different points in the assembly pathway. In serotype K1, the lacunae all appear to contain lipidated polymer (70), and a role can be envisaged for the hydrophobic moiety in overcoming the anionic nature of the polymer and helping form these structures.

KpsF and KpsU are homologs of enzymes involved in synthesis of CMP-Kdo (the activated Kdo donor). Four enzymes (YrbH, KdsA, YrbI, and KdsB) have been identified in the synthesis of CMP-Kdo for LPS biosynthesis. KpsF is a homolog of YrbH, the arabinose-5-phosphate epimerase catalyzing the first step in the pathway (112), and was identified through studies focused on KpsF in *N. meningitidis* (113). KpsU is a homolog of the last enzyme in the pathway, CMP-Kdo synthetase (KdsB) (114), and explains the observation that *E. coli* isolates with group 2 capsules characteristically have elevated synthetase activity at temperatures permissive for capsule formation (i.e., above 20°C) (5). Given the redundancy in KpsF and KpsU, it is perhaps not surprising that neither KpsF nor KpsU enzymes are essential for capsulation, although the amount of capsule in the corresponding K1 mutants is substantially reduced (70, 115). The polymer is lipidated and appears to have a size similar to the wild-type product (70). If the requirement for CMP-Kdo is simply related to the reducing terminal modification, the YrbH and KdsB proteins must provide sufficient CMP-Kdo to sustain some capsule formation and also supply the precursor for the biosynthesis of LPS lipid A core, whose Kdo residues are essential for viability of *E. coli* (reviewed in 20).

A more complex issue is why KpsF and KpsU would have any influence at all on encapsulation in serotype K1 because studies with the closely related K92 and meningococcal serotype B polysialic acids support a direct attachment of the polymer to diacylglycerophosphate, with no Kdo linker (15). Meningococci (unlike *E. coli*) can survive without LPS, allowing the construction of definitive and informative mutants that eliminate all CMP-Kdo synthesis. In the absence of CMP-Kdo, capsule expression is reduced in several different serotypes, regardless of the whether the capsule repeat unit contains Neu5Ac (113). Thus, if Kdo is not present in the linker, it must be involved somewhere in the assembly pathway independent of the structure of the reducing terminal modification. In rationalizing the conflicting structural and biochemical data, Tzeng et al. (113) proposed that the polymer must be removed from the Kdo acceptor prior to its transfer to diacylglycerophosphate. Although there are no data ruling this out, it invokes a rather convoluted and seemingly inefficient assembly scheme for which there are currently no other supporting data. Extended to the *E. coli* systems, it would also require a different mechanism operating in some group 2 capsules depending on the presence (or not) of a reducing terminal Kdo linker. An attractive alternative explanation is that KpsF and KpsU play an additional role independent of their function in CMP-Kdo synthesis (70, 115) (see below).

KpsC and KpsS are cytosolic proteins essential for the appearance of group 2 capsular polymers on the cell surface, and defects lead to the accumulation of intracellular polymer. In initial studies with K5, region 1 deletions (108) and *kpsS* and *kpsC* mutants (111) were proposed to lack lipid modification. This, together with comparable initial data for the meningococcal KpsSC homologs (LipA and LipB) (116), contributed to speculation that KpsS and C are involved in the addition of the diacylglycerophosphate-Kdo modification (6, 67). However, in the K1 system, *kpsS* and *kpsC* mutants accumulate full-length lipidated intracellular polymer (70). The same result was subsequently obtained with meninogococcal *lipAB* mutants, and critically, the structure of the lipid was also definitively determined and shown to be identical to that attached to wild-type polymer (117). Given the high conservation of these homologs (~60% similarity between *E. coli* and meningococci), it seems unlikely that the proteins have different roles in these species. Unless there are cellular components that functionally replace KpsS and KpsC in some (but not all) systems, the majority of available evidence suggests that KpsSC proteins play a role in the export-translocation process that does not involve lipidation per se. Although database searches highlight the wide distribution of KpsS and KpsC homologs in bacteria with group 2-related capsules, they do not provide any clues as to their precise function. Resolution of the exact roles played by KpsS and KpsC is critical for an understanding of the steps involving targeting nascent group 2 (or 3) polymer to the exporter.

In serotype K1 *neuE* mutants, intracellular lacunae are generally located toward one or both poles of the cell (78, 118). NeuE was previously thought to be the initiating GT for polysialic acid synthesis, owing, in part, to its possession of a potential polyisoprenyl-recognition motif in a C-terminal membrane-spanning domain, but this interpretation has to be reevaluated in the light of the intracellular polymer (reviewed in 78, 118). NeuE is the only region 2 gene product that is not essential for polysialic acid biosynthesis.

The properties of the relevant mutants suggest that KpsS, KpsC, KpsF, KpsU, and NeuE may all contribute in some way to a stable enzyme assembly complex, and their absence uncouples the spatial and temporal coupling of biosynthesis and export. In *E. coli* K5, *kpsC*, *kpsS*, and *kpsU* mutants all show a considerable reduction in in vitro GT activity in isolated membranes, yet clearly, the GTs are active in vivo (111, 119). Similarly, in serotype K1, the NeuS polysialyltransferase attaches to membranes in the absence of other Kps proteins except KpsS (80), and the level of in vitro endogenous polysialyltransferase activity is reduced when NeuS is expressed without KpsS and KpsT (120). Roberts and coworkers (121) have performed a systematic analysis of individual mutants in K5 and generated data lending strong support to the concept of a coordinated multienzyme complex. KpsCST proteins all independently target to the inner membrane. In contrast, the KfiAC GTs and the precursor-forming KfiD (UDP-glucose dehydrogenase) only target to the inner membrane in the presence of KpsMT and KpsCS. There is also a hierarchy in the assembly of the complex, with KfiA targeting requiring KfiC, but not KfiB, and KfiB requiring both KfiA and KfiC (90). These data infer complex protein-protein interactions between the various components in the assembly complex. The involvement of NeuE in *E. coli* K1 and the absence of *kpsF* and *kpsU* in group 3 capsule loci (68) suggest that some of the interactions are unique to a particular system. The cytoplasmic lacunae may therefore represent an aberrant site for "dumping" polymers that could not be exported in an appropriate or timely manner.

KpsD and KpsE Mediate Translocation of Group 2 and 3 Capsular Polysaccharides to the Cell Surface

Mutations in *kpsD* (122) and *kpsDE* (111, 123) result in the accumulation of polymer in the periplasm, indicating that their products operate late in the assembly pathway. The corresponding genes from group 3 complement group 2 mutations (74), indicating a conserved function. In ABC transporter-driven efflux pumps, the transporter in the inner membrane is coupled to an outer membrane via a membrane-fusion protein (MFP) or adaptor protein. For example, the TolC outer-membrane channel protein is recruited and coupled to ABC transporters for RND-type drug efflux and type I protein secretion by the putative MFP proteins, AcrA and HlyD, respectively (reviewed in 60, 124). This generates a functional *trans*-envelope complex that provides continuity between the cytoplasm and the cell surface. Assembly of a functional TolC-ABC transporter complex involves a series of molecular interactions and conformational changes that, in the case of hemolysin secretion, are influenced by the presence of the export substrate. An MFP protein was proposed for the group 2 capsule system with KpsE [a representative of the inner-membrane-periplasmic-auxiliary protein-2 family (MPA-2)] providing the candidate (38). KpsE has a tendency to form dimers and higher-order oligomers (125), as expected from studies of the drug efflux MFPs. The bulk of the protein is located in the periplasm (126), and it associates with the outer face of inner membrane via a C-terminal amphipathic α-helix (125, 127). KpsE may be responsible for bridging the periplasm and maintaining the discrete sites at which nascent capsule is translocated to the surface (128).

It is well accepted that the translocation of group 2 capsular polysaccharide must involve some form of channel to facilitate passage of the polymer across the outer membrane, analogous to the role proposed for the Wza outer-membrane protein in group 1 capsule formation (see above). KpsD has been grouped within the OMA family along with Wza from group 1 and 4 capsule systems on the basis of regions of local sequence similarity (38), although the similarity is low and KpsD (unlike Wza) is not a lipoprotein (56). However, the candidacy of KpsD as the outer-membrane "secretin" was obscured by localization studies that placed KpsD in the periplasm (122). It is now apparent that the location of KpsD is dictated by the genetic background. In the presence of KpsE and the envelope protein,

MFP: membrane-fusion protein (or adaptor) protein family

MPA-2 protein: a putative adaptor protein that links an ABC transporter to an outer-membrane channel

Lpp, KpsD appears in the outer membrane (125). As further support for the candidacy of KpsD as the group 2 OMA protein, it is reported that KpsDE can be exchanged with CpxCD (the MPA-2 and OMA proteins from the capsule locus of *Actinobacillus pleuropneumoniae* 5A) with preservation of capsule assembly (105). CpxD shares significant similarity with Wza and other known OMA proteins (56). The presence of KpsD does not signal a unique situation in *E. coli* because a homolog is found in the capsule cluster of *Campylobacter jejuni*, a locus resembling group 3 capsule gene clusters (e.g., K10) and encoding homologs of KpsMTEDFCS (129).

Thus, it seems likely that KpsD fulfills the functional role of OMA proteins and represents a divergent group in the OMA family. If true, the *kps* locus now appears to encode a dedicated OMA protein and putative MFP, consistent with the gene content for other related capsule biosynthesis loci, and those proteins can operate with heterologous ABC transporters (105). If KpsD is indeed a functional equivalent of Wza, structural comparisons will be both interesting and informative.

CONCLUDING REMARKS

Significant progress has been made in identifying the components in capsule biosynthesis and assembly. However, unraveling the underlying molecular mechanisms has been difficult because mutations in many of the components have the same capsule-null phenotype. Furthermore, interpretation of defects at a biochemical level is complicated by potential networks of protein-protein interactions within assembly complexes. The existence of a *trans*-envelope complex would provide a means of coordinating biosynthesis and surface assembly. By creating continuity between the inner membrane and the cell surface, such complexes would overcome the challenges posed by having to cross the periplasm, the peptidoglycan, and the outer membrane. Some of the important individual open questions have been identified above. Clearly, the next step is to determine the structure and function of purified components and resolve the molecular architecture of the complexes. Although biochemical and structural biology methods can be used, the limiting factor is that several of the most important proteins in the systems are integral membrane proteins, for which overexpression has proved difficult, and development of informative activity assays is still a challenge. Current efforts are logically directed at capsule-specific proteins, but capsule assembly occurs within the context of the *E. coli* cell envelope, and it is inevitable that other housekeeping functions will also participate. The recent elegant work of Kahne, Silhavy, and colleagues (130) serves to highlight the relationships and potential interactions between different multiprotein complexes involved in outer-membrane biogenesis. Understanding how the various systems are coordinated to maintain the protective capsule during growth and division of a bacterial cell is a complex and intriguing problem and is of critical importance in virulence.

SUMMARY POINTS

1. *Escherichia coli* capsules are classified into four groups on the basis of genetic and biochemical criteria. The biosynthesis and assembly mechanisms for groups 1 and 4 are closely related and follow Wzy-dependent processes. Group 2 and 3 capsules are both assembled via ABC transporter-dependent pathways.

2. Capsule assembly appears to involve a multiprotein complex spanning the cell envelope, and this may facilitate spatial and temporal coupling of polymer biosynthesis, export, and translocation. The complex consists of core export and translocation

machinery, functioning independent of polymer structure, into which is plugged a serotype-specific polymer biosynthesis system.

3. In the Wzy-dependent mechanism, undecaprenol-linked repeat units are used in a polymerization reaction that requires Wzx and Wzy proteins. Newly synthesized lipid-linked repeat units are exported across the inner membrane in a process requiring Wzx. The growing glycan is transferred from its lipid carrier to the newly exported lipid-linked repeat unit, elongating the polymer in a "block-wise" manner by addition of new repeat units at the reducing terminus. The polymerization reaction is dependent on the Wzy protein.

4. In the ABC transporter-dependent mechanism, processive glycosyltransferase activity extends the nascent glycan in the cytoplasm. Chain extension occurs by addition of residues to the nonreducing terminus, but the nature of the endogenous acceptor for glycan growth is controversial. The polymer is transferred across the inner membrane by the ABC transporter composed of KpsM (transmembrane domain) and KpsT (nucleotide-binding domain). Additional proteins appear to be involved in coupling biosynthesis to export in this system, and other proteins may also participate.

5. Translocation of capsular polysaccharide across the outer membrane requires a member of the OMA family. In group 1 and 4 capsules, this protein is Wza. Wza forms a stable octameric complex. KpsD may fulfill this role in group 2 and 3 capsule assembly systems.

6. In group 1 and 4 capsule assembly, Wza is linked to the inner-membrane biosynthesis process by Wzc (an MPA-1 protein). Wzc forms tetramers and undergoes transphosphorylation of several C-terminal tyrosine residues. Phosphorylation of Wzc and its dephosphorylation by the phosphatase Wzb are both essential for group 1 and 4 capsule assembly, suggesting that Wzc must cycle its phosphorylated state.

7. In group 2 (and possibly group 3) capsules, the periplasmic protein KpsE (MPA-2 protein) may facilitate linkage of the ABC transporter and OMA components of the assembly system. KpsE is a putative membrane-fusion protein, or adaptor protein, and is essential for translocation of polymer to the cell surface.

FUTURE ISSUES

1. What are the mechanisms of action of the characteristic Wzx and Wzy proteins, and how are their activities coordinated in polymerization?

2. What is the nature of the endogenous acceptor for group 2 and 3 capsular polysaccharide biosynthesis, and when is the diacylglycerophosphate anchor added?

3. Do the outer-membrane OMA proteins for groups 1 and 4 and groups 2 and 3 have a conserved function, and what structural features are associated with their coupling to different inner-membrane biosynthesis and export components?

4. If capsule biosynthesis and assembly machinery form a transmembrane complex, as current data suggest, how does this complex cross the stress-bearing peptidoglycan layer without compromising cell wall integrity, and how is its activity coordinated with others involved in membrane biogenesis during cell growth and division?

ACKNOWLEDGMENTS

The author recognizes the contribution of lab members, past and present, to the research and ideas presented here. Helpful and stimulating discussions with Dr. E.R. Vimr are gratefully acknowledged, as are constructive comments on a draft version provided by Dr. Vimr and Dr. A.N. Reid. Our work in this area is generously supported by the Canadian Institutes of Health Research, and C.W. is a recipient of a Canada Research Chair.

LITERATURE CITED

1. Ørskov I, Ørskov F, Jann B, Jann K. 1977. *Bacteriol. Rev.* 41:667–710
2. Rick PD, Silver RP. 1996. In *Escherichia coli and Salmonella: Cellullar and Molecular Biology*, ed. FC Neidhardt, pp. 104–22. Washington, DC: ASM Press
3. Zogaj X, Nimtz M, Rohde M, Bokranz W, Romling U. 2001. *Mol. Microbiol.* 39:1452–63
4. Wang X, Dubey AK, Suzuki K, Baker CS, Babitzke P, Romeo T. 2005. *Mol. Microbiol.* 56:1648–63
5. Jann K, Jann B. 1997. In *Escherichia coli: Mechanisms of Virulence*, ed. M Sussman, pp. 113–43. Cambridge, UK: Cambridge Univ. Press
6. Whitfield C, Roberts IS. 1999. *Mol. Microbiol.* 31:1307–19
7. Hungerer D, Jann K, Jann B, Ørskov F, Ørskov I. 1967. *Eur. J. Biochem.* 2:115–26
8. Goldman RC, White D, Ørskov F, Ørskov I, Rick PD, et al. 1982. *J. Bacteriol.* 151:1210–21
9. Peterson AA, McGroarty EJ. 1985. *J. Bacteriol.* 162:738–45
10. MacLachlan PR, Keenleyside WJ, Dodgson C, Whitfield C. 1993. *J. Bacteriol.* 175:7515–22
11. Peleg A, Shifrin Y, Ilan O, Nadler-Yona C, Nov S, et al. 2005. *J. Bacteriol.* 187:5259–66
12. Amor PA, Whitfield C. 1997. *Mol. Microbiol.* 26:145–61
13. Reid AN, Whitfield C. 2005. *J. Bacteriol.* 187:5470–81
14. Majdalani N, Gottesman S. 2005. *Annu. Rev. Microbiol.* 59:379–405
15. Gotschlich EC, Fraser BA, Nishimura O, Robbins JB, Liu T-Y. 1981. *J. Biol. Chem.* 256:8915–21
16. Schmidt MA, Jann K. 1982. *FEMS Microbiol. Lett.* 14:69–74
17. Jann B, Jann K. 1990. In *Bacterial Capsules*, ed. K Jann, B Jann, pp. 19–42. Berlin: Springer-Verlag
18. Rahn A, Whitfield C. 2003. *Mol. Microbiol.* 47:1045–60
19. Rahn A, Drummelsmith J, Whitfield C. 1999. *J. Bacteriol.* 181:2307–13
20. Raetz CRH, Whitfield C. 2002. *Annu. Rev. Biochem.* 71:635–700
21. Bailey MJ, Hughes C, Koronakis V. 1997. *Mol. Microbiol.* 26:845–51
22. Artsimovitch I, Landick R. 2002. *Cell* 109:193–203
23. Wang L, Curd H, Qu W, Reeves PR. 1998. *J. Clin. Microbiol.* 36:3182–87
24. Stevenson G, Andrianopoulos K, Hobbs M, Reeves PR. 1996. *J. Bacteriol.* 178:4885–93

25. Jayaratne P, Keenleyside WJ, MacLachlan PR, Dodgson C, Whitfield C. 1993. *J. Bacteriol.* 175:5384–94
26. Keenleyside WJ, Bronner D, Jann K, Jann B, Whitfield C. 1993. *J. Bacteriol.* 175:6725–30
27. Valvano MA. 2003. *Front. Biosci.* 8:S452–71
28. Troy FA, Frerman FE, Heath EC. 1971. *J. Biol. Chem.* 246:118–33
29. Drummelsmith J, Whitfield C. 1999. *Mol. Microbiol.* 31:1321–32
30. Whitfield C, Paiment A. 2003. *Carbohydr. Res.* 338:2491–502
31. Price NP, Momany FA. 2005. *Glycobiology* 15:R29–42
32. Wang L, Liu D, Reeves PR. 1996. *J. Bacteriol.* 178:2598–604
33. Pelosi L, Boumedienne M, Saksouk N, Geiselmann J, Geremia RA. 2005. *Biochem. Biophys. Res. Commun.* 327:857–65
34. Daniels C, Vindurampulle C, Morona R. 1998. *Mol. Microbiol.* 28:1211–22
35. McGrath BC, Osborn MJ. 1991. *J. Bacteriol.* 173:649–54
36. Liu D, Cole R, Reeves PR. 1996. *J. Bacteriol.* 178:2102–7
37. Feldman MF, Marolda CL, Monteiro MA, Perry MB, Parodi AJ, Valvano MA. 1999. *J. Biol. Chem.* 274:35129–38
38. Paulsen IT, Beness AM, Saier MHJ. 1997. *Microbiology* 143:2685–99
39. Rick PD, Barr K, Sankaran K, Kajimura J, Rush JS, Waechter CJ. 2003. *J. Biol. Chem.* 278:16534–42
40. Marolda CL, Vicarioli J, Valvano MA. 2004. *Microbiology* 150:4095–105
41. Zhou GP, Troy FA 2nd. 2003. *Glycobiology* 13:51–71
42. Zhou GP, Troy FA 2nd. 2005. *Glycobiology* 15:347–59
43. Schild S, Lamprecht AK, Reidl J. 2005. *J. Biol. Chem.* 280:25936–47
44. Whitfield C, Amor PA, Köplin R. 1997. *Mol. Microbiol.* 23:629–38
45. Dodgson C, Amor P, Whitfield C. 1996. *J. Bacteriol.* 178:1895–902
46. Franco AV, Liu D, Reeves PR. 1996. *J. Bacteriol.* 178:1903–7
47. Morona R, Van den Bosch L, Daniels C. 2000. *Microbiology* 146:1–4
48. Wugeditsch T, Paiment A, Hocking J, Drummelsmith J, Forrester C, Whitfield C. 2001. *J. Biol. Chem.* 276:2361–71
49. Cozzone AJ, Grangeasse C, Doublet P, Duclos B. 2004. *Arch. Microbiol.* 181:171–81
50. Grangeasse C, Doublet P, Cozzone AJ. 2002. *J. Biol. Chem.* 277:7127–35
51. Paiment A, Hocking J, Whitfield C. 2002. *J. Bacteriol.* 184:6437–47
52. Vincent C, Duclos B, Grangeasse C, Vaganay E, Riberty M, et al. 2000. *J. Mol. Biol.* 304:311–21
53. Doublet P, Grangeasse C, Obadia B, Vaganay E, Cozzone AJ. 2002. *J. Biol. Chem.* 277:37339–48
54. Collins RF, Beis K, Clarke BR, Ford RC, Hulley M, et al. 2005. *J. Biol. Chem.* 281:2144–50
55. Daniels C, Morona R. 1999. *Mol. Microbiol.* 34:181–94
56. Drummelsmith J, Whitfield C. 2000. *EMBO J.* 19:57–66
57. Nesper J, Hill CM, Paiment A, Harauz G, Beis K, et al. 2003. *J. Biol. Chem.* 278:49763–72
58. Kostakioti M, Newman CL, Thanassi DG, Stathopoulos C. 2005. *J. Bacteriol.* 187:4306–14
59. Beis K, Collins RF, Ford RC, Kamis AB, Whitfield C, Naismith JH. 2004. *J. Biol. Chem.* 279:28227–32
60. Koronakis V, Eswaran J, Hughes C. 2004. *Annu. Rev. Biochem.* 73:467–89
61. Beis K, Nesper J, Whitfield C, Naismith JH. 2004. *Acta Crystallogr. Sect. D* 60:558–60

61a. Christie PJ, Atmakuri K, Krishnamoorthy V, Jakubowski S, Cascales E. 2005. *Annu. Rev. Microbiol.* 59:451–85
62. Bayer ME, Thurow H. 1977. *J. Bacteriol.* 130:911–36
63. Bayer ME. 1991. *J. Struct. Biol.* 107:268–80
64. Ilan O, Bloch Y, Frankel G, Ullrich H, Geider K, Rosenshine I. 1999. *EMBO J.* 18:3241–48
65. Rahn A, Beis K, Naismith JH, Whitfield C. 2003. *J. Bacteriol.* 185:5882–90
66. Silver RP, Finn CW, Vann WF, Aaronson W, Schneerson R, et al. 1981. *Nature* 289:696–98
67. Roberts IS. 1996. *Annu. Rev. Microbiol.* 50:285–315
68. Barrett B, Ebah L, Roberts IS. 2002. *Curr. Top. Microbiol. Immunol.* 264:137–55
69. Simpson DA, Hammarton TC, Roberts IS. 1996. *J. Bacteriol.* 178:6466–74
70. Cieslewicz M, Vimr E. 1996. *J. Bacteriol.* 178:3212–20
71. Stevens MP, Clarke BR, Roberts IS. 1997. *Mol. Microbiol.* 24:1001–12
72. Rowe S, Hodson N, Griffiths G, Roberts IS. 2000. *J. Bacteriol.* 182:2741–45
73. Deszo EL, Steenbergen SM, Freedberg DI, Vimr ER. 2005. *Proc. Natl. Acad. Sci. USA* 102:5564–69
74. Pearce R, Roberts IS. 1995. *J. Bacteriol.* 177:3992–97
75. Cho JW, Troy FA 2nd. 1994. *Proc. Natl. Acad. Sci. USA* 91:11427–31
76. Lindahl U, Li JP, Kusche-Gullberg M, Salmivirta M, Alaranta S, et al. 2005. *J. Med. Chem.* 48:349–52
77. Troy FA 2nd. 1992. *Glycobiology* 2:5–23
78. Vimr ER, Steenbergen SM. 1993. In *Polysialic Acid from Microbes to Man*, ed. J Roth, U Rutishauser, FA Troy 2nd, pp. 73–91. Basel: Birkhäuser-Verlag
79. Steenbergen SM, Vimr ER. 1990. *Mol. Microbiol.* 4:603–11
80. Steenbergen SM, Wrona TJ, Vimr ER. 1992. *J. Bacteriol.* 174:1099–108
81. McGowen MM, Vionnet J, Vann WF. 2001. *Glycobiology* 11:613–20
82. Steenbergen SM, Vimr ER. 2003. *J. Biol. Chem.* 278:15349–59
83. Charnock SJ, Henrissat B, Davies GJ. 2001. *Plant Physiol.* 125:527–31
84. Shen GJ, Datta AK, Izumi M, Koeller KM, Wong CH. 1999. *J. Biol. Chem.* 274:35139–46
85. Lidholt K, Fjelstad M. 1997. *J. Biol. Chem.* 272:2682–87
86. Finke A, Bronner D, Nikolaev AV, Jann B, Jann K. 1991. *J. Bacteriol.* 173:4088–94
87. Lidholt K, Fjelstad M, Jann K, Lindahl U. 1994. *Carbohydr. Res.* 255:87–101
88. Ninomiya T, Sugiura N, Tawada A, Sugimoto K, Watanabe H, Kimata K. 2002. *J. Biol. Chem.* 277:21567–75
89. Griffiths G, Cook NJ, Gottfridson E, Lind T, Lidholt K, Roberts IS. 1998. *J. Biol. Chem.* 273:11752–57
90. Hodson N, Griffiths G, Cook N, Pourhossein M, Gottfridson E, et al. 2000. *J. Biol. Chem.* 275:27311–15
91. DeAngelis PL, White CL. 2002. *J. Biol. Chem.* 277:7209–13
92. Higa HH, Varki A. 1988. *J. Biol. Chem.* 263:8872–78
93. Pelkonen S, Häyrinen J, Finne J. 1988. *J. Bacteriol.* 170:2646–53
94. Clarke BR, Cuthbertson L, Whitfield C. 2004. *J. Biol. Chem.* 279:35709–18
95. Forsee WT, Cartee RT, Yother J. 2000. *J. Biol. Chem.* 275:25972–78
96. Troy FA, Vijay IK, Tesche N. 1975. *J. Biol. Chem.* 250:156–63
97. Pakkiri LS, Waechter CJ. 2005. *Glycobiology* 15:291–302
98. Rohr TE, Troy FA. 1980. *J. Biol. Chem.* 255:2332–42

99. Rodríguez-Aparicio LB, Reglero A, Ortiz AI, Luengo JM. 1988. *Biochem. J.* 251:589–96
100. Weisgerber C, Troy FA. 1990. *J. Biol. Chem.* 265:1578–87
101. Bliss JM, Silver RP. 1997. *J. Bacteriol.* 179:1400–3
102. Cartee RT, Forsee WT, Yother J. 2005. *J. Bacteriol.* 187:4470–79
103. Davidson AL, Chen J. 2004. *Annu. Rev. Microbiol.* 73:241–68
104. Bliss JM, Silver RP. 1996. *Mol. Microbiol.* 21:221–31
105. Silver RP, Prior K, Nsahlai C, Wright LF. 2001. *Res. Microbiol.* 152:357–64
106. Reizer J, Reizer A, Saier MH Jr. 1992. *Protein Sci.* 1:1326–32
107. Saier MH Jr. 2000. *Mol. Microbiol.* 35:699–710
108. Kröncke K-D, Boulnois G, Roberts I, Bitter-Suermann D, Golecki JR, et al. 1990. *J. Bacteriol.* 172:1085–91
109. Pavelka MS Jr, Hayes SF, Silver RP. 1994. *J. Biol. Chem.* 269:20149–58
110. Bliss JM, Garon CF, Silver RP. 1996. *Glycobiology* 6:445–52
110a. Reyes CL, Chang G. 2005. *Science* 308:1028–31
111. Bronner D, Sieberth V, Pazzani C, Roberts IS, Boulnois GJ, et al. 1993. *J. Bacteriol.* 175:5984–92
112. Meredith TC, Woodard RW. 2003. *J. Biol. Chem.* 278:32771–77
113. Tzeng YL, Datta A, Strole C, Kolli VS, Birck MR, et al. 2002. *J. Biol. Chem.* 277:24103–13
114. Rosenow C, Roberts IS, Jann K. 1995. *FEMS Microbiol. Lett.* 125:159–64
115. Cieslewicz M, Vimr E. 1997. *Mol. Microbiol.* 26:237–49
116. Frosch M, Müller A. 1993. *Mol. Microbiol.* 8:483–93
117. Tzeng YL, Datta AK, Strole CA, Lobritz MA, Carlson RW, Stephens DS. 2005. *Infect. Immun.* 73:1491–505
118. Vimr ER, Kalivoda KA, Deszo EL, Steenbergen SM. 2004. *Microbiol. Mol. Biol. Rev.* 68:132–53
119. Bronner D, Sieberth V, Pazzani C, Smith A, Boulnois G, et al. 1993. *FEMS Microbiol. Lett.* 113:279–84
120. Vimr ER, Aaronson W, Silver RP. 1989. *J. Bacteriol.* 171:1106–17
121. Rigg GP, Barrett B, Roberts IS. 1998. *Microbiology* 144:2905–14
122. Silver RP, Aaronson W, Vann WF. 1987. *J. Bacteriol.* 169:5489–95
123. Pazzani C, Rosenow C, Boulnois GJ, Bronner D, Jann K, Roberts IS. 1993. *J. Bacteriol.* 175:5978–83
124. Eswaran J, Koronakis E, Higgins MK, Hughes C, Koronakis V. 2004. *Curr. Opin. Struct. Biol.* 14:741–47
125. Arrecubieta C, Hammarton TC, Barrett B, Chareonsudjai S, Hodson N, et al. 2001. *J. Biol. Chem.* 276:4245–50
126. Rosenow C, Esumeh F, Roberts IS, Jann K. 1995. *J. Bacteriol.* 177:1137–43
127. Phoenix DA, Brandenburg K, Harris F, Seydel U, Hammerton T, Roberts IS. 2001. *Biochem. Biophys. Res. Commun.* 285:976–80
128. Krönke K-D, Golecki JR, Jann K. 1990. *J. Bacteriol.* 172:3469–72
129. Karlyshev AV, Linton D, Gregson NA, Lastovica AJ, Wren BW. 2000. *Mol. Microbiol.* 35:529–41
130. Wu T, Malinverni J, Ruiz N, Kim S, Silhavy TJ, Kahne D. 2005. *Cell* 121:235–45
131. Matias VR, Al-Amoudi A, Dubochet J, Beveridge TJ. 2003. *J. Bacteriol.* 185:6112–18
132. Whitfield C, Vimr ER, Costerton JW, Troy FA. 1984. *J. Bacteriol.* 159:321–28
133. Reid AN, Whitfield C. 2005. *J. Bacteriol.* 187:5470–81
134. Graham LL, Harris R, Villiger W, Beveridge TJ. 1991. *J. Bacteriol.* 173:1623–33

135. Jann K, Jann B, Schneider KF. 1968. *Eur. J. Biochem.* 5:456–65
136. Choy Y-M, Fehmel F, Frank N, Stirm S. 1975. *J. Virol.* 16:581–90
137. Chakraborty AK, Friebolin H, Stirm S. 1980. *J. Bacteriol.* 141:971–72
138. McGuire EJ, Binkley SB. 1964. *Biochemistry* 3:247–51
139. Fischer W, Schmidt MA, Jann B, Jann K. 1982. *Biochemistry* 21:1279–84
140. Rodriguez ML, Jann B, Jann K. 1988. *Eur. J. Biochem.* 177:117–24
141. Vann WF, Schmidt MA, Jann B, Jann K. 1981. *Eur. J. Biochem.* 116:359–64
142. Schmidt MA, Jann K. 1983. *Eur. J. Biochem.* 131:509–17
143. Vann WF, Jann K. 1979. *Infect. Immun.* 25:85–92
144. Egan W, Liu T-Y, Dorow D, Cohen JS, Robbins JD, et al. 1977. *Biochemistry* 16:3687–92
145. Rodriguez ML, Jann B, Jann K. 1988. *Carbohydr. Res.* 173:243–53
146. Siebarth V, Jann B, Jann K. 1993. *Carbohydr. Res.* 246:219–28
147. Hoffman P, Jann B, Jann K. 1985. *Carbohydr. Res.* 139:261–71
148. Kenne L, Lindberg B, Söderstrom E, Bundle DR, Griffith DW. 1983. *Carbohydr. Res.* 111:289–96
149. Dengler T, Jann B, Jann K. 1986. *Carbohydr. Res.* 150:233–40.
150. Garegg, PJ, Lindberg B, Onn T, Holme T. 1971. *Acta Chem. Scand.* 25:2103–8
151. Andreishcheva EN, Vann WF. 2006. *J. Bacteriol.* 188:1786–97
152. McNulty C, Thompson J, Barrett B, Lord L, Anderson C, Roberts IS. 2006. *Mol. Microbiol.* 59:907–22

RELATED RESOURCE

Luirink J, von Heijne G, Houben E, de Gier JW. 2005. *Annu. Rev. Microbiol.* 59:329–55

NOTE ADDED IN PROOF

While this review was in proof, two papers that provide important new insight into group 2 capsule assembly were published (151, 152):

- Work performed in the laboratory of Vann has identified the minimal complement of gene products necessary for de novo synthesis of polysialic acid in *E. coli* K1 (151). Synthesis requires the precursor (CMP-Neu5Ac), together with the polysialyltransferase (NeuS) and the NeuE and KpsC proteins. The additional presence of KpsS dramatically increases the level of polymer made. These observations now set the stage for deciphering the mechanism of polymer initiation in a reaction that minimally involves NeuES and KpsC.
- Research from Roberts' group has identified a *trans*-envelope complex involved in biosynthesis of the K5 capsule (152). This complex contains the K5 GT enzymes, KpsS, the ABC transporter (KpsMT), the periplasmic KpsE protein, and KpsD. KpsD is confirmed as an outer membrane-spanning protein, thus reinforcing its candidacy as an OMA representative. In an unanticipated observation, KpsS and KpsD colocalize to the poles of the cell and are coincident with the export of nascent K5 polymer. This suggests a requirement for diffusion of polymer across the cell surface during encapsulation. While the universality of this concept has yet to be established, the findings raise intriguing questions concerning capsule growth and its integration with the cell cycle.

Energy Converting NADH:Quinone Oxidoreductase (Complex I)

Ulrich Brandt

Universität Frankfurt, Fachbereich Medizin, Zentrum der Biologischen Chemie, D-60590 Frankfurt am Main, Germany; email: brandt@zbc.kgu.de

Key Words

mitochondria, mitochondrial DNA, iron-sulfur cluster, ubiquinone, proton pumping

Abstract

NADH:quinone oxidoreductase (complex I) pumps protons across the inner membrane of mitochondria or the plasma membrane of many bacteria. Human complex I is involved in numerous pathological conditions and degenerative processes. With 14 central and up to 32 accessory subunits, complex I is among the largest membrane-bound protein assemblies. The peripheral arm of the L-shaped molecule contains flavine mononucleotide and eight or nine iron-sulfur clusters as redox prosthetic groups. Seven of the iron-sulfur clusters form a linear electron transfer chain between flavine and quinone. In most organisms, the seven most hydrophobic subunits forming the core of the membrane arm are encoded by the mitochondrial genome. Most central subunits have evolved from subunits of different hydrogenases and bacterial Na^+/H^+ antiporters. This evolutionary origin is reflected in three functional modules of complex I. The coupling mechanism of complex I most likely involves semiquinone intermediates that drive proton pumping through redox-linked conformational changes.

Contents

INTRODUCTION	70
SUBUNITS OF COMPLEX I	70
Central Subunits	71
Accessory Subunits	73
MODULAR DESIGN OF COMPLEX I	75
The N Module	76
The Q Module	78
The P Module	78
STRUCTURE OF COMPLEX I	79
The Peripheral Arm	79
The Membrane Arm	81
FUNCTION OF COMPLEX I	81
Substrates, Redox Centers, and Inhibitors	82
The Quinone Reducing "Catalytic Core"	83
Proton-Pumping Mechanism	84
MEDICAL ASPECTS OF COMPLEX I	86

NQR: a procaryotic sodium-pumping NADH:quinone oxidoreductase, which is not related to complex I

FMN: flavine mononucleotide

Iron-sulfur cluster: a redox-prosthetic group. In complex I, binuclear (Fe_2S_2) and tetranuclear (Fe_4S_4) clusters are found

INTRODUCTION

As one of the most fundamental metabolic principles, the vast majority of biochemical pathways involves "bound hydrogen" intermediates in the form of NADH, NADPH, or reduced flavoproteins. NADH generated in catabolic pathways is fed into energy converting electron transfer chains via NADH:quinone oxidoreductases. Three enzyme families catalyze this reaction. This review focuses on proton translocating NADH:quinone oxidoreductase: This type of enzyme, usually called complex I, was first described in mitochondria (1), but it is also found in many eubacteria where it is frequently termed NADH dehydrogenase-1 or NDH-1 (2, 3, 4). Moreover, complex I is involved in bacterial photosynthetic electron transport (5, 6). In some prokaryotes, a sodium-pumping NADH:quinone oxidoreductase called NQR is found (7). Its evolution is not related to that of complex I. Finally, mitochondria from plants, many fungi, as well as many bacteria contain up to four so-called alternative NADH dehydrogenases (8). These simple flavoenzymes catalyze the same redox reaction as complex I but do not couple it to proton or sodium translocation. In fermenting yeasts like *Saccharomyces cerevisiae* that do not contain complex I (9), they are the only enzymes that feed reducing equivalents from NADH into the respiratory chain.

Complex I is by far the largest and most complicated enzyme of the respiratory chain. In some organisms, it is a component of so-called respirasomes (10), and the formation of these respiratory supercomplexes seems to be essential for complex I stability (11, 12). Although complex I was first purified from bovine heart mitochondria almost 50 years ago (1), its molecular structure is just emerging, and its mechanism is still elusive. It has become evident in recent years that complex I is involved in numerous pathological conditions and degenerative processes (13, 14). This has resulted in increasing efforts to improve our knowledge about complex I.

SUBUNITS OF COMPLEX I

As for other respiratory chain complexes, prokaryotic complex I represents the minimal form of the enzyme. It comprises 14 subunits that carry all redox centers (15), flavine mononucleotide (FMN), and up to nine iron-sulfur clusters (16, 17); these "central" subunits are sufficient to perform all bioenergetic functions. In addition mitochondrial complex I contains up to 32 additional subunits resulting in a total molecular mass of up to ~1000 kDa (18, 19). The function of these "accessory" subunits is largely unknown, but some of them exhibit remarkable sequence similarities to other cellular proteins. Some insight into the role of the accessory subunits has been obtained from studying the composition of mitochondrial complex I purified from different organisms (20–24).

The identification of the subunits and their assignment to different parts of complex I

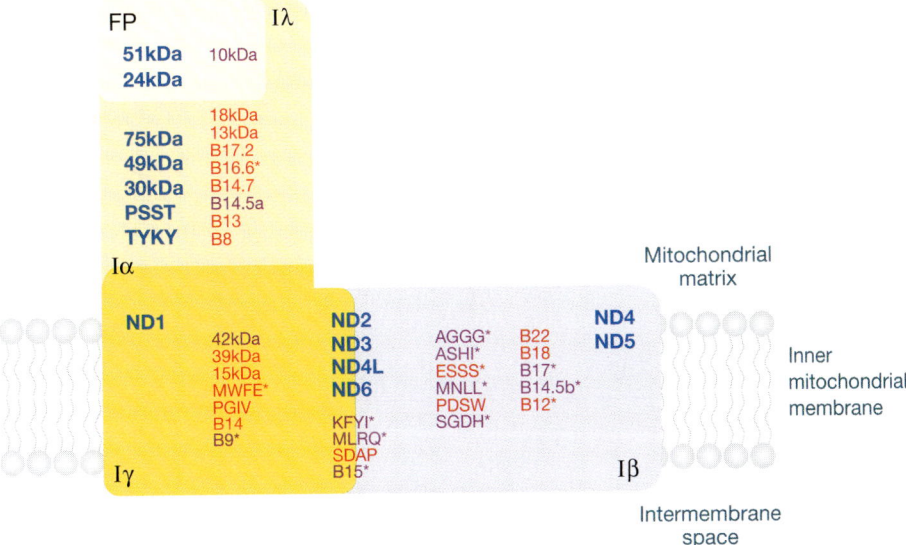

Figure 1

Subcomplexes and subunits of bovine heart complex I. The assignment of subunits to the subcomplexes of complex I was made according to References 18 and 19, and for the central hydrophobic subunits, the figure includes suggestions from Reference 111. Flavoprotein (FP) is part of Iλ (*light yellow*), and Iα essentially is a combination of Iλ and Iγ (*yellow*). Iβ (*blue gray*) forms the major part of the membrane integral arm of complex I. Central subunits are in blue, accessory subunits found in all eukaryotic complexes are in red, and metazoa specific subunits are in purple. Subunits marked with an asterisk are predicted to contain a single transmembrane domain (19).

(**Figure 1**) were achieved largely by using chaotropes and detergents to dissociate the purified bovine complex into various subcomplexes (25, 26) and by analyzing the subunits by gel electrophoresis, Edman degradation, and mass spectrometry (18). For historic reasons, no unified nomenclature has been established for the subunits of complex I. Complex I from bovine heart mitochondria was the first studied and is still the best characterized enzyme (18, 19). Therefore, if not indicated otherwise, the names established for bovine complex I are used in this review for all central subunits and those accessory subunits, analyzed to date, that are found in all mitochondrial complexes. Note that some of the names are directly derived from the molecular mass of the mature bovine heart subunit, which may be different in other species. When possible, SwissProt names are used for proteins not present in the bovine complex.

Central Subunits

Sequence analysis of the 14 central subunits found in prokaryotic and eukaryotic complex I (27) immediately reveals that these proteins fall into two very different categories (**Table 1**): seven subunits are highly hydrophobic and, depending on the algorithm used, are predicted to contain a total of 52–59 transmembrane helices; in contrast, not a single transmembrane segment is consistently predicted for the remaining seven subunits, but they contain all binding motifs for redox prosthetic groups and the substrate NADH. Pointing toward the matrix side in mitochondria and to the cytoplasm in bacteria, the hydrophilic subunits form the peripheral part of the complex, which stands more or less perpendicular on one end of the membrane-embedded part containing the hydrophobic subunits. This results in an L-shaped particle that has been observed for complex I in

Table 1 Central subunits of complex I

Name of central subunit				Transmembrane segments[a]	Gene position[c]	Organisms where subunit is encoded in other compartment[d]
Bos taurus	SwissProt	Homo sapiens	Prokaryotes[b]	a/b		
					Nuclear encoded subunits	
75 kDa	NUAM	NDUFS1	NUOG / NQO3	0/0	2q33-q34	Cryptomonads, various protists
51 kDa	NUBM	NDUFV1	NUOF / NQO1	0/0	11q13	None
49 kDa	NUCM	NDUFS2	NUOD / NQO4	0/0	1q23	Cryptomonads, ciliates, *Trypanosoma*, other protists, plants
30 kDa	NUGM	NDUFS3	NUOC / NQO5	0/0	11p11.11	Cryptomonads, *Trypanosoma*, other protists, some plants
24 kDa	NUHM	NDUFV2	NUOE / NQO2	0/0	18p11.31-p11.2	None
TYKY	NUIM	NDUFS8	NUOI / NQO9	0/0	11q13	Cryptomonads, *Trypanosoma*
PSST	NUKM	NDUFS7	NUOB / NQO6	2/0	19p13.3	Cryptomonads, ciliates
					Mitochondrially encoded subunits	
ND1	NU1M	ND1	NUOH / NQO8	8/8	mt.3307	None
ND2	NU2M	ND2	NUON / NQO14	10/8	mt.4470	*Trypanosoma*
ND3	NU3M	ND3	NUOA / NQO7	3/3	mt.10059	*Chlamydomonas*
ND4	NU4M	ND4	NUOM / NQO13	13/12	mt.10760	None
ND4L	NULM	ND4L	NUOK / NQO11	2/3	mt.10470	*Chlamydomonas*, *Trypanosoma*, ciliates
ND5	NU5M	ND5	NUOL / NQO12	18/14	mt.12337	None
ND6	NU6M	ND6	NUOJ / NQO10	5/5	mt.14149	*Trypanosoma*, ciliates

[a]Transmembrane domains predicted for the bovine subunits using servers (a) http://www.enzim.hu/hmmtop/ (170) and (b) http://www.cbs.dtu.dk/services/TMHMM (171).
[b]The NUO names are used for *Escherichia coli* and *Rhodobacter capsulatus* and NQO names for *Paracoccus denitrificans*, *Thermus thermophilus*, and *Aquifex aeolicus*.
[c]The position of the gene on a human chromosome or the first base in human mitochondrial DNA is given.
[d]Simplified compilation of data, see Reference 33 for detailed analysis and database entries for complete mitochondrial DNA sequences.

different eukaryotic and prokaryotic sources (28–32).

Except for some plants, algae, and protists, the seven hydrophobic subunits are encoded by the mitochondrial genome in most eukaryotes (**Table 1**). However, in some cryptophytic algae such as *Rhodomonas salina* up to 12 subunits are encoded by mitochondria (33). In other organisms, including the green alga *Chlamydomonas reinhardtii*, more than seven central genes have been moved to the nucleus. Only three genes for central complex I subunits (ND1, ND4, and ND5) are always present, and the genes for two subunits (51 kDa and 24 kDa) are never found in the organelle genome (**Table 1**). This variability and the dual origin of the eukaryotic subunits highlight the complexity of evolution of the mitochondrial complex I in different phyla. It should be noted that 11 of the 14 central subunits are also found in chloroplasts from higher plants (34), where they are part of a so-called Ndh complex. This plastidal complex from *Synechocystis* lacks the 75-kDa, 51-kDa, and 24-kDa subunits. Instead, two other subunits (NdhM and NdhN) have been proposed to serve as an alternate electron entry module (35).

Accessory Subunits

Proteomic analyses have been performed for complex I from the mammals *Bos taurus* (18, 19) and *Homo sapiens* (23), the fungus *Neurospora crassa* (24), the yeast *Yarrowia lipolytica* (22), the green alga *C. reinhardtii* (20), and the plants *Arabidopsis thaliana* and *Oryza sativa* (21). These studies have been complemented by genome-wide searches and detailed phylogenetic analyses in other species, including *Mus musculus*, *Drosophila melanogaster*, and *Caenorhabditis elegans* (34). With 46 different subunits, mammalian complex I has the highest number of reported subunits (18, 19). So far, only the sequence of a 10.6-kDa protein has not been determined. Of the remaining 31 accessory subunits of bovine heart complex I (**Figure 1**), 18 proteins have clear orthologues in mitochondria from all other lineages studied (**Table 2**). With the notable exception of the mitochondrial acyl-carrier protein homologue (SDAP), not only the mitochondrial but all nuclear complex I genes are absent in the fermenting yeast *S. cerevisiae*, which lacks complex I (9). Of the remaining 13 accessory subunits, 9 are found in all metazoa analyzed. Two of these proteins seem not to be present in insects (KFYI, B9) and an additional two (MLRQ, MNLL) are missing in *C. elegans*. Some authors have suggested that a 9.5-kDa protein in fungi was orthologous to subunit B9 (34, 36), but this has been challenged by others (20). One accessory subunit (NUXM) is found in fungi, green algae, and higher plants but not in metazoa. Three proteins were identified only in fungal complex I; of the three, subunit NUZM is always present, but subunit NURM and a newly identified 10.4-kDa subunit, found only in *N. crassa*, are not (24). Altogether, in green algae and higher plants, 13 additional proteins were found to be associated with complex I, 4 of these seem to be specific for green algae, and 2 for higher plants (34). The plant-specific subunits comprise four (*A. thaliana*) or five (*C. rheinhardtii*) isoforms of γ-type carbonic anhydrase (37). These proteins form a distinct subcomplex attached to the membrane arm, which confers an additional catalytic function to complex I (38).

As for other respiratory chain complexes, it has been speculated that several accessory subunits are required to organize the biogenesis of eukaryotic complex I, a process that may be complicated by the dual genetic origin of the multiprotein assembly. In bovine complex I, 14 smaller subunits (M_r = 6–19 kDa) classify as single transmembrane domain (STMD) (**Table 2**) proteins. By readily inserting into the mitochondrial inner membrane, these proteins could promote and organize assembly of the mitochondrially encoded subunits by interacting with their highly hydrophobic transmembrane segments and then remain attached to the complex (22). Such a function is likely to depend more on the

Table 2 Accessory subunits of complex I

Bos taurus	SwissProt name	Homo sapiens	Human chromosome[a]	Remarks
39 kDa	NUEM	NDUFA9	12p13.3	Short-chain dehydrogenase, NADPH
15 kDa	NIPM	NDUFS5	1p34.2–p33	Cysteine-rich motif
13 kDa	NUMM	NDUFS6	5p15.33	
AQDQ	NUYM	NDUFS4	5q11.1	Phosphorylation
ESSS	NESM	NDUFB11	Xp11.3	STMD[b], phosphorylation
MWFE[c]	NIMM	NDUFA1	Xq24	STMD, phosphorylation
PDSW	NIDM	NDUFB10	16p13.3	Cysteine-rich motif
PGIV	NUPM	NDUFA8	9q33.2–q34.11	Cysteine-rich motif
SDAP	ACPM	NDUFAB1	16p12.1	Acyl-carrier protein, phosphopantetheine
B22	NI2M	NDUFB9	8q13.3	
B18	NB8M	NDUFB7	19p13.12–p13.11	Cysteine-rich motif
B17.2	N7BM	DAP13	12q22	Nitration
B16.6	NB6M	GRIM19	19p13.2	STMD, proapoptotic factor
B14.7	NUJM	NDUFA11	19p13.3	TIM17/22 family
B14	NB4M	NDUFA6	22q13.2–q13.31	Nitration
B13	NUFM	NDUFA5	7q32	
B12	NB2M	NDUFB3	2q31.3	STMD
B8	NI8M	NDUFA2	5q31	Thioredoxin fold
Metazoa specific subunits				
42 kDa	NUDM	NDUFA10	2q37.3	Phosphorylation
10 kDa	NUOM	NDUFV3	21q22.3	
AGGG	NIGM	NDUFB2	7q34	STMD
ASHI[d]	NIAM	NDUFB8	10q23.2–q23.33	STMD
KFYI[c,e]	NIKM	NDUFC1	4q28.2–q31.1	STMD
MLRQ[c]	NUML	NDUFA4	7p21.3	STMD
MNLL[c]	NINM	NDUFB1	14q32.12	STMD
SGDH	NISM	NDUFB5	3q26.33	STMD
B17	NB7M	NDUFB6	9p21.1	STMD
B15[d]	NB5M	NDUFB4	3q13.33	STMD, nitration
B14.5a	N4AM	NDUFA7	19p13.2	
B14.5b	N4BM	NDUFC2	11q14.1	STMD
B9[c,d,e]	NI9M	NDUFA3	19q13.42	STMD

[a]Map positions were obtained from **www.ncbi.nlm.nih.gov/entrez** in July 2005.
[b]Single transmembrane domain (STMD).
[c]Not found in *C. elegans* (34).
[d]Also found in fungi (34).
[e]Not found in insects (34).

overall structure of the assembly factor than its protein sequence. This may explain why, for as many as eight STMD proteins of bovine complex I, no orthologues could be identified in fungi and plants (**Table 2**).

Four accessory subunits contain several conserved cysteines (**Table 2**), and it was proposed that these proteins may be involved in iron-sulfur cluster assembly (20, 39). Moreover, the NMR structure of the human subunit B8 revealed a Fe_2S_2 ferredoxin fold similar to the one found in thioredoxin (40). This protein is also homologous to the mitochondrial L43/S25 family of ribosomal proteins (34). Finally, it should be noted that a protein recently found to be tightly associated with complex I from *Y. lipolytica* exhibits clear homology to the family of sulfur transferases (41) and confers rhodanese activity to purified complex I (41a).

Subunit B14.7 exhibits some homology to components of the translocase of the inner membrane (42) and is therefore considered a member of the TIM17/22 family (43). It should be noted that so far only two complex I-specific assembly factors CIA30 and CIA84 have been identified in *N. crassa* and that these are not subunits of complex I (44). The assembly process of eukaryotic complex I has been studied in *N. crassa* (45) and humans (46), but the picture emerging from these studies is still rather incomplete and not yet consistent.

Phosphorylation of several accessory subunits of complex I was observed. In initial studies, two proteins were reported to become phosphorylated in a cAMP-dependent fashion. The bigger protein was identified as subunit AQDQ (47, 48), and later the smaller band as subunit MWFE (49). Challenging the initial assignment, it has been demonstrated recently that subunit ESSS, but not subunit AQDQ, becomes phosphorylated. The site of phosphorylation was identified as serine 20, and subunit MWFE is phosphorylated at serine 55 (50). The latter finding is potentially more significant because mutations in subunit MWFE of Chinese hamster cell lines were shown to markedly affect complex I function (51, 52). The third protein that was found to be phosphorylated was the mammal-specific 42-kDa subunit (53). In this subunit, serine 59 of bovine complex I carries the phosphate group. As this residue is missing in man and mouse, a physiological role for this modification seems unlikely. It is worth noting, however, that the 42-kDa subunit has some homology to bacterial deoxyribonucleotide kinase (34).

Subunit B16.6 (54) is identical to GRIM-19 (gene associated with retinoid-interferon-induced mortality), an apoptosis-inducing factor (55, 56). GRIM-19 knockout mice are deficient for complex I assembly (57), but it remains unclear whether the presence of this protein in complex I is directly related to its proapoptotic function. It should be noted that subunit B16.6, but not the characteristic "death domain" of GRIM-19, is conserved in fungi. Subunit B17.2 was first identified as DAP13 (a differentiation-associated protein) by an expression screen in proliferating human erythroid cells (58), but the link to complex I function is unclear. Notably, subunit B17.2 is one of three accessory (**Table 2**) and two central complex I subunits (49 kDa and TYKY) that form tyrosine nitrates when incubating mitochondria with peroxynitrite (59).

Two accessory subunits of complex I carry coenzymes known from fatty acid metabolism and exhibit similarities to enzymes of this metabolic pathway. Complex I from *N. crassa* contains NADPH tightly bound to the 39-kDa subunit (60). The sequence of this protein exhibits homologies to the family of short-chain dehydrogenases and contains a conserved NADPH-binding motif (27, 61). Subunit SDAP is a member of the family of mitochondrial acyl-carrier proteins and carries a serine-linked phosphopantetheine group that in complex I was always found to be loaded with an acyl group (62, 63). Initially this acyl group was identified as 3-hydroxytetradecanoate (64), but it remains unclear whether this assignment is correct (19). Deletion of the gene for the SDAP protein from the genome of *N. crassa* resulted in impaired assembly of complex I and a four-fold increase in mitochondrial lysolipids (65). This suggests that subunit SDAP is directly involved in the fatty acid metabolism of mitochondria, but the exact reactions and pathways have not been elucidated.

MODULAR DESIGN OF COMPLEX I

Rather than just "bound hydrogen," some prokaryotes also utilize H_2 as reducing equivalent. In these organisms, molecular hydrogen is evolved or consumed by specialized enzymes called hydrogenases (66). Remarkably, members of this enzyme class are evolutionarily related to complex I (67–70). Phylogenetic analyses revealed that most of the central

FP: flavoprotein, a subcomplex of complex I, containing the 51-kDa, 24-kDa, and 10-kDa subunits

subunits have evolved from subunits of different types of hydrogenases (**Table 3**). These evolutionary relationships are consistent with a plethora of structural and functional evidence and define three functional modules of complex I: the electron input module or N module that oxidizes NADH, the electron output module or Q module that reduces ubiquinone or in some cases menaquinone, and the P module that translocates the protons across the membrane.

The N Module

The N module is the input module that transfers electrons from NADH via FMN onto a chain of iron-sulfur clusters. It comprises the 75-kDa, 51-kDa, and 24-kDa subunits and has evolved from the α subunit encoded by *hoxF* and in part from the γ subunit encoded by *hoxU* of NAD$^+$-reducing hydrogenases similar to the one in *Alcaligenes eutrophus* (71) (**Table 3**). The N module is identical to the NADH dehydrogenase fragment biochemically prepared from *Escherichia coli* complex I using chaotropic agents (72). In contrast, the so-called flavoprotein (FP), which can be obtained by splitting the peripheral arm of bovine complex I, contains only the 51-kDa, 24-kDa, and 10-kDa subunits (73) (**Figure 1**). Electrons from NADH enter the N module of complex I through FMN, which is noncovalently bound to the 51-kDa subunit. Near the N terminus of the 51-kDa subunit, an NADH-binding motif and a typical βαβ-nucleotide fold are found (27). Binding of NADH to this subunit was demonstrated experimentally using photoreactive NAD$^+$ derivatives (74, 75). FMN serves as a two-to-one electron converter and feeds single electrons into a "wire" composed of iron-sulfur clusters. Near the C terminus of the same protein a conserved iron-sulfur cluster binding motif [CxxCxxC-(x)$_{39}$-C] is found that was assigned experimentally to the tetranuclear iron-sulfur cluster N3 (see Reference 16 for an explanation of the nomenclature of iron-sulfur clusters in complex I; 76, 77). The 24-kDa subunit also contains an iron-sulfur cluster binding motif [CxxxxC-(x)$_{35}$-CxxxC] that binds the binuclear cluster N1a (76). Although the binding motif is conserved, cluster N1a is not detectable by electron paramagnetic resonance (EPR) in fungal complex I (31, 78). The remaining iron-sulfur clusters of the N module, the tetranuclear clusters N4 and N5, and the binuclear cluster N1b are bound in the 75-kDa subunit (16). From the crystal structure of the homologous CpI hydrogenase from *Clostridium pasteurianum* (79) and heterologous expression of the subunit from *Paracoccus denitrificans* (80), it was proposed that cluster N1b is bound by a [C-(x)$_{10}$-CxxC-(x)$_{11–13}$-C] motif, cluster N5 by a [HxxxCxxC-(x)$_5$-C] motif, and cluster N4 by a [CxxCxxC-(x)$_{39–46}$-C] motif. However, although this ligation is found in the hydrogenase structure, the histidine cannot be the fourth ligand of cluster N5 because removing this residue by mutagenesis in *Y. lipolytica* did not affect cluster N5 (81). The EPR spectrum of reduced cluster N5 is only observed at very low temperatures and high microwave power and has been detected only in complex I from bovine heart (16), *Y. lipolytica* (31), and *Rhodobacter sphaeroides* (82). In some bacteria, including *E. coli*, a third tetranuclear cluster called N7 is found; it is bound to a [CxxCxxxC-(x)$_{27}$-C] motif in the part of the 75-kDa subunit not homologous to the γ subunit of NAD$^+$-reducing hydrogenases (70, 72, 80). This cluster was initially believed to be binuclear and therefore was designated N1c, but on the basis of recent evidence, it was renamed cluster N7 (83).

Chloroplasts, blue algae (20), and the archaeon *Archaeoglobus fulgidus* (84) contain enzyme complexes that are related to complex I but lack the N module. In other relatives of complex I from organisms such as *Helicobacter pylori* and *Campylobacter jejuni*, the 51-kDa and 24-kDa subunits are missing (85, 86). These complexes must contain alternate electron input modules specific for other electron donors.

Table 3 Evolutionary relationships between complex I and other enzyme complexes

Subunit symbol	Functional module	Redox prosthetic groups	Homologous subunits in related enzymes					
			NAD⁺-reducing hydrogenase *Alcaligenes eutrophus*	Water-soluble [NiFe] hydrogenase, e.g., *Desulfovibrio fructosovorans*	Membrane-bound [NiFe] hydrogenase *Methanosarcina barkeri*	Hydrogenase 3 (FHL-1) *Escherichia coli*	Hydrogenase 4 (FHL-2) *Escherichia coli*	Antiporter, e.g., *Bacillus subtilis*
75 kDa	N	N1b, N5, N4, (N7)	γ subunit[a]	—	—	—	—	—
51 kDa		FMN, N3	α subunit	—	—	—	—	—
24 kDa		N1a	α subunit	—	—	—	—	—
49 kDa	Q	—	(β subunit)[a]	Large subunit	EchE	HycE	HyfG	—
30 kDa		—	—	—	EchD	HycE	HyfG	—
TYKY		N6a, N6b	—	—	EchF	HycF	HyfH	—
PSST		N2	(δ subunit)[a]	Small subunit	EchC	HycG	HyfI	—
ND1	P	—	—	—	EchB	HycD	HyfC	—
ND2		—	—	—	EchA[b]	HycC[b]	HyfB,D,F[b]	MrpD[b]
ND3		—	—	—	—	—	—	—
ND4		—	—	—	EchA[b]	HycC[b]	HyfB,D,F[b]	MrpD[b]
ND4L		—	—	—	—	—	HyfE?	—
ND5		—	—	—	EchA[b]	HycC[b]	HyfB,D,F[b]	MrpA[b]
ND6		—	—	—	—	—	—	—

[a]Only the first 200 residues of the 75-kDa subunit are homologous to the γ subunit; the homologies to the β and δ subunits are rather weak.
[b]Subunits ND2, ND4, and ND5 are weakly homologous to each other; thus the assignment of the individual subunits of related enzymes is ambiguous.

[NiFe]: nickel-iron center of a hydrogenase that binds hydrogen

The Q Module

The Q module accepts electrons from the iron-sulfur clusters of the N module and transfers them via three more iron-sulfur clusters onto ubiquinone or, in some bacteria, also onto menaquinone. The Q module comprises the 49-kDa, 30-kDa, TYKY, and PSST subunits. The large and small subunits of water-soluble [NiFe] hydrogenases are homologous to the 49-kDa and part of the PSST subunit, respectively (**Table 3**) (67, 68). The 30-kDa subunit is fused to the N terminus of the 49-kDa subunit in *E. coli* (4). Already in membrane-bound type-3 hydrogenase from *E. coli* (67) and Ech hydrogenase from *Methanosarcina barkeri* (87), the small subunit homologue is shortened, which removes the binding folds for two of the three iron-sulfur clusters. This leaves only the domain harboring the proximal cluster intact. The missing iron-sulfur cluster binding domain is replaced by an additional ferredoxin-type subunit that corresponds to the TYKY subunit of complex I (**Table 3**). From the *E. coli* complex I, a so-called connecting fragment representing the Q module can be prepared biochemically (72). Subunit TYKY contains binding motifs for the two tetranuclear iron-sulfur clusters N6a and N6b: [CxxCxxCxxxCP-$(x)_{27}$-CxxCxxCxxxCP] (88). The tetranuclear iron-sulfur cluster N2 is bound to the PSST subunit (89–92). Notably, the first cysteine of the binding motif for the proximal cluster of [NiFe] hydrogenases is missing in subunit PSST. Studies to identify the missing fourth ligand were not successful (91, 93, 94), but modeling studies suggest that the conserved cysteine immediately adjacent to the second cysteine of the hydrogenase motif may be the fourth ligand, resulting in [CC-$(x)_{63}$-C-$(x)_{29}$-CP] as the binding motif for cluster N2 (95). Cluster N2 was shown to interact with an ubisemiquinone species formed during turnover (96). Therefore, cluster N2 is considered as the immediate electron donor to ubiquinone.

A combination of N and Q module subunits is also found in NAD^+-reducing hydrogenase: Proteins of this bacterial enzyme exhibit sequence similarities to subunits of the N module (**Table 3**), and homology of the 49-kDa subunit to the β subunit encoded by *hoxH* and of the PSST subunit of complex I to the δ subunit encoded by *hoxY* were also reported (68). However, the similarities between the two enzymes are rather weak, and the situation may represent an independent combination of related functional modules in complex I and in the NAD^+-reducing hydrogenase.

The P Module

The subunits assigned to the N and Q module are part of the peripheral arm of complex I (**Figure 1**). It follows that the entire electron transfer chain from NADH to ubiquinone is confined to these two modules and that proton pumping must occur in a third, membrane-embedded module. This P module comprises the seven membrane-embedded subunits ND1, ND2, ND3, ND4, ND4L, ND5, and ND6, which are encoded by the mitochondrial genome in most eucaryotes (**Table 1**). Subunits ND2, ND4, and ND5 are homologous to each other, and one representative of this family has already been acquired together with an ND1 homologue at the level of membrane-bound type-3 [NiFe] hydrogenases found in *E. coli* (27) or the archaeon *M. barkeri* (97) (**Table 3**). Two more proteins of the ND2/ND4/ND5 family and an ND4L homologue were identified in the hydrogenase part (Hyd-4, *hyf* operon) of the formate hydrogen lyase (FHL)-2 complex from *E. coli* (98). These proteins are homologous to components of the P module (70). Finally, CO dehydrogenases, e.g., from *Rhodospirillum rubrum*, have a subunit composition similar to membrane-bound [NiFe] hydrogenases and are therefore also related to complex I (97). Remarkably, the ND2/ND4/ND5 family of subunits seems to have evolved from Na^+/H^+ antiporters (27, 99). For example, the MrpA and MrpD proteins encoded by the *mrp* (multiple resistance and pH adaptation) operon of *Bacillus subtilis* show weak homology to these three subunits (**Table 3**). MrpA and MrpD are important for Na^+ resistance and pH homeostasis, and they function as Na^+/H^+ antiporters. Such antiporters are found in many bacteria. The transporter function may have been retained by the related membrane-embedded subunits of complex I. Initially, it was proposed that the transporter subunit found in type-3 hydrogenases was the homologue of ND5 (27), but according to a more recent analysis, it seems impossible to decide just by sequence comparisons which transporter subunit of complex I was derived from the corresponding protein in hydrogenase (99).

STRUCTURE OF COMPLEX I

For a long time, the only direct structural information on complex I was obtained by electron microscopic single particle analysis and two-dimensional crystallography. Most of these studies revealed an L-shaped molecule, comprising a membrane arm and a peripheral arm, which protruded into the mitochondrial matrix or the bacterial cytoplasm (29, 30–32, 100, 101). Only one three-dimensional reconstruction of *E. coli* complex I, obtained at zero ionic strength, showed a horseshoe-shaped particle (101), but this finding could not be reproduced and was interpreted as an artifact by others (102). A horseshoe shape was never reported for eukaryotic complex I. In complex I from *Y. lipolytica*, indications for a bridge or tether connecting the center of the peripheral arm with the end of the membrane arm were found (103), but the significance of this observation remains unclear. Complex I particles exhibit remarkable flexibility (29, 103), and there is evidence of significant redox-dependent conformational changes (104–106).

By combining the results from low- to medium-resolution methods of structure determination with those from biochemical approaches and by exploring structural homologies, our picture of the structural organization of complex I has been improving over recent years. However, only very recently progress toward a high-resolution X-ray structure was made.

The Peripheral Arm

Using crystals of the peripheral arm of complex I from *Thermus thermophilus*, the shape of this soluble part of the complex was solved at ∼4-Å resolution, and the positions of the iron-sulfur clusters were determined (17). Nine iron-sulfur clusters were identified in the structure (**Figure 2**), which matches the number of binding motifs found in the sequences of the central subunits. The deduced relative arrangement of the iron-sulfur clusters is in full agreement with the structural organization of the central subunits as suggested by the modular architecture and the evolutionary origin of complex I discussed above. From the distances between the clusters, it can be concluded that seven of them form an electron "wire" that bridges a distance of 84 Å. Iron-sulfur cluster N1a and the prokaryote-specific cluster N7 were found too far away for direct rapid electron transfer. The electron density of FMN was not identified in the *Thermus* structure, but it was suggested to reside between cluster N1a and N3 (17). This would allow rapid electron exchange with cluster N1a, a feature that may be important for the mechanism of FMN oxidation and reduction.

There is conflicting evidence regarding the critical question of how the peripheral arm of complex I and thus the iron-sulfur clusters have to be oriented relative to the membrane arm. However, it was not possible to align the characteristic Y shape of the peripheral arm of the *T. thermophilus* complex with one of the published three-dimensional models obtained by single particle analysis (29, 30, 101), because none of the the studies shows sufficient structural detail. Only the three-dimensional reconstruction of complex I from *Y. lipolytica* shows a Y shape

Single particle analysis: an electron microscopic method of obtaining three-dimensional images by averaging single molecules

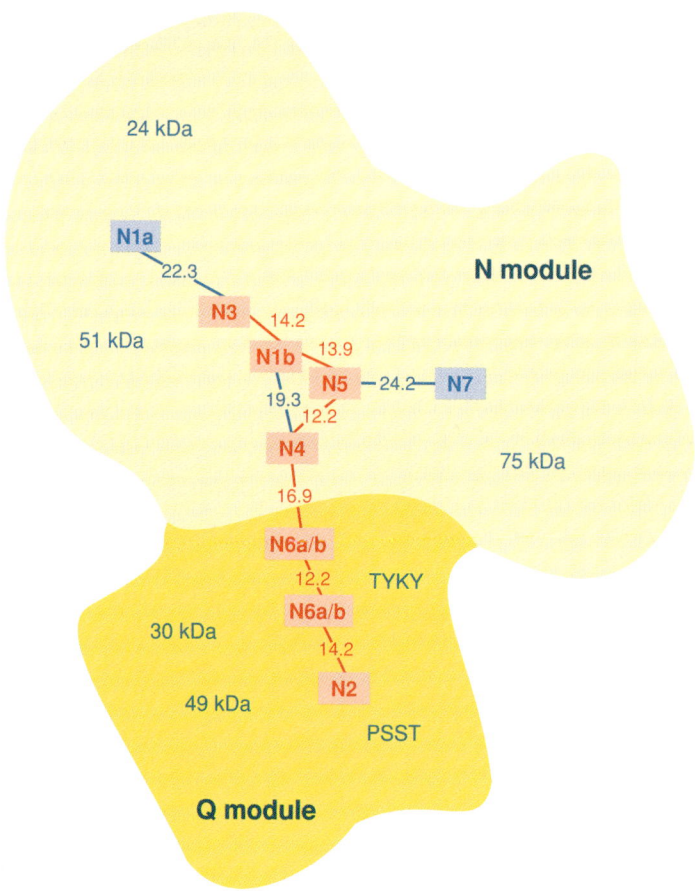

Figure 2
Arrangement of iron-sulfur clusters within the peripheral arm. This cartoon shows the overall shape of the peripheral arm of complex I from *Thermus thermophilus* with the positions of the iron-sulfur clusters (*boxes*) (17). Distances are given in angstroms. Some of the redox centers (*red*) form a contiguous electron transfer chain through the complex. The expected approximate positions of the central subunits (*blue*) are indicated by the subunit names.

for its peripheral arm (M. Radermacher, T. Ruiz, T. Clason, S. Benjamin, U. Brandt, V. Zickermann, submitted) (**Figure 3**). However, placing the prokaryotic substructure into this low-resolution model of a eukaryotic complex I is difficult as the accessory subunits in the periperal arm increase the mass by about 50%. Because cluster N2 is probably the immediate electron donor to hydrophobic ubiquinone, Hinchliffe & Sazanov (17) proposed that this cluster must be the one closest to the membrane. In fact, this view gets some support from data obtained in the *P. denitrificans* complex I, which shows that the hydrophobic ND3 subunit can be cross-linked to the 49-kDa and PSST subunits (107). However, the use of "cholate-washed membranes" may have promoted formation of stacks of membrane fragments. Therefore, the observed cross-links may have occurred between complexes rather than within the same particle. Conversely, it was shown by electron microscopic analysis of *Y. lipolytica* complex I particles decorated with specific monoclonal antibodies that the 49-kDa and 30-kDa subunits reside far away from the membrane arm of complex I (103). This observation gains support from the electron microscopic three-dimensional model of an *E. coli* subcomplex I, which lacked the entire N module (101) and was missing a mass on the inside of the peripheral arm. A similar observation was made with a *Y. lipolytica* subcomplex I, wherein removal of the 51-kDa and 24-kDa subunits resulted in the dissapearence of a portion of the peripheral arm, which normally protrudes near the membrane arm (M. Radermacher and V. Zickermann, personal communication).

Most significantly, no shortening of the peripheral arm was observed in these subcomplexes (**Figure 3**). These results indicate that the interface of the *T. thermophilus* fragment with the membrane arm is not near cluster N2 and that most of the mass of complex I near the membrane is formed by the FP fragment that contains the NADH-binding site. This implies that cluster N2 resides far up in the peripheral arm and that the head group of ubiquinone has to leave the membrane domain via an amphipathic ramp or crevice to reach its electron donor (103). Although it is hard to imagine that nature would have constructed complex I with its electron transfer chain "upside-down," the experimental evidence suggesting this option cannot be ignored. Further studies and a high-resolution structure of the entire complex are needed to resolve this issue.

The Membrane Arm

In most organisms studied to date, the membrane arm of complex I has a length of about 200–230 Å (**Figure 3**). Complex I from *A. thaliana* carries an additional carboanhydrase module that protrudes into the mitochondrial matrix near the middle of the membrane arm (38). Subcomplex Iα, comprising the peripheral arm and part of the membrane arm (see **Figure 1**), contains ND1 and probably ND2 (19, 26). This suggests that these subunits reside at the junction between the peripheral and the membrane arm. Analysis of two-dimensional crystals of a subcomplex lacking subunit ND5 suggests that this protein is located at the distal end of the membrane arm (108). Studies of ND5-deficient mutants in mouse (109) and *C. reinhardtii* (110) support this finding. Subunits ND4 and ND5 can be split off together with subcomplex Iβ placing ND4 in the same region of complex I (25, 26, 111).

The transmembrane topology of the central subunits of the membrane arm has been determined by various approaches for bacterial complex I. Of the transporter-type sub-

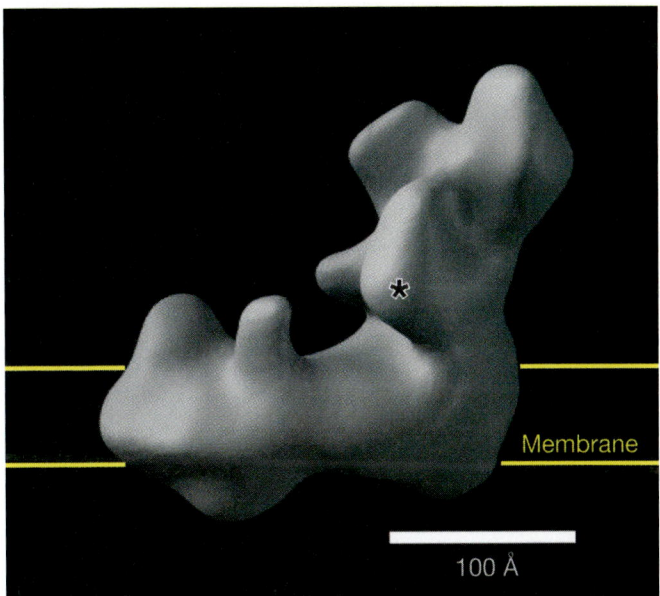

Figure 3

Three-dimensional model of complex I from *Y. lipolytica* (M. Radermacher, T. Ruiz, T. Clason, S. Benjamin, U. Brandt, V. Zickermann, submitted). The approximate position of the membrane is indicated by horizontal lines. The model was obtained by reconstruction from negatively stained single particles using the random conical tilt method. The resolution is approximately 24 Å. The scale bar is 100 Å. The star indicates the mass that is not seen in three-dimensional reconstructions of a subcomplex lacking the 51-kDa and 24-kDa subunits.

units, ND2 and ND4 contain 12, whereas ND5 contains 14 transmembrane segments (99). Subunit ND1 has eight transmembrane helices (112), subunit ND3 (113) and subunit ND4L (114) have three, and subunit ND6 has five (115). According to these studies, the total number of transmembrane domains for the central subunits is 57. Except for subunit ND3, the N terminus of the central hydrophobic subunits was found to reside on the periplasmic side, which corresponds to the intermembrane space in mitochondria.

FUNCTION OF COMPLEX I

Functional analysis of complex I has been limited by the lack of detailed structural information. However, evidence obtained from a great number of biochemical and biophysical studies now excludes many of the

Midpoint potential: the standard electric potential at which 50% of a redox component is reduced. Usually given for pH 7 (E_{m7})

hypothetical mechanisms that have been proposed for redox-linked proton pumping by complex I over the years. In this respect, EPR spectroscopy, the availabilty of numerous inhibitors, and, more recently, the analysis of side-directed mutations have proven to be quite useful.

Substrates, Redox Centers, and Inhibitors

The physiological activity of complex I is the electron transfer from NADH (E_{m7} = −320 mV) to ubiquinone ($E_{m7} \cong$ +90 mV) or, in some cases, to menaquinone ($E_{m7} \cong$ −80 mV); this activity is linked to the formation of an electrochemical membrane potential. Despite a difference in the substrate redox potentials of about 400 mV, the reaction catalyzed by complex I is fully reversible; it was demonstrated a long time ago that in the presence of a protonmotive force mitochondria can transfer electrons from succinate onto NAD^+ (116, 117). There are even indications that, for example, in some photosynthetic bacteria complex I may physiologically operate in reverse (6).

The K_m value of complex I for NADH is in the micromolar range, and weak product inhibition at millimolar concentrations of NAD^+ can be observed (118). In contrast, the K_m of NAD^+ for the reverse reaction is in the micromolar range. This has been taken as an argument for the presence of a second $NADH/NAD^+$-binding site. However, as exemplified by cytochrome c oxidase, this discrepancy may be explained by some kind of "kinetic trapping" mechanism (119). Other arguments that were put forward in favor of two $NADH/NAD^+$-binding sites are the observation that the competitive inhibitor ADP-ribose acts only in the forward reaction (120) and that "transhydrogenase" activities of complex I follow bisubstrate kinetics (121). Because these kinetic issues have to be interpreted with great caution, it is not possible to discuss them in more detail, but considering the linear arrangement of the iron-sulfur clusters in complex I (17), there seems to be no obvious need for a second NADH-binding site.

The primary electron acceptor of complex I is FMN (E_{m7} = −380 mV). Like many flavoproteins, complex I can be inhibited with diphenyleneiodonium by covalent modification of the reduced flavine (122). Two widely used artificial reactions of complex I, which seem to employ only the NADH-binding site and FMN in a ping-pong type mechanism, are the NADH:ferricyanide oxidoreductase (1) and the NADH:hexammineruthenium oxidoreductase (123) activities. Because these reactions are much more robust than the physiological one, they are useful for quantification of complex I, e.g., during purification. The binuclear iron-sulfur cluster N1a (E_{m7} = −370 mV in the bovine enzyme) seems to reside in the vicinity of FMN but is not part of the main electron pathway through complex I (**Figure 2**). A possible reason for this arrangement is that the low-potential redox centers have to cooperate in the two-to-one electron conversion from NADH to the iron-sulfur clusters, a function that is usually attributed exclusively to FMN.

There was a long-standing discussion about the number, sequence, and subunit assignment of the iron-sulfur clusters in complex I. Linear (124) and branched (125) electron transfer pathways were proposed. From the structure of the peripheral arm of *T. thermopilus* (17), it is now clear that seven of the eight iron-sulfur clusters present in all versions of complex I form a linear electron pathway through the protein. Only cluster N1a and the prokaryote-specific cluster N7 seem not to be connected to this chain that "electrically" links FMN to quinone. In fact, with the exception of cluster N2, all redox centers in this chain (clusters N3, N1b, N5, N4, N6a, and N6b) have a midpoint potential of about −250 mV (16, 88) and are, therefore, sometimes called the isopotential iron-sulfur clusters of complex I. Iron-sulfur cluster N2 is the last redox center in the chain (**Figure 1**) and has a more positive and pH-dependent midpoint potential (16, 126). It interacts paramagnetically with two semiquinone species that accumulate during steady-state turnover and differ in their spin relaxation behavior (96, 127). The fast-relaxing species Q_{Nf} is only observed in the presence of a membrane potential, whereas the slow-relaxing species Q_{Ns} is also seen during uncoupled turnover. By analyzing the coupling constants between the two unpaired electrons, a distance of 12 Å between cluster N2 and Q_{Nf} was calculated (127). It cannot be decided, however, whether the two signals reflect two semiquinones bound to separate binding sites or two states of the same molecule. At any rate, the fact that Q_{Nf} is found adjacent to a redox prosthetic group residing in the peripheral arm indicates that at least one ubiquinone-binding site is not deeply buried in the membrane domain of complex I. This is remarkable because the quinone derivatives, such as ubiquinone-9 or -10 that act as natural substrates for complex I, are extremely hydrophobic. Indeed, mutations in the ND1 subunit of the membrane were reported to affect the K_m for hydrophobic quinones (128), and the presence of a ubiquinone-binding motif was discussed for subunits ND4 and ND5 (129). Conversely, complex I also exhibits high activities with comparably hydrophilic ubiquinone derivatives, such as ubiquinone-1 or n-decylubiquinone, which are widely used for complex I assays. The structure of the side chain markedly affects the catalytic potency of quinone derivatives (130), but even rather bulky side chains can be accommodated by complex I (131). This is in line with studies on the binding of hydrophobic inhibitors to complex I, and a plethora of inhibitors that interfere with quinone reduction have been characterized over the years. On the basis of steady-state inhibition studies, they can be divided into three classes (132). However, it was shown by equilibrium-binding studies that they all share one large binding pocket within complex I (133). In addition to the classical inhibitor rotenone, piericidin A and rolliniastatin-1 inhibit complex I in bovine submitochondrial particles in nanomolar concentrations. Rolliniastatin-1 is one of the so-called annonaceous acetogenins, secondary metabolites from *Annonaceae* plants, from which hundreds of derivatives have now been isolated or synthesized (134). Other remarkable compounds that inhibit complex I with lower affinity include the drugs amytal, meperidin, and capsaicin, the vanilloid from peppers. Some complex I inhibitors are used commercially as acaricides (135). Also, several derivatives of ubiquinone (130, 132) and even polyoxyethylene ether detergents, e.g., Triton X-100 or Thesit (136), specifically inhibit complex I.

> **Semiquinone:** a half-reduced radical form of two electron acceptors, such as ubiquinone or flavine mononucleotide

If there is more than one distinct binding site for quinone in complex I, one might envision an arrangement as in the photosynthetic reaction center, where tightly bound Q_A donates electrons directly to the substrate Q_B. However, when complex I is delipidated upon purification, this leads to an almost complete loss of endogenous quinone. Still the inactive delipidated enzyme can be fully reactivated by the addition of phospholipids or reconstitution into liposomes without adding endogenous quinone (137). There have been claims of a covalently bound quinone-like group in complex I (138), but so far, attempts have failed to identify this group. Overall, the number of quinone molecules involved in the mechanism of complex I remains unknown, but as in the possibility of multiple NADH-binding sites, there is no compelling evidence against the simplest assumption that complex I contains a single, but very large, binding domain for its hydrophobic substrate. This notion is in line with the idea, derived from structural studies, of an amphipathic ramp for ubiquinone within the peripheral arm of complex I (103).

The Quinone Reducing "Catalytic Core"

The proximity between iron-sulfur cluster N2 and semiquinone that is only observed in the presence of a membrane potential suggests

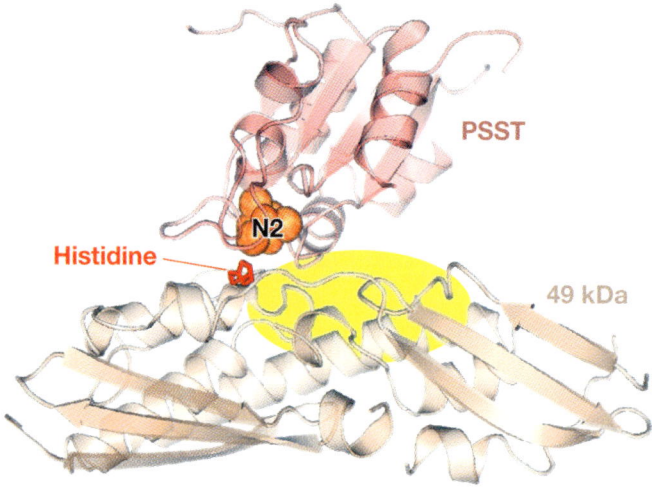

Figure 4

The ubiquinone-reducing catalytic core. The folds of the large and small subunit of water-soluble [NiFe] hydrogenase (Protein Data Bank entry 1FRF), which are conserved in the 49-kDa subunit (*light brown*) and the PSST subunit (*light red*), are shown. Iron-sulfur cluster N2 is shown as a space fill model, and the side chain of the histidine that acts as a redox-Bohr group (*red*) is shown as stick model. This residue is also present in soluble [NiFe] hydrogenases, where it forms a hydrogen bond with the proximal cluster (169). The yellow area indicates the region of the 49-kDa subunit where most of the inhibitor-resistant mutants were identified. It corresponds to the domain around the [NiFe] site in hydrogenase.

Redox-Bohr group: the protonable group controlling the midpoint potential of a pH-dependent redox center

that the 49-kDa and PSST subunits, which have developed from the hydrogen reducing domains of [NiFe] hydrogenase, contain major components of the quinone reducing "catalytic core" of complex I (**Figure 4**) and that reactions occurring there are critical for proton pumping (94). Another property that has stimulated several mechanistic models employing cluster N2 as a critical part of the proton pump (16, 139) is its pronounced pH-dependent midpoint potential or redox-Bohr effect (126). To explore these ideas, residues in the 49-kDa and PSST subunits, which are fully conserved within complex I or even invariant between the orthologous hydrogenase and complex I subunits, were chosen as the targets for mutagenesis (140, 141). Mutations showing functional defects, altered properties of cluster N2 or inhibitor resistance, were identified, clearly demonstrating that a significant part of the fold around the [NiFe]

site and the proximal cluster of hydrogenase is indeed conserved in complex I. As highlighted in **Figure 4**, resistance toward hydrophobic inhibitors was found in a region within the fold that harbors the [NiFe] site in hydrogenase. This region is located at the interface between the 49-kDa and the PSST subunit, which is in line with a report on photoaffinity labeling of subunit PSST (142). Together with other available evidence, this strongly supports the idea that the hydrogen site of hydrogenase was turned into part of the quinone reactive site of complex I, where semiquinone Q_{Nf} is formed as an intermediate during the catalytic cycle. In subunit PSST, two tyrosines near iron-sulfur cluster N2 (143) and as yet unidentified glutamates or aspartates (144) have been reported to exhibit redox-dependent protonations by Fourier transform infrared spectroscopy. Recently, we showed that removing a conserved histidine in the 49-kDa subunit (**Figure 4**) completely abolished the pH dependence of the midpoint potential of cluster N2, thus identifying this residue as the redox-Bohr group associated with this prosthetic group; mutating histidine-226 to methionine in the 49-kDa subunit of *Y. lipolytica* complex I resulted in a pH-independent midpoint potential of -220 mV for cluster N2. Yet complex I was fully active, and proton pumping was not affected (K. Zwicker, A. Galkin, L. Grgic, S. Kerscher, and U. Brandt, submitted).

Proton-Pumping Mechanism

Mitochondrial complex I was shown to pump protons at a stoichiometry of $4 H^+ / 2 e^-$ (145, 146). Pumping of sodium ions has been reported for bacterial complex I from *Klebsiella pneumoniae* (147) and *E. coli* (148), and a stoichiometry of $2 Na^+ / 2 e^-$ was published (149). However, this observation was challenged by other laboratories, and the observed pumping of sodium ions was explained by the presence of an NQR-type NADH dehydrogenase in *K. pneumonia* (150) and by possible secondary H^+/Na^+ antiporter activity in *E. coli* (151).

In fact, the evidence for sodium pumping of complex I by bacterial complex I is not compelling because most of the published rates are very low (147, 148, 152) and are only partially inhibited by complex I inhibitors (149). In addition, considering that these studies used either membranes or a rather crude preparation of complex I, it seems possible that the observed sodium pumping was not due to complex I. Thus, further studies will be required to decide whether complex I is capable of pumping not only protons but also sodium ions.

Many rather different hypothetical mechanisms for the coupling between electron and proton transfers have been proposed over the years (see Reference 139 for a synopsis of early models). As summarized above, recent evidence has imposed a number of critical functional and structural constraints thereby limiting the number of possible mechanistic scenarios. The dichotomy in the structural organization of complex I that places all the redox chemistry into the peripheral arm, although the actual vectorial proton translocation has to occur in the membrane arm, renders all directly coupled mechanisms unlikely. As proposed in a recent model (153), proton uptake on one side and proton release on the other side of the membrane must be linked directly to the redox reaction within complex I. This almost certainly requires that the redox component responsible for proton transfer to be located in the membrane arm. However, even if one follows the interpretation by Hinchliffe & Sazanov (17) that cluster N2 is the redox center closest to the membrane, the potential sensitive semiquinone Q_{Nf} is not buried in the membrane arm of complex I; with a distance of 12 Å from cluster N2, it must reside even then in the head group region of the phospholipids (17, 127). In principle, the second semiquinone Q_{Ns} could extend the electron transfer chain into the membrane, but this is unlikely because this semiquinone species is not affected by the membrane potential.

Overall, available evidence strongly favors the idea of an indirect conformational mechanism of redox-linked proton translocation (154, 155), and indeed redox-dependent conformational changes of complex I have been reported (104–106). Another observation may be important for the understanding of such a conformationally linked mechanism (118): It has been shown in some organisms that complex I may relax into a "deactive" form, exhibiting very slow steady-state turnover. This state of the enzyme can be returned into the "active" form by reduction and reoxidation. Because this transition was found to alter the accessibility of some thiol groups for modifying agents, it may well reflect a functionally important conformational change. To understand a conformational mechanism, one has to identify the precise redox reactions that drive the conformational change, the parts of the complex that transmit the conformational energy and the actual pumping devices in the membrane arm.

Remarkably, the midpoint potentials of all but one of the iron-sulfur clusters that transfer electrons within complex I are on a very similar rather negative level that is only about 70 mV more positive than that of NADH (**Figure 5**). As mentioned above, even turning cluster N2 into another isopotential cluster does not affect proton pumping by complex I (K. Zwicker, A. Galkin, L. Grgic, S. Kerscher, and U. Brandt, submitted). This means that most of the drop in redox potential occurs between cluster N2 and ubiquinone, yet during steady-state turnover the iron-sulfur clusters of complex I are mostly reduced (156). Thus, the reactions generating the conformational power stroke driving proton pumping must occur after the electrons have left cluster N2. This identifies the transiently formed semiquinone species as key components of the pump.

Although at present it is not possible to speculate about the structural components that transmit the conformational change to the membrane arm, there are some clues concerning which subunits of the membrane arm are the actual pumping devices. From their evolutionary origin, the transporter-type

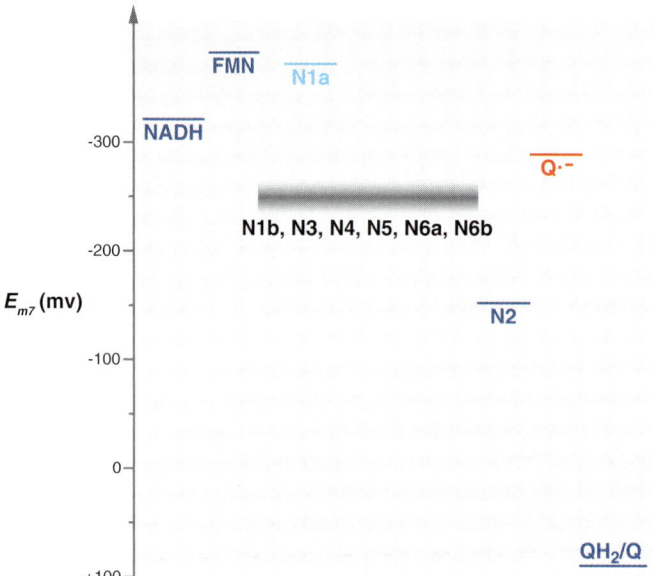

Figure 5

Midpoint potentials of the substrates and redox centers of complex I. The approximate level of the redox midpoint potential of the substrates and the iron-sulfur clusters N1a to N6b is indicated by horizontal bars. Cluster N1a (*blue*) is probably not directly involved in the electron transfer to quinone (Q). Semiquinone (Q$^{·-}$) (*red*) is likely to be a key player in the proton-pumping mechanism.

subunits ND2, ND4, and ND5 are obvious candidates for this function. This notion is supported by the observation that amiloride derivatives, compounds known to inhibit Na$^+$/H$^+$ antiporters, inhibit complex I activity most likely by acting on the ND5 subunit (151, 157, 158). The fact that this subunit is most likely located at the distal end of the membrane arm lends further support to a pumping mechanism involving long-range conformational changes. Like many proton translocating proteins, complex I is also inhibited by the carboxyl-modifying reagent dicyclohexylcarbodiimide (159), suggesting that buried carboxyl groups are involved in proton translocation. Mutagenesis of conserved glutamates and aspartates in several ND subunits of complex I (160, 161) and also of subunit PSST (92) has led to the identification of charged residues that are critical for complex I function, but how they are involved in proton pumping remains unclear.

MEDICAL ASPECTS OF COMPLEX I

About 50% of all mitochondrial disorders affecting the energy metabolism can be traced back to mutations in one of the subunits of complex I (13). The consequences of complex I deficiencies range from selective damage to the optic nerve, as in Leber's hereditary optic neuropathy, to severe encephalomyopathies, as in Leigh syndrome (14). Some of the mutations causing disease in humans have been reconstructed in bacteria (162), *Y. lipolytica* (163) and *N. crassa* (164). Remarkably, chronic application of low doses of complex I inhibitors have been shown to cause a *Morbus Parkinson*-like disease in the rat (165), and complex I dysfunction has been associated with aging phenomena (166). Proteins involved in apoptosis, e.g., GRIM-19 (54) and the apoptosis-inducing factor (167), were demonstrated to be associated with complex I. Moreover, complex I has been shown as a major source of mitochondrial reactive oxygen species (168). Although discussing all the medical implications of complex I goes far beyond the scope of this review, they clearly emphasize the need for a better understanding of the structure and function of this impressive membrane protein complex.

SUMMARY POINTS

1. Proton-pumping NADH:quinone oxidoreductase (complex I) is an L-shaped membrane protein complex containing 14 central subunits that are present in the bacterial and mitochondrial versions of the enzyme.

2. In addition, mitochondrial complex I contains up to 32 accessory subunits, some of which are probably involved in the assembly of the membrane integral multiprotein complex or confer additional functions.

3. The seven hydrophobic central subunits that are usually encoded by mitochondrial DNA reside in the membrane arm of complex I.

4. All known redox centers, flavine mononucleotide and eight to nine iron-sulfur clusters, seven of which form a linear chain, reside in the seven central subunits of the peripheral arm of complex I.

5. Complex I can be subdivided into three functional modules: the electron input module or N module that oxidizes NADH, the electron output module or Q module that reduces quinone, and the P module that pumps protons across the membrane.

6. Central parts of the ubiquinone-reducing catalytic core in the peripheral arm of complex I have evolved from the catalytic domain of [NiFe] hydrogenases.

7. The subunits of the membrane arm most likely to translocate the protons are related to bacterial Na^+/H^+ antiporters.

8. The mechanism of complex I is still unknown, but available evidence suggests that long-range conformational changes linked to the redox chemistry of quinone intermediates drive proton pumping across the membrane.

FUTURE ISSUES TO BE RESOLVED

1. More knowledge is needed concerning the assembly factors and the function of the accessory subunits of mitochondrial complex I.

2. Solving the molecular structure of complex I as a whole will be prerequisite to understand the mechanism of proton pumping and to resolve the current controversy about the localization of cluster N2.

3. Further studies with pure preparations of bacterial complex I are required to test whether in some cases complex I is capable of pumping sodium ions.

ACKNOWLEDGMENTS

This work was supported by SFB472 and SFB628 of the Deutsche Forschungsgemeinschaft. I thank Stefan Dröse, Stefan Kerscher, Hermann Schägger, and Volker Zickermann for carefully reading the manuscript and for helpful discussions.

LITERATURE CITED

1. Hatefi Y, Haavik AG, Griffiths DE. 1962. *J. Biol. Chem.* 237:1676–80
2. Matsushita K, Ohnishi T, Kaback HR. 1987. *Biochemistry* 26:7732–37
3. Yagi T, Yano T, Di Bernardo S, Matsuno-Yagi A. 1998. *Biochim. Biophys. Acta* 1364:125–33
4. Friedrich T. 1998. *Biochim. Biophys. Acta* 1364:134–46
5. Baccarini Melandri A, Zannoni D, Melandri BA. 1973. *Biochim. Biophys. Acta* 314:298–311
6. Dupuis A, Chevallet M, Darrouzet E, Duborjal H, Lunardi J, Issartel JP. 1998. *Biochim. Biophys. Acta* 1364:147–65
7. Hayashi M, Nakayama Y, Unemoto T. 2001. *Biochim. Biophys. Acta* 1505:37–44

8. Kerscher S. 2000. *Biochim. Biophys. Acta* 1459:274–83
9. Ohnishi T, Kawaguchi K, Hagihara B. 1966. *J. Biol. Chem.* 241:1797–806
10. Schägger H, Pfeiffer K. 2000. *EMBO J.* 19:1777–83
11. Schägger H, de Coo R, Bauer MF, Hofmann S, Godinot C, Brandt U. 2004. *J. Biol. Chem.* 279:36349–53
12. Acin-Perez R, Bayona-Bafaluy MP, Fernandez-Silva P, Moreno-Loshuertos R, Perez-Martos A, et al. 2004. *Mol. Cell* 13:805–15
13. Smeitink J, van den Heuvel L, DiMauro S. 2001. *Nat. Rev. Genet.* 2:342–52
14. Smeitink JAM, van den Heuvel LWPJ, Koopman WJH, Nijtmans LGJ, Ugalde C, Willems PHGM. 2004. *Curr. Neurovasc. Res.* 1:29–40
15. Weidner U, Geier S, Ptock A, Friedrich T, Leif H, Weiss H. 1993. *J. Mol. Biol.* 233:109–22
16. Ohnishi T. 1998. *Biochim. Biophys. Acta* 1364:186–206
17. Hinchliffe P, Sazanov LA. 2005. *Science* 309:771–74
18. Carroll J, Fearnley IM, Shannon RJ, Hirst J, Walker JE. 2003. *Mol. Cell. Proteomics* 2:117–26
19. Hirst J, Carroll J, Fearnley IM, Shannon RJ, Walker JE. 2003. *Biochim. Biophys. Acta* 1604:135–50
20. Cardol P, Vanrobaeys F, Devreese B, Van Beeumen J, Matagne RF, Remacle C. 2004. *Biochim. Biophys. Acta* 1658:212–24
21. Heazlewood JL, Howell KA, Millar AH. 2003. *Biochim. Biophys. Acta* 1604:159–69
22. Abdrakhmanova A, Zickermann V, Bostina M, Radermacher M, Schägger H, et al. 2004. *Biochim. Biophys. Acta* 1658:148–56
23. Murray J, Zhang B, Taylor SW, Oglesbee D, Fahy E, et al. 2003. *J. Biol. Chem.* 278:13619–22
24. Marques I, Duarte M, Assuncao J, Ushakova AV, Videira A. 2005. *Biochim. Biophys. Acta* 1707:211–20
25. Finel M, Skehel JM, Albracht SPJ, Fearnley IM, Walker JE. 1992. *Biochemistry* 31:11425–34
26. Sazanov LA, Peak-Chew SY, Fearnley IM, Walker JE. 2000. *Biochemistry* 39:7229–35
27. Fearnley IM, Walker JE. 1992. *Biochim. Biophys. Acta* 1140:105–34
28. Hofhaus G, Weiss H, Leonard K. 1991. *J. Mol. Biol.* 221:1027–43
29. Guenebaut V, Schlitt A, Weiss H, Leonard K, Friedrich T. 1998. *J. Mol. Biol.* 276:105–12
30. Grigorieff N. 1998. *J. Mol. Biol.* 277:1033–46
31. Djafarzadeh R, Kerscher S, Zwicker K, Radermacher M, Lindahl M, et al. 2000. *Biochim. Biophys. Acta* 1459:230–38
32. Peng G, Fritzsch G, Zickermann V, Schägger H, Mentele R, et al. 2003. *Biochemistry* 42:3032–39
33. Gray MW, Lang BF, Cedergren R, Golding GB, Lemieux C, et al. 1998. *Nucleic Acids Res.* 26:865–78
34. Gabaldon T, Rainey D, Huynen MA. 2005. *J. Mol. Biol.* 348:857–70
35. Prommeenate P, Lennon AM, Markert C, Hippler M, Nixon PJ. 2004. *J. Biol. Chem.* 279:28165–73
36. Heinrich H, Azevedo JE, Werner S. 1992. *Biochemistry* 31:11420–24
37. Perales M, Eubel H, Heinemeyer J, Colaneri A, Zabaleta E, Braun HP. 2005. *J. Mol. Biol.* 350:263–77
38. Dudkina NV, Eubel H, Keegstra W, Boekema EJ, Braun HP. 2005. *Proc. Natl. Acad. Sci. USA* 102:3225–29

39. Videira A. 1998. *Biochim. Biophys. Acta* 1364:89–100
40. Brockmann C, Diehl A, Rehbein K, Strauss H, Schmieder P, et al. 2004. *Structure* 12:1645–54
41. Bordo D, Bork P. 2002. *EMBO Rep.* 3:741–46
41a. Abdrakhmanova A, Dobrynin K, Zwicker K, Kerscher S, Brandt U. 2005. *FEBS Lett.* 579:6781–85
42. Wiedemann N, Frazier AE, Pfanner N. 2004. *J. Biol. Chem.* 279:14473–76
43. Carroll J, Shannon RJ, Fearnley IM, Walker JE, Hirst J. 2002. *J. Biol. Chem.* 277:50311–17
44. Küffner R, Rohr A, Schmiede A, Krüll C, Schulte U. 1998. *J. Mol. Biol.* 283:409–17
45. Schulte U. 2001. *J. Bioenerg. Biomembr.* 33:205–12
46. Ugalde C, Janssen RJRJ, Lambert P, van den Heuvel LP, Smeitink JAM, Nijtmans LGJ. 2004. *Hum. Mol. Genet.* 13:659–67
47. Papa S, Sardanelli AM, Cocco T, Speranza F, Scacco SC, Technikova-Dobrova Z. 1996. *FEBS Lett.* 379:299–301
48. Scacco S, Vergari R, Scarpulla RC, Technikova-Dobrova Z, Sardanelli A, et al. 2000. *J. Biol. Chem.* 275:17578–82
49. Raha S, Myint AT, Johnstone L, Robinson BH. 2002. *Free Radic. Biol. Med.* 32:421–30
50. Chen RM, Fearnley IM, Peak-Chew SY, Walker JE. 2004. *J. Biol. Chem.* 279:26036–45
51. Au HC, Seo BB, Matsuno-Yagi A, Yagi T, Scheffler IE. 1999. *Proc. Natl. Acad. Sci. USA* 96:4354–59
52. Yadava N, Houchens T, Potluri P, Scheffler IE. 2004. *J. Biol. Chem.* 279:12406–13
53. Schilling B, Aggeler R, Schulenberg B, Murray J, Row RH, et al. 2005. *FEBS Lett.* 579:2485–90
54. Fearnley IM, Carroll J, Shannon RJ, Runswick MJ, Walker JE, Hirst J. 2001. *J. Biol. Chem.* 276:38345–48
55. Angell JE, Lindner DJ, Shapiro PS, Hofmann ER, Kalvakolanu DV. 2000. *J. Biol. Chem.* 275:33416–26
56. Lufei CC, Ma J, Huang GC, Zhang T, Novotny-Diermayr V, et al. 2003. *EMBO J.* 22:1325–35
57. Huang GC, Lu H, Hao AJ, Ng DCH, Ponniah S, et al. 2004. *Mol. Cell. Biol.* 24:8447–56
58. Gubin AN, Njoroge JM, Bouffard GG, Miller JL. 1999. *Genomics* 59:168–77
59. Murray J, Taylor SW, Zhang B, Ghosh SS, Capaldi RA. 2003. *J. Biol. Chem.* 278:37223–30
60. Schulte U, Haupt V, Abelmann A, Fecke W, Brors B, et al. 1999. *J. Mol. Biol.* 292:569–80
61. Baker ME, Grundy WN, Elkan CP. 1999. *Cell. Mol. Life Sci.* 55:450–55
62. Runswick MJ, Fearnley IM, Skehel JM, Walker JE. 1991. *FEBS Lett.* 286:121–24
63. Sackmann U, Zensen R, Roehlen D, Jahnke U, Weiss H. 1991. *Eur. J. Biochem.* 200:463–69
64. Brody S, Mikolajczyk S. 1988. *Eur. J. Biochem.* 173:353–59
65. Schneider R, Massow M, Lisowsky T, Weiss H. 1995. *Curr. Genet.* 29:10–17
66. Vignais PM, Colbeau A. 2004. *Curr. Issues Mol. Biol.* 6:159–88
67. Böhm R, Sauter M, Böck A. 1990. *Mol. Microbiol.* 4:231–43
68. Albracht SPJ. 1993. *Biochim. Biophys. Acta* 1144:221–24
69. Friedrich T, Weiss H. 1997. *J. Theor. Biol.* 187:529–40
70. Finel M. 1998. *Biochim. Biophys. Acta* 1364:112–21
71. Tran-Betcke A, Warnecke U, Böcker C, Zaborosch C, Friedrich B. 1990. *J. Bacteriol.* 172:2920–29
72. Leif H, Sled VD, Ohnishi T, Weiss H, Friedrich T. 1995. *Eur. J. Biochem.* 230:538–48
73. Galante YM, Hatefi Y. 1978. *Methods Enzymol.* 53:15–21

74. Chen S, Guillory RJ. 1981. *J. Biol. Chem.* 256:8318–23
75. Yagi T, Din TM. 1990. *Biochemistry* 29:5515–20
76. Yano T, Sled VD, Ohnishi T, Yagi T. 1996. *J. Biol. Chem.* 271:5907–13
77. Fecke W, Sled VD, Ohnishi T, Weiss H. 1994. *Eur. J. Biochem.* 220:551–58
78. Wang D-C, Meinhardt SW, Sackmann U, Weiss H, Ohnishi T. 1991. *Eur. J. Biochem.* 197:257–64
79. Peters JW, Lanzilotta WN, Lemon BJ, Seefeldt LC. 1998. *Science* 282:1853–58
80. Yano T, Sklar J, Nakamaru-Ogiso E, Yagi T, Ohnishi T. 2003. *J. Biol. Chem.* 278:15514–22
81. Waletko A, Zwicker K, Abdrakhmanova A, Zickermann V, Brandt U, Kerscher S. 2005. *J. Biol. Chem.* 280:5622–25
82. Sled VD, Friedrich T, Leif H, Weiss H, Fukumori Y, et al. 1993. *J. Bioenerg. Biomembr.* 25:347–56
83. Nakamaru-Ogiso E, Yano T, Ohnishi T. 2005. *J. Biol. Chem.* 280:301–7
84. Brüggemann H, Falinski F, Deppenmeier U. 2000. *Eur. J. Biochem.* 267:5810–14
85. Finel M. 1998. *Trends Biochem. Sci.* 23:412–14
86. Smith MA, Finel M, Korolik V, Mendz GL. 2000. *Arch. Microbiol.* 174:1–10
87. Meuer J, Bartoschek S, Koch J, Künkel A, Hedderich R. 1999. *Eur. J. Biochem.* 265:325–35
88. Rasmussen T, Scheide D, Brors B, Kintscher L, Weiss H, Friedrich T. 2001. *Biochemistry* 40:6124–31
89. Duarte M, Populo H, Videira A, Friedrich T, Schulte U. 2002. *Biochem. J.* 364:833–39
90. Flemming D, Schlitt A, Spehr V, Bischof T, Friedrich T. 2003. *J. Biol. Chem.* 278:47602–9
91. Ahlers P, Zwicker K, Kerscher S, Brandt U. 2000. *J. Biol. Chem.* 275:23577–82
92. Garofano A, Zwicker K, Kerscher S, Okun P, Brandt U. 2003. *J. Biol. Chem.* 278:42435–40
93. Grgic L, Zwicker K, Kashani-Poor N, Kerscher S, Brandt U. 2004. *J. Biol. Chem.* 279:21193–99
94. Kashani-Poor N, Zwicker K, Kerscher S, Brandt U. 2001. *J. Biol. Chem.* 276:24082–87
95. Gurrath M, Friedrich T. 2004. *Proteins* 56:556–63
96. Vinogradov AD, Sled VD, Burbaev DS, Grivennikova VG, Moroz IA, Ohnishi T. 1995. *FEBS Lett.* 370:83–87
97. Hedderich R. 2004. *J. Bioenerg. Biomembr.* 36:65–75
98. Andrews SC, Berks BC, McClay J, Ambler A, Quail MA, et al. 1997. *Microbiology* 143:3633–47
99. Mathiesen C, Hägerhäll C. 2002. *Biochim. Biophys. Acta* 1556:121–32
100. Leonard K, Haiker H, Weiss H. 1987. *J. Mol. Biol.* 194:277–86
101. Böttcher B, Scheide D, Hesterberg M, Nagel-Steger L, Friedrich T. 2002. *J. Biol. Chem.* 277:17970–77
102. Sazanov LA, Carroll J, Holt P, Toime L, Fearnley IM. 2003. *J. Biol. Chem.* 278:19483–91
103. Zickermann V, Bostina M, Hunte C, Ruiz T, Radermacher M, Brandt U. 2003. *J. Biol. Chem.* 278:29072–78
104. Belogrudov G, Hatefi Y. 1994. *Biochemistry* 33:4571–76
105. Yamaguchi M, Belogrudov G, Hatefi Y. 1998. *J. Biol. Chem.* 273:8094–98
106. Mamedova AA, Holt PJ, Carroll J, Sazanov LA. 2004. *J. Biol. Chem.* 279:23830–36
107. Kao MC, Matsuno-Yagi A, Yagi T. 2004. *Biochemistry* 43:3750–55
108. Sazanov LA, Walker JE. 2000. *J. Mol. Biol.* 392:455–64
109. Bai YD, Shakeley RM, Attardi G. 2000. *Mol. Cell. Biol.* 20:805–15
110. Cardol P, Matagne RF, Remacle C. 2002. *J. Mol. Biol.* 319:1211–21
111. Holt PJ, Morgan DJ, Sazanov LA. 2003. *J. Biol. Chem.* 278:43114–20

112. Roth R, Hagerhall C. 2001. *Biochim. Biophys. Acta* 1504:352–62
113. Di Bernardo S, Yano T, Yagi T. 2000. *Biochemistry* 39:9411–18
114. Kao MC, Di Bernardo S, Matsuno-Yagi A, Yagi T. 2002. *Biochemistry* 41:4377–84
115. Kao MC, Di Bernardo S, Matsuno-Yagi A, Yagi T. 2003. *Biochemistry* 42:4534–43
116. Klingenberg M, Slenczka W. 1959. *Biochemistry Z.* 331:486–517
117. Chance B, Hollunger G. 1960. *Nature* 185:666–72
118. Vinogradov AD, Grivennikova VG. 2001. *IUBMB Life* 52:129–34
119. Verkhovsky MI, Morgan JE, Puustinen A, Wikström MKF. 1996. *Nature* 380:268–70
120. Zharova TV, Vinogradov AD. 1997. *Biochim. Biophys. Acta* 1320:256–64
121. Zakharova NV, Zharova TV, Vinogradov AD. 1999. *FEBS Lett.* 444:211–16
122. Majander AS, Finel M, Wikström MKF. 1994. *J. Biol. Chem.* 269:21037–42
123. Sled VD, Vinogradov AD. 1993. *Biochim. Biophys. Acta* 1141:262–68
124. Dutton PL, Moser CC, Sled VD, Daldal F, Ohnishi T. 1998. *Biochim. Biophys. Acta* 1364:245–57
125. Albracht SPJ, de Jong AMP. 1997. *Biochim. Biophys. Acta* 1318:92–106
126. Ingledew WJ, Ohnishi T. 1980. *Biochem. J.* 186:111–17
127. Yano T, Dunham WR, Ohnishi T. 2005. *Biochemistry* 44:1744–54
128. Kurki S, Zickermann V, Kervinen M, Hassinen IE, Finel M. 2000. *Biochemistry* 39:13496–502
129. Fisher N, Rich PR. 2000. *J. Mol. Biol.* 296:1153–62
130. Lenaz G. 1998. *Biochim. Biophys. Acta* 1364:207–21
131. Ohshima M, Miyoshi H, Sakamoto K, Takegami K, Iwata J, et al. 1998. *Biochemistry* 37:6436–45
132. Degli Esposti M. 1998. *Biochim. Biophys. Acta* 1364:222–35
133. Okun JG, Lümmen P, Brandt U. 1999. *J. Biol. Chem.* 274:2625–30
134. Bermejo A, Figadere B, Zafra-Polo MC, Barrachina I, Estornell E, Cortes D. 2005. *Nat. Prod. Rep.* 22:269–303
135. Lümmen P. 1998. *Biochim. Biophys. Acta* 1364:287–96
136. Okun JG, Zickermann V, Zwicker K, Schägger H, Brandt U. 2000. *Biochim. Biophys. Acta* 1459:77–87
137. Dröse S, Zwicker K, Brandt U. 2002. *Biochim. Biophys. Acta* 1556:65–72
138. Friedrich T, Brors B, Hellwig P, Kintscher L, Rasmussen T, et al. 2000. *Biochim. Biophys. Acta* 1459:305–9
139. Brandt U. 1997. *Biochim. Biophys. Acta* 1318:79–91
140. Darrouzet E, Issartel JP, Lunardi J, Dupuis A. 1998. *FEBS Lett.* 431:34–38
141. Kerscher S, Kashani-Poor N, Zwicker K, Zickermann V, Brandt U. 2001. *J. Bioenerg. Biomembr.* 33:187–96
142. Schuler F, Yano T, Di Bernardo S, Yagi T, Yankovskaya V, et al. 1999. *Proc. Natl. Acad. Sci. USA* 96:4149–53
143. Flemming D, Hellwig P, Friedrich T. 2003. *J. Biol. Chem.* 278:3055–62
144. Hellwig P, Scheide D, Bungert S, Mäntele W, Friedrich T. 2000. *Biochemistry* 39:10884–91
145. Wikström MKF. 1984. *FEBS Lett.* 169:300–4
146. Galkin AS, Grivennikova VG, Vinogradov AD. 1999. *FEBS Lett.* 451:157–61
147. Krebs W, Steuber J, Gemperli AC, Dimroth P. 1999. *Mol. Microbiol.* 33:590–98
148. Steuber J, Schmid C, Rufibach M, Dimroth P. 2000. *Mol. Microbiol.* 35:428–34
149. Gemperli AC, Dimroth P, Steuber J. 2002. *J. Biol. Chem.* 277:33811–17
150. Bertsova YV, Bogachev AV. 2004. *FEBS Lett.* 563:207–12

151. Stolpe S, Friedrich T. 2004. *J. Biol. Chem.* 279:18377–83
152. Gemperli AC, Dimroth P, Steuber J. 2003. *Proc. Natl. Acad. Sci. USA* 100:839–44
153. Ohnishi T, Salerno JC. 2005. *FEBS Lett.* 579:4555–61
154. Brandt U, Kerscher S, Dröse S, Zwicker K, Zickermann V. 2003. *FEBS Lett.* 545:9–17
155. Friedrich T. 2001. *J. Bioenerg. Biomembr.* 33:169–77
156. Krishnamoorthy G, Hinkle PC. 1988. *J. Biol. Chem.* 263:17566–75
157. Nakamaru-Ogiso E, Seo BB, Yagi T, Matsuno-Yagi A. 2003. *FEBS Lett.* 14:43–46
158. Steuber J. 2003. *J. Biol. Chem.* 278:26817–22
159. Yagi T. 1987. *Biochemistry* 26:2822–28
160. Kao MC, Di Bernardo S, Perego M, Nakamaru-Ogiso E, Matsuno-Yagi A, Yagi T. 2004. *J. Biol. Chem.* 279:32360–66
161. Kao MC, Nakamaru-Ogiso E, Matsuno-Yagi A, Yagi T. 2005. *Biochemistry* 44:9545–54
162. Zickermann V, Barquera B, Wikström MKF, Finel M. 1998. *Biochemistry* 37:11792–96
163. Kerscher S, Grgic L, Garofano A, Brandt U. 2004. *Biochim. Biophys. Acta* 1659:197–205
164. Duarte M, Schulte U, Ushakova AV, Videira A. 2005. *Genetics* 171:91–99
165. Sherer TB, Betarbet R, Testa CM, Seo BB, Richardson JR, et al. 2003. *J. Neurosci.* 23:10756–64
166. Lenaz G, Bovina C, Castelluccio C, Fato R, Formiggini G, et al. 1997. *Mol. Cell. Biochem.* 174:329–33
167. Vahsen N, Cande C, Briere JJ, Bénit P, Joza N, et al. 2004. *EMBO J.* 23:4679–89
168. Brand MD, Affourtit C, Esteves TC, Green K, Lambert AJ, et al. 2004. *Free Radic. Biol. Med.* 37:755–67
169. Volbeda A, Charon MH, Piras C, Hatchikian EC, Frey M, Fontecilla-Camps JC. 1995. *Nature* 373:580–87
170. Tusnady GE, Simon I. 1998. *J. Mol. Biol.* 283:489–506
171. Krogh A, Larsson B, von Heijne G, Sonnhammer EL. 2001. *J. Mol. Biol.* 305:567–80

RELATED REVIEWS

Hosler JP, Ferguson-Miller S, Mills DA. 2006. *Annu. Rev. Biochem.* 75:165–87
Rasmusson AG, Soole KL, Elthon TE. 2004. *Annu. Rev. Plant Biol.* 55:23–39
Gray MW, Lang BF, Burger G. 2004. *Annu. Rev. Genet.* 38:477–524
Wallace DC. 2005. *Annu. Rev. Genet.* 39:359–407
Rees DC. 2002. *Annu. Rev. Biochem.* 71:221–46

Tyrphostins and Other Tyrosine Kinase Inhibitors

Alexander Levitzki[1] and Eyal Mishani[2]

[1]The Silberman Institute for Life Sciences, Department of Biological Chemistry, The Hebrew University, Givat Ram Campus, Jerusalem 91904, Israel; email: levitzki@vms.huji.ac.il

[2]The Hadassah Hebrew University Hospital, Department of Medical Biophysics and Nuclear Medicine, Ein-Kerem Campus, Jerusalem 91120, Israel; email: mishani@md.huji.ac.il

Key Words

cancer therapy, signal transduction, tyrosine phosphorylation, tyrphostin

Abstract

The development of tyrosine phosphorylation inhibitors has transformed the approach to cancer therapy and is likely to affect other fields of medicine. In spite of the conservation among protein tyrosine kinases (PTKs), one can develop small molecules that block the activity of a narrow spectrum of PTKs and that exhibit much less toxicity than the currently used chemotherapeutic agents. In this review, we discuss principles for inhibiting specific PTKs. We discuss (*a*) the birth of the concept of generating targeted, nontoxic signal transduction inhibitors, (*b*) the potential of substrate-competitive versus the more common ATP-competitive PTK inhibitors, (*c*) the combination of PTK inhibitors with other signal transduction inhibitors to induce apoptosis—the best way to induce the demise of the cancer cell, and (*d*) the potential to utilize PTK inhibitors/tyrphostins to attenuate nonmalignant pathological conditions, such as immune disorders, tissue rejection, and restenosis.

Contents

INTRODUCTION	94
PROTEIN TYROSINE KINASES	95
Signal Transduction Therapy and Protein Tyrosine Kinases	95
TYRPHOSTINS	96
CHRONIC MYELOGENOUS LEUKEMIA, BCR-ABL KINASE, AND ITS INHIBITORS	97
Resistance to STI-571	98
ATP MIMICS VERSUS SUBSTRATE MIMICS	99
EGFR KINASE INHIBITORS	99
Reversible Inhibitors	99
Resistance to Gefitinib and Erlotinib	100
Irreversible Inhibitors	100
EGFR-Directed Tyrphostins in Combination with Other Agents	101
STRUCTURAL CONSIDERATIONS	101
VEGF RECEPTOR TYROSINE KINASE INHIBITORS	102
PDGFR AS A TARGET	103
PDGFR Kinase Inhibitors as Antirestenosis Agents	104
JAK-2 AND JAK-3	104
Jak-2 Inhibitors Combined with Immune Therapy	104
Jak-3 Inhibitors	104
CONCLUSIONS	105

INTRODUCTION

This year marks the sixtieth anniversary of the introduction of the first chemotherapeutic agents against cancer. These agents take advantage of the enhanced cell proliferation that characterizes many cancers and target pathways related to RNA and DNA synthesis. The success of this strategy, although dramatic in some forms of cancers, has been limited. The most prevalent cancers, including lung and colon cancers, breast cancer in women, and prostate cancer in men, tend to respond initially but become refractory to chemotherapy, especially at the metastatic stage. Agents like 5-fluorouracil (5FU) and cisplatin (CDDP) are reasonably effective in the treatment of many cancers, sometimes achieving remarkable results as in the case of testicular cancer. In most cases, however, treatment brings temporary remission with the eventual reemergence of the disease, usually in a more severe, chemotherapy-resistant form. Thus, by the end of the 1970s, cancer therapy had reached an impasse.

There was no real improvement in outlook for cancer patients until the late 1990s. At this time, new therapies finally emerged, resulting from the tide of molecular and biochemical data that began in the late 1970s. A set of genes that code for proteins regulating cell proliferation, cell differentiation, and cell death had been discovered. Mutations that cause aberrant expression of these proteins, their constitutive activation, or both lead to cancer. These genes include oncogenes (genes that cause cancer) and tumor suppressor genes, which encode a number of biochemically different families of proteins such as p53 and Rb. Nuclear oncoproteins derive from transcription factors or nuclear hormone receptors. Cytoplasmic oncoproteins are signal transducers, such as protein kinases and small GTP-binding proteins. Transmembrane proteins are growth factor receptors. Last but not least, many oncoproteins are derivatives of growth factors. Although biochemically distinct, the common denominator of all these proteins is that they function in signal transduction.

The initial interaction of a growth factor receptor with its growth factor leads to activation of the inner, cytoplasmic portion of the receptor. This portion of the receptor can be itself an enzyme or can be linked to an enzyme. Activation of the enzyme portion in turn activates a signal transducer, which is another enzyme, and so the cascade begins to roll. Each of these steps is tightly controlled, and there is not only a mechanism to activate each step but also to quench its activity. In normal transduction, every signaling event is transient. However, if there is a mutation that leads to the appearance of many copies of the receptor or to the production of both the receptor and its growth factor by the same cell, the primary receptor signal becomes amplified and persistent. This leads to abnormally high activity of the whole signaling pathway. Persistent, excess activity of the pathway can also result from a mutation in a downstream element of the signaling pathway that rids the cell of its dependence on the growth factor for activation of the pathway. Another property of cancer cells is genetic instability, leading to chromosome loss, rearrangements, and, consequently, to a reduction in the number of signaling pathways. This loss of redundancy renders the cancer cell highly dependent on those signaling pathways that have become overactive and that give the cell its growth advantage. This very property also makes the cell more susceptible to agents that target these signaling pathways. The recognition of this principle led us to the concept of "signal transduction therapy" (1).

PROTEIN TYROSINE KINASES

Protein tyrosine kinases (PTKs) appear in evolution with multicellular organisms. Yeasts such as *Saccharomyces cerevisiae* or *Candida albicans* do not possess PTKs, whereas Dictostylium, nematodes, and Drosophila do. It appears from what we have learned over the past 25 years that PTKs specialize in communication between cells and within cells. PTKs are involved in embryonic development, metabolism, cell proliferation, angiogenesis, and the immune system. Because PTKs occupy key positions in the function of the multicellular organism, it is not surprising that malfunction of PTKs can lead to disease. Most diseases involve enhanced activity of PTKs owing to various mutations, either in the PTK itself or in the mechanisms of their activation, leading to enhanced, and sometimes persistent, stimulation. Some diseases, such as various forms of diabetes, involve the lack or diminished activity of the PTK [insulin receptor (InsR) in the case of diabetes]. In this review, we discuss the much more common situation—the enhanced activities of PTKs leading to proliferative diseases, including cancers, leukemias, psoriasis, restenosis, and others. PTKs comprise receptor protein tyrosine kinases (RPTKs), which receive signals from outside the cell and transmit it into the cell, and nonreceptor (cellular) PTKs, which do not possess a domain facing outside the cell. RPTKs are glycoproteins spanning the cell membrane where their extracellular domains face outward, interacting with growth factors, triggering the activation of the catalytic intracellular domains, and thereby transmitting the messages. This family of RPTKs consists of ~60 known proteins, such as EGF receptor (EGFR), platelet-derived growth factor receptor (PDGFR), InsR, vascular endothelial growth factor receptors (VEGFRs), and more. Nonreceptor (cellular) PTKs, for example, Abl, the Src family, and the Jak family are activated by upstream signaling molecules, such as receptors of the immune system, G protein–coupled receptors, and RPTKs. In our review, we do not list and discuss all the PTKs (2) but rather focus on the principles of developing inhibitors as manifested through the discussion of representative PTKs.

Signal Transduction Therapy and Protein Tyrosine Kinases

The strategy behind signal transduction therapy is inhibition of the hyperactive signaling pathways on which the cancer cell is dependent for survival and/or proliferation. In 1985, we decided to test the feasibility of this approach on PTKs. After the discovery that pp60Src was a tyrosine kinase, many RPTKs (except the InsR) and cellular (nonreceptor) PTKs, identified in the 1980s, were all found to be associated with various forms of cancer and leukemia. We recognized the immense emerging opportunity to generate low-molecular-weight inhibitors of tyrosine phosphorylation and use them for therapy and the dissection of signal transduction pathways (3). We have since generated many hundreds of PTK inhibitors and coined the term "tyrphostins" (TYRosine PHOSphorylation INhibitors) for such compounds. Initially the work on tyrphostins was met with skepticism because the common belief was that the high degree of conservation of the kinase domain (4) would preclude the possibility of obtaining highly selective molecules. This dogma was reinforced by the observation that all of the newly identified PTK inhibitors obtained from natural sources, including quercetin (5), genistein (6), erbstatin (7), and lavendustin (8), were not selective and also inhibited Ser/Thr kinases (erbstatin inhibits PKC) and other ATP-requiring enzymes (quercetin).

Since its inception, signal transduction therapy has expanded exponentially (1, 9–12). Small molecules that modulate the activity of an aberrant signaling element are entering the clinic in ever increasing numbers. Most of the agents are inhibitors of a signaling protein whose activity is abnormally enhanced. Occasionally, as in the case of p53, an attempt is made to generate agents that activate a signaling element. In addition to small molecules, signal transduction agents include antibodies and recombinant proteins (11, 13). In this review, we focus on PTKs whose enhanced signaling is involved in many cancers as well as in other pathophysiological conditions.

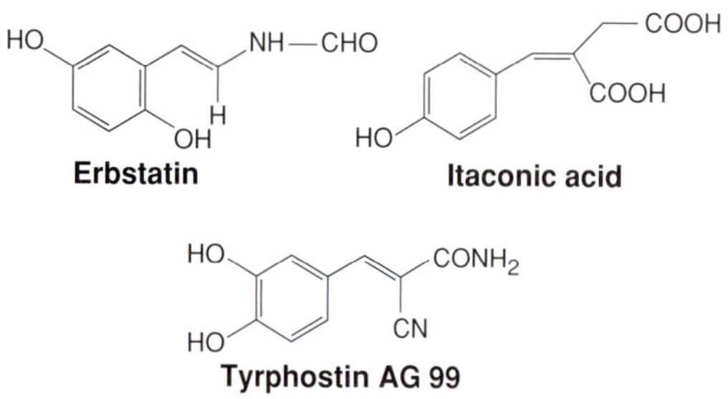

TYRPHOSTINS

The first step in the development of PTK inhibitors began soon after the recognition in the early 1980s that natural compounds, such as quercetin, erbstatin, genistein, and lavendustin A, inhibit the activities of PTKs such as $pp60^{Src}$ and EGFR. Although these natural compounds have rather poor selectivity or mediocre potency, they served as lead compounds for the design and development of synthetic, more potent, and selective PTK inhibitors (tyrphostins). In parallel with the search for natural compounds, our group prepared hydroxyphenyl-containing molecules as tyrosine mimics. We showed that these molecules are competitive inhibitors of the InsR kinase and block insulin-dependent glucose uptake and the antipolytic activity of the hormone (14). The activity of itaconic acid observed by us (14, 15) along with the observations of Umezawa et al. (7) regarding erbstatin triggered the first systematic synthesis of numerous potential PTK inhibitors (tyrphostins) (**Figure 1**) from the family of benzene malononitriles (16, 17). The structure of itaconic acid and erbstatin served as the template for a large number of compounds, many of which demonstrated excellent PTK inhibition, with no significant inhibition of Ser/Thr kinases (16, 17). This initial class of tyrphostins showed for the first time that one can generate a series of compounds that inhibit a particular PTK with minimal toxic effects in cells and in vivo (18). These findings convinced the chief executive officer of Novartis, Daniel Vasella, to move on and develop an inhibitor for the protein tyrosine kinase Bcr-Abl (19, p. 46).

Tyrphostins could be classified into those that were competitive with the substrate and noncompetitive with ATP, ATP competitive (20, 21), bisubstrate competitive (20), and mixed competitive (22). In the mid-1990s,

Figure 1

The origin of early tyrphostins. Itaconic acid was found to be a substrate-competitive inhibitor (14), whereas erbstatin is a mixed-type competitive inhibitor (20). AG 99, one of the early tyrphostins, was designed on the basis of these two inhibitors.

when structure-activity analysis led to bicyclic tyrphostins (21–23) (**Figure 2**), most were found to be ATP competitive but also mixed competitive. Indeed, the main thrust was to develop ATP-competitive kinase inhibitors. This is understandable because the ATP-binding fold is more structurally defined, whereas the substrate-binding domain is more open and is less easy to use for the design of low-molecular-weight inhibitors. The most common chemical entities of ATP mimics are anilinoquinazolines, anilinoquinolines, and anilino-pyridopyrimidines. Although ATP-binding sites are highly conserved among tyrosine kinases, minor differences in kinase domain structures have led to the development of selective PTK inhibitors. Yet, a strong case for substrate-competitive inhibitors can be made (see below).

CHRONIC MYELOGENOUS LEUKEMIA, BCR-ABL KINASE, AND ITS INHIBITORS

The PTK Bcr-Abl kinase causes chronic myelogenous leukemia (CML). In the early chronic phase of the disease, which lasts between three and five years, CML cells depend solely on the kinase activity of Bcr-Abl for their survival. This suggested that inhibiting this enzyme might induce the elimination of the diseased cells from the patient's body. The first potent, selective inhibitors of Bcr-Abl kinase were reported in 1992 (24). One family of inhibitors, AG 957 and its derivatives, is competitive with substrate and noncompetitive with ATP (24) (**Figure 3**). AG 957 induces CML cell death by apoptosis and synergizes with the pro-apoptotic anti-Fas antibody, CH 11 (25). Adaphostin (26) (**Figure 3**), an AG 957 analog, is being developed for clinical use. Another family of Bcr-Abl selective agents, represented by AG 1112 (**Figure 4**) and AG 1318, is ATP competitive (27) and also induces massive apoptosis of K562 (CML) cells (27). In 1996, Druker and the Novartis team (28) reported on a successful Bcr-Abl inhibitor, CGP 57148, later renamed

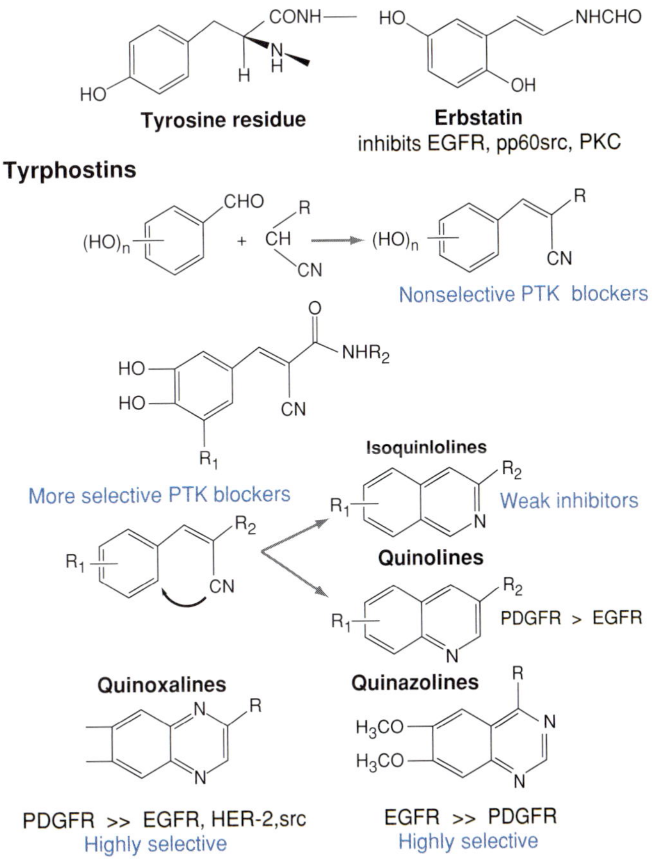

Figure 2

The evolution of tyrphostin/PTK inhibitors. Initially, small molecules tend to be substrate competitive. When more than one aromatic ring is incorporated, compounds tend to be ATP mimics, especially if they possess N atoms in the rings. Thus quinolines, quinazolines, and quinoxalines are ATP competitive or mixed competitive.

STI-571/imatinib mesylate/Gleevec/Glivec (**Figure 4**). The selective inhibition of Bcr-Abl by STI-571 is mediated via interaction between the inhibitor and the amino acids constituting the ATP-binding cleft of the PTK in the inactive state. Clinical trials have demonstrated durable responses in patients in the chronic phase, whereas those with advanced disease relapse.

The fact that STI-571 has only minor side effects and is well tolerated is of great interest and importance. This was initially surprising, given that STI-571 blocks c-Abl, PDGFR, and c-Kit. These kinases play important roles in normal cells, and knockout

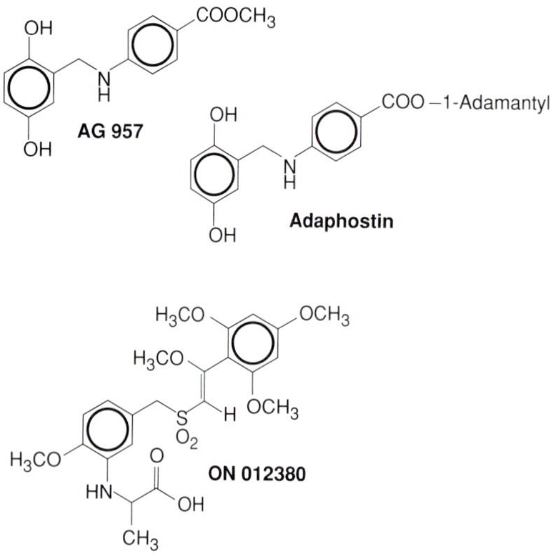

Figure 3
Bcr-Abl substrate-competitive inhibitors. AG957 is the precursor of Adaphostin, whereas ON 012380 is a completely novel structure. See Reference 38.

of their genes is detrimental. The most likely explanation is that healthy cells, which utilize c-Abl, PDGFR, or c-Kit, can get by even when over 90% of these targets are blocked because they can utilize the alternative pathways that all normal cells possess. Conversely, CML cells are dependent on Bcr-Abl for their survival and therefore die when it is blocked. STI-571 is effective also in treating gastrointestinal stromal tumor (GIST) (29), especially in patients who harbor activating mutations in c-Kit in exon 11.

Resistance to STI-571

The most common mechanisms of relapse appear to be mutations of the amino acids involved in the binding of ATP and STI-571 (30–35). A relatively common mutation in STI-571-resistant patients is a single nucleotide change that replaces threonine with isoleucine at position 315 (T315I). Thr 315 forms critical hydrogen bonds with STI-571. Other mechanisms of STI-571 resistance include Bcr-Abl gene amplification (36) and the activation of signaling pathways downstream of or parallel to those of Bcr-Abl (30, 32, 34, 35).

In order to overcome the resistance to STI-571, a number of new inhibitors have been synthesized. The most effective ATP mimics are AMN107 (33) (**Figure 4**) and BMS-354825, which inhibit almost all STI-571-resistant forms of Bcr-Abl (32, 37) but are not effective against the T315I mutant. Interestingly, a substrate-competitive inhibitor of Bcr-Abl, ON012380 (38) (**Figure 3**), was found to bind to all Bcr-Abl kinase mutants, including the T315I mutant, with a similar affinity of ~10 nM. ON012380 synergizes with STI-571, and the combined IC50 for a 1:1 mixture is ~0.1 nM.

Altered Bcr-Abl expression and the emergence of additional signaling pathways are also involved in resistance to STI-571. For example, there is evidence of increased activity of Src family kinases in cells from STI-571-resistant patients (32, 34, 35). This prompted the evaluation of molecules that inhibit both Bcr-Abl and Src activity. Two compounds, PP1 and CGP76030, inhibited Bcr-Abl in a concentration-dependent manner by overlapping binding modes. In contrast to STI-571, PP1 and CGP76030 also blocked cell growth and survival in cells expressing STI-571-resistant Abl mutants by inhibiting the activity of Src family tyrosine kinases (39). These results suggest that the use of Src kinase inhibitors is a potential strategy to prevent or overcome clonal evolution of STI-571 resistance in Bcr-Abl-positive leukemia (39).

Another inhibitor that targets both Src and Bcr-Abl, BMS-354825, is effective in patients who have become resistant to STI-571 (32). Even though it is less specific than STI-571, it seems to be less toxic and more potent than STI-571 in patients (M. Talpaz, unpublished results). This may be because STI-571 binds to the inactive state of the Abl kinase domain, whereas BMS-354825 binds to the active state of the Abl kinase or of the Src kinase (30). In the transformed cell, the Bcr-Abl kinase is permanently in the activated state, whereas in the normal cell, the activated state of c-Abl is achieved transiently under certain biochemical conditions. Hence, the kinase domain in the tumor cell is persistently occupied by the drug. In the normal cell, the inactive kinase is not occupied by the drug and can therefore

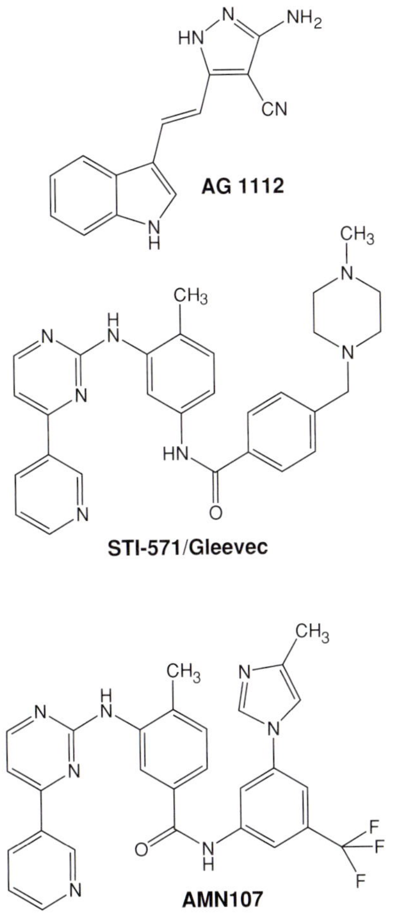

ATP MIMICS VERSUS SUBSTRATE MIMICS

Although ATP mimics are popular, it is advisable to consider seriously the utilization of substrate mimics. They are less likely to hit other targets because the substrate-binding domain is less conserved than the ATP fold. Secondly, a substrate-competitive inhibitor does not need to compete with the high intracellular ATP concentrations, which lead to a requirement for high doses of inhibitor for cellular and in vivo activities. The higher selectivity and the lower doses needed for in vivo activity of substrate-competitive inhibitors markedly reduce toxicity (41). In the case of Ser/Thr kinases, the situation is even more severe because there are five times more Ser/Thr kinases than PTKs. Indeed a PKB/Akt substrate-competitive inhibitor was found to be much superior to an ATP mimic (42). In the case of PTKs, only a few substrate mimics have been made (14, 16, 38, 43, 44).

EGFR KINASE INHIBITORS

In past 10 years, over 25 EGFR inhibitors have been described, many at various stages of preclinical and clinical development. Two have made it to the clinic. Here we discuss the most advanced compounds.

Figure 4

Bcr-Abl ATP-competitive inhibitors. AG 1112 was the first ATP-competitive inhibitor of Bcr-Abl; STI 571/Gleevec was introduced to the clinic in 1999 with great success, and AMN107 is an improved ATP mimic capable of overcoming Gleevec resistance for all Bcr-Abl mutations, except the T315I mutation. ON 012380 (**Figure 3**) overcomes all mutations, including the T315I mutation.

be activated in response to appropriate signals, at least transiently. Had the agent been bound to the inactive state too, the normal cell would have been inhibited more severely, and the toxic effect would have been more apparent. These subtle differences between the two types of compounds may explain, at least partly (40), why BMS-354825 is less toxic than STI-571.

Reversible Inhibitors

Initially, AG1478, an anilinoquinazoline (45, 46), was identified as a potent EFGR kinase inhibitor. A more soluble anilinoquinazoline, Gefitinib, was later developed. AG1478 formulated in captisol remains in clinical development but has not yet reached the clinic (see below). Gefitinib (ZD 1839, Iressa®) (47) (**Figure 5**) is orally available and inhibits the kinase activity of EGFR. Gefitinib binds at the ATP site and is ~100 times less effective in blocking Her-2. Erlotinib [N-(3-ethynylphenyl)-6,7-bis(2-methoxyethoxy)-4-anilinoquinazoline, OSI-774, Tarceva, Genentech, Inc.] is a compound

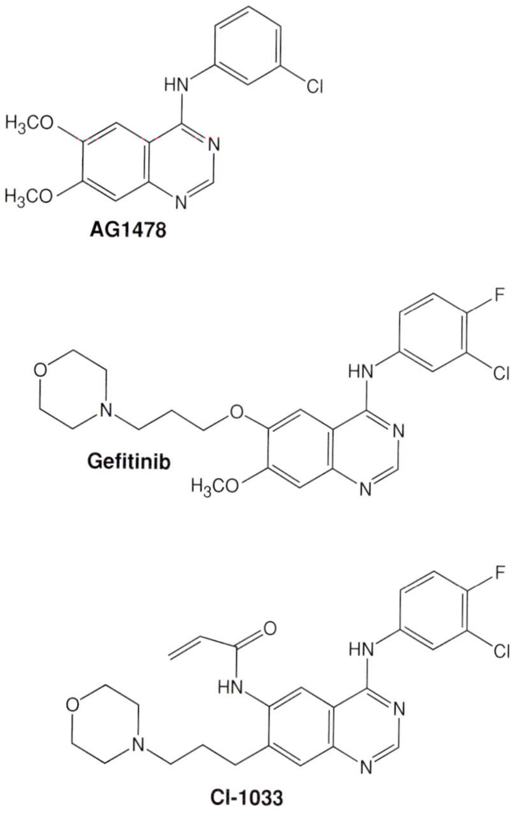

Figure 5
EGFR reversible and irreversible inhibitors.

similar to Gefitinib. Gefitinib has been in the clinic since 2002 for the treatment of non-small cell lung cancer (NSCLC) but is effective only in a small percentage of patients in which EFGR possesses activating mutations in the kinase domain (48–50). Erlotinib yielded results similar to Gefitinib. The small number of patients who respond to these agents carry specific activating mutations in the EGFR kinase domain, and tumor survival seems to depend on the activated form of the EGFR. In all other cases where EGFR overexpression is the hallmark, tumor survival does not appear to depend on EGFR activity, which accounts for inefficacy. In the absence of accurate measurements of EGFR phosphorylation in the human tumor, it is actually not possible to assess whether the failure of Gefitinib and Erlotinib is indeed due to the absence of a survival function of EGFR or to insufficient long-term occupancy of the receptor. The finding that an irreversible EGFR kinase inhibitor is more efficacious as an antitumor agent supports the latter view (51). In view of the weak performance of Gefitinib and Erlotinib, no new reversible EGFR kinase inhibitors are heading for the clinic at present. This attitude may change in view of the recent finding that 10% to 20% of EGFR-overexpressing glioblastoma multiforme patients respond to EGFR kinase inhibitor therapy. It seems that their responses correlate with the expression of PTEN and the mutant $\Delta(2\text{-}7)$EGFR (52). This finding suggests that EGFR kinase inhibitors will find their way to the clinic in cases in which the receptor plays a survival role by itself and/or when the inhibitors can be combined with other signal transduction agents to induce apoptosis selectively in the cancer cell.

Resistance to Gefitinib and Erlotinib

NSCLC patients who initially respond to Gefitinib and Erlotinib become resistant because of secondary mutations in the EGFR (53), in addition to the primary mutations that made them responsive to these inhibitors. Interestingly, the M790T mutation in the EGFR, which confers resistance to Gefitinib and Erlotinib (53), is homologous to the mutations that make Bcr-Abl (T315I), PDGFRα (T674I), and c-Kit (T670I) resistant to STI-571. Current preclinical data show that cell lines expressing the Gefitinib-resistant mutants are inhibited by irreversible EGFR inhibitors, suggesting that these may have future clinical utility (see below). It should be stressed, however, that the primary cause of resistance to EGFR-directed agents is the activity of other signal transduction pathways, such as K-ras activating mutations in lung cancer (54) and the lack of PTEN leading to PKB/Akt hyperactivity (52).

Irreversible Inhibitors

The current reversible EGFR kinase inhibitors are all ATP competitive. Such reversible ATP inhibitors need to compete with high endogenous ATP concentrations within the cell for an extended period of time in order to obtain effective antitumor activity (10, 41). In addition, their rapid clearance from plasma necessitates sustained delivery. The fast washout of reversible inhibitors from the tumor area was demonstrated by positron emission tomography (PET) studies of tumor-bearing animals imaged with fluorine-18-labeled reversible EGFR inhibitors (55). Moreover, fluorine-18-labeled Gefitinib, used to image mice bearing EGFR-overexpressing tumors, did not indicate sustained or significant tumor uptake (56). In order to enhance treatment efficiency, there has been a tremendous effort to develop irreversible EGFR kinase inhibitors on the basis of 4-(phenylamino) quinazoline and quinoline core structures (57–67). The main strategy has been to attach a Michael acceptor functional group to the basic structure of the anilinoquinazoline or quinoline in order to obtain covalent bonding through electrophilic attack of the inhibitor on a cysteine residue in the kinase-binding pocket (58). This strategy was founded on the observation that residues Cys-773 in EGFR and Cys-751 in Her-2 are uniquely positioned within the ATP-binding pocket of the respective kinase. A series of irreversible EGFR and Her-2 inhibitors were synthesized with different Michael acceptor functional groups attached at the 6- and 7-positions of the quinazoline and quinoline rings, and the 6-position yielded the best results. PD168393, which contains the acrylamido functional group at the 6-position, was found to be a potent, irreversible EGFR and Her-2 inhibitor.

Three irreversible inhibitors, CI-1033 (**Figure 5**), HKI 272, and EKB-569, are currently in clinical development. Interestingly, an irreversible EGFR kinase inhibitor is effective against EGFR kinase mutants resistant to Gefitinib (51), suggesting the need for long-term occupancy of the receptor with minimal toxicity (65).

EGFR-Directed Tyrphostins in Combination with Other Agents

Because of the heterogeneity of human tumors, it is likely that a single agent will not have a significant effect over time, but when administered in combination with other agents, it may have much higher impact (9). Fifteen years ago, it was found that the combination of the EGFR-directed tyrphostin, RG 13022, with the anti-EGFR antibody, mAb108, is very effective in inhibiting squamous cell carcinoma grown in vivo (18). The action of the two agents is strongly synergistic, although they target the very same EGFR. Recently, this observation was validated with clinically approved agents. It was shown that the anti-EGFR antibody, Erbitux (mAb 225, Cetuximab), synergizes with Gefitinib to inhibit EGFR-overexpressing tumors generated from A431 cells, grown as xenografts in nude mice (68). In another study, using an EGFR-overexpressing lung cancer xenograft cell model, it was found that Erbitux and Gefitinib or Erlotinib act in an additive manner (69).

A number of reports have appeared on the synergy between Her-2-directed tyrphostins and cytotoxic agents (70) as well as between EGFR kinase inhibitors and cytotoxic agents (71, 72). These data suggest that such combinations may be suitable for the clinic. Nonetheless, Gefitinib failed in a clinical trial during which it was administered together with CDDP (73), even though in preclinical animal experiments synergism was observed.

STRUCTURAL CONSIDERATIONS

The catalytic domains of eukaryotic Ser/Thr and tyrosine kinases are highly conserved in structure and even in amino acid sequence. The kinase domain adopts the bilobate-fold characteristic. The NH_2-terminal lobe of a protein kinase (N-lobe) is formed from mostly β-strands and one conserved α-helix, where the largest COOH-terminal lobe (C-lobe) is mostly α-helical. Although the ATP-binding cleft of GW572016/EGFR is in a relatively closed conformation, the ATP-binding cleft in Erlotinib/EGFR is in a more open one. In another crystal structure study of the kinase domain of Abl in complex with two inhibitors, STI-571 and PD173955 (74), behavior similar to that of EGFR was observed. Although the two inhibitors bind to the canonical ATP-binding site of the kinase domain, because of the differences in size of the two inhibitors their modes of binding differ. STI-571 interacts with a total of 21 amino acid residues, whereas PD173955 interacts with only 11. STI-571 captures a specific inactive conformation of the activation loop of Abl in which the loop mimics bound peptide substrate. In contrast, PD173955 binds to a conformation of Abl in which the conformation of the activation loop resembles that of the active kinase, namely the activation loop is in an extended or open conformation. Furthermore, it has been hypothesized that the reason PD173955 inhibits the Abl kinase more potently than STI-571 is because PD173955 makes fewer contacts with Abl, can bind to multiple conformations of the kinase, and is insensitive to the conformational state of the activation loop. Conversely, STI-571 requires a specific (inactive) conformation of the kinase for productive binding. Detailed knowledge of the crystal structure of Abl-STI-571 has enabled the synthesis of an improved inhibitor of Bcr-Abl, AMN107, which already has been tested and shown to possess better efficacy than STI-571 (33).

From these studies and others, we conclude that different ATP-competitive inhibitors, with a similar basic structure, yet with variations in functional groups, may recognize different conformations of the kinase and produce varied complex enzyme/inhibitor structures. Frequently, even in cases where the structure is known as it is for the Hck-PP1 (75) and Lck-PP2 (76), identifying an improved inhibitor may not be possible (77). Currently, no structural data on PTK complexes with low-molecular-weight substrate-competitive inhibitors are available. The only structure of a substrate-PTK complex involves peptide molecules. As yet, this has not been utilized to generate successful substrate-competitive inhibitors, as was done for the Ser/Thr kinase PKB/Akt (42).

VEGF RECEPTOR TYROSINE KINASE INHIBITORS

The roles of VEGFRs as important mediators in multiple processes of tumor angiogenesis, including increased vascular permeability, endothelial cell migration, proliferation and survival, have been well established (78, 79). Of these receptors, VEGFR-2 is predominantly expressed on endothelial cells and is involved in various aspects of tumor angiogenesis. Consequently, this receptor has become an attractive target for the treatment of multiple types of tumors. Starving the tumor by blocking blood supply is a key strategy to combat cancer. Indeed, many animal models and preclinical experiments have supported this idea. The first demonstration of the possible utility of angiogenesis inhibitors came with the demonstration that certain tyrphostins were able to inhibit VEGFR-2/KDR and to inhibit angiogenesis in vivo (80). Over the next few years, a number of VEGFR kinase inhibitors were synthesized, and a few entered clinical development.

On the basis of the reported crystal structure of the VGEFR-2 kinase domain, a number of small organic molecule families have been developed as potential VEGFR tyrosine kinase inhibitors (81). Among them are 4-anilinoquinazolines (82–84) and 3-substituted indolinones (85–88).

In the 3-substituted indolinone family, SU 11248, which targets VEGFR, PDGFR, Kit, and FLT3, seems to fare much better than its predecessors. This is probably because the compound is multitargeted and hits simultaneously a number of signaling RPTKs involved in tumor blood supply. In mouse xenograft models, SU 11248 exhibited broad and potent antitumor activity. The antitumor activity is combined with antiangiogenic activity, and therefore the compound is very effective in xenograft models of prostate cancer in which PDGFR and angiogenesis play a role. Also, AML (acute myeloid leukemia) patients treated with SU 11248 had reduced Flt3 activity in their blood (89, 90). The Kit kinase is a survival factor for many cases of GIST. The excellent performance of SU 11248 in STI-571-resistant GIST (91) because of its ability to block Kit, including STI-571-resistant Kit, has led to an accelerated path to clinical trials. The future of SU 11248 as an antiangiogenic agent remains to be determined.

Another promising VEGFR-2 kinase inhibitor is BAY 43-9006. Initially, BAY 43-9006 (92) was discovered as an orally available c-Raf (and B-Raf) kinase inhibitor and generated hope that it would be a good agent against metastatic melanoma in which an activating B-Raf mutation is very common. BAY 43-9006 also inhibits VEGFR-2, VEGFR-3, FLT-3, PDGFR, p38, and c-Kit (93). Very quickly it was found that the agent is not clinically effective against melanoma but is rather effective against renal cancer. The clinical utility of BAY 43-9006 appears to be due to its antiantiogenic inhibition of VEGFR-2 (94).

PDGFR AS A TARGET

PDGF receptors and their ligands are involved in a variety of diseases: cancers, atherosclerosis, balloon injury-induced restenosis, pulmonary fibrosis, and others (95–97). The role of PDGF receptors in cancer is hardly restricted to those cases in which PDGF is directly involved in the oncogenic process. PDGF receptors in the tumor stroma regulate tumor interstitial fluid pressure, tumor transvascular transport, and tumor drug uptake (95). Thus, blockade of the PDGFR in every tumor can help drug delivery to the tumor. Furthermore, the involvement of the PDGFR in recruiting blood vessels to the tumor (angiogenesis) makes it a good target for antiangiogenic therapy for many tumors. These findings have induced the research community as well as the pharmaceutical industry to develop agents that block PDGFR signaling. A large number of PDGFR kinase inhibitors with different chemical scaffolds have been synthesized and show efficacy in cells and in vivo, but only STI-571 has entered the clinic. The pyrimidinopyridines were initially developed as PDGFR kinase inhibitors (98) but found their way into the clinic as the Bcr-Abl inhibitors (see above). As a PDGFR inhibitor, STI-571 has shown efficacy in the clinic in the treatment of chronic monomyelocytic leukemia and dermatofibrosarcoma protruberans. STI-571 is also effective in the treatment of those cases of AML in which Flt-3 plays a role and in the treatment of certain cases of GIST driven by mutant c-Kit.

All of the PDGFR inhibitors also target the other members of the PDGF receptor family, c-Kit, and Flt3. The quinoxalines AG 1295, AG 1296 (22) and AGL 2033/43 (99) were found to be highly potent and selective toward the PDGFR and its family members Kit and Flt3. Because STI-571 is the only PDGFR inhibitor that also blocks Bcr-Abl kinase, it presumably has a different mode of binding to the kinase site than the rest of the PDGFR kinase inhibitors. This will be revealed once the structure of the PDGFR kinase domain in complex with the inhibitors is solved. Kinetic/mechanistic studies on the mode of action of the quinoxalines AG 1295 and AG 1296 (22) show it is competitive vis-à-vis ATP in the basal state and is mixed competitive vis-à-vis ATP subsequent to receptor activation (100). This means that AG 1296 binds differently to the kinase domain once the receptor is activated. In the activated state, AG 1296 binds outside the ATP-binding domain but still inhibits the catalytic kinase reaction. This suggests that the quinoxalines bind to unique areas of the PDGFR and may explain their high selectivity (100). In contrast, STI-571 seems to bind to common structural denominators between Bcr-Abl and PDGFR and therefore inhibits both with similar efficacies.

PDGFR Kinase Inhibitors as Antirestenosis Agents

Restenosis is due to the migration of vascular smooth muscle cells (SMC) from the media into the lumen as a result of balloon- and stent-induced injuries. The SMC cross through the injured endothelium and proliferate to generate the neointima, which clogs the blood vessel.

Several growth factors and cytokines are responsible for the process of restenosis, all originating from platelets, mononuclear cells, and endothelial cells. Because PDGF and its receptor are believed to play the major role in the pathogenesis of this process (97), PDGFR kinase inhibitors, AG 1295 (22) and AGL 2043 (99, 101), were examined as antirestenosis agents by local application to the site of injury. Experiments in pigs have shown that when these agents are applied to the site of injury, either by tyrphostin-formulated nanoparticles (102) or by utilizing a tyrphostin eluting stent, balloon-induced or stent-induced stenosis is strongly inhibited (101).

JAK-2 AND JAK-3

Janus kinases (Jaks) are cytoplasmic PTKs that play a crucial role in the initial steps of cytokine signaling. The enhanced activity of Jak-2 and the constitutive activity of Jak-2 fusion proteins generated by oncogenic mutations (103) play key roles in various leukemias, lymphomas, and certain forms of metastatic cancers, such as breast cancer and prostate cancer (104). Jak-2 was therefore identified very early as a potential target for cancer therapy, but no therapeutic agent targeting this kinase has been developed for clinical use. AG 490, which was the first Jak-2 inhibitor generated, was the first to be successfully used for the eradication of a leukemia (105), lymphomas, and a myeloma (106) in animal models. Since then, AG 490 analogs with various degrees of activities in vivo have been generated (107, 108). In all the cases examined, AG 490 and its analogs inhibited the Jak-2/Stat3/5 pathway, which is a key pathway sustaining the oncogenic phenotype.

Jak-2 Inhibitors Combined with Immune Therapy

Because AG 490 does not inhibit the signaling or growth of normal T cells, B cells, and macrophages (105, 106), it was examined in combination with immunotherapy. Immunotherapy by itself has little effect on established tumors but can induce long-term antitumor immunity once tumor volume has been greatly reduced. Reduction in tumor volume can be achieved by treatment with signal transduction inhibitors. Tyrphostins, unlike cytotoxic agents, qualify for combination with immunotherapy because they do not harm the immune system. Indeed, the antitumor efficacy of AG-490 was greatly enhanced by immunotherapy when the tyrphostin was applied first to reduce tumor burden, followed by interleukin (IL)-12. In addition to inducing antitumor immune responses, IL-12 suppresses tumor angiogenesis. These findings (106) suggest that combining nontoxic signaling inhibitors of *JAK*/STAT signaling such as AG-490 with IL-12 cytokine therapy may be extended to other cancers.

Jak-3 Inhibitors

JAK3 is abundantly expressed in lymphoid cells and initiates signaling of IL-2, IL-4, IL-7, IL-9, IL-13, and IL-15. *JAK3* is involved in T-cell activation and proliferation, and it plays a role in leukemia and autoimmune or transplant-induced inflammatory disorders. The selective targeting of *JAK3* in T cells may be clinically beneficial in T-cell-derived pathologic disorders (109), including autoimmune diabetes, allergy, and the rejection of solid organ, pancreatic islet, and bone marrow transplants. A number of Jak-3 inhibitors have already been generated and demonstrated activity in some of these indications (for example, see Reference 110).

CONCLUSIONS

It took 20 years from the inception of the concept of signal transduction therapy until the first effective PTK inhibitors reached the clinic. Signal transduction cancer therapy rests on the finding that it is possible to specifically target and inactivate the overactive kinases upon which the cancer cell depends for its survival. Noncancerous cells are less sensitive to inhibition of a particular kinase because they rely on a larger spectrum of signal transduction pathways for their growth and survival.

The examples we have enumerated in this review highlight several principles for effective signal transduction therapy. Most of the available PTK inhibitors are ATP competitive. We believe that substrate-competitive inhibitors will be more selective because of the greater diversity in substrate-binding domains. Furthermore, ATP mimics require relatively high doses to compete with the high concentrations of intracellular ATP. Irreversibly binding ATP-competitive inhibitors may overcome this difficulty. We therefore believe that more emphasis should be placed on the development of substrate mimics and irreversible ATP mimics as PTK inhibitors. Inhibitors that target the active conformation of a kinase may be more specific for cancer cells, in which the kinase is persistently activated, and may cause fewer side effects to healthy cells, in which the activated state of the kinase is transient.

Most cancers are heterogeneous in nature, and it is therefore unlikely that a single therapeutic moiety will be effective. Even in the case of CML, which is universally associated with Bcr-Abl activity, resistance to STI-571 develops. A promising avenue for treatment of some cancers may be combination of PTK inhibition to reduce tumor load with immunotherapy to induce long-term antitumor immunity.

ACKNOWLEDGMENTS

Research was supported by grants from the European Union, The Israel Cancer Association, Algen Biopharmaceuticals, and the Israel Science Foundation.

LITERATURE CITED

1. Levitzki A. 1994. *Eur. J. Biochem.* 226:1–13
2. Manning G, Whyte DB, Martinez R, Hunter T, Sudarsanam S. 2002. *Science* 298:1912–34
3. Levitzki A. 1990. *Biochem. Pharmacol.* 40:913–18
4. Hanks SK, Quinn AM, Hunter T. 1988. *Science* 241:42–52
5. Graziani Y, Erikson E, Erikson RL. 1983. *Eur. J. Biochem.* 135:583–89
6. Akiyama T, Ishida J, Nakagawa S, Ogawara H, Watanabe S, et al. 1987. *J. Biol. Chem.* 262:5592–95
7. Umezawa H, Imoto M, Sawa T, Isshiki K, Matsuda N, et al. 1986. *J. Antibiot.* 39:170–73
8. Onoda T, Iinuma H, Sasaki Y, Hamada M, Isshiki K, et al. 1989. *J. Nat. Prod.* 52:1252–57
9. Levitzki A. 1992. *FASEB J.* 6:3275–82
10. Levitzki A. 1999. *Pharmacol. Ther.* 82:231–39
11. Klein S, McCormick F, Levitzki A. 2005. *Nat. Rev. Cancer* 5:573–80
12. Levitzki A, Gazit A. 1995. *Science* 267:1782–88
13. Klein S, Levitzki A. 2006. *Curr. Signal Transduct. Ther.* 1:1–12
14. Shechter Y, Yaish P, Chorev M, Gilon C, Braun S, Levitzki A. 1989. *EMBO J.* 8:1671–76

15. Yaish P. 1988. *Synthetic inhibitors for tyrosine kinase*. PhD thesis. The Hebrew Univ., Jerusalem. 125 pp.
16. Yaish P, Gazit A, Gilon C, Levitzki A. 1988. *Science* 242:933–35
17. Gazit A, Yaish P, Gilon C, Levitzki A. 1989. *J. Med. Chem.* 32:2344–52
18. Yoneda T, Lyall RM, Alsina MM, Persons PE, Spada AP, et al. 1991. *Cancer Res.* 51:4430–35
19. Vasella D, Slater R. 2003. In *Magic Cancer Bullet—How a Tiny Orange Pill May Rewrite Medical History*. New York: HarperCollins
20. Posner I, Engel M, Gazit A, Levitzki A. 1994. *Mol. Pharmacol.* 45:673–83
21. Osherov N, Gazit A, Gilon C, Levitzki A. 1993. *J. Biol. Chem.* 268:11134–42
22. Kovalenko M, Gazit A, Bohmer A, Rorsman C, Ronnstrand L, et al. 1994. *Cancer Res.* 54:6106–14
23. Gazit A, Osherov N, Posner I, Bar-Sinai A, Gilon C, Levitzki A. 1993. *J. Med. Chem.* 36:3556–64
24. Anafi M, Gazit A, Gilon C, Ben-Neriah Y, Levitzki A. 1992. *J. Biol. Chem.* 267:4518–23
25. Carlo-Stella C, Regazzi E, Sammarelli G, Colla S, Garau D, et al. 1999. *Blood* 93:3973–82
26. Mow BM, Chandra J, Svingen PA, Hallgren CG, Weisberg E, et al. 2002. *Blood* 99:664–71
27. Anafi M, Gazit A, Zehavi A, Ben-Neriah Y, Levitzki A. 1993. *Blood* 82:3524–29
28. Druker BJ, Tamura S, Buchdunger E, Ohno S, Segal GM, et al. 1996. *Nat. Med.* 2:561–66
29. Blanke CD, Corless CL. 2005. *Cancer Investig.* 23:274–80
30. O'Hare T, Walters DK, Stoffregen EP, Jia T, Manley PW, et al. 2005. *Cancer Res.* 65:4500–5
31. Schindler T, Bornmann W, Pellicena P, Miller WT, Clarkson B, Kuriyan J. 2000. *Science* 289:1938–42
32. Shah NP, Tran C, Lee FY, Chen P, Norris D, Sawyers CL. 2004. *Science* 305:399–401
33. Weisberg E, Manley PW, Breitenstein W, Bruggen J, Cowan-Jacob SW, et al. 2005. *Cancer Cell* 7:129–41
34. Ptasznik A, Nakata Y, Kalota A, Emerson SG, Gewirtz AM. 2004. *Nat. Med.* 10:1187–89
35. Donato NJ, Wu JY, Stapley J, Gallick G, Lin H, et al. 2003. *Blood* 101:690–98
36. Gadzicki D, von Neuhoff N, Steinemann D, Just M, Busche G, et al. 2005. *Cancer Genet. Cytogenet.* 159:164–67
37. Doggrell SA. 2005. *Expert Opin. Investig. Drugs* 14:89–91
38. Gumireddy K, Baker SJ, Cosenza SC, John P, Kang AD, et al. 2005. *Proc. Natl. Acad. Sci. USA* 102:1992–97
39. Warmuth M, Simon N, Mitina O, Mathes R, Fabbro D, et al. 2003. *Blood* 101:664–72
40. Levitzki A, Bohmer FD. 1998. *Anticancer Drug Des.* 13:731–34
41. Levitzki A. 2000. In *Topics in Current Chemistry*, pp. 1–15. Berlin/Heidelberg: Springer-Verlag
42. Livnah N, Levitzki A, Senderovitz H, Yechezkel T, Salitra Y, et al. 2004. *Patent Coop. Treaty WO 204/110337 A2*
43. Blum G, Gazit A, Levitzki A. 2000. *Biochemistry* 39:15705–12
44. Blum G, Gazit A, Levitzki A. 2003. *J. Biol. Chem.* 278:40442–54
45. Ward WH, Cook PN, Slater AM, Davies DH, Holdgate GA, Green LR. 1994. *Biochem. Pharmacol.* 48:659–66
46. Osherov N, Levitzki A. 1994. *Eur. J. Biochem.* 225:1047–53
47. Wakeling AE. 2002. *Curr. Opin. Pharmacol.* 2:382–87
48. Takano T, Ohe Y, Sakamoto H, Tsuta K, Matsuno Y, et al. 2005. *J. Clin. Oncol.* 23:6829–37
49. Paez JG, Janne PA, Lee JC, Tracy S, Greulich H, et al. 2004. *Science* 304:1497–500

50. Lynch TJ, Bell DW, Sordella R, Gurubhagavatula S, Okimoto RA, et al. 2004. *N. Engl. J. Med.* 350:2129–39
51. Kwak EL, Sordella R, Bell DW, Godin-Heymann N, Okimoto RA, et al. 2005. *Proc. Natl. Acad. Sci. USA* 102:7665–70
52. Mellinghoff IK, Wang MY, Vivanco I, Haas-Kogan DA, Zhu S, et al. 2005. *N. Engl. J. Med.* 353:2012–24
53. Pao W, Miller VA, Politi KA, Riely GJ, Somwar R, et al. 2005. *PLoS Med.* 2:e73
54. Pao W, Wang TY, Riely GJ, Miller VA, Pan Q, et al. 2005. *PLoS Med.* 2:e17
55. Bonasera TA, Ortu G, Rozen Y, Krais R, Freedman NM, et al. 2001. *Nucl. Med. Biol.* 28:359–74
56. DeJesus OT, Murali D, Flores LG, Converse AK, Dick DW, et al. 2003. *J. Label. Compd. Radiopharm.* 46 (Suppl. 1):S1–9
57. Fry DW, Bridges AJ, Denny WA, Doherty A, Greis KD, et al. 1998. *Proc. Natl. Acad. Sci. USA* 95:12022–27
58. Tsou HR, Mamuya N, Johnson BD, Reich MF, Gruber BC, et al. 2001. *J. Med. Chem.* 44:2719–34
59. Discafani CM, Carroll ML, Floyd MB Jr, Hollander IJ, Husain Z, et al. 1999. *Biochem. Pharmacol.* 57:917–25
60. Vincent PW, Bridges AJ, Dykes DJ, Fry DW, Leopold WR, et al. 2000. *Cancer Chemother. Pharmacol.* 45:231–38
61. Rabindran SK, Discafani CM, Rosfjord EC, Baxter M, Floyd MB, et al. 2004. *Cancer Res.* 64:3958–65
62. Smaill JB, Rewcastle GW, Loo JA, Greis KD, Chan OH, et al. 2000. *J. Med. Chem.* 43:1380–97
63. Smaill JB, Showalter HD, Zhou H, Bridges AJ, McNamara DJ, et al. 2001. *J. Med. Chem.* 44:429–40
64. Shaul M, Abourbeh G, Jacobson O, Rozen Y, Laky D, et al. 2004. *Bioorg. Med. Chem.* 12:3421–29
65. Mishani E, Abourbeh G, Jacobson O, Dissoki S, Daniel RB, et al. 2005. *J. Med. Chem.* 48:5337–48
66. Wissner A, Overbeek E, Reich MF, Floyd MB, Johnson BD, et al. 2003. *J. Med. Chem.* 46:49–63
67. Denny WA. 2002. *Pharmacol. Ther.* 93:253–61
68. Matar P, Rojo F, Cassia R, Moreno-Bueno G, Di Cosimo S, et al. 2004. *Clin. Cancer Res.* 10:6487–501
69. Huang S, Armstrong EA, Benavente S, Chinnaiyan P, Harari PM. 2004. *Cancer Res.* 64:5355–62
70. Tsai CM, Levitzki A, Wu LH, Chang KT, Cheng CC, et al. 1996. *Cancer Res.* 56:1068–74
71. Nagane M, Levitzki A, Gazit A, Cavenee WK, Huang HJ. 1998. *Proc. Natl. Acad. Sci. USA* 95:5724–29
72. Nagane M, Narita Y, Mishima K, Levitzki A, Burgess AW, et al. 2001. *J. Neurosurg.* 95:472–79
73. Tamura K, Fukuoka M. 2005. *Expert Opin. Pharmacother.* 6:985–93
74. Nagar B, Bornmann WG, Pellicena P, Schindler T, Veach DR, et al. 2002. *Cancer Res.* 62:4236–43
75. Schindler T, Sicheri F, Pico A, Gazit A, Levitzki A, Kuriyan J. 1999. *Mol. Cell* 3:639–48
76. Zhu X, Kim JL, Newcomb JR, Rose PE, Stover DR, et al. 1999. *Struct. Fold. Des.* 7:651–61
77. Karni R, Mizrachi S, Reiss-Sklan E, Gazit A, Livnah O, Levitzki A. 2003. *FEBS Lett.* 537:47–52

78. Dvorak HF. 2003. *Am. J. Pathol.* 162:1747–57
79. Robinson DR, Wu YM, Lin SF. 2000. *Oncogene* 19:5548–57
80. Strawn LM, McMahon G, App H, Schreck R, Kuchler WR, et al. 1996. *Cancer Res.* 56:3540–45
81. McTigue MA, Wickersham JA, Pinko C, Showalter RE, Parast CV, et al. 1999. *Struct. Fold. Des.* 7:319–30
82. Hennequin LF, Thomas AP, Johnstone C, Stokes ES, Plé PA, et al. 1999. *J. Med. Chem.* 42:5369–89
83. Wedge SR, Ogilvie DJ, Dukes M, Kendrew J, Chester R, et al. 2002. *Cancer Res.* 62:4645–55
84. McCarty MF, Wey J, Stoeltzing O, Liu W, Fan F, et al. 2004. *Mol. Cancer Ther.* 3:1041–48
85. Sun L, Tran N, Liang C, Tang F, Rice A, et al. 1999. *J. Med. Chem.* 42:5120–30
86. Sun L, Tran N, Tang F, App H, Hirth P, et al. 1998. *J. Med. Chem.* 41:2588–603
87. Fong TA, Shawver LK, Sun L, Tang C, App H, et al. 1999. *Cancer Res.* 59:99–106
88. Laird AD, Vajkoczy P, Shawver LK, Thurnher A, Liang C, et al. 2000. *Cancer Res.* 60:4152–60
89. O'Farrell AM, Foran JM, Fiedler W, Serve H, Paquette RL, et al. 2003. *Clin. Cancer Res.* 9:5465–76
90. Fiedler W, Serve H, Dohner H, Schwittay M, Ottmann OG, et al. 2005. *Blood* 105:986–93
91. Demetri GD, Desai J, Fletcher JA, Morgan JA, Fletcher CDM, et al. 2004. *SU11248, a multi-targeted tyrosine kinase inhibitor, can overcome imatinib (IM) resistance caused by diverse genomic mechanisms in patients (pts) with metastatic gastrointestinal stromal tumor (GIST).* Presented at ASCO Annu. Meet. Proc., New Orleans
92. Lyons JF, Wilhelm S, Hibner B, Bollag G. 2001. *Endocr. Relat. Cancer* 8:219–25
93. Wilhelm SM, Carter C, Tang L, Wilkie D, McNabola A, et al. 2004. *Cancer Res.* 64:7099–109
94. van Spronsen DJ, de Weijer KJ, Mulders PF, De Mulder RH. 2005. *Anticancer Drugs* 16:709–17
95. Ostman A. 2004. *Cytokine Growth Factor Rev.* 15:275–86
96. Ostman A, Heldin CH. 2001. *Adv. Cancer Res.* 80:1–38
97. Levitzki A. 2004. *Cytokine Growth Factor Rev.* 15:229–35
98. Buchdunger E, Zimmermann J, Mett H, Meyer T, Muller M, et al. 1995. *Proc. Natl. Acad. Sci. USA* 92:2558–62
99. Gazit A, Yee K, Uecker A, Bohmer FD, Sjoblom T, et al. 2003. *Bioorg. Med. Chem.* 11:2007–18
100. Kovalenko M, Ronnstrand L, Heldin CH, Loubtchenkov M, Gazit A, et al. 1997. *Biochemistry* 36:6260–69
101. Banai S, Gertz SD, Gavish L, Chorny M, Perez LS, et al. 2004. *Cardiovasc. Res.* 64:165–71
102. Banai S, Wolf Y, Golomb G, Pearle A, Waltenberger J, et al. 1998. *Circulation* 97:1960–69
103. Kralovics R, Passamonti F, Buser AS, Teo SS, Tiedt R, et al. 2005. *N. Engl. J. Med.* 352:1779–90
104. Verma A, Kambhampati S, Parmar S, Platanias LC. 2003. *Cancer Metastasis Rev.* 22:423–34
105. Meydan N, Grunberger T, Dadi H, Shahar M, Arpaia E, et al. 1996. *Nature* 379:645–48

106. Burdelya L, Catlett-Falcone R, Levitzki A, Cheng F, Mora LB, et al. 2002. *Mol. Cancer Ther.* 1:893–99
107. Grunberger T, Demin P, Rounova O, Sharfe N, Cimpean L, et al. 2003. *Blood* 102:4153–58
108. Gu L, Zhuang H, Safina B, Xiao XY, Bradford WW, Rich BE. 2005. *Bioorg. Med. Chem.* 13:4269–78
109. Cetkovic-Cvrlje M, Uckun FM. 2004. *Arch. Immunol. Ther. Exp.* 52:69–82
110. Changelian PS, Flanagan ME, Ball DJ, Kent CR, Magnuson KS, et al. 2003. *Science* 302:875–78

Break-Induced Replication and Recombinational Telomere Elongation in Yeast

Michael J. McEachern[1] and James E. Haber[2]

[1]Department of Genetics, University of Georgia, Athens, Georgia 30602; email: mjm@uga.edu

[2]Department of Biology and Rosenstiel Center, Brandeis University, Waltham, Massachusetts 02454-9110; email: haber@brandeis.edu

Key Words

homologous recombination, yeast, double-strand break repair

Abstract

When a telomere becomes unprotected or if only one end of a chromosomal double-strand break succeeds in recombining with a template sequence, DNA can be repaired by a recombination-dependent DNA replication process termed break-induced replication (BIR). In budding yeasts, there are two BIR pathways, one dependent on the Rad51 recombinase protein and one Rad51 independent; these two repair processes lead to different types of survivors in cells lacking the telomerase enzyme that is required for normal telomere maintenance. Recombination at telomeres is triggered by either excessive telomere shortening or disruptions in the function of telomere-binding proteins. Telomere elongation by BIR appears to often occur through a "roll and spread" mechanism. In this process, a telomeric circle produced by recombination at a dysfunctional telomere acts as a template for a rolling circle BIR event to form an elongated telomere. Additional BIR events can then copy the elongated sequence to all other telomeres.

Contents

INTRODUCTION 112
REPAIR OF NONTELOMERIC BROKEN ENDS BY BREAK-INDUCED REPLICATION 112
POSSIBLE MOLECULAR MECHANISMS OF BIR 115
GROWTH SENESCENCE AND ITS CONSEQUENCES IN YEAST 116
SURVIVOR FORMATION BY RECOMBINATION-BASED MECHANISMS 119
THE ROLL-AND-SPREAD MODEL OF RTE 123
RTE IN THE PRESENCE OF TELOMERASE 128
RTE AT CHROMOSOMAL ENDS IN OTHER YEAST SPECIES 129
RTE AT YEAST MITOCHONDRIAL DNAs .. 130
SURVIVAL THROUGH FORMATION OF TERMINAL PALINDROMES 130
MAMMALIAN TELOMERE MAINTENANCE IN THE ABSENCE OF TELOMERASE: THE ALT PATHWAY 131

INTRODUCTION

Break-induced replication (BIR), or recombination-dependent DNA replication, appears to play a key role in the repair of stalled and broken replication forks and in the maintenance of eroding telomeres in cells lacking telomerase. In model systems such as *Saccharomyces cerevisiae*, where recombination can be initiated by a specific double-strand break (DSB), BIR occurs when one end of a DSB undergoes strand invasion into a homologous chromosome. Strand invasion is thought to lead to the formation of a unidirectional DNA replication fork that can copy hundreds of kilobase pairs (kb) to the end of a chromosome. One circumstance in which BIR may be particularly important is telomere maintenance by recombination. Although most cells use the telomerase enzyme to maintain telomeres, some cells can instead, or in addition, use recombination. Prominent among these is a subset of human cancers and immortalized cell lines known as ALT (for alternative lengthening of telomere) cells. Yeasts, particularly *S. cerevisiae* and *Kluyveromyces lactis*, have provided much of the information now available on how recombinational telomere elongation (RTE) occurs. RTE in yeast is triggered by circumstances that compromise the protective capping function of telomeres such as telomere shortening and abnormal telomere-binding proteins. RTE can generate different types of lengthened chromosome ends, including elongated telomeric repeat tracts and amplified subtelomeric regions. These different structures arise through Rad51-dependent and Rad51-independent recombination processes that resemble those defined in model systems that study BIR.

REPAIR OF NONTELOMERIC BROKEN ENDS BY BREAK-INDUCED REPLICATION

Some of the earliest models to explain crossing-over during recombination-coupled DNA breakage with the initiation of new DNA replication. Mechanisms called "break-copy" (1) or recombination-dependent DNA replication (2) were invoked to explain the repair and replication of bacteria and bacteriophage. Experiments by Motamedi et al. (3) and by Kuzminov & Stahl (4) with phage λ provided strong evidence that a major pathway to generate crossing-over involves extensive recombination-dependent, break-copy DNA replication. Recombination-dependent replication also appears to be critically important in the re-initiation of DNA replication at a stalled replication fork or in initiating DNA replication in the absence of an origin of replication (5, 6). The experiments of Michel and colleagues (7, 8) illustrated several different pathways whereby stalled replication forks can be restarted by creating a break, so that there is an intact template and a single-ended, partly replicated sister chromatid. Strand invasion and the creation of a new replication fork then allow replication to proceed.

Esposito (9) and his coworkers (10) first reported examples of spontaneous mitotic recombination in budding yeast where there was a nonreciprocal event that extended hundreds of kilobases down a chromosome arm. Voekel-Meiman & Roeder (11) saw similar events promoted by a mitotic "hot spot." The important idea that a broken chromosome end could acquire a new telomere by such a recombination process was provided by Dunn et al. (12), who transformed yeast with a linearized plasmid with one end lacking a telomere but carrying homology to a subtelomeric Y' region. These linear molecules became stable by recombining with an intact, telomere-containing chromosome carrying a Y' sequence adjacent to a telomere. At the same time, Walmsley et al. (13) suggested that normal telomere maintenance could be achieved if one telomeric region used another as a template to extend the end by replication.

The term "break-induced" replication was initially coined to describe events, including replication restart, during which only one end of a DSB was involved in DSB repair (14). A direct demonstration of the replicative nature of BIR repair was provided by Morrow et al. (15), who transformed yeast with a DNA fragment containing an origin of replication and a centromere, and with two oppositely oriented identical DNA segments at the ends (**Figure 1a**). Transformants contained an entirely new chromosome in which both ends of the linear DNA had recombined with the same unique homologous sequence and, both times, had replicated all the way to the chromosome end. Bosco & Haber (16) used HO endonuclease to lop off the end of a

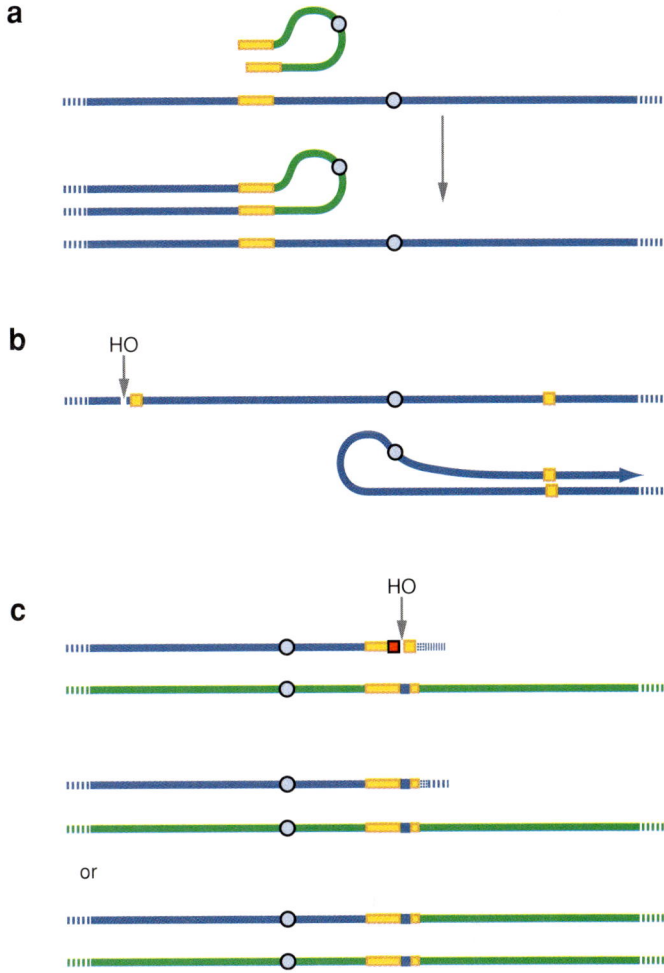

Figure 1

Model systems to study BIR in budding yeast. (*a*) When a linearized DNA fragment containing a centromere and an origin of DNA replication is transformed into yeast, both ends can acquire telomeres by initiating BIR with a unique template sequence. The design of this experiment by Morrow et al. (15) ensures that the new telomeres can not be obtained by a reciprocal exchange placing the intact chromosome end on the fragment and transferring the broken end to the chromosome. (*b*) Cleavage of a cloned HO endonuclease-cutting site on the left arm of yeast chromosome III created a broken chromosome end that was able to initiate BIR with a 70-bp homologous sequence located in the *HMR* locus, 30 kb from its telomere (16). Repair produced a partial isochromosome with a nonreciprocal translocation of the 30-kb distal region onto the opposite chromosome arm. (*c*) BIR initiated in a diploid carrying a truncation of part of the right arm of chromosome, so that HO endonuclease cleavage produced one end with only 46-bp homology to the homologous chromosome (20). About 90% of repair events proceed by Rad51-dependent BIR, with 10% arising by gene conversion (GC).

chromosome in a diploid prior to DNA replication in which the DSB shared extensive homology only centromere-proximal to the DSB and recovered two daughter cells, each of which had become homozygous for the distal region of the template chromosome. Bosco & Haber (16) also showed that BIR could occur between homologous sequences located in generally nonhomologous chromosome arms and could occur when the homology between sequences near the DSB and the distant site was only 70 bp. The DSB was thus repaired by a *RAD52*-dependent recombination event leading to the formation of a 30-kb nonreciprocal translocation (**Figure 1b**).

The genetic requirements of BIR were determined by examining diploids in which there is a single HO-induced DSB in the middle of the right arm of chromosome III. Normally, such a DSB would be repaired by "short patch" gene conversion (GC). A *rad52Δ* diploid shows almost no repair of the broken chromosome; it is simply lost, creating a 2n-1 monosomic derivative. But surprisingly, a *rad51Δ* strain eliminates GCs but still allows BIR to proceed (17). Rad51 is budding yeast's homologue of the bacterial strand exchange protein RecA and is presumed to be the principal agent in promoting strand invasion of an eroding telomere with template sequences. Rad51-mediated recombination is facilitated by the Rad51 paralogs, a Rad55/Rad57 heterodimer, and by the Swi2/Snf2 chromatin-remodeling family member, Rad54. More than 80% of *rad51Δ* colonies derived from single cells give rise to colonies in which at least a sector of cells retained the centromere and left arm of the broken chromosome (by BIR), whereas the other cells in the colony entirely lose the broken chromosome. Similar results are found in *rad54Δ*, *rad55Δ*, and *rad57Δ* mutants (18). Further genetic analysis of this *RAD51*-independent BIR pathway revealed that it is largely dependent on another set of recombination genes: *RAD50*, *MRE11*, *XRS2*, *RAD59*, and *TID1* (*RDH54*) (18). Double mutants, including *rad51Δ rad50Δ*, *rad51Δ rad59Δ*, and *rad54Δ tid1Δ*, fail to repair the DSB more than 90% of the time, leading to chromosome loss. However, still 10% of the cells give rise to sectors that appear to be BIR events by Southern blot and genetic analysis; thus none of these double mutants is as severely defective as a *rad52Δ* strain. This result is reminiscent of a study of spontaneous heteroallelic recombination by Bai & Symington (19) in which a *rad52Δ* strain was threefold more deficient in recombination than a *rad51Δ rad59Δ* double mutant. Thus, there may be a third *RAD52*-dependent pathway.

To examine *RAD51*-dependent BIR, Malkova et al. (20) created a modified diploid in which the target chromosome is truncated such that there is only a 46-bp segment distal to the DSB that is homologous to the template chromosome (**Figure 1c**). This region is too short to permit efficient repair by GC, although 10% of the repair events still occur in this way; the remaining events occur by BIR. *RAD51*-mediated BIR is significantly more efficient than *RAD51*-independent BIR; most colonies derived from single cells suffering a DSB appear to accomplish repair within the first cell cycle. Moreover, the *RAD51*-dependent pathway does not require the distant facilitating sequence that promotes BIR in a *rad51Δ* diploid (21); a large fraction of the repair events are initiated within 3 kb of the DSB.

Because *RAD51*-dependent BIR is highly efficient, it is possible to monitor repair in real time. Several surprises have emerged. First, whereas GC events in a diploid with extensive homology on either side of the DSB occur with little or no activation of the DNA damage checkpoint, there is an extended cell cycle delay when repair occurs by BIR, even when the sequences proximal to the DSB are completely homologous to the template chromosome (20). Second, BIR can take place in G_2-arrested cells when normal replication is completed and when, among other things, the Mcm helicase proteins that are required for elongation of normal DNA replication are largely exported from the nucleus (22). Third,

DSB repair by BIR, monitored by Southern blots, does not appear for about 6 h, whereas the 10% of repair occurring by GC appears in 2 h. This long delay in accomplishing BIR appears to involve a slow step in the formation of a repair replication fork. Chromatin immunoprecipitation experiments show that this slow step occurs after the Rad51 nucleofilament formed on single-stranded (ss) DNA adjacent to the HO cut site has paired with a homologous sequence on the template (N. Tanguy le Gac and J.E. Haber, unpublished observations). Understanding how the repair replication fork assembles in G_2/mitotic (M)-arrested cells remains one of the most important goals of current research. Finally, the rate of repair replication itself, once initiated, is comparable to normal replication because about 100 kb can be replicated in less than 30 min (i.e., at a rate of about 3 kb/min) (20).

A version of BIR has also been studied on a plasmid that contains inverted repeated sequences, one of which is interrupted by an HO endonuclease cleavage site (23). When the homology on both sides of the cleavage site is substantial (several hundred base pairs), recombination proceeds predominantly by Rad51-mediated GC; however, when homology is reduced to 70–100 bp on either side, Rad51-mediated GCs are severely impaired; it appears that to initiate GC, Rad51 requires about 100 bp of homology. With shorter homology, e.g., 33 bp, there is instead a Rad51-independent repair pathway that appears to involve first BIR (to copy sequences to the other, resected end), followed by single-strand annealing (to create a circular, recombined product). The ability of a *RAD51*-independent, *RAD52*-, and *RAD50*-dependent pathway to initiate BIR at very short regions of homology is consistent with one of two recombination pathways that can maintain yeast telomeres in the absence of telomerase by recombination, involving irregular telomeric TG_{1-3} sequences. An interesting result in the plasmid system is that deleting the Sgs1 helicase appears to improve recombination between sequences sharing only short regions of homology (23). This result stands in contrast to the role of Sgs1 in telomere recombination, as discussed below.

Using a transformation assay similar to that of Morrow et al. (15), Davis & Symington (25) demonstrated that there are two different repair mechanisms, one Rad51-dependent and the other Rad51-independent. In the transformation assay, Rad50 is not required for Rad51-independent BIR, whereas deleting Rad50 nearly eliminates Rad51-independent events for a chromosome break (18). This may reflect a difference in how "naked"-transforming DNA is acted upon compared to a break in preexisting chromatin or else a difference reflecting when transformation-initiated recombination occurs in the cell cycle.

POSSIBLE MOLECULAR MECHANISMS OF BIR

BIR events begin as one-ended recombination events either because there is only one free DNA end or because only one of the two ends of the DSB succeeds in strand invasion of a homologous sequence. Precisely what happens is still under active investigation. One possibility is that strand invasion creates a D loop, which then migrates down the template as DNA synthesis occurs (26); this is analogous to the way RNA polymerase copies DNA, displacing a single strand of newly synthesized DNA (**Figure 2a**). Subsequently, the single strand could be filled in a process that might have different requirements than normal lagging-strand DNA synthesis. In this instance, DNA would be conservatively replicated because all the newly synthesized DNA would be associated with the sequences extending from the repaired broken end. Alternatively, the D loop could be transformed into a complete unidirectional replication fork that then migrates down the template chromosome (**Figure 2b**). This results in two semiconservatively replicated molecules and a single Holliday junction that has to be resolved,

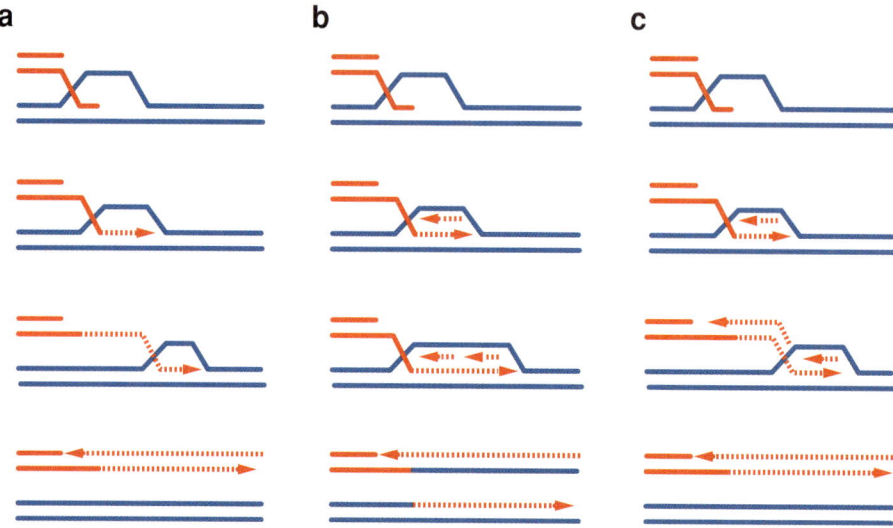

Figure 2

Possible mechanisms of BIR. (*a*) Strand invasion and primer extension leads to the synthesis of a single new strand that is later converted into a double-strand product. (*b*) Strand invasion sets up a unidirectional replication fork that proceeds to the end of the template chromosome. Resolution of a Holliday junction leaves two semiconservative replication products. Note that this process would be accompanied by an apparent sister-chromatid exchange. (*c*) Branch migration of the Holliday junction will displace two newly synthesized strands.

sometimes leading to an apparent mitotic sister-chromatid crossover event. A third version imagines that the replication structure is acted upon by branch-migration enzymes, so that both newly synthesized leading and lagging strands are displaced, and DNA synthesis is conservative if such migration continued to the template's end (**Figure 2*c***). It should be possible to assess how repair replication proceeds by using density-transfer experiments analogous to those used in studying bacterial repair and replication (3).

GROWTH SENESCENCE AND ITS CONSEQUENCES IN YEAST

S. cerevisiae mutants missing one or more of the telomerase components Est1, Est2, Est3, or Tlc1 display a characteristic slow decline in the ability to grow during a span of 50 generations, eventually resulting in the death of most cells (27–29, 30). This phenomenon is termed senescence. Of ∼4300 haploid-viable gene deletions in *S. cerevisiae* screened for telomere length defects, only deletions of the known telomerase components are individually able to generate senescence and survivor phenotypes (31), although some double-mutant combinations also cause senescence (discussed below). Certain alleles of the essential single-strand telomere-binding protein and telomerase recruitment factor, Cdc13, exhibit senescence and survivor formation identical to that of telomerase mutants (32–34). Very similar senescence is also seen in the budding yeast *K. lactis* and in the fission yeast *Schizosaccharomyces pombe* when they are deleted for telomerase components (35, 36). Highly senescent yeast cells appear to be comparable to human cells at crisis when cell death results from overwhelming telomere dysfunction. The term "crisis" is indeed sometimes used to refer to the peak period of cell death in senescing yeast mutants.

Telomeres in *S. cerevisiae* and *K. lactis* are normally a few hundred base pairs in length;

they shorten at a rate of 3 to 5 bp/cell division in the absence of telomerase. The point at which the growth rate is poorest correlates closely with the point at which telomeres are at their shortest size ($<\sim$100 bp) (27, 28, 35). Changing the initial telomere length greatly affects senescence; when telomeres are shorter at the point telomerase is lost, senescence is accelerated; initially longer telomeres significantly delay the onset of senescence (37–40). Reintroduction of telomerase activity will eliminate senescence and again stabilize telomere function (41). Sometimes, however, genome instability can persist after such complementation (42).

Severe telomere shortening in *S. cerevisiae* (and probably in other yeasts) can trigger cell cycle arrest. Cells in early senescence divide more slowly and produce smaller than normal colonies but no significant cell death (27, 43). This cell cycle delay was found to strongly depend on the DNA damage checkpoint genes *MEC1*, *DDC2*, *MEC3*, and *RAD24* (30, 43). Similarly, cells carrying the *cdc13-1* allele, which loses telomere end protection at its restrictive temperature, also experience *RAD9*-dependent G_2/M arrest (44). Activation of Rad53 that occurs during senescence can occur through both Rad9 and Mrc1 (45). Mrc1 is known to be a mediator of the cell cycle arrest that occurs from stalled replication forks but not the arrest brought on by DSBs (46). These results suggest that the cell cycle arrest seen in telomerase-deficient cells has some characteristics of DNA replication stress. One possibility is that telomerase is required to stabilize replication forks when they reach telomeres (45). It should be kept in mind that arrest in telomerase deletion mutants is very asynchronous and might well occur for different reasons in different cells. Tel1, another PI-3 kinase that elicits checkpoint responses primarily to unresected broken chromosomes (47, 48), is not required for arrest in senescing cells lacking telomerase.

There are major differences between an experimentally induced DSB designed to trigger a cell cycle arrest and the shortening telomeres of a telomerase deletion mutant that might contribute to differing DNA damage responses. The number of DNA ends potentially contributing to a cell cycle arrest signal is very different. A single HO endonuclease-induced DSB in the middle of a chromosome arm is sufficient to trigger cell cycle arrest (49); indeed, multiple unrepaired DSBs place cells in an arrested state from which they cannot adapt (50). However, mutations such as *cdc5-ad* that prevent adaptation to a single HO-induced DSB (51) do not significantly change the progress of senescence in budding yeast (D. Toczyski, personal communication). Whether one telomere becomes so short that it loses all telomere character (and the binding of proteins that shield it from triggering the checkpoint) or whether many telomeres each become weak signalers has not been established. Also, as described below, telomeres become recombination prone when they still have a small number of telomeric repeats. The presence of telomere-binding proteins could certainly alter the nature of a DNA damage response. In addition, although induced DSBs are routinely studied using ends that lack homology to other sequences in the cell (and hence can never undergo strand invasion), shortening telomeres normally retain sequence homology to other chromosome ends [in the form of telomeric repeats, subtelomeric elements (such as X and Y'), and subtelomeric gene families]. Thus, it is probably common for shortening telomeres to be repaired by copying sequences from longer telomeres in the same cell. Yet another possibility is that a senescing telomerase deletion mutant could be in a chronic adapted state, permitting cell division in DNA damage circumstances that would arrest an unadapted cell with a single fresh DSB.

Substantial evidence supports the idea that a build up of ssDNA in a cell serves as the signal that triggers cell cycle arrest in the presence of double-stranded breaks or certain other forms of DNA damage (44, 50, 52, 53). The total amount of ssDNA needed to arrest cells with a single DSB or with a stalled replication fork has been estimated to be ∼5 to 10 kb (54), an amount that seems comparable to the 30 to 60 min required to activate cell cycle arrest from a single DSB when the ends are resected at about 4 kb/h (55). Very recent work indicates that large single-stranded tracts of subtelomeric DNA can be present in *S. cerevisiae tlc1Δ rad52Δ* cells (L. Maringele and D. Lydall, personal communication). This directly demonstrates that the processing of telomeric ends to create extensive 3′ overhangs can occur as a result of telomeres being too short. The absence of recombination in these cells presumably allows more ssDNA to accumulate than would otherwise be the case.

The identities of all of the 5′ to 3′ exonucleases that can create ssDNA from unprotected DNA ends are not known. At HO-induced DSBs, the Mre1-Rad50-Xrs2 (MRX) complex is required for full exonuclease activity, but its deletion reduces resection only about twofold (56). The role of the MRX proteins at telomeres is difficult to assess because MRX mutants also exhibit shortened telomeres and because MRX recruits Tel1 (57, 58). Deletion of the exonuclease gene (*EXOI*) permits better growth both of *cdc13-1* cells at a semipermissive temperature (59) and of senescing cells lacking telomerase (60, 61). These results suggest that a diminished degree of 5′ resectioning leads to a lower likelihood of cell cycle arrest in the presence of dysfunctional telomeres or (assuming there is some loss of 3′-ended sequences) better retention of sequences that share homology with other ends.

In addition to triggering cell cycle arrest, dysfunctional telomeres acquire other properties of DSBs. Terminal deletions of chromosomes are very common in diploid *S. cerevisiae* cells lacking telomerase. Hackett & Greider (62) estimated that 6% of such cells suffer a terminal deletion on at least one of the 64 chromosome ends in each cell cycle when telomeres are very short. These deletions likely contain nonreciprocal translocations that are consistent with their having arisen by BIR (42). The terminal deletions in *est1Δ* cells lacking the Est1 protein component of telomerase were ExoI dependent, which is consistent with their being triggered by extensive 5′ to 3′ degradation, although surprisingly cell cycle arrest was not dependent on ExoI. However, chromosomal instability resulting in loss of the terminal 50 kb of chromosome V was observed to continue after diploid cells had been complemented with the *EST1* gene (42), indicating that restoration of telomere function does not always immediately eliminate all DNA damage. In spite of that, fusions between chromosome ends (and the resulting breakage-fusion-bridge cycles) were ruled out as being the primary cause of the truncations; but fusions do sometimes occur in senescing yeast cells (42, 63, 64). Combining a telomerase deletion with any of a variety of deletions of checkpoint or recombination protein often led to an increased appearance of fusions and other gross chromosomal rearrangements (64).

A major expected consequence of 5′ resectioning of dysfunctional telomeres is that the affected chromosome ends should become able to initiate strand invasion and homologous recombination. Consistent with this, *K. lactis* telomerase RNA and MRX mutants with stable short telomeres display highly increased rates of what were described as subtelomeric GCs (65; S. Iyer, L. Harris, and M.J. McEachern, unpublished data). This conversion of one subtelomeric marker by sequences from a homologous subtelomeric region likely resulted from a BIR event because loss of the subtelomeric marker was associated with replacement of telomeric repeats with repeats from another telomere (S. Natarajan and M.J. McEachern, unpublished data).

Gottschling and colleagues developed a telomere capture technique to monitor telomere capping defects (66). In this procedure, a DSB terminating in a short stretch of appropriately oriented telomeric sequence is generated; the assay selects for healing this short end by recombination with a telomere. They reported that short telomere MRX mutants of *S. cerevisiae* failed to increase the rate of telomere capture, although mutations in telomere capping proteins did. This led to the proposal that stable short telomeres in *S. cerevisiae* were not recombinogenic. However, MRX mutants have subsequently been shown to be defective at recombination involving short (<100 bp) nontelomeric sequences (23); thus, MRX may be required for the telomere capture events.

Recombination clearly plays a role in preserving cell growth in senescing telomerase-deleted mutants of both *S. cerevisiae* and *K. lactis* even before survivor formation. This was indicated by the discovery that telomerase deletion mutants also lacking Rad52 or some other recombination proteins senesce more rapidly than mutants lacking only telomerase (27, 37, 67).

SURVIVOR FORMATION BY RECOMBINATION-BASED MECHANISMS

Yeast mutants lacking telomerase that survive beyond the point when most cells die—generally 50 to 100 cell divisions after loss of telomerase—are called postsenescence survivors or frequently just "survivors." Survivors of *S. cerevisiae* and *K. lactis* invariably maintain telomeric repeats at most or all telomeres and usually have chromosome ends that have become elongated relative to those in senescent cells (37, 41, 68). Lineages of senescing cells seldom completely die out. However, the rate of survivor formation can be very low because some senescent yeast colonies with 10^5 to 10^6 cells produce no survivors. A number of factors make the rate of survivor formation difficult to assess. They include the variable and changing growth rates of survivors, plus the fact that multiple events are required for survivor formation. The formation of survivors depends on homologous recombination because virtually all survivors are eliminated when the central recombination gene *RAD52* is deleted.

S. cerevisiae has two recombination-dependent pathways that yield survivors, which are distinguished both by their genetic requirements and by the structures of the telomeres themselves. In the type I pathway, originally described by Lundblad & Blackburn (68), essentially all telomere ends were extended by the acquisition of subtelomeric Y' sequences, which are DNA elements with some features of a degenerate transposable element (**Figure 3b**). The extent of Y' amplification is highly variable but can be greater than 100-fold (68), with the two families of Y' elements (Y' long and Y' short) becoming amplified to different extents within a given survivor. Up to ~10% of total cellular DNA

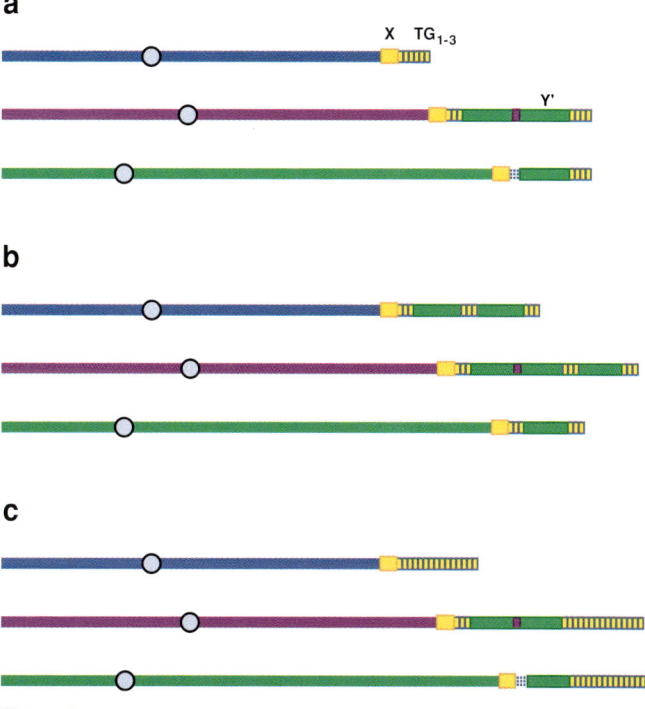

Figure 3

Two types of survivors in budding yeast lacking telomerase. (*a*) Structure of *S. cerevisiae* telomeres. TG_{1-3} telomere sequences are appended either to subtelomeric X or Y' sequences at each end. Some ends have multiple Y' sequences, many of which are separated by short TG_{1-3} tracts. Intrachromosomal recombination between Y' elements can produce extrachromosomal Y' circles, which replicate autonomously. (*b*) Type I survivors lacking telomerase arise in a Rad52- and Rad51-dependent process in which Y' elements, often multiple copies, are proliferated to most chromosome ends. (*c*) Type II survivors arise from a Rad52-dependent, Rad51-independent process in which telomere sequences themselves are elongated.

in survivors can be Y' elements, adding an average of ∼50 kb to each chromosome end. The movement of Y' elements to new chromosome ends is best explained as a recombination event between a deprotected telomere end and similar TG_{1-3} sequences that are sometimes, but not always, found between tandem Y' elements (69) (**Figure 3b**). Alternatively, some recombination may be between Y' sequences themselves. The formation of type I survivors depends on the "canonical" homologous recombination proteins Rad51, Rad54, Rad55, and Rad57 (41, 70). Another feature of type I survivors is that their terminal telomeric repeat arrays are maintained at sizes substantially below those of wild-type cells. This may contribute to the fact that type I survivors are slower growing than either wild-type cells or type II survivors. It is likely that maintenance of telomeric tracts at lengths above a critically short size is more important to the emergence of the improved growth of type I survivors than is the amplification of Y' sequences. How the terminal telomeric arrays of type I survivors are maintained is not clear. One possibility is that strand invasion of a short terminal telomeric tract into a longer telomeric tract followed by a BIR event could lead to an extended telomere somewhat longer than either parental molecule.

The significance of Y' amplification to type I survivors is another unanswered question. The amplification could simply be a by-product of BIR events whose real value to the cell is to moderately elongate the terminal telomeric tracts. Conceivably, Y' amplification could provide some selective advantage. Homogenization of sequences at chromosome ends might, for example, facilitate recombinational repair. Alternatively, the helicase protein encoded by Y' elements and strongly induced in type I survivors might somehow promote survivor formation (71). Y' elements might even contribute directly to protecting the end, much as the HeT-A and TART transposons protect *Drosophila* telomeres that lack typical telomeric repeats (72). Such a selective advantage could contribute to the improved growth of type I survivors relative to senescent cells with similarly short terminal telomeric tracts.

Most chromosomes of type I survivors fail to enter pulsed field gels, and those that do are longer than normal and exhibit a smeared appearance (63). Preliminary evidence suggests that these problems map to the large, telomere-containing terminal *Not*I restriction fragment fragments (E.J. Louis, personal communication). It may be that frequent recombination in the Y' arrays in these survivors, perhaps triggered by chronic telomere capping problems, leads both to rapid variation in the size of Y' arrays as well as to recombination intermediates unable to enter gels. In contrast, chromosomes of type II survivors (described below), even those with an amplified subtelomeric 25-kb fragment that spread to most chromosome ends, do enter pulsed field gels (63; G. Liti and E.J. Louis, personal communication).

A second type of survivor (type II) exhibits elongation of the terminal telomeric repeat tracts (37, 41, 68, 73) (**Figure 3c**). In *S. cerevisiae*, this telomere elongation can be up to 10 kb or more. Type II survivors grow at faster rates than type I survivors. Therefore, even though they arise less frequently than type I survivors in *S. cerevisiae*, they eventually come to dominate in liquid-grown cultures. In *S. cerevisiae*, the RTE that generates type II survivors depends on Rad52 but also on a different set of Rad proteins than type I survivors, namely the MRX complex and Rad59 (41, 57, 67, 70, 73). As would be expected from the analyses of single mutants, survivors are nearly eliminated in a *rad51Δ rad50Δ* double mutant (41, 57, 67, 70, 73). Extremely rarely, survivors can be found in *tlc1 rad50 rad51* mutants (74). These survivors, which required Rad52 for formation, were all type II in terms of telomere structure but presumably represent a minor third pathway of survivor formation.

In general, the genetic requirements for type II RTE are reminiscent of those for Rad51-independent BIR. However, HO-induced BIR in *rad51Δ* strains differs from type II telomere recombination in that the latter needs Sgs1, the yeast helicase related to human Werner syndrome (WRN) and Bloom's syndrome (BLM) (18, 75–78). Because *S. cerevisiae* telomeric sequences are degenerate, it would be expected that a single-stranded resected telomeric end would rarely be a perfect match for another telomeric sequence in a senescing cell. This may explain why mutations in mismatch repair genes facilitate RTE (see below) (79). Because Sgs1 has been shown to prevent homeologous recombination (80, 81), it might also be expected that a deletion of *SGS1* would promote type II RTE, but the opposite is true. It appears that Sgs1 plays a more important role in facilitating type II RTE. One possibility for this role is removing G-quadruplexes or other structures from single-stranded 3′ telomeric tails formed at telomeric ends.

How strand invasion can be accomplished in the absence of Rad51 to generate type II survivors is not clear. Both the MRX complex and Rad59 exhibit strand-annealing activity in vitro [reviewed by Symington (82)]. One possibility could be that one or both of these proteins promotes annealing between a single-stranded telomeric overhang and another partially unwound telomeric terminus. An obvious candidate for such a telomeric unwinding activity is Sgs1. The 3′ end of the telomeric DNA could then be used as a primer for new DNA synthesis and telomere elongation.

The Tel1 and Mec1 checkpoint PI3K proteins are also important to the formation of type II survivors. Single mutants lacking either of these proteins have a reduced ability to form type II survivors, and *tel1 mec1* double mutants can only form type I survivors (83). How the two checkpoint kinases regulate type II repair events is not known. It should be noted that *mec1 tel1* strains are unable to utilize telomerase and undergo senescence (84). Interestingly, a *tel1* mutation partially suppresses the senescence of a telomerase deletion mutant (84). It has been suggested that the role of Tel1 in both telomere maintenance and the DNA damage response involves MRX (47, 85).

Certain mutations in DNA polymerases also can affect survivor formation (83). After senescence, a *tlc1 cdc2-2* (pol δ) mutant grown at a semipermissive temperature could maintain short telomeric tracts and produce somewhat better-growing survivors. However, neither Y′ amplification nor appreciable telomeric tract amplification was observed in these survivors. Polymerase δ may, therefore, be required for synthesizing the long telomeric extensions in both type I and type II survivors. In contrast, a polymerase ϵ mutant at a semipermissive temperature in the absence of telomerase (*tlc1 pol2-18*) showed an accelerated telomere shortening and an accelerated emergence of type II survivors.

Other experiments have shown that the absence of Clb2, yeast's predominant cyclin B, more strongly affects the Rad50-dependent type II mechanism than the Rad51-dependent mechanism (86). Clb2 interacts with Cdc28 (Cdk1), which has recently been shown to promote the 5′ to 3′ resection of double-strand ends that is required for Rad51-dependent homologous recombination (87). It will be interesting to know whether the loss of Clb2 reduces end processing at telomeres.

It is important to keep in mind that the distinction between type I and type II survivors is not always clear. Survivors labeled as type I do not always show extensive Y′ amplification, and survivors labeled as type II may still have some short telomeres or may form from type I survivors that still retain amplified Y′ elements (41).

The survivors formed in *K. lactis* cells lacking telomerase are all type II (37). The absence of subtelomeric blocks of telomeric repeats, such as those that are present in *S. cerevisiae*, prevents the formation of type I-like arrays. Manipulating *K. lactis* telomerase deletion mutants to have a single chromosome end with an extra block of telomeric repeats inserted in subtelomeric sequence will readily lead to survivors with type I-like structures (tandem arrays of the sequence between the original two blocks of telomeric repeats) at multiple chromosome ends (88). The type II RTE of *K. lactis* appears to differ in some ways from that of *S. cerevisiae*. The extent of telomere elongation is often less—seldom more than 2 kb and often much less. These shorter telomeres in *K. lactis* survivors may explain why their growth rate is less stable than *S. cerevisiae* type II survivors. The genetic requirements for RTE in *K. lactis* appear not to be the same as those in *S. cerevisiae*. *RAD59*, the only gene other than *RAD52* studied to date, is not required for *K. lactis* survivor formation (S. Iyer and M.J. McEachern, unpublished data). Also, deletion of *RAD52*, though it severely reduces survivor formation, does not completely eliminate it. Survivors formed in a *rad52* background have modestly elongated telomeres (37). This could be due to *K. lactis* cells simply having fewer telomeres than *S. cerevisiae* (12 vs 32), or it may represent yet another pathway for survivor formation through RTE. At this point, we do not know whether any of the differences between *K. lactis* and *S. cerviseiae* survivors stem from the very different telomeric repeats in the two yeasts (highly heterogeneous in size and sequence in *S. cerevisiae* vs highly uniform 25-bp repeats in *K. lactis*).

A characteristic of postsenescent survivors in both *S. cerevisiae* and *K. lactis* is that they are not stable. With continued growth after their formation, the telomeres of survivors continue to shorten and may initiate additional rounds of recombination-dependent repair and RTE. This is most noticeable with type II survivors. In these cells, it is common for elongated telomeres to shorten gradually over extended periods of growth. Once a telomere again becomes very short, recombination will again lengthen it (37, 73). If all of the telomeres in a survivor that is passaged become critically short more or less simultaneously, the cells can undergo a secondary round of severe growth senescence and survivor formation. However, most survivors do not undergo secondary rounds of senescence that are as severe as the initial senescence. Although the possibility of suppressor mutations cannot be excluded, this probably is a result of heterogeneity in the sizes of telomeres. In *K. lactis* cells lacking telomerase, those that start senescing with telomeres that are both shorter and more uniform in size than those seen in wild-type cells undergo senescence that is more rapid, more uniform, and more severe than senescence seen in cells starting with telomeres of normal length (89). Further evidence that size heterogeneity among telomeres can affect senescence and survivor formation comes from experiments showing that one long telomere can appreciably suppress senescence (see below) (90).

Other factors have also been shown to influence the efficiency and type of survivor formation. Deletion of *EXO1* both reduced the severity of senescence and slowed the emergence of survivors (60, 61). This is consistent with the hypothesis that faster formation of single-stranded 3′ overhangs promotes both cell cycle arrest and one or more forms of telomeric recombination. Deletion of the mismatch repair protein Msh2 increases the rate of survivor formation in both *S. cerevisiae* and *K. lactis* (79). This is likely the result of the mismatch repair system's ability to repress homeologous recombination between sequences containing mismatches (91–94). The recombination events that form survivors are expected to involve mismatched telomeric or subtelomeric sequences that are targets for mismatch repair. Although *K. lactis* has uniform telomeric repeats, polymorphisms are common in the subtelomeric sequence (R element) present at most chromosome ends (95). It is worth noting that Msh2 interacts with ExoI (96), suggesting the alternative possibility that deletion of *MSH2* might act indirectly by freeing up ExoI (60).

Telomere-binding proteins can greatly influence senescence and RTE. Examples are discussed below. A hypothetical factor that might affect RTE in yeast is whether the newly replicated DNA of a telomere is on the leading or the lagging strand. In human cells, these two types of sister telomeres have been shown to be differentially prone to repair events in certain situations (97–99). Some evidence suggests that they are also differentially processed in yeast cells (100).

THE ROLL-AND-SPREAD MODEL OF RTE

Possible mechanisms of RTE include intertelomeric recombination, intratelomeric recombination, and rolling-circle DNA synthesis. Support for a model involving both rolling-circle synthesis and intertelomeric BIR events emerged from experiments with *K. lactis* cells engineered to have telomeres partly or entirely composed of mutationally tagged repeats. The phenotypically silent Bcl mutation creates a restriction site in each copy of the perfect 25-bp *K. lactis* repeat (101, 102). When the wild-type *TER1* gene (encoding the telomerase RNA) is replaced with the *TER1-7C(Bcl)* allele, a cell is produced that has a telomerase with an altered template that synthesizes only Bcl repeats. *TER1-7C(Bcl)* cells, therefore, have all telomeres that have Bcl repeats at their termini but retain wild-type repeats at more internal regions of the telomeres.

Deletion of the *TER1-7C(Bcl)* gene from cells with only a small number of basal wild-type repeats in their telomeres generated cells that, at senescence, sometimes retained terminal Bcl repeats on their shortened telomeres (103). Remarkably, postsenescence survivors that kept any Bcl repeats had telomeres with repeating patterns of wild-type and Bcl repeats (**Figure 4**). A common pattern seemed to be present in most or all telomeres of a particular survivor clone, but different patterns were seen in different survivor clones. This outcome led to the roll-and-spread model (**Figure 5**). According to this hypothesis, a tiny circle of telomeric DNA is occasionally produced by a recombination event at the short telomeres of senescing cells. This circle then serves as a template for rolling-circle DNA synthesis that produces a long tract of telomeric repeats in essentially one step. After one long telomere has been generated, all other telomeres acquire the elongated sequence through BIR events that copy the first long telomere. The repeating patterns seen in survivors with Bcl repeats could arise from

a

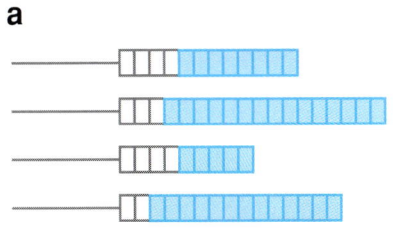

Terminal Bcl repeats from mutant telomerase RNA template

b

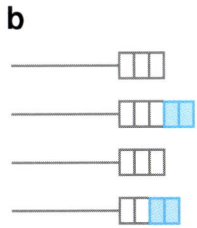

Delete telomerase; gradual telomere shortening

c

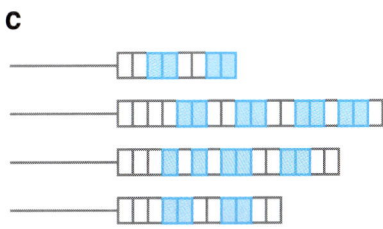

Postsenescence survivors with Bcl repeats in repeating patterns

Figure 4

Patterned telomeres in postsenescence survivors. (*a*) *TER1-7C(Bcl) K. lactis* cells are generated with all but the most centromere-proximal repeats carrying Bcl sequences (*light blue boxes*). (*b*) After deletion of *TER1-7C(Bcl)*, telomeres shorten, and cells senesce. (*c*) Postsenescence survivors, with telomeres lengthened by RTE, that retain at least some Bcl repeats have related repeating patterns of Bcl and wild-type repeats in most or all telomeres.

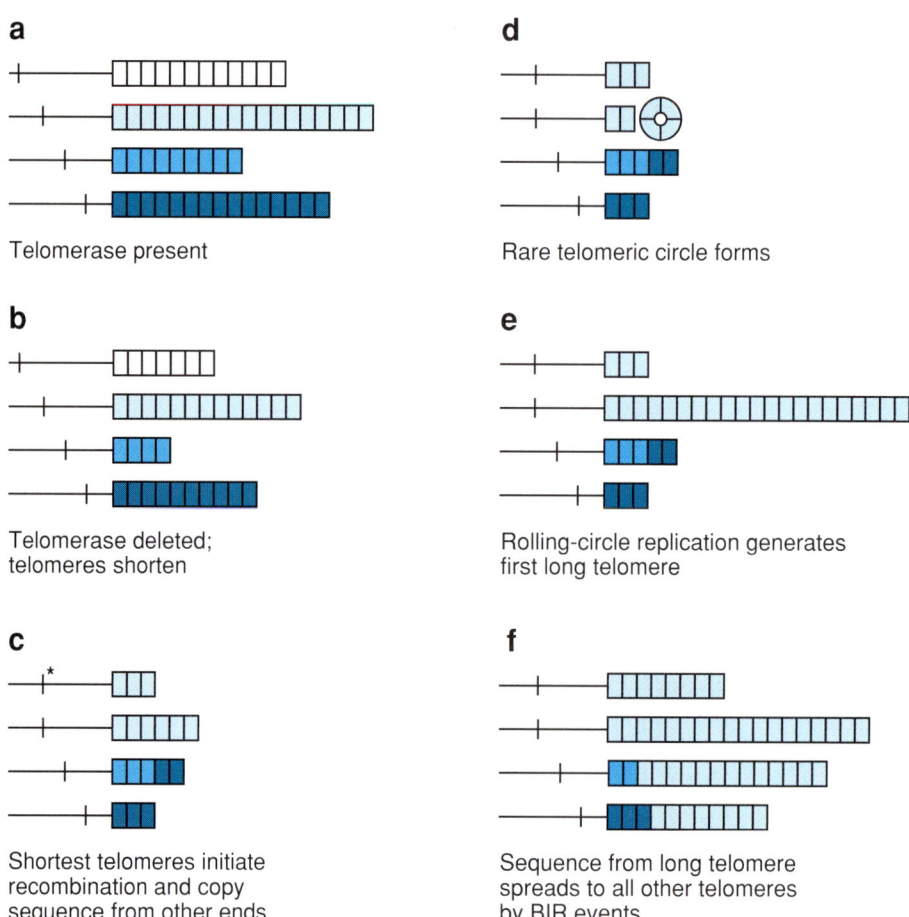

Figure 5

The roll-and-spread model. (*a*) Four telomeres are shown at the point when telomerase is deleted, with the boxes representing telomeric repeats. Telomeric repeats are shown with shading added to illustrate movement of sequences between telomeres. Vertical lines indicate polymorphisms within homologous subtelomeric sequences. Individual telomeres will differ somewhat in size, similar to what occurs when telomerase is present. (*b*) Telomeres gradually shorten in the absence of telomerase. Telomeres that were initially shortest will be the first to become uncapped. (*c*) As particular telomeres become critically short, they become prone to initiating recombination. In this way, a telomere retaining a few telomeric repeats may acquire additional repeats from another telomere by a BIR-like event. Alternatively, the BIR event may result in replacement of both subtelomeric and telomeric DNA with a sequence from another chromosome end (resulting in subtelomeric gene conversion; *asterisk*). This type of telomere-telomere recombination may account for the importance of *RAD52* in early senescence and may generate slightly longer telomeres. Such events may be responsible for maintaining the short terminal telomeric repeat tracts in *S. cerevisiae* type I survivors. (*d*) Formation of a tiny telomeric circle (t-circle), possibly through an intratelomeric deletion (see **Figure 6**). (*e*) Rolling-circle replication with the t-circle serving as a template results in the formation of an appreciably elongated telomere. See **Figure 6** for possible ways this could occur. (*f*) Sequence of the elongated telomere is spread to other telomeres through additional BIR-like events. In the survivors, each of the elongated telomeres contains repeats derived from the sequence of the original t-circle.

a rolling-circle event using a telomeric circle (t-circle) composed of both Bcl and wild-type repeats. Because the telomeres in these survivors had patterns with periodicities as small as 100 bp (103), this predicted that circles of very small size could be both generated and copied by rolling-circle events in $ter1\Delta$ cells.

Several predictions of the roll-and-spread model have been tested and shown to be true in *K. lactis*. One major prediction is that t-circles promote RTE. DNA circles constructed in vitro and composed of a cloned *K. lactis* telomere and an *S. cerevisiae URA3* gene were shown to produce long tandem arrays of the *URA3*-telomere unit at chromosome ends after transformation into *K. lactis*. In $ter1\Delta$ cells with short recombination-prone telomeres, the *URA3*-telomere arrays were always present in many telomeres as early as cells could be examined (~25–30 cell divisions after transformation) (103). In contrast, wild-type transformants usually had only a single telomere with a *URA3*-telomere array. Transformations with mixtures of two *URA3*-telomere circles that differed by only a single restriction site argued that all copies of *URA3*-telomere units originated from a single transforming molecule. These results were interpreted as indicating that wild-type cells could undergo a rolling-circle event that made a single tandem array but that $ter1\Delta$ cells also underwent additional events that copied the arrays onto many other telomeres.

Evidence that very small t-circles can also promote RTE in *K. lactis* came from cotransformation of mostly single-stranded 100-nt t-circles along with plasmid containing an origin of replication into *K. lactis* cells with short recombinogenic telomeres (88). About 1% of transformants had short tandem arrays at their telomeres, which were derived from the minicircle. Interestingly, the degree of telomere elongation observed in these transformants was only hundreds of base pairs, much shorter than the telomere elongation produced by transformation with the 1.6-kb *URA3*-telomere circle but similar to the extent of elongation that occurs in *K. lactis* $ter1\Delta$ survivors. This suggests that rolling-circle replication of a highly constrained small circular template may go through fewer iterations than might be expected on a larger unconstrained circle.

How rolling-circle replication of a t-circle would be primed is not clear. An obvious possibility is that priming is accomplished by a telomeric 3' end after it has annealed to, or strand-invaded, the t-circle (**Figure 6**, bottom left). In this scenario, telomere elongation and rolling-circle replication are combined in the same step. This model predicts that only the C-rich strand of the t-circle would ever be used as a template for replication. However, experiments transforming a heteroduplex 1.6-kb *URA3*-telomere circle into mismatch repair-deficient *K. lactis* cells led to the conclusion that either strand of the circle could be used as a template (88). This may indicate that there can be more than one way for rolling-circle synthesis to be primed. An alternate possibility is that priming occurs extrachromosomally. **Figure 6** shows a few different models of how the priming of rolling-circle events might occur.

A second major prediction of the roll-and-spread model is that tiny t-circles should, at least occasionally, be produced in senescing cells. Although such circles have not been detected in $ter1\Delta$ cells, they are common in some *K. lactis* mutants with long dysfunctional telomeres. The *ter1-16T* mutant (which produces a base change in the Rap1-binding site of the telomeric repeat) contains double-stranded and single-stranded circles of telomeric DNA as small as ~100 bp/nt (104). Only the G strand was found in the single-stranded circles. Although formation of long telomeres in *ter1-16T* was Rad52-independent (and therefore, telomerase driven) (105), the formation of the double-stranded and single-stranded t-circles was Rad52-dependent. Some models of how single-stranded and double-stranded t-circles might form and be used in RTE are shown in **Figure 6**.

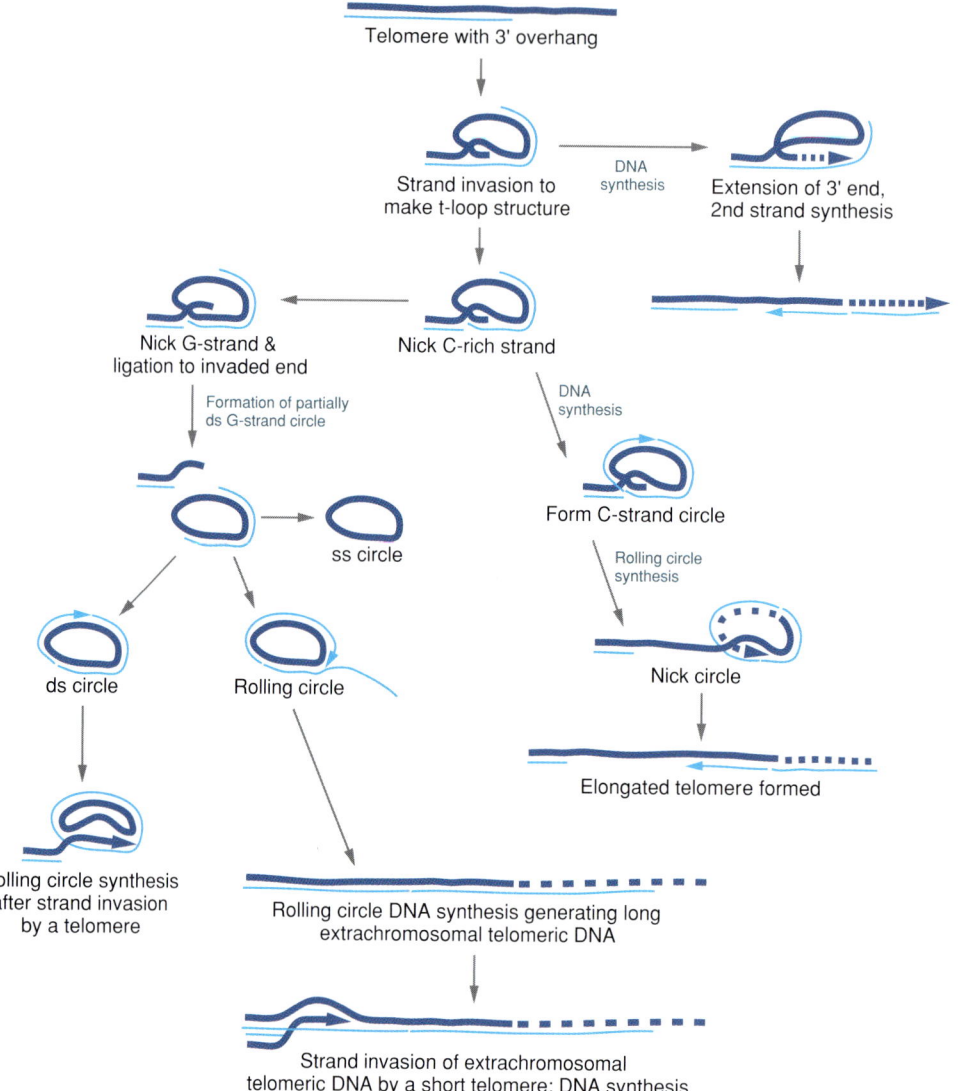

Figure 6

Models of RTE initiated by a t-loop-like intermediate. (*top*) A telomere processed to have a 3′ overhang can undergo intramolecular strand invasion to form a t-loop-like structure. G-rich and C-rich strands (G strand and C strand) are shown as bold and thin lines, respectively. (*left pathway*) A nick in the C strand followed by ligation of the newly cut 5′ end to the strand-invaded 3′ end leads to formation of a G-strand circle with a partial C-strand. This might be processed to be fully single stranded (ss), fully double stranded (ds), or become engaged in rolling-circle DNA synthesis (104). (*middle pathway*) In a second model, a nick in the C strand is extended by DNA synthesis followed by ligation to generate a C-strand circle that may be still base paired to the telomere. The invaded 3′ end could then directly initiate rolling-circle DNA synthesis. If the C-strand circle became nicked, the circle could become part of the extended telomere. In this model, the circle may not be diffusible, and rolling-circle synthesis could be limited to the telomere that formed the circle. (*right pathway*) Extension of the invaded 3′ end to generate a modestly elongated telomere without rolling-circle synthesis. This "rolling-loop" mechanism (67, 103a) would be more limited in the amount of elongation that could be produced in a single event, compared to rolling-circle synthesis.

Topcu et al. (90) found evidence supporting another major prediction of the roll-and-spread model, i.e., most or all elongated telomeres in a postsenescence survivor arise from a single source. *K. lactis* cells containing a single telomere composed entirely of Bcl repeats were first generated by transformation. After the wild-type *TER1* gene was deleted, about 10% of the survivors had Bcl repeats on all of their telomeres, and the remaining survivors did not spread Bcl repeats to other telomeres. These data argue that spreading of sequence from one to all telomeres can occur during survivor formation and is largely an all-or-none phenomenon. The 10% frequency of Bcl spreading is consistent with t-circle formation occurring stochastically from 1 of the 12 telomeres.

A transient intermediate predicted by the roll-and-spread model is a presurvivor cell with only one elongated telomere (**Figure 5**). Construction of *ter1*Δ cells with one long Bcl telomere showed that Bcl repeats now spread to all other telomeres at ∼95% frequency (90). Interestingly, growth senescence was appreciably, albeit temporarily, suppressed. These results suggest that the presence of a single long telomere in a senescing cell bypasses the need for rolling-circle synthesis and promotes earlier spreading and survivor formation.

Among *K. lactis* survivors that spread Bcl repeats to all telomeres, some recombination events involved the spreading of both subtelomeric and telomeric sequences, whereas others spread only Bcl repeats through intertelomeric recombination. From the latter class, it was determined that only telomeres that had dropped below ∼100 bp could initiate recombination with other telomeres (90). These results clearly indicate that complete loss of telomeric repeats is not necessary to permit a telomere from initiating recombination. The relatively sharp transition point between recombination-resistant and recombination-prone states also hints that the transition between the two states may involve an abrupt change in telomere structure, such as the loss of a t loop or disruption of a folded structure.

Experiments with *S. cerevisiae* have also shown that telomeres retaining small amounts of telomeric sequence are also capable of recombining (106). A particular telomere from senescing cells lacking telomerase was examined for changes from its initial sequence. Because *S. cerevisiae* telomeric repeats are heterogeneous in sequence, any changes in the sequence of the telomere in question that occurred from recombination could be readily identified. It was found that telomeres retaining 43 to 226 bp of the original sequence acquired sequence additions of up to 179 bp. The relatively wide range of sizes seen for recombining telomeres in *S. cerevisiae* could conceivably indicate that the heterogeneous *S. cerevisiae* telomeric sequences confer some variability in the size of a minimally functional telomere. For example, not all segments of budding yeast telomeres will bind Rap1, nor are all arrays of Rap1-binding sites equally functional as a telomeric terminus (107).

Some evidence suggests that a roll-and-spread mechanism may also be important to *S. cerevisiae* survivor formation. That type II survivors might form through a rolling-circle mechanism was first suggested from the observation that highly elongated telomeres appeared without obvious intermediate forms (73). Additionally, the simple tendency for individual survivors to generally be characterized as having type I or type II structure favors models wherein spreading of sequence between telomeres occurs by recombination after an initial event has created a suitable template. If formation of type I and type II arrays occurred independently at different telomeres, the 32 telomeres in a haploid *S. cerevisiae* cell would not be expected to all acquire only a type I or only a type II structure. Circles of DNA containing subtelomeric Y' elements (and presumably also telomeric repeat arrays) are detectable even in wild-type *S. cerevisiae* cells (108). Copying such circles could conceivably lead to formation of type I-like arrays. However, individual type I survivors usually amplify both size classes of Y' elements (68). This does not rule out the possibility that type I survivors arise from Y' circles, but it does argue that they do not typically arise from a single sequence source. Clues to how type I arrays form could be found by examining the arrangement of amplified copies of the two types of Y' elements.

Teng and colleagues have proposed that a roll-and-spread mechanism is responsible for *S. cerevisiae* type II survivors. They found that DNA circles, composed of a telomeric repeat tract plus a selectable marker gene (*kan*) and produced by Cre-Lox excision, could lead to formation of tandem arrays of the telomere-Kan^r sequence at telomeres (109). Array formation was greatly reduced in *rad50* and *rad52* mutants but not in a *rad51* mutant. Curiously, wild-type cells formed arrays at a frequency not greatly reduced from that of a telomerase deletion mutant, an observation also true for *K. lactis* (103, 109). Other experiments showed that a *RAD50*-dependent excision of Y' circles was greatly elevated at the point of survivor formation (109). However, the survivors observed in these experiments were type II and thus could not have been derived from the Y' circles. These data are consistent with the possibility that one role of the MRX complex in type II survivor formation is in the formation of t-circles. Consistent with this, t-circle formation in mammalian cells was inhibited in cells lacking Nbs1, a protein that complexes with the mammalian versions of Mre11 and Rad50 (99).

RTE IN THE PRESENCE OF TELOMERASE

Although RTE in yeast is generally associated with mutants lacking components of telomerase, it can sometimes occur when telomerase is still present. *S. cerevisiae* and *S. pombe tel1 mec1* double mutants behave very similarly to telomerase deletion mutants in vivo despite having telomerase that functions in vitro (84). The senescence in these mutants appears because telomerase is unable to extend telomeric ends. *K. lactis* telomerase RNA mutations that do not fully eliminate telomerase's ability to act in vivo can sometimes undergo RTE (M.J. McEachern, unpublished data). Although telomeres in these mutants are generally stable at very short sizes, appreciable jumps in telomere length are sometimes observed. This amplification involves both wild-type repeats and those specified by the tagged telomerase, and consequently, it must arise from recombination. A possible explanation of why telomeres in a cell with a weak telomerase may undergo RTE is that telomeres in such cells may frequently drop below the critical size threshold required to block telomeres from initiating recombination.

The 3' overhangs that form at telomeres are protected in yeast by the Cdc13-Stn1-Ten1 protein complex (44, 110–112). Telomeres become lethally deprotected when this complex is disrupted. When the temperature-sensitive *cdc13-1* allele, unable to grow at 29°C, is combined with various checkpoint mutations, cells exhibited senescent growth—even though telomerase was still active—and then produced survivors (113). All *cdc13-1 mec3* survivors had elongated telomeric repeat tracts but no Y' amplification. Consistent with their being type II survivors, their formation was dependent *RAD50* and *RAD52* but not *RAD51*. These data suggest that RTE was initiated at telomeres that are at or near their normal length but had lost their normal end protection and likely had long single-stranded telomere DNA. These events occurred in the presence of a functional telomerase.

Somewhat similar results arise when *S. cerevisiae cdc13-1 yku70Δ* cells are grown at a semipermissive temperature (74). The *cdc13-1 yku70Δ* double mutants grew poorly and eventually produced faster-growing and more normal-looking cells. These fast-growing survivors had elongated telomeres that were similar in length but more heterogeneous than those of type II survivors. The increased heterogeneity may indicate an increased instability of the telomeres. Further evidence that

Ku influences telomeric recombination came from studies of *tlc1 yku70Δ* double mutants (114). Such cells exhibited a more rapid senescence and were found to have type II telomere structures as soon as DNA from germinating spores could be examined (74). The above data argue that Ku acts as an inhibitor of type II RTE.

An interesting finding is that type II RTE in *cdc13-1 yku70Δ* and *tlc1 yku70Δ* mutants is more dependent upon *RAD51* than upon *RAD50/RAD59* (74). Rad51-dependent recombination events are known to require longer stretches of homology than Rad50-dependent events (23). Alternatively, the diminished requirement for *RAD50* might reflect that RTE occurs independent of t-circles. The absence of type I amplification in the above mutants could simply be due to an increased frequency of type II events. Another possibility is that the short terminal telomeric tracts of type I survivors are too short to permit improved growth in cells with compromised telomere capping.

Another unusual example of RTE is exhibited by *K. lactis* cells carrying a specific mutation in the *STN1* gene (115). The *stn1-M1* mutation leads to a chronic growth defect that resembles a moderately senescent state. These cells show telomerase-independent telomere elongation that is very similar to that seen in human ALT cells, producing a smear of telomeric signal from very short sizes up to the mobility limit in gels. The RTE in this mutant was labeled type IIR for the "runaway elongation" that makes it very different from *K. lactis* cells lacking telomerase, which exhibit much more modest elongation. Telomeres in *stn1-M1* cells are defective at capping as indicated by long single-stranded 3' tails, highly elevated rates of subtelomeric GC, and extreme levels of rapid telomere shortening upon reintroduction of a wild-type *STN1* gene. It was proposed that the runaway RTE arose from telomeres that were able to initiate recombination in a manner that was largely or entirely independent of their length. This could explain why the RTE of *stn1-M1* cells was not self-limiting, as is the case of RTE of *ter1* survivors. It is not yet known whether the type IIR RTE of *stn1-M1* cells involves either t-circles or spreading of sequence from a single telomere source.

A second example of atypical RTE in *K. lactis* involves telomerase deletion mutants constructed to have telomeres composed largely of a particular sequence of mutant repeats (Kpn repeats) (90). The telomeres in such cells become much longer than those of telomerase deletion mutants with only wild-type repeats. The mutant Kpn repeats in question are known to disrupt the negative regulation of telomerase-mediated telomere maintenance (35, 105). This suggests that there may be a common telomeric structure that regulates the access of both telomerase and recombination proteins to the telomeric end.

RTE AT CHROMOSOMAL ENDS IN OTHER YEAST SPECIES

Deletion of the gene encoding the catalytic subunit of telomerase (trt1) from the fission yeast *S. pombe* leads to growth senescence and the emergence of better-growing survivors (36). Although most of these better-growing survivors have circular chromosomes, growth of cultures in liquid media led to the emergence of less common but faster-growing survivors that maintained linear chromosomes. These survivors retained a weak telomeric hybridization signal and showed amplifications and rearrangements of subtelomeric sequences. Whereas deletion of the telomere-binding protein Taz1 from *trt1* mutants produced only survivors with circular chromosomes, deleting trt1 from taz1⁻ mutants with long telomeres resulted in survivors with only linear chromosomes. These survivors retained telomeric repeats and showed fewer signs of subtelomeric rearrangements. The greatly increased rate of formation of survivors retaining telomeric sequences could be due to the presence of long telomeres at the point when telomerase was deleted or to the disrupted telomere function.

A fourth yeast species in which the effects of deleting telomerase have been studied is the diploid opportunistic pathogen *Candida albicans*. The telomeres of *C. albicans* are composed of uniform 23-bp repeats in arrays than are often longer and more heterogeneous than the telomeres of *S. cerevisiae*, *K. lactis*, or *S. pombe*, sometimes ranging up to at least a few kilobases in length (116). Both protein and RNA components of telomerase have been identified from *C. albicans*, and mutants lacking all three genes display evidence of telomere shortening (117; M.J. McEachern, Y. Tzfati, E. Orr, and E. Blackburn, unpublished data). In clones lacking *CaTERT* or *CaEST3*, the shortening was generally progressive over at least 200 cell divisions of growth, but there was a clear ability to generate longer telomeres. These results argue that *C. albicans* telomeres can undergo a type II RTE. The apparent absence of growth defects after telomerase deletion may argue that *C. albicans* telomeres can more readily be extended by recombination than those of other yeasts so far examined. However, as even a single long telomere can appreciably suppress growth senescence and aid survivor formation in *K. lactis*, it would be expected that the long and heterogeneous telomeres of *C. albicans* would also reduce the severity of senescence. Interestingly, deletion of both alleles of the *C. albicans RAD52* gene leads to telomeres that are elongated relative to wildtype cells (118). This could suggest a role for homologous recombination in reducing the size of normal *C. albicans* telomeres. Whether *C. albicans* contains t-circles has not been established.

RTE AT YEAST MITOCHONDRIAL DNAs

Although most mitochondrial DNAs are circular, those of some species, including certain yeasts, are linear in structure (119, 120). The linear mitochondrial DNAs of several yeast species have been shown to terminate in a series of large tandem repeats. Those of *Candida parapsilosis* have been studied most extensively. The mitochondrial telomeres in this species are composed of 738 bp repeats with a 100–120 nucleotide 5′ overhang (121), protected by a mitochondrial telomere-binding protein (122, 123). Two-dimensional gel analysis and electron microscopy has revealed that *C. parapsilosis* mitochondria contain numerous supercoiled DNA circles that appear to be composed of integral numbers of the telomeric repeat (124). Similar t-circles were obtained from *Candida salmanticensis* and *Pichia philodendra*, other species with linear mitochondrial DNA. T-circles were absent in a *C. parapsilosis* strain with circular mitochondrial DNA, arguing that there is a linkage between t-circle formation and linear mitochrondrial DNA structures (125). More recently, small telomeric DNAs that are lasso shaped, with a single-stranded region at the junction between the circle and the tail, have been observed in *C. parapsilosis* (126). The observation of these structures in the absence of psoralen cross-linking suggests that the lasso-shaped molecules are rolling-circle replication intermediates rather than less stable t loops or products of strand invasion. It has been proposed that short conserved palindromic regions in the telomeric repeat are involved in telomeric DNA maintenance (126), perhaps serving as sites of nicking that might initiate rolling-circle synthesis in a manner similar to the replication of certain bacterial phage and plasmids (127, 128).

SURVIVAL THROUGH FORMATION OF TERMINAL PALINDROMES

Recent experiments have shown that *S. cerevisiae* cells lacking both telomerase and recombination can still sometimes produce survivors if they are also mutated to reduce the rate of 5′ end resectioning (129). Maringele and Lydall found that an *exo1Δ* mutation permitted the slow emergence of survivors in both *tlc1Δ rad52Δ* and *mre11Δ yku70Δ* mutants. Deletion of *MRE11*, the nuclease activity that contributes to the generation of 3′ telomeric overhangs, greatly increased the frequency of survivors in a *tlc1Δ rad52Δ exo1Δ* mutant. However, *tlc1Δ rad52Δ mre11Δ* mutants produced no survivors. These data argue that reducing the nucleolytic processing that occurs at shortening permits some type of survivor formation that is independent of both telomerase and recombination.

The survivors that were examined maintained linear chromosomes but suffered large terminal deletions that usually included not only telomeric and subtelomeric DNA but also often kilobases of terminal unique DNA as well. Loss of essential genes in these survivors appears to have been averted at some chromosome ends by the formation of large inverted duplications (palindromes). Sequencing analysis suggests that these palindromes developed from small, naturally occurring inverted repeats as shown in **Figure 7**. It was suggested that once small inverted repeats become single stranded at the end of a degraded chromosome, they could fold into a hairpin with the 3′ end able to act as a primer to initiate DNA synthesis (**Figure 7**). Once normal DNA replication has occurred, the hairpin terminus would lead to the formation of a large terminal palindrome. By extending a chromosomal end, palindrome formation would counteract the gradual loss of sequence from ends that occurs in the absence of telomerase. An interesting implication of these PAL survivors, as they are called, is that reduced 5′ end resectioning may make it possible for yeast cells to grow, albeit very slowly, with chromosome ends often altogether lacking telomeric sequences.

MAMMALIAN TELOMERE MAINTENANCE IN THE ABSENCE OF TELOMERASE: THE ALT PATHWAY

Although many immortalized human cells, including most tumor cells, exhibit a reactivation of the telomerase enzyme that is usu-

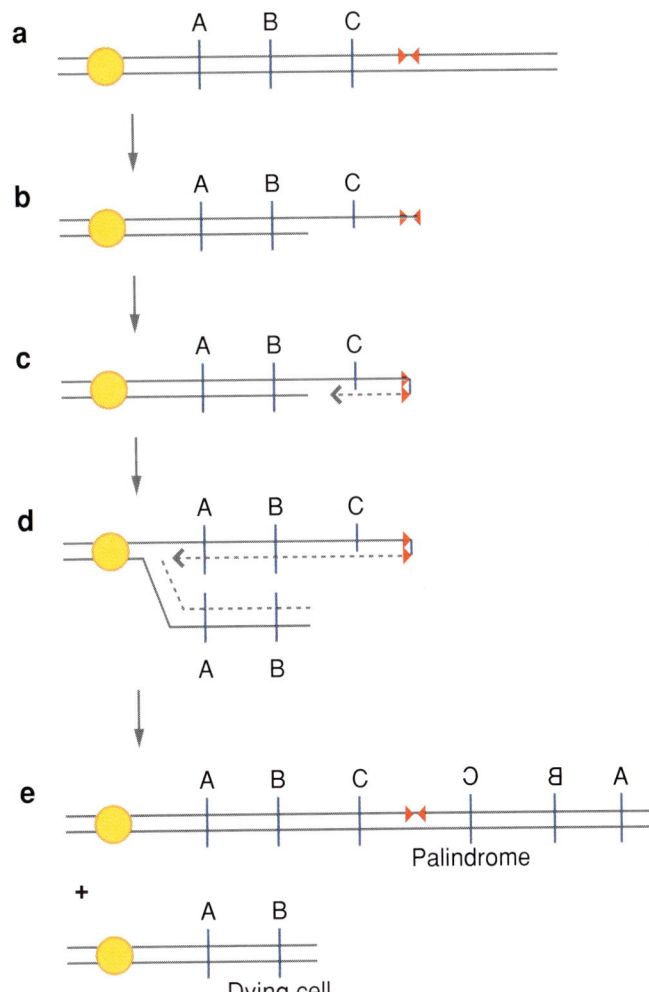

Figure 7
Proposed mechanism for the generation of PAL survivors (based on Reference 129). (*a*) Two strands of a chromosome are shown with a centromere (*circle*). Letters indicate genetic markers, and small red triangles indicate a short inverted repeat (for clarity, depicted on only one strand). (*b*) In the absence of telomerase, Rad52, and ExoI, loss of sequence from the end occurs without repair. Degradation of the 5′ end produces 3′ single-stranded tails. (*c,d*) Single-stranded inverted repeat DNA can form a hairpin in which the 3′ end acts as a primer for DNA synthesis that copies much of the chromosome arm. (*e*) Replication of the hairpin end results in the formation of a large palindrome at the chromosome end of the surviving daughter cell.

ally inactivated in early development, a subset of cell lines and tumor cells can maintain telomeres in the absence of such reactivation (130, 131). These cells are said to have engaged an alternative lengthening of telomeres (ALT) pathway of telomere maintenance. It is tempting to speculate that ALT proceeds by BIR. As yet there are no genetic data to show if this pathway is dependent on the expected cast of recombination proteins, but several papers provide strong evidence that recombination is involved. The first evidence came from Reddel's lab (132), where an ALT cell line was transformed with a selectable marker inserted directly into telomere sequences; subsequently this marker was found to proliferate to other chromosome ends during the growth of the cells. Such events are apparently less frequent in immortalized cells in which telomerase has been reactivated. Whether this represents a nonreciprocal, proliferative (BIR-like) event has not been established. Indeed, recent studies of telomeric regions in ALT cells show that there is a highly elevated level of sister-chromatid exchange, though the relationship to ALT per se is not known (133–136). Sister-chromatid exchange could be intrinsically part of a BIR mechanism that involves a migrating single Holliday junction, as shown in **Figure 2b**.

As noted above, telomere length extension by recombination might involve rolling-circle replication between an exposed telomere end and an extrachromosomal circle. Consistent with this idea is the recent discovery that ALT cells have a high level of such extrachromosomal telomeric circular DNA (99, 137). From their study of the origin of such circles in non-ALT cells expressing a mutant TRF2, Wang et al. (99) inferred that these circles arise by homologous recombination within an intrachromosomal t loop, producing an abruptly shortened telomere and the reciprocal recombination product, an extrachromosomal t-circle. Such circles could act as templates for rolling-circle telomere replication, analogous to events studied in the yeast *K. lactis* (103).

ACKNOWLEDGMENTS

We thank Titia de Lange, David Lydall, Laura Maringele, and Lubo Tomaska for their invaluable comments and Ellen Larson for editorial suggestions. Research in the Haber and McEachern labs was supported by several grants from the National Institutes of Health.

LITERATURE CITED

1. Meselson M, Weigle J. 1961. *Proc. Natl. Acad. Sci. USA* 47:857–68
2. Skalka A. 1974. In *Mechanisms of Recombination*, ed. RF Grell, pp. 431–32. New York: Plenum
3. Motamedi MR, Szigety SK, Rosenberg SM. 1999. *Genes Dev.* 13:2889–903
4. Kuzminov A, Stahl FW. 1999. *Genes Dev.* 13:345–56
5. Kogoma T. 1996. *Cell* 85:625–27
6. Kogoma T. 1997. *Microbiol. Mol. Biol. Rev.* 61:212–38
7. Michel B. 2000. *Trends Biochem. Sci.* 25:173–78
8. Michel B, Grompone G, Flores MJ, Bidnenko V. 2004. *Proc. Natl. Acad. Sci. USA* 101:12783–88
9. Esposito MS. 1978. *Proc. Natl. Acad. Sci. USA* 75:4436–40
10. Esposito MS, Ramirez RM, Bruschi CV. 1994. *Curr. Genet.* 26:302–7
11. Voelkel-Meiman K, Roeder GS. 1990. *Genetics* 126:851–67
12. Dunn B, Szauter P, Pardue ML, Szostak JW. 1984. *Cell* 39:191–201
13. Walmsley RW, Chan CS, Tye BK, Petes TD. 1984. *Nature* 310:157–60
14. Haber JE. 1999. *Trends Biochem. Sci.* 24:271–75
15. Morrow DM, Connelly C, Hieter P. 1997. *Genetics* 147:371–82

16. Bosco G, Haber JE. 1998. *Genetics* 150:1037–47
17. Malkova A, Ivanov EL, Haber JE. 1996. *Proc. Natl. Acad. Sci. USA* 93:7131–36
18. Signon L, Malkova A, Naylor M, Klein H, Haber JE. 2001. *Mol. Cell. Biol.* 21:2048–56
19. Bai Y, Symington LS. 1996. *Genes Dev.* 10:2025–37
20. Malkova A, Naylor ML, Yamaguchi M, Ira G, Haber JE. 2005. *Mol. Cell. Biol.* 25:933–44
21. Malkova A, Signon L, Schaefer CB, Naylor M, Theis JF, et al. 2001. *Genes Dev.* 15:1055–160
22. Labib K, Kearsey SE, Diffley JF. 2001. *Mol. Biol. Cell* 12:3658–67
23. Ira G, Haber JE. 2002. *Mol. Cell. Biol.* 22:6384–92
24. Deleted in proof
25. Davis AP, Symington LS. 2004. *Mol. Cell. Biol.* 24:2344–51
26. Formosa T, Alberts BM. 1986. *Cell* 47:793–806
27. Lundblad V, Szostak JW. 1989. *Cell* 57:633–43
28. Singer MS, Gottschling DE. 1994. *Science* 266:404–9
29. Lendvay TS, Morris DK, Sah J, Balasubramanian B, Lundblad V. 1996. *Genetics* 144:1399–412
30. Ijpma AS, Greider CW. 2003. *Mol. Biol. Cell* 14:987–1001
31. Askree SH, Yehuda T, Smolikov S, Gurevich R, Hawk J, et al. 2004. *Proc. Natl. Acad. Sci. USA* 101:8658–63
32. Nugent CI, Hughes TR, Lue NF, Lundblad V. 1996. *Science* 274:249–52
33. Evans SK, Lundblad V. 1999. *Science* 286:117–20
34. Taggart AK, Teng SC, Zakian VA. 2002. *Science* 297:1023–26
35. McEachern MJ, Blackburn EH. 1995. *Nature* 376:403–9
36. Nakamura TM, Cooper JP, Cech TR. 1998. *Science* 282:493–96
37. McEachern MJ, Blackburn EH. 1996. *Genes Dev.* 10:1822–34
38. McEachern MJ, Blackburn EH. 1997. In *Pezcoller Foundation Symposia No. 8: Genomic Instability and Immortality in Cancer*, ed. E Mihich, L Hartwell, pp. 111–27. New York: Plenum
39. Lingner J, Cech TR, Hughes TR, Lundblad V. 1997. *Proc. Natl. Acad. Sci. USA* 94:11190–95
40. Smolikov S, Krauskopf A. 2003. *Mol. Cell. Biol.* 23:8729–39
41. Teng SC, Zakian VA. 1999. *Mol. Cell. Biol.* 19:8083–93
42. Hackett JA, Feldser DM, Greider CW. 2001. *Cell* 106:275–86
43. Enomoto S, Glowczewski L, Berman J. 2002. *Mol. Biol. Cell* 13:2626–38
44. Garvik B, Carson M, Hartwell L. 1995. *Mol. Cell. Biol.* 15:6128–38
45. Grandin N, Bailly A, Charbonneau M. 2005. *Biol. Cell* 97:799–814
46. Osborn AJ, Elledge SJ. 2003. *Genes Dev.* 17:1755–67
47. Usui T, Ogawa H, Petrini JH. 2001. *Mol. Cell* 7:1255–66
48. Shroff R, Arbel-Eden A, Pilch D, Ira G, Bonner WM, et al. 2004. *Curr. Biol.* 14:1703–11
49. Sandell LL, Zakian VA. 1993. *Cell* 75:729–39
50. Lee SE, Moore JK, Holmes A, Umezu K, Kolodner R, Haber JE. 1998. *Cell* 94:399–409
51. Toczyski DP, Galgoczy DJ, Hartwell LH. 1997. *Cell* 90:1097–106
52. Vaze MB, Pellicioli A, Lee SE, Ira G, Liberi G, et al. 2002. *Mol. Cell* 10:373–85
53. Zou L, Elledge SJ. 2003. *Science* 300:1542–48
54. Sogo JM, Lopes M, Foiani M. 2002. *Science* 297:599–602
55. Pellicioli A, Lee SE, Lucca C, Foiani M, Haber JE. 2001. *Mol. Cell* 7:293–300
56. Ivanov EL, Korolev VG, Fabre F. 1992. *Genetics* 132:651–64
57. Tsukamoto Y, Taggart AK, Zakian VA. 2001. *Curr. Biol.* 11:1328–35

58. Nakada D, Matsumoto K, Sugimoto K. 2003. *Genes Dev.* 17:1957–62
59. Zubko MK, Guillard S, Lydall D. 2004. *Genetics* 168:103–15
60. Maringele L, Lydall D. 2004. *Genetics* 166:1641–49
61. Bertuch AA, Lundblad V. 2004. *Genetics* 166:1651–59
62. Hackett JA, Greider CW. 2003. *Mol. Cell. Biol.* 23:8450–61
63. Liti G, Louis EJ. 2003. *Mol. Cell* 11:1373–78
64. Pennaneach V, Kolodner RD. 2004. *Nat. Genet.* 36:612–17
65. McEachern MJ, Iyer S. 2001. *Mol. Cell* 7:695–704
66. DuBois ML, Haimberger ZW, McIntosh MW, Gottschling DE. 2002. *Genetics* 161:995–1013
67. Chen Q, Ijpma A, Greider CW. 2001. *Mol. Cell. Biol.* 21:1819–27
68. Lundblad V, Blackburn EH. 1993. *Cell* 73:347–60
69. Louis EJ, Naumova ES, Lee A, Naumov G, Haber JE. 1994. *Genetics* 136:789–802
70. Le S, Moore JK, Haber JE, Greider CW. 1999. *Genetics* 152:143–52
71. Yamada M, Hayatsu N, Matsuura A, Ishikawa F. 1998. *J. Biol. Chem.* 273:33360–66
72. Biessmann H, Mason JM. 1997. *Chromosoma* 106:63–69
73. Teng S, Chang J, McCowan B, Zakian VA. 2000. *Mol. Cell* 6:947–52
74. Grandin N, Charbonneau M. 2003. *Mol. Cell. Biol.* 23:3721–34
75. Watt PM, Hickson ID, Borts RH, Louis EJ. 1996. *Genetics* 144:935–45
76. Cohen H, Sinclair DA. 2001. *Proc. Natl. Acad. Sci. USA* 98:3174–79
77. Huang P, Pryde FE, Lester D, Maddison RL, Borts RH, et al. 2001. *Curr. Biol.* 11:125–29
78. Johnson FB, Marciniak RA, McVey M, Stewart SA, Hahn WC, Guarente L. 2001. *EMBO J.* 20:905–13
79. Rizki A, Lundblad V. 2001. *Nature* 411:713–16
80. Myung K, Datta A, Chen C, Kolodner RD. 2001. *Nat. Genet.* 27:113–16
81. Sugawara N, Goldfarb T, Studamire B, Alani E, Haber JE. 2004. *Proc. Natl. Acad. Sci. USA* 101:9315–20
82. Symington LS. 2002. *Microbiol. Mol. Biol. Rev.* 66:630–70
83. Tsai YL, Tseng SF, Chang SH, Lin CC, Teng SC. 2002. *Mol. Cell. Biol.* 22:5679–87
84. Ritchie KB, Mallory JC, Petes TD. 1999. *Mol. Cell. Biol.* 19:6065–75
85. Ritchie KB, Petes TD. 2000. *Genetics* 155:475–79
86. Grandin N, Charbonneau M. 2003. *Mol. Cell. Biol.* 23:9162–77
87. Ira G, Pellicioli A, Balijja A, Wang X, Fiorani S, et al. 2004. *Nature* 431:1011–17
88. Natarajan S, Groff-Vindman C, McEachern MJ. 2003. *Eukaryot. Cell* 2:1115–27
89. McEachern MJ. 2002. In *Telomeres and Telomerases: Cancer and Biology*, ed. G Krupp, R Parwaresch. Eurekah Press, Austin, TX. http://www.eurekah.com
90. Topcu Z, Nickles K, Davis C, McEachern MJ. 2005. *Proc. Natl. Acad. Sci. USA* 102:3348–53
91. Rayssiguier C, Thaler DS, Radman M. 1989. *Nature* 342:396–401
92. Modrich P, Lahue R. 1996. *Annu. Rev. Biochem.* 65:101–33
93. Chen W, Jinks-Robertson S. 1999. *Genetics* 151:1299–313
94. Nicholson A, Hendrix M, Jinks-Robertson S, Crouse GF. 2000. *Genetics* 154:133–46
95. Nickles K, McEachern MJ. 2004. *Yeast* 21:813–30
96. Tishkoff DX, Boerger AL, Bertrand P, Filosi N, Gaida GM, et al. 1997. *Proc. Natl. Acad. Sci. USA* 94:7487–92
97. Bailey SM, Cornforth MN, Kurimasa A, Chen DJ, Goodwin EH. 2001. *Science* 293:2462–65
98. Crabbe L, Verdun RE, Haggblom CI, Karlseder J. 2004. *Science* 306:1951–53

99. Wang RC, Smogorzewska A, de Lange T. 2004. *Cell* 119:355–68
100. Parenteau J, Wellinger RJ. 2002. *Genetics* 162:1583–94
101. Roy J, Fulton TB, Blackburn EH. 1998. *Genes Dev.* 12:3286–300
102. McEachern MJ, Underwood DH, Blackburn EH. 2002. *Genetics* 160:63–73
103. Natarajan S, McEachern MJ. 2002. *Mol. Cell. Biol.* 22:4512–21
103a. Lustig AJ. 2003. *Nat. Rev. Genet.* 4:916–23
104. Groff-Vindman C, Natarajan S, Cesare A, Griffith JD, McEachern MJ. 2005. *Mol. Cell. Biol.* 25:4406–12
105. Underwood DH, Zinzen RP, McEachern MJ. 2004. *Mol. Cell. Biol.* 24:912–23
106. Teixeira MT, Arneric M, Sperisen P, Lingner J. 2004. *Cell* 117:323–35
107. Grossi S, Bianchi A, Damay P, Shore D. 2001. *Mol. Cell. Biol.* 21:8117–28
108. Horowitz H, Haber JE. 1985. *Mol. Cell. Biol.* 5:2369–80
109. Lin CY, Chang HH, Wu KJ, Tseng SF, Lin CC, et al. 2005. *Eukaryot. Cell* 4:327–36
110. Grandin N, Reed SI, Charbonneau M. 1997. *Genes Dev.* 11:512–27
111. Grandin N, Damon C, Charbonneau M. 2001. *EMBO J.* 20:1173–83
112. Pennock E, Buckley K, Lundblad V. 2001. *Cell* 104:387–96
113. Grandin N, Damon C, Charbonneau M. 2001. *EMBO J.* 20:6127–39
114. Nugent CI, Bosco G, Ross LO, Evans SK, Salinger AP, et al. 1998. *Curr. Biol.* 8:657–60
115. Iyer S, Chadha AD, McEachern MJ. 2005. *Mol. Cell. Biol.* 25:8064–73
116. McEachern MJ, Hicks JB. 1993. *Mol. Cell. Biol.* 13:551–60
117. Singh SM, Steinberg-Neifach O, Mian IS, Lue NF. 2002. *Eukaryot. Cell* 1:967–77
118. Ciudad T, Andaluz E, Steinberg-Neifach O, Lue NF, Gow NA, et al. 2004. *Mol. Microbiol.* 53:1177–94
119. Nosek J, Tomaska L, Fukuhara H, Suyama Y, Kovac L. 1998. *Trends Genet.* 14:184–88
120. Nosek J, Tomaska L. 2003. *Curr. Genet.* 44:73–84
121. Nosek J, Dinouel N, Kovac L, Fukuhara H. 1995. *Mol. Gen. Genet.* 247:61–72
122. Tomaska L, Nosek J, Fukuhara H. 1997. *J. Biol. Chem.* 272:3049–56
123. Nosek J, Tomaska L, Pagacova B, Fukuhara H. 1999. *J. Biol. Chem.* 274:8850–57
124. Tomaska L, Nosek J, Makhov AM, Pastorakova A, Griffith JD. 2000. *Nucleic Acids Res.* 28:4479–87
125. Rycovska A, Valach M, Tomaska L, Bolotin-Fukuhara M, Nosek J. 2004. *Microbiology* 150:1571–80
126. Nosek J, Rycovska A, Makhov AM, Griffith JD, Tomaska L. 2005. *J. Biol. Chem.* 280:10840–45
127. Novick RP. 1998. *Trends Biochem. Sci.* 23:434–38
128. Khan SA. 2005. *Plasmid* 53:126–36
129. Maringele L, Lydall D. 2004. *Genes Dev.* 18:2663–75
130. Reddel RR, Bryan TM. 2003. *Lancet* 361:1840–41
131. Scheel C, Poremba C. 2002. *Virchows Arch.* 440:573–82
132. Dunham MA, Neumann AA, Fasching CL, Reddel RR. 2000. *Nat. Genet.* 26:447–50
133. Bailey SM, Brenneman MA, Goodwin EH. 2004. *Nucleic Acids Res.* 32:3743–51
134. Bechter OE, Zou Y, Walker W, Wright WE, Shay JW. 2004. *Cancer Res.* 64:3444–51
135. Bechter OE, Zou Y, Shay JW, Wright WE. 2003. *EMBO Rep.* 4:1138–43
136. Bechter OE, Shay JW, Wright WE. 2004. *Cell Cycle* 3:547–49
137. Cesare AJ, Griffith JD. 2004. *Mol. Cell. Biol.* 24:9948–57

LKB1-Dependent Signaling Pathways

Dario R. Alessi, Kei Sakamoto, and Jose R. Bayascas

Medical Research Council, Protein Phosphorylation Unit, School of Life Sciences, University of Dundee, Dundee DD1 5EH, Scotland; email: d.r.alessi@dundee.ac.uk, k.sakamoto@dundee.ac.uk, j.bayascas@dundee.ac.uk

Key Words

AMPK, cancer, cell growth, Peutz-Jeghers syndrome, polarity, tuberous sclerosis, mTOR, tumor suppressor

Abstract

This review focuses on remarkable recent findings concerning the mechanism by which the LKB1 protein kinase that is mutated in Peutz-Jeghers cancer syndrome operates as a tumor suppressor. We discuss evidence that the cellular localization and activity of LKB1 is controlled through its interaction with a catalytically inactive protein resembling a protein kinase, termed STRAD, and an armadillo repeat-containing protein, named mouse protein 25 (MO25). The data suggest that LKB1 functions as a tumor suppressor by not only inhibiting proliferation, but also by exerting profound effects on cell polarity and, most unexpectedly, on the ability of a cell to detect and respond to low cellular energy levels. Genetic and biochemical findings indicate that LKB1 exerts its effects by phosphorylating and activating 14 protein kinases, all related to the AMP-activated protein kinase. The work described in this review shows how a study of an obscure cancer syndrome can uncover new and important regulatory pathways, relevant to the understanding of multiple human diseases.

Contents

INTRODUCTION 138
EVIDENCE THAT LKB1
 FUNCTIONS AS A TUMOR
 SUPPRESSOR 139
POSTTRANSLATIONAL
 MODIFICATIONS OF LKB1 ... 143
LKB1:STRAD:MO25 COMPLEX .. 144
ROLE OF PSEUDOKINASES 146
ACTIVATION OF THE
 AMP-ACTIVATED PROTEIN
 KINASE BY LKB1 146
TUMOR SUPPRESSOR
 PROPERTIES OF LKB1 AND
 AMPK 149
MEDICAL IMPLICATIONS FOR
 THE TREATMENT OF PJS
 AND OTHER CANCERS 151
LKB1 FUNCTIONS AS A MASTER
 UPSTREAM KINASE 152
MARK AND BRSK KINASES:
 ROLE OF LKB1 IN
 REGULATING CELL
 POLARITY 154
OTHER AMPK-RELATED
 KINASES 156
REGULATION OF
 AMPK-RELATED KINASES 156
CONCLUSIONS AND
 PERSPECTIVE 157

INTRODUCTION

Peutz-Jeghers syndrome (PJS) was first described in 1922 by Dr. Johannes Peutz, a Dutch physician working in Utrecht (1), and further characterized in the 1940s by Dr. Harold Jeghers and coworkers (2, 3). This disease is inherited in an autosomal dominant fashion. Patients with PJS develop benign hamartomatous polyps (overgrowth of differentiated tissues), especially in the gastrointestinal tract, as well as marked cutaneous pigmentation of the mucous membranes. Another key feature of PJS is a greatly increased risk of developing malignant tumors in multiple tissues (4–6); it has been estimated that 93% of PJS patients develop malignant tumor(s) by the age of 43 (7). Estimates for the incidence frequency of PJS vary greatly—from 1 in ~10,000 to 120,000 live births—perhaps because of its rarity and the difficulty in accurately diagnosing this condition (4, 8). Although the majority of patients have a family history of PJS, 10% to 20% of PJS cases result from spontaneous germ line mutations in the *LKB1* gene (9).

In 1997, groundbreaking genetic linkage analysis undertaken by Akseli Hemminki, a graduate student in Lauri Aaltonen's laboratory in Helsinki, revealed that the locus responsible for PJS was located on chromosome 19p13.3 (10). This claim was followed by a frantic multigroup effort in which the cDNAs of 29 genes located in this chromosomal region were sequenced from well-characterized PJS families. This work culminated in 1998 in the discovery that 11 out of the 12 PJS families studied had diverse mutations in a gene encoding a protein kinase that Jun-ichi Nezu, of Chugai Pharmaceuticals, had identified in 1996 in a screen aimed at identifying new kinases and that was termed *LKB1* (11). Nezu deposited the sequence in the database without writing a research paper. LKB1 is highly related to XEEK1, a kinase also identified in 1996 that is expressed in early *Xenopus* embryos and found to be a substrate of the cyclic AMP-dependent protein kinase A (PKA) (12). In parallel studies, another group, which had originally collaborated with the Aaltonen laboratory, also reported mutations in the LKB1 gene in PJS subjects and called this enzyme serine threonine kinase 11 (STK11) (13).

There is only a single isoform of the *LKB1* gene in the human genome, which spans 23 kb and is composed of ten exons, nine of which are coding. The gene is transcribed in the telomere-to-centromere direction. Human LKB1 comprises 433 residues (mouse, LKB1 436), and its catalytic domain (residues 49–309) is somewhat more distantly related to other protein kinases, explaining why it is

Syndrome: signs and symptoms characterizing a particular disease or medical condition

PJS: Peutz-Jeghers syndrome

localized within the center of the human kinome dendrogram (14). The N-terminal and C-terminal noncatalytic regions of LKB1 are not related to any other proteins and possess no identifiable functional domains. LKB1 is expressed at varying levels in all fetal and adult tissues examined (11, 13). Patients with sporadic cancers have also been screened for mutations in the *LKB1* gene, and although their occurrence is relatively rare, an increasing number of mutations are being identified, which are listed in **Figure 1**. Recently, it has been demonstrated that one third of lung adenocarcinomas harbor somatic *LKB1* mutations, supporting a role for LKB1 in the origin of some sporadic tumors (15–17). To date, 144 different mutations in LKB1 have been identified in PJS patients and sporadic cancers, and these are summarized in **Figure 1**. The majority of these mutations result in substantial truncations of the catalytic domain and should impair LKB1 catalytic activity. However, there are also a significant number of point mutations, which are located in the kinase domain and in the C-terminal noncatalytic region (**Figure 1**). Interestingly, no point mutations in the N-terminal noncatalytic region of LKB1 have been identified, indicating that this region of the enzyme is not crucial for tumor suppression (**Figure 1**).

EVIDENCE THAT LKB1 FUNCTIONS AS A TUMOR SUPPRESSOR

A much discussed early finding was that overexpression of wild-type LKB1 caused a G1 cell cycle arrest in HeLa and G361 cells, which do not express LKB1 (18). Catalytically inactive LKB1 mutants, including some of those isolated from PJS patients, failed to inhibit cell growth. Thus, it was assumed that the intrinsic serine-threonine protein kinase catalytic activity of LKB1 was required to block cell division. Heterozygous $LKB1^{+/-}$ mice were viable and displayed no obvious phenotype at birth and early adult life. However, by the age of 45 weeks, most $LKB1^{+/-}$ animals developed polyps in the gastrointestinal tract, predominantly in the glandular stomach, distinct from PJS patients who develop polyps mainly in the small intestine. Differences in preferential sites for intestinal tumor formation in mice and humans have also been observed with adenomatous polyposis coli knockout mice (19). However, a histological analysis revealed that the hamartomatous polyps in $LKB1^{+/-}$ mice were remarkably similar to those found in PJS patients (20–23).

A major question concerning the mechanism of tumor formation in the $LKB1^{+/-}$ animals was whether the polyps still expressed LKB1. Three groups reported that LKB1 levels in polyps from PJS patients had roughly half of the normal amount of LKB1, suggesting that tumor formation resulted from haploinsufficiency of LKB1 (20–22). However, a fourth study reported that three out of twelve polyps isolated from $LKB1^{+/-}$ mice showed loss of the wild-type *LKB1* allele (23). Moreover, the LKB1 protein was not detected in four out of eight polyps from $LKB1^{+/-}$ mice that retained a wild-type *LKB1* allele, suggesting that epigenetic gene inactivation had occurred prior to polyp emergence (23). Thus, these results suggest that total loss of LKB1 expression is required as a precondition of polyp formation. Consistent with this view, $LKB1^{fl/fl}$ hypomorphic mice that express five- to tenfold reduced levels of LKB1 in all tissues examined (except in the brain) have recently been described (24). These mice, at least until the age of 26 weeks, display no signs of developing hamartomas (K. Sakamoto, unpublished observations). In future studies, it would be interesting to study the incidence of hamartoma formation in $LKB1^{+/-}$, $LKB1^{fl/fl}$, and $LKB1^{fl/-}$ mice, which should provide insight into whether it is loss of heterozygosity or a reduction in LKB1 level that is necessary to drive polyposis. A significant number of $LKB1^{+/-}$ mice over the age of 50 weeks developed hepatocellular carcinoma (25). No expression of LKB1 mRNA and protein was observed in these hepatocellular carcinomas,

Mutation: permanent change(s) to the DNA sequence caused by copying errors during cell division and exposure to harmful substances

Protein kinase: a human gene family, which regulates aspects of cell function and, when dysfunctional, causes human disease

Cancer: cells that grow out of control into malignant tumors and replace normal tissue

Carcinoma: a cancerous tumor that begins in tissues covering or lining various body organs

Figure 1

Mutations identified in the human *LKB1* gene in patients with PJS and sporadic cancer showing a schematic representation of their predicted effects on the primary structure of the LKB1 protein. The genomic organization of the coding sequence of the *LKB1* gene is shown on the top, and the functional domains of the protein are shown below with (*a*) stop mutations, in-frame deletions, splicing mutations, and deletions; (*b*) point mutations; (*c*) frameshift mutations. Abbreviations used: NRD, N-terminal regulatory domain; CRD, C-terminal regulatory domain (*white boxes*); Δ, in-frame deletion; fs, frameshift; Ref, reference; X, point mutation. The protein kinase domain (*blue boxes*) and amino acid sequence introduced by the frameshift (*green boxes*) are also indicated.

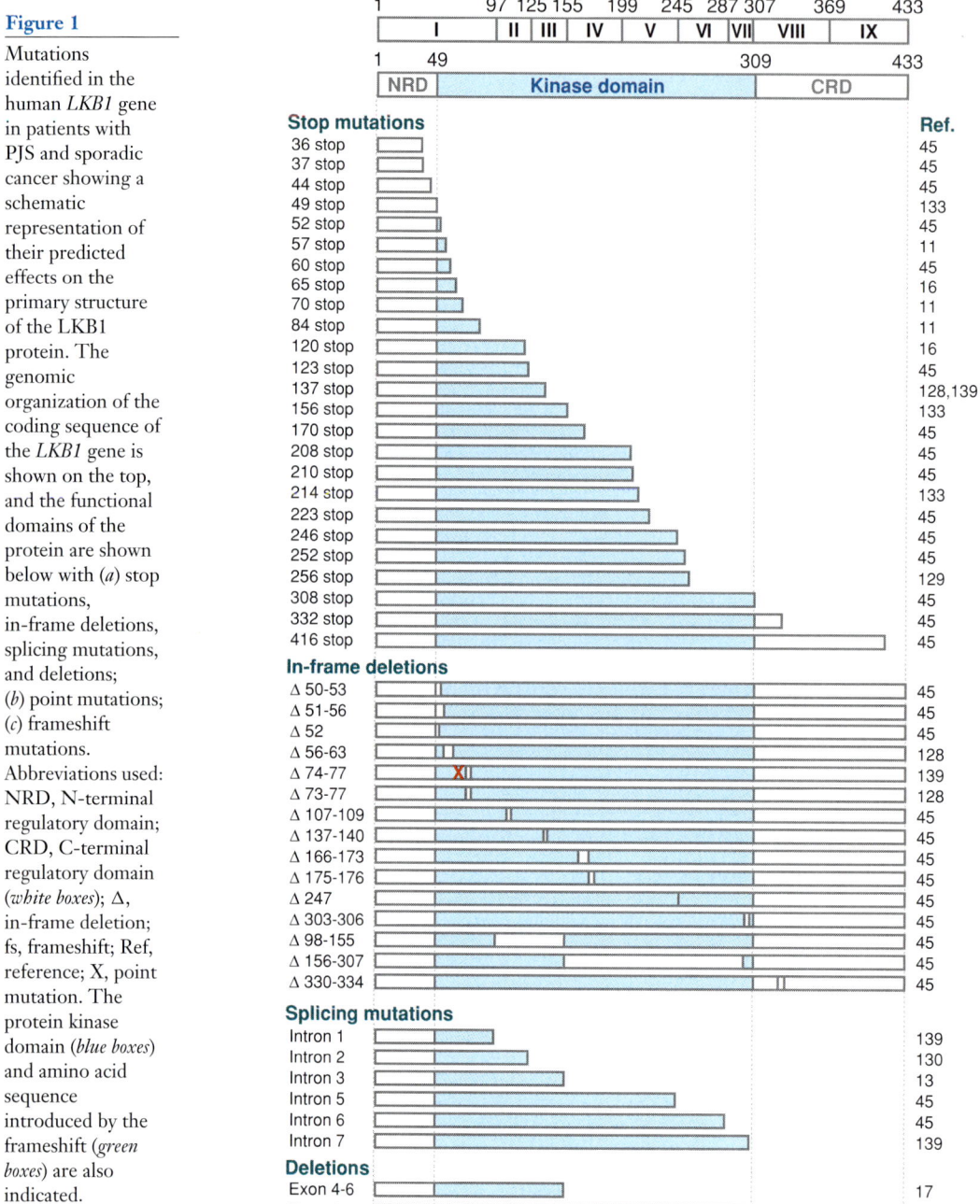

indicating that complete loss of LKB1 expression is required for the development of carcinomas (25).

LKB1 clearly plays an important role in development, as knockout of both *LKB1* alleles results in mid-gestation embryonic lethality at E11.0 (20, 21, 26). $LKB1^{-/-}$ embryos display multiple abnormalities after E8.0, such as defects in neural tube closure, abnormal development of the aorta and hypoplastic first

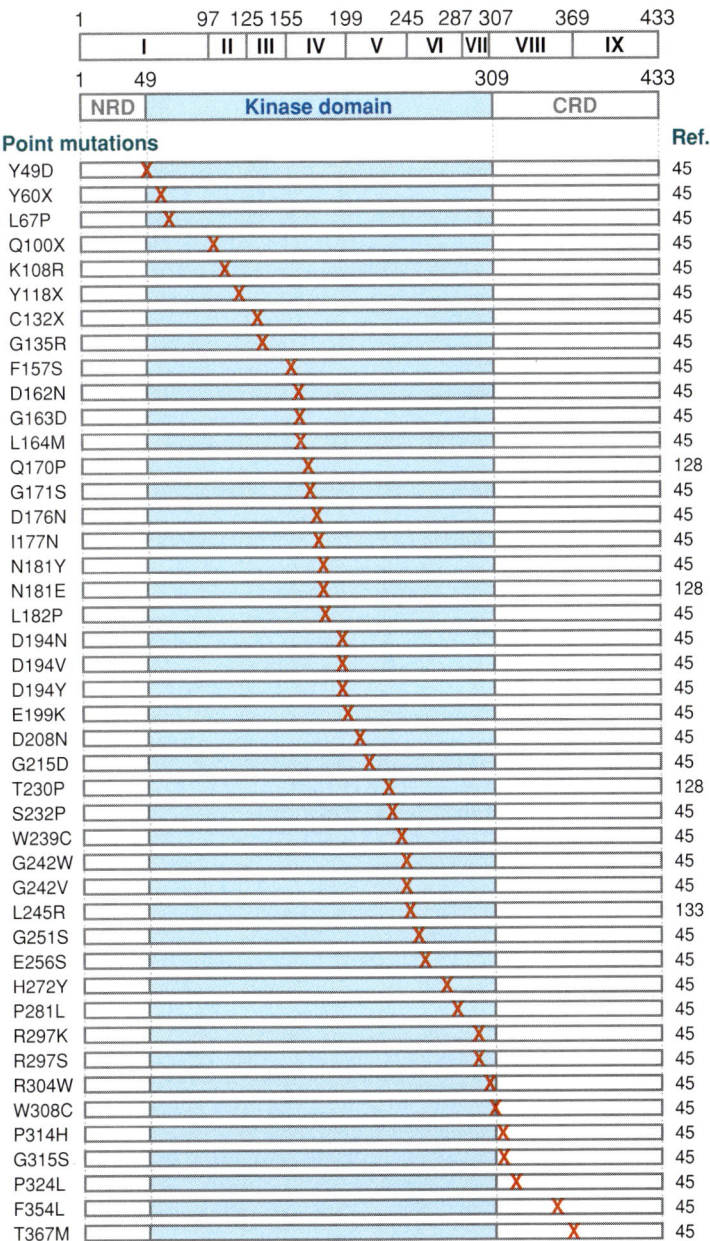

Figure 1

(Continued)

branchial arch, as well as major placental defects in females (26). Vascular endothelial growth factor (VEGF) mRNA levels in $LKB1^{-/-}$ embryos at E8.5 and E9.5 were observed to be elevated, and these embryos displayed abnormally high basal and hypoxia-induced levels of VEGF mRNA, indicating that normally LKB1 might negatively regulate VEGF production and vascular development (26).

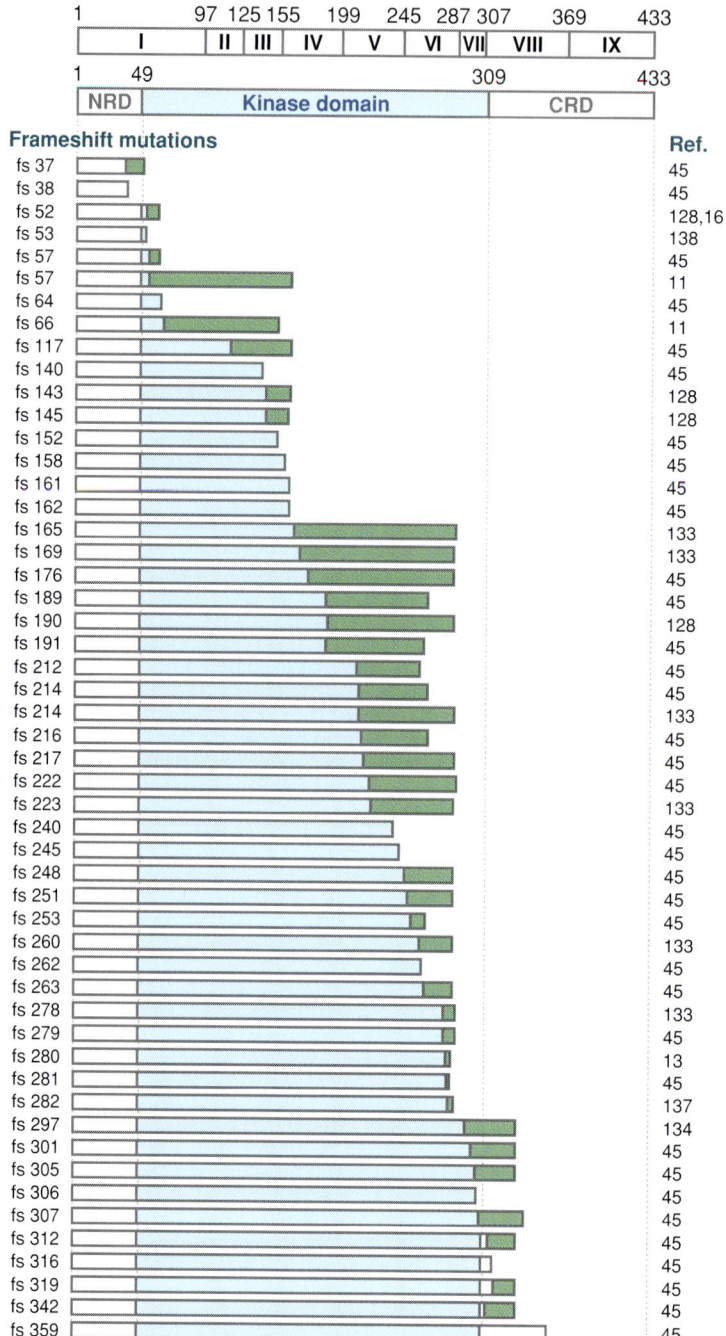

Figure 1

(Continued)

POSTTRANSLATIONAL MODIFICATIONS OF LKB1

As summarized in **Figure 2**, LKB1 is phosphorylated on at least 8 residues. The evidence suggests that LKB1 is phosphorylated at Ser31, Ser325, Thr366, and Ser431 by upstream kinases, whereas LKB1 autophosphorylates itself at Thr185, Thr189, Thr336, and Ser404. Mutation of any of these sites of phosphorylation to either Ala to abolish phosphorylation or to Glu to mimic phosphorylation has thus far not been reported to significantly affect LKB1 catalytic activity in vitro or its cellular localization (27–29). Mutation of Ser431 to either Ala or Glu prevented LKB1 from suppressing the growth of G361 cells in a colony formation assay, suggesting that phosphorylation of this residue is essential for LKB1 to inhibit cell growth (28). Mutation of Thr336, the major autophosphorylation site on LKB1, to a Glu, but not Ala, prevented LKB1 from inhibiting G361 cell growth, indicating that phosphorylation of this residue may somehow inhibit LKB1 tumor suppressor function (27). In contrast, mutation of Ser31, Ser325, or Thr366 had no major effect on the ability of LKB1 to suppress G361 cell growth (27).

The Thr336, Thr366, and Ser431 phosphorylation sites and the residues surrounding these are highly conserved in *Drosophila*, *Xenopus*, and mammalian LKB1, but not in *Caenorhabditis elegans* LKB1 (27). LKB1 is phosphorylated at Ser431 by the p90 ribosomal S6 protein kinase (RSK) and PKA in response to agonists, which trigger the activation of these kinases (28, 30). Therefore, RSK and PKA may regulate cell growth through phosphorylation of LKB1. Phosphorylation of LKB1 at Thr366 is triggered only following exposure of cells to ionizing radiation, and it is likely that the DNA-damage activated ataxia-telangiectasia-mutated (ATM) kinase

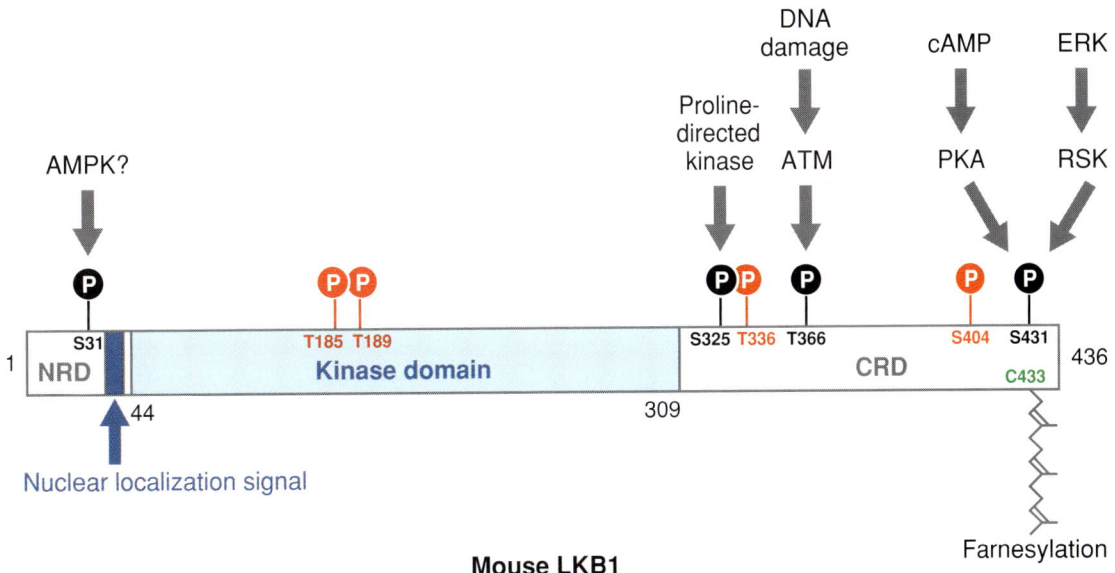

Figure 2

Posttranslational modification sites of the mouse LKB1 protein. Autophosphorylation sites are depicted in red, and the sites phosphorylated by other kinases are in black. The Cys433 farnesylation site is depicted in green. The agonists and upstream protein kinases postulated to phosphorylate each site are indicated. Residues Thr366, Ser404, Ser431, and Cys433 in the mouse sequence correspond to human LKB1 residues Thr363, Thr402, Ser428, and Cys430, respectively. The noncatalytic domains are in white, and the kinase domain is light blue.

AMPK: AMP-activated protein kinase

STRAD: STE20-related adaptor protein

MO25: mouse protein 25

Pseudokinase: comprises ~50 human kinase domains that lack at least one of the conserved catalytic residues and therefore may be catalytically deficient

mediates this phosphorylation in vivo (31). Moreover, after exposure of cells to DNA-damaging ionizing radiation, the N-terminal substrate-binding domain of ATM was observed to associate with LKB1 (32). It would be of interest to explore whether LKB1 could play a role in mediating DNA damage responses in cells. The upstream kinases phosphorylating LKB1 at Ser31 and Ser325 have not yet been characterized. Ser31 lies in a consensus sequence for phosphorylation by AMP-activated protein kinase (AMPK) (27, 33), and it would be interesting to test whether AMPK or one of the AMPK-related kinases could phosphorylate this residue. As Ser325 is followed by a proline, it is likely to be phosphorylated by one of the proline-directed kinases (27).

Collins et al. (30) observed that LKB1 terminated in the amino acid sequence Cys-Lys-Gln-Gln, lying in an optimal consensus sequence for protein prenylation. This motif is conserved in LKB1 homologues in *Xenopus* and *Drosophila*, but not in *C. elegans*. LKB1 expressed in 293 cells was shown to be prenylated by labeling with ^{14}C-mevalonic acid, and mutation of the Cys residue in the prenylation motif (Cys433) abolished prenylation (28, 30). Mass spectrometry analysis revealed that the form of prenylation on LKB1 was farnesylation rather than geranylgeranylation (28). Recent genetic analysis in *Drosophila* indicated that both phosphorylation of the Ser431 residue by PKA and the prenylation of Cys433 are essential for LKB1 to regulate cell polarity (34). However, the mechanism by which prenylation regulates LKB1 is unknown. A mutant of LKB1, in which Cys433 is changed to Ala to prevent prenylation, possessed normal in vitro catalytic activity, was able to suppress G361 cell growth (28), and displayed indistinguishable cellular localization to wild-type LKB1 in HeLa cells (27). Cys433 is located two residues away from Ser431, the site of RSK and PKA modification, but mutation of Ser431 to either Ala or Glu did not affect farnesylation of Cys433. Mutation of Cys433 to Ala also had no effect on phosphorylation of LKB1 at Ser431 in response to agonists that activate RSK and PKA (28).

LKB1:STRAD:MO25 COMPLEX

Through a combination of yeast two-hybrid analysis (35) and affinity purification from mammalian cells (29, 36), LKB1 was found to exist in mammalian cells in a complex with two other proteins, termed STE20-related adaptor (STRAD) and mouse protein 25 (MO25). There are two closely related isoforms of STRAD termed STRADα and STRADβ, which possess high sequence homology to the STE20 family of protein kinases but lack several key catalytic residues required for catalysis and have therefore been classified as pseudokinases. STRADα is inactive and does not autophosphorylate or phosphorylate any substrate tested (35). There are also two isoforms of MO25, known as MO25α and MO25β, that, although closely related to each other, bear no sequence homology to any other protein in the database. *MO25α* was first identified as a gene expressed at the early cleavage stage of mouse embryogenesis (37) and was also noticed in a screen for proteins showing an unusually high degree of evolutionary conservation (38, 39). Complexes of LKB1:STRAD:MO25 can be isolated from mammalian 293 cell expression systems (29) or insect Sf9 cells (G. Kular and D.R. Alessi, unpublished observations) in which the three components are present in a similar stoichiometry, suggesting a relatively high affinity of the individual subunits for each other.

LKB1, when overexpressed on its own in mammalian cells, is localized mainly in the nucleus, although a small fraction was reproducibly found in the cytoplasm (40–42). LKB1 possesses a nuclear localization signal at its N-terminal noncatalytic region (residues 38–43) (**Figure 2**), and mutation of this motif results in LKB1 being located throughout the cell (27, 40). A mutant of LKB1 lacking the nuclear localization signal still retains ability to suppress cell growth (41), suggesting that the cytosolic pool of LKB1 plays an important role in mediating its tumor suppressor properties. This is also consistent with the finding that no point mutations in this region of LKB1 have been reported in human cancers (**Figure 1**). Also supporting this conclusion is the finding that mutants of LKB1, which are unable to enter the nucleus, still suppress cell growth in overexpression studies (41). STRADα and MO25α when expressed alone in cells are both localized throughout the cytoplasm and nucleus, but they are only localized in the cytoplasm and excluded from the nucleus when expressed together. Interestingly, when LKB1 is expressed with STRAD and MO25, it becomes strikingly relocalized in the cytoplasm (29, 35, 43). Point mutants of LKB1 found in human cancers, which are unable to interact with STRAD and MO25, remain localized exclusively within the nucleus when expressed in cells (29, 35, 43). Additional mechanisms may also exist to maintain LKB1 in the cell cytoplasm. The interaction of LKB1 with a protein called LKB1 interacting protein-1 (LIP1) was reported to induce LKB1 cytoplasmic localization in 30% of the cells (44). Whether LIP1 can interact with the heterotrimeric LKB1:STRAD:MO25 complex has not been tested. Prior to the discovery that LKB1 phosphorylated and activated AMPK-related subfamily protein kinases (discussed below), LKB1 was assayed either measuring its autophosphorylation or employing the nonspecific substrate, myelin basic protein, which LKB1 phosphorylates poorly at Thr65 (35). The catalytic activity of LKB1 toward myelin basic protein is increased ~10-fold in the presence of STRAD and MO25 isoforms, indicating that these subunits activated LKB1 (35, 45). Moreover, LKB1 was found to phosphorylate STRADα at Thr329 and Thr419. However, the role that these phosphorylations play is not known because mutation of these sites to either Ala or Asp had no effect on the ability of STRADα to bind, activate, or localize LKB1 in the cytoplasm (35). Furthermore, the Thr329 and Thr419 residues are not conserved in STRADβ (29).

The pseudokinase domain of STRAD isoforms binds directly to the kinase domain of LKB1. There are likely to be multiple sites of interaction between LKB1 catalytic and STRAD pseudokinase domain because 12 PJS point mutations, located within different regions of the LKB1 catalytic domain, abolished interaction with STRAD:MO25 (43). The importance of STRAD is further emphasized by the finding that a non-STRADα-binding mutant of LKB1 was unable to induce a G1 cell cycle arrest when overexpressed in G361 cells, which do not express LKB1 (35). Furthermore, siRNA-mediated knockdown of endogenous STRADα inhibited the ability of wild-type LKB1 to arrest cell growth (35).

MO25α forms a high-affinity interaction with the last three Trp-Glu-Phe amino acids of STRAD. Mutation of any of these three residues abolishes the ability of MO25α to bind STRADα (29). In this regard, MO25 appears to function similarly to many PDZ domains that also recognize the extreme C-terminal residues of their protein-binding partners, although there is no obvious sequence similarity between MO25 and PDZ domains (46). The three-dimensional structure of full-length MO25α in complex with a peptide that encompassed the Trp-Glu-Phe residues of STRADα revealed that MO25α forms an extended α-helical repeat rod-like structure, which was distantly related to the armadillo repeat domain. At its C terminus, MO25α possesses a deep pocket that binds specifically to Trp-Glu-Phe residues of STRADα, and mutation of residues making up this pocket inhibited the binding of MO25α to STRADα (47).

Cotransfection/overexpression studies indicated that the amount of STRADα associated with LKB1 in cells is significantly enhanced by the overexpression of MO25α, and siRNA-mediated knockdown of MO25α reduced the amount of endogenous STRADα associated with LKB1 (29). Interestingly, a STRADα mutant lacking the C-terminal Trp-Glu-Phe residues can still form a complex

Cellular energy: the high cellular ATP level needed to survive and undertake cell division and protein synthesis

with MO25α but only in the presence of LKB1 (29, 43). This indicates that interaction of LKB1 with STRADα creates an additional binding site for either MO25 or LKB1 within the complex, which may explain why MO25α stabilizes the binding of LKB1 to STRADα. Mutational analysis has indicated that Arg240, located on the opposite surface to the Trp-Glu-Phe-binding pocket on MO25α, participates in this second interaction site (43). However, it will be necessary to crystallize the entire LKB1:STRAD:MO25 complex in order to understand in full the molecular mechanism by which these proteins interact.

In addition to interacting with STRAD and MO25 isoforms, a significant pool of cellular LKB1 is associated with a chaperone complex consisting of heat shock protein 90 (Hsp90) and the Cdc37 kinase-specific targeting subunit for Hsp90 (48, 49). The LKB1:STRAD:MO25 and LKB1:Cdc37:Hsp90 are likely to form separate complexes because the immunoprecipitation of STRAD or MO25 results in coimmunoprecipitation of LKB1, but not Cdc37 and Hsp90 (J. Boudeau and D.R. Alessi, unpublished observations). It is possible that Hsp90 and Cdc37 may play a role in enabling LKB1 to assemble into a complex with STRAD and MO25 as it is not possible to simply assemble an active LKB1:STRAD:MO25 complex by mixing the three individual components together in vitro (29). Hsp90 stabilizes and prevents degradation of LKB1 because treatment of cells with Hsp90 inhibitors induces degradation of LKB1 by the proteasome complex (48, 49).

ROLE OF PSEUDOKINASES

Although STRAD is an inactive pseudokinase, it still possesses several conserved motifs found in active protein kinases, including the Gly-rich P-loop motif required for ATP binding to kinases. Interestingly, STRADα was found to bind ATP, ADP, and AMP with relatively high affinity (K_d of ~30–130 μM) (43). Mutation of the conserved Gly residues in the P-loop motif on STRADα abolished its ability to bind ATP, but did not affect the ability of STRADα to activate LKB1 or induce its cytoplasmic localization (43). One possibility is that STRADα evolved from an active protein kinase that was capable of phosphorylating and activating LKB1, and despite losing its catalytic activity, STRADα has retained the ability to bind ATP. Attempts at restoring the catalytic activity of STRADα by mutating residues back to those found in active kinases failed to reactivate STRADα (43). The human genome comprises ~50 pseudokinases (10% of the total number of kinases) that lack one or more of the conserved catalytic residues (14). Interestingly, aside from STRADα, a few of the other mammalian pseudokinases that have been studied also interact with catalytically active kinases. The ErbB3 epidermal growth factor (EGF) receptor pseudokinase, forms heterodimers with other catalytically active members of the ErbB tyrosine kinases, and binding of ErbB3 to these kinases is required for their activation (50, 51). The kinase suppressor of ras (KSR) pseudokinase forms a scaffolding regulatory complex with Raf and regulates signal propagation through the ERK/MAPK pathway (52). The janus kinase (JAK) family of tyrosine kinases possess a pseudokinase domain located next to the catalytically active tyrosine kinase domain. The JAK pseudokinase domain binds to the catalytically active domain and may regulate its activity (53, 54). Thus an emerging theme is that pseudokinases function as key regulators of active protein kinases, and it is likely that interesting information will be learnt from studying the physiological roles of these proteins.

ACTIVATION OF THE AMP-ACTIVATED PROTEIN KINASE BY LKB1

Work carried out over the past 20 years has indicated that the AMPK functions as a master regulator of cellular energy metabolism

(55). AMPK is switched on during situations in which the cellular level of ATP is depleted and the level of AMP is increased, for example, those triggered by various cellular stresses such as hypoxia/ischemia or muscle contraction/exercise. Activation of AMPK restores cellular energy levels by stimulating catabolic pathways, such as glucose uptake and/or glycolysis and fatty acid oxidation, as well as by inhibiting energy-consuming anabolic processes, such as protein synthesis. AMPK is also activated by metformin, the most widely employed antidiabetic drug, which is thought to exert its blood glucose-lowering effects through its ability to activate AMPK in liver and skeletal muscle (56). The mechanism by which metformin or its closely related analog, phenformin, activates AMPK has not yet been fully established, but it may involve inhibition of ATP production via effects on complex I of the mitochondrial respiratory chain (57). AMPK is a heterotrimeric complex comprising a catalytic AMPKα subunit and regulatory AMPKβ and AMPKγ subunits. AMP activates the AMPK complex by binding to a pair of cystathionine-beta-synthase (CBS) domains, located on the AMPKγ subunit (58), thereby stimulating phosphorylation of Thr172 in the T loop of both mammalian AMPKα catalytic subunits, termed AMPKα1 and AMPKα2 (reviewed in 59 and 60). There has been much interest in the identities of the upstream protein kinase(s) that phosphorylate Thr172, but prior to 2004, despite considerable research, there were no obvious candidates for this enzyme(s).

Although there were clear relatives of LKB1 in *Drosophila* (34) and *C. elegans* (61), the genetic studies undertaken on these enzymes did not yield much information on what the molecular targets of LKB1 may be. The breakthrough was unexpectedly made in budding yeast *Saccharomyces cerevisiae*, wherein three kinases termed Elm1, Pak1/Sak1, and Tos3 displayed moderate similarity to LKB1, although they were more closely related to mammalian calmodulin-dependent protein kinase kinase (CAMKK). There was much excitement in 2004, when elegant genetic studies, reported simultaneously by several groups, suggested that Elm1, Pak1/Sak1, and Tos3 were all capable of phosphorylating the T loop of the Snf1, the homolog of mammalian AMPK (62–64). These fascinating findings prompted studies in mammals to investigate whether LKB1 might be capable of activating AMPK. Indeed, it was found that LKB1 efficiently phosphorylated AMPK in vitro specifically at Thr172, the residue that becomes phosphorylated when cellular ATP levels fall (65–67). The ability of LKB1 to activate AMPK was enhanced over 100-fold if it was present with isoforms of STRAD and MO25 in a complex, demonstrating that these subunits are indeed required for the activation of LKB1 (65). Most importantly, it was found that in cell lines, such as HeLa cells that lack LKB1 expression or $LKB1^{-/-}$ mouse embryonic fibroblasts, AMPK could not be activated by a variety of agonists and stresses (65–67). These included phenformin or the drug 5-aminoimidazole-4-carboxamide riboside (AICAR) that activates AMPK in intact cells by being taken up and converted to AICAR-monophosphate, which mimics the effect of AMP on the AMPK system (68). More recently, mice lacking LKB1 expression in skeletal muscle were found to possess vastly reduced AMPKα2 isoform activity and phosphorylation of Thr172 (24). Moreover, AMPKα2 was not stimulated by AICAR, by phenformin, or by muscle contraction in the LKB1-lacking muscle (24). Two of the most studied processes regulated by AMPK in muscle are its ability to induce glucose uptake (69) and to stimulate fatty acid oxidation through phosphorylating acetyl coenzyme A (CoA) carboxylase-2 (70). In LKB1-deficient muscle, phosphorylation of acetyl CoA carboxylase-2 and glucose uptake, triggered by AICAR or muscle contraction, were profoundly inhibited, thereby supporting a role for the LKB1-AMPK pathway in regulating these processes in muscle (24). It was also observed that muscle contraction induced

AICAR: 5-aminioimidazole-4-carboxamide riboside

a markedly greater rise in the AMP:ATP ratio in LKB1-deficient muscle than in wild-type skeletal muscle (24). This provided the first genetic evidence that the LKB1-AMPK pathway is physiologically linked to regulation and maintenance of cellular energy levels. More recently, it has been demonstrated that mice lacking expression of LKB1 in the liver possess greatly reduced AMPK activity, supporting the notion that LKB1 is a key regulator of AMPK in mammalian tissues (70a). Importantly, this study also demonstrated that the antidiabetic drug metformin no longer reduced blood glucose levels in mice that lacked hepatic expression of LKB1, providing the first direct evidence that LKB1-mediated activation of AMPK in the liver is indeed required for the ability of metformin to lower blood glucose levels (70a).

Although LKB1 is not closely related to any other protein kinase, sequence alignments of kinase catalytic domains indicated that LKB1 does display some sequence similarity to the catalytic domain of AMPK. Significantly, several other upstream protein kinases are also homologous to the downstream kinases that they activate. These include cyclin-dependent kinase 7 that activates other CDK isoforms (71), the 3-phosphoinositide-dependent protein kinase 1 (PDK1) that activates AGC family kinases (72), and the WNK (with no lysine kinase) isoforms that are related to the SPAK (STE20/SPS1-related proline-alanine-rich kinase) and OSR1 (oxidative-stress response kinase 1), which they activate (73). These observations may indicate that perhaps the downstream kinases evolved from their upstream regulator or vice versa.

Although much evidence supports the notion that LKB1 is a key regulator of AMPK, it is likely that there are alternative regulators. For example, in LKB1-deficient cells, although AMPK could not be activated by AICAR or phenformin, it possessed significant basal activity and phosphorylation at Thr172 (65, 67). In mouse skeletal muscle lacking LKB1, although the activity of the AMPKα2 isoform of AMPK was virtually abolished, the other isoform AMPKα1 was still significantly active and could be further stimulated with AICAR (24) and phenformin (K. Sakamoto, unpublished results). In cardiac muscle lacking LKB1, AMPKα1 is barely reduced under conditions which AMPKα2 activity is totally ablated (73a). Recent studies have provided compelling evidence that CAMKK isoforms, which, as mentioned above, are more closely related to the yeast Elm1, Pak1/Sak1, and Tos3 activators of Snf1 than LKB1, can function as alternative regulators of AMPK. CAMKK isoforms phosphorylate AMPK in vitro at Thr172 and STO-609, a CAMKK inhibitor [which is not completely specific, but does not inhibit LKB1 (74)], and siRNA-mediated knockdown of CAMKK isoforms inhibit basal AMPK activity in LKB1-deficient cell lines as well as the activation of AMPK, observed in response to agents that elevate Ca^{2+} levels (74–76). CAMKK isoforms are highly expressed in neuronal tissue. K^+-induced depolarization of rat cerebrocortical slices, which increases Ca^{2+} without affecting ATP levels, was found to activate AMPK in a manner that was inhibited by STO-609, suggesting that CAMKK rather than LKB1 regulates AMPK in Ca^{2+}-regulated pathways at least in neuronal tissues (74). It is not yet known whether CAMKK isoforms are involved in phosphorylating the AMPKα1 isoform in the skeletal and cardiac muscle of LKB1-deficient mice. LKB1 can phosphorylate both AMPKα1 and AMPKα2 with similar efficiency in vitro (65). It is possible that the differential cellular localization of AMPK isoforms and upstream activators may play a role in regulating the activation of AMPK isoforms by diverse upstream stimuli.

Thus far, all studies indicate that the LKB1 complex in vitro is not stimulated by AMP, and when the LKB1 complex is assayed after its immunoprecipitation from cell/tissue extracts, it is constitutively active because its activity does not change in cell lines (66, 67, 77) or skeletal and cardiac muscle (73a, 78) under conditions that stimulate AMPK. In vitro studies suggest that binding of AMP to a pair of CBS domains located within the AMPKγ regulatory subunit (58) induces a conformational change that converts AMPK into a better substrate for the LKB1:STRAD:MO25 complex (65). This is reminiscent of the mechanism of activation of protein kinase B (PKB, also designated c-Akt) by the upstream kinase PDK1, in insulin-stimulated cells (72). Insulin does not directly activate PDK1, which like LKB1 is constitutively active, but instead leads to the generation of the lipid second-messenger phosphatidylinositol (3,4,5) trisphosphate (PIP3), which interacts with PKB/Akt and converts it into a better substrate for PDK1 (72).

TUMOR SUPPRESSOR PROPERTIES OF LKB1 AND AMPK

Until the surprising discovery that AMPK activation involved a tumor suppressor, AMPK was mainly considered an enzyme playing a role in controlling metabolic responses that were mainly of relevance to diabetes and exercise physiologists. There was no prior interest in AMPK in the cancer field. On reflection, however, the unique ability of AMPK to directly sense cellular energy places it in an ideal position to ensure that cell division, which is a highly energy-consuming process, only proceeds if cells have sufficient metabolic resources to support a doubling of cell mass.

Recent studies (reviewed in 79) indicate that tuberous sclerosis complex 1 (TSC1) and TSC2 tumor suppressor proteins form a complex and play a key role in controlling cell growth by serving as negative regulators of the mammalian target of rapamycin (mTOR)-signaling pathway and hence of protein synthesis, as summarized in **Figure 3**. The TSC1-TSC2 complex acts as a GTP-activating protein (GAP) on the mTOR activator, Rheb, a small GTPase. Two of the most well characterized downstream effectors of the mTOR pathway are the p70 S6 kinase (S6K) and the eukaryotic initiation factor 4E-binding protein 1 (4E-BP1) translational regulator, but there are likely to be many others (80). In the absence of growth factors the TSC1:TSC2 complex inhibits the mTOR-signaling pathway, and therefore biosynthesis is switched off. Mutations in either TSC1 or TSC2 or other upstream regulators of these proteins, such as the PTEN tumor suppressor, increase the level of the mTOR activator, Rheb-GTP, and lead to the inappropriate stimulation of protein translation and, hence, cell growth (**Figure 3**). In a seminal study, Guan's laboratory (81) reported that AMPK could inhibit cell proliferation under energy starvation conditions by directly phosphorylating TSC2 and enhancing its ability to switch off mTOR-signaling networks through an as yet undefined mechanism. This study also provided evidence that the ability of AMPK to switch off mTOR signaling by phosphorylation of TSC2, protected cells from apoptosis induced by energy deprivation (81). In addition, this study provided the first indication that the tumor suppressor effects mediated by LKB1 might arise from its ability to regulate the mTOR pathway. This notion was further supported by the finding that in LKB1 knockout cells low cellular ATP conditions failed to suppress the mTOR-signaling pathway (82, 83). Furthermore, hamartomatous gastrointestinal polyps derived from LKB1 mutant mice displayed increased S6K activity, leading Shaw and colleagues (83) to conclude that aberrant mTOR activation is likely to be at the nexus of PJS. Overexpression of LKB1 in A549 cells has also been reported to increase the expression of the PTEN tumor suppressor (84), which metabolizes the PIP3 second messenger that activates the mTOR pathway through its ability to activate PKB (**Figure 3**). It is important to investigate whether conversely, PTEN levels are reduced in LKB1-deficient cells and whether this leads to increased PKB and mTOR activity. Interestingly, PTEN is

mTOR: mammalian target of rapamycin

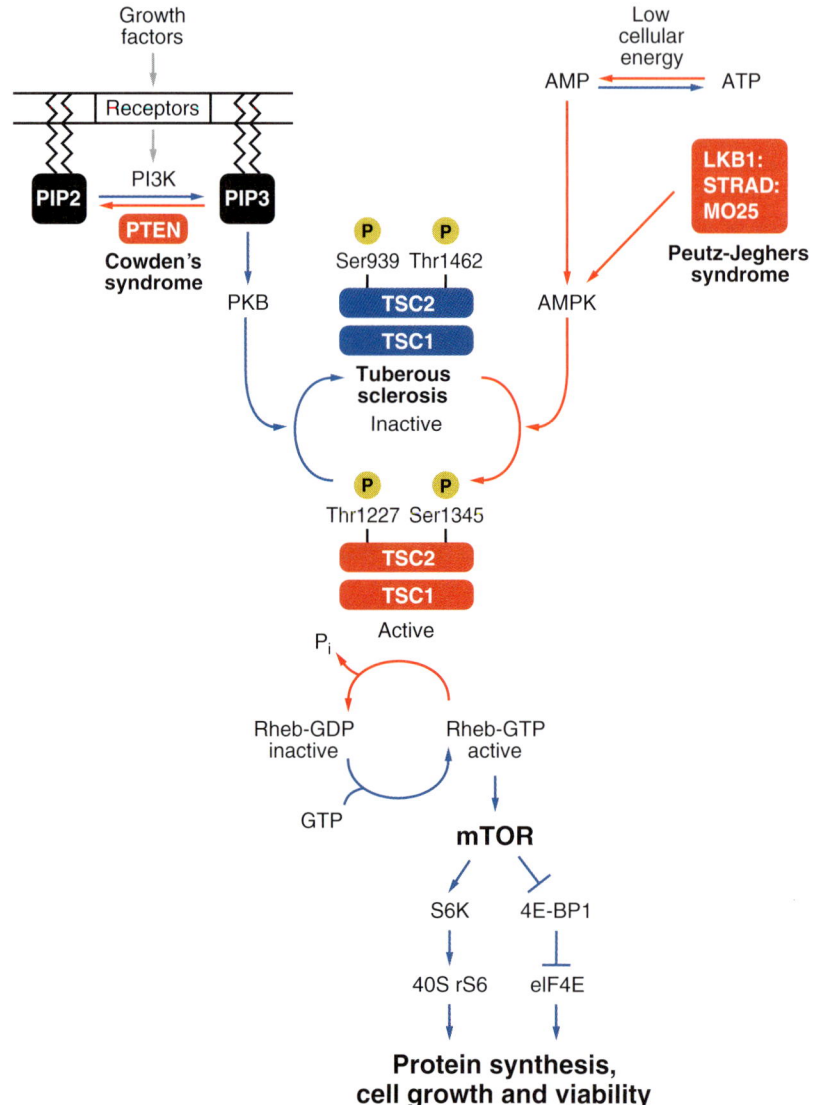

Figure 3

Model describing the regulation of mTOR pathway by the AMPK and the PKB-signaling pathways. Phosphorylation of TSC2 by PKB after growth factor stimulation induces activation of the Rheb GTPase and, hence, the activation of the mTOR pathway promoting protein synthesis, cell growth, and viability. Phosphorylation of TSC2 by AMPK after ATP depletion results in Rheb inactivation and mTOR inhibition. Tumor suppressors that are lost in the indicated human cancer syndromes are indicated (*red*); loss of function of TSC1 or TSC2 in tuberous sclerosis or PTEN in Cowden's syndrome will result in mTOR activation. Loss of LKB1 in PJS prevents the mTOR pathway from becoming inhibited in low-energy conditions. Abbreviations not used in the main text: PIP2, phosphatidylinositol (4,5)-bisphosphate; PI3K, phosphatidylinositol 3-kinase; 40S rS6, 40 S ribosomal S6 protein; and eIF4E, elongation factor 4E.

also mutated in an autosomal dominantly inherited cancer disease, termed Cowden's syndrome, which is phenotypically similar to PJS (85). Moreover, mutations in either the TSC1 or TSC2 genes results in tuberous sclerosis, an inherited syndrome that has some similarities to both PJS and Cowden's syndrome (79). In essence, loss of PTEN, LKB1, TSC1, or TSC2 causes inappropriate activation of the mTOR pathway (**Figure 3**), thereby explaining the link between these tumor suppressors.

It is also possible that activation of AMPK can inhibit cell growth through additional pathways. According to recent reports, glucose deprivation induced AMPK to phosphorylate p53 at Ser15, thereby inducing a cell cycle arrest (86). This study also suggested that persistent activation of AMPK induced cells to undergo a p53-dependent cellular senescence. Overexpression of LKB1 in cell lines that are deficient in LKB1 (G361 or A549) induced expression of several p53-responsive genes, including the p21$^{WAF1/CIP1}$ inhibitor of the cyclin-dependent kinases, further implicating that the p53 pathway is controlled by LKB1. Thus, in addition to its effects on mTOR, there may be additional mechanisms by which LKB1 influences cell division (41, 84).

MEDICAL IMPLICATIONS FOR THE TREATMENT OF PJS AND OTHER CANCERS

The finding that in LKB1-deficient tumors the mTOR-signaling pathway is elevated and not switched off under low-energy conditions suggests that mTOR inhibitors such as rapamycin would be effective at inhibiting growth of those cancer cells (87). Because rapamycin is already used clinically to prevent transplant rejection, it should be possible to treat heterozygous *LKB1*$^{+/-}$ mice and even PJS patients with rapamycin or one of its derivatives to investigate its potential effect in ameliorating the frequency and severity of tumor formation.

Most sporadic cancers are likely to express LKB1 and, hence, should be capable of switching off the mTOR pathway in response to metabolic stresses that activate AMPK. Thus, if it is possible to induce the activation of AMPK with a drug, this might inhibit proliferation of LKB1-expressing cancer cells. Verhoeven and colleagues (88) have recently investigated this idea and reported that treatment of various cancer cell lines with AICAR prevented proliferation and also inhibited DNA synthesis, lipogenesis, and protein translation. It has also been reported that AICAR inhibited proliferation of HepG2 cells (89). These observations further suggest that the AMPK pathway functions as a cellular energy-sensing checkpoint, enabling growth and proliferation of cells to be coupled to the availability of fuel supplies, which could be exploited for cancer treatment and/or prevention.

A mechanism by which AMPK could be activated in human tumors would be to employ the antitype 2 diabetes drug metformin. A recent pilot small-scale epidemiology, population-based case-control study to address whether diabetics prescribed metformin were protected from developing cancer has been undertaken. This suggested that diabetics who take metformin were 23% less likely to get cancer with the reduction in risk rising to 40% for people who have been taking the drug for a longer period of time (90). This was a preliminary study, and the hypothesis needs to be tested in a more rigorous cohort study with a larger number of subjects. Metformin has been utilized for diabetes treatment for nearly 50 years and has very moderate side effects. Thus metformin could be used therapeutically to prevent cancer quite quickly, without the need for lengthy clinical trails that would be necessary to evaluate a new drug. Interestingly, before it was known that metformin activated AMPK, one study reported that metformin markedly inhibited carcinogen-induced pancreatic cancer in hamster (91). It would be interesting to determine whether

Type 2 diabetes: a condition in which resistance to the effects of insulin results in blood with elevated levels of glucose

MARK: microtubule affinity-regulating kinase

SIK: salt-inducible kinase

the striking effect of metformin in cancer that was reported in this study was mediated through the ability of metformin to stimulate AMPK.

Another physiological mechanism by which AMPK is activated in humans is through exercise. Several studies have reported that exercise is beneficial in the prevention of certain cancers. Large-scale clinical trials have suggested that the incidence of recurrence of colon and breast cancer is reduced in patients who undertake long-term exercise regimes (reviewed in 92 and 93). If exercise does have the beneficial outcome of reducing cancer frequency, this effect is likely to be complex and involve many factors, such as reducing whole-body fat content and improving the immune system. However, for the reasons outlined above, it would also be expected that depletion of whole-body ATP levels in tissues induced by exercise would also raise AMPK activity, which would have the potential to inhibit cell growth.

Detailed molecular analysis of the polyps from $LKB1^{+/-}$ mice revealed that the levels of cyclooxygenase-2 (COX-2) protein were significantly increased (22). Induction of COX-2 has been implicated in promotion of tumor formation and progression (94). Makela and colleagues (95) have reported that treatment of $LKB1^{+/-}$ mice with celecoxib, a COX-2 inhibitor, reduced polyp burden by over 50%, an effect that was even more marked if the celecoxib was administered before tumor onset. In a pilot clinical study, celecoxib was also found to reduce polyp burden in two of the six PJS patients who were treated (95). It will be important to undertake similar studies with mTOR inhibitors and metformin in the future.

LKB1 FUNCTIONS AS A MASTER UPSTREAM KINASE

Inspection of the human kinome dendrogram (14) indicated that there is a group of protein kinases highly related to AMPK, termed the AMPK-related kinases (**Figure 4**). The amino acid sequences of the activation loops in these enzymes are homologous to AMPK, which suggested to us that the LKB1:STRAD:MO25 complex might also phosphorylate and activate these enzymes (**Figure 4**). Previous studies indicated that four members of the AMPK-related kinase subfamily (MARK1, MARK2, MARK3, and MARK4—microtubule affinity-regulating kinases), also known as Par-1 or c-TAK1, play roles in regulating cell polarity (96, 97). However, little or no previous research had been performed on the remaining members of the AMPK-related protein kinases (BRSK1/SAD-A, BRSK2/SAD-B, NUAK1/ARK5, NUAK2/SNARK, QIK/SIK2, QSK, SIK, MELK, SNRK, NIM1, TSSK1, TSSK2, TSSK3, TSSK4, SSTK, and HUNK). Evidence to date suggests that 12 of theses AMPK-related kinases (BRSK1, BRSK2, NUAK1, NUAK2, QIK, QSK, SIK, MARK1, MARK2, MARK3, MARK4, and SNRK) are indeed activated over 50-fold by the LKB1-catalyzed phosphorylation of their T loops and that these enzymes are substantially less active when expressed in LKB1-deficient cells (77, 98). As observed for AMPK, the STRAD and MO25 components are essential for LKB1 to phosphorylate every AMPK-related enzyme. These findings suggest that LKB1 functions as a master kinase, similarly to PDK1 (72), to activate many members of the AMPK-related protein kinase subfamily. In addition, these data suggest that the AMPK-related kinases may mediate some of the physiological effects previously ascribed to LKB1 and that one or more of the AMPK-related kinases may themselves function as tumor suppressors.

A useful observation was that the LKB1:STRAD:MO25 complex efficiently phosphorylated a peptide derived from the T loop of NUAK2, termed LKBtide (77), which has been developed into a facile quantitative assay for LKB1 activity (77, 78). Many protein kinases rely on sequences, termed docking sites, lying outside of the catalytic core, which stabilize the interaction between

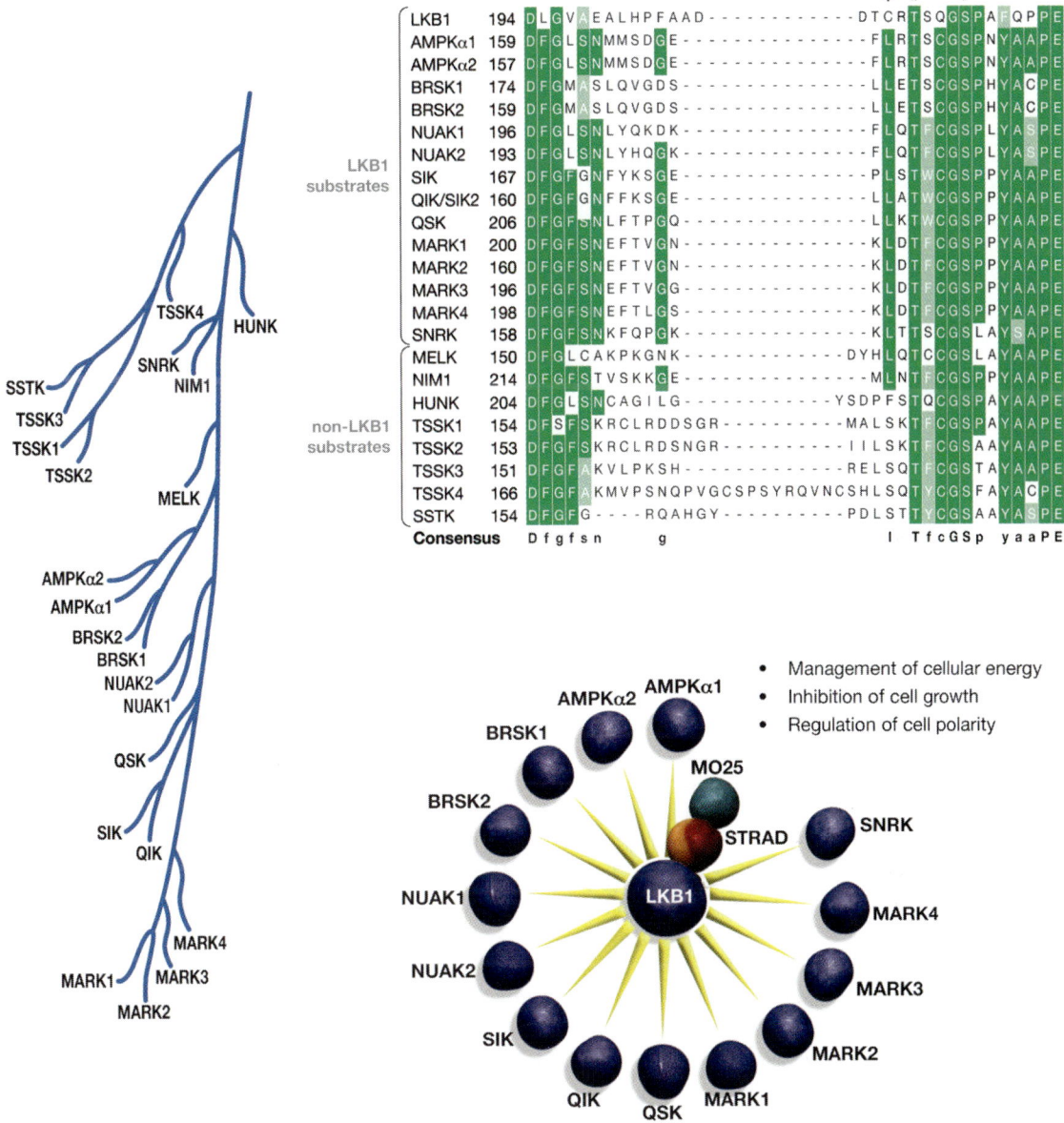

Figure 4

Activation of the AMPK-related kinases by LKB1. (*left panel*) Dendrogram of the AMPK-related protein kinase subfamily (14). (*top right panel*) Multiple sequence alignment of the T-loop sequences of the AMPK-related kinases, including LKB1. The dark green shadow indicates identical residues, and the light green shadow corresponds to residues conserved in more than 50% of the aligned sequences. The T-loop threonine and serine residues are indicated with an asterisk, and the −2 leucine that is thought to be required for LKB1 phosphorylation (67) is marked with an arrowhead. (*bottom right panel*) Diagram illustrating the activation of the two isoforms of AMPK and the 12 AMPK-related kinases by the LKB1:STRAD:MO25 complex.

the kinase and its substrate(s). It is possible that the STRAD and MO25 subunits play a similar role in docking the LKB1 complex to its substrate. However, this role for STRAD and MO25 seems unlikely because these subunits stimulate LKB1 to phosphorylate even the LKBtide peptide, which is probably too short to interact with docking sites. Thus, it would seem that STRAD:MO25 directly activate LKB1.

The MELK, NIM1, and TSSK isoforms of the AMPK-related kinases have the intrinsic ability to autoactivate themselves because they are active and phosphorylated on their T loop when expressed in *E. coli* and therefore may not be controlled by LKB1 (77, 98). Shaw and colleagues (67) have determined the optimal substrate phosphorylation motif for LKB1, employing a degenerate peptide library screen, and found that LKB1 possesses a strong preference to phosphorylate Thr residues, which possess a Leu residue at the −2 position. Interestingly, the 14 AMPK subfamily kinases that are phosphorylated and activated by LKB1 all possesses a Leu residue at the −2 position from the T-loop Thr (**Figure 4**). In contrast, TSSK1, TSSK2, TSSK3, TSSK4, SSTK, and HUNK do not possess such a −2 Leu residue, suggesting one reason why these proteins are not LKB1 substrates. LKB1 itself has a T-loop residue resembling that of AMPK, but it lacks the −2 Leu (**Figure 4**). LKB1 does not autophosphorylate itself at its T-loop residue, nor is the T-loop Thr of LKB1 phosphorylated in cells or required for the activation of LKB1 by binding to STRAD and MO25 (43). LKB1 may have evolved the unusual mechanism of binding to STRAD and MO25 for its activation in order to bypass the requirement that most kinases display for T-loop phosphorylation. The ability of LKB1 to be regulated in a manner different from the downstream kinases that it controls may be the critical property that enables LKB1 to operate as the upstream kinase for the AMPK-related kinase subfamily.

MARK AND BRSK KINASES: ROLE OF LKB1 IN REGULATING CELL POLARITY

The counterpart of mammalian LKB1 in *C. elegans* is termed Par-4 (61), and it was originally identified as a member of the maternally expressed *Par* (partitioning defective) gene family, required for establishing cell polarity during the first cycle of *C. elegans* embryogenesis (99). Maternal lethal mutations in the gene encoding *C. elegans* Par-4 have been shown to affect several aspects of cell polarity (100). Mutation of *Drosophila* LKB1 results in defects in anterior-posterior axis formation and epithelial polarity (34). In a mammalian system, the most spectacular evidence that LKB1 also regulates cell polarization came from the observation that activation of LKB1 in single unpolarized epithelial cells resulted in a dramatic repolarization of these cells in which the actin cytoskeleton was remodelled to form an apical brush border and gap junctions (101). These findings indicate that a major role of mammalian LKB1 is to control cell polarity. Thus, gastrointestinal hamartoma formation in PJS patients could result from inappropriate overgrowths of differentiated cells, which have lost their ability to regulate their polarity as a consequence of LKB1 inactivation.

The mechanisms by which LKB1 mediated its effects on cell polarity were at first unclear. Early genetic studies indicated another Par gene, a kinase termed Par-1, controlled the partitioning of the *C. elegans* zygote (102). More recently, Par-1 homologues in *Drosophila* (34, 103) and *Xenopus* (104) were also shown to play important roles in regulating cell polarity. Taken together these studies strongly indicate that the Par-1 enzymes are master regulators of cell polarity (96). Although the genetic studies suggested a connection between Par-1 and Par-4, the relationship between these two enzymes was not clear, and one study even suggested that Par-1 might lie upstream of Par-4 (34). In humans, there are four closely related

Par-1 like isoforms MARK1, MARK2, MARK3, and MARK4 (96). The demonstration that MARK isoforms are phosphorylated and activated by LKB1 (77, 105), as discussed above, suggests that LKB1/Par-4 regulates cell polarity by activating MARK/Par-1 isoforms.

Numerous mutations that affect only the C-terminal region of LKB1 have been identified in PJS as well as in other tumors (**Figure 1**). It was recently reported that several such C-terminal mutants of LKB1 had reduced ability to polarize epithelial cells in overexpression studies, suggesting that the C-terminal noncatalytic domain of LKB1 may have a role in controlling polarity (106). It is not yet known whether these mutants have a specific defect in their ability to activate MARK isoforms. Two mutations of MARK3 have recently been reported in sporadic colorectal cancer. One is a frameshift mutation that is expected to ablate functional MARK3 expression, whereas the other changes Arg173 in the catalytic domain to a Gln residue (107). The effect that this point mutation has on MARK3 catalytic activity or ability to induce polarization of cells has not yet been investigated.

Recently another kinase, TAO1, purified from pig brain, was identified by its ability to phosphorylate the T-loop Thr residue and thereby activate MARK2 (108). Although TAO1 overexpression also activated MARK2 in vivo (108), no genetic evidence indicating how knockout or knockdown of TAO1 affects the activity of MARK family kinases in vivo has been reported. Such evidence may be difficult to acquire because mammalian cells possess three closely related TAO isoforms (14). However, the fact that MARK3 activity is only moderately reduced in $LKB1^{-/-}$ cells (77), as well as in LKB1-deficient skeletal and cardiac muscle (K. Sakamoto, unpublished results), supports the notion that there are additional regulators of MARK isoforms. It would be interesting to test whether TAO1, or even CAMKK, could regulate MARK3 and other AMPK-related kinases in LKB1-deficient cells or tissues. It should be noted that, unlike LKB1 and CAMKK, TAO1 and LKB1 share no obvious amino acid sequence homology and lie in distinct regions of the human kinase dendrogram (14). TAO1 belongs to the STE20 group of kinases and is therefore related to STRADα and STRADβ, but whether this has any significance is not clear.

In mammals, MARK isoforms also phosphorylate Tau, thereby priming it for hyperphosphorylation by other kinases including glycogen synthase kinase 3 (GSK-3) and Cdk5. These modifications induce the aggregation of Tau into the toxic "neurofibrillary filaments and tangles," which are the deposits found in the brains of patients with Alzheimer's disease (109). Although several kinases phosphorylate Tau at multiple sites to induce the neurofibrillary filaments and tangles, elegant genetic studies performed in *Drosophila* indicate that MARK-mediated phosphorylation of Tau plays a major initiator role in enabling other kinases to phosphorylate Tau (110). These results indicate that drugs that inhibit MARK isoforms or even LKB1 might be useful as therapy for Alzheimer's disease.

The brain-specific kinase 1 (BRSK1 also known as SAD-A) and BRSK2 (SAD-B) that are activated by LKB1 are mainly expressed in the brain and in low levels in the testis (111). Interestingly, the *C. elegans* BRSK homologue was originally found as a kinase that controlled several aspects of presynaptic differentiation, such as presynaptic vesicle clustering and axon termination (112). More recently, mice lacking both the SAD-A and SAD-B kinases were reported to die at birth because of failure of their neurons to polarize and form axons and dendrites (111). These results suggest that the BRSK kinases may play a specific role in regulating polarization of neuronal cells. A key challenge of future research will be to identify the substrates that MARK and BRSK kinases phosphorylate to regulate cell polarity. UV irradiation has also been reported to stimulate BRSK1 activity and cause its translocation from the cytoplasm to the nucleus by an

> **Alzheimer's disease:** a disorder characterized by the progressive and irreversible loss of nerve cells in specific brain areas

unknown mechanisms, suggesting that BRSK isoforms may have a DNA damage checkpoint function (113).

OTHER AMPK-RELATED KINASES

Salt-inducible kinase (SIK) was first cloned from the adrenal glands of rats fed on a high-salt diet (114). SIK mRNA was also induced by membrane depolarization in brain (115), and recent overexpression studies have indicated that SIK may play a role in steroidogenesis (116). The mRNA-expressing QIK (also termed SIK2) was highest in adipose tissue, and in overexpression studies, QIK was reported to phosphorylate the human adaptor protein insulin receptor substrate-1 (IRS1) (117) at the same residue phosphorylated by AMPK (118). Recently, the CREB coactivator protein TORC2 was also found to be phosphorylated at the same sites by QIK and/or AMPK. This modification caused interaction with 14-3-3 proteins and resulted in TORC2 (not related to the mTOR containing complex of the same name) sequestration in the cytoplasm, preventing CREB-dependent gene transcription (70a, 119). Consistent with this, in mice that lack LKB1 expression in the liver, TORC2 was localized within the nucleus and not phosphorylated at the QIK/AMPK sites (70a). QIK activity was also reported to be inhibited by agonists that increased cyclic-AMP levels (119). In other overexpression studies, NUAK1 (ARK5) was shown to suppress apoptosis induced by some stimuli, including nutrient starvation (120). Furthermore, it has been claimed that Akt/PKB phosphorylated NUAK1 at a C-terminal site outside of the catalytic domain, leading to a threefold activation of the enzyme (121). NUAK2 (SNARK) activity was reportedly stimulated by glucose starvation of cells (121, 122). Expression of NUAK2 was also found to be upregulated by the tumor necrosis factor receptor in apoptosis-resistant tumor cell lines (123). Cells with reduced expression of NUAK2 displayed lower motility and invasiveness in response to CD95 engagement, suggesting that NUAK2 plays a role in protecting cancer cells from apoptosis and that inhibitors of NUAK2 might be useful for the treatment of certain apoptosis-resistant tumors (123). RNAi-mediated knockdown of QSK in *Drosophila* cells resulted in mitotic defects that included spindle and chromosome alignment abnormalities, indicating that this enzyme might play a crucial role in regulating cell proliferation (124). SNRK is only expressed in the testis, and its function is unknown, but it may play a role in regulating spermatogenesis and/or sperm motility (98). Testis, in addition to expressing the usual 50-kDa species of LKB1, also expresses a faster migrating 48-kDa species that is likely to represent a C-terminal splice variant (24). Interestingly, male LKB1 hypomorphic mice, which express 10-fold reduced levels of LKB1 and also lack the 48-kDa splice variant of LKB1, are sterile (24). It would be interesting to determine whether SNRK or another AMPK subfamily enzyme was specifically controlled by the 48-kDa splice variant of LKB1 and also to characterize in more detail spermatogenesis and/or sperm motility in LKB1 hypomorphic mice.

REGULATION OF AMPK-RELATED KINASES

The AMPK-related kinases are not activated in response to AICAR, phenformin, or other stresses that stimulate AMPK itself (77, 78). Recently, the tandem affinity purification strategy has been deployed to identify proteins that interact with AMPKα as well as with 12 other AMPK-related kinases activated by LKB1 (125). The AMPKβ and AMPKγ regulatory subunits, required for activation of AMPK by LKB1 in low cellular energy conditions when AMP levels are elevated, were associated only with AMPKα and not with any of the other AMPK-related kinases examined (125). This finding may explain why AMPK-related kinases are not stimulated by AICAR, phenformin, or other stresses, as they lack the subunits to sense AMP concentrations. An important implication of the finding that AMPK-related kinases are not activated by low cellular energy conditions is that the beneficial antidiabetic as well as the antiproliferative effects of these agonists, discussed above, are probably mediated through the activation of AMPK per se, rather than through one of the other AMPK-related kinases. In future studies, it will be important to define the physiological stimuli that regulate each of the AMPK-related kinases and to identify the physiologically relevant targets that these enzymes phosphorylate. The finding that AMPK-related kinases phosphorylate the same set of peptide substrates at different relative rates (77) suggests that these enzymes have distinct substrate preferences and will selectively phosphorylate different substrates in vivo. It has also been recently demonstrated that certain 14-3-3 isoforms can bind directly to the T-loop Thr residue of QSK and SIK after they are phosphorylated by LKB1 (125). Moreover, binding of 14-3-3 to QSK and SIK enhanced their catalytic activity two- to threefold and was also required for the cytoplasmic localization of SIK and for localization of QSK to undefined punctate structures within the cytoplasm (125).

CONCLUSIONS AND PERSPECTIVE

Eighty-three years have now elapsed since Dr. Peutz first described PJS (1). In the last seven years since LKB1 was identified as the causative gene of PJS (11), remarkable progress has been made in the biochemical and genetic analysis of LKB1 in human, mice, and model organisms. These studies indicate that LKB1 functions as a tumor suppressor by not only regulating cell division, but also by controlling polarity and enabling cell growth and division to be coupled to cellular energy levels. The findings outlined in this review also provide unexpected insights for novel therapies for the prevention and treatment of PJS and possibly other cancers. For example, drugs, such as metformin, or exercise that activate AMPK might be used to induce a cell cycle arrest, fooling cancer cells (which express LKB1) into thinking that they lack sufficient energy resources to divide. The finding that cancer cells lacking LKB1 are unable to switch off the mTOR pathway under conditions of low cellular energy suggests that inhibitors of mTOR might be effective at treating PJS patients and other LKB1-deficient tumors, such as lung adenocarcinomas. LKB1 does not represent an obvious drug target itself because drugs that inhibited LKB1 might be expected to induce tumors. However, it could be envisaged that inhibiting LKB1 or AMPK might make cancer cells more susceptible to apoptosis under low-energy conditions, for example hypoxic situations found in tumors or following chemotherapy. The findings that LKB1 hypomorphic mice express markedly lower levels of LKB1 (24) and do not appear to develop detectable tumors by age six months (K. Sakamoto, unpublished findings) may also indicate that a drug that inhibited LKB1 would not cause overt cancer. It would be interesting to test whether the LKB1 hypomorphic animals are protected from developing tumors. Other challenges for future research will be to characterize in greater detail the molecular mechanism by which AMPK-related kinases are controlled and to identify cellular substrates for these enzymes. These studies should provide deeper insight into how the tumor suppressor effects of LKB1 are mediated. Further investigation is also required to understand the roles that phosphorylation, prenylation, and the C-terminal domain of LKB1 play. Apparently, a significant number of inherited forms of PJS found in certain families do not exhibit mutations in the LKB1 gene (126, 127). If this is indeed the case, there could be other causative loci for PJS. Identifying these genes is of crucial importance because these are likely to lie in the same signaling network as LKB1.

SUMMARY POINTS

1. LKB1 is a protein kinase that functions as a tumor supressor, and its mutation in humans causes the rare inherited Peutz-Jeghers cancer syndrome.
2. LKB1 is activated in an unusual manner that involves forming a complex with an inactive pseudokinase termed STRAD and an armadillo repeat scaffolding protein MO25.
3. LKB1 phosphorylates the T loop of at least 14 related protein kinases that belong to the AMPK subfamily, which includes isoforms of AMPK—an important regulator of cellular energy levels—as well as the MARK and BRSK enzymes that control cell polarity.
4. The ability of AMPK to be activated in response to low cellular energy levels depends upon LKB1 in cells and tissues.
5. LKB1 is constitutively switched on in cells, and it is believed that under conditions of low cellular energy and stress, the binding of AMP to a regulatory subunit of AMPK induces a conformational change that permits the phosphorylation and activation of AMPK by LKB1.
6. The LKB1-AMPK pathway functions as a cellular energy-sensing checkpoint, enabling growth and proliferation of cells to be coupled to the availability of fuel supplies, which could be exploited for cancer treatment and/or prevention.
7. Activation of the AMPK by the antidiabetes drug metformin or exercise in humans may reduce the frequency of cancer.

FUTURE ISSUES TO BE RESOLVED

1. The substrates that the AMPK-related kinases phosphorylate and the role that these enzymes play in mediating the tumor suppressor effects of LKB1 are needed to provide insights into the fundamental mechanism by which LKB1 regulates proliferation and polarity as well as other processes that it controls. This information is relevant to the general understanding of cancer and may provide insights for the development of new approaches to anticancer treatments.
2. A three-dimensional structural analysis of the LKB1:STRAD:MO25 complex is needed to provide fundamental information on how LKB1 is activated by binding to a pseudokinase as well as functional insights into how this complex may operate in vivo.
3. The roles that phosphorylation, prenylation, and the C-terminal domain of LKB1 play in regulating its cellular functions should be determined.
4. A putative "second" gene has been postulated to be mutated in a significant number of Peutz-Jeghers syndrome patients, who do not possess obvious mutations in LKB1, and this second gene should be found.

ACKNOWLEDGMENTS

We thank the Association for International Cancer Research, Diabetes UK, the Medical Research Council, the Moffat Charitable Trust, as well as the pharmaceutical companies that support the Division of Signal Transduction Therapy (AstraZeneca, Boehringer-Ingelheim, GlaxoSmithKline, Merck & Co. Inc., Merck KGaA, and Pfizer), which have generously supported our research on LKB1. We also thank Jeremy Thorner for helpful comments on the manuscript.

LITERATURE CITED

1. Peutz JLA. 1921. *Ned. Maandschr. Geneesk.* 10:134–46
2. Jeghers H. 1944. *N. Engl. J. Med.* 231:122–19
3. Jeghers H, McKusick VA, Katz KH. 1949. *N. Engl. J. Med.* 241:992–1005
4. Hemminki A. 1999. *Cell. Mol. Life Sci.* 55:735–50
5. Tomlinson IP, Houlston RS. 1997. *J. Med. Genet.* 34:1007–11
6. Westerman AM, Entius MM, de Baar E, Boor PP, Koole R, et al. 1999. *Lancet* 353:1211–15
7. Giardiello FM, Brensinger JD, Tersmette AC, Goodman SN, Petersen GM, et al. 2000. *Gastroenterology* 119:1447–53
8. Mallory SB, Stough DB 4th. 1987. *Dermatol. Clin.* 5:221–30
9. Boardman LA, Couch FJ, Burgart LJ, Schwartz D, Berry R, et al. 2000. *Hum. Mutation* 16:23–30
10. Hemminki A, Tomlinson I, Markie D, Jarvinen H, Sistonen P, et al. 1997. *Nat. Genet.* 15:87–90
11. Hemminki A, Markie D, Tomlinson I, Avizienyte E, Roth S, et al. 1998. *Nature* 391:184–87
12. Su JY, Erikson E, Maller JL. 1996. *J. Biol. Chem.* 271:14430–37
13. Jenne DE, Reimann H, Nezu J, Friedel W, Loff S, et al. 1998. *Nat. Genet.* 18:38–43
14. Manning G, Whyte DB, Martinez R, Hunter T, Sudarsanam S. 2002. *Science* 298:1912–34
15. Sanchez-Cespedes M, Parrella P, Esteller M, Nomoto S, Trink B, et al. 2002. *Cancer Res.* 62:3659–62
16. Fernandez P, Carretero J, Medina PP, Jimenez AI, Rodriguez-Perales S, et al. 2004. *Oncogene* 23:5084–91
17. Carretero J, Medina PP, Pio R, Montuenga LM, Sanchez-Cespedes M. 2004. *Oncogene* 23:4037–40
18. Tiainen M, Ylikorkala A, Makela TP. 1999. *Proc. Natl. Acad. Sci. USA* 96:9248–51
19. Shibata H, Toyama K, Shioya H, Ito M, Hirota M, et al. 1997. *Science* 278:120–23
20. Jishage KI, Nezu JI, Kawase Y, Iwata T, Watanabe M, et al. 2002. *Proc. Natl. Acad. Sci. USA* 99:8903–8
21. Miyoshi H, Nakau M, Ishikawa TO, Seldin MF, Oshima M, Taketo MM. 2002. *Cancer Res.* 62:2261–66
22. Rossi DJ, Ylikorkala A, Korsisaari N, Salovaara R, Luukko K, et al. 2002. *Proc. Natl. Acad. Sci. USA* 99:12327–32
23. Bardeesy N, Sinha M, Hezel AF, Signoretti S, Hathaway NA, et al. 2002. *Nature* 419:162–67
24. Sakamoto K, McCarthy A, Smith D, Green KA, Hardie DG, et al. 2005. *EMBO J.* 24:1810–20

25. Nakau M, Miyoshi H, Seldin MF, Imamura M, Oshima M, Taketo MM. 2002. *Cancer Res.* 62:4549–53
26. Ylikorkala A, Rossi DJ, Korsisaari N, Luukko K, Alitalo K, et al. 2001. *Science* 293:1323–26
27. Sapkota GP, Boudeau J, Deak M, Kieloch A, Morrice N, Alessi DR. 2002. *Biochem. J.* 362:481–90
28. Sapkota GP, Kieloch A, Lizcano JM, Lain S, Arthur JS, et al. 2001. *J. Biol. Chem.* 276:19469–82
29. Boudeau J, Baas AF, Deak M, Morrice NA, Kieloch A, et al. 2003. *EMBO J.* 22:5102–14
30. Collins SP, Reoma JL, Gamm DM, Uhler MD. 2000. *Biochem. J.* 345 (Part 3):673–80
31. Sapkota GP, Deak M, Kieloch A, Morrice N, Goodarzi AA, et al. 2002. *Biochem. J.* 368:507–16
32. Fernandes N, Sun Y, Chen S, Paul P, Shaw RJ, et al. 2005. *J. Biol. Chem.* 280:15158–64
33. Scott JW, Norman DG, Hawley SA, Kontogiannis L, Hardie DG. 2002. *J. Mol. Biol.* 317:309–23
34. Martin SG, St Johnston D. 2003. *Nature* 421:379–84
35. Baas AF, Boudeau J, Sapkota GP, Smit L, Medema R, et al. 2003. *EMBO J.* 22:3062–72
36. Brajenovic M, Joberty G, Kuster B, Bouwmeester T, Drewes G. 2003. *J. Biol. Chem.* 279:12804–11
37. Miyamoto H, Matsushiro A, Nozaki M. 1993. *Mol. Reprod. Dev.* 34:1–7
38. Karos M, Fischer R. 1999. *Mol. Gen. Genet.* 260:510–21
39. Nozaki M, Onishi Y, Togashi S, Miyamoto H. 1996. *DNA Cell Biol.* 15:505–9
40. Smith CM, Radzio-Andzelm E, Madhusudan, Akamine P, Taylor SS. 1999. *Prog. Biophys. Mol. Biol.* 71:313–41
41. Tiainen M, Vaahtomeri K, Ylikorkala A, Makela TP. 2002. *Hum. Mol. Genet.* 11:1497–504
42. Nezu J, Oku A, Shimane M. 1999. *Biochem. Biophys. Res. Commun.* 261:750–55
43. Boudeau J, Scott JW, Resta N, Deak M, Kieloch A, et al. 2004. *J. Cell Sci.* 117:6365–75
44. Smith DP, Rayter SI, Niederlander C, Spicer J, Jones CM, Ashworth A. 2001. *Hum. Mol. Genet.* 10:2869–77
45. Boudeau J, Sapkota G, Alessi DR. 2003. *FEBS Lett.* 546:159–65
46. Songyang Z, Fanning AS, Fu C, Xu J, Marfatia SM, et al. 1997. *Science* 275:73–77
47. Milburn CC, Boudeau J, Deak M, Alessi DR, van Aalten DM. 2004. *Nat. Struct. Mol. Biol.* 11:193–200
48. Boudeau J, Deak M, Lawlor MA, Morrice NA, Alessi DR. 2003. *Biochem. J.* 370:849–57
49. Nony P, Gaude H, Rossel M, Fournier L, Rouault JP, Billaud M. 2003. *Oncogene* 22:9165–75
50. Holbro T, Beerli RR, Maurer F, Koziczak M, Barbas CF 3rd, Hynes NE. 2003. *Proc. Natl. Acad. Sci. USA* 100:8933–38
51. Berger MB, Mendrola JM, Lemmon MA. 2004. *FEBS Lett.* 569:332–36
52. Roy F, Laberge G, Douziech M, Ferland-McCollough D, Therrien M. 2002. *Genes Dev.* 16:427–38
53. Luo H, Rose P, Barber D, Hanratty WP, Lee S, et al. 1997. *Mol. Cell. Biol.* 17:1562–71
54. Saharinen P, Vihinen M, Silvennoinen O. 2003. *Mol. Biol. Cell* 14:1448–59
55. Hardie DG, Scott JW, Pan DA, Hudson ER. 2003. *FEBS Lett.* 546:113–20
56. Zhou GC, Myers R, Li Y, Chen YL, Shen XL, et al. 2001. *J. Clin. Investig.* 108:1167–74
57. Zou MH, Kirkpatrick SS, Davis BJ, Nelson JS, Wiles WG 4th, et al. 2004. *J. Biol. Chem.* 279:43940–51
58. Scott JW, Hawley SA, Green KA, Anis M, Stewart G, et al. 2004. *J. Clin. Investig.* 113:274–84

59. Hardie DG. 2004. *J. Cell Sci.* 117:5479–87
60. Carling D. 2004. *Trends Biochem. Sci.* 29:18–24
61. Watts JL, Morton DG, Bestman J, Kemphues KJ. 2000. *Development* 127:1467–75
62. Nath N, McCartney RR, Schmidt MC. 2003. *Mol. Cell. Biol.* 23:3909–17
63. Hong SP, Leiper FC, Woods A, Carling D, Carlson M. 2003. *Proc. Natl. Acad. Sci. USA* 100:8839–43
64. Sutherland CM, Hawley SA, McCartney RR, Leech A, Stark MJ, et al. 2003. *Curr. Biol.* 13:1299–305
65. Hawley SA, Boudeau J, Reid JL, Mustard KJ, Udd L, et al. 2003. *J. Biol.* 2:28
66. Woods A, Johnstone SR, Dickerson K, Leiper FC, Fryer LG, et al. 2003. *Curr. Biol.* 13:2004–8
67. Shaw RJ, Kosmatka M, Bardeesy N, Hurley RL, Witters LA, et al. 2004. *Proc. Natl. Acad. Sci. USA* 101:3329–35
68. Corton JM, Gillespie JG, Hawley SA, Hardie DG. 1995. *Eur. J. Biochem.* 229:558–65
69. Winder WW, Hardie DG. 1999. *Am. J. Physiol. Endocrinol. Metab.* 277:E1–10
70. Saha AK, Ruderman NB. 2003. *Mol. Cell. Biochem.* 253:65–70
70a. Shaw RJ, Lamina KA, Vasquez D, Koo SH, Bardeesy N, et al. 2005. *Science* 310:1642–46
71. Lolli G, Johnson LN. 2005. *Cell Cycle* 4:572–77
72. Mora A, Komander D, Van Aalten DM, Alessi DR. 2004. *Semin. Cell. Dev. Biol.* 15:161–70
73. Vitari AC, Deak M, Morrice NA, Alessi DR. 2005. *Biochem. J.* 391:17–24
73a. Sakamoto K, Zarrinpashneh E, Budas GR, Pouleur AC, Dutta A, et al. 2006. *Am. J. Physiol. Endocrinol. Metab.* In press
74. Hawley SA, Pan DA, Mustard KJ, Ross L, Bain J, et al. 2005. *Cell Metab.* 2:9–19
75. Hurley RL, Anderson KA, Franzone JM, Kemp BE, Means AR, Witters LA. 2005. *J. Biol. Chem.* 280:29060–66
76. Woods A, Dickerson K, Heath R, Hong SP, Momcilovic M, et al. 2005. *Cell Metab.* 2:21–33
77. Lizcano JM, Goransson O, Toth R, Deak M, Morrice NA, et al. 2004. *EMBO J.* 23:833–43
78. Sakamoto K, Goransson O, Hardie DG, Alessi DR. 2004. *Am. J. Physiol. Endocrinol. Metab.* 287:E310–17
79. Inoki K, Corradetti MN, Guan KL. 2005. *Nat. Genet.* 37:19–24
80. Pan DJ, Dong JX, Zhang Y, Gao XS. 2004. *Trends Cell Biol.* 14:78–85
81. Inoki K, Zhu T, Guan KL. 2003. *Cell* 115:577–90
82. Corradetti MN, Inoki K, Bardeesy N, DePinho RA, Guan KL. 2004. *Genes Dev.* 18:1533–38
83. Shaw RJ, Bardeesy N, Manning BD, Lopez L, Kosmatka M, et al. 2004. *Cancer Cell* 6:91–99
84. Jimenez AI, Fernandez P, Dominguez O, Dopazo A, Sanchez-Cespedes M. 2003. *Cancer Res.* 63:1382–88
85. Liaw D, Marsh DJ, Li J, Dahia PL, Wang SI, et al. 1997. *Nat. Genet.* 16:64–67
86. Jones RG, Plas DR, Kubek S, Buzzai M, Mu J, et al. 2005. *Mol. Cell* 18:283–93
87. Law BK. 2005. *Crit. Rev. Oncol. Hematol.* 56:47–60
88. Swinnen JV, Beckers A, Brusselmans K, Organe S, Segers J, et al. 2005. *Cancer Res.* 65:2441–48
89. Imamura K, Ogura T, Kishimoto A, Kaminishi M, Esumi H. 2001. *Biochem. Biophys. Res. Commun.* 287:562–67
90. Evans JM, Donnelly LA, Emslie-Smith AM, Alessi DR, Morris AD. 2005. *Br. Med. J.* 330:1304–5

91. Schneider MB, Matsuzaki H, Haorah J, Ulrich A, Standop J, et al. 2001. *Gastroenterology* 120:1263–70
92. Willer A. 2003. *Onkologie* 26:283–89
93. Slattery ML. 2004. *Sports Med.* 34:239–52
94. Dempke W, Rie C, Grothey A, Schmoll HJ. 2001. *J. Cancer Res. Clin. Oncol.* 127:411–17
95. Udd L, Katajisto P, Rossi DJ, Lepisto A, Lahesmaa AM, et al. 2004. *Gastroenterology* 127:1030–37
96. Drewes G. 2004. *Trends Biochem. Sci.* 29:548–55
97. Tassan JP, Le Goff X. 2004. *Biol. Cell* 96:193–99
98. Jaleel M, McBride A, Lizcano JM, Deak M, Toth R, et al. 2005. *FEBS Lett.* 579:1417–23
99. Kemphues KJ, Priess JR, Morton DG, Cheng NS. 1988. *Cell* 52:311–20
100. Morton DG, Roos JM, Kemphues KJ. 1992. *Genetics* 130:771–90
101. Baas AF, Kuipers J, van der Wel NN, Batlle E, Koerten HK, et al. 2004. *Cell* 116:457–66
102. Guo S, Kemphues KJ. 1995. *Cell* 81:611–20
103. Shulman JM, Benton R, St Johnston D. 2000. *Cell* 101:377–88
104. Ossipova O, Bardeesy N, DePinho RA, Green JB. 2003. *Nat. Cell Biol.* 5:889–94
105. Spicer J, Rayter S, Young N, Elliott R, Ashworth A, Smith D. 2003. *Oncogene* 22:4752–56
106. Forcet C, Etienne-Manneville S, Gaude H, Fournier L, Debilly S, et al. 2005. *Hum. Mol. Genet.* 14:1283–92
107. Parsons DW, Wang TL, Samuels Y, Bardelli A, Cummins JM, et al. 2005. *Nature* 436:792
108. Timm T, Li XY, Biernat J, Jiao J, Mandelkow E, et al. 2003. *EMBO J.* 22:5090–101
109. Drewes G, Ebneth A, Preuss U, Mandelkow EM, Mandelkow E. 1997. *Cell* 89:297–308
110. Nishimura I, Yang Y, Lu B. 2004. *Cell* 116:671–82
111. Kishi M, Pan YA, Crump JG, Sanes JR. 2005. *Science* 307:929–32
112. Crump JG, Zhen M, Jin Y, Bargmann CI. 2001. *Neuron* 29:115–29
113. Lu R, Niida H, Nakanishi M. 2004. *J. Biol. Chem.* 279:31164–70
114. Wang Z, Takemori H, Halder SK, Nonaka Y, Okamoto M. 1999. *FEBS Lett.* 453:135–39
115. Feldman JD, Vician L, Crispino M, Hoe W, Baudry M, Herschman HR. 2000. *J. Neurochem.* 74:2227–38
116. Takemori H, Doi J, Horike N, Katoh Y, Min L, et al. 2003. *J. Steroid Biochem. Mol. Biol.* 85:397–400
117. Horike N, Takemori H, Katoh Y, Doi J, Min L, et al. 2003. *J. Biol. Chem.* 278:18440–47
118. Jakobsen SN, Hardie DG, Morrice N, Tornqvist HE. 2001. *J. Biol. Chem.* 276:46912–16
119. Screaton RA, Conkright MD, Katoh Y, Best JL, Canettieri G, et al. 2004. *Cell* 119:61–74
120. Suzuki A, Kusakai G, Kishimoto A, Lu J, Ogura T, Esumi H. 2003. *Oncogene* 22:6177–82
121. Suzuki A, Kusakai G, Kishimoto A, Lu J, Ogura T, et al. 2003. *J. Biol. Chem.* 278:48–53
122. Lefebvre DL, Bai Y, Shahmolky N, Sharma M, Poon R, et al. 2001. *Biochem. J.* 355:297–305
123. Legembre P, Schickel R, Barnhart BC, Peter ME. 2004. *J. Biol. Chem.* 279:46742–47
124. Bettencourt-Dias M, Giet R, Sinka R, Mazumdar A, Lock WG, et al. 2004. *Nature* 432:980–87
125. Al-Hakim AK, Goransson O, Deak M, Toth R, Campbell DG, et al. 2005. *J. Cell Sci.* 118:5661–73
126. Buchet-Poyau K, Mehenni H, Radhakrishna U, Antonarakis SE. 2002. *Cytogenet. Genome Res.* 97:171–78
127. Resta N, Stella A, Susca FC, Di Giacomo M, Forleo G, et al. 2002. *Hum. Mutat.* 20:78–79
128. Amos CI, Keitheri-Cheteri MB, Sabripour M, Wei C, McGarrity TJ, et al. 2004. *J. Med. Genet.* 41:327–33

129. Connolly DC, Katabuchi H, Cliby WA, Cho KR. 2000. *Am. J. Pathol.* 156:339–45
130. Hastings ML, Resta N, Traum D, Stella A, Guanti G, Krainer AR. 2005. *Nat. Struct. Mol. Biol.* 12:54–59
131. Deleted in proof
132. Deleted in proof
133. Schumacher V, Vogel T, Leube B, Driemel C, Goecke T, et al. 2005. *J. Med. Genet.* 42:428–35
134. Shinmura K, Goto M, Tao H, Shimizu S, Otsuki Y, et al. 2005. *Clin. Genet.* 67:81–86
135. Deleted in proof
136. Deleted in proof
137. Tate G, Suzuki T, Mitsuya T. 2003. *Acta Med. Okayama* 57:305–8
138. Trojan J, Brieger A, Raedle J, Roth WK, Zeuzem S. 1999. *Am. J. Gastroenterol.* 94:257–61
139. Wei C, Amos CI, Rashid A, Sabripour M, Nations L, et al. 2003. *J. Histochem. Cytochem.* 51:1665–72

Energy Transduction: Proton Transfer Through the Respiratory Complexes

Jonathan P. Hosler,[1] Shelagh Ferguson-Miller,[2] and Denise A. Mills[2]

[1]Department of Biochemistry, University of Mississippi Medical Center, Jackson, Mississippi 39216; email: jhosler@biochem.umsmed.edu

[2]Department of Biochemistry and Molecular Biology, Michigan State University, East Lansing, Michigan 48824; email: fergus20@msu.edu, millsden@msu.edu

Key Words

cytochrome c oxidase, hydrogen-bonded water

Abstract

A series of metalloprotein complexes embedded in a mitochondrial or bacterial membrane utilize electron transfer reactions to pump protons across the membrane and create an electrochemical potential ($\Delta\mu H^+$). Current understanding of the principles of electron-driven proton transfer is discussed, mainly with respect to the wealth of knowledge available from studies of cytochrome c oxidase. Structural, experimental, and theoretical evidence supports the model of long-distance proton transfer via hydrogen-bonded water chains in proteins as well as the basic concept that proton uptake and release in a redox-driven pump are driven by charge changes at the membrane-embedded centers. Key elements in the pumping mechanism may include bound water, carboxylates, and the heme propionates, arginines, and associated water above the hemes. There is evidence for an important role of subunit III and proton backflow, but the number and nature of gating mechanisms remain elusive, as does the mechanism of physiological control of efficiency.

Contents

INTRODUCTION 166
LONG-RANGE PROTON
    MOVEMENT THROUGH
    PROTEINS 167
CYTOCHROME c OXIDASE: AN
    ELECTRON
    TRANSFER-DRIVEN PUMP ... 168
    Proton Pathways and the Study of
        Proton Movements 168
    Postulated Mechanisms of Proton
        Pumping 170
    Experimental Insights 174
    Computational Insights 178
COMPARISON OF THE PROTON
    MOTIVE COMPLEXES OF
    OXIDATIVE
    PHOSPHORYLATION 181

CcO: cytochrome c oxidase

Respiratory complexes: the membrane-embedded metallocomplexes that make up the respiratory chain used for energy transduction

Energy transduction: change of form of energy such as the electrical potential energy of an electron into a membrane pH gradient

INTRODUCTION

Aerobic organisms have evolved the ability to obtain energy from their environment by extracting electrons from food stuffs and converting the electrical potential energy into a useful chemical form, ATP. Reduced substrates, such as NADH or succinate, donate electrons to a respiratory chain composed of metalloprotein complexes embedded in a mitochondrial or bacterial membrane (**Figure 1**). Three of these complexes have the ability to pump protons across the membrane and create a transmembrane electrochemical potential ($\Delta\mu H^+$) using a series of redox cofactors that achieve a step-wise drop in potential energy. The electrochemical gradient produced is then used for the synthesis of ATP by ATP synthase, also a membrane-inserted complex. In most systems, the ultimate acceptor of electrons in the respiratory chain is oxygen, which is reduced to water, providing a large redox potential drop and maximizing the free energy available for ATP synthesis.

The respiratory system can be "uncoupled"; that is, the electron transfer process can proceed without resulting in net proton pumping or ATP synthesis, either because the pumped protons are leaked back across the membrane by protonophores or by uncoupling proteins that dissipate $\Delta\mu H^+$, leading to the production of heat. This mode of operation is of physiological importance in both plants and animals when there is the need to make heat as well as ATP. There is evidence that the respiratory complex cytochrome c oxidase (CcO) has the ability to control its own pumping efficiency as well by "intrinsic uncoupling" via proton backflow through the protein in response to a buildup of the membrane potential (1) or to signals such as phosphorylation (2). Proton backflow may be one mechanism to prevent the buildup of a high $\Delta\mu H^+$, which can inhibit electron transfer and allow intermediates, such as semiquinones, to persist (3). Semiquinone is known to reduce O_2 directly, ultimately leading to the production of reactive O_2 species (ROS) such as the hydroxyl radical. Mitochondria have sophisticated systems to scavenge ROS, including proteins such as superoxide dismutase, but even low levels of ROS can be damaging. One explanation for the complexity of the mammalian respiratory complexes, in terms of a larger number of subunits compared to the bacterial forms, is that the extra subunits provide protection and stability in the presence of toxic compounds. The far fewer subunits of bacterial complexes, along with a highly homologous catalytic core, make the prokaryotic enzymes excellent model systems for mechanistic studies of energy transduction, especially because they are more amenable to site-directed mutagenesis.

In this review, we discuss the current understanding of the energy transduction mechanism that allows electron transfer energy to be converted to a membrane potential gradient by means of a series of proton pumps. A diversity of chemistries is used by the three proton-pumping complexes, which work in series, each using different segments of the redox potential gradient. In spite of the diversity,

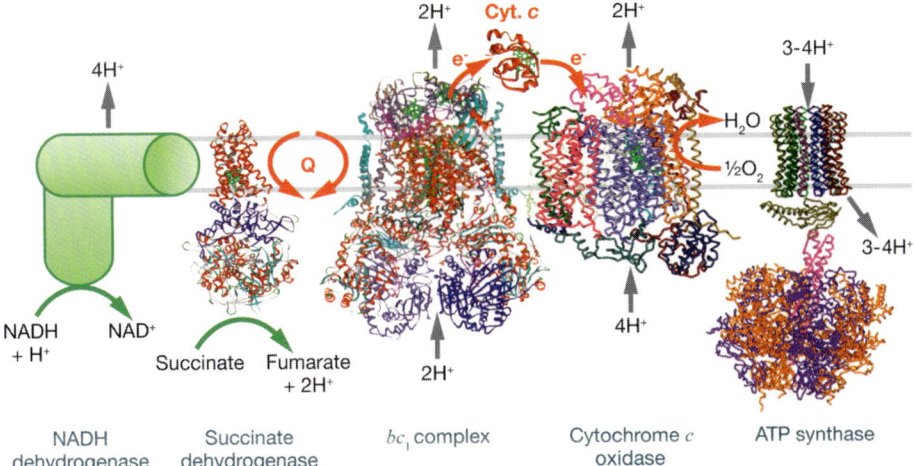

Figure 1
Complexes of the respiratory chain. These include *Escherichia coli* NADH dehydrogenase (145), succinate dehydrogenase 1NEN, bc_1 complex 1PP9, cytochrome *c* oxidase 1V54, and cytochrome *c* 1HRC.

the movement of protons through these proteins has many features in common. The principles of electron-driven proton transfer are illustrated in this review mainly by discussion of the wealth of information available from studies of C*c*O. This complex encompasses a variety of chemistries, including long-range proton transfer, vectorial charge separation, and redox-driven proton pumping. Analogies to the other well-characterized protonmotive system, the bc_1 complex, are also explored.

LONG-RANGE PROTON MOVEMENT THROUGH PROTEINS

Pathways for proton transfer have been identified in the structures of the photosynthetic reaction center, bacteriorhodopsin (4), the bc_1 complex (5), and in C*c*O structures (6). Within the protein, evidence supports the model that protons move rapidly through hydrogen-bonded chains of water and amino acid residues, presumably via a hopping mechanism (7). One mechanism proposes a sequential, isoenergetic exchange of hydrogen bonds and covalent bonds, which allows a proton to be added to one end of a chain of waters while another proton is released from the other end. Proton transfer through proteins may be rapid because a single file of hydrogen-bonded waters, as shown in model channel studies, is capable of moving a proton 40 times faster than proton transfer through bulk water (8). Proton pathways may begin or end with an exchangeable donor or acceptor, such as quinol, quinone, or O_2. At the protein surface, residues that begin or terminate pathways are generally carboxylate or histidine residues and sometimes arginine. The flow of protons, from a donor at the protein surface through a hydrogen-bonded water chain to an acceptor, may be interrupted by a protonatable group, such as a carboxylate residue, that can store a proton and then release it as dictated by the demands of the next acceptor. The waters of the pathways are generally coordinated by polar side chains, but this is not an absolute requirement because molecular dynamics (MD) simulations of cytochrome oxidase and artificial channels (8, 9) predict water chains forming in hydrophobic regions. Not all of the waters that are predicted to play a role in proton transfer are seen in crystal structures—even at high resolution (10, 11), but unseen waters can be modeled into spaces in structures (12). In addition, when many waters are seen in a structure, the proton pathways are not necessarily obvious, e.g., the proton exit pathway of C*c*O. This may relate to the ability of proton pathways to form and dissipate during the catalytic cycle (4). For example, the proton uptake pathway

MD: molecular dynamics

used to reprotonate the Schiff base in bacteriorhodopsin was not observed in the earliest crystal structures, but it was found in the structure of a mutant that mimics a later intermediate of the photocycle (10). Transient proton transfer pathways may be used by proteins to "gate" the flow of protons in order to provide directionality or to control the timing of proton delivery.

CYTOCHROME c OXIDASE: AN ELECTRON TRANSFER-DRIVEN PUMP

Cytochrome c oxidase is the terminal electron acceptor in the electron transfer chain, carrying out the critical tasks of reducing oxygen to water and pumping protons across the membrane. Although it differs significantly in structure, prosthetic groups, and catalytic function from the other members of the chain, the concerted movement of protons and electrons is common to all of the energy-transducing complexes.

Proton Pathways and the Study of Proton Movements

Structural information now exists for two of the three energy-transducing electron transfer complexes of oxidative phosphorylation (**Figure 1**), with crystal structure resolutions often below 3 Å. However, the largest amount of structural, mutational, and kinetic information concerning proton transfer is available for CcO. The bovine CcO structure has been resolved to 1.8 Å (13) along with a considerable amount of well-defined water within the protein. Additionally, there are good crystal structures of the bacterial aa_3-type oxidases from *Paracoccus denitrificans* (14, 15) and *Rhodobacter sphaeroides* (16), with a more recent isotropic crystal (2.35 Å) of the two subunit *R. sphaeroides* enzyme wherein the positions of bound lipids, detergents, and Cd^{2+}, as well as water can be seen clearly (17). Both of these bacterial oxidases are amenable to genetic manipulation leading to the production of numerous mutant forms.

CcO has been well characterized spectroscopically with respect to its multiple metal centers: hemes a and a_3, a dinuclear copper center (Cu$_A$), a type II copper (Cu$_B$), along with a non-redox active Mg and a Ca/Na-binding site (**Figure 2**) (18, 19). Time-resolved spectroscopic studies have shown that electrons from soluble cytochrome c are delivered first to Cu$_A$ in subunit II, then to six-coordinate heme a in subunit I, and then laterally to the heme a_3/Cu$_B$ active site, where O_2 is reduced to water (19).

Two proton uptake pathways lead from the inner surface of cytochrome oxidase (the negative side of the membrane) toward the buried heme a_3-Cu$_B$ active site where O_2 binds to Fe of heme a_3 (**Figure 2**). As yet, clear identification of an exit pathway for pump protons

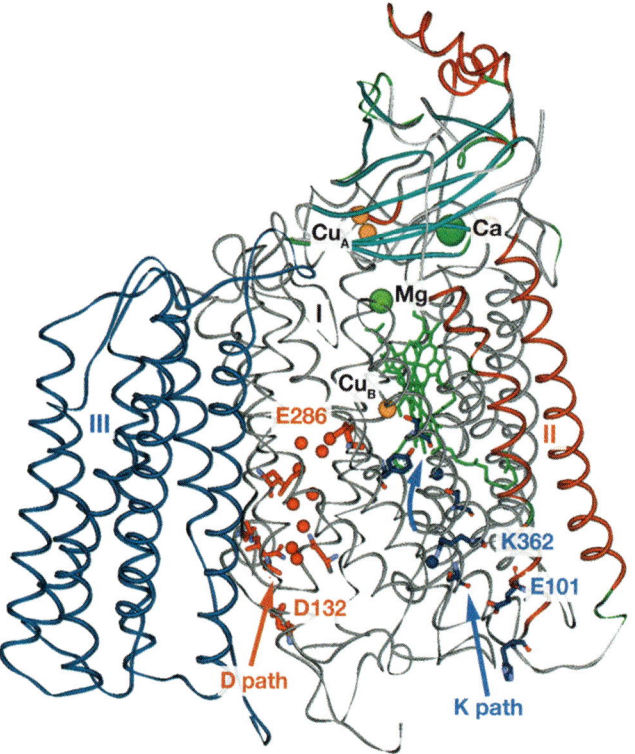

Figure 2

CcO (*R. sphaeroides* numbering in structure 1M56) showing the D path (*red*) and K path (*blue*) with D-path waters (*red spheres*) and K-path waters (*blue spheres*). Heme a and a_3 (*green*) are stick structures with Ca and Mg metals (*green spheres*) and Cu metal (*orange spheres*).

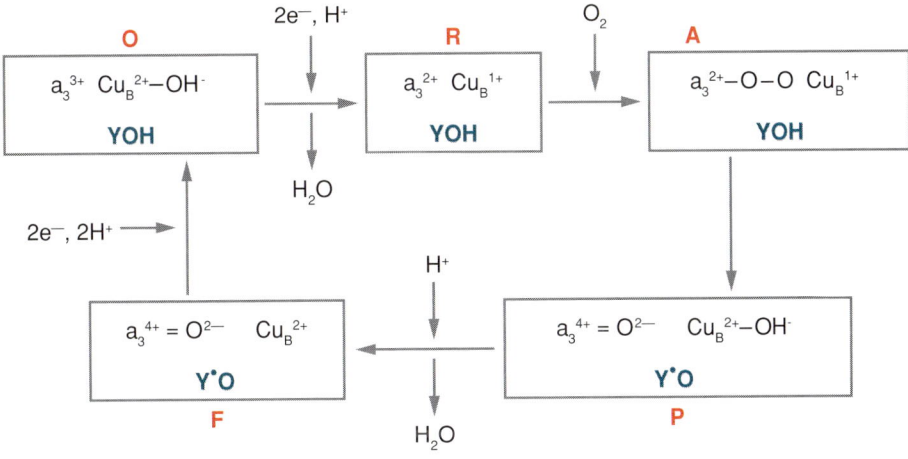

Figure 3
Proposed oxygen reduction reactions at the active site of CcO during steady-state turnover. Abbreviations used are as follows: Y, Y288 (*R. sphaeroides* numbering); R, reduced; A, oxy; P, F, and O are as described in the text. Only substrate protons are indicated. The two phases of the catalytic cycle merge to a certain extent, but the metal reduction phase is essentially O to R, whereas the O_2 reduction phase is A to F.

is lacking. Both proton uptake pathways were identified in the oxidase structures with the help of knowledge gleaned from prior mutagenesis experiments (20–22). The D pathway consists of a series of hydrogen-bonded waters anchored by D132 on the surface and E286 in the interior between hemes a and a_3, approximately 26 Å above D132 (16) (*R. sphaeroides* CcO numbering is used throughout). Substrate protons (those destined to make water) flow from E286 to Cu$_B$, a distance of 10–12 Å, through a short series of waters that are not resolved in the oxidase structures but can be modeled into a hydrophobic cavity between E286 and the active site (23–26). The role of E286 in proton transfer has been extensively examined (27–29). In the native oxidase, E286 appears usually to be in its protonated form because of a high pK_a and rapid reprotonation from the bulk solvent through D132 (30).

The K pathway appears to begin at E101 of subunit II and to sequentially involve S299, K362, T359, the hydroxyl farnesyl of heme a_3 and Y288 of subunit I (**Figure 2**). Computational studies have suggested positions for K pathway waters (31) because only two are resolved in the structures. Side-chain movements of K362 (32), T359 (31), and Y288 (25) have been proposed.

The roles of the D and K pathways are best introduced in the context of the O_2 reduction mechanism (**Figure 3**). Oxygen binds to heme a_3 only after heme a_3 and Cu$_B$ are reduced. This splits the catalytic cycle of CcO into two parts. In the metal reduction phase, electrons flow one at a time from two cytochromes c through Cu$_A$ and heme a to heme a_3/Cu$_B$. Studies indicate the uptake of one proton with each of the two electrons introduced into the active site (33–36); the uptake is argued as a requirement to maintain electrostatic neutrality as the electrons move into the hydrophobic environment (34, 37). Both of the protons of the metal reduction phase may be taken up by the K pathway (35, 38) or the first by the K pathway and the second by the D pathway (39, 40). The initial steps of the subsequent O_2 reduction phase are very rapid, beginning with O_2 binding, moving through O_2 bond scission, with the formation of the **P** oxoferryl state (Fe^{4+}=O Cu$_B$$^{2+}$–OH$^-$) and likely a tyrosine radical. These steps can proceed rapidly because the four electrons (two from Fea_3, one from Cu$_B$, and one from Tyr288) and one

FTIR: Fourier transform infrared

proton to be consumed are already available in the active site. However, the O_2 reduction phase ends with two slower steps that are rate limited by the transfer of substrate protons from the D proton pathway into the active site (27). One proton is required for the **P** to **F** transition, where **F** is postulated to be a protonated form of **P**, and one or two for the **F** to **O** transition (this number depends upon whether the single-turnover reaction or the steady-state cycle is being described). The input of two more electrons (one to Fea_3^{4+} and one to Y288) returns the binuclear center to its oxidized form **O** (**Figure 3**).

Although mutagenesis studies indicate that alteration of the D pathway eliminates proton pumping (41, 42), mutagenesis of the K pathway does not (21); all four of the pump protons appear to be taken up by the D pathway. There is an as yet unidentified path beyond E286 that must take protons to the proton-pumping element(s) and then to a proton exit pathway, leading to the outer surface of the oxidase complex (the positive side of the membrane).

A variety of techniques initially developed to study O_2 reduction and internal electron transfer has been adapted to study proton transfer through C*c*O. The flow-flash technique (43) has provided information on proton uptake (44, 45) and release (46, 47) during the **P** to **F** and **F** to **O** transitions in single-turnover experiments. Laser-induced electron injection has also been used to observe proton release (48) during the **F** to **O** transition. The proton uptake and release events are monitored with pH-sensitive dyes. Electrometric measurements that follow the movement of charge in the protein have been used to deduce proton transfer during the metal reduction phase (49–52) and during the **F** to **O** transition (53, 54). The Mg at the subunit I/II interface can be replaced in vivo with Mn (55) creating a sensitive paramagnetic probe for water and proton movement in the area above the hemes (56–58). The acquisition of high-resolution crystal structures of C*c*O has allowed the prediction of proton

movements by computational methods (23, 59–61). Fourier transform infrared (FTIR) spectroscopy analysis, which has been used to great benefit in studies of bacteriorhodopsin, is beginning to yield information on the more complicated C*c*O enzyme (58, 62–64).

Postulated Mechanisms of Proton Pumping

Protons are known to move rapidly through ordered chains of water, but how the rate and direction of movement is controlled and how it is coordinated to electron transfer is controversial. Mechanistic questions also arise from the fact that protons are moved against an electrochemical gradient, requiring a gating process to prevent the backflow.

Theoretical considerations. Since the discovery of the proton-pumping function of cytochrome oxidase by Wikstrom (65), the theoretical requirements for redox-driven proton pumps have been discussed in detail (66–68). At a minimum, the protein must contain at least one group capable of existing as either a protonated or deprotonated species; this is termed the "pumping element." The pumping element may be a protein residue, a heme propionate, a water, or some combination of these. The pumping element must at one point be accessible to protons from the inner surface of the oxidase (the negative surface of the membrane) and then at a later point be accessible to the outer surface (the positive surface of the membrane) to release the proton. The pumping element may change its pK_a (proton affinity) during the pumping cycle, but this is not absolutely required. There are two absolute requirements. One is a mechanism to couple the free energy of O_2 reduction to the proton pump. The other is for one or more "gating" mechanisms, which may be physical, chemical, or kinetic, that provide directionality to the flow of protons and inhibit a pump proton from dropping back through the electrical gradient to the inner surface of the protein. For full details of the coupling and gating mechanisms proposed for the pumping schemes, discussed below, the reader is directed to the references.

Coupling to electron transfer through Cu_A or heme a. On the basis of a great deal of work on the structure and chemistry of the CcO redox centers, two well-formulated pumping schemes were proposed in the 1980s involving Cu_A and heme a, the centers that transfer electrons to the heme a_3-Cu_B active site. Because the mechanism of pumping is likely to be the same for each of the four pump protons, one attraction for focusing on these metal centers was that they should undergo the same redox chemistry with each transfer of an electron from cytochrome c to O_2. In contrast, the active site is in different conformations at different points in the catalytic cycle. Babcock & Callahan (69) proposed that the alternating strength of a hydrogen bond to the formyl constituent of heme a, during redox cycling of this center, provided the energy for a geometric change of the hydrogen donor. Weakening of the hydrogen bond between the donor side chain and the formyl group allowed transfer of a proton from an input proton pathway to an output proton pathway.

Another mechanism, proposed by Chan and colleagues (70), involved a ligand-switching scheme driven by changes in the coordination geometry of Cu_A upon reduction and oxidation (71). Upon reduction, movement of Cu_A is posited to induce a nearby tyrosine to deprotonate and ligate the metal. The released proton is transferred to a cysteine ligand as the cysteine is displaced from the Cu_A. Relaxation to the oxidized coordination geometry reverses these events. Although ingenious, these schemes lost support with the discovery that the heme-Cu oxidase of *Escherichia coli*, cytochrome bo_3, pumps protons without the benefit of Cu_A or a formyl group on its low-spin heme (72).

A more recent proton-pumping mechanism involving heme a has been proposed by Yoshikawa and colleagues; this is based on their deductions from high-resolution structures of bovine CcO (73) and recent analyses of mutant forms in HeLa cells (13). Comparison of the reduced and oxidized forms shows an alteration in the position of an aspartate residue near the outer surface of subunit I. In the oxidized enzyme, the aspartate is buried in the hydrophobic environment of the protein, its affinity for protons is high, and it is connected to a channel involving water clusters and the formyl group of heme a. In the reduced form, the aspartate is exposed to the outer (positive) surface, its affinity for protons is low, and it is disconnected from its proton source. Proton movement to the aspartate is proposed as coupled to the reduction and oxidation of heme a rather than directly to the chemistry at the active site. However, the critical aspartate and other important residues are not present in CcO of bacteria or plants, implying that the proton-pumping mechanism of CcO is not conserved throughout evolution.

Coupling to electron transfer events at the active site. In the 1990s, the paradigm that CcO activity is primarily regulated by the rate of proton uptake began to emerge from a number of studies. For example, the steps of O_2 reduction seemed to be slowed by the rate of proton delivery (74), and an association was found between proton uptake and the reduction of heme a_3 and Cu_B prior to O_2 binding (75). The intrinsic rates of electron transfer between Cu_A, heme a, and heme a_3/Cu_B were found to be much faster than that of maximum turnover (76, 77), indicating that the electron transfer reactions themselves are not rate determining. With this knowledge, plus comparative analyses of conserved residues in sequences of more than 30 CcO forms, pumping mechanisms emerged, linking the uptake of pump protons to O_2 chemistry at the active site.

In a series of publications, Wikstrom and colleagues (78–80) concluded that proton pumping was restricted to the O_2 reduction phase of the catalytic cycle. In terms of energetics, this was an attractive concept because the free energy available during the O_2 reduction phase appears far greater than that of the metal reduction phase owing to the large redox drop from cytochrome c to O_2 ($\Delta E \sim 580$ mV). In essence, the electron transfer reactions can be viewed as using their energy to create O_2 reduction intermediates with extremely high affinities for protons (high pK_a). The original "histidine cycle" of Wikstrom's group (81) posits that negatively charged O_2 intermediates at the active site provide the driving force for pump proton uptake from the inner surface of the oxidase to the pumping element, a histidine ligand (H334) of Cu_B. The subsequent protonation of the O_2 reduction intermediates by substrate protons provides the electrostatic repulsion that triggers the release of the pump protons from the histidine to the positive surface of the membrane. The histidine cycle mechanism requires that the pump protons are taken up prior to substrate protons in order to ensure the coupling of O_2 reduction to proton pumping. In order to pump two protons in both the **P** to **F** and the **F** to **O** transitions, the model postulates that H334 is bound to Cu_B in its imidazolate form (Im^-) and then dissociates when protonated to the imidazolium form ($ImHH^+$). A critical gating feature is that the pump protons must not be allowed to combine with the O_2 reduction intermediates at the active site, even though it is the electrostatic interaction between the pump protons and the active site charge that drives pump proton uptake.

Shortly following the proposal of the histidine cycle, the acquisition of crystal structures of CcO provided a wealth of information applicable to proton pumping, along with some disappointments. The D and K pathways leading from the inner surface of the oxidase to the active site (**Figure 2**) could be identified with the help of data from prior mutagenesis experiments. However, the plethora of hydrophilic residues and waters above the active site did not reveal an obvious proton exit pathway that would have helped to locate the site of pumping. Further, the D pathway appeared to terminate at an internal glutamate, E286, located between hemes a and a_3, some 10–12 Å from the active site. The water chain that must provide the pathway for substrate protons between E286 and the active site was not resolved in any of the structures.

With the possibility that E286 serves as a branch point that alternately directs protons to the pump or to the active site, it became possible to envision a pumping element that was not a metal ligand at the active site. Rich and colleagues (82) used the structural information to postulate the "glutamate trap," in which the pumping element was proposed as being above E286, with E286 preventing backflow. A more detailed mechanism, based upon structural data indicating conformational changes of E286 depending upon its protonation state, was postulated by Brzezinski & Larsson (83, 84). In this scheme, E286 first donates a substrate proton to either the **P** or **F** oxygen reduction intermediates. Anionic E286 shifts upward toward a cluster of heme a and a_3 propionate groups, closely associated with R481 and R482 (**Figure 4**). The proximity of anionic E286 induces a substantial increase in the pK_a (proton affinity) of this cluster that drives the transfer of a proton from the D pathway (presumably through waters bypassing E286) to the cluster. Reprotonation of E286 from the inner surface of the oxidase allows it to relax to its original position, the pK_a of the accepting cluster drops, and the proton is ejected to the positive surface of the membrane. The movement of E286 is supported by FTIR spectroscopy evidence (28, 64) and computational analyses (23, 59, 85) in addition to the structural data (16).

Another pumping mechanism at a distance from the active site is suggested by recent MD simulations that show movement of a conserved loop that brings W172 within hydrogen-bonding distance of E286 (86). If this movement is reversible, it could form the central element of a proton pump that moves each pump proton from E286 to a heme a_3 propionate via W172.

In 1999, use of the electrometric technique led to the somewhat surprising conclusion that two protons are pumped during the metal reduction phase of the catalytic cycle, but only when this phase immediately follows the O_2 reduction phase (49). Later experiments support this (51, 52). The energy required for pumping in the metal reduction phase may be conserved from the O_2 reduction phase by the retention of hydroxyl ligands with a high pK_a or by virtue of there being significantly more energy available in the metal reduction phase because of a high redox potential for heme a_3/Cu$_B$ during steady-state turnover (50, 52). In a modified form of the histidine cycle (39), E286 alternately directs protons to a dissociated histidine ligand of Cu$_B$ (which binds a single proton in this model) or to O_2 reduction intermediates on heme a_3 and Cu$_B$. As before, the arrival of a substrate proton expels the pump protons from a heme propionate/water cluster to the outer surface of the oxidase. An evolved form of this model proposes specific pathways for the pump and substrate protons deduced from the positions of waters (25) conserved in all of the current CcO structures. Key conserved waters are proposed to be important pumping elements, which account for the number of CcO forms, including engineered mutants (29, 87), that can pump without a carboxyl residue at position 286. This model also invokes a novel gating mechanism, involving the dissociation of the H284-Y288 cross-linked pair from Cu$_B$ in response to the overall charge at the heme a_3-Cu$_B$ center. A hydroxyl ligand on Cu$_B$ provides the driving force for the uptake of pump protons at various stages of the catalytic cycle.

The bacterial protein bacteriorhodopsin is a light-driven proton pump in which light absorption induces isomerization of a bound retinal. Subsequent thermal reisomerization of the retinal drives the protein through a conformational series that pumps a proton across the bacterial membrane. Nine crystal structures, corresponding to different steps in the pumping cycle, have been analyzed to elucidate the mechanism of proton pumping

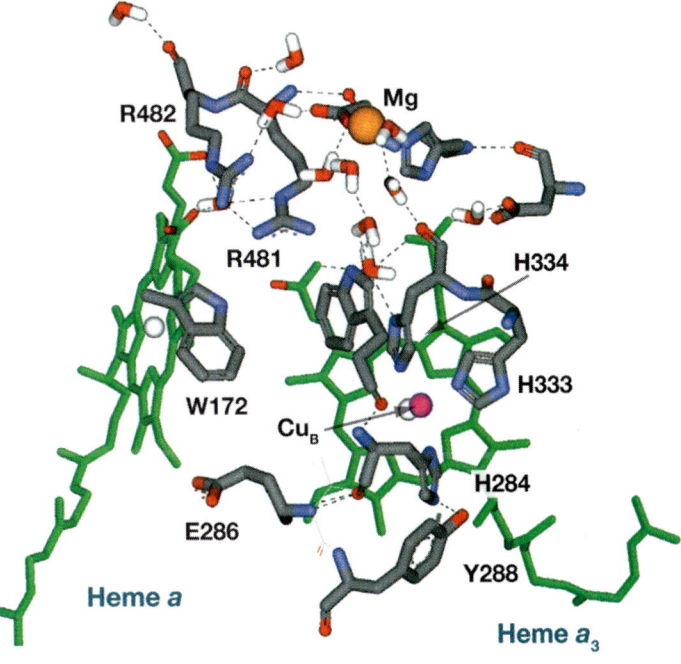

Figure 4

The area surrounding heme a and a_3, including the nonredox Mg, is shown in the *R. sphaeroides* CcO (1M56) (16). The arginine pair interacts closely with the heme propionates. The nitrogen on W172 is close to the heme propionate and away from E286.

(10) along with extensive FTIR spectroscopy analysis (88, 89). One of the findings from this extraordinary work is that the waters required to form connections in the input proton transfer pathway of bacteriorhodopsin are not all in position throughout the catalytic cycle. Rather, the proton pathway dissipates and reforms at various steps. By doing so, directionality is conferred on the pumping process. Transient water chains also provide the required gating function in a new scheme from the Wikstrom group (26). The model invokes the formation and dissipation of short chains of hydrogen-bonded water extending from E286 to either the propionate/arginine/water cluster above E286 or to the heme a_3-Cu$_B$ water center. As in other schemes, the arrival of a substrate proton provides the electrostatic repulsion that expels the pump proton from the propionate/arginine/water cluster to the outside.

C*c*O converts redox energy to a potential gradient by two mechanisms: a proton pump and vectorial redox chemistry. For the latter, electrons come from outside the membrane, and protons are drawn from inside the membrane to accomplish the reduction of O_2 to water at the buried heme a_3-Cu_B center. Here, the charges annihilate as O_2 is reduced to water, resulting in the loss of one negative charge from the outside and one positive charge from the inside, equivalent to the transfer of one proton across the membrane. The addition of the electron transfer-driven proton pump translocates another proton for each electron consumed. Since the introduction of Wikstrom's first histidine cycle, a common theme for proton pumping has emerged, although there is no consensus regarding the details of the process. The system is designed such that the introduction of negative charge (electrons) into the buried heme *a*/heme a_3/Cu_B centers is the driving force for all of the protons taken up from the inner surface of the oxidase. Pump protons are directed to a site where they cannot protonate the O_2 intermediates. Substrate protons both protonate the O_2 intermediates and provide the charge repulsion that expels the pump protons to the outside.

This generalized mechanism provides a framework for efforts to identify the key pumping elements and to understand how rate and efficiency are controlled. Many questions remain. Where is the exit pathway? How is proton pumping gated to inhibit the movement of protons back down the electrical gradient, and how many gates are involved? In what form is free energy provided for pumping during the metal reduction phase? Work on all of these questions, and more, continues.

Experimental Insights

The mitochondrial C*c*O has 13 subunits, but the 3 largest core subunits are highly homologous to those of bacterial C*c*Os, which are usually made up of only four peptides. The simpler bacterial C*c*Os are experimentally more accessible because they can be easily grown and genetically altered. Removal of subunits and site-directed mutagenesis of protein residues have been powerful tools to study the mechanism of proton pumping and metal inhibition.

The effect of subunit III on proton transfer.
Subunit III of C*c*O is a member of the highly conserved, three-subunit "catalytic core" of the enzyme. The subunit is closely associated with subunit I in the membrane (**Figure 2**), and recent work shows that subunit III has a profound effect on the transfer of protons through subunit I (90). The role of this entire subunit can be assessed because subunit III can be removed from C*c*O without strongly inhibiting the initial activity of the enzyme (90–92). Of the two pathways leading from the inner surface of C*c*O, only the D pathway is close to subunit III (**Figure 2**). In fact, the crystal structures of C*c*O show that D132 is located at a junction of subunits I and III such that half of the residues surrounding D132 come from subunit III. In the absence of subunit III, D132 is considerably more exposed to solvent. Single-turnover experiments show large differences in the rate of D pathway proton uptake in the presence and absence of subunit III, from >10,000 to ~350 s-1, respectively, at pH 8 (92). Rapid proton uptake is restored to subunit III-depleted C*c*O at low pH (<6). With steady-state turnover, the overall activity of subunit III-depleted C*c*O is limited by the rate of proton uptake through the D pathway at pH 7 and above. This situation can be exploited for the analysis of D pathway mutants.

Slow proton uptake in the absence of subunit III may result from changes in the pK_a of D132 and/or the loss of proton-collecting groups on the protein surface (92–94). Such a "proton antenna" has been proposed to aid proton uptake in bacteriorhodopsin (95, 96), the photosynthetic reactions center, as well as in C*c*O (6). Surface groups, e.g., carboxylate residues and histidines, are argued to trap protons from the bulk solvent and transfer them along the surface of the protein by a process of release and recapture, effectively increasing the proton concentration near the initial acceptor of a pathway. In C*c*O, single-turnover experiments show that the rate at which E286 is reprotonated from the bulk solvent, via D132 and the D pathway water chain, may be 1000-fold greater than the bimolecular rate constant ($4 \cdot 10^{10}$ M^{-1}s^{-1}) for the diffusion-limited transfer of a buffer proton to D132 (97). This is interpreted as evidence for a proton antenna for the D pathway. By using the rate of O$_2$ reduction by the wild-type and subunit III-depleted oxidases at different pH values (90) to estimate bimolecular rate constants of proton uptake, evidence for the continuous operation of a proton antenna during steady-state turnover can be derived (**Figure 5**). In the absence of subunit III, however, the estimated bimolecular rate constants for proton uptake do not significantly exceed the diffusion-limited value (**Figure 5**); the effect of the postulated proton antenna is clearly diminished. Without subunit III, the number of antenna residues on the inner surface of C*c*O may fall below the threshold required for its operation (98).

In experiments with C*c*O reconstituted into phospholipid vesicles in the absence of uncouplers or ionophores, the rate of O$_2$ reduction is slow. In large part, this is due to the inhibition of proton uptake into the D and K pathways by the membrane potential. Nevertheless, a slow rate (~1/10 V$_{max}$) of "controlled" turnover is observed, apparently supported by proton flow to the active site from the outer surface of the enzyme (99). The removal of subunit III inhibits this backflow of protons, as evidenced by a much slower rate of controlled turnover (100). The effect of removing subunit III is similar, but additive, to inhibiting the backflow pathway with zinc (see the section Metal inhibition of proton uptake, below). Assuming that the backflow proton pathway is actually the normal proton exit pathway operating in reverse, it appears that subunit III facilitates both the uptake and exit of pump protons. This is consistent with the

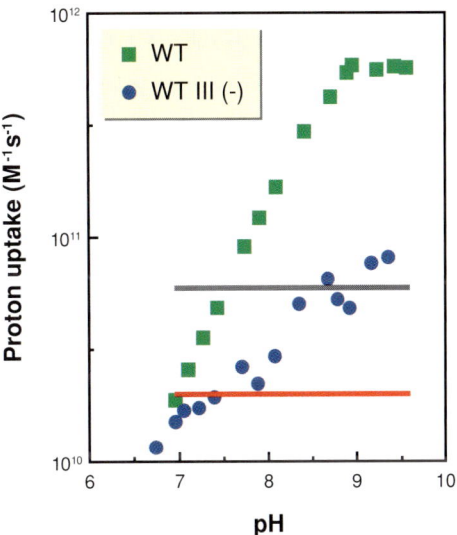

Figure 5

The absence of subunit III decreases the rate of proton uptake into the D pathway to the diffusion limit. Estimated bimolecular rate constants for the uptake of protons into the D pathway during steady-state turnover were calculated from the rates of steady-state activity of wild-type [WT, and without subunit III, WT III (−)] C*c*O. The red and gray lines represent diffusion-limited rate constants of $2-6 \cdot 10^{10}$ M^{-1}s^{-1}, i.e., the range of rate constants expected if the diffusion of buffer to D132 through bulk solvent determines the rate of proton uptake. Rate constants above these limits are suggestive of the function of a proton antenna that increases the local concentration of protons near D132 of the D pathway.

long-known observation that the removal of subunit III decreases the efficiency of proton pumping to approximately half that of C*c*O containing subunit III (101).

Approximately a decade ago, it was noted that turnover of subunit III-depleted C*c*O resulted in the irreversible loss of activity (102). Later characterization showed that this "suicide inactivation" process decreased the catalytic life span of C*c*O (i.e., its total number of turnovers, independent of time) owing to the increased probability that an inactivating structural alteration of the Cu$_B$ center occurs during the catalytic cycle (90, 91). The catalytic life span of subunit III-depleted C*c*O can be less than 0.1% that of the native enzyme. Recent experiments show that suicide inactivation and proton transfer are tightly linked (103). Mutations that inhibit the D pathway, but not the K pathway, induce suicide inactivation. Likewise, inhibition of proton uptake by a membrane potential increases suicide inactivation. Because the D pathway transfers substrate protons after O_2 binds, it is argued that slow proton uptake increases the lifetime of reactive O_2 reduction intermediates (such as heme a_3 oxoferryl forms or a tyrosine radical), which in turn initiate the chemistry of inactivation. Simultaneous inhibition of the proton backflow pathway from the outer surface and the D pathway from the inner surface (using the D132A-R481K mutant described below) greatly increases the probability of suicide inactivation, even in the presence of subunit III. This result suggests a physiological role for proton backflow: Under conditions of a high membrane potential, where proton uptake by the D pathway is strongly inhibited, proton flow to the active site via the backflow pathway helps prevent inactivation of CcO. Measurements of suicide inactivation offer another experimental tool for the analysis of mutants affecting the D and backflow pathways because the catalytic lifetime of CcO is proportional to the rate of proton transfer to the active site.

Metal inhibition of proton uptake. Metal inhibition of voltage-gated proton channels has been well documented (104, 105), and more recently it has been found that proton uptake by the photosynthetic reaction center, the bc_1 complex, and CcO are inhibited by micromolar concentrations of certain metals, including zinc, nickel, and cadmium (1, 106–109). In each case where the metal site has been identified, the coordinating groups include at least one histidine, one or more carboxylate residues, and waters. The sites of metal binding are difficult to determine as it appears that often two of the ligands must be removed in order to prevent metal binding and inhibition (110). In addition, metal binding is pH dependent because protons compete with the metal for binding (1, 104).

Low concentrations of zinc, cadmium, or nickel inhibit CcO catalytic turnover ($K_I \leq$ 5 μM), and inhibition is specific for these three metals (1). The sum of current results suggests the presence of at least three inhibitory binding sites for metal on $R.$ $sphaeroides$ CcO because evidence exists for metal inhibition of proton uptake into the D and K pathways from the inner surface of the complex as well as inhibition of proton backflow from the outer surface.

Single-turnover experiments have shown that zinc slows the uptake of protons about 30-fold from the bulk solvent into the D pathway (109). However, the fact that zinc does not induce suicide inactivation, which is highly sensitive to the rate of proton uptake via the D pathway, indicates that zinc has little effect on D pathway activity during steady-state turnover (103). These two results need not be contradictory if zinc binds more slowly than protons at the D pathway-binding site. In the single-turnover experiment, preincubation with zinc gives it ample time to equilibrate into the proton site for binding at the D pathway, whereas in steady-state turnover, protons may successfully compete with zinc for binding. If the D pathway is not inhibited by zinc during steady-state turnover, it follows that the inhibition of steady-state activity is likely due to inhibition of the K pathway. In support of this, a new crystal structure of $R.$ $sphaeroides$ CcO reveals cadmium binding to residues E101 and H96 of subunit II at the entrance of the K pathway (17).

The fact that low concentrations of zinc further inhibit the slow turnover of CcO in vesicles in the presence of a potential gradient provides strong evidence for a specific proton pathway leading to the active site from the outer surface (1). The proton backflow pathway is inhibited with cadmium or zinc, but not nickel, whereas nickel does inhibit proton uptake from the inner surface. This selectivity argues against inhibition of proton movement through the lipid bilayer by alteration of its permeability by zinc and also argues against metal transfer across the lipid bilayer in the timescale of the experiment. It seems likely that the backflow of protons occurs by reversal of the normal exit pathway for pump protons. Zinc does not inhibit the activity of C*c*O in phospholipid vesicles in the absence of a membrane potential, suggesting that protons moving out of C*c*O successfully compete with the metal for binding or that a membrane potential is required for a conformational change that creates the zinc-binding site. The site of metal binding on the outside remains elusive, but identifying the site should help determine the location of the exit pathway, as in the case of the photosynthetic reaction center proton uptake pathway (111), and help clarify the pumping mechanism.

Mutants affecting proton movement. The mutation of selected amino acid residues has been particularly fruitful in defining the proton uptake pathways. With the advantage of high-resolution structures that resolve waters as well as side chains, it is possible to visualize regions where waters are held and to predict and test the involvement of specific side chains.

D pathway. The ends of the hydrogen-bonded water chain of the D pathway are anchored by D132 on the inner surface of C*c*O and E286, in between hemes a and a_3 (**Figure 2**). Site-directed mutants of D132 and E286 in the aa_3-type oxidases that remove a carboxylate residue from either of these positions strongly inhibit activity and proton uptake (21, 41, 45, 82).

The phenomenon of proton backflow from the outer surface was initially identified in the D132A mutant (41). The O_2 reduction activity of this mutant is accelerated by a membrane potential, presumably because proton backflow is driven by the potential gradient (negative inside) across the membrane. Proton backflow is confirmed by measurement of rapid alkalinization on the outside of D132A vesicles upon initiation of turnover (112, 113). Because the structure of subunit I in the region of the proton backflow/exit pathway is unperturbed in the D132A mutant, it seems likely that proton backflow is a normal phenomenon that occurs with a high membrane potential. This may explain why there is no observation of proton pumping in wild-type C*c*O reconstituted into vesicles under the conditions of a high proton gradient (ΔpH) and transmembrane voltage gradient ($\Delta \Psi$) across the membrane (114).

The O_2 reduction activity of D132A is partially rescued by micromolar concentrations of arachidonic acid, presumably by supplying a carboxylate group in the vicinity of the D pathway (41). The removal of subunit III from D132A further restores activity (99) apparently by exposing an alternative initial acceptor. The addition of arachidonic acid to D132A III (−) restores even more activity (99), but none of these C*c*O forms are capable of proton pumping (90).

A number of alterations of two amides further into the D pathway, N121 and N139, slow proton transfer (115). However, the substitution of N139 with aspartic acid either retains or substantially enhances (42) O_2 reduction activity, but proton pumping is eliminated. This additional protonatable site in the D path is concluded to increase the effective pK_a of E286 from 9.4 to 11.0 (54). Brzezinski and colleagues (116) argue that the resulting decrease in the anionic form of E286 eliminates proton pumping by decreasing the driving force for transfer to the proton-accepting cluster above E286. An alternative explanation is that the extra carboxyl alters the kinetics of proton transfer in the pathway such that the reverse reaction is facilitated to an extent that it successfully competes with proton pumping.

K pathway. The early observation that mutation of K362 inactivated the enzyme and prevented the reduction of heme a_3 (53, 117) strongly supports the concept that the uptake of at least one charge-neutralizing proton via the K pathway is required during the metal reduction phase. The presumed binding site for this proton is a hydroxyl ligand of Cu_B, produced in the previous O_2 reduction phase. The addition of a high concentration of hydrogen peroxide restores reasonable activity

ΔpH: pH gradient across the membrane

$\Delta \Psi$: transmembrane voltage gradient

to K362M, presumably by bypassing the need for a K pathway proton in the metal reduction phase by supplying both oxygen and protons to the active site (38, 112). The mutation of T359 to alanine slows but does not eliminate K pathway proton transfer or CcO activity. The fact that T359A retains efficient proton-pumping activity supports the conclusion that pump protons are not supplied by the K pathway. Controversy remains about the roles of two subunit II residues at the entrance to the K path, E101 and H96, because they appear to affect CcO activity to differing extents depending on the bacterial enzyme that is studied (118).

Exit/backflow pathway. A pair of highly conserved arginine residues located above the hemes, R481 and R482, interact with the D propionates of hemes a and a_3 (**Figure 4**). Arg481 bridges the two hemes by forming a hydrogen bond/ion pair interaction with propionates from each. Alteration of R481 to a lysine residue releases the interaction with the heme a_3 propionate, with the intriguing result of decreased proton backflow. This is seen as inhibition of controlled turnover (plus $\Delta\Psi$) but a normal rate of uncontrolled turnover (119). The closer interaction of R481K with the heme a propionate produces additional effects, including a 20–40 mV decrease in the redox potential of heme a. The double mutation D132A-R481K blocks proton uptake through both the backflow pathway and the D pathway, with the results that O_2 reduction is extremely inhibited (0.25% of native) (119) and that suicide inactivation is rapid (103). The precise role of R481 remains to be determined. The arginine may transfer protons itself, it may influence the protonation state of the propionates of the hemes, or it may help organize a water chain leading from E286 to the Mg center above the hemes (86).

Computational Insights

With the availability of high-resolution crystal structures of proteins and increased computational power, even large proteins can now be simulated in the computer. The movement of protein residues and explicit individual waters can be followed over time in molecular dynamics simulations. Along with electrostatic calculations, these theoretical analyses give insight into possible residue motion and water chain formation, and they suggest new experimental approaches and mechanistic ideas.

Water and proton dynamics. Although crystallography can reveal the positions of side chains and some of the water in a protein, many waters are mobile and hence not observed. For example, in a recent crystal structure of the two-subunit *R. sphaeroides* CcO at 2.35 Å resolution, 110 waters were observed (17), which is less than the calculated content. Yet, the current understanding that we have of energy transduction in the complexes of the respiratory chain critically involves water as part of proton conduction pathways. Thus, the positions of waters and changes in these positions during the reaction cycle are important for understanding proton transfer and pumping mechanisms. Computational methods to simulate water location and mobility are increasingly valuable tools for developing mechanistic models. Programs such as DOWSER (12) or GRID (Molecular Discovery) can be used to fill potential water sites in a structure with no van der Waals overlaps. The number of waters that have been placed within the two-subunit structures of CcO varies from 130 to 755 depending upon selection criteria, such as interaction energies (85, 120, 121).

Using MD simulations, methods have been developed to monitor the formation, nature, and persistence of hydrogen-bonded water chains (23, 31). Simulations are carried out with varying restraints on protein movement; either the complete protein backbone or everything outside the area of interest may be constrained to minimize computational cost. In order to predict the movement of protons themselves through water chains, quantum mechanics calculations have been combined with MD (23, 60, 122).

Simulations of the K pathway. In all C*c*O structures, the K path is found to be fairly "dry," with only two conserved positions for water (25). However, water chains are formed during MD simulations of the two-subunit C*c*O of *R. sphaeroides* after the computational addition of waters (31). Interestingly, rotation of the amino acid side chain of T359 was required for completion of a hydrogen-bonded water chain that reached from S299 to the hydroxyl group of the heme a_3 farnesyl chain. Another water chain was formed from the farnesyl hydroxyl to the active site via Tyr288 despite this being a hydrophobic region. These water chains were formed without any movement of the essential residue K362, which has been suggested to undergo a conformational change (123). However, this may not be inconsistent with the MD simulations, which were run for 2 ns, whereas a full turnover of C*c*O takes a millisecond, and hemes a, a_3, and Cu$_B$ were kept reduced for the duration of the analysis (31).

Simulations of the D pathway. Simulation of water chain formation in the D pathway shows a completely hydrogen-bonded chain from D132 to N139 (**Figure 2**) and then from N139 to E286. There is a break or discontinuity in the water chain around N139 (23). Proton movement through the D pathway varies with the protonation state of E286 at the top of the pathway. Deprotonated E286 causes a proton, introduced as a hydronium at the entrance to the channel, to move through the waters to protonate E286. However, when the simulation is started with protonated E286, the proton is held up near S200 and S201, about halfway along the D path, suggesting a proton trap at this site. Mutation of these serines to alanines inhibits C*c*O activity and slows proton movement in the simulations (23). These studies suggest that the D pathway could accommodate, or trap, a second (86) proton above N139.

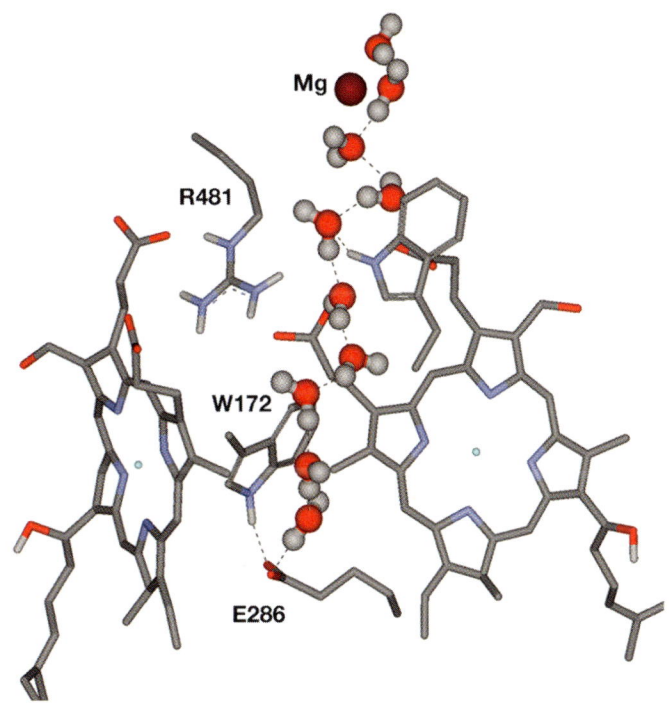

Figure 6

After over a nanosecond of an MD simulation of the *R. sphaeroides* C*c*O structure with added water (31, 86), a chain of hydrogen-bonded waters is clearly seen—stretching from E286 through a hydrophobic cavity to Mg. The glutamate has its carboxyl pointed up and is interacting with W172.

Simulations of the proton exit. Simulation of wild-type C*c*O by MD reveals a possible segment of the exit pathway for pump protons as a continuous hydrogen-bonded water chain extending from E286 to the Mg center (23). Interestingly, the single-file nature of this series of waters is dictated by a narrow hydrophobic channel (**Figure 6**). It is observed that in the R481K mutant a movement of a loop containing the hydrophobic residues leads to the collapse of this water chain. This suggests that significant conformational changes may occur in response to highly conservative single mutations, which in this case could account for the observed phenotype of inhibited proton backflow in R481K (119).

Protons move through C*c*O both for proton pumping across the membrane and for the reduction of O_2 to water. The bifurcation of these two proton streams, at least in the O_2 reduction phase of the catalytic cycle, has often been ascribed to E286, which appears to be capable of some rotational movement based on experimental, crystallographic, and computational methods (16, 26, 27, 59). An alternative to the movement of E286, however, is a conformational change of W172 (86). In MD simulations of wild-type C*c*O, W172 rotates to interact with E286, and such a movement could deliver a proton from E286 to the area above the hemes (**Figure 4**). In the R481K mutant, W172 remains hydrogen bonded to the D propionate of heme a_3.

Electrostatic calculations. Electrostatic calculations are important for estimating the pK_as of amino acid residues that may depend upon the redox states of hemes a, a_3, and Cu_B, and thus aid in predictions of proton movement. Using a continuum dielectric model and Poisson-Boltzmann solution (124), a cluster of 18 residues in C*c*O were calculated to form an electrostatic network (125). The effect of the redox status of the metal centers in subunit I on the pK_a values of these residues was examined. Although K362 in the K pathway did not alter its protonation state with redox state changes, E101 at the entrance to the K pathway did (by 2.6 pH units), even though it is 25 Å from the active site. The redox status of heme a was linked to changes in the arginine pair (R481/R482) above the propionates of hemes a and a_3 and linked with Arg52, which closely interacts with the formyl group of heme a (125, 126). In the D pathway, E286 remained protonated in all redox states. However, later electrostatic analysis of *P. denitrificans* C*c*O suggested that the protonation of E286 was sensitive to redox changes (127).

Using comparative modeling, the *caa*$_3$ oxidase of *Rhodothermus marinus* appears to substitute a tyrosine for E286, as does the *ba*$_3$-type oxidase of *Thermus thermophilus* (127, 128). Calculations suggest that this tyrosine does not undergo a protonation change upon alteration of metal redox status (127). A water associated with the tyrosine at this site may be the protonatable and redox responsive element, opening the possibility that it is water that fulfills this function even in the presence of E286 (25).

Olsson et al. (122) evaluated proton movements using a novel quantum mechanical analysis. From their results, they propose a movement of a proton from within the D pathway to E286 in concert with the movement of a proton from E286 to an acceptor toward the outer surface. Their analysis also supports pumping schemes discussed above (in the Coupling to electron transfer events at the active site section) in that protonation of the D propionate of heme a_3 is followed by protonation of a hydroxyl bound at the heme a_3/Cu_B center. Neutralization of this hydroxyl repulses the proton from the propionate to the outside and causes a high energy barrier for the backflow of protons.

A histidine ligand of Cu_B can be considered to be an unlikely site for protonation because of its nominally high pK_a (122). However, other electrostatic calculations suggest that one or more of the Cu_B ligands can act as a proton loading site (61, 129), allowing for pumping mechanisms that include histidine protonation (see the Coupling to electron transfer events at the active site section, above).

Because of the salt bridges between the arginine pair (R481, R482) and the D propionates of hemes a and a_3, the propionates were predicted to not be part of the proton pump (68). However, mutagenesis (119, 130), computation (86), and FTIR spectroscopy experiments (62, 131) implicate the arginine/propionate cluster as part of the exit route for protons.

The variety of answers coming from different computational approaches suggests that systematic errors continue to be a challenge. Consensus among different analyses on different C*c*O structures could be the hallmark of valuable predictions.

COMPARISON OF THE PROTON MOTIVE COMPLEXES OF OXIDATIVE PHOSPHORYLATION

Little can be said about the pathways of proton transfer through NADH dehydrogenase given the current lack of high-resolution structures. One conundrum is that the metal centers of NADH dehydrogenase are largely or completely located in the extramembrane domain of the complex, although the proton-pumping mechanism must, by definition, be located in the transmembrane domain (132). How the redox reactions are coupled to the pump is a mystery (133).

In the case of the bc_1 complex, a number of structures are available (108, 134–136) and interesting comparative features emerge. The bc_1 complex moves protons from the negative to the positive surface of the membrane using a redox loop mechanism called the Q-cycle. The mechanism of the Q-cycle has been extensively reviewed elsewhere (137–140), but a brief summary follows. Reduced quinol, carrying two electrons and two protons, docks at the ubiquinone (Q_o)-binding site toward the upper (positive) surface of the complex. A concerted two electron oxidation at Q_o sends one electron to a high potential Fe/S center and the other to heme b_L, and the two protons are released to the outer surface. The iron-sulfur protein transfers its electron to soluble cytochrome c, via cytochrome c_1, whereas heme b_L transfers its electron to heme b_H toward the negative or inner surface of the complex. Near heme b_H, there is a second quinone-binding site, ubiquinone (Q_i), that binds oxidized quinone. This quinone is reduced by one electron from b_H and one proton from the interior to produce a temporally stable semiquinone. A second round of quinol oxidation at Q_o reduces another cytochrome c and reduces the semiquinone waiting at Q_i. With the addition of another proton, neutral quinol is released from Q_i into the membrane. When all of the charge movements are accounted for, it can be seen that the Q-cycle moves the equivalent of one proton from the negative to the positive surface of the membrane for each electron delivered to cytochrome c. This is precisely the same charge stoichiometry as in CcO, where only the electrons and protons used to reduce O_2 to water are tallied.

A major difference between CcO and the bc_1 complex is that CcO has evolved a proton pump that moves an additional proton per electron. The bc_1 complex has not evolved a proton pump because there is only sufficient redox energy to drive the Q-cycle in the presence of a transmembrane voltage gradient. With the lack of a pump, there is no requirement for, or observation of, a proton transfer pathway, which extends all of the way through the bc_1 complex. However, the Q_o and Q_i sites are each buried within the transmembrane domain such that proton transfer pathways are required for the exit and uptake of protons. Here the structure of quinones adds complication because the proton-binding carbonyls are separated by the length of the quinone ring. Thus, there exist two proton uptake pathways into Q_i, one leading to each carbonyl. Likewise, there are two exit pathways for the protons released from Q_o. The location of these pathways in the high-resolution yeast bc_1 structure has been elegantly described (5).

The critical reaction in the bc_1 mechanism is the initial oxidation of quinol at Q_o, which is postulated to be a simultaneous reduction of the Fe/S protein and heme b_L that sends the two electrons in two different directions (141, 142). The ability of Q_o to carry out this concerted two-electron reaction is argued to be enhanced by the formation of hydrogen bonds between both quinol hydroxyl groups and two residues that subsequently accept the protons from the quinol (5). These residues are H181 of the Rieske Fe/S protein (yeast numbering) and E272 of cytochrome b. His181 of the Rieske protein is also a ligand of the Fe/S center, so that in accepting a quinol proton, the Fe/S center is acting as a one electron, one proton carrier. In the other proton exit pathway starting with E272, the glutamate rotates nearly 180° after it accepts the quinol proton

Q_o: ubiquinone-binding site toward the outer side of bc_1

Q_i: ubiquinone-binding site toward the inner side of bc_1

in order to pass the proton to a heme propionate of the exit pathway via an intervening water (143). This movement is reminiscent of the postulated movement of deprotonated E286 of CcO up toward the propionates of hemes a and a_3 (23, 83, 85). Interestingly, E272 of the bc_1 complex is located in a conserved loop with the sequence PEWY, and E286 of CcO is located in a conserved sequence of PEVY. Possibly this structural motif is important in allowing glutamate movement.

As in CcO, zinc inhibits the activity of the bc_1 complex at submicromolar concentration (106). A zinc-binding site involving a histidine and an aspartic acid has been observed in the structure of the chicken bc_1 complex at the end of the Q$_o$ exit pathway, approximately 10 Å above E272 (108).

Two proton uptake pathways consisting of hydrogen-bonded water chains coordinated by polar groups lead to the quinone bound at Q$_i$ (144). One of these pathways begins with a glutamate residue, leading to an arginine residue and then onto the quinone via another series of waters. The structure suggests that upon reduction of quinone at Q$_i$, the arginine is deprotonated and then reprotonated by the water chain leading from the glutamate on the inner surface of the complex. This chemistry is similar to that postulated for the D pathway of CcO, where E286 transfers a proton via a water chain to the O$_2$ reduction intermediates and is then reprotonated by the water chain leading from D132 on the inner surface. In yeast, a cardiolipin molecule is well positioned to be the initial proton donor for the second proton uptake pathway leading to Q$_i$. The cardiolipin could transfer protons via water to a lysine and then to the quinone carbonyl via a series of waters. Proton uptake by both the bc_1 complex and CcO may be assisted by structural lipids (5, 90).

If the bc_1 complex keeps pace with the activity of CcO, where $V_{max} \geq 2000$ e$^-$/sec, calculations indicate that a proton antenna would be required at pH values above 7. The cardiolipin at the entrance to one of the proton uptake pathways has been suggested as part of a putative proton-collecting antenna.

In general, it appears that proton movements in CcO and the bc_1 complex have many similarities and that obtaining a high-resolution structure of both has been an important factor in reaching a clearer understanding of the mechanism of coupling to the electron transfer processes.

SUMMARY POINTS

1. Comprehensive knowledge of CcO structure, function, and kinetic/spectroscopic properties create a valuable model for examining the function of an electron transfer-driven proton pump.

2. A central theme has emerged that charge interactions drive both the uptake and the release of pump protons.

3. A cluster of heme propionates, two arginines, and associated waters is likely to be an acceptor site for pump protons en route to the outer surface, regardless of the mechanism of the pump. Similar clusters are involved in the exit of protons from the bc_1 complex.

4. Proton transfer in CcO involves hydrogen-bonded water chains, but experimental results and computational analyses reveal complexities, such as the likely control by side-chain movements, the existence of proton traps, apparent discontinuities in the water chains, and the formation of water chains through both hydrophilic and hydrophobic regions.

5. An increasing understanding of proton transfer mechanisms within CcO has provided the basis for rational models for mechanisms of coupling proton pumping to the transfer of electrons to oxygen.

6. Subunit III of the catalytic core plays a profound role in facilitating the transfer of pump protons through subunit I and may supply a major portion of a proton-collecting antenna, even though it contains no metal centers.

7. The existence of a specific exit pathway for pump protons is supported by evidence for (a) proton backflow from the outer surface of CcO, which is specifically inhibited by Zn/Cd in CcO vesicles; (b) membrane potential-stimulated activity of D-path mutants, which are inhibited in proton uptake from the inner surface; and (c) mutant forms of CcO that are inhibited in proton backflow.

FUTURE ISSUES TO BE RESOLVED

1. Which residues are the critical elements of the electron transfer-driven proton pump of CcO?

2. Is there a defined, controlled exit pathway for pump protons? If so, where is it?

3. What controls the directionality of proton pumping? How many gates are involved? Are the gates physical barriers, e.g., a space through which a proton cannot hop; thermodynamic barriers, e.g., pK_a changes; or kinetic barriers, e.g., dissipation of water chains?

4. Is there physiologically significant variability in the efficiency of proton pumping? If so, is this controlled by regulating the backflow of protons? Does the catalytic core of cytochrome oxidase have a mechanism to regulate its efficiency? Could this be a key function of subunit III or is this only achieved by the additional nuclear-encoded subunits in eukaryotes?

5. How important is the regulation of efficiency of proton pumping in the respiratory chain in the overall process of physiological energy balance? Do signaling systems in the cell, such as phosphorylation, disease states, and aging, play a role in regulating mitochondrial efficiency?

ACKNOWLEDGMENTS

This work was supported by NIH GM 56802 (to J.H.) and NIH GM 26916 (to S. F.-M.). We thank Robert Cukier and Steve Seibold for their input on the computational section.

LITERATURE CITED

1. Mills DA, Schmidt B, Hiser C, Westley E, Ferguson-Miller S. 2002. *J. Biol. Chem.* 277:14894–901
2. Kadenbach B. 2003. *Biochim. Biophys. Acta* 1604:77–94
3. Brand MD, Buckingham JA, Esteves TC, Green K, Lambert AJ, et al. 2004. *Biochem. Soc. Symp.* 71:203–13

4. Lanyi JK. 2004. *Annu. Rev. Physiol.* 66:665–88
5. Hunte C, Palsdottir H, Trumpower BL. 2003. *FEBS Lett.* 545:39–46
6. Adelroth P, Brzezinski P. 2004. *Biochim. Biophys. Acta* 1655:102–15
7. Nagle JF, Tristram-Nagle S. 1983. *J. Membr. Biol.* 74:1–14
8. Dellago C, Naor MM, Hummer G. 2003. *Phys. Rev. Lett.* 90:105902
9. Wu YJ, Voth GA. 2003. *Biophys. J.* 85:864–75
10. Schobert B, Brown LS, Lanyi JK. 2003. *J. Mol. Biol.* 330:553–70
11. Luecke H, Schobert B, Richter H-T, Cartailler J-P, Lanyi JK. 1999. *J. Mol. Biol.* 291:899–911
12. Zhang L, Hermans J. 1996. *Proteins* 24:433–38
13. Tsukihara T, Shimokata K, Katayama Y, Shimada H, Muramoto K, et al. 2003. *Proc. Natl. Acad. Sci. USA* 100:15304–9
14. Iwata S, Ostermeier C, Ludwig B, Michel H. 1995. *Nature* 376:660–69
15. Ostermeier C, Harrenga A, Ermler U, Michel H. 1997. *Proc. Natl. Acad. Sci. USA* 94:10547–53
16. Svensson-Ek M, Abramson J, Larsson G, Tornroth S, Brzezinski P, Iwata S. 2002. *J. Mol. Biol.* 321:329–39
17. Qin L. 2005. PhD thesis. *X-ray crystallographic studies of cytochrome c oxidase from Rhodobacter sphaeroides*. Mich. State Univ., East Lansing. 205 pp.
18. Gennis R, Ferguson-Miller S. 1995. *Science* 269:1063–64
19. Ferguson-Miller S, Babcock G. 1996. *Chem. Rev.* 96:2889–907
20. Hosler JP, Ferguson-Miller S, Calhoun MW, Thomas JW, Hill J, et al. 1993. *J. Bioenerg. Biomembr.* 25:121–36
21. Fetter JR, Qian J, Shapleigh J, Thomas JW, García-Horsman JA, et al. 1995. *Proc. Natl. Acad. Sci. USA* 92:1604–8
22. Gennis RB. 1998. *Science* 280:1712–13
23. Cukier RI. 2004. *Biochim. Biophys. Acta* 1656:189–202
24. Tashiro M, Stuchebrukhov AA. 2005. *J. Phys. Chem. B* 109:1015–22
25. Sharpe MA, Qin L, Ferguson-Miller S. 2005. In *Biophysical and Structural Aspects of Bioenergetics*, ed. M Wikstrom, pp. 26–54. Cambridge: R. Soc. Chem.
26. Wikstrom M, Verkhovsky MI, Hummer G. 2003. *Biochim. Biophys. Acta* 1604:61–65
27. Adelroth P, Karpefors M, Gilderson G, Tomson FL, Gennis RB, Brzezinski P. 2000. *Biochim. Biophys. Acta* 1459:533–39
28. Lubben M, Prutsch A, Mamat B, Gerwert K. 1999. *Biochemistry* 38:2048–56
29. Aagaard A, Gilderson G, Mills DA, Ferguson-Miller S, Brzezinski P. 2000. *Biochemistry* 39:15847–50
30. Namslauer A, Brzezinski P. 2004. *FEBS Lett.* 567:103–10
31. Cukier RI. 2005. *Biochim. Biophys. Acta* 1706:134–46
32. Brändén M, Sigurdson H, Namslauer A, Gennis RB, Adelroth P, Brzezinski P. 2001. *Proc. Natl. Acad. Sci. USA* 98:5013–18
33. Rich PR. 1995. *Aust. J. Plant Physiol.* 22:479–84
34. Rich P, Meunier B, Mitchell R, Moody R. 1996. *Biochim. Biophys. Acta* 1275:91–95
35. Forte E, Scandurra F, Richter O-MH, D'Itri E, Sarti P, et al. 2004. *Biochemistry* 43:2957–63
36. Parul D, Palmer G, Fabian M. 2005. *Biochemistry* 44:4562–71
37. Artzatbanov VY, Konstantinov A, Skulachev VP. 1978. *FEBS Lett.* 87:180–85
38. Vygodina TV, Pecoraro C, Mitchell D, Gennis R, Konstantinov AA. 1998. *Biochemistry* 37:3053–61

39. Wikstrom M, Jasaitis A, Backgren C, Puustinen A, Verkhovsky MI. 2000. *Biochim. Biophys. Acta* 1459:514–20
40. Ruitenberg M, Kannt A, Bamberg E, Ludwig B, Michel H, Fendler K. 2000. *Proc. Natl. Acad. Sci. USA* 97:4632–36
41. Fetter JR, Sharpe M, Qian J, Mills D, Ferguson-Miller S, Nicholls P. 1996. *FEBS Lett.* 393:155–60
42. Pawate AS, Morgan J, Namslauer A, Mills D, Brzezinski P, et al. 2002. *Biochemistry* 41:13417–23
43. Gibson Q, Greenwood C. 1963. *Biochem. J.* 86:541–54
44. Oliveberg M, Hallén S, Nilsson T. 1991. *Biochemistry* 30:436–40
45. Adelroth P, Ek MS, Mitchell DM, Gennis RB, Brzezinski P. 1997. *Biochemistry* 36:13824–29
46. Nilsson T, Hallén S, Oliveberg M. 1990. *FEBS Lett.* 260:45–47
47. Flaxen K, Gilderson G, Adelroth P, Brzezinski P. 2005. *Nature* 437:286–89
48. Zaslavsky D, Sadoski RC, Rajagukguk S, Geren L, Millett F, et al. 2004. *Proc. Natl. Acad. Sci. USA* 101:10544–47
49. Verkhovsky MI, Jasaitis A, Verkhovskaya ML, Morgan JE, Wikstrom M. 1999. *Nature* 400:480–83
50. Wikstrom M, Verkhovsky MI. 2002. *Biochim. Biophys. Acta* 1555:128–32
51. Ruitenberg M, Kannt A, Bamberg E, Fendler K, Michel H. 2002. *Nature* 417:99–102
52. Bloch D, Belevich I, Jasaitis A, Ribacka C, Puustinen A, et al. 2004. *Proc. Natl. Acad. Sci. USA* 101:529–33
53. Konstantinov AA, Siletsky S, Mitchell D, Kaulen A, Gennis RB. 1997. *Proc. Natl. Acad. Sci. USA* 94:9085–90
54. Siletsky SA, Pawate AS, Weiss K, Gennis RB, Konstantinov AA. 2004. *J. Biol. Chem.* 279:52558–65
55. Hosler JP, Espe MP, Zhen Y, Babcock GT, Ferguson-Miller S. 1995. *Biochemistry* 34:7586–92
56. Florens L, Schmidt B, McCracken J, Ferguson-Miller S. 2001. *Biochemistry* 40:7491–97
57. Schmidt B, McCracken J, Ferguson-Miller S. 2003. *Proc. Natl. Acad. Sci. USA* 100:15539–42
58. Schmidt B, Hillier W, McCracken J, Ferguson-Miller S. 2004. *Biochim. Biophys. Acta* 1655:248–55
59. Pomes R, Hummer G, Wikstrom M. 1998. *Biochim. Biophys. Acta* 1365:255–60
60. Xu JC, Voth GA. 2005. *Proc. Natl. Acad. Sci. USA* 102:6795–800
61. Popovic DM, Stuchebrukhov AA. 2005. *J. Phys. Chem. B* 109:1999–2006
62. Behr J, Hellwig P, Mantele W, Michel H. 1998. *Biochemistry* 37:7400–6
63. Iwaki M, Puustinen A, Wikstrom M, Rich PR. 2003. *Biochemistry* 42:8809–17
64. Nyquist RM, Heitbrink D, Bolwien C, Gennis RB, Heberle J. 2003. *Proc. Natl. Acad. Sci. USA* 100:8715–20
65. Wikstrom MK. 1977. *Nature* 266:271–73
66. Wikstrom M, Krab K, Saraste M. 1981. *Annu. Rev. Biochem.* 50:623–55
67. Malmstrom BG. 1985. *Biochim. Biophys. Acta* 811:1–12
68. Popovic DM, Stuchebrukhov AA. 2004. *FEBS Lett.* 566:126–30
69. Babcock GT, Callahan PM. 1983. *Biochemistry* 22:2314–19
70. Gelles J, Blair DF, Chan SI. 1986. *Biochim. Biophys. Acta* 853:205–36
71. Chan SI, Li PM. 1990. *Biochemistry* 29:1–12
72. Puustinen A, Finel M, Virkki M, Wikström M. 1989. *FEBS Lett.* 249:163–67

73. Tsukihara T, Aoyama H, Yamashita E, Tomizaki T, Yamaguchi H, et al. 1995. *Science* 269:1069–74
74. Verkhovsky MI, Morgan JE, Wikstrom M. 1995. *Biochemistry* 34:7483–91
75. Mitchell R, Mitchell P, Rich PR. 1992. *Biochim. Biophys. Acta* 1101:188–91
76. Winkler JR, Malmstrom BG, Gray HB. 1995. *Biophys. Chem.* 54:199–209
77. Verkhovsky MI, Morgan JE, Wikstrom M. 1992. *Biochemistry* 31:11860–63
78. Wikstrom M. 1989. *Nature* 338:776–78
79. Babcock GT, Wikstrom M. 1992. *Nature* 356:301–9
80. Wikstrom M. 2004. *Biochim. Biophys. Acta* 1655:241–47
81. Wikström M, Bogachev A, Finel M, Morgan JE, Puustinen A, et al. 1994. *Biochim. Biophys. Acta* 1187:106–11
82. Rich PR, Junemann S, Meunier B. 1998. *J. Bioenerg. Biomembr.* 30:131–38
83. Brzezinski P, Larsson G. 2003. *Biochim. Biophys. Acta* 1605:1–13
84. Brzezinski P. 2004. *Trends Biochem. Sci.* 29:380–87
85. Hofacker I, Schulten K. 1998. *Proteins* 30:100–7
86. Seibold SA, Mills DA, Ferguson-Miller S, Cukier RI. 2005. *Biochemistry* 44:10475–85
87. Gilderson G, Aagaard A, Gomes CM, Adelroth P, Teixeira M, Brzezinski P. 2001. *Biochim. Biophys. Acta* 1503:261–70
88. Braiman MS, Bousché O, Rothschild KJ. 1991. *Proc. Natl. Acad. Sci. USA* 88:2388–92
89. Garczarek F, Brown LS, Lanyi JK, Gerwert K. 2005. *Proc. Natl. Acad. Sci. USA* 102:3633–38
90. Hosler JP. 2004. *Biochim. Biophys. Acta* 1655:332–39
91. Bratton M, Pressler M, Hosler J. 1999. *Biochemistry* 38:16236–45
92. Gilderson G, Salomonsson L, Aagaard A, Gray J, Brzezinski P, Hosler J. 2003. *Biochemistry* 42:7400–9
93. Marantz Y, Nachliel E, Aagaard A, Brzezinski P, Gutman M. 1998. *Proc. Natl. Acad. Sci. USA* 95:8590–95
94. Marantz Y, Einarsdottir OO, Nachliel E, Gutman M. 2001. *Biochemistry* 40:15086–97
95. Gutman M, Nachliel E. 1990. *Biochem. Biophys. Acta* 1015:391–414
96. Checover S, Marantz Y, Nachliel E, Gutman M, Pfeiffer M, et al. 2001. *Biochemistry* 40:4281–92
97. Namslauer A, Aagaard A, Katsonouri A, Brzezinski P. 2003. *Biochemistry* 42:1488–98
98. Georgievskii Y, Medvedev ES, Stuchebrukhov AA. 2002. *Biophys J.* 82:2833–46
99. Mills DA, Tan Z, Ferguson-Miller S, Hosler J. 2003. *Biochemistry* 42:7410–17
100. Mills DA, Ferguson-Miller S. 2003. *FEBS Lett.* 545:47–51
101. Wilson KS, Prochaska LJ. 1990. *Arch. Biochem. Biophys.* 282:413–20
102. Haltia T, Saraste M, Wikstrom M. 1991. *EMBO J.* 10:2015–21
103. Mills DA, Hosler JP. 2005. *Biochemistry* 44:4656–66
104. Cherny V, DeCoursey T. 1999. *J. Gen. Physiol.* 114:819–38
105. DeCoursey T. 2003. *Physiol. Rev.* 83:475–579
106. Link T, von Jagow G. 1995. *J. Biol. Chem.* 270:25001–6
107. Axelrod H, Abresch E, Paddock M, Okamura M, Feher G. 2000. *Proc. Natl. Acad. Sci. USA* 97:1542–47
108. Berry E, Zhang Z, Bellamy H, Huang L. 2000. *Biochim. Biophys. Acta* 1459:440–48
109. Aagaard A, Namslauer A, Brzezinski P. 2002. *Biochim. Biophys. Acta* 1555:133–39
110. Paddock ML, Adelroth P, Chang C, Abresch EC, Feher G, Okamura MY. 2001. *Biochemistry* 40:6893–902
111. Paddock M, Graige M, Feher G, Okamura M. 1999. *Proc. Natl. Acad. Sci. USA* 96:6183–88

112. Mills DA, Ferguson-Miller S. 1998. *Biochim. Biophys. Acta* 1365:46–52
113. Zhen Y, Mills D, Hoganson CW, Lucas RL, Shi W, et al. 1999. In *Frontiers of Cellular Bioenergetics: Molecular Biology, Biochemistry and Physiopathology*, ed. S Papa, F Guerrieri, JM Tager, pp. 157–78. New York: Kluwer Acad./Plenum
114. Proteau G, Wrigglesworth JM, Nicholls P. 1983. *Biochem. J.* 210:199–205
115. Pfitzner U, Hoffmeier K, Harrenga A, Kannt A, Michel H, et al. 2000. *Biochemistry* 39:6756–62
116. Namslauer A, Pawate AS, Gennis RB, Brzezinski P. 2003. *Proc. Natl. Acad. Sci. USA* 100:15543–47
117. Hosler JP, Shapleigh JP, Mitchell DM, Kim Y, Pressler M, et al. 1996. *Biochemistry* 35:10776–83
118. Richter O-MH, Durr K, Kaant A, Ludwig B, Scandurra F, et al. 2005. *FEBS J.* 272:404–12
119. Mills DA, Geren L, Hiser C, Schmidt B, Durham B, et al. 2005. *Biochemistry* 44:10457–65
120. Olkhova E, Hutter MC, Lill MA, Helms V, Michel H. 2004. *Biophys. J.* 86:1873–89
121. Zheng XH, Medvedev DM, Swanson J, Stuchebrukhov AA. 2003. *Biochim. Biophys. Acta* 1557:99–107
122. Olsson MH, Sharma PK, Warshel A. 2005. *FEBS Lett.* 579:2026–34
123. Adelroth P, Gennis RB, Brzezinski P. 1998. *Biochemistry* 37:2470–76
124. Tiede DM, Vashishta A-C, Gunner MR. 1993. *Biochemistry* 32:4515–31
125. Kannt A, Roy C, Lancaster D, Michel H. 1998. *Biophys. J.* 74:708–21
126. Lancaster CRD. 2003. *FEBS Lett.* 545:52–60
127. Soares CM, Baptista AM, Pereira MM, Teixeira M. 2004. *J. Biol. Inorg. Chem.* 9:124–34
128. Chen Y, Hunsicker-Wang L, Pacoma RL, Luna E, Fee JA. 2005. *Protein Expr. Purif.* 40:299–318
129. Kannt A, Lancaster CR, Michel H. 1998. *Biophys. J.* 74:708–21
130. Brändén G, Brändén M, Schmidt B, Mills DA, Ferguson-Miller S, Brzezinski P. 2005. *Biochemistry* 44:10466–74
131. Behr J, Michel H, Mantele W, Hellwig P. 2000. *Biochemistry* 39:1356–63
132. Friedrich T, Abelmann A, Brors B, Guenebaut V, Kintscher L, et al. 1998. *Biochem. Biophys. Acta* 1365:215–19
133. Brandt U, Kerscher S, Drose S, Zwicker K, Zickermann V. 2003. *FEBS Lett.* 545:9–17
134. Xia D, Yu C-A, Kim H, Xia J-Z, Kachurin AM, et al. 1997. *Science* 277:60–66
135. Iwata S, Lee JW, Okada K, Lee JK, Iwata M, et al. 1998. *Science* 281:64–71
136. Lange C, Hunte C. 2002. *Proc. Natl. Acad. Sci. USA* 99:2800–5
137. Crofts AR. 2004. *Annu. Rev. Physiol.* 66:689–733
138. Crofts AR, Hacker B, Barquera B, Yun CH, Gennis RB. 1992. *Biochim. Biophys. Acta* 1101:162–65
139. Trumpower BL. 2002. *Biochim. Biophys. Acta* 1555:166–73
140. Osyczka A, Moser CC, Dutton PL. 2005. *Trends Biochem. Sci.* 30:176–82
141. Junemann S, Heathcote P, Rich PR. 1998. *J. Biol. Chem.* 273:21603–7
142. Darrouzet E, Moser CC, Dutton PL, Daldal F. 2001. *Trends Biochem. Sci.* 26:445–51
143. Crofts AR, Barquera B, Gennis RB, Kuras R, Guergova-Kuras M, Berry EA. 1999. *Biochemistry* 38:15807–26
144. Hunte C, Koepke J, Lange C, Rossmanith T, Michel H. 2000. *Struct. Fold. Des.* 8:669–84
145. Holt P, Morgan DJ, Sazanov LA. 2003. *J. Biol. Chem.* 278:43114–20

The Death-Associated Protein Kinases: Structure, Function, and Beyond

Shani Bialik and Adi Kimchi

Department of Molecular Genetics, Weizmann Institute of Science, Rehovot, 76100 Israel; email: shanib@weizmann.ac.il, adi.kimchi@weizmann.ac.il

Key Words

autophagy, DAP kinase, DRP-1, programmed cell death, ZIP kinase

Abstract

Death-associated protein kinase (DAPk) is the founding member of a newly classified family of Ser/Thr kinases, whose members not only possess significant homology in their catalytic domains, but also share cell death–associated functions. The realization that *DAPk* is a tumor suppressor gene, whose expression is lost in multiple tumor types, has spurred a flurry of interest in the kinase family and produced an impressive body of literature concerning its function, regulation, and connection to disease. The DAPk family has been linked to several cell death–related signaling pathways, and functions other than cell death have also been proposed. This review presents a thorough structural analysis of the kinases, discusses methods of regulation, clarifies their cellular targets and functions, and shows how these functions are integrated. Although many gaps in our knowledge still remain, the data generated to date can be combined to delineate a place for the DAPk family within the general cell death-signaling network.

Contents

- INTRODUCTION 190
- DEATH IS A FAMILY MATTER: DAPk AND RELATED PROTEINS 191
  - Structural Features of DAPk Family Members 191
  - Mechanisms of Regulation of the DAPk Family 196
- DOWNSTREAM OF DAPk: A CLOSE LOOK AT DAPk FUNCTION 197
  - Death-Associated Protein Kinases: How DAPk Earned Its Name .. 198
  - Molecular Dissection of the Death Pathways Induced by the DAPk Family 199
  - Death-Related Effects on the Cytoskeleton 201
  - Cross Talk with Other Kinases 203
  - Nonapoptotic Functions 203
  - The DAPk Signaling Network, an Overview 204

INTRODUCTION

Death-associated protein kinase (DAPk) is a Ca^{2+}/calmodulin (CaM)-regulated Ser/Thr kinase that mediates cell death. Increased DAPk activity, due to overexpression of the kinase, leads to pronounced death-associated cellular changes, which include membrane blebbing, cell rounding, detachment from extracellular matrix, and the formation of autophagic vesicles (1–8). Furthermore, DAPk activity is necessary for the induction of cell death by multiple death signals, including those generated by death receptors, cytokines, matrix detachment, and oncogene-induced hyperproliferation (1–4, 6, 7, 9).

DAPk acts as a tumor suppressor largely because of its ability to sensitize cells to many of the apoptotic signals that are encountered as a cell undergoes tumorigenesis.

CaM: calmodulin

DAPk: Death-associated protein kinase

DAPk suppresses cellular transformation at early stages of tumor development (4) and also inhibits later metastatic events, as shown in a mouse model of tumor metastasis (3). Interestingly, embryonic fibroblasts derived from DAPk −/− mice have a greater spontaneous immortalization rate than their wild-type (WT) counterparts (H. Berissi, T. Raveh, and A. Kimchi, unpublished data). This indicates that lack of DAPk also interferes with cellular senescence. These data have motivated numerous researchers to examine the status of DAPk in human primary tumors. The majority of these studies found a significant loss of DAPk expression in a large variety of tumor types, mainly owing to DNA methylation. **Supplementary Table 1** summarizes these findings (follow the Supplemental Material link from the Annual Reviews home page at http://www.annualreviews.org). The tumor screens combined with the previously mentioned experimental data on the tumor suppressive functions of DAPk suggest that DAPk may play a causative role in tumor development, and assays of its expression levels may serve as a diagnostic and prognostic tool to evaluate disease severity, progression, metastatic rate, and recurrence. Furthermore, DAPk, which is highly abundant in the brain, has been linked to diseases associated with neuronal injury and may serve as a target for therapeutic intervention in the treatment of neurodegeneration. The reader is referred to a recent review for an in-depth discussion of the contribution that loss and gain of DAPk function has to cancer and neuronal disease, respectively, as well as the potential strategies available to modulate DAPk's function (10).

Over the past few years, several groups have joined the effort to decipher DAPk's cellular function, focusing on areas that include its biochemical properties, regulation, and target substrates. From these studies, it has become apparent that DAPk has multiple functions and is linked to several cell death–related signaling pathways. Furthermore, it is now recognized that DAPk is the prototype of a family of closely related kinases, all of which have been associated with cell death. The purposes of this review are to provide a comprehensive overview of the recent literature on DAPk and its family members, to resolve some discrepancies that have arisen, and ultimately, to place DAPk in the proper context of cell death signaling.

DEATH IS A FAMILY MATTER: DAPk AND RELATED PROTEINS

DAPk belongs to a family of related death kinases, all of which share significant sequence and functional homology (11–14). The family includes two closely related homologues of DAPk: ZIPk [ZIP kinase, also known as Dlk (DAP-like kinase) or DAPk3] (11, 12) and DRP-1 (DAPk-related protein 1, also known as DAPk2) (13, 14). The human genes share 83% and 80% identity at the amino acid level, respectively, with DAPk's catalytic domain. More distantly related are human DRAK-1 and DRAK-2 (DAPk-related apoptosis-inducing protein kinase-1 and -2), whose kinase domains are only 50% identical to DAPk (15). Phylogenetically, the DAPk family is most closely related to the family of CaM-regulated kinases, in particular to myosin light chain kinase (MLCK), which shares 44% identity within the corresponding catalytic domain. DAPk orthologues exist in rodent (16) and in *Caenorhabditis elegans* (13) but not in *Drosophila* or lower organisms. In contrast, ZIPk and DRP-1 are only present in mammals (Y. Shoval and A. Kimchi, unpublished data). An orthologue of DRAK-2, but not DRAK-1, exists in the rodent and is expressed almost exclusively in lymphoid tissue (17). For the most part, the DRAKs have not been well characterized, although a mouse knockout of DRAK-2 suggests that the protein negatively regulates T-cell activation, with no apparent role in apoptosis (17). This review therefore focuses on the three closer death-associated relatives, DAPk, ZIPk, and DRP-1.

Structural Features of DAPk Family Members

DAPk, DRP-1, and ZIPk are grouped together into one kinase subfamily because of the high degree of conservation within their common catalytic domains. However, outside this region, the kinases vary greatly in size and structure (**Figure 1**). This section compares the various structural domains of the kinase family and discusses how these domains affect the function and localization of each kinase.

ZIPk: ZIP kinase
DRP-1: DAPk-related protein 1
MLCK: myosin light chain kinase

The catalytic domain: structure and regulation. Each DAPk family member contains at its N terminus a catalytic domain composed of the typical 11 subdomains found in all Ser/Thr kinases. The X-ray crystal structure of DAPk's catalytic domain has been resolved to an impressive 1.5 Å and has provided structural hints as to DAPk's mechanism of activation, its interactions with substrates, and its potential inhibitor design (18 and reviewed in 10). Of note, the presence of two clusters of acidic amino acids at the proposed substrate-binding site suggests that complementary interactions with basic residues near the substrate's core phosphorylation site may play a role in substrate recognition. In fact, many proposed substrates of the DAPk family possess two to three basic residues just N-terminal to the phosphorylated Ser or Thr (see **Table 1**), and mutation of such a basic cluster in an optimized DAPk peptide substrate reduced phosphorylation efficiency (19). Another outstanding feature of the structure is a highly ordered, positively charged loop, enriched in basic residues, that is positioned on the upper lobe of the catalytic domain (18). The 12 amino acids (aa 45–57) that comprise this loop are a conserved feature of the DAPk family members and are thus referred to as "the fingerprint" of the family (13). Of note, mutation of the basic residues in the loop did not affect the K_m of a peptide substrate (19), indicating it is not directly involved in substrate binding, but rather may be

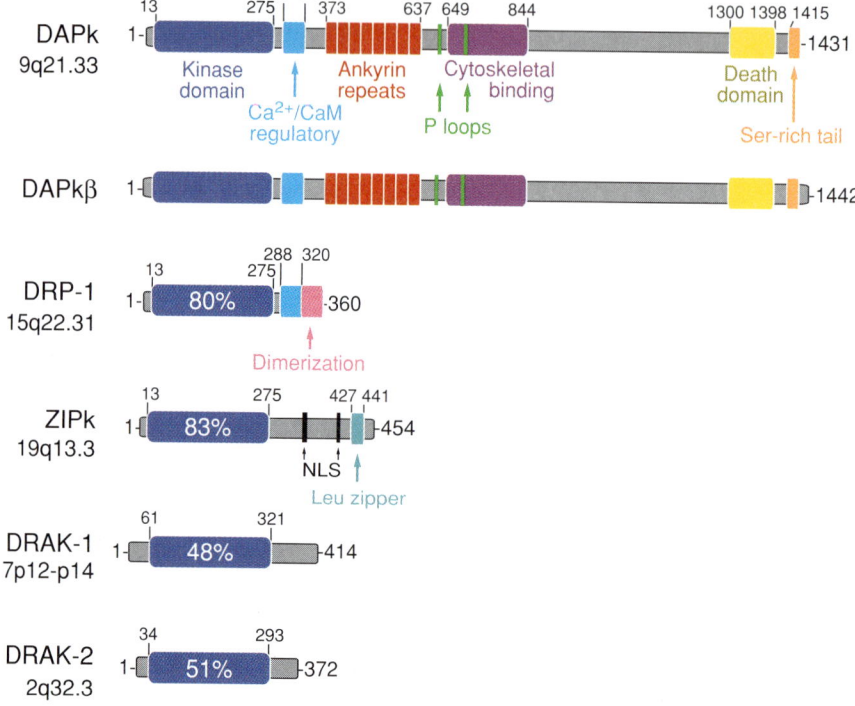

Figure 1
The DAPk family. Shown are the protein domains of the members of the DAPk family. The numbers above the proteins demarcate the amino acid position of each domain. The quantities within the kinase domains indicate the degree of amino acid identity to the kinase domain of DAPk. Chromosomal designations for the human orthologues are indicated. Abbreviation: NLS, nuclear localization signal.

aa: amino acids

invoved in other functions, as detailed below in the section on Mechanisms of Regulation.

The kinase domains of DAPk and DRP-1 are followed by a CaM autoregulatory/binding segment, which serves to suppress catalytic activity by binding to the catalytic cleft, and functioning as a pseudo-substrate (20, 21; for a detailed discussion, also see 10 and 22). In addition, this domain undergoes autophosphorylation at Ser308, an inhibitory event that reduces its affinity to CaM and may further stabilize its docking within the substrate-binding site (20, 21). Hence, activation of DAPk and DRP-1 requires two events: First, as in all CaM-dependent kinases, binding of Ca^{2+}-activated CaM to the autoregulatory/CaM-binding segment pulls this domain out from the catalytic cleft (1, 13, 14). Second, and unique to DAPk and DRP-1, dephosphorylation of Ser308 increases the affinity for CaM and, as a consequence, promotes catalytic activity at low CaM levels (20, 21). Interestingly, a low level of catalytic activity is detected upon Ser308 dephosphorylation of DAPk even in the absence of CaM (21). In DRP-1's case, dephosphorylation of Ser308 promotes homodimerization, which is also critical for CaM binding and full functional activation in cells (20). In support of this model, deletion of the CaM-binding domain from either DAPk or DRP-1, or substitution of Ser308 to Ala, generates constitutively active kinases, which exhibit greater Ca^{2+}-independent catalytic activity in vitro and stronger killing potential in vivo (1, 13, 14, 20, 21).

The third member, ZIPk, lacks the CaM regulatory domain, and thus its regulation differs substantially from DAPk and DRP-1. Recent work has demonstrated that the catalytic activity of ZIPk may be regulated through positive autophosphorylation. ZIPk undergoes autophosphorylation on multiple sites both in vitro and in vivo, and three of these were mapped by a combination of phosphopeptide analysis and mutagenesis to the catalytic domain (Thr180, Thr225, Thr265) (23). Individual mutation of these residues

Table 1 Substrates of DAPk family members and proposed function of phosphorylation

Substrate	Kinase	Sequence	In vivo[a]	Proposed function
DAPk	DAPk	FAARKKWKQS^{308}VRL	Yes	Inhibits catalytic activity
DRP-1	DRP-1	QYVRRRWKLS^{308}FSI	Yes	Inhibits catalytic activity
ZIPk	ZIPk	GNEFKNIFGT^{180}PEF PFLGETKQET^{225}LTN VKDPKRRMT^{265}IAQ RKPERRRLKT^{299}TRL RLKEYT^{306}IKSHS^{311}SL	Yes	Enhances catalytic activity Localization
ZIPk	DAPk	RKPERRRLKT^{299}TRL RLKEYTIKS^{309}HS^{311}S^{312}L PPNNS^{318}YADFERFS^{326}K	—	Localization and oligomerization
MLC	DAPk ZIPk DRP-1	KKRPQRATS^{19}NVF and/or KKRPQRAT^{18}S^{19}NVF	Yes Yes —	Activates myosin, membrane blebbing
MYPT (MBS)	ZIPk	RQARQSRRST^{697}QGV	—	Activates myosin
CPI-17	ZIPk	AQRLGKRVLS^{12}KLQ GLQKRHARVT^{38}VKY	—	Activates myosin
p21Cip1	ZIPk	DSQGRKRRQT^{145}SMT	Yes	Protein stability
Mdm2[b]	ZIPk	STSSRRRAIS^{166}ETE	—	p53 regulation
Par-4	ZIPk	—	—	
Histone H3	ZIPk	RTKQTARKST^{11}GGK	—	Mitosis
H2A, H4		—	—	
ASF/SF2	ZIPk	—	—	Splicing
Syntaxin-1A	DAPk	SGIIMDSSIS^{188}KQA	Yes[c]	Synaptic vesicle/membrane fusion
α-, β-tubulin	ZIPk	—	—	Cytokinesis
CaMKK	DAPk	GSRREERSLS^{511}APG	Yes	Inhibits activation

[a]Indicates confirmation of phosphorylation in vivo by exogenous or endogenous kinase.
[b]Phosphorylated as peptide but not in the context of full-length protein.
[c]Indirect in vitro proof of in vivo phosphorylation.

to Ala greatly reduced the catalytic activity of ZIPk toward exogenous substrate in vitro and suppressed its ability to induce membrane blebbing and cell rounding in vivo (23). Of note, Thr180 is a conserved residue that lies within the universal kinase activation loop, and phosphorylation of this loop is often an important step in activation of the catalytic activity of many protein kinases (24). Although phosphorylation of Thr180 suggests that ZIPk may be subjected to a regulatory mechanism common to many kinases, such as CaM-regulated kinase I and IV (25), it is not known what signals, if any, regulate ZIPk's autophosphorylation. Interestingly, all three Thr residues are conserved in DAPk and DRP-1. Yet, there is no evidence to suggest that either DAPk or DRP-1 are positively regulated through phosphorylation of the catalytic domain. Moreover, the crystal structure of DAPk suggests that phosphorylation of the activation loop is unnecessary, because unlike many other kinases which display autoinhibited forms, DAP kinase's catalytic domain is constitutively in the active "closed" conformation even in the absence of a peptide substrate (18).

The extracatalytic domains. Beyond the common kinase domain, the family members

TNF: tumor necrosis factor

TGF: transforming growth factor

differ in structure. DAPk, at 160 kDa, contains a large C-terminal extension with multiple functional domains. A series of eight ankyrin repeats follows the catalytic domain, after which is a region that has been shown to direct the kinase to the actin cytoskeleton (1, 26). Two putative P loops of unknown function reside at aa 639–646 and 695–702 of DAPk, respectively, the second of which partially overlaps the cytoskeletal-interacting domain. Studies of exogenous and endogenous DAPk for the most part agree that DAPk is an actin-associated protein kinase. This was demonstrated by imaging live or fixed HeLa cells expressing a GFP-DAPk fusion protein, which localized to actin stress fibers and cortical actin filaments (26). Consistent with this localization, actin coimmunoprecipitates with exogenous DAPk in an almost stoichiometric manner (G. Shohat and A. Kimchi, unpublished data). Endogenous DAPk can be solubilized in mild detergent extraction buffers only upon pretreatment of cells with latrunculin B, an actin-depolymerizing agent (1). Additionally, a recent report indicated that upon induced formation of stress fibers in serum-stimulated NIH3T3 cells, endogenous DAPk localized to the actin stress fibers in the central region of the cell but not at the cell periphery (27).

The ankyrin repeats, a protein-protein interaction motif common to cytoskeletal proteins, are also necessary for the proper localization of the kinase to the actin stress fibers (26). In their absence, DAPk is sequestered to focal adhesions and does not induce cell death morphology. Further evidence in support of an important functional role for the ankyrin repeats is that a peptide from this region (aa 451–498) was isolated as a dominant-interfering fragment of DAPk death-promoting activity (28). Thus, presumably, the ankyrin repeats mediate an interaction with some second factor that is critical for localization and/or function. The ankyrin repeats were shown to interact by a yeast two-hybrid screen with a second ankyrin repeat-containing protein called DAPk-interacting protein 1 (DIP-1), and association of the full-length proteins was confirmed in vivo (29). DIP-1, an E3 ubiquitin ligase, was capable of ubiquitinating DAPk in vitro and in vivo upon co-overexpression with ubiquitin. Overexpression of DIP-1 led to a modest, somewhat unconvincing decline in endogenous protein levels, although this was more pronounced when large quantities of DIP-1's RING domains were used instead. However, the authors did not assess whether DAPk is ubiquitinated by endogenous DIP-1 or actually regulated at the level of protein stability. In fact, DIP-1 is orthologous to the *Drosophila* Mind bomb and mouse Mind bomb-1, which have been shown to mediate ubiquitination and subsequent endocytosis of ligands of Notch (30–32). Thus, DIP-1/Mind bomb's ubiquitin ligase activity can promote activities unrelated to proteasome-mediated degradation of its targets, suggesting that the functional significance of the DIP-1/DAPk interaction may be unrelated to protein stability and is, as of yet, unknown.

The C terminus of DAPk contains a death domain, followed by a 17-aa tail rich in Ser residues, a feature common to other death domain-containing proteins (33). Both the death domain and the Ser-rich tail of DAPk are critical regulatory elements; both were isolated as dominant-negative-interfering fragments of DAPk function (28). Expression of the death domain protected 293 cells from death by tumor necrosis factor (TNF)-α and Fas, and Hep3B cells from transforming growth factor (TGF)-β-induced apoptosis (2, 7). Similarly, ceramide-induced death of primary hippocampal neurons was inhibited by administration of a sythentic peptide corresponding to the Ser-rich tail (9). Deletion of the tail generated a gain-of-function mutant that exhibited a greater killing potential without any effect on in vitro kinase activity (28). Thus, the C-terminal tail negatively regulates the cellular functions of DAPk. In contrast, deletion of the death domain attenuated DAPk's killing ability in some settings (2, 27). This suggests that the death domain competes with the full-length kinase by sequestering a critical death domain-interacting partner.

Two proteins have been shown to independently interact with DAPk's death domain. The first is the extracellular signal-regulated kinase (ERK), which binds a docking site within DAPk's death domain (34). As will be discussed in more detail below, ERK phosphorylates DAPk on Ser735 within the cytoskeletal binding region, a catalytically stimulating modification that presumably depends on the interaction with DAPk's death domain (34). DAPk's death domain also mediates homotypic interactions with the UNC5H2 dependence receptor (35), a death domain-containing protein that induces cell death when unbound to its ligand, netrin-1. Interestingly, the death domains of DAPk and UNC5H2 are more closely related to each other than to other death domains, such as that of Fas (33), which may explain the selectivity of the DAPk/UNC5H2 death domain interaction. Death induced by expression of unliganded UNC5H2 was partially attenuated by the absence of DAPk in cells derived from knockout mice, indicating that the interaction between the two proteins is critical for UNC5H2's death effects (35). Netrin-1 binding to the receptor had no effect on its association with DAPk; however, removal of the ligand led to dephosphorylation of DAPk at Ser308 (35). Thus, the death domain mediates two interactions that serve, in different circumstances, to activate DAPk. Deletion of the death domain from DAPk, or competition with excess death domain, would disrupt these interactions and thus interfere with DAPk activation and cellular funtion.

Interestingly, an alternatively spliced isoform of DAPk, DAPkβ, which possesses a unique 12-aa extension following the Ser-rich tail, has been identified in the mouse (**Figure 1**) (36). An antibody raised to this unique C terminus recognizes a protein of the correct molecular weight in HeLa cell extract (37), suggesting that this second isoform exists in humans as well, although it has not been observed in human databases (Y. Shoval and A. Kimchi, unpublished observations). This longer kinase was reported to have anti-apoptotic activity and blocked TNFα-induced killing in HeLa and MDCK cells (36), suggesting that the C-terminal extension augments the inhibitory nature of the tail region. It should be noted, however, that DAPkβ functioned identically to the prodeath human DAPk when tested by independent investigators (S. Bialik and A. Kimchi, unpublished data). Considering that Jin et al. (36) did not see any difference in the two isoforms' in vitro kinase activity, or in their ability to facilitate TNFα-induced phosphorylation of an apoptosis-relevant substrate, myosin II regulatory light chain (MLC), the anti-apoptotic nature of DAPk requires further substantiation.

DRP-1, a 42-kDa protein, contains a 40-aa extension C-terminal to the CaM regulatory domain. This region is essential for homodimerization of the protein, a feature necessary for its functional activity and, in particular, for its ability to be regulated by CaM (13, 20). DRP-1 lacks any of the cytoskeletal features of DAPk, and consistent with this, exogenous DRP-1 was shown to be a soluble, cytoplasmic protein (13). In addition, Flag-tagged DRP-1 accumulated within double-membrane-enclosed autophagic vesicles (5). However, as there is a lack of efficient antibodies that recognize the endogenous protein, it is not known whether the endogenous DRP-1 behaves similarly.

The 52-kDa ZIPk possesses a C-terminal leucine zipper motif (11, 12). This domain mediates homodimerization and interactions between ZIPk and additional leucine zipper-containing proteins, such as ATF4, AATF, and rat CDC5 (12, 20, 23, 38, 39). There are also several regions that resemble nuclear localization signals (NLSs), two of which have been shown to be functional within the exogenous protein (40). Curiously, although ZIPk itself lacks DAPk's death domain and cytoskeletal-interacting region, it binds Par-4, a 38-kDa

ERK: extracellular signal-regulated kinase

MLC: myosin II regulatory light chain

IFN: interferon

death domain-containing protein that directly interacts with actin filaments (41, 42). Thus, the ZIPk/Par-4 complex can be thought of as a bipartite structural mimic of DAPk. These structural features of ZIPk predict both cytoplasmic and nuclear distributions, and in fact, overexpressed ZIPk has been observed either in the cytoplasm, the nucleus, or both (11, 12, 43, 44). In one system, coexpression of Par-4 shifted the localization of ZIPk from the nucleus to the cytoplasm, where the two proteins colocalized to actin filaments (41). The nuclear distribution of ZIPk has been reported as both diffuse and speckled, and associations with PML (promyelocytic leukemia protein) nuclear bodies (44), chromatin (12), centrosomes, centromeres during mitosis, and the contractile ring during cytokinesis have been observed (45, 46). The discrepancies among the data were initially attributed to the inherent problematic nature of studies using overexpressed protein and/or to reflect differences among the tested cell lines. Yet, immunostaining of endogenous rat ZIPk in REF-2A cells revealed both a diffuse nuclear distribution as well as an association with actin stress fibers, indicating that ZIPk can maintain both nuclear and cytoplasmic localizations within a single cell (47). Interestingly, in addition to full-length ZIPk, a smaller-molecular-weight form is present in lysates from these cells. This is similar in size to a 32-kDa ZIPk observed in bovine smooth muscle, which was isolated in association with myofibers (48). This small ZIPk is most likely a cellular proteolytic fragment of the full-length protein because there is no experimental evidence or sequence data to support the presence of alternate splicing (49). The small form was isolated as an active kinase, so it is predicted to terminate just beyond the catalytic domain and to lack the C-terminal portion of the extracatalytic region, including the NLS and the Leu zipper. In fact, it is reminiscent of a previously described artificial C-terminal truncation of the rat protein that localized to the actin cytoskeleton when overexpressed (40). Thus, the dual localization of ZIPk in these cells may reflect alternate behaviors of the two protein forms. Other posttranslation events, such as phosphorylation of ZIPk by DAPk, may also change its intracellular localization, as detailed below in the next section.

Mechanisms of Regulation of the DAPk Family

DAPk, DRP-1, and ZIPk are all ubiquitously expressed in numerous adult mouse and rat tissue (11, 12, 14, 36, 50, 51). DAPk is particularly abundant in the adult and embryonic brain, especially the hippocampus. The constitutive presence of these potentially lethal proteins in normal tissue necessitates tight regulation, which, on the one hand, maintains their silence in growing cells but, on the other hand, facilitates efficient activation in response to death signals. Most of the activation mechanisms characterized so far in DAPk and DRP-1 influence catalytic activity by targeting the CaM autoregulatory/binding segment. As stated above, relief from the inhibition by this domain requires binding of Ca^{2+}-activated CaM and dephosphorylation of Ser308. In fact, dephosphorylation of both DAPk and DRP-1 at Ser308 has been observed in vivo in response to certain death stimuli, such as C_6-ceramide, TNF-α, inhibition of mitochondrial respiration, unliganded UNC5H2, and interferon (IFN)-γ (20, 21, 35, 52; G. Shohat, and A. Kimchi, unpublished data). Thus, two important cellular factors are predicted to activate DAPk and DRP-1: elevation in intracellular Ca^{2+}, which is often observed during different sceranios of programmed cell death, and a death-regulated phosphatase that specifically dephosphorylates Ser308. This combination should confer the required specificity to allow activation of these kinases in limited circumstances and not in response to every local spike in intracellular Ca^{2+}. Clearly, identification of the putative phosphatase and its mode of activation during cell death would greatly contribute to our understanding of how these kinases are regulated in vivo.

Additional phosphorylation events, involving ERK and mitogen-activated protein kinase (MAPK) signaling cascades, have been shown to regulate DAPk activity. As mentioned above, ERK phosphorylates DAPk on Ser735 both in vitro and in vivo (34). Increased phosphorylation on Ser735 in vivo correlated with enhanced phosphorylation of a DAPk substrate, MLC, and increased killing activity by overexpressed DAPk. Interestingly, this was attributed to a direct enhancement of in vitro catalytic activity, although the mechanism by which phosphorylation of a residue within the distant cytoskeletal interacting domain can affect catalysis is not clear. A second study has recently demonstrated that a downstream effector of ERK, the p90 ribosomal S6 kinase (RSK), inhibited exogenous DAPk cellular activity by phosphorylation of Ser289 within the CaM-autoregulatory/binding segment (53). These two studies suggest that the same upstream signals can lead to reverse effects on DAPk activity. It is not known, in fact, whether these phosphorylations occur simultaneously; perhaps they are sequential modifications that act to transiently turn DAPk activity on or off. Alternatively, they may represent different adaptations to subtle variations in the upstream signaling events, which may require alternative cellular responses. Importantly, it is not yet known whether MAPK/ERK signaling modifies endogenous DAPk and whether phosphorylation at either Ser289 or Ser735 serves to regulate DAPk activity in response to physiologic life and death decisions.

Under certain circumstances, particularly in cells with minimal DAPk protein expression, transcriptional and/or translational mechanisms may be an important means of activation of the kinase. *DAPk* gene expression increases in response to TGFβ (7) and to stimuli that activate p53, such as DNA-damaging agents and oncogene expression (54). In fact, the *DAPk* promoter contains both a TGFβ-response element, which is activated, at least in part, by the Smad transcription factor family (7) and a p53-binding element (54). Other death triggers, such as IFN, C_6-ceramide, and expression of transforming oncogenes, induce DAPk expression as well (4, 9, 55).

Less is known about the mechanisms that activate ZIPk in response to stress signals. Recent work, however, has suggested that the cellular localization of ZIPk may be a critical determinant in ZIPk's cell death–promoting activity. Intial studies of ZIPk subcellular distribution, described in detail above, correlated ZIPk's killing activity with its presence in the cytosol (11, 12, 40, 43). The cytoplasmic distribution may be regulated through phosphorylation of ZIPk by DAPk. DAPk and ZIPk form a complex, which enables phosphorylation of DAPk on ZIPk, but not vice versa (43). Importantly, the interaction has been confirmed at the level of the endogenous kinases. Mutational analysis of the interacting domains indicated that it is mediated by the catalytic domains, with the fingerprint basic loop contributing to the heterodimerization (43). DAPk phosphorylates ZIPk on multiple sites, including Thr299, along with five serines clustered within the region spanning aa 309–326. Significantly, a ZIPk phosphorylation mimetic, in which the target residues were mutated to Asp, was preferentially retained in the cytosol and displayed higher cell-killing activity (23, 43). Interestingly, Thr299 is also a site of ZIPk autophosphorylation (23). This ZIPk/DAPk interaction suggests a means by which a death signal can be transferred from one kinase to another in a catalytic amplification loop.

DOWNSTREAM OF DAPk: A CLOSE LOOK AT DAPk FUNCTION

DAPk's primary function is, as its name suggests, to regulate cell death. There is, in fact, a strong body of evidence that shows that DAPk is an essential component of different cell death signaling pathways. As discussed above, endogenous DAPk undergoes activation in response to various death stimuli, and death

MAPK: mitogen-activated protein kinase

Type II (autophagic) cell death: the result of autodigestion of intracellular organelles, distinguished by the accumulation of double-membrane-enclosed autophagocytic vesicles

MEF (or REF): mouse (or rat) embryonic fibroblast

stimuli have been associated with an increase in DAPk catalytic activity (6, 52). ZIPk and DRP-1 are mostly guilty by association with their more famous relative DAPk. The evidence to support ZIPk's and DRP-1's roles as cell killers is rather circumstantial and is based mainly on their ability to promote cell death–related morphologies when overexpressed (5, 11, 13, 14, 23, 40, 41, 43, 56). Further investigation into their functions at the level of the endogenous kinases is required. This is especially demanded for ZIPk, which has been additionally shown to interact with and/or phosphorylate several nuclear factors, such as histone H3 and p21^{WAF1} (see **Table 1**), suggesting potential roles in transcription, splicing, chromatin modification, and cell cycle control (11, 38, 39, 45, 46, 57). Several functions unrelated to cell death have also been ascribed to DAPk. The sections below describe and evaluate the various functional arms of the kinases, focusing mainly on DAPk, for which there is more substantial data.

Death-Associated Protein Kinases: How DAPk Earned Its Name

DAPk was originally isolated in an unbiased antisense-based genetic screen for genes whose protein products were necessary for IFN-γ-induced death in HeLa cells (58). IFNs are multifunctional cytokines, which can, depending on the cell setting, exhibit antiproliferative, immunomodulatory, differentiation- or death-promoting properties (55). In the HeLa cell system used in the screen, IFN-γ led to a slow but very efficient cell death, classified as Type II (see below) (5, 58). The fact that a reduction in DAPk expression led to increased cell survival in long-term clonal viability assays was the first clue that DAPk is a death-promoting kinase. These DAPk antisense-expressing HeLa cells were also resistant to cell death induced by Fas (2). Numerous studies have since demonstrated that the dominant-negative DAPk K42A mutant, or interfering fragments such as the death domain or C-terminal tail segment, reduced the extent of cell death in response to multiple triggers, such as Fas and TNFα (2), ceramide signaling (6, 9), TGF-β (7), and expression of the UNC5H2 dependence receptor in the absence of its ligand (35). More elegant approaches to elimination of DAPk function involved generation of a mouse deleted of *DAPk* through gene targeting (4; and D. Gozuacik, T. Raveh, and A. Kimchi, unpublished data). Although no defects were observed in developmental cell death, DAPk's necessity for cell death was observed when primary cells derived from the knockout mouse were subjected to external stresses. For example, DAPk −/− hippocampal neurons were much more resistant than their WT counterparts to ceramide-induced apoptosis, triggered either directly by administration of C$_6$-ceramide or indirectly by activation of p75 neurotrophin receptors with nerve growth factor (NGF) (9). In addition, mouse embryonic fibroblasts (MEFs) derived from DAPk −/− mice showed decreased induction of p53 and p19ARF and, consequently, decreased levels of apoptosis in response to the hyperproliferative signals generated by forced expression of oncogenes such as c-myc and E2F (4). Likewise, UNC5H2-mediated cell death was partially attenuated in immortalized DAPk −/− MEFs (35). In the intact animal, retinal ganglion cells from DAPk −/− mice showed increased survival (79% vs 56% in matched controls) following administration of glutamate (59). Furthermore, renal tubular cells in which the DAPk catalytic domain was deleted by homologous recombination were more resistant to apoptosis in a mouse model of chronic obstructive uropathy (60). All of these studies prove independently the necessity of DAPk for cell death.

Still further evidence emerges from the cellular phenotype obtained by ectopic expression of DAPk, which has shown that DAPk is sufficient to induce cell death in various cell lines (1, 2, 4, 5, 7, 8). Furthermore, expression of DAPk enhanced ceramide- and UNC5H2-induced death (6, 35). Although constitutively active forms of the kinase (e.g., DAPk lacking the CaM-autoregulatory/binding segment) were used in many of these studies, the WT form of the kinase is also capable of inducing cell death, albeit to lower levels (1). In fact, the degree of cell death correlates with the levels of protein expressed; low levels, such as those achieved upon stable, inducible expression of DAPk, do not lead to an apparent cell death phenotype (36; O. Cohen and A. Kimchi, unpublished observations). Such low amounts of protein, albeit more accurately reflecting the physiologic situation, are more likely to be subject to the same regulatory control that keeps endogenous DAPk inactive in the absence of an activating stimulus. Expression of particularly high levels of DAPk, however, may bypass the inhibitory regulation and result in the production of active kinase that is capable of inducing cell death. Thus, although such experiments deal with nonphysiologic protein levels, they very likely reflect the behavior of the endogenous kinase upon activation by a death signal.

Molecular Dissection of the Death Pathways Induced by the DAPk Family

Closer inspection of the molecular and cellular events that occur as a consequence of DAPk activation or that are blocked by loss of DAPk function have linked DAPk to both caspase-dependent (Type I) and caspase-independent cell death (4, 5, 7). ZIPk, too, was shown to induce two distinct morphologic stages of cell death within one cell type: The first stage was not blocked by caspase inhibitors or Bcl-x_L, but a later terminal stage had all the features of Type I apoptosis (61). The emerging model is that the DAPk family members can be linked to various molecular pathways depending on the cellular setting, which culminate in different forms of cell death. Importantly, the functional death effects of DAPk and its family members require catalytic activity; catalytically inactive K42A mutants of DAPk, DRP-1, or ZIPk fail to induce a death phenotype (1, 11, 13, 14). Thus, phosphorylation of some specific substrate(s) is required for the death-inducing effects of the DAPk family (see **Table 1**). The main challenges now are to dissect the death pathways at the molecular level, to elucidate the specific substrates within each pathway, and to determine the factors that enable one pathway to predominate over another.

One cellular setting in which DAPk was linked to Type I (apoptotic) cell death is primary embryonic fibroblasts with a functional p53-signaling pathway. In WT, but not p53-null primary REFs and MEFs, activated DAPk triggered caspase-dependent Type I apoptosis (4). Furthermore, in the absence of DAPk, p53 was only partially upregulated in response to proliferative signals generated by oncogene expression, indicating that DAPk is an upstream activator of p53 along this particular pathway (4). In fact, DAPk expression induced p53 and its response genes in both transformed and primary fibroblasts (4, 8). This is particularly interesting, considering that DAPk itself is a transcriptional target of p53 (54), which indicates a signaling feedback loop in which DAPk and p53 can activate each other. These experiments show that p53 activation molecularly links DAPk to the classical death pathway involving mitochondrial-based activation of the caspase cascade.

DAPk's upregulation of p53 and subsequent induction of Type I cell death require the presence of p19ARF (4), an inhibitor of Mdm2, which normally promotes the ubiquitin-dependent degradation of p53 (62). DAPk can phosphorylate p19ARF in vitro, although this phosphorylation is not efficient (T. Raveh, S. Kahan-Reef, and A. Kimchi, unpublished observations). Intriguingly, ZIPk was shown to interact directly with Mdm2 and phosphorylated peptide derivatives of Mdm2 on Ser166 (57). Phosphorylation of the full-length protein, however, was extremely weak, and no kinetic parameters were presented to evaluate its efficiency. Thus, although p19ARF and Mdm2

Type I (apoptotic) cell death: involves caspase-mediated dismantling of the cell and nucleus into residual apoptotic bodies that are engulfed by neighboring cells

THE DUAL NATURE OF AUTOPHAGY

The manner in which autophagy contributes to the overall death outcome is controversial. At times, both apoptotic and autophagic processes contribute to cell death and may even be regulated by the same signaling molecules (64, 80, 81); although with some signals, Type II cell death is only observed when the more predominant apoptotic pathway is blocked (82). Autophagy, however, is also used by the cell to maintain homeostasis. The controlled breakdown of cellular components provides an emergency nutrient supply during times of cell stress, and removal of damaged organelles, such as mitochondria with perturbed membrane potential, can fend off more serious damage (63, 83). In these conditions, inhibition of autophagy actually enhances apoptotic cell death (84–86). Thus, the accumulation of autophagic vesicles in any death scenario may be causative to cell lethality or may be a futile attempt at rescue. The dual nature of autophagy may, in fact, explain why the death-inducing DAPk has occasionally shown anti-apoptotic behavior (36, 37). It is unclear when and how autophagy shifts from its cytoprotective role to a "point of no return." DAPk may be critical in this decision or may be an intrinsic component of the autophagic machinery in both circumstances.

are alluring substrates, there is little physiologic evidence to support the connection between DAPk and p53 in this manner. An additional means of activating p53 independently of p19ARF may derive from DAPk's ability to inhibit integrin signaling (8) (see the Death-Related Effects on the Cytoskeleton section below for further discussion). Expression of DAPk in NIH3T3 cells (which lack an intact p19ARF locus), grown without serum on fibronectin, led to loss of adhesion and induction of p53. Forced activation of integrins blocked DAPk's effects on both adherence and p53. Thus, at least in this system, p53 activation and, presumably, subsequent activation of the mitochondria-dependent apoptotic pathway were responses to DAPk's inhibition of integrin-mediated adhesion. Yet the links between DAPk and the integrins, as well as between the integrins and p53, are still undefined.

At times, DAPk can be connected to the classical apoptotic route in a p53-independent manner. For example, treatment of p53-null Hep3B hepatoma cells with TGF-β led to induction of DAPk expression and subsequent mitochondrial-dependent apoptosis, which was significantly attenuated by DAPk inactivation (7). Ectopic expression of DAPk in these cells to levels similar to those induced by TGFβ triggered Type I apoptosis, including DNA fragmentation. Thus, several roads link DAPk to mitochondrial-based caspase activation processes.

In other cellular settings DAPk, DRP-1, and ZIPk induce an alternate type of programmed cell death, referred to as Type II or autophagic cell death, which occurs independently of caspases (5, 43). The most prominent feature of this type of cell death is the formation of double-membrane-enclosed, intracellular autophagic vesicles that consume organelles and other cytoplasmic components (63). In fact, overexpression of the DAPk family members leads, in some cells, to the appearance of autophagic vesicles, extensive cell rounding and membrane blebbing, nuclear condensation without DNA degradation, and no measurable loss of mitochondrial membrane potential or release of cytochrome c (5). Furthermore, knockdown of DAPk by antisense RNA and inactivation of DRP-1 function by use of a dominant-negative kinase have demonstrated that the two proteins are necessary for autophagy, induced either by IFN-γ in HeLa cells or by steroid withdrawal and amino acid starvation of MCF-7 cells, respectively (5). The involvement of DAPk in Type II cell death may explain why some studies failed to see DAPk's death-inducing function; these studies relied on assays for activation of caspases and caspase-dependent apoptotic events, such as DNA fragmentation, and would have missed the autophagic component of DAPk expression (36).

The studies discussed in this section indicate that DAPk is capable of regulating both Type I apoptotic and Type II autophagic cell deaths, depending on the cell system and specific death signal. The extent to which DAPk contributes to Type I death, especially, depends on other signaling pathways, such as p53. Thus, in Hep3B cells, Type I death induced by TGF-β, but not by UV irradiation, required functional DAPk (7). Other studies showed that the presence of DAPk was necessary for Type II death by IFN-γ but not for Type I apoptotic death by TNFα (5, 37). Interestingly, it is now recognized that some classical apoptotic stimuli, such as etoposide and ceramide, can lead to autophagy (64, 65). DAPk may be particularly important for these signals. Considering this, it is important when analyzing DAPk-mediated death to assess the type of death with specific markers for individual death-associated events, such as autophagosome formation, mitochondrial depolarization and permeabilization, caspase activation, and DNA fragmentation, rather than rely on measurements of overall cell viability or morphologic changes that are common to both pathways.

Death-Related Effects on the Cytoskeleton

One of the most prominent features of DAPk family-induced death is the effect on the cytoskeleton. This is manifested in multiple ways, including cell rounding, loss of matrix attachment, and membrane blebbing. Expression of DAPk, DRP-1, or ZIPk leads to cell rounding and the formation of spherical membrane blebs at the cell surface, with ultimate loss of adherence to the matrix (1, 5, 8, 11, 13, 14). In more adherent cell lines, DAPk and ZIPk have been shown to induce the extrusion of lamellipodia-like protrusions and remodeling of actin stress fibers into ring-like structures at the periphery of the cell (26, 41, 56). Importantly, expression of DRP-1 K42A or the death domain of DAPk blocked the extent of membrane blebbing during TNFα-induced apoptosis, even though this did not block overall cell death (5). Thus, the DAPk family is not only sufficient to induce blebbing but is also necessary in some cases for the blebbing phenomenon that accompanies cell death.

Membrane blebbing is a common morphologic feature of apoptosis, brought about by increased myosin contractility, resulting from phosphorylation of MLC, in conjunction with weakening of the structural integrity of the cortical actin network owing to proteolysis of many of its components (66–68). Consistent with their high homology to MLCK, DAPk and ZIPk are capable of phosphorylating MLC in vitro, both in isolation and as part of the intact myosin II molecule (1, 12, 26, 36, 69, 70). For DAPk, this phosphorylation occurs mainly on Ser19, which is critical for myosin II activation, and to a lesser degree on Thr18 (26, 27, 36), whereas ZIPk phosphorylates both sites efficiently (69, 70). DRP-1 also possesses MLC kinase activity, although this has so far been demonstrated only in vitro with isolated myosin light chain (13, 14).

Importantly, both DAPk and ZIPk are well positioned within the cell to encounter MLC as a substrate. DAPk, in particular, is found in close proximity to myosin within the blebbing cell, including a ring-like structure at the periphery of the cell body and the base of the blebs (26), and can bind to recombinant GST-MLC in vitro (27). As stated above, ZIPk, under certain conditions, can localize to actin filaments (40, 41, 47), where it encounters myosin and leads to enhanced MLC phosphorylation (42). Significantly, expression of mouse DAPk in TNFα-treated MDCK cells led to enhanced phosphorylation of endogenous MLC (36). Furthermore, in 293T, HeLa, and NIH3T3 cells, expression of DAPk alone was sufficient to phosphorylate endogenous MLC, whereas expression of ZIPk in HeLa cells induced its diphosphorylation, implying that both DAPk and ZIPk are capable of phosphorylating the light chain in vivo (26, 27, 56). DAPk- and ZIPk-mediated phosphorylation of MLC was not affected by the presence of Y27632, an inhibitor of the Rho-activated kinase (ROCK) (26, 47), indicating that MLC phosphorylation is not

ROCK: Rho-activated kinase

an indirect result of activation of ROCK by these kinases. At least in the case of ZIPk, the morphologic changes and rearrangements of the actin cytoskeleton that accompany its overexpression were directly attributed to myosin phosphorylation because they were blocked by expression of a form of MLC that can not be phosphorylated (56). Conversely, expression of the dominant-negative DAPk K42A or siRNA to ZIPk led to reductions in the basal levels of MLC phosphorylation (27, 47), indicating that the endogenous kinases also phosphorylate myosin. Thus, although not excluding the existence of additional substrates, phosphorylation of MLC and activation of myosin at the stress fibers may account for the contractile forces that result in membrane blebbing and/or protrusion formation.

ZIPk, like ROCK, can also negatively regulate the myosin phosphatase responsible for reverting MLC to its inactive, dephosphorylated form. In smooth muscle, ZIPk (in its full-length or truncated form) interacts with and phosphorylates the myosin-binding subunit (MBS) MYPT1 of the smooth muscle myosin phosphatase (48, 49), although not nearly as efficiently as it phosphorylates MLC (70, 71). This phosphorylation, which occurs on Thr697, inactivates the phosphatase. ZIPk can also phosphorylate CPI-17, an inhibitor of the phosphatase whose function is enhanced by phosphorylation (72). Thus, ZIPk can lead to increased phosphorylation of MLC, either through direct phosphorylation on Ser19 and Thr18 or indirectly by inhibiting the activity of the myosin phosphatase. It should be noted, however, that in vivo the direct route may be more prominent because knockdown of ZIPk in NIH3T3 cells by siRNA, which effectively reduced MLC phosphorylation, had little effect on the phosphorylation of the MBS (47).

DAPk and ZIPk can now be added to a growing list of kinases that phosphorylate MLC, which includes MLCK and ROCK, both of which have been implicated in apoptotic membrane blebbing (66, 67, 73). From kinetic analysis alone, MLCK is the best candidate (K_m, 5–12 µM; K_{cat}, 21–62 s^{-1}). ZIPk, too, is an adequate MLC kinase, with a less effective K_{cat} (0.74 s^{-1}), but a lower K_m (1.3 µM) (56). DAPk, in contrast, is much less efficient at phosphorylating either the intact MLC or a peptide derived from the phosphorylation site (19, 34, 36). ROCK is also a relatively inefficient MLC kinase, and it is believed to exert its effects on MLC phosphorylation preferentially through phosphorylation of the myosin phosphatase (74). Yet, in vivo the cell may purposely utilize the different kinetic properties of the various MLC kinases to achieve different contractile activities (75). For example, ROCK and MLCK were shown to have differential spatial roles in 3T3 cells, with both kinases directly phosphorylating MLC in different regions of the actin network (75). Furthermore, ZIPk is unique in that it efficiently produces diphosphorylated MLC (69, 70), which is differentially distributed within the cell and may have different effects on myosin assembly and contractility, compared to monophosphorylated MLC (47). In motile fibroblasts, ZIPk was shown to be responsible for a major part of the MLC diphosphorylation, particularly at central stress fibers, whereas ROCK activity appeared to be primarily directed at regulation of the myosin phosphatase (47). Furthermore, in vivo local concentrations of substrate, kinase, and other regulators/activators may influence overall kinetics. For example, phosphorylation of DAPk by ERK reduces DAPk's K_m toward MLC to a very respectable 10–17 µM (34). Thus, it remains plausible that DAPk and ZIPk act as direct MLC kinases in vivo.

In addition to its effects on myosin contraction, DAPk also affects adhesion of the cells to the extracellular matrix (ECM) (8). This may be a result of indirect changes in the contractile nature of the actin cytoskeleton or, perhaps, of direct phosphorylation of a novel substrate involved in adhesion. DAPk-expressing 293T, NIH3T3, or MCF10A mammary epithelial cells exhibited decreased adherence to fibronectin or laminin; conversely, the dominant-negative K42A mutant led to enhanced spreading and attachment (8). The loss of adherence was attributed to inhibition of integrin signaling; decreased Tyr phosphorylation of Fak and paxillin was observed upon DAPk expression, and the anti-adhesion effects could be blocked by forcibly activating β1 integrins (8). For cells that are dependent on ECM signaling for viability, DAPk's ability to interfere with integrin function may directly contribute to cell death. In this regard, the anti-adhesion property of DAPk correlated with its ability to induce an anoikis-like apoptotic death in NIH3T3 or MCF10A cells plated on fibronectin in the absence of serum and/or growth factors, and forced activation of integrin signaling attenuated DAPk's apoptotic-inducing properties in these cells (8). Conversely, for cells that are capable of anchorage-independent growth, DAPk's effects on integrin activity may not be a direct cause of death but may account for the reduced adhesion and cell rounding observed upon activation of DAPk (26).

Cross Talk with Other Kinases

In addition to the direct consequences of DAPk activity on cell morphology, DAPk can also regulate several kinase cascades that may impact cell viability. DAPk interacts with CaM-regulated kinase kinase (CaMKK)β and phosphorylates it on Ser511 with an impressive K_m of 3 μM (76). Significantly, expression of the DAPk catalytic domain in neuroblastoma cells led to enhanced phosphorylation of endogenous CaMKK on Ser511. This phosphorylation site is located adjacent to the CaM regulatory domain, and preincubation with DAPk attenuated CaMKK's ability to undergo CaM-dependent autophosphorylation in vitro, which is important for activation of the kinase. Although this suggests a mechanism by which DAPk can inactivate CaMKK and, consequently, the survival pathways that lie downstream, there is no physiological evidence that DAPk regulates CaMKK signal transduction in vivo nor that such cross talk is critical to DAPk's cell-killing functions.

More substantial evidence supports a role for DAPk as a negative regulator of the pro-survival signals from the MAPK/ERK pathway. In addition to ERK's ability to phosphorylate and activate DAPk, DAPk can exert its influence on ERK by promoting its cytoplasmic retention, through some unknown mechanism, effectively sequestering it away from nuclear substrates that may be critical for cell survival (34). In this manner, DAPk and ERK cooperate in a feedback mechanism whereby ERK activates DAPk, which then promotes death by shutting down the ERK survival pathway. As an example, phorbol-12-myristate-13-acetate (PMA) treatment of D2 erythroblasts in suspension, but not adherent cells, led to cell death. In the nonadherent population, the DAPk/ERK interaction predominated, resulting in cytoplasmic accumulation of ERK, whereas DAPk and ERK did not interact to any appreciable degree in the adherent cells. It is not known what secondary signals, such as those generated by cell detachment, promote the DAPk/ERK interaction. Furthermore, it remains to be determined whether inactivation of the ERK survival pathway directly contributes to DAPk's death effects or whether it serves as a means to silence the opposition so that DAPk can execute the cell through direct phosphorylation of its substrates in the various functional arms described above.

Nonapoptotic Functions

Until now, DAPk and its relatives have been presented as death-promoting kinases. However, events such as membrane blebbing and loss of adhesion, although often associated with death, are not restricted to dead-end paths. For example, blebbing can occur in such diverse processes as mitosis, cell spreading, and differentiation of primary neurons (77). Furthermore, DAPk's ability to mediate MLC phosphorylation may impact additional myosin-dependent events, such as

ECM: extracellular matrix

CaMKK: CaM-regulated kinase kinase

cytokinesis. In support of this, DAPk-expressing MEFs and D122 Lewis lung carcinoma cells that did not die exhibited defects in cytokinesis, which resulted in multinucleation (4, 78). Also, exogenous ZIPk colocalized with phosphorylated MLC to the contractile ring during cytokinesis (45). Enhanced myosin contractility due to phosphorylation of MLC by DAPk was also shown to affect stress fiber formation. Expression of DAPk in serum-starved NIH3T3 cells promoted stress fiber formation, blocking the dissolution of stress fibers that occurs in response to serum starvation (27). This was attributed to MLC phosphorylation, as an inhibitor of myosin, and a MLC mutated at Ser19 blocked this phenotype. Conversely, DAPk was shown to be necessary for serum-induced stress fiber formation in NIH3T3 cells, and increases in DAPk catalytic activity were observed following serum stimulation (27). Similarly, depletion of ZIPk by siRNA in NIH3T3 fibroblasts perturbed stress fiber formation and also led to a reduction in focal adhesions (47). This actually contrasted with DAPk, whose overexpression in serum-stimulated cells, rather than depletion, led to focal adhesion disassembly (27). The combination of these effects on contractility, stress fibers, and adhesion may also influence cell migration. Fibroblasts with reduced ZIPk expression exhibited impaired migration toward a chemotactic factor on fibronectin-coated surfaces, a phenotype attributed to the altered stress fiber structure resulting from decreased myosin phosphorylation (47). Thus, the consequences of DAPk/ZIPk's cytoskeletal activities are important in a more global sense, and the DAPk family members may display more general cytoskeletal phenomena, some of which are specifically recruited during cell death.

In this regard, the DAPk family has been shown to phosphorylate substrates that are not a priori linked to apoptosis. DAPk interacts via its C terminus with syntaxin-1A, a component of the SNARE complex that mediates docking and fusion of synaptic vesicles with the plasma membrane (51). Significantly, DAPk phosphorylates syntaxin-1A in a Ca^{2+}-dependent manner on Ser188. Identification of a nonapoptotic substrate involved in neuronal synaptic transmission may account for DAPk's abundance in adult hippocampal neurons. Phosphorylated syntaxin-1A had reduced binding to Munc18-1, a regulator that prevents formation of the SNARE complex (51). Although this could be a possible mechanism by which DAPk regulates vesicle docking or fusion, no effect on the assembly or stability of the SNARE complex was seen, and no functional relevance of this phosphorylation was demonstrated. However, an RNA-interference-based screen of the human kinome for regulators of endocytosis identified DAPk and DRP-1 as kinases that were necessary for clathrin-mediated endocytosis (79). Knockdown of either gene resulted in the accumulation of early/late endosomes underneath the cell membrane. At the same time, caveolae-mediated and lipid raft-mediated endocytosis was enhanced. This study functionally links DAPk to vesicular trafficking, although it does not address the mechanism or specific stage involved. The two studies together suggest that DAPk may be a general regulator of vesicle fusion. Furthermore, under the right circumstances, DAPk's ability to modulate membrane trafficking may be recruited to mediate related death processes such as autophagy, which involves numerous membrane fusion events, from the starting point when membrane components are recruited to form vesicles to the final stages in which early autophagosomes fuse with lysosomes (63). Thus, the physical and functional interactions of DAPk with components of the endocytotic and exocytotic pathways may be an enlightening line of research into multiple DAPk functions.

The DAPk Signaling Network, an Overview

Although gaps remain in our understanding of the DAPk family signaling network, and certain models await further verification, a schematic map can now be assembled, which includes both upstream regulators and functional arms that emanate from the active kinases (**Figure 2**). For simplicity's sake, we will refer to DAPk as the representative of the family. DAPk can be activated by several mechanisms (**Figure 2a**), including binding of Ca^{2+}-activated CaM, phosphorylation of Ser735 by ERK, and dephosphorylation of Ser308 by an unknown phosphatase, which can be activated by several death signals. Other triggers lead to increases in DAPk expression through p53- and/or Smad-mediated transcriptional upregulation of the *DAPk* gene. Once activated, DAPk can trigger a range of death responses leading to multiple phenotypes (**Figure 2b**, *I–V*). DAPk predominantly triggers the formation of autophagic vesicles, which results in a slow Type II death (I). The substrates responsible for this phenotype are unknown, although syntaxin or other related proteins that are involved in vesicle fusion are possible candidates. Autophagy can be accompanied by cytoskeletal rearrangements and global contractility (II), most likely caused by direct phosphorylation of cytoskeletal substrates, including, but not necessarily restricted to, MLC. Through these contractile changes, and/or phosphorylation of unknown substrates, DAPk inactivates integrins and blocks their signaling, leading to decreased adhesion to cell matrix and, consequently, cell rounding (III). In the proper signaling context, alternate pathways can also be initiated. For example, in cells where ECM provides an essential survival signal and p53 is present, inhibition of integrins can additionally lead to anoikis, a caspase-dependent apoptotic pathway initiated by loss of matrix adherence (IV). p53 also serves to link DAPk to the caspase-dependent, mitochondrial-based death pathway through the p19ARF regulatory loop, although the identity of the exact in vivo substrate in this pathway is not yet known. Once activated, p53 triggers a caspase-dependent Type I death process involving the hallmarks of apoptosis, such as DNA degradation, and cellular fragmentation, as well as membrane blebbing and cell rounding (V). This pathway can also be activated by DAPk independently of p53. In addition, DAPk can potentially promote death by shutting off survival pathways that function in parallel, such as those activated by ERK and CaMKK, the former through cytoplasmic retention of ERK, and the latter through direct phosphorylation of CaMKK. The choice of which of these pathways will be followed will ultimately depend on the cell context, the initial substrates that DAPk encounters, and the presence or abundance of factors that interact with or modulate these substrates and the signaling pathways that they regulate. Certainly, many additional layers of cross interaction among the functional arms are still to be uncovered.

In addition to the multiple signaling pathways emanating from DAPk, an additional level of complexity arises from the presence of multiple family members. There seems to be cross talk among the family members because DAPk, DRP-1, and ZIPk are all capable of interacting through their respective kinase domains and because DAPk can trans phosphorylate ZIPk (43; G. Shani, L. Marash, H. Berissi, and A. Kimchi, unpublished data). Furthermore, the levels of killing achieved by coexpression of low, nonfatal quantities of DAPk and WT ZIPk, but not the nonphosphorylatable ZIPk mutant, were much greater than the additive effect of expressing either one alone, indicating that the two synergize to induce cell killing (43). This suggests a hierarchical relationship among the kinases, with perhaps the most downstream kinase, ZIPk, acting as the effector kinase. Alternatively, the family members may collaborate in parallel to activate a common pathway, mediated by phosphorylation of shared substrates. In this case, the cross talk between DAPk and ZIPk may reflect an internal amplification loop within the overall network to ensure maximal phosphorylation of the common

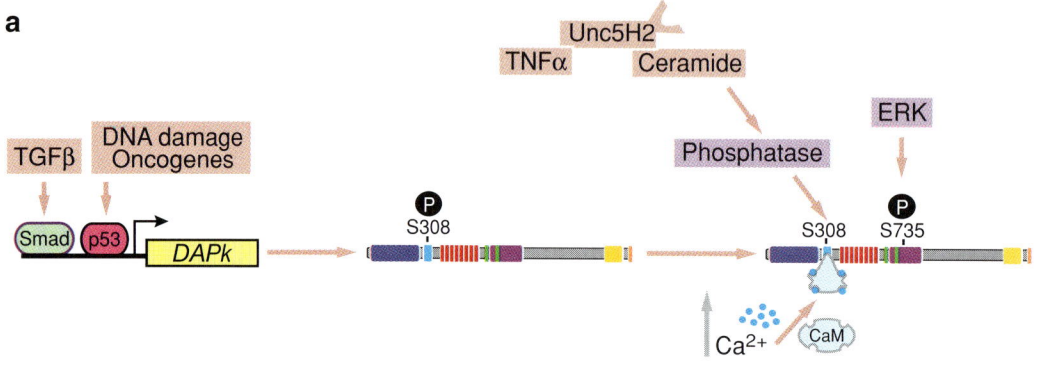

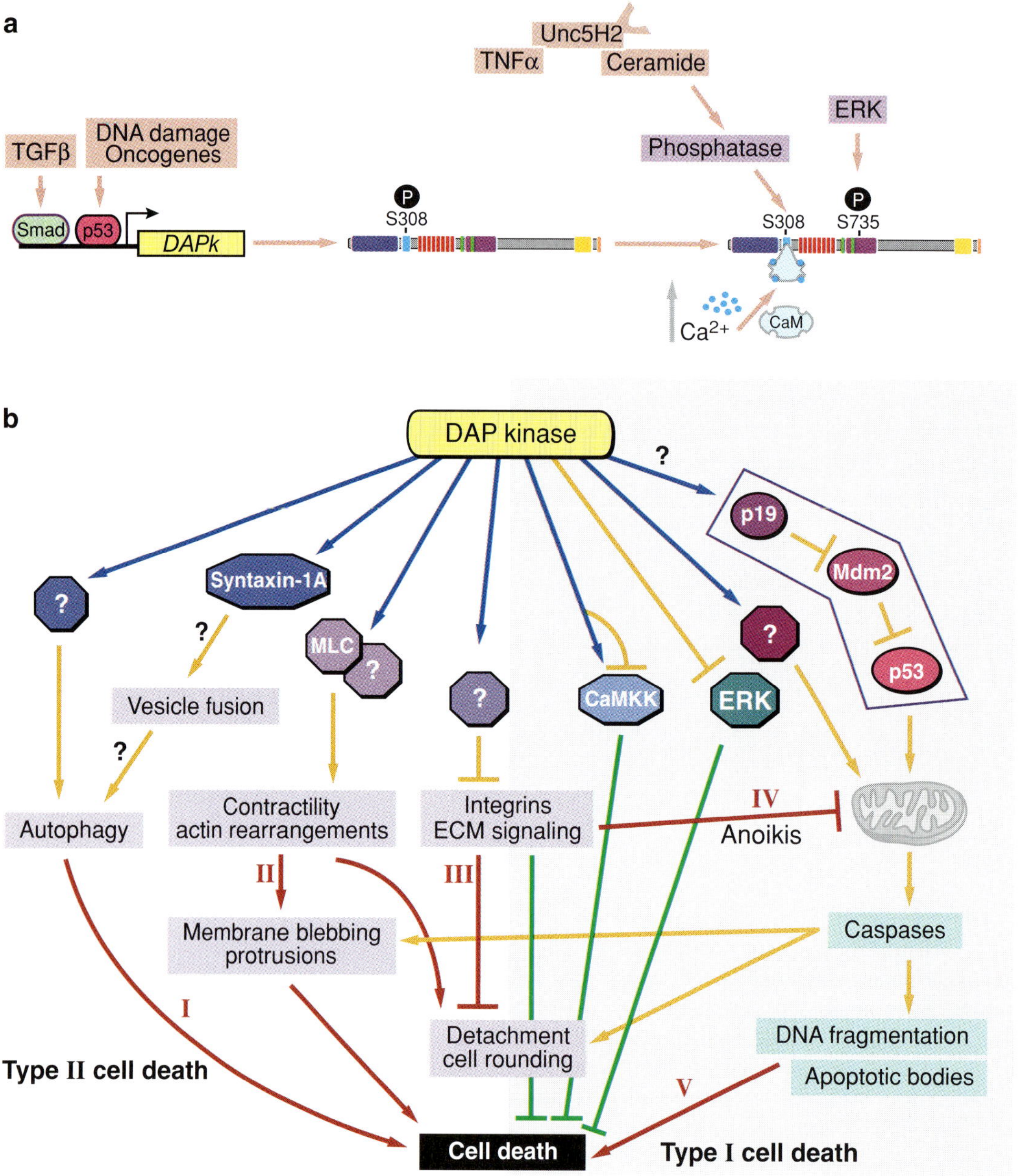

Figure 2

(*a*) Regulation of DAPk. DAPk is regulated by multiple signals, either at the level of transcription or at the protein level. Transcriptional activators include Smad and p53, which are each activated by death signals. Other signals lead to changes in the phosphorylation status of Ser308 and Ser735 or to increases in Ca^{2+}, which enables binding of CaM, serving to activate DAPk's catalytic activity. (*b*) The DAPk death signaling network. DAPk regulates numerous functional arms through direct phosphorylation of multiple substrates (*blue arrows*). These pathways result in different phenotypes (*red arrows*, I–V), culminating in cell death. Additional signaling events (*yellow arrows*) and survival pathways (*green arrows*) are indicated. Caspases can also contribute to the membrane blebbing and cell rounding phenotypes through cleavage of cytoskeletal proteins and activation of the MLC kinase, ROCK. See the text for further details.

substrates. A third possibility is that each kinase phosphorylates a unique set of substrates, leading to different cellular outcomes, with coordinated activity among the three kinases to ensure the complete global phenotype. These issues, as well as the complete in vivo substrate profile of each kinase, still await resolution.

SUMMARY POINTS

1. DAPk shares structural and functional homology with a family of death-related kinases, including DRP-1 and ZIPk.
2. The DAPk family members regulate both Type II autophagic cell death and Type I apoptotic death, depending on the cellular setting.
3. DAPk activity leads to multiple phenotypes, including membrane blebbing, autophagosome formation, and loss of adhesion, which results from phosphorylation of specific substrates, such as MLC.
4. Functions and substrates unrelated to cell death have also been identified, underscoring the diversity of the functional arms that this kinase family mediates.

FUTURE ISSUES TO BE RESOLVED

1. Elucidation of the complete substrate profiles and specificities for DAPk, DRP-1, and ZIPk is needed. Although attempts have been made to predict DAPk substrates using a positional-scanning peptide library (19), the optimal sequence obtained actually does not match the phosphorylation sites of some of the known DAPk substrates, and thus this method is a less than adequate tool. New approaches to substrate identification are necessary, and these need to be followed by rigorous assessment of the functional consequences of substrate phosphorylation.
2. The cross talk among the DAPk family members requires further elucidation, as does the assignment of specific roles to each kinase.
3. It is essential to develop methods of gene deletion and/or knockdown, individually and in combination, in order to assess the function and regulation of the endogenous kinases and their substrates in vivo.
4. An understanding of how each kinase contributes to the development of diseases such as cancer and neurodegeneration is necessary. Although this has been the subject of much interest for DAPk, the contribution of DRP-1 and ZIPk to tumorigenesis, or to other pathological disorders, is not known, and analysis of epigenetic modifications of their loci has not yet been attempted.

ACKNOWLEDGMENTS

This work was supported by grants from the European Union (LSHB-CT-2004-511983) and by the Center of Excellence grant from the Flight Attendant Medical Research Institute (FAMRI). A.K. is the incumbent of Helena Rubinstein Chair of Cancer Research.

LITERATURE CITED

1. Cohen O, Feinstein E, Kimchi A. 1997. *EMBO J.* 16:998–1008
2. Cohen O, Inbal B, Kissil JL, Raveh T, Berissi H, et al. 1999. *J. Cell Biol.* 146:141–48
3. Inbal B, Cohen O, Polak-Charcon S, Kopolovic J, Vadai E, et al. 1997. *Nature* 390:180–84
4. Raveh T, Droguett G, Horwitz MS, DePinho RA, Kimchi A. 2001. *Nat. Cell Biol.* 3:1–7
5. Inbal B, Bialik S, Sabanay I, Shani G, Kimchi A. 2002. *J. Cell Biol.* 157:455–68
6. Yamamoto M, Hioki T, Ishii T, Nakajima-Iijima S, Uchino S. 2002. *Eur. J. Biochem.* 269:139–47
7. Jang CW, Chen CH, Chen CC, Chen JY, Su YH, Chen RH. 2002. *Nat. Cell Biol.* 4:51–58
8. Wang WJ, Kuo JC, Yao CC, Chen RH. 2002. *J. Cell Biol.* 159:169–79
9. Pelled D, Raveh T, Riebeling C, Fridkin M, Berissi H, et al. 2002. *J. Biol. Chem.* 277:1957–61
10. Bialik S, Kimchi A. 2004. *Semin. Cancer Biol.* 14:283–94
11. Kawai T, Matsumoto M, Takeda K, Sanjo H, Akira S. 1998. *Mol. Cell. Biol.* 18:1642–51
12. Kogel D, Plottner O, Landsberg G, Christian S, Scheidtmann KH. 1998. *Oncogene* 17:2645–54
13. Inbal B, Shani G, Cohen O, Kissil JL, Kimchi A. 2000. *Mol. Cell. Biol.* 20:1044–54
14. Kawai T, Nomura F, Hoshino K, Copeland NG, Gilbert DJ, et al. 1999. *Oncogene* 18:3471–80
15. Sanjo H, Kawai T, Akira S. 1998. *J. Biol. Chem.* 273:29066–71
16. Sakagami H, Kondo H. 1997. *Brain Res. Mol. Brain Res.* 52:249–56
17. McGargill MA, Wen BG, Walsh CM, Hedrick SM. 2004. *Immunity* 21:781–91
18. Tereshko V, Teplova M, Brunzelle J, Watterson DM, Egli M. 2001. *Nat. Struct. Biol.* 8:899–907
19. Velentza AV, Schumacher AM, Weiss C, Egli M, Watterson DM. 2001. *J. Biol. Chem.* 276:38956–65
20. Shani G, Henis-Korenblit S, Jona G, Gileadi O, Eisenstein M, et al. 2001. *EMBO J.* 20:1099–113
21. Shohat G, Spivak-Kroizman T, Cohen O, Bialik S, Shani G, et al. 2001. *J. Biol. Chem.* 276:47460–67
22. Shohat G, Spivak-Kroizman T, Eisenstein M, Kimchi A. 2002. *Eur. Cytokine Netw.* 13:398–400
23. Graves PR, Winkfield KM, Haystead TA. 2005. *J. Biol. Chem.* 280:9363–74
24. Nolen B, Taylor S, Ghosh G. 2004. *Mol. Cell* 15:661–75
25. Soderling TR. 1999. *Trends Biochem. Sci.* 24:232–36
26. Bialik S, Bresnick AR, Kimchi A. 2004. *Cell Death Differ.* 11:631–44
27. Kuo JC, Lin JR, Staddon JM, Hosoya H, Chen RH. 2003. *J. Cell Sci.* 116:4777–90
28. Raveh T, Berissi H, Eisenstein M, Spivak T, Kimchi A. 2000. *Proc. Natl. Acad. Sci. USA* 97:1572–77
29. Jin YJ, Blue EK, Dixon S, Shao ZL, Gallagher PJ. 2002. *J. Biol. Chem.* 277:46980–86
30. Koo BK, Lim HS, Song R, Yoon MJ, Yoon KJ, et al. 2005. *Development* 132:3459–70
31. Lai EC, Roegiers F, Qin XL, Jan YN, Rubin GM. 2005. *Development* 132:2319–32
32. Koo BK, Yoon KJ, Yoo KW, Lim HS, Song R, et al. 2005. *J. Biol. Chem.* 280:22335–42
33. Feinstein E, Kimchi A, Wallach D, Boldin M, Varfolomeev E. 1995. *Trends Biochem. Sci.* 20:342–44
34. Chen CH, Wang WJ, Kuo JC, Tsai HC, Lin JR, et al. 2005. *EMBO J.* 24:294–304
35. Llambi F, Calheiros-Lourenco F, Gozuacik D, Guix C, Pays L, et al. 2005. *EMBO J.* 24:1192–201

36. Jin Y, Blue EK, Dixon S, Hou L, Wysolmerski RB, Gallagher PJ. 2001. *J. Biol. Chem.* 276:39667–78
37. Jin Y, Gallagher PJ. 2003. *J. Biol. Chem.* 278:51587–93
38. Engemann H, Heinzel V, Page G, Preuss U, Scheidtmann KH. 2002. *Nucleic Acids Res.* 30:1408–17
39. Page G, Lodige I, Kogel D, Scheidtmann KH. 1999. *FEBS Lett.* 462:187–91
40. Kogel D, Bierbaum H, Preuss U, Scheidtmann KH. 1999. *Oncogene* 18:7212–18
41. Page G, Kogel D, Rangnekar V, Scheidtmann KH. 1999. *Oncogene* 18:7265–73
42. Vetterkind S, Illenberger S, Kubicek J, Boosen M, Appel S, et al. 2005. *Exp. Cell Res.* 305:392–408
43. Shani G, Marash L, Gozuacik D, Bialik S, Teitelbaum L, et al. 2004. *Mol. Cell. Biol.* 24:8611–26
44. Kawai T, Akira S, Reed JC. 2003. *Mol. Cell. Biol.* 23:6174–86
45. Preuss U, Bierbaum H, Buchenau P, Scheidtmann KH. 2003. *Eur. J. Cell Biol.* 82:447–59
46. Preuss U, Landsberg G, Scheidtmann KH. 2003. *Nucleic Acids Res.* 31:878–85
47. Komatsu S, Ikebe M. 2004. *J. Cell Biol.* 165:243–54
48. MacDonald LA, Borman MA, Muranyi A, Somlyo AV, Hartshorne DJ, Haystead TA. 2001. *Proc. Natl. Acad. Sci. USA* 98:2419–24
49. Endo A, Surks HK, Mochizuki S, Mochizuki N, Mendelsohn ME. 2004. *J. Biol. Chem.* 279:42055–61
50. Yamamoto M, Takahashi H, Nakamura T, Hioki T, Nagayama S, et al. 1999. *J. Neurosci. Res.* 58:674–83
51. Tian JH, Das S, Sheng ZH. 2003. *J. Biol. Chem.* 278:26265–74
52. Shang T, Joseph J, Hillard CJ, Kalyanaraman B. 2005. *J. Biol. Chem.* 280:34644–53
53. Anjum R, Roux PP, Ballif BA, Gygi SP, Blenis J. 2005. *Curr. Biol.* 15:1762–67
54. Martoriati A, Doumont G, Alcalay M, Bellefroid E, Pelicci PG, Marine JC. 2005. *Oncogene* 24:1461–66
55. Chawla-Sarkar M, Lindner DJ, Liu YF, Williams BR, Sen GC, et al. 2003. *Apoptosis* 8:237–49
56. Murata-Hori M, Fukuta Y, Ueda K, Iwasaki T, Hosoya H. 2001. *Oncogene* 20:8175–83
57. Burch LR, Scott M, Pohler E, Meek D, Hupp T. 2004. *J. Mol. Biol.* 337:115–28
58. Deiss LP, Feinstein E, Berissi H, Cohen O, Kimchi A. 1995. *Genes Dev.* 9:15–30
59. Schori H, Yoles E, Wheeler LA, Raveh T, Kimchi A, Schwartz M. 2002. *Eur. J. Neurosci.* 16:557–64
60. Yukawa K, Hoshino K, Kishino M, Mune M, Shirasawa N, et al. 2004. *Int. J. Mol. Med.* 13:515–20
61. Kogel D, Reimertz C, Mech P, Poppe M, Fruhwald MC, et al. 2001. *Br. J. Cancer* 85:1801–8
62. Sherr CJ, Weber JD. 2000. *Curr. Opin. Genet. Dev.* 10:94–99
63. Gozuacik D, Kimchi A. 2004. *Oncogene* 23:2891–906
64. Feng ZH, Zhang H, Levine AJ, Jin S. 2005. *Proc. Natl. Acad. Sci. USA* 102:8204–9
65. Daido S, Kanzawa T, Yamamoto A, Takeuchi H, Kondo Y, Kondo S. 2004. *Cancer Res.* 64:4286–93
66. Mills JC, Stone NL, Erhardt J, Pittman RN. 1998. *J. Cell Biol.* 140:627–36
67. Sebbagh M, Renvoize C, Hamelin J, Riche N, Bertoglio J, Breard J. 2001. *Nat. Cell Biol.* 3:346–52
68. Mills JC, Stone NL, Pittman RN. 1999. *J. Cell Biol.* 146:703–8
69. Murata-Hori M, Suizu F, Iwasaki T, Kikuchi A, Hosoya H. 1999. *FEBS Lett.* 451:81–84
70. Niiro N, Ikebe M. 2001. *J. Biol. Chem.* 276:29567–74

71. Kiss E, Muranyi A, Csortos C, Gergely P, Ito M, et al. 2002. *Biochem. J.* 365:79–87
72. MacDonald JA, Eto M, Borman MA, Brautigan DL, Haystead TA. 2001. *FEBS Lett.* 493:91–94
73. Coleman ML, Sahai EA, Yeo M, Bosch M, Dewar A, Olson MF. 2001. *Nat. Cell Biol.* 3:339–45
74. Bresnick AR. 1999. *Curr. Opin. Cell Biol.* 11:26–33
75. Totsukawa G, Yamakita Y, Yamashiro S, Hartshorne DJ, Sasaki Y, et al. 2000. *J. Cell Biol.* 150:797–806
76. Schumacher AM, Schavocky JP, Velentza AV, Mirzoeva S, Watterson DM. 2004. *Biochemistry* 43:8116–24
77. Hagmann J, Burger MM, Dagan D. 1999. *J. Cell. Biochem.* 73:488–99
78. Raveh T, Kimchi A. 2001. *Exp. Cell Res.* 264:185–92
79. Pelkmans L, Fava E, Grabner H, Hannus M, Habermann B, et al. 2005. *Nature* 436:78–86
80. Mills KR, Reginato M, Debnath J, Queenan B, Brugge JS. 2004. *Proc. Natl. Acad. Sci. USA* 101:3438–43
81. Pyo JO, Jang MH, Kwon YK, Lee HJ, Jun JI, et al. 2005. *J. Biol. Chem.* 280:20722–29
82. Shimizu S, Kanaseki T, Mizushima N, Mizuta T, Arakawa-Kobayashi S, et al. 2004. *Nat. Cell Biol.* 6:1221–28
83. Priault M, Salin B, Schaeffer J, Vallette FM, di Rago JP, Martinou JC. 2005. *Cell Death Differ.* 12:1613–21
84. Kuma A, Hatano M, Matsui M, Yamamoto A, Nakaya H, et al. 2004. *Nature* 432:1032–36
85. Boya P, Gonzalez-Polo RA, Casares N, Perfettini JL, Dessen P, et al. 2005. *Mol. Cell. Biol.* 25:1025–40
86. Lum JJ, Bauer DE, Kong M, Harris MH, Li C, et al. 2005. *Cell* 120:237–48

Mechanisms for Chromosome and Plasmid Segregation

Santanu Kumar Ghosh, Sujata Hajra, Andrew Paek, and Makkuni Jayaram

Section of Molecular Genetics and Microbiology, University of Texas at Austin, Austin, Texas 78712-0612; email: ghoshsk@mail.utexas.edu, hajras@mail.utexas.edu, apaek@email.arizona.edu, jayaram@icmb.utexas.edu

Key Words

genome partitioning, cohesion, condensing, *par* loci, Rep proteins, STB

Abstract

The fundamental problems in duplicating and transmitting genetic information posed by the geometric and topological features of DNA, combined with its large size, are qualitatively similar for prokaryotic and eukaryotic chromosomes. The evolutionary solutions to these problems reveal common themes. However, depending on differences in their organization, ploidy, and copy number, chromosomes and plasmids display distinct segregation strategies as well. In bacteria, chromosome duplication, likely mediated by a stationary replication factory, is accompanied by rapid, directed migration of the daughter duplexes with assistance from DNA-compacting and perhaps translocating proteins. The segregation of unit-copy or low-copy bacterial plasmids is also regulated spatially and temporally by their respective partitioning systems. Eukaryotic chromosomes utilize variations of a basic pairing and unpairing mechanism for faithful segregation during mitosis and meiosis. Rather surprisingly, the yeast plasmid 2-micron circle also resorts to a similar scheme for equal partitioning during mitosis.

Contents

INTRODUCTION.................. 212
CHROMOSOME SEGREGATION
  IN BACTERIA................... 213
  Orientation and Dynamics of the
    Bacterial Nucleoid............. 213
  The Driving Force for DNA
    Segregation: A Replication
    Factory?..................... 213
  Cellular Addresses for
    Chromosomal Loci............ 215
  Do Nascent Duplexes Stay
    Associated During
    Replication?.................. 215
CHROMOSOME SEGREGATION
  IN EUKARYOTES............... 216
  Remembering Replication and
    Counting Chromosomes....... 217
  Molecular Basis for Chromosome
    Cohesion 217
  Separase: Not Just a Protease 220
  Pairing and Unpairing During
    Meiosis...................... 220
CONDENSINS AND
  CHROMOSOME
  SEGREGATION................. 222
  Chromosome Condensation and
    Bacterial DNA Segregation.... 223
  Mechanism of DNA Compaction
    by Condensin................. 224
  Roles for Condensin and Cohesin
    not Directly Related to
    Chromosome Segregation..... 224
SEGREGATION OF BACTERIAL
  PLASMIDS..................... 225
  Partitioning Systems of Low-Copy
    Plasmids..................... 226
  Mechanisms of Plasmid
    Partitioning.................. 227
  Centromeres and Par Proteins in
    the Segregation of Bacterial
    Chromosomes................ 229
PLASMID SEGREGATION IN
  YEAST 230
  The Cohesin Complex and Yeast
    Plasmid Partitioning.......... 231
  The Mitotic Spindle Promotes
    Recruitment of Cohesin by the
    Yeast Plasmid 231
  Models for Yeast Plasmid
    Segregation 232
THE CYTOSKELETON,
  DYNAMIC PROTEIN RINGS,
  COILS, AND OSCILLATORS
  IN BACTERIAL CELL
  DIVISION AND
  CHROMOSOME
  SEGREGATION................. 233
TOPOISOMERASES,
  SITE-SPECIFIC
  RECOMBINASES AND DNA
  TRANSLOCASES: FAITHFUL
  CHROMOSOME
  SEGREGATION AND
  GENOME MAINTENANCE ... 233
ADDENDUM..................... 235

INTRODUCTION

The faithful duplication and transmission of genetic information to daughter cells is a fundamental attribute of life. Yet, the process is fraught with spatial, topological, mechanical, and mechanistic challenges. Genomes present in multiple copies, i.e., certain bacterial plasmids, for example, can propagate stably by random segregation. They may maintain their steady-state copy number by appropriate replication control. Single-copy (or low-copy) genomes such as bacterial chromosomes and certain bacterial plasmids have to utilize active partitioning mechanisms to prevent missegregation. The task of faithful genome segregation is further magnified in bacteria and viruses with segmented genomes and in haploid eukaryotes. Diploid eukaryotes have the additional task, during mitosis, of guarding against a daughter cell receiving a pair of sister chromatids instead of a pair of chromosome homologues. Conversely, during meiosis, they must hold chromosome sisters together, segregate homologues during the first division, and subsequently segregate sisters during the second division.

In this review, our goal is to summarize the logic of evolutionary solutions to some of the problems outlined above. We rely primarily on *Escherichia coli* and *Saccharomyces cerevisiae* (the budding yeast) as model systems to illustrate general principles and unifying themes. Occasionally, we turn to other systems as well, especially when they harbor specialized mechanisms. Supplemental movies of plasmid and chromosome segregation and protein oscillations related to bacterial cell division are available. (Follow the Supplemental Material link from the Annual Reviews home page at **http://www.annualreviews.org**.)

CHROMOSOME SEGREGATION IN BACTERIA

The once held view that the bacterial nucleoid is an amorphous entity, perhaps the "last refuge of entropy," is no longer valid (1). The emerging picture reveals the nucleoid as a well-organized moiety within which chromosome replication and partitioning are carried out with remarkable spatial and temporal regulation. Some of the essential ingredients for accomplishing this feat are (*a*) the localization of replication to one or a small number of fixed sites, (*b*) the force from replication that directs daughter chromosomes away from each other, (*c*) the condensation of replicated regions of the chromosome at defined cell positions, and (*d*) the action of localized molecular machines to insure chromosome separation and movement away from the division septum.

Orientation and Dynamics of the Bacterial Nucleoid

The bacterial nucleoid has a finite orientation, with well-defined cellular locations for the replication origin and terminus (2, 3). Under slow-growth conditions, the origin (*oriC*) and terminus (*terC*) of the circular chromosome in a newborn *E. coli* cell are located at the nucleoid borders proximal to the old and new cell poles, respectively (4). They migrate to the mid-cell region, as if in preparation for the start of bidirectional DNA replication (**Figure 1a**) (5, 6). The duplicated *oriC*s move away from each other toward opposite poles. After completion of replication, chromosome segregation and cell division, the termini will be situated close to the new poles. Under fast-growth conditions, when the generation time is equal to or less than the replication time, the *E. coli* cell responds by increasing the frequency of *oriC* firing. Chromosome dynamics are also adjusted accordingly (**Figure 1b**). When cells divide every 20 min, a newborn will contain four *oriC*s. Of these, the two localized near the poles continue to more or less hold their position during cell growth. The other two move toward each other as replication proceeds and reach the mid-zone in the predivision cell. At this stage, the duplicated termini occupy the one-fourth and three-fourths positions or the mid-zones of the would-be daughter cells. After division, the choreography of coordinated replication and DNA movement repeats itself.

The Driving Force for DNA Segregation: A Replication Factory?

What provides the motive force for the directed chromosome movement? The earlier notion that the duplicated DNA molecules are attached to the membrane and move passively as the cell elongates (7) has been largely discounted by contrary evidence. The measured rate of chromosome movement in *Bacillus subtilis*, for example, is much faster than that of cell elongation: 0.17 μm min^{-1} versus 0.011 to 0.025 μm min^{-1} (8). In principle, an abundant and powerful cellular motor such as RNA polymerase could drive DNA translocation, provided the transcription machinery is constrained within the cell.

oriC: replication origin of bacterial chromosome

terC: replication terminus of bacterial chromosome

Consistent with this idea, an inhibitor of RNA polymerase prevents the separation of newly replicated duplex regions near the replication origin in B. subtilis (9). The biased orientation of transcription units away from the origin can impart the proper directionality to the segregation force.

Using similar reasoning, chromosome segregation may be promoted by physically constraining the replication machinery itself. A stationary replisome (also called a replication factory) located at the mid-cell position may function as a DNA pump, ingesting the parental duplex and extruding the daughter duplexes (**Figure 1c**). In the "extrusion-capture" model (10), the spooling effect produced by the replisome pushes the newly replicated molecules away from each other to be captured and positioned by an anchoring mechanism (**Figure 1c**). This positional information may be provided by the replication origin itself or sequences in its proximity in conjunction with a cognate protein or proteins. The pushing force generated by the replisome may be complemented and reinforced by the pulling force provided by proteins that condense DNA (discussed below). However, further assistance from a motor protein such as FtsK of E. coli (11, 12) is required to dispatch the nascent DNA emerging from the replisome toward the cell pole before it collapses into a random coil.

B. subtilis sporulation provides another example of active DNA transfer in a vectorial fashion. The RacA protein, which binds to short repeated sequences within a centromere-like region spanning the origin, is responsible for capturing sister chromosomes and anchoring them to opposite ends of the cell (13). During asymmetric patterning into a large mother cell and a smaller prespore cell, the sporulation septum traps only about 30% of the chromosome, spanning the origin and centromere, in the latter compartment (14, 15). The rest is pumped into the prespore by the SpoIIIE motor protein, a Maxwell's demon acting at the septum. The ATPase activity of SpoIIIE, stimulated by double-stranded

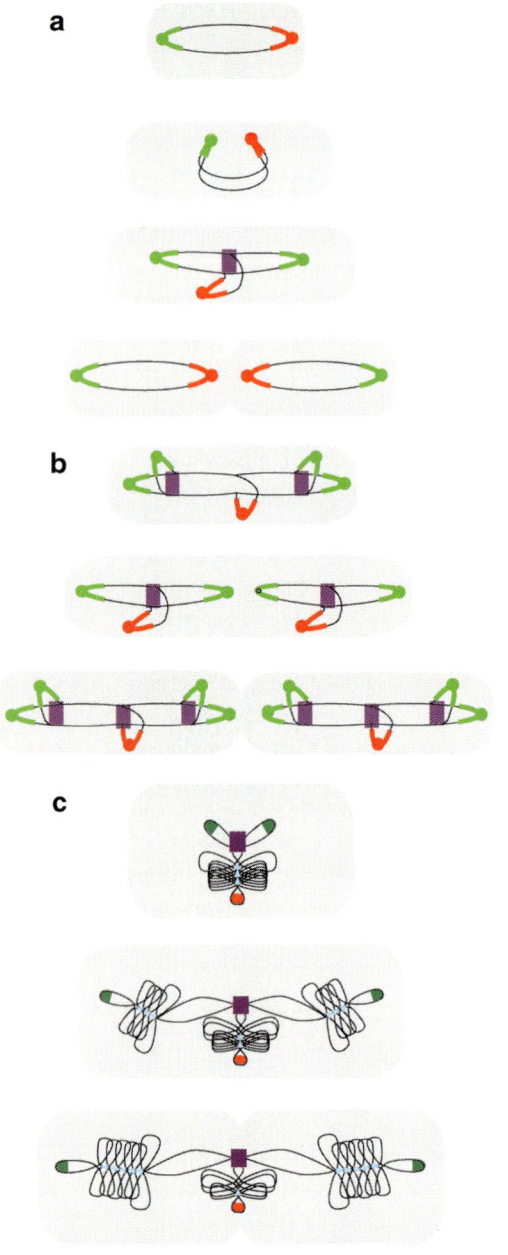

Figure 1

Dynamics of chromosome replication in E. coli. The movements of the replication origin (*green*) and terminus (*red*) in slow- (*a*) and fast- (*b*) growing E. coli cells are schematically shown. The squares (*purple*) indicate the replication factories. In the "extrusion and capture" model for chromosome segregation (*c*), the nascent duplexes emerging from the replication factory are ejected in opposite directions. The origins are captured near cell poles, and the rest of the DNA is condensed into the nucleoid. The diagram is adapted from Draper & Gober (5) with permission.

DNA, permits it to track along DNA (or to translocate DNA if the protein is stationary) (16). SpoIIIE is synthesized during vegetative phase as well and appears to enhance the fidelity of chromosome segregation by sweeping away any DNA trapped at the site of septum closure.

Cellular Addresses for Chromosomal Loci

Perhaps the most striking case of replication-coupled and ordered distribution of nascent duplexes in "future" daughter cells came from a sophisticated analysis carried out recently in *Caulobacter crescentus* by Viollier et al. (17). Among a set of 112 chromosomal loci examined by fluorescence tagging, each one had a specific subcellular address, and all were arrayed in a linear order along the long axis of the cell. Genes closest to the origin localized near the origin-proximal cell pole, those farthest from it localized to the terminus-proximal cell pole, and the rest lined up in between in a follow-the-leader fashion (**Figure 2**). Furthermore, time-lapse microscopy revealed that, as the origin and a set of 10 selected loci in the origin-proximal half of the chromosome were duplicated, the nascent DNA segments corresponding to each of them moved in a chronological order to their final destinations. Because replication was bidirectional, chromosome order in a resting cell and chromosome dynamics in a dividing cell were directly correlated to the temporal order in which each locus was duplicated. The spatial specification of loci likely reflected an intrinsic structure of the chromosome itself (1) or, perhaps, one conferred on it by an underlying subcellular foundation.

The organizational unit of bacterial chromosomes, the topological domain, is likely of the order of 10 kbp (18, 19). Given the resolution limit of the Viollier et al. (17) assays, the colinearity of loci in *Caulobacter* may be less perfect than meets the eye; deviations from linearity within domains would not have been

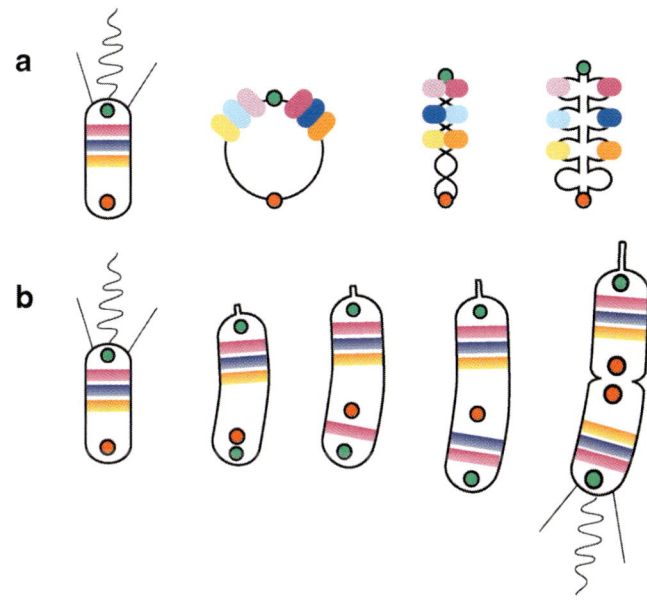

Figure 2

Ordered chromosome organization and segregation in *C. crescentus*. (*a*) The replication origin (*green*), the terminus (*red*) and the loci in between (shown in different colors) have a linear order within the *Caulobacter* nucleoid. This order could derive from a plectonemic or side-by-side configuration of the chromosome (1, 17). (*b*) The migration of the origin and the rest of the chromosome during replication also follows the same linear order.

revealed (1). Nevertheless, the almost obsessive neatness with which the *Caulobacter* cell lays down its DNA (which is a thousand times longer than itself) suggests strong evolutionary selection for this pattern of genome organization. Perhaps this arrangement is important not only for ordered replication but also for spatial and temporal control of gene expression. Let entropy find its last refuge elsewhere; certainly, not in the nucleoid!

Do Nascent Duplexes Stay Associated During Replication?

Whether sister duplexes part their ways immediately or stay together for a while during bacterial chromosome replication has been an issue of debate. The notion of cohesion came about with the observations of Sunako et al. (20) on the segregation timings of the origin, the terminus, and several loci in between,

Cohesion: the pairing of sister chromatids by the cohesin complex

Cohesin: the multiprotein complex that mediates chromosome pairing

marked by fluorescence in situ hybridization (FISH). In synchronously replicating *E. coli* chromosomes, sister copies of loci in the *oriC*-proximal half of the chromosome appear to stay together until late in replication and separate from each other at about the same time as those of loci near the terminus. The process is reminiscent of cohesion of sister chromatids followed by their separation in eukaryotes. Obviously, this segregation pattern challenges the idea of a stationary replication factory and the mechanism of DNA extrusion and capture. A subsequent experiment combining flow cytometry and fluorescence microscopy seems to indicate that little time elapses before duplicated copies of the original, marked by green fluorescent protein, separate from each other (21), thus apparently contradicting the cohesion model.

Bates & Kleckner (22) revisited the cohesion problem using a novel and highly efficient method for cell synchronization. They found that the *oriC* and *terC* sequences behave as functionally independent domains from the rest of the *E. coli* chromosome with respect to segregation. The replicated *oriC* sequences did not immediately separate. Furthermore, even after the *oriC*s split, other sister loci remained colocalized (cohesed) until a large portion of the chromosome had been replicated. They then separated in a coordinated single-step event, presumably triggered by a concerted loss of cohesion, and moved apart to form a bi-lobed nucleoid. This en masse separation led to a reorganization of *oriC* and *terC* sequences. The net result of these dynamics was to establish the *ori*-out/*ter*-in conformation that signified the twofold symmetry of chromosome segregation (**Figure 3**; also **Figure 1**).

Because *E. coli* does not possess a molecular equivalent of the eukaryotic cohesin complex to hold sisters together, the pairing is likely mediated through catenane links. Although the sister cohesion mode of segregation does not absolutely exclude a stationary replication factory, the data from Bates & Kleckner (22) argue against a long-lived factory.

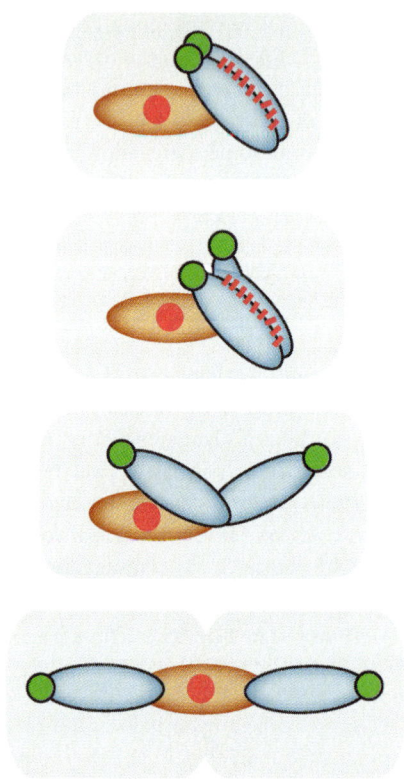

Figure 3

Cohesion between sister duplexes during chromosome replication and segregation in *E. coli*. In this highly schematized representation of the nucleoid, replication origin (*green circles*) and the terminus (*red circles*) are shown. Duplicated and unduplicated portions of the chromosome are colored blue and red, respectively. The sequential steps in chromosome segregation involve splitting of the paired origins, subsequent unpairing of sisters, and reorientation of replicated and unreplicated regions of the chromosome (22).

CHROMOSOME SEGREGATION IN EUKARYOTES

The principles that govern the mechanics of chromosome segregation in eukaryotes and prokaryotes are quite different. The primary infrastructure for moving eukaryotic chromosomes is the tubulin-based cytoskeletal apparatus, the spindle. Attachment of individual chromosomes to the spindle is mediated by microtubules connected to the kinetochore, an elaborate protein assembly organized at the centromere (23). A multisubunit protein complex called cohesin holds replicated sister chromatids together, so that their kinetochores can be attached to the spindle in the opposite orientations during mitotic chromosome segregation (24, 25). Furthermore, a surveillance mechanism ensures that wrong connections are undone and indeed corrected (26, 27). (The spindle-kinetochore attachments are managed quite differently during meiosis.) To avoid the entanglement and self-guillotining of segregating chromosomes, catenane linkages resulting from replication are resolved by topoisomerase II (28), and each chromatid is kept away from its sister by DNA condensation promoted by the condensin complex (29). Finally, the disassembly of cohesin by proteolytic cleavage of one of its subunits causes chromosome sisters to separate and be pulled away toward opposite cell poles (30).

The logic underlying the complex events of eukaryotic mitosis is simple, yet highly effective: Keep the sister duplexes formed by replication together and give the cell enough time to complete quality control tests before they are allowed to split asunder. A general analogy, albeit a bit contrived, is made to the solution devised by two blind shoppers who realize that the socks they purchased have all been placed in one shopping bag (34). (Of course, by a curious and convenient coincidence, the number of pairs and the different colors they bought happen to be identical, so the bag contains two pairs of a given color.) Using a pair of scissors, they cut the label holding together each pair, and distribute the individual socks into two separate bags. At the end, each man's bag would contain a full set of socks and no mismatched ones. Indeed, the pairing and unpairing strategy greatly simplifies the equal segregation challenge.

Condensin: the multiprotein complex that compacts chromosomes

Remembering Replication and Counting Chromosomes

The pairing of sister chromatids by the cohesin complex serves several useful purposes during eukaryotic chromosome segregation. It provides the cell with a memory of replication events. It permits a pair of sister chromatids to be distinguished from the corresponding pair of homologue sisters. And it is an efficient mechanism for counting the products of replication in a binary fashion. The time interval between the duplication (S) and segregation (M) phases of the eukaryotic cell cycle offers the opportunity for checkpoint mechanisms to postpone the commitment to chromosome segregation until DNA damages have been repaired, replication events completed, and spindle integrity and bi-oriented spindle attachment of sister chromosomes verified (**Figure 4**). These checkpoints either block entry into mitosis by stopping S-phase progression, arresting cells in G2, or prevent the onset of anaphase by arresting them in metaphase (31–33).

Molecular Basis for Chromosome Cohesion

The cohesin complex in the budding yeast is composed of four primary subunits: Smc1p, Smc3p, Scc1p/Mcd1p, and Scc3p (**Figure 5**). In addition, a fifth protein, Pds5p, may associate with the complex after its assembly on chromosomes and is required only for the maintenance of sister chromatid cohesion (34). The Smc subunits belong to the SMC (structural maintenance of chromosomes) family of proteins that are conserved in bacteria, archaea, and eukaryotes and that play important roles in chromosome condensation, cohesion, and repair (35, 36). The functional equivalents of the non-Smc subunits of cohesin are distributed throughout eukaryotes. Mitotic and meiotic cohesins may differ in their non-Smc components. For example, the meiosis-specific counterpart of Scc1p in *S. cerevisiae* is Rec8p; the corresponding mitotic and meiotic proteins in *Schizosaccharomyces pombe* are Rad21p and Rec8p, respectively. In mammals, there exists a meiotic version of Smc1 termed Smc1-beta.

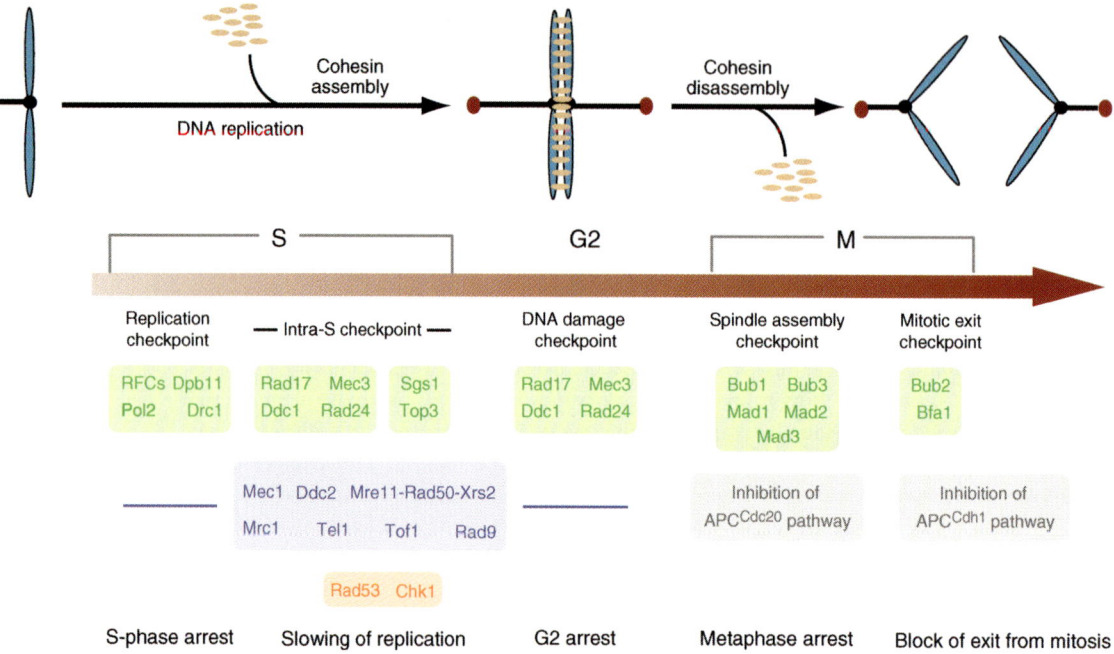

Figure 4

Cell cycle checkpoints and mitotic chromosome segregation in *S. cerevisiae*. Chromosomes duplicated during S phase are paired with their sisters by the cohesin complex and remain so until the disassembly of cohesin during anaphase. Sister kinetochores are attached to the spindle in a bipolar fashion. Following cohesin disassembly, the pulling force exerted by the microtubules dispatches the separated sisters toward opposite cell poles. Checkpoints operating during the cell cycle along with the responsible protein factors are indicated (RFC, replication factor C). The S, G2, and M checkpoints take advantage of the interval between chromosome duplication and anaphase to ensure that replication is complete, DNA damages are repaired, the spindle is functional, and sister chromatids are bi-oriented before permitting chromosome segregation to proceed.

The Smc1 and Smc3 proteins are intramolecularly folded through their long coiled-coil arms via a hinge region to form V shaped molecules (**Figure 5**), thus bringing their N- and C-terminal globular domains into juxtaposition (37). As a result, the Walker A motif from the N-terminal domain and the Walker B motif from the C-terminal domain are united to form an ATP-binding domain, which is structurally related to the nucleotide-binding domain of the ABC (ATP-binding cassette) family of proteins (38). The Scc1 and Scc3 subunits are likely associated with the globular heads and may form a closed protein ring (39). It is possible that sister chromatids are held captive topologically within this ring (24). Consistent with this topological restraining mechanism, release of sister-to-sister cohesion can be effected not only by the cleavage of Scc1 (as is the norm during the cell cycle) but also by artificial cleavage of the coiled-coil region of Smc3p (40). Although the notion of sister chromatids being held together within the embrace of the cohesin ring is quite appealing, the proof for it is not absolute.

Centromeres are strong cohesin-binding sites, with additional sites distributed at approximately 10 kb intervals along the budding yeast chromosomes (41). Yeast kinetochores serve as enhancers to promote cohesion assembly at pericentric regions that cover tens

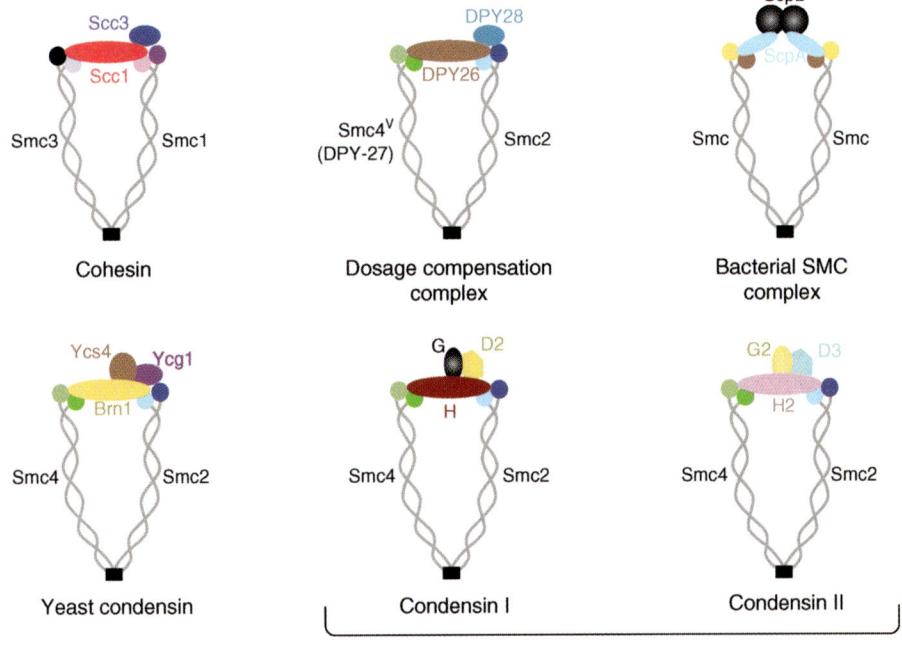

Figure 5

Architecture of the cohesin, condensin, and related complexes. The general structures of cohesin, condensin, and related complexes (*Drosophila* dosage compensation and bacterial SMC complexes) are schematically shown. Their common V-shaped backbone is provided by the hinge and coiled-coil regions of the SMC proteins, whose globular head domains associate with regulatory subunits. In the metazoan condensins, the condensin-associated proteins are CAP-H (or H2), CAP-G (or G2), and CAP-D2 (or D3).

of kilobases (42). Chromosomal cohesin loading is dependent on a separate complex consisting of the Scc2 and Scc4 proteins (43). The establishment of cohesion, but not its maintenance, is dependent on the Ctf7 protein (44). Although the cohesin complex may associate with chromosomal loci in late G1 phase, cohesion appears to occur concomitantly with the passage of replication forks. An advancing fork is believed to pause at a cohesin-binding site and exchange the resident polymerase with polymerase σ as a prerequisite for establishment of cohesion (45). This step appears to require participation of the processivity clamp proliferating cell nuclear antigen and an alternate form of the clamp loader replication factor C. Cohesin loaded at chromosomal sites may translocate to more permanent locations that represent sites of convergent transcription (46). This unexpected flexibility is suggestive of a protein ring that can slide along DNA when pushed by a protein machine.

In the budding yeast, once cohesion between sister chromatids has been established, it lasts until the onset of anaphase when the cysteine protease Esp1 (separase), released from its sequestered state in association with securin, cleaves Scc1p at two target sites (47). The timely cleavage of Scc1p is controlled, at least in part, through its phosphorylation by the polo-like kinase Cdc5p (48). This straightforward "cut and separate" mechanism may be an oversimplified picture. Additional regulatory steps must operate because deletion of the securin gene (*pds1*Δ), together with

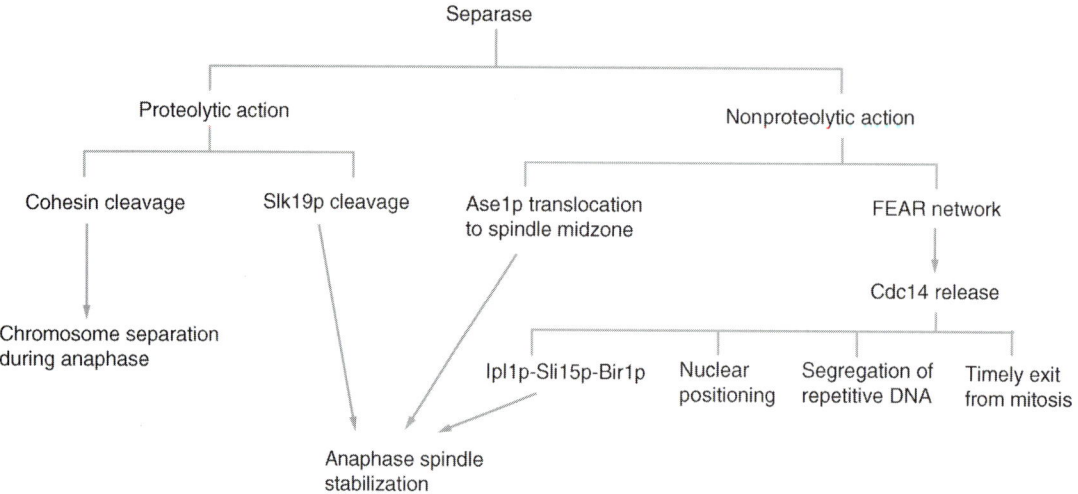

Figure 6

Multiple functions of separase. In addition to cleaving the Scc1p/Mcd1p subunit of the cohesin complex (and Slk19p of the FEAR complex), separase is involved in a number of steps that ensure the orderly progression of M phase and the exit from mitosis. Nearly all of these functions are mediated through the Cdc14 phosphatase.

overexpression of Cdc5p, does not advance the timing of Scc1p cleavage significantly. In *Xenopus* egg extracts, separase is phosphorylated by the cyclin-dependent kinase Cdk1; a dephosphorylation step, in addition to the destruction of securin, is likely required for separase activation (49). Unlike the budding yeast, higher eukaryotes remove cohesin in two distinct waves (50, 51). The first occurs during prophase and causes the dissociation of cohesin bound to chromosome arms. The process is independent of protease cleavage but is dependent on phosphorylation by the polo and aurora kinases. The second occurs in anaphase and primarily removes centromere-bound cohesin by proteolysis. Perhaps protective molecule(s) present at the centromere can locally block the action of the prophase pathway.

Separase: Not Just a Protease

Separase function in chromosome segregation goes beyond its proteolytic role in the cleavage of Scc1p (**Figure 6**). In a protease-independent manner, separase functions as part of the FEAR (fourteen early anaphase release) network to release Cdc14 phosphatase from the nucleolus, where it is anchored by Net1p (52). Although chromosome segregation at a global level can be achieved by cohesin disassembly alone, repeated DNA in budding yeast (rDNA and telomeres) fail to segregate when the FEAR pathway and release of Cdc14p are blocked (53, 54). This cohesin-independent cohesion of rDNA is resolved in a step (which requires condensin) that probably does not involve DNA compaction per se. Although this last-to-be-broken bonding between sisters is likely mediated by catenane links, and they are disjoined by topo II (54), the issue is as yet unsettled (53). In addition to facilitating the "last act" of sister chromatid segregation, separase contributes toward stabilizing and orienting the spindle, and it helps set the stage for exit from mitosis by counteracting Cdk1p activity in multiple ways (54–56) (**Figure 6**).

Pairing and Unpairing During Meiosis

The basic mechanism of pairing and unpairing has to be suitably modified spatially and

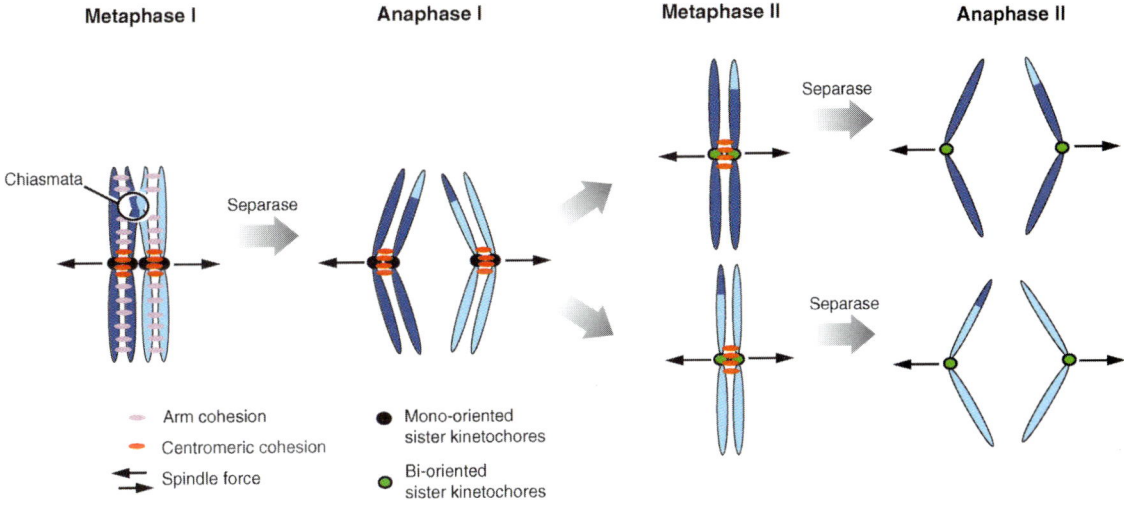

Figure 7

Chromosome pairing and unpairing in meiosis. During meiosis I, sisters become paired by cohesin, and DNA exchange between homologues takes place. Unlike in mitosis, sister kinetochores attach to the spindle in monopolar fashion; attachment of the homologue kinetochores is bipolar. Following resolution of crossovers and cohesin disassembly along the arms (but not at centromeres), the homologues segregate. During meiosis II, sister kinetochores bi-orient on the spindle, and cohesin removal at the centromeres triggers their segregation.

temporally to meet the demands of meiotic chromosome segregation. In order to produce haploid gametes, the meiotic cell must perform two rounds of segregation following a single round of chromosome duplication (**Figure 7**). During meiosis I, sisters must stay together, and homologues that have undergone recombination events must segregate. This is accomplished by preserving kinetochore cohesion between sisters and ensuring their monopolar (syntellic) attachment to the spindle. Arm cohesion, by contrast, is destroyed coincidently with the resolution of crossovers or chiasmata. In budding yeast, Spo13p, a meiosis-specific centromere-associated protein, facilitates the recruitment of the monopolin complex, consisting of monopolin (Mam1p) and the nucleolar proteins Csm1p and Lrs4p, that is essential for monopolar attachment and segregation of homologues (57, 58). During meiosis II, the rules of mitosis, bi-orientation of sister kinetochores and dissolution of kinetochore cohesion, come into play to segregate sisters.

Because the same enzyme, separase, appears to be responsible for cohesin cleavage during meiosis I and meiosis II in worms, yeasts, and mice (59–62), an important question is how the differentiation between centromere cohesion and arm cohesion is accomplished in the two cases. In principle, the existence of a meiosis I-specific protector of centromeric cohesin could take care of the problem. The search for the protector in the fission yeast has identified shugoshin (Sgo1p), guardian angel in Japanese, which localizes to the centromere in a Bub1 kinase-dependent manner and shields cohesin's Rec8p from cleavage by separase (63, 64). In the absence of Sgo1p, centromeric cohesion cannot be sustained during meiosis I, and sister chromosomes segregate randomly. The Sgo1p of the budding yeast is not only required for proper meiosis I, but also plays a role in sister chromatid segregation during meiosis II (and mitosis as well) (65). The fission yeast has a paralogue Sgo2p that is required for faithful chromosome segregation during mitosis (63). Sgo2p is also expressed during both meiotic divisions, and *sgo2* mutants display high incidence of precocious equational division (segregation of sisters). Perhaps Sgo2p contributes to monopolar orientation as well as protection of centromeric cohesion during meiosis I. How does it then function during meiosis II (and mitosis), during which segregation demands bipolar orientation and loss of centromeric cohesion? The activity of Sgo2p may be regulated differentially to perform different roles under the two distinct contexts. The same argument may apply to Sgo1p in budding yeast, which does not have the Sgo2p counterpart and acts during mitosis, mieosis I, and meiosis II. Shugoshins have been identified in fungi, plants, flies, and vertebrates (63, 64) by virtue of conserved architectural motifs—an N-terminal putative coiled-coil, a central region rich in charged and hydroxylated residues, and a C-terminal basic motif—rather than by high amino acid sequence conservation. Perhaps these proteins have evolved by convergence or have coopted similar peptide modules to perform mitotic and meiotic functions in their specific biological contexts. It will not be surprising if, in vertebrate mitosis, a shugoshin-like protein mediates the retention of cohesin at centromeres during the prophase removal of arm-bound cohesin.

CONDENSINS AND CHROMOSOME SEGREGATION

The condensin complex is architecturally quite similar to the cohesin complex (**Figure 5**) but plays a very different role in chromosome segregation. Two Smc proteins, Smc2p and Smc4p in the budding yeast, bent at their hinge regions provide the V-shaped structural frame of condensin. The coiled-coil arms of the V end in twin-lobed globular ATPase heads, with which additional subunits interact to perhaps form a protein ring, are analogous to how ring closure has been proposed to take place for the cohesin complex (29, 66). Vertebrate cells have two condensin complexes, each with a distinct set of regulatory subunits. Condensin II acts during prophase, in the early step of chromosome compaction. Condensin I engages a chromosome after breakdown of the nuclear envelope and cooperates with condensin II to organize highly condensed and fully resolved metaphase chromosomes. By contrast, there is only one condensin complex, corresponding to condensin I, in the budding yeast. The relatively small sizes of individual yeast chromosomes as well as the lack of nuclear membrane breakdown during fungal mitosis may account for this evolutionary parsimony. However, the correlation between genome size and the presence or absence of condensin II is not perfect. For example, all condensin II genes are present in the unicellular red algae *Cyanidioschyzon merolae*, whose genome size is comparable to that of the budding yeast (29). Although condensin I is conserved from yeast to humans, the nematodes *Caenorhabditis elegans* and *Caenorhabditis briggsae* are exceptions to this rule. Perhaps their holocentric chromosome structure, with multiple centromeres arrayed along the entire length of chromosomes, may override the requirement for condensin I (67). Or, these nematodes may have adapted and retooled a primordial condensin I to perform dosage compensation that equalizes the expression of X-linked genes in the two sexes (68). Despite the similarity in their subunit composition, condensin and cohesin display distinct arm conformations with characteristic hinge angles when examined by electron microscopy (39). This structural difference may be important in their functional specializations.

Although the spatial and temporal controls of condensin action during the cell cycle may differ widely among organisms from the budding yeast to vertebrates, they are nevertheless directed toward the common goal of organizing and maintaining metaphase chromosome architecture that is conducive for unimpeded segregation. Condensation of the rDNA array in the budding yeast can be divided into two distinct cell cycle–regulated pathways (69). The first one spanning G2 to M is a multi-step process in which an early "random coil" state of condensation matures by "alignment" and "resolution" into a high-order state of condensation in a cohesin-dependent manner (**Figure 8**). It has been suggested that cohesin bound at its cognate sites along a chromosome may provide boundary marks or anchoring points that assist condensin to assemble a uniform array of condensed DNA domains. The second one is a post-metaphase process that is independent of cohesin but requires the Ipl1 (aurora) kinase, one of whose targets is the condensin subunit Ycg1 (69). The reason for orchestrating chromosome condensation in two separate steps is not immediately obvious. Perhaps the second phase of condensation, after cohesin has already been disassembled from chromosomes, may provide an additional safeguard against sister chromatid bridging and/or the trapping of chromosome laggards in the plane of cytokinesis.

Condensin's chromosome compacting role is central to chromosome segregation during meiosis as well. In the budding yeast, condensin subunits localize along the axial core of pachytene chromosomes (70). In condensin mutants, Zip1p, a component of the central element, is improperly localized, leading to defective assembly of the synaptonemal complex. Furthermore, lack of condensin function adversely affects the processing of meiotic double-strand breaks, pairing of homologues, resolution of recombination-mediated homologue linkages during meiosis I, and segregation of sister chromatids during meiosis II. The requirement for condensin during meiosis I and II has been demonstrated in *Arabidopsis* and *C. elegans* as well (71, 72). In contrast to the budding yeast situation, the association of condensin with chromosomes and the process of chromosome restructuring in the worm occur only after exit from pachytene. Additionally, the non-Smc components of the dosage compensation complex are required for meiotic chromosome segregation (73), but they

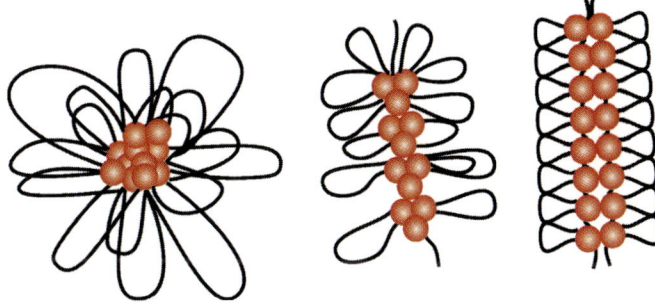

Figure 8

The stepwise condensation of budding yeast rDNA locus in the G2-M cell cycle window. The early intermediate in rDNA condensation by the condensin complex is a random coil (*left*), which undergoes partial alignment (*center*) before proceeding to the highly organized state of condensation (*right*). The process is dependent on the cohesin complex, which binds at centromeres and along chromosome arms with fairly uniform periodicity. The cohesin-bound sites may thus provide anchoring points for arranging uniform loops of condensed DNA. The diagram is adapted from Lavoie et al. (69).

do not seem to play a role in mitotic chromosome segregation.

Chromosome Condensation and Bacterial DNA Segregation

DNA condensation plays an important role in prokaryotic chromosome segregation as well (74, 75). Unlike eukaryotes, bacteria usually possess a single SMC protein or a distantly related, although functionally equivalent, protein such as MukB, the active core of the MukBEF complex of *E. coli*. Mutations in *smc* of *B. subtilis* or *C. crescentus* or in *mukB* of *E. coli* result in poorly condensed chromosomes that segregate abnormally (76–78). The segregation defect caused by the lack of *mukB* function can be suppressed by increasing the negative superhelical density of the chromosome, implying collaborative roles for plectonemic supercoiling and DNA condensation in promoting bacterial chromosome segregation (79, 80). In the extrusion-capture model for replication and segregation, the MukB protein may pull a newly replicated duplex poleward by supercoiling it into a more condensed form. The observation that *mukB* mutants are hypersensitive to inhibitors of gyrase

(81) strengthens the argument that DNA supercoiling by the two proteins acts additively in condensing daughter duplexes away from each other.

Two auxiliary subunits of the SMC complex, ScpA and ScpB, have been discovered in archaea and gram-positive bacteria (82, 83). ScpA shows similarity to the Scc1 subunit of eukaryotic cohesin (84), and its binding to the SMC head domains is stabilized by ScpB (85). In *B. subtilis*, *scpA* and *smc* mutations produce nearly identical phenotypes, including high-frequency anucleate cells and chromosome guillotining. The SMC complexes in archaea and bacteria thus share architectural and functional similarities with eukaryotic condensins.

Mechanism of DNA Compaction by Condensin

In vitro experiments have yielded important clues on plausible modes of condensin action, although the precise mechanism of DNA compaction is not fully understood (29, 86). In the presence of ATP and condensin, DNA knots formed by topoisomerase II are chiral, with exclusively + crossings (**Figure 9a**). This result fits a model in which condensin stabilizes global positive writhe in DNA. Recent data using atomic force microscopy suggest a loop fastener mechanism for condensin action (**Figure 9b**; middle). Here, the hinge region of the Smc proteins bind to one region of DNA, and in an ATP-dependent opening and closing reaction, the non-Smc subunits clamp the V to enclose a DNA loop. Electron spectroscopic imaging favors an "oriented gyre" (**Figure 9b**; right) because two stacked supercoils appear to be captured by a single condensin molecule. A single DNA molecule assay using magnetic tweezers demonstrates highly reversible and dynamic compaction of DNA by condensin I. The reaction requires ATP hydrolysis and may proceed by a looping mechanism.

Stable binding of double-stranded DNA by bacterial Smc dimers involves the engagement of their catalytic head domains through two bound ATP molecules (87, 88). This DNA association may occur either through an intramolecular "embrace" mechanism or an intermolecular "hand-in-hand" interaction between adjacent Smc dimers (**Figure 9c**). The non-SMC subunits ScpA and ScpB may exert either negative or positive regulation on DNA association, depending on whether they stabilize ring closure before or after the DNA has been engaged by the Smc dimer.

Roles for Condensin and Cohesin not Directly Related to Chromosome Segregation

Aside from their critical role in chromosome segregation, condensin and cohesin complexes participate in control of gene expression globally in a chromosome-wide manner, regionally within chromosomal domains, and locally for individual genes (86). Examples include transcriptional control by the condensin-like dosage compensation complex (see **Figure 5**) during hermaphrodite (XX) development in nematodes, repression of the silent mating cassettes *HML* and *HMR* in the budding yeast and regulation of the *Drosophila Abd-B* gene that specifies body segment identity in the posterior of the fly. Cohesin and related complexes such as the Smc5-Smc6 complex are also crucial for promoting repair of DNA damages (89–91). The Smc5p and Smc6p functions are required in the budding yeast for the proper segregation of repeated loci, rDNA, and telomeres. Presumably, they assist resolution of the X-shaped recombination intermediates formed during sister chromatid exchange (92). The meiotic recombination protein Rad50p, which forms part of the MRX (Mre11p, Rad50p, Xrs2p) double-strand break repair and telomere maintenance complex in the budding yeast (MRN complex in humans; N = Nbs1), is a structural relative of SMC proteins, displaying the globular ATPase head domains and two coiled-coil

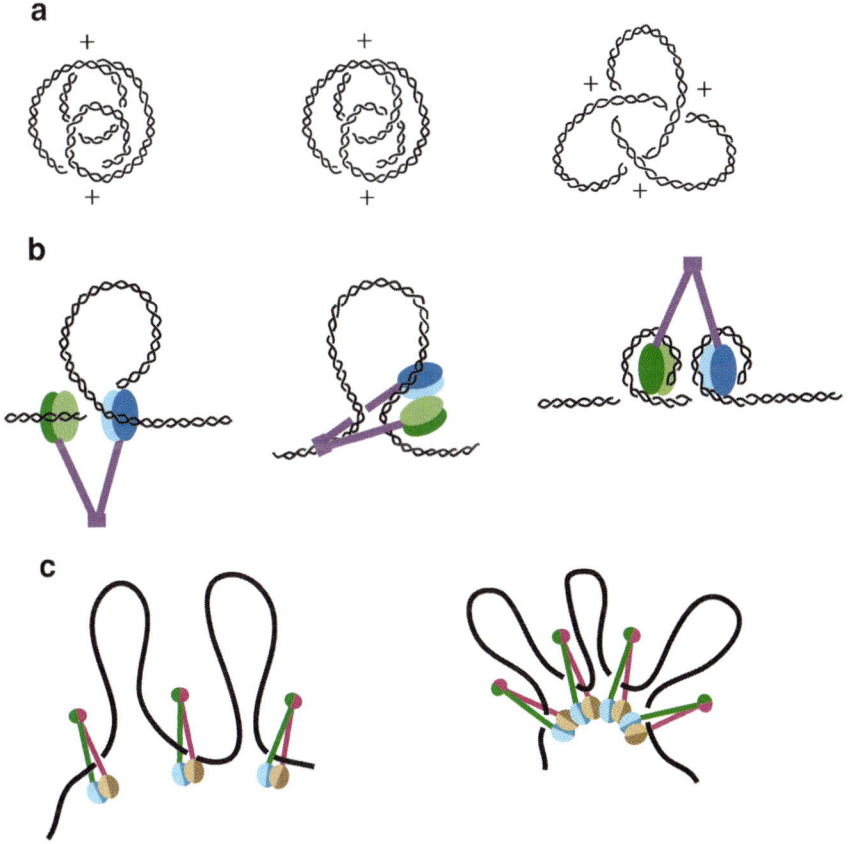

Figure 9

Plausible mechanisms of DNA compaction by condensin. (*a*) Knots generated by topoisomerase II in presence of condensin and ATP have exclusive + crossings, suggesting that condensin induces global positive writhe in DNA (*b, left*). Strand passage by topoiosmerase II between two overlying positively writhed DNA loops will generate a (+) trefoil (a simple knot with three crossings) as shown in panel *a*. (*b*) Other plausible modes of DNA condensation include a loop fastener mechanism (*middle*) or a twin-looped oriented gyre mechanism (*right*). (*c*) Engagement of the catalytic sites by two bound ATP molecules stabilizes the closed configuration of the bacterial SMC dimer and promotes stable encircling of DNA by the coiled-coil arms. Catalytic site engagement may occur by an intramolecular "embrace" mechanism (*left*) or an intermolecular "hand-in-hand" mechanism (*right*).

regions connected by a "zinc-hook" hinge motif (93).

The condensin and cohesin complexes are more than just chromosome compactors and glue, respectively, devoted to chromosome segregation alone (86). By combining pairs of SMC or SMC-like proteins with alternative sets of regulatory components, evolution has formulated solutions to multiple problems related to organization, integrity, accessibility, and transmission of genomes.

SEGREGATION OF BACTERIAL PLASMIDS

Depending on copy number, stable propagation of a bacterial plasmid can be accomplished via random segregation or active partitioning. Provided the steady-state copy number is reasonably high, and there is no barrier to free diffusion, random segregation will suffice. The probability of a plasmid-free cell arising at any given cell division will be

par locus:
partitioning locus of bacterial plasmids

Par proteins:
partitioning proteins of bacterial plasmids

quite low. Furthermore, control mechanisms operating at the replication level can correct for any deviation from the normal copy number because of unequal segregation. By contrast, unit-copy or low-copy number plasmids must rely on an active partitioning system to avoid missegregation. Certain plasmids also possess an intricate postsegregational killing mechanism to eliminate competition from plasmid-free cells formed as a result of missegregation events. The "addiction module" harbored by such plasmids codes for a long-lived toxin and a short-lived antidote (94–96). Cells that lose the plasmid are condemned to death as they inherit the toxin without the benefit of the antidote. Whereas the stable toxins are normally protein moieties, the unstable antidotes may be either proteins or small antisense RNA molecules that prevent translation of the toxin mRNAs.

Partitioning Systems of Low-Copy Plasmids

Unit-copy or low-copy bacterial plasmids such as P1, F, and R1 encode partitioning genes (*par*) whose protein products act in conjunction with a centromere-like locus to facilitate their faithful segregation (97, 98). To avoid possible confusion by different nomenclatures used in differing systems, we present a simplified picture of their unifying features in **Figure 10a**. In general, the *par* locus specifies two *trans*-acting proteins encoded within an operon and a *cis*-acting "centromere-like" element. The polypeptide products of the upstream and downstream *par* genes are generalized here as ParA and ParB, respectively. The centromere *parC* (*sopC* in F, *parS* in P1, and *parC* in R1) contains iterons, multiple copies of a sequence element that is characteristic of each individual plasmid. ParA (SopA in F, ParA in P1, and ParM in R1) is an ATPase that can associate with the centromere, usually assisted by its partner ParB (SopB in F, ParB in P1, and ParR in R1). ParB normally binds directly to *parC* to form a "prepartitioning" complex that then recruits ParA to complete the partitioning complex. The location of *parC* with respect to the operon is variable. For example, *parC* may harbor the promoter to the operon or may reside distal to it and downstream of the protein-coding sequences. Some of the *par* loci are characterized by multiple direct repeats upstream and downstream of the *par* genes, and the two sets may independently or cooperatively serve the centromere-like function. An extensive search of sequence databases using partitioning ATPases of well-studied plasmid systems has revealed novel organization patterns for plasmid partitioning loci (99). On the basis of these additional data, plasmid partitioning systems have been divided into five representative classes. Nevertheless, all of these systems reflect variations of a common theme.

The ParA ATPases can be divided into two groups, the Walker type and the actin-like ATPases (100, 101). Four of the five partitioning classes (that include P1 and F) harbor ATPases that display Walker A, Walker B, and A′ motifs, typifying nucleotide-Mg^{2+} binding, plus a fourth motif whose function is not known. The ATPases of the other partitioning group (that includes R1) contain ATP-binding motifs characteristic of the actin-like ATPase family. There is strong evidence that the *par* loci are under strict autoregulation, presumably because the correct stoichiometry of the Par proteins is important for partitioning. The ParA protein of P1 (and SopA of F) can associate with operator sequences in the *par* promoter region using an N-terminal DNA-binding domain to downregulate transcription; the extent of transcriptional repression is augmented by the ParB (or SopB) protein acting as a corepressor (102–105). In the case of F, maximal repression also requires the centromere, suggesting a role for DNA looping in this process (106). In vitro studies indicate that P1 ParA interacts with the centromere-ParB partition complex in the ATP-associated form and binds to the *par* promoter more strongly in the ADP-associated form (107). Thus, the dual function of ParA, as

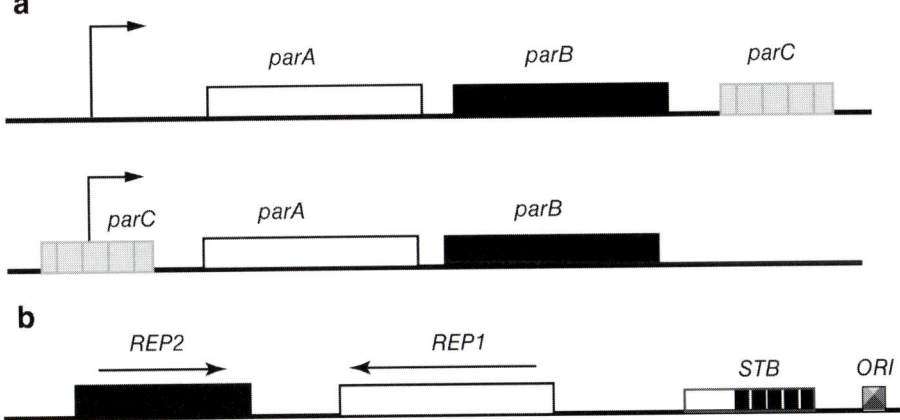

Figure 10

Organization of partitioning loci in bacterial and yeast plasmids. (*a*) In this simplified representation of bacterial *par* loci, *parA* and *parB* are the genes coding for the partitioning proteins, and *parC* is the centromere-like locus consisting of iterated sequence elements. ParA proteins are ATPases that harbor typical Walker motifs or belong to the actin-like family of ATPases. The ParB proteins in general bind to *parC* and recruit ParA to the partitioning complex. The bent arrows indicate the site of initiation and direction of transcription. (*b*) The yeast plasmid partitioning system also consists of two plasmid-coded proteins Rep1p and Rep2p and the *cis*-acting locus *STB*. There is no functional correspondence between the partitioning proteins of the bacterial and yeast plasmids. The *STB* element can be functionally divided into two subloci. The one proximal to the replication origin (*ORI*) contains approximately six iterations of a consensus 65-bp repeat element. The distal-*STB* region is important in maintaining the active configuration of *STB* and contains the termination site for two transcripts directed toward the origin.

a partitioning agent or as a transcription regulator, appears to be subject to allosteric regulation by the bound form of the nucleotide ligand. In the R1 plasmid, the ParM protein (equivalent to ParA ATPase) does not appear to participate in regulating the *par* operon, leaving this function entirely to the ParR protein (equivalent to ParB) (108).

Mechanisms of Plasmid Partitioning

Our current understanding of how the partitioning proteins assist plasmid segregation is primarily phenomenological. In general, the partitioning complex, formed by association of the Par proteins with the centromere, specifies plasmid localization within the bacterial cell and provides an important spatial determinant for proper segregation (99). The identities of host factors involved in bacterial plasmid segregation have remained elusive. Although the *E. coli* integration host factor protein promotes the assembly of the P1 partitioning apparatus by increasing the affinity of ParB for its centromere, it is not essential for the partitioning process (109). The P1 and F plasmids are normally present at the center of a newborn cell and stay there until they are replicated (110, 111). The duplicated copies then quickly move to the one-fourth and three-fourths cell positions (112), which mark the midpoints of the future daughter cells. An occasional second round of replication prior to cell division can result in a daughter cell containing two copies of the plasmid. Time-lapse fluorescence microscopy has provided a more refined and better resolved perspective of P1 localization and dynamics in live *E. coli* cells (113). A focus containing one or more plasmid molecules is captured at the cell center just prior to replication, and foci of nascent plasmid molecules are ejected bidirectionally along the long axis of the cell before cell division (**Figure 11***a*). Plasmid

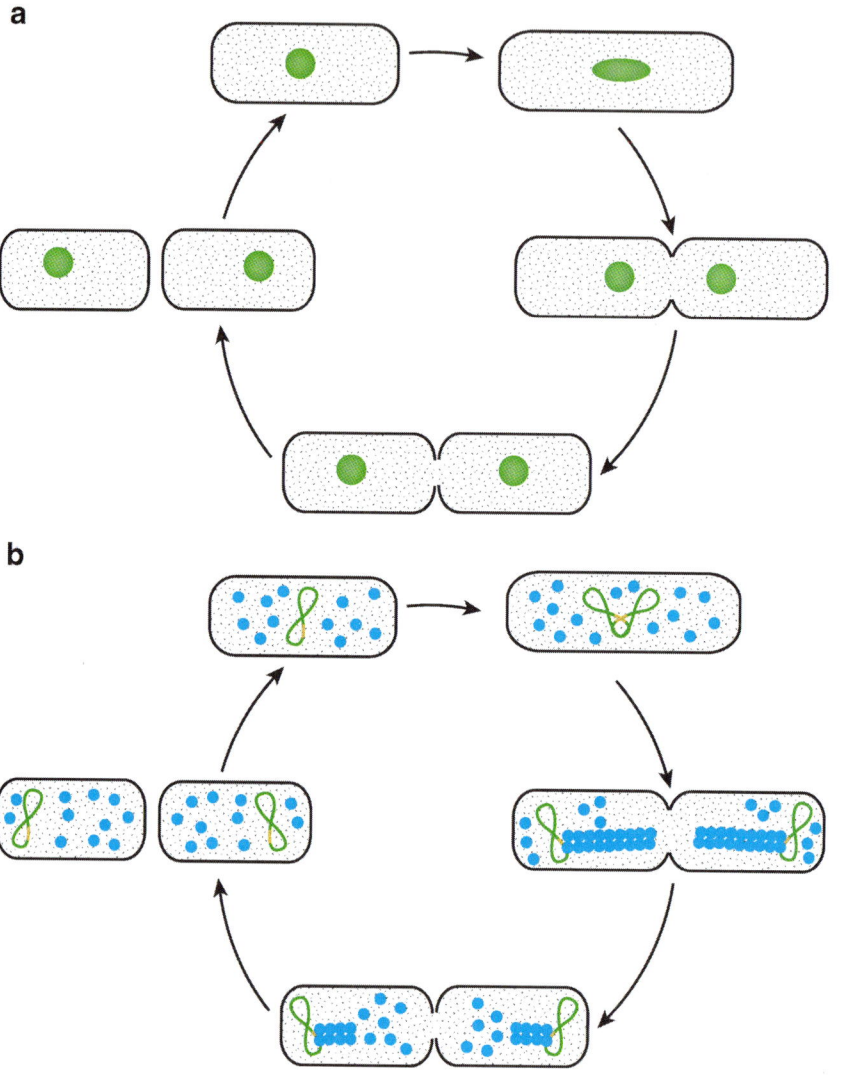

Figure 11

Segregation of P1 and R1 plasmids in *E. coli*. (*a*) The focus of P1 plasmid, which is free to diffuse in a newborn cell, localizes to the mid-cell position just before replication. The foci of duplicated plasmid molecules are extruded in opposite directions. (*b*) The R1 plasmid (*green*) has a pole-proximal location in the newborn cell. The plasmid translocates to the cell center for replication. ParM (*blue*)-ATP polymerization into filaments at the ParR-bound *parC* loci (*gold*) transports the replicated plasmid molecules toward opposite cell poles. The filament depolymerizes, starting at the tail end, to generate ParM-ADP. The figure is patterned after diagrams from Li & Austin (113) and Gerdes and coworkers (122).

copies are free to move, associate, and dissociate in the newborn daughter cell before they are captured at the cell center to start a new replication and segregation cycle. At the mid-cell position (or the one-fourth or three-fourths positions of a cell about to divide), there does not appear to be a single "partitioning center" to which plasmids harboring distinct partitioning systems are tethered. Pairs of compatible plasmids such as F and RK2 do reside close to each other but are not necessarily colocalized (114). They also differ in their segregation timings. Although the importance of accurate plasmid localization in association with the partitioning complex for efficient segregation is well documented (111, 115), the subsequent mechanisms remain obscure.

An analysis of in vivo plasmid topology, along with a related set of cell biological observations, suggests that P1 or F plasmid molecules are physically restrained within the partitioning complex or at their cellular tethering site and may need to be released from this state prior to segregation (105, 115, 116). ParB/SopB is likely responsible for plasmid pairing or grouping through its association with the centromere sequence; perhaps it is up to ParA/SopA to actively unpair/ungroup the replicated molecules. Consistent with this notion, certain types of mutations in ParA/SopA cause "worse than random" segregation, as if the copy number has been lowered because the plasmids have been "glued" together (117, 118). Furthermore, as revealed by time-lapse assays, ParB is required for tethering the P1 plasmid focus at mid-cell, whereas the ParA ATPase is essential for the active longitudinal (and bidirectional) ejection of foci after replication (119).

The most well-understood bacterial plasmid segregation system, with respect to mechanism, is that harbored by the R1 plasmid (120). The plasmid shows a dynamic pattern of localization within the *E. coli* cell. Although it is positioned near the cell pole in a newborn cell, it moves to the central position during cell growth (for duplication by the replisome), and the daughter plasmids move back to their pole-proximal location (121). The foci formed by the ParA ATPase of R1 (the ParM protein) are coincident with the plasmid and show identical dynamics. The actin-like ParM protein can oligomerize into filaments in vivo and in vitro (122). Filament formation in vitro is dependent on ATP and Mg^{2+}, whereas filament disassembly requires ATP hydrolysis. The ParB protein of R1 (ParR) bound to the centromere *parC* can mediate the pairing of two plasmid molecules in vitro (123). The ParR-*parC* complex is essential for ParM polymerization in vivo and for filament formation in vitro at low ParM concentrations. Furthermore, in vivo the plasmids are located at the distal tips of the ParM filament, as it extends from the cell center toward the poles (124). These findings provide the basis for the R1 segregation pathway described below and illustrated in **Figure 11b**.

Following duplication of R1, the daughter plasmids are held together at *parC* by the bound ParR. The ParR-*parC* complex provides the nucleation site for the polymerization of ParM-ATP. As the filament grows, the plasmids are unpaired and forced apart from each other toward the cell poles. It has been suggested that the plasmid transport is mediated by a bundle of protofilaments, perhaps each one connected to one of the ten ParR-binding sites within *parC*. The filament pushes the plasmids to their native locations within would-be daughter cells. Depolymerization of the filament at the tail end would release ParM-ADP monomers into the cytoplasm. They have to be recharged by nucleotide exchange for the next round of plasmid segregation in the subsequent cell cycle.

Centromeres and Par Proteins in the Segregation of Bacterial Chromosomes

Partitioning systems of bacterial plasmids are simple in organization yet efficient in function. It is natural to wonder whether bacterial chromosomes also harbor partitioning loci that are functionally similar to their plasmid counterparts. Indeed, almost all of them do with the exception, perhaps, of certain γ-proteobacteria, including *E. coli* and *Haemophilus influenzae* (97, 125). The ParA and ParB homologues of *B. subtilis* are Soj and Spo0J, respectively, and there are eight centromere iterons (*parS*) in the origin-proximal region of the chromosome (126). Although the cellular localization patterns of Spo0J in

Rep1p, Rep2p: partitioning proteins of the yeast 2-micron plasmid

***STB*:** partitioning locus of the yeast 2-micron plasmid

***FRT*:** Flp recombination target

the presence or absence of Soj are generally consistent with those expected of a centromere system, origin localization and dynamics are not affected in the absence of Spo0J (8). However, deletion of the *spo0j* locus does result in a small but significant fraction of anucleate cells during vegetative growth (127). The role of the Soj-Spo0J-*parS* system appears to be rather subtle during mitotic chromosome segregation, whereas it is more prominent during sporulation. Genetic and cytological analyses reveal a partial functional overlap between Soj and the RacA protein in bringing *oriC* to the DivIVA protein located at the pole of the forespore (128). The *par*-system of *C. crescentus* is constituted by the ParA and ParB proteins together with the origin-proximal ParB-binding sites (*parS*). The ParAB proteins are not only important in chromosome segregation but also serve a checkpoint-like function in linking chromosome segregation to cell division and/or DNA replication (129, 130). In other systems that have been investigated, *Pseudomonas putida* and *Streptomyces coelicolor*, the Par proteins are required for proper chromosome segregation during sporulation or in cells transitioning from exponential growth to stationary phase (131, 132). Lack of Par functions appears to have no effect or only modest effects on chromosome segregation during vegetative growth, although overproduction of Par proteins leads to an increased frequency of anucleate cells in *P. putida* (131, 133).

Is a partitioning locus truly absent in *E. coli*, or is there such a novel locus that is yet to be characterized? A recent analysis, based on the assumption that the centromere sequence would be the first to migrate away from the site of replication toward the cell pole, has identified a potential centromere at 89′ on the *E. coli* chromosome, close to *oriC* at 84′ and probably a second one at 79′ (134). Interestingly, a 25-bp sequence *migS*, located at the putative centromeric region at 89′, serves as a *cis*-acting element that directs the bipolar positioning of replicated *oriC*s (135). Future work will decide whether *migS* and associated sequences (perhaps) represent an authentic chromosomal centromere and, if so, what the corresponding Par proteins are.

PLASMID SEGREGATION IN YEAST

The 2-micron plasmid, nearly ubiquitously present in *Saccharomyces* yeast strains, is a multicopy extrachromosomal element that resides in the nucleus and propagates itself stably in host cell populations (136). The plasmid does not confer any advantage to its host, nor does it impose any obvious disadvantage at its steady-state copy number of 40–60 molecules per cell. The chromosome-like stability of the plasmid (a loss rate of 10^{-5} to 10^{-4} per cell division) is conferred by the combined action of a plasmid partitioning system and a plasmid amplification system. The former ensures equal or roughly equal distribution of replicated plasmid molecules to daughter cells; the latter corrects any decrease in copy number caused by a rare missegregation event. The partitioning system consists of two plasmid-coded proteins Rep1p and Rep2p and a *cis*-acting partitioning locus *STB* (**Figure 10***b*). The amplification system consists of the plasmid-coded Flp site-specific recombinase and a pair of *FRT* (Flp recombination target) sites present on the plasmid genome in head-to-head orientation. Here, we discuss the mechanism of action of the partitioning system, and we outline the amplification process below.

It may seem paradoxical that a plasmid with as high a copy number as 40 to 60 would require a partitioning system. Random segregation should be eminently suitable for plasmid propagation especially because a copy number can be corrected by the amplification system. As it turns out, the plasmid exists as a tight-knit cluster of dynamic foci within the yeast nucleus, and these foci stay together throughout the cell cycle (137). The Rep1 and Rep2 proteins form an integral part of the plasmid cluster by their association with *STB*. The cluster, which likely includes host proteins as well, is also the entity in segregation (137, 138), thus reducing the copy number effectively to unity. Hence, the operation of a partitioning system makes sense.

The 2-micron circle partitioning system appears to resemble bacterial plasmid partitioning systems in its general organization in that both consist of two protein components that act in conjunction with a *cis*-acting centromere-like locus (**Figure 10**). However, there are no functional similarities between the two. A number of recent experiments suggest that the Rep/*STB* system is a clever molecular device to channel components of chromosome segregation toward plasmid partitioning. The plasmid-chromosome connection in segregation has been suspected from observations indicating that the dynamics and segregation kinetics of a fluorescence-tagged reporter plasmid are remarkably similar to those of a similarly tagged chromosome (137). Furthermore, conditional mutations that affect segregation of chromosomes appear to affect the 2-micron plasmid in quite a similar manner. In the nonpermissive state, the plasmid cluster appears to missegregate in tandem with the bulk of the chromosomes. This comissegregation phenotype is lost when either of the Rep proteins is mutated or the *STB* locus is removed from the plasmid, thus identifying the partitioning system as the potential agent that couples segregation of the plasmid with that of the chromosomes.

The Cohesin Complex and Yeast Plasmid Partitioning

A breakthrough in elucidating the possible mechanism of 2-micron circle partitioning came with the discovery that the yeast cohesin complex is recruited specifically to the *STB* locus in a Rep1p-Rep2p-dependent manner (139). Cohesin associates with the plasmid during S phase in synchrony with chromosomal cohesin-binding sites, and the lifetime of this associated state is the same for the plasmid and the chromosomes. The disassembly of cohesin by Scc1p/Mcd1p cleavage during anaphase is as critical for plasmid segregation as it is for chromosome segregation. When a noncleavable version of Scc1p is overexpressed from an inducible promoter, the replicated plasmids do not split into two separate clusters, just as sister chromatids fail to separate. These findings imply that the cohesin complex plays similar functional roles during chromosome segregation and plasmid segregation. We propose that, concomitant with plasmid replication, cohesin holds sister clusters together until dissolution of the cohesin bridge annuls this union, and they are segregated in a one-to-one fashion (**Figure 12**).

The chromatin structure at the *STB* locus, which is dependent on the RSC2 chromatin remodeling complex, is important in equal plasmid segregation (140). In an $rsc2\Delta$ yeast strain, the 2-micron plasmid is lost at a high rate. Lack of Rsc2p blocks Rep1p association with *STB* and consequently cohesin recruitment by the plasmid (141, 142). Chromatin immunoprecipitation (ChIP) assays suggest that the functional nucleosome organization at *STB* is reset de novo during each cell cycle between the late G1 phase and early S phase. During this window, the Rep proteins transiently dissociate from *STB* and reassociate with it shortly afterwards, in time for cohesin recruitment. It is likely that the recycling of the Rep proteins, assembly of the functional chromatin architecture at *STB*, and the acquisition of cohesin by the plasmid are tightly coordinated with DNA replication.

The Mitotic Spindle Promotes Recruitment of Cohesin by the Yeast Plasmid

The yeast mitotic spindle promotes 2-micron circle segregation in quite an unconventional manner (143). Normally, the plasmid cluster has a specific nuclear address close to the spindle pole, and preparations of chromosome spreads reveal the presence of the plasmid. It is not clear whether plasmids are directly attached to the chromosomes or are anchored

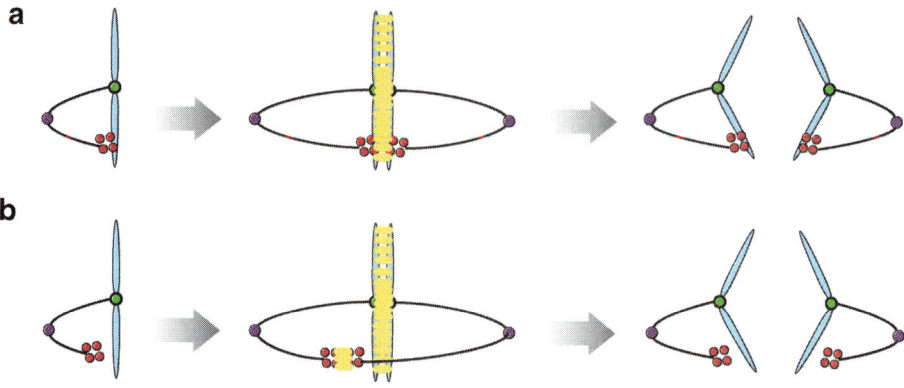

Figure 12

Models for segregation of the yeast 2-micron plasmid. (*a*) In this hitchhiking model, the plasmid cluster is tethered to a chromosome. Cohesin assembly at the *STB* locus during DNA replication keeps the duplicated clusters paired, facilitating the attachment of the second cluster to the sister chromosome. Following disassembly of cohesin, each of the two segregating sisters carries with it one plasmid cluster. (*b*) In the chromosome-independent model, plasmid segregation is still dependent on cohesin-mediated pairing and unpairing but takes place without chromosome assistance. The model does not specify how the driving force for plasmid segregation is derived. Recent findings demonstrate that precise nuclear localization of the plasmid and cohesin acquisition by it are dependent on the nuclear spindle. Perhaps the plasmid segregates in a spindle-associated fashion.

in some way to the same subcellular structures as chromosomal domains. When spindle integrity is disrupted by microtubule depolymerizing agents or by specific tubulin mutations, the plasmid cluster becomes less compact and loses its precise nuclear localization. Concomitantly, chromosome spreads fail to display the plasmid, and cohesin assembly at *STB* does not occur. A role for the mitotic spindle in cohesin acquisition is quite surprising because neither centromeric cohesion nor chromosome arm cohesion is dependent on the spindle. Plasmid molecules replicate normally in the absence of spindle, just as chromosomes do; when the spindle is allowed to reform, cohesin associates with *STB*. However, this postreplicative association is not effective in plasmid segregation, presumably because replication-dependent one-to-one pairing of sister clusters is a mandatory step in partitioning. Chromosomes, in contrast, having been cohesed in a replication-dependent manner, go on to form bipolar spindle attachment and segregate normally. It is only through spindle disassembly or through mutations of the partitioning system itself that we have been able to uncouple 2-micron circle segregation from chromosome segregation.

Models for Yeast Plasmid Segregation

Two general models under consideration for plasmid segregation in yeast are outlined in **Figure 12**. According to the hitchhiking model, the plasmid cluster is tethered to one of the chromosomes, perhaps with the assistance of the Rep proteins bound at *STB*. DNA replication and cohesin-mediated DNA bridging result in a pair of sister clusters, linked to each other as well as attached to sister chromatids. Subsequent cohesin disassociation will permit the clusters to hitchhike in opposite directions on the sister chromosomes. Stable propagation by attachment to host chromosomes has been demonstrated for bovine papilloma and Epstein-Barr viruses, whose genomes have an episomal existence (144). In the chromosome-independent model, cohesin still mediates the one-to-one segregation of the plasmid clusters but without physical linkage between plasmid and chromosome. How the unpaired plasmid clusters move away from each other is not specified in the model. Perhaps this movement occurs in association with the spindle or with a subnuclear entity that is partitioned evenly between daughter cells.

In both bacterial plasmids and the yeast plasmid, precise cellular localization mediated through their respective partitioning systems appears to be an important spatial determinant for proper segregation. Positioning of the 2-micron circle is critically dependent on the integrity of the mitotic spindle. The role of host factors in specifying the locations of bacterial plasmids is not understood. Subsequent events in partitioning are quite different for the bacterial and yeast plasmid systems. As suggested by the example of the R1 plasmid, the *par* system appears to be functionally autonomous and mediates segregation apparently independently of the chromosome segregation machinery. The yeast plasmid, by contrast, utilizes its partitioning system to gain access to the highly efficient pathway that faithfully segregates the chromosomes of its host.

gation) underneath the cell membrane. Targeted and programmed inactivation of MreB in *Caulobacter* specifically affects the poleward migration of *oriC*, with no effect on DNA replication per se or the segregation of the remainder of the chromosome (146). The role of the ParM helical filament in the segregation of the R1 plasmid was discussed above. The ring structure formed by the tubulin-like FtsZ protein to mark the bacterial division site (whose assembly and disassembly is regulated by the collaborative action of several proteins) may itself be a helix with a highly compressed pitch (147). The location of the ring is specified through a dual-protein (MinC-MinD) harmonic oscillator, set up by the MinCDE system, that sweeps from cell pole to cell pole in a helical wave (148; J. Lutkenhaus, unpublished results). Similarly, oscillations by partitioning ATPases may help localize plasmids or chromosomal replication origins within bacterial cells (124). Rod-shaped bacteria appear to have evolved a system of molecular spirals and oscillators for maintaining cell shape, mediating cell growth, and specifying important cellular landmarks.

THE CYTOSKELETON, DYNAMIC PROTEIN RINGS, COILS, AND OSCILLATORS IN BACTERIAL CELL DIVISION AND CHROMOSOME SEGREGATION

The participation of cytoskeletal structures, such as a dynamic scaffolding to regulate cell shape and movement, chromosome segregation, and cell division (once thought to be an exclusive feature of the eukaryotic cell), has been unveiled in bacteria as well. All three signature elements of the eukaryotic cytoskeleton, actin, tubulin, and intermediate filaments, play important roles in cellular organization and dynamics in bacteria (145). The actin-like Mre/Mbl proteins form extended continuously moving helical filaments (which are well suited for transporting cellular components and structures and, perhaps, for promoting chromosome segre-

TOPOISOMERASES, SITE-SPECIFIC RECOMBINASES AND DNA TRANSLOCASES: FAITHFUL CHROMOSOME SEGREGATION AND GENOME MAINTENANCE

Topoisomerases are indispensable in overcoming the topological barriers to replication and in completely undoing the inevitable catenane linkages between daughter duplexes. In *E. coli*, DNA gyrase is primarily responsible for removing positive supercoils ahead of the fork, whereas topo IV, a type II topoisomerase, almost exclusively resolves precatenane and catenane links between nascent duplexes (3, 149). Topo III, a type I topoisomerase, may also contribute to precatenane resolution by acting at the single-stranded DNA present at the replication fork (150).

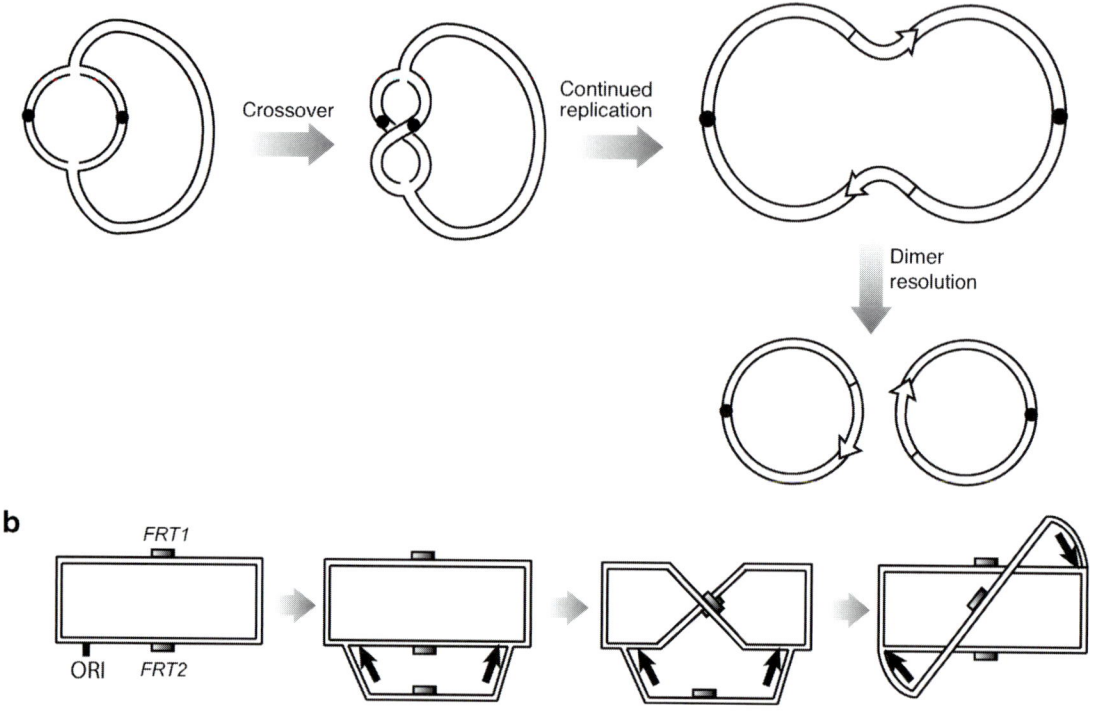

Figure 13

Site-specific recombination promotes genome segregation or copy-number maintenance. (*a*) In *E. coli*, resolution of a chromosome dimer resulting from homologous recombination is mediated by the XerC/XerD site-specific recombinase. Coordination of recombination and cell division is mediated through the FtsK motor protein that localizes at the division septum and also activates the XerD subunit of the recombinase. (*b*) Copy-number amplification of the 2-micron yeast plasmid is triggered by a replication-coupled recombination event mediated by the Flp site-specific recombinase. During bidirectional replication, the origin-proximal Flp recombination target site 2 (*FRT2*) is duplicated before the distal *FRT1*. Recombination between one of the duplicated sites and the unduplicated site inverts the direction of one of the replication forks and triggers the amplification process.

Whereas decatenation is critical in the segregation of both circular and linear chromosomes, circularity further complicates matters for a unit copy genome. A single crossover or an odd number of crossovers during homologous recombination will result in a chromosome dimer that has to be resolved into monomers (**Figure 13***a*). This reaction is mediated in *E. coli* by a tyrosine family site-specific recombinase XerC/XerD, and by its relatives in other bacteria (151). The location of the recombination target site *dif* near the replication terminus on the *E. coli* chromosome helps coordinate the recombination event with the act of cell division. A central player in this regulation is FtsK, a DNA-translocating ATPase that localizes to the division septum (11, 12, 152, 153). There are multiple roles for FtsK in bacterial chromosome segregation (152, 154): restraining *terC* regions of chromosome dimers at the mid-cell position, promoting synapsis of *dif* sites by XerC/XerD, recruiting topo IV and XerC/XerD to the division septum, activating decatenation and chromosome dimer resolution, and clearing DNA away from the constricting septum.

Site-specific recombination serves a different purpose for the yeast plasmid in the context of genome maintenance, namely, copy-number amplification. The highly asymmetric location of the plasmid replication origin with respect to the head-to-head *FRT* sites is a clever design for increasing copy number without violating the ban on more than one replication initiation event per cell cycle. As suggested by Futcher (155), an appropriately timed recombination reaction during bidirectional plasmid replication can invert one of the forks with respect to the other (**Figure 13b**). The two unidirectional forks chasing each other around the circular template will give rise to an amplicon comprised of multiple tandem copies of the plasmid. A second recombination event may terminate amplification by restoring bidirectionality to the forks. Individual copies of the plasmid can be resolved from the amplicon by homologous recombination or by Flp-mediated site-specific recombination.

The actions of XerC/XerD and Flp, both tyrosine recombinases, have starkly different physiological consequences. The former neutralizes the threat posed by homologous recombination to equal chromosome segregation; the latter negates potential erosion in plasmid copy number caused by missegregation. Thus, two related recombination systems contribute to genome maintenance and integrity via chemically similar yet biologically distinct pathways.

ADDENDUM

We have listed below some of the features of plasmid and chromosome segregation that have come to light or been brought to our notice since submission of this chapter.

1. There have been new revelations on the role of ParA ATPases in proper positioning and segregation of bacterial plasmids (and perhaps chromosomes as well). Partitioning mechanisms based on dynamic protein filaments, first demonstrated for the R1 plasmid, appear to be more widespread. The partitioning ATPase SopA of the F plasmid polymerizes into filaments in vitro in an ATP-dependent manner and elongates at a rate consistent with plasmid separation in vivo (156). Furthermore, SopA undergoes cycles of polymerization/depolymerization inside the cell and shuttles back and forth between nucleoprotein complexes consisting of SopB associated with the plasmid partitioning locus *sopC*. Dynamic polymerization of SopA is likely the driving force for F plasmid separation.

2. Although the mechanism by which oscillations of filament forming ATPases mediate plasmid segregation is unknown, a simple model based on the outward force exerted by ParM filaments on the R plasmid has been posited (157). The assumption is that the ATPase oscillation occurs over the distance of the nucleoid, whose borders provide toeholds for the oscillating protein. According to the model, the force exerted on a single plasmid focus over an oscillation cycle will localize it to the center of the cell. Under the same force, two plasmid foci will be dispatched to quarter cell positions.

3. The structure of the partitioning protein ParB of plasmid P1 has been solved (158). The rotational freedom of the DNA binding helix-turn-helix modules of ParB about a flexible linker explains how the protein recognizes the A and B boxes despite their complex arrangements within a sharply bent *parS* locus.

4. It has been brought to our attention (J. Pogliano, personal communication) that the model for P1 plasmid segregation described here (113) may not represent the whole picture. The notion that replicated plasmid molecules

are held at cell center until just before septation/division is not consistent with time-lapse assays conducted by Gordon et al. (159).

5. Simultaneous tracking of pairs of genetic loci and divisome proteins in *E. coli* provides new insights into how DNA replication, chromosome segregation, chromosome organization, and cell division are related to each other. Observations on different loci in the replication termination region (*ter*) suggest an asymmetric pattern of segregation of leading and lagging strand templates following their duplication (160).

6. Further evidence has been provided in support of the notion that the cohesin complex topologically entraps sister chromatids by forming a ring around them (161). Circular minichromosomes isolated from yeast nuclei retain some tightly bound cohesin, and the association between DNA and protein can be relieved by cleavage of either a cohesin component by a protease or the minichromosome by a restriction enzyme.

SUMMARY POINTS

1. Within the bacterial nucleoid, chromosome replication is subjected to strict spatial and temporal controls. In *C. crescentus*, chromosomal loci display a remarkably ordered organization. The directed movement of daughter duplexes as they are formed, likely coupled with DNA compaction, recreates the original chromosome order in newborn cells.

2. Eukaryotic chromosome segregation is facilitated by two architecturally similar protein assemblies, the cohesin and condensin complexes. Cohesin is central to a sister chromatid pairing and unpairing mechanism that retains replication memory, distinguishes chromosome sisters from homologues, and accommodates equational or reductional modes of segregation. Condensin promotes ordered intramolecular compaction of DNA and prevents chromosome entanglement during segregation.

3. Single-copy or low-copy bacterial plasmids harbor partitioning proteins and partitioning loci that function collaboratively to impose spatial and temporal controls, which ensure efficient and faithful plasmid segregation.

4. The multicopy yeast plasmid, because of its clustered organization, is effectively a single-copy entity in segregation. The Rep-*STB* partitioning system promotes a one-to-one segregation of duplicated plasmid clusters via the cohesion-mediated pairing and unpairing strategy, which is analogous to that utilized by the yeast chromosomes.

5. Cytoskeletal structures play a central role in prokaryotic genome segregation, as illustrated by the importance of the FtsZ ring in specifying the bacterial cell division septum, the ParM filament in R1 plasmid partitioning, and the MreB filament in *Caulobacter* chromosome segregation.

6. DNA topoisomerases, translocases, and site-specific recombinases contribute to genome segregation and integrity by promoting topological unlinking, chromosome dimer resolution, and copy-number maintenance.

FUTURE ISSUES TO BE RESOLVED

1. The generality of the impeccable order in organization and segregation revealed for the *Caulobacter* genome needs to be verified in other bacterial systems.

2. With the exception of the R1 plasmid system, investigations on the partitioning of bacterial plasmids such as P1 and F call for a deliberate shift in emphasis from phenomenology to mechanism.

3. The demonstrated roles for actin-like filaments in R1 plasmid and *Caulobacter* chromosome segregation demand incisive studies on the involvement of the bacterial cytoskeleton in plasmid and chromosome segregation mechanisms in general.

4. The question whether cohesin-mediated segregation of the yeast 2-micron plasmid occurs in a chromosome-tethered fashion or not awaits resolution.

5. The advent of single-molecule studies is likely to shed new light on the role of DNA motor proteins, condensins, and cohesins in ordered, rapid, and large-scale transport of chromosomes during segregation.

ACKNOWLEDGMENTS

We thank S. Austin, D.J. Sherratt, W. Margolin, J. Pogliano, and L. Shapiro for providing the movies on bacterial chromosome/plasmid segregation. We acknowledge critical comments on sections of this chapter from D. Bates, T. Hirano, W. Margolin, J. Pogliano, and M. Thanbichler. We have been supported over the years by the National Institutes of Health, National Science Foundation, Robert F. Welch Foundation, Council for Tobacco Research, Texas Higher Education Coordinating Board, and Human Frontiers in Science Program.

LITERATURE CITED

1. Breier AM, Cozzarelli NR. 2004. *Proc. Natl. Acad. Sci. USA* 101:9175–76
2. Lau IF, Filipe SR, Soballe B, Okstad OA, Barre FX, Sherratt DJ. 2003. *Mol. Microbiol.* 49:731–43
3. Sherratt DJ. 2003. *Science* 301:780–85
4. Niki H, Hiraga S. 1998. *Genes Dev.* 12:1036–45
5. Draper GC, Gober JW. 2002. *Annu. Rev. Microbiol.* 56:567–97
6. Gordon GS, Wright A. 2000. *Annu. Rev. Microbiol.* 54:681–708
7. van Helvoort JM, Woldringh CL. 1994. *Mol. Microbiol.* 13:577–83
8. Webb CD, Graumann PL, Kahana JA, Teleman AA, Silver PA, Losick R. 1998. *Mol. Microbiol.* 28:883–92
9. Dworkin J, Losick R. 2002. *Proc. Natl. Acad. Sci. USA* 99:14089–94
10. Lemon KP, Grossman AD. 2001. *Genes Dev.* 15:2031–41
11. Pease PJ, Levy O, Cost GJ, Gore J, Ptacin JL, et al. 2005. *Science* 307:586–90
12. Saleh OA, Bigot S, Barre FX, Allemand JF. 2005. *Nat. Struct. Mol. Biol.* 12:436–40
13. Ben-Yehuda S, Fujita M, Liu XS, Gorbatyuk B, Skoko D, et al. 2005. *Mol. Cell* 17:773–82
14. Sharpe ME, Errington J. 1996. *Mol. Microbiol.* 21:501–9
15. Wu LJ, Errington J. 1997. *EMBO J.* 16:2161–69
16. Bath J, Wu LJ, Errington J, Wang JC. 2000. *Science* 290:995–97

17. Viollier PH, Thanbichler M, McGrath PT, West L, Meewan M, et al. 2004. *Proc. Natl. Acad. Sci. USA* 101:9257–62
18. Higgins NP. 1999. In *Organization of the Prokaryotic Genome*, ed. RL Charlebois, pp. 189–202. Washington, DC: ASM Press
19. Pettijohn DE. 1996. In *Escherichia coli and Salmonella*, ed. FC Neidhardt, pp. 158–66. Washington, DC: ASM Press
20. Sunako Y, Onogi T, Hiraga S. 2001. *Mol. Microbiol.* 42:1233–41
21. Li Y, Sergueev K, Austin S. 2002. *Mol. Microbiol.* 46:985–96
22. Bates D, Kleckner N. 2005. *Cell* 121:899–911
23. Hauf S, Watanabe Y. 2004. *Cell* 119:317–27
24. Nasmyth K. 2002. *Science* 297:559–65
25. Uhlmann F. 2004. *Exp. Cell Res.* 296:80–85
26. Musacchio A, Hardwick KG. 2002. *Nat. Rev. Mol. Cell Biol.* 3:731–41
27. Tanaka TU. 2002. *Curr. Opin. Cell Biol.* 14:365–71
28. Wang JC. 2002. *Nat. Rev. Mol. Cell Biol.* 3:430–40
29. Hirano T. 2005. *Curr. Biol.* 15: R265–75
30. Nasmyth K. 2005. *Cell* 120:739–46
31. Kolodner RD, Putnam CD, Myung K. 2002. *Science* 297:552–57
32. Bartek J, Lukas C, Lukas J. 2004. *Nat. Rev. Mol. Cell Biol.* 5:792–804
33. Taylor SS, Scott MI, Holland AJ. 2004. *Chromosome Res.* 12:599–616
34. Nasmyth K. 2001. *Annu. Rev. Genet.* 35:673–745
35. Hagstrom KA, Holmes VF, Cozzarelli NR, Meyer BJ. 2002. *Genes Dev.* 16:729–42
36. Jessberger R. 2003. *IUBMB Life* 55:643–52
37. Haering CH, Lowe J, Hochwagen A, Nasmyth K. 2002. *Mol. Cell* 9:773–88
38. Lowe J, Cordell SC, van den Ent F. 2001. *J. Mol. Biol.* 306:25–35
39. Anderson DE, Losada A, Erickson HP, Hirano T. 2002. *J. Cell Biol.* 156:419–24
40. Gruber S, Haering CH, Nasmyth K. 2003. *Cell* 112:765–77
41. Laloraya S, Guacci V, Koshland D. 2000. *J. Cell Biol.* 151:1047–56
42. Weber SA, Gerton JL, Polancic JE, DeRisi JL, Koshland D, Megee PC. 2004. *PLoS Biol.* 2:E260
43. Ciosk R, Shirayama M, Shevchenko A, Tanaka T, Toth A, Nasmyth K. 2000. *Mol. Cell* 5:243–54
44. Toth A, Ciosk R, Uhlmann F, Galova M, Schleiffer A, Nasmyth K. 1999. *Genes Dev.* 13:320–33
45. Carson DR, Christman MF. 2001. *Proc. Natl. Acad. Sci. USA* 98:8270–75
46. Lengronne A, Katou Y, Mori S, Yokobayashi S, Kelly GP, et al. 2004. *Nature* 430:573–78
47. Uhlmann F. 2001. *EMBO Rep.* 2:487–92
48. Alexandru G, Uhlmann F, Mechtler K, Poupart MA, Nasmyth K. 2001. *Cell* 105:459–72
49. Stemmann O, Zou H, Gerber SA, Gygi SP, Kirschner MW. 2001. *Cell* 107:715–26
50. Gimenez-Abian JF, Sumara I, Hirota T, Hauf S, Gerlich D, et al. 2004. *Curr. Biol.* 14:1187–93
51. Waizenegger IC, Hauf S, Meinke A, Peters JM. 2000. *Cell* 103:399–410
52. Stegmeier F, Visintin R, Amon A. 2002. *Cell* 108:207–20
53. D'Amours D, Stegmeier F, Amon A. 2004. *Cell* 117:455–69
54. Sullivan M, Higuchi T, Katis VL, Uhlmann F. 2004. *Cell* 117:471–82
55. Higuchi T, Uhlmann F. 2005. *Nature* 433:171–76
56. Ross KE, Cohen-Fix O. 2004. *Dev. Cell* 6:729–35
57. Katis VL, Matos J, Mori S, Shirahige K, Zachariae W, Nasmyth K. 2004. *Curr. Biol.* 14:2183–96

58. Rabitsch KP, Petronczki M, Javerzat JP, Genier S, Chwalla B, et al. 2003. *Dev. Cell* 4:535–48
59. Buonomo SB, Clyne RK, Fuchs J, Loidl J, Uhlmann F, Nasmyth K. 2000. *Cell* 103:387–98
60. Kitajima TS, Miyazaki Y, Yamamoto M, Watanabe Y. 2003. *EMBO J.* 22:5643–53
61. Siomos MF, Badrinath A, Pasierbek P, Livingstone D, White J, et al. 2001. *Curr. Biol.* 11:1825–35
62. Terret ME, Wassmann K, Waizenegger I, Maro B, Peters JM, Verlhac MH. 2003. *Curr. Biol.* 13:1797–802
63. Kitajima TS, Kawashima SA, Watanabe Y. 2004. *Nature* 427:510–17
64. Rabitsch KP, Gregan J, Schleiffer A, Javerzat JP, Eisenhaber F, Nasmyth K. 2004. *Curr. Biol.* 14:287–301
65. Katis VL, Galova M, Rabitsch KP, Gregan J, Nasmyth K. 2004. *Curr. Biol.* 14:560–72
66. Hirano T. 2004. *Cell Cycle* 3:26–28
67. Maddox PS, Oegema K, Desai A, Cheeseman IM. 2004. *Chromosome Res.* 12:641–53
68. Blackwell TK, Walker AK. 2002. *Genes Dev.* 16:769–72
69. Lavoie BD, Hogan E, Koshland D. 2004. *Genes Dev.* 18:76–87
70. Yu HG, Koshland DE. 2003. *J. Cell Biol.* 163:937–47
71. Chan RC, Severson AF, Meyer BJ. 2004. *J. Cell Biol.* 167:613–25
72. Siddiqui NU, Stronghill PE, Dengler RE, Hasenkampf CA, Riggs CD. 2003. *Development* 130:3283–95
73. Lieb JD, Capowski EE, Meneely P, Meyer BJ. 1996. *Science* 274:1732–36
74. Pogliano K, Pogliano J, Becker E. 2003. *Curr. Opin. Microbiol.* 6:586–93
75. Strunnikov A. 2006. *Plasmid.* 55:135–44
76. Britton RA, Lin DC, Grossman AD. 1998. *Genes Dev.* 12:1254–59
77. Jensen RB, Shapiro L. 1999. *Proc. Natl. Acad. Sci. USA* 96:10661–66
78. Niki H, Jaffe A, Imamura R, Ogura T, Hiraga S. 1991. *EMBO J.* 10:183–93
79. Holmes VF, Cozzarelli NR. 2000. *Proc. Natl. Acad. Sci. USA* 97:1322–24
80. Sawitzke JA, Austin S. 2000. *Proc. Natl. Acad. Sci. USA* 97:1671–76
81. Weitao T, Nordstrom K, Dasgupta S. 1999. *Mol. Microbiol.* 34:157–68
82. Mascarenhas J, Soppa J, Strunnikov AV, Graumann PL. 2002. *EMBO J.* 21:3108–18
83. Soppa J, Kobayashi K, Noirot-Gros MF, Oesterhelt D, Ehrlich SD, et al. 2002. *Mol. Microbiol.* 45:59–71
84. Schleiffer A, Kaitna S, Maurer-Stroh S, Glotzer M, Nasmyth K, Eisenhaber F. 2003. *Mol. Cell* 11:571–75
85. Volkov A, Mascarenhas J, Andrei-Selmer C, Ulrich HD, Graumann PL. 2003. *Mol. Cell. Biol.* 23:5638–50
86. Hagstrom KA, Meyer BJ. 2003. *Nat. Rev. Genet.* 4:520–34
87. Hirano M, Hirano T. 2004. *EMBO J.* 23:2664–73
88. Lammens A, Schele A, Hopfner KP. 2004. *Curr. Biol.* 14:1778–82
89. Harvey SH, Sheedy DM, Cuddihy AR, O'Connell MJ. 2004. *Mol. Cell. Biol.* 24:662–74
90. Lehman AR. 2005. *DNA Repair* 4:309–14
91. Strom L, Lindroos HB, Shirahige K, Sjogren C. 2004. *Mol. Cell* 16:1003–15
92. Torres-Rosell J, Machin F, Farmer S, Jarmuz A, Eydmann T, et al. 2005. *Nat. Cell Biol.* 7:412–19
93. Shin DS, Chahwan C, Huffman JL, Tainer JA. 2004. *DNA Repair* 3:863–73
94. Engelberg-Kulka H, Glaser G. 1999. *Annu. Rev. Microbiol.* 53:43–70
95. Gerdes K, Gultyaev AP, Franch T, Pedersen K, Mikkelsen ND. 1997. *Annu. Rev. Genet.* 31:1–31

96. Greenfield TJ, Ehli E, Kirshenmann T, Franch T, Gerdes K, Weaver KE. 2000. *Mol. Microbiol.* 37:652–60
97. Gerdes K, Moller-Jensen J, Jensen RB. 2000. *Mol. Microbiol.* 37:455–66
98. Hiraga S. 2000. *Annu. Rev. Genet.* 34:21–59
99. Funnell BE, Slavcev RA. 2004. In *Plasmid Biology*, ed. BE Funnell, PJ Phillips, pp. 81–103. Washington, DC: ASM Press
100. Bork P, Sander C, Valencia A. 1992. *Proc. Natl. Acad. Sci. USA* 89:7290–94
101. Koonin EV. 1993. *J. Mol. Biol.* 229:1165–74
102. Friedman SA, Austin SJ. 1988. *Plasmid* 19:103–12
103. Mori H, Mori Y, Ichinose C, Niki H, Ogura T, et al. 1989. *J. Biol. Chem.* 264:15535–41
104. Hayes F, Radnedge L, Davis MA, Austin SJ. 1994. *Mol. Microbiol.* 11:249–60
105. Hirano M, Mori H, Onogi T, Yamazoe M, Niki H, et al. 1998. *Mol. Gen. Genet.* 257:392–403
106. Yates P, Lane D, Biek DP. 1999. *J. Mol. Biol.* 290:627–38
107. Bouet JY, Funnell BE. 1999. *EMBO J* 18:1415–24
108. Jensen RB, Dam M, Gerdes K. 1994. *J. Mol. Biol.* 236:1299–309
109. Funnell BE. 1991. *J. Biol. Chem.* 266:14328–37
110. Gordon GS, Sitnikov D, Webb CD, Teleman A, Straight A, et al. 1997. *Cell* 90:1113–21
111. Niki H, Hiraga S. 1997. *Cell* 90:951–57
112. Onogi T, Miki T, Hiraga S. 2002. *J. Bacteriol.* 184:3142–45
113. Li Y, Austin S. 2002. *Plasmid* 48:174–78
114. Ho TQ, Zhong Z, Aung S, Pogliano J. 2002. *EMBO J.* 21:1864–72
115. Erdmann N, Petroff T, Funnell BE. 1999. *Proc. Natl. Acad. Sci. USA* 96:14905–10
116. Edgar R, Chattoraj DK, Yarmolinsky M. 2001. *Mol. Microbiol.* 42:1363–70
117. Fung E, Bouet JY, Funnell BE. 2001. *EMBO J.* 20:4901–11
118. Libante V, Thion L, Lane D. 2001. *J. Mol. Biol.* 314:387–99
119. Li Y, Dabrazhynetskaya A, Youngren B, Austin S. 2004. *Mol. Microbiol.* 53:93–102
120. Moller-Jensen J, Borch J, Dam M, Jensen RB, Roepstorff P, Gerdes K. 2003. *Mol. Cell* 12:1477–87
121. Jensen RB, Gerdes K. 1999. *EMBO J.* 18:4076–84
122. Moller-Jensen J, Jensen RB, Lowe J, Gerdes K. 2002. *EMBO J.* 21:3119–27
123. Jensen RB, Lurz R, Gerdes K. 1998. *Proc. Natl. Acad. Sci. USA* 95:8550–55
124. Gerdes K, Moller-Jensen J, Ebersbach G, Kruse T, Nordstrom K. 2004. *Cell* 116:359–66
125. Yamaichi Y, Niki H. 2000. *Proc. Natl. Acad. Sci. USA* 97:14656–61
126. Lin DC, Grossman AD. 1998. *Cell* 92:675–85
127. Ireton K, Gunther NW 4th, Grossman AD. 1994. *J. Bacteriol.* 176:5320–29
128. Wu LJ, Errington J. 2003. *Mol. Microbiol.* 49:1463–75
129. Mohl DA, Easter J Jr, Gober JW. 2001. *Mol. Microbiol.* 42:741–55
130. Mohl DA, Gober JW. 1997. *Cell* 88:675–84
131. Godfrin-Estevenon AM, Pasta F, Lane D. 2002. *Mol. Microbiol.* 43:39–49
132. Kim HJ, Calcutt MJ, Schmidt FJ, Chater KF. 2000. *J. Bacteriol.* 182:1313–20
133. Lewis RA, Bignell CR, Zeng W, Jones AC, Thomas CM. 2002. *Microbiology* 148:537–48
134. Fekete RA, Chattoraj DK. 2005. *Mol. Microbiol.* 55:175–83
135. Yamaichi Y, Niki H. 2004. *EMBO J.* 23:221–33
136. Jayaram M, Mehta S, Uzri D, Voziyanov Y, Velmurugan S. 2004. *Prog. Nucleic Acids Res. Mol. Biol.* 77:127–72
137. Velmurugan S, Yang XM, Chan CS, Dobson M, Jayaram M. 2000. *J. Cell Biol.* 149:553–66
138. Scott-Drew S, Wong CM, Murray JA. 2002. *Cell Biol. Int.* 26:393–405

139. Mehta S, Yang XM, Chan CS, Dobson MJ, Jayaram M, Velmurugan S. 2002. *J. Cell Biol.* 158:625–37
140. Wong MC, Scott-Drew SR, Hayes MJ, Howard PJ, Murray JA. 2002. *Mol. Cell. Biol.* 22:4218–29
141. Huang J, Hsu JM, Laurent BC. 2004. *Mol. Cell* 13:739–50
142. Yang XM, Mehta S, Uzri D, Jayaram M, Velmurugan S. 2004. *Mol. Cell. Biol.* 24:5290–303
143. Mehta S, Yang XM, Jayaram M, Velmurugan S. 2005. *Mol. Cell. Biol.* 25:4283–98
144. McBride AA, McPhillips MG, Oliveira JG. 2004. *Trends Microbiol.* 12:527–29
145. Michie KA, Löwe J. 2006. *Annu. Rev. Biochem.* 75:467–92
146. Gitai Z, Dye NA, Reisenauer A, Wachi M, Shapiro L. 2005. *Cell* 120:329–41
147. Margolin W. 2003. *Curr. Biol.* 13:R16–18
148. Lutkenhaus J. 2002. *Curr. Opin. Microbiol.* 5:548–52
149. Espeli O, Marians KJ. 2004. *Mol. Microbiol.* 52:925–31
150. Nurse P, Levine C, Hassing H, Marians KJ. 2003. *J. Biol. Chem.* 278:8653–60
151. Barre F-X, Sherratt DJ. 2002. In *Mobile DNA II*, ed. NL Craig, R Craigie, M Gellert, AM Lambowitz, pp. 149–61. Washington, DC: ASM Press
152. Massey TH, Aussel L, Barre FX, Sherratt DJ. 2004. *EMBO Rep.* 5:399–404
153. Weiss DS. 2004. *Mol. Microbiol.* 54:588–97
154. Espeli O, Lee C, Marians KJ. 2003. *J. Biol. Chem.* 278:44639–44
155. Futcher AB. 1986. *J. Theor. Biol.* 119:197–204
156. Lim GE, Derman AI, Pogliano J. 2005. *Proc. Natl. Acad. Sci. USA* 102:17658–63
157. Ebersbach G, Gerdes K. 2005. *Annu. Rev. Genet.* 39:453–79
158. Schumacher MA, Funnell BE. 2005. *Nature* 438:516–19
159. Gordon S, Rech J, Lane D, Wright A. 2004. *Mol. Microbiol.* 51:461–69
160. Wang X, Possoz C, Sherratt DJ. 2005. *Genes Dev.* 19:2367–77
161. Ivanov D, Nasmyth K. 2005. *Cell* 122:849–60

Chromatin Modifications by Methylation and Ubiquitination: Implications in the Regulation of Gene Expression

Ali Shilatifard

Saint Louis University School of Medicine and the Saint Louis University Cancer Center, St. Louis, Missouri 63104; email: shilatia@slu.edu

Key Words

histone methylation, histone ubiquitination, epigenetic regulation, MLL, transcriptional elongation

Abstract

It is more evident now than ever that nucleosomes can transmit epigenetic information from one cell generation to the next. It has been demonstrated during the past decade that the posttranslational modifications of histone proteins within the chromosome impact chromatin structure, gene transcription, and epigenetic information. Multiple modifications decorate each histone tail within the nucleosome, including some amino acids that can be modified in several different ways. Covalent modifications of histone tails known thus far include acetylation, phosphorylation, sumoylation, ubiquitination, and methylation. A large body of experimental evidence compiled during the past several years has demonstrated the impact of histone acetylation on transcriptional control. Although histone modification by methylation and ubiquitination was discovered long ago, it was only recently that functional roles for these modifications in transcriptional regulation began to surface. Highlighted in this review are the recent biochemical, molecular, cellular, and physiological functions of histone methylation and ubiquitination involved in the regulation of gene expression as determined by a combination of enzymological, structural, and genetic methodologies.

Contents

INTRODUCTION 244
THE BASIC CHEMISTRY AND SITES OF POSTTRANSLATIONAL MODIFICATIONS ON HISTONES BY METHYLATION 245
THE ENZYMATIC MACHINERY INVOLVED IN HISTONE METHYLATION 246
THE ROLE OF METHYLATION OF HISTONE H3 ON LYSINE 9 IN THE INITIATION AND MAINTENANCE OF HETEROCHROMATIC SILENCING 248
HISTONE H3 LYSINE 4 METHYLATION BY COMPASS AND ITS MAMMALIAN HOMOLOGUE, THE MLL COMPLEX 251
THE ROLE OF SET2 IN METHYLATING LYSINE 36 OF HISTONE H3 253
HISTONE H3 LYSINE 27 METHYLATION, H2A LYSINE 119 UBIQUITINATION, REGULATION OF POLYCOMB-GROUP SILENCING, AND X-CHROMOSOME INACTIVATION 254
NON-SET DOMAIN-CONTAINING LYSINE-SPECIFIC HISTONE METHYLTRANSFERASES 256
HISTONE ARGININE METHYLTRANSFERASES 258
HISTONE MONOUBIQUITINATION IN SIGNALING FOR HISTONE METHYLATION AND THE REGULATION OF GENE EXPRESSION 259
HISTONE METHYLATION AND TRANSCRIPTIONAL MEMORY 260
DEMETHYLATING HISTONES.. 261
FUTURE DIRECTIONS.......... 263

INTRODUCTION

PcG: Polycomb group

Su(var): suppression of position-effect variegation

Eukaryotic DNA, which is several meters long, must remain functional when packaged into chromatin (1–5). We still have much to learn about the molecular mechanisms required for the packaging of the genetic information and how the RNA polymerase II (RNAPII) machinery and its regulatory factors access packaged DNA sequences. Several factors, including DNA methylation, histone modifications, and small nuclear RNAs, have been implicated in the regulation of transcription from chromatin. This mode of regulation has been referred to as "epigenetic regulation," which denotes an inherited state of gene regulation that is independent of the genetic information encoded within DNA itself (6–8).

Several classes of proteins required for proper gene expression through the control of chromatin structures have been identified. Two protein families, the trithorax (TRX) group and the Polycomb group (PcG), have been shown to play opposing roles in this process (9–11). These two chromatin-associated classes of proteins function by activating and repressing transcription, respectively. Both classes of proteins contain a 130– to 140–amino acid motif called the SET domain (12, 13). This domain takes its name from the *Drosophila* proteins Su(var)3–9, Enhancer of zeste [E(z)], and trithorax (SET) (14, 15) and has recently been shown to be involved in methylating histones within chromatin (16–19). The SET domain is found in a variety of chromatin-associated proteins. The genes encoding for these proteins can also mutate and/or translocate to form fusions with other proteins, resulting in the pathogenesis of hematological malignancies (20–22).

It has been known for some time now that proteins can be posttranslationally modified via the enzymatic addition of methyl groups from the donor *S*-adenosylmethionine (SAM) to proteins on either carboxyl groups or the nitrogen atoms in the N-terminal and side-chain positions (23). The addition of methyl esters on the carboxyl group of proteins is potentially reversible; however, posttranslational modifications by methylation occurring on nitrogen atoms in the N-terminal and/or side-chain positions of proteins are generally considered very stable and in some forms irreversible. Nevertheless, such posttranslational modifications of proteins by methylation have wide-ranging effects, including transcriptional regulation, protein targeting, signal transduction, RNA metabolism, and modulation of enzymatic activity, as well as roles in several behavioral phenomena, such as chemotaxis (17–19, 23–25). Despite the critical function of protein methylation in the regulation of biology, we know very little about the exact molecular mechanism of protein methylation. Because this review focuses on the process of histone methylation and the consequences of this modification on the regulation of gene expression, the discussion henceforth is limited to histone methylation.

THE BASIC CHEMISTRY AND SITES OF POSTTRANSLATIONAL MODIFICATIONS ON HISTONES BY METHYLATION

It was demonstrated that histones can be methylated either on their arginine or lysine residues (26–29). The lysine residue of histones can be methylated on the ε-nitrogen by either the SET domain- or non-SET domain-containing lysine histone methyltransferases (KHMTase). As shown in **Figure 1a**, lysine methylation can occur in mono-, di-, or trimethylated forms. The arginine residue in histone proteins, however, can only be mono- or dimethylated. The dimethylation of the arginine residue can be found in either symmetric or asymmetric configurations (**Figure 1b**).

Initial investigations taking advantage of metabolic labeling and bulk sequencing provided a large body of evidence that residues within histones are methylated. However, the first experimental evidence supporting a link between histone methylation and transcriptional regulation was not reported until recently (16, 30–33). Studies during the past decade have provided evidence that chromatin is highly modified posttranslationally in several different ways and that such modifications play pivotal roles in the regulation of gene expression.

Initial investigations demonstrated that chromatin appears as a series of "beads on a string," with the beads being the individual nucleosomes (1, 5). Since the discovery of the beads on a string, it has been revealed that each nucleosome consists of eight core histone proteins (two each of H3, H4, H2A, and H2B), which are wrapped by 147 base pairs of DNA in a left-handed superhelix, forming the intact nucleosome (34). Extending away from the core of the nucleosome are the histone tails. Histone tails can be modified and are available for interactions with DNA and/or other proteins. It has been demonstrated that histone tails are the site of interaction for diverse classes of enzymatic machinery capable of covalently modifying the tails through acetylation, phosphorylation, sumoylation, ubiquitination, and methylation. **Figure 2** demonstrates the known sites and enzymatic machinery involved in the modification of histones by methylation and ubiquitination. For reviews on other modifications of histones, please see References 35–38.

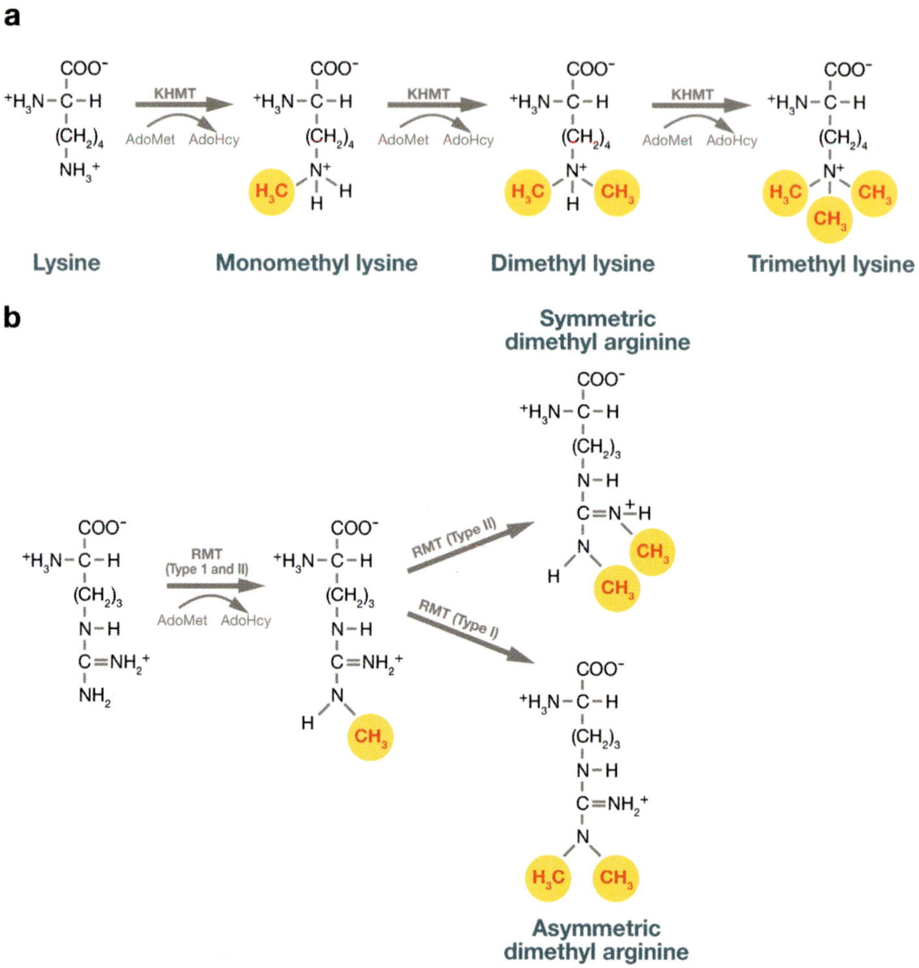

Figure 1

The chemistry of methylation on lysine and arginine residues of histones. (*a*) The lysine residues on histones can be mono-, di-, and trimethylated by histone methyltransferases (HMTases) such as the Set1/COMPASS or its human homologue, the MLL complex. Recent studies have demonstrated the presence of multiple roles for the different forms of lysine methylation (64). (*b*) The arginine residues on histones can be mono- and dimethylated as well. Type I and II protein arginine methyltransferases catalyze asymmetric and symmetric dimethylation, respectively.

THE ENZYMATIC MACHINERY INVOLVED IN HISTONE METHYLATION

The process of histone methylation was described many years ago, but the biological significance of this modification and its role in the regulation of gene expression had remained elusive. The attachment of methyl groups from the donor SAM to histone proteins occurs predominantly on specific lysine or arginine residues on histones H3 and H4 (**Figure 2**). Initial mass spectrometric studies demonstrated that histone lysine residues are mono-, di-, or trimethylated in vivo. Recent biochemical studies have confirmed this observation and have demonstrated that the ε-amino group of histone lysines residues can accept up to three methyl groups, therefore

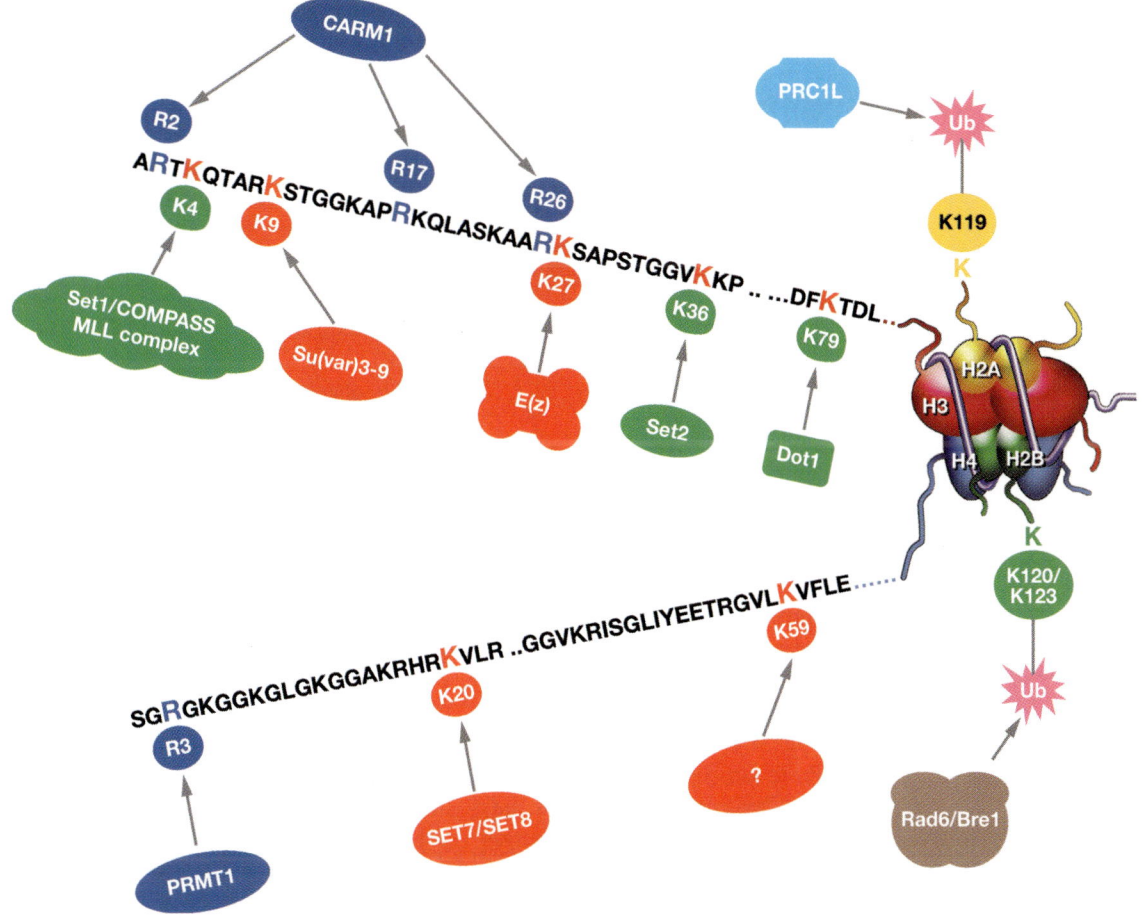

Figure 2

The known enzymatic machinery involved in the methylation of lysine and arginine residues and ubiquitination of the lysine residues of histones. The N-terminal amino acid sequences of histones H3 and H4 are shown along with the positions of specific methylation sites and the known enzymatic machinery responsible for the corresponding modification. Ubiquitination of the lysine residues on histone H2B and H2A by Rad6/Bre1 and the polycomb repressive complex 1-like (PRC1-L) is shown.

supporting the idea that histone lysine residues can be mono-, di- or trimethylated (**Figure 1***a*). As will be discussed later, although the enzymatic machinery capable of removing mono- and dimethylated histones has been described, there are no known enzymes that can remove a methyl group from a trimethylated histone. Therefore, histone methylation (specifically histone trimethylation) is considered a much more stable mark in comparison to other modes of histone modification such as phosphorylation, ubiquitination, or acetylation.

During the past few years, remarkable progress has been made in identifying the enzymatic machinery involved in the posttranslational modification of histones by methylation. These enzymes have been grouped into several classes, including (*a*) the lysine-specific SET domain-containing histone methyltransferases (HMTases) involved in methylation of lysines 4, 9, 27, and 36 of histone

PEV: position-effect variegation

H3 and lysine 20 of histone H4; (*b*) non-SET domain-containing lysine methyltransferases involved in methylating lysine 79 of histone H3; and (*c*) arginine methyltransferases involved in methylating arginine 2, 17, and 26 of histone H3 as well as arginine 3 of histone H4.

THE ROLE OF METHYLATION OF HISTONE H3 ON LYSINE 9 IN THE INITIATION AND MAINTENANCE OF HETEROCHROMATIC SILENCING

In metazoan development, different cells within an organism become committed to different fates, partly through heritable, quasi-stable changes in gene expression. Several families of proteins were initially genetically characterized to play a fundamental role in the process of development and segmentation. Two such families of proteins include the products of the *trithorax* (*trx*) group and the *Polycomb* (*Pc*) groups of genes (9–11). TrxG and PcG group gene products are chromatin-associated proteins required for transcription of the clustered homeotic genes in the *Bithorax* and *Antennapedia* gene complexes, and these gene products are known to function by activating and repressing transcription, respectively (9–11). A common feature of the Trx and PcG group proteins is the presence of a 130– to 140–amino acid motif called the SET domain (12–13). Since the completion of genome sequencing from several organisms including humans, it has been shown that the SET domain is found in a variety of chromatin-associated proteins. Some of the first SET domain-containing HMTases to be identified included the products of *Su(var)3–9* (suppressor of position-effect variegation) in *Drosophila*, its homologue *Clr4* (cryptic locus regulator) in *Schizosaccharomyces pombe*, and *SUV39H1* and *SUV39H2* in humans (16). This class of enzymes was initially demonstrated to be required for the proper formation of heterochromatin, which in higher eukaryotic organisms is characterized by histone hypoacetylation and the methylation of histone H3 on lysine 9 (6). The products of each of these genes are responsible for the catalysis of histone H3 K9 methylation in their respective species. Previous studies from several laboratories genetically demonstrated that Su(var)3–9 was an effective modifier of position-effect variegation (PEV) in *Drosophila*, suggesting a direct role for this factor in heterochromatin formation (15).

The phenomenon of PEV (defined as a variegation caused by the inactivation of a gene in some cells through its abnormal juxtaposition with heterochromatin) was first described by Muller under the label of "eversporting displacement" (39). This phenomenon was attributed by Muller to either chromosomal instability or an effect of chromosomal position and interaction with local gene products. For decades now, PEV has provided the scientific community with a critical entry point for understanding the role that histone proteins play in the formation and maintenance of heterochromatin (6, 40). Initial genetic studies demonstrated that when a transcriptionally active euchromatic gene is brought near the pericentric heterochromatin, the gene becomes silenced. Such an alteration in the pattern of gene expression for an active gene has been proposed to be caused by the spreading of heterochromatin into the active gene, resulting in its inactivation. However, not all *Drosophila* cells inactivate a euchromatic gene when juxtaposed to the pericentric heterochromatin (**Figures 3** and **4**).

Because PEV can be measured, genetic screens searching for modifiers of PEV were initiated many years ago, resulting in the identification of both enhancers and suppressors of PEV (6, 15, 40–42). Mutations in the suppressor of variegation genes—known as *Su(var)* genes—resulted in the identification of factors such as Su(var)2–5 or HP1 (a chromodomain-containing protein) (41, 42) and Su(var)3–9, which we now know encodes a histone H3

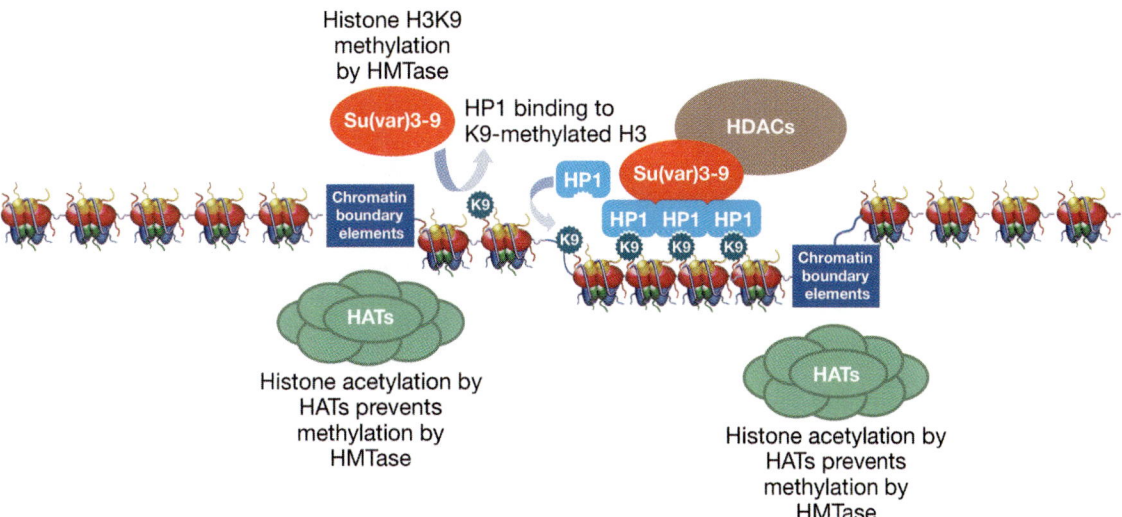

Figure 3

The stepwise model for the formation of heterochromatin assembly. Histone deacetylation by the histone deacetylase complexes (HDACs) allows for the methylation of histone H3 on lysine 9 to take place by the Su(var)3–9 family of HMTase. The HP1 protein can recognize and bind to the lysine 9-methylated histone H3 and continue the assembly of heterochromatin. The progression of HP1 binding in the heterochromatin can be stopped by "boundary elements," which are considered to be potential sites for the recruitment of histone acetyltransferases that would prevent the methylation of K9 of histone H3 by the Su(var)3–9 family of HMTase.

lysine 9 methyltransferase (15, 16). Biochemical studies have demonstrated that the chromodomain of HP1 can specifically recognize histone H3 methylated at lysine 9 (43–45). This recognition of lysine 9-methylated H3 by HP1, in part, is required for the establishment and maintenance of heterochromatin (**Figure 3**).

An important question in the field has been how Su(var)3–9 recognizes regions of chromatin to be silenced, subsequently targeting them for methylation of histone H3 at lysine 9 to assemble heterochromatin. A recent surprising discovery implicated repetitive DNA elements and RNA interference (RNAi) machinery in recruiting the S. pombe homologue of Su(var)3–9, the Clr4 protein, to the centromeric heterochromatic region (8, 46–48). The centromeric repeats are transcribed bidirectionally, resulting in the production of noncoding double-stranded RNA, which is then processed into interfering RNA or short heterochromatic RNA (shRNA) by the RNAi machinery (8, 46) (**Figure 4**). These observations have suggested that production of shRNA in heterochromatic regions is involved in the recruitment of histone H3 lysine 9 methyltransferase machinery to establish lysine 9 methylation, resulting in the recruitment of HP1 (8, 46). Once HP1 is recruited to heterochromatin, the process of heterochromatin formation is initiated. The spreading of heterochromatin occurs through self-association of HP1 with other HP1 molecules and the use of its chromoshadow domain to recruit additional histone H3 lysine 9 methyltransferase machinery, further catalyzing histone H3 lysine 9 methylation and the recruitment of more HP1 (8, 49–50) (**Figure 4**). Although in vitro studies have demonstrated that the interaction of HP1 with methylated chromatin results in the repression of transcription (51), it is not clear at this time how these events lead to gene silencing in vivo. Furthermore, we do not know how the spreading of heterochromatin is regulated.

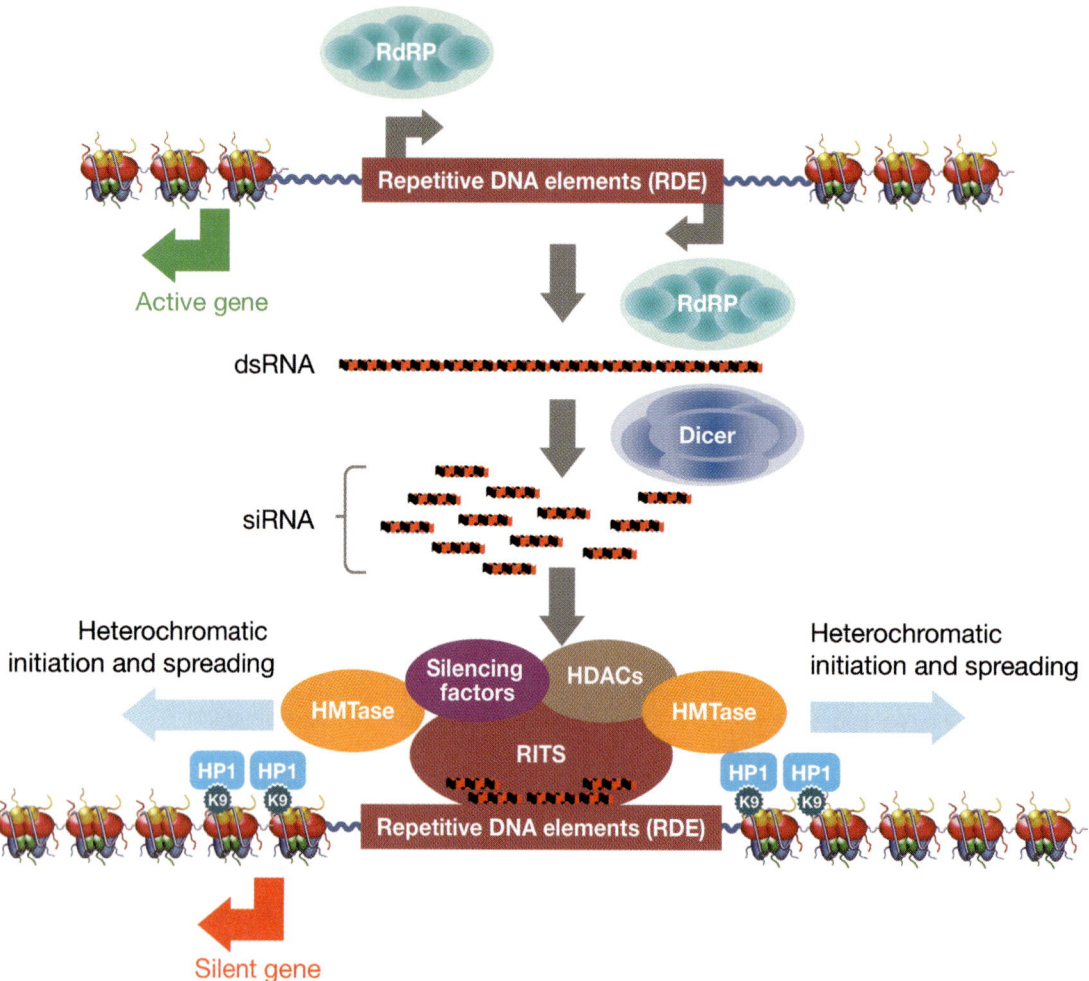

Figure 4

The role of RNA interference (RNAi) in heterochromatin assembly. The repetitive DNA sequences in the heterochromatin serve as templates for the synthesis of double-stranded RNA (dsRNA) via the enzymatic activity of the RNA-dependent RNA polymerase (RdRP). Also, the transcription from external and internal promoters with opposite orientation can result in the formation of dsRNA as well. The enzymatic activity of Dicer, a component of the RNAi machinery, is required for the processing of the dsRNA for the generation of the small interfering RNAs (siRNAs). The localization of the histone modifying complexes is directed by the siRNAs and the newly described RITS (RNA-induced initiation of transcriptional gene silencing) complex. The chromodomain-containing protein Chp1 within RITS and the HP1 protein can bind to the lysine 9-methylated histone H3. The further assembly and spreading of heterochromatin then follows resulting from the combined activities of HATs, HDACs, HMTase, and HP1, as described in **Figure 3**.

Both of these points pose very important questions for future studies in the field.

The mechanism described above for the establishment and maintenance of heterochromatin in higher eukaryotic organisms is now very well accepted; however, in budding yeast, there are no histone H3 lysine 9 methyltransferases. As will be discussed below, other modes of histone modification are used for the proper establishment and maintenance of heterochromatic silencing in yeast *Saccharomyces cerevisiae* (19, 52–54).

There are several reported HMTases capable of methylating histone H3 at lysine 9. Because the degree of lysine 9 methylation correlates with distinct chromatin regions, it appears that the mechanism and machinery for histone H3 methylation at lysine 9 are different in euchromatin versus heterochromatin (55–57). In an in vitro reaction, both the HMTases Suv39h1 and G9a can methylate histone H3. However, in vivo, the HMTase G9a mediates the methylation of histone H3 on lysine 9 in euchromatin, and Suv39h1 mediates the trimethylation of histone H3 on lysine 9 in constitutive pericentric heterochromatin (56–57). These enzymes also seem to display different patterns of localization on chromatin as well as different specificity for various types of histone H3 lysine 9 methylation. The G9a protein is the major histone H3 lysine 9 dimethylase in euchromatin, whereas Suv39h1 and Clr4 appear to be the major histone H3 lysine 9 trimethylases in pericentric heterochromatin (44, 58–59).

HISTONE H3 LYSINE 4 METHYLATION BY COMPASS AND ITS MAMMALIAN HOMOLOGUE, THE MLL COMPLEX

Following the report on the role of the SET domain of Su(var)3–9 as a HMTase (16), and owing to the completion of the genome sequencing of several different organisms including humans, the field of histone methylation has rapidly progressed toward the discovery of roles of other SET domain-containing proteins as HMTases. The human *MLL* gene, which encodes a SET domain-containing protein, was cloned over 15 years ago on the basis of its translocation properties associated with the pathogenesis of several different forms of hematological malignancies, including acute myeloid leukemia (AML) (20–22, 60). MLL is a 3968–amino acid protein consisting of an N-terminal A-T hook DNA-binding domain, a DNA methyltransferase-like domain with several continuous zinc fingers near the center of the molecule, and a SET domain at the C terminus (20–22). Chromosomal translocations involving the *MLL* gene occur in approximately 80% of infants with either AML or acute lymphoblastic leukemia (ALL). They also occur in approximately 5% of adult patients with AML, and up to 10% with ALL (20–22, 61). Although the cDNA for *MLL* was cloned in the early 1990s, it was not until recently that we learned about the biochemical properties and enzymatic activity of MLL and its macromolecular complexes. The yeast *S. cerevisiae* Set1 protein was noted several years ago to be highly related to the MLL protein (32). Because yeast is a great model organism for biochemical and genetic studies, characterization of the biochemical and biological properties of the Set1 and its macromolecular complex in yeast was initiated to learn more about MLL (32). These biochemical studies resulted in the identification of the Set1-containing complex, which is called COMPASS (complex proteins associated with Set1) (32). COMPASS contains the MLL-related Set1 protein and seven other polypeptides, several of which contain WD domains found in other trithorax-related complexes. Work from several laboratories has shown that COMPASS associates with the early elongating RNAPII via the polymerase II-associated factor 1 (Paf1) complex to methylate lysine 4 of histone H3 within the early body of a transcribed gene (19, 31–33, 62–63) (**Figure 5**). Unlike other SET domain-containing proteins, Set1 is not active by itself and requires the presence of other components of COMPASS for its full H3 lysine 4 methyltransferase activity. Set1 was initially identified as a gene product required for the proper regulation of telomere-associated gene silencing, and similar to Set1, several components of COMPASS are required for telomeric silencing as well (31, 32). These subunits appear to be the same ones required for histone methylation by COMPASS,

COMPASS: complex proteins associated with Set1

MLL: myeloid/lymphoid or mixed lineage leukemia

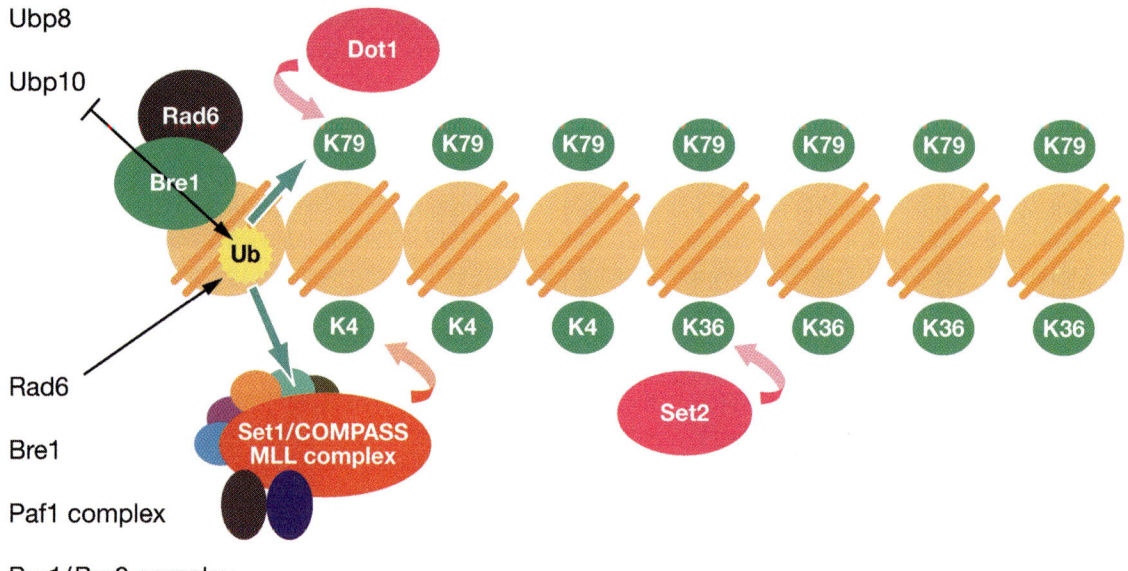

Figure 5

Schematic representation of the molecular machinery required for proper histone H3 methylation on lysines 4, 36, and 79. The Rad6/Bre1 complex is required for the monoubiquitination of histone H2B on lysine 123. Via an unknown mechanism, histone H2B monoubiquitination signals for the activation of histone H3 methylation on lysines 4 and 79 by COMPASS and Dot1p, respectively. Other factors such as the Paf1 complex and the Bur1/Bur2 complex have been demonstrated to be required for proper histone H2B monoubiquitination. The monoubiquitinated histone H2B is deubiquitinated by the action of deubiquitinating enzymes, such as Ubp8 and Ubp10. The enzymatic removal of ubiqutin from the monoubiquitinated histone H2B can negatively regulate the methylase activities of COMPASS and Dot1p.

therefore linking telomere-associated gene expression to histone tail methylation (31). Recent studies from our laboratory have demonstrated that several components of COMPASS, namely Cps40 and Cps60, are required for specific histone H3 lysine 4 trimethylation (64). The loss of Cps40 and/or Cps60 does not affect the recruitment of COMPASS to chromatin, indicating that the loss of H3 lysine 4 trimethylation is due to the overall conformational changes in the complex and/or shifts in the active site of Set1 itself. Furthermore, the loss of histone H3 lysine 4 trimethylation has very little effect on telomere-associated gene silencing, indicating that perhaps the threshold of mono- or dimethylation on lysine 4 of H3 is essential to maintain telomeric silencing in yeast (64). Overall, this study has indicated the presence of multiple roles for different forms of histone methylation by COMPASS.

On the basis of the studies performed in yeast and the homology between the Set domain of Set1 and MLL, MLL was tested for histone H3 lysine 4-specific methyltransferase activity. Similar to yeast Set1, it was demonstrated that MLL's SET domain is a histone H3 lysine 4-specific methyltransferase whose activity is stimulated with acetylated H3 peptides (65, 66). Also, a leukemogenic MLL fusion protein that activates *Hox* gene expression appears to have no effect on histone methylation, further supporting the presence of a distinct mechanism for gene regulation by MLL and MLL chimeras found in translocations associated with leukemia.

In a separate study identifying macromolecular complexes associated with menin, the tumor suppressor protein, a product of the *MEN1* gene, Meyerson and colleagues (67) demonstrated that the mammalian MLL2 exists in a COMPASS-like complex. Since then, several human MLL- and MLL2-containing complexes have been reported in the literature, all of which are found in COMPASS-like complexes (67–69). The MLL-containing complexes are also HMTases that methylate histone H3 on lysine 4 (67–69). Most interestingly, a subclass of human tumors derived from mutations in menin lacks HMTase activity (67). Furthermore, similar to yeast COMPASS, this menin/MLL-containing complex is associated with serine 5-phosphorylated RNAPII, and as anticipated, the complex is recruited to *Hoxc6* and *Hoxc8*.

The compositional and functional conservation between the MLL and Set1 complexes establishes that a highly conserved, ancient molecular machinery for the modification of histone H3 on its fourth lysine by methylation is required for the proper regulation of gene expression. This finding emphasizes the generality and significance of the information obtained from yeast in defining the molecular role of histone methylation by the yeast MLL-like complex, COMPASS. As discussed below, the activity of COMPASS in yeast is highly regulated via the recruitment of this complex to the transcribing RNAPII by elongation factors, and also via histone monoubiquitination by the Rad6/Bre1 complex. Studies in yeast have now set the stage for analyzing the role of such factors in the regulation of the methyltransferase activity of MLL and its complex in humans.

THE ROLE OF SET2 IN METHYLATING LYSINE 36 OF HISTONE H3

The Set2 protein was originally identified in a genetic screen by Johnston and colleagues as a factor involved in transcriptional repression in budding yeast. On the basis of the homology in its Set domain, Set2 was then purified and shown to have HMTase activity (70) (**Figure 5**). Employing biochemical and genetic approaches, Allis and colleagues (70) demonstrated that the HMTase activity associated with Set2 is specific for lysine 36 of histone H3. Set2 is the only SET domain-containing protein in *S. cerevisiae* that is capable of methylating lysine 36 of histone H3. Furthermore, Set2's methyltransferase activity is necessary for the repression of *GAL4* basal expression. In vivo studies have also demonstrated that Set2 is required for the maintenance of the low basal expression of the *GAL4* gene in *S. cerevisiae* (70, 71).

In the quest for defining the molecular mechanism of Set2 in the regulation of gene expression, several laboratories set out to purify to homogeneity macromolecular complexes associated with Set2. Such endeavors have resulted in the copurification of Set2 with RNAPII (72–76). These studies collectively demonstrated that Set2 associates with serine 2-phosphorylated RNAPII (the elongating form of RNAPII). In support of this observation, the deletion of approximately 10 heptapeptide repeats of the C-terminal domain of RNAPII results in a significant global loss of histone H3 lysine 36 methylation. Chromatin immunoprecipitation studies have also demonstrated that Set2 is recruited within the coding regions of transcriptionally active genes. Furthermore, enzymatic activity of the CTK kinase, which is required for the RNAPII C-terminal domain phosphorylation, is also required for the Set2-dependent lysine 36 methylation of histone H3 (74, 76). However, because not all genes are methylated on lysine 36 of histone H3, there appears to be a gene-selective targeting mechanism for Set2.

Interestingly, the copurification of RNAPII with Set2 was not detected when Set2 was purified based on its HMTase activity (70). However, tagging Set2 and its subsequent purification showed the interaction of Set2 with subunits of RNAPII (72–76). This in part may indicate that there are free forms of Set2 within cells that do not associate with RNAPII and therefore are

PRC1: polycomb repressive complex 1

not recruited to chromatin. The polymerase free form of Set2 may also work on other substrates in addition to histone H3.

It was recently demonstrated that a novel domain called SRI (for <u>S</u>et2 <u>R</u>pb1 <u>I</u>nteracting) exists in the C terminus of Set2 (77). This domain is required for the interaction of Set2 with RNAPII, and the SRI domain of Set2 binds specifically to RNAPII CTD repeats that are doubly modified on serine 2 and serine 5 by phosphorylation. Because the SRI domain is required for the interaction of Set2 with RNAPII, its deletion results in the loss of histone H3 lysine 36 methylation (77). This finding, along with studies performed in several other laboratories, indicates that the recruitment and interaction of Set2 with RNAPII are required for establishing K36 methylation on chromatin (19, 78). The future identification of the higher eukaryotic homologues of Set2 and their roles in the regulation of gene expression and development will shed further light on the importance of the specific role of histone H3 lysine 36 methylation in development.

HISTONE H3 LYSINE 27 METHYLATION, H2A LYSINE 119 UBIQUITINATION, REGULATION OF POLYCOMB-GROUP SILENCING, AND X-CHROMOSOME INACTIVATION

As described above, the trithorax group (TrxG) and the PcG families of proteins in *Drosophila* have provided a great model for studying the molecular mechanism of how heritable transcriptional states are maintained during development. Detailed genetic and biochemical studies first suggested that PcG and TrxG proteins provided transcriptional memory through alterations of chromatin structure. As described in previous sections, it was recently demonstrated for the mammalian MLL (the homologue of the *Drosophila* trithorax protein) that this class of proteins functions as histone H3 lysine 4 methyltransferases (21, 22). The PcG proteins are essential for the maintenance of the silenced state of homeotic genes. Recent biochemical and genetic studies have shown that the PcG proteins are also HMTases that function in at least two distinct macromolecular complexes. These include the polycomb repressive complex 1 (PRC1) and the E(z) ESC, Enhancer of Zeste [E(Z)] protein complex. As discussed below, PcG complexes are also HMTases, and part of their silencing function is mediated by the HMTase activity. Furthermore, in addition to a role in Hox gene silencing, the HMTase activity of PcG protein complexes is required for X inactivation.

Initial biochemical studies demonstrated that histone H3 is methylated on lysine 27. It was originally reported that the G9a is a HMTase capable of methylating both lysine 9 and 27 of histone H3 (79). Recent studies from several laboratories have now shown that the SET domain within the E(Z) protein can methylate lysine 27 of histone H3 within nucleosomes (80–81). In *Drosophila* embryos, the purified ESC-E(Z) is found in a large macromolecular complex. The four major components of this complex include the ESC, E(Z), SU(Z)12, and NURF-55. Histone H3 lysine 27 methyltransferase activity associated with this complex can be reconstituted from these four subunits, and mutations in the E(Z) SET domain disrupting the methyltransferase activity results in the repression of *Hox* gene expression (80). The human homologue of this complex, the EED-EZH2 complex, is also capable of methylating histone H3 on lysine 27 (81). Chromatin immunoprecipitation in human cells has demonstrated that methylation of histone H3 on lysine 27 is dependent on E(Z) binding at an Ultrabithorax (Ubx) polycomb response element. Also, the level of Ubx repression correlates with H3 lysine 27 methylation, perhaps through facilitating the binding of the polycomb component of the PRC1 complex to lysine 27-methylated histone H3 (81). The functional conservation between *Drosophila* and human PcG proteins in methylating histone H3 on lysine 27 has resulted in the development of a model for PcG-mediated gene silencing. In this model, histone H3 lysine 27 methylation facilitates the binding of polycomb, a component of the PRC1 complex, to histone H3 through its chromodomain, which is required for the regulation of silencing by the PcG complex (81).

It has been demonstrated that mice homozygous for an *eed* mutation (embryonic ectoderm development, a member of the mouse PcG of genes) are defective in the maintenance of X-chromosome silencing in extraembryonic, but not embryonic, tissues, and the levels of Eed Ezh2 and Eed are enriched on the inactive X chromosome in the trophoblast stem cells. Therefore, a role for histone H3 lysine 27 methylation in X-chromosome inactivation has also been suggested (82, 83). The process of dosage compensation in mammals is attained by transcriptional silencing of one X chromosome in female cells (84). Proper X inactivation is a multistage process requiring the concerted action of several factors and involving the choice of the active X chromosome, the initiation of silencing on the inactive X chromosome, and the maintenance of the inactive X chromosome throughout the life of the cell (84). The Xist RNA, which is transcribed exclusively from the inactive X chromosome in female somatic cells, plays a role at every stage of X-chromosome inactivation (84). The Xist RNA remains in the nucleus and is found associated with the inactive X chromosome. Given the role for the PcG complex and its *Drosophila* counterpart, the ESC-E(Z) complex in the methylation of H3 on lysine 27, a possible role was tested for this histone H3 modification in X-chromosome inactivation. It has now been demonstrated that the recruitment of the Eed-Ezh2 complex to the inactive X chromosome takes place during initiation of X inactivation and is accompanied by histone H3 lysine 27 methylation. The recruitment of this complex and modification of histone H3 are dependent on Xist RNA; however, this process is independent of the silencing function of PcG proteins (85–87).

Overall, methylation of histone H3 at lysine 27 exhibits some functional similarities to that of lysine 9. First, both lysines are found within the sequence of ARKS in histone H3; however, they require different enzymatic machinery for their methylation (**Figure 2**). This, perhaps, indicates that the enzymatic machinery recognizing these sites uses sequences outside of the ARKS or that other epigenetic information is required for their specificity. Supporting this hypothesis, lysine 9 methylation and lysine 27 methylation represent different degrees and distribution of methylation on chromatin. For example, in pericentric heterochromatin, monomethylated lysine 27 is found along with trimethylated lysine 9 (88). However, it appears that lysine 27 methylation is a characteristic of the inactive X chromosome during the initial stage of X inactivation (85–87). At the same time, some level of dimethylated, but not trimethylated, lysine 9 can be found in the inactive X chromosome (89–91). However, the exact physiological role for histone H3 lysine 9 methylation in X-chromosome inactivation is not clear at this time. Given that Suv39h double-null mouse embryonic fibroblasts still maintain some level of histone H3 lysine 9 dimethylation on the inactive X, a different HMTase may be involved in this process.

Another modification of histone that has been linked to polycomb silencing, and X-chromosome inactivation is the modification of histone H2A via ubiquitination (92, 93). It was demonstrated recently that ubiquitinated H2A is found on the inactive X chromosome in females and that its presence is correlated with the recruitment of the PcG proteins belonging to the PRC1, referred to as PRC1-like (PRC1-L). PRC1-L was purified to homogeneity and was found to ubiquitinate histone H2A within the nucleosomes at lysine

Dot1: disruptor of telomeric silencing 1

119 (92). Consistent with its role in regulating gene expression, it was shown that the removal of the Ring protein in tissue culture cells via RNAi resulted in the loss of H2A ubiquitination and the derepression of Ubx. In embryonic stem cells, which are null for Ring1B, an extensive depletion of global ubiquitinated H2A levels has been observed. However, on inactive X chromosomes, ubiquitinated H2A is maintained in either Ring1A or Ring1B null cells, but not in double knockouts (93). These studies have now linked H2A ubiquitination to X inactivation and polycomb silencing. However, the relationship between ubiquitination on lysine 119 of histone H2A and methylation of lysine 27 of histone H3 and their exact molecular role in polycomb silencing and X-chromosome inactivation are not clear at this time.

To define the downstream targets of the PcG complexes in mammalian cells, Farnham and colleagues (94) have taken advantage of RNA expression arrays and CpG-island DNA arrays. The siRNA-mediated removal of Suz12 enabled researchers to identify a number of genes whose expression was also altered. Employing this technology, Farnham and colleagues have demonstrated that the PRC complex colocalizes to the target promoters with Suz12 and that its recruitment coincides with the methylation of histone H3 on lysine 27. However, it is still not yet clear how PRC complexes are directed to their loci, as no site-specific DNA-binding factor has been isolated in the PRC complexes. Furthermore, even identifying a gene as being regulated by a PRC does not provide information regarding the site of recruitment of the PRC complex to that gene, because the PRC-specific element could be located a long distance away from the transcription start site. Identifying a large number of target genes for the PRC complex will provide further information for the identification of such common elements for PRC recruitment to chromatin. However, it is also feasible that other chromatin modifications and epigenetic information may play a role in this process.

NON-SET DOMAIN-CONTAINING LYSINE-SPECIFIC HISTONE METHYLTRANSFERASES

The enzyme Dot1 (disruptor of telomeric silencing 1) is the only HMTase identified so far that lacks the characteristic SET domain and can methylate the lysine residue of histones. Dot1 was initially discovered by Gottschling and colleagues in a high-copy suppressor screen while they were searching for factors that affect telomeric-associated gene silencing (95). Several groups have reported that Dot1 is capable of methylating histone H3 within the nucleosomes exclusively at lysine 79 (96, 97) (**Figures 2** and **5**). Dot1 methylates approximately 90% of histone H3 found in the chromatin (96). The N terminus of Dot1 contains its active site, which is linked to the C-terminal domain by a loop that also serves as part of the AdoMet-binding site. The loop, and thus the SAM-binding site, is highly conserved between Dot1 and other AdoMet-binding proteins. Unlike other SET domain-containing enzymes that modify lysine residues in the histone tails, methylation by Dot1 takes place at a site in the histone H3 core (96–98). Based on the crystal structure of the nucleosome reported by Luger and colleagues (34), the lysine 79 residue of histone H3 is located on the accessible surface of the outside (the "top" and "bottom") of the nucleosome core and does not contact other histones or DNA.

Both HMTase activity of Dot1 and lysine 79 of histone H3 are required for the establishment of telomeric-associated gene silencing. The loss of Dot1 or mutations in the lysine 79 of histone H3 abolish silencing and reduce Sir2p and Sir3p association with silenced regions. Because of this observation, it has been proposed that at silent domains Sir proteins interact with histone H3 that is hypomethylated at Lys79. In this model, methylation of histone H3 at lysine 79 in bulk chromatin prevents the binding of Sir proteins to chromatin at weak protosilencers.

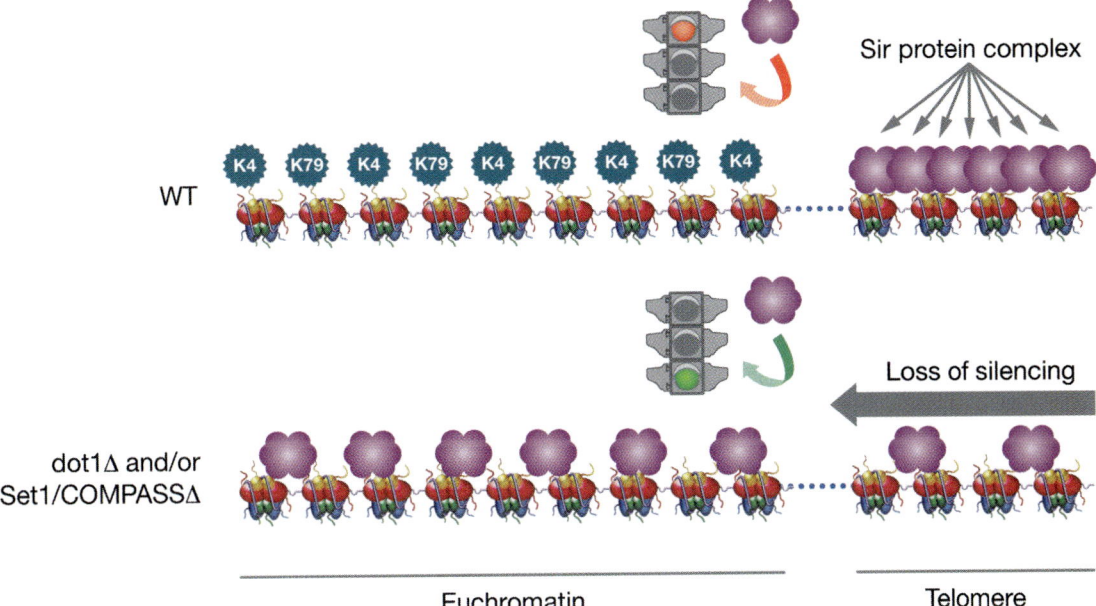

Figure 6
A model for the role of histone H3 lysine 4 and 79 methylation in the regulation of telomere-associated gene silencing. In the wild-type (WT) cells, the Sir protein complex is recruited to telomeres and perhaps to other silent domains within chromatin via recruitment by specific DNA-binding proteins such as Rap1. The Sir complex can interact with hypomethylated histone H3 that is found within the silent chromatin domains. Methylation of histone H3 on lysines 4 and 79 within the euchromatin prevents the binding of the Sir complex to chromatin at positions with single binding sites for Rap1p (known as weak protosilencers). In the absence of Dot1 and/or Set1/COMPASS (or factors required for their enzymatic activity such as Rad6/Bre1 or the Paf1 complex), the Sir complex binding to euchromatin becomes promiscuous. Such promiscuous binding of the Sir complex to euchromatin at weak protosilencer sites results in the reduction of the concentration of Sir proteins found normally in silent domains and, therefore, results in the loss of silencing.

However, in the absence of methylated histone at lysine 79, Sir protein binding to chromatin becomes promiscuous and can bind to chromatin at weak protosilencers (**Figure 6**). This promiscuous binding by the Sir proteins in the absence of H3-methylated K79 results in reduced availability of Sir proteins that normally interact with the silent domains, leading to the misregulation of gene expression at silent chromatin. Telomere-associated gene silencing as a result of histone H3 lysine 4 methylation by COMPASS appears to also follow the same model as silencing associated with H3 K79 methylation.

In a yeast two-hybrid screen, human DOT1 was recently reported to interact with the AF10 protein (98). AF10 is one of the fusion partners of MLL involved in the pathogenesis of leukemia (20–22). Although MLL-Dot1 translocations have not been reported in patients suffering from AML, direct fusion of Dot1 to MLL results in the immortalization of myeloid progenitor cells. Most importantly, mutations effecting Dot1 methyltransferase activity resulted in the loss of the immortalization by the MLL-Dot1 chimera. This study may indicate that the methylation of histone H3 on lysine 79 is required for the proper regulation of *Hox* gene expression and that constitutively active MLL-Dot1 may misregulate the transcription of MLL-regulated genes such as the *Hoxa9* locus.

Development of chemical inhibitors modulating Dot1's HMTase activity will be instrumental in testing such models and perhaps could play an important role for targeted therapy of MLL-AF10-associated leukemia.

HISTONE ARGININE METHYLTRANSFERASES

Methylation of arginine residues has been identified on many cytosolic and nuclear proteins. This posttranslational modification of arginine has been implicated in a variety of cellular processes, such as RNA processing, transcription, cellular signaling, and DNA repair. Although protein arginine methylation is involved in many cellular processes, this section of the review concentrates only on the role of this class of enzymes in histone methylation. For a most recent detailed review on the role of protein arginine methyltransferases (PRMTs) in other cellular processes, please see References 25 and 99.

As early as the 1960s, it was demonstrated that the arginine residues within proteins were methylated (100). The positively charged arginine can mediate hydrogen bonding and amino-aromatic interactions. Its posttranslational modifications by methylation can occur on its nitrogen and result in the addition of one or two methyl groups to the guanidino nitrogen atoms of arginine. As with the SET domain-containing HMTases, PRMTs catalyze the transfer of methyl groups from S-adenosyl-L-methionine to the guanidino nitrogens of arginine residues (25, 99). Recent studies have identified three distinct forms of methylation that occur on arginine residues on histone tails. These include N^G-mono-methylarginine, $N^G N^G$-symmetric dimethylarginine (in which both guanido nitrogens are methylated), and $N^G N'^G$-asymmetric dimethylarginine (in which only one guanido nitrogen receives two methyl groups) (**Figure 1**).

CARM1 is a PRMT that can methylate histone H3 at arginine 2, 17, and 26 (101). This posttranslational modification by CARM1 has been shown to enhance transcriptional activation by nuclear receptors (102). It has also been demonstrated that the methyltransferase activity of CARM1 and its association with p160 coactivators are required for its coactivator function with nuclear receptors. These studies demonstrate that the recruitment of CARM1 and the subsequent modification of arginine residues on histone H3 by methylation are indispensable parts of the transcriptional activation process. Another PRMT involved in histone methylation is the product of the *PRMT1* gene. PRMT1 has been shown to methylate histone H4 at arginine 3 both in vitro and in vivo (102–104). This enzymatic activity of PRMT1 has been shown to be required for transcriptional activation by nuclear receptors (103).

Recent transient transfection studies have demonstrated that multiple coactivators capable of histone-modifying activities can cooperate synergistically. For example, the modification of histone H3 by CARM1 can cooperate with the arginine methylation of histone H4 by PRMT1 (105). In the same respect, CARM1 activity can be synergized with other histone-modifying machinery such as CBP, pCAF, and p300, which are involved in histone acetylation (106, 107). Not yet clear are the exact molecular mechanisms by which the modification of arginine residues within nucleosomes contributes to chromatin remodeling and transcriptional activation. Not only can the histone tails be modified by such enzymatic machinery, but the tails are also available for additional intermolecular interactions. Histone arginine methylation and/or acetylation can play a role in the disruption of nucleosome stability or internucleosomal interactions. Recent studies defining the role of PRMTs in the regulation of gene expression have emphasized the intricate details and the importance of the histone-modifying machinery in transcriptional regulation. Such studies have also brought about the understanding that the pattern of such posttranslational modifications can cooperate. Future studies identifying other possible histone and nonhistone substrates for these protein arginine methyltransferases, involved in transcriptional regulation, promise to shed more light on the complexity of this process.

HISTONE MONOUBIQUITINATION IN SIGNALING FOR HISTONE METHYLATION AND THE REGULATION OF GENE EXPRESSION

Biochemical screens geared toward identifying the molecular machinery required for histone H3 methylation by COMPASS have been instrumental in dissecting the molecular pathways for the functional regulation of several HMTases. A functional proteomic screen approach called GPS (global proteomic analysis in *S. cerevisiae*) was developed to define the molecular mechanism of histone H3 methylation by COMPASS (108). In GPS, extracts of each of the nonessential yeast genes were initially tested via Western analysis using antibodies specific to lysine 4-methylated histone H3 to identify factors required for proper histone H3 methylation by COMPASS. By testing each of the nonessential yeast gene deletion mutants for defects in methylation of histone H3 on lysine 4, it was first shown that histone H2B monoubiquitination by Rad6 is required for histone methylation by COMPASS (109) (**Figure 5**). Other studies searching for factors involved in telomere-associated gene silencing also resulted in the observation that histone monoubiquitination is required for histone methylation (110).

Interestingly, similar to COMPASS, it was reported that histone methylation by the non-SET domain enzyme Dot1 requires histone monoubiquitination by Rad6 (111–113). Collectively, these studies provided evidence for the existence of a "cross-talk" pathway for histone tail modifications (**Figure 5**). However, it is not clear at this time how ubiquitination on lysine 123 of histone H2B results in the activation of the catalytic activity of both COMPASS and Dot1p.

All E2 ubiquitin-conjugating enzymes require the presence of an E3 ligase to provide substrate specificity for the enzyme. Several E3 ligases, such as Ubr1, Ubr2, and Rad18, were demonstrated to function with Rad6, but none of these E3 ligases are required for either histone monoubiquitination or methylation (109, 110). Via the GPS screen, the Ring finger protein Bre1 was identified as the E3 ligase that is required for monoubiquitination of histone H2B, histone H3 methylation by COMPASS and Dot1p, and the association of Rad6 with chromatin (111). This study also demonstrated that the Rad6/Bre1 complex can be purified biochemically and that mutations affecting their interactions result in the loss of histone monoubiquitination and, therefore, methylation (111).

The GPS screen and other biochemical studies have also identified the role of the Paf1 complex in the regulation of histone ubiquitination and methylation (114, 115). Initial studies from several laboratories demonstrated that the Paf1 complex is associated with the elongating form of RNAPII (19, 78, 116–120). Later, it was demonstrated that the Paf1 complex is required for histone monoubiquitination and, therefore, methylation by playing a role in the recruitment of factors such as COMPASS to the transcribing polymerase (121, 122). The Paf1 complex was also demonstrated to play a role as a "platform" for the association of COMPASS and perhaps other HMTases with the elongating form of Pol II, therefore linking transcriptional elongation to histone methylation for the first time (19, 114). The Paf1 complex appears to be required for the functional activation of Rad6/Bre1 in histone monoubiquitination via an unknown mechanism (121). Recently, it was demonstrated that Rad6/Bre1 may also associate with elongating RNAPII and monoubiquitinate histone H2B on the body of a transcribed gene (123). However, given the low abundance of UbH2B in

GPS: global proteomic analysis in *S. cerevisiae*

comparison to that of methylated lysine 79 and/or lysine 4 of histone H3, and owing to the absence of antibodies specifically recognizing monoubiquitinated H2B, the directness of such observations has not been tested. Other factors playing a role in the regulation of histone H2B monoubiquitination by Rad6/Bre1 have recently been identified via GPS. These include the serine/threonine protein kinase Bur1 and its divergent cyclin Bur2 complex, which function in the regulation of histone H2B monoubiquitination via the phosphorylation of Rad6 and the recruitment of the Paf1 complex (124).

Ubiquitination is a reversible process, and several very exciting studies have recently demonstrated that monoubiquitinated histone H2B can be deubiquitinated by the enzyme Ubp8 (125, 126). Because Ubp8 is a component of the SAGA histone acetyltransferase, it has been proposed that the Rad6-catalyzed monoubiquitination of histone H2B is followed by the recruitment of SAGA to the ubiquitinated nucleosome and subsequent deubiquitination of histone H2B, which is required to initiate transcription. In support of this observation, mutations affecting Ubp8 led to a rise in global histone H2B ubiquitination and a decrease in the transcription of SAGA-regulated genes (125, 126).

Another deubiquitinating enzyme, Ubp10/DOT4, which was originally isolated by Gottschling and colleagues in a screen for high-copy disruptors of telomeric silencing in yeast (95), also targets monoubiquitinated histone H2B for deubiquitination (127). However, this enzyme exhibits reciprocal Sir2-dependent preferential localization proximal to telomeres and also localizes to the rDNA locus. Comparative studies of Ubp10 and Ubp8 functions have demonstrated that the deubiquitination activities involved in telomeric-associated gene silencing are functions specific to Ubp10. This study indicates that such deubiquitinating enzymes have distinct functions in the regulation of gene expression via the targeting of histone H2B deubiquitination (**Figure 5**).

HISTONE METHYLATION AND TRANSCRIPTIONAL MEMORY

Histone H3 lysine 4 methylation by COMPASS in yeast and its homologue, the MLL complex, in mammalian cells have been linked to transcriptionally active genes (22, 128, 129). Initially, Kouzarides and colleagues (128) demonstrated that transcriptionally active coding regions are enriched by histone H3 trimethylated at lysine 4. Detailed chromatin immunoprecipitation with antibodies against mono-, di-, and trimethylated histone H3 lysine 4, coupled with DNA array analysis, demonstrated a close connection between RNAPII transcriptional activity and levels of histone H3 lysine 4 methylation both for COMPASS and, now, for the MLL complex (19, 114, 115, 130). Such studies have demonstrated that the methylation of histone H3 on lysine 4 is necessary for transcription and is a specific mark for transcriptionally active genes in eukaryotic organisms.

GPS studies performed in our laboratory as well as studies performed in Struhl's laboratory demonstrated the requirement for the elongation machinery of the Paf1 complex in the regulation of both the activity of Rad6/Bre1 in the monoubiquitination of histone H2B and the recruitment of COMPASS to the early elongating RNAPII (19, 114, 115). These studies have linked transcriptional elongation by RNAPII to histone methylation. Related to this conclusion, analysis of the distribution of both histone H3 lysine 4 methylation and Set1/COMPASS throughout the transcriptionally active genes has demonstrated that they are confined to the 5′ end of transcribed regions in yeast (114, 115). When the distribution of these factors was analyzed on the *GAL10* gene during and after activation, it was demonstrated that the occupancy of Set1/COMPASS and levels of H3 lysine 4 methylation at the 5′ coding region rose rapidly upon activation. When the gene is switched off, Set1/COMPASS occupancy falls rapidly, similar to transcription by RNAPII; however, the levels of histone

H3 lysine 4 di- and trimethylation fall relatively slowly. Our analysis of the bulk lysine 4-trimethylated histone H3 under regulated Paf1 expression also demonstrated that the loss of trimethylation on histone H3 is a very slow process. Collectively, these observations have resulted in the proposal of a "short-term memory" model for lysine 4 methylation of histone H3 (19, 114, 115) (**Figure 7**). The methylation of histone H3 on lysine 4 by COMPASS is observed to be associated with early elongating RNAPII, and the methylation of histone H3 on lysine 36 by Set2 appears to be associated in the body of a transcriptionally active gene. Therefore, in the memory model, histone H3 methylation informs the cell of the transcription status of a given gene. The pattern of methylation can indicate that the transcription of a given gene has occurred in the recent past but is not necessarily happening at the present time. Furthermore, the pattern of methylation can inform the cell how far the RNAPII has transcribed through a given gene. Histone methylation only lasts for a portion of an individual cell cycle, so this modification cannot be faithfully transmitted to all daughter cells. Thus, histone H3 methylation could provide the cell with a memory for recently transcribed genes that is mechanistically distinct from the epigenetic inheritance that occurs in position-effect variegation and transcriptional silencing.

DEMETHYLATING HISTONES

Several of the many known covalent modifications affecting histone tails, such as phosphorylation, ubiquitination, and acetylation, have all been shown to be reversible. Therefore, if modification of histone tails by phosphorylation, ubiquitination and/or acetylation influences gene expression, its removal may have the opposite effect. In this way, cells can rapidly respond to such regulatory modifications. Histone modification by methylation, however, has been considered to be a fairly stable and irreversible mark on histones. This has partly been due to a number of early ob-

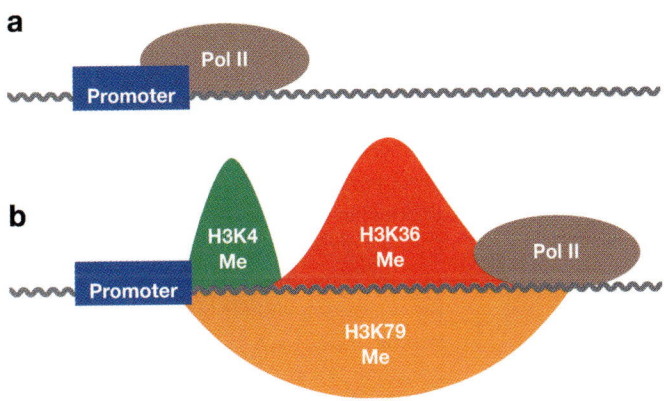

Figure 7
Histone methylation and transcriptional memory. (*a*) RNA polymerase II (RNAPII) associated within the preinitiation complex. (*b*) The transcribing RNA polymerase II. Histone H3 lysine 4 methylation is associated within the early body of transcribed genes. Histone H3 lysine 36 methylation is associated within the open reading frames (ORFs) of transcribed genes. Modification of histone by methylation appears to be a relatively stable mark, so it has been proposed that the pattern of methylation can serve as a mark of transcriptional memory. As discussed in the text, this mode of memory is described as a noninheritable memory that is a mark for a recently transcribed gene. Because the pattern of localization of histone H3 on lysine 79 appears to be broad, this modification is shown throughout the open reading frame.

servations that indicated the half-life of histones and methyl-lysine residues within them are the same (131, 132). Furthermore, the lack of identification of histone demethylases and the stability of the methyl marks on histones have led to the dogma that once a methyl group is added to a histone, it cannot be removed via an active mechanism and will remain on the chromatin until a natural histone turnover or until DNA replication replaces the modified histone with an unmodified one. This stability of histone methylation is in line with the observation of the role of histone methylation found at centromeric heterochromatin for heterochromatic silencing; however, several studies have indicated that an active turnover mechanism for methyl groups on histone tails may exist.

For some time now, histone replacement has received much attention as a possible mechanism for a response-mediated removal of a methylated histone from chromatin (133–135). In this model, the methylated histone can be replaced with an unmodified version on chromatin. In support of this model, the expression of core histones is coordinately up-regulated at the onset of the S phase, which is consistent with histone deposition during DNA replication. In addition to a replication-dependent deposition, Ahmad & Henikoff (133) have clearly demonstrated the presence of a replication-independent mechanism for the deposition of the histone H3.3 variant at active rDNA arrays (133). This same phenomenon seems to hold true for the *HSP70* gene in *Drosophila* (136). Upon heat shock induction, the *HSP70* gene can rapidly lose histone H3 and acquire the H3.3 variant. This replacement seems to require the process of active transcription. In support of the generality of this model for the whole genome, the histone H3.3 variant appears to be enriched in the open reading frames of all active genes, implicating the presence of a histone deposition mechanism that is linked to transcription elongation (136). However, the role of the RNAPII general elongation factors such as ELLs, Elongins and DSIF in this process have not yet been tested.

In addition to replication-dependent and replication-independent mechanisms, two classes of enzymes were recently reported to be required for either the inhibition of histone methylation or demethylation of histone tails. The first report was the identification of PADI4 as a histone deiminase that antagonizes arginine methylation on histones (137–139). It was demonstrated that either free or monomethylated arginine can be cleaved at the guanidine C-N bond by the arginine deiminase PADI4. The by-products of such reaction are citrulline and methylammonium. Although PADI4 is capable of deiminating free and monomethylated arginine, dimethylation of arginines prevents deimination by PADI4. This conversion prevents arginine methylation by the HMTase CARM1. On the basis of this observation, it has been proposed that the deimination of arginine residues on histones can be reversed either by a distinct enzymatic activity yet to be characterized or by histone replacement. Because dimethylation of arginine cannot be reversed by PADI4 or the deiminating pathway, the discovery of PADI4 still does not address how cells deal with methylated arginine once these residues are methylated. Therefore, it is still possible that methylated arginine can be removed either by histone replacement or by another enzymatic activity capable of directly demethylating arginine residues on histones. Further experimentation in this area will define the molecular mechanism of the removal of dimethylated arginine from histones and the way cells respond once the arginine residues on histones are demethylated.

Histone lysine methylation, which has been shown to have important roles in epigenetic silencing, has been regarded as a very stable modification of histones. However, the first evidence for a histone lysine demethylase that reverts an activating methyl mark (lysine 4 methylated on histone H3) was recently reported (140, 141). The KIAA0601 (BHC110) protein, which is a riboflavin-binding protein and a member of a flavin adenine dinucleotide (FAD)-dependent enzyme superfamily, was initially reported to be a component of the Co-REST and other repressor complexes that also contain histone deacetylase complex HDACs (142–145). Recombinant BHC110/LSD1 (lysine-specific demethylase 1) shares extensive sequence homology to metabolic FAD-dependent amine oxidases. Recombinant LSD1 can catalyze the amine oxidation of methylated histone H3 lysine 4 to generate unmodified lysine and formaldehyde. Considerable in vitro evidence, using recombinant enzymes and various histone peptides, was reported. This seems to be the likely mechanism for the demethylation of lysine 4-methylated histone H3. However, in the reported study, it was demonstrated that the oxidation of aminomethyl requires the presence of the cofactor FAD and a protonated nitrogen. Therefore, LSD1 can only demethylate mono- or dimethylated lysines and not trimethylated lysine 4 or other trimethylated lysines.

The discovery of the role of LSD1 as a histone demethylase is very exciting. The exquisite selectivity of this enzyme for mono- and dimethylated H3 lysine 4 in vitro on a free histone tail in the face of the broad mechanism of amine oxidation indicates that perhaps other enzymes may be required for the demethylation of histone H3 methylated on lysines 9, 36, or 79. Also, because this enzyme has been reported to be part of a larger macromolecular complex, it is feasible that its interacting partners may play a role in the substrate section, such as nucleosomes in vivo. Analysis of this macromolecular complex in demethylating free histone and histone within the nucleosome, as well as kinetic comparative studies on the role of free and complexed LSD1 in histone demethylation, promises to shed further light on the role of this protein in histone demethylation.

The identification of the enzymatic machinery involved in either the prevention of methylation or demethylation of histones represents a landmark discovery in the rapidly moving field of histone modifications. Such pioneering studies also demonstrate that no modifications of histones last forever. This observation is a testimony to the dynamic nature of histone modifications in the regulation of gene expression. Future studies in defining the roles of PADI4 and LSD1, their macromolecular complexes, and homologues in other organisms with amenable genetics and biochemistry will yield more milestone discoveries in this field.

FUTURE DIRECTIONS

A large body of studies from many laboratories during the past five years has demonstrated that histone methylation on both lysine and arginine residues of histones are undoubtedly involved at many levels in the regulation of gene expression, signal transduction, and development. However, the precise mechanisms by which histone methylation contributes to these physiological processes are mostly unresolved. For example, it is not clear at this time how cells decipher the histone methylation signal. A few classes of proteins, such as HP1, have been identified to bind methylated histone tail (H3K9 for HP1) and to translate the signal for silencing. However, there are no known factors to bind to the modified histone H3 on lysines 27, 36, 79, trimethylated lysine 4, or monoubiquitinated H2B. We also do not know much about other nonhistone substrates for the identified methyltransferases. Furthermore, the question of reversibility of histone methylation remains for the most part unresolved for trimethylated histones. Future investigations addressing these questions are needed to understand the exact molecular mechanism and biological ramification of histone methylation in the regulation of gene expression.

SUMMARY POINTS

1. "Epigenetic information," which is a form of the inherited state of gene regulation that lies outside of the DNA sequence itself, was shown to be required for the proper regulation of gene expression. Several factors including DNA methylation, small nuclear RNAs, and histone modifications (such as histone methylation) are required for proper epigenetic regulation.

2. Histone methylation is found on several lysine and arginine residues on histones and is associated with various biological processes ranging from transcriptional activation and regulation of gene expression to epigenetic silencing via heterochromatin assembly.

3. Lysine residues within histones can be mono-, di-, or trimethylated. It has recently been demonstrated that different forms of histone methylation may have multiple roles in the regulation of gene expression.

4. Unlike histone acetyltransferases, histone lysine methyltransferases are very dedicated enzymes with each enzymatic machinery devoted to methylation of a specific lysine residue on a specific histone. For example, the Set1/COMPASS or MLL class of HMTases can specifically methylate only the fourth lysine of histone H3, whereas Su(var)3-9 class of methyltransferases are specific for the methylation of lysine 9 of histone H3.

5. Histone H2B can be monoubiquitinated by enzymatic action of Rad6/Bre1. This modification of H2B by ubiquitination is a regulatory mark in signaling for histone methylation by COMPASS.

6. Histone H3 methylation on lysine 27 and histone H2A ubiquitination on lysine 119 are required for PcG gene silencing and X-chromosome inactivation.

7. The differential pattern of histone methylation on the open reading frame of a transcribed gene (histone H3 lysine 4 methylation on the early body and H3 lysine 36 methylation on the body of transcribed genes) can play a role as a noninheritable mark for the "transcriptional memory" of recently transcribed genes.

8. Once lysine residues within histones are modified by methylation, such modifications are considered to be stable. However, a few sets of enzymes have been identified to be capable of either inhibiting the methylation of arginine residues or demethylating lysine residues within the histones.

FUTURE ISSUES TO BE RESOLVED

1. The molecular mechanism of "cross talk" between histone H2B monoubiquitination by Rad6/Bre1 and histone H3 methylation by COMPASS should be defined.

2. How the methylation of histone H3 K27 and ubiquitination of histone H2A lysine 119 can result in the regulation of polycomb group gene silencing and X-chromosome inactivation have not been delineated, and whether ubiquitination of histone H2A is required for histone H3 K27 methylation by E(z) needs to be determined.

3. The identification and characterization of histone demethylating and/or replacement machineries that are capable of removing methyl groups from lysines 4, 9, 27, 36, 79 and arginines 2, 17, and 26 of histone H3, as well as methylated residues within histone H4 would be extremely informative.

4. Characterization of the HMTase activity of MLL and its chimeras found in leukemia and the determination of whether the HMTase activity of MLL or its chimeras are required for the pathogenesis of leukemia will be of great clinical significance.

ACKNOWLEDGMENTS

I would like to thank Elizabeth Torno for her editorial assistance in preparing this review. Given the limitation in space and the broad nature of the field of chromatin modifications, I apologize to many of my colleagues whose work was not described or referenced in this review. The work in the author's laboratory is supported by grants from the National Institutes of Health (2R01CA089455 and 1R01GM069905) and the American Cancer Society. The author is a Scholar of the Leukemia and Lymphoma Society.

LITERATURE CITED

1. Kornberg RD, Thomas JO. 1974. *Science* 184:865–68
2. Kornberg RD. 1974. *Science* 184:868–71
3. Rill RL, Shaw BR, Van Holde KE. 1978. *Methods Cell Biol.* 18:69–103
4. van Holde KE. 1989. *Chromatin*, pp. 111–48. New York: Springer-Verlag
5. Kornberg RD, Lorch Y. 1999. *Cell* 98:285–94
6. Richards EJ, Elgin SC. 2002. *Cell* 108:489–500
7. Ahmad K, Henikoff S. 2002. *Cell* 111:281–84
8. Grewal SI, Moazed D. 2003. *Science* 301:798–802
9. Pirrotta V. 1998. *Cell* 93:333–36
10. Mahmoudi T, Verrijzer CP. 2001. *Oncogene* 20:3055–66
11. Orlando V. 2003. *Cell* 112:599–606
12. Stassen MJ, Bailey D, Nelson S, Chinwalla V, Harte PJ. 1995. *Mech. Dev.* 52:209–23
13. Jenuwein T, Laible G, Dorn R, Reuter G. 1998. *Cell. Mol. Life Sci.* 54:80–93
14. Jones RS, Gelbart WM. 1993. *Mol. Cell. Biol.* 13:6357–66
15. Tschiersh B, Hofmann A, Krauss V, Dorn R, Korge G, Reuter G. 1994. *EMBO J.* 13:3822–31
16. Rea S, Eisenhaber F, O'Carroll D, Strahl BD, Sun ZW, et al. 2000. *Nature* 406:593–99
17. Zhang Y, Reinberg D. 2001. *Genes Dev.* 15:2343–60
18. Jenuwein T, Allis CD. 2001. *Science* 293:1074–80
19. Gerber M, Shilatifard A. 2003. *J. Biol. Chem.* 278:26303–6
20. Rowley JD. 1998. *Annu. Rev. Genet.* 32:495–519
21. Hess JL. 2004. *Trends Mol. Med.* 10:500–7
22. Tenney K, Shilatifard A. 2005. *J. Cell. Biochem.* 95:429–36
23. Clarke S. 1993. *Curr. Opin. Cell Biol.* 5:977–83
24. Springer MS, Goy MF, Adler J. 1979. *Nature* 280:279–84
25. Bedford MT, Richard S. 2005. *Mol. Cell* 18:263–72
26. Murray K. 1964. *Biochemistry* 3:10–15
27. DeLange RJ, Fambrough DM, Smith EL, Bonner J. 1969. *J. Biol. Chem.* 244:319–34
28. Patterson BD, Davies DD. 1969. *Biochem. Biophys. Res. Commun.* 34:791–94
29. Gershey EL, Haslett GW, Vidali G, Allfrey VG. 1969. *J. Biol. Chem.* 244:4871–77
30. Chen D, Ma H, Hong H, Koh SS, Huang SM, et al. 1999. *Science* 284:2174–77
31. Krogan NJ, Dover J, Khorrami S, Greenblatt JF, Schneider J, et al. 2002. *J. Biol. Chem.* 277:10753–55
32. Miller T, Krogan NJ, Dover J, Erdjument-Bromage H, Tempst P, et al. 2001. *Proc. Natl. Acad. Sci USA* 98:12902–7
33. Briggs SD, Bryk M, Strahl BD, Cheung WL, Davie JK, et al. 2001. *Genes Dev.* 15:3286–95
34. Luger K, Mader AW, Richmond RK, Sargent DF, Richmond TJ. 1997. *Nature* 389:251–60

35. Vermaak D, Ahmad K, Henikoff S. 2003. *Curr. Opin. Cell Biol.* 15:266–74
36. Rusche LN, Kirchmaier AL, Rine J. 2003. *Annu. Rev. Biochem.* 72:481–516
37. Kouzarides T. 2000. *EMBO J.* 19:1176–79
38. Workman JL, Kingston RE. 1998. *Annu. Rev. Biochem.* 67:545–79
39. Muller HJ. 1930. *J. Genet.* 22:299–334
40. Weiler KS, Wakimoto BT. 1995. *Annu. Rev. Genet.* 29:577–605
41. James TC, Elgin SC. 1986. *Mol. Cell Biol.* 6:3862–72
42. James TC, Eissenberg JC, Craig C, Dietrich V, Hobson A, Elgin SC. 1989. *Eur. J. Cell Biol.* 50:170–80
43. Lachner M, O'Carroll D, Rea S, Mechtler K, Jenuwein T. 2001. *Nature* 410:116–20
44. Nakayama J, Rice JC, Strahl BD, Allis CD, Grewal SI. 2001. *Science* 292:110–13
45. Jacobs SA, Khorasanizadeh SA. 2002. *Science* 295:2080–83
46. Volpe TA, Kidner C, Hall IM, Teng G, Grewal SI, et al. 2002. *Science* 297:1833–37
47. Allshire R. 2002. *Science* 297:1818–19
48. Reinhart BJ, Bartel DP. 2002. *Science* 297:1831
49. Noma K, Allis CD, Grewal SI. 2001. *Science* 293:1150–55
50. Khorasanizadeh S. 2004. *Cell* 116:259–72
51. Loyola A, LeRoy G, Wang YH, Reinberg D. 2001. *Genes Dev.* 15:2837–51
52. van Leeuwen F, Gafken PR, Gottschling DE. 2002. *Cell* 109:745–56
53. van Leeuwen F, Gottschling DE. 2002. *Curr. Opin. Cell Biol.* 14:756–62
54. Ng HH, Feng Q, Wang H, Erdjument-Bromage H, Tempst P, et al. 2002. *Genes Dev.* 16:1518–27
55. Maison C, Bailly D, Peters AHFM, Quivy JP, Roche D, et al. 2002. *Nat. Genet.* 30:329–34
56. Rice JC, Briggs SD, Ueberheide B, Barber CM, Shabanowitz J, et al. 2003. *Mol. Cell* 12:1591–98
57. Peters AH, O'Carroll D, Scherthan H, Mechtler K, Sauer S, et al. 2001. *Cell* 107:323–37
58. Tachibana M, Sugimoto K, Fukushima T, Shinkai Y. 2001. *J. Biol. Chem.* 276:25309–17
59. Tachibana M, Sugimoto K, Nozaki M, Ueda J, Ohta T, et al. 2002. *Genes Dev.* 16:1779–91
60. Ziemin-van der Poel S, McCabe NR, Gill HJ, Espinosa R III, Patel Y, et al. 1991. *Proc. Natl. Acad. Sci. USA* 88:10735–39
61. Daser A, Rabbitts TH. 2004. *Genes Dev.* 18:965–74
62. Nagy PL, Griesenbeck J, Kornberg RD, Cleary ML. 2002. *Proc. Natl. Acad. Sci. USA* 99:90–94
63. Roguev A, Schaft D, Shevchenko A, Pijnappel WW, Wilm M, et al. 2001. *EMBO J.* 20:7137–48
64. Schneider J, Wood A, Lee JS, Schuster R, Dueker J, et al. 2005. *Mol. Cell* 19:849–56
65. Milne TA, Briggs SD, Brock HW, Martin ME, Gibbs D, et al. 2002. *Mol. Cell* 10:1107–17
66. Nakamura T, Mori T, Tada S, Krajewski W, Rozovskaia T, et al. 2002. *Mol. Cell* 10:1119–28
67. Hughes CM, Rozenblatt-Rosen O, Milne TA, Copeland TD, Levine SS, et al. 2004. *Mol. Cell* 13:587–97
68. Yokoyama A, Wang Z, Wysocka J, Sanyal M, Aufiero DJ, et al. 2004. *Mol. Cell. Biol.* 24:5639–49
69. Dou Y, Milne TA, Tackett AJ, Smith ER, Fukuda A, et al. 2005. *Cell* 121:873–85
70. Strahl BD, Grant PA, Briggs SD, Sun ZW, Bone JR, et al. 2002. *Mol. Cell. Biol.* 22:1298–306
71. Landry J, Sutton A, Hesman T, Min J, Xu RM, et al. 2003. *Mol. Cell. Biol.* 23:5972–78
72. Li J, Moazed D, Gygi SP. 2002. *J. Biol. Chem.* 277:49383–88

73. Li B, Howe L, Anderson S, Yates JR 3rd, Workman JL. 2003. *J. Biol. Chem.* 278:8897–903
74. Xiao TJ, Hall H, Kizer KO, Shibata Y, Hall MC, et al. 2003. *Genes Dev.* 17:654–63
75. Schaft D, Roguev A, Kotovic KM, Shevchenko A, Sarov M, et al. 2003. *Nucleic Acids Res.* 31:2475–82
76. Krogan NJ, Kim M, Tong A, Golshani A, Cagney G, et al. 2003. *Mol. Cell. Biol.* 23:4207–18
77. Kizer KO, Phatnani HP, Shibata Y, Hall H, Greenleaf AL, et al. 2005. *Mol. Cell. Biol.* 25:3305–16
78. Hampsey M, Reinberg D. 2003. *Cell* 113:429–32
79. Tachibana M, Sugimoto K, Fukushima T, Shinkai Y. 2001. *J. Biol. Chem.* 276:25309–17
80. Muller J, Hart CM, Francis NJ, Vargas ML, Sengupta A. 2002. *Cell* 111:197–208
81. Cao R, Wang LJ, Wang HB, Xia L, Erdjument-Bromage H, et al. 2002. *Science* 298:1039–43
82. Wang JB, Mager J, Chen YJ, Schneider E, Cross JC, et al. 2001. *Nat. Genet.* 28:371–75
83. Mak W, Baxter J, Silva J, Newall AE, Otte AP, et al. 2002. *Curr. Biol.* 12:1016–20
84. Avner P, Heard E. 2001. *Nat. Rev. Genet.* 2:59–67
85. Plath K, Fang J, Mlynarczyk-Evans SK, Cao R, Worringer KA, et al. 2003. *Science* 300:131–35
86. Silva J, Mak W, Zvetkova I, Appanah R, Nesterova TB, et al. 2003. *Dev. Cell* 4:481–95
87. Plath K, Talbot D, Hamer KM, Otte AP, Yang TP, et al. 2004. *J. Cell Biol.* 167:1025–35
88. Peters AH, Kubicek S, Mechtler K, O'Sullivan RJ, Derijck AA, et al. 2003. *Mol. Cell* 12:1577–89
89. Boggs BA, Cheung P, Heard E, Spector DL, Chinault AC, et al. 2002. *Nat. Genet.* 30:73–76
90. Peters AHFM, Mermoud JE, O'Carroll D, Pagani M, Schweizer D, et al. 2002. *Nat. Genet.* 30:77–80
91. Heard E, Rougeulle C, Arnaud D, Avner P, Allis CD, et al. 2001. *Cell* 107:727–38
92. Wang H, Wang LJ, Erdjument-Bromage H, Vidal M, Tempst P, et al. 2004. *Nature* 431:873–78
93. de Napoles M, Mermoud JE, Wakao R, Tang YA, Endoh M, et al. 2004. *Dev. Cell* 7:663–76
94. Kirmizis A, Bartley SM, Kuzmichev A, Margueron R, Reinberg D, et al. 2004. *Genes Dev.* 18:1592–605
95. Singer MS, Kahana A, Wolf AJ, Meisinger LL, Peterson SE, et al. 1998. *Genetics* 150:613–32
96. van Leeuwen F, Gafken PR, Gottschling DE. 2002. *Cell* 109:745–56
97. Ng HH, Feng Q, Wang HB, Erdjument-Bromage H, Tempst P, et al. 2002. *Genes Dev.* 16:1518–27
98. Okada Y, Feng Q, Lin YH, Jiang Q, Li YQ, et al. 2005. *Cell* 121:167–78
99. Lee DY, Teyssier C, Strahl BD, Stallcup MR. 2005. *Endocr. Rev.* 26:147–70
100. Paik WK, Kim S. 1967. *Biochem. Biophys. Res. Commun.* 29:14–20
101. Schurter BT, Koh SS, Chen D, Bunick GJ, Harp JM, et al. 2001. *Biochemistry* 40:5747–56
102. Chen D, Ma H, Hong H, Koh SS, Huang S-M, et al. 1999. *Science* 284:2174–77
103. Wang HB, Huang Z-Q, Xia L, Feng Q, Erdjument-Bromage H, et al. 2001. *Science* 293:853–57
104. Strahl BD, Briggs SD, Brame CJ, Caldwell JA, Koh SS, et al. 2001. *Curr. Biol.* 11:996–1000
105. Koh SS, Chen DG, Lee Y-H, Stallcup MR. 2001. *J. Biol. Chem.* 276:1089–98
106. Lee Y-H, Koh SS, Zhang X, Cheng XD, Stallcup MR. 2002. *Mol. Cell. Biol.* 22:3621–32
107. Daujat S, Bauer UM, Shah V, Turner B, Berger S, Kouzarides T. 2002. *Curr. Biol.* 12:2090–97

108. Schneider J, Dover J, Johnston M, Shilatifard A. 2004. *Methods Enzymol.* 377:227–34
109. Dover J, Schneider J, Tawiah-Boateng MA, Wood A, Dean K, et al. 2002. *J. Biol. Chem.* 277:28368–71
110. Sun ZW, Allis CD. 2002. *Nature* 418:104–8
111. Wood A, Krogan NJ, Dover J, Schneider J, Heidt J, et al. 2003. *Mol. Cell* 11:267–74
112. Briggs SD, Xiao TJ, Sun ZW, Caldwell JA, Shabanowitz J, et al. 2002. *Nature* 418:498
113. Ng HH, Xu RM, Zhang Y, Struhl K. 2002. *J. Biol. Chem.* 277:34655–57
114. Krogan NJ, Dover J, Wood A, Schneider J, Heidt J, et al. 2003. *Mol. Cell* 11:721–29
115. Ng HH, Robert F, Young RA, Struhl K. 2003. *Mol. Cell* 11:709–19
116. Mueller CL, Jaehning JA. 2002. *Mol. Cell. Biol.* 22:1971–80
117. Pokholok DK, Hannett NM, Young RA. 2002. *Mol. Cell* 9:799–809
118. Porter SE, Washburn TM, Chang M, Jaehning JA. 2002. *Eukaryot. Cell* 1:830–42
119. Squazzo SL, Costa PJ, Lindstrom DL, Kumer KE, Simic R, et al. 2002. *EMBO J.* 21:1764–74
120. Krogan NJ, Kim M, Ahn SH, Zhong G, Kobor MS, et al. 2002. *Mol. Cell. Biol.* 22:6979–92
121. Wood A, Schneider J, Dover J, Johnston M, Shilatifard A. 2003. *J. Biol. Chem.* 278:34739–42
122. Ng HH, Dole S, Struhl K. 2003. *J. Biol. Chem.* 278:33625–28
123. Xiao TJ, Kao CF, Krogan NJ, Sun ZW, Greenblatt JF, et al. 2005. *Mol. Cell. Biol.* 25:637–51
124. Wood AN, Schneider J, Dover J, Johnstone M, Shilatifard A. 2005. *Mol. Cell* 20:589–99
125. Henry KW, Wyce A, Lo WS, Duggan LJ, Emre NC, et al. 2003. *Genes Dev.* 17:2648–63
126. Daniel JA, Torok MS, Sun ZW, Schieltz D, Allis CD, et al. 2004. *J. Biol. Chem.* 279:1867–71
127. Emre NC, Ingvarsdottir K, Wyce A, Wood A, Krogan NJ, et al. 2005. *Mol. Cell* 17:585–94
128. Santos-Rosa H, Schneider R, Bannister AJ, Sherriff J, Bernstein BE, et al. 2002. *Nature* 419:407–11
129. Bernstein BE, Humphrey EL, Erlich RL, Schneider R, Bouman P, et al. 2002. *Proc. Natl. Acad. Sci. USA* 99:8695–700
130. Guenther MG, Jenner RG, Chevalier B, Nakamura T, Croce CM, et al. 2005. *Proc. Natl. Acad. Sci. USA* 102:8603–8
131. Byvoet P, Shepherd GR, Hardin JM, Noland BJ. 1972. *Arch. Biochem. Biophys.* 148:558–67
132. Duerre JA, Lee CT. 1974. *J. Neurochem.* 23:541–47
133. Ahmad K, Henikoff S. 2002. *Mol. Cell* 9:1191–200
134. McKittrick E, Gafken PR, Ahmad K, Henikoff S. 2004. *Proc. Natl. Acad. Sci. USA* 101:1525–30
135. Goll MG, Bestor TH. 2002. *Genes Dev.* 16:1739–42
136. Schwartz BE, Ahmad K. 2005. *Genes Dev.* 19:804–14
137. Bannister AJ, Schneider R, Kouzarides T. 2002. *Cell* 109:801–6
138. Cuthbert GL, Daujat S, Snowden AW, Erdjument-Bromage H, Hagiwara T, et al. 2004. *Cell* 118:545–53
139. Wang Y, Wysocka J, Sayegh J, Lee YH, Perlin JR, et al. 2004. *Science* 306:279–83
140. Shi YJ, Lan F, Matson C, Mulligan P, Whetstine JR, et al. 2004. *Cell* 119:941–53
141. Kubicek S, Jenuwein T. 2004. *Cell* 119:903–6
142. Lunyak VV, Burgess R, Prefontaine GG, Nelson C, Sze SH, et al. 2002. *Science* 298:1747–52
143. Hakimi MA, Bochar DA, Chenoweth J, Lane WS, Mandel G, et al. 2002. *Proc. Natl. Acad. Sci. USA* 99:7420–25

144. Hakimi MA, Dong Y, Lane WS, Speicher DW, Shiekhattar R. 2003. *J. Biol. Chem.* 278:7234–39
145. Shi YJ, Sawada J, Sui GC, Affar EB, Whetstine JR, et al. 2003. *Nature* 422:735–38

RELATED REVIEWS

Sauve AA, Wolberger C, Schramm VL, Boeke JD. 2006. *Annu. Rev. Biochem.* 75:435–65
Spector DL. 2003. *Annu. Rev. Biochem.* 2003. 72:573–608
Shilatifard A, Conaway RC, Conaway JW. 2003. *Annu. Rev. Biochem.* 72:693–715

Structure and Mechanism of the Hsp90 Molecular Chaperone Machinery

Laurence H. Pearl and Chrisostomos Prodromou

Section of Structural Biology, Institute of Cancer Research, Chester Beatty Laboratories, London SW3 6JB, United Kingdom; email: laurence.pearl@icr.ac.uk, chris.prodromou@icr.ac.uk

Key Words

conformational change, protein crystallography, ATPase, protein-protein interaction

Abstract

Heat shock protein 90 (Hsp90) is a molecular chaperone essential for activating many signaling proteins in the eukaryotic cell. Biochemical and structural analysis of Hsp90 has revealed a complex mechanism of ATPase-coupled conformational changes and interactions with cochaperone proteins, which facilitate activation of Hsp90's diverse "clientele." Despite recent progress, key aspects of the ATPase-coupled mechanism of Hsp90 remain controversial, and the nature of the changes, engendered by Hsp90 in client proteins, is largely unknown. Here, we discuss present knowledge of Hsp90 structure and function gleaned from crystallographic studies of individual domains and recent progress in obtaining a structure for the ATP-bound conformation of the intact dimeric chaperone. Additionally, we describe the roles of the plethora of cochaperones with which Hsp90 cooperates and growing insights into their biochemical mechanisms, which come from crystal structures of Hsp90 cochaperone complexes.

Contents

INTRODUCTION AND
    PERSPECTIVES 272
HSP90 STRUCTURE AND
    CONFORMATION 273
    N-Terminal Domain—Nucleotide
        and Drug Binding 273
    Middle Segment—Client Protein
        Binding and Catalytic Loop 274
    C-Terminal
        Domain—Dimerization 275
    A Conformationally Coupled
        ATPase Cycle 276
    Structure of the ATP-Bound
        Conformation of Hsp90 279
COCHAPERONES AND
    COMPLEXES 281
    TPR-Domain Cochaperones 281
    ATPase Regulation by
        Cochaperones 283
    ATPase Arrest by Cdc37 284
    ATPase Activation by Aha1 285
    Conformation-Dependent Binding
        of P23/Sba1 286
CLIENT PROTEIN ACTIVATION 289

INTRODUCTION AND PERSPECTIVES

The 90-kDa heat shock proteins (Hsp90s) are a widespread family of molecular chaperones found in bacteria and all eukaryotes, but they are apparently absent from the archaea. Many eukaryotes possess multiple Hsp90 homologues, including endoplasmic reticulum, mitochondrial, and chloroplast-specific isoforms. Although the bacterial HtpG proteins are typically nonessential (1), eukaryotes require a functional cytoplasmic Hsp90 for viability under all conditions tested (2). This requirement for a functional Hsp90 has facilitated the use of yeast as an effective reporter for analyzing structure/function relationships in Hsp90 itself (3–6), as well as a powerful tool for identifying and characterizing the plethora of cochaperone proteins with which Hsp90 collaborates functionally (7–11).

Unlike other well-characterized molecular chaperones, such as GroEL and Hsp70, Hsp90 displays a clear but eclectic selectivity for the type of "client" proteins it chaperones. Its broad clientele includes proteins as structurally and functionally different as telomerase (12), the actin organizer N-WASP (13), and nitric oxide synthase (14) as well as a range of nuclear hormone receptors (reviewed in Reference 15) and an ever-growing selection of protein kinases (reviewed in Reference 16). Even with Hsp90 clients that are functionally closely related, there can be a high degree of specificity, with close homologues having radically different dependence on Hsp90 for their activation. The molecular basis for Hsp90's specificity for client proteins is largely unknown. For nuclear steroid hormone receptors, which are probably the most studied clients in terms of their association with Hsp90, recruitment to the Hsp90 system involves a prior interaction with the Hsp70/Hsp40 chaperone system, which is then coupled to Hsp90 via the bridging cochaperone Hop/Sti1 (15, 17). It is not clear, however, what proportion of specificity for a client receptor comes from each of the collaborating chaperone systems. For protein kinases, which constitute the largest coherent client group, specificity probably resides not with Hsp90 itself but with the Cdc37 adaptor protein, which binds Hsp90 and protein kinase clients (16, 18). However, the puzzle remains because the origins of Cdc37's specificity are also unknown. Adaptor-mediated specificity has been observed with other clients, such as the recruitment of the p21[WAF1/CIP1] cyclin-dependent kinase inhibitor by WISp39 (19), but it is far from clear that it is a widespread phenomenon.

Perhaps the least well-understood aspect of the Hsp90 system is what association with Hsp90 does to or for a bound client protein that facilitates its biological function. Although Hsp90 will stick to selected denatured proteins in vitro (20), this appears to be a nonspecific phenomenon largely divorced from Hsp90's biological interaction with authentic clients in vivo. However, a domain from at least one bona fide Hsp90 client, p53, has been shown to be unfolded in complex with Hsp90 (21), and this is also likely to be true of Hsp90's interaction with p21$^{WAF1/CIP1}$, which is an inherently unstructured protein (22). Protein kinases by contrast appear to interact with Hsp90 in a largely folded, structured, and active state (for example 23–25). Clearly the physical nature of the client protein when bound to Hsp90 is inextricably linked to issues of client-protein specificity, and both of these will only be resolved by the crystal structure of an authentic Hsp90-client complex. However, the technical challenge this presents should not be underestimated.

HSP90 STRUCTURE AND CONFORMATION

Crystallization of a full-length Hsp90 was first reported nearly 10 years ago (26). However, well-ordered diffraction-quality crystals capable of yielding an atomic resolution structure remain elusive, although some progress is being made (see Reference 27 and below). In the interim, an enormous amount has been learned from structural studies of domains and large subconstructs, so that atomic structures for almost all segments of the Hsp90 structure are now known (**Figure 1**).

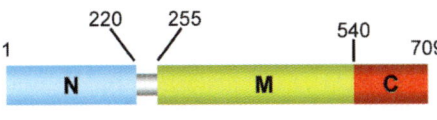

Figure 1

Schematic of the domain structure of yeast Hsp90, defined by limited proteolysis and structural studies. The segment linking the N and M regions of the molecule is highly variable in length and composition between species.

N-Terminal Domain—Nucleotide and Drug Binding

The first breakthrough into Hsp90 structure came from the identification of an N-terminal ∼25-kDa domain, which was readily released from the protein by limited proteolysis. Crystal structures of this domain from human (28) and yeast (29) Hsp90s revealed a two-layer α/β sandwich structure (CATH Code: 3.30.565.10) in which the helices delimit a pocket extending from the surface to the buried face of the highly twisted antiparallel β-sheet. In the human structure, this pocket was found to be the binding site for the macrocyclic antitumor agent geldanamycin, whose binding to Hsp90 had been shown to disrupt productive complexes with protein kinase and steroid hormone receptor clients (30–33). In the absence of any mechanistic model for Hsp90 function at the time, inhibition by geldanamycin was erroneously attributed to competitive blockade by the macrocycle of a putative binding site for an unfolded client protein (28). The yeast structure was also subject to some misinterpretation, with formation of an antiparallel β-ribbon between C-terminal β-strands from adjacent molecules in the crystal, which were taken to represent a functional dimerization interface (29).

The true function of the N-terminal domain and its pocket only became apparent with the recognition of limited sequence homology between Hsp90 and two classes of ATP-dependent DNA-manipulating proteins—the type II topoisomerases and the MutL mismatch repair proteins (34)—and determination of a cocrystal structure of the yeast Hsp90 N terminus and ATP/ADP (35) (**Figure 2a**). This landmark structure revealed the N-terminal pocket as the binding site for adenine nucleotides, identifying key conserved functional residues and explaining the action of geldanamycin and related compounds as competitive inhibitors of ATP binding (**Figure 2b**). Involvement of ATP in Hsp90 function had been suggested several years earlier, but it was associated

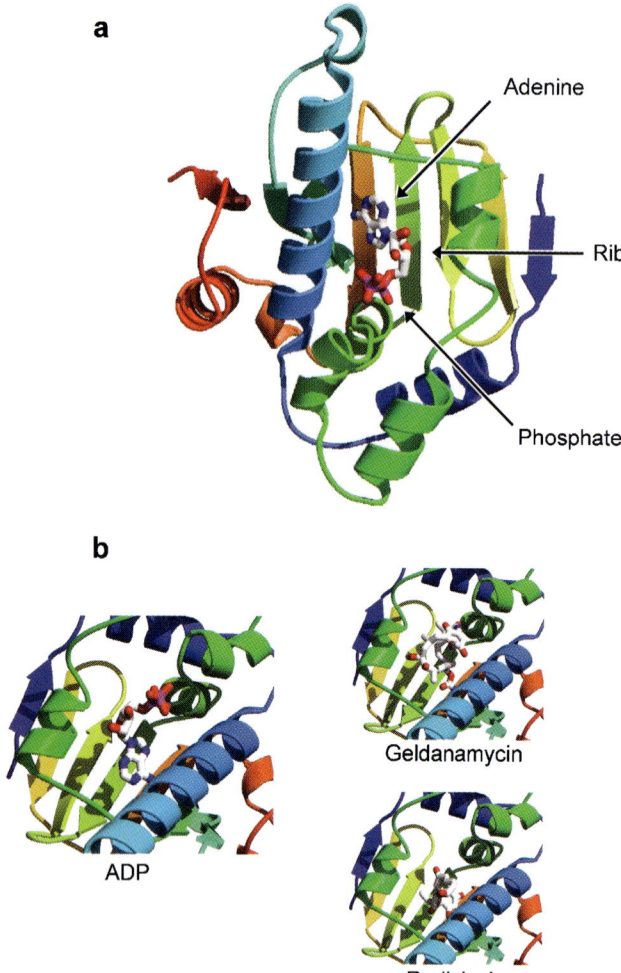

Figure 2

(*a*) Crystal structure of the N domain of yeast Hsp90, showing the binding site for ATP/ADP. The protein is shown as a secondary structure cartoon rainbow (blue-red) from the N to the C terminus of the construct. (*b*) Comparison of cocrystal structures of yeast Hsp90 N domain and bound ADP, with complexes of the antibiotics geldanamycin and radicicol. These compounds are potent competitive inhibitors of Hsp90s ATPase activity, and geldanamycin derivatives are currently in clinical trials as antitumor agents.

with apparent autophosphorylation (36–39), suggesting contamination with protein kinases, although subsequent biochemical analysis seemed to show conclusively that Hsp90 lacked significant ADP/ATP affinity (40). With the realization that geldanamycin was a highly specific competitive inhibitor of ATP binding to Hsp90, it became possible to measure the low inherent ATPase of Hsp90s accurately, without interference by contamination kinases or other more active ATPases. The identification of key residues in the N-terminal structure made it possible to demonstrate unequivocally, by mutagenesis in engineered yeast strains, that the biological function of Hsp90 was absolutely dependent on its ability to both bind and hydrolyze ATP (4, 5). The recognition of related N-terminal domain folds and ATPase activities unites Hsp90 with MutL and type II toposimerases, along with bacterial histidine kinases and branched-chain alpha-ketoacid dehydrogenase kinase, into a distinct superfamily, the GHKL ATPases (41). Crystal structures have subsequently been determined for N-terminal domains from the mammalian endoplasmic reticulum Hsp90 isoform GRP94 (42) and, as part of a larger construct, for *Escherichia coli* HtpG (43).

Middle Segment—Client Protein Binding and Catalytic Loop

Following from structural analysis of the yeast and human N-terminal domains, the crystal structure of a proteolytically resistant middle segment from yeast Hsp90 was determined (3) (**Figure 3a**). This structure consists of a large αβα domain at the N terminus of the construct connecting to a small αβα domain at the C terminus via a series of short α-helices in a tight coil. Consistent with the structural homology found between the N-terminal domains, the larger of the two αβα domains has a fold similar to equivalent regions of MutL and DNA gyrase B (Gyr B), although there is considerable difference between loop regions implicated in DNA interaction in MutL and GyrB. The smaller of the two domains is unique to Hsp90, and although it has architectural similarity to classic αβα domains, its fold is novel. Analogy with GyrB and MutL and a variety of mutagenesis studies (3, 44) implicate the middle segment as a major site for client protein interactions, with a conserved hydrophobic patch centered on Trp 300 and an unusual amphipathic protrusion formed by residues 327–340, with one exposed hydrophobic side and one positively charged side, being particularly important (**Figure 3b**). Again by analogy with other GHKL proteins, it was anticipated that the middle segment also contributed a key catalytic lysine residue that would interact with the γ-phosphate of an ATP molecule bound in the N-terminal domain. Topologically, this residue was expected to lie in the loop extending from the end of a strand in the central β-sheet of the larger αβα domain, but no conserved lysine is present in this Hsp90 region. Instead, a conserved arginine, Arg 380, was found by mutagenesis to be essential for ATPase activity in vitro and for the essential functions of Hsp90 in yeast in vivo, suggesting that it plays the same role of polarizing the β-γ phosphodiester bond, as the lysines do in the other GHKL ATPases. However, at least in the isolated Hsp90 middle segment, this connecting loop is highly structured with Arg 380 involved in a short α-helix, so that substantial remodeling of this segment would be required for Arg 380 to fulfill its catalytic role.

C-Terminal Domain—Dimerization

More recently the structure of the C-terminal domain from the *E. coli* Hsp90 homologue HtpG has been determined (27) (**Figure 4**). Consistent with previous biochemical observations (45) that the C-terminal domain is essential for Hsp90 dimerization, the HtpG C-terminal structure is a dimer of a small mixed α/β domain. The dimer interface is formed by a pair of helices at the C-terminal end of the domain packing together to create a four-helix bundle. Compared to the rest of the protein, the C-terminal domain of HtpG is much more divergent from the equivalent region of the eukaryotic Hsp90s, with a lower sequence similarity and two small deletions relative to the eukaryotic sequences between secondary structural elements. Most significantly, the HtpG C-terminal domain lacks a >35-residue extreme C-terminal segment, which provides the MEEVD motif implicated in binding to cochaperones with tetratricopeptide repeat (TPR) domains (46, 47). Nonetheless, the overall fold of the C-terminal domains in eukaryotic and bacterial Hsp90s is likely to be very similar, and in the present absence of a

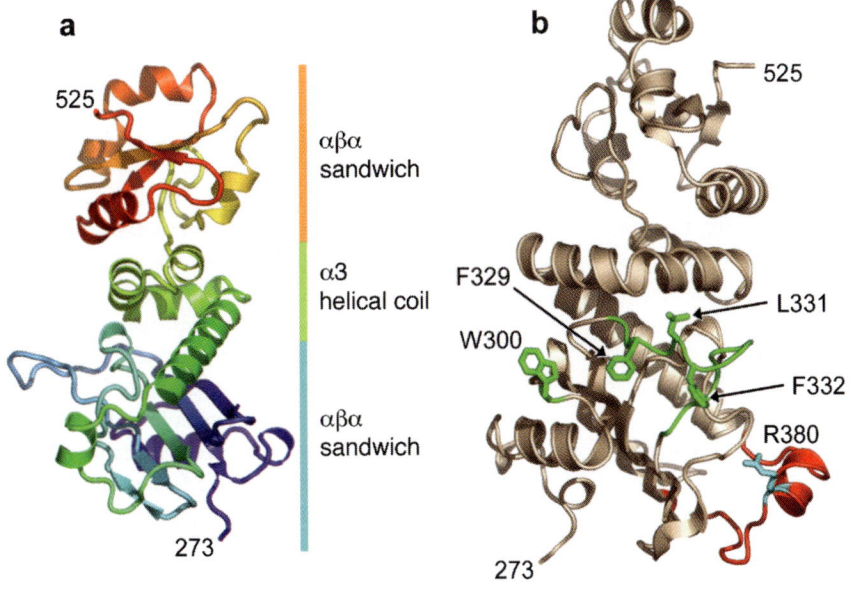

Figure 3

(*a*) Crystal structure of the middle segment of yeast Hsp90, showing the three subdomains. (*b*) As in (*a*) but rotated ∼90° around the vertical, with the hydrophobic residues implicated in client protein binding (*green*) and the catalytic loop bearing the essential residue Arg 380 (*red*).

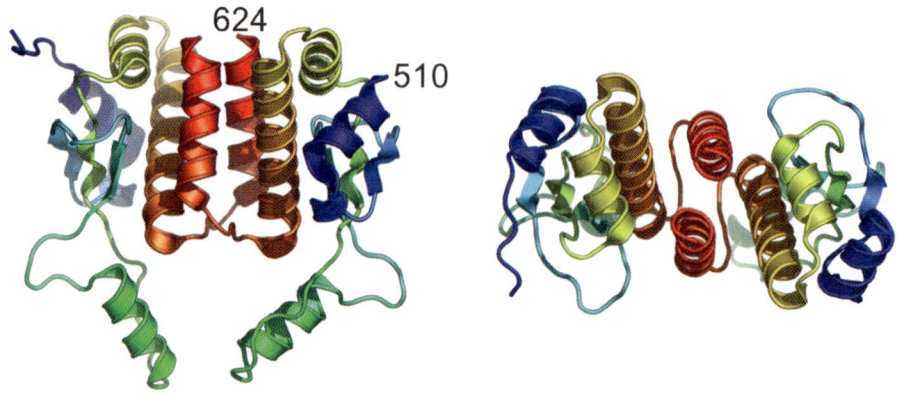

Figure 4
Crystal structure of the dimeric C-terminal domain of the *E. coli* Hsp90 homologue, HtpG (*left*) and viewed down the local pseudotwofold symmetry axis (*right*). This region corresponds to residues 549–674 of the yeast protein and lacks the extreme C terminus of ~30 residues that provides the binding site for TPR domain cochaperones in eukaryotic Hsp90s.

structure for a eukaryotic example, the HtpG C domain provides a good working model.

With structures of the N-terminal, middle, and C-terminal segments, the structure of Hsp90 is known—at least in pieces. The only remaining region of unknown structure is the very poorly conserved segment that links the N-terminal and middle segments of the protein. This segment, which is absent from the bacterial HtpG and mitochondrial proteins, varies between 30–70 residues in different species and isoforms, and it consists almost entirely of runs of charged residues, both acidic and basic. Although there has been some suggestion of functional involvement by this segment (48), its removal has no significant effect on the inherent ATPase activity of Hsp90 in vitro (6), nor does it noticeably impair the essential functions of yeast Hsp90 in vivo (49).

A Conformationally Coupled ATPase Cycle

The unequivocal demonstration that binding and hydrolysis of ATP are key to Hsp90 function begs the question as to how this ATPase cycle changes Hsp90's interactions with its client proteins to favor their activation. Conformational changes of Hsp90 in the presence of ATP had been previously observed, and it was interpreted generally in terms of a structural transition from an open hydrophobic state to a more closed conformation (38, 50, 51). Some structural insight into these changes were provided by biochemical and mutational analysis of Hsp90, which showed that the N-terminal domains were transiently associated in the ATP-bound state, so that the ATPase cycle was intimately coupled to opening and closing of a molecular clamp (6, 52)—a biochemical mechanism it shares with the GHKL ATPase Gyr B (53). A more direct view of this conformational change has come from rotary shadowing-electron microscopy of Hsp90, which showed an essentially linear antiparallel arrangement of the monomers dimerized via the C terminus in the absence of ATP but a ring-shaped structure in the presence of ATP, consistent with a parallel arrangement with dimerization at the C-terminal and N-terminal domains of the protein (54). By contrast, a low-resolution crystal structure of full-length HtpG (27) shows a "V"-shaped structure dimerized at the C terminus but with a parallel arrangement of the monomers. The different conformation observed in the microscopy and the crystal structure may well reflect the considerable flexibility of the ATP-unbound Hsp90 molecule, which is unlikely to have a single fixed conformation.

Association of the N-terminal domains has been demonstrated by several different methods (6, 52, 54), using Hsp90 protein from yeast, chicken, and mammalian sources. The current model for conformational coupling of the ATPase cycle suggests that N-terminal dimerization is required for ATP hydrolysis, and as such, the two halves of Hsp90 cooperate to achieve a conformation of the Hsp90 dimer that is able to hydrolyze ATP. The rate-limiting step to achieving this conformation is a complex structural process in which several regions of Hsp90 undergo interdependent and coupled rearrangements, which are required for either active site in the dimer to be catalytically active (55). This is broadly consistent with a kinetic analysis of human Hsp90, which found no evidence for formal cooperativity in the classic sense between the protomers in an Hsp90 dimer in hydrolysis of ATP (56). Interestingly, the measured rate of hydrolysis for a construct lacking a C-terminal constitutive dimerization interface in that study was found to be one tenth of that observed in the dimerized protein, consistent with the current model based on the yeast protein (6), clearly showing that the rate of hydrolysis in one monomer is affected by the presence of the other monomer. Although the possibility of a significant mechanistic difference between yeast and mammalian proteins cannot be dismissed, the very high degree of conservation between them and the ability of the human enzyme to complement a yeast Hsp90 defect (57) make this most unlikely, but comparable kinetic analysis of the yeast protein would be worthwhile.

Implicit in the notion of a functionally coupled ATPase cycle is the idea that binding of ATP stabilizes a "tense" conformational and functional state of Hsp90 that is dependent on the presence of the γ-phosphate of the ATP, so that subsequent hydrolysis to ADP destabilizes that conformation and allows the Hsp90

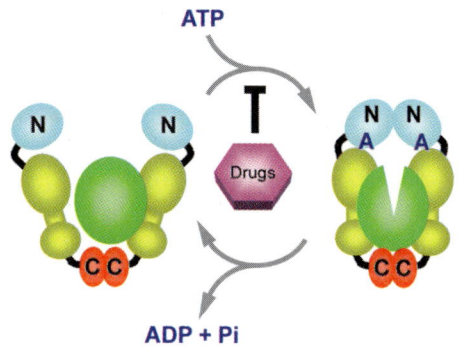

Figure 5

Schematic of the current model for the gross conformational changes that accompany binding and hydrolysis of ATP by Hsp90 on the basis of structural and biochemical evidence. The N-, middle, and C-terminal regions in the Hsp90 dimer (colored *cyan*, *yellow*, and *red*, respectively) are shown; a client protein (*green*) is able to bind in the absence of ATP and undergoes some change of state during the passage through the "tense" ATP-bound conformation of the chaperone. Inhibition of ATP binding by drugs, such as geldanamycin, blocks client protein activation.

dimer to relax to a default state (**Figure 5**). The currently accepted model for that tense ATP-bound state of the Hsp90 dimer is similar to the comparable state of the DNA gyrase B and MutL dimers, in which the two N-terminal domains are in intimate proximity and each is tightly packed against the middle segment of the same chain. This allows the key γ-phosphate-binding residues from the middle segment to project into the top of the nucleotide-binding pocket in the N-terminal domain and provide the missing catalytic interaction. Modeling of an ATP-bound Hsp90 dimer based on its structural homology to the MutL and GyrB structures (3, 6) indicates clearly that several changes in the conformation of structural elements in the N-terminal domain and middle segment would be required to allow formation of the tightly packed interfaces observed in the other dimeric GHKL ATPases.

Within the N-terminal domain, the most significant change required is the remodeling of a contiguous segment of structure (residues

94–125 in yeast Hsp90), known as the "lid," which must move to prevent steric overlap with its equivalent in the other monomer and with the region of the middle segment from the same monomer that interacts with the N-terminal domain. In MutL and DNA gyrase B, the equivalent lid segment undergoes a disorder-order transition during ATP binding (58, 59), with a GxxGxG motif at the C-terminal end of the lid segment wrapping around the triphosphate segment of the bound ATP. In Hsp90, at least in the isolated N-terminal domain, the lid segment is a fully α-helical hairpin in the absence of any ligand (28, 29). For the Hsp90 lid to adopt a similar conformation to the lid in GyrB or MutL and position its GxxGxG motif in the same way, it must restructure from its observed position, effectively folding over the ATP-binding pocket. Mutation of lid residues, whose environment would be changed significantly, had significant effects on the ATPase activity of Hsp90, which was consistent with this mechanistic model (6) (**Figure 6**). However, this predicted conformational change has not been observed so far in crystallographic studies of the isolated Hsp90 N-terminal domain on binding of nucleotides or inhibitors (60). Some conformational change in the lid segment has been observed in crystal structures of the isolated N-terminal domain of the endoplasmic reticulum Hsp90 isoform, Grp94, on binding of ADP or ATP (61). This involves a relatively small rigid body shift of the two helices in the lid along with the first helix of the domain, which is quite different from the lid closure predicted by analogy with other GHKL ATPases. The biological significance of this change in conformation is far from clear because it is only observed in one of the two crystallographically independent molecules in the crystal asymmetric unit, whereas the lid in the other becomes disordered. Furthermore, in the shifted position, the lid helices are involved in the major lattice contact of the crystal, so that this conformation may be favored by crystallization. Although Grp94 might conceivably operate by a distinct mechanism from Hsp90 and other GHKL ATPases as has been proposed (62), the total conservation of the GxxGxG motif in Grp94s suggests that this is unlikely.

In addition to restructuring of the lid segment within the N-terminal domain, the current unified GHKL model for the ATP-bound conformation of Hsp90 requires that the middle segment and N-terminal domain come into close association, so that a catalytic loop from the middle segment is able to project into the top of the nucleotide pocket and deliver a putative key residue (Arg 380) to interact with the γ-phosphate of ATP. This proposal has not been supported by a recent crystal structure of the contiguous N-terminal

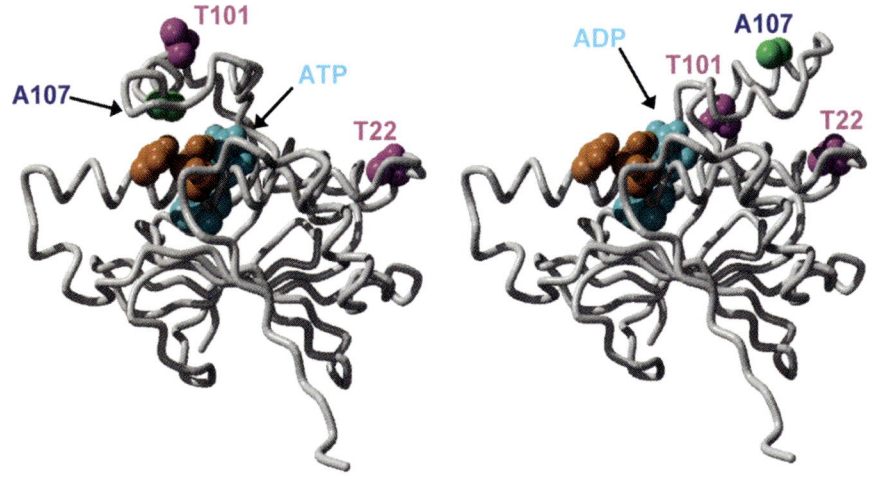

Figure 6
Model of the conformational switch in the "lid" segment (residues 94–125) in the N domain when going from the ADP-bound state (*right*) to the ATP-bound state (*left*) on the basis of biochemical analysis of mutations in residues Thr 22, Thr 101, and Ala 107.

and middle segment of the *E. coli* HtpG protein in the presence of ADP (43), which adopt a quite different relative position than expected, such that the HtpG equivalent to the proposed catalytic loop in eukaryotic Hsp90s is too far away from the bound nucleotide to perform its proposed role (**Figure 7**). Compared with the apo structure, binding of nucleotide appears to engender a disordering of the lid segment in HtpG, which also conserves the GxxGxG motif. Although the N-middle construct lacks the C-terminal dimerization domain and is predominantly monomeric in solution, pairs of molecules come in to close proximity within the crystals and have been interpreted as possibly biologically significant associations. The arrangement of pairs of HtpG N-middle protein molecules is substantially different between the apo and ADP-bound crystal structures, but in both cases, they are arranged in a head-to-tail antiparallel orientation, not parallel as expected. This arrangement is difficult to reconcile with the clear parallel arrangement of the dimerized C-terminal domains and the V-shaped parallel low-resolution crystal structure for intact apo (27).

Structure of the ATP-Bound Conformation of Hsp90

At the time of writing, a structure has been determined at ~3.8 Å for the full yeast Hsp90, which is intact save for a shortened charged linker, in complex with a nonhydrolyzable ATP analogue (AMPPNP) and with the cochaperone p23/Sba1 (M.M.U. Ali, S.M. Roe, C. Vaughan, P. Meyer, B. Panaretou, P.W. Piper, C. Prodromou, L.H. Pearl, unpublished data) (**Figure 8***a*). Although the fine detail of the middle-segment catalytic loop and the N-terminal domain lid segment are not fully defined at this resolution, the overall conformation of these loops and the arrangement of the domains in the complex is unambiguous. This Hsp90 structure clearly shows a parallel dimer arrangement in which the two N-terminal domains are in close proximity to

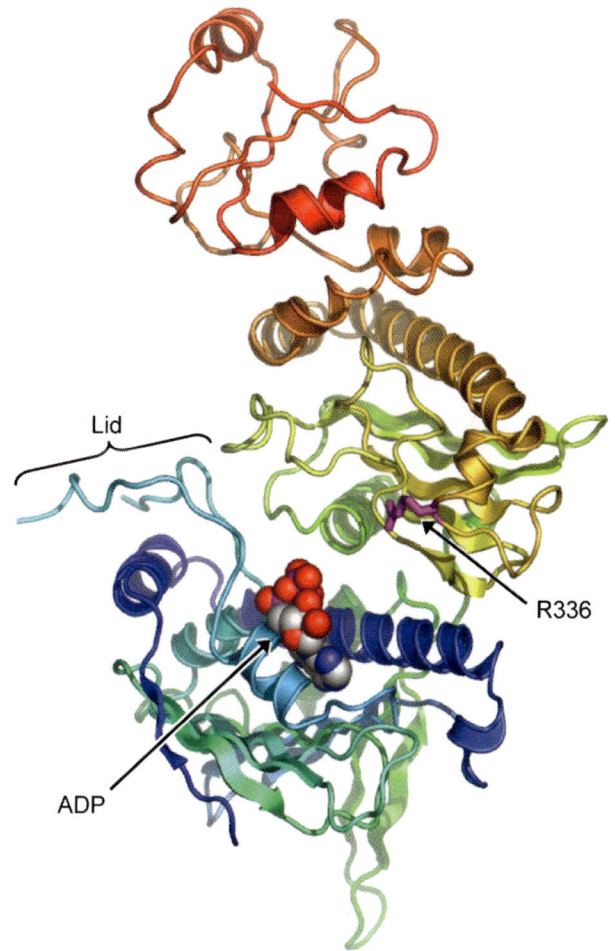

Figure 7

Crystal structure of the N domain and middle segment of *E. coli* HtpG with bound ADP. The distance between the putative catalytic residue Arg 336 (equivalent to Arg 380 in yeast Hsp90) and the ADP is too far for this to represent an active conformation. The "lid" segment in this structure is largely disordered and projects out of the N domain.

each other and their respective middle segments. Several important differences are evident between the structures of the N-terminal domain and middle segment in isolation and their structures in the full Hsp90 dimer. First, the N-terminal β-strand (residues 1–9) from each monomer crosses over to hydrogen bond to the edge of the main β-sheet in the N-terminal domain of the other monomer with concomitant movement of the first α-helix—an arrangement also seen in the MutL and GyrB dimers (**Figure 8***a,b*). Second,

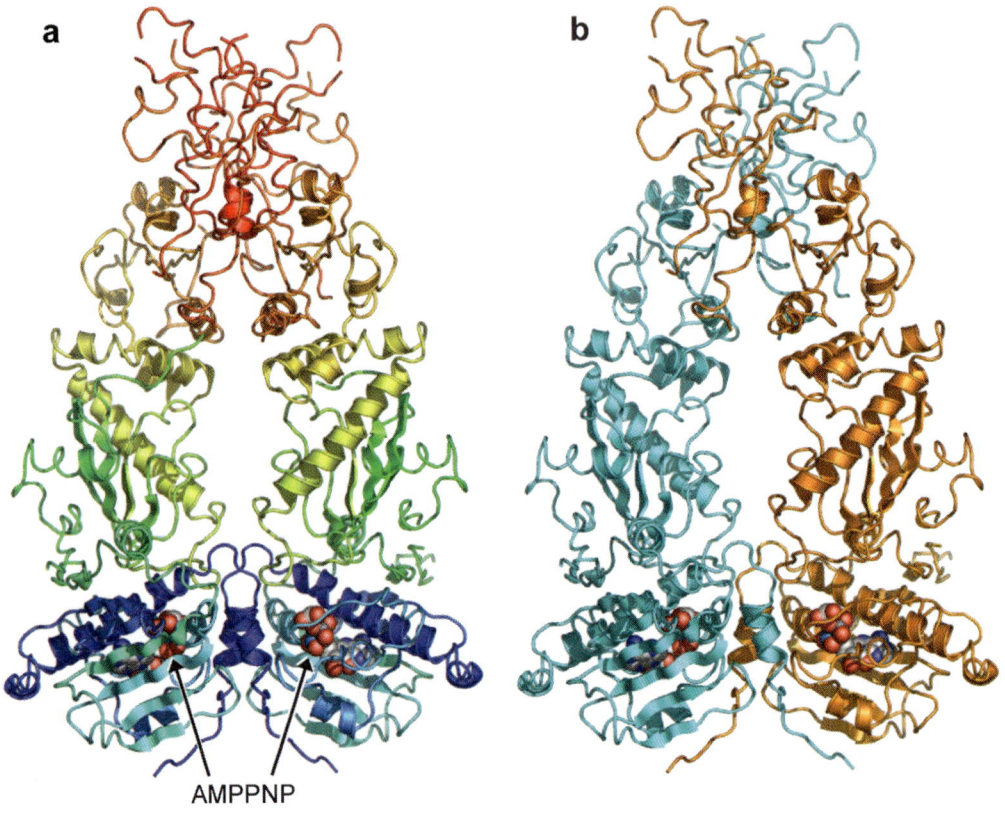

Figure 8

(*a*) Medium/low-resolution crystal structure of full-length yeast Hsp90 in the ATP-bound closed state, with the separate chains (*rainbow colored*). The AMPPNP molecules bound in each N domain are indicated (*arrows*). The P23/Sba1 cochaperone that is also present in the crystals has been omitted for clarity, but see **Figure 13b**, which is as in (*a*) but with the two chains in different colors. The "domain swap" of the N-terminal strand and the overall helical twist of the monomers from the N to C termini can be seen.

the lid segment from residues 94–125 is completely displaced from its position packed over residues 11–22 and 124–127 in the isolated N-terminal domain structure and folds back much as predicted over the mouth of the nucleotide pocket, with the 118-GxxGxG-123 motif at the C-terminal end of the lid wrapped around the β- and γ-phosphates, and more distal parts of the lid interact with the ribose sugar of the bound ATP (**Figure 9**). The observed change in conformation is radically different from that observed in the isolated Grp94 N-terminal domain (61).

In the middle segment, the catalytic loop (residues 370–390), which forms a short helix, unravels and extends down toward the mouth of the nucleotide-binding pocket in the N-terminal domain. The tip of the loop, bearing the putative catalytic residue Arg 380, then occupies the same site as the tip of the lid segment in the isolated N-terminal domain. Mobility of this loop was noted in different crystal forms of the middle domain (3), and promotion of this conformational change was suggested as a means for Hsp90 activation by the Aha1 cochaperone (see below). Further rearrangements in the middle segment involving the projecting amphipathic loop and interaction with a projecting helix from the C-terminal domain are evident but cannot be fully interpreted at the present resolution. The relative juxtaposition of the N and middle domains in the full Hsp90-P23-AMPPNP complex is radically different from that in the C-terminal deleted HtpG structure (43), which is most likely to be an artifact of crystallization in the absence of the C-terminal domain.

The architecture of the full Hsp90 in the ATP-bound state reveals a high degree of interchain and interdomain contacts. Between the N-terminal domains, in addition to the contacts made by the domain-swapped N-terminal strands, there is a substantial hydrophobic interface involving mutual symmetric interaction of the side chains of Gln 14, Leu 15, Leu 18, Thr 95, Ile 96, and Ala 97 at the N-terminal hinge point of the lid and Phe 120 from the C-terminal hinge point. Although Thr 95, Ile 96, and Ala 97 are exposed to solvent in the isolated N-terminal domain structure, Leu 15 and Leu 18 are substantially buried by the lid segment in its "open" form. The exposure of these residues when the lid folds over the nucleotide pocket in the ATP-bound conformation generates a new hydrophobic patch, whose burial stabilizes association of the N-terminal domains, as previously suggested (6). The middle segments themselves make no direct mutual contact, but each interacts with the N-terminal domain of the other monomer in addition to the substantial interface it makes with its own N-terminal domain. This interstrand interdomain contact involves Leu 376 and Leu 378 on one side of the middle segment catalytic loop in a hydrophobic interaction with the side chains of Thr 22, Val 23, and Tyr 24 in the loop connecting the first and second α-helix of the N-terminal domain. Adjacent to this an additional polar interaction occurs between the side chains of Asn 386 to form the middle segment and Ser 25 in the N-terminal domain. The involvement of Thr 22 in this interaction is particularly interesting because a Thr22Ile mutation had previously been isolated as a *ts* allele in yeast Hsp90 in vivo (63) and found to have an activating ef-

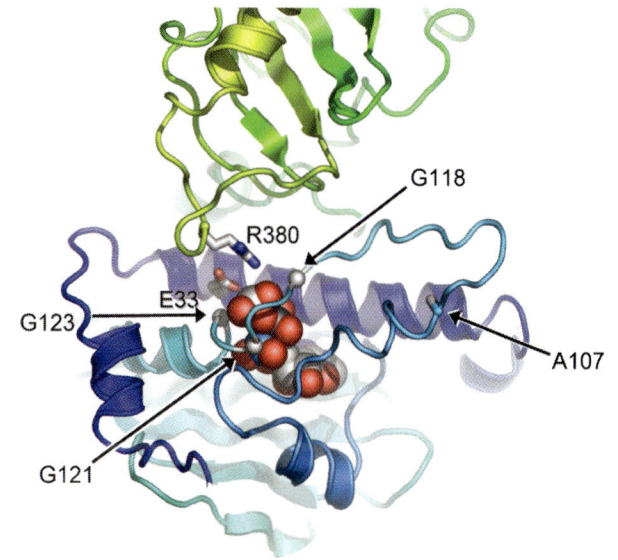

Figure 9

Close-up of the interactions around the AMPPNP in the full-length Hsp90 structure. As predicted from biochemical and mutagenesis studies, the lid segment, bearing residue A107, folds over the nucleotide, with the 118-GxxGxG-123 motif wrapping around the phosphates. Movement of the lid allows access by the catalytic loop from the middle segment with Arg 380 coming into contact with the γ-phosphate of the AMPPNP.

fect on Hsp90s ATPase in vitro (6), although the mechanism of this activation was not clear at the time.

COCHAPERONES AND COMPLEXES

Hsp90 functions in vivo as the core component of a dynamic set of multiprotein complexes, involving a plethora of collaborating proteins or cochaperones of which more than 12 have been identified. Defining the biological roles and biochemical mechanisms of these cochaperones and defining the combinations in which they simultaneously interact with Hsp90 are two of the major challenges in the field.

TPR-Domain Cochaperones

The largest definable class of cochaperones and the first to be identified are those that possess one or more TPR domains, a helical coil structure formed by concatenation of

∼34 residue helical hairpins (64), which binds to an MEEVD motif at the extreme C terminus of Hsp90 (46, 47, 65–69). TPR domains are very widespread, and the presence of a TPR domain in a protein does not automatically implicate it as an Hsp90-binding protein. Indeed, in at least one known Hsp90-associating TPR-domain protein, Sgt1, the TPR domain is not required for binding (70). A closely related motif IEEVD occurs at the C terminus of some Hsp70 chaperone family members, and several TPR-domain cochaperones are able to bind to either Hsp90 or Hsp70. Hsp90-binding TPR-domain proteins are an eclectic set, with little obvious biochemical similarity other than their TPR domains. WISp39, for example, appears to act as a specificity factor for recruitment of a client protein (19), and Hop/Sti1/p60 (71) or the outer mitochondrial membrane receptor Tom70 (72) facilitates recruitment of Hsp90 complexes to other multiprotein systems. Some TPR-domain cochaperones incorporate additional enzymatic functionality; thus, the immunophilins Cyp40 and FKBPs are peptidyl-prolyl isomerases (73, 74), PP5 is a Ser/Thr protein phosphatase (75), and CHIP is an E3/E4-ubiquitin ligase (76). Apart from CHIP, where the ubiquitin ligase activity is required for proteasome-targeted destruction of client proteins (77), the biological role of TPR-associated enzymatic functions is far from clear, and at least in the case of the yeast cyclophilin Cpr7, the growth defect associated with a null allele can be effectively suppressed by the TPR domain alone (78).

The structural basis for specific recruitment of TPR-domain cochaperones has been defined by crystallographic and biochemical studies of TPR domains from Hop/Sti1 bound to the C-terminal EEVD motifs of Hsp90 and Hsp70 (79, 80) (**Figure 10**). Hop/Sti1 possesses three TPR domains with distinct properties; the most N-terminal domain binds the C terminus of Hsp70 exclusively, the central TPR binds the C terminus of Hsp90 exclusively, and the last domain appears to bind neither. Hsp90 and Hsp70 C-terminal peptides both bind to their cognate Hop/Sti1 TPR domain in essentially similar extended conformations, with the side chain and α-carboxylates of the C-terminal aspartate making polar interactions with residues lining the TPR domain's

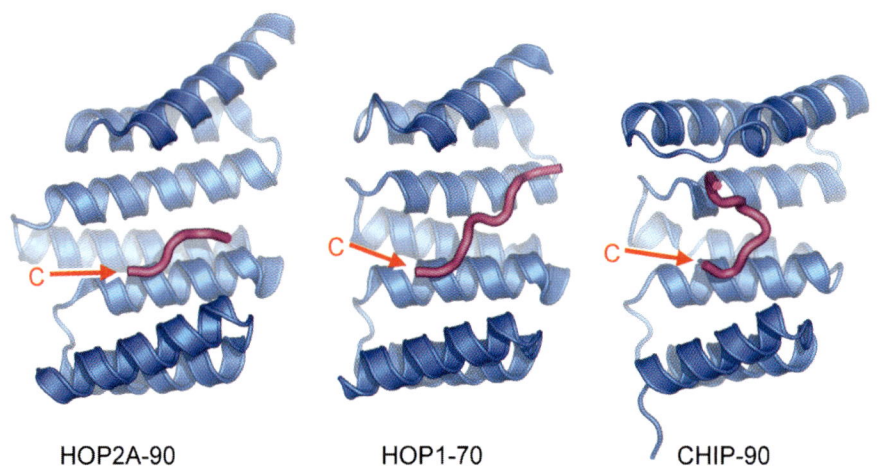

Figure 10

Comparison of the binding of C-terminal EEVD-containing peptides from Hsp90 and Hsp70 to TPR domains of cochaperones. In Hop/Sti1, different TPR domains (2A and 1) bind the C terminus of Hsp90 (*left*) and Hsp70 (*middle*) with the peptide in an extended conformation. The single TPR domain of CHIP (*right*) binds either chaperone C terminus but in a curled conformation in which only the common (M/I)EEVD sequence is in contact with the TPR domain.

concave groove, and the penultimate valine makes hydrophobic interactions in both cases. However, the sequences upstream of the common EEVD motif are substantially different between the two chaperones, (DDTSRMEEVD in yeast Hsp90, GSGPTIEEVD in yeast Ssa1/Hsp70), so that the different TPR domains provide specific and exclusive interactions for one or the other sequence.

Although the presence of multiple selective TPR domains allows Hop/Sti1 to bind Hsp90 and Hsp70 simultaneously, in other cases, notably CHIP, a single TPR domain facilitates exclusive binding of either Hsp90 or Hsp70 (81, 82). Structural studies of CHIP bound to a C-terminal Hsp90 peptide (83) show an interaction between the TPR domain and the common EEVD motif similar to that between this motif and the Hop/Sti1 TPR domain. However, instead of an extended conformation, the CHIP TPR-bound peptide makes a tight turn upstream of this position, facilitated by insertion of the Met (or Ile in Hsp70) into a hydrophobic pocket in the base of the TPR-domain groove. This hoists the upstream and chaperone-specific residues out of the TPR domain, thereby allowing either Hsp90 or Hsp70 peptides to be bound with comparable affinity.

As a dimer, Hsp90 possesses two copies of the C-terminal MEEVD TPR-binding sequence, each of which can in principle bind a different TPR-domain cochaperone. However, several cochaperones, such as Hop/Sti1 and CHIP, are themselves dimers (84, 85) and simultaneously occupy both TPR-binding sites on Hsp90, competitively excluding other TPR-domain cochaperones, as has been experimentally observed (66, 75, 86). Hop/Sti1 can also bind Hsp70, which is primarily a monomeric protein, as well as the dimeric Hsp90. However, the stoichiometry of the Hsp70-Hop interaction is dictated by the presence or absence of Hsp90, so that two Hsp70s bind to a Hop dimer in the absence of Hsp90, but only one Hsp70 binds to the Hop-Hsp90 dimer-dimer complex (87).

ATPase Regulation by Cochaperones

The biological roles of the many cochaperones in facilitating Hsp90-dependent client protein activation are far from clear. However, some understanding of the nature of their interactions with Hsp90 and their possible biochemical mechanisms is being gained. Three different cochaperones, Hop/Sti1, Cdc37, and p23, have been found to have an inhibitory effect on the ATPase cycle of Hsp90 (11, 84, 88), and two, Aha1 and Cpr6, have an activating effect (11, 89). The common ability of Hop/Sti1 and Cdc37, which are implicated in recruitment of client proteins to the Hsp90 complex, to arrest the ATPase cycle suggests that this is an important process in facilitating the loading of client protein at the start of the ATPase-coupled chaperone cycle. Hop/Sti1 binds to Hsp90 via interaction of its middle TPR domain with the C-terminal MEEVD peptide of Hsp90 (see above). However, mutagenesis studies indicated a secondary interaction with the N-terminal nucleotide-binding domain of Hsp90 (46), and spectroscopic studies showed perturbation of the circular dichroism spectrum of the Hsp90-bound ATP-competitive inhibitor geldanamycin, suggesting a possible direct interaction of Hop/Sti1 with the nucleotide-binding pocket. However, kinetic analysis shows that inhibition by Hop/Sti1 is noncompetitive with respect to ATP (90) and, instead, probably involves stabilizing a conformational state of Hsp90 in which the ATPase cycle cannot progress. The nature of this conformational state, and how Hop/Sti1 stabilizes it, is at present unknown. The TPR-domain immunophilin Cpr6 is able to relieve ATPase inhibition by Hop/Sti1, competitively displacing it from the C-terminal TPR-binding sites of Hsp90 (84). However, in low-salt conditions, Cpr6 displays an ability to stimulate ATPase activity by ~twofold (11). This ability did not depend on the presence of the peptidyl-prolyl isomerase domain, and the mechanism by which it occurs is unknown.

ATPase Arrest by Cdc37

In contrast to Hop/Sti1, the mechanism of Hsp90 ATPase inhibition by the cochaperone Cdc37 is well understood at the molecular level. Cdc37 was first identified as pp50, a phosphoprotein component of an Hsp90 complex with the viral oncogene protein tyrosine kinase v-Src (91). Subsequently, p50 was identified as the mammalian homologue of the yeast Cdc37 gene product (92, 93), required for formation of active cyclin-dependent kinase complexes (94). Cdc37 is associated with a wide range of Hsp90-dependent protein kinases (reviewed in Reference 16) and acts as a specific adaptor or scaffold protein, binding client protein kinases via its N-terminal region and Hsp90 via its central and C-terminal domains (95–97). Although competition studies suggested that the binding site for Cdc37 on Hsp90 overlapped with that for TPR-domain cochaperones in the C terminus of the protein (95), domain dissection identified the N-terminal nucleotide-binding domain of Hsp90 as the primary binding site and allowed the crystallization and structure determination of a core N-Hsp90-C-Cdc37 complex (98).

The C-terminal segment of Cdc37 in the core complex consists of a large six-helix domain at the N-terminal end of the structure (residues 148–245) connecting via a long helix (residues 246–286) to a small 3-helix domain (residues 292–347), which is less well ordered than the rest of the structure (**Figure 11***a*). The topology of the helices in the large helical domain is unusual, and no similar fold could be identified in databases of known protein structures. Within the crystals a small but highly specific homodimeric interface was observed, which involved residues from the large domain and connecting helix and was consistent with biochemical studies that show Cdc37 has distinct dimerization interfaces in different domains of the protein (88, 99). The interaction between Cdc37 and Hsp90 consists of a substantially hydrophobic interface involving the large helical domain of Cdc37, which packs against the surface of the lid segment of the Hsp90 N-terminal domain. The hydrophobic core is reinforced by a network of polar interactions, including a hydrogen bond contact in which the guanidinium side chain of Arg 167 of Cdc37 inserts into the mouth of the nucleotide-binding pocket

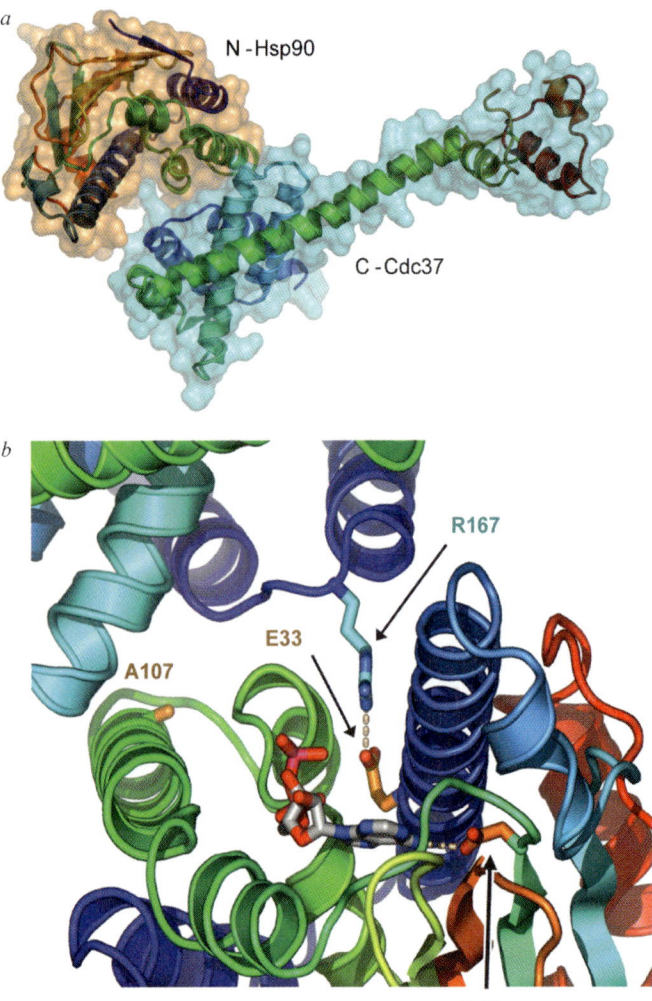

Figure 11

(*a*) Crystal structure of the complex between the N-terminal domain of Hsp90 (*orange surface*) and the C-terminal region of Cdc37 (*blue surface*). (*b*) Cdc37 binds onto the surface of the "lid" segment bearing Ala107 in its open conformation and inserts an arginine residue into the mouth of the nucleotide-binding pocket to interact with the catalytic residues Glu 33.

of N-Hsp90 to interact with the side chain of the catalytic residue Glu 33 (**Figure 11b**).

From the structure, it is evident that Cdc37 arrests the Hsp90 ATPase cycle by blocking key structural rearrangements of Hsp90 that are required to achieve the ATP-triggered transactivating interaction seen in the full-length Hsp90 structure. First, the interaction of Cdc37 with the Hsp90 lid segment locks it in the open conformation and prevents it closing over a bound ATP molecule. Second, the arrangement of C-Cdc37 and N-Hsp90 molecules in the crystals suggests that a Cdc37 dimer would simultaneously bind between N-terminal domains of an Hsp90 dimer, holding them apart. Third, Cdc37 sits over the open mouth of the nucleotide pocket in the N-terminal domain, engaging the catalytic residue Glu 33 in a polar interaction and blocking access to the middle segment and the catalytic loop bearing Arg 380.

ATPase Activation by Aha1

Aha1 was originally identified as a homologue of Hch1, a yeast protein whose overexpression suppressed an Hsp90 temperature-sensitive allele *E381K* (100). Unlike Hch1, which is restricted to simple yeasts, Aha1 homologues are widespread throughout eukaryotes with at least two human homologues. Subsequent analysis revealed that in yeast and human cells Aha1 is coregulated with Hsp90 and known Hsp90 cochaperones, is involved in client protein activation by the Hsp90 system in vivo, interacts directly with Hsp90, and activates the basal ATPase activity of Hsp90 more potently than Hch1 does (11, 101). At the sequence level, apart from the N-terminal ~150 residues, which are homologous to the full Hch1 sequence, Aha1 shows no homology to any other protein family, nor does optimal sequence threading suggest any structural relationships that might indicate how Aha1 activates Hsp90. Domain dissection of Hsp90 and Aha1 (3, 11, 101) revealed a core interaction in which the N terminus of Aha1 (equivalent to all of Hch1) bound to the middle segment of Hsp90 with comparable affinity to the interaction of the intact proteins. The unique C-terminal 200 residues of Aha1 were unable to bind to Hsp90 or stimulate the ATPase activity in isolation, but their presence in the intact Aha1 protein did increase the ATPase activation of yeast Hsp90 by ~threefold over that elicited by the isolated Hch1 or the N-terminal part of Aha1 in isolation.

The basis for the recruitment of Aha1 to the Hsp90 machinery, and for the activation mechanism of Hsp90 by the common Aha1/Hch1 core domain, was revealed by a cocrystal structure of the middle segment of yeast Hsp90 and the N-terminal domain (residues 1–153) of yeast Aha1 (102) (**Figure 12a**). N-Aha1 has an elongated cylindrical structure, formed by an N-terminal α-helix leading into a four-stranded meandering antiparallel β-sheet, followed by a C-terminal α-helix. Within the Hsp90-Aha1 complex, the two interacting molecules are orientated with their long axes parallel, so that N-Aha1 interacts with all three subdomains of the elongated Hsp90 middle segment. The majority of the contacts between Hsp90 and Aha1 are polar, including an extensive hydrogen bonding and ion-pair network involving four basic residues (lysines 387, 390, 394, and 398), exposed on the surface of a long α-helix in the large subdomain of Hsp90, and four acidic residues (Asp 53, Asp 68, Glu 97, and Asp 101) from Aha1. The only significant nonpolar contact between the two protein molecules involves a hydrophobic patch on Aha1, which packs against a similar patch on Hsp90, formed by the side chains of Leu 315, and Val 391, and the side chain of Ile 388, which lies within the catalytic loop segment of Hsp90, containing Arg 380. In the N-Aha1 complex, this segment of the Hsp90 polypeptide chain undergoes a substantial change in conformation and increase in order, compared to its state in the free Hsp90 middle segment structure. In particular, Lys 387 and Ile 388 at the C-terminal end of the catalytic loop move by >15 Å and >8 Å, respectively, and become incorporated into the beginning of the long

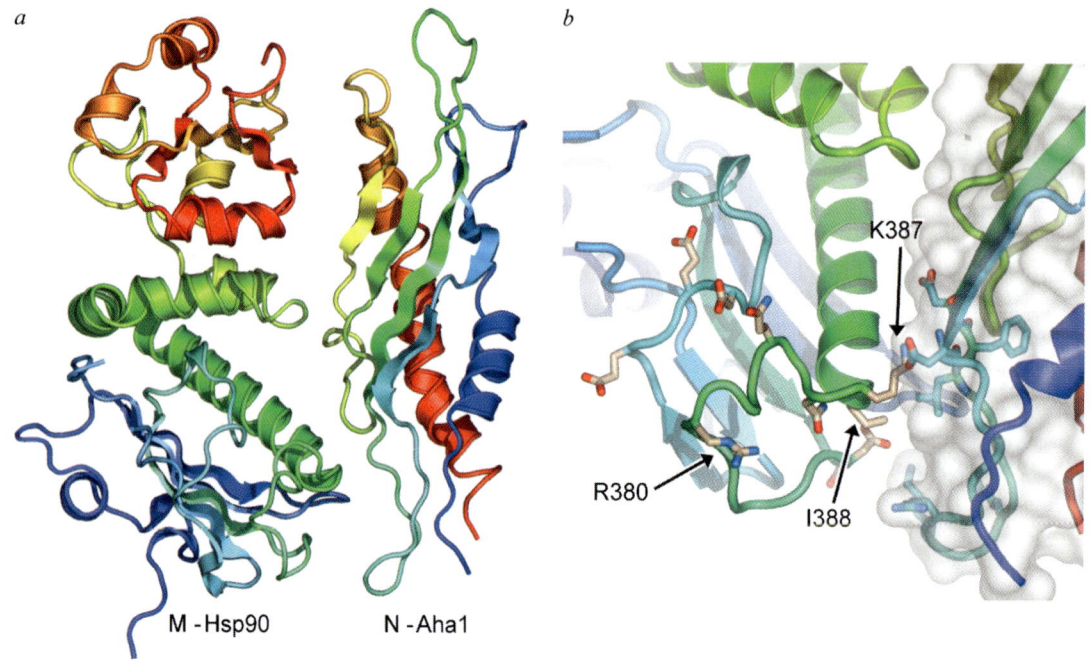

Figure 12

(*a*) Crystal structure of the complex between the middle segment of Hsp90 and the N-terminal domain of the Hsp90 activator cochaperone Aha1. (*b*) Close-up of the interaction between Aha1 and Hsp90, which centers on residues at the C-terminal end of the catalytic loop bearing Arg 380. Movement of Lys 387 and Ile 388 into the interface after Aha1 binding melts the secondary structure of the catalytic loop in the absence of Aha1 and frees Arg 380 to interact with ATP in the N-terminal domain.

α-helix that follows the catalytic loop as they contact Aha1. Movement of these and adjacent residues has the effect of destabilizing the short α-helix in the catalytic loop and releasing Arg 380 from the hydrogen-bonding interactions with Glu 353 and Glu 355 in the unliganded Hsp90 structure (**Figure 12***b*). Thus, binding of N-Aha1/Hch1 facilitates a conformational switch in the middle segment catalytic loop that helps it move from a retracted and inactive conformation toward the extended conformation and enables it to contact the ATP bound in the N-terminal domain. By lowering the energetic barrier to this essential conformational change, N-Aha1/Hch1 accelerates the overall rate of the ATPase cycle. How the additional 200 residues of the C-terminal part of Aha1 provide a further rate enhancement is not yet understood.

Conformation-Dependent Binding of P23/Sba1

The p23 cochaperone was first identified as a component of Hsp90-steroid hormone nuclear receptor complexes, purified in the presence of molybdate (103, 104). Subsequently p23 has been found associated with a wide range of Hsp90-client complexes (105–107), and although its precise role in client protein activation is unclear, it appears to be involved at a late stage in the chaperone cycle, enhancing the release of active client protein (108, 109). Binding of p23 or its yeast homologue Sba1 to Hsp90 is much stronger in the presence of nonhydrolyzable ATP analogues, or molybdate, which together with ADP probably mimics an ATP hydrolysis transition state (10, 104). Subsequent studies suggest that p23 binding to Hsp90 is not facilitated by

ATP binding by Hsp90 as such but is by the N-terminally associated conformation of Hsp90 that ATP binding engenders (6, 52). Quantitative analysis using isothermal titration calorimetry shows that the interaction between Sba1 and Hsp90 is ~70-fold weaker in the absence of an ATP analogue than in its presence (K_d ~2 μM) (55).

Similar to Hop/Sti1 and Cdc37, p23/Sba1 also exerts an inhibitory effect on the ATPase activity of Hsp90 (11, 55, 89, 110), but it is unable to inhibit the reaction past ~50%. One possible explanation for this behavior comes from the observation that p23/Sba1 only binds to an N-terminally associated Hsp90 dimer that has already bound ATP and is therefore committed to ATP hydrolysis. Stabilization of Hsp90 in this conformation by bound p23/Sba1 may slow the release of ADP and phosphate products but cannot fully prevent it. An alternative explanation comes from the surprising observation by both circular dichroism spectroscopy and isothermal titration calorimetry that Sba1 preferentially binds Hsp90 in a 1:2 stoichiometry and with a second lower affinity binding at higher molar ratio (55, 110). Thus, binding of a single p23/Sba1 may be fully inhibitory to one active site in an Hsp90 dimer but exerts a negative cooperativity via generation of an asymmetric conformation that disfavors binding of a second p23/Sba1 molecule, resulting in effective half-of-sites activity for the Hsp90-p23/Sba1 complex.

The structural basis for the conformation-dependent binding of p23/Sba1 was revealed by the crystal structure of yeast Sba1 bound to Hsp90 in the presence of the nonhydrolyzable ATP analogue AMPPNP (M.M.U. Ali, S.M. Roe, C. Vaughan, P. Meyer, B. Panaretou, P.W. Piper, C. Prodromou, L.H. Pearl, unpublished data) (**Figure 13a**). Although the resolution of the structure is limited to ~3.8 Å, the location of p23/Sba1 relative to Hsp90 within the complex is unambiguously defined, and many of the sites of interaction between the two proteins are clearly revealed. Although binding of a second Sba1 is less favorable than the first (55, 110), within the crystals, Hsp90 and Sba1 form a 1:1 stoichiometric complex, with crystallographically independent Sba1 molecules binding on either side of the Hsp90 dimer, and the complete $(Hsp90)_2$-$(Sba1)_2$ complex constituting the crystallographic asymmetric unit. Because both Sba1 molecules are involved in crystal contacts, the added stability of the lattice probably overcomes the negative cooperativity of the Sba1-Hsp90 interaction and favors the symmetric loading of Sba1 molecules.

Each Sba1 molecule packs into a broad groove formed at the interface of the two Hsp90 N domains in the dimer, making contact with both via distinct patches on the Sba1 surface (**Figure 13b**). One interface involves residues at the tips of β-hairpins formed by residues 31–37 and 85–91 of Sba1 in substantially polar interactions with residues on the exposed surface of the first α-helix (12–21) and the β-hairpin loop from residues 151–155 of Hsp90. The interface with the other N-terminal domain involves a broad loop from Sba1, formed by residues 113–118, which (along with residues 13–16 from the N-terminal strand) pack against the exposed surface of the lid segment of Hsp90 N domain (residues 94–125) in its closed ATP-bound conformation. A third segment of Sba1 (residues 120–125) at the end of the globular core defined by the crystal structure of isolated mammalian p23 (111) binds close to the interstrand interdomain interface between residues 23–27 from the N-terminal domain of one Hsp90 monomer and the ends of the catalytic loop from residues 376–386 of the middle segment of the other monomer. Residues from this segment of Sba1, the most conserved region of the molecule, make multiple polar and hydrophobic contacts with Hsp90 that bridge the two Hsp90 domains and reinforce their interaction. Additional structure for Sba1 is visible beyond residue 125 and appears to include a helical segment that packs against the outer face of the middle segment of Hsp90, and it may interact

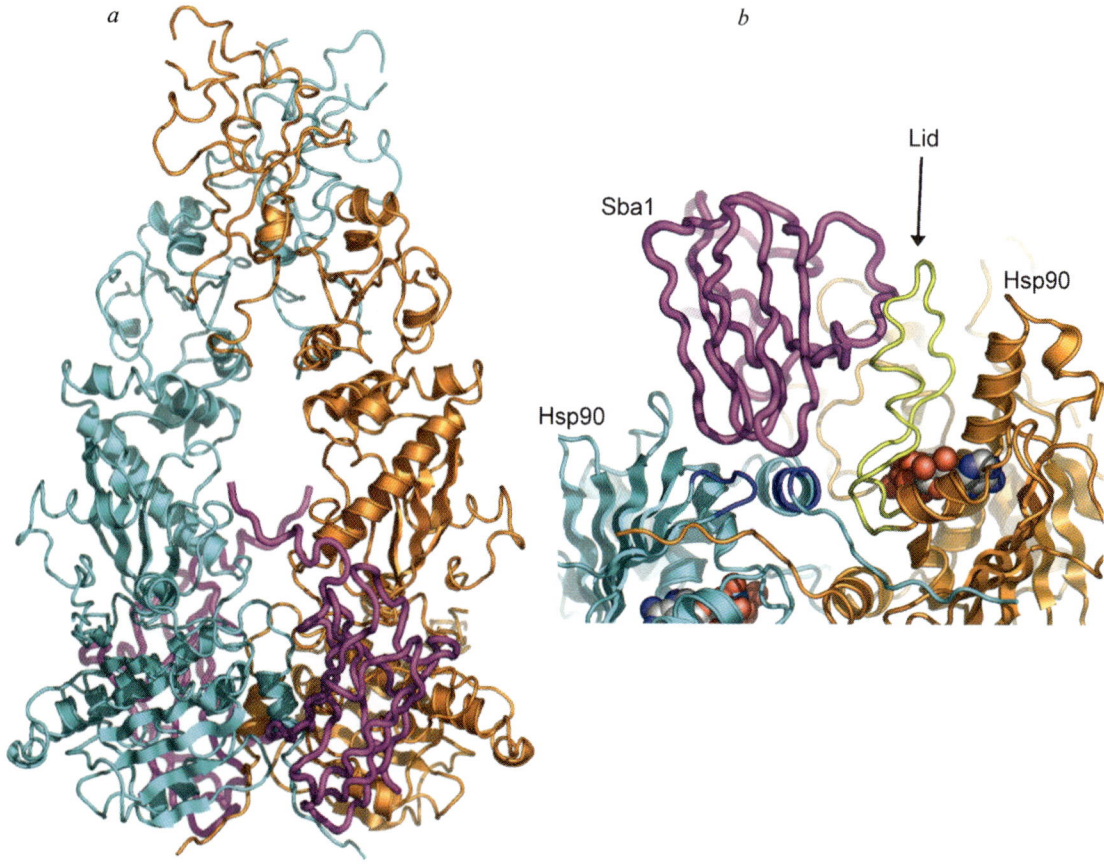

Figure 13

(*a*) Medium/low-resolution crystal structure of a full-length yeast Hsp90 in the ATP-bound closed state and in complex with p23/Sba1. The two chains of the Hsp90 dimer are colored individually, and both molecules of Sba1 (*purple*) are shown as "worms." (*b*) Close-up showing the multiple points of interaction made by each Sba1, bridging the two chains of the Hsp90 dimer in its ATP-bound, lid-closed conformation. The interacting patch (*dark blue*) involving residues 14–19 and 151–155 from one N domain (*cyan*) is shown, and the lid segment (*yellow*) from the other N domain (*orange*) is also shown.

with residues at the end of the 370–390 catalytic loop, in a manner similar to Aha1, although the detailed side-chain interactions cannot be distinguished at the present resolution. None of the patches of interaction between Sba1 and Hsp90 involve the significant burial of a hydrophobic surface or extensive charge neutralization that would give significant individual affinity, but in combination, these patches add a substantial degree of interaction. Because p23/Sba1, unlike Hsp90, is a relatively rigid single domain structure, the full set of interactions required for tight binding can only occur when Hsp90 is in the correct conformation so that the interaction sites are presented in the correct three-dimensional configuration. Thus, tight binding of p23/Sba1 requires lid closure, association of the N-terminal domains, and docking of the middle domains against the mouth of the nucleotide-binding pocket, all of which are contingent on ATP binding. In doing so, p23/Sba1 provides bridging interactions that favor this ATP-bound closed conformation, stabilizing it to some extent even after ATP hydrolysis has occurred, and extending the lifetime of the conformational state required for client protein activation.

CLIENT PROTEIN ACTIVATION

The process by which association with the Hsp90 chaperone machinery and progress through its ATPase-coupled chaperone cycle assists the activation, stabilization, or assembly of client proteins is the central question in the field. Although many of the biochemical mechanisms that accompany that process are being defined, the process itself remains a mystery. The mystery is deepened further by the paradox of Hsp90's selectivity for its client proteins, which argues for a specific mechanism, set against the structural unrelatedness of the broad range of Hsp90 clients, which argues for a generic mechanism. There is now a general consensus that Hsp90 is not involved in protein folding or unfolding in a manner similar to chaperones such as GroEL or Hsp70, but rather it promotes subtle changes in conformation or domain arrangement in its substantially folded client proteins, which often (but not always) facilitate interaction with a ligand molecule. The best example of this is provided by the steroid hormone receptors, whose interaction with the Hsp90 system (via the Hsp70 system) converts them from a low-affinity state to a structure that can bind their cognate ligand and dimerize (112). Structurally, this is likely to involve the transient stabilization of a conformation of the unliganded and less compact steroid hormone receptor in which the otherwise buried hydrophobic cavity is opened to allow ingress of the ligand (113). However, the direct contribution of Hsp90 itself to this structural change is far from clear because the minimal reconstituted system for maturation of steroid hormone receptors requires Hsp70 and Hsp90 coupled via Hop/Sti1 (114).

Activation of protein kinases by the Hsp90 system has been far less well characterized. Although cell-free systems have been described that allow exploration of the binding of kinases to the Hsp90/Cdc37 system and Hsp90-bound kinase complexes have been isolated (115–118), no Hsp90-dependent in vitro kinase activation system comparable to that developed for steroid hormone receptors has been described. Again, unlike the steroid hormone receptors wherein Hsp90 association facilitates a ligand-binding event, it is far from clear what change in state occurs in kinases. For Hsp90 clients, such as the cyclin-dependent kinase Cdk4 or the haem-regulated eIF2α kinase (HRI), binding of a ligand (cyclin D and haem, respectively) is certainly required for full activation, but that is not the case for other kinase clients such as Raf-1, PKB/Akt, or Lck. For many, if not all Hsp90 client kinases, phosphorylation of the activation segment, which is often disordered or locked in an inactive conformation in the unphosphorylated state, is a key event in generation of a catalytically active enzyme. Hsp90 could conceivably facilitate kinase activation by changing the conformation of the activation segment to permit access to the active site of an activating protein kinase. However, when PKB/Akt is isolated as part of an Hsp90-Cdc37 complex, it is already fully active and presumably phosphorylated (25), so this again cannot be a general mechanism. Intriguingly, binding to Hsp90-Cdc37 substantially protects phosphorylated and active PKB/Akt from dephosphorylation and inactivation (44). This suggests yet another possible role for Hsp90 association in sequestering a pool of phosphorylated kinases and maintaining them in an activated state until required. Given the broad range of Hsp90 clients in general, it is entirely possible that Hsp90 plays different roles in activation of different protein kinases. Clearly, a crystal structure of a protein kinase bound to an Hsp90 complex is ultimately the only way in which this key question will be effectively addressed.

SUMMARY POINTS

1. Despite the conformational flexibility of Hsp90 and difficulties in obtaining a crystal structure of the intact dimeric chaperone, crystallographic studies of individual domains, coupled with biochemical and mutagenesis, have provided considerable insight into its mechanism of action.

2. The current model for the ATPase-coupled molecular clamp model for the chaperone cycle of Hsp90, in which ATP-bound N-terminal domains are transiently associated, has come under criticism recently, but it is fully vindicated by the medium/low-resolution crystal structure of a full-length Hsp90-p23/Sba1-AMPNP complex.

3. The ATPase cycle of Hsp90 is regulated by a range of cochaperones, which selectively bind to and stabilize different conformational states of the chaperone dimer. The mechanisms of action of three of these, Cdc37, Aha1, and p23/Sba1, can now be understood from crystal structures of Hsp90-cochaperone complexes.

4. The nature of the changes engendered in Hsp90 clients by association with Hsp90 and passage through the ATPase-coupled chaperone cycle remain unknown.

FUTURE ISSUE TO BE RESOLVED

Despite considerable progress in understanding the structure of Hsp90, its conformational coupling to the ATPase cycle, and its regulation by cochaperones, the key question as to how progress through this cycle facilitates activation of the broad range of client proteins remains largely unanswered.

ACKNOWLEDGMENT

We are very grateful to our colleagues Maruf Ali, Philippe Meyer, Mark Roe, Cara Vaughan, and Minghao Zhang and to our collaborators Peter Piper, Barry Panaretou, Giuliano Siligardi, and Paul Workman for their considerable and superb contributions to our studies of the Hsp90 molecular chaperone system. Our work in this field benefits from the generous support of a Programme Grant from the Wellcome Trust and infrastructural support from the Institute of Cancer Research and Cancer Research UK.

LITERATURE CITED

1. Versteeg S, Mogk A, Schumann W. 1999. *Mol. Gen. Genet.* 261:582–88
2. Borkovich KA, Farrelly FW, Finkelstein DB, Taulien J, Lindquist S. 1989. *Mol. Cell. Biol.* 9:3919–30
3. Meyer P, Prodromou C, Hu B, Vaughan C, Roe SM, et al. 2003. *Mol. Cell* 11:647–58
4. Obermann WMJ, Sondermann H, Russo AA, Pavletich NP, Hartl FU. 1998. *J. Cell Biol.* 143:901–10
5. Panaretou B, Prodromou C, Roe SM, O'Brien R, Ladbury JE, et al. 1998. *EMBO J.* 17:4829–36
6. Prodromou C, Panaretou B, Chohan S, Siligardi G, O'Brien R, et al. 2000. *EMBO J.* 19:4383–92

7. Dey B, Lightbody JJ, Boschelli F. 1996. *Mol. Biol. Cell* 7:1405–17
8. Duina AA, Marsh JA, Gaber RF. 1996. *Yeast* 12:943–52
9. Chang HCJ, Nathan DF, Lindquist S. 1997. *Mol. Cell. Biol.* 17:318–25
10. Fang Y, Fliss AE, Rao J, Caplan AJ. 1998. *Mol. Cell. Biol.* 18:3727–34
11. Panaretou B, Siligardi G, Meyer P, Maloney A, Sullivan JK, et al. 2002. *Mol. Cell* 10:1307–18
12. Holt SE, Aisner DL, Baur J, Tesmer VM, Dy M, et al. 1999. *Genes Dev.* 13:817–26
13. Park SJ, Suetsugu S, Takenawa T. 2005. *EMBO J.* 24:1557–70
14. Garcia-Cardena G, Fan R, Shah V, Sorrentino R, Cirino G, et al. 1998. *Nature* 392:821–24
15. Pratt WB, Toft DO. 2003. *Exp. Biol. Med.* 228:111–33
16. Pearl LH. 2005. *Curr. Opin. Genet. Dev.* 15:55–61
17. Kimmins S, MacRae TH. 2000. *Cell Stress Chaperones* 5:76–86
18. Grammatikakis N, Lin J-H, Grammatikakis A, Tsichlis PN, Cochran BH. 1999. *Mol. Cell. Biol.* 19:1661–72
19. Jascur T, Brickner H, Salles-Passador I, Barbier V, El Khissiin A, et al. 2005. *Mol. Cell* 17:237–49
20. Wiech H, Buchner J, Zimmermann R, Jakob U. 1992. *Nature* 358:169–70
21. Rudiger S, Freund SMV, Veprintsev DB, Fersht AR. 2002. *Proc. Natl. Acad. Sci. USA* 99:11085–90
22. Fink AL. 2005. *Curr. Opin. Struct. Biol.* 15:35–41
23. Wartmann M, Davis RJ. 1994. *J. Biol. Chem.* 269:6695–701
24. Chen GQ, Cao P, Goeddel DV. 2002. *Mol. Cell* 9:401–10
25. Basso AD, Solit DB, Chiosis G, Giri B, Tsichlis P, Rosen N. 2002. *J. Biol. Chem.* 277:39858–66
26. Prodromou C, Piper PW, Pearl LH. 1996. *Proteins: Struct. Funct. Genet.* 25:517–22
27. Harris SF, Shiau AK, Agard DA. 2004. *Structure* 12:1087–97
28. Stebbins CE, Russo AA, Schneider C, Rosen N, Hartl FU, Pavletich NP. 1997. *Cell* 89:239–50
29. Prodromou C, Roe SM, Piper PW, Pearl LH. 1997. *Nat. Struct. Biol.* 4:477–82
30. Whitesell L, Mimnaugh EG, De Costa B, Myers CE, Neckers LM. 1994. *Proc. Natl. Acad. Sci. USA* 91:8324–28
31. Schulte TW, Blagosklonny MV, Ingui C, Neckers L. 1995. *J. Biol. Chem.* 270:24585–88
32. Smith DF, Whitesell L, Nair SC, Chen S, Prapapanich V, Rimerman RA. 1995. *Mol. Cell. Biol.* 15:6804–12
33. Chavany C, Mimnaugh E, Miller P, Bitton R, Nguyen P, et al. 1996. *J. Biol. Chem.* 271:4974–77
34. Bergerat A, de Massy B, Gadelle D, Varoutas PC, Nicolas A, Forterre P. 1997. *Nature* 386:414–17
35. Prodromou C, Roe SM, O'Brien R, Ladbury JE, Piper PW, Pearl LH. 1997. *Cell* 90:65–75
36. Nadeau K, Sullivan MA, Bradley M, Engman DM, Walsh CT. 1992. *Protein Sci.* 1:970–79
37. Nadeau K, Das A, Walsh CT. 1993. *J. Biol. Chem.* 268:1479–87
38. Csermely P, Kajtar J, Hollosi M, Jalsovszky G, Holly S, et al. 1993. *J. Biol. Chem.* 268:1901–7
39. Csermely P, Miyata Y, Schnaider T, Yahara I. 1995. *J. Biol. Chem.* 270:6381–88
40. Jakob U, Scheibel T, Bose S, Reinstein J, Buchner J. 1996. *J. Biol. Chem.* 271:10035–41
41. Dutta R, Inouye M. 2000. *Trends Biochem. Sci.* 25:24–28
42. Soldano KL, Jivan A, Nicchitta CV, Gewirth DT. 2003. *J. Biol. Chem.* 278:48330–38

43. Huai Q, Wang HC, Liu YD, Kim HY, Toft D, Ke HM. 2005. *Structure* 13:579–90
44. Sato S, Fujita N, Tsuruo T. 2000. *Proc. Natl. Acad. Sci. USA* 97:10832–37
45. Minami Y, Kimura Y, Kawasaki H, Suzuki K, Yahara I. 1994. *Mol. Cell. Biol.* 14:1459–64
46. Chen SY, Sullivan WP, Toft DO, Smith DF. 1998. *Cell Stress Chaperones* 3:118–29
47. Young JC, Obermann WMJ, Hartl FU. 1998. *J. Biol. Chem.* 273:18007–10
48. Scheibel T, Siegmund HI, Jaenicke R, Ganz P, Lilie H, Buchner J. 1999. *Proc. Natl. Acad. Sci. USA* 96:1297–302
49. Louvion JF, Warth R, Picard D. 1996. *Proc. Natl. Acad. Sci. USA* 93:13937–42
50. Grenert JP, Sullivan WP, Fadden P, Haystead TAJ, Clark J, et al. 1997. *J. Biol. Chem.* 272:23843–50
51. Sullivan W, Stensgard B, Caucutt G, Bartha B, McMahon N, et al. 1997. *J. Biol. Chem.* 272:8007–12
52. Chadli A, Bouhouche I, Sullivan W, Stensgard B, McMahon N, et al. 2000. *Proc. Natl. Acad. Sci. USA* 97:12524–29
53. Kampranis SC, Bates AD, Maxwell A. 1999. *Proc. Natl. Acad. Sci. USA* 96:8414–19
54. Maruya M, Sameshima M, Nemoto T, Yahara I. 1999. *J. Mol. Biol.* 285:903–7
55. Siligardi G, Hu B, Panaretou B, Piper PW, Pearl LH, Prodromou C. 2004. *J. Biol. Chem.* 279:51989–98
56. McLaughlin SH, Ventouras LA, Lobbezoo B, Jackson SE. 2004. *J. Mol. Biol.* 344:813–26
57. Piper PW, Panaretou B, Millson SH, Trumana A, Mollapour M, et al. 2003. *Gene* 302:165–70
58. Ban C, Junop M, Yang W. 1999. *Cell* 97:85–97
59. Lewis RJ, Singh OM, Smith CV, Skarzynski T, Maxwell A, et al. 1996. *EMBO J.* 15:1412–20
60. Roe SM, Prodromou C, O'Brien R, Ladbury JE, Piper PW, Pearl LH. 1999. *J. Med. Chem.* 42:260–66
61. Immormino RM, Dollins DE, Shaffer PL, Soldano KL, Walker MA, Gewirth DT. 2004. *J. Biol. Chem.* 279:46162–71
62. Dollins DE, Immormino RM, Gewirth DT. 2005. *J. Biol. Chem.* 280:30438–47
63. Nathan DF, Lindquist S. 1995. *Mol. Cell. Biol.* 15:3917–25
64. Das AK, Cohen PTW, Barford D. 1998. *EMBO J.* 17:1192–99
65. Radanyi C, Chambraud B, Baulieu EE. 1994. *Proc. Natl. Acad. Sci. USA* 91:11197–201
66. Owens-Grillo JK, Czar MJ, Hutchison KA, Hoffmann K, Perdew GH, Pratt WB. 1996. *J. Biol. Chem.* 271:13468–75
67. Owens-Grillo JK, Stancato LF, Hoffmann K, Pratt WB, Krishna P. 1996. *Biochemistry* 35:15249–55
68. Carrello A, Ingley E, Minchin RF, Tsai S, Ratajczak T. 1999. *J. Biol. Chem.* 274:2682–89
69. Ratajczak T, Carrello A. 1996. *J. Biol. Chem.* 271:2961–65
70. Lee YT, Jacob J, Michowski W, Nowotny M, Kuznicki J, Chazin WJ. 2004. *J. Biol. Chem.* 279:16511–17
71. Smith DF, Sullivan WP, Marion TN, Zaitsu K, Madden B, et al. 1993. *Mol. Cell. Biol.* 13:869–76
72. Young JC, Hoogenraad NJ, Hartl FU. 2003. *Cell* 112:41–50
73. Smith DF, Baggenstoss BA, Marion TN, Rimerman RA. 1993. *J. Biol. Chem.* 268:18365–71
74. Dolinski K, Muir S, Cardenas M, Heitman J. 1997. *Proc. Natl. Acad. Sci. USA* 94:13093–98
75. Silverstein AM, Galigniana MD, Chen M-S, Owens-Grillo J, Chinkers M, Pratt WB. 1997. *J. Biol. Chem.* 272:16224–30

76. Jiang J, Ballinger CA, Wu Y, Dai Q, Cyr DM, et al. 2001. *J. Biol. Chem.* 276:42938–44
77. Murata S, Minami Y, Minami M, Chiba T, Tanaka K. 2001. *EMBO Rep.* 2:1133–38
78. Duina AA, Marsh JA, Kurtz RB, Chang H-CJ, Lindquist S, Gaber RF. 1998. *J. Biol. Chem.* 273:10819–22
79. Scheufler C, Brinker A, Bourenkov G, Pegoraro S, Moroder L, et al. 2000. *Cell* 101:199–210
80. Brinker A, Scheufler C, von der Mulbe F, Fleckenstein B, Herrmann C, et al. 2002. *J. Biol. Chem.* 277:19265–75
81. Ballinger CA, Connell P, Wu Y, Hu Z, Thompson LJ, et al. 1999. *Mol. Cell. Biol.* 19:4535–45
82. Connell P, Ballinger CA, Jiang J, Wu Y, Thompson LJ, et al. 2001. *Nat. Cell Biol.* 3:93–96
83. Zhang M, Windheim M, Roe SM, Peggie M, Prodromou C, et al. 2005. *Mol. Cell* 20:525–38
84. Prodromou C, Siligardi G, O'Brien R, Woolfson DN, Regan L, et al. 1999. *EMBO J.* 18:754–62
85. Nikolay R, Wiederkehr T, Rist W, Kramer G, Mayer MP, Bukau B. 2004. *J. Biol. Chem.* 279:2673–78
86. Ratajczak T, Hlaing J, Brockway MJ, Hahnel R. 1990. *J. Steroid Biochem.* 35:543–53
87. Hernandez MP, Sullivan WP, Toft DO. 2002. *J. Biol. Chem.* 277:38294–304
88. Siligardi G, Panaretou B, Meyer P, Singh S, Woolfson DN, et al. 2002. *J. Biol. Chem.* 277:20151–59
89. McLaughlin SH, Smith HW, Jackson SE. 2002. *J. Mol. Biol.* 315:787–98
90. Richter K, Muschler P, Hainzl O, Reinstein J, Buchner J. 2003. *J. Biol. Chem.* 278:10328–33
91. Brugge JS. 1986. *Curr. Top. Microbiol. Immunol.* 123:1–22
92. Stepanova L, Leng XH, Parker SB, Harper JW. 1996. *Genes Dev.* 10:1491–502
93. Perdew GH, Wiegand H, Vanden Heuvel JP, Mitchell C, Singh SS. 1997. *Biochemistry* 36:3600–7
94. Gerber MR, Farrell A, Deshaies RJ, Herskowitz I, Morgam DO. 1995. *Proc. Natl. Acad. Sci. USA* 92:4651–55
95. Silverstein AM, Grammatikakis N, Cochran BH, Chinkers M, Pratt WB. 1998. *J. Biol. Chem.* 273:20090–95
96. Shao J, Grammatikakis N, Scroggins BT, Uma S, Huang WJ, et al. 2001. *J. Biol. Chem.* 276:206–14
97. Shao J, Irwin A, Hartson SD, Matts RL. 2003. *Biochemistry* 42:12577–88
98. Roe SM, Ali MM, Meyer P, Vaughan CK, Panaretou B, et al. 2004. *Cell* 116:87–98
99. Roiniotis J, Masendycz P, Ho S, Scholz GM. 2005. *Biochemistry* 44:6662–69
100. Nathan DF, Vos MH, Lindquist S. 1999. *Proc. Natl. Acad. Sci. USA* 96:1409–14
101. Lotz GP, Lin H, Harst A, Obermann WMJ. 2003. *J. Biol. Chem.* 278:17228–35
102. Meyer P, Prodromou C, Liao CY, Hu B, Roe SM, et al. 2004. *EMBO J.* 23:511–19
103. Johnson JL, Toft DO. 1994. *J. Biol. Chem.* 269:24989–93
104. Johnson JL, Toft DO. 1995. *Mol. Endocrinol.* 9:670–78
105. Cox MB, Miller CA 3rd. 2004. *Cell Stress Chaperones* 9:4–20
106. Forsythe HL, Jarvis JL, Turner JW, Elmore LW, Holt SE. 2001. *J. Biol. Chem.* 276:15571–74
107. Munoz MJ, Bejarano ER, Daga RR, Jimenez J. 1999. *Genetics* 153:1561–72
108. Freeman BC, Felts SJ, Toft DO, Yamamoto KR. 2000. *Genes Dev.* 14:422–34
109. Young JC, Hartl FU. 2000. *EMBO J.* 19:5930–40

110. Richter K, Walter S, Buchner J. 2004. *J. Mol. Biol.* 342:1403–13
111. Weaver AJ, Sullivan WP, Felts SJ, Owen BA, Toft DO. 2000. *J. Biol. Chem.* 275:23045–52
112. Pratt WB, Galigniana MD, Morishima Y, Murphy PJ. 2004. *Essays Biochem.* 40:41–58
113. Steinmetz AC, Renaud JP, Moras D. 2001. *Annu. Rev. Biophys. Biomol. Struct.* 30:329–59
114. Dittmar KD, Hutchison KA, Owens-Grillo JK, Pratt WB. 1996. *J. Biol. Chem.* 271:12833–39
115. Stancato LF, Chow Y-H, Hutchison KA, Perdew GH, Jove R, Pratt WB. 1993. *J. Biol. Chem.* 268:21711–16
116. Prince T, Matts RL. 2004. *J. Biol. Chem.* 279:39975–81
117. Scroggins BT, Prince T, Shao J, Uma S, Huang W, et al. 2003. *Biochemistry* 42:12550–61
118. Zhao Q, Boschelli F, Caplan AJ, Arndt KT. 2004. *J. Biol. Chem.* 279:12560–64

Biochemistry of Mammalian Peroxisomes Revisited

Ronald J.A. Wanders and Hans R. Waterham

Department of Clinical Chemistry and Pediatrics, Laboratory Genetic Metabolic Disease, Academic Medical Center, University of Amsterdam, 1105 AZ Amsterdam, The Netherlands; email: r.j.wanders@amc.uva.nl, h.r.waterham@amc.uva.nl

Key Words

fatty acid oxidation, plasmalogens, reactive oxygen species, genetic diseases

Abstract

In this review, we describe the current state of knowledge about the biochemistry of mammalian peroxisomes, especially human peroxisomes. The identification and characterization of yeast mutants defective either in the biogenesis of peroxisomes or in one of its metabolic functions, notably fatty acid beta-oxidation, combined with the recognition of a group of genetic diseases in man, wherein these processes are also defective, have provided new insights in all aspects of peroxisomes. As a result of these and other studies, the indispensable role of peroxisomes in multiple metabolic pathways has been clarified, and many of the enzymes involved in these pathways have been characterized, purified, and cloned. One aspect of peroxisomes, which has remained ill defined, is the transport of metabolites across the peroxisomal membrane. Although it is clear that mammalian peroxisomes under in vivo conditions are closed structures, which require the active presence of metabolite transporter proteins, much remains to be learned about the permeability properties of mammalian peroxisomes and the role of the four half ATP-binding cassette (ABC) transporters therein.

Contents

INTRODUCTION 296
PEROXISOMAL PROTEINS 297
   When to Call a Protein a
      Peroxisomal Protein? 297
   Strategies to Identify Putative
      Peroxisomal Proteins 300
   Strategies to Demonstrate the
      Peroxisomal Localization of
      Proteins 300
SELECTED METABOLIC
   PATHWAYS 301
   Oxygen Metabolism, Reactive
      Oxygen Species, and Reactive
      Nitrogen Species Metabolism .. 301
   Ether-Phospholipid Biosynthesis . 302
   Peroxisomal Fatty Acid
      Beta-Oxidation 305
   Peroxisomal Fatty Acid
      Alpha-Oxidation 311
   Glyoxylate Metabolism 312
   Amino Acid Catabolism 313
   Pentose Phosphate Pathway 313
   Polyamine Oxidation 313
   Miscellaneous Peroxisomal
      Enzyme Activities 314
   Isoprenoid and Cholesterol
      Metabolism 314
BIOCHEMISTRY OF HUMAN
   PEROXISOMAL DISORDERS .. 315
MOUSE MODELS FOR
   PEROXISOMAL DISORDERS .. 316
PEROXISOMAL METABOLITE
   TRANSPORT 319
   Permeability Properties of
      Peroxisomes 319
   The Intraperoxisomal pH 322
   Peroxisomal ABC Transporters 322
   Peroxisomal ATP Transporter 324
   Other Putative Peroxisomal
      Transporters 325
CONCLUDING REMARKS 325

INTRODUCTION

Peroxisomes belong to the microbody family, with glyoxysomes and glycosomes as the other members, and represent a class of ubiquitous and essential cell organelles characterized by the presence of a proteinaceous matrix surrounded by a single membrane. Since this topic was last reviewed in this journal in 1992 (1), our knowledge about the biochemistry of peroxisomes has increased substantially for a number of different reasons. First, the identification and characterization of yeast mutants defective in peroxisome biogenesis have allowed the resolution of the principal features of peroxisome biogenesis, which includes the targeting of peroxisomal matrix proteins via one of two distinct peroxisomal targeting signals (PTS1 and PTS2). This knowledge has been used to perform computer-based searches for proteins that contain a PTS1 or a PTS2 notably in the yeast *Saccharomyces cerevisiae* and to identify the corresponding mammalian orthologues, using homology probing. This strategy, together with more classical approaches, such as protein purification, has led to the identification of a series of peroxisomal proteins. Second, the recognition of a large class of genetic diseases in man, in which either peroxisome biogenesis per se or a certain peroxisomal function is defective, has provided new insights into the metabolic role of peroxisomes in humans. As a result of these combined studies, it is now clear that fatty acid (FA) beta-oxidation is a general feature of virtually all types of peroxisomes. In addition, peroxisomes in higher eukaryotes, including humans, catalyze a number of additional peroxisomal functions not shared by peroxisomes in lower eukaryotes, including ether phospholipid biosynthesis, FA alpha-oxidation, and glyoxylate detoxification.

Another major step forward has been the discovery that, in contrast to the long-held view that the peroxisomal membrane is freely permeable to low-molecular-weight substances, peroxisomes are in fact closed structures under in vivo conditions. This requires the presence of peroxisomal membrane proteins, which allow the specific entrance and exit of metabolites. Some progress has been made in this respect with the identification of a set of half ATP-binding cassette (ABC) transporters, possibly involved in transmembrane FA transport.

In this review, we present the current state of knowledge about (*a*) the biochemistry of peroxisomes, with special emphasis on the enzymology and metabolic functions of mammals, notably in humans; (*b*) the transport properties of mammalian peroxisomal membranes; and (*c*) the biochemistry in patients suffering from different peroxisomal disorders and corresponding mouse models.

PEROXISOMAL PROTEINS

It has become clear that peroxisomes are involved in a variety of metabolic pathways, which implies the presence of a large number of proteins in the peroxisomal matrix. Below we define criteria for the peroxisomal localization of proteins and discuss some commonly used strategies to demonstrate the peroxisomal localization for a given protein.

When to Call a Protein a Peroxisomal Protein?

It is estimated that mammalian peroxisomes contain some 50 different enzyme activities, many of which have been attributed to established peroxisomal proteins (**Table 1**). In addition, a number of peroxisomal proteins have been reported without a known catalytic activity. This includes the peroxisomal protein PeP, encoded by *FNDC5*, with no significant homology to any known protein except for a short stretch of amino acids containing the fingerprint of the fibronectin type III superfamily (2). In addition, a number of peroxisomal proteins have been identified with a catalytic activity of unknown function, as, for example, the peroxisomal nudix hydrolase Nudt7 (3). Furthermore, some enzyme activities are not yet linked to a specific peroxisomal protein. These activities may be catalyzed by new, yet unidentified peroxisomal proteins or may be a side reaction of a known protein, as shown, for instance, for 2,4-dienoyl-coenzyme A (CoA) reductase, which also catalyzes the NADPH-dependent reduction of retinal to retinol (4), at least in vitro, and the peroxisomal enzyme D-bifunctional protein, which has abundant 17 beta-estradiol dehydrogenase activity under in vitro, but not in vivo, conditions. The latter data further imply that the identification of a certain enzyme activity in peroxisomes under in vitro conditions does not necessarily imply that peroxisomes also catalyze this activity under in vivo conditions.

Most enzyme activities listed in **Table 1** are unique to peroxisomes, but some are shared with other subcellular compartments, including the mitochondria and cytosol. Such a multiple subcellular localization may be due to the existence of different isoforms of a protein targeted to different subcellular sites, as, for example, is the case for the enzymes involved in FA beta-oxidation in mammals (peroxisomal and mitochondrial), or the presence of multiple targeting signals within the same protein, as, for example, shown for 3-hydroxy-3-methylglutaryl-CoA lyase (5) and alpha-methylacyl-CoA racemase (AMACR) (6–8). These enzymes are located in both peroxisomes and mitochondria.

Conclusive evidence for the peroxisomal localization of a certain protein and/or its activity requires the actual identification and characterization of the protein and the demonstration of its physical presence inside the organelles. In this respect, it should be noted that the peroxisomal localization of a certain protein may be species dependent. Furthermore, the finding of a peroxisomal localization in one species does not necessarily mean that its homologue is also peroxisomal in other species. A well-documented example of this is alanine:glyoxylate aminotransferase (AGT), which is peroxisomal in humans, rabbits, and guinea pigs, has a dual peroxisomal and mitochondrial location in rats and

FA: fatty acid

Table 1 List of peroxisomal enzymes and other peroxisomal proteins from humans and their peroxisomal targeting sequences (PTS1 or PTS2)

Peroxisomal (enzyme) protein	Gene symbol	Enzyme symbol	EC number	Human locus	Targeting signal PTS1 or PTS2	Targeting sequence
Peroxisomal beta-oxidation						
Acyl-CoA oxidase 1 (palmitoyl-CoA oxidase)	*ACOX1*	ACOX1	1.3.3.6	17q25	PTS1	-SKL
Acyl-CoA oxidase 2 (branched-chain acyl-CoA oxidase)	*ACOX2*	ACOX2	1.3.3.6	3p14.3	PTS1	-SKL
Acyl-CoA oxidase 3 (pristanoyl-CoA oxidase)	*ACOX3*	ACOX3	1.3.3.6	4p15.3	PTS1	-SKL
L-bifunctional protein (peroxisomal multifunctional enzyme 1)	*EHHADH*	LBP/MFP1	1.1.1.35; 5.3.3.8; 4.2.1.17	3q26.3–3q28	PTS1	-SKL
D-bifunctional protein (peroxisomal multifunctional enzyme 2)	*HSD17B4*	DBP/MFP2	4.2.1.-; 1.1.1.35	5q2	PTS1	-AKL
Peroxisomal beta-ketothiolase 1 (straight-chain thiolase)	*ACAA1*	—	2.3.1.16	3p23–p22	PTS2	-RLQVVLGHL
Peroxisomal beta-ketothiolase 2 (branched-chain thiolase)	*SCP2*	SCP2	—	1p32	PTS1	-AKL
Alpha-methylacyl-CoA racemase	*AMACR*	AMACR	5.1.99.4	5p13.2–q11.1	PTS1	-(K)ASL
Carnitine acetyltransferase	*CRAT*	CAT	2.3.1.7	9q34.1	PTS1	-AKL
Carnitine octanoyltransferase	*CROT*	COT	—	7q21.1	PTS1	-THL
Delta3,5-, delta2,4-dienoyl-CoA isomerase	*ECHI*	—	—	19q13.1	PTS1	-SKL
Peroxisomal 2,4-dienoyl-CoA reductase 2	*DECR2*	—	—	16p13.3	PTS1	-AKL
Peroxisomal 3,2-trans-enoyl-CoA isomerase	*PEC1*	—	—	6p24.3	PTS1	-SKL
Very-long-chain acyl-CoA synthetase	*SLC27A2*	VLCS	6.2.1.-	15q21.2	PTS1	-LKL
Acyl-CoA thioesterase 2	*PTE1*	—	3.1.1.2	20q12–q13.1	PTS1	-SKL
Acyl-CoA thioesterase 1B	*PTE2*	—	3.1.1.2	14q24.3	PTS1	-SKV
Peroxisomal *trans*-2-enoyl-CoA reductase (NADPH)	*PECR*	—	—	2q35	PTS1	-AKL
Peroxisomal alpha-oxidation						
Phytanoyl-CoA 2-hydroxylase	*PHYH/PAHX*	PHYH/PAHX	1.14.11.18	10pter-p11.2	PTS2	-RLQIVLGHL
2-Hydroxyphytanoyl-CoA lyase	*HPCL2*	HPCL2	—	3p25.1	PTS1	-(R)SNM
Plasmalogen biosynthesis						
Dihydroxyacetone phosphate acyltransferase	*GNPAT*	DHAPAT	2.3.1.42	1q42.11–42.3	PTS1	-AKL
Alkyldihydroxyacetone phosphate synthase	*AGPS*	ADHAPS	2.5.1.26	2q31	PTS2	-RLRVLSGHL
Fatty acyl-CoA reductase 1	*MLSTD2*	FAR1	—	11p15.2	—	—
Fatty acyl-CoA reductase 2	*MLSTD1*	FAR2	—	12p11.22	—	—

Glyoxylate metabolism						
Alanine:glyoxylate aminotransferase	AGXT	AGT	2.6.1.44; 2.6.1.55	2q36-q37	PTS1	-KKL
Lysine metabolism						
Peroxisomal sarcosine oxidase/L-pipecolate oxidase	PIPOX	PIPOX	—	17q11.2	PTS1	-AHL
Oxygen metabolism						
Catalase	CAT	CAT	1.11.1.6	11p13	PTS1	-(K)ANL
Peroxiredoxin V (PMP20)	PRDX5	PROX5/PMP20	1.11.1.7	11q13	PTS1	-SQL
D-amino acid oxidase	DAO	DAOX	1.4.3.3	12q24	PTS1	-SHL
D-aspartate oxidase	DDO	DASPOX	1.4.3.1	6q21	PTS1	-(K)SNL
Glycolate oxidase (hydroxyacid oxidase 1)	HAO1	GOX/HAO1	1.1.3.15	20p12	PTS1	-SKI
Hydroxyacid oxidase 2	HAO2	HAO2	1.1.3.15	1p13.3–p13.1	PTS1	-SRL
Hydroxyacid oxidase 3	HAO3	HAO3	1.1.3.15	—	PTS1	-SRL
Epoxide hydrolase	EPHX2	EPH2	3.3.2.3	8p21-p12	PTS1	-SKM
Glutathione S-transferase class Kappa	GSTK1	GSTK1	—	7	PTS1	-ARL
Polyamine metabolism						
N^1-acetylspermine/spermidine oxidase	PAOX	PAO	—	10q26.3	PTS1	-(R)PRL
Additional (enzyme) proteins						
Malonyl-CoA decarboxylase	MLYCD	—	4.1.1.9	16q12	PTS1	-SKL
3-Hydroxy-3-methylglutaryl-CoA lyase	HMGCL	HL	4.1.3.4	1p36.1-p35	PTS1 (+ MTS)	-CKL
Isocitrate dehydrogenase ($NADP^+$-linked)	IDH1	IDH1	1.1.1.42	2q33.3	PTS1	-AKL
Nudix hydrolase specific for CoA	NUDT7	NUDT7	—	16q23.1	PTS1	-SRL
Insulin-degrading enzyme	IDE	IDE	3.4.24.56	10q23-q25	PTS1	-AKL
Serine hydrolase like	SERHL	—	—	22q13.2–q13.31	—	—
Lon protease	LONP	—	—	16q12.1	PTS1	-SKL
Nudix-type motif 19 (Roswell-Park complex 2)	D7RP2e	RP2p	—	19q13.11	PTS1	-SHL
Trim37	TRIM37	—	—	6p21.3	—	—
PeP	FNDC5	FNDC5/PeP	—	1p35.1	PTS1	-SKI
PMP22	PXMP2	PMP22	—	12q24.33	—	—

marmosets, and is mitochondrial in cats (9). Another example is the localization of the FA beta-oxidation pathway, which is exclusively peroxisomal in yeast but shows a dual peroxisomal and mitochondrial localization in mammals and plants (10).

Strategies to Identify Putative Peroxisomal Proteins

Both forward and reverse genetics approaches have been employed to identify putative peroxisomal proteins. In short, the forward genetics approach involves the conventional purification of a protein, which exhibits a certain enzyme activity assumed to be peroxisomal, followed by the generation of specific antibodies against the purified protein, which can be used to demonstrate its actual subcellular localization. Furthermore, the encoding cDNA and corresponding gene can be identified by a degenerated polymerase chain reaction approach that is based on a partial amino acid sequence of the purified protein. The reverse genetics approach became popular because of the availability of the genomic sequences of various organisms, including yeasts, mice, and humans, in conjunction with the increased knowledge of functional domains within amino acid sequences (e.g., catalytic sites and targeting signals), allowing selective database searches for genes encoding putative peroxisomal proteins with certain activities. This *in silico* strategy has been very successful in identifying novel putative peroxisomal proteins [e.g., proteins with consensus peroxisomal targeting signals (11)] as well as orthologues (i.e., functional homologues identified on the basis of significant sequence similarity) of proteins determined to be peroxisomal in other species, a strategy also referred to as homology probing (12, 13).

Strategies to Demonstrate the Peroxisomal Localization of Proteins

The most important criterion for the peroxisomal localization of a certain protein remains that the protein should be shown as physically present inside peroxisomes. This is true particularly for candidate peroxisomal proteins identified by the reverse genetics approach, because the presence of a consensus PTS in the amino acid sequence, i.e, a PTS1-consensus sequence: (S/A/C)-(K/R/H)-(L/M) or a PTS2-consensus sequence: (R/K)-(L/I/V)-X_5-(Q/H)-(L/I/V), does not necessarily mean that the protein is truly peroxisomal. For example, although both pristanoyl-CoA oxidase and the bile acid conjugating (BACAT) enzyme harbor the same C-terminal PTS1-like SQL tripeptide, only pristanoyl-CoA oxidase is peroxisomal and BACAT cytosolic (14, 15). Also, despite the presence of a consensus PTS1 (-SRL), phosphomevalonate kinase was recently demonstrated to be a cytosolic protein (16).

Several approaches can be employed to demonstrate a peroxisomal localization. These include conventional subcellular fractionation studies by which tissues or cells are first homogenized and separated by differential centrifugation into organelle-enriched fractions that are further fractionated by density gradient centrifugation. The fractions of these gradients are then analyzed by specific enzyme activity measurements and/or immunoblot analysis, using specific antibodies to determine the profile of the protein of interest in the gradient in comparison to the profile of established marker proteins/enzymes for the various subcellular organelles. As activity measurements may not always be specific because enzymes often can handle multiple substrates, the combined approach with specific antibodies is highly recommended.

The peroxisomal localization of soluble proteins can also be studied through controlled permeabilization of cellular membranes with digitonin. Because digitonin permeabilizes cellular membranes forming a complex with cholesterol and organelle membranes contain lower levels of cholesterol than the plasma membrane, this treatment leads to differential leakage of proteins. This leakage can be assessed by determining the release of the protein of interest in comparison to that of established marker proteins using specific enzyme activity measurements and/or immunoblot analyses using specific antisera (see, for examples, References 16, 17, and 18).

Conclusive evidence for the peroxisomal localization of a protein can be obtained by in situ (immuno) microscopical techniques, including immunoelectron microscopy, immunofluorescence microscopy, and immunohistochemistry. Because the outcome of these techniques relies heavily on the quality of antisera used to detect the protein, it is essential to obtain antisera that are highly specific and exclusively recognize the protein of interest under native conditions. The importance of performing immunolocalization studies to establish the peroxisomal localization of a certain protein using specific antibodies has been shown on numerous occasions. This is exemplified by recent studies by Yokota and coworkers (19), showing that the NADP-linked isocitrate dehydrogenase encoded by *IDH1* is predominantly, if not exclusively, peroxisomal, whereas subcellular fractionation studies, based on differential centrifugation of homogenates, revealed that the enzyme is predominantly cytosolic and only partially peroxisomal (19).

The functionality of putative PTS1 or PTS2 motifs in the amino acid sequence of a protein can be tested by reporter studies in which the protein or portions thereof are fused to a reporter protein such as green fluorescent protein or specific epitopes (e.g., myc, HA). This allows easy detection of the protein constructs upon expression in cells. It should be noted, however, that the observation that a certain (truncated) amino acid sequence is capable of targeting a reporter to peroxisomes does not necessarily imply that it also functions as a true PTS in the authentic protein if the latter has not been shown to actually reside in peroxisomes by other means. Indeed, a putative C-terminal PTS1 may not be functional when, for example, the PTS1 is hidden within the three-dimensional structure of the protein or when, in addition, an N-terminal mitochondrial targeting sequence is present in the same protein. Furthermore, great care should be taken with the interpretation of results obtained with reporter studies because in most cases these involve overexpression, which may introduce unpredictable artifacts (16, 20). It should be noted that the specific context of a putative PTS is important for targeting, by increasing the affinity between the PTS-containing peptide and its receptor (21–24). For instance, it has been shown for catalase, which ends in –ANL, a weak PTS1, that the lysine present in the (–4) position of catalase (-KANL) greatly stimulates binding to the PTS1-receptor (22). Optimized amino acid residues at positions (–4) and (–5) can enhance affinities by at least two orders of magnitude (23).

SELECTED METABOLIC PATHWAYS

Below we describe the role of peroxisomes in selected metabolic pathways.

Oxygen Metabolism, Reactive Oxygen Species, and Reactive Nitrogen Species Metabolism

Peroxisomes harbor a number of oxidases that reduce O_2 to H_2O_2 (25). The H_2O_2 produced can be disposed of via several enzymes, including catalase, glutathione peroxidase, and peroxiredoxin V (PMP20). The decomposition of H_2O_2 by catalase may occur catalytically ($2H_2O_2 \rightarrow O_2 + 2H_2O$) or peroxidatically ($H_2O_2 + AH_2 \rightarrow A + 2H_2O$), in which the conversion of one molecule H_2O_2 to two molecules of H_2O is coupled to the oxidation of different hydrogen donors (AH_2), such as ethanol, methanol, formaldehyde, formate, and nitrite. In addition to catalase, peroxisomes also contain glutathione peroxidase activity (26). A third peroxisomal enzyme that removes H_2O_2 is PMP20 (27), which exhibits thiol-specific antioxidant activity.

Apart from H_2O_2, peroxisomal enzymes also generate other reactive species, including superoxide anions. One source of superoxide anions is xanthine oxidoreductase, which can exist in two forms, including a dehydrogenase and oxidase form (the latter form generates superoxide anions). Angermuller et al. (28) were the first to identify xanthine oxidase activity in the core, but not in the matrix, of peroxisomes. Recent studies in rat liver in which use was made of improved methods applied to unfixed cryostat sections have shown that xanthine oxidase is not only present in the core of peroxisomes but also in the peroxisomal matrix (29). Furthermore, xanthine oxidase appears to be the predominant, if not exclusive, form of xanthine oxidase in peroxisomes, whereas in the cytosol, the reverse is true with xanthine dehydrogenase predominating over xanthine oxidase.

Inactivation of superoxide anions is brought about by superoxide dismutases. Several reports have shown the presence of Cu/Zn-SOD (30, 31) and Mn-SOD activities (32) in peroxisomes, although the proteins responsible for these activities remain to be identified. It has also been claimed recently that peroxisomes contain inducible nitric oxide (NO) synthase activity (33), which, if true, would be an important intraperoxisomal generator of NO species. Together with superoxide anions, NO would generate peroxynitrite, a highly reactive species. Interestingly, peroxiredoxin V, which has been localized to peroxisomes (see **Table 1**) as well as to mitochondria and cytosol, was recently shown to exhibit potent peroxynitrite reductase activity (34).

Peroxisomes also contain epoxide hydrolase activity (35–37). Epoxides are a group of highly reactive molecules of both exogenous and endogenous origin. Some of the most potent carcinogenic and mutagenic compounds only become active when transformed into their epoxides. Because they are very electrophilic, they easily react with nucleophilic groups such as lipids containing unsaturated FAs, DNA, RNA, and proteins. Epoxides, which can be synthesized endogenously, include epoxides of prostaglandins, leukotriens, arachidonic acid, cholesterol, and unsaturated FAs. One single gene coding for a protein with a weak PTS1 signal (SKI) has been identified, which gives rise to a bicompartmental distribution of epoxide hydrolase in both the peroxisomes and cytosol (36). According to others, however, the peroxisomal epoxide hydrolase is different from that present in other compartments (37).

Finally, peroxisomes also contain glutathione S-transferase (GST) activity (38). GSTs catalyze the conjugation of electrophilic substrates to glutathione but, in addition, have reduced glutathione-dependent peroxidase and isomerase activities. The GST identified in peroxisomes belongs to the kappa family. The true function of this GSTK1 remains to be identified, however. The enzyme shows reactivity with 1-chloro-2,4-dinitrobenzene as well as with cumene hydroperoxide and 15-S-hydroperoxy-5,8,11,13-eicosatetraenoic acid (38).

Ether-Phospholipid Biosynthesis

Ether phospholipids may occur in two forms, including (*a*) plasmanyl-phospholipids and (*b*) plasmenyl-phospholipids (plasmalogens) with a 1-O-alkyl and 1-O-alk-1^1-enyl ether bond, respectively, and usually contain ethanolamine or choline as head group with ethanolamine predominating in humans (fourfold). In humans, plasmalogens make up some 18% of total phospholipid mass and show a cell- and tissue-specific distribution. High levels of plasmenyl ethanolamine occur in brain, heart, lung, kidney, spleen, skeletal muscle, and testis, whereas high levels of plasmenyl choline occur in heart and skeletal muscle with very low levels in all other tissues. Macrophages and neutrophils contain not only high plasmenyl ethanolamine levels, but also significant levels of the saturated ether-phospholipid plasmanyl choline, which is used by these cells for the production of platelet-activating factor (1-O-alkyl-2-acyl-*sn*-glycero-3-phosphocholine). The *sn*-1 position of plasmalogens is occupied predominantly by C16:0, C18:0, and C18:1 fatty alcohols, whereas the *sn-2* position of ether phospholipids usually contains polyunsaturated FAs.

Peroxisomes and the enzymology of ether-phospholipid biosynthesis. The first committed step in the biosynthesis of ether-linked glycerolipids is the formation of the ether linkage by the enzyme alkyldihydroxyacetone phosphate synthase (alkyl-DHAP synthase or ADHAPS) (**Figure 1a**) encoded by the *AGPS* gene. In this reaction, the ester-linked FA of acyl-DHAP is replaced by a fatty alcohol with an ether bond, forming alkyl-DHAP. The reaction proceeds by a ping-pong mechanism: acyl-DHAP first binds to the enzyme, followed by release of the FA, resulting in an activated enzyme-DHAP complex, which then reacts with a fatty alcohol to produce alkyl-DHAP. Alkyl-DHAP synthase is an established peroxisomal enzyme (39–41), which can react with a range of fatty alcohols, including saturated (C10:0 to C18:0) as well as mono- (C18:1) and polyunsaturated (C18:2 and C18:3) alcohols, and this contrasts markedly to the fatty alcohols found in plasmalogens, which are C16:0, C18:0, and C18:1 only (see below). Alkyl-DHAP synthases have been identified in various eukaryotic species and represent one of the few peroxisomal enzymes equipped with a PTS2-targeting sequence in all organisms except *Caenorhabditis elegans*. In this organism, the enzyme contains a PTS1, which is in line with the notion that the PTS2 pathway is missing in *C. elegans* (42).

The two substrates of alkyl-DHAP synthase, i.e., acyl-DHAP and a long-chain fatty alcohol, are also generated by peroxisomes (**Figure 1b**). Acyl-DHAP is synthesized from DHAP and an acyl-CoA ester by the peroxisomal enzyme dihydroxyacetone phosphate acyltransferase (DHAPAT) encoded by *GNPAT*. The enzyme can handle only a small range of acyl-CoAs, including saturated (C14:0 and C16:0) and unsaturated (C18:1) acyl-CoAs (43), shows a broad pH optimum between 7.0 and 9.0, and is membrane associated, with its catalytic site exposed to the peroxisome interior. All DHAPAT amino acid sequences known from different eukaryotic species contain a PTS1 sequence (44, 45). DHAPAT is crucial for plasmalogen synthesis because DHAPAT-deficient human and Chinese hamster ovary (CHO) cell lines are unable to synthesize plasmalogens. Interestingly, acyl-DHAP can also be synthesized outside peroxisomes by other acyltransferases including microsomal G3PAT (46), but this acyl-DHAP is not available for

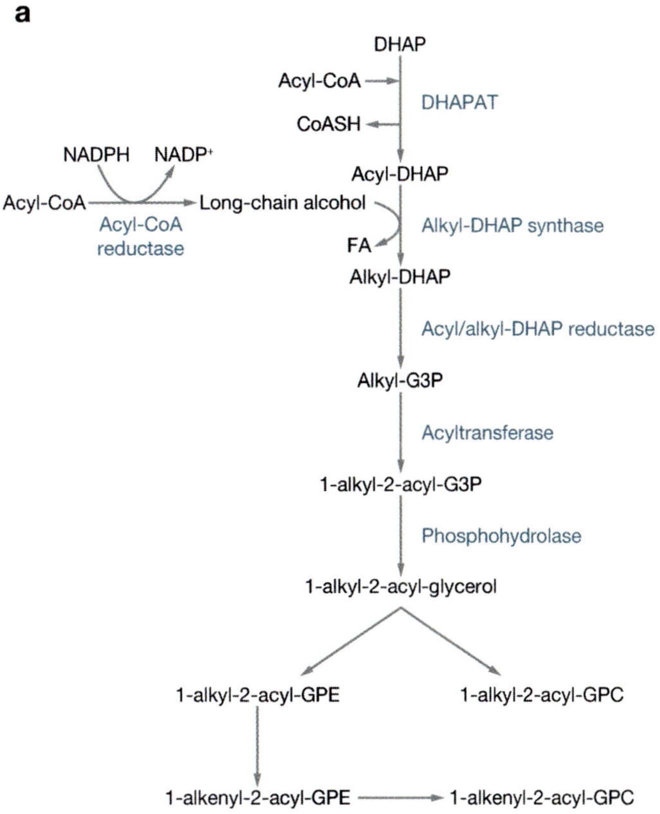

Figure 1

(*a*) Schematic representation of the steps involved in the biosynthesis of PC (phosphatidylcholine) and PE (phosphatidylethanolamine) plasmalogens and (*b*) the topology of the enzymes involved in the biosynthesis of plasmalogens. Abbreviations used: AADHAPR, acylalkyl-dihydroxyacetone phosphate reductase; ADHAPS, alkyl-DHAP synthase; DHAPAT, dihydroxyacetone phosphate acyltransferase; FAR, fatty acyl-CoA reductase; G3PDH, glycerol-3-phosphate dehydrogenase; and VLCS, very-long-chain acyl-CoA synthetase.

CHO: Chinese hamster ovary

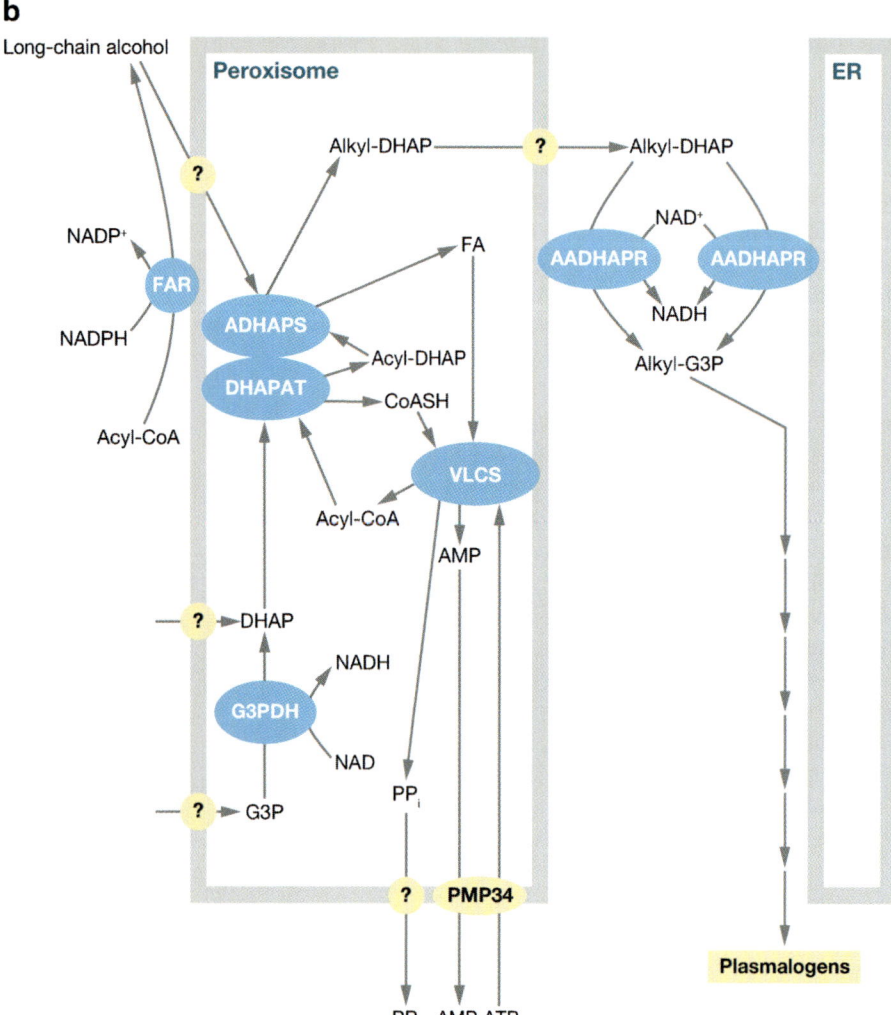

Figure 1
(Continued)

peroxisomal alkyl-DHAP synthase, most likely because acyl-DHAP synthesized outside peroxisomes is unable to traverse the peroxisomal membrane (**Figure 1b**). Alkyl-DHAP synthase and DHAPAT form a 210-kDa heterotrimeric complex within peroxisomes (44, 47) (**Figure 1b**). DHAPAT is only stable inside peroxisomes and when present in the 210-kDa complex, whereas alkyl-DHAP synthase is stable in peroxisomes and active even in the absence of DHAPAT (48).

Although some of the long-chain fatty alcohols required for the alkyl-DHAP synthase reaction may come from dietary sources, the bulk is synthesized from acyl-CoAs. The two consecutive reductions required to transform acyl-CoAs into the corresponding alcohols (acyl-CoA → aldehyde → alcohol) are catalyzed by the same fatty acyl-CoA reductase (FAR), which does not release the intermediate aldehyde and has NADPH as the preferred cosubstrate. In developing rat brain, a long-chain acyl-CoA reductase activity was identified as reactive with C16:0-, C18:0-, and C18:1-CoA only (49). The enzyme was specifically localized in the peroxisomal membrane with its catalytic site exposed to the cytosol (49) (see **Figure 1b**).

On the basis of a specific substrate spectrum, it was speculated that this enzyme is responsible for the virtually exclusive presence of C16:0, C18:0, and C18:1-alk(en)yl chains at the *sn*-1 position of ether phospholipids. Recently, the identification of two acyl-CoA reductases, called FAR1 and FAR2, with different substrate specificities and tissue distributions was reported (50). Both FAR1 and FAR2 are peroxisomal membrane proteins, but they lack clear transmembrane-spanning regions as well as PTS1 or PTS2 sequences. FAR1 is reactive with saturated and unsaturated acyl-CoAs of 16–18 carbons, whereas FAR2 prefers saturated acyl-CoAs of 16 or 18 carbon atoms only. FAR1 expression was identified in many mouse tissues, with the highest level in the preputial gland, a modified sebaceous gland. FAR2 expression was more restricted in distribution and most abundant in the eyelid, which contains wax-laden meibomian glands. Both FAR1 and FAR2 expression was observed in the brain, a tissue rich in ether lipids. These findings suggest that fatty alcohol synthesis in mammals is accomplished by two FAR enzymes (FAR1 and FAR2), encoded by *MLSTD2* and *MLSTD1*, respectively, and expressed at high levels in tissues known to synthesize wax esters and ether lipids (50).

The last contribution of peroxisomes to ether-phospholipid biosynthesis is the reduction of alkyl-DHAP, generated by alkyl-DHAP synthase, to alkyl-G3P. The responsible enzyme acyl/alkyl-DHAP reductase is membrane bound, faces the cytosol both in peroxisomes and in the endoplasmatic reticulum, and preferentially reacts with NADPH rather than NADH (51). All subsequent steps occur in the endoplasmic reticulum (**Figure 1*b***) (see References 46 and 52 for recent reviews).

The physiological role of ether phospholipids, including plasmalogens, has not been established with certainty. They have been implicated in membrane dynamics, intracellular signaling, cholesterol transport and metabolism, oxidative stress, and polyunsaturated FA metabolism (46, 53–58). The fact that isolated deficiencies of DHAPAT and alkyl-DHAP synthase in humans are associated with severe clinical abnormalities and early death (see the section on peroxisomal disorders) indicates that ether phospholipids are essential for life.

Peroxisomal Fatty Acid Beta-Oxidation

In contrast to most other functions of peroxisomes, which may vary between different species and within specific cell types in a single species, FA beta-oxidation is a universal property of peroxisomes in most, if not all, organisms. In yeast and plants, peroxisomes are the sole site of FA beta-oxidation, whereas in higher eukaryotes beta-oxidation may occur in both mitochondria and peroxisomes, following a mechanism involving dehydrogenation, hydration, dehydrogenation again, and thiolytic cleavage as depicted in the panels of **Figure 2*a,b*** for mitochondrial beta-oxidation and peroxisomal beta-oxidation, respectively. Although similar in mechanism, mitochondrial and peroxisomal beta-oxidation fulfill different functions, as concluded from the usually severe but different clinical signs and symptoms associated with inherited defects in either mitochondrial (59) or peroxisomal beta-oxidation (60).

FAs destined for beta-oxidation may originate from outside the cell or result from intracellular breakdown of lipids, for instance in lysosomes. Extracellular FAs probably enter cells by a saturable mechanism mediated by candidate proteins, such as the plasma membrane $FABP_{PM}$, the FA translocase (FAT/CD36), as well as one or more members of the FA transport protein (FATP) family of molecules, which have been hypothesized to harbor both FA transport as well as acyl-CoA synthetase activity (61–63). FAs generated within the cell may be activated by one of the acyl-CoA synthetase enzymes (Acs1-6), which activate different FAs with unique efficiencies (64). Once activated, FAs cannot repartition back into the membrane

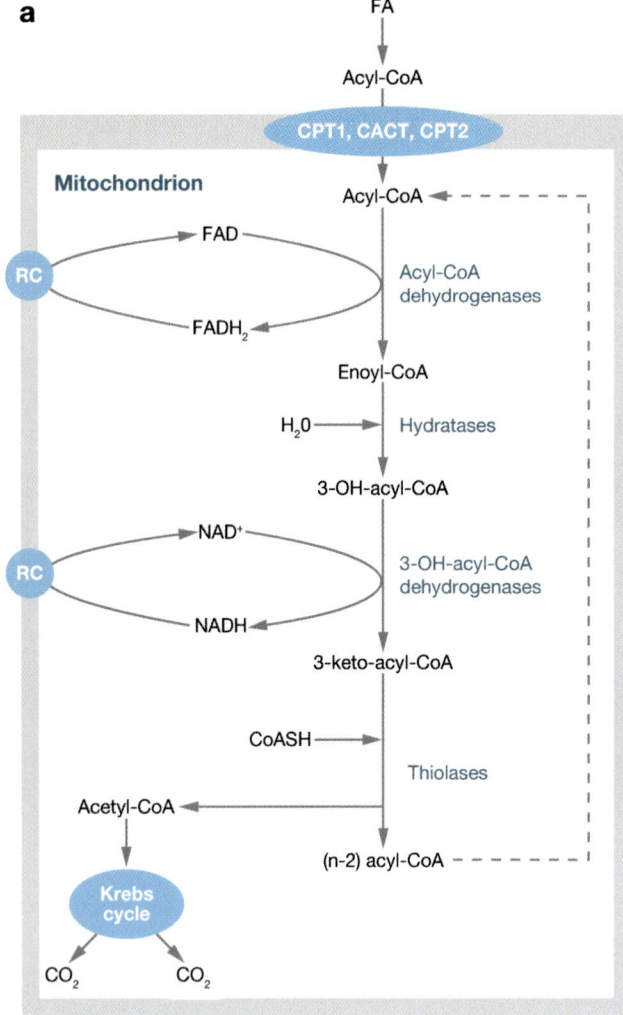

Figure 2
Schematic representation of the mitochondrial and peroxisomal beta-oxidation systems in humans. (*a*) In mitochondria, the FADH$_2$ and NADH, generated in the first and third steps of beta-oxidation, are directly reoxidized by the respiratory chain (RC), (*b*) whereas in peroxisomes, molecular oxygen is the electron acceptor in the first step of beta-oxidation, resulting in the formation of H$_2$O$_2$, which is reconverted into O$_2$ by catalase. The NADH generated in the third step of peroxisomal beta-oxidation is reoxidized via a NAD(H)-redox shuttle, involving the cytosolic and peroxisomal isoforms of malate dehydrogenase in *S. cerevisiae* and lactate dehydrogenase in higher eukaryotes.

VLCFA:
very-long-chain fatty acid

cytosol. The major differences between peroxisomal and mitochondrial beta-oxidation include different substrate specificities and transport of substrates and products of beta-oxidation across the membrane (see References 65 and 66 for reviews).

Different substrate specificities. Short- and medium-chain FAs are exclusively and long-chain FAs are predominantly beta-oxidized in mitochondria, whereas very-long-chain FAs (VLCFAs), notably 26:0, can only be handled by peroxisomes. Other substrates handled only by peroxisomes are (*a*) pristanic acid (2,4,6,10-tetramethylpentadecanoic acid), derived from dietary sources such as pristanic acid itself or its precursor phytanic acid, which is converted to pristanic acid by alpha-oxidation; (*b*) di- and trihydroxycholestanoic acid (DHCA and THCA), produced from cholesterol in the liver and converted to chenodeoxycholic and cholic acid, respectively, after one cycle of beta-oxidation in the peroxisome; (*c*) long-chain dicarboxylic acids, produced by omega-oxidation of long-chain monocarboxylic acids; (*d*) certain polyunsaturated FAs, including tetracosahexaenoic acid (C24:6), which undergoes one cycle of beta-oxidation in peroxisomes to produce docosahexaenoic acid (C22:6); (*e*) certain prostaglandins and leukotrienes; (*f*) some xenobiotics; and (*g*) vitamins E and K.

Transport of substrates and products of beta-oxidation across the membrane. In the case of mitochondria, long-chain FAs (LCFAs) enter the mitochondrial space via the carnitine cycle (**Figure 2***a*), whereas short- and medium-chain FAs enter directly in their protonated form. For peroxisomes, the situation is less clear, but a carnitine-mediated import mechanism has been ruled out (66). As discussed below, the FAs destined for beta-oxidation in peroxisomes probably enter peroxisomes as acyl-CoA esters. Oxidation of FAs in peroxisomes generates a number of acyl-CoA esters, including (*a*) medium-chain

because of their decreased hydrophobicity. The activation also ensures low unesterified FA levels in the cell, thereby maintaining a concentration gradient that is favorable for the entry of more unesterified FAs into the

b

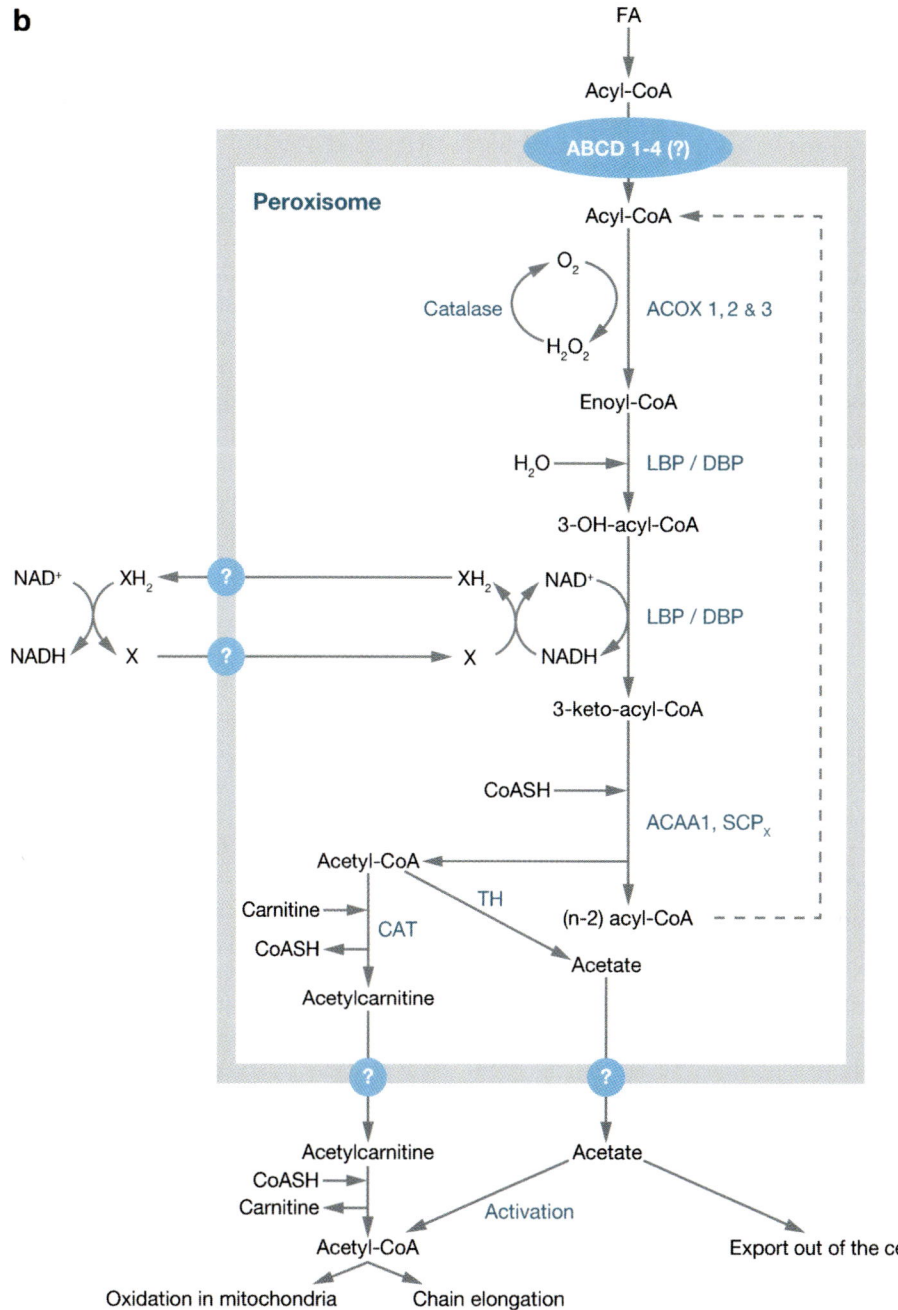

Figure 2
(*Continued*)

acyl-CoAs, e.g., 4,8-dimethylnonanoyl-CoA in the case of pristanoyl-CoA beta-oxidation; (*b*) proprionyl-CoA, and (*c*) acetyl-CoA. The fate of each of these products may vary among different organs and cell types. In principle, there are different ways in which these CoA esters can be metabolized further (**Figure 2*b***). First, different acyl-CoAs can be converted into the corresponding carnitine esters via the peroxisomal enzymes carnitine

LCFA: long-chain fatty acid

acetyltransferase and carnitine octanoyltransferase, as encoded by *CRAT* and *CROT*, respectively, followed by export from the peroxisomes and uptake into the mitochondrion via the carnitine acylcarnitine carrier as was shown for acetyl-CoA and propionyl-CoA (67) and 4,8-dimethylnonanoyl-CoA (68) in cultured skin fibroblasts. Furthermore, acyl-CoA esters may be hydrolyzed within the peroxisome by one of the peroxisomal acyl-CoA thioesterases (69), yielding the free acid and CoA (**Figure 2b**). In hepatocytes, the thioesterase route is very active, with acetate as a major product of the acetyl-CoA units produced in peroxisomes (70). Recent studies on the peroxisomal and mitochondrial oxidation of FAs have shown that there is no detectable transfer of peroxisomal acetyl-CoA units to the mitochondrion for oxidation to CO_2 and H_2O, at least in perfused rat hearts (71). This implies that, in contrast to the liver, peroxisomal FA oxidation in the heart is not accompanied by the hydrolysis of acetyl-CoA and release of acetate. The exact fate of the acetyl-CoA units produced in heart peroxisomes remains to be determined. Recent studies in HepG2 cells have shown that the acetyl-CoA units generated in liver peroxisomes are not only converted into acetate (70) but are also used for chain elongation (72) (see **Figure 2b**).

Enzymology of the peroxisomal beta-oxidation system. Saturated unbranched and 2-methyl-branched FAs are the only FAs that can undergo direct beta-oxidation. In contrast, other FAs, such as mono- and polyunsaturated FAs, 3-methyl branched-chain FAs, and 2-hydroxy FAs, first need to undergo remodeling before they become substrate for peroxisomal beta-oxidation (**Figure 3**). The first step of beta-oxidation in mammalian peroxisomes is catalyzed by different acyl-CoA oxidases, with important differences between the rat and human. Extrahepatic peroxisomes in the rat contain two acyl-CoA oxidases, including palmitoyl-CoA oxidase (ACOX1) and pristanoyl-CoA oxidase (ACOX3), whereas liver peroxisomes contain an additional cholestanoyl-CoA oxidase (ACOX2), specifically reacting with the CoA esters of the bile acid intermediates DHCA and THCA (73). Rat ACOX1 is active with CoA esters of straight-chain mono- and dicarboxylic FAs, prostaglandins, VLC FAs, and xenobiotics, whereas rat ACOX3 is active with 2-methyl-branched-chain acyl-CoAs, such as pristanoyl-CoA, but also handles long and very-long straight-chain acyl-CoAs (73). In the rat, only ACOX1 is inducible by peroxisome proliferators. Interestingly, human peroxisomes contain only two oxidases; the first one is palmitoyl-CoA oxidase, the counterpart of rat ACOX1, with similar substrate spectrum and molecular characteristics (74). The second human peroxisomal oxidase is the branched-chain acyl-CoA oxidase, active with 2-methyl-branched compounds, such as pristanoyl-CoA and the CoA esters of DHCA and THCA, as well as straight-chain acyl-CoAs, including the CoA esters of VLCFAs and dicarboxylic acids (74). Cloning of the cDNA for human branched-chain acyl-CoA oxidase revealed that it is the homologue of the rat liver-specific cholestanoyl-CoA oxidase (75). In contrast to the rat enzyme, however, human branched-chain acyl-CoA oxidase is ubiquitously present in all tissues. Remarkably, the gene coding for the homologue of rat ACOX3 was also identified in the human genome, but both immunoblot and northern blot analyses failed to identify its expression in human tissues (76). It was therefore speculated that the gene was only expressed under certain, for instance developmental, conditions. Interestingly, abundant expression of this pristanoyl-CoA oxidase was found recently in human prostate tissue as well as in some prostate cancer cell lines (77).

Human, rat, and mouse peroxisomes contain two distinct bifunctional proteins that display both enoyl-CoA hydratase and 3-hydroxy-acyl-CoA dehydrogenase activities and catalyze the conversion of 2-*trans*-enoyl-CoAs to 3-ketoacyl-CoAs. L-bifunctional

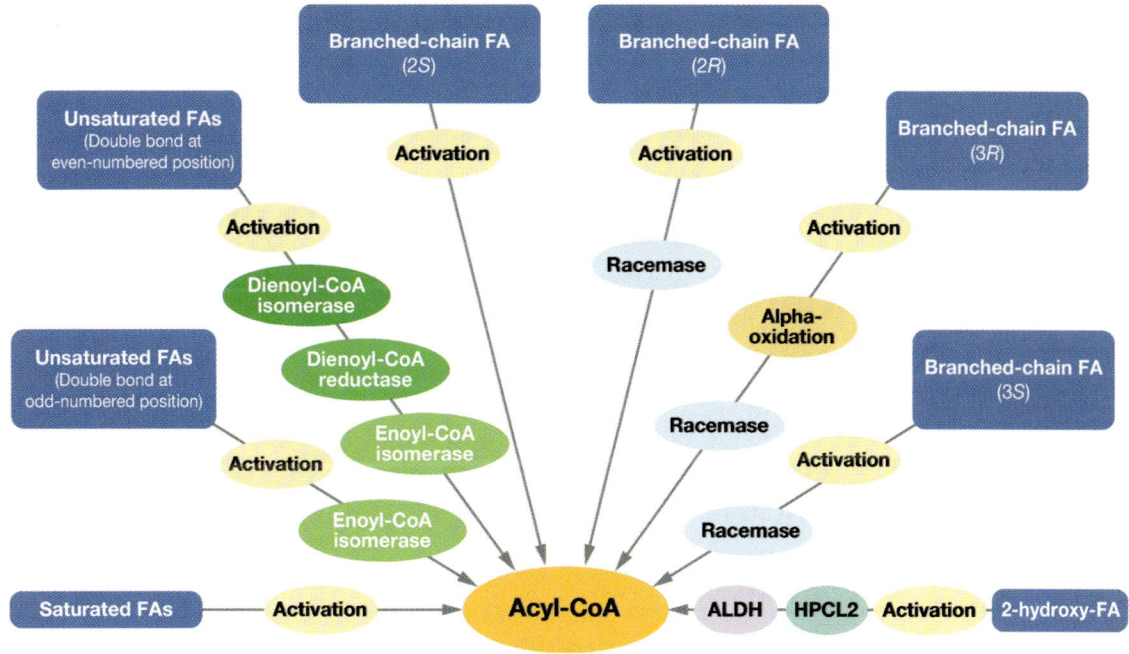

Figure 3

Schematic description of the enzymes required for the conversion of different FAs into the corresponding acyl-CoA esters. Abbreviations used: ALDH, aldehyde dehydrogenase; HPCL2, 2-hydroxyphytanoyl-CoA lyase.

protein (LBP) forms and dehydrogenates L-3-hydroxyacyl-CoAs, and D-bifunctional protein (DBP) forms and dehydrogenates D-3-hydroxyacyl-CoAs. Alternative names for LBP and DBP are multifunctional enzymes I and II (perMFE-I and -II) (78), multifunctional proteins 1 and 2 (MFP-1 and -2) (79), and L- and D-peroxisomal bifunctional enzyme (L-PBE and D-PBE) (80). It is well established now that DBP is the main, if not exclusive, enzyme involved in the beta-oxidation of VLCFAs, pristanic acid, as well as DHCA and THCA (**Figure 4**). Substrate specificity studies have shown that both LBP and DBP react with straight-chain enoyl-CoAs, whereas only DBP reacts with the enoyl-CoA esters of pristanic acid, DHCA, and THCA (78, 79, 81–84). The importance of DBP in the beta-oxidation of all these compounds has become clear through the identification and characterization of patients with a deficiency of DBP (85) and by the generation of a DBP (MFP-2)

knockout mouse (86). The physiological role of LBP remains unclear, although recent studies indicated that it may be the primary enzyme involved in dicarboxylic acid oxidation (87).

Both bifunctional proteins show very little sequence homology and are structurally very different. The N-terminal part of LBP contains the enoyl-CoA hydratase activity and the C-terminal part, the 3-hydroxyacyl-CoA dehydrogenase activity. In addition, LBP also harbors $\Delta 3, \Delta 2$-enoyl-CoA isomerase activity. In contrast, the N-terminal domain of DBP is responsible for the 3-hydroxyacyl-CoA dehydrogenase activity, the central part contains the enoyl-CoA hydratase activity, and the C-terminal domain, sterol carrier protein (SCP) 2 activity (88).

Mammalian peroxisomes also contain multiple peroxisomal thiolases. Mouse and rat liver peroxisomes contain three different 3-oxoacyl-CoA thiolases, including

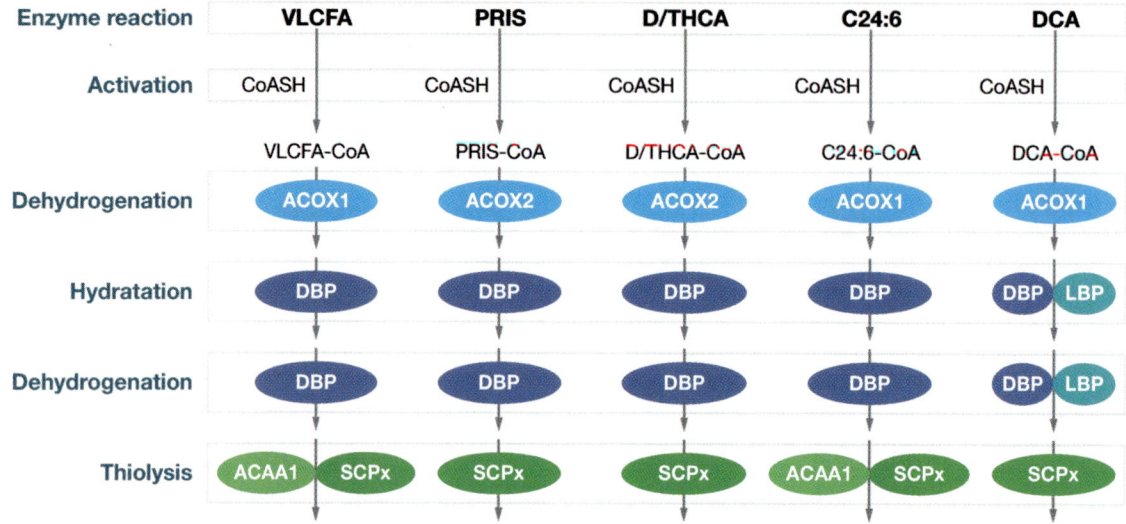

Figure 4

Overview depicting the involvement of the different peroxisomal beta-oxidation enzymes in the peroxisomal beta-oxidation of VLCFAs, pristanic acid (PRIS), DHCA, THCA, tetracosahexaenoic acid (C24:6), and long-chain dicarboxylic acids (DCA). Abbreviation: CoASH, free unesterified coenzyme A.

(*a*) 3-oxoacyl-CoA thiolase A, (*b*) 3-oxoacyl-CoA thiolase B, and (*c*) SCP-2/3-oxoacyl-CoA thiolase (SCPx). The constitutively expressed thiolase A and inducible thiolase B have a virtually identical substrate spectrum, which is active toward short-, medium-, long-, and very-long-chain 3-oxoacyl-CoAs, and are involved in the peroxisomal beta-oxidation of straight-chain FAs (89). SCPx is active toward medium-, long-, and very-long-chain 3-oxoacyl-CoAs, and it is also reactive toward the 3-oxoacyl-CoA species of 2-methyl branched-chain FAs, such as pristanic acid and the bile acid intermediates DHCA and THCA (90–93). Human peroxisomes only contain two thiolases; these include a straight-chain 3-oxoacyl-CoA thiolase, encoded by *ACAA1*, and the SCPx, encoded by *SCP2*, which is essential for the oxidation of 2-methyl branched-chain FAs, i.e., pristanic acid, DHCA, and THCA (94). The involvement of the different beta-oxidation enzymes in the oxidation of VLCFAs, pristanic acid, DHCA, THCA, C24:6, and long-chain dicarboxylic acids is shown in **Figure 4**.

The enzymes described above are necessary and sufficient for the beta-oxidation of straight-chain saturated FAs as well as alpha-methyl branched-chain FAs with the methyl group in the (2S)-configuration. However, auxiliary enzymes are needed for the beta-oxidation of (2R)-methyl branched-chain FAs and unsaturated FAs (**Figure 3**). Oxidation of (2R)-methyl branched-chain FAs requires the active participation of the peroxisomal enzyme 2-methylacyl-CoA racemase, which is capable of converting (2R)- into (2S)-branched-chain acyl-CoAs (95–97). Interestingly, a single gene (*AMACR*) codes for a protein equipped with both a mitochondrial and peroxisomal targeting signal thus explaining its bicompartmental presence in both peroxisomes and mitochondria (6–8). Both the peroxisomal and mitochondrial AMACRs are required for the oxidation of pristanic acid (7).

Peroxisomes also contain the full enzymatic machinery to remove the double bonds in mono- and polyunsaturated FAs. FAs with a double bond at an even-numbered position require the subsequent action of two auxiliary enzymes, including 2,4-dienoyl-CoA reductase and Δ^3,Δ^2-enoyl-CoA isomerase, to produce the corresponding saturated acyl-CoA esters. The 2,4-dienoyl-CoA reductase and Δ^3,Δ^2-enoyl-CoA isomerase in peroxisomes are different from their mitochondrial counterparts and the products of distinct genes, i.e., the *DECR2* (98) and *PEC1* (99) genes (see **Table 1**). Oxidation of FAs with a double bond at an odd-numbered position may proceed via two different pathways. Pathway one only involves the Δ^3,Δ^2-enoyl-CoA isomerase, whereas pathway two requires the active participation of three auxiliary enzymes, including a distinct $\Delta^{3,5},\Delta^{2,4}$-dienoyl-CoA isomerase, $\Delta^{2,4}$-dienoyl-CoA reductase, and Δ^3,Δ^2-enoyl-CoA isomerase. After its prior identification in mitochondria, He et al. (100) detected $\Delta^{3,5},\Delta^{2,4}$-dienoyl-CoA isomerase activity in peroxisomes. Subsequent studies have shown that a single gene *ECH1* codes for a protein with both a mitochondrial and peroxisomal targeting signal, thus explaining the bicompartmental distribution of $\Delta^{3,5},\Delta^{2,4}$-dienoyl-CoA isomerase in both mitochondria and peroxisomes (101).

Peroxisomal Fatty Acid Alpha-Oxidation

FAs with a methyl-group at the carbon 3 position are not a substrate for beta-oxidation but must first undergo alpha-oxidative decarboxylation to produce the corresponding (n-1) FA, with the methyl-group at the 2 position, which then can undergo beta-oxidation. In the past decade, it has become clear that, in contrast to FA beta-oxidation, FA alpha-oxidation is confined to peroxisomes and only accepts acyl-CoA esters as substrate (**Figure 5**). Most studies on alpha-oxidation have been performed with the physiological substrate phytanic acid (3,7,11,15-tetramethylhexadecanoic acid), which is known to accumulate in different genetic disorders including Refsum disease (see below). Activation of phytanic acid to its CoA-ester can occur either outside the peroxisome by the enzyme long-chain acyl-CoA synthetase (102), present at the cytosolic face of the peroxisome (103), or inside the peroxisome by the enzyme very-long-chain acyl-CoA synthetase, a peripheral peroxisomal membrane protein equipped with a PTS1-like signal (**Figure 5**) (104). Subsequently, phytanoyl-CoA is converted into 2-hydroxyphytanoyl-CoA by the enzyme phytanoyl-CoA 2-hydroxylase, which belongs to the family of 2-oxoglutarate-dependent oxygenases, the largest known family of nonheme metal-dependent oxidizing enzymes (see References 105 and 106 for reviews). Conversion of phytanoyl-CoA to 2-hydroxyphytanoyl-CoA is stoichiometrically coupled to the decarboxylation of 2-oxoglutarate into succinate and CO_2, after which one of the oxygen atoms of the dioxygen (O_2) molecule is incorporated into the carboxyl group of succinate and the other in the 2-hydroxy group of 2-hydroxyphytanoyl-CoA. The primary sequences of a number of phytanoyl-CoA 2-hydroxylases from different species have become available in recent years with little overall sequence identity. All phytanoyl-CoA 2-hydroxylases contain the 2-His-1-carboxylate motif and the RXS motif responsible for the binding of iron and 2-oxoglutarate, respectively (107, 108). Site-directed mutagenesis studies have established the identity of His-175, Asp-177, and His-264 as the iron-binding ligands in the human phytanoyl-CoA 2-hydroxylase (109).

All hydroxylases identified so far contain PTS2 sequences except for the *C. elegans* 2-hydroxylase, which contains a PTS1-like sequence (-RSNL) in accordance with the absence of the PTS2 pathway in *C. elegans* (42). The next enzyme in the pathway, i.e., 2-hydroxyphytanoyl-CoA lyase, converts 2-hydroxyphytanoyl-CoA into pristanal and formyl-CoA, is a peroxisomal matrix enzyme of four identical 63-kDa subunits, and contains a PTS1-like sequence (-SNM) (110). At neutral pH, formyl-CoA is split into formate and free CoA nonenzymatically

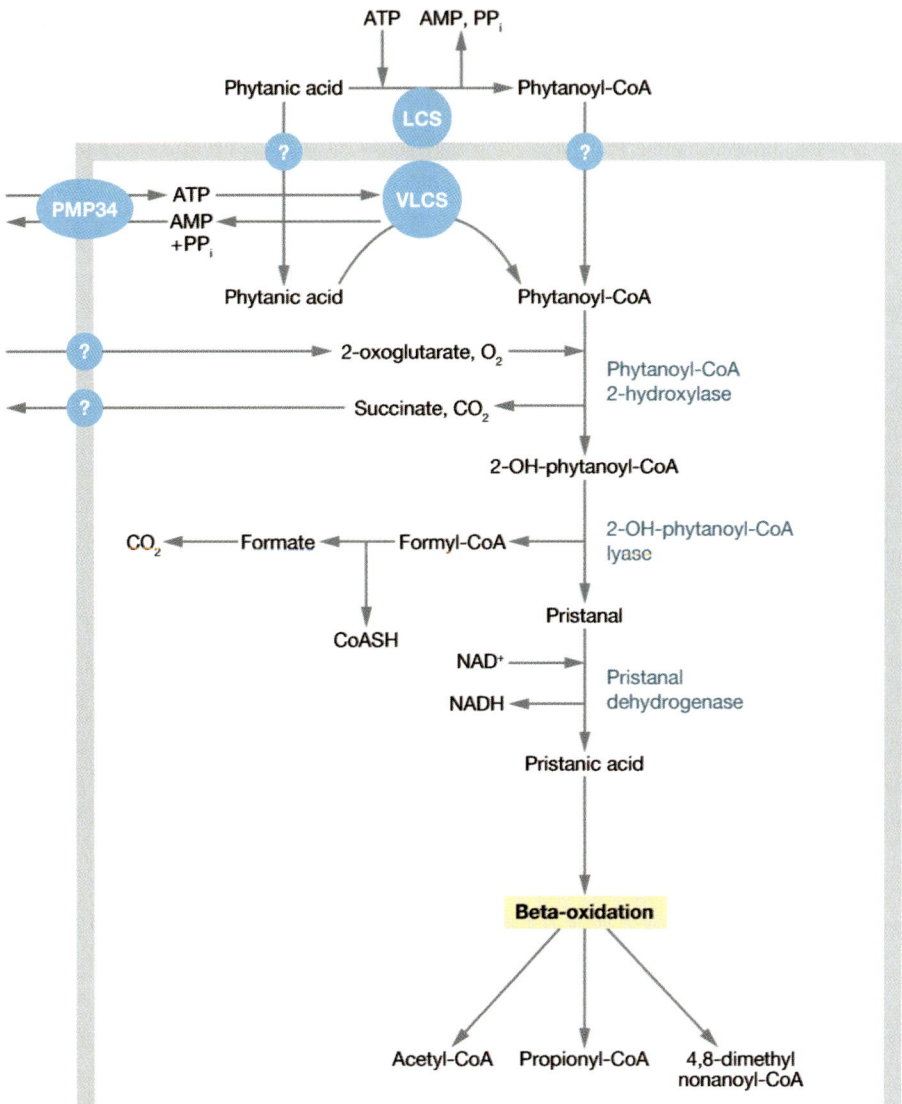

Figure 5
Schematic illustration and topology of the enzymes involved in the phytanic acid alpha-oxidation pathway in mammalian peroxisomes.

(111). Pristanal is converted into pristanic acid via a still poorly defined peroxisomal aldehyde dehydrogenase (112). Finally, pristanic acid is activated to pristanoyl-CoA, probably via the enzyme VLCS (104, 113). Pristanoyl-CoA undergoes three cycles of beta-oxidation in the peroxisome to produce 4,8-dimethylnonanoyl-CoA, which is then transported to the mitochondria for full oxidation (68). Recent studies have shown that peroxisomes also catalyze the oxidation of 2-hydroxy FAs, which are first activated to the respective CoA esters, followed by cleavage into the corresponding aldehyde and formyl-CoA by 2-hydroxyphytanoyl-CoA lyase (114).

Glyoxylate Metabolism

In humans, the enzyme alanine:glyoxylate aminotransferase (AGT) is exclusively expressed in liver peroxisomes and converts glyoxylate generated in peroxisomes into glycine using alanine as the primary amino group donor. This prevents the conversion of glyoxylate into the toxic metabolite oxalate, which can be catalyzed by various dehydrogenases and oxidases, including lactate dehydrogenase. In the case of AGT deficiency, as in patients with hyperoxaluria type 1, glyoxylate accumulates and is converted into oxalate, which precipitates in the liver as well as in other organs, including the kidneys, ultimately causing kidney failure and loss. It has been postulated that glyoxylate, which accumulates in peroxisomes, diffuses across the peroxisomal membrane and is converted into oxalate by cytosolic lactate dehydrogenase. Such a scenario is unlikely, however, because under these conditions the cytosolic enzyme glyoxylate reductase is expected to convert cytosolic glyoxylate to glycolate. Therefore, it is more likely that glyoxylate, which accumulates in peroxisomes, is converted into oxalate by peroxisomal enzymes, including glycolate oxidase and lactate dehydrogenase (115). Interestingly, hyperoxaluria type 1 may not only be caused by a functional deficiency of peroxisomal AGT but also by mislocalization of AGT to mitochondria (116). Under the latter conditions, the AGT, mislocalized to mitochondria, is catalytically active, but glyoxylate generated in peroxisomes cannot reach the AGT in mitochondria and instead is converted into oxalate, giving rise to hyperoxaluria type 1 (9).

Amino Acid Catabolism

Mammalian peroxisomes contain D-amino acid oxidase, which oxidizes the D-isomers of neutral and basic amino acids, as well as D-aspartate oxidase (117), which oxidizes the D-isomers of acidic amino acids. The oxidation of amino acids by both enzymes yields the corresponding keto acids, ammonia, and hydrogen peroxide. Peroxisomes are also involved in the oxidation of some L-amino acids, e.g., L-lysine, which may be degraded to L-2-amino adipic acid either via the saccharopine pathway or via the L-pipecolate pathway. L-pipecolate oxidase, which oxidizes L-pipecolate to Δ1-piperideine-6-carboxylate, is a peroxisomal enzyme identified in human (118) and monkey liver (119), has been purified and cloned (120, 121), and is a typical PTS1 protein. L-pipecolate accumulates in tissues and body fluids of patients who lack peroxisomes, emphasizing the importance of the L-pipecolate pathway in humans. Lysine, hydroxylysine, and tryptophan can be converted to glutaryl-CoA, for which a peroxisomal glutaryl-CoA oxidase has been described in rat and man. However, because the glutaryl-CoA oxidase activity copurifies with palmitoyl-CoA oxidase activity (73), the existence of a separate glutaryl-CoA oxidase is questionable.

Pentose Phosphate Pathway

In the pentose phosphate pathway, NADPH is generated when glucose 6-phosphate is oxidized to ribose-5-phosphate. Although the pentose phosphate pathway has always been assumed to be cytosolic, approximately 10% of the total activity of the two pentose phosphate pathway enzymes, glucose-6-phosphate dehydrogenase and 6-phosphogluconate dehydrogenase, is peroxisomal (122). It is proposed that these two enzymes provide intraperoxisomal NADPH as needed, for example, for the 2,4-dienoyl-CoA reductase reaction. As discussed below, peroxisomes also contain a different system that provides intraperoxisomal NADPH, i.e., via the 2-oxoglutarate/isocitrate NADP(H) shuttle. Because peroxisomes lack any of the subsequent enzymes of the pentose phosphate pathway, the intraperoxisomal generation of NADPH by the glucose 6-phosphate and 6-phosphogluconate dehydrogenases would require import of glucose 6-phosphate and export of ribulose 5-phosphate.

Polyamine Oxidation

Spermine (SPM) and spermidine (SPD) are required for numerous fundamentally important cellular processes, and the levels of these polyamines are under tight control. The main mechanism by which spermine and spermidine are degraded involves transformation of SPM and SPD into N^1-acetyl-SPM and N^1-acetyl-SPD by the cytosolic enzyme acetyl-CoA:SPD/SPM N^1-acetyltransferase, followed by oxidation of the N^1-acetylated polyamines by a peroxisomal N^1-acetylated polyamine oxidase. This enzyme converts N^1-acetyl-SPM to SPD and 3-acetamidopropanal and N^1-acetyl-SPD to putrescine and 3-acetamidopropanal. The same enzyme also oxidizes SPM to SPD and 3-aminopropanal, although very inefficiently. Most cells also contain a cytosolic spermine oxidase, which oxidizes SPM to SPD and 3-aminopropanal but does not react with N^1-acetyl-SPM and N^1-acetyl-SPD. The mouse, bovine, and human peroxisomal N^1-acetylated polyamine oxidases (123, 124) all have a typical PTS1 sequence (**Table 1**). The fate of the products of the peroxisomal polyamine oxidase reaction is not clear. It may well be that the 3-acetamidopropanal is further metabolized within peroxisomes because peroxisomes harbor both aldehyde (125) and alcohol dehydrogenase (126) activities.

Miscellaneous Peroxisomal Enzyme Activities

Peroxisomes have been said to contain a number of different enzyme activities for which the responsible enzymes and genes have remained unknown as well as proteins with unknown functions (**Table 1**). Indeed, rat liver peroxisomes contain a clofibrate-inducible alcohol:NAD$^+$ oxidoreductase activity, with tetradecanol showing the highest catalytic efficiency, i.e. V_{max}/K_m (126). Furthermore, a clofibrate-inducible aldehyde dehydrogenase activity was identified in rat liver peroxisomes with nonanal as the substrate with the highest catalytic efficiency (125). Peroxisomes also contain pristanal dehydrogenase activity (112) and retinal reductase activity (4). As discussed above, the peroxisomal 2,4-dienoyl-CoA reductase is responsible for the latter activity (4). Peroxisomes also contain a *trans*-2-enoyl-CoA reductase, which may play a role in chain elongation (127). Because of the observation that phytol is converted into phytanic acid via the reduction of phytenoyl-CoA into phytanoyl-CoA, it was recently hypothesized that the enoyl-CoA reductase identified by Das et al. (127) may mediate this reduction step (128).

Recent studies have shown that one or more members of the nudix hydrolase family are present in peroxisomes (3). One of these enzymes, called NUDT7, catalyzes the hydrolysis of CoA and its derivatives, and its function may be to eliminate oxidized CoA from peroxisomes and/or to regulate the levels of CoASH and acyl-CoAs in this organelle in response to metabolic demands. Recently, we performed proteomic analysis of mouse peroxisomes and identified several new peroxisomal proteins, including a novel nudix hydrolase designated RP2p and encoded by the *D7RP2e* gene. RP2p is a CoA diphosphatase with activity toward CoASH, oxidized CoASH, and a wide range of acyl-CoA esters (R. Ofman and R.J.A. Wanders, submitted for publication). Finally peroxisomes also contain NAD-linked glycerol-3-phosphate dehydrogenase activity (122, 129), which may play a role in the provision of DHAP for the DHA-PAT reaction (see **Figure 1b**).

Isoprenoid and Cholesterol Metabolism

Many studies performed from the 1950s to the 1980s showed that eight of the nine enzymes of the first part of the isoprenoid biosynthesis pathway, involved in the conversion of acetyl-CoA to farnesyl pyrophosphate, are cytosolic (130). The exception is 3-hydroxy-3-methylglutaryl-CoA reductase, which had been localized to the endoplasmic reticulum (ER), as are the enzymes involved specifically in cholesterol synthesis. Since 1985, however, a series of reports have claimed that many of the enzymes (or the reactions they catalyze) of the first part of the pathway are partly, mainly, or even exclusively located in peroxisomes (131). Moreover, several enzymes involved in cholesterol synthesis were reported to be colocalized in peroxisomes and in the ER. This has led to the rather generally accepted view that peroxisomes would be directly involved in isoprenoid and cholesterol biosynthesis.

Different observations have been used to support the claim that peroxisomes would be involved in isoprenoid/cholesterol biosynthesis. Among these are the decreased activities of several isoprenoid biosynthetic enzymes observed in postmortem liver homogenates of patients, who suffered from a fatal peroxisome biogenesis disorder (PBD) and thus lacked functional peroxisomes (see References 131 and 132 for reviews). Recent studies, however, showed that these decreased activities result from inactivation owing to the bad condition and/or preservation of the livers rather than from mislocalization to the cytosol (17). This corresponds to the finding of normal activities and protein levels of these enzymes in cultured primary skin fibroblasts of PBD patients and in liver of the *pex5* knockout mouse (18, 133), which constitutes a well-defined model for human PBDs (134).

Conflicting data have been published on the de novo cholesterol synthesis rate in peroxisome-deficient fibroblasts and CHO cells. Although a few groups reported decreased rates (135–137), others found normal or even increased rates in such cells (138–140), indicating that the loss of peroxisomes per se does not affect the enzyme activities or de novo cholesterol biosynthesis.

Also, with respect to the subcellular localization of isoprenoid biosynthetic enzymes, conflicting data have been published indicating either a cytosolic or a peroxisomal localization of the proteins. It should be noted, however, that in most cases the claim of a peroxisomal (co)localization was based on (*a*) the finding of only (very) low amounts of proteins in peroxisomal fractions obtained after subcellular fractionation of rat liver tissue, (*b*) immunocytochemical studies using antisera of undefined specificity, and/or (*c*) the results of overexpression studies with tagged proteins or (portions of) proteins fused to reporter proteins in cell lines. As discussed above, the latter studies can be informative but may be full of pitfalls and can only support, but never fully replace, studies aimed at determining the subcellular localization under physiological conditions. Indeed, the subcellular localization under physiological conditions and after overexpression of the three human isoprenoid biosynthetic enzymes mevalonate kinase, phosphomevalonate kinase, and mevalonate pyrophosphate decarboxylase, which were previously claimed to be predominantly peroxisomal, was reinvestigated in great detail using a variety of biochemical and microscopical techniques. The results of these studies unambiguously pointed to an exclusive cytosolic localization of these enzymes with no indication of even a partial peroxisomal localization (16–18).

When combining all available data with emphasis on studies of the subcellular localization of authentic, nonengineered proteins under physiological conditions, the conclusion must be that there is little, if any, evidence for a direct peroxisomal involvement in the biosynthesis of isoprenoids and cholesterol.

PBD: peroxisome biogenesis disorder

BIOCHEMISTRY OF HUMAN PEROXISOMAL DISORDERS

The importance of peroxisomes for normal mammalian development and growth is underlined by the existence of a group of inherited diseases in humans, the peroxisomal disorders, which can be classified into two groups, including (*a*) the PBDs and (*b*) the single peroxisomal enzyme deficiencies. The fatal cerebro-hepato-renal syndrome, in short Zellweger syndrome (ZS), is the prototype of the first group and is characterized by the complete absence of peroxisomes. The underlying basis for the inability to synthesize peroxisomes in ZS has been resolved in recent years and involves mutations in at least 12 different *PEX* genes (141–143).

The absence of peroxisomes has major consequences for most of the metabolic pathways in which peroxisomes are involved. This is evident from the biochemical abberrations observed in ZS patients; these abberrations range from the accumulation of substrates normally handled by peroxisomes, e.g., VLCFAs, pristanic acid, phytanic acid, DHCA, THCA, and pipecolic acid, to a shortage of end products of peroxisomal metabolism, e.g., plasmalogens, cholic and chenodeoxycholic acid, and docosahexaenoic acid (see **Table 2**). From these and other observations, it has become clear that the sequestration of (or parts of) certain metabolic pathways in peroxisomes is essential for efficient substrate channeling and to protect the cell against toxic metabolites generated in peroxisomes, e.g., reactive oxygen species, reactive nitrogen species, and glyoxylate. The abnormalities observed in ZS patients are caused by the deficiency of most, but not all, peroxisomal enzymes destined for peroxisomes. Indeed, most peroxisomal enzymes are unstable in the cytosol and are rapidly degraded, whereas a few peroxisomal enzymes, such as catalase (144) and alanine:glyoxylate aminotransferase (145), are assembled correctly into active multimers (tetramer and dimer, respectively) and are stable in the cytosol.

Most peroxisomal disorders belong to group two, which can be subdivided further into distinct subgroups, depending upon which peroxisomal function is impaired (**Table 2**). Virtually all peroxisomal enzyme deficiencies are associated with severe clinical aberrations. Remarkably, defects in enzymes within the same metabolic pathway may result in different phenotypes (**Table 2**). This is especially true for the disorders of peroxisomal beta-oxidation (**Table 2**). For example, patients with DBP deficiency are severely affected with clinical signs and symptoms resembling those observed in ZS patients. In patients with DBP deficiency, the peroxisomal beta-oxidation of all major substrates is impaired, resulting in the accumulation of VLCFAs, pristanic acid, DHCA, and THCA in tissues and plasma (85, 146). Conversely, patients with AMACR deficiency show a mild clinical phenotype resembling Refsum disease (see below) and only accumulate pristanic acid, DHCA, and THCA (147, 148). Another example is X-linked adrenoleukodystrophy, which affects boys who develop normally for the first few years of life and then rapidly deteriorate, followed by early death. In X-linked adrenoleukodystrophy, only VLCFAs accumulate (68, 149).

Other single peroxisomal enzyme deficiencies are rhizomelic chondrodysplasia punctata type 2 and 3, caused by mutations in *GNPAT* and *AGPS*, respectively, which encode the peroxisomal enzymes DHAPAT and ADHAPS, respectively. Patients affected by these deficiencies show markedly lowered plasmalogen levels, which is in line with the notion that both enzymes play an indispensable role in ether-phospholipid biosynthesis (52). Refsum disease also belongs to group 2 and is caused by mutations in the gene encoding phytanoyl-CoA hydroxylase; as a result, phytanic acid gradually accumulates and reaches toxic levels later in life (150, 151). Another disorder belonging to group 2 is hyperoxaluria type 1, caused by mutations in the alanine:glyoxylate aminotransferase encoding gene (*AGXT*). When glyoxylate accumulates in peroxisomes, oxalate is formed, which precipitates as calcium oxalate in tissues, including the kidney. This explains the patients' loss of kidney function over time. Acatalasaemia is the last single peroxisomal enzyme deficiency, caused by mutations in the catalase-encoding gene, and is associated with an increased tendency to develop oral gangrene in otherwise asymptomatic patients (152).

MOUSE MODELS FOR PEROXISOMAL DISORDERS

Because the number of patients with specific peroxisomal defects is rather limited and the majority of defects lead to early death,

Table 2 Biochemistry of human peroxisomal disorders[a]

Groups	Peroxisomal disorder	Mutant gene	Biochemical abnormalities						Clinical signs and symptoms
			VLCFA	PRIS	PHYT	D/THCA	PL		
Group 1 (peroxisome biogenesis defects)	Zellweger spectrum disorders (ZS, NALD, IRD)	PEX1,2,3,5, 6,10,13,14, 16,19,26	↑	↑[b]	↑[b]	↑	↑		ZS, NALD, and IRD represent a spectrum of disease severity with ZS being the most and IRD the least severe disorder. Common to ZS, NALD, and IRD are liver disease, variable neuro-developmental delay, retinopathy, and perceptive deafness. ZS patients are usually hypotonic from birth and die before one year of age, whereas NALD patients show neonatal onset hypotonia and seizures, and they have progressive white matter disease, usually dying in late infancy. IRD patients may survive beyond infancy, and some may even reach adulthood
	RCDP type 1	PEX7	N	N	↑[b]	N	↑		Patients have a disproportionally short stature primarily affecting the proximal parts of the extremities. Other symptoms include typical facial abnormalities, congenital contractures, ocular aberrations, severe growth deficiency, and mental retardation
Group 2 (single peroxisomal enzyme deficiencies)	X-linked adrenoleukodystrophy (X-ALD)	ABCD1	↑	N	N	N	N		Two major forms, including childhood cerebral adrenoleukodystrophy (CCALD) and adrenomyeloneuropathy (AMN); in the severe form (CCALD), normal development until six years of age, followed by rapid deterioration and death within two years

(Continued)

Table 2 (Continued)

Groups	Peroxisomal disorder	Mutant gene	Biochemical abnormalities					Clinical signs and symptoms
			VLCFA	PRIS	PHYT	D/THCA	PL	
	Acyl-CoA oxidase deficiency (ACOX1 deficiency)	ACOX1	↑	N	N	N	N	Hypotonia, early onset seizures, hearing loss, retinopathy, neurological abnormalities
	D-Bifunctional protein deficiency/multifunctional protein 2 deficiency (DBP/MFP2 deficiency)	HSD17B4	↑	↑[b]	↑[b]	↑	N	Craniofacial abnormalities; neurological disturbances; Zellweger-like phenotype, including neuronal migration defect
	2-Methyl-acyl-CoA racemase deficiency (AMACR deficiency)	AMACR	N	↑[b]	↑[b]	↑	N	Slow, progressive loss of vision; neurological deterioration; in some patients marked hepatopathy
	RCDP2 (DHAPAT deficiency)	GNPAT	N	N	N	N	→	Severe growth retardation, mental retardation, rhizomelia, early death
	RCDP3 (ADHAPS deficiency)	AGPS	N	N	N	N	→	Severe growth retardation, mental retardation, rhizomelia, early death
	Refsum disease (phytanoyl-CoA hydroxylase deficiency)	PAHX/PHYH	N	N	↑	N	N	Loss of vision, cerebellar ataxia, anosmia, ichtyosis, cardiac problems
	Hyperoxaluria type 1 (AGT deficiency)	AGXT	N	N	N	N	N	Progressive loss of kidney function
	Acatalasaemia	CAT	N	N	N	N	N	Increased tendency to develop oral gangrene

[a]Abbreviations: ZS, Zellweger syndrome; NALD, neonatal adrenoleukodystrophy; IRD, infantile Refsum disease; RCDP, rhizomelic chondrodysplasia punctata; X-ALD, X-linked adrenoleukodystrophy; VLCFA, very-long-chain FAs; PRIS, pristanic acid; PHYT, phytanic acid; D/THCA, di- and trihydroxycholestanoic acid; PL, plasmalogens; N, normal; ↑, elevated; ↓, decreased.

[b]Levels may vary from normal to elevated because phytanic acid and pristanic acid are derived from exogenous (dietary) sources only.

most biochemical studies on these defects have been done on cells of such patients, in particular primary skin fibroblasts. Although valuable and informative, these studies do not show how and why a given peroxisomal dysfunction leads to the specific pathophysiology associated with these defects. To obtain more insight into these pathophysiological consequences, a number of mouse models have been generated; the biochemical and phenotypical characteristics of these models are summarized in **Table 3**. All of these mice were generated through targeted gene disruption.

Most mouse models, for which currently a human peroxisomal disorder is known, show biochemical and phenotypical defects similar to those observed in the corresponding human disorders. Moreover, studies on these mouse models have led to the identification of additional defects that could not be readily investigated in patients or their cells, including ossification and neuronal migration defects.

A few mouse knockouts have been generated that create peroxisomal enzyme deficiencies for which no human peroxisomal disorder has been identified (see **Table 3**). The biochemical and phenotypical characterization of these mouse models may aid in the recognition of their possible human counterparts. Interestingly, it appears that, at least in mice, the absence of certain peroxisomal enzymes including L-bifunctional enzyme (80), thiolase-B (153), and catalase (154) does not result in noticeable biochemical and phenotypical abnormalities, making the specific physiological function of these enzymes difficult to discern. It should be noted, however, that such enzymes may be required only for certain specific conditions, as exemplified by the *Scp2* (−/−) mouse. In this mouse model, severe biochemical and phenotypical defects became apparent only after feeding it the phytanic acid precursor phytol, leading to the accumulation of pristanic acid. From this observation and additional findings in the *Amacr* (−/−) mouse (155), one may predict that the possible human counterpart of SCP deficiency presents only later in life, as is the case with Refsum disease, which presents only after gradual accumulation of phytanic acid to toxic levels over time.

Another aspect that should be mentioned is the short life span of mice in comparison to humans, which may not allow enough time for the development of certain phenotypical abnormalities. One example is the observation that *Abcd1* (−/−) mice, a mouse model for X-linked adrenoleukodystrophy (X-ALD), only develop mild neurological and behavioral abnormalities later in life (> 15 months of age), and these abnormalities resemble those observed in adrenomyeloneuropathy, a milder variant of X-ALD (156).

The generation of additional mouse models for selected peroxisomal proteins has been reported for which no detailed information on the biochemical and phenotypical characteristics is available yet. These include mice with targeted disruptions of the genes encoding phytanoyl-CoA hydroxylase (a model for Refsum disease), peroxisomal ABC proteins, i.e., the 70-kDa peroxisomal membrane protein (PMP70), ALD-related protein (ALDRP), PMP70-related protein (PMP70R) (no human diseases known), and alanine:glyoxylate aminotransferase (a model for hyperoxaluria type 1).

PEROXISOMAL METABOLITE TRANSPORT

Correct execution of the many metabolic functions of peroxisomes requires the transport of a large variety of metabolites across the peroxisomal membrane. In recent years, much has been learned about the permeability properties of peroxisomes as discussed below.

Permeability Properties of Peroxisomes

Because after isolation mammalian peroxisomes are freely permeable to low-molecular-weight compounds and peroxisomal enzymes do not show structure-linked latency, it has long been assumed that the peroxisomal

Table 3 Mouse models of peroxisome biogenesis and peroxisome function[a]

Disrupted gene	Deficient (enzyme) protein	Corresponding human disease	Biochemical phenotype					Clinical characteristics	References
			VLCFA	PRIS	PHYT	D/THCA	PL		
Pex2	Pex2	Zellweger spectrum disorder (ZS/NALD/IRD)	↑	↑	↑	↑	↓	Intrauterine growth retardation, severe hypotonia, neonatal death, delayed neuronal migration in CNS, cerebellar abnormalities with reduced Purkinje cell development	(192, 193)
Pex5	Pex5	Zellweger spectrum disorder (ZS/NALD/IRD)	↑	↑	↑	↑	↓	Low birth weight, hypotonia, poor feeding, neuronal migration defect, neonatal death	(134, 194)
Pex13	Pex13	Zellweger spectrum disorder (ZS/NALD/IRD)	↑	↑	↑	↑	↓	Low birth weight, hypotonia, poor feeding, neuronal migration defect, neonatal death	(195)
Pex7	Pex7	RCDP type 1	N	N	↑	N	↓	Intrauterine growth retardation, severe hypotonia, delayed ossification of distal bone elements, dwarfism, delayed neuronal migration	(196)
Pex11α	Pex11α	—	N	N	N	N	N	No phenotypic abnormalities	(197)
Pex11β	Pex11β	Zellweger spectrum disorder (ZS/NALD/IRD)	N	N	N	N	N	Intrauterine growth retardation, hypotonia, developmental delay, neonatal death, impaired neuronal migration	(198)
Gnpat	Dhapat	RCDP type 2	N	N	N	N	↓	Intrauterine growth retardation, hypotonia, male infertility, defects in eye development, cataract, optic nerve hypoplasia, prenatal death of *Dhapat* (−/−) embryos	(199)
Acox1	Acox1	Acyl-CoA oxidase deficiency	↑	N	N	N	N	Viable, but infertile; retarded postnatal growth; microvesicular steatosis; focal cell death; inflammatory reactions; liver tumors at later age (> 15 months)	(200)

Gene	Protein	Disease						Description	Ref.
Hsd17B4	Dbp/Mfp2	D-Bifunctional protein deficiency	↑		↑	↑	N	Normal birth weight, dramatic growth retardation, up to 30% die before postnatal day 12, male infertility, no neuronal migration defect	(86)
Scp2	Peroxisomal thiolase 2 (Scpx)	—	N	↑	↑	↑	N	No phenotypic abnormalities, phytol feeding induces weight loss, neurological abnormalities, and early death within three weeks of birth	(94)
Abcd1	Aldp	Adrenomyeloneuropathy	↑	N	N	N	N	No apparent phenotype; however, beyond age 15 months, late-onset neurological and behavioral abnormalities, axonal loss in the spinal cord, and slower nerve conduction	(156, 201–203)
Ehhadh	Lbp/Mfp1	—	N	N	N	N	N	No phenotypic abnormalities	(204)
Slc27a2	Vlcs	—	N	N	N	N	N	No phenotypic abnormalities	(176)
Amacr	Amacr	AMACR deficiency	N	↑	N–↑	↑	N	No phenotypic abnormalities, intolerance to phytol with liver disease and early death	(155)
mThB	Thiolase B	—	N	N	N	N	N	No phenotypic abnormalities	(153)
Cat	Catalase	Acatalasemia	N	N	N	N	N	No phenotypic abnormalities except for increased susceptibility to trauma-induced dysfunction of brain mitochondria	(154)

[a]See **Table 2** for abbreviations.

membrane does not constitute a permeability barrier to small molecules, at least in mammals. More recent studies, notably in the yeast *S. cerevisiae* and partly confirmed in mammalian cells, however, revealed that in vivo the peroxisomal membrane is impermeable to small metabolites, which implies the existence of peroxisomal metabolite carriers. Hence, it seems plausible that the in vitro permeability is due to the disruption of protein/membrane structures as a consequence of the cell fractionation methods used for their isolation. It should be noted that structure-linked latency of peroxisomal enzymes has been observed in other members of the microbody family, including glyoxysomes and glycosomes. On the basis of pulse-labeling experiments in *Trypanosoma brucei*, it was concluded that the glycosomal membrane is poorly permeable to glycolytic intermediates with the exception of glycerol-3-phosphate, which was postulated to be transported via a specific translocator (157). Recent studies by Antonenkov et al. (158, 159) have shown that under properly controlled conditions several peroxisomal enzymes do show structure-linked latency in peroxisomes from rat liver.

The Intraperoxisomal pH

One of the first indications that the peroxisomal membrane may form a closed structure in vivo was the demonstration of a pH gradient across the membrane in yeast, which could be dissipated by uncouplers (160). More recently, a peroxisomal pH gradient was also shown in human cells (161). It remains unclear whether this gradient is due to an active proton-translocating protein, the consequence of a high intraperoxisomal metabolic activity, or both. Moreover, there is no agreement concerning the orientation of the proton gradient because some groups reported a lower (162), whereas others found a higher or similar intraperoxisomal pH in comparison to the cytosolic pH (163), despite the fact that similar methodologies and cell systems were used. This also makes it difficult to determine whether the proton gradient is used as a driving force for metabolite transport, e.g., FA transport (163), or is a mere consequence of metabolite transport (164). Further insight into a possible physiological function of a peroxisomal proton gradient may come from in vitro studies with reconstituted peroxisomal membrane proteins aimed at determining the properties, characteristics, and requirements of peroxisomal metabolite carriers and shuttle systems, which were identified recently.

Peroxisomal ABC Transporters

Mammalian peroxisomes contain four different half ABC transporters, including the adrenoleukodystrophy protein (ALDP) (or ABCD1), the ALDRP (or ABCD2), PMP70 (or ABCD3), and the PMP70R (or ABCD4) (165). Yeast peroxisomes only contain two half ABC transporters, Pxa1p and Pxa2p, which are most similar to mammalian ALDP and PMP70, respectively (166–170).

The ABC transporters constitute a large family of structurally similar proteins that couple ATP hydrolysis to substrate transport (165). Functional ABC transporters are composed of two structurally similar hydrophobic halves, each comprising six transmembrane alpha helices and one conserved hydrophilic nucleotide-binding domain. In the case of a full transporter, the two halves are encoded by one gene, whereas in the case of a half transporter, they are encoded by two genes. Half transporters need to homo- or heterodimerize to become functional ABC transporters.

The four mammalian proteins display specific but often overlapping expression patterns in different cell types and tissues, raising the possibility that different dimeric combinations of the half transporters may promote the transport of different substrates (171). Several studies have addressed the issue of dimerization of the peroxisomal half ABC transporters. One study examined the possible interactions between ALDP, ALDPR, and PMP70 by expression of the hydrophilic C-terminal halves of these proteins in the yeast two-hybrid system and by coimmunoprecipitation experiments using mouse 3T3 cells, which only express ALDP and PMP70 endogenously, or 3T3 cells genetically modified to also express human ALDP or mouse ALDRP (172). Although it was concluded that each of these three peroxisomal ABC half transporters can dimerize with itself or with a related partner, the physiological relevance of these results remains unclear given that in the coimmunoprecipitation experiments one of the heterodimeric partners involved an overexpressed protein. Another study involved coimmunoprecipitation experiments of c-myc epitope-tagged human ALDP, PMP70, and ALDRP synthesized by in vitro transcription and translation, which revealed ALDP homodimers and heterodimers of ALDP with PMP70 and ALDRP (173). The fact that the interactions between these integral membrane proteins were studied in vitro in the complete absence of (peroxisomal) membranes makes the physiological relevance of these data questionable. Other experiments performed with purified rat liver peroxisomes revealed the formation of a stable complex of ALDP and PMP70 with other peroxisomal membrane proteins in the presence of ATP, but the results did not establish whether this involves homo- or heterodimers (174). Most recently, protein purification and coimmunoprecipitation experiments performed with mouse liver peroxisomes did not reveal the existence of heterodimers between ALDP with PMP70 or ALDRP, suggesting that at least the vast majority of mouse ALDP and mouse PMP70 occurs as homodimer (175).

The two half ABC transporters in yeast most probably function as heterodimers, as indicated when disruption of either of the two encoding genes in *S. cerevisiae* resulted in impaired growth on oleic acid (166–169). This conclusion is supported by coimmunoprecipitation experiments (166).

The function of the peroxisomal half ABC transporters appears most clear for yeast. In yeast, LCFAs, such as oleate, are activated into their corresponding CoA esters in the cytosol by an LCFA-CoA synthetase and then transported across the peroxisomal membrane by Pxa1p/Pxa2p heterodimers to become substrate for peroxisomal beta-oxidation (169). Also, medium-chain FAs (MCFAs) are substrate for peroxisomal beta-oxidation in yeast, but these enter the peroxisome as free FAs via a different pathway and are activated to their corresponding CoA esters inside the peroxisomes by a peroxisomal MCFA-CoA synthetase (169). When the peroxisomal MCFA-CoA synthetase is relocated to the cytosol by removal of its PTS1, the MCFAs are activated in the cytosol and then also become substrate for the Pxa1p/Pxa2p complex (169). This latter observation supports the concept that the Pxa1p/Pxa2p complex accepts activated acyl-CoA esters and not free FAs as substrate.

The functions of the four mammalian half ABC transporters remain to be established. Most of the available information concerns the possible function of ALDP, which is defective in the peroxisomal disorder X-ALD. As discussed above, peroxisomes are involved in the oxidative degradation of VLCFAs, which need to enter the peroxisomes to become substrate for the beta-oxidation system. In cultured skin fibroblasts of patients with X-ALD, the VLCFA beta-oxidation is reduced by 60% to 80% but can be restored by transfection with ALDP cDNA, indicating a role of ALDP in the VLCFA beta-oxidation. In analogy to the putative role of Pxa1p/Pxa2p in the transmembrane transport of acyl-CoA esters in *S. cerevisiae* (169), it has been postulated that ALDP transports the CoA esters of VLCFAs across the peroxisomal membrane. This hypothesis requires the extra- rather than intraperoxisomal activation of VLCFAs. Although VLCFAs can be activated both by intra- as well as extraperoxisomal synthetases, studies by Heinzer et al. (176) in the *Vlcs* (−/−) mouse have shown that VLCS, the intraperoxisomal VLCFA-activating enzyme

(104, 113), does not contribute to VLCFA beta-oxidation, which supports the notion that VLCFAs are activated in the cytosol of human cells similar to the activation of LCFAs in the cytosol of S. cerevisiae cells.

On the basis of the observation that PMP70 overexpression in CHO cells leads to a two- to threefold increase in the rate of palmitic acid beta-oxidation, it has been suggested that PMP70 is involved in the transport of LCFAs across the peroxisomal membrane (177). Conversely, overexpression of PMP70 in mouse 3T3 cells reduced the oxidation rate of VLCFA by 30% to 40%. From studies in a PMP70 knockout mouse, PMP70 also has been implicated in the transport of 2-methyl (branched-chain) acyl-CoA esters, including pristanoyl-CoA, DHC-CoA, and THC-CoA (178). Because overexpression of ALDRP corrects the accumulation of VLCFAs in vitro in X-ALD fibroblasts (179, 180) as well as in vivo in the *Abcd1* (−/−) mouse (181), it has been suggested that ALDRP has a substrate specificity that overlaps with ALDP (179–181). So far, there is no indication for a physiological function of PMP70R.

Peroxisomal ATP Transporter

The best-characterized peroxisomal membrane protein involved in metabolite transport is the peroxisomal adenine nucleotide transporter Ant1p of S. cerevisiae, which has orthologues in other organisms, including PMP47 in the yeast *Candida boidinii* and PMP34 in mammals (182, 183). Ant1p and its orthologues are members of the large mitochondrial carrier family, which comprises at least 35 different, but structurally similar, proteins involved in the transport of various solutes (184). Although most members of this family are located in the mitochondrial membrane, Ant1p and its orthologues are exclusively located in the peroxisomal membrane. Ant1p is most similar to the mitochondrial ADP/ATP exchanger, but topology studies indicated a reverse orientation of Ant1p in the peroxisomal membrane in comparison with the mitochondrial ADP/ATP exchanger (185, 186).

Disruption of the *ANT1* gene in S. cerevisiae results in impaired growth on MCFAs, but growth on LCFAs remained unaffected, suggesting that a transport step specific for MCFA beta-oxidation is impaired in the *ant1* deletion strain (187). Because MCFA beta-oxidation in yeast requires the intraperoxisomal ATP-dependent activation of the MCFAs into their corresponding CoA esters, this mutant phenotype suggested that Ant1p may be involved in peroxisomal ATP import. This was confirmed by the observation of a markedly reduced (ATP-dependent) peroxisomal luciferase activity in intact cells of a transformed *ant1* deletion strain when compared with transformed wild-type cells (187). Conclusive evidence for the role of Ant1p as ATP transporter has come from elegant transport studies with purified Ant1p protein functionally reconstituted in liposomes (164). These studies indicate that Ant1p may catalyze both uniport and exchange of adenine nucleotides. In the reconstituted system, these processes involved proton movement and, in the case of nucleotide hetero-exchange transport, were proton compensated and electroneutral (164, 169). Ant1p is able to exchange ATP for ADP, but also AMP, which clearly differs from the substrate specificity of the mitochondrial ADP/ATP exchanger. Such an ATP/AMP exchange may be required for the import into the peroxisomal lumen of cytosolic ATP in exchange for peroxisomal AMP that is generated from ATP during the intraperoxisomal activation of MCFAs (164, 169). Functional complementation of the *ant1* deletion strain by human PMP34 and reconstitution experiments in liposomes of purified human PMP34 confirmed that this orthologous peroxisomal membrane protein also functions as an adenine nucleotide transporter and also is required for the provision of intraperoxisomal ATP (188).

Other Putative Peroxisomal Transporters

With our current knowledge of peroxisomal metabolism and the various enzymes, substrates, metabolites, and cofactors involved, and knowing that in vivo the peroxisomal membrane constitutes a permeability barrier, one can postulate the existence of various transport proteins and shuttles in the peroxisomal membrane, which function in substrate import, product export, or cofactor regeneration. Evidence for the existence of these comes from genetic studies performed with *S. cerevisiae*, which are based on the strict dependence on peroxisomes for growth on FAs of this yeast. The genetic approach is based on the principle that a specific block in peroxisomal membrane metabolite transport in a constructed or chemically induced yeast mutant should only lead to a metabolic defect in intact cells but not in lysates of such cells. This approach has led to the postulation of a malate/oxaloacetate transport shuttle in the peroxisomal membrane of yeast involved in the regeneration of intraperoxisomal NAD^+ by mediating the transfer of electrons from peroxisomal NADH to cytosolic NAD^+, which is essential for the beta-oxidation of saturated FAs (189). Similarly, the occurrence of an isocitrate/2-oxoglutarate transport shuttle has been postulated for the regeneration of intraperoxisomal NADPH by mediating the transfer of electrons from cytosolic NADPH to peroxisomal $NADP^+$, which is essential for the beta-oxidation of unsaturated FAs (189). Furthermore, there is evidence that peroxisomal export of acetyl-CoA, the end product of FA beta-oxidation in yeast, occurs in the form of either acetyl-carnitine, presumably mediated by an acetyl-carnitine/carnitine transport shuttle similar to that in mitochondria, or as glyoxylate cycle intermediates, or both (190). Finally, recent studies with bovine kidney peroxisomal membrane proteins reconstituted in liposomes revealed the existence of a peroxisomal (pyro)phosphate transport system (191).

Conclusive evidence for the occurrence of additional transport or shuttle systems in the peroxisomal membrane of yeast (and higher organisms) awaits the actual demonstration of transport activities in peroxisomal membranes and the identification of the protein components responsible for these activities. Ways to achieve this are through the determination of transport properties of purified peroxisomal membrane proteins reconstituted in liposomes and via detailed characterization of the protein components of purified peroxisomal membranes (a proteomics approach) to identify proteins with sequence similarity to known transport proteins. Alternatively, these components may be identified through the physiological characterization of constructed yeast deletion strains that lack genes coding for candidate peroxisomal transport proteins identified by database searches.

CONCLUDING REMARKS

In the past several years, great advances have been made in our understanding of the biochemistry of mammalian peroxisomes. Both the generation of specific yeast mutants and the recognition of ZS and related peroxisomal disorders as diseases of the peroxisome have had equally important roles in these advances. Despite all these achievements, many of the fine details of the metabolic functions of peroxisomes remain to be resolved, particularly the mechanisms involved in the transmembrane transport of metabolites. Progress in this field can only be made if candidate transport proteins, either purified from peroxisomes or expressed, can be reconstituted in artificial liposomes, thereby allowing true transport studies. This is especially important to resolve the functional characteristics of the four mammalian half ABC transporters, including ALDP, which has been proposed as an acyl-CoA ester transporter. We hope that this review helps to further the field.

ACKNOWLEDGMENTS

In this review, we tried to give an overview of the current state of the biochemistry of peroxisomes. We apologize to our colleagues whose important contributions to this field have not been included because of the limited space available and the limited number of references permitted. The authors' work was financially supported by European Union grants QLG1-CT-2001-01277, Mouse models of peroxisomal diseases; QLG3-CT-2002-00696, Refsum's disease: diagnosis, pathology, and treatment; LSHG-CT-2004-512018, Peroxisomes in health and disease; and also by two NWO (Dutch Organisation for Scientific Research) grants, including NWO-CW-99008 and NWO-MW-901-03-097. The authors gratefully acknowledge Mrs. Maddy Festen for expert preparation of the manuscript and Mr. Jos Ruiter for expert preparation of the figures.

LITERATURE CITED

1. van den Bosch H, Schutgens RBH, Wanders RJA, Tager JM. 1992. *Annu. Rev. Biochem.* 61:157–97
2. Ferrer-Martinez A, Ruiz-Lozano P, Chien KR. 2002. *Dev. Dyn.* 224:154–67
3. Gasmi L, McLennan AG. 2001. *Biochem. J.* 357:33–38
4. Lei Z, Chen W, Zhang M, Napoli JL. 2003. *Biochemistry* 42:4190–96
5. Ashmarina LI, Rusnak N, Miziorko HM, Mitchell GA. 1994. *J. Biol. Chem.* 269:31929–32
6. Kotti TJ, Savolainen K, Helander HM, Yagi A, Novikov DK, et al. 2000. *J. Biol. Chem.* 275:20887–95
7. Ferdinandusse S, Denis S, Ijlst L, Dacremont G, Waterham HR, Wanders RJA. 2000. *J. Lipid Res.* 41:1890–96
8. Amery L, Fransen M, De Nys K, Mannaerts GP, Van Veldhoven PP. 2000. *J. Lipid Res.* 41:1752–59
9. Birdsey GM, Lewin J, Cunningham AA, Bruford MW, Danpure CJ. 2004. *Mol. Biol. Evol.* 21:632–46
10. Kunau WH, Dommes V, Schulz H. 1995. *Prog. Lipid Res.* 34:267–342
11. Emanuelsson O, Elofsson A, von Heijne G, Cristobal S. 2003. *J. Mol. Biol.* 330:443–56
12. Gould SJ, Valle D. 2000. *Trends Genet.* 16:340–45
13. Sacksteder KA, Gould SJ. 2000. *Annu. Rev. Genet.* 34:623–52
14. Hunt MC, Yang YZ, Eggertsen G, Carneheim CM, Gafvels M, et al. 2000. *J. Biol. Chem.* 275:28947–53
15. Vanhooren JC, Fransen M, de Bethune B, Baumgart E, Baes M, et al. 1996. *Eur. J. Biochem.* 239:302–9
16. Hogenboom S, Tuyp JJM, Espeel M, Koster J, Wanders RJA, Waterham HR. 2004. *J. Cell. Sci.* 117:631–39
17. Hogenboom S, Tuyp JJM, Espeel M, Koster J, Wanders RJA, Waterham HR. 2004. *J. Lipid Res.* 45:697–705
18. Hogenboom S, Tuyp JJM, Espeel M, Koster J, Wanders RJA, Waterham HR. 2004. *Mol. Genet. Metab.* 81:216–24
19. Yoshihara T, Hamamoto T, Munakata R, Tajiri R, Ohsumi M, Yokota S. 2001. *J. Histochem. Cytochem.* 49:1123–31
20. Yokota S, Kamijo K, Oda T. 2000. *Histochem. Cell Biol.* 114:433–46
21. Motley A, Lumb MJ, Oatey PB, Jennings PR, De Zoysa PA, et al. 1995. *J. Cell. Biol.* 131:95–109
22. Purdue PE, Lazarow PB. 1996. *J. Cell. Biol.* 134:849–62

23. Maynard EL, Gatto GJ Jr, Berg JM. 2004. *Proteins* 55:856–61
24. Swinkels BW, Gould SJ, Subramani S. 1992. *FEBS Lett.* 305:133–36
25. Schrader M, Fahimi HD. 2004. *Histochem. Cell Biol.* 122:383–93
26. Singh AK, Dhaunsi GS, Gupta MP, Orak JK, Asayama K, Singh I. 1994. *Arch. Biochem. Biophys.* 315:331–38
27. Yamashita H, Avraham S, Jiang S, London R, Van Veldhoven PP, et al. 1999. *J. Biol. Chem.* 274:29897–904
28. Angermuller S, Bruder G, Volkl A, Wesch H, Fahimi HD. 1987. *Eur. J. Cell Biol.* 45:137–44
29. Frederiks WM, Vreeling-Sindelarova H. 2002. *Acta Histochem.* 104:29–37
30. Keller GA, Warner TG, Steimer KS, Hallewell RA. 1991. *Proc. Natl. Acad. Sci. USA* 88:7381–85
31. Dhaunsi GS, Gulati S, Singh AK, Orak JK, Asayama K, Singh I. 1992. *J. Biol. Chem.* 267:6870–73
32. Singh I. 1996. *Ann. NY Acad. Sci.* 804:612–27
33. Stolz DB, Zamora R, Vodovotz Y, Loughran PA, Billiar TR, et al. 2002. *Hepatology* 36:81–93
34. Dubuisson M, Vander SD, Clippe A, Etienne F, Nauser T, et al. 2004. *FEBS Lett.* 571:161–65
35. Waechter F, Bentley P, Bieri F, Staubli W, Volkl A, Fahimi HD. 1983. *FEBS Lett.* 158:225–28
36. Arand M, Knehr M, Thomas H, Zeller HD, Oesch F. 1991. *FEBS Lett.* 294:19–22
37. Pahan K, Smith BT, Singh I. 1996. *J. Lipid Res.* 37:159–67
38. Morel F, Rauch C, Petit E, Piton A, Theret N, et al. 2004. *J. Biol. Chem.* 279:16246–53
39. Hajra AK, Bishop JE. 1982. *Ann. NY Acad. Sci.* 386:170–82
40. Singh H, Beckman K, Poulos A. 1993. *J. Lipid Res.* 34:467–77
41. Hardeman D, van den Bosch H. 1989. *Biochim. Biophys. Acta* 1006:1–8
42. Motley AM, Hettema EH, Ketting R, Plasterk R, Tabak HF. 2000. *EMBO Rep.* 1:40–46
43. Ofman R, Wanders RJA. 1994. *Biochim. Biophys. Acta* 1206:27–34
44. Thai TP, Heid H, Rackwitz HR, Hunziker A, Gorgas K, Just WW. 1997. *FEBS Lett.* 420:205–11
45. Ofman R, Hettema EH, Hogenhout EM, Caruso U, Muijsers AO, Wanders RJA. 1998. *Hum. Mol. Genet.* 7:847–53
46. Nagan N, Zoeller RA. 2001. *Prog. Lipid Res.* 40:199–229
47. Biermann J, Just WW, Wanders RJA, van den Bosch H. 1999. *Eur. J. Biochem.* 261:492–99
48. de Vet EC, Ijlst L, Oostheim W, Wanders RJA, van den Bosch H. 1998. *J. Biol. Chem.* 273:10296–301
49. Burdett K, Larkins LK, Das AK, Hajra AK. 1991. *J. Biol. Chem.* 266:12201–6
50. Cheng JB, Russell DW. 2004. *J. Biol. Chem.* 279:37789–97
51. Datta SC, Ghosh MK, Hajra AK. 1990. *J. Biol. Chem.* 265:8268–74
52. Brites P, Waterham HR, Wanders RJA. 2004. *Biochim. Biophys. Acta* 1636:219–31
53. Zoeller RA, Morand OH, Raetz CR. 1988. *J. Biol. Chem.* 263:11590–96
54. Morand OH, Zoeller RA, Raetz CR. 1988. *J. Biol. Chem.* 263:11597–606
55. Zoeller RA, Lake AC, Nagan N, Gaposchkin DP, Legner MA, Lieberthal W. 1999. *Biochem. J.* 338:769–76
56. Mandel H, Sharf R, Berant M, Wanders RJA, Vreken P. 1998. *Biochem. Biophys. Res. Commun.* 250:369–73
57. Maeba R, Ueta N. 2003. *J. Lipid Res.* 44:164–71

58. Munn NJ, Arnio E, Liu D, Zoeller RA, Liscum L. 2003. *J. Lipid Res.* 44:182–92
59. Rinaldo P, Matern D, Bennett MJ. 2002. *Annu. Rev. Physiol.* 64:477–502
60. Wanders RJA. 2004. *Mol. Genet. Metab.* 83:16–27
61. Glatz JFC, Storch J. 2001. *Curr. Opin. Lipidol.* 12:267–74
62. Stahl A, Gimeno RE, Tartaglia LA, Lodish HF. 2001. *Trends Endocrinol. Metab.* 12:266–73
63. DiRusso CC, Li H, Darwis D, Watkins PA, Berger J, Black PN. 2005. *J. Biol. Chem.* 280:16829–37
64. Watkins PA. 1997. *Prog. Lipid Res.* 36:55–83
65. Mannaerts GP, Van Veldhoven PP. 1996. *Ann. NY Acad. Sci.* 804:99–115
66. Wanders RJA, Vreken P, Ferdinandusse S, Jansen GA, Waterham HR, et al. 2001. *Biochem. Soc. Trans.* 29:250–67
67. Jakobs BS, Wanders RJA. 1995. *Biochem. Biophys. Res. Commun.* 213:1035–41
68. Verhoeven NM, Roe DS, Kok RM, Wanders RJA, Jakobs C, Roe C. 1998. *J. Lipid Res.* 39:66–74
69. Hunt MC, Alexson SE. 2002. *Prog. Lipid Res.* 41:99–130
70. Leighton F, Bergseth S, Rortveit T, Christiansen EN, Bremer J. 1989. *J. Biol. Chem.* 264:10347–50
71. Bian F, Kasumov T, Thomas KR, Jobbins KA, David F, et al. 2005. *J. Biol. Chem.* 280:9265–71
72. Wong DA, Bassilian S, Lim S, Lee WNP. 2004. *J. Biol. Chem.* 279:41302–9
73. Van Veldhoven PP, Vanhove G, Assselberghs S, Eyssen HJ, Mannaerts GP. 1992. *J. Biol. Chem.* 267:20065–74
74. Vanhove GF, Van Veldhoven PP, Fransen M, Denis S, Eyssen HJ, et al. 1993. *J. Biol. Chem.* 268:10335–44
75. Baumgart E, Vanhooren JC, Fransen M, Marynen P, Puype M, et al. 1996. *Proc. Natl. Acad. Sci. USA* 93:13748–53
76. Vanhooren JC, Marynen P, Mannaerts GP, Van Veldhoven PP. 1997. *Biochem. J.* 325(Pt. 3):593–99
77. Zha S, Ferdinandusse S, Hicks JL, Denis S, Dunn TA, et al. 2005. *Prostate* 63:316–23
78. Qin YM, Poutanen MH, Helander HM, Kvist AP, Siivari KM, et al. 1997. *Biochem. J.* 321:21–28
79. Dieuaide M, Novikov DK, Carchon H, Van Veldhoven PP, Mannaerts GP. 1996. *Ann. NY Acad. Sci.* 804:680–81
80. Qi Ch, Zhu Y, Pan J, Usada N, Maeda N, et al. 1999. *J. Biol. Chem.* 274:15775–80
81. Leenders F, Tesdorpf JG, Markus M, Engel T, Seedorf U, Adamski J. 1996. *J. Biol. Chem.* 271:5438–42
82. Jiang LL, Kurosawa T, Sato M, Suzuki Y, Hashimoto T. 1997. *J. Biochem.* 121:506–13
83. Qin YM, Haapalainen AM, Conry D, Cuebas DA, Hiltunen JK, et al. 1997. *Biochem. J.* 328:377–82
84. Dieuaide-Noubhani M, Asselberghs S, Mannaerts GP, Van Veldhoven PP. 1997. *Biochem. J.* 325:367–73
85. Wanders RJA. 2004. *Am. J. Med. Genet.* 126A:355–75
86. Baes M, Huyghe S, Carmeliet P, Declercq PE, Collen D, et al. 2000. *J. Biol. Chem.* 275:16329–36
87. Ferdinandusse S, Denis S, van Roermund CWT, Wanders RJA, Dacremont G. 2004. *J. Lipid Res.* 45:1104–11
88. Qin YM, Marttila MS, Haapalainen AM, Siivari KM, Glumoff T, Hiltunen JK. 1999. *J. Biol. Chem.* 274:28619–25

89. Antonenkov VD, Van Veldhoven PP, Waelkens E, Mannaerts GP. 1999. *Biochim. Biophys. Acta* 1437:136–41
90. Seedorf U, Brysch P, Engel T, Schrage K, Assmann G. 1994. *J. Biol. Chem.* 269:21277–83
91. Wanders RJA, Denis S, Wouters F, Wirtz KW, Seedorf U. 1997. *Biochem. Biophys. Res. Commun.* 236:565–69
92. Antonenkov VD, Van Veldhoven PP, Waelkens E, Mannaerts GP. 1997. *J. Biol. Chem.* 272:26023–31
93. Bunya M, Maebuchi M, Kamiryo T, Kurosawa T, Sato M, et al. 1998. *J. Biochem.* 123:347–52
94. Seedorf U, Raabe M, Ellinghaus P, Kannenberg F, Fobker M, et al. 1998. *Genes Dev.* 12:1189–201
95. Schmitz W, Albers C, Fingerhut R, Conzelmann E. 1995. *Eur. J. Biochem.* 231:815–22
96. Schmitz W, Conzelmann E. 1997. *Eur. J. Biochem.* 244:434–40
97. Schmitz W, Helander HM, Hiltunen JK, Conzelmann E. 1997. *Biochem. J.* 326:883–89
98. Fransen M, Van Veldhoven PP, Subramani S. 1999. *Biochem. J.* 340:561–68
99. Geisbrecht BV, Zhang D, Schulz H, Gould SJ. 1999. *J. Biol. Chem.* 274:21797–803
100. He XY, Shoukry K, Chu C, Yang J, Sprecher H, Schulz H. 1995. *Biochem. Biophys. Res. Commun.* 215:15–22
101. Filppula SA, Yagi AI, Kilpelainen SH, Novikov D, FitzPatrick DR, et al. 1998. *J. Biol. Chem.* 273:349–55
102. Watkins PA, Howard AE, Gould SJ, Avigan J, Mihalik SJ. 1996. *J. Lipid Res.* 37:2288–95
103. Miyazawa S, Hashimoto T, Yokota S. 1985. *J. Biochem.* 98:723–33
104. Steinberg SJ, Wang SJ, Kim DG, Mihalik SJ, Watkins PA. 1999. *Biochem. Biophys. Res. Commun.* 257:615–21
105. Wanders RJA, Jansen GA, Lloyd MD. 2003. *Biochim. Biophys. Acta* 1631:119–35
106. Mukherji M, Schofield CJ, Wierzbicki AS, Jansen GA, Wanders RJA, et al. 2003. *Prog. Lipid Res.* 42:359–76
107. Mukherji M, Chien W, Kershaw NJ, Clifton IJ, Schofield CJ, et al. 2001. *Hum. Mol. Genet.* 10:1971–82
108. Prescott AG, Lloyd MD. 2000. *Nat. Prod. Rep.* 17:367–83
109. Searls T, Butler D, Chien W, Mukherji M, Lloyd MD, Schofield CJ. 2005. *J. Lipid Res.* 46:1660–67
110. Foulon V, Antonenkov VD, Croes K, Waelkens E, Mannaerts GP, et al. 1999. *Proc. Natl. Acad. Sci. USA* 96:10039–44
111. Croes K, Van Veldhoven PP, Mannaerts GP, Casteels M. 1997. *FEBS Lett.* 407:197–200
112. Jansen GA, van den Brink DM, Ofman R, Draghici O, Dacremont G, Wanders RJA. 2001. *Biochem. Biophys. Res. Commun.* 283:674–79
113. Uchiyama A, Aoyama T, Kamijo K, Uchida Y, Kondo N, et al. 1996. *J. Biol. Chem.* 271:30360–65
114. Foulon V, Sniekers M, Huysmans E, Asselberghs S, Mahieu V, et al. 2005. *J. Biol. Chem.* 280:9802–12
115. Baumgart E, Fahimi HD, Stich A, Volkl A. 1996. *J. Biol. Chem.* 271:3846–55
116. Danpure CJ, Purdue PE, Fryer P, Griffiths S, Allsop J, et al. 1993. *Am. J. Hum. Genet.* 53:417–32
117. Van Veldhoven PP, Brees C, Mannaerts GP. 1991. *Biochim. Biophys. Acta* 1073:203–8
118. Wanders RJA, Romeyn GJ, Schutgens RBH, Tager JM. 1989. *Biochem. Biophys. Res. Commun.* 164:550–55
119. Mihalik SJ, McGuinness M, Watkins PA. 1991. *J. Biol. Chem.* 266:4822–30

120. Dodt G, Kim DG, Reimann SA, Reuber BE, McCabe K, et al. 2000. *Biochem. J.* 345:487–94
121. Ijlst L, de Kromme I, Oostheim W, Wanders RJA. 2000. *Biochem. Biophys. Res. Commun.* 270:1101–5
122. Antonenkov VD. 1989. *Eur. J. Biochem.* 183:75–82
123. Vujcic S, Liang P, Diegelman P, Kramer DL, Porter CW. 2003. *Biochem. J.* 370:19–28
124. Wu T, Yankovskaya V, McIntire WS. 2003. *J. Biol. Chem.* 278:20514–25
125. Antonenkov VD, Pirozhkov SV, Panchenko LF. 1985. *Eur. J. Biochem.* 149:159–67
126. Sakuraba H, Noguchi T. 1995. *J. Biol. Chem.* 270:37–40
127. Das AK, Uhler MD, Hajra AK. 2000. *J. Biol. Chem.* 275:24333–40
128. van den Brink DM, van Miert JNI, Dacremont G, Rontani JF, Wanders RJA. 2005. *J. Biol. Chem.* 280:26838–44
129. Gee R, McGroarty E, Hsieh B, Wied DM, Tolbert NE. 1974. *Arch. Biochem. Biophys.* 161:187–93
130. Goldstein JL, Brown MS. 1990. *Nature* 343:425–30
131. Kovacs WJ, Olivier LM, Krisans SK. 2002. *Prog. Lipid Res.* 41:369–91
132. Kovacs WJ, Krisans S. 2003. *Adv. Exp. Med. Biol.* 544:315–27
133. Vanhorebeek I, Baes M, Declercq PE. 2001. *Biochim. Biophys. Acta* 1532:28–36
134. Baes M, Gressens P, Baumgart E, Carmeliet P, Casteels M, et al. 1997. *Nat. Genet.* 17:49–57
135. Hodge VJ, Gould SJ, Subramani S, Moser HW, Krisans SK. 1991. *Biochem. Biophys. Res. Commun.* 181:537–41
136. Mandel H, Getsis M, Rosenblat M, Berant M, Aviram M. 1995. *J. Lipid Res.* 36:1385–91
137. Aboushadi N, Engfelt WH, Paton VG, Krisans SK. 1999. *J. Histochem. Cytochem.* 47:1127–32
138. Appelkvist EL, Venizelos N, Zhang Y, Parmryd I, Hagenfeldt L, Dallner G. 1999. *Pediatr. Res.* 46:345–50
139. Malle E, Oettl K, Sattler W, Hoefler G, Kostner GM. 1995. *Eur. J. Clin. Investig.* 25:59–67
140. Van Heusden GP, van Beckhoven JR, Thieringer R, Raetz CR, Wirtz KW. 1992. *Biochim. Biophys. Acta* 1126:81–87
141. Weller S, Gould SJ, Valle D. 2003. *Annu. Rev. Genomics Hum. Genet.* 4:165–211
142. Shimozawa N, Tsukamoto T, Nagase T, Takemoto Y, Koyama N, et al. 2004. *Hum. Mutation* 23:552–58
143. Wanders RJA, Waterham HR. 2005. *Clin. Genet.* 67:107–33
144. Wanders RJA, Strijland A, van Roermund CWT van den, Bosch H, Schutgens RBH, et al. 1987. *Biochim. Biophys. Acta* 923:478–82
145. Wanders RJA, van Roermund CWT, Westra R, Schutgens RBH, van der Ende MA, et al. 1987. *Clin. Chim. Acta* 165:311–19
146. van Grunsven EG, van Berkel E, Ijlst L, Vreken P, de Klerk JB, et al. 1998. *Proc. Natl. Acad. Sci. USA* 95:2128–33
147. Ferdinandusse S, Denis S, Clayton PT, Graham A, Rees JE, et al. 2000. *Nat. Genet.* 24:188–91
148. Setchell KD, Heubi JE, Bove KE, O'Connell NC, Brewsaugh T, et al. 2003. *Gastroenterology* 124:217–32
149. Moser HW, Loes DJ, Melhem ER, Raymond GV, Bezman L, et al. 2000. *Neuropediatrics* 31:227–39
150. Wanders RJA, Jansen GA, Skjeldal OH. 2001. *J. Neuropathol. Exp. Neurol.* 60:1021–31
151. Wierzbicki AS, Lloyd MD, Schofield CJ, Feher MD, Gibberd FB. 2002. *J. Neurochem.* 80:727–35

152. Eaton JW, Mouchou MA. 1995. In *The Metabolic and Molecular Basis of Inherited Disease*, ed. CR Scriver, AL Beaudet, WS Sly, D Valle, pp. 2371–83. New York: McGraw-Hill
153. Chevillard G, Clemencet MC, Latruffe N, Nicolas-Frances V. 2004. *Biochimie* 86:849–56
154. Ho YS, Xiong Y, Ma W, Spector A, Ho DS. 2004. *J. Biol. Chem.* 279:32804–12
155. Savolainen K, Kotti TJ, Schmitz W, Savolainen TI, Sormunen RT, et al. 2004. *Hum. Mol. Genet.* 13:955–65
156. Pujol A, Hindelang C, Callizot N, Bartsch U, Schachner M, Mandel JL. 2002. *Hum. Mol. Genet.* 11:499–505
157. Visser N, Opperdoes FR, Borst P. 1981. *Eur. J. Biochem.* 118:521–26
158. Antonenkov VD, Sormunen RT, Hiltunen JK. 2004. *J. Cell Sci.* 117:5633–42
159. Antonenkov VD, Sormunen RT, Hiltunen JK. 2004. *Am. J. Physiol. Cell Physiol.* 287:C1623–35
160. Nicolay K, Veenhuis M, Douma AC, Harder W. 1987. *Arch. Microbiol.* 147:37–41
161. Dansen TB, Wirtz KW, Wanders RJA, Pap EH. 2000. *Nat. Cell Biol.* 2:51–53
162. Lasorsa FM, Scarcia P, Erdmann R, Palmieri F, Rottensteiner H, Palmieri L. 2004. *Biochem. J.* 381:581–85
163. van Roermund CWT, De Jong M, Ijlst L, Van Marle J, Dansen TB, et al. 2004. *J. Cell Sci.* 117:4231–37
164. Palmieri L, Rottensteiner H, Girzalsky W, Scarcia P, Palmieri F, Erdmann R. 2001. *EMBO J.* 20:5049–59
165. Dean M, Annilo T. 2005. *Annu. Rev. Genomics Hum. Genet.* 6:123–42
166. Shani N, Valle D. 1996. *Proc. Natl. Acad. Sci. USA* 93:11901–6
167. Shani N, Sapag A, Valle D. 1996. *J. Biol. Chem.* 271:8725–30
168. Shani N, Watkins PA, Valle D. 1995. *Proc. Natl. Acad. Sci. USA* 92:6012–16
169. Hettema EH, van Roermund CWT, Distel B, van den Berg M, Vilela C, et al. 1996. *EMBO J.* 15:3813–22
170. Swartzman EE, Viswanathan MN, Thorner J. 1996. *J. Cell. Biol.* 132:549–63
171. Langmann T, Mauerer R, Zahn A, Moehle C, Probst M, et al. 2003. *Clin. Chem.* 49:230–38
172. Liu LX, Janvier K, Berteaux-Lecellier V, Cartier N, Benarous R, Aubourg P. 1999. *J. Biol. Chem.* 274:32738–43
173. Smith KD, Kemp S, Braiterman LT, Lu JF, Wei HM, et al. 1999. *Neurochem. Res.* 24:521–35
174. Tanaka AR, Tanabe K, Morita M, Kurisu M, Kasiwayama Y, et al. 2002. *J. Biol. Chem.* 277:40142–47
175. Guimaraes CP, Domingues P, Aubourg P, Fouquet F, Pujol A, et al. 2004. *Biochim. Biophys. Acta* 1689:235–43
176. Heinzer AK, Watkins PA, Lu JF, Kemp S, Moser AB, et al. 2003. *Hum. Mol. Genet.* 12:1145–54
177. Imanaka T, Aihara K, Takano T, Yamashita A, Sato R, et al. 1999. *J. Biol. Chem.* 274:11968–76
178. Jimenez-Sanchez G, Hebron KJ, Silva-Zolezzi I, Mihalik S, Watkins P, et al. 2000. *Am. J. Hum. Genet.* 67:65 (Abstract)
179. Kemp S, Wei HM, Lu JF, Braiterman LT, McGuinness MC, et al. 1998. *Nat. Med.* 4:1261–68
180. Netik A, Forss-Petter S, Holzinger A, Molzer B, Unterrainer G, Berger J. 1999. *Hum. Mol. Genet.* 8:907–13
181. Pujol A, Ferrer I, Camps C, Metzger E, Hindelang C, et al. 2004. *Hum. Mol. Genet.* 13:2997–3006

182. Wylin T, Baes M, Brees C, Mannaerts GP, Fransen M, Van Veldhoven PP. 1998. *Eur. J. Biochem.* 258:332–38
183. McCammon MT, Dowds CA, Orth K, Moomaw CR, Slaughter CA, Goodman JM. 1990. *J. Biol. Chem.* 265:20098–105
184. Kunji ER. 2004. *FEBS Lett.* 564:239–44
185. Honsho M, Fujiki Y. 2001. *J. Biol. Chem.* 276:9375–82
186. McCammon MT, McNew JA, Willy PJ, Goodman JM. 1994. *J. Cell Biol.* 124:915–25
187. van Roermund CWT, Drissen R, van den Berg M, Ijlst L, Hettema EH, et al. 2001. *Mol. Cell. Biol.* 21:4321–29
188. Visser WF, van Roermund CWT, Waterham HR, Wanders RJA. 2002. *Biochem. Biophys. Res. Commun.* 299:494–97
189. van Roermund CWT, Elgersma Y, Singh N, Wanders RJA, Tabak HF. 1995. *EMBO J.* 14:3480–86
190. van Roermund CWT, Hettema EH, van den Berg M, Tabak HF, Wanders RJA. 1999. *EMBO J.* 18:5843–52
191. Visser WF, van Roermund CW, Ijlst L, Hellingwerf KJ, Wanders RJA, Waterham HR. 2005. *Biochem. J.* 389:717–22
192. Faust PL, Hatten ME. 1997. *J. Cell. Biol.* 139:1293–305
193. Faust PL, Su HM, Moser A, Moser HW. 2001. *J. Mol. Neurosci.* 16:289–97
194. Gressens P, Baes M, Leroux P, Lombet A, Van Veldhoven P, et al. 2000. *Ann. Neurol.* 48:336–43
195. Maxwell M, Bjorkman J, Nguyen T, Sharp P, Finnie J, et al. 2003. *Mol. Cell. Biol.* 23:5947–57
196. Brites P, Motley AM, Gressens P, Mooyer PA, Ploegaert I, et al. 2003. *Hum. Mol. Genet.* 12:2255–67
197. Li X, Baumgart E, Dong GX, Morrell JC, Jimenez-Sanchez G, et al. 2002. *Mol. Cell. Biol.* 22:8226–40
198. Li X, Baumgart E, Morrell JC, Jimenez-Sanchez G, Valle D, Gould SJ. 2002. *Mol. Cell. Biol.* 22:4358–65
199. Rodemer C, Thai TP, Brugger B, Kaercher T, Werner H, et al. 2003. *Hum. Mol. Genet.* 12:1881–95
200. Fan CY, Pan J, Chu R, Lee D, Kluckman KD, et al. 1996. *J. Biol. Chem.* 271:24698–710
201. Lu JF, Lawler AM, Watkins PA, Powers JM, Moser AB, et al. 1997. *Proc. Natl. Acad. Sci. USA* 94:9366–71
202. Kobayashi K, Kobayashi H, Ueda M, Honda Y. 1997. *Exp. Eye Res.* 64:719–26
203. Forss-Petter S, Werner H, Berger J, Lassmann H, Molzer B, et al. 1997. *J. Neurosci. Res.* 50:829–43
204. Qi C, Zhu Y, Pan J, Usuda N, Maeda N, et al. 1999. *J. Biol. Chem.* 274:15775–80

Protein Misfolding, Functional Amyloid, and Human Disease

Fabrizio Chiti[1] and Christopher M. Dobson[2]

[1]Dipartimento di Scienze Biochimiche, Università degli Studi di Firenze, I-50134 Firenze, Italy; email: fabrizio.chiti@unifi.it

[2]Department of Chemistry, University of Cambridge, Cambridge, CB2 1EW, United Kingdom; email: cmd44@cam.ac.uk

Key Words

aggregation mechanism, Alzheimer, Parkinson, prion, protein aggregation

Abstract

Peptides or proteins convert under some conditions from their soluble forms into highly ordered fibrillar aggregates. Such transitions can give rise to pathological conditions ranging from neurodegenerative disorders to systemic amyloidoses. In this review, we identify the diseases known to be associated with formation of fibrillar aggregates and the specific peptides and proteins involved in each case. We describe, in addition, that living organisms can take advantage of the inherent ability of proteins to form such structures to generate novel and diverse biological functions. We review recent advances toward the elucidation of the structures of amyloid fibrils and the mechanisms of their formation at a molecular level. Finally, we discuss the relative importance of the common main-chain and side-chain interactions in determining the propensities of proteins to aggregate and describe some of the evidence that the oligomeric fibril precursors are the primary origins of pathological behavior.

Contents

- INTRODUCTION 334
- THE ROLE OF AMYLOID-LIKE STRUCTURES IN DISEASE AND IN NORMAL BIOLOGY .. 335
  - Many Human Diseases Are Associated with Protein Aggregation 336
  - Formation of Amyloid Fibrils Is Sometimes Exploited by Living Systems 339
  - Amyloid Structures Can Serve as Nonchromosomal Genetic Elements 339
- THE STRUCTURES OF AMYLOID FIBRILS 341
  - High-Resolution Structural Studies Using Solid-State NMR 341
  - High-Resolution Structural Studies Using X-ray Crystallography ... 343
  - Other Approaches to Defining the Structural Properties of Amyloid Fibrils 343
  - Similarities and Differences in Fibrillar Structures from Various Systems 344
  - The Polymorphism of Amyloid Fibrils 345
- MECHANISMS OF AMYLOID FIBRIL FORMATION 346
  - Amyloid Formation Occurs via a Nucleated Growth Mechanism .. 347
  - Oligomers Preceding Amyloid Fibril Formation: Structured Protofibrils 347
  - Oligomers Preceding Fibril and Protofibril Formation: Unstructured Aggregates 348
  - Aggregation of Globular Proteins Can Occur via Partial Unfolding 349
  - Aggregation of Globular Proteins Can Occur via Formation of Native-Like Oligomers 349
  - A Multitude of Conformational States Is Accessible to Polypeptide Chains 350
- THE INFLUENCE OF SEQUENCE ON AMYLOID FORMATION 352
  - Hydrophobicity, Charge, and Secondary Structure Propensities Strongly Influence Amyloid Formation 353
  - The Amino Acid Sequence Affects Fibril Structure and Aggregation Rate 353
  - Unfolded Regions Play Critical Roles in Promoting the Aggregation of Partially Folded States 355
  - Variations in Fibrillar Structure Can Be Reconciled by Common Determinants of the Aggregation Process 356
- THE PATHOGENESIS OF PROTEIN DEPOSITION DISEASES 357
  - The Search Is on for the Causative Agents of Protein Aggregation Diseases 357
  - The Toxicity of Prefibrillar Aggregates Results from their Misfolded Nature 358
- PERSPECTIVES 359

INTRODUCTION

Writing a review focused on protein misfolding and the diseases with which it is related is both an exciting and a challenging activity. This is in part because recent interest in this topic has led to an explosion in the number of papers published across a broad spectrum of disciplines, and in part because many of the pathological features of the different diseases, and the characteristics of the proteins with which they are associated, appear at first sight to be quite diverse. Despite this diversity, it is increasingly evident from the experimental data emerging from a wide range of studies that there are some, perhaps many, common features in the underlying physicochemical and biochemical origins of the various disorders and, indeed, of the cases in which similar processes contribute positively to biological function. It has been one of our primary objectives during the writing of this article to explore the extent to which such common features can provide the foundation on which to develop a deeper understanding of the various phenomena associated with protein misfolding and its consequences. Fortunately, within the past year or two, a variety of excellent reviews and books has appeared on the more specific features of many aspects of this complex subject, such as the two-volume book entitled *Protein Misfolding, Aggregation and Conformational Diseases* (1).

To provide a framework on which to build this article, we first describe the variety of human diseases that are now thought to arise from the misfolding of proteins, particularly those, perhaps the majority, in which misfolding results in the formation of highly organized and generally intractable thread-like aggregates termed amyloid fibrils. We point out, however, that in addition living organisms can take advantage of the inherent ability of proteins to form such structures to generate novel and diverse biological functions. Second, we describe the dramatic advances that have recently been made toward the elucidation of the structures of amyloid fibrils at a molecular level and emphasize that our knowledge of these structures is no longer limited to the notion of a fibrillar morphology and an ordered "cross-β" arrangement of the polypeptide chains of which they are composed. We then describe the progress that is being made toward understanding the mechanism of aggregation and toward identifying the nature of key intermediates in the aggregation process. Finally, we discuss some of the important ideas that are emerging about the pathogenesis of the various protein deposition diseases and show that, in at least some cases, the prefibrillar aggregates, rather than the mature and stable fibrils into which they convert, are the likely origins of pathological behavior.

From the evidence that emerges from such considerations, we have tried to pull together the various threads of this complex subject in an attempt to identify both the common features of the various disorders and the differences that lead to their individual identities. We also try to show that, in delving into the general phenomenon of protein misfolding, considerable light can be shed on the origins of some of the most debilitating and increasingly common diseases that affect humanity as well as on the strategies that are likely to be most effective for their prevention and treatment.

THE ROLE OF AMYLOID-LIKE STRUCTURES IN DISEASE AND IN NORMAL BIOLOGY

A broad range of human diseases arises from the failure of a specific peptide or protein to adopt, or remain in, its native functional conformational state. These pathological conditions are generally referred to as *protein misfolding (or protein conformational) diseases*. They include pathological states in which an impairment in the folding efficiency of a given protein results in a reduction in the quantity of the protein that is available to play its normal role. This reduction can arise as the result of one of several posttranslational processes, such as an increased probability of degradation via the quality control system of the endoplasmic reticulum, as occurs in cystic fibrosis (2), or the improper trafficking of a protein, as seen in early-onset emphysema (3). The largest group of misfolding diseases, however, is associated with the conversion of specific peptides or proteins from their soluble functional states ultimately into highly organized fibrillar aggregates. These structures are

Protein misfolding: the conversion of a protein into a structure that differs from its native state

Amyloid fibrils: protein aggregates having a cross-β structure and other characteristics, e.g., specific dye-binding

Protein deposition disease: any pathological state associated with the formation of intracellular or extracellular protein deposits

Amyloidosis: any pathological state associated with the formation of extracellular amyloid deposits

TEM: transmission electron microscopy

AFM: atomic force microscopy

ThT: thioflavin T

CR: Congo red

Protofilaments: the constituent units of amyloid fibrils. They should not be confused with protofibrils

generally described as amyloid fibrils or plaques when they accumulate extracellularly, whereas the term "intracellular inclusions" has been suggested as more appropriate when fibrils morphologically and structurally related to extracellular amyloid form inside the cell (4). For simplicity, however, we shall describe all such species as amyloid fibrils in this article. It is also becoming clear that fibrillar species with amyloid characteristics can serve a number of biological functions in living organisms, provided they form under controlled conditions. Perhaps the most fascinating of these functions lies in the ability of such structures to serve as transmissible genetic traits distinct from DNA genes.

Many Human Diseases Are Associated with Protein Aggregation

A list of known diseases that are associated with the formation of extracellular amyloid fibrils or intracellular inclusions with amyloid-like characteristics is given in **Table 1**, along with the specific proteins that in each case are the predominant components of the deposits. The diseases can be broadly grouped into neurodegenerative conditions, in which aggregation occurs in the brain, non-neuropathic localized amyloidoses, in which aggregation occurs in a single type of tissue other than the brain, and nonneuropathic systemic amyloidoses, in which aggregation occurs in multiple tissues (**Table 1**).

Some of these conditions, such as Alzheimer's and Parkinson's diseases, are predominantly sporadic (labeled ^{c} in **Table 1**), although hereditary forms are well documented. Other conditions, such as the lysozyme and fibrinogen amyloidoses, arise from specific mutations and are hereditary (labeled ^{d} in **Table 1**). In addition to sporadic (85%) and hereditary (10%) forms, spongiform encephalopathies can also be transmissible (5%) in humans as well as in other mammals. It has also been found that intravenous injection or oral administration of preformed fibrils from different sources can result in accelerated AA amyloidosis in mice subjected to an inflammatory stimulus (5, 6). It has therefore been postulated that an environment enriched with fibrillar material could act as a risk factor for amyloid diseases (6). Similarly, injection of the recombinant mouse prion protein in the form of amyloid-like fibrils has been reported to generate disease in mice that express the prion protein (7).

The extracellular proteinaceous deposits found in patients suffering from any of the amyloid diseases have a major protein component that forms the core and then additional associated species, including metal ions, glycosaminoglycans, the serum amyloid P component, apolipoprotein E, collagen, and many others (8, 9). Ex vivo fibrils, representing the amyloid core structures, can be isolated from patients, and closely similar fibrils can also be produced in vitro using natural or recombinant proteins; in this case, mildly denaturing conditions are generally required for their rapid formation, at least for proteins that normally adopt a well-defined folded structure (see below).

The fibrils can be imaged in vitro using transmission electron microscopy (TEM) or atomic force microscopy (AFM). These experiments reveal that the fibrils usually consist of a number (typically 2–6) of protofilaments, each about 2–5 nm in diameter (10). These protofilaments twist together to form rope-like fibrils that are typically 7–13 nm wide (10, 11) or associate laterally to form long ribbons that are 2–5 nm thick and up to 30 nm wide (12–14). X-ray fiber diffraction data have shown that in each individual protofilament the protein or peptide molecules are arranged so that the polypeptide chain forms β-strands that run perpendicular to the long axis of the fibril (11). The fibrils have the ability to bind specific dyes such as thioflavin T (ThT) and Congo red (CR) (15), although the specificity of binding of CR to amyloid fibrils and the resulting green birefringence under

Table 1 Human diseases associated with formation of extracellular amyloid deposits or intracellular inclusions with amyloid-like characteristics

Disease	Aggregating protein or peptide	Number of residues[a]	Native structure of protein or peptide[b]
Neurodegenerative diseases			
Alzheimer's disease[c]	Amyloid β peptide	40 or 42[f]	Natively unfolded
Spongiform encephalopathies[c,e]	Prion protein or fragments thereof	253	Natively unfolded (residues 1–120) and α-helical (residues 121–230)
Parkinson's disease[c]	α-Synuclein	140	Natively unfolded
Dementia with Lewy bodies[c]	α-Synuclein	140	Natively unfolded
Frontotemporal dementia with Parkinsonism[c]	Tau	352–441[f]	Natively unfolded
Amyotrophic lateral sclerosis[c]	Superoxide dismutase 1	153	All-β, Ig like
Huntington's disease[d]	Huntingtin with polyQ expansion	3144[g]	Largely natively unfolded
Spinocerebellar ataxias[d]	Ataxins with polyQ expansion	816[g,h]	All-β, AXH domain (residues 562–694); the rest are unknown
Spinocerebellar ataxia 17[d]	TATA box-binding protein with polyQ expansion	339[g]	α+β, TBP like (residues 159–339); unknown (residues 1–158)
Spinal and bulbar muscular atrophy[d]	Androgen receptor with polyQ expansion	919[g]	All-α, nuclear receptor ligand-binding domain (residues 669–919); the rest are unknown
Hereditary dentatorubral-pallidoluysian atrophy[d]	Atrophin-1 with polyQ expansion	1185[g]	Unknown
Familial British dementia[d]	ABri	23	Natively unfolded
Familial Danish dementia[d]	ADan	23	Natively unfolded
Nonneuropathic systemic amyloidoses			
AL amyloidosis[c]	Immunoglobulin light chains or fragments	∼90[f]	All-β, Ig like
AA amyloidosis[c]	Fragments of serum amyloid A protein	76–104[f]	All-α, unknown fold
Familial Mediterranean fever[c]	Fragments of serum amyloid A protein	76–104[f]	All-α, unknown fold
Senile systemic amyloidosis[c]	Wild-type transthyretin	127	All-β, prealbumin like
Familial amyloidotic polyneuropathy[d]	Mutants of transthyretin	127	All-β, prealbumin like
Hemodialysis-related amyloidosis[c]	β2-microglobulin	99	All-β, Ig like
ApoAI amyloidosis[d]	N-terminal fragments of apolipoprotein AI	80–93[f]	Natively unfolded
ApoAII amyloidosis[d]	N-terminal fragment of apolipoprotein AII	98[i]	Unknown
ApoAIV amyloidosis[c]	N-terminal fragment of apolipoprotein AIV	∼70	Unknown
Finnish hereditary amyloidosis[d]	Fragments of gelsolin mutants	71	Natively unfolded
Lysozyme amyloidosis[d]	Mutants of lysozyme	130	α+β, lysozyme fold
Fibrinogen amyloidosis[d]	Variants of fibrinogen α-chain	27–81[f]	Unknown
Icelandic hereditary cerebral amyloid angiopathy[d]	Mutant of cystatin C	120	α+β, cystatin like

(Continued)

Table 1 (*Continued*)

Disease	Aggregating protein or peptide	Number of residues[a]	Native structure of protein or peptide[b]
Nonneuropathic localized diseases			
Type II diabetes[c]	Amylin, also called islet amyloid polypeptide (IAPP)	37	Natively unfolded
Medullary carcinoma of the thyroid[c]	Calcitonin	32	Natively unfolded
Atrial amyloidosis[c]	Atrial natriuretic factor	28	Natively unfolded
Hereditary cerebral haemorrhage with amyloidosis[d]	Mutants of amyloid β peptide	40 or 42[f]	Natively unfolded
Pituitary prolactinoma	Prolactin	199	All-α, 4-helical cytokines
Injection-localized amyloidosis[c]	Insulin	21 + 30[j]	All-α, insulin like
Aortic medial amyloidosis[c]	Medin	50[k]	Unknown
Hereditary lattice corneal dystrophy[d]	Mainly C-terminal fragments of kerato-epithelin	50–200[f]	Unknown
Corneal amylodosis associated with trichiasis[c]	Lactoferrin	692	α+β, periplasmic-binding protein like II
Cataract[c]	γ-Crystallins	Variable	All-β, γ-crystallin like
Calcifying epithelial odontogenic tumors[c]	Unknown	~46	Unknown
Pulmonary alveolar proteinosis[d]	Lung surfactant protein C	35	Unknown
Inclusion-body myositis[c]	Amyloid β peptide	40 or 42[f]	Natively unfolded
Cutaneous lichen amyloidosis[c]	Keratins	Variable	Unknown

[a]Data refer to the number of residues of the processed polypeptide chains that deposit into aggregates, not of the precursor proteins.
[b]According to Structural Classification Of Proteins (SCOP), these are the structural class and fold of the native states of the processed peptides or proteins that deposit into aggregates prior to aggregation.
[c]Predominantly sporadic, although in some cases hereditary forms associated with specific mutations are well documented.
[d]Predominantly hereditary, although in some cases sporadic forms are documented.
[e]Five percent of the cases are transmitted (e.g., iatrogenic).
[f]Fragments of various lengths are generated and have been reported to be present in ex vivo fibrils.
[g]Lengths shown refer to the normal sequences with nonpathogenic traits of polyQ.
[h]Length shown is for ataxin-1.
[i]The pathogenic mutation converts the stop codon into a Gly codon, extending the 77-residue protein by 21 additional residues.
[j]Human insulin consists of two chains (A and B, with 21 and 30 residues, respectively) covalently linked by disulfide bridges.
[k]Medin is the 245–294 fragment of human lactadherin.

cross-polarized light has recently been questioned (16, 17).

The proteins found as intractable aggregates in pathological conditions do not share any obvious sequence identity or structural homology to each other. Considerable heterogeneity also exists as to secondary structure composition or chain length (**Table 1**). Interestingly, some amyloid deposits in vivo and fibrils generated in vitro have both been found to include higher-order assemblies, including highly organized species known as spherulites, which can be identified from a characteristic Maltese cross pattern when observed under cross-polarized light (18, 19). Such species are also observed in preparations of synthetic polymers, such as polyethylene, a finding consistent with the idea that amyloid fibrils have features analogous to those of classical polymers.

Formation of Amyloid Fibrils Is Sometimes Exploited by Living Systems

An increasing number of proteins with no link to protein deposition diseases has been found to form, under some conditions in vitro, fibrillar aggregates that have the morphological, structural, and tinctorial properties that allow them to be classified as amyloid fibrils (20, 21). This finding has led to the idea that the ability to form the amyloid structure is an inherent or generic property of polypeptide chains, although, as we discuss below, the propensity to form such a structure can vary dramatically with sequence. This generic ability can increasingly be seen to have been exploited by living systems for specific purposes, as some organisms have been found to convert, during their normal physiological life cycle, one or more of their endogenous proteins into amyloid fibrils that have functional rather than disease-associated properties. A list of such proteins is reported in **Table 2**.

One particularly well-studied example of functional amyloid is that of the proteinaceous fibrils formed from the protein curlin that are used by *Escherichia coli* to colonize inert surfaces and mediate binding to host proteins. Consistent with the characteristics of other amyloid structures, these fibrils are 6–12 nm in diameter, possess extensive β-sheet structure, as revealed by circular dichroism (CD) spectroscopy, and bind to CR and ThT (22). A second example involves the filamentous bacterium *Streptomyces coelicolor* that produces aerial hyphae, which allow its spores to be dispersed efficiently; a class of secreted proteins called chaplins has been identified in the hyphae of this organism with the ability to form amyloid fibrils that act cooperatively to bring about aerial development (23). All these systems have extremely highly regulated assembly processes; generation of the bacterial *curli*, for example, involves several proteins, including one that nucleates a different protein to form fibrils.

As well as these examples from bacteria, the formation of functional amyloid-like structures has recently been observed in a mammalian system. The melanosomes, lysosome-related organelles that differentiate in melanocytes to allow the epidermal production of the melanin pigment, are characterized by intralumenal fibrous striations upon which melanin granules form. This fibrous material, sharing significant analogies with amyloid fibrils, is assembled from the intralumenal domain of the membrane protein Pmel17 that is proteolyzed by a proprotein convertase (24). This result is a direct indication that even in higher organisms amyloid formation can be physiologically useful for specific and specialized biological functions, provided it is regulated and allowed to take place under highly controlled conditions.

Amyloid Structures Can Serve as Nonchromosomal Genetic Elements

As we discussed in the previous paragraph, it is clear that living systems can utilize the amyloid structure as the functional state of some specific proteins. It is also clear, however, that nature has selected, or at the very least has not selected against, some proteins that can exist within normally functioning biological systems in both a soluble conformation and in an aggregated amyloid-like form. Remarkably, this phenomenon has resulted in the latter state being self-perpetuating, infectious, and inheritable as a non-Mendelian nonchromosomal genetic trait (25). Proteins with such behavior are called prions and are listed in **Table 2**. Although the only endogenous mammalian protein so far recognized to have such properties is associated with the group of invariably fatal and transmissible diseases, the heritable conformational changes of prion proteins from some other organisms have, in some cases, been found beneficial.

The prion proteins from *Saccharomyces cerevisiae*, including Ure2p and Sup35p, give rise to distinct phenotypes when adopting either one or the other forms of the

Functional amyloid: an amyloid structure found to have a beneficial function in living systems

CD: circular dichroism

Table 2 Proteins forming naturally nonpathological amyloid-like fibrils with specific functional roles

Protein	Organism	Function of the resulting amyloid-like fibrils	References
Curlin	*Escherichia coli* (bacterium)	To colonize inert surfaces and mediate binding to host proteins	22
Chaplins	*Streptomyces coelicolor* (bacterium)	To lower the water surface tension and allow the development of aerial hyphae	23
Hydrophobin[a] EAS	*Neurospora crassa* (fungus)	To lower the water surface tension and allow the development of aerial hyphae	23a
Proteins of the chorion of the eggshell[b]	*Bombyx mori* (silkworm)	To protect the oocyte and the developing embryo from a wide range of environmental hazards	23b
Spidroin	*Nephila edulis* (spider)	To form the silk fibers of the web	23c
Intralumenal domain of Pmel17	*Homo sapiens*	To form, inside melanosomes, fibrous striations upon which melanin granules form	24
Ure2p (prion)	*Saccharomyces cerevisiae* (yeast)	To promote the uptake of poor nitrogen sources ([URE3])	25
Sup35p (prion)	*Saccharomyces cerevisiae* (yeast)	To confer new phenotypes ([PSI+]) by facilitating the readthrough of stop codons on mRNA	26–28
Rnq1p (prion)	*Saccharomyces cerevisiae* (yeast)	Not well understood ([RNQ+], also known as [PIN+], phenotype)	28a
HET-s (prion)	*Podospora anserina* (fungus)	To trigger a complex programmed cell death phenomenon (heterokaryon incompatibility)	31, 32
Neuron-specific isoform of CPEB (prion)	*Aplisia californica* (marine snail)	To promote long-term maintenance of synaptic changes associated with memory storage	30

[a]Other proteins from this class, collectively called hydrophobins, have been found to play similar roles in other species of filamentous fungi.
[b]Suggested to form amyloid-like fibrils in vivo, although amyloid formation has only been observed in vitro.

protein (soluble or fibrillar). These proteins are not related to each other, although they do have some characteristics in common, such as the presence of a globular domain and an unstructured portion of the sequence and the high occurrence of glutamine and asparagine residues in the unstructured domain. The polymerization-mediated inactivation of Sup35p, a protein involved in the termination of mRNA translation, confers a wide variety of novel phenotypes ([PSI+]) by facilitating the readthrough of stop codons (26–28). The aggregation of Ure2p destroys its ability to bind and sequester the transcription factor Gln3p; this results in the activation of a series of genes involved in the uptake of poor nitrogen sources (25). The resulting yeast cells [URE3] can grow on media that, for example, lack uracil but contain its precursor ureidosuccinate (25). Although the low natural occurrence of [URE3] and [PSI+] strains suggest that the corresponding phenotypes are not generally beneficial (29), they can still be advantageous under particular environmental circumstances.

In the marine snail *Aplysia californica*, a neuron-specific isoform of cytoplasmic polyadenylation element-binding protein (CPEB) has also been found to exist in a soluble and a self-perpetuating prion form (30). The prion form was found to be more active than the soluble form in stimulating translation of CPEB-regulated mRNA. From this finding, the suggestion was made that the polymerization of the protein could be essential for the long-term maintenance of synaptic changes associated with memory storage (30). Finally, the polymerization of the HET-s protein from *Podospora anserina* is involved in a controlled programmed cell death phenomenon termed heterokaryon incompatibility (31, 32).

It is evident, even from the relatively few examples that have been studied in detail so far, that the aggregation of proteins into amyloid-like structures can generate a number of extremely diverse biological functions. The presence of many other sequences in the genomes of different organisms with the characteristics of prions suggests that there may yet be surprises in store for us when their properties are investigated.

THE STRUCTURES OF AMYLOID FIBRILS

For many years the only structural information about amyloid fibrils came from imaging techniques such as TEM, and more recently AFM, and from X-ray fiber diffraction (10, 11, 33). Despite the structural insight given by these techniques, as outlined above, one of the most common statements in the introductory sections of papers in this field until about three years ago was to the effect that "amyloid fibrils cannot be characterized in detail at the molecular level because they are not crystalline yet they are too large to be studied by solution NMR spectroscopy." The situation has changed dramatically recently as a result of major progress in the application of solid-state NMR (SSNMR) spectroscopy to preparations of amyloid fibrils (34–36) and of successes in growing nano- or microcrystals of small peptide fragments that have characteristics of amyloid fibrils yet are amenable to single crystal X-ray diffraction analysis (37, 38).

High-Resolution Structural Studies Using Solid-State NMR

Using SSNMR, in conjunction with computational energy minimization procedures, Tycko and coworkers (34, 39, 40) have put forward a structure of the amyloid fibrils formed from the 40-residue form of the amyloid β peptide ($A\beta_{1-40}$) at pH 7.4 and 24°C under quiescent conditions. In this structure, each $A\beta_{1-40}$ molecule contributes a pair of β-strands, spanning approximately residues 12–24 and 30–40, to the core region of the fibrils (**Figure 1a**). These strands, connected by the loop 25–29, are not part of the same β-sheet, however, but participate in the formation of two distinct β-sheets within the same protofilament (**Figure 1a**). The different Aβ molecules are stacked on to each other, in a parallel arrangement and in register, at least from residue 9 to 39 (39, 40). By invoking additional experimental constraints, such as the diameter of the protofilaments observed using TEM, and the mass per unit length, measured by means of scanning transmission electron microscopy (STEM) (34, 41), it has been suggested that a single protofilament is composed of four β-sheets separated by distances of ∼10 Å (**Figure 1a**).

Support for key elements of this proposed structure comes from experiments of site-directed spin labeling coupled to electron paramagnetic resonance (SDSL-EPR) (42). The values of the inverse central line width in the EPR spectra for a series of labeled residues indicate that the segments of the $A\beta_{1-42}$ molecule corresponding to residues 13–21 and 30–39 are highly structured in the fibrils, parallel and in register. High flexibility and exposure to the solvent of the N-terminal region, in contrast to considerable structural rigidity detected for the remainder of the sequence, are also suggested by experimental strategies that use hydrogen-deuterium exchange methods in conjunction with mass spectrometry (43), limited proteolysis (44), and proline-scanning mutagenesis (45).

SSNMR, in conjunction with site-directed fluorescence labeling and an ingenious hydrogen/deuterium exchange protocol applied previously to probe the regions of β2-microglobulin fibrils that are involved in persistent structure (46), has led to identification of the regions of the C-terminal fragment of HET-s that are involved in the core of the fibril (36). In the proposed structure, each molecule contributes four β-strands, with strands one and three forming the same parallel β-sheet and with strands two and four

SSNMR: solid-state nuclear magnetic resonance

STEM: scanning transmission electron microscopy

SDSL-EPR: site-directed spin labeling coupled to electron paramagnetic resonance

Aβ: amyloid β peptide

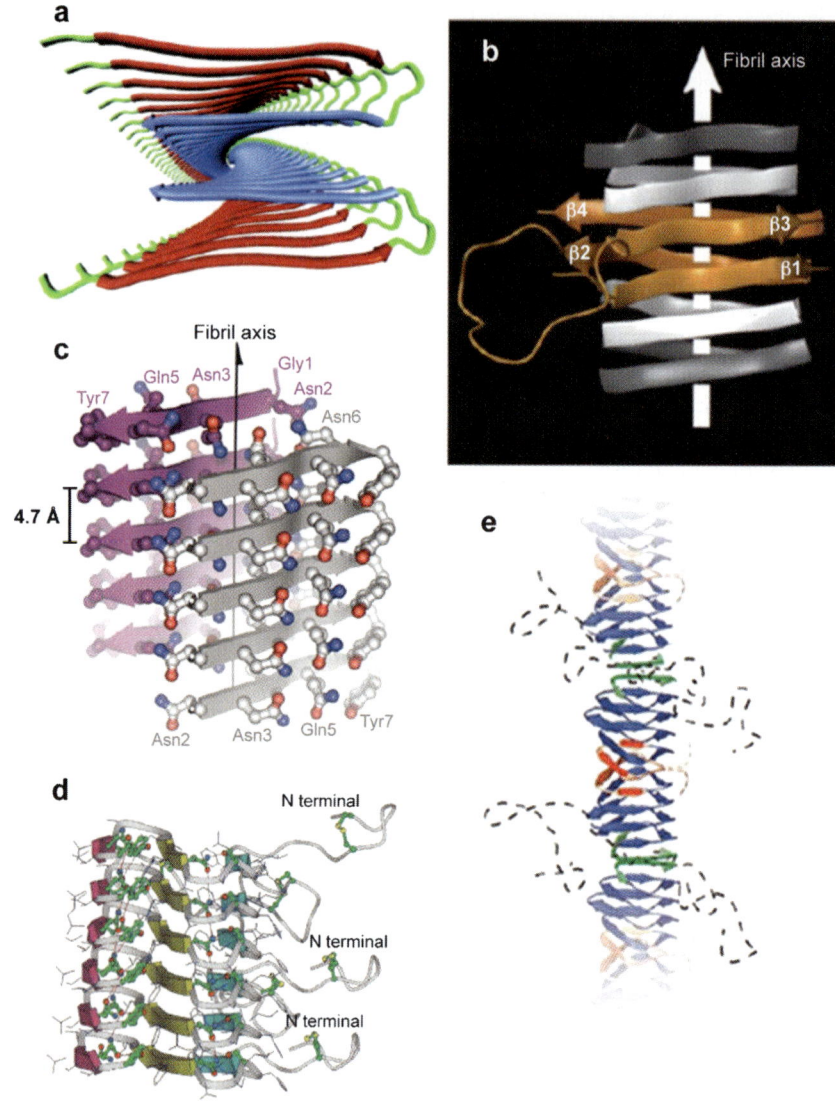

Figure 1

Recent three-dimensional structural models of fibrillar aggregates from different sources. (*a*) The protofilament of Aβ viewed down the long axis of the fibril. Reprinted with permission (177), copyright (2003) American Chemical Society. The segments 12–24 (*red*) and 30–40 (*blue*) are shown. (*b*) The fibril from the C-terminal domain 218–289 of the fungal prion protein HET-s [reproduced with permission (36)]. The ribbon diagram shows the four β-strands (*orange*) (residues 226–234, 237–245, 262–270, and 273–282) and the long loop between β2 and β3 from one molecule. Flanking molecules along the fibril axis (*gray*) are shown. (*c*) Atomic structure of the microcrystals assembled from the GNNQQNY peptide [reproduced with permission (38)]. Each β-strand is a peptide molecule. (*d*) The protofilament from amylin [reprinted with permission from Elsevier (51)]. Green, yellow, and pink β-strands indicate residues 12–17, 22–27, and 31–37, respectively. The unstructured N-terminal tail is shown on the right of the panel along with the disulfide bridge between Cys2 and Cys7. (*e*) The fibril from the NM region of Sup35p [reproduced with permission (52)]. The colored ribbons indicate residues 25–38 (*red*), 39–90 (*blue*), and 91–106 (*green*). The unstructured regions 1–20 (*red dashed lines*) and 158–250 (*black dashed lines*) are shown.

forming another parallel β-sheet ∼10 Å away (**Figure 1***b*).

Advances in SSNMR techniques that enable specific internuclear distances and torsion angles to be measured have also allowed the structure of a 11-residue fragment of transthyretin within an amyloid-like fibril to be defined in atomic detail (35, 47). This study shows that the peptide adopts an extended β-strand within the fibrils. Most importantly, however, this pioneering study reveals that the molecules within the fibrils possess a degree of uniformity, even at the level of the side-chain torsion angles, that has previously only been associated with crystalline materials. Because this regularity is reflected in the very narrow resonance lines in the SSNMR spectra, we can anticipate that complete atomic-level structures will soon begin to emerge for a range of systems, transforming our understanding of this facet of the amyloid phenomenon.

High-Resolution Structural Studies Using X-ray Crystallography

The remarkable achievement of inducing a peptide derived from Sup35p (GNNQQNY) and another with sequence KFFEAAAKKFFE, to form three-dimensional crystals that possess key characteristics of amyloid fibrils, has allowed both the structure of the peptides and the way the molecules could be packed together to be determined with unprecedented resolution (37, 38). In the case of the Sup35p fragment, the crystal consists of pairs of parallel β-sheets in which each individual peptide molecule contributes a single β-strand (**Figure 1***c*). The stacked β-strands are parallel and in register in both sheets. The two sheets interact with each other through the side chains of Asn2, Gln4, and Asn6 to such a degree that water is excluded from the region between them. The remaining side chains on the outer faces of the sheets are hydrated and more distant from the next pair of β-sheets, suggesting that this less intimate interaction could represent a crystal contact rather than a feature of the fibrillar state.

A particularly significant aspect of these structures determined with X-ray or SSNMR is that they are strikingly similar to proposals from cryo-electron microscopy (EM) analysis of the amyloid fibrils formed from an SH3 domain and from insulin, in which the electron density maps were interpreted as arising from pairs of relatively flat untwisted β-sheets (48, 49). Such similarities suggest that many amyloid fibrils could have core structures that have very similar features, which are primarily dictated by the intrinsic conformational preferences of polypeptide chains. The specific nature of the side-chain packing, including such characteristics as the alignment of adjacent strands and the separation of the sheets (50), however, provides an explanation for the occurrence of variations in the details of the structures for specific types of fibril. Hopefully, these pioneering X-ray and SSNMR studies may represent the first entries in a new database of structures similar to the current Protein Data Bank.

Other Approaches to Defining the Structural Properties of Amyloid Fibrils

As discussed above, SSNMR and X-ray crystallography have recently made major contributions to our knowledge of the structures of amyloid fibrils. Considerable progress in this quest has also come from other approaches, typically involving the combination of data from a number of different biophysical experiments (13, 51, 52). One example is the three-dimensional structure proposed for amyloid fibrils from amylin (**Figure 1***d*) (51). The polypeptide chains were configured within the fibrils on the basis of a cross-β structure, deduced from X-ray diffraction data along with measurements of the protofilament diameter and mass per unit length, determined using TEM and STEM, respectively (53, 54). Additional constraints were provided by evidence of a parallel and in register arrangement of the β-strands formed by adjacent molecules from SDSL-EPR data (55) and by evidence of the high propensity of various amylin segments to form fibrils when dissected from the rest of the sequence (51). In the resulting model, the N-terminal "tail" (residues 1–11) is unstructured, and residues 12–17, 22–27, and 31–37 form β-strands in a serpentine arrangement, contributing to different β-ribbons in the protofilament (**Figure 1d**).

In another particularly elegant example, detailed structural information on the fibrils, formed from the NM region (residues 1–250) of the yeast prion protein Sup35p, was also obtained by combining a variety of experimental strategies (52). Carefully chosen residues spaced along the fragment of the protein were mutated so as to generate 37 variants, each having a single cysteine residue at a desired position in the molecule. The variants were then labeled with fluorescent probes. The wavelength maximum and total emission intensity of the fluorescent probes were then used to provide information about the degree of burial from solvent of the various residues and about the distances between probes attached to different molecules within the fibrils. Dimeric constructs were also generated for each variant by covalently linking the free thiol group of one molecule to the same group in a second molecule, either directly by a disulfide bridge or by the insertion of a linker. The ability, or lack of ability, of such dimers to form fibrils was used to estimate the distances between corresponding regions of the sequence from adjacent molecules in the fibrils.

Taken together, these complementary sets of data allowed a model to be defined that describes the molecular structure of the filaments (52). In this structure (**Figure 1e**), two segments of the N domain, corresponding to residues 25–38 and 91–106 (colored green and red in **Figure 1e**, respectively), interact with the corresponding regions in other molecules to form a "head-to-head" and "tail-to-tail" arrangement. The large central region of the sequence between these two segments (blue in **Figure 1e**) is folded in such a way that it forms only intramolecular interactions. The C-terminal region of the N domain and the proximal portion of the M domain (residues 107–157) are also structured within the fibrils, whereas the N-terminal region (residues 1–20) and the distal end of the M domain (residues 158–250) appear to be structurally heterogeneous and solvent exposed (dashed lines in **Figure 1e**).

Finally, although detailed structural models have not yet been proposed, much has been learned about the characteristics of other types of fibrils through similar approaches. This has led, for example, to the identification of regions of the polypeptide chain that are associated with an ordered structure in α-synuclein and tau fibrils using SDSL-EPR (56, 57). It was also possible to determine the most structured regions in α-synuclein as well as in both straight and curly fibrils from β2-microglobulin using hydrogen-deuterium exchange (46, 58, 58a), limited proteolysis (59, 59a), and SSNMR (59b). In addition, from X-ray fiber diffraction studies a cylindrical β-sheet model for fibrils from a poly-Gln peptide and the exon-1 peptide of huntingtin has been proposed (60). The polyglutamine fibrils are of particular interest because of the possibility that the additional array of hydrogen-bonding interactions involving the side chains results in a structure significantly different from that of the classical amyloid fibrils. Evidence that this situation can arise comes from the absence of the 10 Å reflection in the X-ray fiber diffraction patterns of these systems.

Similarities and Differences in Fibrillar Structures from Various Systems

Comparison of the information about the structural properties of various fibrillar systems, discussed in the previous three paragraphs, allows us to draw a number of tentative conclusions about their similarities and differences. Different fibrils clearly have many properties in common, including the canonical cross-β structure and the frequent presence of repetitive hydrophobic or polar interactions along the fibrillar axis. The ubiquitous presence of a cross-β structure strongly supports the view that the physicochemical properties of the polypeptide chain are the major determinants of the fibrillar structure in each case. Moreover, several of the proposed structures, despite very different sequences of their component polypeptides, suggest that the core region is composed of two to four sheets that interact closely with each other. An interesting feature of these sheets is that they appear to be much less twisted than expected from the analysis of the short arrays of β-strands that form β-sheets in globular protein structures. This feature was first proposed from cryo-EM and has been supported by Fourier transform infrared (FTIR) analyses (48, 61).

Nevertheless, it is clear that there are significant differences in detail attributable to the influence of the side chains on the structures adopted by the various systems. These appear to include the lengths of the β-strands and whether they are arranged in a parallel or antiparallel arrangement within each sheet; the lengths and conformational properties of the loops, turns, and other regions that are not included within the core structure; and the number of β-sheets in the protofilament. It is clear that the fraction of the residues of a polypeptide chain that are incorporated in the core structure can vary substantially (e.g., from all the residues of the 7-mer peptide to only about 13% of the residues in the full-length HET-s) and that the exact spacing between the β-sheets varies with factors such as the steric bulk of the side chains that are packed together in the core (50). In addition, the presence of disulfide bonds in proteins such as insulin may perturb the way in which the sheets can stack together (49). In cases such as the polyglutamine sequences, other interactions between the side chains may generate larger perturbations of the structure to generate such motifs as β-helices (60), which are also seen under similar circumstances in the structures of globular proteins.

The structure that will normally be adopted in the fibrils will be the lowest in free energy and/or the most kinetically accessible. What is clear, therefore, is that the interactions of the various side chains with each other and with solvent are crucial in determining the variations in the fibrillar architecture even though the main-chain interactions determine the overall framework within which these variations can occur. In other words, the interactions and conditions (see below) involving the side chains in a given sequence can tip the balance between the alternative "variations on a common theme" arrangements of a polypeptide "polymer" chain in its fibrillar structure. Such a situation contrasts with that pertaining to the native structures of the highly selected protein molecules, which are able to fold to unique structures that are significantly more stable for a given sequence than any alternatives.

FTIR: Fourier transform infrared

The Polymorphism of Amyloid Fibrils

Even before the molecular structures of amyloid fibrils began to emerge, it was clear that significant morphological variation can exist between different fibrils formed from the same peptide or protein (12, 48, 49, 54). Evidence is now accumulating that such variations in morphology is linked to heterogeneity in molecular structure, i.e., in the structural positioning of the polypeptide chains within the fibrils. One example of such heterogeneity involves the peptide hormone glucagon, wherein fibrils formed at different temperatures (25°C or 50°C) are morphologically distinct; measurements of CD and FTIR spectroscopy reveal differences in the secondary structure adopted by the constituent peptide molecules (14). A particularly important study in this regard addresses the origin of the marked differences in the morphology of $A\beta_{1-40}$ fibrils that can be observed in TEM studies of samples prepared under agitation or quiescent conditions; differences in the SSNMR spectra recorded from the different preparations provide clear evidence that this polymorphism is linked to differences in molecular structure (62).

Another example of conformational variability involves fibrils formed from the yeast prion protein Ure2p, where two independent studies came to somewhat different conclusions about the fibril structure. Both studies find that the globular C-terminal domain maintains a largely native-like structure. However, in one case, it appears that the fibrils possess a cross-β core involving only the N-terminal domains, each arranged in a serpentine fashion and forming a series of consecutive strands and loops (63–65). The parallel and in register stacking of serpentines from different molecules then forms the cross-β core with the C-terminal globular units decorating it (64). In the other study, the C- and N-terminal domains of the protein appear to interact with each other, and these fibrils do not have the characteristic 4.7-Å reflection typical of a cross-β structure (17, 66, 67). These apparently conflicting reports are likely to reflect structural differences in the fibrils, probably caused by the slightly different conditions used to prepare them.

Conformational polymorphism has also been found in other yeast prion proteins and is of particular significance because of the light it sheds on the existence of "strains" of mammalian prions and on the nature of the crucial barriers to infectivity that limit transmissibility between species (68). Efficiency of interspecies prion transmission decreases as the sequences of the infectious prions diverge, probably because each prion sequence can give rise to a limited number of misfolded conformations, which have low cross-seeding efficiency. However, a strain conformation of Sup35p has recently been identified that allows transmission from *S. cerevisiae* to the highly divergent *Candida albicans* (68). Similarly, mammalian PrP_{23-144} fibrils from different species vary in morphology and secondary structure, and these differences appear to be controlled by one or two residues in a critical region of the polypeptide sequence (69).

In all of these cases, preformed seeds can propagate their morphology and structure as well as overcome sequence- or condition-based structural preferences, resulting in fibrils that inherit the characteristics of the template (14, 62, 68, 69). These results show that each protein sequence can form a spectrum of structurally distinct fibrillar aggregates and that kinetic factors can dictate which of these alternatives is dominant under given circumstances. Of the many possible conformations that could be present in the amyloid core for a given protein, the specific ones that play this role will depend simply on the thermodynamic and, in many cases, the kinetic factors that are dominant under those circumstances. By contrast, natural globular proteins have been selected by evolution to fold into one specific three-dimensional structure, and the complex free-energy landscapes associated with their sequences have a single and well-defined minimum, under physiological conditions, corresponding to the native state.

MECHANISMS OF AMYLOID FIBRIL FORMATION

The full elucidation of the aggregation process of a protein requires the identification of all the conformational states and oligomeric structures adopted by the polypeptide chain during the process and the determination of the thermodynamics and kinetics of all the conformational changes that link these different species. It also implies characterizing each of the transitions in molecular detail and identifying the residues or regions of the sequence that promote the various aggregation steps. The identification and characterization of oligomers preceding the formation of well-defined fibrils is of particular interest because of an increasing awareness that these species are likely to play a critical role in the pathogenesis of protein deposition diseases.

Amyloid Formation Occurs via a Nucleated Growth Mechanism

It is widely established that amyloid fibril formation has many characteristics of a "nucleated growth" mechanism. The time course of the conversion of a peptide or protein into its fibrillar form (measured by ThT fluorescence, light scattering, or other techniques) typically includes a lag phase that is followed by a rapid exponential growth phase (70–73). The lag phase is assumed to be the time required for "nuclei" to form. Once a nucleus is formed, fibril growth is thought to proceed rapidly by further association of either monomers or oligomers with the nucleus.

Such a nucleated growth mechanism has been well studied both experimentally and theoretically in many other contexts, most notably for the process of crystallization of both large and small molecules (74). As with many other processes dependent on a nucleation step, including crystallization, addition of preformed fibrillar species to a sample of a protein under aggregation conditions ("seeding") causes the lag phase to be shortened and ultimately abolished when the rate of the aggregation process is no longer limited by the need for nucleation (70, 71). It has been shown also that changes in experimental conditions, or certain types of mutations, can also reduce or eliminate the length of the lag phase, again assumed to result from a situation wherein nucleation is no longer rate limiting (72, 73, 75). The absence of a lag phase, therefore, does not necessarily imply that a nucleated growth mechanism is not operating, but it may simply be that the time required for fibril growth is sufficiently slow relative to the nucleation process and that the latter is no longer the slowest step in the conversion of a soluble protein into the amyloid state. Although fibrils do not appear to a significant extent during the lag phase, it is increasingly clear that this stage in fibril formation is an important event in which a variety of oligomers form, including β-sheet-rich species that provide nuclei for the formation of mature fibrils.

The efficiency of preformed fibrils to promote further aggregation through a seeding mechanism decreases dramatically as the sequences diverge (68, 76, 76a). Using a number of immunoglobulin domains sharing different degrees of sequence identity, it was shown that coaggregation between different types of domain is not detectable if the sequence identity is lower than ∼30% to 40% (76). A bioinformatics analysis of consecutive homologous domains in large multimodular proteins shows that such domains almost exclusively have sequence identities of less than 40%, suggesting that such low sequence identities could play a crucial role in safeguarding proteins against aggregation (76).

Oligomers Preceding Amyloid Fibril Formation: Structured Protofibrils

The past decade has seen very substantial efforts directed toward identifying, isolating, and characterizing the oligomeric species that are present in solution prior to the appearance of fibrils, both because of their likely role in the mechanism of fibril formation and because of their implication as the toxic species involved in neurodegenerative disorders. We focus initially on amyloid formation by the Aβ peptide because this has been widely studied owing to its links with Alzheimer's disease. Aggregation of this peptide is preceded by the formation of a series of metastable, nonfibrillar species that can be visualized using AFM and TEM (33, 77–79). Some appear to be spherical beads of 2–5 nm in diameter. Others appear to be beaded chains with the individual beads again having a diameter of 2–5 nm and seeming to assemble in linear and curly chains. Yet others appear as annular structures, apparently formed by the circularization of the beaded chains. All of these aggregates, which have been termed protofibrils by the authors who first observed them (33, 77–79), should not be confused with the

---

Oligomers: clusters of small numbers of protein or peptide molecules without a fibrillar appearance

Protofibrils: protein aggregates of isolated or clustered spherical beads 2–5 nm in diameter with β-sheet structure

protofilaments that are the constituent units of mature fibrils. Protofibrils from Aβ can bind CR and ThT (79), contain an extensive β-sheet structure (79), and, in the form of the smaller spherical species, are made up of ∼20 molecules (80). A first exciting attempt to determine the structure of Aβ protofibrils was published using proline-scanning mutagenesis (81).

Analogous spherical and chain-like protofibrillar structures have been observed for many other systems, including α-synuclein (82), amylin (80), the immunoglobulin light chain (83), transthyretin (84), polyQ-containing proteins (80), β2-microglobulin (85), equine lysozyme (86), the *Sulfolobus solfataricus* acylphosphatase (Sso AcP) (87), and an SH3 domain (87a). These species are generally characterized by extensive β-structure and sufficient structural regularity to bind ThT and CR. The exciting finding that a specific antibody can bind to protofibrillar species from different sources, but not to their corresponding monomeric or fibrillar states, suggests that such soluble amyloid oligomers have some important common structural elements (88).

Data have been reported showing that in some cases protofibrils can be on-pathway to fibrils (33, 71). In other cases, they appear to be off-pathway (85, 89). It has been reported that the transition from the protofibrillar to the fibrillar state of the peptide 109–122 of the Syrian hamster prion protein occurs concomitantly with the alignment of β-strands within sheets in which the strands are initially misaligned (89a). Such an aligment involves detachment and re-annealing of the strands, but may also occur through an internal structural reorganization within the sheets, depending on conditions (89b). Regardless of the precise role played by protofibrils in the overall process of fibril formation, the elucidation of their mechanism of formation and of their structures is extremely important, not least because these species could be the primary toxic agents involved in neurodegenerative disorders.

Oligomers Preceding Fibril and Protofibril Formation: Unstructured Aggregates

Following the isolation and characterization of protofibrils, studies based on photo-induced cross-linking of unmodified proteins (PICUP) began to identify other oligomeric species that appeared to precede their formation (90, 91). Both the 40 and 42 residue forms of Aβ have been shown to exist as soluble oligomers in rapid equilibrium with the corresponding monomeric forms. These oligomers appear to be composed of 2–4 and 5–6 molecules for $A\beta_{1-40}$ and $A\beta_{1-42}$, respectively, and CD measurements suggest that they are relatively disorganized (91). Interest in these low-molecular-weight oligomers has been particularly intense as species of this type have also been detected in the brains of Alzheimer's disease patients (92) and in the lysates and conditioned media of cultured cells expressing the amyloid β protein precursor (93, 94).

The NM region of the yeast prion Sup35p has been shown to form "structurally fluid" oligomers rapidly, and these oligomers only later convert to species with extensive β-structures that are capable of nucleating fibril formation (71). Such a conversion has been found to be facilitated by the covalent dimerization of NM molecules when residues in the "head" region of N (residues 25–38) are cross-linked (52). Moreover, if the fluid oligomers are maintained under oxidizing conditions, intermolecular disulfide bridges are found to form more easily for variants in which cysteine residues are introduced into the head region of N rather than elsewhere. These results indicate that the interaction of the head regions of two N molecules nucleates the formation of an amyloid-like structure within the aggregates (52).

Similar behavior has been observed for the aggregation of denatured yeast phosphoglycerate kinase at low pH using dynamic light scattering and far-UV CD spectroscopy (95). β-sheet structure is increasingly stabilized as the aggregates grow in size. When a critical mass is reached, the oligomers associate with each other to form short, curly protofibrils that are similar in appearance to those observed with Aβ and α-synuclein (95). Moreover, unfolding of the SH3 domain from the bovine phosphatidylinositol 3′ kinase at pH 3.6 results in the rapid formation of a broad distribution of unstructured oligomers that subsequently convert into thin, curly, ThT-binding protofibrils (87a). All these experimental results, along with computer simulations carried out using simple polyalanine peptides (96), suggest that structured protofibrillar species can form from the reorganization or assembly of small and relatively disorganized oligomers that are formed rapidly after the initiation of the aggregation process.

Aggregation of Globular Proteins Can Occur via Partial Unfolding

So far we have discussed systems that are largely unstructured prior to the aggregation process. It is generally believed that globular proteins need to unfold, at least partially, to aggregate into amyloid fibrils (21, 97, 98). Evidence supporting this hypothesis comes from a large body of experimental data. It is clear, for example, that globular proteins have an increased propensity to aggregate under conditions that promote their partial unfolding, such as high temperature, high pressure, low pH, or moderate concentrations of organic solvents (85, 99–102). In addition, for some familial forms of disease in which the proteins involved in aggregation normally adopt folded conformations (see **Table 1**), there is clear evidence that a destabilization of the native structure, resulting in an increase in the population of nonnative states, is the primary mechanism through which natural mutations mediate their pathogenicity (103–105).

A strong correlation between a decreased conformational stability of the native state and an increased propensity to aggregate into amyloid-like structures has also been shown in vitro for nondisease-associated proteins (100, 106). Remarkably, aggregation of human lysozyme and HypF-N can be initiated by a population of less than 1% of a partially folded state that is in equilibrium with the native conformation (104, 107). Conversely, the binding of ligands and other species, such as antibodies, that stabilize the native state can decrease dramatically the propensity of proteins to aggregate (108–111). Such observations have inspired an extensive search of potential pharmaceutical compounds for the treatment of the diseases associated with transthyretin through specific binding to the tetrameric native state of the protein (109).

Aggregation of Globular Proteins Can Occur via Formation of Native-Like Oligomers

Although the "conformational change hypothesis" is undoubtedly the most appropriate way to describe the formation of amyloid fibrils by many globular proteins, recent observations have suggested that in some cases the major conformational change associated with amyloid aggregation may not take place until after the initial aggregation step. Formation of amyloid fibrils by insulin at low pH, for example, is preceded by an oligomerization step in which a native-like content of α-helical structure is almost completely retained, and aggregates with a morphology reminiscent of amyloid protofibrils and with a high content of β-structure appear only later in the process (112). In addition, within a group of variants of the protein S6 from *Thermus thermophilus*, no significant correlation was found between the rate of fibril formation under conditions in which a quasi-native state was populated prior to aggregation and the unfolding rate or conformational stability (73). Similarly, the native state of the pathogenic variant of ataxin-3, the protein associated with spinocerebellar ataxia type-3, does not appear to be significantly destabilized, leading to the proposal that the pathway for fibril formation can be distinct from that of unfolding (113).

Details of the manner in which aggregation under these conditions can take place has come from studies of the aggregation of Sso AcP. These studies have shown that unfolding of the protein can be two orders of magnitude slower than the formation of amyloid protofibrils when the protein is placed under conditions in which the native state is thermodynamically more stable than the dominant partially unfolded state (87). The first event in the aggregation of Sso AcP under these conditions is the formation of oligomers that do not bind to ThT or CR and, remarkably, not only have a native-like topology but also retain enzymatic activity (114). These native-like oligomers then undergo structural reorganization to form amyloid protofibrils that have extensive β-structure, bind ThT and CR, but are not enzymatically active. The fact that protofibril formation is also faster than the rate of disaggregation of the initially formed oligomers shows that dissolution of the latter followed by renucleation cannot be the dominant process giving rise to the structural conversion.

In the case of Ure2p, a mechanism of the type observed for Sso AcP appears to give rise to a situation wherein a native-like conformation is even retained in the fibrils themselves under some conditions (17, 66, 67). The significant propensity of native or native-like structures to aggregate is not surprising if we consider that there is a multitude of conformers even in the native ensemble of a globular protein (115). Some of these conformers will be only transiently populated but could be significant for aggregation just as they are for the hydrogen exchange of their main-chain amide groups.

Finally in this section, despite their apparent differences, there are in fact substantial similarities between the fundamental mechanism of aggregation described here for folded proteins and that of natively unfolded systems, such as Aβ and Sup35p NM. In both cases, the polypeptide molecules assemble first into species that can have characteristics far from those of the final aggregates but similar to those of the precursor structures, whether natively unfolded or natively folded. The initial aggregates then transform into species that are not yet fibrillar in their morphologies but have other properties characteristic of amyloid-like structures, notably β-sheet structure and binding to CR and ThT. Clearly, fully or partially unfolded states of globular proteins are generally more susceptible to aggregation than the native states. Nevertheless, in some situations, particularly those close to physiological, the much higher populations of the latter can result in their playing an important role in initiating an aggregation process that could be significant on the very slow timescales of the amyloid disorders.

A Multitude of Conformational States Is Accessible to Polypeptide Chains

The differing features of the aggregation processes, described in the previous paragraphs, reveal that polypeptide chains can adopt a multitude of conformational states and interconvert between them on a wide range of timescales. The network of equilibria, which link some of the most important of such states both inside and outside the cell, is schematically illustrated in **Figure 2**. Following biosynthesis on a ribosome, a polypeptide chain is initially unfolded. It can then populate a wide distribution of conformations, each of which contains little persistent structure, as in the case of natively unfolded proteins, or fold to a unique compact structure, often through one or more partly folded intermediates. In such a conformational state, the protein can remain as a monomer or associate to form oligomers or higher aggregates, some of which are functional with characteristics far from those of amyloid structures, such as in actin, myosin, and microtubules. Sooner or later, the vast majority of proteins will be degraded, usually under very carefully

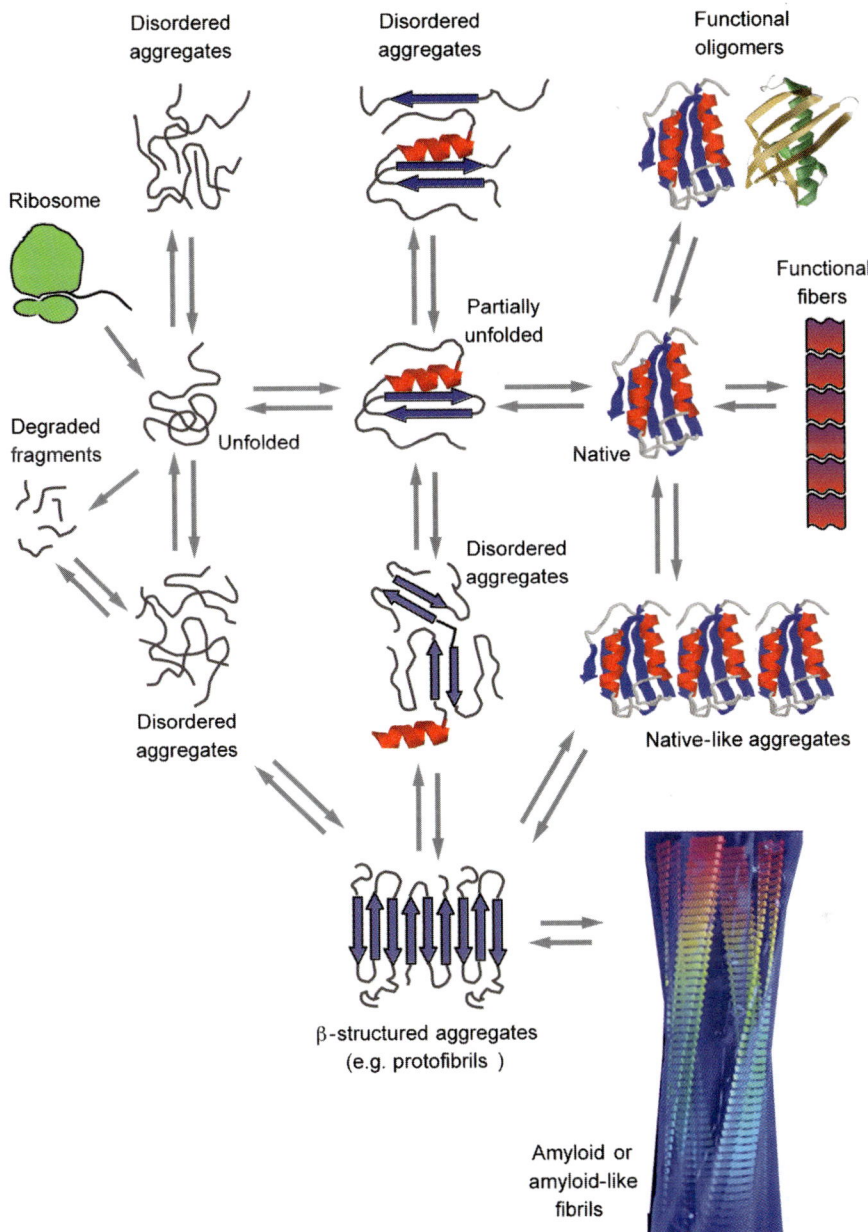

Figure 2

A schematic representation of some of the many conformational states that can be adopted by polypeptide chains and of the means by which they can be interconverted. The transition from β-structured aggregates to amyloid fibrils can occur by addition of either monomers or protofibrils (depending on protein) to preformed β-aggregates. All of these different conformational states and their interconversions are carefully regulated in the biological environment, much as enzymes regulate all the chemistry in cells, by using machinery such as molecular chaperones, degradatory systems, and quality control processes. Many of the various states of proteins are utilized functionally by biology, including unfolded proteins and amyloid fibrils, but conformational diseases will occur when such regulatory systems fail, just as metabolic diseases occur when the regulation of chemical processes becomes impaired.

controlled conditions and as a part of normal biochemical processes, with their amino acids often being recycled.

This description of normal functional behavior, honed by millions of years of evolution, is, however, only part of the story. Fully or partially unfolded ensembles on the pathways to their functional states (or generated as the result of stress, chemical modification, or genetic mutation) are particularly vulnerable to aggregation (**Figure 2**). Peptides and proteins that are natively unfolded, as well as fragments of proteins generated by proteolysis and unable to fold in the absence of the remainder of the polypeptide chain, can also aggregate under some circumstances, for example, if their concentrations become elevated. Some of the initial amorphous aggregates simply dissociate again, but others may reorganize to form oligomers with the germ of an amyloid structure, including the spherical, chain-like, and annular amyloid protofibrils observed for many systems. In order to generate long-range order in such structures, a critical number of molecules must be present such that the favorably enthalpic terms associated with their regular stacking can most effectively offset the accompanying loss of configurational entropy.

The structured polypeptide aggregates can then sometimes grow into mature fibrils by further self-association or through the repetitive addition of monomers. Proteins that adopt a folded structure under physiological conditions can also aggregate under some circumstances. This latter type of protein can either unfold, fully or partially, and aggregate through the mechanism described above or they can oligomerize prior to such a substantial conformational change. In the latter process, a structural reorganization to give amyloid-like assemblies occurs later and may in some cases be promoted by the existence of intermolecular contacts within native-like aggregates.

Every state of a polypeptide molecule, except the unique native state of globular proteins wherein the side chains pack together in a unique manner, is a broad ensemble of often diverse conformations. It is not surprising, therefore, that even the fibrillar end products of aggregation processes are characterized by morphological and structural diversity, representing variations on a common theme. Under most conditions in living systems, misfolding and aggregation of proteins are intrinsic side effects of the conformational transitions essential to the functioning of the organism. Formation of aggregates is normally inhibited by molecular chaperones and degradation processes as well as being disfavored by the amino acid sequences that are carefully selected by evolution to inhibit aggregation. But under some circumstances, as we discuss below, these aggregation processes can escape from the host of natural defenses and then give rise to pathogenic behavior.

THE INFLUENCE OF SEQUENCE ON AMYLOID FORMATION

We have stressed that amyloid formation results primarily from the properties of the polypeptide chain that are common to all peptides and proteins. We have seen, however, that the sequence influences the relative stabilities of all the conformational states accessible to a given molecule, most notably the native state, and will thereby contribute to the susceptibility of a given polypeptide chain to convert into amyloid fibrils. Moreover, it is clear that polypeptide chains with different sequences can form amyloid fibrils at very different rates, even when these processes occur from fully or partially unfolded states. We start the exploration of this topic with a description of the determinants of the aggregation of those unfolded polypeptide chains that can broadly be described as unstructured, i.e., having no significant elements of persistent or cooperative structure. By considering these systems, we can examine how the properties of the sequence influence its intrinsic aggregation behavior rather than affect the stability of a given protein fold.

Hydrophobicity, Charge, and Secondary Structure Propensities Strongly Influence Amyloid Formation

One important determinant of the aggregation of an unfolded polypeptide chain is the hydrophobicity of the side chains. Amino acid substitutions within regions of the sequence that play a crucial role (e.g., if they are in the region that nucleates aggregation) in the behavior of the whole sequence can reduce (or increase) the aggregation propensity of a sequence when they decrease (or increase) the hydrophobicity at the site of mutation (116–118). Moreover, there is evidence that protein sequences have evolved to avoid clusters of hydrophobic residues; for example, groups of three or more consecutive hydrophobic residues are less frequent in natural protein sequences than would be expected in the absence of evolutionary selection (119).

Another property likely to be a key factor in protein aggregation is charge, as a high net charge either globally or locally may hinder self-association (120, 121). For example, the effects of single amino acid substitutions were investigated on the propensity of AcP denatured in trifluoroethanol to aggregate (120). Although mutations decreasing the positive net charge of the protein resulted in an accelerated formation of β-sheet containing aggregates able to bind CR and ThT, mutations increasing the net charge resulted in the opposite effect. Further indications of the importance of charge in protein aggregation come from observations that aggregation of polypeptide chains can be facilitated by interactions with macromolecules, which exhibit a high compensatory charge (50, 122–125).

Comparison of large data sets of natively unfolded and natively folded proteins has shown that the former have a lower content of hydrophobic residues and a higher net charge than the latter (126). These properties undoubtedly contribute to maintaining the aggregation propensity of natively unfolded proteins sufficiently low to avoid the formation of aggregates under normal physiological conditions despite the fact that all, or at least the very large majority, of the side chains are accessible for intermolecular interactions.

In addition to charge and hydrophobicity, a low propensity to form α-helical structure and a high propensity to form β-sheet structure are also likely to be important factors encouraging amyloid formation (45, 50, 101, 117, 127, 128). Patterns of alternating hydrophilic and hydrophobic residues have been shown to be less frequent in natural proteins than expected on a random basis, suggesting that evolutionary selection has reduced the probability of such sequence patterns that favor β-sheet formation (127). Furthermore, it has been suggested that the high conservation of proline residues in a fibronectin type III superfamily and of glycine residues in AcPs can be rationalized on the grounds that such residues have a low propensity to form β-structure and hence inhibit aggregation (129, 130).

The Amino Acid Sequence Affects Fibril Structure and Aggregation Rate

The demonstration that the various physicochemical factors described in the previous paragraph are important determinants of the formation of amyloid structure by unfolded polypeptide chains has proved to be of great value in understanding the mechanism of aggregation at a molecular level. For example, changes in the rate of aggregation of unfolded AcP following a series of mutations were used to generate a phenomenological equation, based on physicochemical principles, that is able to rationalize these rates in a robust manner (131). This expression was, remarkably, found to rationalize just as well similar data for a whole series of other unstructured peptides and proteins (**Figure 3***a*). This finding also provides compelling evidence for the close similarity of the principles underlying the aggregation behavior of different polypeptide molecules.

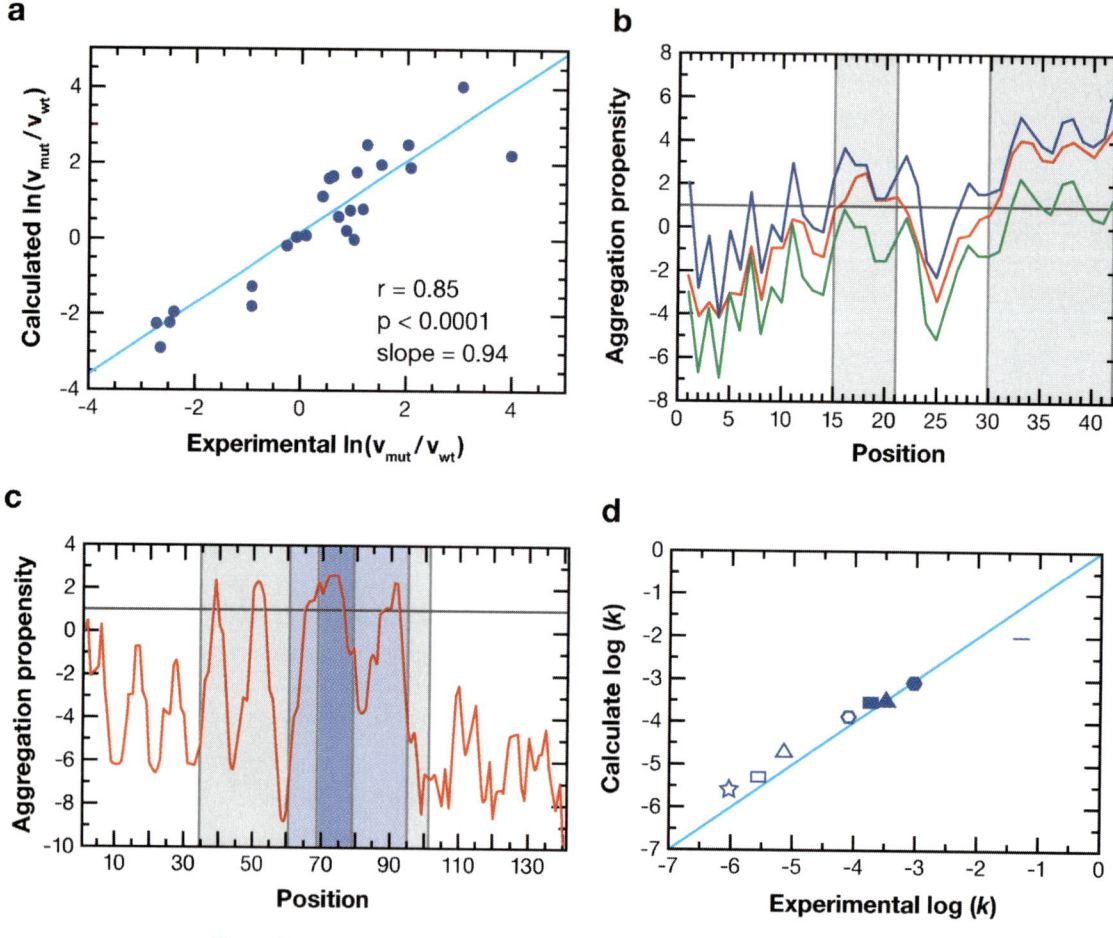

Figure 3

(*a*) Predicted versus experimental changes of aggregation rate following mutation. The mutations are in a variety of different unstructured polypeptide chains, such as Aβ, α-synuclein, amylin, and tau. Reproduced with permission (131). (*b*) Aggregation propensity profile (*red line*) for Aβ. The gray areas indicate regions of the sequence found experimentally to form and stabilize the amyloid cross-β core structure. A horizontal line at a propensity of 1 is drawn to highlight the aggregation-promoting regions that have values above this line. (*c*) Aggregation propensity profile (*red line*) for α-synuclein. The large region of the protein thought to be structured in the fibrils (*pale gray*) (56) is shown and includes all the peaks in the profile. The highly amyloidogenic NAC region (*light blue*) (178) and the 69–79 region (*dark blue*), found to be a particularly amyloidogenic segment within the NAC region (179) and containing the most prominent peak in the profile, are shown. The figures shown in panels *b* and *c* are reprinted with permission from Elsevier (134). (*d*) Predicted versus experimental aggregation rates (*k*) for a number of unstructured systems. The data points refer to Aβ$_{1-40}$ (Θ), ABri (□), denatured transthyretin (Δ), amylin (ο), AChE (■), unfolded PrP (σ), unfolded human AcP (λ), and unfolded *E. coli* HypF-N (–). The straight line has a slope of 1 and indicates the ideal correlation between theory and experiment. Reprinted with permission from Elsevier (136).

A related approach, which considers additional factors such as the changes in the number of aromatic side chains, exposed surface area, and dipole moment upon mutation, has also been shown to predict the effect of mutations on the aggregation rates of a wide variety of polypeptide chains (132). The success of these rather straightforward relationships between the rates of aggregation of unfolded polypeptide chains and simple physicochemical factors is strong support for the idea that such aggregation reflects the situation wherein a polypeptide chain behaves as a simple "polymer." Such behavior contrasts with the process of folding a globular protein for which the rates of folding are closely coupled to the specific structures of the highly evolved native states associated with individual sequences.

As described above, only a fraction of the residues of even the most highly amyloidogenic proteins are found in the core structure of the fibrils. In addition, conservative mutations have an effect on amyloid formation only when they are located in specific regions of the sequence (45, 117). Our increasing knowledge of such effects is beginning to lead to an understanding of the factors that cause specific segments of the sequence, rather than others, to form the characteristic cross-β structure. By extension of the analysis that led to the recognition of the links between aggregation propensities and the physicochemical characteristics of the constituent amino acid residues of a polypeptide chain, new algorithms have been developed to identify the regions of the sequence that are likely to promote aggregation within an unstructured polypeptide chain (133, 134). The outcome of both these approaches is a plot of aggregation propensity as a function of residue number, similar to the hydropathy profiles introduced to predict the regions of sequences that span the lipid bilayer in membrane proteins (135). The success of this type of approach is illustrated particularly well by the very good agreement between the regions of the sequence predicted to promote the aggregation of the Aβ peptide and α-synuclein and the regions found experimentally to form and stabilize the fibril core and/or to play a primary role in fibril formation (**Figure 3***b,c*) (134).

In a similar type of approach, multiple regression analysis has generated an equation that includes in the aggregation predictions the effects of extrinsic factors, such as the concentration of protein as well as the pH and ionic strength of the solution in which it is located, in addition to the intrinsic factors associated with the amino acid sequence (136). This equation reproduces the experimentally determined aggregation rates, which span five orders of magnitude, from the unstructured states of a set of nonhomologous protein sequences (**Figure 3***d*). In an exciting development, two reports have appeared recently that point to the validity of these concepts for in vivo situations, at least in bacteria (137, 138). The expenditure of effort and ingenuity in devising new methods to make quantitative analyses of such aggregation behavior in living organisms is therefore likely to bring rich rewards.

Unfolded Regions Play Critical Roles in Promoting the Aggregation of Partially Folded States

Although the key regions of the sequence that promote fibril formation by an unfolded peptide or protein can now be broadly identified using relatively simple physicochemical parameters, the aggregation of polypeptides that contain significant levels of persistent secondary structure and long-range interactions will be influenced by additional factors. For example, the yeast prion proteins Sup35p, Ure2p, and HET-s all have unstructured and globular domains in their soluble states. In each of these three cases, the region forming the cross-β core and responsible for the prion properties has been found to be the domain that is unstructured in the soluble form of the protein (63, 139–141).

Further insights into this issue have again come from studies of proteins other than those found to form functional or pathogenic amyloid structures in vivo. An approach using limited proteolysis has shown that in the partially unfolded state adopted by AcP in the presence of moderate concentrations of trifluoroethanol, the regions of the sequence found to promote amyloid aggregation are flexible and/or solvent exposed in addition to having an intrinsically high propensity to aggregate (142). Other regions that are not involved in aggregation despite having high propensities to aggregate were found to be at least partially buried in residual structure, whereas other solvent-exposed regions not involved in the aggregation process possess a low propensity to form amyloid fibrils. Similarly, the ease with which apomyoglobin converts to fibrils under different solution conditions correlates with the degree of denaturation, suggesting that fibrils assemble by association of unfolded polypeptide segments rather than by the docking of preformed structured elements (143).

Even α-synuclein, the protein associated with Parkinson's disease and assumed to have no significant structural preferences, has recently been shown to possess some significant long-range interactions between the negatively charged C-terminal region and the central amyloidogenic NAC region (144, 145). Structural perturbations that destabilize the interactions between these two portions of the protein molecule appear to increase the exposure of the amyloidogenic NAC region. Such perturbations include the presence of positively charged ions able to interact with the C terminus, a decrease in pH that reduces the net charge of the C-terminal region, and deletion of the C terminus; all result in a more rapid aggregation reaction (125, 146). Although it is clear that partial neutralization of the negative charge of α-synuclein will stimulate aggregation on a purely electrostatic argument, pairs of variants with a similar net charge but opposite signs (for example +3 and −3) aggregate more rapidly when the NAC region is unprotected (146).

Variations in Fibrillar Structure Can Be Reconciled by Common Determinants of the Aggregation Process

Our ability to rationalize, and particularly to predict, important features of the process of amyloid assembly emphasizes in a dramatic manner that common traits are dominant in the aggregation behavior of different peptides and proteins. Although the structural analysis of fibrils at the level of specific residues (described above) highlights differences in the details of the manner in which individual molecules are incorporated into the fibrils, the fact remains that the generic cross-β structure and the frequent presence of stabilizing rows of hydrophobic interactions that run along the fibril axis (apart from important exceptions such as the fibrillar species associated with polyQ traits where additional side-chain hydrogen-bonding interactions are undoubtedly important) indicate the presence of common features in the aggregation of polypeptide chains. This commonality explains our ability to predict, often with a high degree of success, the regions involved in the formation of the amyloid core and the effect of mutations in this process.

Unlike the extreme dependence of the evolved native fold on protein sequence, it is unlikely that a single arrangement of a given chain in the amyloid core structure provides unique stability relative to all other arrangements. As noted above, this conclusion also means that the specific regions of a sequence found in such structures can vary with solution conditions, that there can be subtle differences in the manner in which a given polypeptide sequence is arranged in a cross-β core structure even under essentially identical conditions, and that the details of the resulting structures may be determined by kinetic rather than thermodynamic factors. This lack of a single unique structure, coupled with the extremely high degree of repetitive order within individual fibrils, may be the origin of the strain phenomena observed in both yeast and mammalian prions. Another important facet of this topic is that chemical modifications, for example those induced by physiologically formed metabolites (147, 148), or interactions with small molecules or metal ions (149) may play a much more important role in the aggregation process than might be imagined, e.g., by perturbing the thermodynamics of kinetics sufficiently to alter the details of the resulting amyloid structure.

THE PATHOGENESIS OF PROTEIN DEPOSITION DISEASES

The presence of highly organized and stable fibrillar deposits in the organs of patients suffering from protein deposition diseases led initially to the reasonable postulate that this material is the causative agent of the various disorders. This view was later reinforced by a number of observations; for example, amyloid fibrils formed from the Aβ peptide were found to be toxic to cultured neuronal cells (150, 151) and to cause both membrane depolarization and alterations in the frequency of their action potentials (152). Moreover, Aβ fibrils were shown to cause neuronal loss and microglial activation when injected into the cerebral cortex of aged rhesus monkeys (153). However, more recent findings have raised the possibility that precursors to amyloid fibrils, such as low-molecular-weight oligomers and/or structured protofibrils, are the real pathogenic species, at least in neuropathic diseases. Here we describe some of the most compelling evidence supporting this view, starting again from the well documented Aβ case.

The Search Is on for the Causative Agents of Protein Aggregation Diseases

The severity of cognitive impairment in Alzheimer's disease correlates with the levels of low-molecular-weight species of Aβ, including small oligomers, rather than with the amyloid burden (154–156). In addition, transgenic mice show deficits in cognitive impairment, cell function, and synaptic plasticity well before the accumulation of significant quantities of amyloid plaques (157, 158). Similarly, phenotypic changes reminiscent of Alzheimer's disease precede amyloid plaque formation, or occur in their absence, in transgenic *Drosophila* expressing Aβ_{1-42} and Aβ_{1-40} (159, 160).

Further evidence comes from the finding that a single injection of a monoclonal anti-Aβ antibody does not reduce amyloid deposits in the brains of transgenic mice expressing Aβ_{1-42}, but it does reverse the associated memory loss, perhaps as a result of enhanced peripheral clearance and/or sequestration of soluble forms of the Aβ peptide (161). Genetic evidence also supports the theory that the precursor aggregates, as opposed to mature fibrils, are the pathogenic species: The aggressive "Arctic" (E693G) mutation of the amyloid β precursor protein, associated with a heritable early-onset manifestation of Alzheimer's disease, has been found in vitro to enhance protofibril, but not fibril, formation (162).

A similar scenario concerning the toxicity of early aggregates also holds for Parkinson's disease, a neurodegenerative condition associated with the formation of intracellular fibrillar deposits, notably Lewy bodies, in the dopaminergic neurons of the *substantia nigra*. In this disease, those dopaminergic neurons that survive, whether or not they contain Lewy bodies, show no quantifiable differences in viability (163, 164). Furthermore, mutations associated with juvenile Parkinson's disease or early-onset forms of Parkinsonism give rise to early neuronal degeneration in the absence of the accumulation of Lewy bodies (165). Overexpression of α-synuclein in transgenic flies or rats does not result in neuronal loss concomitant with the formation of detectable intracellular deposits (166, 167). By contrast, transgenic mice with nonfibrillar deposits of α-synuclein in various regions of the brain are characterized by substantial motor deficiencies and losses of dopaminergic neurons (168).

It is increasingly evident that prefibrillar aggregates from peptides and proteins other than Aβ and α-synuclein can either be toxic to cells or perturb their function. Early, nonfibrillar aggregates of transthyretin have been found toxic to neuronal cells under conditions in which the native tetramer and the mature fibrils are not (169). Consistent with this finding, symptoms of familial amyloid polyneuropathy appear when transthyretin is deposited in an aggregated but nonfibrillar form that does not stain with CR (169). Reixach and coworkers (170) have found that such toxicity originates from low-molecular-weight oligomers of transthyretin of up to ~100 kDa in size.

The likelihood that such behavior is much more general is suggested by the finding that prefibrillar forms of the nondisease-related HypF-N from *E. coli*, the SH3 domain from bovine phosphatidylinositol 3' kinase, lysozyme from horse, and apomyoglobin from sperm whale are also highly toxic to cultured fibroblasts and neurons, whereas the monomeric native states and the amyloid-like fibrils (all formed in vitro) displayed very little, if any, toxicity (171, 171a, 171b). Interestingly in this context, the most highly infective form of the mammalian prion protein has been identified as an oligomer of about 20 molecules, indicating that such small aggregates are the most effective initiators of transmissible spongiform encephalopathies (172).

The Toxicity of Prefibrillar Aggregates Results from their Misfolded Nature

The reason why prefibrillar aggregates are toxic to cells, and hence appear to be the most likely culprits for the origins of at least some of the protein deposition diseases, is now at the front line of research in this field. A wide variety of biochemical, cytological, and physiological perturbations has been identified following the exposure of neurons to such species, both in vivo and in vitro. A detailed description of all of the reported effects is beyond the scope of the present review, and indeed, it is still too early to draw definitive conclusions about the similarities or differences of the effects of particular types of aggregates in different diseases.

Despite differences in the specific mechanisms of pathogenic behavior giving rise to distinct diseases, it is clear that the conversion of a protein from its soluble state into oligomeric forms will invariably generate a wide distribution of nonnative species, the populations of which will vary with sequence, time, and conditions. It seems likely that all of these inherently "misfolded" species will be toxic to some degree because they will inevitably expose on their surfaces an array of groups that are normally buried in globular proteins or dispersed in highly unfolded peptides or proteins. Small aggregates have a higher proportion of residues on their surfaces than larger aggregates, including mature amyloid fibrils, and therefore are likely in general to have a higher relative toxicity. In the crowded and highly organized environment of a living organism, the nonnative character of misfolded oligomers is particularly likely to trigger aberrant events resulting from their inappropriate interactions with cellular components, such as membranes, small metabolites, proteins, or other macromolecules. Such events will, in some situations, lead to the malfunctioning of crucial aspects of the cellular machinery, whether it is axonal transport, oxidative stress, ion balance, sequestration of essential proteins, or a combination of disparate factors, ultimately leading to apoptosis or other forms of cell death.

Although the natural defenses against misfolded proteins will act to sequester and neutralize such species, and/or inhibit their formation, it is inevitable that these mechanisms will sometimes be overwhelmed (98, 173). Such situations include mutations that dramatically increase aggregation rates, as in familial diseases; ingestion of preformed aggregates that are able to seed more extensive aggregation, as in prion diseases; or the age-related decline of chaperone and ubiquitin/proteasome responses, as in sporadic forms of diseases. Oligomer-mediated cytotoxicity is a key issue in neuropathic protein deposition diseases, although the question arises as to whether a similar mechanism is central in the pathogenesis of nonneuropathic diseases. Systemic amyloidoses are often associated with accumulation of large quantities (even kilograms in some cases) of amyloid deposits in the affected tissues and organs (174). Undoubtedly, the impairment and disruption of tissue architecture, caused by these deposits in vital organs, are major features of these diseases and could well be the most important factors in the pathogenesis of at least some of these nonneuropathic degenerative conditions (174). Patients can have mechanical problems in carrying out even routine everyday tasks. Examples include difficulties in swallowing when amyloid accumulation occurs in the tongue and in moving because of extreme pain when amyloid accumulation occurs in joints. However, suggestions that early oligomeric species could have a more important role than fibril accumulation in the pathogenesis of nonneuropathic amyloidoses have been put forward (169, 174a). The elucidation of the mechanism of tissue damage by amyloid fibril proteins is undoubtedly an important issue in therapeutic approaches, although the optimum strategy must be to prevent aggregation or even production of the amyloidogenic protein before it can generate any potential damaging deposits.

PERSPECTIVES

Despite the complexity of the protein aggregation process, the findings described above show that dramatic progress in its elucidation has been made in recent years. This progress relates particularly to our understanding of the nature and significance of amyloid formation and to how this process relates to the normal and aberrant behavior of living organisms. Increasingly sophisticated techniques are now being applied to elucidate the "amyloid phenomenon" in ever greater detail. Of special significance is the manner in which a wide variety of ideas from across the breadth of the biological, physical, and medical sciences is being brought together to probe important unifying principles. Much, of course, still remains to be discovered, but we are personally optimistic that the investigation of an increasing number of proteins both in vitro and in vivo will shed new light on the relationships between protein folding and misfolding as well as on the manner in which the multitude of different states accessible to proteins are regulated and interact with each other and with other cellular components. In addition, even our present understanding of the mechanism of amyloid formation is leading to more reliable methods of early diagnosis and to more rational therapeutic strategies that are either in clinical trials or approaching such trials (175, 176). Thus, despite the rapidity with which diseases of the type discussed here are increasingly afflicting the human populations of the modern world, there are grounds for optimism that present progress in understanding their nature and origins will lead, in the not too distant future, to the beginnings of widely applicable and effective means to combat their spread and their debilitating consequences.

SUMMARY POINTS

1. A variety of human diseases is now thought to be associated with the formation of highly organized and generally intractable thread-like aggregates termed amyloid or amyloid-like fibrils.

2. Living organisms can take advantage of the inherent ability of proteins to form such structures to generate novel and diverse biological functions.

3. Dramatic advances have recently been made toward the elucidation of the structures of amyloid fibrils at a molecular level.

4. Amyloid fibril formation is preceded by formation of a wide range of aggregates such as unstructured oligomers and structured protofibrils.

5. We are now able to rationalize some of the issues regarding the molecular mechanism of amyloid formation, e.g., identify the regions of the sequence that form and stabilize the fibril core and/or play a primary role in fibril formation.

6. At least in some cases, prefibrillar aggregates, rather than the mature fibrils into which they convert, are the likely origins of pathological behavior. Despite obvious differences in detail, the pathogenic nature of these species lies in the exposure of groups that are normally buried in a folded protein or dispersed in an unfolded ensemble.

FUTURE ISSUES TO BE RESOLVED

1. Although considerable progress has been made in the elucidation of amyloid fibril properties at a molecular level, very little is yet known about the structure of the amyloid protofibrils and unstructured aggregates that precede their formation and are likely to play a key role in the pathogenesis of protein deposition diseases.

2. Present research has been remarkably successful in providing a framework for understanding the fundamental nature of protein aggregation. The challenge now is to explore in more detail the links between these largely structural principles and the cellular and animal environments in which aggregation takes place.

3. The precise origin of the pathogenic nature of the amyloid deposits and their precursors remains elusive in each pathological condition associated with formation of these species.

4. The rational design of successful therapeutic strategies requires further characterization of the processes of amyloid formation occuring in vivo and of the interaction of the resulting aggregates with the various components of living organisms.

LITERATURE CITED

1. Uversky VN, Fink AL. 2006. *Protein Misfolding, Aggregation and Conformational Diseases*, Vols. 1, 2. Kluwer Academic/Plenum
2. Amaral MD. 2004. *J. Mol. Neurosci.* 23:41–48
3. Lomas DA, Carrell RW. 2002. *Nat. Rev. Genet.* 3:759–68
4. Westermark P, Benson MD, Buxbaum JN, Cohen AS, Frangione B, et al. 2005. *Amyloid* 12:1–4
5. Lundmark K, Westermark GT, Nystrom S, Murphy CL, Solomon A, Westermark P. 2002. *Proc. Natl. Acad. Sci. USA* 99:6979–84
6. Lundmark K, Westermark GT, Olsen A, Westermark P. 2005. *Proc. Natl. Acad. Sci. USA* 102:6098–102
7. Legname G, Baskakov IV, Nguyen HO, Riesner D, Cohen FE, et al. 2004. *Science* 305:673–76
8. Hirschfield GM, Hawkins PN. 2003. *Int. J. Biochem. Cell. Biol.* 35:1608–13
9. Alexandrescu AT. 2005. *Protein Sci.* 14:1–12

10. Serpell LC, Sunde M, Benson MD, Tennent GA, Pepys MB, Fraser PE. 2000. *J. Mol. Biol.* 300:1033–39
11. Sunde M, Blake C. 1997. *Adv. Protein Chem.* 50:123–59
12. Bauer HH, Aebi U, Haner M, Hermann R, Muller M, Merkle HP. 1995. *J. Struct. Biol.* 115:1–15
13. Saiki M, Honda S, Kawasaki K, Zhou D, Kaito A, et al. 2005. *J. Mol. Biol.* 348:983–98
14. Pedersen JS, Dikov D, Flink JL, Hjuler HA, Christiansen G, Otzen D. 2005. *J. Mol. Biol.* 355:501–23
15. Nilsson MR. 2004. *Methods* 34:151–60
16. Khurana R, Uversky VN, Nielsen L, Fink AL. 2001. *J. Biol. Chem.* 276:22715–21
17. Bousset L, Redeker V, Decottignies P, Dubois S, Le Marechal P, Melki R. 2004. *Biochemistry* 43:5022–32
18. Jin LW, Claborn KA, Kurimoto M, Geday MA, Maezawa I, et al. 2003. *Proc. Natl. Acad. Sci. USA* 100:15294–98
19. Krebs MRH, MacPhee CE, Miller AF, Dunlop LE, Dobson CM, Donald AM. 2004. *Proc. Natl. Acad. Sci. USA* 101:14420–24
20. Stefani M, Dobson CM. 2003. *J. Mol. Med.* 81:678–99
21. Uversky VN, Fink AL. 2004. *Biochim. Biophys. Acta* 1698:131–53
22. Chapman MR, Robinson LS, Pinkner JS, Roth R, Heuser J, et al. 2002. *Science* 295:851–55
23. Claessen D, Rink R, de Jong W, Siebring J, de Vreugd P, et al. 2003. *Genes Dev.* 17:1714–26
23a. Mackay JP, Matthews JM, Winefield RD, Mackay LG, Haverkamp RG, et al. 2001. *Structure* 9:83–91
23b. Iconomidou VA, Vriend G, Hamodrakas SJ. 2000. *FEBS Lett.* 479:141–45
23c. Kenney JM, Knight D, Wise MJ, Vollrath F. 2002. *Eur. J. Biochem.* 269:4159–63
24. Berson JF, Theos AC, Harper DC, Tenza D, Raposo G, Marks MS. 2003. *J. Cell Biol.* 161:521–33
25. Chien P, Weissman JS, DePace AH. 2004. *Annu. Rev. Biochem.* 73:617–56
26. Eaglestone SS, Cox BS, Tuite MF. 1999. *EMBO J.* 18:1974–81
27. True HL, Lindquist SL. 2000. *Nature* 407:477–83
28. True HL, Berlin I, Lindquist SL. 2004. *Nature* 431:184–87
28a. Giaever G, Chu AM, Ni L, Connelly C, Riles L, et al. 2002. *Nature* 418:387–91
29. Nakayashiki T, Kurtzman CP, Edskes HK, Wickner RB. 2005. *Proc. Natl. Acad. Sci. USA* 102:10575–80
30. Si K, Lindquist S, Kandel ER. 2003. *Cell* 115:879–91
31. Coustou V, Deleu C, Saupe S, Begueret J. 1997. *Proc. Natl. Acad. Sci. USA* 94:9773–78
32. Saupe SJ. 2000. *Microbiol. Mol. Biol. Rev.* 64:489–502
33. Harper JD, Lieber CM, Lansbury PT Jr. 1997. *Chem. Biol.* 4:951–59
34. Petkova AT, Ishii Y, Balbach JJ, Antzutkin ON, Leapman RD, et al. 2002. *Proc. Natl. Acad. Sci. USA* 99:16742–47
35. Jaroniec CP, MacPhee CE, Astrof NS, Dobson CM, Griffin RG. 2002. *Proc. Natl. Acad. Sci. USA* 99:16748–53
36. Ritter C, Maddelein ML, Siemer AB, Luhrs T, Ernst M, et al. 2005. *Nature* 435:844–48
37. Makin OS, Atkins E, Sikorski P, Johansson J, Serpell LC. 2005. *Proc. Natl. Acad. Sci USA* 102:315–20
38. Nelson R, Sawaya MR, Balbirnie M, Madsen AO, Riekel C, et al. 2005. *Nature* 435:773–78

39. Antzutkin ON, Balbach JJ, Leapman RD, Rizzo NW, Reed J, Tycko R. 2000. *Proc. Natl. Acad. Sci. USA* 97:13045–50
40. Balbach JJ, Petkova AT, Oyler NA, Antzutkin ON, Gordon DJ, et al. 2002. *Biophys. J.* 83:1205–16
41. Goldsbury CS, Wirtz S, Muller SA, Sunderji S, Wicki P, et al. 2000. *J. Struct. Biol.* 130:217–31
42. Torok M, Milton S, Kayed R, Wu P, McIntire T, et al. 2000. *J. Biol. Chem.* 277:40810–15
43. Kheterpal I, Zhou S, Cook KD, Wetzel R. 2000. *Proc. Natl. Acad. Sci. USA* 97:13597–601
44. Kheterpal I, Williams A, Murphy C, Bledsoe B, Wetzel R. 2001. *Biochemistry* 40:11757–67
45. Williams AD, Portelius E, Kheterpal I, Guo JT, Cook KD, et al. 2004. *J. Mol. Biol.* 335:833–42
46. Hoshino M, Katou H, Hagihara Y, Hasegawa K, Naiki H, Goto Y. 2002. *Nat. Struct. Biol.* 9:332–36
47. Jaroniec CP, MacPhee CE, Bajaj VS, McMahon MT, Dobson CM, Griffin RG. 2004. *Proc. Natl. Acad. Sci. USA* 101:711–16
48. Jimenez JL, Guijarro JI, Orlova E, Zurdo J, Dobson CM, et al. 1999. *EMBO J.* 18:815–21
49. Jimenez JL, Nettleton EJ, Bouchard M, Robinson CV, Dobson CM, Saibil HR. 2002. *Proc. Natl. Acad. Sci. USA* 99:9196–201
50. Fandrich M, Dobson CM. 2002. *EMBO J.* 21:5682–90
51. Kajava AV, Aebi U, Steven AC. 2005. *J. Mol. Biol.* 348:247–52
52. Krishnan R, Lindquist SL. 2005. *Nature* 435:765–72
53. Makin O, Serpell LC. 2004. *J. Mol. Biol.* 335:1279–88
54. Goldsbury CS, Cooper GJ, Goldie KN, Muller SA, Saafi EL, et al. 1997. *J. Struct. Biol.* 119:17–27
55. Jayasinghe SA, Langen R. 2004. *J. Biol. Chem.* 279:48420–25
56. Der-Sarkissian A, Jao CC, Chen J, Langen R. 2003. *J. Biol. Chem.* 278:37530–35
57. Margittai M, Langen R. 2004. *Proc. Natl. Acad. Sci. USA* 101:10278–83
58. Yamaguchi K, Katou H, Hoshino M, Hasegawa K, Naiki H, Goto Y. 2004. *J. Mol. Biol.* 338:559–71
58a. Del Mar C, Greenbaum EA, Mayne L, Englander SW, Woods VL Jr. 2005. *Proc. Natl. Acad. Sci. USA* 102:15477–82
59. Monti M, Principe S, Giorgetti S, Mangione P, Merlini G, et al. 2002. *Protein Sci.* 11:2362–69
59a. Miake H, Mizusawa H, Iwatsubo T, Hasegawa M. 2002. *J. Biol. Chem.* 2002. 277:19213–19
59b. Heise H, Hoyer W, Becker S, Andronesi OC, Riedel D, Baldus M. 2005. *Proc. Natl. Acad. Sci. USA* 102:15871–76
60. Perutz MF, Finch JT, Berriman J, Lesk A. 2002. *Proc. Natl. Acad. Sci. USA* 99:5591–95
61. Zandomeneghi G, Krebs MR, McCammon MG, Fandrich M. 2004. *Protein Sci.* 13:3314–21
62. Petkova AT, Leapman RD, Guo Z, Yau WM, Mattson MP, Tycko R. 2005. *Science* 307:262–65
63. Baxa U, Taylor KL, Wall JS, Simon MN, Cheng NQ, et al. 2003. *J. Biol. Chem.* 278:43717–27
64. Kajava AV, Baxa U, Wickner RB, Steven AC. 2004. *Proc. Natl. Acad. Sci. USA* 101:7885–90
65. Baxa U, Cheng NQ, Winkler DC, Chiu TK, Davies DR, et al. 2005. *J. Struct. Biol.* 150:170–79

66. Bousset L, Thomson NH, Radford SE, Melki R. 2002. *EMBO J.* 17:2903–11
67. Bousset L, Briki F, Doucet J, Melki R. 2003. *J. Struct. Biol.* 141:132–42
68. Tanaka M, Chien P, Yonekura K, Weissman JS. 2005. *Cell* 121:49–62
69. Jones EM, Surewicz WK. 2005. *Cell* 121:63–72
70. Naiki H, Hashimoto N, Suzuki S, Kimura H, Nakakuki K, Gejyo F. 1997. *Amyloid* 4:223–32
71. Serio TR, Cashikar AG, Kowal AS, Sawicki GJ, Moslehi JJ, et al. 2000. *Science* 289:1317–21
72. Uversky VN, Li J, Souillac P, Millett IS, Doniach S, et al. 2002. *J. Biol. Chem.* 277:11970–78
73. Pedersen JS, Christensen G, Otzen DE. 2004. *J. Mol. Biol.* 341:575–88
74. Chayen NE. 2005. *Prog. Biophys. Mol. Biol.* 88:329–37
75. Fezoui Y, Teplow DB. 2002. *J. Biol. Chem.* 277:36948–54
76. Wright CF, Teichmann SA, Clarke J, Dobson CM. 2005. *Nature* 438:878–81
76a. Krebs MR, Morozova-Roche LA, Daniel K, Robinson CV, Dobson CM. 2004. *Protein Sci.* 13:1933–38
77. Harper JD, Wong SS, Lieber CM, Lansbury PT Jr. 1997. *Chem. Biol.* 4:119–25
78. Walsh DM, Lomakin A, Benedek GB, Condron MM, Teplow DB. 1997. *J. Biol. Chem.* 272:22364–72
79. Walsh DM, Hartley DM, Kusumoto Y, Fezoui Y, Condron MM, et al. 1999. *J. Biol. Chem.* 274:25945–52
80. Kayed R, Sokolov Y, Edmonds B, McIntire TM, Milton SC, et al. 2004. *J. Biol. Chem.* 279:46363–66
81. Williams AD, Sega M, Chen ML, Kheterpal I, Geva M, et al. 2005. *Proc. Natl. Acad. Sci. USA* 102:7115–20
82. Conway KA, Harper JD, Lansbury PT Jr. 2000. *Biochemistry* 39:2552–63
83. Ionescu–Zanetti C, Khurana R, Gillespie JR, Petrick JS, Trabachino LC, et al. 1999. *Proc. Natl. Acad. Sci. USA* 96:13175–79
84. Quintas A, Vaz DC, Cardoso I, Saraiva MJM, Brito RMM. 2001. *J. Biol. Chem.* 276:27207–13
85. Gosal WS, Morten IJ, Hewitt EW, Smith DA, Thomson NH, Radford SE. 2005. *J. Mol. Biol.* 351:850–64
86. Malisauskas M, Zamotin V, Jass J, Noppe W, Dobson CM, Morozova-Roche LA. 2003. *J. Mol. Biol.* 330:879–90
87. Plakoutsi G, Taddei N, Stefani M, Chiti F. 2004. *J. Biol. Chem.* 279:14111–19
87a. Bader R, Bamford R, Zurdo J, Luisi BF, Dobson CM. 2006. *J. Mol. Biol.* 356:189–208
88. Kayed R, Head E, Thompson JL, McIntire TM, Milton SC, et al. 2003. *Science* 300:486–89
89. Morozova-Roche LA, Zamotin V, Malisauskas M, Ohman A, Chertkova R, et al. 2004. *Biochemistry* 43:9610–19
89a. Petty SA, Adalsteinsson T, Decatur SM. 2005. *Biochemistry* 44:4720–26
89b. Petty SA, Decatur SM. 2005. *Proc. Natl. Acad. Sci. USA* 102:14272–77
90. Bitan G, Lomakin A, Teplow DB. 2001. *J. Biol. Chem.* 276:35176–84
91. Bitan G, Kirkitadze MD, Lomakin A, Vollers SS, Benedek GB, Teplow DB. 2003. *Proc. Natl. Acad. Sci. USA* 100:330–35
92. Roher AE, Chaney MO, Kuo YM, Webster SD, Stine WB, et al. 1996. *J. Biol. Chem.* 271:20631–35
93. Podlisny MB, Ostaszewski BL, Squazzo SL, Koo EH, Rydell RE, et al. 1995. *J. Biol. Chem.* 270:9564–70

94. Walsh DM, Tseng BP, Rydel RE, Podlisny MB, Selkoe DJ. 2000. *Biochemistry* 39:10831–39
95. Modler AJ, Gast K, Lutsch G, Damaschun G. 2003. *J. Mol. Biol.* 325:135–48
96. Nguyen HD, Hall CK. 2004. *Proc. Natl. Acad. Sci. USA* 101:16180–85
97. Kelly JW. 1998. *Curr. Opin. Struct. Biol.* 8:101–6
98. Dobson CM. 1999. *Trends Biochem. Sci.* 24:329–32
99. Guijarro JI, Sunde M, Jones JA, Campbell ID, Dobson CM. 1998. *Proc. Natl. Acad. Sci. USA* 95:4224–28
100. Chiti F, Taddei N, Bucciantini M, White P, Ramponi G, Dobson CM. 2000. *EMBO J.* 19:1441–49
101. Villegas V, Zurdo J, Filimonov VV, Aviles FX, Dobson CM, Serrano L. 2000. *Protein Sci.* 9:1700–8
102. Ferrao-Gonzales AD, Souto SO, Silva JL, Foguel D. 2000. *Proc. Natl. Acad. Sci. USA* 97:6445–50
103. Raffen R, Dieckman LJ, Szpunar M, Wunschl C, Pokkuluri PR, et al. 1999. *Protein Sci.* 8:509–17
104. Canet D, Last AM, Tito P, Sunde M, Spencer A, et al. 2002. *Nat. Struct. Biol.* 9:308–15
105. Hammarstrom P, Jiang X, Hurshman AR, Powers ET, Kelly JW. 2002. *Proc. Natl. Acad. Sci. USA* 99:16427–32
106. Ramirez-Alvarado M, Merkel JS, Regan L. 2000. *Proc. Natl. Acad. Sci. USA* 97:8979–84
107. Marcon G, Plakoutsi G, Canale C, Relini A, Taddei N, et al. 2005. *J. Mol. Biol.* 347:323–35
108. Chiti F, Taddei N, Stefani M, Dobson CM, Ramponi G. 2001. *Protein Sci.* 10:879–86
109. Sacchettini JC, Kelly JW. 2002. *Nat. Rev. Drug Discov.* 1:267–75
110. Dumoulin M, Last AM, Desmyter A, Decanniere K, Canet D, et al. 2003. *Nature* 424:783–88
111. Ray SS, Nowak RJ, Brown RH Jr, Lansbury PT Jr. 2005. *Proc. Natl. Acad. Sci. USA* 102:3639–44
112. Bouchard M, Zurdo J, Nettleton EJ, Dobson CM, Robinson CV. 2000. *Protein Sci.* 9:960–67
113. Chow MK, Ellisdon AM, Cabrita LD, Bottomley SP. 2005. *J. Biol. Chem.* 279:47643–51
114. Plakoutsi G, Bemporad F, Calamai M, Taddei N, Dobson CM, Chiti F. 2005. *J. Mol. Biol.* 351:910–22
115. Lindorff-Larsen K, Best RB, Depristo MA, Dobson CM, Vendruscolo M. 2005. *Nature* 433:128–32
116. Otzen DE, Kristensen O, Oliveberg M. 2000. *Proc. Natl. Acad. Sci. USA* 97:9907–12
117. Chiti F, Taddei N, Baroni F, Capanni C, Stefani M, et al. 2002. *Nat. Struct. Biol.* 9:137–43
118. Wurth C, Guimard NK, Hecht MH. 2002. *J. Mol. Biol.* 319:1279–90
119. Schwartz R, Istrail S, King J. 2001. *Protein Sci.* 10:1023–31
120. Chiti F, Calamai M, Taddei N, Stefani M, Ramponi G, Dobson CM. 2002. *Proc. Natl. Acad. Sci. USA* 99:16419–26
121. Schmittschmitt JP, Scholtz JM. 2003. *Protein Sci.* 12:2374–78
122. Konno T. 2001. *Biochemistry* 40:2148–54
123. Goers J, Uversky VN, Fink AL. 2003. *Protein Sci.* 12:702–7
124. Giasson BI, Forman MS, Higuchi M, Golbe LI, Graves CL, et al. 2003. *Science* 300:636–40
125. Fernandez CO, Hoyer W, Zweckstetter M, Jares-Erijman EA, Subramaniam V, et al. 2004. *EMBO J.* 23:2039–46

126. Uversky VN. 2002. *Protein Sci.* 11:739–56
127. Broome BM, Hecht MH. 2000. *J. Mol. Biol.* 296:961–68
128. Kallberg Y, Gustafsson M, Persson B, Thyberg J, Johansson J. 2001. *J. Biol. Chem.* 276:12945–50
129. Steward A, Adhya S, Clarke J. 2002. *J. Mol. Biol.* 318:935–40
130. Parrini C, Taddei N, Ramazzotti M, Degl'Innocenti D, Ramponi G, et al. 2005. *Structure* 13:1143–51
131. Chiti F, Stefani M, Taddei N, Ramponi G, Dobson CM. 2003. *Nature* 424:805–8
132. Tartaglia GG, Cavalli A, Pellarin R, Caflisch A. 2004. *Protein Sci.* 13:1939–41
133. Fernandez-Escamilla AM, Rousseau F, Schymkowitz J, Serrano L. 2004. *Nat. Biotechnol.* 22:1302–6
134. Pawar AP, DuBay KF, Zurdo J, Chiti F, Vendruscolo M, Dobson CM. 2005. *J. Mol. Biol.* 350:379–92
135. Kyte J, Doolittle RF. 1982. *J. Mol. Biol.* 157:105–32
136. DuBay KF, Pawar AP, Chiti F, Zurdo J, Dobson CM, Vendruscolo M. 2004. *J. Mol. Biol.* 341:1317–26
137. Ignatova Z, Gierasch LM. 2004. *Proc. Natl. Acad. Sci. USA* 101:523–28
138. Calloni G, Zoffoli S, Stefani M, Dobson CM, Chiti F. 2005. *J. Biol. Chem.* 280:10607–13
139. Paushkin SV, Kushnirov VV, Smirnov VN, Ter-Avanesyan MD. 1996. *EMBO J.* 15:3127–34
140. Masison DC, Maddelein ML, Wickner RB. 1997. *Proc. Natl. Acad. Sci. USA* 94:12503–8
141. Balguerie A, Dos Reis S, Ritter C, Chaignepain S, Coulary-Salin B, et al. 2003. *EMBO J.* 22:2071–81
142. Monti M, di Bard BLG, Calloni G, Chiti F, Amoresano A, et al. 2004. *J. Mol. Biol.* 336:253–62
143. Fandrich M, Forge V, Buder K, Kittler M, Dobson CM, Diekmann S. 2003. *Proc. Natl. Acad. Sci. USA* 100:15463–68
144. Bertoncini CW, Jung YS, Fernandez CO, Hoyer W, Griesinger C, et al. 2005. *Proc. Natl. Acad. Sci. USA* 102:1430–35
145. Dedmon MM, Lindorff-Larsen K, Christodoulou J, Vendruscolo M, Dobson CM. 2005. *J. Am. Chem. Soc.* 127:476–77
146. Hoyer W, Cherny D, Subramaniam V, Jovin TM. 2004. *Biochemistry* 43:16233–42
147. Nilsson MR, Driscoll M, Raleigh DP. 2002. *Protein Sci.* 11:342–49
148. Zhang Q, Powers ET, Nieva J, Huff ME, Dendle MA. 2004. *Proc. Natl. Acad. Sci. USA* 101:4752–57
149. Bush AI. 2003. *Trends Neurosci.* 26:207–14
150. Pike CJ, Walencewicz AJ, Glabe CG, Cotman CW. 1991. *Brain Res.* 563:311–14
151. Lorenzo A, Yankner BA. 1994. *Proc. Natl. Acad. Sci. USA* 91:12243–47
152. Hartley DM, Walsh DM, Ye CP, Diehl T, Vasquez S, et al. 1999. *J. Neurosci.* 19:8876–84
153. Geula C, Wu CK, Saroff D, Lorenzo A, Yuan M, Yankner BA. 1998. *Nat. Med.* 4:827–31
154. Lue LF, Kuo YM, Roher AE, Brachova L, Shen Y, et al. 1999. *Am. J. Pathol.* 155:853–62
155. McLean CA, Cherny RA, Fraser FW, Fuller SJ, Smith MJ, et al. 1999. *Ann. Neurol.* 46:860–66
156. Wang J, Dickson DW, Trojanowski JQ, Lee VM. 1999. *Exp. Neurol.* 158:328–37
157. Moechars D, Dewachter I, Lorent K, Reverse D, Baekelandt V, et al. 1999. *J. Biol. Chem.* 274:6483–92
158. Larson J, Lynch G, Games D, Seubert P. 1999. *Brain Res.* 840:23–35
159. Iijima K, Liu HP, Chiang AS, Hearn SA, Konsolaki M, Zhong Y. 2001. *Proc. Natl. Acad. Sci. USA* 101:6623–28

160. Crowther DC, Kinghorn KJ, Miranda E, Page R, Curry JA, et al. 2005. *Neuroscience* 132:123–35
161. Dodart JC, Bales KR, Gannon KS, Greene SJ, DeMattos RB, et al. 2002. *Nat. Neurosci.* 5:452–57
162. Nilsberth C, Westlind-Danielsson A, Eckman CB, Condron MM, Axelman K, et al. 2001. *Nat. Neurosci.* 4:887–93
163. Gertz HJ, Siegers A, Kuchinke J. 1994. *Brain Res.* 637:339–41
164. Bergeron C, Petrunka C, Weyer L. 1996. *Am. J. Pathol.* 148:273–79
165. Kitada T, Asakawa S, Hattori N, Matsumine H, Yamamura Y, et al. 1998. *Nature* 392:605–8
166. Auluck PK, Chan HY, Trojanowski JQ, Lee VM, Bonini NM. 2002. *Science* 295:865–68
167. Lo Bianco C, Ridet JL, Schneider BL, Deglon N, Aebischer P. 2002. *Proc. Natl. Acad. Sci. USA* 99:10813–18
168. Masliah E, Rockenstein E, Veinbergs I, Mallory M, Hashimoto M, et al. 2000. *Science* 287:1265–69
169. Sousa MM, Cardoso I, Fernandes R, Guimaraes A, Saraiva MJ. 2001. *Am. J. Pathol.* 159:1993–2000
170. Reixach N, Deechongkit S, Jiang X, Kelly JW, Buxbaum JN. 2004. *Proc. Natl. Acad. Sci. USA* 101:2817–22
171. Bucciantini M, Giannoni E, Chiti F, Baroni F, Formigli L, et al. 2002. *Nature* 416:507–11
171a. Sirangelo I, Malmo C, Iannuzzi C, Mezzogiorno A, Bianco MR, et al. 2004. *J. Biol. Chem.* 279:13183–89
171b. Malisauskas M, Ostman J, Darinskas A, Zamotin V, Liutkevicius E, et al. 2005. *J. Biol. Chem.* 280:6269–75
172. Silveira JR, Raymond GJ, Hughson AG, Race RE, Sim VL, et al. 2005. *Nature* 437:257–61
173. Soti C, Pal C, Papp B, Csermely P. 2005. *Curr. Opin. Cell. Biol.* 17:210–15
174. Pepys MB. 2001. *Philos. Trans. R. Soc. London Ser. B* 356:203–10
174a. Merlini G, Bellotti V. 2003. *N. Engl. J. Med.* 349:583–96
175. Cohen FE, Kelly JW. 2003. *Nature* 426:905–9
176. Merlini G, Westermark P. 2004. *J. Intern. Med.* 255:159–78
177. Tycko R. 2002. *Biochemistry* 42:3151–59
178. Han HY, Weinreb PH, Lansbury PT Jr. 1995. *Chem. Biol.* 2:163–69
179. El-Agnaf OMA, Irvine GB. 2002. *Biochem. Soc. Trans.* 30:559–65

Obesity-Related Derangements in Metabolic Regulation

Deborah M. Muoio[1] and Christopher B. Newgard[1,2]

[1]Sarah W. Stedman Nutrition and Metabolism Center and Departments of Pharmacology and Cancer Biology, Medicine, and [2]Biochemistry, Duke University Medical Center, Durham, North Carolina 27704; email: muoio@duke.edu, newga002@mc.duke.edu

Key Words

diabetes, lipid metabolism, mitochondria, pancreas, pyruvate cycling, skeletal muscle

Abstract

An epidemic surge in the incidence of obesity has occurred worldwide over the past two decades. This alarming trend has been triggered by lifestyle habits that encourage overconsumption of energy-rich foods while also discouraging regular physical activity. These environmental influences create a chronic energy imbalance that leads to persistent weight gain in the form of body fat and a host of other abnormalities in metabolic homeostasis. As adiposity increases, so does the risk of developing comorbidities such as diabetes, hypertension, and cardiovascular disease. The intimate association between obesity and systemic metabolic dysregulation has inspired a new area of biochemistry research in which scientists are seeking to understand the molecular mechanisms that link chronic lipid oversupply to tissue dysfunction and disease development. The purpose of this chapter is to review recent findings in this area, placing emphasis on lipid-induced functional impairments in the major peripheral organs that control energy flux: adipose tissue, the liver, skeletal muscle, and the pancreas.

Contents

INTRODUCTION 368
LIVER AND ADIPOSE TISSUE AS THE DRIVERS OF METABOLIC DYSFUNCTION 370
LIPID-INDUCED FUNCTIONAL IMPAIRMENT OF RESPONDER PERIPHERAL TISSUES 374
    Skeletal Muscle 374
    Pancreatic Islets 385
CONCLUDING REMARKS 393

TAG: triacylglycerols

INTRODUCTION

As we enter the new millennium, 65% of adults in the United States are overweight [body mass index (BMI) > 25 kg/m^2], and 30% are obese (BMI > 30 kg/m^2) (1), reflecting similar trends in many other countries around the world. For most of human evolution, the ability to store nutrients in the form of esterified lipids [triacylglycerols (TAG)] during times of food abundance has constituted a survival advantage for times of famine and/or energy deficit. In more recent times, this "thrifty" fuel economy has been challenged by overconsumption of energy-dense foods and reduced physical activity, leading to dysfunction of major tissues and organs and alarming increases in the incidence of obesity-related diseases such as diabetes, hypertension, and cardiovascular disease.

These events have led to a new emphasis on understanding of mechanisms by which chronic exposure of tissues to elevated concentrations of lipids and other nutrients may contribute to tissue dysfunction and disease development. The purpose of this chapter is to review recent findings in this area, with particular focus on changes in metabolic function induced in muscle, the liver, adipose tissue, and pancreatic islets by obesity and chronic lipid exposure.

To understand the biological perturbations created by chronic lipid exposure, one must first appreciate the contributions of different tissues to normal lipid homeostasis. In this regard it is important to note that dietary lipids are critical for normal biological functions, including energy production, membrane biosynthesis, covalent modification of proteins, and intracellular signaling. We therefore briefly review the distribution and metabolism of lipids in the fed and fasted states.

A typical meal contains lipids mainly in the form of triglycerides, with some free fatty acids, cholesterol, and other sterols. Unlike dietary carbohydrates, which are cleaved to monomeric sugars such as glucose and fructose and absorbed directly into the blood through the gastrointestinal tract, lipids enter the blood primarily via the lymphatic system as chylomicron lipoprotein particles. Digestion of dietary lipids begins in the stomach with partial hydrolysis of triglycerides and the formation of large fat globules that also contain free fatty acids, phospholipids, and sterols. As these globules enter the intestinal lumen, they are mixed with bile salts and hydrolyzed by pancreatic lipases to form monoacylglycerols and long-chain free fatty acids, which then enter enterocytes via fatty acid transporter proteins. Free fatty acids are then used as substrates for esterification of lysophosphatidic acid or monoacylglycerol, leading to formation of diacylglycerol and triacylglycerol. Fatty acids are also used to form cholesterol esters. Finally, triacylglycerols and cholesterol esters are packaged with apolipoprotein B to form lipoprotein particles, primarily chylomicrons, as well as a smaller number of very low-density lipoprotein (VLDL) particles, which then enter the circulation via the lymph (2).

As chylomicrons enter the circulation, they interact with lipoprotein lipases, which are located on the luminal surface of capillary endothelial cells, resulting in hydrolysis of triglycerides to free fatty acids and monoacylglycerol. The released nonesterified fatty acids (NEFA) then enter cells or bind to albumin for distribution in the circulation. Transport of fatty acids into cells can occur to some extent via diffusion across the lipid bilayer of the plasma membrane, but the predominant mechanism is protein mediated. Two major classes of fatty acid transporter proteins have been described. CD36 (also known as FAT and GPIV) is an integral membrane glycoprotein found on the surface of a variety of cells, including adipocytes and oxidative muscle fibers (3). Mice lacking CD36 expression exhibit a 60% decrease in fatty acid uptake in the heart, skeletal muscle, and adipose tissue (4). A family of fatty acid transporter proteins (FATPs) 1–6 has also been identified with no structural homology to CD36 (3, 5). Some, but not all, studies of these proteins demonstrate that their overexpression enhances cellular uptake of fatty acids. FATP family members are homologous to long-chain fatty acid–coenzyme A (CoA) synthetases and catalyze CoA activation of long-chain fatty acids to varying degrees (3). Thus it remains unclear to what extent increases in fatty acid uptake in FATP-expressing cells are a result of intrinsic fatty acid transport activities of these proteins or are secondary to the enhanced rate of CoA activation. Knockout of FATP1, which is primarily expressed in skeletal muscle, results in prevention of fat-induced insulin resistance (6), whereas knockout of FATP4 results in disrupted skin function and early death (7).

Once transported into cells and activated to their CoA esters, the metabolic fate of fatty acids varies according to cell/tissue type. In humans or other mammals that are physically fit and near their ideal body weight, the primary fate of fatty acyl-CoAs in adipose tissue is re-esterification to triglycerides, whereas the major fate of fatty acids in skeletal and cardiac muscle is oxidation to provide ATP for muscle contraction. In the fed state, the liver helps to distribute lipid by esterification of fatty acids into triglycerides and packaging into VLDL for export into the circulation and ultimate storage of fatty acids in the adipose triglyceride pool. Pancreatic islets are equipped for both oxidation and esterification of exogenous fatty acids, but in addition, fatty acids serve both as enablers and potentiators of fuel-stimulated insulin secretion in these cells.

In fasted or other catabolic states (e.g., exercise), chylomicron levels decline, and the adipose tissue becomes the major source of fatty acids. The fall in insulin-to-glucagon ratio and increased catecholamine levels associated with fasted/catabolic states cause activation of hormone-sensitive lipase, resulting in hydrolysis of triglycerides in adipose tissue and release of glycerol and free fatty acids into the circulation. Coincident with the increase in fatty acid supply, exercise or fasting stimulates fatty acid oxidation, while simultaneously decreasing glucose utilization, via the following series of events. Glycolytic flux is decreased in liver via phosphorylation of the bifunctional protein (6-phosphofructo-2-kinase/fructose-2,6-bisphosphatase) by cyclic AMP–dependent protein kinase, resulting in reduced levels of fructose-2,6-bisphosphate and lower 6-phosphofructo-1-kinase activity. This in turn leads to reduced flux of glucose into mitochondrial pathways and export of mitochondrial citrate to the cytoplasm. Cytoplasmic citrate is used to synthesize acetyl-CoA and malonyl-CoA. Fasting causes phosphorylation and activation of 5'AMP kinase (AMPK), which in turn phosphorylates and inhibits acetyl-CoA carboxylase (ACC), the enzyme that converts acetyl-CoA to malonyl-CoA. The decline in malonyl-CoA concentrations induced by the combined effects of reduced ACC activity and citrate supply has two major consequences: suppression of de novo lipogenesis and stimulation of fatty acid oxidation due to the removal of malonyl-CoA inhibition of carnitine palmitoyltransferase 1 (CPT1). Finally, stimulation of fatty acid oxidation leads to further impairment of glucose oxidation, via accumulation of acetyl-CoA, NADH, and ATP, all of which inhibit pyruvate dehydrogenase (PDH) by allosteric mechanisms, while stimulating pyruvate

AMPK: 5'AMP kinase

CPT1: carnitine palmitoyltransferase 1

Adipokines: peptide hormones and cytokines produced and secreted by adipocytes, and which in turn regulate fuel utilization and storage in other peripheral tissues

dehydrogenase kinases, resulting in phosphorylation and further inhibition of PDH. The net effect of these changes is to increase the rate of fat oxidation and to decrease the rate of glucose oxidation. Increased rates of fatty acid oxidation in the liver also help to stimulate gluconeogenesis via allosteric activation of pyruvate carboxylase (PC) and generation of ATP for this energy-expensive process.

The interplay of various tissues in control of lipid homeostasis is perturbed in animals or humans that consume high-fat diets. With ingestion of such diets and onset of obesity, triglycerides begin to be stored at sites other than adipose tissue, including skeletal muscle, the heart, kidney, and liver. These changes are often associated with chronic elevations in circulating free fatty acids and triglycerides (hyperlipidemia). This has led to the widely accepted notion that obesity-associated tissue dysfunction, including insulin resistance and cell death, is a direct consequence of chronic exposure of tissues to elevated lipids and consequent accumulation of toxic by-products of lipid metabolism. The central goal of this chapter is to provide a critical review of studies on lipid-induced tissue dysfunction and to contrast these pathophysiologic responses with those that occur in response to physiologic states (e.g., fasting or exercise) in which circulating and tissue lipids are appropriately increased.

LIVER AND ADIPOSE TISSUE AS THE DRIVERS OF METABOLIC DYSFUNCTION

All of the major organs considered in this chapter are directly susceptible to functional impairment in response to chronic exposure to elevated lipids. However, in the setting of overnutrition and obesity in whole animals or humans, there is increasing evidence these responses are hierarchical, with adipose tissue and the liver acting as the primary or "driver" organs for systemic metabolic dysregulation, whereas skeletal muscle and pancreatic islets serve as the secondary or "responder" systems.

It is now appreciated that adipocytes play a critical metabolic regulatory role via their capacity to produce a number of endocrine hormones with potent modulatory effects on food intake, energy balance, and metabolic homeostasis. Another key event in the development of obesity-related metabolic dysfunction appears to be the failure of adipocytes to adequately sequester excess lipids, resulting in their redistribution to other organs and tissues. Finally, there is growing evidence that obesity and excess lipids activate production of inflammatory mediators by adipose tissue and the liver (and/or immune cells resident in those tissues), contributing to metabolic dysregulation in responder tissues such as skeletal muscle. This section reviews these aspects of adipose tissue and liver biology and the evidence that they play a primary role in systemic metabolic dysfunction. **Figure 1** provides a schematic view of the interplay of the liver, adipose tissue, skeletal muscle, and pancreatic islets in lipid-induced metabolic function.

Once considered a passive energy reservoir, adipose tissue is now recognized as an important endocrine organ that informs the brain and peripheral tissues of changes in whole-body energy status. The endocrine function of the adipocyte came to light with the hallmark discovery of leptin, a hormone synthesized and secreted by adipocytes that controls body weight through actions on both feeding behavior and energy expenditure. In 1994, Friedman and colleagues identified leptin as the mutated gene in homozygous *ob/ob* mice, which exhibit an obesity syndrome characterized by severe adiposity, hyperphagia, hypothermia, hyperlipidemia, hyperinsulinemia, and insulin resistance (8, 9). Leptin replacement in *ob/ob* mice restores energy balance by acting on central and peripheral receptors that mediate changes in feeding behavior and systemic fuel metabolism (8, 9). The ensuing decade of research has revealed a network of circulating adipokines that signal changes in adipose tissue energy status to other metabolic organs that control fuel consumption and redistribution. These

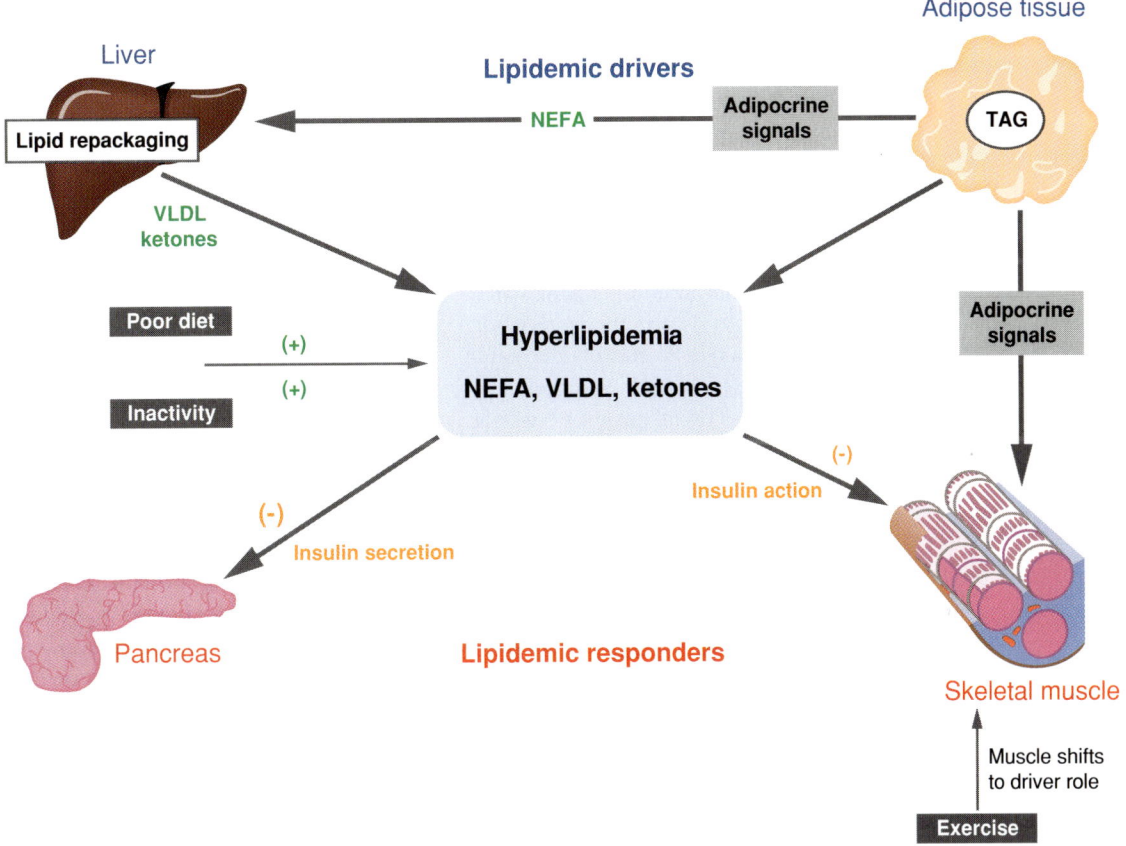

Figure 1

Obesity-related perturbations in systemic metabolic control. Under conditions of chronic energy overload (poor diet and inactivity) the adipocyte eventually fails to adequately sequester excess fuel, leading to uncontrolled lipolysis, elevated circulating nonesterified free fatty acids (NEFA) and aberrant production/secretion of adipokines. Concurrently, the liver repackages or transforms excess fuel into very low-density lipoprotein (VLDL) and ketones. Whereas the adipose tissue and liver operate as the primary drivers of systemic hyperlipidemia, skeletal muscle and pancreas respond to the metabolic cues by attempting to adjust energy flux and substrate selection (e.g., by elevating uptake and metabolism of lipid fuels). These adaptations allow appropriate handling of acute elevations in circulating lipids that occur in exercise and fasting; chronic diet-induced hyperlipidemia causes dysregulated mitochondrial metabolism in the responder tissues, contributing to functional impairments in insulin action and secretion, respectively. Abbreviation: TAG, triacylglycerols.

include peptide hormones such as adiponectin (also Acrp30) and resistin and proinflammatory cytokines such as interleukin (IL)-6 and tumor necrosis factor α (TNFα). Many of these adipocyte-secreted hormones, which appear to function interdependently, have been found to correlate with diabetes and/or insulin sensitivity (10–12) and to regulate both lipid and glucose metabolism (13). For example, the two best-characterized antidiabetic adipokines, leptin and adiponectin, have been shown to decrease TAG synthesis, promote fatty acid catabolism, and enhance insulin action in both skeletal muscle and the liver. Although information on the signaling mechanisms that mediate these actions is still unfolding, growing evidence indicates both leptin and adiponectin target AMPK (14, 15), a fuel-sensing serine kinase that plays a central role in regulating cellular energy metabolism.

With the exception of rare genetic cases, leptin levels increase with obesity as a result of peripheral leptin resistance (8, 9, 16). Impaired leptin action has been linked to increased expression of suppressor of cytokine signaling-3 (17). In contrast to leptin, circulating levels of adiponectin are reduced in a variety of obese and insulin-resistant states. Low adiponectin levels, together with leptin resistance, are thought to contribute to systemic abnormalities in both lipid and glucose homeostasis (18). Conversely, weight loss and improved metabolic control have been associated with a decline in serum leptin, improved leptin sensitivity, and increased adiponectin levels. In addition to perturbations relating to leptin and adiponectin, obesity-induced glucose intolerance is associated with increased adipocyte production of insulin-desensitizing molecules such as resistin, IL-6, TNFα, and retinol-binding protein 4 (RBP-4), all of which have been shown to induce insulin resistance in both muscle and the liver (10, 19–21). Accumulating data suggest the spectrum of hormones secreted by a given adipocyte depends on the size of the fat cell. Whereas small adipocytes preferentially secrete insulin-sensitizing hormones (leptin and adiponectin), enlarged adipocytes, characteristic of obesity, secrete higher levels of insulin-desensitizing hormones such as resistin and TNFα.

The adverse impact of insufficient lipid storage in adipose depots is clearly evidenced by the severe metabolic turmoil caused by lipoatrophy. An understanding of operative mechanisms of metabolic dysregulation in lipodystrophic syndromes has been facilitated by the creation of several strains of transgenic mice with ablation or loss of function of white adipose tissue. This has been achieved by the expression of various molecules [attenuated diphtheria toxin, a dominant negative regulator of normal adipose development (A-ZIP/F), or a truncated form of the sterol regulatory binding protein-1c (SREPB-1c)] under control of the adipose-specific promoter aP2 (22–24). Animals with ablated white adipose tissue have severe hepatic and muscle insulin resistance, occurring in concert with large increases in triglyceride stores in both tissues (25). Moreover, transplantation of normal fat tissue into aP2-A-ZIP/F mice restores insulin sensitivity (26). The restoration of insulin sensitivity appears to be mediated in part by a redistribution of fat from the liver and muscle to the adipose depot and via endocrine factors produced by the transplanted fat tissue. Most evidence points to leptin as the key endocrine mediator in these studies. Thus, leptin infusion ameliorates insulin resistance in aP2-SREBP-1c and aP2-A-ZIP/F mice (27, 28), whereas transplantation of fat from leptin-deficient mice into aP2-A-ZIP/F mice fails to improve insulin sensitivity (29). Furthermore, leptin administration to humans with severe lipodystrophy partially reverses their severe insulin resistance and hyperlipidemia (30).

Other mechanisms of interorgan communication for control of fuel homeostasis have emerged in recent years. For example, it is well known that expression of the insulin-regulated glucose transporter 4 (GLUT-4) is strongly depressed in adipose tissue but is much less reduced in skeletal muscle in animals and humans with type 2 diabetes (31). Because skeletal muscle accounts for approximately 80% of glucose disposal in the postprandial state, the diabetes-associated reduction in adipose GLUT-4 did not at first seem highly relevant to metabolic dysregulation. However, subsequent studies showed that mice with adipose-specific knockout of GLUT-4 have impaired insulin sensitivity in muscle and liver (31). The impairment in insulin action is only apparent in tissues in situ and not in excised tissue samples, implying participation of a blood-borne hormone or metabolite that mediates the effect. Whereas the impaired muscle and hepatic insulin sensitivity could not be linked to changes in leptin or free fatty acid levels (31), a subsequent study has demonstrated that mice deficient in adipose GLUT-4 have elevated levels of RBP-4 in blood that is apparently produced in part by adipose tissue. Furthermore, experimental elevation of circulating RBP-4 levels in normal mice by infusion or transgenic expression causes insulin resistance (21). Interestingly, food deprivation (fasting) also causes a form of insulin resistance and is associated with a decrease in adipose GLUT-4 expression (32). This raises the possibility that regulation of adipose production of insulin antagonists such as RBP-4, TNFα, and resistin may actually be a mechanism originally designed to prevent hypoglycemia in the fasted state, which with the advent of overnutrition and senescence in modern life has been subverted to create pathophysiology.

The liver also plays a primary or driver role in whole-organism metabolic control. For example, in rats fed a high-fat diet, hepatic expression of malonyl-CoA decarboxylase (MCD) ameliorates whole-animal and hepatic insulin resistance (16). Surprisingly, hepatic MCD expression also caused near-complete reversal of severe muscle insulin resistance, as evidenced by metabolic and signaling assays (16). MCD affects lipid partitioning by degrading malonyl-CoA to acetyl-CoA, thereby relieving inhibition of CPT1, the enzyme that regulates entry of long-chain fatty acyl-CoAs (LC-CoAs) into the mitochondria for fatty acid oxidation. In addition, malonyl-CoA is the immediate precursor for de novo lipogenesis. To gain insight into lipid-derived metabolites that might participate in the cross talk between the liver and muscle in the regulation of insulin sensitivity, metabolic profiling of 36 acyl-carnitine species was performed by tandem mass spectrometry. These studies revealed a unique decrease in the concentration of one lipid-derived metabolite, β-OH-butyrylcarnitine, in muscle of MCD-overexpressing animals that is likely a result of a change in intramuscular ketone metabolism (16). These studies have revealed a mechanism by which the liver controls muscle fuel metabolism via modulation of circulating free fatty acid levels. (These and subsequent studies are discussed in more detail in the section below on mechanisms of lipid-mediated impairment of skeletal muscle function.) Another example of the profound effects of altered lipid partitioning in control of whole-animal metabolic status comes from studies of animals deficient in stearoyl-CoA desaturase-1 activity in liver. This enzyme catalyzes the conversion of saturated fatty acids (e.g., C 16:0, C18:0) to monounsaturated fatty acids (C16:1, C18:1). Knockout of stearoyl-CoA desaturase-1 in *ob/ob* mice reverses obesity and insulin resistance in these animals (33, 34). This effect appears to be mediated by enhanced rates of oxidation of saturated versus unsaturated LC-CoAs. There is also evidence to suggest stearoyl-CoA desaturase-1 deficiency results in activation of AMPK, thereby further enhancing overall rates of fatty acid oxidation (35).

Finally, there is growing evidence that adipose tissue and the liver play important roles in the regulation of insulin sensitivity via inflammatory mechanisms (36). At high doses, salicylates (aspirin) reverse insulin resistance and hyperlipidemia in obese rodents while suppressing activation of the NF-κB transcription factor (37, 38). Subsequently, it has been demonstrated that high-fat diets or obesity result in activation of NF-κB and its transcriptional targets in the liver. Overexpression of a constitutively active version of the NF-κB activating kinase, IkB kinase catalytic subunit β (IKK-β), in liver of normal rodents to a level designed to mimic the effects of high-fat feeding results in liver and muscle insulin resistance and diabetes (20). In addition, both high-fat feeding and IKK-β overexpression increase expression of proinflammatory cytokines such as IL-6, IL-1β, and TNFα in the liver, and lead to increased levels of these molecules in blood. Systemic antibody-mediated neutralization of IL-6 in these models partially restores insulin sensitivity (20). Interestingly, mice with IKK-β knockout in the liver are protected from diet-induced

PPAR: peroxisome proliferator-activated receptor

impairment of hepatic insulin action but still develop muscle and adipose insulin resistance (39). In contrast, mice with IKK-β knockout in myeloid cells are protected against diet-induced insulin resistance in all tissues (39). These findings suggest the primary mediator of the inflammatory response to elevated lipids may be macrophages that reside within the liver and adipose depots. The precise biochemical mechanisms that link excess lipids and inflammatory responses remain undefined. One intriguing possibility is that excess lipids may trigger stress responses in the endoplasmic reticulum (ER) (40). Thus, markers of ER stress are elevated in the liver and adipose tissue of genetic or diet-induced forms of obesity, and this in turn is linked to activation of the c-jun amino-terminal kinases, which are known to interfere with insulin signaling via serine phosphorylation of insulin receptor substrate-1. Moreover, genetic manipulations that relieve ER stress also confer resistance against diet-induced metabolic dysfunction. The question of whether obesity-induced disturbances in ER function stem from chronic lipid overload, the anabolic pressures of hyperinsulinemia, cytokine-induced signaling, mitochondrial dysfunction, and/or other pathophysiological assaults now awaits further investigation. In this regard, it is interesting to note that several of the enzymes responsible for processing excess lipid (e.g., enzymes of lipid esterification) are integral membrane proteins that reside in the ER.

LIPID-INDUCED FUNCTIONAL IMPAIRMENT OF RESPONDER PERIPHERAL TISSUES

Skeletal Muscle

This section provides a current view of the molecular and metabolic networks that regulate glucose-fatty acid interactions in skeletal muscle. Emphasis centers on likely sites of obesity-associated metabolic dysfunction, as well as new insights into the long-standing debate over whether muscle insulin resistance links to high or low rates of fat oxidation. Finally, we will consider potential mechanisms that connect metabolic wellness to habitual exercise, a physiological paradigm in which muscle becomes a primary "driver" of energy homeostasis.

Physiological regulation of skeletal muscle lipid metabolism. Before attempting to dissect the pathophysiological consequences of lipid overload, it is necessary to summarize normal adaptive responses that occur when skeletal muscle is presented with a large influx of fatty acids. This section therefore introduces several key molecular and metabolic components of lipid regulation and handling in skeletal muscle.

Lipid-induced transcriptional reprogramming of muscle. Under conditions of lipid influx major metabolic reprogramming occurs at a transcriptional level, which is mediated in large part by a family of lipid-activated nuclear hormone receptors known as the peroxisome proliferator-activated receptors (PPARs) (41–43). In 1990, Issemann & Green (41) identified PPARα as the steroid hormone receptor responsible for mediating the metabolic effects of a class of drugs known as the peroxisome proliferators. Two other subtypes, PPARβ/δ and PPARγ, were subsequently cloned (reviewed in 42). The three subtypes have distinct tissue distributions that reflect their discrete but overlapping functions. PPARα is expressed most abundantly in tissues such as skeletal muscle, the heart, and the liver, where it plays a key role in regulating pathways of β-oxidation (43). PPARγ, the target of the insulin-sensitizing thiazolidinediones, is expressed primarily in adipose tissue where it activates programs of adipocyte differentiation and lipogenesis (44). Recent studies demonstrated that muscle-specific deletion of PPARγ in mice resulted in whole-body insulin resistance, suggesting the low levels of this receptor in muscle are physiologically important (45). PPARβ/δ, the most ubiquitous and least characterized of these receptors, has been shown to regulate both fatty acid oxidation and cholesterol efflux, apparently sharing many duties with PPARα (42, 46). Recent findings also suggest PPARδ may participate in the adaptive metabolic and histologic (fiber-type switching) response of skeletal muscle to endurance exercise (47).

In both animal and muscle cell culture models, pharmacological activation of either PPARα or PPARδ results in the robust induction of lipid-regulatory genes, including several associated with lipid trafficking, interorgan lipid transport and cholesterol efflux, fatty acid metabolism, β-oxidative enzymes, uncoupling proteins (UCPs), and cross talk between lipid and glucose substrates (42, 46). Upregulation of a similar set of genes occurs in response to physiological and pathophysiological circumstances that raise circulating fatty acids, such as obesity, diabetes, overnight starvation, high-fat feeding, and acute exercise (42, 46, 48). Studies in PPARα-null mice indicate the alpha subtype is essential for regulating both constitutive and inducible expression of β-oxidative genes in the liver and heart (43). However, skeletal muscles from the PPARα-null mice exhibited surprisingly modest perturbations in lipid metabolism and retained their ability to upregulate at least some PPAR-target genes in response to starvation and exercise. The lack of a severe phenotype in skeletal muscle may be a result of functional redundancies between PPARα and PPARδ (42, 46).

PPARγ coactivator 1 and fuel homeostasis. In the context of this discussion it is necessary to recognize the emergent role of the PPARγ coactivators (PGC1) as master regulators of mitochondrial function and energy homeostasis. In 1998, Puigserver and colleagues (49) identified PGC1α as a PPARγ interacting protein responsible for regulating mitochondrial genesis in brown fat. Subsequent studies identified a second isoform (PGC1β) and determined that both proteins are expressed ubiquitously and function as promiscuous coactivators of most nuclear hormone receptors, as well as several other transcription factors (50). These proteins are thereby capable of regulating several distinct biological programs. In skeletal muscle, PGC1α has been shown to stimulate mitochondrial biogenesis via coactivation of the nuclear respiratory factor (51) and to regulate genes involved in oxidative phosphorylation through interactions with estrogen-related receptor α (52). PGC1α also coactivates the PPARs (50), thereby regulating pathways of lipid metabolism, and myocyte enhancer factor-2 (51), a muscle-specific transcription factor involved in fiber-type programming. PGC1α is more abundant in red/oxidative muscle and is induced by exercise, whereas its expression is decreased both by inactivity and chronic high-fat feeding (53, 54). In contrast, PGC1β mRNA levels are unaltered by these manipulations.

Regulation by the CPT1-malonyl-CoA system. Upon entering cells, fatty acids are activated to acyl-CoA by one of several acyl-CoA synthetase isozymes. The subsequent transfer of long-chain acyl-CoAs into the mitochondria for β-oxidation requires the carnitine shuttle system. CPT1, which spans the outer mitochondrial membrane, catalyzes the initial and pace-setting step in this process. The acylcarnitines formed by CPT1 traverse the inner membrane via a specific translocase coupled to CPT2, which regenerates acyl-CoA upon transporting the fatty acyl groups into the mitochondrial matrix. The two CPT1 isoforms (CPT1α in liver and CPT1β in heart/skeletal muscle) exhibit distinct kinetic properties; however, both are inhibited by malonyl-CoA, whereas the other CPT family members are not.

Physiological alterations in malonyl-CoA concentrations correlate inversely with changes in β-oxidation. For example, starvation and exercise decrease tissue malonyl-CoA levels and increase fat oxidation (55), whereas, conversely, carbohydrate feeding increases malonyl-CoA production in conjunction with low fat oxidation (56).

UCP: uncoupling protein

PGC1: PPARγ coactivators

Lipotoxicity:
lipid-induced tissue dysfunction resulting from persistent elevations in the supply, storage, and metabolism of lipid fuels

IMTG:
intramuscular triacylglycerol

Paradoxically, however, concentrations of malonyl-CoA measured in skeletal muscle (1–4 µM) (55) should completely inhibit CPT1 activity at all times. This is because the muscle isoform of CPT1 is approximately 100 times more sensitive to malonyl-CoA than its counterpart in liver. The question of how fatty acid oxidation proceeds in muscle despite constitutively high malonyl-CoA levels has remained unresolved. One theory suggests malonyl-CoA is compartmentalized into physically and/or functionally discrete pools synthesized by the distinct ACC1 (cytosolic) and ACC2 (mitochondrial) (57) isozymes. Thus, the pool that regulates mitochondrial CPT1 enzyme activity might be controlled by ACC2, independent of total cellular concentrations. Additionally, evidence of a malonyl-CoA-insensitive CPT1 activity in red skeletal muscle has been reported (58).

The CPT1/malonyl-CoA system is primarily controlled by three factors that impact ACC activity: glucose availability, insulin status, and the activity of AMPK, which increases in response to energy stress (e.g., a high AMP/ATP ratio) (59). Whereas glucose and insulin favor high ACC activity, AMPK phosphorylates and inactivates both isoforms of the enzyme. There is also evidence to suggest AMPK activates MCD (59), leading to malonyl-CoA degradation and disinhibition of CPT1. Muscle AMPK is activated by exercise, leptin, and adiponectin, and by the antidiabetic agent metformin, and is thus presumed to play a major role in mediating the therapeutic effects of these manipulations (18).

Mechanisms of lipid-induced insulin resistance in skeletal muscle. The lipotoxicity model of insulin resistance predicts that increased levels of bioactive lipid intermediates oppose glucose homeostasis by blocking insulin signal transduction. Indeed, numerous studies have reported a strong correlation between impaired insulin action and intramuscular accumulation of specific lipid species. This section reviews current theories to explain how lipids disrupt glucose disposal, and furthermore explores the widely held view that maladaptive changes in muscle fat oxidation play a central role in driving glucose intolerance.

Candidate mediators of insulin resistance. In both animals and humans, elevated intramuscular TAG (IMTG) storage is well recognized as a common feature of obese and/or insulin-resistant states (60, 61). Disease-associated accumulation of IMTG droplets has been demonstrated by biochemical and histological analyses as well as by computed tomography and magnetic resonance spectroscopy. Both cross-sectional and prospective studies have found that IMTG content correlates more closely with systemic insulin resistance than with other important factors such as BMI, waist-to-hip ratio, and total adiposity (62). Most researchers agree, however, that the TAG themselves are unlikely to play a direct role in causing the associated insulin resistance. Rather, these lipid reserves may provide a source of fuel that competes with glucose as an oxidative substrate and/or contributes to the synthesis of other lipid-derived entities that directly interfere with insulin signal transduction (63).

Using ^{13}C and ^{31}P magnetic resonance spectroscopy to assess metabolic changes in human skeletal muscle, lipid infusion has been shown to inhibit insulin-stimulated glucose transport activity in association with intracellular accumulation of LC-CoAs and diacylglycerol (DAG), lipid molecules with known signaling actions (64). These findings triggered a flurry of research activity that focused heavily on LC-CoA and DAG as candidate mediators of insulin resistance. In many cases, muscle levels of these lipids have been shown to correlate positively with IMTG and negatively with insulin sensitivity. This has been observed in cross-sectional studies of both humans and rodents as well as in several experimental and transgenic models of lipid-induced insulin resistance (61, 63, 65).

It has also been suggested that these lipid molecules can exert a direct, negative impact on insulin signal transduction by activating inhibitory signaling cascades (63, 65). Insulin signal transduction depends on a tyrosine phosphorylation cascade that begins with autoactivation of the insulin receptor tyrosine kinase, followed by tyrosine phosphorylation of proximal targets such as the insulin receptor substrate 1 (66). At least some forms of insulin resistance appear to be mediated by a serine kinase cascade that targets the insulin receptor and/or its downstream signaling partners (66–69). In contrast to tyrosine phosphorylation, which serves to propagate the signal, serine phosphorylation acts as a blockade that prevents the signal from reaching its final destination.

Acute lipid infusion in rodents has been shown to cause marked muscle insulin resistance in association with LC-CoA accumulation, increased phosphorylation of protein kinase C-θ, and phosphorylation of insulin receptor substrate 1 on serine 307 (65). Phosphorylation at serine 307 impairs insulin receptor–mediated tyrosine phosphorylation of insulin receptor substrate-1 and consequently inhibits its association with phosphatidylinositol-3-kinase. This in turn is thought to impair phosphorylation and activation of distal components of the pathway, such as protein kinase B (AKT) and glycogen synthase kinase, thus ultimately perturbing insulin-mediated GLUT-4 translocation and glycogen synthesis. Further investigations have since identified several inhibitory serine phosphorylation sites in multiple protein members of the insulin signaling cascade. In vitro assays suggest both LC-CoAs and DAG can activate several serine kinases, including protein kinase C-θ, c-jun amino-terminal kinases, and IKK-β. Moreover, inhibition of these kinases by pharmacological or genetic interventions has been shown to reverse insulin resistance in rodents (37–39, 70). Thus, strong evidence suggests this family of stress-induced serine kinases participates in the etiology of insulin resistance, possibly in response to elevated levels of nonoxidative lipid metabolites.

Still, definitive evidence that the aforementioned lipid molecules are directly responsible for insulin resistance has yet to surface. Their associations with both the serine kinases and impaired insulin signaling could reflect coincidence rather than causation. In fact, there are a number of instances in which the relationship between IMTG, LC-CoAs, and insulin resistance has not held true. For example, muscles from exercise-trained subjects are highly insulin sensitive despite IMTG levels similar to or even higher than those found in association with obesity and diabetes (60). Likewise, both TAG stores and insulin sensitivity are higher in muscle composed of predominantly red (type I) myofibers compared with white (type II) myofibers (60, 71). Furthermore, exercise intervention in type 2 diabetic patients has been shown to improve insulin sensitivity without a corresponding decrease in muscle LC-CoA levels (72). Notably, a hallmark of both exercise-trained and type I skeletal muscle is their enhanced mitochondrial density and performance (73), thus implying these properties might confer protection against lipid surplus. Consistent with this notion, skeletal muscle oxidative capacity (assessed by citrate synthase activity) has been shown to be a better predictor of insulin sensitivity than either IMTG or LC-CoA content (74).

The exercise paradigm is not the only one that has challenged the LC-CoA theory. Additionally, muscles from mice lacking the heart/muscle-specific fatty acid binding proteins (H-FABP) were found to be protected against high-fat diet–induced insulin resistance, despite the presence of LC-CoA levels comparable with those found in wild-type muscles (75). Interestingly, deletion of H-FABP resulted in dramatic decreases in muscle fat oxidation and TAG storage. Similarly, insulin-mediated phosphorylation of AKT-1 and glycogen synthase kinase-3 in muscle of high-fat fed rats could be restored by a maneuver that tended to increase rather than decrease LC-CoA levels (16). Instead, improved insulin signaling corresponded with a sharp decrease in a mitochondrial-derived lipid metabolite (16). These observations could indicate the LC-CoAs are merely markers of other cellular events more directly connected to glucose homeostasis.

Ceramide, a sphingolipid derived from the condensation of palmitoyl-CoA and serine, has emerged as another attractive candidate mediator of insulin resistance (reviewed in 76). Similar to other lipid species (IMTG, LC-CoA and DAG), muscle levels of ceramide increase in association with obesity and diabetes, both in animals and humans. Moreover, diabetogenic agents (such as NEFA, TNFα, and glucocorticoids) have been shown to stimulate ceramide accumulation, either by increasing its palmitoyl-CoA precursor and/or inducing serine palmitoyltransferase, the enzyme that catalyzes the initial and rate-limiting step in de novo ceramide biosynthesis (76–78). Conversely, exercise-mediated improvements in insulin sensitivity have been shown to correlate with reduced ceramide levels in muscle (79).

Cell-based experiments have demonstrated that ceramide inhibits insulin-stimulated glucose uptake, GLUT-4 translocation, and/or glycogen synthesis (80, 81). These effects are associated with impaired insulin-stimulated activation of insulin receptor substrate-1, phosphatidylinositol-3-kinase, and/or AKT (82, 83). Moreover, both chemically and genetically mediated suppression of ceramide synthesis prevented palmitate-induced inactivation of AKT in C_2C_{12} muscle cells. Conversely, pharmacological antagonists of ceramide degradation both mimicked and exacerbated the insulin inhibitory properties of palmitate. However, this mechanism has been found to be inoperative when muscle is exposed to lipid mixtures consisting of high mono- or polyunsaturated fatty acids (65). Therefore, the role of ceramide in mediating lipid-induced insulin desensitization appears specific to saturated fatty acids. It is also noteworthy that much of the evidence linking ceramide to fatty acid–induced insulin resistance comes from studies in which cells were exposed to palmitate without carnitine supplementation. Under these conditions palmitate can be particularly cytotoxic because (when given alone) it serves as a relatively poor substrate for de novo TAG synthesis, and additionally, its use as an oxidative substrate is severely impaired by carnitine insufficiency. Thus, the experimental conditions described in some of these reports may have fueled an artifactual rise in cellular palmitoyl-CoA and ceramide levels. Further studies are necessary to determine whether this pathway is invoked by a more physiological lipid environment.

Obesity-related impairments in substrate utilization. Fatty acids and glucose constitute the primary oxidative fuels that support skeletal muscle contractile activity. In the context of normal physiology, this tissue exhibits a robust capacity to transition between the two substrates. Thus, skeletal muscle adapts or "responds" to a given set of metabolic and neurohumoral cues by adjusting fuel selection to match both energy supply and demand. The proposal that elevated fatty acid oxidation inhibits glycolysis and glucose oxidation was first presented by Randle et al. (84) in 1963 as the "glucose-fatty acid cycle." Principal elements of this model hold that (*a*) provision of lipid fuels (fatty acids or ketones) promotes fatty acid oxidation and inhibits glucose metabolism, (*b*) the inhibitory effects of lipid fuels on glucose oxidation are mediated via inhibition of hexokinase, phosphofructokinase, and pyruvate dehydrogenase, and (*c*) these lipid-induced changes in metabolic regulation lead to diminished insulin-stimulated glucose transport (85). Conversely, in 1977 McGarry et al. (86) demonstrated that high glucose concentrations suppress fatty acid oxidation via malonyl-CoA-mediated inhibition of CPT1. This pathway represents a near-exact complement to the mechanism described by Randle and is thus often referred to as the "reverse glucose-fatty acid cycle."

In more recent years the CPT1-malonyl-CoA "partnership" has been featured as a key constituent of the lipotoxicity paradigm (87), in which elevated levels of malonyl-CoA and impaired fatty acid catabolism are thought to encourage cytosolic accumulation of "toxic" lipid species that disrupt glucose control. Fitting with model, muscle insulin senstivity correlates positively with high AMPK activity, which in turn favors inactivation of ACC, enhanced fat oxidation and diminished lipid accumulation. This model also proposes that the malonyl-CoA/CPT1 axis might be disrupted with obesity and/or diabetes. Consistent with this notion, muscle malonyl-CoA concentrations are elevated in several (but not all) models of rodent obesity, occurring in conjunction with increased IMTG levels and accumulation of LC-CoAs (18, 59). Furthermore, knockout mice lacking ACC2 have decreased muscle malonyl-CoA levels, increased β-oxidation, and protection against diet-induced obesity and insulin resistance (88). In light of these findings, it was surprising that several genetic models of diminished AMPK activity failed to exhibit defects in muscle insulin sensitivity and glucose transport/metabolism (89). However, these studies did not evaluate the impact of impaired AMPK function on tissue malonyl-CoA levels.

In humans, the relationship between malonyl-CoA and insulin resistance is less clear. Although several laboratories have shown human muscle malonyl-CoA content increases in response to a hyperinsulinemic-euglycemic clamp, in association with decreased fat oxidation (90, 91), basal levels were found to be similar in lean, obese, and type 2 diabetic subjects (92). Moreover, fat oxidation rates during hyperinsulinemic conditions were actually increased in diabetic subjects compared with control subjects, despite similarly high levels of malonyl-CoA (64, 65). Thus, whereas the malonyl-CoA/CPT1 axis plays a key role in regulating muscle lipid oxidation, it is unclear whether disturbances in this system are an essential component of insulin resistance.

Although in many ways Randle and McGarry held similar views regarding the fundamental role of lipids in mediating glucose intolerance, Randle's hypothesis that insulin resistance stems from increased muscle fatty acid oxidation is apparently at odds with the model put forth by McGarry and others (59, 65) predicting that obesity-associated increases in malonyl-CoA antagonize fat oxidation, thereby causing insulin-desensitizing lipids to accumulate. Adding further confusion to this subject, a survey of the literature reveals a number of inconsistent reports describing either increased or decreased muscle fat oxidation in association with obesity, thus seeming to support both possibilities. Perhaps neither is entirely correct or incorrect. To reconcile these discrepancies the concept of "metabolic inflexibility" has been proposed, holding that muscles from obese and insulin-resistant mammals lose their capacity to switch between glucose and lipid substrates (93). In support of this idea, fat oxidation (assessed by respiratory quotient across the leg) in obese and type 2 diabetic subjects compared with lean subjects is greater in the postprandial state (simulated by hyperinsulinemic, euglycemic clamp) but depressed in the postabsorptive state (94). Thus, whereas control subjects adjusted muscle substrate selection in response to a changing nutrient supply, the insulin-resistant subjects did not. Several other groups have similarly demonstrated that the increase in fatty acid oxidation that normally occurs in response to fasting, exercise, or β-adrenergic stimulation is diminished in obese and/or diabetic subjects (reviewed in 95).

Also at odds with a model in which obesity and insulin resistance link to impaired β-oxidation are a number of reports showing that both disorders are accompanied by increased expression of several PPAR-targeted genes involved in lipid catabolism (96–98). Likewise, both acute and chronic lipid

> **Metabolic inflexibility:** the inability of muscle tissue to appropriately adjust substrate selection in response to metabolic and neurohormonal cues

Incomplete fat oxidation: partial oxidation of fatty acids resulting in the production of β-oxidative intermediates but not the end product of CO_2

TCA: tricarboxylic acid

exposure cause insulin resistance in association with the induction of lipid trafficking and β-oxidative genes (46, 53, 99). Consistent with these changes in transcript levels, muscles from obese Zucker rats (100) or rats fed a high-fat diet (101) exhibit increased fatty acid oxidation when compared to the respective non-obese control groups. Interestingly, isolated mitochondria from (high-fat) diet-induced obese rats were found to have increased rates of incomplete fat oxidation, but without a coincident change in complete oxidation to CO_2 (53). Perhaps this distinction is one that bears on the aforementioned discrepancies.

Complete oxidation of fatty acids to CO_2 requires not only their entry into mitochondria and processing through β-oxidation, but also the coordinated regulation of downstream metabolic pathways such as the tricarboxylic acid (TCA) cycle and electron transport chain (ETC) (**Figure 2**). In the event that rates of β-oxidation exceed energy demand, excess acetyl-CoA and other acyl-CoA intermediates can be converted back to acylcarnitine by CPT2 and another key member of the acyltransferase family, carnitine acetyltransferase, which resides within the mitochondrial matrix and exhibits high specificity for short chain acyl-CoAs (102). Because these enzymes catalyze the freely reversible exchange of CoA for carnitine, they play an important role in maintaining an adequate mitochondrial pool of free CoA. The acylcarnitine esters formed by this reaction are thought to exit the matrix through the carnitine acylcarnitine translocase. Upon entering the cytoplasmic compartment, which lacks acyltransferase activity, these metabolites can either be recycled for use as an oxidative substrate, or exported from the tissue into the general circulation. Blood levels and urinary excretion of specific acylcarnitine species increase dramatically in cases of inherited mitochondrial disorders, consistent with a critical detoxifying role for this system.

Whereas mass spectrometry-based measurement of acylcarnitines has long been used as a tool for detecting inborn metabolic disorders, more recent studies have used the technology to detect incomplete β-oxidation in animal models of insulin resistance. These studies found that muscle levels of fatty acylcarnitine metabolites, measured in the fed state, were abnormally high in obese compared to lean rats (16, 53). Similar acylcarnitine profiles have been observed in muscle from obese Zucker diabetic fatty (ZDF) rats compared to lean controls. Moreover, rats fed a standard chow diet decreased muscle production of acylcarnitines during the transition from the fasted to the fed conditions, whereas in comparison, rats on the high-fat diet exhibited little or no change. Finally, the same study showed that a three-week exercise intervention in mice fed on a chronic high-fat diet lowered muscle acylcarnitine levels, in association with increased PGC1α expression, enhanced mitochondrial performance and restoration of glucose tolerance (53).

The specific role of PGC1α in permitting an efficient lipid-induced substrate switch was tested in studies that compared adaptive metabolic responses in rat L6 myocytes expressing typically low levels of PGC1α to those occuring when the gene was overexpressed (53). Similar to muscle mitochondria from high-fat fed rats, L6 myocytes exposed to increasing fatty acid concentrations exhibited disproportionate increases in the rates of incomplete (assessed by measuring incorporation of the label from ^{14}C oleate into acid-soluble β-oxidative intermediates) relative to complete (label incorporation into CO_2) β-oxidation of fatty acids. Overexpression of PGC1α in lipid-cultured L6 cells caused production of $^{14}CO_2$ to increase and maintain pace with production of $[^{14}C]$-labeled acid-soluble β-oxidative intermediates. In other words, the ratio of complete to incomplete β-oxidation was dramatically increased by PCG1α expression. Consistent with these functional assessments, cDNA microarray analyses showed that fatty acid exposure in the context of low PGC1α activity resulted in the induction of classic PPAR-targeted genes involved in lipid trafficking

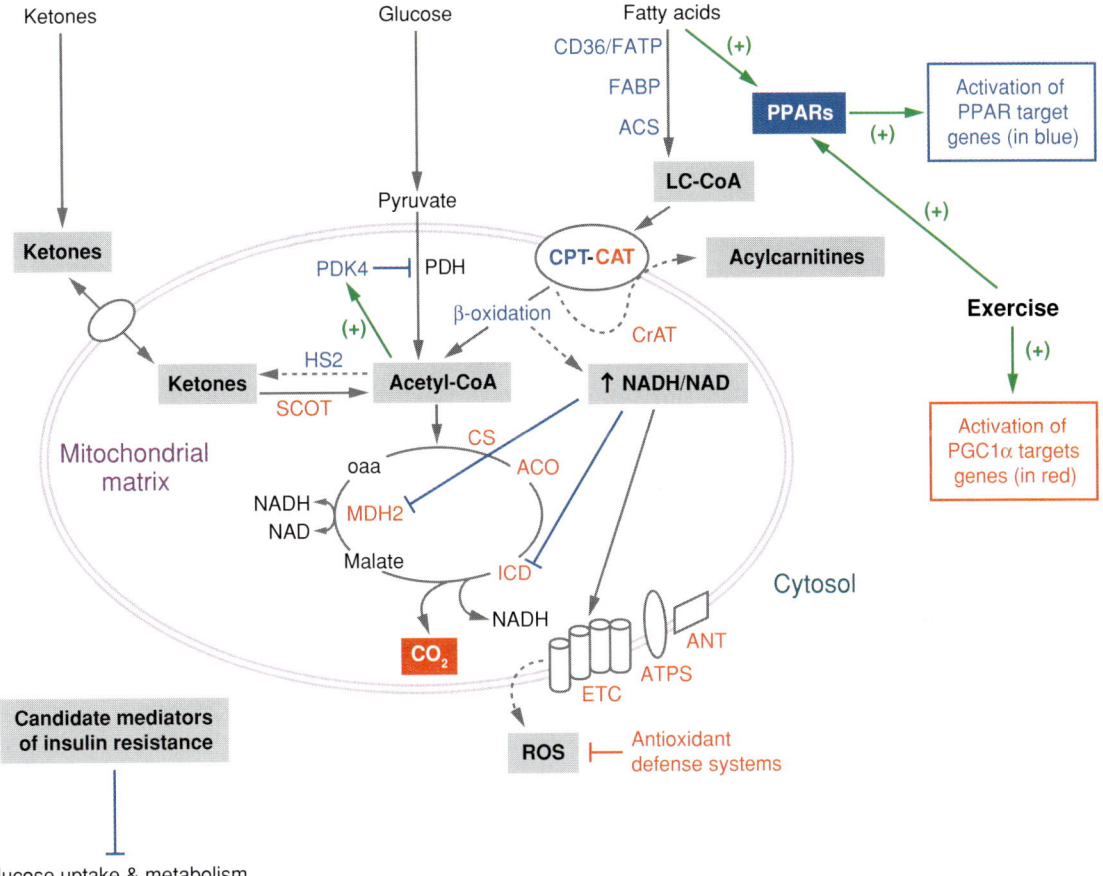

Figure 2

Lipid-induced mitochondrial dysfunction in skeletal muscle. During conditions of overnutrition and inactivity, fatty acid influx and peroxisome proliferator-activated receptor (PPAR)-mediated activation of target genes (*blue*) enhance β-oxidation without an accompanying increase in tricarboxylic acid (TCA) cycle flux. Mitochondrial performance is further perturbed by a rising NADH/NAD ratio and depletion of both free coenzyme A (CoA) and carnitine. As a result, metabolic by-products of incomplete fatty acid oxidation (*gray boxes*) may trigger insulin inhibitory stress-activated serine kinases. Exercise and/or activation of PPARγ coactivator 1 α (PGC1α) combat lipid stress by coordinating ligand-induced PPAR activity with the remodeling of downstream metabolic pathways (*red*), thereby favoring complete fatty acid oxidation to CO_2. Abbreviations: ACO, aconitase; ANT, adenine nucleotide transporter, ATPS, ATP synthase; CPT-CAT, carnitine palmitoyltransferase system; CrAT, carnitine acetyltransferase; CS, citrate synthase; ETC, electron transport chain; FABP, fatty acid binding proteins; FATP, fatty acid transport proteins; HS2, mitochondrial HMG-CoA synthase; ICD, isocitrate dehydrogenase; LC-CoAs, long-chain fatty acyl-CoAs; MDH, malate dehydrogenase; PDH, pyruvate dehydrogenase; PDK4, pyruvate dehydrogenase kinase 4; ROS, reactive oxygen species; SCOT, succinyl-CoA oxoacid transferase.

and β-oxidation, but with little or no change in other downstream pathways that regulate respiratory capacity. In contrast, high PGC1α expression enabled the coordinated induction of β-oxidative enzymes with equally important downstream targets (e.g., TCA cycle, ETC, and NADH shuttle systems) (**Figure 2**). These findings imply that PGC1α enables tighter coupling between β-oxidation and the TCA cycle.

Taken together, these metabolic studies underscore several important points. First, the accumulation of fatty acylcarnitines in muscle of obese/insulin-resistant rats implies increased rather than decreased rates of mitochondrial fatty acid uptake and β-oxidation. Second, experiments in isolated mitochondria from high-fat rats suggest that PPAR-mediated increases in β-oxidative activity exceeded the capacity of the TCA cycle to fully oxidize the incoming acetyl-CoA. This supports the idea that assessment of complete fat oxidation via measurement of CO_2 production provides only a partial view of lipid catabolism. Lastly, the acylcarnitine profiles from fed and fasted rats suggested that mitochondria from obese animals were unable to appropriately adjust mitochondrial fatty acid influx in response to nutritional status, thus supporting the observation of metabolic inflexibility in humans (94). An important question remaining is whether the high rates of fatty acid catabolism in the obese state are insufficient to compensate for increased lipid delivery, thereby allowing excess cytosolic fatty acids to impair insulin signaling, or alternatively, whether persistently high rates of mitochondrial β-oxidation directly contribute to the development of insulin resistance.

Mitochondrial malfunction: a unifying hypothesis? Whereas considerable controversy still surrounds the question of whether oxidative or nonoxidative lipid metabolites act as the primary mediators of insulin resistance, a universal theme emanating from both models is the concept of mitochondrial malfunction (103). This is not surprising given mitochondria serve as both the metabolic powerhouse of the cell and the primary site of β-oxidation. In recent years a growing number of studies have pointed to a strong and potentially causative connection between muscle mitochondrial dysfunction and the development of both acquired and inherited forms of insulin resistance (103–107). For example, age-related muscle insulin resistance is accompanied by decreased mitochondrial oxidative activity and lower rates of ATP synthesis, in association with increased IMTG and LC-CoA content (108). Another investigation used magnetic resonance spectroscopy to demonstrate that mitochondrial function was compromised in insulin-resistant offspring of parents with type 2 diabetes compared with age- and weight-matched controls (107). They additionally reported that insulin resistance and reduced mitochondrial function were associated with a lower proportion of type 1 muscle fibers relative to type 2 muscle fibers. Other groups have likewise found the severity of insulin resistance in obese and type 2 diabetes subjects correlated with reduced oxidative enzyme capacity, measured in muscle homogenates, and reduced activity of the ETC, assessed in isolated mitochondria (109, 110). Furthermore, transmission electron microscopy revealed smaller mitochondria in muscle from obese and type 2 diabetic subjects compared with lean controls (109).

In several reports, the mitochondrial abnormalities observed in association with obesity and/or diminished insulin sensitivity correlated with decreased muscle expression and activity of PGC1α (53, 54, 103, 104, 111). Similarly, polymorphisms in PGC1α are associated with increased risk of diabetes (112). Conversely, conditions in which muscle exhibits both high PGC1α levels and high oxidative capacity (red fiber type and exercise training) are the same circumstances in which the correlation between lipid content (IMTG and LC-CoA) and insulin resistance is disrupted. Together, these observations again point to the probability that the accumulated lipids seen in obese and diabetic subjects are merely markers of suboptimal mitochondrial function. Moreover, recent findings suggest habitual exercise combats the potentially toxic effects of lipid oversupply via the enhancement of mitochondrial performance as well as mitochondrial number (53), adaptations that are at least partly attributable to the activation of PGC1α.

What underlying mechanisms could be responsible for linking mitochondrial dysfunction to insulin resistance? One theory presented predicts that reduced oxidative capacity forces preferential partitioning of lipids away from mitochondrial catabolism and toward cytosolic accumulation of IMTG, LC-CoA, DAG, and ceramide (103) (discussed above). However, at least in the case of diet-induced insulin resistance, the weight of evidence indicates early stage defects in fuel homeostasis are associated with increased mitochondrial uptake and oxidation of fatty acids. We therefore call attention to alternative mechanisms that center on the concept of lipid-induced mitochondrial stress, which conceivably could trigger the same inhibitory serine kinases shown to antagonize insulin signaling (**Figure 3**).

First, we consider a modified rendition of the Randle cycle in which accelerated incomplete fatty acid oxidation, resulting from a mismatch between β-oxidation and TCA/ETC activity, disrupts the energy milieu of the mitochondria by perturbing redox status (e.g., increasing the NADH/NAD ratio) and by depleting free CoA and carnitine to critically low levels. There are in fact several lines of evidence that link increased β-oxidation to insulin resistance. Throughout this discussion we cite examples of muscle-specific transgenic mouse models that display increased rates of β-oxidation and exhibit coincident impairments in insulin sensitivity and vice versa (4, 6, 98). In the muscle-specific PPARα transgenic mice, both local and systemic glucose intolerance were observed in association with marked induction of several lipid-oxidative genes (98). Notably, the diabetic phenotype of these animals was reversed by administration of the CPT1 inhibitor oxfenicine, further implicating a role for excessive β-oxidation. Equally supportive are data from studies using PPARα-null mice. These animals exhibit decreased muscle fat oxidation, increased circulating NEFA, and elevated TAG content in several peripheral tissues (43, 46, 98). Remarkably, however, despite this severe lipid-dysregulated phenotype, the PPARα-null mice were protected against diet-induced glucose intolerance (98, 113). These animals represent one of the few models in which elevated NEFA actually accompany improved insulin sensitivity, thus implying fatty acids are less destructive when β-oxidation is limited.

Consistent with this theory, several studies have shown that CPT1 inhibitors improve systemic insulin sensitivity, both in animals and humans (114–116). Even more intriguing, salicylates, which are thought to improve insulin sensitivity in high-fat fed rodents via their anti-inflammatory actions, can also inhibit fatty acid oxidation (117). At first glance these results may seem to contradict the finding that transgenic overexpression of ACCβ in mice conferred protection against diabetes; however, whole-body transgenic manipulations affect systemic-wide changes in lipid dynamics. Thus, similar to many pharmacological therapies, the metabolic impact on the driver organs (the liver and adipose tissue) may override the potential negative effects in muscle. Finally, a growing number of investigations have reported that L-carnitine supplementation also improves whole-body insulin sensitivity (118). One might anticipate that the therapeutic actions of L-carnitine occur as a result of increased fat oxidation. On the contrary, the supplement has been shown to enhance glucose oxidation (118), in association with increased PDH activity and improved mitochondrial respiratory function (118–122). Thus, because carnitine also participates in the export of acetyl- and acyl-CoAs from the mitochondrial matrix, the surplus carnitine may be functioning to rid the organelle of excess lipids, thereby restoring homeostatic control of the mitochondrial microenvironment.

Is there a mitochondrial-derived lipid signal that couples incomplete β-oxidation to insulin resistance? The possibility that the acylcarnitine esters might function in some sort of signaling capacity cannot be discounted. These metabolites have been shown to alter

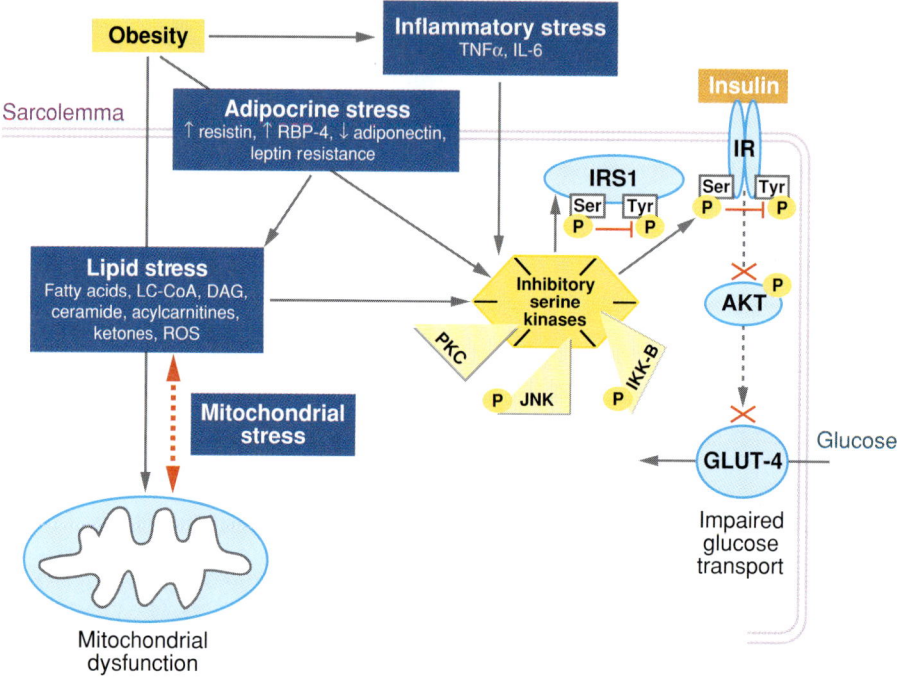

Figure 3

Obesity-related metabolic dysfunction in skeletal muscle. Obesity is associated with a host of metabolic, inflammatory, and adipocrine stresses that combine to mount a full-scale attack on both mitochondrial function and insulin signaling. These stresses converge to activate a network of inhibitory serine kinases that interfere with insulin signal transduction and hence glucose homeostasis. Abbreviations: AKT, protein kinase B; DAG, diacylglycerol; GLUT-4, insulin-regulated glucose transporter 4; IKK-β, IkB kinase catalytic subunit β; IL-6, interleukin-6; IRS1, insulin receptor substrate 1; JNK, c-jun amino-terminal kinases; LC-CoA, long-chain acyl-CoA; PKC, protein kinase C; RBP-4, retinol-binding protein 4; ROS, reactive oxygen species; Ser, serine; TNFα, tumor necrosis factor-α; Tyr, tyrosine.

ROS: reactive oxygen species

calcium signaling in the heart (123), and they could conceivably participate in processes related to protein acetylation/acylation. Alternatively, recent findings have implicated intramuscularly synthesized β-OH-butyrate, a ketone, as a novel candidate-mediator of insulin resistance (16). In this study, fatty acid surplus increased muscle expression of the mitochondrial isoform of hydroxymethylglutaryl (HMG)-CoA synthase, a ketogenic enzyme typically thought of as a liver-specific protein. Perhaps the induction of this gene in skeletal muscle reflects a bleed-off pathway intended to accommodate the lipid-mediated mismatch between β-oxidation and TCA cycle activity (53). Interestingly, the literature supports a strong inverse correlation between ketogenesis and glucose tolerance (84, 124–126). The potential role of intramuscularly derived ketones as local modulators of insulin action now warrants further investigation.

Another attractive and highly plausible candidate signal is one that stems from lipid-induced oxidant stress. In addition to their principal tasks of substrate oxidation and ATP generation, mitochondria are also a primary source of reactive oxygen species (ROS), and they play major roles in antioxidant defense (127). Catabolism of fatty acids via β-oxidation produces reducing equivalents (NADH and $FADH_2$) that are subsequently oxidized by the ETC to generate ATP. This process of coupled respiration also results in the production of ROS, in the form of superoxide and hydrogen peroxide. The generation of these specific ROS has been shown to increase when electron transport is slowed by a high proton gradient across the inner mitochondrial membrane (128) (a condition that might be expected when rates of β-oxidation exceed energy demand). ROS can react rapidly with DNA, protein, and lipids, thereby inflicting oxidative damage to these cellular constituents.

Fatty acids in particular are highly susceptible to oxidative attacks and thus are easily transformed to highly reactive lipid peroxides. If not sufficiently managed, these lipid by-products can in turn cause further harm to DNA and protein targets residing within the mitochondria, ultimately prohibiting the organelle from performing at an optimal level. Furthermore, oxidant stress is known to activate several of the stress-induced kinases and transcription factors implicated in insulin resistance, including the mitogen-activated kinases, c-jun amino-terminal kinases, protein kinase C, IKK-β, and NFkB (reviewed in 129). It is therefore not difficult to imagine how a vicious feed-forward cycle could result in rapid deterioration of cellular energy homeostasis. In this light, and given that mitochondrial integrity is critical to cell survival, it could be argued that adaptations favoring increased IMTG and lowered fat oxidation, typically seen in late or more severe obesity (61, 130, 131), might operate to protect muscle mitochondria from lipid-induced insults.

Importantly, mitochondria are equipped with several antioxidant defense mechanisms, including the Mn-dependent superoxide dismutase, catalase, peroxiredoxins, and the glutathione systems (127). Additionally, there is growing support for the notion that the fatty acid–inducible UCPs (UCP2 and UCP3 in muscle) play a key role in combating oxidative stress by dissipating the proton gradient across the inner mitochondrial membrane (128, 132). Notably, many of these defense systems have been shown to be induced by both PGC1α and exercise (133, 134). Likewise, cross-sectional studies have found that muscle lipid peroxidation is lower in endurance-trained subjects compared with either obese or insulin-resistant subjects, despite higher rates of lipid uptake, storage, and catabolism in the latter (135). Thus, the lipid-tolerant phenotype of exercise-trained and/or type I muscle fibers may relate to a resident mitochondrial population that is better prepared to cope with oxidant stress.

In closing, although the precise molecular mechanisms that link fatty acid oversupply to muscle insulin resistance are yet unfolding, many important clues have emerged in recent years. It is clear that ingestion of high-fat diets and obesity contribute to changes in adipose and liver-derived hormones, cytokines, and metabolites that impair insulin action in muscle. Increased delivery of lipids to muscle in normal physiological cycles of fasting or exercise induces a program of increased fatty acid oxidation, which is appropriate for sparing glucose and energy requirements, respectively. However, chronic lipid oversupply seems to induce a "disconnect" between β-oxidation of fatty acids and the further oxidation of acetyl-CoA in the TCA cycle. Whereas more work is required to fully understand how these metabolic regulatory changes impinge upon insulin action in muscle tissue, recent findings support the provocative possibility that insulin desensitization depends on a signal or set of signals that emanate from events occurring within the mitochondrial compartment.

Pancreatic Islets

As is true for skeletal muscle, the effects of lipids on pancreatic islet function, growth, and survival are complex. When applied to islets in an acute fashion, fatty acids stimulate insulin secretion in a glucose-dependent manner. These effects of fatty acids are dependent on their chain length and degree of saturation with long-chain saturated fatty acids (C16:0, C18:0) exhibiting the highest potency (136). A further, recently emergent complexity is that fatty acids may potentiate

GSIS:
glucose-stimulated insulin secretion

Pyruvate cycling:
pyruvate carboxylase-catalyzed exchange of pyruvate with TCA cycle intermediates

insulin secretion either by generation of messengers during metabolism or via interaction with cell-surface G protein–coupled receptors (137, 138), or some combination of both mechanisms. Fatty acids also have chronic effects on islet cell function and survival. Thus, total depletion of islet triglyceride stores in rats with experimental hyperleptinemia (139) or by treatment with the antilipolytic agent nicotinic acid (140) results in loss of insulin secretion in response to glucose and many other secretagogues, whereas acute provision of fatty acids to the perfused pancreas of hyperleptinemic rats (85) or infusion of lipids into nicotinic acid–treated rats (141) rapidly restores secretory function. Conversely, the dysfunctional islets in animal models of obesity and type 2 diabetes are lipid-laden, with the rise in stored triglyceride preceding the onset of β-cell dysfunction (142). Moreover, chronic exposure of rodent islets, human islets, or insulinoma cell lines to fatty acids causes impairment of glucose-stimulated insulin secretion (143–146). However, this effect of lipid exposure may depend on concomitant hyperglycemia (147, 148), and in the in vivo setting it may require genetic mutations that predispose to lipid damage.

To provide context for discussion of the mechanisms of lipid-induced impairment of β-cell function, we first provide a brief review of the current understanding of the glucose-stimulated insulin secretion (GSIS) process in normal islets. Glucose enhances insulin secretion not by binding to a receptor on the surface of β-cells, but via its metabolism to generate second messengers that trigger exocytosis of insulin-containing secretory granules (149, 150). Classically, much attention has been focused on ATP production (and increases in the ATP:ADP ratio) as a by-product of glucose metabolism that regulates insulin secretion via closure of ATP-sensitive K$^+$ (K$_{ATP}$) channels that reside at the β-cell plasma membrane, resulting in turn in membrane depolarization, activation of voltage-gated Ca^{2+} channels, and Ca^{2+}-mediated stimulation of granule exocytosis (149–152). However, it has become increasingly apparent that this mechanism is inadequate to fully explain the entire process of GSIS. Thus, islets treated with elevated concentrations of K$^+$ and diazoxide, an agent that opens K$_{ATP}$ channels, are still capable of significant insulin secretion in response to changes in glucose concentrations, described as occurring via a "K$_{ATP}$ channel-independent pathway" (153, 154). These findings strongly imply the involvement of glucose-derived second messengers other than ATP and ADP in control of GSIS.

More recent work suggests nonoxidative pathways of mitochondrial glucose metabolism may be the source of second messengers that synergize with changes in ATP and ADP levels to regulate insulin secretion. For example, subclones of the rat insulinoma cell line (INS)-1 (155) have been isolated and characterized as having either robust or weak GSIS (156). Comprehensive analysis of mitochondrial metabolism of ^{13}C glucose by NMR-based mass isotopomer analysis has revealed no differences in pyruvate dehydrogenase-catalyzed entry of pyruvate into the TCA cycle in robustly versus poorly glucose responsive clones. In contrast, glucose responsiveness is found to be tightly correlated with pyruvate carboxylase-catalyzed anaplerotic influx of pyruvate to the TCA cycle as oxaloacetate and further participation of oxaloacetate in various pyruvate cycling pathways (**Figure 4a**) (157, 158). Pyruvate cycling is enabled in β-cells by a high level of PC expression and very low or absent expression of phosphoenolpyruvate carboxykinase (159, 160). Thus in β-cells, unlike the liver, anaplerotic influx of pyruvate does not lead to formation of phosphoenolpyruvate and gluconeogenesis, but rather results in resynthesis of pyruvate.

The strong correlation between pyruvate cycling pathways and the capacity for GSIS has focused recent attention on metabolic by-products of pyruvate cycling pathways as stimulus/secretion coupling factors. Prominent candidates that have been investigated include glutamate, malonyl-CoA, long-chain

a

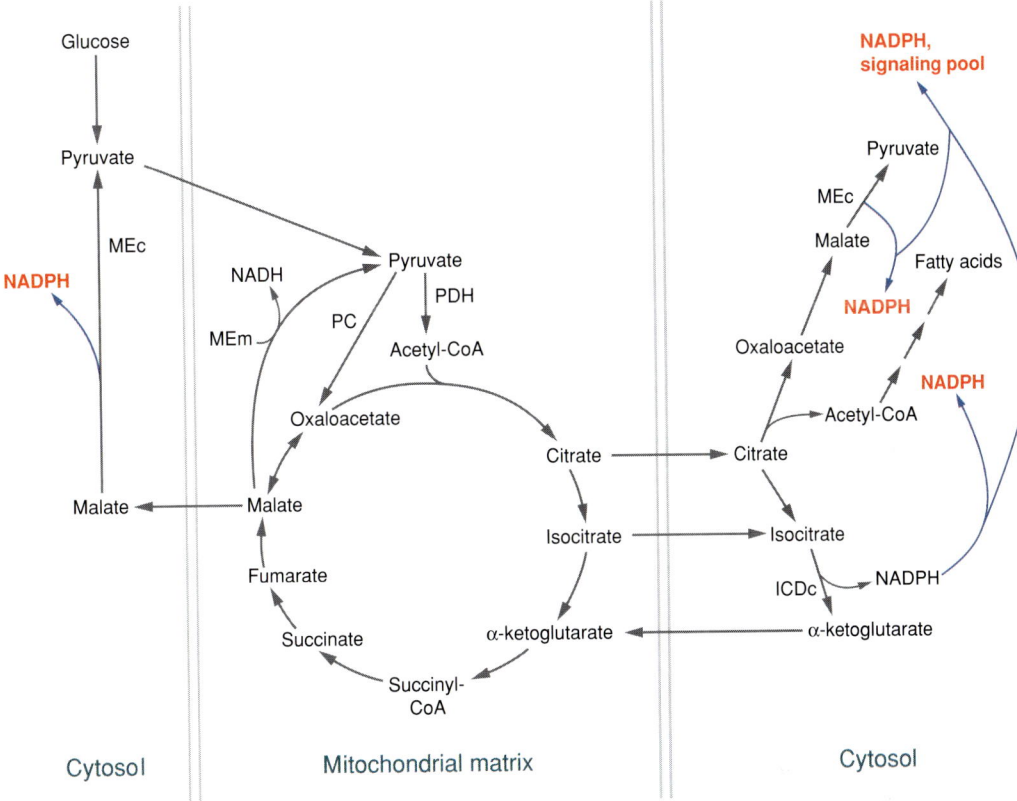

Figure 4

Mechanisms of glucose-stimulated insulin secretion and impairment by lipids. (*a*) Role of pyruvate cycling pathways in glucose-stimulated insulin secretion (GSIS). During GSIS, pyruvate derived from glucose can enter mitochondrial metabolic pathways either via pyruvate dehydrogenase (PDH) or pyruvate carboxylase (PC). Pyruvate that enters via PC serves an anaplerotic function and allows for egress of the tricarboxylic acid (TCA) cycle intermediates malate, citrate, and isocitrate to the cytosol, where they are recycled to pyruvate. As detailed in the text, NADPH has emerged as an attractive candidate stimulus-secretion coupling factor that would be produced by the pyruvate cycling enzymes cytosolic malic enzyme (MEc) or cytosolic isocitrate dehydrogenase (ICDc). The mechanism by which glucose-induced increases in NAPDH stimulate exocytosis of insulin-containing secretory granules is unknown but could be via service as a cofactor for de novo lipogenesis or via binding to specific signaling proteins. (*b*) Proposed mechanism by which chronic exposure of pancreatic islets to elevated free fatty acids leads to loss of the glucose-induced increment in pyruvate cycling activity and impairment of GSIS. In this model, exposure of islets to excess long-chain fatty acids (LCFA) results in increased expression of enzymes of fatty acid oxidation, resulting in enhanced production of acetyl-coenzyme A (CoA) in the cytosol. This high level of acetyl-CoA results in increased PC activity via allosteric activation of the enzyme, resulting in enhanced pyruvate cycling activity at basal glucose. This increase in basal cycling activity obliterates the normal glucose-induced rise in pyruvate cycling activity (15). Recent studies also suggest GPR40 receptors may contribute to lipid-induced impairment of β-cell function (161). Abbreviations: ACS, acyl-CoA synthetase; CPT, carnitine palmitoyltransferase; MEm, mitochondrial malic enzyme.

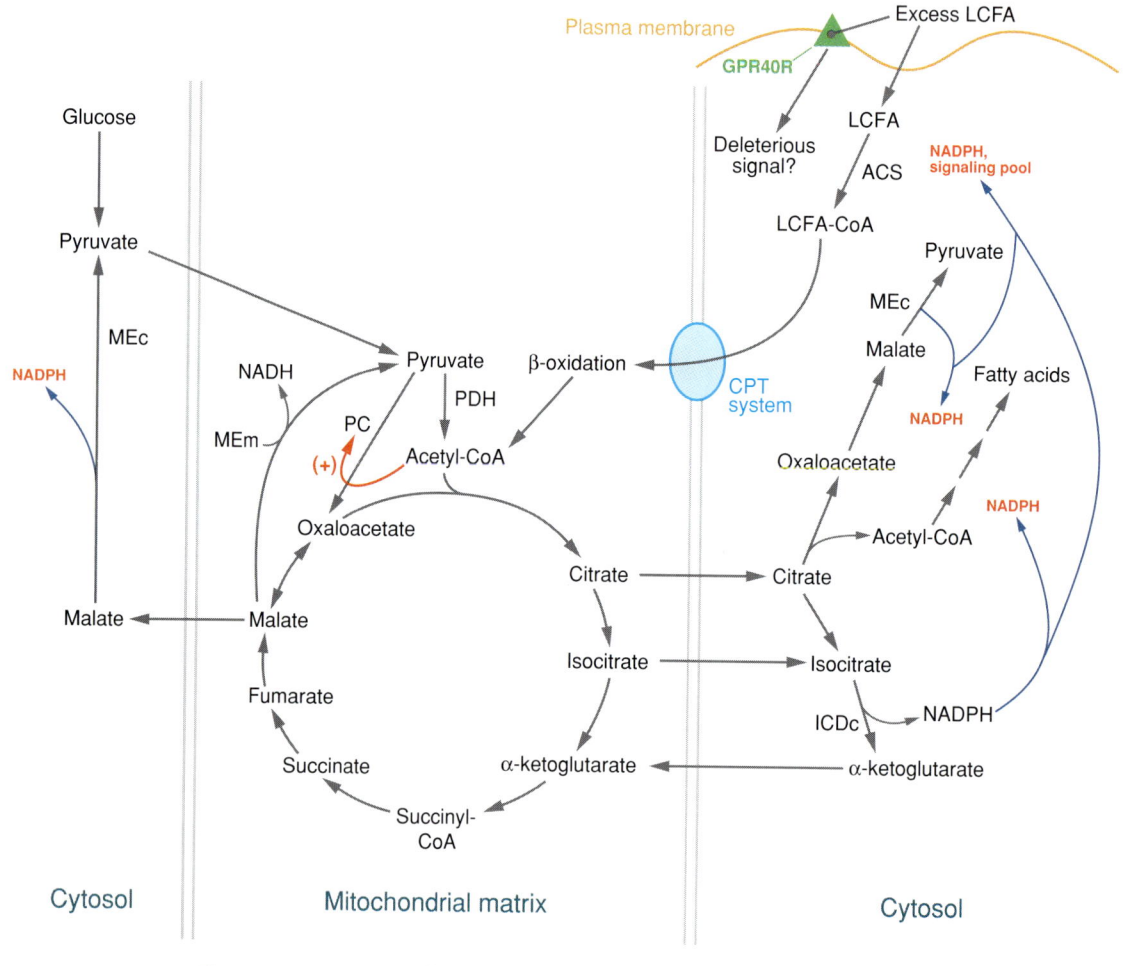

Figure 4

(Continued)

acyl-CoAs, and NAPDH. It was proposed that glucose-induced increases in cytosolic glutamate concentrations, achieved by conversion of α-ketoglutarate to glutamate by glutamate dehydrogenase, would stimulate insulin granule exocytosis (162). However, a number of subsequent studies by other groups have failed to establish any clear causal relationship between changes in glutamate concentration and insulin release (157, 163–165). The existence of a familial form of hyperinsulinism caused by an activating mutation of glutamate dehydrogenase seemed to buttress the glutamate signaling hypothesis (166), but it has subsequently been revealed that these mutations enhance glutamate oxidation rather than accumulation (167).

The "malonyl-CoA/long-chain acyl-CoA signaling pathway" was proposed as a mediator of GSIS on the basis of a rapid glucose-induced increase in the levels of malonyl-CoA (preceding insulin secretion) in β-cells (168, 169) and because this metabolite has a strong inhibitory effect on CPT1 activity (87).

Inhibition of CPT1 is predicted to suppress fatty acid oxidation and increase lipid esterification pathways, and possibly LC-CoA pools. However, subsequent work has clearly shown that expression of malonyl-CoA decarboxylase in insulinoma cell lines and islets blocks the glucose-induced rise in malonyl-CoA with no effect on GSIS (170–172). It has been suggested that an effect of MCD-mediated suppression of malonyl-CoA can be revealed by inclusion of fatty acids in the secretion media (172), but this has not been observed in all laboratories (171). Manipulation of malonyl-CoA and fatty acyl-CoA levels has also been attempted in the context of studies of lipid-induced impairment of GSIS, as discussed in more detail below. Thus, at the present time, the weight of evidence argues against a role of glutamate or malonyl-CoA in regulation of GSIS.

At present, the most attractive candidate second messenger for GSIS derived from pyruvate cycling pathways appears to be NADPH. As seen in **Figure 4**, all of the possible cycling pathways (including the pyruvate/malate cycle, the pyruvate/citrate cycle, and the pyruvate/isocitrate cycle) will result in NADPH production in the cytosol, either via malic enzyme or cytosolic NADP-dependent isocitrate dehydrogenase (ICDc). Evidence supporting a signaling role for NADPH includes the following:

1. Glucose causes dose-dependent increases in NADPH and the NADPH:NADP ratio in rodent islets and β-cell lines, whereas NADH and NADH levels are not as well correlated (173, 174).
2. Addition of NAPDH to patch-clamped β-cells stimulates exocytosis as measured by increases in cell capacitance (173).
3. Suppression of ICDc expression by small interfering RNA (siRNA) technology reduces NADPH and the NAPDH:NADP ratio, and it strongly impairs GSIS in β-cell lines and primary rat islets (174).

The mechanisms by which changes in the cytosolic NAPDH:NADP ratio regulate insulin granule exocytosis remain to be established (159). One hypothesis is that cellular redox acceptor proteins such as glutaredoxin (GRX) or thiaredoxin may be involved. Consistent with this idea, GRX is expressed at higher levels in β-cells than in non-β-cells in the pancreatic islets. Moreover, coaddition of GRX and NADPH to patch-clamped β-cells results in augmentation of granule exocytosis relative to NAPDH alone, whereas thiaredoxin antagonizes the stimulatory effect of NADPH (173). Although these initial findings are intriguing, they await confirmation via manipulation of GRX expression in intact islets and β-cell lines, and the molecular links between NADPH, GRX, and the secretory apparatus also remain to be established. Another hypothesis for a role of NADPH is its participation in de novo lipogenesis as an obligate cofactor in the fatty acid synthetase reaction (159). Indeed, suppression of the NADPH:NADP ratios by delivery of siRNA for ICDc results in reduced incorporation of radiolabeled glucose into β-cell lipids (174). However, direct manipulation of pathways that convert TCA cycle intermediates to fatty acids has yielded conflicting results, such that no definitive conclusions can yet be advanced on the possible role of such pathways in GSIS.

We now turn our attention to a discussion of the mechanism(s) of β-cell failure in obesity and type 2 diabetes, with a specific focus on the deleterious effects of chronic exposure of β-cells to elevated lipid levels. A well-studied model of development of β-cell dysfunction has been the ZDF rat. These animals have a point mutation in the leptin receptor, leading to hyperphagia, obesity, insulin resistance, and, by age 11–12 weeks, frank diabetes (hyperglycemia) (142). Shortly after birth, animals are lean, and the islets are of a normal size and cellular architecture. As insulin resistance appears, ZDF rats are initially able to mount a compensatory response, manifested by an increase in β-cell mass secondary to

ICDc: NADP-dependent isocitrate dehydrogenase

β-cell hyperplasia. However, as the insulin resistance worsens, islet cell mass decreases in concert with the appearance of fibrotic lesions between the islet cells. Simultaneous with the loss of β-cell mass comes a loss of β-cell function, involving an initial increase in basal insulin secretion followed by a loss of glucose's ability to stimulate insulin secretion above the now elevated basal rate of insulin release (142). Importantly, recent studies indicate the changes in β-cell mass and function observed in ZDF rats appear to closely mimic changes that occur in the progression to human diabetes (175).

The foregoing β-cell deterioration in ZDF rats is accompanied by a large accumulation of islet triglycerides, and the rise in stored lipids immediately precedes the onset of β-cell dysfunction (142). Moreover, exposure of isolated pancreatic islets or β-cell lines to elevated levels of fatty acids causes impairment of GSIS and, with more prolonged exposures, triggers β-cell apoptosis (143–146). However, prolonged feeding of a high-fat diet to normal rats (176) or mice (177) or diabetes-resistant female ZDF rats (178)—although resulting in obesity, hyperlipidemia, and severe insulin resistance—does not result in dramatic β-cell decompensation. Thus, in these models, the enlarged islet mass is sustained, and GSIS does not fail. Similarly, obese humans can manifest severe insulin resistance for many years without progression to frank diabetes, probably because of prolonged β-cell compensation. A recent study has shown that acute lipid infusion causes dramatic impairment of β-cell function in subjects with a family history of type 2 diabetes compared to no functional defect in controls with no family history (179). These findings suggest that in a nonpredisposing genetic environment, increased lipid supply may actually contribute to increased β-cell mass and function. However, in a predisposing genetic environment, chronic lipid exposure eventually leads to β-cell failure. Below we attempt to delineate those biochemical adaptations to chronic lipid exposure that constitute part of the normal compensatory response from those that may participate in the development of β-cell dysfunction.

One of the first hypotheses to emerge in an attempt to explain the link between hyperlipidemia and β-cell dysfunction was the simple idea that β-cell functional impairment is directly linked to TAG accumulation. Support for this idea came from studies showing that TAG content is inversely correlated with glucose responsiveness (144, 180) and that lipid-induced impairment of β-cell function and TAG accumulation requires coexposure to elevated glucose concentrations (147, 148, 181, 182). Also, adenovirus-mediated expression of diacylglycerol acyltransferase in rat islets causes TAG accumulation and coincident impairment of GSIS (183). More recently, a direct test of the role of TAG in β-cell dysfunction has been achieved via adenovirus-mediated expression of MCD in lipid-exposed β-cell lines and islets from ZDF rats (158). As discussed above, this enzyme catalyzes the degradation of malonyl-CoA to acetyl-CoA, thereby relieving allosteric inhibition of CPT1 by the former metabolite and "shunting" lipids away from synthesis/esterification pathways and toward mitochondrial oxidation. Exposure of the robustly glucose-responsive insulinoma (INS-1)-derived cell line 832/13 to 0.5-mM oleate:palmitate (2:1, complexed to bovine serum albumin) resulted in a large increase in TG content and a strong impairment of GSIS. Treatment of lipid-cultured cells with the virus containing the MCD cDNA prevented TAG accumulation but did not prevent lipid-induced impairment of GSIS. Similarly, expression of MCD in islets from ZDF rats significantly decreased their TAG levels without restoring GSIS. Thus, TAG accumulation per se does not explain lipid-induced impairment of GSIS, suggesting instead that this serves as a marker of another fat-mediated pathophysiologic event. Indeed, TAG storage has even been proposed to represent a protective mechanism against lipotoxicity, based on finding that palmitate, a saturated long-chain fatty acid, causes more potent inhibition of

---

INS-1: insulinoma cell line

GSIS and stimulation of apoptosis than the monounsaturated fatty acid oleate, whereas oleate is the preferred substrate for lipid esterification and TAG synthesis (184). It remains possible that MCD expression failed to remove lipid-derived metabolites other than TG that are the real causal agent for β-cell impairment. For example, the fatty acid–derived metabolite ceramide has been linked both to impairment of insulin gene expression (185) and increased rates of β-cell apoptosis (186) in lipid-cultured β-cells.

A second potential mechanism of lipid-induced β-cell failure involves uncoupling protein 2 (UCP2) and the formation of ROS. Interest in this area was stimulated by studies showing that UCP2-deficient mice have enhanced glucose tolerance, whereas overexpression of UCP2 results in impaired GSIS (187). Furthermore, UCP2 expression is increased in islets of *ob/ob* mice, and breeding of UCP (−/−) mice with *ob/ob* mice results in the restoration of first-phase insulin secretion and normalization of blood glucose levels (187). Subsequently, several laboratories have shown that exposure of islets or insulinoma cell lines to elevated fatty acid levels increases UCP2 expression (188–190), whereas lipid-induced impairment of β-cell function is prevented in islets of UCP2 (−/−) mice (191). There is also evidence that lipid-induced increases in UCP2 expression are mediated by SREBP1c, whose effects are antagonized by AMPK (190, 192). Thus in the fasted state or in islets cultured at low glucose, AMPK is active and SREBP1c levels are relatively low. In the fed state or in response to an increase in medium glucose, SREBP1c is induced and AMPK activity is decreased, leading to increased UCP2 expression. Indeed overexpression of SREBP1c in islets increases UCP2 expression and impairs GSIS, an effect that can be partially reversed by siRNA-mediated silencing of UCP2 expression (190). Finally, a recent study provides evidence for regulation of UCP2 expression by Sirt1, the mammalian ortholog of Sir 2, an NAD-dependent histone deacetylase linked to longevity and metabolic reprogramming in yeast and flies. Overexpression of Sirt1 specifically in β-cells results in decreased UCP2 expression, improved glucose tolerance, and increased insulin secretion (193).

Although the foregoing data seem to establish an important role for UCP2 in control of β-cell function and lipid-induced β-cell failure, several important questions remain. Compared with UCP1, which is expressed in brown fat and clearly involved in thermogenesis, the uncoupling/thermogenic effects of UCP2 are relatively modest (194), causing uncertainty as to the specific metabolic effects of increased UCP2 expression in the β-cell. There is evidence that UCP2 overexpression mimics the effects of chronic palmitate exposure by reducing glucose-stimulated mitochondrial membrane potential via its mitochondrial proton leak activity, resulting in impaired ATP production during glucose stimulation (195, 196). It has also been reported that palmitate increases ROS production in normal islets, and this response is not observed in UCP2 (−/−) islets (195). However, knockout of UCP2 causes an increase in ROS production, an effect seemingly at odds with the enhanced insulin secretion in these islets. Moreover, UCP2 proton leak activity appears to be activated by a specific ROS, superoxide, and impaired GSIS seen in islets with chronic exposure to elevated glucose levels (25 mM) or in *ob/ob* islets can be partially corrected by the lowering of intracellular superoxide levels by overexpression of Mn-dependent superoxide dismutase (197). Finally, another recent study presents evidence against the involvement of oxidative stress in lipid-mediated β-cell functional impairment (198). This study shows no increase in ROS (peroxide) or reactive nitrogen species in response to 72 h of exposure of rat islets to 16.7-mM glucose + 0.5-mM palmitate or oleate. Moreover, the addition of antioxidants such as N-acetyl-cysteine or pyridoxamine failed to correct lipid impairment of GSIS in these studies. It should also be noted that UCP2 expression is increased to only a modest extent (generally in the range of 30%–100%) by chronic exposure of islets to elevated fatty acids. Thus, whereas the functional impact of genetic manipulation of UCP2 expression or chemical activation of its proton leak activity is clear, it remains uncertain if UCP2 activation is an essential component of β-cell failure in diabetes. In particular, if UCP2 expression is specifically increased by fatty acids, why do islets in normal rodents exhibit sustained compensation for insulin resistance induced by high-fat feeding, even in the face of hyperlipidemia? This issue requires further investigation.

Another possible effect of the prolonged exposure of islets to increased levels of fatty acids could reside at the level of mitochondrial fuel metabolism and generation of secretion coupling factors via pyruvate cycling pathways. Until recently, it has been difficult to develop a clear picture from a survey of the literature in the field owing to a number of significant disagreements. For example, one group has reported that islets exposed to fatty acids experience a reduction in PDH activity in concert with a fall in glucose oxidation, leading to the suggestion that a glucose–fatty acid (Randle) cycle is operative in such cells (144, 180, 199). However, studies from two other laboratories failed to demonstrate significant lipid-induced impairment of PDH activity in INS-1 cells (199) or rat islets (200). In the Randle hypothesis, a rise in citrate is suggested to slow glycolytic flux via inhibition of 6-phosphofructo-1-kinase activity. However, citrate levels are reported to be either unchanged (199) or decreased (201) and phosphofructokinase activity to be increased (202) in various β-cell preparations following lipid exposure. More recently, it has been reported that long-term exposure of MIN-6 mouse insulinoma cells to fatty acids results in a reduction of the levels of PC protein (203). These authors also suggested the consequence of such a lowering of PC might be a reduction in "malate-pyruvate shuttle flux," but this conclusion appeared to be based solely on a decrease in NAD(P)H autofluoresence in fat-cultured cells, rather than any direct measurement of a metabolic pathway. In contrast, another group has reported no change in PC V_{max} in fat-cultured islets and further speculated that malate-pyruvate shuttle flux would be increased rather than decreased because of a 60% rise in intracellular pyruvate concentrations (200). The same group has also reported an increase in PC V_{max} in islets from nondiabetic Zucker fatty rats and has suggested this would lead to increased pyruvate cycling, thereby possibly explaining the enhanced insulin secretion of such islets that compensates for insulin resistance (204). However, this conclusion was based on static measurement of enzyme activities and concentrations of selected metabolic intermediates rather than any direct measurement of metabolic flux.

In light of this confusion, more comprehensive methods of metabolic analysis were required. The application of ^{13}C NMR to analyze pathways of pyruvate metabolism in β-cell mitochondria led to the discovery that the activity of PC-catalyzed pyruvate cycling pathways can be used to distinguish robustly glucose-responsive from poorly glucose-responsive INS-1-derived cell lines, as discussed above (157). These methods have recently been applied in conjunction with more traditional biochemical assays to gain insight into lipid-induced metabolic changes in β-cells (158). These studies reveal that chronic lipid exposure causes a rise in oxygen consumption at 3-mM glucose that occurs in concert with an increase in endogenous substrate oxidation. Furthermore, the normal glucose-induced increment in pyruvate cycling activity is completely ablated in lipid-cultured cells as a result of a large increase in cycling at basal glucose levels. Interestingly, these changes occur in the absence of any significant change in the rate of ^{13}C glucose oxidation in lipid-cultured versus control cells, at either 3- or 12-mM glucose. The coordinate increase in oxygen consumption and endogenous fuel oxidation at 3-mM glucose observed in lipid-cultured cells is probably a result of an increase in fatty acid oxidation. This follows from the obvious increase in supply of this substrate in fat-cultured cells and is consistent with prior studies demonstrating increased expression of enzymes of lipid oxidation in β-cells in response to chronic lipid exposure (205, 206). Acetyl-CoA generated from lipid oxidation is known to influence pyruvate metabolism via its capacity to activate PC. It would appear that this mechanism is retained in β-cells, given the large increase in PC-catalyzed pyruvate cycling activity that occurs at basal glucose in response to chronic lipid culture (158). This rise in basal pyruvate cycling activity eliminates the normal glucose-induced increment in cycling activity, with the prediction that by-products of cycling pathways such as NAPDH (discussed above) would no longer increase in response to glucose. A schematic summary of these ideas is presented in **Figure 4b**. This model may also explain why overexpression of malonyl-CoA decarboxylase in islets or β-cell lines cultured in normal medium (lacking exogenous fatty acids) have completely normal GSIS (170, 172, 207), in that such cells will have only limited endogenous lipid stores for oxidation, acetyl-CoA production, and activation of PC.

Further evidence for the potential importance of dysregulated pyruvate cycling in mediating lipid-induced β-cell failure comes from studies with a membrane permeant malate ester. Malate is an intermediate common to all of the potential pyruvate cycles shown in **Figure 4a**, suggesting its provision to dysfunctional islets and β-cell could overcome the lipid-induced impairment in insulin secretion. Remarkably, addition of dimethylmalate to lipid-cultured 832/13 cells or lipid-laden ZDF islets during static incubation insulin secretion assays caused dramatic improvement in GSIS (158). In conjunction with the clear change in pyruvate cycling evoked by lipid culture, these data provide strong support for the idea that lipid-induced impairment of GSIS is caused at least in part by alteration of the metabolic fate of pyruvate. As for the case of UCP2, however, one must still explain the enormous capacity of islets from normal animals to compensate for systemic insulin resistance, even in the face of hyperlipidemia. As discussed above, the damaging effects of lipids may require concomitant hyperglycemia (147). This would imply that in obesity and overnutrition, transient elevations of glucose levels above normal values may occur as the β-cells struggle to deliver enough insulin to clear the large amounts of glucose and other metabolic fuels coming from the diet. As these episodes increase in frequency and severity, the full brunt of lipid-induced impairment of β-cell function is realized. A recent study also suggests the GPR40 receptor, first described as involved in fatty acid potentiation of GSIS, may also contribute to loss of GSIS in rodents fed on a high-fat diet (161) (see **Figure 4b**). Clearly, further studies are required to unravel the relative contributions and potential synergies of elevated levels of glucose and fatty acids in the alteration of pyruvate metabolism, and the possible contribution of cell-surface fatty acid receptors to this and other mechanisms of functional deterioration of the β-cell.

CONCLUDING REMARKS

In this chapter, we attempt to develop two broad concepts related to mechanisms of obesity and high-fat diet–induced metabolic dysregulation. First, we summarize studies that argue for an important primary role for adipose tissue and the liver in the disruption of metabolic fuel homeostasis. This includes a review of the important roles these tissues play in the production of hormones (adiponectin, leptin), cytokines (IL-6, TNFα), and metabolites (free fatty acids and VLDL) that contribute to impaired insulin action and secretion in skeletal muscle and islets, respectively. The second key concept is that the loss of function in the responder tissues, skeletal muscle, and pancreatic islets may be mediated in part by increased rather than decreased rates of fatty acid oxidation, in a failed attempt to compensate for lipid overload. In the case of muscle, metabolic dysregulation occurs when TCA cycle enzymes are unable to keep pace with the increased rate of β-oxidation, whereas in islets, increased fatty acid oxidation may stimulate pyruvate cycling to an extent that obviates the normal glucose-induced increment in this pathway. We hope these views stimulate continued research toward a complete understanding of obesity-related metabolic dysfunction.

SUMMARY POINTS

1. Strong evidence argues for an important primary or driver role for adipose tissue and the liver in the disruption of metabolic fuel homeostasis.
2. The failure of the adipocyte to adequately sequester excess fuel may ultimately precipitate the metabolic syndrome by causing systemic hyperlipidemia.
3. Lipid oversupply to peripheral tissues leads to excessive uptake, storage, and metabolism of fatty acids, eventually leading to functional impairments.
4. Lipid-induced dysfunction in the responder tissues, skeletal muscle, and pancreatic islets may be mediated in part by increased, rather than decreased, rates of fatty acid oxidation in a failed attempt to compensate for lipid overload.
5. In the case of muscle, metabolic dysregulation may occur when TCA cycle enzymes are unable to keep pace with the increased rates of β-oxidation.
6. In the pancreatic islets, increased fatty acid oxidation may stimulate pyruvate cycling to an extent that obviates the normal glucose-induced increment in this pathway.

FUTURE ISSUES TO BE RESOLVED

1. Is mitochondrial stress and/or failure the principal cause of lipid-induced tissue dysfunction?
2. Do specific fatty acid species play distinct roles in causing or combating lipid-induced tissue dysfunction, and if so, what underlying biochemical and molecular mechanisms are responsible for these distinctions?
3. Can pharmacological activation of PGC1α and/or other targets that enhance muscle TCA cycle activity mimic the metabolic benefits of exercise training?

ACKNOWLEDGMENTS

We thank the many members of our laboratories for their contributions to the concepts put forth in this review. We are grateful for the following support relevant to this article: National Institutes of Health grants PO1-DK-58398 (C.B.N.), RO1-DK46492 (C.B.N.), and KO1-DK067200 (D.M.M.), Takeda Pharmaceuticals (C.B.N.), the American Diabetes Association (D.M.M.), and GlaxoSmithKline (D.M.M.).

LITERATURE CITED

1. Stein CJ, Colditz GA. 2004. *J. Clin. Endocrinol. Metab.* 89:2522–25
2. Shi YG, Burn P. 2004. *Nat. Rev. Drug Discov.* 3:695–710
3. Hajri T, Abumrad NA. 2002. *Annu. Rev. Nutr.* 22:383–415
4. Coburn CT, Knapp FF Jr, Febbraio M, Beets AL, Silverstein RL, Abumrad NA. 2000. *J. Biol. Chem.* 275:32523–29
5. Hirsch D, Stahl A, Lodish HF. 1998. *Proc. Natl. Acad. Sci. USA* 95:8625–29
6. Kim JK, Gimeno RE, Higashimori T, Kim HJ, Choi H, et al. 2004. *J. Clin. Investig.* 113:756–63
7. Herrmann T, van der Hoeven F, Grone HJ, Stewart AF, Langbein L, et al. 2003. *J. Cell Biol.* 161:1105–15
8. Zhang Y, Proenca R, Maffei M, Barone M, Leopold L, Friedman JM. 1994. *Science* 372:425–32
9. Pelleymounter MA, Cullen MJ, Baker MB, Hecht R, Winters D, et al. 1995. *Science* 269:540–43
10. Steppan CM, Bailey ST, Bhat S, Brown EJ, Banerjee RR, et al. 2001. *Nature* 409:307–12
11. Uysal KT, Wiesbrock SM, Marino MW, Hotamisligil GS. 1997. *Nature* 389(6651):610–14
12. Yamauchi T, Kamon J, Waki H, Terauchi Y, Kubota N, et al. 2001. *Nat. Med.* 7(8):941–46
13. Moller DE. 2000. *Trends Endocrinol. Metab.* 11(6):212–17
14. Tomas E, Tsao TS, Saha AK, Murrey HE, Zhang CC, et al. 2002. *Proc. Natl. Acad. Sci. USA* 99:16309–13
15. Minokoshi Y, Kim YB, Peroni OD, Fryer LG, Muller C, et al. 2002. *Nature* 415:339–43
16. An J, Muoio DM, Shiota M, Fujimoto Y, Cline GW, et al. 2004. *Nat. Med.* 10:268–74
17. Bjorbaek C, Elmquist JK, Frantz JD, Shoelson SE, Flier JS. 1998. *Mol. Cell* 1:619–25
18. Ruderman NB, Saha AK, Kraegen EW. 2003. *Endocrinology* 144:5166–71
19. Uysal KT, Wiesbrock SM, Marino MW, Hotamisligil GS. 1997. *Nature* 389:610–14
20. Cai DS, Yuan MS, Frantz DF, Melendez PA, Hansen L, et al. 2005. *Nat. Med.* 11:183–90
21. Yang Q, Graham TE, Mody N, Preitner F, Peroni OD, et al. 2005. *Nature* 436:356–62
22. Ross SR, Graves RA, Spiegelman BM. 1993. *Genes Dev.* 7:1318–24
23. Moitra J, Mason MM, Olive M, Krylov D, Gavrilova O, et al. 1998. *Genes Dev.* 12:3168–81
24. Shimomura I, Hammer RE, Richardson JA, Ikemoto S, Bashmakov Y, et al. 1998. *Genes Dev.* 12:3182–94
25. Reitman ML, Gavrilova O. 2000. *Int. J. Obes. Relat. Metab. Disord.* 24(Suppl. 4):S11–14
26. Gavrilova O, Marcus-Samuels B, Graham D, Kim JK, Shulman GI, et al. 2000. *J. Clin. Investig.* 105:271–78
27. Shimomura I, Hammer RE, Ikemoto S, Brown MS, Goldstein JL. 1999. *Nature* 401:73–76
28. Ebihara K, Ogawa Y, Masuzaki H, Shintani M, Miyanaga F, et al. 2001. *Diabetes* 50:1440–48
29. Colombo C, Cutson JJ, Yamauchi T, Vinson C, Kadowaki T, et al. 2002. *Diabetes* 51:2727–33
30. Oral EA, Simha V, Ruiz E, Andewelt A, Premkumar A, et al. 2002. *N. Engl. J. Med.* 346:570–78
31. Abel ED, Peroni O, Kim JK, Kim YB, Boss O, et al. 2001. *Nature* 409:729–33
32. Sivitz WI, Desautel SL, Kayano T, Bell GI, Pessin JE. 1989. *Nature* 340:72–74
33. Ntambi JM, Miyazaki M, Stoehr JP, Lan H, Kendziorski CM, et al. 2002. *Proc. Natl. Acad. Sci. USA* 99:11482–86

34. Cohen P, Miyazaki M, Socci ND, Hagge-Greenberg A, Liedtke W, et al. 2002. *Science* 297:240–43
35. Dobrzyn P, Dobrzyn A, Miyazaki M, Cohen P, Asilmaz E, et al. 2004. *Proc. Natl. Acad. Sci. USA* 101:6409–14
36. Wellen KE, Hotamisligil GS. 2005. *J. Clin. Investig.* 115:1111–19
37. Yuan M, Konstantopoulos N, Lee J, Hansen L, Li ZW, et al. 2001. *Science* 293:1673–77
38. Kim JK, Kim YJ, Fillmore JJ, Chen Y, Moore I, et al. 2001. *J. Clin. Investig.* 108:437–46
39. Arkan MC, Hevener AL, Greten FR, Maeda S, Li ZW, et al. 2005. *Nat. Med.* 11:191–98
40. Ozcan U, Cao Q, Yilmaz E, Lee AH, Iwakoshi NN, et al. 2004. *Science* 306:457–61
41. Issemann I, Green S. 1990. *Nature* 347:645–50
42. Gilde AJ, Van Bilsen M. 2003. *Acta Physiol. Scand.* 178:425–34
43. Leone TC, Weinheimer CJ, Kelly DP. 1999. *Proc. Natl. Acad. Sci. USA* 96:7473–78
44. Rosen ED, Sarraf P, Troy AE, Bradwin G, Moore K, et al. 1999. *Mol. Cell. Biol.* 4:611–17
45. Norris AW, Chen L, Fisher SJ, Szanto I, Ristow M, et al. 2003. *J. Clin. Investig.* 112:608–18
46. Muoio DM, MacLean PS, Lang DB, Li S, Houmard JA, et al. 2002. *J. Biol. Chem.* 277:26089–97
47. Wang YX, Zhang CL, Yu RT, Cho HK, Nelson MC, et al. 2004. *PLoS Biol.* 2:e294
48. Yechoor VK, Patti ME, Saccone R, Kahn CR. 2002. *Proc. Natl. Acad. Sci. USA* 99:10587–92
49. Puigserver P, Wu ZD, Park CW, Graves R, Wright M, Spiegelman BM. 1998. *Cell* 92:829–39
50. Puigserver P, Spiegelman BM. 2003. *Endocr. Rev.* 24:78–90
51. Lin J, Wu H, Tarr PT, Zhang CY, Wu ZD, et al. 2002. *Nature* 418:797–801
52. Mootha VK, Handschin C, Arlow D, Xie X, St Pierre J, et al. 2004. *Proc. Natl. Acad. Sci. USA* 101:6570–75
53. Koves TR, Li P, An J, Akimoto T, Slentz D, et al. 2005. *J. Biol. Chem.* 280:33588–98
54. Sparks LM, Xie H, Koza RA, Mynatt R, Hulver MW, et al. 2005. *Diabetes* 54:1926–33
55. Chien D, Dean D, Saha AK, Flatt JP, Ruderman NB. 2000. *Am. J. Physiol. Endocrinol. Metab.* 279:E259–55
56. Saha AK, Vavvas D, Kurowski TG, Apazidis A, Witters L, et al. 1997. *Am. J. Physiol. Endocrinol. Metab.* 272:E641–48
57. Abu-Elheiga L, Oh W, Kordari P, Wakil SJ. 2003. *Proc. Natl. Acad. Sci. USA* 100:10207–12
58. Kim JY, Koves TR, Yu GS, Gulick T, Cortright RN, et al. 2002. *Am. J. Physiol. Endocrinol. Metab.* 282:E1014–22
59. Saha AK, Ruderman NB. 2003. *Mol. Cell. Biochem.* 253:65–70
60. Goodpaster BH, Kelley DE. 2002. *Curr. Diabetes Rep.* 2:216–22
61. Hulver MW, Berggren JR, Cortright RN, Dudek RW, Thompson RP, et al. 2003. *Am. J. Physiol. Endocrinol. Metab.* 284:E741–47
62. Krssak M, Falk PK, Dresner A, DiPietro L, Vogel SM, et al. 1999. *Diabetologia* 42:113–16
63. Shulman GI. 2000. *J. Clin. Investig.* 106:171–76
64. Griffin ME, Marcucci MJ, Cline GW, Bell K, Barucci N, et al. 1999. *Diabetes* 48:1270–74
65. Yu CL, Chen Y, Cline GW, Zhang DY, Zong HH, et al. 2002. *J. Biol. Chem.* 277:50230–36
66. Saltiel AR, Pessin JE. 2002. *Trends Cell Biol.* 12:65–71
67. Hirosumi J, Tuncman G, Chang LF, Gorgun CZ, Uysal KT, et al. 2002. *Nature* 420:333–36

68. Perseghin G, Petersen K, Shulman GI. 2003. *Int. J. Obes. Relat. Metab. Disord.* 27(Suppl. 3):S6–11
69. Shoelson SE, Lee J, Yuan M. 2003. *Int. J. Obes. Relat. Metab. Disord.* 27(Suppl. 3):S49–52
70. Kim JK, Fillmore JJ, Sunshine MJ, Albrecht B, Higashimori T, et al. 2004. *J. Clin. Investig.* 114:823–27
71. Muoio DM, Dohm GL, Tapscott EB, Coleman RA. 1999. *Am. J. Physiol. Endocrinol. Metab.* 276:E913–21
72. Bruce CR, Kriketos AD, Cooney GJ, Hawley JA. 2004. *Diabetologia* 47:23–30
73. Koves TR, Noland RC, Bates AL, Henes ST, Muoio DM, Cortright RN. 2005. *Am. J. Physiol. Cell Physiol.* 288:C1074–82
74. Bruce CR, Anderson MJ, Carey AL, Newman DG, Bonen A, et al. 2003. *J. Clin. Endocrinol. Metab.* 88:5444–51
75. Erol E, Cline GW, Kim JK, Taegtmeyer H, Binas B. 2004. *Am. J. Physiol. Endocrinol. Metab.* 287:E977–82
76. Summers SA, Nelson DH. 2005. *Diabetes* 54:591–602
77. Meyer SG, de Groot H. 2003. *Biochim. Biophys. Acta* 1643:1–4
78. Linn SC, Kim HS, Keane EM, Andras LM, Wang E, Merrill AH Jr. 2001. *Biochem. Soc. Trans.* 29:831–35
79. Helge JW, Dobrzyn A, Saltin B, Gorski J. 2004. *Exp. Physiol.* 89:119–27
80. Chavez JA, Knotts TA, Wang LP, Li G, Dobrowsky RT, et al. 2003. *J. Biol. Chem.* 278:10297–303
81. Schmitz-Peiffer C, Craig DL, Biden TJ. 1999. *J. Biol. Chem.* 274:24202–10
82. Chavez JA, Holland WL, Bar J, Sandhoff K, Summers SA. 2005. *J. Biol. Chem.* 280:20148–53
83. Chavez JA, Summers SA. 2003. *Arch. Biochem. Biophys.* 419:101–9
84. Randle PJ, Garland PB, Hales CN, Newsholme EA. 1963. *Lancet* 1:785–89
85. Frayn KN. 2003. *Biochem. Soc. Trans.* 31:1115–19
86. McGarry JD, Mannaerts GP, Foster DW. 1977. *J. Clin. Investig.* 60:265–70
87. McGarry JD. 2002. *Diabetes* 51:7–18
88. Abu-Elheiga L, Matzuk MM, Abo-Hashema KAH, Wakil SJ. 2001. *Science* 291:2613–16
89. Viollet B, Andreelli F, Jorgensen SB, Perrin C, Flamez D, et al. 2003. *Biochem. Soc. Trans.* 31:216–19
90. Ruderman NB, Cacicedo JM, Itani S, Yagihashi N, Saha AK, et al. 2003. *Biochem. Soc. Trans.* 31:202–6
91. Rasmussen BB, Holmback UC, Volpi E, Morio-Liondore B, Paddon-Jones D, Wolfe RR. 2002. *J. Clin. Investig.* 110:1687–93
92. Bavenholm PN, Kuhl J, Pigon J, Saha AK, Ruderman NB, Efendic S. 2003. *J. Clin. Endocrinol. Metab.* 88:82–87
93. Kelley DE, Mandarino LJ. 2000. *Diabetes* 49:677–83
94. Kelley DE, Goodpaster B, Wing RR, Simoneau JA. 1999. *Am. J. Physiol. Endocrinol. Metab.* 277:E1130–41
95. Blaak EE. 2004. *Proc. Nutr. Soc.* 63:323–30
96. Yechoor VK, Patti ME, Saccone R, Kahn CR. 2002. *Diabetes* 51:A258
97. Way JM, Harrington WW, Brown KK, Gottschalk WK, Sundseth SS, et al. 2001. *Endocrinology* 142:1269–77
98. Finck BN, Bernal-Mizrachi C, Han DH, Coleman T, Sambandam N, et al. 2005. *Cell Metab.* 1:133–44
99. de Fourmestraux, V, Neubauer H, Poussin C, Farmer P, Falquet L, et al. 2004. *J. Biol. Chem.* 279:50743–53

100. Turcotte LP, Swenberger JR, Zavitz TM, Yee AJ. 2001. *Diabetes* 50:1389–96
101. Oakes ND, Cooney GJ, Camilleri S, Chisholm DJ, Kraegen EW. 1997. *Diabetes* 46:1768–74
102. Ramsay RR, Naismith JH. 2003. *Trends Biochem. Sci.* 28:343–46
103. Lowell BB, Shulman GI. 2005. *Science* 307:384–87
104. Patti ME, Butte AJ, Crunkhorn S, Cusi K, Berria R, et al. 2003. *Proc. Natl. Acad. Sci. USA* 100:8466–71
105. Patti ME, Butte A, Cusi K, Kohane I, Landaker EJ, et al. 2001. *Diabetes* 50:A247
106. Iossa S, Mollica MP, Lionetti L, Crescenzo R, Tasso R, Liverini G. 2004. *Diabetes* 53:2861–66
107. Petersen KF, Dufour S, Befroy D, Garcia R, Shulman GI. 2004. *N. Engl. J. Med.* 350:664–71
108. Petersen KF, Befroy D, Dufour S, Dziura J, Ariyan C, et al. 2003. *Science* 300:1140–42
109. Kelley DE, He J, Menshikova EV, Ritov VB. 2002. *Diabetes* 51:2944–50
110. Ritov VB, Menshikova EV, He J, Ferrell RE, Goodpaster BH, Kelley DE. 2005. *Diabetes* 54:8–14
111. Mootha VK, Lindgren CM, Eriksson KF, Subramanian A, Sihag S, et al. 2003. *Nat. Genet.* 34:267–73
112. Ling C, Poulsen P, Carlsson E, Ridderstrale M, Almgren P, et al. 2004. *J. Clin. Investig.* 114:1518–26
113. Guerre-Millo M, Rouault C, Poulain P, Andre J, Poitout V, et al. 2001. *Diabetes* 50:2809–14
114. Deems RO, Anderson RC, Foley JE. 1998. *Am. J. Physiol. Regul. Integr. Comp. Physiol.* 274:R524–28
115. Giannessi F, Pessotto P, Tassoni E, Chiodi P, Conti R, et al. 2003. *J. Med. Chem.* 46:303–9
116. Zarain-Herzberg A, Rupp H. 2002. *Expert Opin. Investig. Drugs* 11:345–56
117. Deschamps D, Fisch C, Fromenty B, Berson A, Degott C, Pessayre D. 1991. *J. Pharmacol. Exp. Ther.* 259:894–904
118. Mingrone G. 2004. *Ann. NY Acad. Sci.* 1033:99–107
119. Ames BN, Liu J. 2004. *Ann. NY Acad. Sci.* 1033:108–16
120. Kumaran S, Subathra M, Balu M, Panneerselvam C. 2005. *Exp. Aging Res.* 31:55–67
121. Kumaran S, Savitha S, Anusuya DM, Panneerselvam C. 2004. *Mech. Ageing Dev.* 125:507–12
122. Kumaran S, Subathra M, Balu M, Panneerselvam C. 2004. *Chem. Biol. Interact.* 148:11–18
123. Yamada KA, Kanter EM, Newatia A. 2000. *J. Cardiovasc. Pharmacol.* 36:14–21
124. Oakes ND, Kennedy CJ, Jenkins AB, Laybutt DR, Chisholm DJ, Kraegen EW. 1994. *Diabetes* 43:1203–10
125. Singh BM, Krentz AJ, Nattrass M. 1993. *Diabetes Res. Clin. Pract.* 20:55–62
126. Krentz AJ, Singh BM, Hale PJ, Robertson DA, Nattrass M. 1992. *Diabetes Res.* 20:51–60
127. Andreyev AY, Kushnareva YE, Starkov AA. 2005. *Biochemistry* 70:200–14
128. Schrauwen P, Hesselink MK. 2004. *Diabetes* 53:1412–17
129. Chakraborti S, Chakraborti T. 1998. *Cell. Signal.* 10:675–83
130. Thyfault JP, Kraus RM, Hickner RC, Howell AW, Wolfe RR, Dohm GL. 2004. *Am. J. Physiol. Endocrinol. Metab.* 287:E1076–81
131. Hulver MW, Berggren JR, Carper MJ, Miyazaki M, Ntambi JM, et al. 2005. *Cell Metab.* 2:251–61
132. Hesselink MK, Mensink M, Schrauwen P. 2003. *Obes. Res.* 11:1429–43
133. McArdle F, Spiers S, Aldemir H, Vasilaki A, Beaver A, et al. 2004. *J. Physiol.* 561:233–44

134. Hollander J, Fiebig R, Gore M, Bejma J, Ookawara T, et al. 1999. *Am. J. Physiol. Regul. Integr. Comp. Physiol.* 277:R856–62
135. Russell AP, Gastaldi G, Bobbioni-Harsch E, Arboit P, Gobelet C, et al. 2003. *FEBS Lett.* 551:104–6
136. Stein DT, Stevenson BE, Chester MW, Basit M, Daniels MB, et al. 1997. *J. Clin. Investig.* 100:398–403
137. Itoh Y, Kawamata Y, Harada M, Kobayashi M, Fujii R, et al. 2003. *Nature* 422:173–76
138. Briscoe CP, Tadayyon M, Andrews JL, Benson WG, Chambers JK, et al. 2003. *J. Biol. Chem.* 278:11303–11
139. Koyama K, Chen G, Wang MY, Lee Y, Shimabukuro M, et al. 1997. *Diabetes* 46:1276–80
140. Stein DT, Esser V, Stevenson BE, Lane KE, Whiteside JH, et al. 1996. *J. Clin. Investig.* 97:2728–35
141. Dobbins RL, Chester MW, Stevenson BE, Daniels MB, Stein DT, McGarry JD. 1998. *J. Clin. Investig.* 101:2370–76
142. Lee Y, Hirose H, Zhou YT, Esser V, McGarry JD, Unger RH. 1997. *Diabetes* 46:408–13
143. Unger RH. 1995. *Diabetes* 44:863–70
144. Zhou YP, Grill VE. 1994. *J. Clin. Investig.* 93:870–76
145. Milburn JL Jr, Hirose H, Lee YH, Nagasawa Y, Ogawa A, et al. 1995. *J. Biol. Chem.* 270:1295–99
146. Segall L, Lameloise N, Assimacopoulos-Jeannet F, Roche E, Corkey P, et al. 1999. *Am. J. Physiol. Endocrinol. Metab.* 277:E521–28
147. Poitout V, Robertson RP. 2002. *Endocrinology* 143:339–42
148. Briaud I, Harmon JS, Kelpe CL, Segu VB, Poitout V. 2001. *Diabetes* 50:315–21
149. Newgard CB, McGarry JD. 1995. *Annu. Rev. Biochem.* 64:689–719
150. Newgard CB, Matschinsky FM. 2001. In *Handbook of Physiology*, ed. J Jefferson, A Cherrington, 2:125–52. London: Oxford Univ. Press
151. Johnson RD, Manske DD. 1977. *Pestic. Monit. J.* 11:116–31
152. Cook DL, Hales CN. 1984. *Nature* 311:271–73
153. Gembal M, Detimary P, Gilon P, Gao ZY, Henquin JC. 1993. *J. Clin. Investig.* 91:871–80
154. Komatsu M, Schermerhorn T, Noda M, Straub SG, Aizawa T, Sharp GW. 1997. *Diabetes* 46:1928–38
155. Asfari M, Janjic D, Meda P, Li G, Halban PA, Wollheim CB. 1992. *Endocrinology* 130:167–78
156. Hohmeier HE, Mulder H, Chen G, Henkel-Rieger R, Prentki M, Newgard CB. 2000. *Diabetes* 49:424–30
157. Lu D, Mulder H, Zhao PY, Burgess SC, Jensen MV, et al. 2002. *Proc. Natl. Acad. Sci. USA* 99:2708–13
158. Boucher A, Lu DH, Burgess SC, Telemaque-Potts S, Jensen MV, et al. 2004. *J. Biol. Chem.* 279:27263–71
159. MacDonald MJ, Fahien LA, Brown LJ, Hasan NM, Buss JD, Kendrick MA. 2005. *Am. J. Physiol. Endocrinol. Metab.* 288:E1–15
160. MacDonald MJ. 1995. *J. Biol. Chem.* 270:20051–58
161. Steneberg P, Rubins N, Bartoov-Shifman R, Walker MD, Edlund H. 2005. *Cell Metab.* 1:245–58
162. Maechler P, Wollheim CB. 1999. *Nature* 402:685–89
163. MacDonald MJ, Fahien LA. 2000. *J. Biol. Chem.* 275:34025–27
164. Yamada S, Komatsu M, Sato Y, Yamauchi K, Aizawa T, Hashizume K. 2001. *Endocr. J.* 48:391–95

165. Bertrand G, Ishiyama N, Nenquin M, Ravier MA, Henquin JC. 2002. *J. Biol. Chem.* 277:32883–91
166. Stanley CA, Lieu YK, Hsu BY, Burlina AB, Greenberg CR, et al. 1998. *N. Engl. J. Med.* 338:1352–57
167. MacDonald MJ. 2003. *Metabolism* 52:993–98
168. Corkey BE, Glennon MC, Chen KS, Deeney JT, Matschinsky FM, Prentki M. 1989. *J. Biol. Chem.* 264:21608–12
169. Prentki M, Vischer S, Glennon MC, Regazzi R, Deeney JT, Corkey BE. 1992. *J. Biol. Chem.* 267:5802–10
170. Antinozzi PA, Segall L, Prentki M, McGarry JD, Newgard CB. 1998. *J. Biol. Chem.* 273:16146–54
171. Mulder H, Lu DH, Finley J, An J, Cohen J, et al. 2001. *J. Biol. Chem.* 276:6479–84
172. Roduit R, Nolan C, Alarcon C, Moore P, Barbeau A, et al. 2004. *Diabetes* 53:1007–19
173. Ivarsson R, Quintens R, Dejonghe S, Tsukamoto KV, Renstrom E, Schuit FC. 2005. *Diabetes* 54:2132–42
174. Ronnebaum SM, Burgess SC, Sherry AD, Becker TC, Newgard CB, Jensen MV. 2005. *Diabetes* 54(Suppl. 1):A423 (abstract)
175. Butler AE, Janson J, Bonner-Weir S, Ritzel R, Rizza RA, Butler PC. 2003. *Diabetes* 52:102–10
176. Lee Y, Wang MY, Kakuma T, Wang ZW, Babcock E, et al. 2001. *J. Biol. Chem.* 276:5629–35
177. Hull RL, Kodama K, Utzschneider KM, Carr DB, Prigeon RL, Kahn SE. 2005. *Diabetologia* 48:1350–58
178. Corsetti JP, Sparks JD, Peterson RG, Smith RL, Sparks CE. 2000. *Atherosclerosis* 148:231–41
179. Kashyap S, Belfort R, Gastaldelli A, Pratipanawatr T, Berria R, et al. 2003. *Diabetes* 52:2461–74
180. Zhou YP, Ling ZC, Grill VE. 1996. *Metabolism* 45:981–86
181. Jacqueminet S, Briaud I, Rouault C, Reach G, Poitout V. 2000. *Metabolism* 49:532–36
182. Briaud I, Kelpe CL, Johnson LM, Tran PO, Poitout V. 2002. *Diabetes* 51:662–68
183. Kelpe CL, Johnson LM, Poitout V. 2002. *Endocrinology* 143:3326–32
184. Listenberger LL, Han X, Lewis SE, Cases S, Farese RV Jr, et al. 2003. *Proc. Natl. Acad. Sci. USA* 100:3077–82
185. Kelpe CL, Moore PC, Parazzoli SD, Wicksteed B, Rhodes CJ, Poitout V. 2003. *J. Biol. Chem.* 278:30015–21
186. Shimabukuro M, Zhou YT, Levi M, Unger RH. 1998. *Proc. Natl. Acad. Sci. USA* 95:2498–502
187. Zhang CY, Baffy G, Perret P, Krauss S, Peroni O, et al. 2001. *Cell* 105:745–55
188. Lameloise N, Muzzin P, Prentki M, Assimacopoulos-Jeannet F. 2001. *Diabetes* 50:803–9
189. Li LX, Skorpen F, Egeberg K, Jorgensen IH, Grill V. 2002. *Endocrinology* 143:1371–77
190. Medvedev AV, Robidoux J, Bai X, Cao WH, Floering LM, et al. 2002. *J. Biol. Chem.* 277:42639–44
191. Joseph JW, Koshkin V, Saleh MC, Sivitz WI, Zhang CY, et al. 2004. *J. Biol. Chem.* 279:51049–56
192. Yamashita T, Eto K, Okazaki Y, Yamashita S, Yamauchi T, et al. 2004. *Endocrinology* 145:3566–77
193. Moynihan KA, Grimm AA, Plueger MM, Bernal-Mizrachi E, Ford E, et al. 2005. *Cell Metab.* 2:105–17

194. Brand MD, Esteves TC. 2005. *Cell Metab.* 2:85–93
195. Joseph JW, Koshkin V, Zhang CY, Wang J, Lowell BB, et al. 2002. *Diabetes* 51:3211–19
196. Chan CB, Saleh MC, Koshkin V, Wheeler MB. 2004. *Diabetes* 53(Suppl. 1):S136–42
197. Winzell MS, Svensson H, Enerback S, Ravnskjaer K, Mandrup S, et al. 2003. *Diabetes* 52:2057–65
198. Moore PC, Ugas MA, Hagman DK, Parazzoli SD, Poitout V. 2004. *Diabetes* 53:2610–16
199. Zhou YP, Grill VE. 1995. *Diabetes* 44:394–99
200. Liu YQ, Tornheim K, Leahy JL. 1999. *Diabetes* 48:1747–53
201. Liu YQ, Tornheim K, Leahy JL. 1998. *Diabetes* 47:1889–93
202. Liu YQ, Tornheim K, Leahy JL. 1998. *J. Clin. Investig.* 101:1870–75
203. Iizuka K, Nakajima H, Namba M, Miyagawa J, Miyazaki J, et al. 2002. *Biochim. Biophys. Acta* 1586:23–31
204. Liu YQ, Jetton TL, Leahy JL. 2002. *J. Biol. Chem.* 277:39163–68
205. Assimacopoulos-Jeannet F, Thumelin S, Roche E, Esser V, McGarry JD, Prentki M. 1997. *J. Biol. Chem.* 272:1659–64
206. Zhou YT, Shimabukuro M, Wang MY, Lee Y, Higa M, et al. 1998. *Proc. Natl. Acad. Sci. USA* 95:8898–903
207. Mount LA, Antunes JL. 1975. *J. Neurosurg.* 42:189–93

Cold-Adapted Enzymes

Khawar Sohail Siddiqui and Ricardo Cavicchioli

School of Biotechnology and Biomolecular Sciences, The University of New South Wales, Sydney, NSW 2052, Australia; email: k.siddiqui@unsw.edu.au, r.cavicchioli@unsw.edu.au

Key Words

enzyme activity, enzyme stability, protein structure, protein flexibility, psychrophile

Abstract

By far the largest proportion of the Earth's biosphere is comprised of organisms that thrive in cold environments (psychrophiles). Their ability to proliferate in the cold is predicated on a capacity to synthesize cold-adapted enzymes. These enzymes have evolved a range of structural features that confer a high level of flexibility compared to thermostable homologs. High flexibility, particularly around the active site, is translated into low-activation enthalpy, low-substrate affinity, and high specific activity at low temperatures. High flexibility is also accompanied by a trade-off in stability, resulting in heat lability and, in the few cases studied, cold lability. This review addresses the structure, function, and stability of cold-adapted enzymes, highlighting the challenges for immediate and future consideration. Because of the unique properties of cold-adapted enzymes, they are not only an important focus in extremophile biology, but also represent a valuable model for fundamental research into protein folding and catalysis.

Contents

INTRODUCTION: ADAPTATION
    TO THE COLD 404
ACTIVITY OF COLD-ADAPTED
    ENZYMES 405
    Cold-Adapted Enzymes Have a
        High Reaction Rate 405
    Can k_{cat} Be Improved? 407
    Effect of Viscosity on Activity
        and Activation Energy 408
    Optimization of K_m in
        Cold-Adapted Enzymes........ 408
STABILITY OF COLD-ADAPTED
    ENZYMES 410
    Unfolding of Cold-Adapted
        Enzymes 410
    Kinetic Stability 411
    Conformational Stability.......... 412
    Cold Denaturation 414
ACTIVITY-STABILITY
    RELATIONSHIP................. 415
CONCEPT OF FLEXIBILITY 417
    Global Versus Local Flexibility.... 418
FLEXIBILITY AND
    STRUCTURAL
    ADAPTATION 419
    Hydrophobic Interactions......... 421
    Surface Hydrophilicity 423
    Electrostatic Interactions 423
    Secondary Structure Elements 425
    Surface Loops 426
    Other Factors 426

INTRODUCTION: ADAPTATION TO THE COLD

Life, particularly microbial life, has evolved the capacity to proliferate in a broad range of different thermal environments. Thermal adaptation, particularly to extremes, has also limited the range of temperatures any individual organism may tolerate. This is illustrated for hot environments by a member of the *Archaea*, which was isolated from a deep sea hydrothermal vent and is capable of and restricted to growth at temperatures between 85°C and 121°C (1). At the opposite thermal extreme, psychrophilic (cold-adapted) microorganisms have been described that are capable of metabolizing in snow and ice at −20°C, and numerous psychrophilic isolates have been characterized by their ability to proliferate at ≤0°C and are restricted to <30°C (2–5). Clearly, mechanisms have evolved that provide the ability, and restrict the scope, of an organism to adapt to a particular thermal environment.

The largest proportion of biomass on Earth is generated at cold temperatures (≤5°C). This is mainly due to the contribution of vast numbers of microorganisms in the world's oceans, although the cold biosphere extends to permanently cold alpine regions, caves, the upper atmosphere, and polar regions, in addition to seasonally cold environments (2, 3, 5–7). Representatives of organisms from the three domains of life (*Bacteria*, *Archaea*, *Eucarya*), including bacteria, yeast, archaea, algae, fungi, lichens, moss, plants, invertebrates, and fish, have been isolated and characterized from these cold environments (2, 5).

All these organisms are at thermal equilibrium with their environment, and all components of their cells must be suitably adapted to the cold (5). Their phylogenetic diversity underscores the potential mechanistic diversity that may have evolved to enable cold adaptation, and to some degree, cell-specific adaptation strategies have been identified. For example, antifreeze proteins are a feature of Antarctic and Arctic fish, although they are not typically found in psychrophilic microorganisms. What has emerged as a general feature from a number of decades of research is that organisms that live in permanently cold environments harbor enzymes that function effectively in the cold (3–5, 8–10).

Genomic (e.g., 11) and proteomic (e.g., 12, 13) studies of psychrophiles are now emerging, and the potential biotechnological applications of cold-adapted enzymes have been extensively considered (e.g., 5, 14, 15).

This review focuses on the structure, function, and stability of cold-adapted enzymes and is not restricted to enzymes from microorganisms. However, with the exception of fish, most cold-adapted enzymes have been sourced from prokaryotes. The α-amylase from *Pseudoalteromonas haloplanktis* (AHA) is referred to frequently, as it is the most comprehensively studied cold-adapted enzyme.

ACTIVITY OF COLD-ADAPTED ENZYMES

How enzymes avoid a decrease in their activity as a consequence of a drop in temperature from 37°C to 0°C is the crux of cold adaptation and is discussed below.

Cold-Adapted Enzymes Have a High Reaction Rate

The rate of all reactions including enzymatic reactions is described by the Arrhenius equation (16)

$$k_{cat} = A\kappa e^{-E_a/RT}, \quad 1.$$

where k_{cat} is the enzyme reaction rate, which increases with an increase in absolute temperature (T) and a decrease in activation energy (E_a), A is the preexponential factor, κ is the dynamic transmission coefficient (generally assumed to be 1), and R is the universal gas constant (8.314 J mol^{-1} K^{-1}).

According to Equation 1, at very low temperatures (0°–4°C), insufficient kinetic energy is available in the system to overcome reaction barriers. Psychrophilic organisms have evolved several strategies to compensate for the very slow metabolic rates that would occur at low temperatures as a result of this kinetic effect. These include an energetically expensive strategy of increasing enzyme concentration (17, 18), seasonal expression of isoenzymes in fish and nematodes (19–21), and the evolution of enzymes in which reaction rates tend to become temperature independent and approach diffusion control (10). The majority of cold-adapted enzymes are characterized by a shift in apparent T_{opt} (optimum temperature of activity) to a low temperature with a concomitant decrease in stability. Moreover, they tend to exhibit a high-reaction rate (up to 10-fold higher k_{cat} compared to heat-stable homologs) by decreasing the activation free-energy ($\Delta G^{\#}$) barrier between the ground state (substrate) and the transition state (TS$^{\#}$) (10). The $\Delta G^{\#}$ is composed of two components (22):

$$\Delta G^{\#} = \Delta H^{\#} - T\Delta S^{\#}, \quad 2.$$

where $\Delta H^{\#}$ is the change in activation enthalpy, $\Delta S^{\#}$ is the change in activation entropy and T is the absolute temperature.

According to transition-state theory (TST), k_{cat} is related to temperature and thermodynamic activation parameters (10, 16, 23–25) by the following equation:

$$k_{cat} = (k_B T/h)e^{-\Delta G^{\#}/RT}, \quad 3.$$

where k_B is the Boltzman constant (1.38 × 10^{-23} J K^{-1}) and h is the Planck constant (6.63 × 10^{-34} J s). Almost all cold-adapted enzymes studied to date (**Table 1**) have a low $\Delta H^{\#}$. As a result, the reaction rate tends to be less dependent on temperature, and a high-reaction rate (k_{cat}) is maintained at low temperature (16, 24).

To consider the effects of $\Delta S^{\#}$ and $\Delta H^{\#}$ on k_{cat}, the value of $\Delta G^{\#}$ from Equation 2 was used in Equation 3 to give Equation 4:

$$k_{cat} = (k_B T/h)e^{-\{(\Delta H^{\#}/RT)+(\Delta S^{\#}/R)\}}. \quad 4.$$

From Equation 4, in order for k_{cat} to increase at low temperatures, either $\Delta S^{\#}$ has to increase, or $\Delta H^{\#}$ has to decrease. As described above, in cold-adapted enzymes, a decrease in $\Delta H^{\#}$ is observed to increase k_{cat}. This decrease in $\Delta H^{\#}$ is structurally accomplished by a reduction in the number of enthalpy-related interactions that need to be broken during transition-state formation. This is likely to generate enhanced flexibility of the active site in cold-adapted enzymes (see the Concept of Flexibility section, below). As a consequence of active-site flexibility, the ground-state enzyme-substrate (ES) complex resides in a

---

AHA: the α-amylase from *Pseudoalteromonas haloplanktis* is a multidomain, monomeric, Ca^{2+}- and Cl^{-}-dependent enzyme

Thermodynamic activation parameter (#): a measure of the energy between the ground state and the transition state

Flexibility: the sum of the fluctuations of interconverting protein conformations

Table 1 Kinetic and thermodynamic activation parameters of catalysis of selected thermally adapted enzymes

Enzyme/organism[a]	k_{cat}, min^{-1}[b]	K_m, mM[c]	T_{opt} (°C)	$\Delta H^{\#}$ (kJmol^{-1})	$\Delta S^{\#}$ (Jmol^{-1}K^{-1})	$\Delta G^{\#}$ (kJmol^{-1})	References
AHA[P]	17640 (10°C)	—	28	35	−81	58	(72)
	41820 (25°C)	0.23 (pNME)					
PPA[M]	5820 (10°C)	—	54	46	−43	59	(72)
	17460 (25°C)	0.06 (pNME)					
α-Amylase[T] Bacillus amyloliquefaciens	840 (10°C)	—	84	70	27	63	(72)
Cellulase, endo[P] Pseudoalteromonas haloplanktis	10.8 (4°C)	6 (pNPC)	45	46	−92	71	(118)
Cellulase, endo[T] Erwinia chrysanthemi	0.6 (4°C)	1.5 (pNPC)	55	66	−45	78	(118)
Trypsin[P] Gadus morhua	240 (25°C)	0.08 (amide)	50	32	−57	49	(152)
Trypsin[M] Bovine	120 (25°C)	0.7 (amide)	47	53	−25	60	(152)
Subtilisin[P] Bacillus. sp. TA41	1920 (5°C)	0.026 (amide)	40	36	−92	62	(130)
Subtilisin[T] Bacillus subtilis	1000 (5°C)	0.006 (amide)	—	46	−70	66	(26, 130)
Phosphatase, alkaline[P] Pandalus borealis	48740 (37°C)	0.13 (pNPP)	40	35	−10	37.7	(83)
Phosphatase, alkaline[P] (chemically modified) P. borealis	97363 (37°C)	0.85 (pNPP)	40	28	−25	35.8	(83)
GTPase (EF-2)[P] Methanococcoides burtonii	2.0 (40°C)	0.02 (GTP)	—	75	−35	86	(24)
GTPase (EF-2)[M] Methanosarcina thermophila	3.7 (40°C)	0.03 (GTP)	—	88	13	84	(24)
Isocitrate DH[P] Colwellia maris	1.0 (15°C)	0.062 (isocitrate)	20	6.5	−189	61	(35)
Isocitrate DH[M] Azotobacter vinelandii	1.2 (15°C)	0.008 (isocitrate)	42	14.7	−157	60	(35)
Glutamate DH[P] Chaenocephalus aceratus (ice-fish)	228 (5°C)	2.0 (glutamate)	30	34	−114	65	(84)
Glutamate DH[M] Bovine	54 (5°C)	0.87 (glutamate)	30	59	−32	68	(84)
Chitobiase[P] Arthrobacter sp.TAD20	5880 (15°C)	0.027 (NPAG)	—	45	−51	60	(66)
Chitobiase[M] Serratia marcescens	1080 (15°C)	0.038 (NPAG)	—	72	28	64	(66)
Chitanase A[P] Arthrobacter sp.TAD20	102 (15°C)	—	—	60	−31	69	(34)

(Continued)

Table 1 (Continued)

Enzyme/organism[a]	k_{cat}, min^{-1}[b]	K_m, mM[c]	T_{opt} (°C)	$\Delta H^{\#}$ (kJmol^{-1})	$\Delta S^{\#}$ (Jmol^{-1}K^{-1})	$\Delta G^{\#}$ (kJmol^{-1})	References
Chitanase A^M Serratia marcescens	235 (15°C)	—	—	74	25	67	(34)
Xylanase^P Pseudoalteromonas haloplanktis	30930 (10°C) 74820 (25°C) 99000 (35°C)	2.8 % (xylan)	35	21	−116	54	(76)
Xylanase^M Streptomyces sp. S38	3600 (10°C)	—	50	58	−7	60	(76)
Xylanase, LAX^T Scopulariopsis sp.	95000 (40°C) 79000 (35°C)[d]	3.0 % (xylan)	50	66	28	58	(36)

[a]Codes and abbreviations used: ^{P}, psychrophile; ^{M}, mesophile; ^{T}, thermophile; pNME, p-nitrophenyl-α-D-maltoheptaoside-4,6-O-ethylidene; pNPC, p-nitrophenyl-β-D-cellobioside; pNPP, p-nitrophenyl phosphate; GTP, guanosine triphosphate; DH, dehydrogenase; NPAG, p-nitrophenyl-acetyl glucosamine; LAX, least acidic xylanase isoenzyme.
[b]Values in parentheses represent the temperature at which the enzyme assay was performed.
[c]Values in parentheses represent the substrate used in the enzyme assay.
[d]Calculated from published data (36, 76) using Q_{10} of 2.4 for LAX^T and 1.32 for xylanase^P.

wider distribution of conformational states than the activated enzyme-transition-state complex. Depending on the reaction, $\Delta S^{\#}$ can be negative or positive as this term also includes contributions from the redistribution of water molecules. However, the key point is that the difference in the activation entropy between an enzyme from a mesophile and a psychrophile is always negative (16, 25–27).

The gain in k_{cat} would be massive if the decrease in $\Delta H^{\#}$ was not accompanied by a decrease in $\Delta S^{\#}$ (16). It has been shown theoretically that maintaining a constant $\Delta S^{\#}$ and decreasing $\Delta H^{\#}$ by 20 kJ mol^{-1} would result in a ∼50,000-fold increase in k_{cat} at 15°C (16). However, in practice, such a large increase in k_{cat} is not observed in enzymes from psychrophiles because of an enthalpy-entropy compensation (28–30). Enthalpy-entropy compensation implies that a decrease in $\Delta H^{\#}$ accompanied by a decrease in $\Delta S^{\#}$ produces an overall small change in $\Delta G^{\#}$.

Can k_{cat} Be Improved?

The activity of enzymes from psychrophiles measured at their environmental temperature (0°–4°C) is in general lower than that for homologous enzymes from mesophiles at their environmental temperature (∼37°C). This has been shown for AHA (10, 25, 31) and a xylanase (32). Although the enzymes from the psychrophiles are indeed active at their environmental temperature, it appears that higher activities might be achievable and that adaptation might therefore be considered incomplete (10, 25, 33).

For a limited number of enzymes, the k_{cat} of an enzyme from a thermophile is higher or nearly comparable to the k_{cat} (at the same temperature) for the enzyme from a psychrophile (Table 1). These include a GTPase (24), chitanase (34), isocitrate dehydrogenase (35), and xylanase (36). In these cases, the relatively high activity of the enzyme from the thermophile is entropically driven (higher and/or positive $\Delta S^{\#}$). These findings imply that the k_{cat} of a cold-adapted enzyme could be enhanced by simultaneously decreasing $\Delta H^{\#}$ and increasing $\Delta S^{\#}$. This may be achieved by an indirect increase in the entropy (∼40 J K mol^{-1}) of the system via water dislocation (37, 38). In glucanases (e.g., xylanase), a chain of well-ordered water molecules carpets the length of the active site. Some water molecules are displaced when substrate (ground state) or activated substrate (transition state) binds to the active site (39). If more water molecules are released upon binding of the transition state to the enzyme than upon binding of the ground-state substrate, then there will be considerable entropic benefit for the formation of an enzyme-transition-state complex (38, 40–44). A decrease in $\Delta S^{\#}$ may be kept to a minimum in cold-adapted enzymes by manipulating the active site to enable the activated substrate to displace a greater number of water molecules than the substrate. It has also been suggested that this may be achieved by adding a small amount of rigidity to part of the active site, while maintaining $\Delta H^{\#}$ (16).

Effect of Viscosity on Activity and Activation Energy

TST (Equation 3) is based on the solvent being an ideal dilute substance treated as a heat bath, where it may be applicable to enzyme-catalyzed reactions only at low-solution concentrations (45). High viscosity at low temperatures may particularly limit the usefulness of Equation 3 (46). A modified TST equation that takes into account the viscosity of the medium has recently been proposed (45). At low temperatures, the dynamic transmission factor (κ) in Equation 1 cannot be assumed to be 1:

$$\kappa = \sqrt{1 + (\eta'/\eta_1)^2} - (\eta'/\eta_1) \qquad 5.$$

and

$$\eta_1 = (m\omega_b/3\pi a \eta_w), \qquad 6.$$

where η' is the relative viscosity (viscosity of solution divided by the viscosity of water, η_w), and ω_b, a, and m are the vibrational energy, radius, and mass of a spherical particle, respectively. η_1 is the dimensionless unknown parameter, which can be determined by fitting the experimental data to Equation 1.

When η_1 is 10 or above, κ is close to 1, and the TST (Equation 3) can be used. For smaller values of η_1, dramatic changes take place in the rate and E_a/RT. For example, when η_1 is 0.01, the rate is reduced by a factor of 1/200 and E_a/RT is reduced by −5.3 from its value for TST. When η_1 is 0.001, the corresponding numbers are 1/2000 and −7.6, respectively. For determining $\Delta G^{\#}$ between the ES complex and the transition state of a cold-adapted enzyme at low temperature and high viscosity, Equation 7 can be used (45):

$$\Delta G^{\#} = -RT \ln(h k_{cat}/k_B T \kappa). \qquad 7.$$

Optimization of K_m in Cold-Adapted Enzymes

The strength of the ES interactions may decrease (electrostatic interactions) or increase (hydrophobic interactions) with increasing temperature. The former are derived exothermically, whereas the latter are formed endothermically, within a temperature range of 0°–30°C (46). However, both types of interactions are affected by changes in water structure that occur as a function of temperature (47). For example, the overall free-energy change (ΔG_{Total}) that occurs during the formation of an ionic bond is complicated by the effect of temperature on the dielectric constant of water.

$$\Delta G_{Total} = \Delta G_{Desolvation} + \Delta G_{Electrostatic} + \Delta G_{Protein}, \qquad 8.$$

where $\Delta G_{Desolvation}$ is the energy penalty of desolvating the two charged species, $\Delta G_{Electrostatic}$ is the favorable Coulombic and van der Waals interaction between two oppositely charged ionic residues, and $\Delta G_{Protein}$ is the free-energy change describing the interactions of the ion pair with other amino acid side chains in the surrounding region of the protein (48).

In cold-adapted enzymes, K_m will be determined by the contributions of the various bond types involved in the ES interactions (49). Enzymes with low K_m have more negative binding energy (ΔG_{ES}) than those with higher K_m, in accordance with the relationship (50) in Equation 9:

$$\Delta G_{ES} = -RT \ln(1/K_m). \qquad 9.$$

As a result of higher ES affinity, the reaction falls deeper into a thermodynamic well (more negative ΔG_{ES}) from where it has to climb in order to form a transition state (20, 25, 26, 37). An enzyme can increase k_{cat} by lowering $\Delta G^{\#}$. This can be achieved by stabilizing the activated substrate in the transition state or by destabilizing the ES complex (increasing K_m) in line with Equation 9 (51). There is experimental evidence that the transition-state, but not the ground-state, substrate forms strong electrostatic bonds and strong H-bonds (hydrogen bonds) with the active site (37, 52). For glucanases, binding of the ground-state substrate is mainly afforded by hydrophobic interactions, whereas the activated substrate (charged oxocarbonium ion) is polar and binds electrostatically (53). As a result, compared to the ground-state substrate, enzyme affinities for the activated substrate are expected to increase much more steeply with decreasing temperature (38). The high k_{cat} of cold-adapted enzymes is consistent with low temperatures favoring transition-state binding.

The majority of cold-adapted enzymes (**Table 1**) have a higher k_{cat} and K_m than their thermostable counterparts, with the exception of enzymes, which work at a [S] close to K_m (e.g., rate-limiting enzymes involved in metabolic pathways) (26). In addition to those listed in **Table 1**, a malate dehydrogenase (54), β-lactamase (55), DNA ligase (56), and aspartate aminotransferase (57) have similar characteristics. This relationship is well illustrated for α-amylases, wherein mutants of AHA tended to exhibit proportional decreases in k_{cat} and K_m (9, 25, 58).

Higher K_m and k_{cat} are also characteristics of lactate dehydrogenase (LDH-A$_4$) enzymes from cold-water fish (4, 20, 59, 60). The higher K_m in cold-adapted LDH-A$_4$ results in a decrease in ΔG_{ES} with a concomitant decrease in the energy of activation required to form the transition state, thereby increasing k_{cat} (20, 59, 60). Consistent with this, mutagenesis studies have shown that E_a changes without a shift in $K_m^{pyruvate}$ when mutations are distant from the LDH-A$_4$ active site, whereas K_m always varies inversely to E_a in mutants that have changes in or near the active site (61). Similar findings have been reported for isocitrate dehydrogenase (**Table 1**) (35).

To facilitate substrate binding at a low-energy cost, the active site of cold-adapted enzymes tends to be larger and more accessible to the substrate. In citrate synthase, this type of architecture (see the section Flexibility and Structural Adaptation, below) is achieved by replacing amino acids that have bulky groups with those that have smaller side chains and by increasing the length of loops around the active site (62). In addition to increasing K_m, these structural features have been found to cause a reduction in substrate specificity in an elastase (8) and alcohol dehydrogenase (63).

It is clear that the majority of cold-adapted enzymes have a higher K_m than more stable homologs. However, some cold-adapted enzymes have a lower K_m than thermostable homologs. For secreted enzymes from marine microorganisms, the requirement for a low K_m may relate to the need to scavenge substrates that are at low concentrations in the environment (9, 54, 64, 65). In a chitobiase from an Antarctic marine bacterium, k_{cat} is 8-fold higher and K_m 25-fold lower than a homolog from a mesophile (66). The chitobiase is a multidomain enzyme with a relatively rigid substrate-binding cleft, but it has a relatively flexible region surrounding the catalytic site (see the Concept of Flexibility section, below). The low K_m was attributed to the replacement of two Trp residues (mesophile) by two polar residues in the substrate-binding site, with the ionic interactions responsible for high-substrate affinity at a low temperature (5°C) (66).

Comparatively few cold-adapted enzymes have been examined from the *Archaea* (67, 68), although comparative genomic studies have been performed with two Antarctic representatives (11). Studies of a GTPase [elongation factor 2 (EF-2)] from the Antarctic archaeon, *Methanococcoides burtonii*, and a closely related thermophile, *Methanosarcina thermophila*, have shown that cold adaptation involves a decrease in K_m (69, 70). However, the studies also demonstrated that the

CD: circular dichroism

DSC: differential scanning calorimetry

TUG-GE: transverse urea-gradient gel electrophoresis

physiological activity of the enzyme was affected by interaction with partner proteins (ribosomes) and by the thermally regulated levels of intracellular solutes (68). These findings highlight that, although inherent properties of individual proteins make essential contributions to adaptation, cellular factors can play an important role and should be considered when interpreting the mechanisms and effectiveness of cold adapation.

STABILITY OF COLD-ADAPTED ENZYMES

Because of their highly flexible structures (see the section Global Versus Local Flexibility, below), cold-adapted enzymes unfold at low to moderate temperatures with a concomitant reduction in kinetic and conformational stabilities. This is discussed below.

Unfolding of Cold-Adapted Enzymes

A number of spectrophotometric [fluorescence and circular dichroism (CD)], calorimetric [differential scanning calorimetry (DSC)], and electrophoretic [transverse urea-gradient gel electrophoresis (TUG-GE)] methods have been employed to study unfolding/folding transitions and measure kinetic and conformational stabilities of enzymes (71–73). Most studies of cold-adapted enzymes using these techniques have been carried out on multidomain proteins and include chitobiase (66), dihydrofolate reductase (74), ornithine carbamoyltransferase (75), phosphoglycerate kinase (64), xylanase (32, 76), DNA ligase (77), β-galactosidase (65), EF-2 (70), and AHA (58, 71, 72, 78–80).

Heat-induced unfolding of large multidomain proteins (including cold-adapted proteins) tends to be kinetically driven as a result of the usually irreversible nature of unfolding. AHA is presently the only example of a cold-adapted enzyme that displays fully reversible unfolding (71). When subjected to unfolding at 20°C or above, AHA shows reversible unfolding (71, 72, 79). This is illustrated by 100% recovery of ΔH_{cal} (the total amount of heat absorbed during unfolding) during a second DSC scan following the initial thermal denaturation (71). The ratio of ΔH_{cal} to $\Delta H_{vantHoff}$ (cooperativity of unfolding determined from the slope of the transition) is 1, implying true two-state unfolding (58).

Cooperative unfolding is typically associated with small-molecular-weight enzymes, which have a tightly packed core (81). AHA is a large enzyme (~50 kDa), and cooperative unfolding appears to be due to the small number of interactions between structural elements that preserve the native state (58). Simultaneous disruption of the limited number of interactions is likely to facilitate two-state unfolding (58).

In pancreatic porcine α-amylase (PPA) and more stable mutants of AHA, two-state unfolding does not occur (or partially occurs). Non-two-state unfolding can be attributed to increased ionic interactions, which increase the frequency of intramolecular mismatches during folding. In a large number of AHA mutants, the rate of thermal inactivation has been found to be directly proportional to the extent of reversibility (58). An N12R replacement produces an R12 to D15 salt bridge in domain A, which is not present in the wild-type AHA, and the mutant exhibits a 75% reduction in reversibility (58).

Using TUG-GE it was shown that at 3°C and 12°C, AHA unfolds reversibly and sequentially showing two transitions (79). The transition that unfolded at lower urea concentration was identified as belonging to the active-site region (79). Cooperative unfolding of structures forming the active site (parts of domain A and domain B) was shown to precede independent unfolding of other more stable regions of the protein (parts of domain A and domain C) (79). By constructing chimeric enzymes from thermolabile and thermostable domains of isocitrate dehydrogenase enzymes, the substrate-binding region of the cold-adapted enzyme was found to be the most flexible region (with high K_m) from which unfolding initiated (35). The findings

Table 2 Kinetic and thermodynamic activation parameters of thermal inactivation of cold-adapted enzymes compared to mesophilic and thermophilic homologs

Enzyme[a]	T (°C)	$t_{1/2}$ (min)	$\Delta H^{\#}$ (kJ mol^{-1})	$T\Delta S^{\#}$ (kJ mol^{-1})	$\Delta G^{\#}$ (kJ mol^{-1})	Reference
AHA[P] (N12R mutant)[b]	43	0.23 (inact)	721	635	86	(72)
		0.23 (denat)	459	373	85	
PPA[M]	60	0.23 (inact)	640	550	90	(72)
		0.23 (denat)	354	264	90	
α-Amylase[T] *Bacillus amyloliquefaciens*	80	0.23 (inact)	245	149	96	(72)
		0.23 (denat)	310	215	96	
DNA ligase[P] *Pseudoalteromonas haloplanktis*	35	0.43 (denat)	422	337	85	(77)
DNA ligase[M] *Escherichia coli*	35	124 (denat)	162	63	99	(77)
GTPase (EF-2)[P] *Methanococcoides burtonii*	46.5	1.7 (denat)	282	190	92	(24)
GTPase (EF-2)[T] *Methanosarcina thermophila*	46.5	92 (denat)	474	372	102	(24)
Glycosyl hydrolase[P] (family 8) *Pseudoalteromonas haloplanktis*	65	0.008 (inact)	458	376	82	(76)
		0.007 (denat)	414	332	82	
Glycosyl hydrolase[T] (family 8) *Clostridium thermocellum*	65	963 (inact)	305	190	115	(76)
		608 (denat)	345	232	113	
Phosphatase, Alkaline[P] *Pandalus borealis*	50	11 (inact)	210	112.5	98	(83)
Phosphatase, Alkaline[M] Bovine	65	1 (inact)	202	106	96	(83)
Glutamate dehydrogenase[P] *Chaenocephalus aceratus*	52	3 (inact)	745	666	80	(84)
Glutamate dehydrogenase[M] Bovine	52	16 (inact)	565	483	82	(84)
Trypsin[P] Atlantic cod	60	21 (inact)	122	19	103	(153, 154)
Trypsin[M] Bovine	60	106 (inact)	172	66	107	(153, 154)

[a] Codes and abbreviations used: [P], psychrophile; [M], mesophile; [T], thermophile; inact, loss in activity determined by enzyme inactivation; denat, enzyme denaturation determined by DSC.
[b] Data for irreversibly unfolding N12R mutant were included in lieu of data for reversibly unfolding wild type.

for AHA and isocitrate dehydrogenase illustrate that instability of the active-site region is a feature of these heat-labile enzymes, and it will be valuable to examine this experimentally in a broader range of cold-adapted enzymes.

Kinetic Stability

Kinetic stability can be followed by enzyme inactivation (k_{inact}) or denaturation (k_{denat}) (73, 82). The majority of cold-adapted enzymes have a half-life ($t_{1/2}$) of <12 min at 50°C, and some, such as a DNA ligase (77), denature at temperatures as low as 35°C (**Table 2**). For kinetic stability, it is important to consider the magnitude of the free-energy change ($\Delta G^{\#}$) between the folded (active) state and the transition state:

$$F \underset{}{\overset{K}{\rightleftharpoons}} TS^{\#} \overset{k}{\to} D, \qquad \text{Scheme 1.}$$

Kinetic stability: a measurement of how rapidly an enzyme unfolds irreversibly at a given temperature

Conformational stability: Gibbs free-energy (ΔG) change between folded and unfolded states of a protein, which are in equilibrium

where F is the folded enzyme, K is the equilibrium constant and k is the first-order rate constant for the conversion of TS$^{\#}$ to the denatured state (D).

In order to increase the rate of thermal unfolding, cold-adapted enzymes require a decreased $\Delta G^{\#}$ as shown in Equation 10.

$$\Delta G^{\#} = -RT \ln K. \qquad 10.$$

In accordance with Equations 2 and 10, the reduced thermostability of cold-adapted enzymes could be caused by low $\Delta H^{\#}$ of the folded form [implying a reduced number of noncovalent interactions that need to be broken to reach transition state (TS$^{\#}$)] or by increased $\Delta S^{\#}$ of the unfolded form (implying a higher disordered TS$^{\#}$) (25, 83).

The thermodynamic activation parameters of the cold-adapted enzymes, glutamate dehydrogenase (84), glycosyl hydrolase (76), AHA (72), and DNA ligase (77) (**Table 2**), indicate that low thermostability has resulted from increased disorder of the transition state (high $T\Delta S^{\#}$). It is important to note that the decreased entropy of thermostable enzymes (**Table 2**) may also arise from the hydration of nonpolar groups during unfolding. Water may form ordered structures around hydrophobic side chains and decrease the entropy of the system (77, 83).

The reason cold-adapted enzymes have higher $\Delta H^{\#}$ than homologs from mesophiles and thermophiles (**Table 2**) is as follows: The enthalpic contribution to thermolability may reflect the higher cooperativity of unfolding that arises from the lower number of interactions that need to be disrupted. Cold-adapted enzymes unfold in a narrow temperature range, thereby producing a steep slope in an Arrhenius plot and generating high values of $\Delta H^{\#}$ (25, 77).

For EF-2 (24) and a chemically modified alkaline phosphatase (83), low thermostability was achieved through a decrease in $\Delta H^{\#}$. The reduced $\Delta H^{\#}$ of the two enzymes from the psychrophiles may result from a reduction in interactions that need to be broken in order to reach TS$^{\#}$ and subsequent collapse of the structure. To establish the relevance of these findings, it will be valuable to determine activation parameters of denaturation for a larger range of cold-adapted enzymes and their thermostable homologs. The cold-adapted citrate synthase from *Arthrobacter* sp. has a large number of ion pairs compared to thermostable homologs (85), making it a good candidate for determining the importance of enthalpic contributions to its kinetic stability.

Conformational Stability

Conformational stability is typically determined for small, single-domain enzymes that undergo reversible unfolding (Scheme 2) (71, 73).

$$F \overset{K}{\rightleftharpoons} U, \qquad \text{Scheme 2.}$$

where U is the unfolded enzyme.

ΔG is a measure of the thermodynamic stability of a protein and is most easily calculated from the determination of K at varying temperatures, or concentrations of a denaturant, in accordance with Equation 10. For most proteins, the value of ΔG is small (e.g., 20–60 kJ mol^{-1}), equivalent to a few noncovalent interactions, and thus proteins appear to be only moderately stable at in vivo temperatures.

The thermodynamic parameters of conformational stability (**Table 3**) illustrate that enzymes from psychrophiles are thermolabile compared to those from mesophiles and thermophiles. The importance of enthalpic contributions to conformational stability is also reflected in site-directed mutants, which produce additional intramolecular interactions. For example, in an N150D (extra salt bridge with K190) and V196F (two extra aromatic interactions with Y82 and F198) AHA double mutant, T_m (melting temperature) and ΔH_{cal} increase by 2.4°C and 243 kJ mol^{-1}, respectively (58).

Conformational stabilities may be evaluated using the Gibbs-Helmholtz equation:

Table 3 Thermodynamic parameters of conformational stability of cold-adapted enzymes compared to homologs from mesophiles and thermophiles

Thermally adapted enzymes[a]	[denat]$_{1/2}$ (M)[b]	$\Delta G_{(H_2O)}$ (kJ mol^{-1})[b]	T_m (°C)[c]	Trans[c]	ΔH_{cal} (kJ mol^{-1})[c]	Reference
AHA[P]	0.9 (20°C)	15.5 (20°C)	44[F] 61[I]	1	895[F] 1272[I]	(72)
PPA[M]	2.6 (20°C)	29 (20°C)	66[F] 82[I]	2	1234[F] 1280[I]	(72)
α-Amylase[T] *Bacillus amyloliquefaciens*	6.0 (20°C)	100 (20°C)	86[F] 86[I]	3	1787[F] 1452[I]	(72)
DNA ligase[P] *Pseudoalteromonas haloplanktis*	1.5 (18°C)	27 (18°C)	33	1	193	(77)
DNA ligase[M] *Escherichia coli*	2.5 (25°C)	31 (25°C)	52,54	1st 2nd Σ	193 615 1059	(77)
DNA ligase[T] *Thermus scotoductus*	5.8 (25°C)	60.4 (25°C)	92, 96, 101	1st 2nd 3rd Σ	301, 1009, 419 1729	(77)
Glycosyl hydrolase[P] (family 8) *Pseudoalteromonas haloplanktis*		Irreversible	54	—	1058	(76)
Glycosyl hydrolase[T] (family 8) *Clostridium thermocellum*		Irreversible	81.4	—	1247	(76)
Dihydrofolate reductase[P] *Moritella profunda*	1.6 (15°C)	13.4 (15°C)	45.5	—	209 (40°C)	(74)
Dihydrofolate reductase[M] *Escherichia coli*	3.1 (15°C)	25 (15°C)	52	—	247 (46°C)	(74, 155)
Dihydrofolate reductase[P] *Thermotoga maritima*	5.5 (15°C)	144 (15°C)	81	—	466 (60°C)	(74, 156)

[a]Codes and abbreviations used: P, psychrophile; M, mesophile; T, thermophile; trans, transitions inferred from deconvolution of DSC thermograms; F, free enzyme; I, enzyme-competitive inhibitor (acarbose) complex; ΔH_{cal}, enthalpy change determined by DSC; T_m, melting temperature; [denat]$_{1/2}$, guanidinium-HCl or urea concentration at which $\Delta G = 0$.
[b]Used fluorescence spectrometry.
[c]Used differential scanning calorimetry (DSC).

$$\Delta G(T) = \Delta H_m(1 - T/T_m) + \Delta C_p(T - T_m) - T\Delta C_p \ln(T/T_m), \quad 11.$$

where $\Delta G(T)$ is the free energy required to disrupt the folded state and is 0 at T_m, ΔH_m is the enthalpy change at T_m, and ΔC_p is the heat capacity at constant pressure.

Protein aggregation that is caused by irreversible denaturation (particularly for multidomain enzymes) can become negligible in the presence of sulfobetaine (72). Under these conditions, k is assumed to be small, and the unfolding transition can be treated as obeying equilibrium thermodynamics (see Scheme 3).

$$F \underset{}{\overset{K}{\rightleftharpoons}} U^{\#} \overset{k}{\rightarrow} D. \quad \text{Scheme 3.}$$

Applying Equation 11 and Scheme 3, stability curves can be generated by plotting ΔG versus T (**Figure 1**) (73, 74). In principle, the low stability of cold-adapted enzymes may be attained through a shift in the stability curve (**Figure 1**). The low stability of AHA (lowest ΔG/residue for any reversibly unfolding protein studied to date) arises from a global

Stability curve: plot of ΔG versus T. The temperature at the peak of the stability curve is T_{max}

Table 4 Thermodynamic properties deduced from stability curves

| Thermodynamic parameters[a] | α-Amylase[b] | | | | | | Dihydrofolate reductase[c] | | |
| | (0°C) | | | (37°C) | | | (40°C) | | |
	P	M	T	P	M	T	P	M	T
T_{max} (°C)	20	30	33	20	30	33	15	18	42.5
ΔG (kJ mol^{-1})	10.5	25	91	17	67	155	3	13	50
ΔH (kJ mol^{-1})	−664	−985	−1000	649	322	310	—	—	—
$T\Delta S$ (kJ mol^{-1})	−675	−1010	−1091	632	251	159	—	—	—

[a]Abbreviations used: P, psychrophile; M, mesophile; T, thermophile; T_{max}, peak of stability curve.
[b]Reference 72.
[c]Reference 74.

collapse of the stability curve in which both ΔG and T_m decrease (**Table 4**) (33, 71, 72). For thermostable homologs, T_{max} (maximal temperature) lies to the left-hand side of the growth temperature optima of their respective mesophilic and thermophilic hosts (74). This indicates their structure is stabilized enthalpically, and unfavorable entropic contributions contribute to normal enzyme mobility. The stability curves for AHA show that T_{max} lies to the right-hand side of the environmental temperature for *P. haloplanktis*, indicating that in the cold the structure of AHA is stabilized entropically and destabilized enthalpically (72). This implies that increased flexibility of AHA mainly involves hydration of polar and nonpolar groups (72) (see the section Flexibility and Structural Adaptation, below). Other than AHA, this type of stability curve has only been constructed for dihydrofolate reductase, and similar conclusions were reached about its conformational stability (74).

Cold Denaturation

At temperatures below T_{max}, cold denaturation commences and becomes complete below 0°C at the T_m where the curve crosses the x-axis (**Figure 1**). The stability curves of AHA (72) (**Table 4**) and dihydrofolate

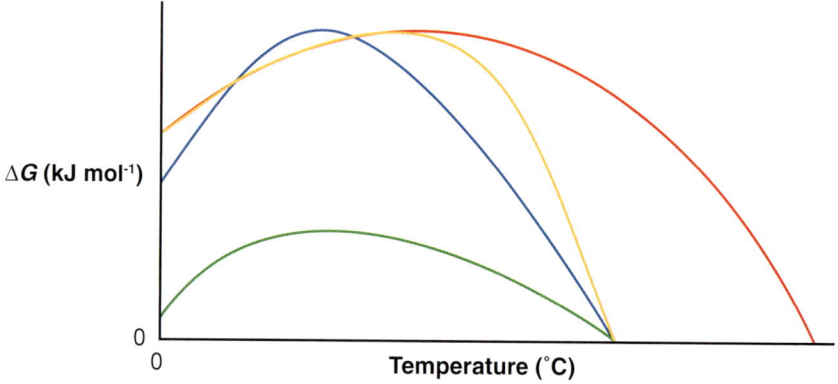

Figure 1
Hypothetical stability curves depicting possible ways to achieve low stability. The stability curve for a cold-adapted enzyme may be shifted to a lower temperature (*blue line*) compared to that of a thermostable homolog (*red line*). The gradient of the heat denaturation side of the curve may be increased while maintaining a constant T_{max} and ΔG (T_{max}) (*orange line*). Global collapse of the curve may occur, corresponding to both a lower ΔG and T_m (*green line*).

reductase (74) (**Table 4**) show that the cold-adapted enzymes are more cold labile than their thermostable homologs, with AHA predicted to become cold denatured at approximately $-5°C$ (72). In the presence of urea, cold denaturation of AHA was found to occur at $3°C$, with the catalytic domain being more cold labile than the rest of the protein (79). These findings demonstrate that the region containing the active site is more heat (see the section Unfolding of Cold-Adapted Enzymes) and cold labile in AHA.

There is little experimental data examining cold denaturation. The hydration of polar and nonpolar groups has been implicated in cold denaturation (72, 86). This is consistent with enzymes from psychrophiles tending to have a less packed hydrophobic core and fewer ionic interactions (see the section Flexibility and Structural Adaptation, below). A good candidate for assessing the importance of ionic interactions is the citrate synthase from *Arthrobacter*, which has a large number of ionic interactions (compared to thermostable homologs), which have been implicated in enhancing stability at low temperatures (47, 62, 85, 87).

ACTIVITY-STABILITY RELATIONSHIP

Cold-adapted enzymes are characterized by high flexibility, thermolability, and specific activity at low temperatures (3, 9, 10), whereas thermostable homologs have low activity and are structurally rigid at low temperatures (88). These apparently mutually exclusive properties indicate that an activity-stability trade-off exists in cold-adapted enzymes. A list of enzymes that exhibit intrinsically high activity at low temperatures (green font), high stability (blue font), or low activity or stability (red font) are compiled in **Table 5**. The table highlights the properties of enzyme variants (genetically or chemically modified) that were designed to increase the stability of a cold-adapted enzyme while maintaining its high activity or to increase the low temperature activity of a thermostable enzyme while maintaining its high stability. For many variants, a gain in activity (green font) is accompanied by a decrease in stability (red font), and an increase in stability (blue font) is accompanied by a decrease in activity (red font), illustrating an activity-stability trade-off (33). There are numerous examples of this in the literature, demonstrating that this is a general principle (33). However, there are a number of examples highlighted in **Table 5** wherein the activity-stability trade-off relationship has been defied, e.g., Lipase B (89, 90) and shrimp alkaline phosphatase (83).

It is noteworthy that activity measurements are typically performed using small synthetic substrates. Subtilisin mutants have been generated that demonstrate high activity toward small substrates but no enhancement for casein (91). It remains to be determined whether the mutant or modified enzymes demonstrate the same trend toward naturally occuring substrates or whether their substrate specificity has also been modified (90).

In phosphoglycerate kinase, the active site is formed by a flexible, heat-labile catalytic domain that is brought together with a rigid, stable substrate-binding domain through a hinge region (64). A similar separation of catalytic and substrate-binding domains is present in a cold-adapted chitobiase (66). These examples illustrate that the stability and activity of an enzyme can involve distinct regions of the protein (59, 72).

Heat lability of cold-adapted enzymes may reflect the need for high flexibility in the active-site region rather than the absence of selective pressure for stability (25). This view is consistent with the ability to engineer increased thermostability without compromising activity (33, 92) and with the proposal that cold-adapted enzymes evolved from a more thermostable ancestor (93). In this context, it is interesting to consider why the catalytic domains of cold-adapted enzymes have evolved flexibility, whereas the noncatalytic domains have retained relative stability (9).

Table 5 Changes in activity and stability parameters following genetic or chemical modification

Enzyme, organism, reference[a]	Method: changes[b,c]	Activity			Stability	
		k_{cat} (sec^{-1})	K_m (μM)	T_{opt} (°C)	$t_{1/2}$ (min)	T_m (°C)
AHA (58)	SDM: ND, VF	697 ↓25°C 642	234 ↓(maltose)$_7$ 174	—	5.1 ↓45°C 22	44 ↓DSC 46.4
AHA (72)	SDM: KR, ND, VF, QI, TV	697 ↓25°C 308	234 ↓(maltose)$_7$ 66	—	—	44 ↓DSC 49.4
AHA (151)	SDM: QC, AC	697 ↓25°C 340	234 ↓(maltose)$_7$ 112	—	5.1 ↓45°C 2.5	44 ↓DSC 40, 44
Triosephosphate isomerase *Vibrio marinus* (139)	SDM: AS	7000 ↓(10°C) 4667	1900 ↓(Gluc-3-P) 4800	—	10 ↓(25°C) 27	41 ↓DSC 46
β-Glucosidase *Paenibacillus polymyxa* (157)	RM: EK, MI, NK	13 ↓30°C 35	900 ↓pNPGP 1400	—	2 ↓50°C >90	—
Citrate synthase *Arthrobacter* DS2-3R (158)	SDM: KL, AR	18 ↓20°C 8	7 ↓OA 19	32°C ↓27°C	0% left ↓45°C 42% left	—
Lipase B *Candida antarctica* (89)	RM: VI, AE	1.4 ↓25°C 32	5.7 ↓DiFMU 13	—	8 ↓70°C 211	57.7 ↓CD 52.1
Lipase B *Candida antarctica* (90)	CM: DAP-Dex	100% ↓30°C 160%	—	—	18 ↓70°C 168	—
Alkaline phosphatase *Pandalus borealis* (83)	CM: Hydrophobic ↓COO$^-$	156 ↓10°C 15	260 ↓pNPP 60	40°C ↓ 40°C	11 ↓50°C 182	—
Subtilisin *Bacillus* TA39 (159)	SDM: TD	80 ↓15°C 230	27 ↓amide 30	—	6 ↓50°C 60	—
Subtilisin S41 ↓S3–2G7 *Bacillus* TA41 (92)	RM: SI, ST x 2, KP, RA, KE, NI	57 ↓(30°C) 160	192 ↓(amide) 210	55 ↓(amide) 65	1 ↓(60°C) 450	50 ↓CD 72
Subtilisin S3–2G7 ↓8–4A9 *Bacillus* TA41 (160)	RM: ND, QH, ST x 5, SI x 2, KP x 2, RA, DE, KE, SA, NI	179 ↓(30°C) 185	210 ↓(amide) 170	65 ↓(amide) 75	450 ↓(60°C) 1082	72 ↓CD 75
Subtilisin *Bacillus sphaericus* (161)	RM: TA, KR, DN, SF	16 ↓(10°C) 104	280 ↓(amide) 190	—	14 ↓(70°C) 4	—

(Continued)

Table 5 (Continued)

Enzyme, organism, reference[a]	Method: changes[b,c]	Activity			Stability	
		k_{cat} (sec^{-1})	K_m (µM)	T_{opt} (°C)	$t_{1/2}$ (min)	T_m (°C)
Esterase *Bacillus subtilis* (162)	RM: IV, LM, LF, HY, AV, MV, YF, GE, IT	720 ↓(30°C) 470	1900 ↓(pNB) 600	45 ↓(pNB) 58	—	52.5 ↓DSC 66.5
3-Isopropylmalate dehydrogenase *Thermus thermophilis* (163)	RM: VM	2.4 ↓40°C 17.7	1.2 ↓IPM 1.2	—	10 ↓87°C 10	87 ↓CD 87
β-Glucosidase *Pyrococcus furiosus* (164)	RM: NS	13 ↓20°C 49	500 ↓pNPGP 1000	98 ↓85	100% left ↓ 106°C, 1h 1% left	—
Xylose isomerase *Thermus thermophilus* (165)	RM: EG	47 ↓60°C 412	3400 ↓xylose 29,000	80°C ↓90°C	95% left ↓ 70°C, 55h 55% left	—
Ornithine carbamoyltransferase *Pyrococcus furiosus* (166)	RM: AD, EG	370 ↓30°C 2900	100 ↓ornithine 2000	—	600 ↓75°C 14	—

[a]Meaning of colored fonts: green font, increased k_{cat} and K_m characteristic of cold-adapted enzymes; blue font, increased stability characteristic of thermostable enzymes; red font, decreased activity or stability.
[b]Symbols and abbreviations used: ↓, change following modification; SDM, site-directed mutagenesis; CM, chemical modification; RM, randomized mutagenesis; $t_{1/2}$, half-life at specified temperature; T_m, melting temperature; Gluc-3-P, glucose-3-phosphate; pNPGP, p-nitrophenyl-β-D-glucopyranoside; OA, oxaloacetate; pNPP, p-nitrophenyl phosphate; pNB, p-nitrophenyl butyrate; IPM, isopropylmalate; DiFMU, 6, 8-difluoro-4-methylumbelliferyl octanoate; DAP-Dex, dialdehyde polysaccharide-dextran (40 kDa).
[c]Changes represented by dual, single amino acid codes to identify amino acids from wild types (first amino acid) and mutants (second amino acid).

CONCEPT OF FLEXIBILITY

The low stability and high activity of cold-adapted enzymes at low temperatures implies a flexibile enzyme structure. Amino acids involved in catalysis are conserved between cold and thermostable homologs, illustrating that causes of flexibility must reside in other parts of the enzyme (59, 94)

Two types of flexibility can be considered: static and dynamic (33, 95). Static flexibility can be approximated by techniques that provide the average mobility of amino acid side chains, such as B- (temperature) factors (estimate of occupancy derived from electron density maps averaged over a population and time) from X-ray structures (96) and hydrogen/deuterium (H/D) exchange. H/D exchange measures an amino acid's accessibility to solvent and is proportional to the exchange rate of the enzyme's protons with deuterium of the solvent (10, 33, 60, 97).

Based on B-factors from a large set of high-resolution structures, amino acids have been classified as rigid (WYFCIVHLMA) or flexible (GTRSNQDPEK), with flexibility increasing from W to K (98). The average B-factor for a cold-adapted malate dehydrogenase was reported to be lower than for a homolog from a thermophile (54). However, the B-factors were twofold higher for regions involved in substrate and cofactor binding. In

Static flexibility: the number and structural variety of possible enzyme conformations

Dynamic flexibility: a measure of how rapidly structures within an enzyme interconvert between conformations

a separate study, B-factors were reported to be invariant in homologs from psychrophiles, mesophiles, and thermophiles (33). When H/D exchange was compared between LDH-A_4 from cold- and warm-water-adapted fish, no difference was observed at 2°C (indicating equivalent flexibility) (60). However, at 23°C, the H/D exchange rate was higher for LDH-A_4 from the cold-water fish, implying higher flexibility of the cold-adapted enyzme. For a cold-adapted 3-isopropylmalate dehydrogenase, the H/D exchange was not supportive of it having a high level of static flexibility (97).

Dynamic flexibility is measured by dynamic fluorescence quenching (72, 73, 77) and proteolytic nicking (99). For a flexible enzyme, the quencher (e.g., acrylamide) will infiltrate the inner part of the structure and decrease Trp fluorescence, thereby providing a high index of permeability (72, 73). Permeability provides a measure of all conformational conversions averaged over a timescale that is sufficient to enable the quencher to diffuse into the enzyme (72). Applied to α-amylases (72), DNA ligases (77), and Ca^{2+}-Zn^{2+} proteases (33, 100), protein permeability was found to be highest for the cold-adapted enzymes. Proteolytic nicking has been used to identify flexible regions in chimeric isocitrate dehydrogenases that were constructed from cold-adapted and thermostable homologs (35). The study confirmed that the components derived from the psychrophile contained more flexible regions (more nicking) than those of the mesophile.

The flexibility of cold-adapted enzymes has also been examined by indirect means. The X-ray structures of an elastase (101), citrate synthase (62), and AHA (72) have been predicted to have enhanced substrate accessibility conferred by a highly flexible active-site region. These studies predict that improved accessibility afforded by enhanced flexibility should not only reduce the energy required to access large substrates but also decrease substrate specificity.

Global Versus Local Flexibility

A number of studies have addressed whether cold-adapted proteins have flexibility throughout their structure (global flexibility) or whether they have distinct regions of local flexibility. It has been argued that global flexibility may promote high activity and low stability in cold-adapted enzymes; however, it may also enhance incorrect folding (58). Studies of wild-type and mutant forms of AHA, which demonstrate that the enzyme is able to undergo two-state reversible unfolding, support the view that the structure of AHA is uniformly flexible (10, 58, 71, 72). Molecular dynamic simulations of a protease from a mesophile (savinase) that has been engineered to contain a loop structure that is present in a protease from a psychrophile (subtilisin) also support the view that a gain in psychrophilic character is accompanied by an overall increase in flexibility (102).

Support for local flexibility comes from studies that propose that thermal unfolding starts from the most flexible loop present on the protein surface (103), from extremities of the protein (8, 25), or from the active site (79). In the case of AHA, data supporting global (10, 58, 71, 72) or local flexibility (72, 79) have been described (also see Unfolding of Cold-Adapted Enzymes). Conclusions drawn reflect the methodology (DSC, CD, fluorescence, TUG-GE, X-ray B-factors, H/D exchange) and experimental conditions (temperature, pH, absence/presence of urea, Ca^{2+}, Cl^-), as well as a developing understanding gained from wild-type, genetically manipulated, and chemically modified variants.

Consistent with discussions in previous sections (Activity-Stability Relationship and Concept of Flexibility), in order to attain sufficient activity at low temperatures, the active-site cleft may need to be more flexible than peripheral parts of the enzyme. In support of this, loop structures around active sites have been reported to have enhanced flexibility, resulting in increased K_m (35, 59, 104, 105). Cold-adapted enzymes that appear to have enhanced flexibility of the active site include a protein-tyrosine phosphatase (106), uracil-DNA glycosylase (107, 108), chitobiase (66), and phosphoglycerate kinase (64).

FLEXIBILITY AND STRUCTURAL ADAPTATION

Considerable effort has been directed toward defining the essential structural features of a protein that describe the thermal characteristics of its activity and stability. The first X-ray structure for a cold-adapted protein appeared in 1994 (116), and 15 structures are presently in the Protein Data Bank (**Table 6**). The structural basis of cold adaptation has been inferred from comparisons of X-ray structures (**Table 6**) with homology models of proteins from mesophiles and thermophiles. As discussed above in Optimization of K_m in Cold-Adapted Enzymes, the activity and stability of a candidate enzyme can be greatly affected by its physiological environment. It is also noteworthy that the majority of studies have been performed on cold-adapted enzymes from marine organisms. To some degree (particularly for secreted enzymes), compositional and structural charcteristics may reflect halophilic as well as psychrophilic adaptation (8).

The comparison data sets used influence the nature of conclusions drawn about structural adaptation. X-ray structures of citrate synthase have been compared from phylogenetically and biochemically distinct members of *Bacteria*, *Eucarya*, and *Archaea* (85, 87). In the absence of crystal structure information, some modeling studies have purposefully chosen phylogenetically closely related organisms (e.g., 10, 117, 118). Thomas & Cavicchioli (117) found that despite choosing a phylogenetically coherent group of organisms, and a protein (EF-2) that is highly conserved, it was difficult to identify amino acid changes that were obviously linked to thermal adaptation. Because the changes that are critical for thermal adaptation are hidden amid those produced by genetic drift and other effectors of natural selection, it can be beneficial to maximize thermal differences and minimize phylogenetic differences in comparative data sets (6, 117), particularly if X-ray structures are not available.

An alternative approach to comparisons of single-protein sets is to identify statistically valid trends from the analysis of large data sets. A comparative genomics study has been performed on archaea spanning the growth temperature range 0°C to 110°C (11). Principle component analysis revealed that proteins from the cold-adapted archaea had a higher content of noncharged polar amino acids, particularly Gln and Thr, and a lower content of hydrophobic amino acids, particularly Leu. Using threading and homology modeling, 1111 modeled protein structures were constructed. Analysis of the models from the cold-adapted archaea showed a strong tendency in the solvent-accessible area for more Gln, Thr, and hydrophobic residues as well as fewer charged residues. Future statistical analyses of genome data sets should enhance the ability to identify the most abundant and significant structural characteristics of cold adaptation.

FOLDING FUNNEL MODEL

On the basis of kinetic, biophysical, and structural data, a "folding funnel model" has been proposed for AHA and other cold-adapted enzymes (72). In this model, a cold-adapted enzyme in its native state consists of a large population of conformations with low-energy barriers between conformations (4, 33, 109). This type of energy landscape would promote conversion between conformations and result in high structural flexibility. The increased flexibility would cause the cold-adapted enzyme to spend more time in conformations that are not optimal for substrate binding and result in high K_m (4). The low-energy barriers across a spectrum of microstates would result in an overall decreased activation energy and a high k_{cat}. It would be insightful to test this model by measuring the activity and activation energies of individual molecules (110–115) of sets of cold-adapted enzymes and their thermostable homologs.

Table 6 Structural and compositional elements characteristic of cold-adapted enzymes deduced from X-ray structure comparisons

Cold-adapted enzyme and organism	Decreased core hydrophobicity	Less Ile	Increased surface hydrophobicity	Less total charged residues	Increased surface hydrophilicity	More His	High (Glu + Asp)/(Lys + Arg) ratio	Less hydrogen bonds	Low Arg/Lys ratio
1. AHA[a,b]	+	+	+	−	+	+	+	0	+
2. Citrate synthase *Arthrobacter* sp.	+	+	+	−	+	+	+	+	−
3. Metalloprotease *Pseudomonas* sp.	+	X	0	0	X	+	0	0	−
4. Xylanase *Pseudoalteromonas haloplanktis*	X	X	+	+	X	X	+	0	+
5. Malate dehydrogenase *Aquaspirillium arcticum*	0	−	0	+	X	0	+	−	+
6. Triosephosphate isomerase *Vibrio marinus*	0	+	0	0	X	−	+	X	0
7. Uracil-DNA glycosylase *Gadus morhua*	+	0	X	0	X	0	0	−	−
8. Adenylate kinase *Bacillus globisporus*	+	X	+	0	0	−	+	+	+
9. Pepsin *Gadus morhua*	X	+	+	+	+	+	+	X	+
10. Elastase *Salmo salar*	X	−	X	−	−	−	−	X	+
11. Trypsin *Salmo salar*	+	+	X	−	+	+	+	+	−
12. Subtilisin *Vibrio* sp.	+	+	−	0	X	0	+	−	−
13. Alkaline phosphatase *Pandalus borealis*	X	−	+	−	+	0	+	X	+
14. Cellulase, catalytic module *Pseudoalteromonas haloplanktis*	X	X	+	+	+	X	+	+	+

[a] Codes used: +, presence of structural or compositional feature; −, absence of structural or compositional feature; 0, structural or compositional feature not appreciably different between cold-adapted enzyme and thermostable homolog; X, not examined; N, not present in structure.

[b] Cold-adapted enzyme structure compared to homolog from: 1,7,9–11,13, mammal; 2, hyperthermophilic *Pyrococcus furiosus*; 3, mesophilic *Pseudomonas aeruginosa*; 4, endoglucanase from *Clostridium thermocellum*; 5, thermophilic *Thermus flavus*; 6, mesophilic *E. coli*; 8, thermophilic *Bacillus stearothermophilus*; 12, mesophilic fungal proteinase K; 14, mesophilic *Erwinia chrysanthemi*.

Experimental verification and empirical determination of important structural features have been examined through rational design and screening of variants derived by genetic or chemical modification (73). Three X-ray structures of mechanistic mutants of AHA (with the amino acid change and Protein Data Bank numbers that follow: K300R, 1JD7; K300Q, 1JD9; D174N, 1KXH) have been solved. However, no structures are available for mutants of any cold-adapted enzymes with altered thermal properties.

Structural features implicated in the flexibility of cold-adapted enzymes have been extensively documented (e.g., 3, 5, 8, 26, 49, 60, 119). The following sections describe key findings from a large number of studies.

Table 6 (Continued)

Less aromatic interactions	Less salt bridges	Weaker intersubunit/ interdomain contacts	More/ longer loops	More Gly	Less Pro (loops)	More Pro (α-helices)	More Met	Less metal-binding sites and/or lower affinity	Less disulphide bridges	References/ PDB code
+	+	+	−	−	+	−	+	+	+	(122) 1AQM
X	−	+	+	−	+	+	+	N	N	(62) 1A59
X	0	X	−	+	0	N	+	+	N	(100) 1H71, 1G9K
X	+	X	X	−	−	−	X	N	N	(125) 1H12, 1H13, 1H14
X	−	+	−	0	−	0	0	N	N	(54) 1B8P
−	0	X	0	+	−	−	−	N	N	(139) 1AW1, 1AW2
X	+	X	X	0	−	−	0	N	N	(124) 1OKB
X	+	N	X	−	+	+	−	0	N	(123) 1S3G
X	X	X	−	+	+	X	−	X	−	(167) 1AM5
X	+	X	−	+	+	X	+	X	−	(168) 1ELT
−	+	+	−	−	−	X	+	N	−	(126, 169) 2TBS
X	0	X	X	+	−	X	−	−	−	(132, 170) 1SH7
X	X	X	X	+	+	X	−	X	0	(127) 1K7H
+	+	X	+	+	+	X	X	N	N	(134) 1TVN, 1TVP

Hydrophobic Interactions

The major factors contributing to the structural flexibility and thermolability of a cold-adapted protein are the nature and magnitude of interactions between hydrophobic residues (core hydrophobicity) and between hydrophobic residues and solvent water molecules (surface hydrophobicity). These hydrophobic interactions are discussed below.

Core hydrophobicity. The buried amino acids in cold-adapted enzymes tend to be smaller and less hydrophobic than in homologs from mesophiles and thermophiles (**Table 6**). Van der Waals interactions are weak, very short range, and extremely distance sensitive. Therefore, the distance between hydrophobic groups in the interior of a protein will determine the enthalpic contribution to stabilization (8, 86, 120). Cold-adapted enzymes will be destabilized owing to reduced van der Waals interactions and increased movement of internal groups. Because of its branching and size, Ile can pack more efficiently inside the core and stabilize a protein. Fewer Ile residues are present in the inner core of cold-adapted trypsins (121), citrate synthase (62), and AHA (122) (**Table 6**). In an adenylate kinase from a thermophile, Ile is present at postion 26. In the cold-adapted homolog, the presence of Thr26 appears to increase the solvent accessibility of the core residue, Val21, from 23 Å^2 to 42 Å^2 (123). In a cold-adapted uracil-DNA glycosylase (124) and the catalytic domain of a cellulase (118), decreased hydrophobicity of the core leads to an increase in both the number and volume of internal cavities. The thermolability of the uracil-DNA glycosylase is consistent with location of the cavities being near the active site and the DNA-binding loop (124).

The solubilities of hydrophobic side chains in water are minimal at 20°C, and therefore hydrophobic interactions are strongest at room temperature (47). A study of 31 proteins that unfold reversibly found that ∼3/4 had maximum stability around room temperature (47). Although the study only included one protein from a psychrophile, these findings illustrate the important role that hydrophobic interactions in the core of a protein play in enhancing protein stability at low to moderate temperatures.

Surface hydrophobicity. The surfaces of cold-adapted enzymes tend to have a higher proportion of hydrophobic (nonpolar) residues (**Table 6**). This was demonstrated by the X-ray structures of glyceraldehyde-3-phosphate dehydrogenase (119), citrate synthase (62), AHA (122), xylanase (125), trypsin (126), shrimp alkaline phosphatase (127), and adenylate kinase (123). A similar trend was observed in large-scale modeling and structural studies, which demonstrated that the mean fraction of the solvent-accessible surface (11) or buried surface (119) that is hydrophobic was higher in cold-adapted proteins.

The solvent-exposed area of hydrophobic residues is 7854 and 4929 Å^2 in citrate synthase from the psychrophile and hyperthermophile, respectively (62). However, in the oligomeric form of citrate synthase, the intersubunit hydrophobicity is decreased compared to thermophilic homologs (85). In particular, the X-ray structures reveal that Ile clusters at the subunit interface are absent in the cold-adapted enzyme, whereas tightly packed hydrophobic clusters are present in the homolog from a hyperthermophile (62).

For adenylate kinase, the exposed hydrophobic surface of the cold-adapted enzyme is increased by 483 Å^2 and equates to a ΔG of 4-11 kJ mol^{-1} (123). In the cold-adapted enzyme, Thr replaces Met at position 179 of a thermophilic homolog. Thr179 is unable to maintain hydrophobic interactions with a distant part of the protein, resulting in a marked increase in solvent accessibility of Val21. Because Val21 is part of the hydrophobic core, the Thr179Met replacement appears to be important for generating flexibility of adenylate kinase (123).

Flexibility in shrimp alkaline phosphatase is likely to be enhanced by hydrophobic residues (55% of the protein surface), which are arranged in clusters (127). Nonpolar amino acids constitute 32% and 26% of the surface of DNA ligases from psychrophilic and thermophilic bacteria, respectively (77). Hydrophobic surfaces also appear to be enriched in order to destabilize the active site of the cold-adapted DNA ligase (77). A similar finding was reported for a protein-tyrosine phosphatase (106).

Hydrophobic surface residues will destabilize a protein structure because of the decreased entropy of water molecules, which form cage-like structures around nonpolar residues. However, at low temperatures, the entropy gain is reduced owing to the decreased mobility of the released water molecules (47, 87, 120, 128). This implies that cold-adapted enzymes may gain flexibility from, and have a greater capacity to tolerate, increased surface hydrophobicity (11). Cold adaptation appears to involve a strategic arrangement of the total exposed and buried nonpolar fraction, leading to increased exposure of the nonpolar residues to water, with a concomittant reduction in the packing of the hydrophobic core (119).

Surface Hydrophilicity

An increase in surface charge, particularly a negative charge, has been described for cold-adapted trypsins (121, 126, 129), β-lactamase (55), malate dehydrogenase (54), subtilisin (130–132), citrate synthase (62), and cellulase (118). For shrimp alkaline phosphatase, the negative surface charge is particularly high (−80), with a patch of positive charge located near the active site (127).

The dielectric constant of water increases from 55.5 Debye (ε_r) at 100°C to 88 ε_r at 0°C (87). At low temperatures (approaching 0°C), the energetic cost of disrupting H-bond networks is very high because of the high viscosity and high surface tension of water (87). In cold-adapted enzymes, the energetic cost may be offset by surface-charged or polar amino acids interacting with water molecules of a high dielectric constant (60, 130), thereby enabling proper solvation and maintaining flexibility (87). Better solvent interactions with positively charged His ($pK_a \approx 7$) may also play a role in improving flexibility at low temperatures (**Table 6**). The imidazole ring of His has high $\Delta H_{ionization}$, and a decrease in temperature favors its protonation (133).

In addition to improving solvent interactions, the localization of acidic residues in surface patches may produce charge-charge repulsions causing destabilization of overall protein structure. The catalytic and cellulase-binding modules of a cellulase from *P. haloplanktis* are connected by an unusually long linker (109 residues). The linker contains 23 acidic amino acids and lacks positively charged residues. Charge repulsion of acidic residues is likely to create a high level of flexibility in the linker region and is proposed to be a major structural feature of cold adaptation in this enzyme (134). In the case of shrimp alkaline phosphatase (127), the positively charged surface may direct the negatively charged substrate to the active site.

In contrast to the above studies, a genomic study reported that cold adaptation involves a decrease in the mean fraction of solvent-accessible and buried surface that is charged (11). This trend may reflect characteristics of proteins from *Archaea*. However, similar trends have been identified using genome sequences of *Bacteria* (135). Increasing surface hydrophilicity in proteins from both psychrophiles and thermophiles have been reported (136). The increased surface charge in thermostable proteins was linked to an ability to form networks of salt bridges (48), which is in contrast to the interaction with water molecules in cold-adapted enzymes (136). As ionic interactions become stronger with decreasing temperature, there appears to be a minimization of their number that allows cold-adapted proteins to retain flexibility at low temperatures.

As addressed above (see the section Flexibility and Structural Adaptation, above), the marine origin of many of the cold-adapted enzymes, which have been studied, may explain some of the charged (particularly anionic) characteristics (8). An extreme example of halophilic adaptation is the overrepresentation of acidic amino acids in the genomes of haloarchaea (137).

Electrostatic Interactions

Intramolecular, noncovalent, electrostatic interactions are of prime importance in maintaining secondary and tertiary structure. These interactions occur with polar and charged amino acids and mainly comprise H-bonds, salt bridges, and aromatic interactions. Electrostatic interactions are effective over long distances and, as a result, can be difficult to predict, particularly in homology models.

H-bonds. H-bonds are the most abundant type of noncovalent interactions, and 4–12 kJ mol^{-1} is required to break a single bond (138). Because the total difference in stabilization energy between homologs from a psychrophile and mesophile may be as little as 40–50 kJ mol^{-1}, it is clear that this may be accounted for by a few critically placed H-bonds. Cold adaptation of a triosephosphate isomerase was linked to an Ala replacement of a Ser; Ser is expected to confer thermostability by forming two additional intramolecular H-bonds (139).

The interpretation of H-bonds in homology models is not reliable because H-bonds are highly directional. No H-bond differences were identified from comparative analyses of the X-ray structures of AHA, citrate synthase, and malate dehydrogenase (54, 62, 122) (**Table 6**). However, cold-adapted enzymes may have fewer interdomain or intersubunit H-bonds (8). In one study, a cold-adapted uracil-DNA glycosylase (124) was reported to have three more H-bonds (219 H-bonds) than its counterpart from a mesophile (216 H-bonds).

Arginine-mediated interactions. Many cold-adapted enzymes have a reduced Arg/Lys ratio, including β-galactosidase (65), chitobiase (66), phosphoglycerate kinase (64), and the majority of enzymes listed in **Table 6**. In a comparative genomics study, protein models generated from cold-adapted *Archaea* were found to have fewer charged residues (Arg + Lys + Glu) on their solvent-accessible surface (11). However, a number of cold-adapted enzymes (e.g., citrate synthase, metalloprotease, trypsin, subtilisin) (**Table 6**) have a higher Arg/Lys ratio compared to thermostable homologs. A cold-adapted uracil-DNA glycosylase has three more Arg residues than a homolog from a mesophile (124). In this enzyme, Arg residues are located mostly near its surface. Unless they participate in ionic interactions, they are likely to form interactions with water, thereby imparting flexibility to the overall enzyme structure (8, 124).

Arginine residues are generally thought to enhance enzyme thermostability more than Lys residues by facilitating a greater number of electrostatic (up to two salt bridges and five H-bond) interactions through their guanidino group (55, 56, 58, 140, 141). Experimental support for the general thermostabilizing effect of Arg over Lys comes from studies on a range of mammalian enzymes that were reported as stabilized by the introduction of guanidinium groups (140). In a cold-adapted protein-tyrosine phosphatase, Met147 is present in a loop that is a part of the active site. The homolog from a mesophile contains Arg at an equivalent location in the structure (Arg73) and is able to make five additional H-bonds. It is likely that the Met replacement imparts flexibility to the active site of the cold-adapted enzyme (106). AHA has only 13 Arg and 13 Lys residues compared to 28 Arg and 19 Lys residues in PPA (78, 142). As a result, there is a reduction in Arg-mediated salt bridges (−9), amino-aromatic interactions (−11), and H-bonds (−22) (25, 122). The stability of AHA has been found to increase by converting the side chains of Lys residues to homoarginine (142a). This is consistent with Lys replacing Arg as a means of generating flexibility in AHA.

Aromatic interactions. The aromatic rings in Trp, Tyr, and Phe have a dipole because of the partial negative charge on the face of the ring caused by the π-electron cloud and by a partial positive charge on the C-H edges (25). This polarity permits favorable interactions between aromatic rings at right angles to each other (aromatic-aromatic interactions) or between aromatic rings and the side chains of Arg and Lys (aromatic-amino interactions). Aromatic interactions may therefore promote thermostabilization through an enthalpic contribution.

Cold-adapted subtilisin has a general lack of aromatic interactions in contrast to 11 interactions identified on the surface of a thermophilic homolog (130). In a β-lactamase from a mesophile, two Trp aromatic-aromatic interactions are present in comparison to a homolog from a psychrophile and appear to be important for thermostabilization (55). In AHA, a reduced number of aromatic-amino, but not aromatic-aromatic, interactions appear to be important for maintaining flexibility (25).

Salt bridges. A salt bridge is defined as an ion pair with a distance of 2.5–4.0 Å between charged nonhydrogen atoms (143) and contributes 12–21 kJ mol^{-1} of stabilization to an enzyme (138). The formation cost of a single ion pair is very high on the surface of an enzyme (138). However, ion-pair networks are more stable owing to the reduction by half of the desolvation penalty (138). In addition to these effects, because of water's high dielectric constant at low temperatures, ion pairs become destabilizing (see the section Optimization of K_m in Cold-Adapted Enzymes, above) and may therefore enhance flexibility of cold-adapted enzymes at low temperatures (48, 87).

Several comparative studies based on X-ray structure data have reported that the number of salt bridges is lower in cold-adapted enzymes (e.g., 119), and this is reflected in the majority of those listed in **Table 6**. Cold-adapted subtilisin has two salt bridges compared with five and ten salt bridges in the homologs from mesophiles and thermophiles, respectively (130). The number of salt bridges (particularly the Arg-mediated ones) in AHA is considerably less than in PPA and may contribute to AHA's thermolability and ability to unfold reversibily (25, 58, 122).

Citrate synthase from a psychrophile is reported to have more intrasubunit salt bridges than a homolog from a hyperthermophile, 27 versus 42, respectively (62). However, the cold-adapted enzyme has fewer intersubunit interactions than a homolog from a mesophile (62). A reduced number of intersubunit and ion-pair networks also appears to be important for the heat lability of a malate dehydrogenase from a psychrophile (54). In a genomics study, the number of predicted salt bridges was found to decrease significantly for enzymes from psychrophiles for some protein models (e.g., Cpn60); however, unlike the trends for solvent accessibility for charged and hydrophobic residues, the trend for salt bridges was not a general feature (11). In association with the data assembled in **Table 6**, these studies indicate that cold adaptation correlates more strongly with a reduced number of interdomain and intersubunit interactions, rather than the number of salt bridges per se.

The importance of the strategic placement of salt bridges is well illustrated in citrate synthase, where the absence of an intersubunit R375-E48 salt bridge appears to make a critical contribution to its low stability (62). An ion pair may be considered critical if the bridge is between residues that are distantly located (>10 residues) and if the ion pair cannot form in a more thermolabile homolog because of amino acid replacement or the ion pair's distance apart (≥6 Å) (123). This type of critical replacement appears to make an important contribution to the thermolability of a cold-adapted adenylate kinase (123).

Secondary Structure Elements

The elements of secondary structures (α-helix and β-sheets) define how the tertiary structure of a protein will fold. As they play a central role in the structure and function of a protein, only minor changes in secondary structure can be tolerated. The stability of an α-helix depends on the intrahelical H-bonds and the side-chain-to-side-chain interactions. Helix termini lack intrahelical interactions. The properties of Ncap (N-terminal residue of the helix) and Ccap residues can also affect the stability of an enzyme (144). In a helix, the peptide units tend to align parallel to the helical axis, and their cumulative effect is to create a helix macro-dipole (approximately +0.5 unit charge at the N terminus and −0.5 at the C terminus). As a result, residues toward the N terminus of a helix tend to be negatively charged, and those toward the C terminus tend to be positively charged.

The amino acid differences between cold-adapted enzymes and thermostable homologs create subtle differences in the properties of α-helices (8). In AHA compared to PPA, negatively charged Ncap residues decrease from four to two, and positively charged Ccap residues decrease from six to three (25). The weakened charge-dipole interactions in AHA may generate flexibility (25). In the case of cold-adapted triosephosphate isomerase, flexibility may be promoted by a higher level of positive charge near the Ncap and negative charge near the Ccap (145).

When located in the center of α-helices, Pro residues cause at least two H-bonds to be lost and, owing to their restricted degrees of freedom, can cause destabilization by acting as a helix breaker (146, 147). However, when present at the Ncap, they may induce helix formation (146). A number of cold-adapted enzymes have a higher number of Pro residues in α-helices (**Table 6**). For example, a cold-adapted citrate synthase has additional Pro residues in the center of two α-helices (62).

Lacking a side chain and the conformational freedom of Gly residues, they may destabilize α-helices (147, 148). This capacity does not seem to play a role in cold adaptation because no significant trends involving Gly residues have been identified in comparative studies (8).

Surface Loops

The majority of loops that connect secondary structure elements are found on the enzyme's surface. Many catalytic pockets are surrounded by loop structures and generate flexibility of the active site (8). For reasons similar to those described in the Secondary Structure Elements section above, the side chain of a Pro residue is covalently bound to the N atom of the peptide backbone, thereby restricting the rotation about N-C$_\alpha$ bonds and reducing conformational flexibility of loop structures (56). In contrast, Gly residues in loops increase conformational freedom of the native state (60).

The number of Pro residues in loops is lower in cold-adapted subtilisin, malate dehydrogenase, trypsin, alkaline phosphatase, DNA ligase and AHA, compared to their respective mesophilic and thermophilic homologs (**Table 6**). In a cold-adapted 3-isopropylmalate dehydrogenase, Ala and Ser replace two Pro residues that are present in two loops in thermostable homologs and in one loop near the active site (143). In a cold-adapted citrate synthase, a loop connecting two α-helices is nine residues longer than in thermostable enzymes and contains more charged residues (62, 149). The cold-adapted enzyme also contains eight fewer Pro residues in loops compared with the homolog from a hyperthermophile.

The flexibility of LDH-A$_4$ from an Antarctic fish appears to be enhanced by a loop possessing 2 additional Gly residues that are 10 residues apart (60). However, a cold-adapted citrate synthase has 7 less Gly residues (a total of 22) than a homolog from a hyperthermophile; the number of Gly residues is the same in α-helices but is reduced by 7 in loops of the cold-adapted enzyme (62). It appears that the relative abundance of Gly residues in loops is only partially explained by the contribution they make to the overall flexibility of the enzyme.

A cold-adapted triosephosphate isomerase has more and longer loops connecting α-helices and β-sheets than homologs from mesophiles and thermophiles (139, 145). Longer surface loops increase the possible amplitude of the movement between secondary structures and may decrease enzyme stability.

Other Factors

Methionine content is higher in a number of cold-adapted enzymes (**Table 6**). Met residues may confer flexibility because of their high degree of freedom and because they lack branching and charge or dipole interactions (117). Alternatively, a high Met content, identified in cold-adapted trypsins from fish, may reflect halophilic adaptation (8, 121).

Metals can stabilize an enzyme by simultaneously bridging numerous secondary structures or domains. Many Ca^{2+}-binding, cold-adapted enzymes are characterized by low binding constants. The Ca^{2+}-binding affinity is 2000-fold lower in AHA compared to PPA (78), and a cold-adapted subtilisin from a *Bacillus* spp. has a lower Ca^{2+}-binding affinity than a thermostable homolog from *Thermoactinomycetes vulgaris* (130). A cold-adapted subtilisin from *Vibrio* sp. has one more Ca^{2+}-binding site compared to the *T. vulgaris* enzyme (132). The significance of the larger number of Ca^{2+}-binding sites in the *Vibrio* sp. enzyme will become clearer when the affinity constants are calculated.

The absence of disulfide bridges in a cold-adapted alkaline phosphatase has been suggested to confer flexibility (150). In contrast, three disulfide bridges are present in a cold-adapted subtilisin-like serine protease compared to none in the homolog from a thermophile (132). Four disulfide bridges are present in AHA compared to five in PPA (78). Introduction of a fifth disulfide bridge in AHA resulted in increased conformational stability and decreased activity (151). Removal (by chemical modification) of all disulfide bonds in AHA led to stabilization of the least stable region of the enzyme, including the active site, and a decrease in activity (80). In AHA, the disulfide bridges appear to prevent the active site from developing ionic interactions, thereby promoting a localized destabilization to preserve activity.

SUMMARY POINTS

1. Insight into cold adaptation has occurred through studies performed on 15 X-ray structures, biochemical and biophysical analysis of activity and stability of numerous enzymes, and compositional and structural (homology modeling) analyses of genome sequence data from a limited number of psychrophiles. Studies have been performed on cold-adapted enzymes from members of the *Bacteria*, *Archaea*, and *Eucarya*, representing wild-type, mutant, and chemically modified forms of native and recombinant proteins.

2. A range of structural features correlates with enzyme cold adaptation. However, no structural feature is present in all cold-adapted enzymes, and no structural features always correlate with cold adaptation.

3. Proteins from psychrophiles are thermolabile and more flexibile than their counterparts from thermophiles. Many cold-adapted proteins have regions of local flexibility, particularly around the active site. The high local flexibility is translated into a reduction in $\Delta H^{\#}$, high k_{cat}, and, in the majority of cases, high K_m.

4. The high activity and low stability of cold-adapted enzymes underlie a general principle of activity-stability trade-off. The outcomes of mutagenesis and chemical modification studies indicate, to some degree, that variants of naturally occuring cold-adapted enzymes can defy this rule. Seeking out ways to optimize enzyme cold adaptation will help promote a mechanistic understanding and provide improved products for commercial exploitation.

FUTURE ISSUES TO BE RESOLVED

1. Observational trends derived from compositional and structural inference need to be experimentally probed, using a range of biophysical methods, and structures determined for mutants to verify predictions.

2. In addition to increasing the number of single-enzyme comparisons of phylogenetically closely related and thermally divergent models, a significant increase in genome sequence data for psychrophiles is required in order to identify broad, robust trends in thermal adaptation. These should include a number of nonmarine organisms (e.g., psychrophiles from freshwater lakes) to ensure that halophilic and psychrophilic adaptation can be distinguished.

3. In order to more fully determine the characteristics of enzymes from psychrophiles, attention should be paid to physiological and environmental factors, including the use of natural substrates.

4. Cold denaturation needs to be experimentally defined for a larger number of cold-adapted enzymes. The generation of stability curves of sets of thermally adapted enzymes at temperatures between the T_m of cold denaturation and the T_m of heat denaturation would be useful for probing the mechanisms of cold unfolding.

ACKNOWLEDGMENTS

This work was supported by the Australian Research Council. Thanks are extended to Charles Gerday, Michael Danson, Kevin Barrow, Neil Saunders, and Torsten Thomas for critical appraisal of the manuscript and to Marilyn Katrib for editorial assistance.

LITERATURE CITED

1. Kashefi K, Lovley DR. 2003. *Science* 301:934
2. Margesin R, Schinner F, eds. 1999. *Cold-Adapted Organisms—Ecology, Physiology, Enzymology and Molecular Biology*. Berlin: Springer-Verlag. 416 pp.
3. Feller G, Gerday C. 2003. *Nat. Rev. Microbiol.* 1:200–8
4. Somero GN. 2004. *Comp. Biochem. Physiol. B* 139:321–33
5. Cavicchioli R, Siddiqui KS. 2004. In *Enzyme Technology*, ed. A Pandey, C Webb, CR Soccol, C Larroche, pp. 615–38. New York, NY: Springer Science
6. Sheridan PP, Panasik N, Coombs JM, Brenchley JE. 2000. *Biochim. Biophys. Acta* 1543:417–33
7. Wainwright M, Wickramasinghe NC, Narlikar JV, Rajaratnam P. 2003. *FEMS Microbiol. Lett.* 218:161–65
8. Smalas AO, Leiros HK, Os V, Willassen NP. 2000. *Biotechnol. Annu. Rev.* 6:1–57
9. D'Amico S, Claverie P, Collins T, Georlette D, Gratia E, et al. 2002. *Philos. Trans. R. Soc. London Ser. B* 357:917–25
10. Georlette D, Blaise V, Collins T, D'Amico S, Gratia E, et al. 2004. *FEMS Microbiol. Rev.* 28:25–42
11. Saunders NF, Thomas T, Curmi PM, Mattick JS, Kuczek E, et al. 2003. *Genome Res.* 13:1580–88
12. Goodchild A, Raftery M, Saunders NF, Guilhaus M, Cavicchioli R. 2004. *J. Proteome Res.* 3:1164–76
13. Goodchild A, Saunders NF, Ertan H, Raftery M, Guilhaus M, et al. 2004. *Mol. Microbiol.* 53:309–21

14. Cavicchioli R, Siddiqui KS, Andrews D, Sowers KR. 2002. *Curr. Opin. Biotechnol.* 13:253–61
15. Margesin R, Schinner F. 1999. *Biotechnological Applications of Cold-Adapted Organisms.* Berlin: Springer-Verlag. 338 pp.
16. Lonhienne T, Gerday C, Feller G. 2000. *Biochim. Biophys. Acta* 1543:1–10
17. Crawford DL, Powers DA. 1992. *Mol. Biol. Evol.* 9:806–13
18. Devos N, Ingouff M, Loppes R, Matagne RF. 1998. *J. Phycol.* 34:655–60
19. Baldwin J, Hochachka PW. 1970. *Biochem. J.* 116:883–87
20. Somero GN. 1995. *Annu. Rev. Physiol.* 57:43–68
21. Jagdale GB, Gordon R. 1997. *Comp. Biochem. Physiol. A* 118:1151–56
22. Low PS, Bada JL, Somero GN. 1973. *Proc. Natl. Acad. Sci. USA* 70:430–32
23. Eyring H, Stearn AE. 1939. *Chem. Rev.* 24:253–70
24. Siddiqui KS, Cavicchioli R, Thomas T. 2002. *Extremophiles* 6:143–50
25. Feller G. 2003. *Cell. Mol. Life Sci.* 60:648–62
26. Feller G, Gerday C. 1997. *Cell. Mol. Life Sci.* 53:830–41
27. Gerday C, Aittaleb M, Arpigny JL, Baise E, Chessa JP, et al. 1997. *Biochim. Biophys. Acta* 1342:119–31
28. Barnes R, Vogel H, Gordon I. 1969. *Proc. Natl. Acad. Sci. USA* 62:263–70
29. Low PS, Somero GN. 1974. *Comp. Biochem. Physiol. B* 49:307–12
30. Somero GN. 2003. *Comp. Biochem. Physiol. B* 136:577–91
31. Feller G, Lonhienne T, Deroanne C, Libioulle C, Van Beeumen J, et al. 1992. *J. Biol. Chem.* 267:5217–21
32. Collins T, Meuwis MA, Stals I, Claeyssens M, Feller G, et al. 2002. *J. Biol. Chem.* 277:35133–39
33. Zecchinon L, Claverie P, Collins T, D'Amico S, Delille D, et al. 2001. *Extremophiles* 5:313–21
34. Lonhienne T, Baise E, Feller G, Bouriotis V, Gerday C. 2001. *Biochim. Biophys. Acta* 1545:349–56
35. Watanabe S, Yasutake Y, Tanaka I, Takada Y. 2005. *Microbiology* 151:1083–94
36. Afzal AJ, Ali S, Latif F, Rajoka MI, Siddiqui KS. 2005. *Appl. Biochem. Biotechnol.* 120:51–70
37. Fersht A. 1999. *Structure and Mechanism in Protein Science.* New York: Freeman. 631 pp.
38. Wolfenden R, Snider MJ. 2001. *Acc. Chem. Res.* 34:938–45
39. Harris GW, Jenkins JA, Connerton I, Pickersgill RW. 1996. *Acta Crystallogr. D* 52:393–401
40. Lienhard GE. 1973. *Science* 180:149–54
41. Snider MJ, Lazarevic D, Wolfenden R. 2002. *Biochemistry* 41:3925–30
42. Licht SS, Lawrence CC, Stubbe J. 1999. *Biochemistry* 38:1234–42
43. Loftfield RB, Eigner EA, Pastuszyn A, Lovgren TN, Jakubowski H. 1980. *Proc. Natl. Acad. Sci. USA* 77:3374–78
44. Fan YX, McPhie P, Miles EW. 2000. *Biochemistry* 39:4692–703
45. Siddiqui KS, Bokhari SA, Afzal AJ, Singh S. 2004. *IUBMB Life* 56:403–7
46. Georlette D, Bentahir M, Claverie P, Collins T, d'Amico D. 2001. In *Physics and Chemistry Basis of Biotechnology*, ed. M De Cuyper, JWM Bulte, pp. 177–96. Dordrecht, Neth.: Kluwer Acad.
47. Kumar S, Tsai CJ, Nussinov R. 2002. *Biochemistry* 41:5359–74
48. Kumar S, Ma B, Tsai CJ, Nussinov R. 2000. *Proteins* 38:368–83
49. Arpigny JL, Feller G, Davail S, Genicot S, Narinx E, et al. 1994. In *Advances in Comparative and Environmental Physiology*, ed. R Gilles, pp. 269–95. Berlin: Springer-Verlag

50. Tsuruta H, Aizono Y. 2003. *J. Biochem.* 133:225–30
51. Carlow DC, Short SA, Wolfenden R. 1996. *Biochemistry* 35:948–54
52. Miller BG, Wolfenden R. 2002. *Annu. Rev. Biochem.* 71:847–85
53. Strynadka NC, James MN. 1991. *J. Mol. Biol.* 220:401–24
54. Kim SY, Hwang KY, Kim SH, Sung HC, Han YS, Cho Y. 1999. *J. Biol. Chem.* 274:11761–67
55. Feller G, Zekhnini Z, Lamotte-Brasseur J, Gerday C. 1997. *Eur. J. Biochem.* 244:186–91
56. Georlette D, Jonsson ZO, Van Petegem F, Chessa J, Van Beeumen J, et al. 2000. *Eur. J. Biochem.* 267:3502–12
57. Birolo L, Tutino ML, Fontanella B, Gerday C, Mainolfi K, et al. 2000. *Eur. J. Biochem.* 267:2790–802
58. D'Amico S, Gerday C, Feller G. 2001. *J. Biol. Chem.* 276:25791–96
59. Fields PA, Somero GN. 1998. *Proc. Natl. Acad. Sci. USA* 95:11476–81
60. Fields PA. 2001. *Comp. Biochem. Physiol. A* 129:417–31
61. Fields PA, Houseman DE. 2004. *Mol. Biol. Evol.* 21:2246–55
62. Russell RJ, Gerike U, Danson MJ, Hough DW, Taylor GL. 1998. *Structure* 6:351–61
63. Tsigos I, Velonia K, Smonou I, Bouriotis V. 1998. *Eur. J. Biochem.* 254:356–62
64. Bentahir M, Feller G, Aittaleb M, Lamotte-Brasseur J, Himri T, et al. 2000. *J. Biol. Chem.* 275:11147–53
65. Hoyoux A, Jennes I, Dubois P, Genicot S, Dubail F, et al. 2001. *Appl. Environ. Microbiol.* 67:1529–35
66. Lonhienne T, Zoidakis J, Vorgias CE, Feller G, Gerday C, Bouriotis V. 2001. *J. Mol. Biol.* 310:291–97
67. Cavicchioli R, Thomas T, Curmi PM. 2000. *Extremophiles* 4:321–31
68. Thomas T, Cavicchioli R. 2002. *Curr. Protein Pept. Sci.* 3:223–30
69. Thomas T, Cavicchioli R. 2000. *J. Bacteriol.* 182:1328–32
70. Thomas T, Kumar N, Cavicchioli R. 2001. *J. Bacteriol.* 183:1974–82
71. Feller G, d'Amico D, Gerday C. 1999. *Biochemistry* 38:4613–19
72. D'Amico S, Marx JC, Gerday C, Feller G. 2003. *J. Biol. Chem.* 278:7891–96
73. Cavicchioli R, Curmi PM, Siddiqui KS, Thomas T. 2006. In *Extremophiles-Methods in Microbiology*, Vol. 35, ed. FA Rainey, A Oren, pp.395–436. London: Elsevier-Acad.
74. Xu Y, Feller G, Gerday C, Glansdorff N. 2003. *J. Bacteriol.* 185:5519–26
75. Xu Y, Feller G, Gerday C, Glansdorff N. 2003. *J. Bacteriol.* 185:2161–68
76. Collins T, Meuwis MA, Gerday C, Feller G. 2003. *J. Mol. Biol.* 328:419–28
77. Georlette D, Damien B, Blaise V, Depiereux E, Uversky VN, et al. 2003. *J. Biol. Chem.* 278:37015–23
78. Feller G, Payan F, Theys F, Qian M, Haser R, Gerday C. 1994. *Eur. J. Biochem.* 222:441–47
79. Siddiqui KS, Feller G, D'Amico S, Gerday C, Giaquinto L, Cavicchioli R. 2005. *J. Bacteriol.* 187:6197–205
80. Siddiqui KS, Poljak A, Guilhaus M, Feller G, D'Amico S, et al. 2005. *J. Bacteriol.* 187:6206–12
81. Privalov PL. 1992. In *Protein Folding*, ed. TE Creighton, pp. 83–126. New York: Freeman
82. Siddiqui KS, Saqib AA, Rashid MH, Rajoka MI. 2000. *Enzyme Microb. Technol.* 27:467–74
83. Siddiqui KS, Poljak A, Cavicchioli R. 2004. *Cell. Mol. Biol.* 50:657–67
84. Ciardiello MA, Camardella L, Carratore V, di Prisco G. 2000. *Biochim. Biophys. Acta* 1543:11–23
85. Bell GS, Russell RJ, Connaris H, Hough DW, Danson MJ, Taylor GL. 2002. *Eur. J. Biochem.* 269:6250–60

86. Makhatadze GI, Privalov PL. 1994. *Biophys. Chem.* 51:291–304
87. Kumar S, Nussinov R. 2004. *ChemBioChem.* 5:280–90
88. Eijsink VG, Bjork A, Gaseidnes S, Sirevag R, Synstad B, et al. 2004. *J. Biotechnol.* 113:105–20
89. Zhang NY, Suen WC, Windsor W, Xiao L, Madison V, Zaks A. 2003. *Protein Eng.* 16:599–605
90. Siddiqui KS, Cavicchioli R. 2005. *Extremophiles* 9:471–76
91. Taguchi S, Komada S, Momose H. 2000. *Appl. Environ. Microbiol.* 66:1410–15
92. Miyazaki K, Wintrode PL, Grayling RA, Rubingh DN, Arnold FH. 2000. *J. Mol. Biol.* 297:1015–26
93. Marshall CJ. 1997. *Trends Biotechnol.* 15:359–64
94. D'Amico S, Gerday C, Feller G. 2000. *Gene* 253:95–105
95. Tang KE, Dill KA. 1998. *J. Biomol. Struct. Dyn.* 16:397–411
96. Halle B. 2002. *Proc. Natl. Acad. Sci. USA* 99:1274–79
97. Svingor A, Kardos J, Hajdu I, Nemeth A, Zavodszky P. 2001. *J. Biol. Chem.* 276:28121–25
98. Smith DK, Radivojac P, Obradovic Z, Dunker AK, Zhu G. 2003. *Protein Sci.* 12:1060–72
99. Siddiqui KS, Rangarajan M, Hartley BS, Kitmitto A, Panico M, et al. 1993. *Biochem. J.* 289:201–8
100. Aghajari N, Van Petegem F, Villeret V, Chessa JP, Gerday C, et al. 2003. *Proteins* 50:636–47
101. Aittaleb M, Hubner R, Lamotte-Brasseur J, Gerday C. 1997. *Protein Eng.* 10:475–77
102. Tindbaek N, Svendsen A, Oestergaard PR, Draborg H. 2004. *Protein Eng. Des. Sel.* 17:149–56
103. Clarke J, Fersht AR. 1993. *Biochemistry* 32:4322–29
104. Brandsdal BO, Heimstad ES, Sylte I, Smalas AO. 1999. *J. Biomol. Struct. Dyn.* 17:493–506
105. Merz A, Yee MC, Szadkowski H, Pappenberger G, Crameri A, et al. 2000. *Biochemistry* 39:880–89
106. Tsuruta H, Mikami B, Aizono Y. 2005. *J. Biochem.* 137:69–77
107. Moe E, Leiros I, Riise EK, Olufsen M, Lanes O, et al. 2004. *J. Mol. Biol.* 343:1221–30
108. Olufsen M, Smalas AO, Moe E, Brandsdal BO. 2005. *J. Biol. Chem.* 280:18042–48
109. Tsai CJ, Kumar S, Ma B, Nussinov R. 1999. *Protein Sci.* 8:1181–90
110. Xue Q, Yeung ES. 1995. *Nature* 373:681–83
111. Craig DB, Arriaga EA, Wong JCY, Lu H, Dovichi NJ. 1996. *J. Am. Chem. Soc.* 118:5245–53
112. Edman L, Foldes-Papp Z, Wennmalm S, Rigler R. 1999. *Chem. Phys.* 247:11–22
113. Lu HP, Xun L, Xie XS. 1998. *Science* 282:1877–82
114. Shoemaker GK, Juers DH, Coombs JM, Matthews BW, Craig DB, et al. 2003. *Biochemistry* 42:1707–10
115. Lee AI, Brody JP. 2005. *Biophys. J.* 88:4303–11
116. Uppenberg J, Hansen MT, Patkar S, Jones TA. 1994. *Structure* 2:293–308
117. Thomas T, Cavicchioli R. 1998. *FEBS Lett.* 439:281–86
118. Garsoux G, Lamotte J, Gerday C, Feller G. 2004. *Biochem. J.* 384:247–53
119. Gianese G, Bossa F, Pascarella S. 2002. *Proteins* 47:236–49
120. Privalov PL, Gill SJ. 1988. *Adv. Protein Chem.* 39:191–234
121. Leiros HK, Willassen NP, Smalas AO. 1999. *Extremophiles* 3:205–19
122. Aghajari N, Feller G, Gerday C, Haser R. 1998. *Protein Sci.* 7:564–72
123. Bae E, Phillips GN Jr. 2004. *J. Biol. Chem.* 279:28202–8

124. Leiros I, Moe E, Lanes O, Smalas AO, Willassen NP. 2003. *Acta Crystallogr. D* 59:1357–65
125. Van Petegem F, Collins T, Meuwis MA, Gerday C, Feller G, et al. 2003. *J. Biol. Chem.* 278:7531–39
126. Smalas AO, Heimstad ES, Hordvik A, Willassen NP, Male R. 1994. *Proteins* 20:149–66
127. de Backer M, McSweeney S, Rasmussen HB, Riise BW, Lindley P, et al. 2002. *J. Mol. Biol.* 318:1265–74
128. Tsai CJ, Maizel JV Jr, Nussinov R. 2002. *Crit. Rev. Biochem. Mol. Biol.* 37:55–69
129. Leiros HK, Willassen NP, Smalas AO. 2000. *Eur. J. Biochem.* 267:1039–49
130. Davail S, Feller G, Narinx E, Gerday C. 1994. *J. Biol. Chem.* 269:17448–53
131. Narinx E, Davail S, Feller G, Gerday C. 1992. *Biochim. Biophys. Acta* 1131:111–13
132. Arnorsdottir J, Kristjansson MM, Ficner R. 2005. *FEBS. J.* 272:832–45
133. Bhattacharya S, Lecomte JT. 1997. *Biophys. J.* 73:3241–56
134. Violot S, Aghajari N, Czjzek M, Feller G, Sonan GK, et al. 2005. *J. Mol. Biol.* 348:1211–24
135. Methe BA, Nelson KE, Deming JW, Momen B, Melamud E, et al. 2005. *Proc. Natl. Acad. Sci. USA* 102:10913–18
136. Spassov VZ, Karshikoff AD, Ladenstein R. 1995. *Protein Sci.* 4:1516–27
137. Kennedy SP, Ng WV, Salzberg SL, Hood L, DasSarma S. 2001. *Genome Res.* 11:1641–50
138. Vieille C, Zeikus GJ. 2001. *Microbiol. Mol. Biol. Rev.* 65:1–43
139. Alvarez M, Zeelen JP, Mainfroid V, Rentier-Delrue F, Martial JA, et al. 1998. *J. Biol. Chem.* 273:2199–206
140. Cupo P, El-Deiry W, Whitney PL, Awad WM Jr. 1980. *J. Biol. Chem.* 255:10828–33
141. Mrabet NT, Van den Broeck A, Van den Brande I, Stanssens P, Laroche Y, et al. 1992. *Biochemistry* 31:2239–53
142. Chessa JP, Feller G, Gerday C. 1999. *Can. J. Microbiol.* 45:452–57
142a. Siddiqui KS, Poljak A, Guilhaus M, De Francisci D, Curmi PMG, et al. 2006. *Proteins* In press
143. Wallon G, Lovett ST, Magyar C, Svingor A, Szilagyi A, et al. 1997. *Protein Eng.* 10:665–72
144. Kapp GT, Richardson JS, Oas TG. 2004. *Biochemistry* 43:3814–23
145. Rentier-Delrue F, Mande SC, Moyens S, Terpstra P, Mainfroid V, et al. 1993. *J. Mol. Biol.* 229:85–93
146. Richardson JS, Richardson DC. 1988. *Science* 240:1648–52
147. O'Neil KT, DeGrado WF. 1990. *Science* 250:646–51
148. Aurora R, Creamer TP, Srinivasan R, Rose GD. 1997. *J. Biol. Chem.* 272:1413–16
149. Gerike U, Danson MJ, Russell NJ, Hough DW. 1997. *Eur. J. Biochem.* 248:49–57
150. Asgeirsson B, Nielsen BN, Hojrup P. 2003. *Comp. Biochem. Physiol. B* 136:45–60
151. D'Amico S, Gerday C, Feller G. 2002. *J. Biol. Chem.* 277:46110–15
152. Asgeirsson B, Fox JW, Bjarnason JB. 1989. *Eur. J. Biochem.* 180:85–94
153. Venkatesh R, Sundaram PV. 1998. *Ann. NY Acad. Sci.* 864:512–16
154. Venkatesh R, Sundaram PV. 1998. *Protein Eng.* 11:691–98
155. Ionescu RM, Smith VF, O'Neill JC Jr, Matthews CR. 2000. *Biochemistry* 39:9540–50
156. Dams T, Jaenicke R. 1999. *Biochemistry* 38:9169–78
157. Gonzalez-Blasco G, Sanz-Aparicio J, Gonzalez B, Hermoso JA, Polaina J, et al. 2000. *J. Biol. Chem.* 275:13708–12
158. Gerike U, Danson MJ, Hough DW. 2001. *Protein Eng.* 14:655–61
159. Narinx E, Baise E, Gerday C. 1997. *Protein Eng.* 10:1271–79

160. Wintrode PL, Miyazaki K, Arnold FH. 2001. *Biochim. Biophys. Acta* 1549:1–8
161. Wintrode PL, Miyazaki K, Arnold FH. 2000. *J. Biol. Chem.* 275:31635–40
162. Giver L, Gershenson A, Freskgard PO, Arnold FH. 1998. *Proc. Natl. Acad. Sci. USA* 95:12809–13
163. Suzuki T, Yasugi M, Arisaka F, Yamagishi A, Oshima T. 2001. *Protein Eng.* 14:85–91
164. Lebbink JH, Kaper T, Bron P, van der Oost J, de Vos WM. 2000. *Biochemistry* 39:3656–65
165. Lonn A, Gardonyi M, van Zyl W, Hahn-Hagerdal B, Otero RC. 2002. *Eur. J. Biochem.* 269:157–63
166. Roovers M, Sanchez R, Legrain C, Glansdorff N. 2001. *J. Bacteriol.* 183:1101–5
167. Karlsen S, Hough E, Olsen RL. 1998. *Acta Crystallogr. D* 54:32–46
168. Berglund GI, Smalas AO, Hansen LK, Willassen NP. 1995. *Acta Crystallogr. D* 51:393–94
169. Toyota E, Ng KK, Kuninaga S, Sekizaki H, Itoh K, et al. 2002. *J. Mol. Biol.* 324:391–97
170. Arnorsdottir J, Smaradottir RB, Magnusson OT, Thorbjarnardottir SH, Eggertsson G, Kristjansson MM. 2002. *Eur. J. Biochem.* 269:5536–46

The Biochemistry of Sirtuins

Anthony A. Sauve,[1] Cynthia Wolberger,[2] Vern L. Schramm,[3] and Jef D. Boeke[4]

[1]Department of Pharmacology, Weill Medical College of Cornell University, New York, New York 10021; email: aas2004@med.cornell.edu

[2]Department of Biophysics and Biophysical Chemistry, Howard Hughes Medical Institute, Johns Hopkins University, Baltimore, Maryland 21205; email: cwolberg@jhmi.edu

[3]Department of Biochemistry, Albert Einstein College of Medicine, Yeshiva University, Bronx, New York 10465; email: vern@aecom.yu.edu

[4]Department of Molecular Biology and Genetics, Johns Hopkins University, Baltimore, Maryland 21205; email: jboeke@jhmi.edu

Key Words

Sir2, deacetylase, longevity, O-acetyl-ADP-ribose, NAD^+, gene silencing

Abstract

Sirtuins are a family of NAD^+-dependent protein deacetylases widely distributed in all phyla of life. Accumulating evidence indicates that sirtuins are important regulators of organism life span. In yeast, these unique enzymes regulate gene silencing by histone deacetylation and via formation of the novel compound $2'$-O-acetyl-ADP-ribose. In multicellular organisms, sirtuins deacetylate histones and transcription factors that regulate stress, metabolism, and survival pathways. The chemical mechanism of sirtuins provides novel opportunities for signaling and metabolic regulation of protein deacetylation. The biological, chemical, and structural characteristics of these unusual enzymes are discussed in this review.

Contents

INTRODUCTION AND BRIEF
   HISTORY OF SIRTUINS 436
THE DIVERSE BIOLOGY OF
   THE SIRTUIN FAMILY: *SIR2*
   GENE FAMILIES IN
   BACTERIA, FUNGI, AND
   METAZOANS 437
   Biological Effects of
      O-Acetyl-ADP-Ribose as
      a Potential Effector Molecule .. 441
   NAD^+ Metabolism and Sirtuin
      Activity 442
   Sirtuins, Caloric Restriction,
      and Aging 444
   Sirtuin Activities: Deacetylase or
      ADP-Ribosyl Transferase 445
CHEMICAL MECHANISM OF
   SIRTUINS 445
   ADP-Ribosyl Transferase
      Chemistry 445
   Products and Solvent Labeling
      Patterns 446
   Mechanistic Explanations of Sir2
      Activity 448
   Regulation by NAD^+ and
      Nicotinamide 452
   Partition Model of Imidate
      Reactivity 453
   Nicotinamide Antagonists as
      Sirtuin Activators 454
STRUCTURAL PROPERTIES OF
   SIRTUINS 454
   Structure of the Conserved
      Catalytic Core of Sirtuins 454
   Binding of Acetylated Peptide 455
   NAD^+ Binding 457
   Structural Insights into Enzyme
      Mechanism 458
   Multimerization of Sirtuins 459
SMALL MOLECULE
   MODULATORS OF SIRTUINS 460
   Activators of Sirtuins 460
   Inhibitors of Sirtuins 461
CONCLUSION 461

INTRODUCTION AND BRIEF HISTORY OF SIRTUINS

Sir2 is the founding member of a large and diverse family of protein-modifying enzymes known as sirtuins, which regulate key pathways throughout biology: in eubacteria, archaea, eukaryotes, and even viruses (1). Consistent with the role of yeast Sir2 in silencing transcription, this enzyme is a histone deacetylase (2–4). Unlike previously characterized histone deacetylases, however, which catalyze the simple hydrolysis of acetyllysine (5, 6), Sir2 was shown to deacetylate lysine residues in a novel chemical reaction that consumes nicotinamide adenine dinucleotide (NAD^+), releasing nicotinamide, O-acetyl ADP ribose (AADPR), and the deacetylated substrate. The basic chemistry and enzymology of this reaction are summarized in **Figure 1**. The seemingly baroque and energetically costly chemistry of sirtuins permits their enzymatic activity to be regulated by a variety of metabolites and may enable the enzyme to serve other chemical functions in addition to deacetylation. As studies continue to uncover the roles that members of the sirtuin family play in important biological processes, such as life span regulation (7–11), fat mobilization in human cells (12), insulin secretion (13), cellular response to stress (11, 14, 15), axonal degeneration (16), basal transcription factor activity (17), regulating enzyme activity (18), rDNA recombination (19–21), and switching between morphological states in *Candida* (22) and apoptosis (23, 24), there has been substantial progress in uncovering the chemical and structural details of these fascinating enzymes.

The discovery of Sir2 emerged from studies of how yeast cell type, known as mating type, is regulated. Mating type in the yeast *Saccharomyces cerevisiae* is determined by a single locus known as *MAT*, which can contain either of two alleles, *MAT***a** or *MAT*α. *MAT***a** cells can mate only with *MAT*α cells; the products of such matings are *MAT***a**/α diploid cells, which themselves cannot mate. Additionally,

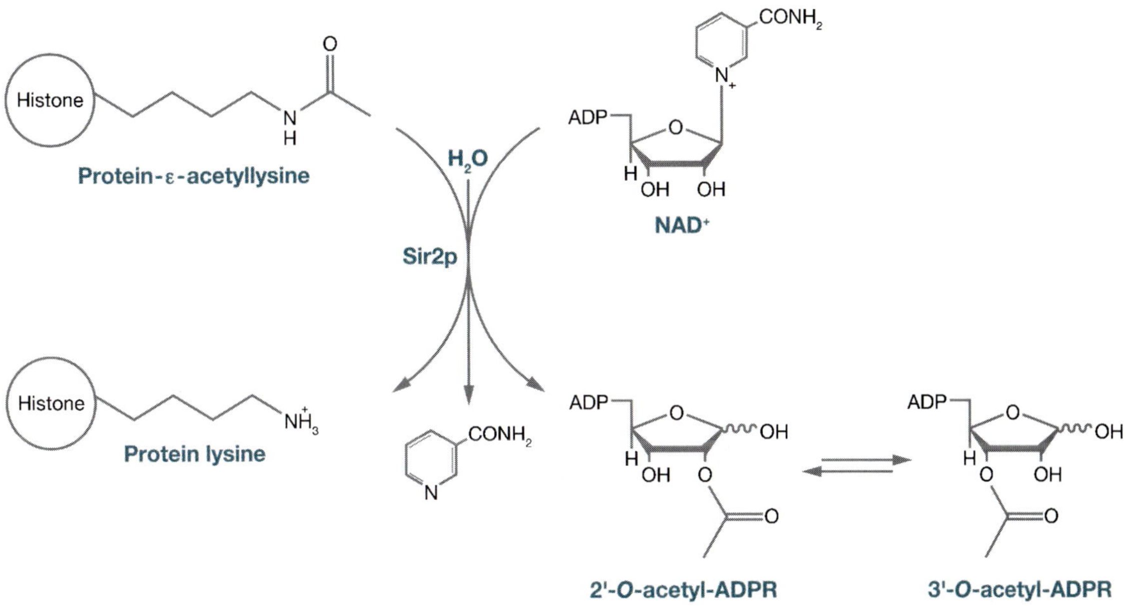

Figure 1

The Sir2 reaction. Deacetylation of protein acetyllysines catalyzed by Sir2p. Acetyl-group transfer to the ADP-ribose (ADPR) moiety of NAD^+ occurs via Sir2p chemistry to form 2'-O-acetyl-ADPR. 3'-O-acetyl-ADPR is formed nonenzymatically after release of 2'-O-acetyl-ADPR from the enzyme.

yeast also harbor, on the same chromosome as the *MAT* locus, but distinct from it, two "silent" copies of the mating-type information known as *HMR*a and *HML*α (collectively referred to as the *HM* loci), which encode the a and α information, respectively, but which are transcriptionally silent. The *SIR2* gene was first identified in yeast via a mutation, originally called *mar1-1*. Klar and colleagues (25) isolated and analyzed a spontaneous mutant with genetic properties indicating a defect in maintaining the silent state of the *HM* loci. The interpretation of the complex panoply of phenotypes conferred by these early *sir* mutants played a pivotal role in not only establishing the phenomenon of transcriptional silencing per se, but also helping to define the pivotal cassette model for yeast mating-type interconversion. Similar mutants were isolated by several other groups (26, 27), and these mutants define the four *SIR* genes, now known to function collectively in maintaining silencing at the *HM* loci, at all yeast telomeres (28), and in the rDNA (21, 29, 30). In all three types of silent loci, the histone deacetylase Sir2p, the only Sir protein with an enzymatic activity, is required for all three types of silencing. The other three Sir proteins play active roles in silencing only at the *HM* loci and telomeres. This is reflected by the fact that Sir2p can interact with different molecular partners; Sir3p and Sir4p interact with Sir2p in the *HM* loci and telomeres, whereas Net1p/Cfi1p interacts with Sir2p in the rDNA to form a completely distinct complex with roles in cell cycle control, as well as silencing (31, 32).

THE DIVERSE BIOLOGY OF THE SIRTUIN FAMILY: *SIR2* GENE FAMILIES IN BACTERIA, FUNGI, AND METAZOANS

Sirtuins are found in virtually all organisms in all three domains of life. Bacterial genomes and most archaea encode only a single sirtuin, although there are exceptions, such as the archeon *Archaeoglobus fulgidus*, which encodes two sirtuins. Eukaryotes typically have multiple sirtuins, with yeast containing four sirtuins (Hst1–4) in addition to Sir2p, and humans containing seven (SirT1–7) sirtuins. The presence of multiple sirtuin species encoded in eukaryotic genomes naturally invites speculation as to the biological roles of the different types of sirtuins. Phylogenetic analyses by Frye (33, 34), which analyzed sequence conservation in the ∼250 amino acid core domain, have organized the sirtuins into five main classes. Class I is eukaryote specific and includes yeast Sir2 and the four Hst proteins, as well as mammalian SirT1, SirT2, and SirT3. Class II includes mammalian SirT4, as well as a number of related eukaryotic sirtuins, but also includes a number of bacterial proteins. Class III is mostly composed of bacterial sirtuins (both archaeal and eubacterial), but also includes mammalian SirT5 and a number of other eukaryotic orthologs. Class IV includes mammalian SirT6 and SirT7 and is also eukaryote specific. Finally, Class U contains sequences unique to gram-positive eubacteria. A more extensive analysis of additional sequences (W. Hawse & C. Wolberger, unpublished data) suggests that modest revisions to this scheme are required to reflect more accurately the increased sequence diversity of data now available. There are no obvious biological correlations between the sirtuin classes and specific biological functions. For example, histone and tubulin deacetylases are found in Class I, as is a mitochondria-localized deacetylase.

Some sirtuins, including yeast Sir2 and its orthologs, contain significant extensions that flank the conserved enzymatic core at the N and C termini. Human SirT1, for example, has ∼240 amino acid N- and C-terminal extensions. There is some evidence that these extensions may promote binding to specific substrates (35). In the case of Sir2, there is good evidence that these regions contain binding sites for Sir4, which forms a complex with Sir2 in the cell (36). Among the human sirtuins, all six proteins have N-terminal extensions of 30 or more residues relative to a bacterial sirtuin, which presumably represents the core catalytic subunit. It is conceivable that these terminal domains bind to distal sites on substrate macromolecular complexes and do mechanical work as a consequence of the catalysis, as large conformational changes are thought to accompany the catalytic cycle (37). In this connection, it is interesting that sirtuin substrates include at least two large fibrous macromolecular complexes, chromatin (yeast Sir2) and tubulin (human SirT2) (38).

The various sirtuins have very diverse substrates (see **Table 1**). These can be broken down into three major overlapping classes: transcriptional, apoptosis regulating, and metabolic regulating. In general, transcriptional downregulation by sirtuins is associated with deacetylation of histones, at least in microorganisms, and there is extensive evidence that this deacetylation leads to transcriptional downregulation in the case of silencing in *S. cerevisiae*. Histones H3 and H4 are physiological substrates for yeast Sir2 and Hst1 proteins, and there is extensive evidence that K16 in histone H4 is a critical residue for transcriptional silencing, apparently by controlling binding of Sir3p and Sir4p to histone H4 N termini (37, 39, 40). The Sir2p protein also shows specificity in vitro for deacetylation of the K16 residue (2).

In another well-studied yeast, *Schizosaccharomyces pombe*, sirtuins play roles in the transcriptional silencing of telomeres and cryptic mating loci (41), and also of centromeres (42). However, in *S. pombe* (and in most eukaryotes, but not in *S. cerevisiae*), a separate form of posttranscriptional silencing, the RNAi (RNA interference) system, operates as well. Remarkably, the RNAi system is also intimately involved in establishing the sites of transcriptional silencing, mediated through *Su(var3-9)*-dependent methylation of K9 in histone H3; this modification creates a site for the major heterochromatin protein 1 (HP1) (43). It is interesting that potent silencing in *S. cerevisiae* can occur in the absence of the RNAi system,

Table 1 Sirtuin substrates

Substrate[a]	Residue(s)	Reference	Role of sirtuin; comments	Sirtuin class
Salmonella enterica cobB (sirtuin)				U
Acs (acetyl-CoA synthetase)	K609	(18)	Activates catalysis	
Sulfolobus sulfotaricus Sir2				U
Alba	K16	(91)	Antitranscriptional Enhanced DNA binding	
Saccharomyces cerevisiae Sir2				I
Histone H3	K9/14	(2)	Antitranscriptional	
Histone H4	K16		Antitranscriptional	
Schizosaccharomyces pombe Sir2				I
Histone H3	K16	(92)	Antitranscriptional	
Histone H4	K9			
Homo sapiens SirT1				I
Histone H1	K26	(46)	Antitranscriptional	
Histone H3	K9	(46)	Antitranscriptional	
Histone H4	K16	(46)	Antitranscriptional	
p53	K317/370	(48)	Anti-apoptotic	
	K382	(24)		
		(48)		
p300	K102/1024	(54)	Antitranscriptional	
FOXO3a	K242/259/271/290/569	(23)	Antitranscription Anti-apoptotic	
		(15)		
RelA/p65 (NFκB)	K310	(80)	Pro-apoptotic	
FOXO1	K242/245/262	(93)		
FOXO4	ND	(94)	Protranscriptional	
HIV Tat	K50	(95)	Protranscriptional	
PGC-1α	K77/144/183/253/270/277/320	(56)	Protranscriptional and antitranscriptional	
	K412/441/450/757/778			
PCAF	ND	(96)		
MyoD	K99/102/104	(96)		
Ku70	K539/542	(11)	Anti-apoptotic	
SIRT2				I
α-tubulin	K40	(38)		
SIRT6				IV
SIRT6	ND	(90)	Auto-ADP ribosylation reported	
Mus musculus Sir2α				I
TAF(I)68	ND	(17)		
Trypanosoma brucei TbSIR2RP1				I
Histone H2A	ND	(89)	Both deacetylation and ADP ribosylation reported; increased DNA repair?	
Histone H2B	ND	(89)	Both deacetylation and ADP ribosylation reported; increased DNA repair?	

[a]Abbreviations: cob, cobalamin biosynthesis; FOXO, forkhead box, type O; PGC, PPAR-gamma coactivator; PCAF, p300/CBP-associated factor; ND, residue identity not determined.

K9 methylation, and HP1 and its associated machinery; this suggests that lysine deacetylation by sirtuins may represent an ancestral form of silencing, whereas histone methylation and RNAi and DNA methylation represent more advanced, eukaryote-specific evolutionary embellishments. Accordingly, in *S. pombe*, prior deacetylation of K9 by a sirtuin appears to be required for subsequent lysine methylation (44).

In metazoans, if and how sirtuins generally downregulate gene expression is far less clear, and there is no direct evidence that histone deacetylation in vivo by SirT1, the closest homolog of Sir2p, globally controls transcription. In fact, SirT1 is not necessary for maintaining global silencing of the genome in embryonic stem cells (45). One recent publication (46) did, however, report that SirT1 can carry out local deacetylation of chromatin quite similar to that seen in silent loci in *S. cerevisiae*. In this study, SirT1 was tethered to a GAL4 binding site, where it also deacetylated histone H1, the linker histone associated with heterochromatin, which is absent from *S. cerevisiae* (46). In addition, a native chicken transcription factor appears to recruit SIRT1 to specific target DNA sites, at which it apparently deacetylates histones (47).

Whether or not histone deacetylation is a major mechanism of SirT1 action in mammalian cells, a variety of mammalian transcription factors have been identified as Sir2 substrates, and in some cases deacetylation of these transcription factors is associated with transcriptional activation rather than repression. In most cases, these transcription factors control genes related to growth, cell cycle, and/or apoptosis control. A good example is p53, which carries several acetylation sites (along with a number of other modifications) in its C-terminal domain; several of these are excellent substrates for SirT1 (24, 48). The precise role of the acetylation as well as the relative contributions of sirtuins and NAD^+-independent deacetylases to its deacetylation are debated. The consensus is that deacetylation of the protein is required for its degradation, and thus deacetylation of p53 by sirtuins is predicted to lead to its destruction by an MDM2-dependent ubiquitin-mediated pathway. The p53 protein activates the transcriptional program that promotes apoptosis, and thus decreases in p53 abundance reduce apoptosis. However, deacetylation of p53 may also play more direct roles in apoptosis, as many additional functions beyond serving as a transcriptional regulator have been ascribed to p53. SirT1 is localized to PML bodies, where p53 and other apoptotic proteins also reside (49). Mouse embryo fibroblasts lacking SirT1 have unlimited replicative life span, and although p53 is highly acetylated, it is rendered less stable owing to upregulation of p19Arf (50). Hyperacetylation of p53 occurs in *sirt1* knockout animals (51), providing compelling evidence that the SirT1/p53 interaction is meaningful in vivo.

FOXO (Forkhead box type O) proteins are another important set of transcription factors regulated by SirT1. The proteins are orthologous to the Dauer-formation (Daf)-16 protein that controls longevity in the nematode *Caenorhabditis elegans* (52) and are part of the insulin/insulin-like growth factor-1 pathway. Interestingly, in the nematode, excess Sir2 activity extends life span and this requires Daf-16 (7). This Dauer-formation protein is part of a specialized pathway that allows long-term survival in response to stress in the nematode; extension of life span is indeed Daf-16 dependent (7). Both pro- and antitranscriptional effects of sirtuins on the homologous mammalian FOXO factors have been noted (**Table 1**) (53), indicating that the connection between Sir2 and these factors is not an artifact of the specialized pathway. As is the case with p53, SirT1-mediated deacetylation is anti-apoptotic and promotes cell survival. An important member of this family, FOXO3a, is deacetylated by SirT1, as is the p300 acetyltransferase that acetylates it (23, 54).

A third class of sirtuin substrates comprises a series of proteins intimately involved with carbon metabolism and its regulation.

A unique example is the eubacterial acetyl-CoA (coenzyme A) synthetase enzyme, which is regulated directly by cobB, the corresponding sirtuin that removes an acetyl residue from a critical lysine in the active site, thereby activating it for catalysis (55). Interestingly, this protein is part of a larger superfamily of proteins, including nonribosomal peptide synthetases and luciferases, which suggests that potentially these can all be activated in the same way. Thus far this protein family is the only instance of an enzyme activity directly regulated by lysine acetylation.

Two mammalian regulatory proteins deserve special mention in the metabolic context, PPARγ and its coactivator PGC-1α. A critical mechanism controlling energy metabolism in metazoans is their ability to store fat in specialized adipose tissues. In these tissues, PPARγ regulates fat storage by controlling the expression of adipogenic-specific genes. However, how PPARγ is regulated is not totally understood. Fatty acids are mobilized from white fat during food withdrawal by activating lipolysis, and it has been suggested that SirT1 plays a role in this process (12). Insulin treatment of pre-adipocytes results in adipogenesis, and under these conditions, SirT1 levels rise. In this context, overexpression of SirT1 leads to reduced fat accumulation (12). The effect is mediated by repression of PPARγ through its negative regulatory subunit NcoR, a transcriptional corepressor recruited by PPARγ and to which SirT1 binds (12). Finally, SirT1 was observed bound to chromatin near PPARγ-regulated genes only in food-deprivation conditions, suggesting that lipolysis was activated. This binding suggests that SirT1 may exert its effect on transcription in this context by targeting histone proteins.

PGC-1α, in partnership with various transcription factors, controls genes for gluconeogenesis, as well as other pathways such as mitochondrial biogenesis, thermogenesis in brown fat, and fiber-type switching in skeletal muscle (55). PGC-1α was shown to form a specific complex with SirT1 and other factors, and this resulted in deacetylation of specific lysines in PGC-1α. This led to upregulation of the gluconeogenic pathway and downregulation of the glycolytic pathway, presumably in response to changes in NAD^+ or a related metabolite (56). This is a good example of SirT1 having both and pro- and antitranscriptional effects in a single study and, presumably, as a consequence of deacetylation of a single substrate.

Biological Effects of O-Acetyl-ADP-Ribose as a Potential Effector Molecule

In most literature to date, the biological effect of sirtuins is presumed to derive from the removal of an acetyl group from lysine side chains. The Sir2 protein is capable of deacetylating multiple lysines in the N termini of histones H3 and H4, although K16 is known from genetic experiments to be a key substrate (57). However hypoacetylation is observed at these other sites in histones H3 and H4, at telomeres, and at *HM* loci (58). Perplexingly, mutation of Sas2, the major K16 acetyltransferase, fails to suppress a knockout of Sir2 (59, 60), suggesting an additional silencing function for Sir2 beyond its deacetylating activity. However, there was an early investigation into the possibility that another sirtuin reaction product, O-AADPR, may have biological activity as a novel signaling molecule (61).

Evidence for the importance of O-AADPR as a biological effector molecule is limited but accumulating. The first demonstration of its potential came from the observation that microinjection of starfish oocytes with AADPR could delay or block embryonic cell division (61). If AADPR is indeed a signaling molecule, it would then be likely that cells have a mechanism for breaking this metabolite down when it is no longer required. A family of enzymes called nudix hydrolases can cleave pyrophosphate bonds, and some of these enzymes degrade AADPR, as well as ADPR (62). Interestingly, these studies also identified two other activities specific for degradation of AADPR, an esterase and an acetyltransferase. Structural studies show that AADPR can form a complex with the sirtuin that produced it, suggesting that AADPRs can be potent biological inhibitors of sirtuins (63, 64). A more clear-cut case for a biological role for these molecules is in yeast silencing, in which Liou and colleagues (37) showed dramatic evidence that AADPR stimulates formation of the Sir2p/Sir3p/Sir4p complex by recruiting Sir3p from a Sir3p homopolymeric state into a preformed Sir2p/Sir4p complex. Although the binding site of AADPR is not known, an intriguing possibility is the AAA ATPase domain of Sir3p, which lacks the catalytic residues found in other AAA ATPases, but presumably retains the nucleotide binding site that could potentially bind AADPR. There is also some evidence that macroH2A, a variant histone incorporated into regions silenced by X-chromosome inactivation, may also bind AADPR (65), albeit no more avidly than ADP-ribose. The full implications of the signaling potential of AADPR are likely to be highly significant, but remain to be clarified.

NAD$^+$ Metabolism and Sirtuin Activity

Because sirtuins are absolutely dependent on NAD$^+$, the abundance of free NAD$^+$ and its biosynthetic and breakdown products in cells are very relevant to the enzymatic activity of sirtuins. There are two main routes to NAD$^+$ biosynthesis in yeast and mammals (**Figure 2**): a de novo kynurenine pathway, derived from tryptophan as a precursor, and a salvage pathway that utilizes nicotinamide, produced from NAD$^+$ by sirtuins as well as ADP-ribosyltransferases and polymerases or exogenous nicotinic acid (66). Finally, Bieganowski & Brenner (67) recently identified a novel pathway to NAD$^+$ in yeast and humans, starting from nicotinamide riboside, which can be provided externally. Another important recent finding is that mammalian cells have a fundamentally different salvage pathway than the one known in bacteria and yeast (68). In yeast, nicotinamide is deaminated by Pnc1p (pyrazin-amide/nicotinamide hydrolase), converting it to nicotinic acid. Nicotinic acid is then converted to NaMN (nicotinic acid mononucleotide) by nicotinic acid phosphoribosyltransferase. In mammalian cells, however, nicotinamide is converted directly to nicotinamide mononucleotide (NMN) by Nampt (nicotinamide phosphoribosyltransferase). Increased expression of Nampt in response to various stresses elevates cellular NAD$^+$ levels, which in turn regulate catalytic activity of Sir2 (69). Recent evidence suggests that changes in NAD$^+$ metabolites can have tissue-specific effects: For example, NAD$^+$ increases in nuclear neurons prevent axonal degeneration in a SirT1-dependent manner. (16). Mammalian de novo biosynthesis is also organized differently than in plants and prokaryotes (70, 71).

Studies from the Sinclair and Smith labs have implicated nicotinamide as a potent in vivo inhibitor of Sir2p (14, 72–74). Exogenously added nicotinamide phenocopies a *sir2* mutant, providing a clear demonstration of potent nicotinamide effects on Sir2p in vivo and simultaneously defining an important pharmacologic tool for probing sirtuin involvement in a variety of pathways in yeast and mammalian cells. In addition, overexpression of Pnc1p activates Sir2p in vitro and in vivo, indicating that levels of nicotinamide naturally produced by sirtuins (and perhaps other enzyme activities) are sufficient to inhibit Sir2p; elevation of Pnc1p activity increases sirtuin activity. Although changes in the activity of this NAD$^+$ salvage pathway enzyme may affect Sir2p activity through an alteration of nuclear NAD$^+$ levels, further studies implicate the change in nuclear nicotinamide concentrations as the primary cause of the change in Sir2-mediated silencing (74). This view of Sir2p regulation was recently supported by chemical interference of nicotinamide with a nicotinamide isostere called isonicotinamide, which enhances Sir2-mediated silencing (75). Interestingly, the Pnc1 enzyme itself is directly regulated by various forms

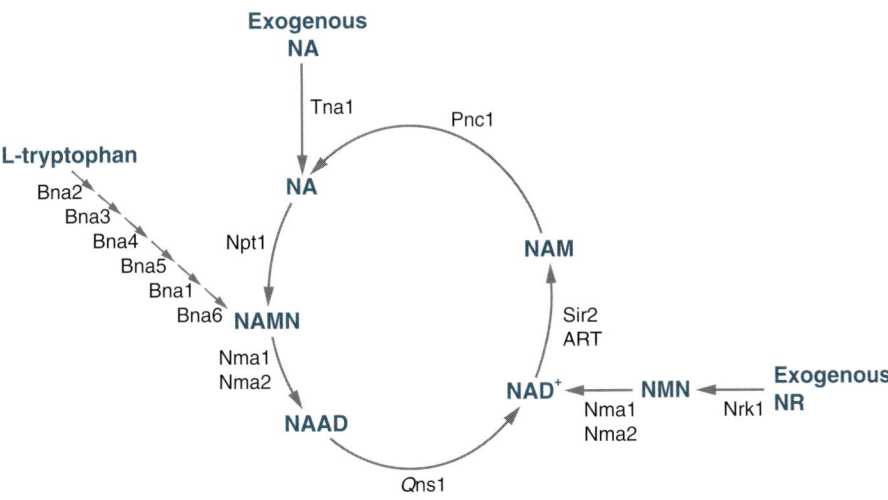

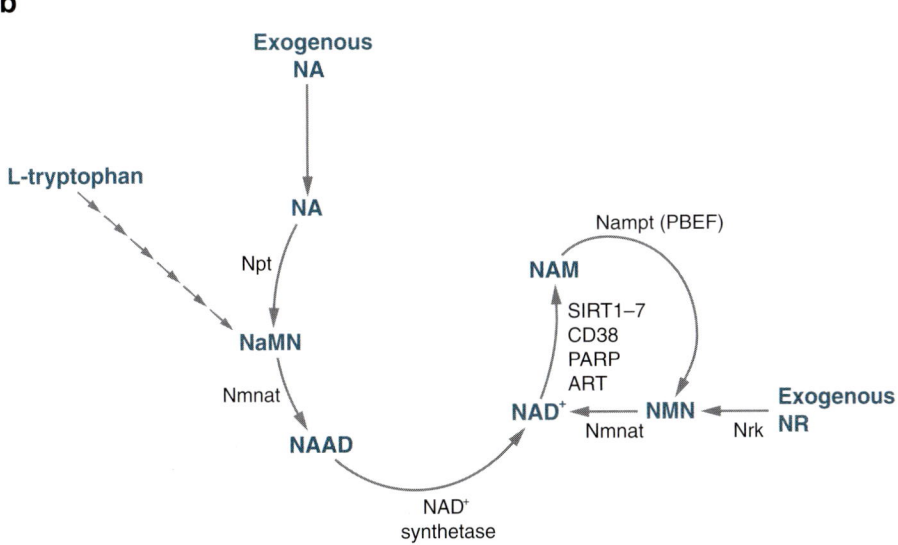

Figure 2

(*a*) NAD⁺ biosynthetic pathway in yeast. (*b*) NAD⁺ biosynthetic pathway in mammals. Mammals notably lack an enzyme to degrade nicotinamide (NAM) to nicotinic acid (NA) (Pnc1p), but compensate by producing an enzyme known as Nampt [nicotinamide phosphoribosyltransferase; also known as pre-B cell colony-enhancing factor (PBEF)], which can directly convert nicotinamide to nicotinamide mononucleotide (NMN). Abbreviations: ART, ADP-ribosyl transferase; Nmnat, nicotinamide mononucleotide adenylyltransferase; Nrk, nicotinamide riboside kinase; Pnc1, pyrazin-amide/nicotinamide hydrolase; Npt, nicotinic acid phosphoribosyltransferase; NAAD, nicotinic acid adenine dinucleotide; NaMN, nicotinic acid mononucleotide; NR, nicotinamide riboside; PARP, poly-ADP-polymerases; Bna, biosynthesis of nicotinic acid; Tna, transport of nicotinic acid; Nma, nicotinic acid mononucleotide adenylyltransferase; Qns, glutamine-dependent NAD⁺ synthetase.

of stress, indicating that there is potentially a cellular signaling system that telegraphs various stress responses to the cell via chemomodulation of Sir2p activity (14, 74). Remarkably, recent studies in mammalian cells established that the enzyme Nampt, which carries out a completely different enzymatic, nicotinamide-consuming reaction process in mammalian cells, performs and regulates the analogous nicotinamide-scavenging function, although the topology of this NAD^+ salvage pathway is rather different from the yeast pathway (69) (**Figure 2**).

Sirtuins, Caloric Restriction, and Aging

The complex and controversial relationships between Sir2 proteins, caloric restriction (CR), and aging have been reviewed in depth (76, 77) and are covered more briefly here. Accumulated evidence suggests that sirtuins are involved in promoting longevity, particularly longevity associated with CR regimens, in several organisms. Two key early findings supporting this are the discovery that extra doses of sirtuins promote longevity in *C. elegans* (7) and the correlation of yeast mother cell longevity (a somewhat specialized model for longevity) with the activity of Sir2 and other Sir proteins (78). When yeast cells divide asymmetrically, the mother cells (the larger of the two cells) can divide only a finite number of times, typically 20–30 times. Mutants lacking Sir2 have a shortened life span in that they divide a fewer number of times. Prematurely aging *sir2* mutant mother cells were shown to accumulate extrachromosomal rDNA circles, which accumulate because rDNA recombination is disregulated in the *sir2* mutant (20, 21, 77). Subsequent studies showed that CR could be mimicked in yeast by a simple glucose starvation regimen. Such treatment led to transcriptional changes, which suggests metabolism shifted in favor of respiration, as opposed to the normal fermentative yeast lifestyle, and extended mother cell longevity in a Sir2-dependent manner (79). Experiments suggest that in these cells, $NAD^+/NADH$ ratios are increased, and this may increase sirtuin activity due to relief of inhibition by NADH, providing a possible mechanism for the life span extension. Because the absolute levels of free pyridine nucleotides in nuclei are difficult to determine, it is difficult to rigorously evaluate this possibility. Alternatively, the activation of yeast Pnc1 owing to CR-induced stress would lead to a decrease in nicotinamide and the activation of Sir2 activity. Although either or both of these mechanisms may account for the effect of CR on Sir2 activity, it remains to be shown why an increase in Sir2 activity increases yeast life span and whether this mechanism is relevant to higher organisms.

Pharmacologic approaches have also provided interesting support for the notion that elevated Sir2 activity promotes longevity. Treatment of yeast cells with resveratrol, an agonist of Sir2 (at least in certain in vitro assays) extends mother cell longevity in a Sir2-dependent manner (10). Experiments in human tissue culture cells (80), *C. elegans*, and *Drosophila* (81, 82) suggest that resveratrol (and related polyphenolic compounds) can increase longevity in very diverse organisms. These studies are confounded by the fact that resveratrol only significantly activates human SirT1 and not yeast Sir2p in vitro, and only in a nonphysiologic assay. The SirT1 activation requires a fluorophore in the substrate absent from native substrates (83). Another study confirmed the SirT1 specificity and alleged a general lack of evidence for in vivo stimulation of Sir2p activity; that is, the strength of its silencing phenotypes was not enhanced by resveratrol treatment, nor was mother cell longevity enhanced in two different strain backgrounds tested (84). From this and other studies (78), it is clear that there can be multiple mechanisms of life span extension, some Sir2p dependent and some Sir2p independent, and that these mechanisms vary between yeast strain backgrounds, which are genetically diverse. One recent study implicated the Sir2p paralog Hst2p in Sir2p-independent mother cell life span extension (85). Returning to resveratrol and related compounds, these tend to be bioactive in a number of signaling pathways unrelated to Sir2p, which means biological results in response to these compounds must be interpreted cautiously.

Although the specifics of the resveratrol connection to Sir2p remain controversial, there is an elegant theoretical underpinning for a connection. Molecules like resveratrol, produced in abundance by plants undergoing a variety of environmental stresses, may actually have been exploited as an interspecies signal that allows other organisms feeding on the plant material to detect environmental stressors, referred to as xenohormesis (86).

Sirtuin Activities: Deacetylase or ADP-Ribosyl Transferase

The NAD^+-dependent deacetylase activity is well established for a multitude of sirtuin enzymes. Catalytic activity has been demonstrated in vitro with purified proteins and specific sites of deacetylation have been identified (**Table 1**). However, the initial report of enzymatic activity in yeast Sir2 described weak mono-ADP-ribosyl transferase (ART) activity (87). For sirtuins, this function had been anticipated on the basis of indirect genetic experiments in *Salmonella typhimurium* in which cobB, a bacterial sirtuin, was shown to complement a deletion of a mononucleotide phosphoribosyltransferase in the cobalamin biosynthesis pathway (88). With the discovery of the more robust NAD^+-dependent deacetylase activity in yeast Sir2 and other sirtuins, the ART activity was attributed to nonenzymatic labeling. However, two recent publications (89, 90) provide evidence for the transfer of radioactive phosphate from NAD^+ to a target protein(s) in vitro that is highly elevated in two specific sirtuins, a specific trypanosomal sirtuin that both deacetylates and transfers labels from NAD^+ to histones (89) and the human SirT6 protein (90), which autolabels with ^{32}P-labeled NAD^+ but lacks deacetylase activity. It is assumed in all these cases that the reaction is an ADP-ribosylation reaction, although only limited evidence for this has been provided. In one case (90), evidence is presented that an antibody specific for ADPR recognizes the products. As of this writing, a specific site of modification, or the nature of the modified amino acid, has yet to be characterized.

The striking biochemical activities of sirtuins make it clear that they are an important class of protein-modifying enzymes. Their requirement for a high-energy cofactor, NAD^+, ties them directly to a cellular metabolic states. Could there be more to the sirtuin reaction than the simple removal (or addition) of a small chemical modification on proteins? Could they actually be involved in more complex biochemical reactions that directly effect changes in the structure of large macromolecular complexes such as chromatin? The expenditure of a high-energy bond during the deacetylation reaction is consistent with such a function. Alternatively, recent results from the Moazed lab (37) suggest that the Sir2-deacetylation reaction product *O*-AADPR, rather than enzyme movements, may be sufficient to promote or stabilize the formation of a silent chromatin structure.

CHEMICAL MECHANISM OF SIRTUINS

The unexpected and chemically unnecessary involvement of NAD^+ in the deacetylation reaction catalyzed by sirtuins has led to extensive investigation into the chemical mechanism. The NAD^+ requirement sets sirtuin chemistry apart from all known deacetylases but links the sirtuins to the ribosyl transferases.

ADP-Ribosyl Transferase Chemistry

Sirtuins belong to a widely distributed group of enzymes called ADP-ribosyl transferases, which catalyze transfer of the ADPR group of NAD^+ to nucleophiles (97). Members of this group of enzymes include ARTs

(98), which posttranslationally modify nucleophilic protein side-chain residues with the ADPR moiety (a group of enzymes that include the bacterial toxins and mammalian Arg ADP-ribosyl transferases), the poly-ADP-polymerases (PARP) (99), which synthesize ADPR polymers onto proteins, and the ADP-ribosyl cyclases/NAD$^+$ glycohydrolases (100), which hydrolyze NAD$^+$ and also form the signaling molecule cyclic-ADPR (cADPR) from NAD$^+$. The synthesis of cADPR is accomplished by a cyclization reaction via intramolecular ADP ribosylation of the N1 position of adenine (100). This enzyme can also synthesize nicotinic acid adenine dinucleotide (NAADP) from NADP via a nicotinic acid/nicotinamide base-exchange reaction (100).

The sirtuin NAD$^+$-dependent deacetylation reaction is not an ADP-ribosyl transferase reaction in the traditional sense. However, several reactions that are catalyzed by sirtuin enzymes involve ribosyl transfer. For example, the first reaction characterized for a sirtuin was a ribosyltransfer reaction catalyzed by cobB (88), a Sir2 homologue found in bacteria in which 5,6-dimethylbenzimidazole is substituted for nicotinic acid of NaMN to form the metabolite α-ribose-5,6-benzimidazole (88). This reaction accomplishes synthesis of a necessary metabolite on the pathway of de novo cobalamin biosynthesis (88). The discovery of this reaction not only provided a clear example of ribosyltransferase activity for this family of enzymes, but led to the eventual discovery of NAD$^+$-dependent deacetylase and ADP-ribosyl transferase activities of sirtuins. Observations that sirtuins (archaean, yeast, mouse, and human) catalyze a nicotinamide base-exchange reaction, in which radiolabeled nicotinamide is exchanged into NAD$^+$, demonstrate that sirtuins are ADP-ribosyl transferases (101–104). This reaction is diagnostic for reversible formation of an ADPR intermediate formed on the enzyme and is a well-known feature of ADP-ribosyl-cyclase/NAD$^+$ glycohydrolase enzymes that form an ADPR intermediate during catalysis (105, 106). Sirtuins therefore use ADP-ribosyl transfer as an integral feature of their catalytic mechanisms. The apparent coupling of ADP-ribosyl transferase chemistry with protein deacetylation is a remarkable feature that provides unique biochemical opportunities for signaling and metabolic regulation. The chemistry, mechanisms, and structural characteristics that illuminate the novelty of these catalysts are discussed below.

Products and Solvent Labeling Patterns

In 2000, three independent laboratories reported sirtuin NAD$^+$-dependent deacetylation (2–4). The role of NAD$^+$ in this reaction was originally undetermined, but a tight coupling of NAD$^+$ consumption with amide deacetylation confirmed stoichiometric requirement of NAD$^+$ in the deacetylation reaction (101). A novel product, AADPR, was later identified as the acetyl acceptor, formed via acetyltransfer to NAD$^+$ with hydrolysis of the nicotinamide bond (107, 108). The stoichiometric involvement of water was established by ^{18}O isotope labeling of AADPR derived from Sir2 reactions conducted in ^{18}O water (109).

The definitive mechanism of Sir2 catalysis was provided by the complete identification of the AADPR reaction product. This previously unknown compound was first detected by mass spectrometry and high-pressure liquid chromatography (HPLC) and mistakenly identified as β-1′-AADPR (107). Independent synthesis of 1′-AADPR and comparison to Sir2-derived AADPR determined them to be different (109). Sir2-derived AADPR was fully characterized by mass spectrometry and NMR and identified as an equilibrating mixture of 2′- and 3′-AADPR (109, 110), in which only the 2′-AADPR isomer was the immediate enzymatic product (109, 110). The full stoichiometry of sirtuin deacetylation is shown in **Figure 1**. This NAD$^+$-dependent deacetylation reaction is catalyzed by enzymes from

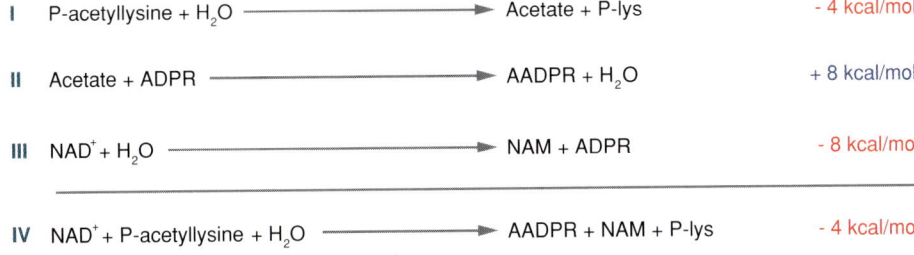

Figure 3

A reaction sum approach to describe components of the Sir2 reaction. Reaction I is the overall reaction for amide hydrolysis catalyzed by histone deacetylases (HDACs, Class I and Class II). Reaction II is the ester synthesis of AADPR from acetate and ADPR. Reaction III is hydrolysis of NAD^+ to nicotinamide (NAM) and APDR. Reaction IV is the overall deacetylation reaction catalyzed by sirtuin enzymes. Values for the energy changes are drawn from literature sources (6, 111).

archaea, eubacteria, yeast, mouse, and human sources (2–4, 101, 102, 107–110).

The stoichiometric involvement of NAD^+ is kinetically and thermodynamically unnecessary to achieve protein deacetylation. Widely distributed Class I and Class II HDACs catalyze deacetylation using only water as a cosubstrate (Reaction I; **Figure 3**), with a favorable free energy change and an equilibrium constant in favor of products on the order of 500 at pH 7.5 (6, 111). One rationale for the involvement of NAD^+ in sirtuin deacetylation chemistry is the synthesis of the novel reaction product AADPR. Sirtuin deacetylation catalysis couples amide hydrolysis to ester synthesis (AADPR synthesis). Amide-to-ester acyltransfer is unfavorable (Reaction I and II) (111), but hydrolysis of NAD^+ (Reaction III) can provide a favorable driving force for the overall sirtuin reaction (Reaction IV). The free energy change for the sirtuin deacetylation reaction is comparable to Reaction I alone; therefore there is no energetic reason to favor coupling of NAD^+ consumption with deacetylation. A complete stoichiometry establishes three reactants and three products for sirtuin deacetylation reactions. The mechanistic basis for protein deacetylation as catalyzed by sirtuins is entirely unprecedented. The biological logic of consuming NAD^+ in deacetylation reactions may be rooted in the advantage of forming a diffusible, second messenger AADPR and/or in the creation of a regulatory couple that links protein deacetylation to cellular metabolism. Insight into the chemistry of these enzymes has been obtained by a series of mechanistic observations presented below.

Base exchange. Sirtuins catalyze the exchange of radiolabeled nicotinamide into NAD^+, thereby displaying reversible loss of the high-energy nicotinamide-ribose bond, a characteristic of enzymes that form the covalent ADPR intermediate such as that formed on the ADP-ribosylcyclase CD38 (105, 112). Properties of this ADPR intermediate include a sufficient life span to permit exchange of nicotinamide into the active site and preservation of the energy of the original nicotinamide-ribose bond of NAD^+ to permit reversible reactivity of the intermediate with nicotinamide (105). An important feature of the exchange reaction catalyzed by sirtuins is that it occurs only when an acetylated protein or peptide substrate is present in the reaction mixture (101, 102). The requirement of peptide in this process has been interpreted in two ways. One proposal is that peptide binding facilitates an allosteric change in enzyme structure that enables the reaction of NAD^+ with an enzyme side-chain nucleophile, thereby forming an enzyme-stabilized ADPR intermediate analogous to that observed in reactions of CD38/NAD+ glycohydrolase (101, 102, 107). Alternatively, it has been proposed that the peptide is required to participate chemically in order to generate the exchangeable ADPR intermediate (109). X-ray structures support a close approach of the peptide acetylamide and the NAD+ molecule, which would enable formation of an ADPR-peptidyl intermediate (102, 113).

Oxygen labeling. Reactions of Sir2 in ^{18}O water show that the mixture of 2'- and 3'-AADPR is ^{18}O labeled in the carbonyl oxygen of the acetyl group, confirming that the original carbonyl double bond of the acetyllysine side chain of substrate is cleaved as a consequence of sirtuin deacetylation catalysis and replaced by a new double bond with an oxygen derived from solvent (109) (**Figure 4**). The cleaved O of acetylamide bonds to C1' of ADPR, as indicated by ^{16}O labeling of the C1' hydroxyl of AADPR determined for Sir2 reactions conducted in ^{18}O water (109) (**Figure 4**). The observed oxygen isotope-labeling pattern supports the concept that an ADP-ribose(C1')-oxygen (acetyl) bond is formed during sirtuin deacetylation catalysis. This covalent bond between C1' of ADPR and the carbonyl oxygen from the acetyllysine provides the hydroxyl that is bonded at the C1' position in AADPR (109).

NMR studies. Reactions conducted at low temperature under conditions favorable to observation of a single turnover product formation indicate that Sir2 enzymes form 2'-O-AADPR (96, 97). The failure to observe α-1'-O-AADPR suggests that this compound fails to form during enzymatic turnover (96). Spontaneous interconversion of 2'-O-AADPR and 3'-O-AADPR has been observed by NMR and by HPLC when there is an excess of either regioisomer (96, 97). These species rapidly establish equilibrium in solution, and HPLC analyses of Sir2 reactions always form a mixture of both isomers (109, 110). The biological significance of this interconversion of acetyl isomers is unknown, but the chemical phenomenon of acetyl migration between vicinal hydroxyls is well known in carbohydrate chemistry. 2'- and 3'-O-AADPR are chemically unstable to hydrolytic breakdown at physiological pH and form acetate and ADPR (101, 107–110).

Mechanistic Explanations of Sir2 Activity

Three distinct chemical pathways have been proposed to rationalize the chemical reactivities of the sirtuin enzymes. These alternative reaction explanations are discussed in light of their utility to explain experimental data available for sirtuin chemical behavior.

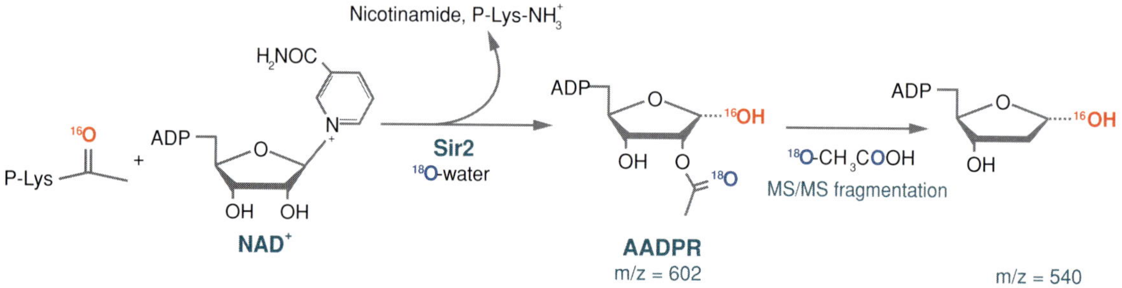

Figure 4

Reaction of sirtuins in ^{18}O water. Mass spectrometry (MS) analysis of AADPR (in negative detection mode) reveals that AADPR is labeled by one oxygen atom derived from solvent (m/z = 602), compared to AADPR, which is formed in ^{16}O water (m/z = 600). Fragmentation observed in MS/MS reveals that ^{18}O from solvent is incorporated into the acetyl moiety of the product. This result also shows that the acetyl oxygen of acetyllysine is transferred to the C1' position of the product.

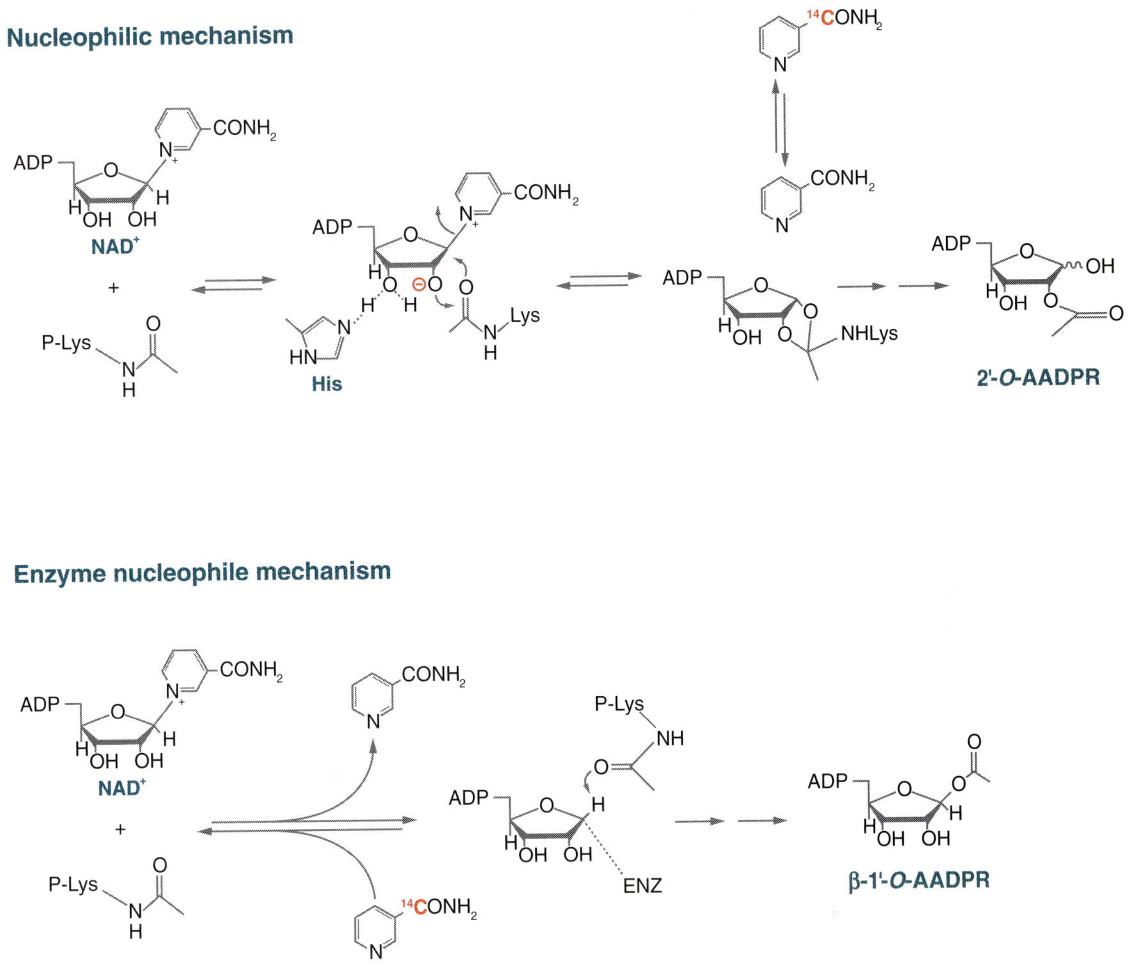

Nucleophilic mechanism

Enzyme nucleophile mechanism

Figure 5
Proposed mechanisms to account for deacetylation and base-exchange reactions catalyzed by Sir2 enzymes. The nucleophilic mechanism proposes that a nucleophilic attack on the amide by the 2′-hydroxyl group of NAD⁺ initiates deacetylation chemistry. According to this mechanism, base exchange occurs at the same time or later than the 2′-hydroxyl nucleophilic attack. A conserved histidine at the active site could be involved in activating the hydroxyl for this chemistry. Base exchange could occur if the intermediate is formed reversibly. The enzyme nucleophile mechanism proposes that an enzyme side-chain nucleophile is able to react with NAD⁺ to stabilize an enzyme-bonded ADPR species that can reverse to account for base-exchange chemistry. This mechanism has been shown to operate for ADP-ribosyl cyclases. A prediction of this mechanism is the formation of β-1′-O-AADPR as the acetyltransfer product.

Nucleophilic attack mechanism. A nucleophilic mechanism has been proposed in which the 2′-hydroxyl of NAD⁺ acts as the initiating nucleophile in the reaction that attacks the acetylamide group, leading to formation of a 2′-acetylated NAD⁺ as a reaction intermediate (108). Such chemistry could be facilitated by a proximal active site histidine that is proposed to act as a base catalyst for deprotonation of the 2′-hydroxyl (**Figure 5**). In several X-ray crystal structures of sirtuins, this universally conserved histidine is coordinated to the 3′-hydroxyl of the ribose of NAD$^+$, consistent with a catalytic role to activate the 2′-hydroxyl for nucleophilic attack (102, 113). This proposal predicts that acetyltransfer occurs in the catalytic mechanism of sirtuins first and that loss of nicotinamide occurs second (**Figure 5**). Several independent observations have discredited this mechanism. First, the acetylated NAD$^+$ species has not been observed. Second, mutagenesis of the highly conserved catalytic His residue disables deacetylation catalysis but not base exchange (102), suggesting that deacetylation/acetyltransfer reaction occurs downstream of nicotinamide-bond cleavage chemistry (102, 109), in direct contradiction to the proposed mechanism. Finally, the proposed mechanism is eliminated by observations that 2′-deoxy-NAD$^+$ compounds retain activity as base-exchange substrates but cannot complete deacetylation chemistry (104). These 2′-deoxy and mutagenesis studies support the concept that the deacetylation catalysis requires the 2′-hydroxyl and that the histidine provides an activation of nucleophilicity. However, it is apparent that the base-exchange process is independent of the presence of the 2′-hydroxyl. Consequently, the nucleophilic step in deacetylation catalysis on Sir2 enzymes occurs after nicotinamide loss in a step subsequent to the base-exchange reaction.

Enzyme nucleophile mechanism. Another mechanism proposed for Sir2-deacetylation catalysis invokes the precedent of a well-known ADP-ribosyl transferase mechanism to explain the observed base-exchange pathway. In this proposal, NAD$^+$ is attacked by a catalytic site amino acid nucleophile to form a stable ADPR intermediate (**Figure 5**) (101, 102, 107). This reaction pathway is analogous to the well-studied reaction mechanism of CD38, in which a Glu nucleophile reacts with NAD$^+$ to covalently stabilize an ADPR intermediate that can be reversed by reaction with nicotinamide (112). Analogously, the sirtuin-reactive intermediate can reform NAD$^+$ in the presence of nicotinamide to complete the base-exchange mechanism (102, 107). Although attractive as a mechanism, a crystal structure of NAD$^+$ complexed to an *A. fulgidus* enzyme provided no evidence for a nucleophile in the vicinity of the α-carbon of NAD$^+$ (102). Also problematic for this proposed mechanism is the prediction that the acetyltransfer product would necessarily be formed with retention of configuration to produce β-1′-AADPR (107, 108) and not the observed 2′-AADPR product.

Adpr-peptidyl-imidate mechanism. A mechanism first proposed by the authors rationalizes the chemistry observed for sirtuins (109). This mechanism is called the ADPR-peptidyl-imidate mechanism of sirtuin catalysis. Inherent to this mechanism is that base-exchange and deacetylation pathways are united by a common reaction intermediate known as the ADPR-peptidyl-imidate Complex III (109), which is formed by electrophilic capture of the acetyl oxygen in an ADP-ribosyl transfer reaction (**Figure 6**). This chemical step accounts for the requirement of peptide in base exchange (101, 102, 109), the observed lack of an enzyme nucleophile in X-ray structures (102), the requirement of a C1′-O (acetyl) bond-forming step in Sir2 catalysis (109), and the viability of base exchange in the absence of the active site His residue (102) or when 2′-OH is removed from substrate (104). The proposed capture of an acetyl oxygen by ADPR requires a highly electrophilic ribosylating agent, given the weakness of the amide as a nucleophile (109). However, ADPR-transfer reaction mechanisms have been determined previously for the bacterial toxins by kinetic isotope effects and computational studies (114). These studies show that the enzymes stabilize well-developed oxacarbenium-ion transition states that are highly reactive with nucleophiles (105, 114). A potently reactive oxacarbenium ion is proposed to account for imidate formation (Complex II) in sirtuin ADP-ribosyl transfer reactions (109), but

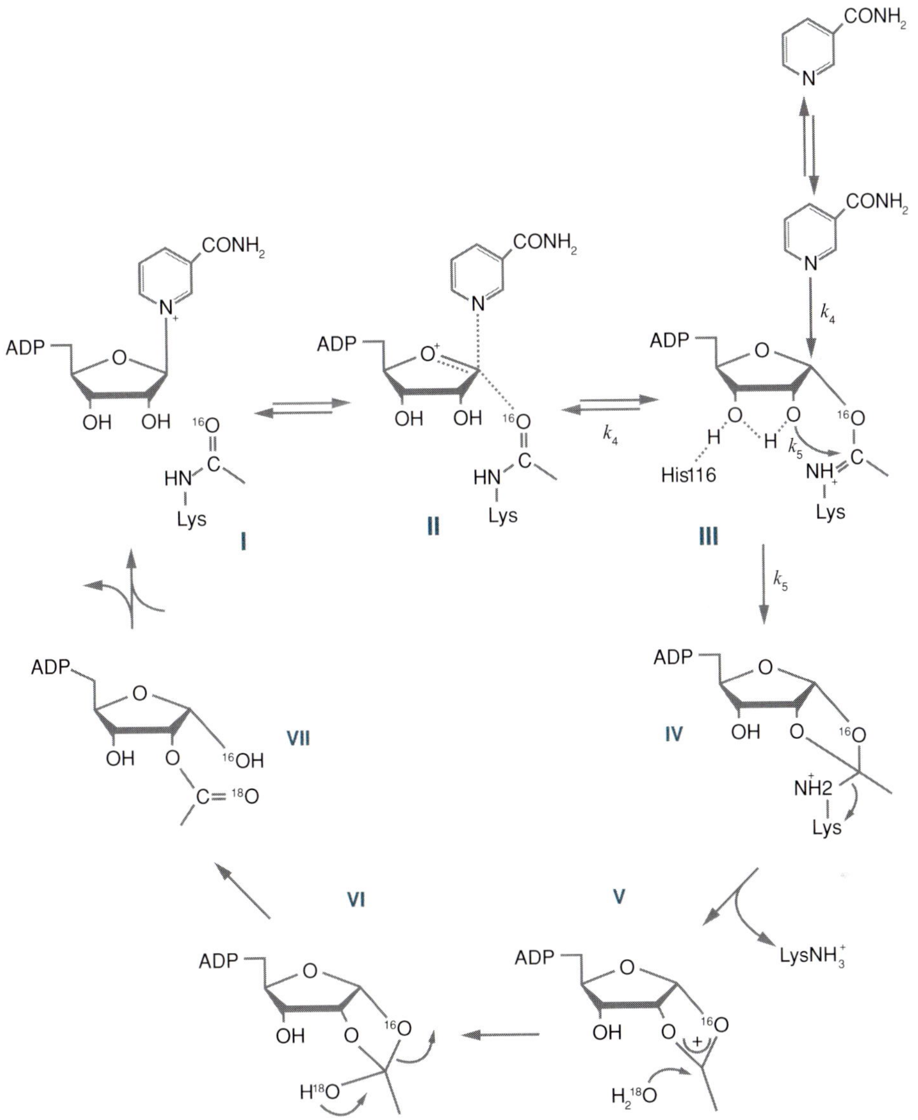

Figure 6

Peptidyl-imidate mechanism of sirtuin deacetylation. The reaction of sirtuins, as proposed to occur via the electrophilic capture of the peptide acyloxygen by an enzyme-stabilized oxacarbenium-ion transition state (Complex II) to form an ADPR-peptidyl imidate (Complex III). The imidate complex has a long enough lifetime to equilibrate nicotinamide in the active-site nicotinamide pocket and can reform NAD$^+$ by reaction reversal to account for base-exchange reactivity. Subsequent nucleophilic attack on the imidate by the 2′-hydroxyl forms an intermediate (Complex IV), which subsequently eliminates to form a proposed oxonium species (Complex V), which can capture water to form a second tetrahedral intermediate (Complex VI). This species is proposed to decompose via elimination of the C1′ hydroxyl group to generate the 2′-O-AADPR as the final product on the enzyme (Complex VII). This mechanism accounts for ^{18}O labeling of the acetyl group and ^{16}O labeling of C1′ hydroxyl as observed by reactions conducted in ^{18}O water.

convincing evidence of this transition state is not yet available. The ADPR transfer step that forms the imidate is written as a reversible step to account for nicotinamide reactivity through base exchange to reform NAD$^+$ (109).

Subsequent breakdown of the intermediate can occur via attack of the imidate by the ribose 2′-hydroxyl, with subsequent attack by water to complete deacetylation chemistry (**Figure 6**, Complexes III–VII) (103, 104, 109). This order of two sequential nucleophilic attacks (2′-hydroxyl then water) completes deacetylation, with release of 2′-AADPR as the enzymatic product, as observed experimentally. The alternative order of attacks (water then 2′-hydroxyl) would predictably yield an α-1′-AADPR compound (104, 105, 109) prior to 2′-AADPR formation. To date, there has been no evidence for an α-1′-AADPR compound, even at low-temperature studies of Sir2 catalysis (109).

Direct characterization of any single intermediate in the sirtuin deacetylation pathway is lacking. Convincing evidence for Complexes II–VI (**Figure 6**) has not been obtained, nor have any of the putative transition states of sirtuin catalysis been clearly identified by kinetic isotopic methods. Either result would provide deeper insight into the nature of the chemical transitions that occur on these enzymes. Despite the lack of direct evidence for the structural identities of intermediates in the Sir2 reaction pathway, the imidate mechanism best explains all available experimental data.

Regulation by NAD$^+$ and Nicotinamide

The regulation of sirtuins is debated, although it is clear that sirtuins are sensitive to perturbations in NAD$^+$ levels and energy metabolism. Regulatory mechanisms involving NAD$^+$ and nicotinamide from the sirtuin reaction mechanism include modulation of activity by substrate requirement of NAD$^+$. In support of this view, genetic modifications in yeast and in mammalian cells that alter NAD$^+$ levels appear to change sirtuin activity. Kinetic studies of Sir2 have determined K_m values for NAD$^+$ in the range of 150–500 μM (10, 72, 83, 84, 103, 104, 116), consistent with physiological control of activity by NAD$^+$ levels. Regulation by NAD$^+$ is especially relevant in stressed mammalian cells, such as those exposed to genotoxicity or trauma, where NAD$^+$ levels may plummet as a consequence of PARP activation (117).

Sirtuins are inhibited noncompetitively (versus both NAD$^+$ and acetyllysine substrates) by the nicotinamide product of the sirtuin deacetylation reaction. K_i values for this inhibition have been reported in the range 50–150 μM (72, 101, 103, 104). Although measurements of physiological nicotinamide concentrations are scarce in the literature, new measurements of nicotinamide in yeast suggest that endogenous nicotinamide levels are inhibitory to Sir2 deacetylase activity (cellular concentrations of 10–150 μM) (75). Nicotinamide inhibition of sirtuins is a consequence of the base-exchange reaction common to sirtuins (103, 104). This mechanism of nicotinamide inhibition is supported by the finding that the K_m for nicotinamide exchange and the K_i for nicotinamide inhibition of deacetylation are similar for evolutionarily diverse sirtuins (archaea, yeast, mouse, and human) (75, 103, 104, 115). Moreover, nicotinamide isosteres that are weak base-exchange substrates are correspondingly weak as deacetylation inhibitors (75, 103, 104).

The unification of nicotinamide base-exchange reactivity and nicotinamide inhibition of the deacetylation reaction is a consequence of the ADPR-imidate mechanism of sirtuin deacetylation. According to this mechanism, nicotinamide equilibrates at the active site of an ADPR-peptidyl imidate intermediate downstream of NAD$^+$ and peptide reaction on the enzyme. Nicotinamide reaction with the intermediate causes reaction reversal and depletes the intermediate from the enzyme (90, 91). Because the base-exchange and deacetylation processes compete for the imidate intermediate to complete their respective reactions, base-exchange depletion of the intermediate necessarily inhibits deacetylation catalysis (103, 104). Accordingly, equivalence for the base-exchange parameter K_m and the inhibition constant K_i is predicted and observed (75, 103, 104, 115). Nicotinamide does not compete for binding with either NAD$^+$ or peptide in this mechanism, and noncompetitive inhibition occurs with respect to both substrates.

The imidate intermediate complex formed on sirtuin enzymes behaves as a metabolic sensor for nicotinamide (75, 103, 104, 115). A long lifetime for the covalent intermediate [estimated to be in excess of 100 ms at 25°C (103, 115)] permits full equilibration of intracellular nicotinamide at the active site. When the intermediate reacts backward to reform substrates, deacetylation of sirtuin is inhibited. The reversible reactivity of the intermediate introduces a novel mechanism of metabolic control on sirtuin reactivity that functions as a reaction checkpoint or switch (103). This nicotinamide checkpoint/switch reverses or forwards NAD$^+$-initiated chemistry depending on the status of nicotinamide concentration within the cell. In summary, the sirtuin catalytic mechanism provides reaction steps (imidate formation) that can be increased by metabolite concentration (NAD$^+$) or reversed by another metabolite (nicotinamide) in the NAD$^+$ biosynthetic pathway. Consequently, the mechanism of sirtuins integrates independent and counteracting inputs of the NAD$^+$ biosynthetic pathway to regulate protein deacetylation. This catalytic control mechanism appears to be unique among enzymatic reactions.

Partition Model of Imidate Reactivity

Nicotinamide reactivity in the base-exchange pathway inhibits sirtuin deacetylation. The ADPR-imidate partitions between these pathways and causes unusual properties of nicotinamide inhibition. Surprisingly, under conditions where nicotinamide concentration fully saturates the base-exchange reaction, the deacetylation rate is not completely inhibited (103). An extreme case is the *A. fulgidus* Sir2Af2 enzyme, where at saturation of the base-exchange reaction, the deacetylation rate retains 75% of the uninhibited rate (103). For yeast Sir2p, the uninhibited rate is reportedly 5% to 35% depending on the identity of peptide substrate (103, 104, 115), and for mouse Sir2α, the residual deacetylation rate is 5% (103).

To rationalize partial inhibition by nicotinamide, Sauve & Schramm (103) argued that the nucleophilic reactivities of the intermediate leading to base exchange and deacetylation are sterically independent. This possibility can be appreciated by considering that the nicotinamide attacks the imidate from the β-face of the ribosyl sugar, whereas 2'-hydroxyl attacks the imidate from the α-face (**Figure 6**). These nucleophilic attacks can proceed independently. Thus, it was proposed that the residual deacetylation rates in the presence of saturating nicotinamide are determined by the partitioning of the intermediate to deacetylation, as predicted by the relative rate constants (k_4 and k_5) (**Figure 6**) for base exchange and deacetylation, respectively (103). As an illustrative case, this model of nicotinamide inhibition predicts that k_5 is greater than k_4 for the Sir2Af2 enzyme because deacetylation chemistry is only modestly inhibited by nicotinamide when the base-exchange reaction is saturated (103). Measurements show that the uninhibited steady-state deacetylation rate (k_{cat}) is faster than the steady-state base-exchange rate (k_{cat}) (103), consistent with $k_5 > k_4$ for this enzyme (103). Conversely, yeast Sir2p, mouse Sir2α, and human SIRT2 are more potently inhibited by nicotinamide, and the steady-state nicotinamide base-exchange rate k_{cat} is faster than the steady-state deacetylation rate k_{cat} (103, 104). These data indicate that nicotinamide reactivity exceeds 2'-hydroxyl reactivity ($k_4 > k_5$) on these enzymes (103). The biochemical and structural basis for nicotinamide sensitivity is still not well understood. It is apparent that nicotinamide sensitivity is better developed in yeast and mammalian sirtuins. This supports the hypothesis that nicotinamide is an endogenous regulator of sirtuins in yeast and mammals.

Nicotinamide Antagonists as Sirtuin Activators

Nicotinamide inhibition of sirtuin deacetylation arises from nicotinamide reactivity with the imidate intermediate. It was proposed that a small molecule could fit into the nicotinamide pocket and selectively inhibit base exchange but not deacetylation (103). The deacetylation reaction could proceed as usual because the 2′-hydroxyl remains unhindered by ligand binding to the nicotinamide site. A ligand with these properties is predicted to be a nicotinamide antagonist that can relieve nicotinamide inhibition of sirtuin deacetylation activity (75, 103). A small molecule, isonicotinamide, was identified, which inhibits nicotinamide base exchange but does not inhibit the deacetylation reaction (75). Observed values for K_i (deacetylation) and K_m (base exchange) both increased in the presence of isonicotinamide, indicating that isonicotinamide inhibition of nicotinamide base exchange weakens nicotinamide inhibition of the enzyme (75). By virtue of its ability to displace nicotinamide from the imidate, isonicotinamide can increase deacetylation activity of yeast Sir2 when inhibitory nicotinamide is present (**Figure 7**) (75). The ability of isonicotinamide to antagonize nicotinamide suggests that it may be able to activate Sir2 in cells to provide evidence of nicotinamide regulation of sirtuins. In yeast, isonicotinamide is a potent stabilizer of heterochromatin formation at *TEL*, *rDNA*, and *HM* genetic loci, as determined by both negative and positive growth selections (75). Isonicotinamide binding to Sir2 is weak (at millimolar concentrations), but the biological effects of this compound have provided chemical evidence for nicotinamide regulation of Sir2 activity in wild-type yeast cells, in agreement with results obtained by genetic studies (72–74). The potent inhibition of mammalian enzymes by nicotinamide suggests that molecules with properties similar to isonicotinamide could provide a pharmacological approach to activating sirtuins in humans.

STRUCTURAL PROPERTIES OF SIRTUINS

The unprecedented chemical reaction catalyzed by sirtuins implicated a unique arrangement of acetyllysine peptide and NAD^+ at the catalytic sites of these enzymes. The structural data reported here are from several structures of sirtuins from lower organisms, with structures from human sirtuins still mostly undetermined. The mammalian sirtuin X-ray structures are of great interest given the rapid effort to describe their biological and biochemical functions.

Structure of the Conserved Catalytic Core of Sirtuins

There are approximately 250 amino acids in the sirtuin's catalytic core domain, which is conserved among virtually all organisms (**Figure 8**). A significant amount of information on the structure of the catalytic core and its binding to ligands comes from crystallographic studies of the archaeal sirtuins, Sir2Af1 (89) and Sir2Af2 (113, 118, 119), human SirT2 (120), yeast Hst2 (121–123), and bacterial sirtuins cobB (64) and Sir2Tm (119). The sirtuin fold consists of two characteristic domains (89, 102, 113, 118–123) (**Figure 9**). The larger domain adopts the classic pyridine dinucleotide binding fold, or Rossman fold, which is commonly found in proteins that bind $NAD^+/NADH$ or $NADP^+/NADPH$ (124). The small domain is composed of residues from two insertions within the Rossmann fold, one comprising a zinc-binding module that contains a structural zinc atom coordinated by four invariant cysteines, and the other forming a helical module that includes a flexible loop. The position of the

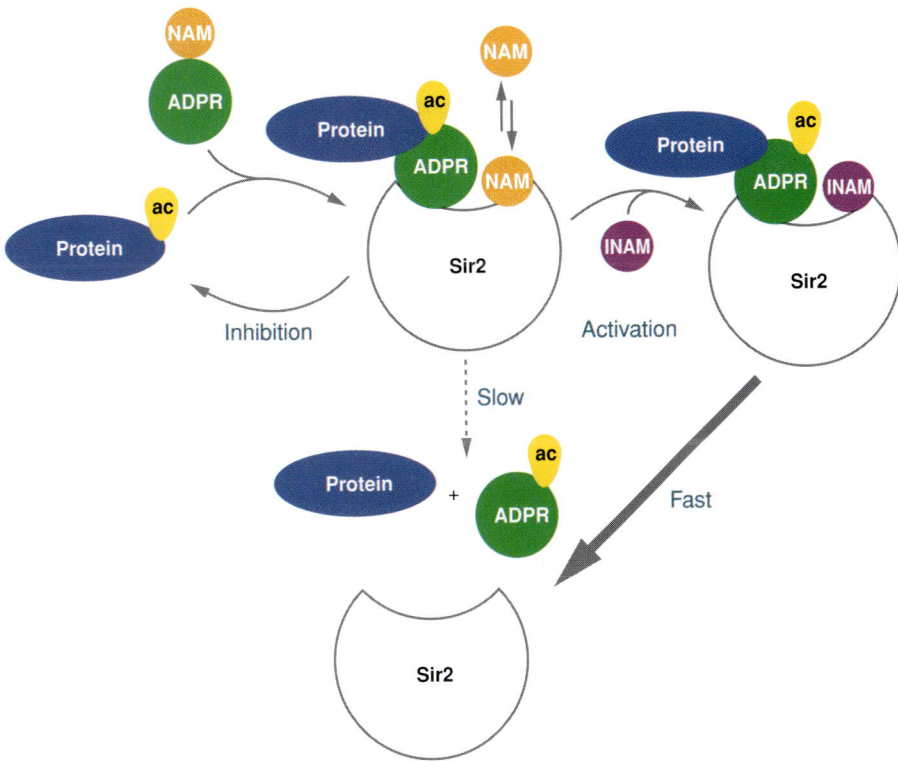

Figure 7

Activation mechanism of sirtuins by antagonism of nicotinamide. As shown, sirtuins catalyze reaction of NAD$^+$, shown as NAM-ADPR, and a protein acetyllysine group, the acetyl group labeled as ac, to form an imidate intermediate (*center top*) that releases nicotinamide (NAM). NAM can rebind to reverse the reaction through the base-exchange mechanism (*left arrow labeled inhibition*), or the imidate complex can proceed downward to complete the deacetylation reaction (*dashed arrow*). In the presence of nicotinamide, the deacetylation reaction is slow because of depletion of the imidate by nicotinamide reactivity. Isonicotinamide (INAM) can bind the nicotinamide pocket and prevent reaction reversal. The INAM-bound imidate complex can complete deacetylation more efficiently (*thick arrow*), leading to activation of sirtuin catalytic activity in cells.

small domain relative to the large domain varies among different sirtuin structures and appears to be influenced by ligand binding as well as contact with other proteins. The protein and NAD$^+$ cosubstrates bind in a cleft between the two domains of the sirtuin fold.

Binding of Acetylated Peptide

As first seen in the structure of Sir2Af2 bound to an acetylated peptide (113), sirtuins bind their peptide substrates primarily through β-sheet interactions between main chain atoms of the peptide and backbone atoms in the large and small flanking domains. This gives rise to a three-stranded enzyme-substrate β-sheet, with the central substrate peptide strand flanked by a β strand from the large and small domains (**Figure 10**). The β-sheet-like interactions with the peptide backbone fix the N- to C-terminal orientation of the substrate backbone and extend roughly two residues to either side of the acetyllysine. There are additional contacts between some of the peptide side chains and residues on the surface of the enzyme. The acetyllysine side-chain inserts

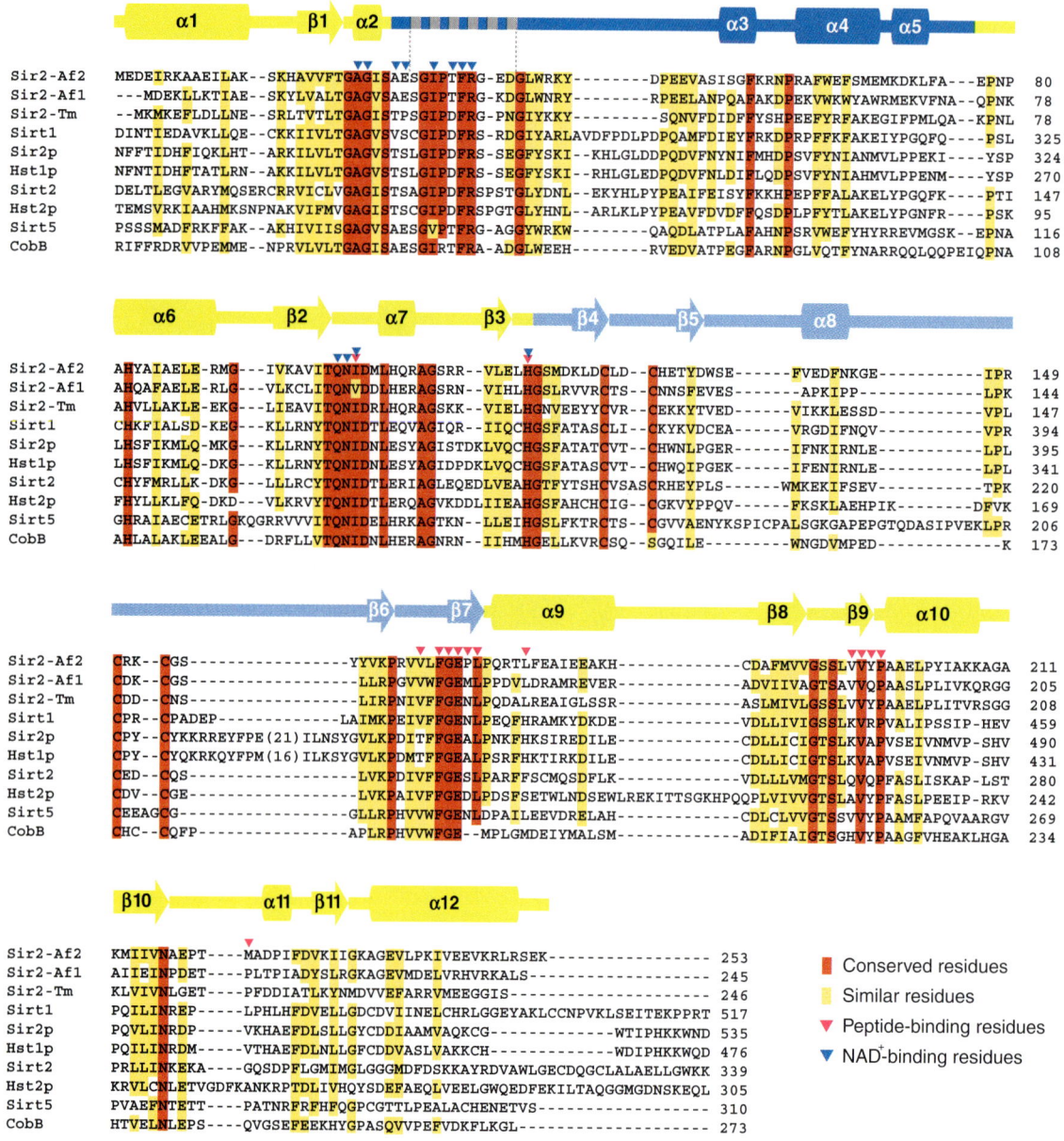

Figure 8

Multiple-sequence alignment of the sirtuin core domain. Red highlighting indicates sequence identity, and yellow highlighting indicates chemically similar residues. The coloring of secondary structural elements, indicated at the top of the sequence, follows the color scheme used to indicate structural domains in **Figure 9**.

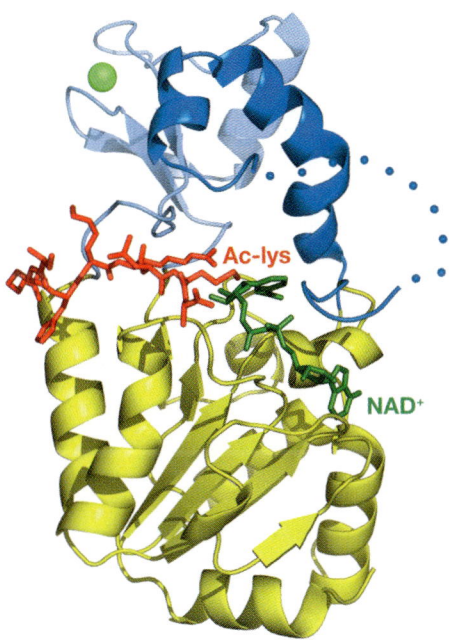

Figure 9

Structure of a sirtuin bound to acetylated peptide and nicotinamide adenine dinucleotide (NAD$^+$). The Rossmann-fold domain is colored yellow. The zinc-binding module (*light blue*) and the helical module (*royal blue*), each arising from an insertion within the Rossmann fold, associate to form a single domain. The acetylated peptide (*red*) and NAD$^+$ (*green*) bind in the cleft that separates these two domains.

into a largely hydrophobic tunnel that extends into the cleft and is anchored by a hydrogen bond between the Nε of the acetyllysine and a backbone carbonyl (**Figure 10**). The aliphatic side chain lies atop His118 (Sir2Af2 numbering), an invariant residue required for catalytic activity (87), and is also contacted by invariant side chain Phe165. Structures of yeast Hst2 (121, 123) and bacterial cobB (64) sirtuins bound to acetylated histone H4 peptide show essentially the same mode of binding for the acetyllysine and for the immediately flanking residues in the peptide.

The predominance of protein backbone interactions in peptide binding suggests that sirtuins are designed to bind to unstructured peptide substrates that can thus form the β-sheet interactions observed in all crystal structures. Although the predominance of peptide backbone interactions suggests that there is limited sequence specificity in binding of the peptide substrate within the enzyme cleft, detailed studies of the substrate specificity of other sirtuins indicate that these enzymes discriminate among different acetylated substrates (115). This specificity may arise, in part, from residues on the surface of the enzyme that contact side chains in the peptide substrate, although the effect appears to be relatively modest (113). It is likely that the binding of sirtuins to their biological targets is determined in large part by domains that lie N- or C-terminal to the catalytic core, or by other proteins.

NAD$^+$ Binding

NAD$^+$ binds in the cleft between the sirtuin large and small domains, immediately adjacent to the acetyllysine binding tunnel. The adenine base and the adjacent ribose bind to the Rossmann-fold domain, forming a fixed set of contacts (**Figure 9**). By contrast, the nicotinamide ring and the adjacent ribose and phosphates are free to adopt a variety of positions in the absence of peptide (102, 113). It is only when the acetyllysine binding tunnel is occupied, either by acetyllysine (K. Hoff, J. Avalos, K. Sens, and C. Wolberger, manuscript submitted) or by a fortuitously bound PEG (polyethylene glycol) molecule (118), that the nicotinamide is buried within a conserved region, dubbed the C pocket (102). In this conformation, the carboxamide of the nicotinamide hydrogen bonds with an invariant aspartic acid (Asp103 of Sir2Af2), as well as with the backbone carbonyl of Ile102. These interactions rotate the carboxamide 30° out of the plane of the nicotinamide ring, in contrast with the coplanar arrangement seen in the structure of free nicotinamide (119). The nicotinamide ring forms additional van der Waals interactions with other residues within the C pocket. Interestingly, a number of the invariant residues that comprise the front portion of the C pocket and contact nicotinamide

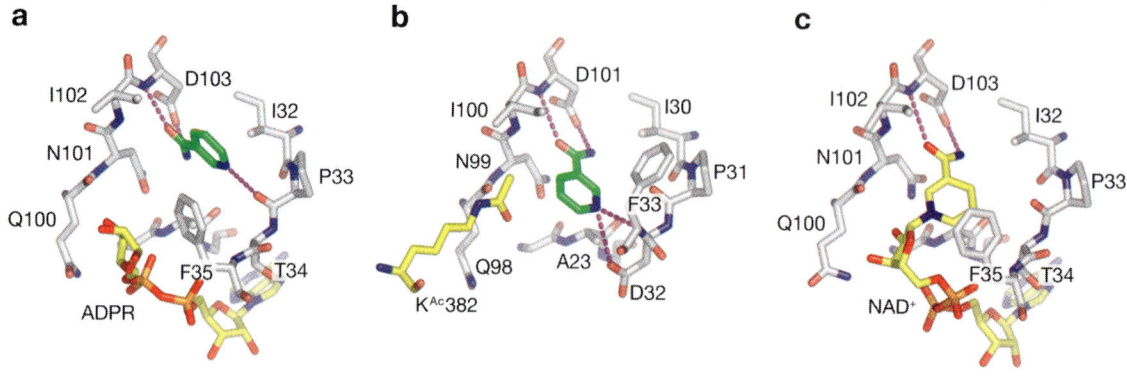

Figure 10

Binding of free nicotinamide in the C pocket of sirtuins. (*a*) Binding of free nicotinamide in the C pocket of Sir2Af2 and (*b*) of Sir2Tm. (*c*) Nicotinamide moiety of NAD$^+$ bound in the C pocket of Sir2Af2.

lie in the flexible loop, which is disordered in some structures that lack NAD$^+$ (113, 122).

Structural Insights into Enzyme Mechanism

Proposals for the structural basis of the NAD$^+$-dependent deacetylation reaction have come from several different studies of sirtuins bound to substrates and products. As mentioned above, isotope-labeling experiments suggest that the deacetylation reaction begins with cleavage of the nicotinamide glycosidic bond and an attack by the carbonyl oxygen of acetyllysine on the C1$'$ of the nicotinamide ribose (109). Two distinct mechanisms have been proposed for the nicotinamide step of the reaction. A dissociative, SN1-like reaction has been proposed on the basis of the structure of a ternary complex containing yeast Hst2, acetylated peptide, and carba-NAD$^+$, a nonreactive analog that contains cyclopentane in place of the nicotinamide ribose (123). In that structure, the acetyl oxygen hydrogen bonds to the 2$'$- and 3$'$-hydroxyls of the nicotinamide ribose. The orientation of the ribose does not allow access by the acetyl oxygen to the C1$'$ of the ribose, thereby precluding a direct nucleophilic attack. Instead, Zhao and colleagues (123) propose that a highly ordered water in the vicinity of the ribose stabilizes formation of a positively charged oxacarbenium ion, thereby promoting cleavage of the glycosidic bond. The ribose would then have to undergo a significant rotation and translation for the C1$'$ of the oxacarbenium to react with the acetyl oxygen. Although a dissociative mechanism has been proposed for NAD$^+$ cleavage by bacterial ADP-ribosylating toxins, those enzymes contain a negatively charged glutamate that can stabilize the oxacarbenium. It remains problematic that any water molecule in the vicinity of an oxacarbenium ion would be expected to react immediately, thereby aborting the subsequent steps that lead to lysine deacetylation.

An associative, SN2-like reaction mechanism was proposed on the basis of the structure of Sir2Af2 bound to NAD$^+$ and an acetyllysine mimic, polyethylene glycol (118). In this structure, the nicotinamide ribose adopts a different conformation from that of the cyclopentane of carba-NAD$^+$. Importantly, the sugar ring is oriented with its α face exposed to the acetyl oxygen, which could permit a direct nucleophilic attack by the acetyl oxygen on the C1$'$ of the ribose. Burial of the charged nicotinamide ring in a largely hydrophobic environment and rotation of the carboxamide ring out of plane would decrease the energetic barrier to cleaving the glycosidic bond. However, the oxygen of the amide is still a very weak nucleophile, and an SN2 mechanism for an ADP-ribosyl transferase would be unprecendented, as all other reaction coordinates of this type involve SN1 mechanisms with migration of the ribosyl cation. A more recent structure of a ternary complex containing wild-type Sir2Tm bound to NAD$^+$ and acetylated peptide (K. Hoff, J. Avalos, K. Sens, and C. Wolberger, manuscript submitted) provides further insights into the true nature of the Michaelis complex. As in the structure of Sir2Af2 with NAD$^+$ and an acetyllysine mimic, the ribose is oriented with its α face exposed to the acetyl oxygen, although the absolute ribose conformation is somewhat different from that in the PEG complex. The distance between the acetyl oxygen and the C1$'$ is 3.2 Å, as predicted for nucleophilic displacement by electrophile migration, but somewhat long for an SN2-like nucleophilic attack. The ribose 2$'$- and 3$'$-hydroxyls hydrogen bond to the acetyl oxygen, albeit in a different relative orientation from that observed in the carba-NAD$^+$ structure (123). Additionl hydrogen bonds between the 2$'$- and 3$'$-hydroxyls of the nicotinamide ribose and the catalytic histidine (His116 in Sir2Tm) may further help to position NAD$^+$ in a conformation that is primed for catalysis. In particular, the proximal phosphate oxygen is within van der Waals contact distance (3.1–3.7 Å) of several of the ribose atoms, which could help stabilize a developing negative charge on the ribose predicted by an SN1 mechanism. Following cleavage of the glycosidic bond, the catalytic histidine would need to remain within hydrogen-bonding distance of the 3$'$-hydroxyl in order to carry out the proposed deprotonation of the ribose and promote the rearrangements that lead to migration of the acetyl group from the C1$'$ to the 2$'$ and 3$'$ positions (**Figure 6**). Indeed, the same hydrogen bond with the ribose is seen in the structure of Hst2 bound to the product, O-AADPR (64), suggesting that this interaction could be maintained throughout the deacetylation reaction.

Structural studies have also provided insights into how nicotinamide acts as a non-competitive inhibitor by reacting with the enzyme-bound O-alkylimidate intermediate, reforming NAD$^+$ and releasing the acetylated peptide in a base-exchange reaction (103, 104). As shown in structural studies of the Sir2Af2 and Sir2Tm enzymes (119), free nicotinamide can bind in the C pocket when the enzyme is simultaneously in complex with peptide, ADP-ribose, or NAD$^+$ that is in a nonproductive conformation, thereby leaving the C pocket free to bind nicotinamide (**Figure 10**). The carboxamide of free nicotinamide forms the same hydrogen bonds with the enzyme as when NAD$^+$ binds in the productive conformation. However, the ring in free nicotinamide adopts the low-energy conformation (125, 126) and is nearly coplanar with the carboxamide. While the hydrogen-bonding interactions anchor the carboxamide to the enzyme, the ring is able to adopt a number of different conformations. It has been suggested that only a subset of the conformations of nicotinamide bound in the C pocket are in a position to react with the O-alkyl imidate intermediate (119). The relevance of the C pocket to the inhibitory base-exchange reaction is supported by experiments on Sir2Tm enzyme in which the invariant C pocket, aspartic acid, in which hydrogen bonds with the amino group of nicotinamide, is changed to asparagine. This substitution enables the mutant to catalyze the inhibitory base-exchange reaction with nicotinic acid, which does not inhibit the wild-type enzyme. The glutamic acid to glutamine substitution also enables the enzyme to catalyze NAAD$^+$-dependent deacetylation, indicating that the C pocket plays dual roles in deacetylation and base exchange.

Multimerization of Sirtuins

There is evidence that at least one sirtuin, yeast Hst2, may regulate its own activity by homo-oligomerization. Crystals of the Hst2 apoenzyme contain crystallographic trimers that arise from contacts with neighboring monomers made by N- and C-terminal residues that lie outside the conserved sirtuin core (64). A seven-residue amino acid extension binds in the substrate binding cleft of a neighboring molecule, inserting the N-terminal methionine into the acetyllysine binding tunnel, while an additional C-terminal helix not found in other sirtuins mediates additional contacts with the catalytic domain. Hst2 forms trimers in solution with a dissociation constant that falls roughly in the physiological range of Hst2 concentration, suggesting that the crystallographic trimer may also be relevant in the cell. Trimerization is dependent upon the presence of the seven-residue amino acid extension. The potential significance of the observed interaction is that the N-terminal residues inhibit enzymatic activity, raising the K_m for peptide by approximately twofold, whereas the additional C-terminal residues raise the K_m for NAD^+ by approximately twofold. This provides a potential auto-inhibitory mechanism, which can be exploited by the cell to activate Hst2, by either competing for the inhibitory N-terminal peptide or binding to another protein that relieves the inhibitory effect of the C-terminal helix. Most sirtuins contain additional residues or domains flanking the catalytic domain that could similarly autoregulate enzymatic activity.

SMALL MOLECULE MODULATORS OF SIRTUINS

A number of small molecule agents have been identified that either inhibit or activate sirtuin enzymes. These chemical reagents have been utilized to examine the biological activities of sirtuins in diverse organisms including yeast, flies, and mammals. Examples of these molecules are briefly described.

Activators of Sirtuins

Small molecule activators of enzymes are uncommon, but two distinct examples have been found for sirtuins. These activators include polyphenols, exemplified by resveratrol, and another type of compound that can antagonize nicotinamide inhibition of sirtuin deacetylation activity. (See the Chemical Mechanism of Sirtuins, above, for discussion of nicotinamide antagonists).

Resveratrol and plant polyphenols. The polyphenol sirtuin activators were discovered through a high-throughput screen using a fluorophore-linked peptide substrate as a reaction reporter. This screen used low concentrations of the fluorophore-linked substrate and also used low NAD^+ concentrations with SirT1 as the enzyme. These conditions were designed so that compounds that provide activation could be discovered. A set of well-known polyphenol compounds with known biological and anti-oxidant activities scored high in this assay. Among these compounds were resveratrol, quercetin, and a set of other known catechins produced by plants such as tea and grapes (resveratrol is abundant in red wine). The activation effects of these compounds in the assay were two- to eightfold, with the best activation observed for resveratrol. Kinetic studies indicated that resveratrol reduced K_m values for substrates (by two- to eightfold), most notably for peptides. Subsequent studies of the biological responses of cells treated with resveratrol showed effects consistent with stimulation of SirT1 activity. These effects included increased resistance of cells to apoptosis (10), reduced adipogenesis in fat-producing cells (consistent with SirT1 antagonism of PPARγ) (12), and increased stress resistance of neurons in a SirT1-dependent mechanism (16). Problematic to resveratrol is a lack of specificity for SirT1 in vivo, as the compound is unstable in cells to oxidation (127) and is also a potent inhibitor of cyclooxygenases and other enzymes (127). This lack of specificity complicates interpretation of the effects observed for resveratrol in cells.

Recent studies indicating that resveratrol does not activate SirT1 activity if peptide substrates do not contain a fluorophore are problematic for this intriguing molecule as an activator of sirtuins (83, 84). Two independent laboratories investigated resveratrol activity in activating SirT1 activity for peptide substrates analogous to those used in the original assays with and without fluorophore. These studies showed that fluorophore peptides have significantly higher K_m values than the underivatized peptides in the absence of resveratrol, leading to the conclusion that resveratrol stabilizes binding of these modified substrates to the enzyme. Underivatized peptides did not bind better in the presence of resveratrol, so no activation effects were observed with these more physiologic substrates (83, 84). Given that resveratrol effects in vivo are consistent with potent SirT1 activation, these in vitro results suggest that the mechanism of resveratrol (and other polyphenols) and its physiologic impact on SirT1 activity is poorly understood.

Inhibitors of Sirtuins

Several inhibitors of the sirtuin enzymes have been discovered. The most available is the product molecule nicotinamide. Problematic to the use of this compound as an inhibitor is the ability of nicotinamide to be directly incorporated into metabolism. In addition, all sirtuins studied thus far exhibit some sensitivity to nicotinamide inhibition. Thus, nicotinamide is likely to inhibit all sirtuins in a cell or organism to some extent. This consequence is likely to cause difficulty attributing a biological effect to the specific interaction of nicotinamide with a given sirtuin target. The need for specific sirtuin inhibitors and the characterization of their effects on individual sirtuin enzymes is apparent.

Sirtinol. The compound sirtinol was discovered by a high-throughput screen to identify compounds that would antagonize yeast silencing in a yeast-plating assay (128). This compound includes an imine function constructed from 2-hydroxy-1-napthaldehyde condensed with an aromatic amine. This derivative is a micromolar inhibitor of the yeast Sir2p enzyme and is also a potent inhibitor of the human sirtuin SirT2 (128). There is no evidence for inhibition of SirT1. It is not clear where sirtinols bind on Sir2 enzymes.

Splitomicins. Like sirtinol, splitomicins were among the first compounds identified in a library screening to discover disruptors of gene silencing in yeast (129). Both sirtinol and splitomicins incorporate a 1,2-substituted-napthalene nucleus as a central feature of the molecule. Splitomicins also have a third fused lactone ring (129–131). The parent molecule is partially saturated and inhibits Sir2 specifically. The compound dehydrosplitomicin exhibits specificity for the sirtuin Hst1p and is a very weak inhibitor of Sir2p (130, 131). This latter compound has been used to show that the Hst1p enzyme regulates NAD^+ biosynthesis in yeast (124). Splitomicin and dehydrosplitomicin are not potent inhibitors of human sirtuins. However, a splitomicin-like compound called HR73 was recently reported to inhibit human SirT1 (95). A full characterization of the inhibitory properties of this compound against SirT1 has not been reported.

CONCLUSION

The comparative lack of specific and potent inhibitors of mammalian sirtuin enzymes poses an obstacle to our understanding of their activities in cells. The use of sirtinol and splitomicin in studies investigating sirtuin action in mammalian cells raises concerns because the action sites have not been determined, and there is no evidence that they potently inhibit SirT1. These issues highlight the need for additional efforts to discover inhibitors that target this intriguing class of enzymes (132).

DISCLOSURE STATEMENT

Under a licensing agreement between Sirtris Pharmaceuticals, Inc., Albert Einstein College of Medicine of Yeshiva University, and the Johns Hopkins University, J. Boeke, V.L. Schramm, and A.A. Sauve are entitled to a share of royalty received by the universities on sales of future products, which may be related to compounds described in this article. J. Boeke, V.L. Schramm, A.A. Sauve, and the universities own Sirtris Pharmaceuticals, Inc., stock, which is subject to certain restrictions under each university policy. The terms of this arrangement are being managed by the Albert Einstein College of Medicine and Johns Hopkins University in accordance with conflict of interest policies at each institution. Aspects of A.A. Sauve's potential conflict of interest are managed through Weill Medical College of Cornell University in accord with university policy. Dr. Wolberger is on the scientific advisory board of Sirtris.

ACKNOWLEDGMENTS

Work on this review was supported in part by the National Institutes of Health.

LITERATURE CITED

1. Miller ES, Heidelberg JF, Eisen JA, Nelson WC, Durkin AS, et al. 2003. *J. Bacteriol.* 185:5220–33
2. Imai S, Armstrong CM, Kaeberlein M, Guarente L. 2000. *Nature* 403:795–800
3. Landry J, Sutton A, Tafrov ST, Heller RC, Stebbins J, et al. 2000. *Proc. Natl. Acad. Sci. USA* 97:5807–11
4. Smith JS, Brachmann CB, Celic I, Kenna MA, Muhammad S, et al. 2000. *Proc. Natl. Acad. Sci. USA* 97:6658–63
5. Marmorstein R. 2001. *Structure* 9:1127–33
6. Hernick M, Fierke CA. 2005. *Arch. Biochem. Biophys.* 433:71–84
7. Tissenbaum HA, Guarente L. 2001. *Nature* 410:227–30
8. Kaeberlein M, Andalis AA, Fink GR, Guarente L. 2002. *Mol. Cell. Biol.* 22:8056–66
9. Kaeberlein M, McVey M, Guarente L. 1999. *Genes Dev.* 13:2570–80
10. Howitz KT, Bitterman KJ, Cohen HY, Lamming DW, Lavu S, et al. 2003. *Nature* 425:191–96
11. Cohen HY, Miller C, Bitterman KJ, Wall NR, Hekking B, et al. 2004. *Science* 305:390–92
12. Picard F, Kurtev M, Chung N, Topark-Ngarm A, Senawong T, et al. 2004. *Nature* 429:771–76
13. Moynihan KA, Grimm AA, Plueger MM, Bernal-Mizrachi E, Ford E, et al. 2005. *Cell Metab.* 2:105–17
14. Anderson RM, Bitterman KJ, Wood JG, Medvedik O, Sinclair DA. 2003. *Nature* 423:181–85
15. Brunet A, Sweeney LB, Sturgill JF, Chua KF, Greer PL, et al. 2004. *Science* 303:2011–15
16. Araki T, Sasaki Y, Milbrandt J. 2004. *Science* 305:1010–13
17. Muth V, Nadaud S, Grummt I, Voit R. 2001. *EMBO J.* 20:1353–62
18. Starai VJ, Celic I, Cole RN, Boeke JD, Escalante-Semerena JC. 2002. *Science* 298:2390–92
19. McMurray MA, Gottschling DE. 2003. *Science* 301:1908–11
20. Gottlieb S, Esposito RE. 1989. *Cell* 56:771–76
21. Smith JS, Boeke JD. 1997. *Genes Dev.* 11:241–54
22. Perez-Martin J, Uria JA, Johnson AD. 1999. *EMBO J.* 18:2580–92
23. Motta MC, Divecha N, Lemieux M, Kamel C, Chen D, et al. 2004. *Cell* 116:551–63

24. Vaziri H, Dessain SK, Ng-Eaton E, Imai SI, Frye RA, et al. 2001. *Cell* 107:149–59
25. Klar AJ, Fogel S, Macleod K. 1979. *Genetics* 93:37–50
26. Rine J, Strathern JN, Hicks JB, Herskowitz I. 1979. *Genetics* 93:877–901
27. Haber JE, George JP. 1979. *Genetics* 93:13–35
28. Aparicio OM, Billington BL, Gottschling DE. 1991. *Cell* 66:1279–87
29. Bryk M, Banerjee M, Murphy M, Knudsen KE, Garfinkel DJ, et al. 1997. *Genes Dev.* 11:255–69
30. Fritze CE, Verschueren K, Strich R, Easton Esposito R. 1997. *EMBO J.* 16:6495–509
31. Straight AF, Shou W, Dowd GJ, Turck CW, Deshaies RJ, et al. 1999. *Cell* 97:245–56
32. Visintin R, Hwang ES, Amon A. 1999. *Nature* 398:818–23
33. Frye RA. 2000. *Biochem. Biophys. Res. Commun.* 273:793–98
34. Frye RA. 1999. *Biochem. Biophys. Res. Commun.* 260:273–79
35. Cuperus G, Shafaatian R, Shore D. 2000. *EMBO J.* 19:2641–51
36. Moazed D, Kistler A, Axelrod A, Rine J, Johnson AD. 1997. *Proc. Natl. Acad. Sci. USA* 94:2186–91
37. Liou GG, Tanny JC, Kruger RG, Walz T, Moazed D. 2005. *Cell* 121:515–27
38. North BJ, Marshall BL, Borra MT, Denu JM, Verdin E. 2003. *Mol. Cell* 11:437–44
39. Hecht A, Laroche T, Strahl-Bolsinger S, Gasser SM, Grunstein M. 1995. *Cell* 80:583–92
40. Carmen AA, Milne L, Grunstein M. 2002. *J. Biol. Chem.* 277:4778–81
41. Freeman-Cook LL, Gomez EB, Spedale EJ, Marlett J, Forsburg SL, et al. 2005. *Genetics* 169:1243–60
42. Freeman-Cook LL, Sherman JM, Brachmann CB, Allshire RC, Boeke JD, et al. 1999. *Mol. Biol. Cell* 10:3171–86
43. Bernstein E, Allis CD. 2005. *Genes Dev.* 19:1635–55
44. Shankaranarayana GD, Motamedi MR, Moazed D, Grewal SI. 2003. *Curr. Biol.* 13:1240–46
45. McBurney MW, Yang XF, Jardine K, Bieman M, Th'ng J, Lemieux M. 2003. *Mol. Cancer Res.* 1:402–9
46. Vaquero A, Scher M, Lee D, Erdjument-Bromage H, Tempst P, et al. 2004. *Mol. Cell* 16:93–105
47. Senawong T, Peterson VJ, Avram D, Shepherd DM, Frye RA, et al. 2003. *J. Biol. Chem.* 278:43041–50
48. Luo JY, Nikolaev AY, Imai S, Chen DL, Su F, et al. 2001. *Cell* 107:137–48
49. Langley E, Pearson M, Faretta M, Bauer UM, Frye RA, et al. 2002. *EMBO J.* 21:2383–96
50. Chua KF, Mostoslavsky R, Lombard DB, Pang WW, Franco S, et al. 2005. *Cell Metab.* 2:67–76
51. Cheng HL, Mostoslavsky R, Saito S, Manis JP, Gu YS, et al. 2003. *Proc. Natl. Acad. Sci. USA* 100:10794–99
52. Kenyon C. 2001. *Cell* 105:165–68
53. Giannakou ME, Partridge L. 2004. *Trends Cell Biol.* 14:408–12
54. Bouras T, Fu MF, Sauve AA, Wang F, Quong AA, et al. 2005. *J. Biol. Chem.* 280:10264–76
55. Puigserver P. 2005. *Int. J. Obes. Relat. Metab. Disord.* 29(Suppl. 1):S5–9
56. Rodgers JT, Lerin C, Haas W, Gygi SP, Spiegelman BM, et al. 2005. *Nature* 434:113–18
57. Park EC, Szostak JW. 1990. *Mol. Cell. Biol.* 10:4932–34
58. Suka N, Suka Y, Carmen AA, Wu J, Grunstein M. 2001. *Mol. Cell* 8:473–79
59. Kimura A, Umehara T, Horikoshi M. 2002. *Nat. Genet.* 32:370–77
60. Suka N, Luo K, Grunstein M. 2002. *Nat. Genet.* 32:378–83
61. Borra MT, O'Neill FJ, Jackson MD, Marshall B, Verdin E, et al. 2002. *J. Biol. Chem.* 277:12632–41

62. Rafty LA, Schmidt MT, Perraud AL, Scharenberg AM, Denu JM. 2002. *J. Biol. Chem.* 277:47114–22
63. Chang JH, Kim HC, Hwang KY, Lee JW, Jackson SP, et al. 2002. *J. Biol. Chem.* 277:34489–98
64. Zhao KH, Chai XM, Marmorstein R. 2004. *J. Mol. Biol.* 337:731–41
65. Kustatscher G, Hothorn M, Pugieux C, Scheffzek K, Ladurner AG. 2005. *Nat. Struct. Mol. Biol.* 12:624–25
66. Llorente B, Dujon B. 2000. *FEBS Lett.* 475:237–41
67. Bieganowski P, Brenner C. 2004. *Cell* 117:495–502
68. Preiss J, Handler P. 1958. *J. Biol. Chem.* 233:488–92
69. Revollo JR, Grimm AA, Imai S. 2004. *J. Biol. Chem.* 279:50754–63
70. Katoh A, Hashimoto T. 2004. *Front. Biosci.* 9:1577–86
71. Penfound T, Foster JW. 1996. In *Escherichia Coli and Salmonella: Cellular and Molecular Biology*, ed. FC Neidhardt, pp. 721–30. Washington, DC: ASM Press
72. Bitterman KJ, Anderson RM, Cohen HY, Latorre-Esteves M, Sinclair DA. 2002. *J. Biol. Chem.* 277:45099–107
73. Sandmeier JJ, Celic I, Boeke JD, Smith JS. 2002. *Genetics* 160:877–89
74. Gallo CM, Smith DL Jr, Smith JS. 2004. *Mol. Cell. Biol.* 24:1301–12
75. Sauve AA, Moir R, Schramm VL, Willis I. 2005. *Mol. Cell* 17:595–601
76. Guarente L, Picard F. 2005. *Cell* 120:473–82
77. Sinclair DA, Guarente L. 1997. *Cell* 91:1033–42
78. Kaeberlein M, Kirkland KT, Fields S, Kennedy BK. 2004. *PLoS Biol.* 2:e296
79. Lin SJ, Kaeberlein M, Andalis AA, Sturtz LA, Defossez PA, et al. 2002. *Nature* 418:344–48
80. Yeung F, Hoberg JE, Ramsey CS, Keller MD, Jones DR, et al. 2004. *EMBO J.* 23:2369–80
81. Wood JG, Rogina B, Lavu S, Howitz K, Helfand SL, et al. 2004. *Nature* 430:686–89
82. Rogina B, Helfand SL. 2004. *Proc. Natl. Acad. Sci. USA* 101:15998–6003
83. Borra MT, Smith BC, Denu JM. 2005. *J. Biol. Chem.* 280:17187–95
84. Kaeberlein M, McDonagh T, Heltweg B, Hixon J, Westman EA, et al. 2005. *J. Biol. Chem.* 280:17038–45
85. Lamming DW, Latorre-Esteves M, Medvedik O, Wong SN, Tsang FA, et al. 2005. *Science* 309:1861–64
86. Lamming DW, Wood JG, Sinclair DA. 2004. *Mol. Microbiol.* 53:1003–9
87. Tanny JC, Dowd GJ, Huang J, Hilz H, Moazed D. 1999. *Cell* 99:735–45
88. Tsang AW, Escalante-Semerena JC. 1998. *J. Biol. Chem.* 273:31788–94
89. Garcia-Salcedo JA, Gijon P, Nolan DP, Tebabi P, Pays E. 2003. *EMBO J.* 22:5851–62
90. Liszt G, Ford E, Kurtev M, Guarente L. 2005. *J. Biol. Chem.* 280:21313–20
91. Bell SD, Botting CH, Wardleworth BN, Jackson SP, White MF. 2002. *Science* 296:148–51
92. Shankaranarayana GD, Motamedi MR, Moazed D, Grewal SI. 2003. *Curr. Biol.* 13:1240–46
93. Daitoku H, Hatta M, Matsuzaki H, Aratani S, Ohshima T, et al. 2004. *Proc. Natl. Acad. Sci. USA* 101:10042–47
94. van der Horst A, Tertoolen LG, de Vries-Smits LM, Frye RA, Medema RH, et al. 2004. *J. Biol. Chem.* 279:28873–79
95. Pagans S, Pedal A, North BJ, Kaehlcke K, Marshall BL, et al. 2005. *PLoS Biol.* 3:210–20

96. Fulco M, Schiltz RL, Iezzi S, King MT, Zhao P, et al. 2003. *Mol. Cell* 12:51–62
97. Ueda K, Hayaishi O. 1985. *Annu. Rev. Biochem.* 54:73–100
98. Corda D, Di Girolamo M. 2003. *EMBO J.* 22:1953–58
99. Smith S. 2001. *Trends Biochem. Sci.* 26:174–79
100. Lee HC. 2001. *Annu. Rev. Pharmacol. Toxicol.* 41:317–45
101. Landry J, Slama JT, Sternglanz R. 2000. *Biochem. Biophys. Res. Commun.* 278:685–90
102. Min JR, Landry J, Sternglanz R, Xu RM. 2001. *Cell* 105:269–79
103. Sauve AA, Schramm VL. 2003. *Biochemistry* 42:9249–56
104. Jackson MD, Schmidt MT, Oppenheimer NJ, Denu JM. 2002. *J. Biol. Chem.* 278:50985–98
105. Sauve AA, Schramm VL. 2004. *Curr. Med. Chem.* 11:807–26
106. Kaplan N. 1960. In *The Enzymes*, ed. P Boyer, 3:105–64. New York: Academic. 2nd ed.
107. Tanner KG, Landry J, Sternglanz R, Denu JM. 2000. *Proc. Natl. Acad. Sci. USA* 97:14178–82
108. Tanny JC, Moazed D. 2001. *Proc. Natl. Acad. Sci. USA* 98:415–20
109. Sauve AA, Celic I, Avalos J, Deng H, Boeke JD, Schramm VL. 2001. *Biochemistry* 40:15456–63
110. Jackson MD, Denu JM. 2002. *J. Biol. Chem.* 277:18535–44
111. Carpenter FH. 1960. *J. Am. Chem. Soc.* 82:1111–22
112. Sauve AA, Deng H, Angelletti RH, Schramm VL. 2000. *J. Am. Chem. Soc.* 122:7855–59
113. Avalos JL, Celic I, Muhammad S, Cosgrove MS, Boeke JD, Wolberger C. 2002. *Mol. Cell* 10:523–35
114. Scheuring J, Schramm VL. 1998. *Biochemistry* 40:15456–63
115. Borra MT, Langer MR, Slama JT, Denu JM. 2004. *Biochemistry* 43:9877–87
116. Schmidt MT, Smith BC, Jackson MD, Denu JM. 2004. *J. Biol. Chem.* 279:40122–29
117. Zhang J. 2003. *BioEssays* 8:808–14
118. Avalos JL, Boeke JD, Wolberger C. 2004. *Mol. Cell* 13:639–48
119. Avalos JL, Bever KM, Wolberger C. 2005. *Mol. Cell* 17:855–68
120. Finnin MS, Donigian JR, Pavletich NP. 2001. *Nat. Struct. Biol.* 8:621–25
121. Zhao KH, Chai XM, Marmorstein R. 2003. *Structure* 11:1403–11
122. Zhao KH, Chai XM, Clements A, Marmorstein R. 2003. *Nat. Struct. Biol.* 10:864–71
123. Zhao KH, Harshaw R, Chai XM, Marmorstein R. 2004. *Proc. Natl. Acad. Sci. USA* 101:8563–68
124. Rossmann MG, Argos P. 1978. *Mol. Cell. Biochem.* 21:161–82
125. Bell CE, Yeates TO, Eisenberg D. 1997. *Protein Sci.* 6:2084–96
126. Olsen RA, Liu L, Ghaderi N, Johns A, Hatcher ME, Mueller LJ. 2003. *J. Am. Chem. Soc.* 125:10125–32
127. Frémont L. 2000. *Life Sci.* 66:663–73
128. Grozinger CM, Chao ED, Blackwell HE, Moazed D, Schreiber SL. 2001. *J. Biol. Chem.* 276:38837–43
129. Bedalov A, Gatbonton T, Irvine WP, Gottschling DE, Simon JA. 2001. *Proc. Natl. Acad. Sci. USA* 98:15113–18
130. Hirao M, Posokony J, Nelson M, Hruby H, Jung M, et al. 2003. *J. Biol. Chem.* 278:52773–82
131. Posokony J, Hirao M, Stevens S, Simon JA, Bedalov A. 2004. *J. Med. Chem.* 47:2635–44
132. Posakony J, Hirao M, Bedalov A. 2004. *Comb. Chem. High Throughput Screen.* 7:661–68

Dynamic Filaments of the Bacterial Cytoskeleton

Katharine A. Michie and Jan Löwe

Medical Research Council Laboratory of Molecular Biology, Cambridge CB2 2QH, UK; email: kmichie@mrc-lmb.cam.ac.uk, jyl@mrc-lmb.cam.ac.uk

Key Words

FtsZ, MreB, ParM, tubulin, actin

Abstract

Bacterial cells contain a variety of structural filamentous proteins necessary for the spatial regulation of cell shape, cell division, and chromosome segregation, analogous to the eukaryotic cytoskeletal proteins. The molecular mechanisms by which these proteins function are beginning to be revealed, and these proteins show numerous three-dimensional structural features and biochemical properties similar to those of eukaryotic actin and tubulin, revealing their evolutionary relationship. Recent technological advances have illuminated links between cell division and chromosome segregation, suggesting a higher complexity and organization of the bacterial cell than was previously thought.

Contents

INTRODUCTION	468
THE TUBULIN HOMOLOGUES	468
FtsZ Protein	468
BtubA/B	475
THE ACTIN HOMOLOGUES	476
Actin	476
ParM	477
MreB	479
FtsA	482
INTERMEDIATE FILAMENT HOMOLOGUE	483
Crescentin	483
WALKER A CYTOSKELETAL ATPASE—A NEW FAMILY OF CYTOSKELETAL PROTEINS?	483

Cytoskeleton: the proteinaceous filaments that provide intracellular organization

Protofilament: a linear structural precursor assembled from protein that is able to assemble into a larger superstructure

INTRODUCTION

Research over the past two decades has uncovered the existence of a well-developed bacterial cytoskeleton. This revolution in the understanding of bacterial cell structure and dynamics has come about largely through the development of fluorescent labeling techniques for bacterial cells and the availability of complete genome sequences, together with structural and biochemical studies. We now know that proteinaceous filaments encircle and wind helically around the inside of the cell, providing intracellular organization and cytoskeletal functions reminiscent of those in eukaryotic cells. Although the discovery of a well-organized bacterial cytoskeleton caused much excitement, another surprise came when sequence analysis and/or the three-dimensional structural determination of many of the proteins involved revealed their homology to eukaryotic cytoskeletal proteins actin, tubulin, and those comprising intermediate filaments. The eukaryotic proteins perform a myriad of important functions, including establishing cell shape, providing mechanical strength, contributing to cell locomotion, assisting in the intracellular transport of organelles, as well as bringing about chromosome separation during mitosis and meiosis. It is becoming evident that the bacterial homologues have analogous or overlapping functions.

This review presents the current understanding of the most well-characterized bacterial cytoskeletal proteins with a focus on their biochemistry (for reviews on the cell biology of the bacterial cytoskeleton, see References 1–3). In particular, we focus on proteins whose assembly into dynamic filaments is regulated by cycles of nucleotide binding and hydrolysis. An intermediate filament homologue is also discussed; however, intermediate filaments are not dynamic and are not found in all eukaryotic cells—being required mainly for mechanical strength.

THE TUBULIN HOMOLOGUES

Tubulin is an indispensable eukaryotic cytoskeletal protein involved in many mechanical cellular processes (examples include chromosome segregation during mitosis and vesicular transport). Heterodimeric αβ-tubulin forms protofilaments that combine to make dynamic microtubules. Microtubules form "scaffolds" that motor proteins such as dynein and kinesin are able to track along.

FtsZ Protein

The *ftsZ* gene (4) encodes a guanosine triphosphatase (GTPase) (5, 6) that is essential for cell division (7, 8). FtsZ was suggested to be a cytoskeletal protein (9) and was predicted to be a homologue of tubulin on the basis of a short sequence motif of seven amino acids (10–12). It is a highly conserved protein that is found in virtually all eubacteria and archaea (13), with a few exceptions (14–16). FtsZ is also present in some chloroplasts and mitochondria (17, 18).

FtsZ in vivo. FtsZ is the first protein known to localize to the mid-cell position prior to septum invagination during cell division, remaining positioned as a ring structure (the Z ring) at the leading edge of the constricting division septum (9). Time-lapse images of cells expressing FtsZ–green fluorescent protein (GFP) fusions show the Z ring contracting while septum constriction occurs (19). Under some circumstances, FtsZ has been observed to form helical structures in cells (20–25), leading to the suggestion that the mature Z ring is a compressed helix (26).

FtsZ is an abundant protein [with estimates of between 3200 and possibly 15,000 molecules per *Escherichia coli* cell (27, 28) and 5000 for *Bacillus subtilis* (29)]. In vitro assembly of FtsZ and its homology to tubulin (discussed below) has led to the idea that FtsZ assembles into linear polymers called protofilaments. A filament formed by end-to-end association of FtsZ monomers (40 Å in length) would require fewer than 500 monomers to span the circumference of a typical bacterial cell with a 1-μm diameter, but it is likely that the Z ring is more complex in structure than a single protofilament. Instead, the Z ring may contain several of these protofilaments associated (or bundled) together to form the Z-ring superstructure. Consistent with this, fluorescence recovery after photobleaching (FRAP) experiments indicate that an average of 30% of the total cellular FtsZ is assembled into the Z ring, more than enough to comprise a Z ring of several FtsZ filaments (30).

FtsZ displays a dynamic localization in that it exchanges between the Z ring and the cytoplasmic pool on a timescale of seconds (30, 31). Time-lapse movies of FtsZ assemblies in vivo have shown FtsZ rearranging from helical structures to rings and vice versa (21, 22, 26). However, as yet, there is no detailed information regarding the large-scale structure of FtsZ polymers in vivo, and this significantly limits our understanding of FtsZ action.

FtsZ self-assembles in vitro. FtsZ protofilaments have been assembled in vitro in a nucleotide-dependent manner (32, 33). Assembly with either nonhydrolyzable GMPCPP [guanylyl-(α,β)-methylene diphosphate] or with GDP indicates that nucleotide hydrolysis is not required for assembly (34, 35). Filaments of FtsZ have been studied by negative stain electron microscopy (EM), and a wide variety of polymer morphologies have been observed (including tubules, sheets, asters, straight and curved protofilaments, and minirings) (32, 34, 36–40). The wide range of conditions favoring FtsZ polymerization suggests that FtsZ could assemble unassisted into a polymer in vivo; however, the diversity of superstructures observed for FtsZ in vitro indicate that some (or all) of the filament types are probably artifactual. The question is which, if any, are representative of the in vivo situation. FtsZ polymers show highly dynamic and flexible behavior in vitro (41–43), and a recent analysis by atomic force microscopy (42) has shown that the dynamic FtsZ filaments continuously rearrange. Also, end-to-end joining of FtsZ filaments and depolymerization of FtsZ from within the middle of filaments has been observed in vitro (42).

There are a large number of in vitro FtsZ filament morphologies, but at low concentrations, FtsZ assembles into apparently single protofilaments. Coupled with the high abundance of FtsZ in the cell, it is thought that the in vivo form of FtsZ is likely to be composed of linear filaments of FtsZ laterally associated with a defined topology. Our attempt to understand the Z-ring structure in vivo is complicated by the presence of accessory and regulatory proteins. At least eight proteins (FtsA, ZipA, ZapA, EzrA, Noc, SlmA, MinC, and SulA) affect FtsZ assembly either by direct or indirect interactions in vivo. Some of these proteins are restricted to a limited number of organisms (for example ZipA is only present in the γ subdivision of the gram-negative bacteria) and are not well conserved, implying their mechanisms have evolved independently. These accessory proteins may inhibit Z-ring assembly, assist in the correct positioning of the Z ring, or directly effect the dynamics of the Z-ring structure. Very little is known about how these proteins exert their molecular effect on FtsZ [with the exception

FRAP: fluorescence recovery after photobleaching

Lateral interaction: any interaction between protofilaments

of SulA, whose mode of action is to titrate away monomeric FtsZ by binding to one of the polymerization interfaces (44)], and their biochemistry is beyond the scope of this review. (The reader is referred to References 45 and 46 for a review of these proteins.) Interestingly, current structural and sequence analysis of these proteins has failed to reveal homology to any of the eukaryotic tubulin accessory proteins. It is possible that the accessory proteins evolved after the evolutionary split of prokaryotes and eukaryotes: this is consistent with the highly extended evolutionary histories speculated for the large divergence in eukaryotic and bacterial actins and tubulins (47, 48).

Subunit structure. Although FtsZ's primary sequence identity to tubulin is low (10% to 18%), it is now generally accepted that FtsZ is a true prokaryotic homologue of tubulin because the three-dimensional tertiary structure revealed that the two proteins share the same fold (**Figure 1**) and both proteins assemble into remarkably similar protofilaments (49–52). FtsZ comprises two domains (52), reflecting thermal denaturation profiles of FtsZ from *Methanococcus jannaschii* and *E. coli*, which both show a clear two-step unfolding process, whereby the domains first separate from each other and then unfold completely (53, 54). It has been suggested that these two domains were derived from two separate proteins earlier in evolution because the N-terminal domain has a Rossmann fold similar to that of many ATPases, and the C-terminal domain is homologous to the family of chorismate mutase-like proteins (52). The N-terminal domain contains the central helix H7 and is essentially the nucleotide-binding domain. Compared with tubulin, this domain of FtsZ has an extra helix (called H0) that protrudes from the N terminus and shows high variability across FtsZ sequences (49). The C-terminal domain of FtsZ shows much less conservation than the N terminus when compared with tubulin. The C-terminal domain is also considerably shorter. In both proteins, this domain contains some residues impor-

tant for GTP hydrolysis (51), which occurs during assembly into filaments (see below). It is interesting to note that FtsZ also has a conserved hydrophobic pocket similar to the taxol-binding pocket in tubulin at the interface of the N- and C-terminal domains, immediately adjacent to helix H7 and the active site. In tubulin, this binding pocket lies on the inside face of microtubules in a region thought to mediate lateral contacts (55). Filling this pocket with an as yet unknown natural accessory protein may help stabilize the filaments in the "straight," assembly-competent conformation (see below). It is thought that the natural binding partner for this pocket in tubulin might be a microtubule-associated protein (MAP) (56), although this is controversial (57). Regardless of the eukaryotic substrate, FtsZ has a homologous pocket, and whether proteins (such as Zap, ZipA, and FtsA) with functions analogous to MAPs that might stabilize FtsZ assembly bind in a similar fashion has yet to be determined.

Filament structure. Semicontinuous tubulin-like protofilaments of FtsZ have been successfully crystallized (52) (**Figure 1**, right), providing us with insight into how FtsZ may assemble into protofilaments. Not unexpectedly, the data suggest that FtsZ assembles in an orientation very similar to that observed for polymerized tubulin, with each FtsZ monomer maintaining head-to-tail interactions. These head-to-tail interactions are referred to as longitudinal contacts and are the basis of protofilament formation. All other interactions are referred to as lateral and function to bring protofilaments together. Lateral interactions may play important roles in Z-ring nucleation, assembly, regulation, and disassembly. Although the regions required for lateral tubulin-tubulin interactions within microtubules are known, the corresponding regions in FtsZ are quite different, with little conservation observed in the relatively short loop regions of FtsZ (51), which is consistent with the notion that FtsZ and tubulin do not

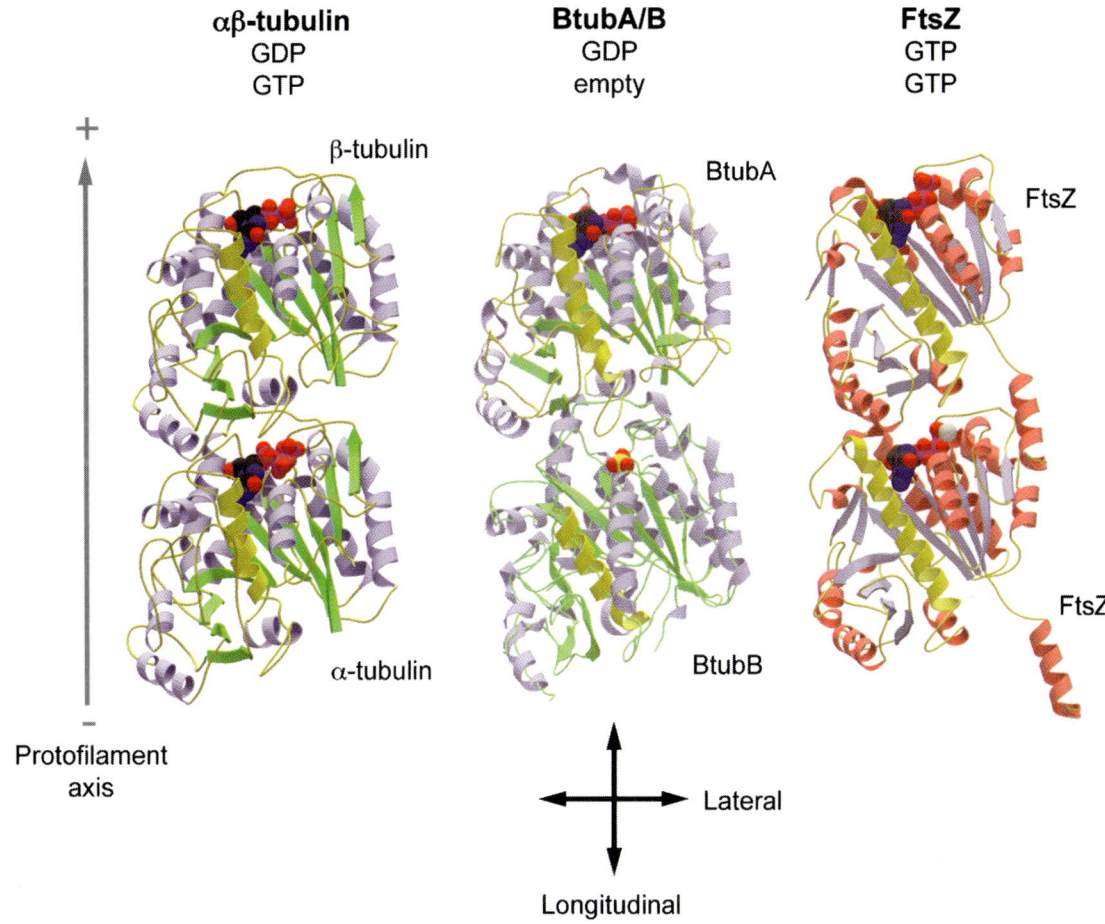

Figure 1
Structures of the α/β-tubulin heterodimer (*left*), the BtubA/BtubB heterodimer (*center*) and FtsZ dimer (*right*), showing the position of the nucleotide at the dimer interface, the conservation of fold, and the axis of protofilament extension (*up the page*). Lateral interactions between protofilaments could be formed at all or any of the interfaces perpendicular to the longitudinal axis of protofilament assembly. (*left*) The α/β-tubulin heterodimer observed in tubulin zinc sheets [Protein Data Bank (PDB) entry 1JFF] (172). (*center*) BtubA/BtubB heterodimer from *Prosthecobacter dejongeii* (PDB entry 2BTQ) (88). (*right*) FtsZ dimer obtained from nucleotide-free FtsZ from *Methanococcus jannaschii* soaked in MgGTP (PDB entry 1W5A) (52).

share similar lateral interactions or accessory proteins.

Active site. The GTPase active site is formed by the association of two FtsZ monomers, with the catalytic T7 loop [or synergy loop (58)] in the C-terminal domain of one monomer inserting into the nucleotide-binding pocket of the N-terminal domain of the adjacent molecule (**Figure 1**, right), thereby leading to association-dependent activation of the GTPase activity (51, 52, 59). Catalysis occurs by the polarization of a water molecule hydrogen bonded to two conserved aspartate side chains (*M. jannaschii* residues 235 and 238) within the T7 loop, promoting nucleophilic attack on the γ-phosphate and thus hydrolysis of GTP. A further

Longitudinal interaction: an interaction responsible for the formation of protofilaments

Dynamic instability: the switching of biological protein polymers between phases of steady elongation and rapid shortening

Isodesmic assembly: all intersubunit contacts are equivalent

contribution to the polarization of the γ-phosphate is provided by a magnesium ion coordinated by glutamine (*M. jannaschii* residue 75), several water molecules, and the α,β phosphates. Thus, GTP hydrolysis requires Mg^{2+}, consistent with in vitro observation (5, 6, 12).

What can we learn about FtsZ from tubulin? Although FtsZ is similar to tubulin in structure, there are some important differences between the two proteins. Microtubules are comprised of α- and β-tubulin, which form a tight α/β heterodimer in solution. The α/β heterodimer has a nonhydrolyzed, nonexchanging GTP bound at the dimer interface. In contrast, most bacteria have only one isoform of FtsZ (60–62), and thus each subunit interface is equivalent.

Tubulin assembly characteristics. Microtubules are assembled from α/β heterodimers joined end-to-end so that the α and β isoforms of tubulin alternate, with a second, hydrolyzable GTP molecule between each heterodimer subunit. After assembly, the nucleotide within the subunit cannot exchange, nor can subunits of tubulin within the filament exchange with the cytoplasmic supply. In addition to the longitudinal interactions between α/β heterodimers, extensive lateral associations between protofilaments further stabilize microtubule assembly, and a complete microtubule is comprised of 13 parallel protofilaments. Tubulin protofilaments have a distinct polarity because of the head-to-tail association of tubulin subunits, and owing to the alternating α and β forms of tubulin, β-tubulin is always present at one end (designated the plus end or fast-growing end), and α-tubulin is present at the other end (designated the minus end).

Microtubules exhibit dynamic instability, enabling them to disassemble rapidly in vivo. GTP hydrolysis drives the protofilaments toward a bent or curved polymeric state that is incompatible with the geometry of the microtubule wall. Kinetic stability of the microtubule is maintained by a GTP-bound cap that restrains and stabilizes the polymer in the straight conformation. If this cap is hydrolyzed, the tubulin filaments can adopt the curved or bent morphology, resulting in spontaneous disassembly of the filament. Thus, the state of the GTP cap determines how microtubules switch between states of rapid growth and rapid shrinkage (63).

Isodesmic versus cooperativity assembly. Both actin and tubulin assemble via cooperative mechanisms in which nucleation is a rate limiting step (**Figure 2**). As a consequence, polymer assembly can be controlled by providing specific nucleation sites at a defined time and place in the cell. For an isodesmic assembly mechanism (**Figure 2**, top), polymer will rapidly assemble and disassemble at all places in the cell, and the cell must provide stabilization factors (to assemble the polymer into a defined structure) and topological information (to position it correctly). Research

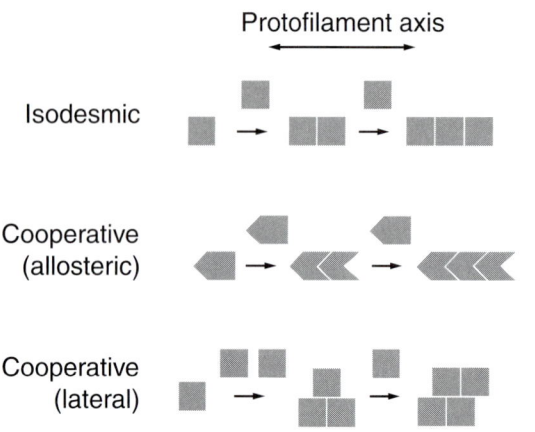

Figure 2

Schematic description of isodesmic and cooperative assembly mechanisms. The isodesmic model (*top*) assumes all subunit additions are equivalent, and thus the likelihood of assembly is directly proportional to the concentration of protein. Cooperative assembly (subdivided into allosteric and lateral) occurs when subunit additions are not equivalent. In the case of allosteric cooperative assembly (*center*), the nucleation of a dimer results in a conformational change within the dimer, which increases the affinity for the next subunit to bind. In the case of lateral cooperative assembly (*bottom*), assembly of more than two molecules is stabilized by a third molecule at a different interface; in the case of tubulin and FtsZ, this is known as a lateral protofilament interaction.

into FtsZ assembly has so far failed to establish conclusively the kinetics and mechanism of FtsZ assembly. Significant factors limiting our understanding are the wide variation in FtsZ behavior under different experimental conditions and the consequent uncertainty about which of the observed behaviors may be relevant in vivo.

Isodesmic assembly occurs when linear multimers form in which each bond has an identical contact, and nucleation is just as favored as filament extension (**Figure 2**, top). Cooperative assembly occurs when multimers of protein only become stable after forming an unfavored but defined nucleus, wherein the bonds between the subunits are relatively weak and initiation is difficult. Once the unfavored nucleation step occurs, the filament can extend rapidly because subunits in the larger structure are stabilized by multiple bonds formed with other adjoining subunits (**Figure 2**, middle and bottom). There are three characteristics typical of cooperative assembly: a critical concentration for assembly, a lag in the assembly kinetics at low protein concentration, and a distribution of subunits at equilibrium into two distinct populations of monomers and very long polymers.

A cooperative assembly model for FtsZ protofilaments is now generally accepted for three reasons. First, both FtsZ assembly and GTPase activity have a critical concentration (64, 65). Second, FtsZ assembly shows an ~1 s lag of assembly even at very high protein concentrations, suggesting an "activation" step. This might be the time required to release GDP and bind GTP (66). Third, a size limit and relatively homogeneous protofilament length for FtsZ have been reported (67).

Results from analytical ultracentrifugation and scanning transmission electron microscopy suggest that the FtsZ protofilament is formed by a single chain of FtsZ monomers associated head to tail (68, 69). However, if the FtsZ molecules are rigid and if there is no communication between the two binding faces on a single monomer, then linear protofilament formation cannot be cooperative.

There are several models that attempt to reconcile the observed association into single protofilaments with the data showing cooperative assembly. It is possible that, initially, isodesmic assembly into single protofilaments occurs, followed by cooperative association of protofilaments (such that the cooperative kinetics overwhelm the effects of the isodesmic assembly), giving rise to an apparent overall cooperative assembly. However, this model is unlikely to be true because cooperative assembly is still observed at low concentrations when no bundling or grouping of protofilaments is detected (64). Mingorance et al. (42) have suggested that the isodesmic assembly of protofilaments is followed by a stabilizing cyclization event that they argue would show overall cooperative kinetics. They support this hypothesis with images of cyclic single protofilaments formed in vitro; however, the biological relevance of the observed small ring protofilaments is unclear.

Finally, if we abandon the assumption that FtsZ is a rigid molecule and suggest that binding nucleotide or another subunit could induce a conformational change, then it is also possible that FtsZ exhibits cooperativity within a single protofilament. In this case, the initial dimerization process would cause a conformational change that increases the affinity of the next molecule to bind. Such a model could explain the observed cooperative kinetics reported for single protofilaments of FtsZ.

Nucleotide exchange. The structural data for the FtsZ filament strongly suggests that there are major differences in the solvent accessibility of the nucleotide pocket of FtsZ compared with tubulin; however, this data was obtained from crystals in which the nucleotide was soaked in and may not represent the bona fide nucleotide-bound state of the protofilament (52). Within tubulin protofilaments, the nucleotide-binding pockets are occluded, and nucleotide exchange is prohibited.

This characteristic is essential for the dynamic instability of tubulin polymers and provides the stabilizing GTP cap.

By contrast, the FtsZ nucleotide-binding pocket is partially exposed, potentially allowing nucleotide exchange, which could make nucleotide hydrolysis the rate-limiting step in filament disassembly (52, 70). This potential for nucleotide exchange may have important implications for FtsZ assembly and dynamics. If significant nucleotide exchange can occur, FtsZ filaments would not experience dynamic instability because the high ratio of GTP to GDP in the cell (71) would ensure that every molecule in the filament is in the GTP-bound state.

Data suggesting that FtsZ is unable to exchange nucleotides and has a GTP cap have been reported (41, 72), consistent with the tubulin paradigm of dynamic instability. However, other strong data suggest that nucleotide exchange does occur, thus excluding the possibility of dynamic instability (70, 73). Some attempts to determine the nucleotide state within filaments have suggested that the majority of FtsZ in protofilaments is bound to GTP (70, 73), although other data suggest the contrary (74). The conflicting data most probably arise from the different conditions used in the various assays, such as Ca^{2+} concentrations (discussed below) and bundling states of the filaments that may sterically inhibit nucleotide exchange. In particular, the initial form of the soluble protein may be entirely monomeric in some investigations but include dimers and larger oligomers in other cases. Some mutants with reduced GTPase activity are able to support cell division, although with much slower dynamics (30, 75, 76), suggesting that either GTPase activity is modulated in vivo, or GTPase activity is much higher than required to support division. The observed rates of FtsZ assembly appear to differ greatly between different organisms (77), but these differences may also arise from varying experimental conditions.

The presence of Ca^{2+} reduces FtsZ's GTPase activity in vitro (40, 43, 78) and increases bundling of the protofilaments (40, 43, 78, 79). Ca^{2+} has also been reported to reduce the exchange of nucleotide (73). This effect on bundling observed with Ca^{2+} may simply be an indirect effect of a decrease in GTPase activity. The decreased GTPase activity would lead to an increased stability of filaments. Long filaments should be present for a longer period of time, and this should result in conditions promoting lateral association. It is also possible that bundling may sterically inhibit nucleotide hydrolysis. Alternatively, increased bundling may be due to a more specific mechanism involving a currently unknown Ca^{2+}-binding site (40, 80).

Conformational change. It is thought that nucleotide hydrolysis brings about the disassembly of microtubules by a mechanism linked to a conformational change (81–83). In this hypothesis, the GDP-bound form of tubulin adopts a bent or curved form [observed experimentally (81)] that destabilizes lateral bonds in the microtubule, resulting in the peeling back of curved filaments from the end of a microtubule and the eventual disassembly of the filaments by heterodimer dissociation. Bending occurs at all interfaces in the tubulin protofilament (83), even between the two subunits of each heterodimer where the GTP is never hydrolyzed; thus bending need not be absolutely linked to nucleotide hydrolysis (see below).

Early work observed that FtsZ protofilament disassembly was concomitant with rapid nucleotide hydrolysis (33). FtsZ in the GTP-bound state predominantly forms straight protofilaments (32, 36), whereas GDP-bound FtsZ forms curved protofilaments (37, 65, 69, 84, 85). This observation has led to the suggestion that nucleotide hydrolysis causes a conformational change in the FtsZ filament that may be involved in converting the chemical energy of nucleotide hydrolysis into mechanical energy for constricting the Z ring (35). It has been predicted, on the basis of in silico modeling, that the T3 loop in the nucleotide-binding site will undergo a large conformational change upon the GDP-to-GTP transition (86); however, no large conformational changes have been observed in the various cocrystal structures of FtsZ with GDP or GTP (52). It is important to note that FtsZ filaments formed in vitro do not always show consistent morphologies with the type of bound nucleotide (42). It has been suggested that the hydrolysis of GTP to GDP simply produces a repulsive electrostatic effect with the loss of the γ phosphate and that this chemical repulsion is the mechanism behind disassembly (52, 68).

FtsZ function and disassembly in vivo may be regulated by the combination of a conformational change and an electrostatic effect. However, the oversimplified "spring" model in analogy to the GDP-depolymerization of tubulin is unlikely to reflect the real mechanism of FtsZ function in vivo. The filament energy formed by large superstructures (where the potential binding surfaces of the subunits are large) is very significant and can overcome almost all other molecular effects. For example, although FtsZ may prefer to adopt a bent conformation when bound to GDP in a free protofilament, the binding energy that accompanies favorable lateral association of protofilaments in a certain filamentous form of FtsZ might be large enough to restrain the FtsZ protofilament so that it is straight. Again, only elucidation of the in vivo superstructure of FtsZ will allow us to determine which of the behaviors observed in vitro is biologically relevant.

BtubA/B

Two tubulin homologues (BtubA and BtubB) have been identified recently in the *Prosthecobacter* bacterial genus, and both show a closer relationship to eukaryotic tubulin than to FtsZ (87). BtubA is 31% to 35% identical and BtubB is 34% to 37% identical to α- and β-tubulin, respectively but only 8% to 11% identical to FtsZ. These proteins do not exist in most bacterial species, and their low divergence from eukaryotic tubulin suggests that they are a product of a distant horizontal gene transfer (88). The cellular function of these two proteins is unknown; however, they assemble in vitro, and it has been speculated that BtubA/BtubB may contribute to the elongated spindle shape of *Prosthecobacter* (89).

The structure of BtubA bound to GTP closely resembles the structure of tubulin (**Figure 1**, middle), including the long loops responsible for lateral interactions in microtubules and the large helix-loop-helix domain (often referred to as the third C-terminal domain) that forms the outer surface of microtubules (88). The third C-terminal domain is absent from FtsZ (49). As in both tubulin and FtsZ, the N-terminal domain (which provides loops T1–T6 for nucleotide binding) is separated from the second domain by the central helix T7. Similarly, the second domain provides the T7 loop (which deviates substantially from both tubulin and FtsZ) that activates nucleotide hydrolysis in the protofilament when inserted into the active site of the adjacent subunit.

BtubA and BtubB form a weak heterodimer in vitro (88). A point of interest is that both BtubA and BtubB are able to refold in vitro without the help of chaperones, which is similar to some FtsZs and unlike normal tubulins (88). The crystal structure of BtubA/BtubB shows a tubulin-like heterodimer, with both intra- and interdimer bends evident. It is not possible to describe BtubA or BtubB as being analogous to either α- or β-tubulin because both BtubA and BtubB have mixed characteristics of the two tubulin forms (88).

In vitro self-assembly. BtubA does not self-assemble in vitro (regardless of nucleotide presence), whereas a His-tagged BtubB (with any guanine nucleotide supplied) assembles into rings that appear to be one subunit thick (89). Interestingly, mixtures of BtubB and BtubA assemble into bundled linear protofilaments. Sontag et al. (89) observed bundles of BtubA/BtubB comprising 4–7 protofilaments that were similar to double protofilaments as well as bundles of straight double protofilaments that twist, as shown by Schlieper et al. (88). It was not possible to determine the parallel/antiparallel arrangement of BtubA and BtubB protofilaments within these structures; however, the monomer repeat is 41.6 Å, a value that lies between the monomer spacings for FtsZ and tubulin, suggesting the filaments are similar (88). Filament bundles with hollow tubular profiles 40 nm in diameter (thicker than the 25 nm of microtubules) have also been observed (89). Analysis of BtubA/BtubB protofilaments indicates an equimolar ratio of each protein within the filaments, further supporting the idea that these proteins assemble in a fashion similar to αβ-tubulin whereby the BtubA/BtubB subunits alternate within the polymer (88, 89).

Little is known about the kinetics of BtubA/BtubB assembly, but critical concentrations for both GTPase activity and assembly have been observed, suggesting a cooperative assembly mechanism (89). Protofilament assembly is reversible; the filaments assemble relatively quickly and, over time, disassemble because of GTP consumption (88).

The role of nucleotide hydrolysis. BtubA and BtubB are able to bind one molecule of guanine nucleotide each (89); however, the rates of GTPase activity differ significantly between the two proteins (0.40 mol GTP per min per mol for BtubB and 0.13 mol GTP per min per mol for BtubA). When mixed together in equimolar amounts, the GTPase activity of the combined proteins is higher than that of either protein alone, suggesting the direct interaction of BtubA and BtubB (89).

THE ACTIN HOMOLOGUES

Besides tubulin, actin is the other essential and ubiquitous eukaryotic cytoskeletal protein. Actin forms double-helical thin filaments composed of two strands. Actin filaments form the "tracks" that myosin (a motor protein) is able to move along.

Actin

Actin is the prototypical member of a superfamily of ATPases that includes hexokinase and Hsp70. The actin family is very diverse in sequence and in function, showing a conserved fold related to ATPase activity. In a landmark paper in 1992, it was reported that the actin family also includes three bacterial proteins: ParM (StbA), MreB (and relatives), and FtsA. These proteins were identified as actin homologues (showing higher similarity to actin and Hsp70 than to the hexokinases) because they contain five conserved sequence motifs related to nucleotide binding and hydrolysis (90).

The actin fold is comprised of two large domains (named I and II). These two domains can be divided into two subdomains (A and B), and the larger subdomains (designated IA and IIA) share a common fold consisting of a five-strand ß-sheet surrounded by three α-helices. The two smaller subdomains (IB and IIB) show a wider variability in size and structure across the actin family, bestowing some of the properties unique to each protein. The two major domains of actin can rotate with respect to one another, and between the two domains lies a highly conserved ATP-binding pocket. Proteins of the actin family bind ATP, normally in association with Mg^{2+} or Ca^{2+}, and coordinating Asp residues are important for nucleotide hydrolysis. Unlike Hsp70 and hexokinase, actin assembles in vivo into a dynamic helical polymer (called F-actin) with cooperative assembly characteristics. Within the filaments, actin assembles in a head-to-tail arrangement; thus, similar to tubulin, it has a distinct asymmetry. Actin shows structural changes upon polymerization (91) and exhibits the characteristic termed treadmilling. Treadmilling occurs when the two ends of a filament have different affinities for polymerization. New subunits assemble at the preferred end, and after nucleotide hydrolysis and phosphate release, subunits dissociate from the nonpreferred end thus leading to a flux of subunits though the filament. When the rates of assembly and disassembly are equivalent, the filament maintains a constant length. Along with the rate of treadmilling, filament growth and shrinkage are controlled by the rates of monomer addition and dissociation. Actin dynamics are regulated by a range of accessory proteins that affect the assembly, disassembly, and rearrangement of actin filaments in vivo.

ParM

ParM (previously StbA) is one of three components required for the correct partitioning of R1 low-copy number plasmids in *E. coli*. The two other components of the *par* system are *parC* and ParR. *parC* is a centromere-like sequence of DNA that contains the R1 *par* promoter sequence. ParR is a repressor protein that binds to the *parC* locus. ParM interacts with the ParR-*parC* complex (92) and apparently functions as a primitive mitosis-like spindle to move newly replicated plasmids to opposite poles of the cell (93).

The crystal structure of ParM confirmed the original sequence-based assignment of an actin fold (94), although ParM does show significant differences in loop, helix, and sheet arrangement within domains IB, IA, and IIB (**Figure 3**). The subdomain IB lacks a helix present in both MreB (see below) and actin. This subdomain also shows a longer loop (quite different from the equivalent Dnase I–binding loop of actin) as well as an unusual insertion of a strand from domain IA that is not observed for any other member of the actin family. Other differences include the replacement of a β-sheet with a helix-loop-helix

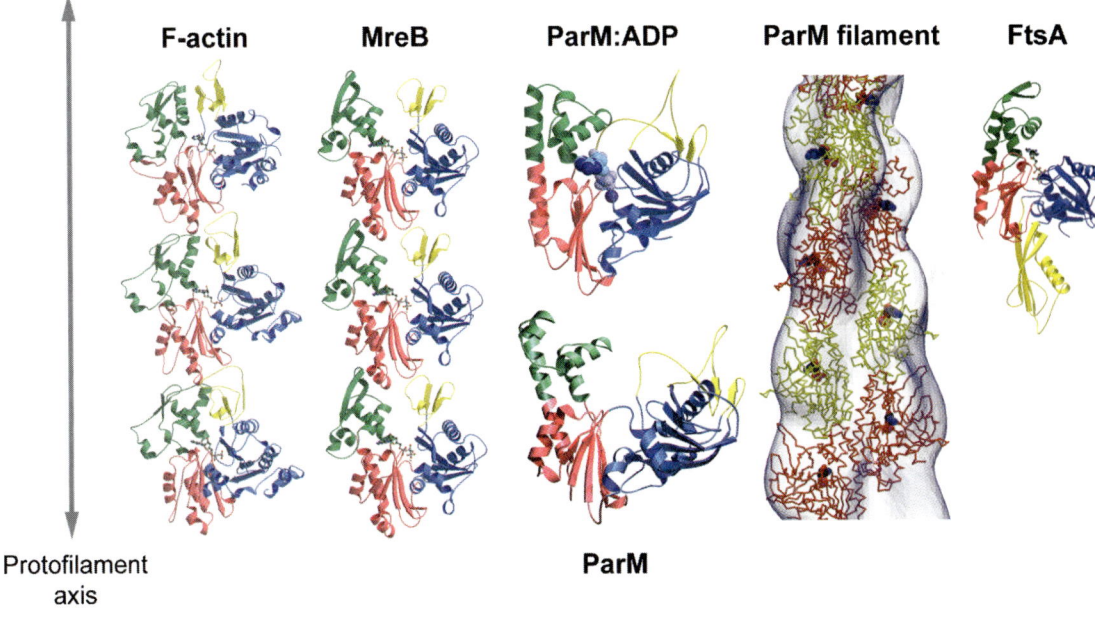

Figure 3

(*left*) Structures of F-actin filaments (PDB entry 1YAG) (91), (*second from the left*) MreB filaments from *Thermotoga maritima* (PDB entry 1JCE) (112), (*center*) ParM in the "open" and "closed" conformations, (*second from the right*) ParM filament (94), and (*right*) FtsA, showing the position of the nucleotide within the interdomain cleft, the conservation of fold, and the axis of the protofilament extension (*arrow*). (*center, top*) Note the closed conformation (PDB entry 1MWM) and (*bottom*) the open conformation of the apo and ADP-bound forms of ParM, both from *E. coli* plasmid R1 (94). The conformational change is predicted for all actin homologues. (*right*) AMPPNP-bound FtsA from *T. maritima* (PDB entry 1E4G) (136).

motif at the top of subunit IIB and the absence of a helix in subdomain IIA. These structural differences occur in the equivalent regions of actin that are involved in protofilament contacts. The three-dimensional structures of ADP-bound and apo forms of ParM revealed a ∼25° conformational difference between domains I and II that closes the interdomain cleft (**Figure 3**). This conformational change is facilitated by the connecting helix H5 that acts as a mechanical hinge. Like actin, ParM is able to assemble into filaments, and the change in conformation is thought to be linked to the assembly characteristics of ParM filaments.

ParM filaments. In vitro, ParM self-assembles in the presence of ATP, ATPγS, AMPPNP, or ADP (93, 95), forming long double-helical filaments that gently twist with a crossover of 300 Å (shorter than the variable actin crossover that averages 360 Å). The helical nature of these filaments is thought to increase the overall strength of the filament and reduce the propensity to form stable lateral interactions, making these filaments less likely to bundle. Modeling ParM filaments using the crystal structure and three-dimensional reconstruction of ParM filaments from negative stain EM, coupled with our understanding of MreB and actin filaments, suggests that ParM assembles in a head-to-tail orientation (**Figure 3**) (94).

Kinetic analysis of ParM assembly in vitro also shows clear differences from actin. ParM filaments are believed to nucleate via a nucleation-condensation mechanism involving three monomers, as expected for a two-stranded helical polymer. ParM shows a 300-fold faster rate of nucleation to that of F-actin (95). Actin has a significant kinetic challenge to spontaneously nucleate, requiring specific nucleation factors that assist filament assembly in vivo. No such factors are apparently required for ParM. Although the in vitro rate of assembly ($5.3 \pm 1.3 \mu M^{-1} s^{-1}$) of ParM filaments is similar to actin, ParM shows some interesting characteristics. ParM filaments display symmetrical, bidirectional polymerization, but actin assembles unidirectionally. The rate of assembly of ParM filaments in both directions is equal, whereas the disassembly of ParM filaments is unidirectional and catastrophic (shortening results in rapid, complete filament disassembly with a disassembly rate of 64 ± 20 s^{-1}). This dynamic instability is also a feature of tubulin assembly but not actin. Interestingly, in vitro filaments of ParM assemble to a fairly constant length of 1.5 μm (95).

In vitro ParM shows cooperative ATPase activity (92). Disassembly of ParM filaments requires nucleotide hydrolysis (ADP-ParM filaments are extremely unstable with a critical concentration of ∼100 μM compared to ∼2 μM for ATP-ParM), and as a consequence of this, mutants deficient in ATPase activity show hyperstability, both in vitro and in vivo. The dynamic instability of ParM filaments is an integral part of ParM function, and ATPase mutants form stable filaments in vivo that are not dynamic and do not support plasmid partitioning (92, 93). It is assumed that ParM filament disassembly occurs via a mechanism similar to that postulated for F-actin: The energy released by nucleotide hydrolysis might invoke a large conformational change, which in turn weakens the intramolecular bonds between subunits, promoting filament disassembly. The conformational change seen in the two crystal structures of ParM containing ADP but without a nucleotide might represent such a mechanism (**Figure 3**) (94).

Finally, the spontaneous disassembly of ADP-ParM filaments differs from ADP-actin filaments, which requires severing factors, such as cofilin, to promote disassembly. Monomeric ADP-ParM subunits dissociate from filaments at a rate ∼100 times faster than ADP-actin, and actin requires the nucleotide exchange factor profilin to achieve the same rate of ADP dissociation from monomers as ParM achieves alone. Recent work by Garner et al. (95) indicates that like actin, ParM filaments are stabilized by a cap of ATP-bound monomers.

In vivo filaments—how does ParM really work? Immunofluorescence microscopy revealed that ParM assembles into dynamic pole-to-pole axial filaments that are essential for plasmid partitioning (93). The intracellular expression of ParM produces ~15,000–18,000 molecules per cell (93). When assembled into filaments, this should be enough ParM to form a filament 15–20 times the cell length. So, it is certainly possible that the ParM filament is comprised of several parallel filaments.

Most *par* systems of plasmid partitioning only comprise three components. ParM is unable to assemble without the other components ParR and *parC* (93), and plasmid replication is required for ParM filament formation (96). The binding of ParM to ParR-*parC* is ATP dependent, and this stimulates ParM's ATPase activity. Møller-Jensen et al. (96) have observed plasmids attached to each end of the ParM filament. As the ParM filament extends in length, so does the distance between the plasmids. These data imply that ParM may provide the mechanical force required to push the plasmids to the poles of the cell, analogous to the mitotic apparatus in eukaryotic cells. It was proposed that ParR-*parC* complex functions as a nucleation point for ParM polymerization (96).

Garner et al. (95) propose that, at cellular concentrations of ParM, spontaneous nucleation and filament elongation would occur throughout the cell. Thus, rather than nucleation being a regulatory point in the mechanism, only filaments that manage to locate and bind to plasmid DNA (the ParR-*parC* interaction is known to stabilize ParM filaments below the steady-state critical concentration) would be protected from dynamic instability. Only when both ends are stabilized by binding to the ParR/plasmid complex, would segregation occur. It has been proposed that bidirectional elongation of ParM filaments stabilized by the interaction with the ParR-*parC* complex drives plasmid segregation.

Questions still remain unanswered. How is the ParM filament extended in vivo if the plasmids are bound to each end? It has been suggested that the DNA might be propelled by a treadmilling mechanism of ParM filaments, and the addition of new subunits to the ends of filaments may move the plasmid forward. It has also been speculated that in an unidentified mechanism ParM filaments might serve as a track on which motor proteins carry the plasmids to their destination. No potential motor proteins have been identified as yet. Another interesting issue is raised by the attachment of plasmids to each end of the ParM filament. As described above, ParM assembles into filaments with an inherent polarity. A mechanism whereby asymmetrical depolymerization occurs from a symmetrically assembled filament is difficult to reconcile. Either the two ends of the filament are chemically different, and the plasmid attachment sites are nonidentical, or the in vivo filament of ParM consists of filaments aligned in an antiparallel manner thus forming equivalent ends.

Finally, an axial filament that crosses the mid-cell site must be disassembled prior to cell division or somehow severed. In cells over expressing ParM with a deficient ATPase, hyperstable filaments form, and cell division is blocked. How is disassembly regulated? It is possible that once the plasmids meet the poles of the cell, they attach to the membrane or some kind of target, releasing the ParR-*parC* interaction and causing ParM filament to become unstable and disassemble.

MreB

MreB is encoded in a cluster of genes involved in determining cell shape formation although its precise role(s) are still unknown (97–100). Some bacteria contain several related MreB-like genes. For example *B. subtilis* encodes MreB, Mbl, and MreBH, with each one showing a similar degree of sequence identity (101, 102). In *B. subtilis*, MreB appears to be required for the control of cell diameter (102, 103), and Formstone & Errington (98) suggest that MreB specifically functions to restrain cell diameter. Mbl in *B. subtilis* is specifically involved in cell elongation and is required for the helical insertion of the peptidoglycan necessary for growth of some rod-shaped cells (104). In *Caulobacter crescentus*, MreB depletion causes abnormal lemon-shaped cell morphology (instead of the normal rod-shaped cell) with defects in cell wall integrity (105). Analysis of the conservation of MreB is complicated because the phylogenic assignment of these proteins is difficult owing to their similarities. It is interesting to note that those organisms that do not have MreB often have Mbl homologues, which may eventually turn out to be MreB (or vice versa).

Because MreB shows homology with actin and ParM, it was expected that MreB would form some kind of filament capable of mechanical work. On the basis of such an idea, hypothesis-driven research has implicated MreB in chromosome segregation for several organisms, including *B. subtilis* (101, 103, 106), *E. coli* (107), and *C. crescentus* (108), and data from the latter organism are the most convincing. MreB's involvement in chromosome segregation has been investigated with the aid of a small-molecule inhibitor called A22 (whose specific target is MreB) (108, 109). Treatment of *C. crescentus* cells with A22 causes a specific, rapid, and reversible disruption of MreB function. Studies involving the administration of A22 at specific times in the cell cycle revealed that MreB played an important role in the segregation of the origin-proximal loci of the *C. crescentus* chromosome. It appears that segregation requires at least two separate mechanisms, the first being an MreB-dependent separation, whereby a region near the origin is initially segregated, followed by a second mechanism that is independent of MreB, whereby the rest of the chromosome follows the origin (108, 110). In addition to a role in cell shape determination and chromosome segregation, MreB is also thought to mediate cell polarity in *C. crescentus* (111). Sequence analysis of MreB initially revealed similarities to FtsA (97), and MreB was predicted to have an ATPase fold similar to actin and hsp70 in the paper by Bork et al. (90).

MreB structure. The MreB crystal structure shows that MreB has a conserved actin fold comprising the two domains (I and II) with a nucleotide-binding site in the interdomain cleft between them (**Figure 3**) (112). Typical of other members of the actin family, the smaller domains, IB and IIB, show more diversity when compared to those of actin. These smaller domains, however, show the same topology as those of actin, suggesting a closer relationship of MreB to actin than Hsp70, FtsA, or hexokinase because the topology of these domains differs. Significant differences between MreB and actin are evident within the helix H8 loop. In actin, this region contains specific sequence insertions required for subunit-to-subunit interactions. These sequences are absent in MreB (112, 113).

MreB filament. All biochemical studies have been performed on MreB from *Thermotoga maritima* (a hyperthermophilic eubacterium) because MreB from most mesophilic organisms is difficult to handle. An advantage of this is that the data obtained about MreB are directly comparable. MreB assembles in vitro into straight and curved protofilaments in the presence of ATP (112, 114). Ring-like structures and filament bundling have also been reported; however, the biological implications of these structures have yet to be determined. Filamentous bundles of MreB formed in vitro have an increased rigidity, increasing the overall strength of the filaments, which may be very important to their function particularly if they are part of a mechanical apparatus (114).

No high-resolution data of the in vivo MreB filament are currently available, although crystals containing protofilaments of MreB have provided us with an atomic resolution insight into the self-association of MreB monomers (**Figure 3**). MreB assembles into filaments similar to that of F-actin, in that the subunit repeat, structure, and subunit orientation are approximately the same (91, 112). However, F-actin shows axial rotation (or twist) within the filaments (F-actin filaments can be described as two twisted protofilaments), but only a small number of MreB filaments show a slight axial rotation (112).

The critical concentration for assembly of MreB (~3 nM) is much lower than the critical concentration of F-actin (~0.25 μM) (114, 115), implying that MreB has a higher affinity for other MreB monomers and MreB filaments than actin does for other actin monomers and filaments. This also suggests that MreB nucleation is a much more favorable process than actin nucleation, and it has been proposed that MreB polymerization occurs either without a nucleation phase or with an extremely short-lived nucleation step (114). Rapid nucleation is also suggested by the almost instantaneous assembly observed in the presence of ATP even at very low MreB concentration (114). Although actin requires accessory proteins to assist with in vivo assembly, the ease of MreB nucleation suggests that MreB may be more kinetically tuned to have less need for regulatory or accessory proteins to assist in filament assembly. Consistent with this, MreB shows faster polymerization rates than actin.

The role of nucleotide hydrolysis with respect to MreB assembly dynamics is poorly understood. Comparison of the MreB nucleotide-binding site with that of the ATP-bound actin indicates that most of the active-site residues are in the same position, with the exception of some of the residues that bind the γ-phosphate. Free phosphate is released in solutions of self-assembling MreB after ATP addition. A time lag between phosphate release and MreB polymerization is the basis for the suggestion that ATP hydrolysis might occur after MreB monomers incorporate into filaments (114). Currently, there is no data about the stability of MreB filaments in vitro, and the study of the disassembly of MreB filaments with respect to nucleotide hydrolysis should be very interesting.

MreB in vivo. In a variety of organisms, MreBs have been observed assembling into varied structures, i.e., rings in *Rhodobacter sphaeroides* (116), helical structures in *B. subtilis* (102, 117) and *E. coli* (107, 118), as well as bands and helical structures in *C. crescentus* (105). Time-lapse images and FRAP experiments have revealed that MreB and Mbl filaments are dynamic in vivo (101, 111, 117), although the biological significance for this is unknown.

Molecular function. How might MreB proteins function in vivo and what might be their roles? It has been speculated that the helical filaments of MreB-like proteins might provide positional information for the localization of the wall-synthesizing enzymes, the penicillin-binding proteins (PBPs), thereby controlling cell wall morphogenesis and thus cell shape (104, 105, 119). In the 1970s, experimental observations of *B. subtilis* mutants with helical cell morphology led to the suggestion that cell wall elongation occurs by helical peptidoglycan insertion (120). The suggestions that MreB-like proteins may distribute factors that affect the organization and mechanics of the cell wall and that the filament structures themselves may directly contribute to the mechanics of the cell wall (102, 104) provide a possible molecular mechanism for helical peptidoglycan growth. Results from *C. crescentus* revealed MreB localizes in an FtsZ-dependent manner to the mid-cell position during cell division, leading to the proposal that MreB directs the switch from cell wall elongation to septum extension during division (105).

MreB's involvement in chromosome segregation may arise from an indirect function of MreB, or MreB might attach to an (as yet) unknown bacterial centromere (108, 119). Finally, it is possible that the functions of MreB proteins provide physical markers in the cell to which other essential processes are linked, thus giving rise to the apparent involvement in chromosome segregation, polarity determination, and cell shape definition. Recently, it has been shown that MreC and MreD are linked to cell wall synthesis, and it has been proposed that the Mre proteins (MreB, Mbl, MreC and MreD) provide the link between intracellular organization and the extracellular cell wall synthetic machinery (121, 122). It may not be generally appreciated that filaments of MreB, MreC, MreD, and the peptidoglycan-synthesizing PBPs comprise a system vaguely analogous to the machinary used to position cellulose synthase in plants. This plant machinary utilizes cytosolic helical microtubules to provide information about where cellulose (rather than peptidoglycan in the case of MreB) is helically inserted on the outside of the cell (123).

WACA: Walker A cytoskeletal ATPase

FtsA

FtsA was one of the first cell division proteins to be identified (124), and sequence analysis indicated that FtsA might belong to the actin superfamily (90). This caused some excitement because FtsA interacts directly with FtsZ (125–131). In vivo FtsA localizes to the Z ring and also to FtsZ helical structures (23) in an FtsZ-dependent manner (20, 23, 29). Its C terminus (containing a conserved amphipathic helix) has a role in interacting with FtsZ and in interacting with the membrane, suggesting that FtsA tethers the Z ring to the cell membrane (132, 133). Consistent with this, a defined FtsA/FtsZ ratio is required for normal cell division to occur (134, 135).

Structure determination of FtsA revealed an actin fold (less conserved than ParM or MreB) showing some differences particularly in the topology of the two small subdomains (136), with the small subdomain located on the opposite side of domain I when compared to actin (**Figure 3**) (136). This arrangement shows no homology with any known structure.

The actin superfamily is a diverse family united by a common ATPase domain, and self-assembly is an exception rather than the rule. FtsA's conserved ATPase fold, including a conserved catalytic ATP-binding pocket and its preferential ability to bind ATP suggest that an intrinsic part of FtsA's function is to hydrolyze ATP. However, large differences in its enzymatic activity have been observed. Although ATPase activity has been reported for *B. subtilis* FtsA (29), no activity has been detected for FtsA from *Streptococcus pneumoniae* (127). It is possible that ATP hydrolysis by FtsA is regulated in vivo by a conformational change evoked by some form of protein-protein interaction.

The observation that FtsA localizes to the Z ring in vivo, with its similarities to actin, indicates that FtsA itself might self-assemble into a polymer. Generally, attempts to study the self-assembly of FtsA from a variety of organisms has yielded largely negative results; however, very recently *S. pneumoniae* FtsA was observed to polymerize in vitro into bent and bundled long corkscrew-like helices, composed of paired protofilaments (127). The filaments were highly stable, requiring both adenosine nucleotide and magnesium for initial assembly, and showed no dynamic behavior (127). Because FtsA is unable to assemble into a detectable superstructure in vivo in the absence of FtsZ, it is possible that FtsA may require in vivo accessory proteins to assemble, explaining the different self-assembly potentials observed. The high stability of the FtsA filaments reported for *S. pneumoniae* FtsA hints at a more complicated in vivo scenario. Filaments that assemble spontaneously and that are extremely stable would need to be very carefully controlled by the cell. It is possible that with an external activator, FtsA filamentation or self-interaction, becomes reversible, as in the case of the Walker A cytoskeletal ATPase (WACA) family of proteins (see below). Very small changes in vivo may be significant in transforming a monomeric protein into a multimeric assembly. Alternatively, FtsA does not self-assemble but plays a role in tethering the Z ring to the membrane and in stabilizing the ring structure. This possibility is supported by the recent observation that a conserved amphipathic helix in the

C terminus of FtsA is essential for targeting FtsA to the membrane and to the Z ring (132).

INTERMEDIATE FILAMENT HOMOLOGUE

Intermediate filaments are a class of cytoskeletal elements in eukaryotes that are often expressed tissue-specifically. They are comprised of five different filament structures formed from various forms of keratins, lamins, and other specialized proteins. Examples include filensin (which is found in the lens of the eye), the keratins that are expressed in epithelial cells, and the lamins (which are required for nuclear envelope integrity).

Crescentin

Crescentin has been postulated to be a bacterial homologue of intermediate filament proteins (IFs) (137). Its amino acid sequence has a distinct seven-residue repeat that is predicted to form coiled-coil structures. Because of the dominating coiled-coil repeat, sequence comparisons are unreliable, but crescentin shares some important overall features with eukaryotic IF proteins. Analysis has revealed that the domain organization of crescentin is similar to animal IF proteins, suggesting that crescentin probably is a prokaryotic homologue of IFs.

Crescentin is required for determining the vibrioid or helical shape of *C. crescentus* cells. In vivo immunofluorescence microscopy and deconvolution analysis revealed that crescentin localizes as a continuous pole-to-pole helical filament along one side of the cell (137). In vitro purified crescentin is able to assemble into filaments with a width of about 10 nm. Remarkably, these filaments assemble spontaneously (in the absence of any energy source or cofactor) similar to IFs.

The correlation that crescentin is involved in determining cell shape and that it forms long filaments within the cell suggest that crescentin filaments assemble and somehow (directly or indirectly) associate with the cytoplasmic membrane specifically on one side of the cell. Furthermore, if the shape and helicity of the filament was somehow applied to the cell, a vibrioid or helical cell shape might be formed. In support of this, in stationary-phase cultures of *C. crescentus*, the cells become filamentous but also helical (138). Vibrioid cells are shorter than the helical pitch of the filament and are simply curved. In cells treated with cephalexin (which disrupts normal peptidoglycan synthesis), the intracellular localization of the crescentin filament was gradually disrupted. So, the function of crescentin is somehow linked to the biosynthesis of peptidoglycan, and it seems likely that its function must also be coordinated with the cell cycle. This poses an interesting question. Because crescentin does not require cofactors or nucleotide, and apparently assembles independently, how is this filament regulated? When a cell is dividing, how is the crescentin filament disassembled to allow daughter cell separation? It seems most likely that cofactors must be present to control crescentin assembly and disassembly.

IF: intermediate filament

WALKER A CYTOSKELETAL ATPASE—A NEW FAMILY OF CYTOSKELETAL PROTEINS?

From the discussion above it is clear that prokaryotes possess a cytoskeleton composed of classical actin, tubulin and possibly intermediate filament-like proteins. However, the spatial organization of bacterial cells also relies on a further group of proteins that has no known direct counterpart in the cytoplasm of eukaryotes. We propose a new subclass of proteins, called the WACA proteins, which are required for the spatial regulation of chromosome partitioning and cell division. The WACA proteins belong to a large and functionally diverse family of ATPases that have a conserved deviant Walker A motif and dimerize in an ATP-dependent manner (139). Although these deviant Walker A proteins are structurally homologous, their functions differ significantly, and it has been proposed that these proteins may be molecular switches (140). Recently, nitrogenase (an archetypal deviant Walker A ATPase) was shown to undergo conformational changes thought to control electron transfer processes. It was suggested that such conformational changes might be used to achieve directed motion, thus bestowing possible mechanical roles on the deviant Walker A ATPases (141). The WACA proteins are a specific subset of the deviant Walker A ATPases that have evolved the specialized function of forming ATP-induced, surface-dependent polymers (140), which might be considered as an additional component of the prokaryotic cytoskeleton.

The WACA family of proteins is comprised of MinD and the ParA/Soj plasmid and chromosome partitioning proteins, including SopA and ParF [reviewed by Hiraga (142)]. These proteins share extensive sequence homology and a similar three-dimensional structure (140, 143) (**Figure 4**). MinD is involved in the processes of Z-ring positioning during cell division, and ParA and Soj have roles in the processes of chromosome segregation, transcription, and organization of plasmids and chromosomes. All deviant Walker A proteins form dimers and are able to bind and hydrolyze ATP. The ATPase activity and dynamic behavior of the WACA proteins are regulated by the interaction of the WACA with an activation protein (144–146). MinD's ATPase activity is modulated by an interaction with MinE, ParB contains a small, N-terminal peptide known to activate ParA, and similarly Spo0J has a 20-amino acid N-terminal tail that activates Soj's ATPase (140, 147).

Typically the WACA proteins show dynamic behavior in vivo. They all have time-dependent localization patterns in the cell, with ParA alternating between nucleoids (148), MinD (from *E. coli*) oscillating from cell pole to cell pole (149), and Soj moving from pole to pole or nucleoid to nucleoid (150, 151). A notable difference between the WACA proteins is in the period of their oscillation. Although the Min system shows a regular fast oscillation of roughly a minute (152), both ParA and Soj show irregular,

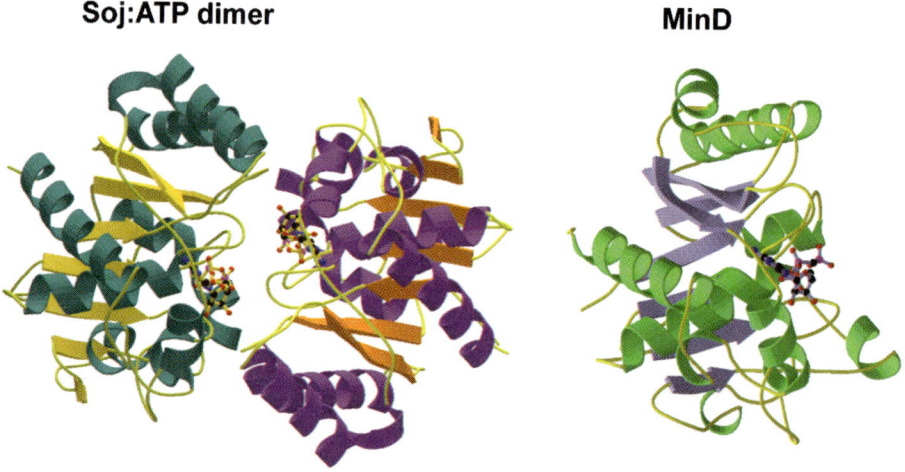

Figure 4

Structures of the Walker A cytoskeletal ATPase (WACA proteins). (*left*) Crystal structure of the MgATP-induced Soj (D44A mutant) dimer from *Thermus thermophilus* (PDB entry 2BEK) (140). (*right*) Structure of MinD (binding AMPPCP, a slowly hydrolyzable ATP analogue) from *Pyrococcus furiosus* (PDB entry code 1G3R) (143).

erratic "jumping" with periodicities of minutes, sometimes extending up to an hour (148, 150, 151).

The mechanism behind MinD oscillation is probably best understood. Discovered several years ago in *E. coli* (153), the MinCDE proteins help place the septum in the middle of the cell by inhibiting cell division at the cell poles. ATP-bound MinD is tethered to the membrane (154) by a C-terminal amphipathic helix that binds to the phospholipid bilayer (155, 156), similar to the amphipathic helix required for FtsA's membrane interaction (see below) (132). ATP-bound MinD interacts with MinC [the cell division inhibitor that directly inhibits FtsZ-ring formation (157)], recruiting MinC to the membrane. MinE is also able to bind to the ATP-bound MinD via a short activating peptide (at its N terminus). This interaction enhances MinD's ATPase turnover (158), and the ADP-bound MinD is released from the membrane. Thus, a self-organizing oscillating system is generated because MinD is more likely to rebind the membrane where there is no MinE and where there is already some MinD bound (on the other side of the cell). Thus, MinD oscillation is closely linked to ATP hydrolysis, and consistent with this, MinD mutants deficient in ATPase activity show a reduced oscillation rate (158). Several mathematical models of this system have been produced and faithfully reproduce properties observed in living cells (159–162). It is possible that related oscillatory mechanisms may be used by ParA (163) and Soj (164) to position plasmids and chromosome origins, respectively. It is interesting to note that MinD in *B. subtilis* has not been observed to oscillate but appears to be tethered to the cell poles by another protein called DivIVA. It is surprising that homologous proteins performing identical functions have such different mechanisms of action.

The WACA proteins show other interesting similarities consistent with cytoskeletal elements. Both MinD (118) and ParA (163) have been shown to assemble into dynamic helical structures in vivo. In the case of MinD, these helices were similar to but different in helical pitch and general spread along the length of the cell from those found for MreB (102), and it is thought that the MinD helices are independent of MreB filaments (118). The helical structures of ParA were also observed in the absence of MinD, indicating that ParA filaments are not simply interacting with MinD assemblies. The potential relationship of ParA with MreB has not been explored, but it is possible that the ParA structures are independent of other known helical filaments.

In vitro filaments of MinD have been observed; however, their formation is strongly dependent on the presence of both phospholipid and ATP (154, 165). Similarily, Soj assembly also has a dependence on ATP and DNA, and in their presence, Soj forms nucleoprotein filaments in a cooperative manner (140). Filaments have also been reported for ParF (a ParA homologue) with a requirement for ATP (166). The in vitro filaments described for Soj (140), MinD (165), and ParF (166) are not dynamic by themselves, but this simply reflects the fact that in contrast to FtsZ, the WACA subclass of proteins have specialized activator proteins that are required for ATP hydrolysis.

WACA proteins show an unusual characteristic whereby they bind to the entire surface area of their substrate. In vitro MinD coats phospholipid vesicles with high density; something that could be explained by surface-assisted polymerization (154), and Soj binds to DNA coating it completely (167). The process of surface-assisted polymerization can explain Soj/Spo0J oscillation along the nucleoid, similar to membrane binding of MinD and subsequent displacement by MinE, leading to pole-to-pole oscillation (146, 168).

An important unresolved question is how these proteins travel through the cell. Simple diffusion is one possibility; however, polymerization-depolymerization dynamics is another. Because there is not enough MinD in the cell to cover the whole membrane and helical structures have been detected in vivo for the proteins, the second alternative seems more likely to us at the moment.

Dynamic behavior (fuelled by nucleotide turnover) seems to be generated by a set of two proteins (the WACA subclass and their ATPase-activating counterparts) in these positioning systems. In contrast, dynamic behavior is maintained by a single component in the case of FtsZ. However, it is interesting to note that it has been speculated that FtsZ protein arises from the fusion of a GTP-binding protein with its activator protein (52). Thus, in order to evolve, a dynamic system biology required two components: a nucleotide-binding protein and an activation domain; however, the two components need not be separate.

SUMMARY POINTS

1. Bacterial cells contain a variety of dynamic filamentous proteins that bring about spatial and temporal organization analogous to the eukaryotic cytoskeleton.

2. Many of these proteins are related to actin and tubulin by an extended evolutionary history. These proteins retain the actin and tubulin protein fold, and they are able to form filaments in vitro and in vivo.

3. The tubulin homologues include FtsZ and a pair of cotranscribed proteins called BtubA and BtubB.

4. The actin homologues include ParM (a plasmid partitioning protein), MreB, and FtsA.

5. A single protein that forms filaments homologous to intermediate filaments has been identified in *C. crescentus* and is thought to regulate cell shape. Surprisingly, it has some sequence similarity and domain arrangement that are analogous to IFs but also assembles in the absence of a nucleotide or cofactors.

6. We propose that a subclass of the deviant Walker A ATPases [named Walker A cytoskeletal ATPases (WACAs)] have important, dynamic roles in organizing bacterial cells during cell division and plasmid/chromosome partitioning. These proteins should be categorized as a new class of bacterial cytoskeletal proteins.

7. Great leaps forward in our understanding of bacterial cellular organization have been facilitated by modern technologies (such as in vivo fluorescent tagging of proteins), and the discovery of the existence of cytoskeletal scaffolds has provided the means to speculate on the molecular mechanisms by which cell division, cell shape, and chromosome segregation are executed within bacterial cells. It seems likely that many of the mechanisms that have evolved in bacteria share similarities to the molecular processes that regulate cell shape, cell division, chromosome segregation, and possibly endo- and exocytosis in eukaryotes.

8. Currently, our understanding of the molecular processes behind cell division, cell shape determination, and plasmid/chromosome segregation is restricted by our inability to determine the in vivo forms of the superstructures, formed by the bacterial cytoskeletal proteins. It is expected that in time many accessory proteins with roles in assisting the assembly, disassembly, and regulation of such superstructures will be identified.

FUTURE ISSUES TO BE RESOLVED

1. The first reports of dynamic filaments observed in bacteria caused quite a stir; however, with hindsight, the requirement for such in vivo assemblies is not surprising. It has always been perplexing that bacteria maintain their many and varied shapes in the absence of cytoskeletal elements, and the mechanisms involved in chromosome segregation and cytokinesis seemed hard to fathom without the existence of some form of scaffold or cytoskeletal organization. Various models have been proposed that can explain both chromosome segregation and cytokinesis in the absence of cytoskeletal elements (169, 170). Although bacteria clearly possess proteins that form cytoskeletal elements, the molecular mechanisms by which these cytoskeletal proteins may affect cellular processes are unknown, and much biochemical information is required to resolve this.

2. Why are dynamic assemblies a feature of the cytoskeleton? First, the dynamic nature of these filaments allows for their rapid reorganization. These dynamic structures have inherent flexibility, perfect for bringing about mechanical work and also for adapting to changing cell shape and size. Furthermore, dynamic instability greatly assists in the regulation of filament nucleation. Any filaments that might form without defined initiation directed by the cell will disassemble (171).

3. Biology has typically supplied accessory proteins to modulate the behavior of polymerizing proteins inside eukaryotic cells. In bacteria, the cytoskeletal proteins are apparently self-assembling proteins, and for most, no accessory proteins have been identified, with the exception of FtsZ. In the future, it is likely that many more accessory proteins will be identified that have roles in the specific nucleation, localization, disassembly, or stabilization of the various filaments. So far, all the bacterial proteins likely to be involved in accessory roles for the filamenting cell division, cell shape, and plasmid/chromosome partitioning proteins show no similarities to accessory proteins from eukaryotes.

4. Because many of these filamentous proteins are likely to assemble in vivo into structures with a thickness greater than a single molecule, the formation of lateral interactions between protofilaments and their regulation have important consequences for understanding filament assembly and function. Currently, little is known about any of the lateral interactions of any of the bacterial cytoskeletal proteins, and this limitation severely impedes our understanding the molecular mechanisms behind the processes of cell shape determination, cell division, and possibly chromosome segregation.

5. Identification of the mechanisms behind filament nucleation and assembly, coupled with an understanding of the in vivo filament form, should help us answer fundamental questions, such as how the asymmetry of the crescentin filament is established. Many of these cytoskeletal proteins assemble into helical structures in vivo; however, in vitro they form straight filaments. How is the helicity established?

6. Both actin and tubulin filaments show clear polarity that controls the direction of motor proteins and enables spatial organization in the eukaryotic cell. So far, there is little data indicating that any of these cytoskeletal homologues provide polarity in bacteria (except for MreB in *C. crescentus*). Although the proteins assemble in a

head-to-tail fashion and could provide polarity, little is known about any in vivo consequence of this. No motor proteins that might interact with the bacterial cytoskeleton have been identified.

7. Many of these cytoplasmic bacterial cytoskeletal proteins apparently convey structural direction to the peptidoglycan, thus adaptor proteins that traverse the membrane should exist to pass information from the cytoplasm to the peptidoglycan synthesizing machinery in the periplasmic space. Recent data implicates MreC and MreD with these roles in the case of cell shape determination (122). It seems likely that similar molecular links will assist in the correct positioning and assembly of the septum synthesizing apparatus during cell division and also in the coupling of cell division with chromosome/plasmid segregation. These molecular links should reveal much about how the processes of cell shape, cytokinesis, and chromosome segregation are brought about.

ACKNOWLEDGMENTS

K.A. Michie acknowledges support from UNESCO-L'Oreal and ESF/MRC EuroDYNA.

LITERATURE CITED

1. Carballido-Lopez R, Errington J. 2003. *Trends Cell Biol.* 13:577–83
2. Errington J, Daniel RA, Scheffers DJ. 2003. *Microbiol. Mol. Biol. Rev.* 67:52–65
3. Møller-Jensen J, Löwe J. 2005. *Curr. Opin. Cell Biol.* 17:75–81
4. Lutkenhaus JF, Wolf-Watz H, Donachie WD. 1980. *J. Bacteriol.* 142:615–20
5. de Boer P, Crossley R, Rothfield L. 1992. *Nature* 359:254–56
6. RayChaudhuri D, Park JT. 1992. *Nature* 359:251–54
7. Beall B, Lutkenhaus J. 1991. *Genes Dev.* 5:447–55
8. Dai K, Lutkenhaus J. 1991. *J. Bacteriol.* 173:3500–6
9. Bi EF, Lutkenhaus J. 1991. *Nature* 354:161–64
10. Bermudes D, Hinkle G, Margulis L. 1994. *Microbiol. Rev.* 58:387–400
11. Erickson HP. 1995. *Cell* 80:367–70
12. Mukherjee A, Dai K, Lutkenhaus J. 1993. *Proc. Natl. Acad. Sci. USA* 90:1053–57
13. Margolin W, Wang R, Kumar M. 1996. *J. Bacteriol.* 178:1320–27
14. Glass JI, Lefkowitz EJ, Glass JS, Heiner CR, Chen EY, Cassell GH. 2000. *Nature* 407:757–62
15. Stephens RS, Kalman S, Lammel C, Fan J, Marathe R, et al. 1998. *Science* 282:754–59
16. Kawarabayasi Y, Hino Y, Horikawa H, Yamazaki S, Haikawa Y, et al. 1999. *DNA Res.* 6:83–101
17. Beech PL, Nheu T, Schultz T, Herbert S, Lithgow T, et al. 2000. *Science* 287:1276–79
18. Osteryoung KW, Vierling E. 1995. *Nature* 376:473–74
19. Sun Q, Margolin W. 1998. *J. Bacteriol.* 180:2050–56
20. Addinall SG, Bi E, Lutkenhaus J. 1996. *J. Bacteriol.* 178:3877–84
21. Ben-Yehuda S, Losick R. 2002. *Cell* 109:257–66
22. Grantcharova N, Lustig U, Flardh K. 2005. *J. Bacteriol.* 187:3227–37
23. Ma X, Ehrhardt DW, Margolin W. 1996. *Proc. Natl. Acad. Sci. USA* 93:12998–3003

24. Mileykovskaya E, Sun Q, Margolin W, Dowhan W. 1998. *J. Bacteriol.* 180:4252–57
25. Stricker J, Erickson HP. 2003. *J. Bacteriol.* 185:4796–805
26. Thanedar S, Margolin W. 2004. *Curr. Biol.* 14:1167–73
27. Rueda S, Vicente M, Mingorance J. 2003. *J. Bacteriol.* 185:3344–51
28. Lu C, Stricker J, Erickson HP. 1998. *Cell Motil. Cytoskelet.* 40:71–86
29. Feucht A, Lucet I, Yudkin MD, Errington J. 2001. *Mol. Microbiol.* 40:115–25
30. Stricker J, Maddox P, Salmon ED, Erickson HP. 2002. *Proc. Natl. Acad. Sci. USA* 99:3171–75
31. Anderson DE, Gueiros-Filho FJ, Erickson HP. 2004. *J. Bacteriol.* 186:5775–81
32. Mukherjee A, Lutkenhaus J. 1994. *J. Bacteriol.* 176:2754–58
33. Mukherjee A, Lutkenhaus J. 1998. *EMBO J.* 17:462–69
34. Löwe J, Amos LA. 2000. *Biol. Chem.* 381:993–99
35. Lu C, Reedy M, Erickson HP. 2000. *J. Bacteriol.* 182:164–70
36. Bramhill D, Thompson CM. 1994. *Proc. Natl. Acad. Sci. USA* 91:5813–17
37. Erickson HP, Taylor DW, Taylor KA, Bramhill D. 1996. *Proc. Natl. Acad. Sci. USA* 93:519–23
38. Löwe J, Amos LA. 1999. *EMBO J.* 18:2364–71
39. Oliva MA, Huecas S, Palacios JM, Martin-Benito J, Valpuesta JM, Andreu JM. 2003. *J. Biol. Chem.* 278:33562–70
40. Yu XC, Margolin W. 1997. *EMBO J.* 16:5455–63
41. Chen Y, Erickson HP. 2005. *J. Biol. Chem.* 280:22549–54
42. Mingorance J, Tadros M, Vicente M, Gonzalez JM, Rivas G, Velez M. 2005. *J. Biol. Chem.* 280:20909–14
43. Mukherjee A, Lutkenhaus J. 1999. *J. Bacteriol.* 181:823–32
44. Cordell SC, Robinson EJ, Löwe J. 2003. *Proc. Natl. Acad. Sci. USA* 100:7889–94
45. Löwe J, van den Ent F, Amos LA. 2004. *Annu. Rev. Biophys. Biomol. Struct.* 33:177–98
46. Romberg L, Levin PA. 2003. *Annu. Rev. Microbiol.* 57:125–54
47. Doolittle RF. 1995. *Philos. Trans. R. Soc. London Ser. B* 349:235–40
48. Doolittle RF, York AL. 2002. *BioEssays* 24:293–96
49. Löwe J, Amos LA. 1998. *Nature* 391:203–6
50. Nogales E, Wolf SG, Downing KH. 1998. *Nature* 391:199–203
51. Nogales E, Downing KH, Amos LA, Löwe J. 1998. *Nat. Struct. Biol.* 5:451–58
52. Oliva MA, Cordell SC, Löwe J. 2004. *Nat. Struct. Mol. Biol.* 11:1243–50
53. Andreu JM, Oliva MA, Monasterio O. 2002. *J. Biol. Chem.* 277:43262–70
54. Santra MK, Panda D. 2003. *J. Biol. Chem.* 278:21336–43
55. Nogales E, Whittaker M, Milligan RA, Downing KH. 1999. *Cell* 96:79–88
56. Kar S, Fan J, Smith MJ, Goedert M, Amos LA. 2003. *EMBO J.* 22:70–77
57. Santarella RA, Skiniotis G, Goldie KN, Tittmann P, Gross H, et al. 2004. *J. Mol. Biol.* 339:539–53
58. Erickson HP. 1998. *Trends Cell Biol.* 8:133–37
59. Scheffers DJ, de Wit JG, den Blaauwen T, Driessen AJ. 2002. *Biochemistry* 41:521–29
60. Faguy DM, Doolittle WF. 1998. *Curr. Biol.* 8:R338–41
61. Gilson PR, Beech PL. 2001. *Res. Microbiol.* 152:3–10
62. Margolin W, Long SR. 1994. *J. Bacteriol.* 176:2033–43
63. Desai A, Mitchison TJ. 1997. *Annu. Rev. Cell Dev. Biol.* 13:83–117
64. Caplan MR, Erickson HP. 2003. *J. Biol. Chem.* 278:13784–88
65. Huecas S, Andreu JM. 2004. *FEBS Lett.* 569:43–48
66. Chen Y, Bjornson K, Redick SD, Erickson HP. 2005. *Biophys. J.* 88:505–14

67. Gonzalez JM, Velez M, Jimenez M, Alfonso C, Schuck P, et al. 2005. *Proc. Natl. Acad. Sci. USA* 102:1895–900
68. Rivas G, Lopez A, Mingorance J, Ferrandiz MJ, Zorrilla S, et al. 2000. *J. Biol. Chem.* 275:11740–49
69. Romberg L, Simon M, Erickson HP. 2001. *J. Biol. Chem.* 276:11743–53
70. Romberg L, Mitchison TJ. 2004. *Biochemistry* 43:282–88
71. Neuhard J, Nygaard P. 1987. *Escherichia coli and Salmonella Typhimurium: Cellular and Molecular Biology*. Washington, DC: Am. Soc. Microbiol. Press
72. Scheffers DJ, den Blaauwen T, Driessen AJ. 2000. *Mol. Microbiol.* 35:1211–19
73. Mingorance J, Rueda S, Gomez-Puertas P, Valencia A, Vicente M. 2001. *Mol. Microbiol.* 41:83–91
74. Scheffers DJ, Driessen AJ. 2002. *Mol. Microbiol.* 43:1517–21
75. Lu C, Stricker J, Erickson HP. 2001. *BMC Microbiol.* 1:7
76. Phoenix P, Drapeau GR. 1988. *J. Bacteriol.* 170:4338–42
77. White EL, Ross LJ, Reynolds RC, Seitz LE, Moore GD, Borhani DW. 2000. *J. Bacteriol.* 182:4028–34
78. Marrington R, Small E, Rodger A, Dafforn TR, Addinall SG. 2004. *J. Biol. Chem.* 279:48821–29
79. Esue O, Tseng Y, Wirtz D. 2005. *Biochem. Biophys. Res. Commun.* 333:508–16
80. Santra MK, Beuria TK, Banerjee A, Panda D. 2004. *J. Biol. Chem.* 279:25959–65
81. Chretien D, Fuller S, Karsenti E. 1995. *J. Cell Biol.* 129:1311–28
82. Ravelli RB, Gigant B, Curmi PA, Jourdain I, Lachkar S, et al. 2004. *Nature* 428:198–202
83. Wang HW, Nogales E. 2005. *Nature* 435:911–15
84. Lu C, Erickson HP. 1999. *Cell. Struct. Funct.* 24:285–90
85. Huecas S, Andreu JM. 2003. *J. Biol. Chem.* 278:46146–54
86. Diaz JF, Kralicek A, Mingorance J, Palacios JM, Vicente M, Andreu JM. 2001. *J. Biol. Chem.* 276:17307–15
87. Jenkins C, Samudrala R, Anderson I, Hedlund BP, Petroni G, et al. 2002. *Proc. Natl. Acad. Sci. USA* 99:17049–54
88. Schlieper D, Oliva MA, Andreu JM, Löwe J. 2005. *Proc. Natl. Acad. Sci. USA* 102:9170–75
89. Sontag CA, Staley JT, Erickson HP. 2005. *J. Cell Biol.* 169:233–38
90. Bork P, Sander C, Valencia A. 1992. *Proc. Natl. Acad. Sci. USA* 89:7290–94
91. Holmes KC, Popp D, Gebhard W, Kabsch W. 1990. *Nature* 347:44–49
92. Jensen RB, Gerdes K. 1997. *J. Mol. Biol.* 269:505–13
93. Møller-Jensen J, Jensen RB, Löwe J, Gerdes K. 2002. *EMBO J.* 21:3119–27
94. van den Ent F, Møller-Jensen J, Amos LA, Gerdes K, Löwe J. 2002. *EMBO J.* 21:6935–43
95. Garner EC, Campbell CS, Mullins RD. 2004. *Science* 306:1021–25
96. Møller-Jensen J, Borch J, Dam M, Jensen RB, Roepstorff P, Gerdes K. 2003. *Mol. Cell* 12:1477–87
97. Doi M, Wachi M, Ishino F, Tomioka S, Ito M, et al. 1988. *J. Bacteriol.* 170:4619–24
98. Formstone A, Errington J. 2005. *Mol. Microbiol.* 55:1646–57
99. Levin PA, Margolis PS, Setlow P, Losick R, Sun D. 1992. *J. Bacteriol.* 174:6717–28
100. Varley AW, Stewart GC. 1992. *J. Bacteriol.* 174:6729–42
101. Soufo HJD, Graumann PL. 2004. *EMBO Rep.* 5:789–94
102. Jones LJ, Carballido-Lopez R, Errington J. 2001. *Cell* 104:913–22
103. Soufo HJD, Graumann PL. 2003. *Curr. Biol.* 13:1916–20
104. Daniel RA, Errington J. 2003. *Cell* 113:767–76
105. Figge RM, Divakaruni AV, Gober JW. 2004. *Mol. Microbiol.* 51:1321–32

106. Soufo HJD, Graumann PL. 2005. *BMC Cell Biol.* 6:10
107. Kruse T, Møller-Jensen J, Lobner-Olesen A, Gerdes K. 2003. *EMBO J.* 22:5283–92
108. Gitai Z, Dye NA, Reisenauer A, Wachi M, Shapiro L. 2005. *Cell* 120:329–41
109. Iwai N, Nagai K, Wachi M. 2002. *Biosci. Biotechnol. Biochem.* 66:2658–62
110. Bates D, Kleckner N. 2005. *Cell* 121:899–911
111. Gitai Z, Dye N, Shapiro L. 2004. *Proc. Natl. Acad. Sci. USA* 101:8643–48
112. van den Ent F, Amos LA, Löwe J. 2001. *Nature* 413:39–44
113. Galkin VE, VanLoock MS, Orlova A, Egelman EH. 2002. *Curr. Biol.* 12:570–75
114. Esue O, Cordero M, Wirtz D, Tseng Y. 2005. *J. Biol. Chem.* 280:2628–35
115. Nishida E, Sakai H. 1983. *J. Biochem.* 93:1011–20
116. Slovak PM, Wadhams GH, Armitage JP. 2005. *J. Bacteriol.* 187:54–64
117. Carballido-Lopez R, Errington J. 2003. *Dev. Cell* 4:19–28
118. Shih YL, Le T, Rothfield L. 2003. *Proc. Natl. Acad. Sci. USA* 100:7865–70
119. Kruse T, Gerdes K. 2005. *Trends Cell Biol.* 15:343–45
120. Mendelson NH. 1976. *Proc. Natl. Acad. Sci. USA* 73:1740–44
121. Kruse T, Bork-Jensen J, Gerdes K. 2005. *Mol. Microbiol.* 55:78–89
122. Leaver M, Errington J. 2005. *Mol. Microbiol.* 57:1196–209
123. Burk DH, Ye ZH. 2002. *Plant Cell* 14:2145–60
124. Donachie WD, Begg KJ, Lutkenhaus JF, Salmond GP, Martinez-Salas E, Vincente M. 1979. *J. Bacteriol.* 140:388–94
125. Descoteaux A, Drapeau GR. 1987. *J. Bacteriol.* 169:1938–42
126. Din N, Quardokus EM, Sackett MJ, Brun YV. 1998. *Mol. Microbiol.* 27:1051–63
127. Lara B, Rico AI, Petruzzelli S, Santona A, Dumas J, et al. 2005. *Mol. Microbiol.* 55:699–711
128. Ma X, Sun Q, Wang R, Singh G, Jonietz EL, Margolin W. 1997. *J. Bacteriol.* 179:6788–97
129. Ma X, Margolin W. 1999. *J. Bacteriol.* 181:7531–44
130. Wang X, Huang J, Mukherjee A, Cao C, Lutkenhaus J. 1997. *J. Bacteriol.* 179:5551–59
131. Yan K, Pearce KH, Payne DJ. 2000. *Biochem. Biophys. Res. Commun.* 270:387–92
132. Pichoff S, Lutkenhaus J. 2005. *Mol. Microbiol.* 55:1722–34
133. Pla J, Dopazo A, Vicente M. 1990. *J. Bacteriol.* 172:5097–102
134. Dai K, Lutkenhaus J. 1992. *J. Bacteriol.* 174:6145–51
135. Dewar SJ, Begg KJ, Donachie WD. 1992. *J. Bacteriol.* 174:6314–16
136. van den Ent F, Löwe J. 2000. *EMBO J.* 19:5300–7
137. Ausmees N, Kuhn JR, Jacobs-Wagner C. 2003. *Cell* 115:705–13
138. Wortinger MA, Quardokus EM, Brun YV. 1998. *Mol. Microbiol.* 29:963–73
139. Koonin EV. 1993. *J. Mol. Biol.* 229:1165–74
140. Leonard TA, Butler PJ, Löwe J. 2005. *EMBO J.* 24:270–82
141. Tezcan FA, Kaiser JT, Mustafi D, Walton MY, Howard JB, Rees DC. 2005. *Science* 309:1377–80
142. Hiraga S. 1992. *Annu. Rev. Biochem.* 61:283–306
143. Cordell SC, Löwe J. 2001. *FEBS Lett.* 492:160–65
144. Ma L, King GF, Rothfield L. 2004. *Mol. Microbiol.* 54:99–108
145. Zhou H, Schulze R, Cox S, Saez C, Hu Z, Lutkenhaus J. 2005. *J. Bacteriol.* 187:629–38
146. Leonard TA, Møller-Jensen J, Löwe J. 2005. *Philos. Trans. R. Soc. London Ser. B* 360:523–35
147. Radnedge L, Youngren B, Davis M, Austin S. 1998. *EMBO J.* 17:6076–85
148. Ebersbach G, Gerdes K. 2001. *Proc. Natl. Acad. Sci. USA* 98:15078–83
149. Raskin DM, de Boer PA. 1999. *J. Bacteriol.* 181:6419–24
150. Marston AL, Errington J. 1999. *Mol. Cell* 4:673–82
151. Quisel JD, Lin DC, Grossman AD. 1999. *Mol. Cell* 4:665–72

152. Raskin DM, de Boer PA. 1999. *Proc. Natl. Acad. Sci. USA* 96:4971–76
153. de Boer PA, Crossley RE, Rothfield LI. 1989. *Cell* 56:641–49
154. Hu Z, Gogol EP, Lutkenhaus J. 2002. *Proc. Natl. Acad. Sci. USA* 99:6761–66
155. Szeto TH, Rowland SL, Rothfield LI, King GF. 2002. *Proc. Natl. Acad. Sci. USA* 99:15693–98
156. Zhou H, Lutkenhaus J. 2003. *J. Bacteriol.* 185:4326–35
157. Hu Z, Mukherjee A, Pichoff S, Lutkenhaus J. 1999. *Proc. Natl. Acad. Sci. USA* 96:14819–24
158. Hu Z, Lutkenhaus J. 2001. *Mol. Cell* 7:1337–43
159. Howard M, Rutenberg AD, de Vet S. 2001. *Phys. Rev. Lett.* 87:278102
160. Kruse K. 2002. *Biophys. J.* 82:618–27
161. Meinhardt H, de Boer PA. 2001. *Proc. Natl. Acad. Sci. USA* 98:14202–7
162. Drew DA, Osborn MJ, Rothfield LI. 2005. *Proc. Natl. Acad. Sci. USA* 102:6114–18
163. Ebersbach G, Gerdes K. 2004. *Mol. Microbiol.* 52:385–98
164. Marsh JW, Taylor RK. 1999. *J. Bacteriol.* 181:1110–17
165. Suefuji K, Valluzzi R, RayChaudhuri D. 2002. *Proc. Natl. Acad. Sci. USA* 99:16776–81
166. Barilla D, Rosenberg MF, Nobbmann U, Hayes F. 2005. *EMBO J.* 24:1453–64
167. Leonard TA, Butler PJ, Löwe J. 2004. *Mol. Microbiol.* 53:419–32
168. Doubrovinski K, Howard M. 2005. *Proc. Natl. Acad. Sci. USA* 102:9808–13
169. Lemon KP, Grossman AD. 2000. *Mol. Cell* 6:1321–30
170. Woldringh CL, Mulder E, Huls PG, Vischer NOE. 1991. *Res. Microbiol.* 142:309–20
171. Mitchison T, Kirschner M. 1984. *Nature* 312:237–42
172. Löwe J, Li H, Downing KH, Nogales E. 2001. *J. Mol. Biol.* 313:1045–57

The Structure and Function of Telomerase Reverse Transcriptase

Chantal Autexier[1] and Neal F. Lue[2]

[1]Bloomfield Center for Research in Aging, Lady Davis Institute for Medical Research, Sir Mortimer B. Davis Jewish General Hospital, and Department of Anatomy and Cell Biology and Department of Medicine, McGill University, Montreal, Quebec, Canada; email: chantal.autexier@mcgill.ca

[2]Department of Microbiology and Immunology, W. R. Hearst Microbiology Research Center, Weill Medical College of Cornell University, New York, New York 10021; email: nflue@med.cornell.edu

Key Words

telomere, polymerase, ribonucleoprotein, processivity

Abstract

The structure and integrity of telomeres are essential for genome stability. Telomere dysregulation can lead to cell death, cell senescence, or abnormal cell proliferation. The maintenance of telomere repeats in most eukaryotic organisms requires telomerase, which consists of a reverse transcriptase (RT) and an RNA template that dictates the synthesis of the G-rich strand of telomere terminal repeats. Structurally, telomerase reverse transcriptase (TERT) contains unique and variable N- and C-terminal extensions that flank a central RT-like domain. The enzymology of telomerase includes features that are both similar to and distinct from those characteristic of other RTs. Two distinguishing features of TERT are its stable association with the telomerase RNA and its ability to repetitively reverse transcribe the template segment of RNA. Here we discuss TERT structure and function; its regulation by RNA-DNA, TERT-DNA, TERT-RNA, TERT-TERT interactions, and TERT-associated proteins; and the relationship between telomerase enzymology and telomere maintenance.

Contents

INTRODUCTION 494
STRUCTURAL FEATURES AND
    EVOLUTION 496
ENZYMATIC PROPERTIES OF
    TERTS AND THEIR
    STRUCTURAL
    DETERMINANTS 499
    Catalysis 499
    Elongation and Processivity 500
    TERT and Nucleotide Addition
        Processivity 501
    TERT and Repeat Addition
        Processivity 502
    RNA Binding 502
    Template Utilization 503
    Nuclease Activity 504
    Dimerization/Multimerization 505
    TERT-Associated Proteins 507
THE RELATIONSHIP BETWEEN
    TELOMERASE
    ENZYMOLOGY AND
    TELOMERE
    MAINTENANCE 509
CONCLUSIONS 511

INTRODUCTION

Eukaryotic telomeres are special nucleoprotein structures located at the ends of linear chromosomes. In most organisms, telomeres consist of short, reiterated sequences as well as proteins that interact directly or indirectly with these sequences (reviewed in 1–4). Telomeres protect chromosomal termini against fusion, degradation, and other inappropriate reactions and promote proper partitioning of chromosomes during mitosis and meiosis. Functional telomeres are also required to sustain cell proliferation. Significant telomere loss in cultured somatic cells is accompanied by a cessation in cell division and by global metabolic changes.

Though crucial for many cellular functions, telomeres are subject to a variety of transactions that result in sequence loss. Incomplete duplication of telomeres by the conventional replicative machinery, for example, leads to loss of DNA, a phenomenon often referred to as the "end replication problem" (5–7). The ability of nucleases to process and degrade terminal DNAs may also contribute to telomere attrition (reviewed in 1, 8). Although recombination has been demonstrated as one means of replenishing telomere DNA, the major offsetting activity against telomere erosion in most organisms is provided by a cellular reverse transcriptase (RT) called telomerase (reviewed in 9–11). Telomerase replenishes telomeres by extending the telomere strand with the 3' end, thus enabling other polymerases to synthesize the complementary strand. Telomerase is active in unicellular organisms and required for indefinite proliferation of the cell population. In certain multicellular organisms, including humans, telomerase is strongly repressed in normal somatic tissues but expressed in highly proliferative ones, including ovaries, testis, and hematopoietic tissues (reviewed in 12). Remarkably, telomerase is also strongly upregulated in most cancer cells, a feature that accounts for their proliferative capacity. It is not surprising that telomerase has attracted considerable attention as a plausible target for cancer therapy (13).

First identified in ciliated protozoa, telomerase is now known to be almost universally conserved in eukaryotes. It is an obligate ribonucleoprotein (RNP) whose catalytic function depends minimally on two components: the TERT (telomerase reverse transcriptase) protein and telomerase RNA (known as TR, TER, or TERC). The telomerase RNA contains a short segment, which encodes the cognate telomere repeat, and this segment serves as the template for reverse transcription by TERT (**Figure 1**). The primer for reverse transcription is provided by the 3'-OH group of the terminal nucleotide at telomeres. Although sharing many properties of "conventional" RTs, TERT also exhibits unique features; the primary ones are its utilization

---

Telomere: a special nucleoprotein complex located at the ends of chromosomes that protects the termini from fusion, degradation, and recombination

RT: reverse transcriptase

Telomerase: a ribonucleoprotein complex responsible for extending the G-rich strand of telomere repeats

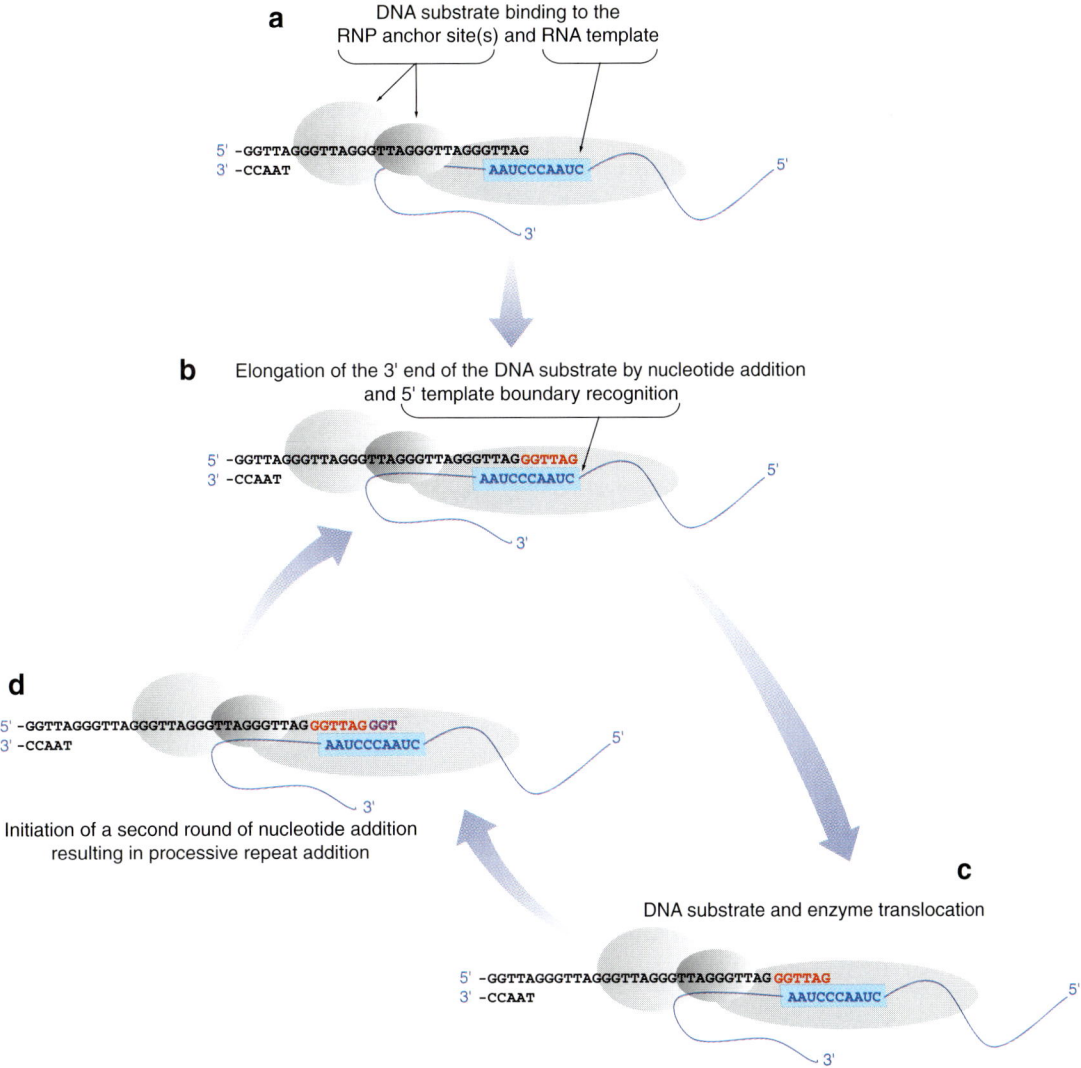

Figure 1.

Model for processive elongation by telomerase. Processive extension of DNA by telomerase requires a number of steps, as illustrated here for the human enzyme. (*a*) First, the telomeric DNA is recognized by the telomerase ribonucleoprotein (RNP), consisting of at least the TERT protein, including the anchor site(s) (*shades of gray*) and the RNA subunit (depicted in *blue* and as unstructured here for simplicity). The 3′ end of the DNA forms a hybrid with the RNA template (*boxed in blue*), whereas the more 5′ region of the DNA is postulated to interact with the "template-proximal" and "template-distal" anchor site(s). (*b*) Next, template-directed addition of nucleotides to the 3′ end of the DNA occurs sequentially until the 5′ end of the template is reached (nucleotide addition processivity). Added nucleotides (*red*) are displayed. (*c*) Telomerase undergoes the translocation reaction and repositions the 3′ end of the DNA in concert with recognition of the 3′ template boundary. (*d*) Another round of nucleotide addition is initiated. Added nucleotides (*purple*) are displayed. Reiterative translocation and nucleotide addition result in the addition of multiple repeats. Repeat addition is regulated by RNA-DNA, TERT-DNA, TERT-RNA, and TERT-TERT interactions and possibly by an intrinsic nuclease activity as well as telomerase-associated proteins. This model is based on numerous studies performed with *Tetrahymena*, human, and yeast telomerases (60, 89–92, 97–99, 182). See text for more details.

Telomerase Reverse Transcriptase (TERT): the catalytic component of the telomerase complex

of a template embedded in a large RNA and its ability to add multiple complements of the template through repeated cycles of extension and "translocation" reactions (see below) (**Figure 1**). Considerable efforts in the past few years have been directed toward understanding the structure and function of this unusual and medically important RT, and insights on its mechanisms of action have emerged from genetic and biochemical analysis of the protein component in diverse systems, including ciliates, yeasts and mammals. The goals of this review are to summarize the unifying themes on TERT that have emerged from available studies and to point out intriguing biological variations between different systems. The structure and function of telomerase RNA and the regulation of telomerase by telomere-associated proteins have also received intensive analysis. These topics have been reviewed recently (2, 14) and are only discussed here in relation to the enzymology and mechanisms of TERT.

STRUCTURAL FEATURES AND EVOLUTION

The gene encoding the TERT protein was initially cloned from a budding yeast (*Saccharomyces cerevisiae*) and subsequently from a ciliated protozoa (*Euplotes aediculatus*) (15, 16). Sequence comparison between these two proteins and prototypical RTs immediately suggested common as well as distinct features of the TERT family members. Consistent with their reverse transcription activity, all seven of the universally conserved RT motifs can be discerned within a central region of the TERT proteins (**Figure 2**). However, the telomerase proteins are also distinguished by a rather large insertion between two of the conserved motifs named A and B'. In the commonly deployed analogy of a "right hand" for nucleic acid polymerases, the two motifs are positioned within the "palm" and "fingers" domain, and the insertion is presumed to be an elaboration of the fingers structure and has been referred to as the insertion in fingers domain (IFD) (see below) (17, 18). In addition to the central RT-like domain, TERT also possesses a large, $\sim$ 400 amino acid N-terminal extension (NTE) and a short, $\sim$ 150–200 amino acid C-terminal extension (CTE) (**Figure 2**). With few exceptions, this general scheme has more or less withstood the subsequent identification and characterization of more than 40 TERTs or TERT-like proteins from all phyla of the eukaryotic kingdom, including fungi, plants, protozoa, and mammals (11, 16, 19–28). Sequence alignment of these

Figure 2

Structure and organization of TERTs. The organization of TERT is illustrated for the *Tetrahymena thermophila* (tTERT), *Saccharomyces cerevisiae* (ScEst2p), *Homo sapiens* (hTERT), *Plasmodium falciparum* (PfTERT), and *Caenorhabditis elegans* (CeTERT) proteins in comparison with HIV-1 RT. Features of TERTs include the RT motifs (1, 2, A, B', C, D, and E) and the telomerase-specific T motif and N- and C-terminal extensions (NTE, CTE) (16, 19–22, 24, 27, 28). The NTE is interrupted by a linker of variable length. Alternative nomenclature for overlapping N-terminal regions and motifs, as defined in different studies, is indicated underneath the ScEst2p and hTERT diagrams. The GQ, CP, and QFP motifs were identified on the basis of multiple sequence alignments (101). The I-A, DAT, I-B, II and III, RNA interaction domains 1 (RID1) and RID2 in hTERT were defined functionally (46, 109). Unigenic evolution led to the characterizations of regions I-IV in *S. cerevisiae* TERT (44). The T2 of tTERT falls within the GQ motif (83). IFD designates a telomerase-specific insertion in the fingers domain of RT, between conserved motifs A and B'. RT motifs are drawn in red/orange. High-affinity RNA-binding domains are indicated in blue. N-terminal domains, including some implicated in low-affinity RNA interactions, are illustrated in green. The CP2 and VSR motifs are only conserved in ciliate and vertebrate TERTs, respectively. The PfTERT and CeTERT were included in the figure because of their unusual features: large and small size and the lack of certain conserved motifs. Regions of hTERT implicated in binding to other proteins are also illustrated. Schematics are drawn only approximately to scale.

Fingers and palm
Thumb
1 2 A B' C D E
HIV-1 RT 66 kDa
Linker
TERTs
NTE
NTE
RT motifs
CTE
tTERT 133 kDa
T2 CP2 CP T 1 2 A B' C D E
ScEst2p 103 kDa
GQ CP QFP T 1 2 A B' C D E
I II III IV
IFD
hTERT 127 kDa
GQ CP QFP T 1 2 A B' C D E
RID1
VSR RID2
I-A N-DAT 1-B
II III
p23 foldasome
c-Abl
TEP1
14-3-3
NoLD/PinX1
KIP
MKRN1
PfTERT 280 kDa
GQ QFP T 1 2 A B' C D E
CeTERT 66 kDa
1 2 A B' C D E

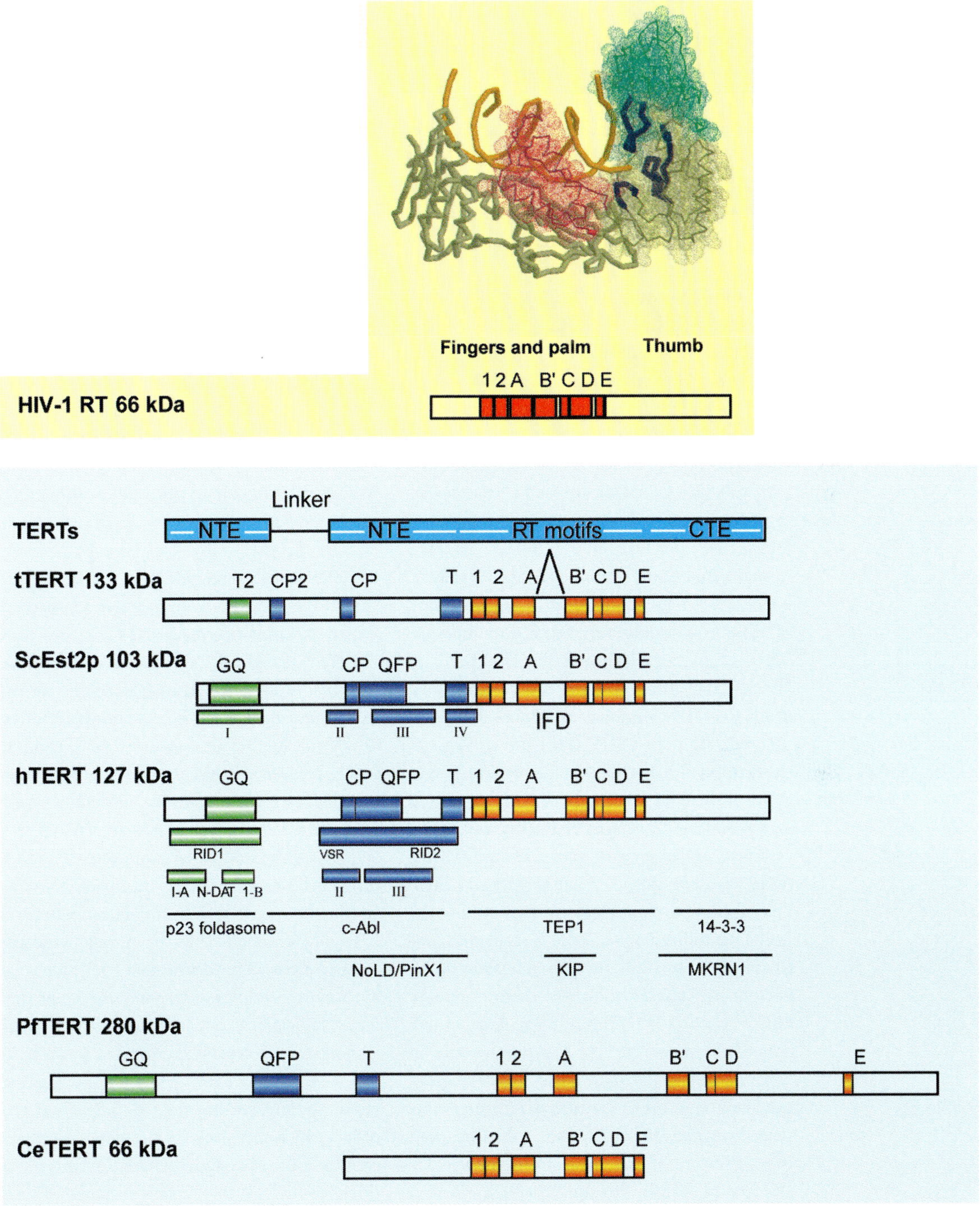

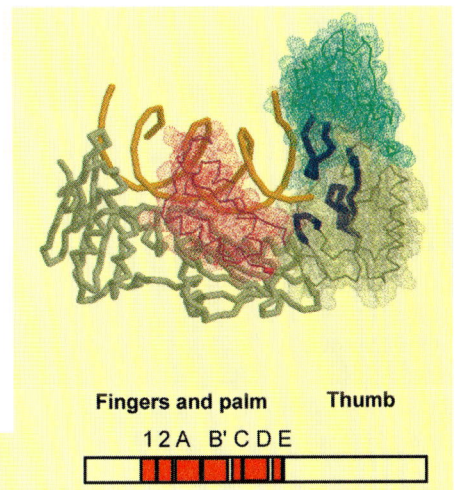

Processivity: the propensity of polymerase to add successive nucleotides to the substrate without termination or dissociation

family members has additionally revealed the existence of telomerase-specific motifs (e.g., GQ, CP, QFP, and T) within TERT-specific regions (reviewed in 10).

That all the conserved telomerase motifs are functionally important has been amply borne out by extensive mutagenesis of TERT in multiple systems (see below). However, the domain organization of a few TERTs is exceptional, and their unusual features offer interesting lessons on the diverse strategies that can be employed by TERT to accomplish essential goals. One exception relates to the CTE of TERTs. Although this domain is almost universally conserved, several roundworm members (*Caenorhabditis elegans*, *Caenorhabditis briggsae*, and *Caenorhabditis remanei*) appear to lack this structure entirely (29) (**Figure 2**). As described below, current evidence implicates CTE in promoting telomerase processivity and in regulating telomerase localization but not in essential catalytic functions. Thus, loss of CTE may be compensated by acquisition or optimization of other functionally analogous domains. Another interesting variation concerns a putative "linker" that interrupts the most N-terminal motif of TERT [known as GQ or RNA interaction domain 1 (RID1)] from the rest of the protein. This is the most diverse region of the protein in terms of sequence and ranges in size from about 20 amino acids (in *Encephalitozoon cuniculi*) to greater than 500 amino acids (in *Plasmodium* spp.) (30, 31) (**Figure 2**). Such diversity suggests a great deal of conformational flexibility between the N-terminal motif and the rest of the protein, consistent with current hypotheses about the distinctive role of this motif in DNA interaction (see below). The *Plasmodium* TERTs are noteworthy in light of not only their large linker region, but also many other insertions between conserved motifs, such that they are more than twice the size of other family members (31) (**Figure 2**). These insertions, which are rich in hydrophilic tracts, have been observed in numerous *Plasmodium* proteins and may be related to the high A+T content of the genome

(32). Although their significance is not entirely clear, their presence in TERT testifies to the tolerance of the protein to insertions. Thus, to summarize, TERT family members can be unified by a description of important conserved motifs, yet TERTs display impressive variability in arranging the motifs within the polypeptide. Interestingly, whereas most organisms appear to possess a single TERT gene, the ciliated protozoan *Euplotes crassus* was recently shown to contain as many as three TERTs, whose expression is restricted to specific developmental stages and is correlated with telomerase of distinct enzymatic properties (33).

The wide-spread nature of the protein family members also hints at its ancient origin (arguably in the earliest eukaryotes) and raises fascinating questions about its evolutionary relationship to other RT families (e.g., retroviral and retrotransposon RTs). Comparison of the RT motifs can be used to gauge the evolutionary distances between the different RT families (34). Such exercises have led to the conclusion that TERTs are most closely related to RTs of the non-LTR (long terminal repeat) retrotransposons and retrons of mitochondria (35, 36). This evolutionary kinship is further underscored by the superficial similarity between the mechanisms of non-LTR retrotransposons and telomerase. In both cases, the templates are encoded by a stably associated RNA component, and the primers for reverse transcription are provided by the DNAs that are subject to extension. A key difference is that the retrotransposons require a cleavage reaction (carried out by an autonomous nuclease) to generate the primer. That *Drosophila*, one of the few organisms lacking telomerase, uses retrotransposons to cap telomeres provokes further speculations about their relationships (37, 38). Two scenarios can be imagined and indeed have been proposed for the origins of TERTs and retrotransposon RTs. In the first scenario, TERT is of the more ancient origin and gave rise to retrotransposons through the acquisition of an alternative RNA and a nuclease capable of cleaving the target DNA (39, 40). In the second scenario, the retrotransposons are of the more ancient origin and were co-opted by cells for an essential function in chromosome maintenance during or shortly after the transition from circular to linear genomes. One plausible argument for the second scenario is based on the reasoning that because retrons are present in prokaryotes, they are likely to predate the origin of eukaryotes when TERT presumably came into existence. However, this argument is compromised by the mobile nature of retroelements, which can conceivably be transferred laterally to prokaryotes following their origin in eukaryotes. Further phylogenetic analysis seems unlikely to resolve these arguments. Perhaps a more detailed understanding of the structural and mechanistic basis of these two classes of enzymes would prove informative.

ENZYMATIC PROPERTIES OF TERTS AND THEIR STRUCTURAL DETERMINANTS

The cloning of genes encoding TERTs from ciliates, yeasts, and mammals led to the identification of conserved RT and telomerase-specific motifs as discussed in the previous section (**Figure 2**). A multitude of genetic and biochemical studies have provided insights on their mechanisms and function. As emphasized earlier, the two major features that distinguish TERT from other RTs are its tight association with telomerase RNA and its ability to repetitively reverse transcribe a short segment of the RNA (the template segment), thereby adding multiple copies of the telomeric sequence repeats onto the same DNA substrate (repeat addition processivity) (**Figure 1**). These features are predicated on a complex set of RNA-DNA, TERT-DNA, TERT-RNA, and TERT-TERT interactions and may be regulated as well by telomerase-associated proteins and an intrinsic nuclease activity. The contributions of specific TERT structures to these interactions and activities and other important aspects of telomerase enzymology are the focus of the following sections.

Catalysis

All polymerases are believed to use a two-metal mechanism to mediate the chemistry of nucleotide transfer (41). Principal to this mechanism are conserved metal-binding Asp or Glu residues near the active site, which can be visualized in multiple crystallographic models of different polymerases, including HIV-1 RT. The identification of three Asp residues in RT motifs A and C of TERT that align well with comparable residues in retroviral RTs immediately suggests that telomerase also employs the two-metal mechanism of catalysis (16). Indeed, as expected, Ala substitution of the Asp residues abolished telomerase activity in vitro and telomere maintenance in vivo (16, 42). Nevertheless, in the absence of a high-resolution structure, it could be argued that loss of activity may be due to reasons other than loss of catalysis. Even in the case of HIV-1 RT, where crystallographic structures are available, there has been little direct evidence that the metals in the active site are crucial. In this respect, a recent study of HIV-1 RT active-site mutations proved quite informative. In this study, an essential Asp residue was changed to Asn rather than Ala, and this led not to a complete loss of activity but to a rather different binding preference for the magnesium and the manganese ion (43). As a consequence, the catalytic rate of the enzyme in the presence of different metal ions was dramatically altered, thereby providing direct support for the two-metal mechanism of catalysis. A comparable mutation in TERT has yet to be characterized and may prove informative.

The RT domain of TERT is not the only domain essential for telomerase activity; many extra-RT regions contribute to the activity as measured by standard in vitro assays. Not surprisingly, some extra-RT domains are essential because they help to integrate the RNA component into the complex (see below). A potentially interesting interspecies difference between TERTs in terms of enzyme activity is the role of the most N- and C-terminal regions of the protein. Initial studies suggested that the N terminus of human and of *Tetrahymena* TERT are quite important for enzyme activity, in contrast to the analogous region of yeast TERT (44–48). Similarly, the C terminus of human and of *Tetrahymena* TERTs were demonstrated to be critical for activity, whereas the same region of yeast TERT was shown to be dispensable (44, 45, 47, 49–52). The species-specific requirements for the most N- and C-terminal portions of TERT are interesting given the lack of strong sequence conservation in these regions. However, because of the potential effects of primer substrates and assay protocols on enzyme activity (53, 54), more studies are necessary to determine if the apparent discrepancies are in fact reflective of real differences.

In the case of telomerase, further questions concerning the mechanism of catalysis have arisen because of the long-standing observation that mutations in the RNA subunit can largely abrogate enzyme activity (55). Although some mutations can be explained by their impact on protein-binding or template utilization (see below), others provoke speculations on a potential role for the RNA in catalysis. It was suggested, for example, that telomerase may have evolved from an ancient ribozyme, but that significant catalytic function was later transferred to an acquired protein component (56). In this scenario, the RNA may continue to provide functional groups that contribute to the covalent chemistry of the nucleotide transfer reaction. However, a recent study demonstrated that in the presence of adequate concentrations of manganese, the TERT proteins from yeast and humans can catalyze nucleotide addition without the telomerase RNA component (57). The reaction, however, is still strictly dependent on the catalytic Asp residues within the RT domain. Thus, the effect of RNA mutations on telomerase activity is more plausibly explained by their impact on RNA-protein interaction and/or RNA-directed substrate recognition, rather than by their impact directly on active-site chemistry.

Elongation and Processivity

The processive nature of telomerase was first observed for the *Tetrahymena* enzyme and subsequently in humans (58, 59). In a typical 60-minute in vitro reaction, products greater than 2000 nucleotides can be synthesized by the *Tetrahymena* enzyme (60). Reiterative extension of telomeric DNA is postulated to entail (*a*) binding of the telomeric DNA to the RNA template and to other domains of the RNP [e.g., the anchor site(s)]; (*b*) successive addition of nucleotides to the 3' end of the DNA (nucleotide addition processivity); (*c*) upon reaching the 5' template boundary, translocation and repositioning of the 3' end of the DNA in concert with recognition of the 3' template boundary; and (*d*) subsequent rounds of elongation (**Figure 1**). Although *Tetrahymena* and human telomerase efficiently synthesize long products in vitro, telomerases from mouse and most yeasts (with the notable exception of *Saccharomyces castelli*) are typically nonprocessive and generate only short products in standard reactions (61–64). Numerous factors that influence telomerase processivity have been identified including temperature, substrate (primer and dNTP) concentrations, primer sequence, and G-quadruplex-interacting agents (59, 65–68). Processivity may also be modulated by proteins that interact with telomerase or telomeric DNA. Studies of *E. crassus* telomerase, for example, revealed that developmentally regulated association of proteins with telomerase is correlated with changes in repeat addition processivity (69, 70). More recently, both human Pot1 and *E. aediculatus* telomerase subunit p43 were shown to modulate enzyme activity and processivity in vitro (71, 72). Determinants of processivity have also been mapped to template and nontemplate regions of the

RNA component (68, 73–81). The discussion in the next two sections focuses on structural elements of the TERT protein that have been implicated in processivity control.

TERT and Nucleotide Addition Processivity

A role for TERT in regulating nucleotide addition processivity has received compelling support. The theme that has emerged from a multitude of analyses is a high degree of mechanistic conservation between TERT and prototypical RTs with regard to the determinants of nucleotide addition processivity. Previous studies of retroviral RTs have implicated conserved RT motifs 1, 2, B′, C, and E as well as the "thumb" domain in processivity regulation. With few exceptions, analogous structures in TERT have been demonstrated to have similar roles in processivity control.

Three conserved RT motifs in TERT, including two in the putative "fingers" domain (motifs 1 and 2) and one at the interface between the "palm" and "thumb" domain (motif E), are clearly important for optimal processivity. Many point mutations in the yeast and *Tetrahymena* motifs engendered the expected impairment in successive nucleotide additions (51, 82, 83). Importantly, at conserved positions, the exact same residues are required to optimize the processivity of both telomerase and retroviral RTs, indicative of a high degree of mechanistic conservation. For example, conversion of two residues in motif E of yeast TERT (Est2p) to amino acids present at comparable positions of HIV-1 RT resulted in an increase in the processivity of yeast telomerase (51). As revealed by the crystal structure of an HIV-1 RT transcribing complex, the conserved RT motifs make contact with either the nucleotide triphosphate (motifs 1 and 2) or the primer terminus (motif E), and their roles in processivity regulation can be easily rationalized by contributions to substrate binding (84).

Analysis of conserved motif C revealed a somewhat more complicated picture with regard to mechanistic conservation. This motif harbors two invariant Asp residues that are essential for metal binding and polymerase chemistry. The two residues immediately preceding the Asp pairs in retroviral RTs are typically large hydrophobic amino acids but are more variable in TERTs. Replacing the Leu at position 813 (two residues prior to the two conserved aspartic acids) of *Tetrahymena* TERT by a Tyr that is characteristic of retroviral RTs resulted in an enzyme with increased processivity (85). In contrast, substitution of the two residues preceding the Asp pair in Est2p by Tyr and Thr caused a significant reduction in processivity (51). These results suggest that the effects of motif C on processivity are mediated by its overall conformation and chemistry rather than individual amino acids. In the crystal structure of HIV-1 RT, the Tyr, which is two amino acids prior to the Asp pairs, makes contact with the penultimate nucleotide in the DNA primer, suggesting one plausible mechanism for influencing processivity (84). Of interest would be to substitute the residue immediately preceding the two Asp amino acids of TERT with a Met, which is present in HIV-1 RT and interacts with the primer terminus. Although mutating this residue in HIV-1 RT does not appear to affect polymerization function, fidelity of DNA synthesis is compromised (86).

Another set of residues implicated in nucleotide addition processivity falls within the C-terminal domain. Despite weak conservation of sequences, multiple mutations in this domain of both yeast and human TERT impair nucleotide addition processivity, consistent with functional conservation (53, 87). By analogy with prototypical RTs, this domain of TERT was proposed to constitute the "thumb" domain of telomerase (51, 53). This notion was further supported by effects of C-terminal mutations on K_m and telomerase-substrate stability as well as by demonstration of a weak nucleic acid-binding activity for this domain of yeast TERT (87).

Est2p: ever shorter telomeres 2 protein

TERT and Repeat Addition Processivity

Because repeat addition processivity is a unique biochemical attribute of telomerase, it is natural to suspect that telomerase-specific structures are responsible for this activity. Indeed, the three groups of TERT residues that have been demonstrated to promote repeat addition appear to be largely confined to telomerase-specific motifs, although they fall both within and outside of the RT domain. These residues are discussed separately in the following paragraphs.

First, as pointed out above, in comparison with prototypical RTs, the RT domain of TERT harbors a large insertion between the conserved motifs A and B′, also known as IFD (**Figure 2**). In yeast telomerase, a specific mutation in this domain leads to defects in repeat addition and an apparent reduction in the K_m of telomerase for DNA substrates (18). Yeast telomerase, however, exhibits naturally a very low level of repeat addition in vitro. Thus, it would be worthwhile to evaluate the role of comparable residues in the more processive telomerases, such as those from *Tetrahymena* and humans.

A second group of residues implicated in repeat addition falls within the C-terminal domain of hTERT. As noted before, this domain of TERT was proposed to constitute the "thumb" domain of telomerase (**Figure 2**). Although many mutations in the C-terminal domain of yeast and human TERT appear to impair primarily nucleotide addition processivity, several deletion mutations in this domain of hTERT also reduce repeat addition, suggesting an additional function for the thumb domain of human telomerase (53).

The last group of residues that contribute to repeat addition processivity has been mapped to the hTERT RID1 and the homologous N-GQ domain of yeast TERT (54, 78) (**Figure 2**). Multiple studies showed that deletions and point mutations in this domain selectively impair repeat addition processivity and can reduce the ability of telomerase to extend partially or completely nontelomeric primers (50, 54, 78, 88). The apparent K_m of the mutants for telomeric DNA was also reduced, again suggesting a contribution for the RID1/N-GQ domain to DNA binding.

Related to the role of TERT in promoting repeat addition is a classic question in telomerase enzymology, namely the identity of the anchor site (**Figure 1**). Early functional analysis of ciliate and human telomerase led to the proposal that telomerase contains a second, template-independent DNA substrate-binding site (89–91). The existence of this site, also known as anchor site, was deduced from the influence of 5′ DNA substrate sequences on the repeat addition processivity, binding affinity, and catalytic rate of the enzyme (90, 92–95). Subsequent cross-linking experiments performed with enzymes derived from *E. aediculatus* and *S. cerevisiae* identified a TERT-sized protein as being responsible for physical contact with the 5′ region of DNA (95, 96). Additional studies suggest that the anchor site may be functionally divisible into a template-proximal and a template-distal site and that only the former may be necessary for repeat addition processivity (88, 90–92, 94, 97–100). The contribution of the RID1/N-GQ domain to repeat addition processivity and DNA substrate recognition suggests that this domain may constitute the template-proximal anchor site (47, 78, 88, 101, 102). This notion is supported by recent proteolytic mapping of the DNA-cross-linking site on yeast TERT (54). The RID1 of hTERT is known to be a site of interaction with the pseudoknot-template region of telomerase RNA, which is also implicated in promoting repeat addition processivity (76, 78). The relative contribution and mutual influence of the protein and RNA domains in anchor site function would appear to warrant further investigation.

RNA Binding

Because telomerase requires both the TERT and the RNA component for the synthesis of telomere DNA repeats, stable assembly of these two components is essential for enzyme activity. Two domains of TERT have been demonstrated to mediate interaction with telomerase RNA. A relatively high-affinity RNA-binding domain corresponds to the C-terminal part of NTE, which encompasses three conserved telomerase-specific motifs named CP, QFP, and T (30, 44) (**Figure 2**). This part of NTE has been shown to interact with the telomerase RNA in yeast, humans, and *Tetrahymena* (30, 44, 45, 47, 48, 52). Surprisingly, although this domain apparently constitutes the high-affinity RNA-binding structure in all TERTs, it seems to interact with different targets in the telomerase RNA in different organisms (**Figure 3**). In yeast, a region encompassing a potential stem-loop/pseudoknot structure has been identified as essential for TERT binding (103–106). In humans, a conserved CR4-CR5 region distant from the template appears to be the target site for the high-affinity domain (also known as RID2) (45, 52, 78, 107). In *Tetrahymena*, TERT binding requires a few nucleotides on the 5′ side of the RNA template (template binding/boundary element) as well as stem II, and TERT-binding does not appear to involve the conserved pseudoknot (45, 108). That a conserved protein domain should recognize apparently diverse RNA structures is not easily rationalized. One potential explanation invokes the region of the protein N-terminal to the CP motif. The high-affinity RNA-binding domain of yeast, human, and *Tetrahymena* TERT appears to extend into this poorly conserved region, which exhibits species-specific features [e.g., a CP2 motif that is ciliate specific (83) and a VSR motif that is mammalian specific (109)]. Thus, the highly conserved motifs may form a common scaffold to which species-specific structures are attached to mediate recognition of different RNA targets. In addition to promoting the stable assembly of telomerase RNP, the *Tetrahymena* RNA-binding domain has an additional role in helping to define the 5′ boundary of the template; weakening the interaction results in significant reverse transcription beyond the normal template boundary (110). It would be interesting to determine if comparable mutations in other TERTs can also impact on the template boundary definition.

A second, lower-affinity interaction with RNA is apparently mediated by the RID1/N-GQ domain (also known as Nterm in *Tetrahymena*) (44, 78, 109, 111). Here, too, the RNA target appears to be distinct in different organisms. The *Tetrahymena* RNA target encompasses a stem-loop (helix IV and associated loop) and a segment immediately 3′ of the template known as the template recognition element (TRE) (111, 112). The human target appears to reside in the pseudoknot-template domain (78), whereas the yeast target has not been defined. As discussed above and below, the RID1/N-GQ domain has been shown to promote repeat addition processivity and enzyme complementation, and further studies are necessary to determine the role of RNA interaction in these functions.

Template Utilization

Regulation of template usage and template 5′ and 3′ boundary recognition is essential for synthesis of proper telomere sequences and ultimately for the maintenance of telomere structure and integrity. Proper 5′ and 3′ template boundary definition is also essential for repeat addition processivity. Template usage in *Tetrahymena* telomerase is dependent on the TRE sequences 3′ of the template (112) (**Figure 3**). A comparable element has not been identified in other telomerases. TRE was recently shown to affect the interaction of telomerase RNA with an N-terminal domain of *Tetrahymena* TERT, hinting at a plausible role for TERT in template recognition (111).

In all the cases that have been analyzed thus far, the definition of the telomerase 5′ template boundary depends critically on particular RNA sequences or structures (**Figure 3**). For example, in yeast, a conserved stem (helix I) immediately 5′ of the template is essential to prevent aberrant reverse transcription beyond

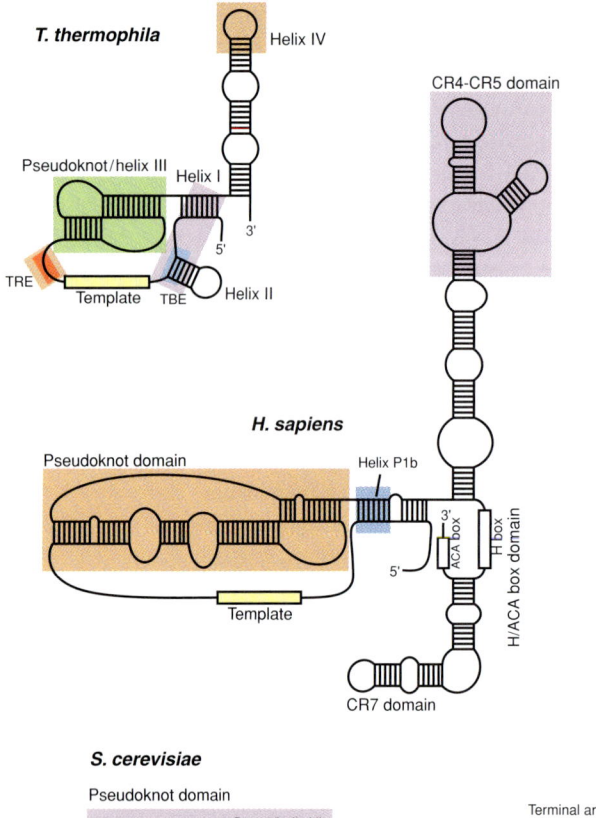

Figure 3

Structures of telomerase RNAs. Secondary structures of *T. thermophila*, *H. sapiens*, and *S. cerevisiae* telomerase RNAs are illustrated on the basis of published studies (103, 105, 106, 183, 184). Template regions (*yellow*), main TERT-binding regions (*highlighted in purple boxes*) (45, 47, 52, 78, 103, 107, 185), and template boundary regulating elements (*boxed in blue*) (74, 110, 113–115, 124) are indicated. The template boundary element (TBE) in *T. thermophila* overlaps with the main TERT-binding region (83, 110). Low-affinity TERT-binding sites in helix IV and the template recognition element (TRE) in *Tetrahymena* and in the pseudoknot/template domain in humans have also been identified (47, 52, 77, 78, 107, 111, 185). These regions are illustrated in light brown. Several structures have been proposed for the yeast telomerase RNA pseudoknot region, only one of which is presented here (103, 105, 106). Dimethyl sulphate-based footprinting analysis suggests that the yeast pseudoknot structure may be in equilibrium with other conformational state(s) (186). The yeast RNA is unusually large and contains, in addition to the central core presented in the figure, several arms that interact with Est1p, Ku, and other proteins. These remaining parts are schematically represented by lines interrupted with slashes.

the normal template boundary (113, 114). A similarly positioned stem (the P1b helix) was also implicated in human telomerase template boundary definition (115). In both instances, the structure of the RNA helix rather than the precise sequence was shown to be the necessary feature. In contrast, template boundary definition in *Tetrahymena* telomerase is mediated by a specific RNA sequence located immediately 5′ to the boundary (74). This sequence is also required for interaction between the RNA and a high-affinity RNA-binding domain of TERT, suggesting a role for the TERT protein. Subsequent studies indicate that all *Tetrahymena* TERT motifs involved in RNA binding (including CP2, CP, and T) contribute to a proper 5′ template boundary definition (83, 110). One possibility then is that this protein-RNA interaction constitutes a steric block to the movement of the RT domain. Although it is plausible to imagine that paired RNA helices (in yeast and humans) and an RNA-bound protein domain (in *Tetrahymena*) may act as efficient steric blocks, for this mechanism to prevail, the TERT protein itself must lack the ability to displace the interfering RNA strand or protein, in contrast to some retroviral RTs, which are capable of strand displacement synthesis (116).

In hTERT, mutations in RID1 and near the C terminus were found to affect 5′ template usage and incorporation of noncognate nucleotides 5′ of the template (117). In *Tetrahymena* telomerase, alterations in 5′ and 3′ template usage have been reported for mutations in several RT motifs and CP2 (83). The mechanistic basis for the effects of these mutations remains to be elucidated.

Nuclease Activity

Telomerase preparations derived from ciliates, yeast, and humans have been reported to contain a nuclease activity (62, 63, 90, 118–120). That the nuclease is intrinsic to telomerase is supported by a substantial body of evidence. First, the cleavage activity copurifies with the primer extension activity over multiple chromatographic steps (121, 122). Second, cleavage products can be observed using telomerases that are reconstituted by the expression of the two core components (TERT and the telomerase RNA) in rabbit reticulocyte lysate (27, 119, 120). Third, telomerase RNA mutations (some located within the template region) caused specific alterations in the cleavage pattern (75, 123). Lastly, mutations in certain TERT residues were shown to promote nucleolytic cleavage of the primer in preference to direct elongation (82, 83). Interestingly, these TERT mutations map to conserved RT motifs 1 and 2 and are in close proximity to the nucleotide-binding pocket. Together, the specific effects of the TERT and RNA template mutations suggest that cleavage may be mediated by the same site that catalyzes nucleotide transfer. In all studies performed to date, detection of the cleavage activity has relied on elongation (and labeling) of a cleaved DNA. Further characterization of the telomerase nuclease activity requires an assay that is independent of product elongation, especially if the same structural determinants are involved in both activities.

One proposed function for the telomerase-associated nuclease is to enhance the fidelity of telomere synthesis (90, 118–120, 122, 124). In support of this notion, nontelomeric sequences can be removed from the substrates, and cleavage is stimulated by primer-template mismatches. Moreover, cleavage is often found to occur at the 5′ end of the template, which may mitigate against read-through reverse transcription (90, 118, 119, 121). However, in contrast to the classical proof-reading nucleases, there is no compelling evidence that mismatched nucleotides are preferentially removed by the nucleolytic activity of telomerase. An alternative function for cleavage is to rescue paused, stalled, or arrested complexes by generating a substrate that can be processively elongated (90, 118–122). A precedent for such a function has been documented for the nuclease activity of RNA polymerase (125). Needless to say, in order to test any of the hypotheses on the role of cleavage, it would be extremely useful to have separation-of-function mutations that selectively impair the nucleolytic activity of telomerase.

Dimerization/Multimerization

Both TERT and the telomerase RNA have been shown capable of forming dimeric/multimeric complexes. For ease of discussion, we refer to the complexes as dimers even though their stoichiometry, in most cases, has not been determined. Dimerization has been investigated using endogenous or recombinant RNPs, or separately, the TERT or telomerase RNA component. Evidence for physical and functional dimer formation has been reported for human, yeast, and *E. crassus* telomerases (46, 50, 95, 109, 126–129). Conversely, *Tetrahymena* telomerase was shown rather unequivocally to function as a monomer (130). Thus, dimerization does not appear to be a universal feature of telomerase.

Complementation studies suggest that the N terminus of one hTERT molecule (RID1) can act in *trans* to restore the activity of a truncated hTERT containing the RID2-RT-CTE region (50, 78, 109). This head-to-tail interaction is consistent with the presence of a flexible and protease-sensitive linker between RID1 and the RID2-RT-CTE region (54, 101). Physical association between RID1 and RT-CTE in the absence of telomerase RNA can be detected in vitro (78). However, interactions between other TERT regions and domains have also been reported, as summarized in **Table 1**. For example, a region of hTERT that overlaps with RID2 (301–538) has been shown to bind a C-terminal fragment that spans motif E and CTE (914–1132) (127). In addition, two small fragments of *E. crassus* TERT (186–354 and 755–857) were found to interact with full-length EcTERT, but could not form homo or heterodimers with each other (128). Interestingly, both of these minimal fragments can interact with larger

Table 1 Summary of reported interactions between TERT regions or domains

TERT	Fragments[a]	Physical interaction[b]	Reconstituted activity	Reference
hTERT	1–350 + 1–1132	+	+	50
hTERT	1–300 + 301–1132	n.d.[c]	+	50
hTERT	1–927 + 928–1132	n.d.	−	50
hTERT	301–538 + 1–1132	+	n.d.	127
hTERT	914–1132 + 1–1132	+	n.d.	127
hTERT	928–1132 + 1–1132	−	n.d.	127
hTERT	301–538 + 914–1132	+	n.d.	127
hTERT	914–1132 + 914–1132	−	n.d.	127
hTERT	1–200 + 1–595	+	n.d.	78
hTERT	1–200 + 1–946	+	n.d.	78
hTERT	1–200 + 595–946	+	n.d.	78
hTERT	1–200 + 595–1132	+	n.d.	78
hTERT	1–200 + 946–1132	+	n.d.	78
EcTERT	186–354 + 1–1032	+	n.d.	128
EcTERT	755–857 + 1–1032	+	n.d.	128
EcTERT	1–354 + 186–354	+	n.d.	128
EcTERT	755–857 + 755–1032	+	n.d.	128
EcTERT	186–354 + 755–857	−	n.d.	128
EcTERT	186–354 + 186–354	−	n.d.	128
EcTERT	755–857 + 755–857	−	n.d.	128

[a]TERT fragment boundaries are indicated by amino acid positions. hTERT and EcTERT are 1132 and 1032 amino acids in total length.
[b]Experiments were generally performed by immunoprecipitation of one tagged fragment and visualization of the other fragment via labeling with ^{35}S-methionine or by Western blot analysis against a different tag. TERT fragments were expressed in vitro using coupled transcription and translation systems in mammalian cells or in insect cells.
[c]Abbreviation: n.d., not determined.

N- and C-terminal fragments in a manner that is indicative of head-to-tail, head-to-head, and tail-to-tail interactions (128). The physiologic significance of these interactions remains to be determined.

What is the function of telomerase dimerization? Several hypotheses have been advanced. One, proposed on the basis of the parallel extension model, postulates that dimerization may allow telomerase to coordinate the extension of two independent substrates such as a pair of sister chromatids (10, 95, 126). Another hypothesis, proposed on the basis of the template-switching model, invokes the passing of the DNA substrate between the two active sites of a dimeric enzyme; such that following one round of synthesis, the 3′ end of the DNA is transferred to the 3′ end of the alternate template for continued elongation (10, 126). As a consequence, processive elongation is predicted to require alternate utilization of the two templates. This prediction is not supported by a recent study involving an obligatory human telomerase dimer containing two different RNA template sequences (79). Another processivity-related hypothesis postulates two specialized functions for each of the protomers, with one supplying the template and the other providing the anchor site to maintain substrate-enzyme interactions during translocation (10, 79, 126).

Why are some telomerases functional as monomers and others as dimers? Assuming that dimeric telomerases are more active, one possibility is that high levels of the enzyme in organisms such as *Tetrahymena* may obviate the requirement for dimerization (130). However, this is difficult to reconcile with the dimeric nature of telomerases

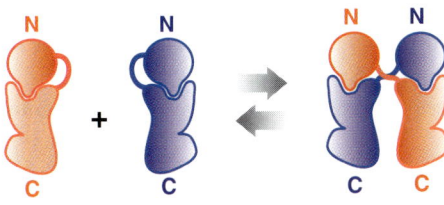

Figure 4.
Model for domain swapping in a dimeric TERT complex. A speculative model for the conversion of telomerase between the monomeric and dimeric state postulates the existence of an interface between the N- and C-terminal domain of the polypeptide and a flexible linker that connects the domains. In the monomeric form, the interface is formed between the two domains of the same polypeptide. Disruption of the interface followed by intermolecular pairing between the domains can lead to the formation of a dimeric complex.

that are also abundant, such as the *E. crassus* telomerase. Another possibility is that both monomers and dimers exist but have specialized functions during different stages of the cell cycle, or different stages of the life cycle in ciliates, or at different subcellular locations (128, 130, 131). Can telomerase interconvert between the monomeric and dimeric state? It is not necessary to invoke a new set of interactions if domain swapping for TERT can occur (**Figure 4**). For example, as noted above, a functional TERT monomer may harbor interactions between an N- and a C-terminal domain. Disruption of the intramolecular interaction followed by intermolecular pairing between the N terminus of one and the C terminus of another monomer could result in a functional telomerase dimer. Analysis of domain-swapped proteins revealed that most swapped domains are at either the N or C terminus and that a conformationally flexible hinge region is often critical for swapping (132, 133). The existence in TERT of independently folded N- and C-terminal domains, interdomain interaction, and the hinge region is all supported by experimental evidence (50, 78, 87, 101, 109, 134).

TERT-Associated Proteins

A large number of telomerase-associated proteins have been identified in ciliates, yeast, and vertebrates. However, few are common to all telomerases, and the functional significance of many of these interacting proteins in telomere metabolism has not been established (reviewed in 2, 11). In this section, we restrict our discussion to proteins that are likely to interact directly with TERT, including the p23 chaperone, TEP1, 14-3-3, c-Abl, Ku, hEST1, KIP, PinX1, and MKRN1 (**Figure 2**). These proteins are believed to regulate telomerase assembly, posttranslational modification, localization, and enzymatic function.

One of the first TERT interacting proteins to be identified was the p23 chaperone (135). p23 was identified in a yeast two-hybrid screen using an N-terminal fragment of hTERT (amino acids 1–195) as the bait. p23 is part of a foldosome, commonly associated with protein complex assembly, and is required, along with Hsp90, for the efficient assembly of active telomerase both in vitro and in vivo. Contrary to the association of chaperones with most proteins, Hsp90 and p23 remain stably associated with an active telomerase following assembly and may therefore contribute additional regulatory functions at telomeres (136).

Two covalent modifications have been reported for mammalian TERT: phosphorylation and ubiquitination. With regard to phosphorylation, multiple kinases including PKCα, PKCζ, and Akt, have been reported to act on hTERT (reviewed in 137). The available data largely support phosphorylation as a positive regulator of telomerase, although the effect of most phosphorylation events on telomere lengths has not been determined. In contrast, a recent study revealed an inhibitory effect of c-Abl kinase-mediated phosphorylation on telomerase activity (138). The functional relevance of this phosphorylation event was supported by the finding of increased telomerase activity and telomere lengths in cAbl-/- cells. The specificity of interaction between c-Abl and hTERT is apparently conferred by a c-Abl SH3-binding motif in hTERT (amino acids 308–316). In the case of ubiquitination, the only known modifying enzyme is a ubiquitin ligase named MKRN1. MKRN1 interacts with a C-terminal fragment of hTERT (amino acids 946–1132) in two-hybrid assays and coimmunoprecipitates with full-length hTERT (139). Overexpression of MKRN1 promotes the degradation of hTERT via ubiquitination and, as a consequence, causes a decrease in telomerase activity as well as in telomere lengths.

In addition to covalent modification, control of localization is another possible posttranslational mechanism for telomerase regulation. Several regions of the TERT protein have been implicated in nucleolar localization, including amino acids 1–15 and a nucleolar localization domain (NoLD) mapped to amino acids 326–620 (140, 141). However, the relevant interacting partners for nucleolar localization have not been identified. One protein that has been demonstrated to regulate hTERT intracellular localization is the 14-3-3 protein, which has previously been implicated in the localization of many other proteins. 14-3-3 was shown to be an hTERT-interacting protein in two-hybrid assays using the C-terminal 129 amino acids of hTERT as the bait (142). 14-3-3 apparently enhances the nuclear localization of hTERT by inhibiting the binding of CRM1 to a nuclear export signal motif in hTERT. Although this regulation is expected to enhance telomerase function in vivo, its effect on telomere lengths and structure has not been determined.

Additional evidence for regulation of TERT localization to telomeres came from studies of the DAT (dissociates activities of telomerase) mutations. These mutations, positioned near the N and C terminus of TERT, have no apparent impact on telomerase activity in vitro yet abolish the ability of the protein to maintain telomeres in vivo (44, 46, 49, 101, 143). On the basis of their phenotypes, the DAT mutations have been postulated to affect the association with regulatory proteins or to influence the recruitment of telomerase to telomere ends. For instance, the region encompassing the yeast DAT mutants are known to be important for association of telomerase with Est3p, a regulatory subunit (131). The telomere maintenance defects of several hTERT N-terminal DAT mutants are partially rescued by fusion to the telomere-binding proteins hTRF2 or hPot1, consistent with a role for the mutated sequences in the recruitment of telomerase to telomeres (144, 145). Two recent studies of hTERT, with the use of selected DNA primers in direct primer extension assays, however, revealed processivity and DNA synthesis defects of several DAT mutants (88, 102). These studies thus point to the possibility that the telomere maintenance defects of the DAT mutants may be due to impairment of enzymatic function rather than loss of recruitment.

Five other proteins, namely Ku, hEST1, TEP1, KIP, and PinX1, may regulate TERT activity through direct binding. The human heterodimeric Ku complex was initially linked to telomere maintenance through its interaction with telomeric DNA and the telomere-repeat-binding proteins TRF1 and TRF2 (146). More recent studies suggest that the complex can also independently associate with hTERT and human telomerase RNA (147, 148). Interestingly in yeast, Ku binds specifically to a stem-loop in the telomerase RNA and may promote the localization of telomerase to telomere termini (149, 150). Whether the interactions between Ku and human telomerase components serve a similar function is not yet known.

hEST1A and hEST1B are both human homologs of the yeast telomerase subunit Est1p, whose essential functions in telomere maintenance have been extensively characterized (151, 152). Yeast Est1p mediates the recruitment of telomerase to telomere ends by binding to both telomerase RNA and Cdc13p, a telomere end-binding protein (153–155). Est1p also apparently serves a postrecruitment or activating function (154,

156). Direct interaction between Est1p and yeast TERT has not been detected. In contrast, both hEST1A and hEST1B were shown to bind hTERT independently of the RNA subunit when expressed in vitro (157, 158). The human proteins also associate specifically with telomerase activity in cell extracts. However, thus far, only hEST1A has been functionally linked to telomere maintenance and protection in vivo; overexpression of hEST1A results in telomere uncapping and elevated levels of chromosome fusion (158) and, in combination with hTERT overexpression, can provoke substantial telomere elongation (157).

TEP1 was first identified as a telomerase-associated protein that interacts specifically with mammalian telomerase RNA (159, 160). Subsequent binding studies show that TEP1 can also interact with the RT domain of hTERT in an RNA-independent fashion (47). The functional relevance of these interactions remains in doubt given the predominantly cytoplasmic localization of TEP1 and the failure of TEP1 deletion to alter either telomerase activity or telomere length in vivo (161–163).

KIP was initially characterized as a DNA-PKcs-binding protein and subsequently shown to interact with a fragment of hTERT (amino acids 762–855) in two-hybrid assays (164). Consistent with a positive regulatory function, overexpression of KIP in tumor cells was shown to increase telomerase activity and telomere lengths. KIP binding to DNA-PKcs and hTERT appear to be mutually exclusive, suggesting that a competition between the two interactions may influence the functions of hTERT and DNA-PKcs at the telomeres.

Human PinX1 was first identified as a TRF1 interacting protein and a negative regulator of telomerase enzyme activity and telomere length (165). PinX1 binds hTERT, primarily to the RNA interaction site between amino acids 326 and 620. PinX1 can also bind telomerase RNA, although only in the presence of hTERT (166). Although the yeast homolog of human PinX1 was originally thought to mediate exclusively nucleolar functions (167), a more recent analysis suggests that the yeast protein, too, may negatively regulate telomerase function, most likely by sequestering yeast TERT in an inactive, RNA-free complex (168). As PinX1 is, to date, one of the few regulators of telomerase that binds directly to the RNP, its mechanism of inhibition would appear to warrant further analysis.

THE RELATIONSHIP BETWEEN TELOMERASE ENZYMOLOGY AND TELOMERE MAINTENANCE

What is the relationship between telomerase activity and telomere lengths? Studies that seek to correlate the levels of telomerase to telomere lengths have resulted in disparate findings. Some observations argue for a strong dependence of telomere length on telomerase activity. For example, a threshold level of enzyme activity is necessary to avert progressive telomere shortening and senescence. In normal human somatic cells, a very low level of telomerase activity can be detected but is clearly insufficient to counteract telomere loss (169, 170). Likewise, a significant level of ectopic telomerase activity was shown to be required for immortalization and transformation (171). In addition, beyond the threshold level, telomeres are maintained stably around an average length that correlates positively with the levels of telomerase RNP and activity (172). In mouse ES cells, knocking out just one TERT allele is sufficient to cause telomere attrition (173). Similarly, the amount of yeast telomerase RNP can be reduced to different extents by various mutations in the RNA-binding domain of Est2p. The yeast strains modified in this fashion exhibit a good correlation between the telomerase RNP level and equilibrium telomere lengths (30). Furthermore, altering the processivity of telomerase through mutations in TERT (as judged by in vitro assays) resulted in corresponding changes in telomere lengths, implying an important role for the elongation propensity of telomerase in vivo (51). Notwithstanding

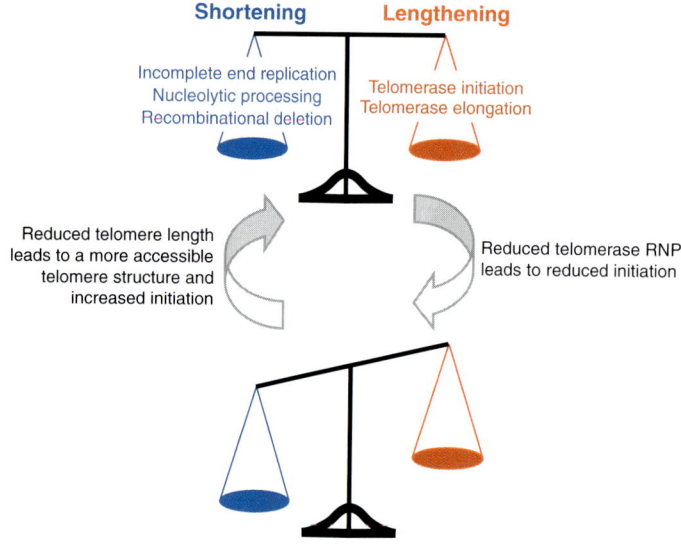

Figure 5.
A model for the dependence of telomere length on telomerase activity. Telomere lengths are at equilibrium when the average loss (owing to incomplete replication, nucleolytic degradation, and recombination) is counterbalanced by telomerase-mediated extension, which is a function of both initiation frequency and processivity. As illustrated in the diagram, perturbing the balance by decreasing the level of telomerase leads to an imbalance in the average lengthening and shortening rates, which causes telomere lengths to decrease. This in turn produces a more accessible telomere structure, which increases the initiation frequency of telomerase and reestablishes homeostasis at a shorter average length.

eres enable the binding of larger numbers of telomere recognition proteins and their partners. This longer and larger telomere complex in turn has a greater propensity to sequester the ends into an inaccessible conformation. The second insight came from recent analysis of in vivo telomere extension in yeast within a single cell cycle at a specific terminus (179). This analysis confirmed the notion that short telomeres are more susceptible to extension and further indicated that the length of tracts added is not dependent on the initial telomere length. Thus, although the elongation propensity of telomerase is likely influenced by a variety of factors as discussed above, this propensity does not appear to be regulated by telomere lengths.

Given these premises, the homeostatic telomere length can be modeled by a simple equation. At equilibrium, the average shortening rate (SR), which is dictated by incomplete end replication and the length of G-strand overhangs (but presumably not by telomere length), would be balanced by average extension, which is the product of the probability of initiation (P_i) and the average length of extension (E), such that

$$SR = P_i(L) \times E.$$

Notably as discussed above, P_i is negatively correlated with telomere length (L). This model can readily account for most of the aforementioned observations (see **Figure 5** for an illustration). For example, in the case of a reduction in telomerase RNP level, P_i is reduced at standard wild-type telomere length simply because of the law of mass action, whereas E is unaffected. Given that SR is greater than the product of P_i (L) and E, telomeres will gradually shorten. However, during the shortening process, P_i will gradually increase until the shortening and lengthening rate is equivalent again. If the RNP level is reduced to a level that cannot be balanced by increase in the accessibility of telomere, then unremitting telomere attrition will ensue, thus explaining the threshold requirement for telomerase in telomere maintenance.

these apparent correlations, many findings argue against a strict relationship between telomerase activity and telomere lengths. Reconstitution of telomerase activity in primary human cells leads generally to telomere maintenance, but a clear correlation between activity levels and telomere length is difficult to establish (174–176). Survey of large numbers of tumor cell lines or samples also failed to disclose a correspondence between the two variables (177, 178).

How does one reconcile these apparently disparate observations? Two insights that have emerged from recent studies may be particularly relevant. First, it is almost certain that a feedback loop exists to regulate the accessibility of telomere ends to telomerase as a function of telomere length, such that longer telomeres are more refractory to extension (2). Specifically, it is thought that longer telom-

When telomerase is compromised in terms of processivity, E is reduced, and new equilibrium is reached only when P_i is increased sufficiently because of the opening of the telomere structure following length reduction. In the case of a very severe reduction in processivity, even the maximal P_i (presumably close to 1) may be incapable of balancing the lengthening and shortening rate, thus resulting in unremitting telomere attrition.

How then, does one explain the lack of close correlation between telomere length and telomerase activity in cancer cells and in primary cells expressing ectopic TERT? A reasonable hypothesis is that there is variability in the telomere nucleoprotein structures (as well as the presence of telomerase recruitment factors), such that the dependence of P_i on telomere length varies in different cell populations. It is not surprising, then, that the best correlation between telomerase enzymatic function and telomere length is observed in genetically well-defined systems using defined mutations that perturb telomerase levels or activities. Clearly, a future challenge is to define quantitatively the link between the key parameters that govern the relationship between telomerase activity, telomere structure, and telomere lengths.

CONCLUSIONS

It is evident that structure and function analyses of TERT have revealed much about the workings of this unusual RT. Perhaps not surprisingly, the experimental findings have generally been interpreted with an emphasis on the structural and biochemical attributes of TERT that distinguish it from conventional RTs (typically used to designate retroviral RTs). This is rightly so given the many unique features that have been uncovered. Yet, the potential for analyses of TERT to provide general insights on RT mechanisms should also not be overlooked. For example, even within the conserved RT motifs, there are clear preferences in the TERT family at selected positions for certain amino acid residues. By analyzing the consequences of substitutions and interpreting the results in light of the biological requirement, one can reasonably hope to gain appreciation of the multiple ways a basic polymerase scaffold is modified to yield enzymes of diverse biological functions (e.g., turning out numerous retroviral particles with relatively little regard for mutations vs adding a few nucleotides to the chromosome ends). For another example, retrotransposon RTs are also known to form dimers, but the basis for dimerization is not yet well understood. It is not implausible that a better understanding of TERT dimerization may inform studies of retrotransposons as well.

In terms of future biochemical analysis of TERT, two key challenges lie ahead. One is to obtain a high-resolution structure(s) of the protein or protein domains. Although many of the existing biochemical findings can be interpreted in light of the crystal structures of retroviral RTs, many involve telomerase-specific residues that can only be understood with a TERT structure. Needless to say, such a structure(s) would also be invaluable in motivating additional mutagenesis and mechanistic experiments. Another key challenge is to devise in vitro assays that more accurately recapitulate aspects of in vivo regulation. Information regarding in vivo regulation has come from studies of yeast and human telomerases, and many participating players have been identified (2, 180). By exploiting such knowledge (e.g., obtaining adequate quantities of relevant recombinant proteins), one is in a better position to reconstitute telomerase regulation in vitro. For example, a particularly interesting and critical regulation in yeast appears to entail the recruitment of telomerase to telomere ends through an interaction between a telomerase subunit (Est1p) and a telomere protein (Cdc13p) (153, 155). A significant advance in this regard would be to reconstitute a positive stimulatory effect of Cdc13p or Est1p in vitro. This type of system could then allow detailed investigation of the function of TERT domains and motifs in a regulated, more physiologic assay. Ultimately, it will be necessary to analyze the mechanisms of TERT in the context of a purified reconstituted in vitro system that accurately mimics the regulation of telomere extension in vivo.

The discovery of TERT close to 10 years ago represented a milestone in the analysis of telomerase. Studies over the ensuing years have only strengthened interests in this unusual and medically relevant polymerase. Insights on TERT mechanisms are anticipated to lead to many applications in medicine, not the least of which is the development of anticancer therapeutics (13, 181). It can be stated with some confidence that analysis of this protein will continue to occupy investigators for years to come and will impact on many areas of biology.

SUMMARY POINTS

1. TERT is conserved in evolution and represents a family of special RTs dedicated to maintaining eukaryotic telomeres.
2. TERT comprises a central RT-like domain flanked by N- and C-terminal telomerase-specific motifs, all of which are required for its function.
3. The basic nucleotide transfer and elongation mechanism of TERT is nearly identical to conventional RTs.
4. The ability of telomerase to mediate multiple cycles of template copying and translocation (repeat addition processivity) is predicated on telomerase-specific structures located both within and outside of the RT domain, including an N-terminal domain that represents the classically defined anchor site.
5. Several conserved telomerase motifs located N-terminal to the RT domain mediate recognition of apparently diverse RNA structures in different organisms.
6. The level and processivity of telomerase are two variables that govern the equilibrium lengths of telomeres in vivo.

FUTURE ISSUES TO BE RESOLVED

1. How do the conserved RT and telomerase-specific motifs mediate diverse aspects of telomerase enzymology at the atomic level?
2. What are the features of telomerase that promote unpairing of the RNA-DNA duplex during and after each cycle of polymerization?
3. How do conserved and essential features of telomerase RNA (aside from the template) promote/influence TERT function?
4. How does telomere structure regulate at the molecular level the propensity of telomerase to initiate telomere extension?

ACKNOWLEDGMENTS

We are especially indebted to Vicki Lundblad, Joachim Lingner, and Lea Harrington for insightful comments on this manuscript as well as to many colleagues for communicating results prior to publication. We thank Tara Moriarty and members of our laboratories for helpful critiques and suggestions, and we thank Sylvain Huard for help with the preparation of **Figure 3**. Work in our labs is supported by grants from the Canadian Institutes of Health Research and a Boehringer Ingelheim (Canada) Young Investigator Award (C.A.) and from the National Institutes of Health (N.F.L). C.A. is a recipient of the Fonds de la Recherche en Santé du Québec Chercheur-boursier Award and N.F.L. is a recipient of the Irma Hirschl Monique Weill-Caulier Career Development Award.

LITERATURE CITED

1. Chakhparonian M, Wellinger RJ. 2003. *Trends Genet.* 19:439–46
2. Smogorzewska A, de Lange T. 2004. *Annu. Rev. Biochem.* 73:177–208
3. Ferreira MG, Miller KM, Cooper JP. 2004. *Mol. Cell* 13:7–18
4. Blackburn EH. 2001. *Cell* 106:661–73
5. Lingner J, Cooper JP, Cech TR. 1995. *Science* 269:1533–34
6. Watson JD. 1972. *Nat. New Biol.* 239:197–201
7. Olovnikov AM. 1973. *J. Theor. Biol.* 41:181–90
8. Lansdorp PM. 2005. *Trends Biochem. Sci.* 30:388–95
9. Henson JD, Neumann AA, Yeager TR, Reddel RR. 2002. *Oncogene* 21:598–610
10. Kelleher C, Teixeira MT, Forstemann K, Lingner J. 2002. *Trends Biochem. Sci.* 27:572–79
11. Harrington L. 2003. *Cancer Lett.* 194:139–54
12. Dong CK, Masutomi K, Hahn WC. 2005. *Crit. Rev. Oncol. Hematol.* 54:85–93
13. Shay JW, Wright WE. 2002. *Cancer Cell* 2:257–65
14. Chen JL, Greider CW. 2004. *Trends Biochem. Sci.* 29:183–92
15. Lendvay TS, Morris DK, Sah J, Balasubramanian B, Lundblad V. 1996. *Genetics* 144:1399–412
16. Lingner J, Hughes TR, Shevchenko A, Mann M, Lundblad V, Cech TR. 1997. *Science* 276:561–67
17. Cristofari G, Lingner J. 2003. *Cell* 113:552–54
18. Lue NF, Lin YC, Mian IS. 2003. *Mol. Cell. Biol.* 23:8440–49
19. Nakamura TM, Morin GB, Chapman KB, Weinrich SL, Andrews WH, et al. 1997. *Science* 277:955–59
20. Harrington L, Zhou W, McPhail T, Oulton R, Yeung DS, et al. 1997. *Genes Dev.* 11:3109–15
21. Meyerson M, Counter CM, Eaton EN, Ellisen LW, Steiner P, et al. 1997. *Cell* 90:785–95
22. Counter CM, Meyerson M, Eaton EN, Weinberg RA. 1997. *Proc. Natl. Acad. Sci. USA* 94:9202–7
23. O'Reilly M, Teichmann SA, Rhodes D. 1999. *Curr. Opin. Struct. Biol.* 9:56–65
24. Kilian A, Bowtell DD, Abud HE, Hime GR, Venter DJ, et al. 1997. *Hum. Mol. Genet.* 6:2011–19
25. Greenberg RA, Allsopp RC, Chin L, Morin GB, DePinho RA. 1998. *Oncogene* 16:1723–30
26. Martin-Rivera L, Herrera E, Albar JP, Blasco MA. 1998. *Proc. Natl. Acad. Sci. USA* 95:10471–76
27. Collins K, Gandhi L. 1998. *Proc. Natl. Acad. Sci. USA* 95:8485–90
28. Bryan TM, Sperger JM, Chapman KB, Cech TR. 1998. *Proc. Natl. Acad. Sci. USA* 95:8479–84
29. Malik HS, Burke WD, Eickbush TH. 2000. *Gene* 251:101–8
30. Bosoy D, Peng Y, Mian IS, Lue NF. 2003. *J. Biol. Chem.* 278:3882–90

31. Figueiredo LM, Rocha EPC, Mancio-Silva L, Prevost C, Hernandez-Verdun D, Scherf A. 2005. *Nucleic Acids Res.* 33:1111–22
32. Gardner MJ, Tettelin H, Carucci DJ, Cummings LM, Aravind L, et al. 1998. *Science* 282:126–32
33. Karamysheva Z, Wang LB, Shrode T, Bednenko J, Hurley LA, Shippen DE. 2003. *Cell* 113:565–76
34. Xiong Y, Eickbush TH. 1990. *EMBO J.* 9:3353–62
35. Eickbush TH. 1997. *Science* 277:911–12
36. Nakamura TM, Cech TR. 1998. *Cell* 92:587–90
37. Pardue ML, Danilevskaya ON, Traverse KL, Lowenhaupt K. 1997. *Genetica* 100:73–84
38. Pardue ML, DeBaryshe PG. 2003. *Annu. Rev. Genet.* 37:485–511
39. Luan DD, Korman MH, Jakubczak JL, Eickbush TH. 1993. *Cell* 72:595–605
40. Malik HS, Burke WD, Eickbush TH. 1999. *Mol. Biol. Evol.* 16:793–805
41. Steitz TA. 1999. *J. Biol. Chem.* 274:17395–98
42. Weinrich SL, Pruzan R, Ma L, Ouellette M, Tesmer VM, et al. 1997. *Nat. Genet.* 17:498–502
43. Bolton EC, Mildvan AS, Bockc JD. 2002. *Mol. Cell* 9:879–89
44. Friedman KL, Cech TR. 1999. *Genes Dev.* 13:2863–74
45. Lai CK, Mitchell JR, Collins K. 2001. *Mol. Cell. Biol.* 21:990–1000
46. Armbruster BN, Banik SSR, Guo CH, Smith AC, Counter CM. 2001. *Mol. Cell. Biol.* 21:7775–86
47. Beattie TL, Zhou W, Robinson MO, Harrington L. 2000. *Mol. Biol. Cell* 11:3329–40
48. Bryan TM, Goodrich KJ, Cech TR. 2000. *Mol. Cell* 6:493–99
49. Banik SSR, Guo CH, Smith AC, Margolis SS, Richardson DA, et al. 2002. *Mol. Cell. Biol.* 22:6234–46
50. Beattie TL, Zhou W, Robinson MO, Harrington L. 2001. *Mol. Cell. Biol.* 21:6151–60
51. Peng Y, Mian IS, Lue NF. 2001. *Mol. Cell* 7:1201–11
52. Bachand F, Autexier C. 2001. *Mol. Cell. Biol.* 21:1888–97
53. Huard S, Moriarty TJ, Autexier C. 2003. *Nucleic Acids Res.* 31:4059–70
54. Lue NF. 2005. *J. Biol. Chem.* 280:26586–91
55. Chan SR, Blackburn EH. 2004. *Philos. Trans. R. Soc. London Ser. B* 359:109–21
56. Blackburn EH. 1999. In *The RNA World*, ed. R Gesteland, T Cech, J Atkins, pp. 609–36. Cold Spring Harbor, NY: Cold Spring Harbor Lab. Press
57. Lue NF, Bosoy D, Moriarty TJ, Autexier C, Altman B, Leng SY. 2005. *Proc. Natl. Acad. Sci. USA* 102:9778–83
58. Greider CW, Blackburn EH. 1985. *Cell* 43:405–13
59. Morin GB. 1989. *Cell* 59:521–29
60. Greider CW. 1991. *Mol. Cell. Biol.* 11:4572–80
61. Prowse KR, Avilion AA, Greider CW. 1993. *Proc. Natl. Acad. Sci. USA* 90:1493–97
62. Cohn M, Blackburn EH. 1995. *Science* 269:396–400
63. Lue NF, Peng Y. 1997. *Nucleic Acids Res.* 25:4331–37
64. Fulton TB, Blackburn EH. 1998. *Mol. Cell. Biol.* 18:4961–70
65. Bosoy D, Lue NF. 2004. *Nucleic Acids Res.* 32:93–101
66. Sun D, Lopez-Guajardo CC, Quada J, Hurley LH, Von Hoff DD. 1999. *Biochemistry* 38:4037–44
67. Maine IP, Chen SF, Windle B. 1999. *Biochemistry* 38:15325–32
68. Hardy CD, Schultz CS, Collins K. 2001. *J. Biol. Chem.* 276:4863–71
69. Greene EC, Shippen DE. 1998. *Genes Dev.* 12:2921–31

70. Bednenko J, Melek M, Greene EC, Shippen DE. 1997. *EMBO J.* 16:2507–18
71. Lei M, Zaug AJ, Podell ER, Cech TR. 2005. *J. Biol. Chem.* 280:20449–56
72. Aigner S, Cech TR. 2004. *RNA* 10:1108–18
73. Autexier C, Greider CW. 1994. *Genes Dev.* 8:563–75
74. Autexier C, Greider CW. 1995. *Genes Dev.* 9:2227–39
75. Ware TL, Wang H, Blackburn EH. 2000. *EMBO J.* 19:3119–31
76. Chen JL, Greider CW. 2003. *EMBO J.* 22:304–14
77. Lai CK, Miller MC, Collins K. 2003. *Mol. Cell* 11:1673–83
78. Moriarty TJ, Marie-Egyptienne DT, Autexier C. 2004. *Mol. Cell. Biol.* 24:3720–33
79. Rivera MA, Blackburn EH. 2004. *J. Biol. Chem.* 279:53770–81
80. Ly H, Blackburn EH, Parslow TG. 2003. *Mol. Cell. Biol.* 23:6849–56
81. Mason DX, Goneska E, Greider CW. 2003. *Mol. Cell. Biol.* 23:5606–13
82. Bosoy D, Lue NF. 2001. *J. Biol. Chem.* 276:46305–12
83. Miller MC, Liu JK, Collins K. 2000. *EMBO J.* 19:4412–22
84. Huang H, Chopra R, Verdine GL, Harrison SC. 1998. *Science* 282:1669–75
85. Bryan TM, Goodrich KJ, Cech TR. 2000. *J. Biol. Chem.* 275:24199–207
86. Pandey VN, Kaushik N, Rege N, Sarafianos SG, Yadav PNS, Modak MJ. 1996. *Biochemistry* 35:2168–79
87. Hossain S, Singh S, Lue NF. 2002. *J. Biol. Chem.* 277:36174–80
88. Moriarty TJ, Ward RJ, Taboski MAS, Autexier C. 2005. *Mol. Biol. Cell* 16:3152–61
89. Harrington LA, Greider CW. 1991. *Nature* 353:451–54
90. Collins K, Greider CW. 1993. *Genes Dev.* 7:1364–76
91. Morin GB. 1991. *Nature* 353:454–56
92. Lee M, Blackburn EH. 1993. *Mol. Cell. Biol.* 13:6586–99
93. Melek M, Davis B, Shippen DE. 1994. *Mol. Cell. Biol.* 14:7827–38
94. Wang H, Blackburn EH. 1997. *EMBO J.* 16:866–79
95. Prescott J, Blackburn EH. 1997. *Genes Dev.* 11:2790–800
96. Hammond PW, Lively TN, Cech TR. 1997. *Mol. Cell. Biol.* 17:296–308
97. Lue NF, Peng Y. 1998. *Nucleic Acids Res.* 26:1487–94
98. Baran N, Haviv Y, Paul B, Manor H. 2002. *Nucleic Acids Res.* 30:5570–78
99. Collins K. 1999. *Annu. Rev. Biochem.* 68:187–218
100. Wallweber G, Gryaznov S, Pongracz K, Pruzan R. 2003. *Biochemistry* 42:589–600
101. Xia JQ, Peng Y, Mian IS, Lue NF. 2000. *Mol. Cell. Biol.* 20:5196–207
102. Lee SR, Wong JMY, Collins K. 2003. *J. Biol. Chem.* 278:52531–36
103. Lin J, Ly H, Hussain A, Abraham M, Pearl S, et al. 2004. *Proc. Natl. Acad. Sci. USA* 101:14713–18
104. Chappell AS, Lundblad V. 2004. *Mol. Cell. Biol.* 24:7720–36
105. Zappulla DC, Cech TR. 2004. *Proc. Natl. Acad. Sci. USA* 101:10024–29
106. Dandjinou AT, Levesque N, Larose S, Lucier JF, Elela SA, Wellinger RJ. 2004. *Curr. Biol.* 14:1148–58
107. Mitchell JR, Collins K. 2000. *Mol. Cell* 6:361–71
108. Licht JD, Collins K. 1999. *Genes Dev.* 13:1116–25
109. Moriarty TJ, Huard S, Dupuis S, Autexier C. 2002. *Mol. Cell. Biol.* 22:1253–65
110. Lai CK, Miller MC, Collins K. 2002. *Genes Dev.* 16:415–20
111. O'Connor CM, Lai CK, Collins K. 2005. *J. Biol. Chem.* 280:17533–39
112. Miller MC, Collins K. 2002. *Proc. Natl. Acad. Sci. USA* 99:6585–90
113. Seto AG, Umansky K, Tzfati Y, Zaug AJ, Blackburn EH, Cech TR. 2003. *RNA* 9:1323–32
114. Tzfati Y, Fulton TB, Roy J, Blackburn EH. 2000. *Science* 288:863–67

115. Chen JL, Greider CW. 2003. *Genes Dev.* 17:2747–52
116. Fisher TS, Darden T, Prasad VR. 2003. *J. Mol. Biol.* 325:443–59
117. Moriarty TJ, Marie-Egyptienne DT, Autexier C. 2005. *RNA* 11:1448–60
118. Melek M, Greene EC, Shippen DE. 1996. *Mol. Cell. Biol.* 16:3437–45
119. Huard S, Autexier C. 2004. *Nucleic Acids Res.* 32:2171–80
120. Oulton R, Harrington L. 2004. *Mol. Biol. Cell* 15:3244–56
121. Greene EC, Bednenko J, Shippen DE. 1998. *Mol. Cell. Biol.* 18:1544–52
122. Niu HW, Xia JQ, Lue NF. 2000. *Mol. Cell. Biol.* 20:6806–15
123. Bhattacharyya A, Blackburn EH. 1997. *Proc. Natl. Acad. Sci. USA* 94:2823–27
124. Prescott J, Blackburn EH. 1997. *Genes Dev.* 11:528–40
125. Cramer P. 2004. *Curr. Opin. Genet. Dev.* 14:218–26
126. Wenz C, Enenkel B, Amacker M, Kelleher C, Damm K, Lingner J. 2001. *EMBO J.* 20:3526–34
127. Arai K, Masutomi K, Khurts S, Kaneko S, Kobayashi K, Murakami S. 2002. *J. Biol. Chem.* 277:8538–44
128. Wang LB, Dean SR, Shippen DE. 2002. *Nucleic Acids Res.* 30:4032–39
129. Ly H, Xu LF, Rivera MA, Parslow TG, Blackburn EH. 2003. *Genes Dev.* 17:1078–83
130. Bryan TM, Goodrich KJ, Cech TR. 2003. *Mol. Biol. Cell* 14:4794–804
131. Friedman KL, Heit JJ, Long D, Cech TR. 2003. *Mol. Biol. Cell* 14:1–13
132. Bennett MJ, Eisenberg D. 2004. *Structure* 12:1339–41
133. Liu Y, Eisenberg D. 2002. *Protein Sci.* 11:1285–99
134. Jacobs SA, Podell ER, Wuttke DS, Cech TR. 2005. *Protein Sci.* 14:2051–58
135. Holt SE, Aisner DL, Baur J, Tesmer VM, Dy M, et al. 1999. *Genes Dev.* 13:817–26
136. Forsythe HL, Jarvis JL, Turner JW, Elmore LW, Holt SE. 2001. *J. Biol. Chem.* 276:15571–74
137. Cong YS, Wright WE, Shay JW. 2002. *Microbiol. Mol. Biol. Rev.* 66:407–25
138. Kharbanda S, Kumar V, Dhar S, Pandey P, Chen C, et al. 2000. *Curr. Biol.* 10:568–75
139. Kim JH, Park SM, Kang MR, Oh SY, Lee TH, et al. 2005. *Genes Dev.* 19:776–81
140. Yang YH, Chen YH, Zhang CY, Huang H, Weissman SM. 2002. *Exp. Cell Res.* 277:201–9
141. Etheridge KT, Banik SSR, Armbruster BN, Zhu YS, Terns RM, et al. 2002. *J. Biol. Chem.* 277:24764–70
142. Seimiya H, Sawada H, Muramatsu Y, Shimizu M, Ohko K, et al. 2000. *EMBO J.* 19:2652–61
143. Counter CM, Hahn WC, Wei W, Caddle SD, Beijersbergen RL, et al. 1998. *Proc. Natl. Acad. Sci. USA* 95:14723–28
144. Armbruster BN, Etheridge KT, Broccoli D, Counter CM. 2003. *Mol. Cell. Biol.* 23:3237–46
145. Armbruster BN, Linardic CM, Veldman T, Bansal NP, Downie DL, Counter CM. 2004. *Mol. Cell. Biol.* 24:3552–61
146. Downs JA, Jackson SP. 2004. *Nat. Rev. Mol. Cell Biol.* 5:367–78
147. Chai WH, Ford LP, Lenertz L, Wright WE, Shay JW. 2002. *J. Biol. Chem.* 277:47242–47
148. Ting NSY, Yu YP, Pohorelic B, Lees-Miller SP, Beattie TL. 2005. *Nucleic Acids Res.* 33:2090–98
149. Stellwagen AE, Haimberger ZW, Veatch JR, Gottschling DE. 2003. *Genes Dev.* 17:2384–95
150. Fisher TS, Taggart AKP, Zakian VA. 2004. *Nat. Struct. Mol. Biol.* 11:1198–205
151. Taggart AKP, Zakian VA. 2003. *Curr. Opin. Cell Biol.* 15:275–280
152. Lundblad V. 2003. *Curr. Biol.* 13:R439–41

153. Pennock E, Buckley K, Lundblad V. 2001. *Cell* 104:387–96
154. Evans SK, Lundblad V. 2002. *Genetics* 162:1101–15
155. Bianchi A, Negrini S, Shore D. 2004. *Mol. Cell* 16:139–46
156. Taggart AKP, Teng SC, Zakian VA. 2002. *Science* 297:1023–26
157. Snow BE, Erdmann N, Cruickshank J, Goldman H, Gill RM, et al. 2003. *Curr. Biol.* 13:698–704
158. Reichenbach P, Hoss M, Azzalin CM, Nabholz M, Bucher P, Lingner J. 2003. *Curr. Biol.* 13:568–74
159. Harrington L, McPhail T, Mar V, Zhou W, Oulton R, et al. 1997. *Science* 275:973–77
160. Nakayama J-I, Saito M, Nakamura H, Matsuura A, Ishikawa F. 1997. *Cell* 88:875–84
161. Liu Y, Snow BE, Hande MP, Baerlocher G, Kickhoefer VA, et al. 2000. *Mol. Cell. Biol.* 20:8178–84
162. Kickhoefer VA, Liu Y, Kong LB, Snow BE, Stewart PL, et al. 2001. *J. Cell Biol.* 152:157–64
163. Kickhoefer VA, Stephen AG, Harrington L, Robinson MO, Rome LH. 1999. *J. Biol. Chem.* 274:32712–17
164. Lee GE, Yu EY, Cho CH, Lee J, Muller MT, Chung IK. 2004. *J. Biol. Chem.* 279:34750–55
165. Zhou XZ, Lu KP. 2001. *Cell* 107:347–59
166. Banik SSR, Counter CM. 2004. *J. Biol. Chem.* 279:51745–48
167. Guglielmi B, Werner M. 2002. *J. Biol. Chem.* 277:35712–19
168. Lin J, Blackburn EH. 2004. *Genes Dev.* 18:387–96
169. Broccoli D, Young JW, de Lange T. 1995. *Proc. Natl. Acad. Sci. USA* 92:9082–86
170. Masutomi K, Yu EY, Khurts S, Ben-Porath I, Currier JL, et al. 2003. *Cell* 114:241–53
171. Hamad NM, Banik SSR, Counter CM. 2002. *Oncogene* 21:7121–25
172. Swiggers SJ, Nibbeling HA, Zeilemaker A, Kuijpers MA, Mattern KA, Zijlmans JM. 2004. *Exp. Cell Res.* 297:434–43
173. Liu Y, Snow BE, Hande MP, Yeung D, Erdmann NJ, et al. 2000. *Curr. Biol.* 10:1459–62
174. Franco S, MacKenzie KL, Dias S, Alvarez S, Rafii S, Moore MAS. 2001. *Exp. Cell Res.* 268:14–25
175. Bodnar AG, Ouellette M, Frolkis M, Holt SE, Chiu C-P, et al. 1998. *Science* 279:349–52
176. MacKenzie KL, Franco S, May C, Sadelain M, Moore MAS. 2000. *Exp. Cell Res.* 259:336–50
177. Hastie ND, Dempster M, Dunlop MG, Thompson AM, Green DK, Allshire RC. 1990. *Nature* 346:866–68
178. Yan P, Benhattar J, Coindre JM, Guillou L. 2002. *Int. J. Cancer* 98:851–56
179. Teixeira MT, Arneric M, Sperisen P, Lingner J. 2004. *Cell* 117:323–35
180. Evans SK, Lundblad V. 2000. *J. Cell Sci.* 113(Pt. 19):3357–64
181. Damm K, Hemmann U, Garin-Chesa P, Hauel N, Kauffmann I, et al. 2001. *EMBO J.* 20:6958–68
182. Lue NF. 2004. *BioEssays* 26:955–62
183. Romero DP, Blackburn EH. 1991. *Cell* 67:343–53
184. Chen JL, Blasco MA, Greider CW. 2000. *Cell* 100:503–14
185. Chen JL, Opperman KK, Greider CW. 2002. *Nucleic Acids Res.* 30:592–97
186. Forstemann K, Lingner J. 2005. *EMBO Rep.* 6:361–66

Relating Protein Motion to Catalysis

Sharon Hammes-Schiffer and Stephen J. Benkovic

Department of Chemistry, Pennsylvania State University, University Park, Pennsylvania 16802; email: shs@chem.psu.edu, sjb1@psu.edu

Key Words

protein conformational motions, enzyme catalysis, dihydrofolate reductase, liver alcohol dehydrogenase

Abstract

This review examines the linkage between protein conformational motions and enzyme catalysis. The fundamental issues related to this linkage are probed in the context of two enzymes that catalyze hydride transfer, namely dihydrofolate reductase and liver alcohol dehydrogenase. The extensive experimental and theoretical studies addressing the role of protein conformational changes in these enzyme reactions are summarized. Evidence is presented for a network of coupled motions throughout the protein fold that facilitate the chemical reaction. This network is comprised of fast thermal motions that are in equilibrium as the reaction progresses along the reaction coordinate and that lead to slower equilibrium conformational changes conducive to the chemical reaction.

Contents

INTRODUCTION 520
DIHYDROFOLATE
   REDUCTASE 520
   Mutagenesis of DHFR 522
   Conformational Changes
      in DHFR 526
   Theoretical Studies of Motion
      in DHFR 528
LIVER ALCOHOL
   DEHYDROGENASE 530
   Conformational Changes and
      Mutagenesis in LADH 532
   Theoretical Studies of Motion in
      LADH 533
CONCLUSIONS 535

DHFR: dihydrofolate reductase

LADH: liver alcohol dehydrogenase

H_2F: 7,8-dihydrofolate

H_4F: 5,6,7,8-tetrahydrofolate

INTRODUCTION

Rather than present another review that initially recites the historical record that underpins our understanding of how enzymes execute their catalytic process, we presume that background on the part of the reader and choose instead to commence with a series of questions that shape contemporary research efforts. Partial answers to these questions are just emerging through a combination of experimental and theoretical investigations. We have recently written a perspective on enzyme catalysis that traced the progression of hypotheses from the early "lock & key" model (1) to those of more recent vintage [e.g., induced fit (2–5), preorganization of active sites (6, 7), and near-attack conformations of bound substrates (8)], all of which require conformational flexibility within the enzyme and substrate (9). For the present review, our starting point is based on the well-supported observations that protein structures exhibit dynamical fluctuations on a wide range of timescales (10–22). These timescales include bond vibrations on the 10–100 fs timescale, rotations of side chains at the protein surface on the 10–100 ps timescale, hinge bending at domain interfaces on the 100 ps to 10 ns timescale, water structure reorganization on the 10 ns timescale, and rotation of medium-sized side chains in the protein interior on the micro- to millisecond timescale. The fundamental issues are whether there is indeed a linkage between these fluctuations of enzymatic structures and their catalytic function and, if so, what is the nature of this linkage.

Because of the extensive experimental and theoretical data sets required to interrogate this issue, we will deliberately limit our discussion to two enzymes that catalyze one of the most elementary transfer processes, namely, hydride transfer. The two enzymes are dihydrofolate reductase (DHFR) and liver alcohol dehydrogenase (LADH). Both enzymes have been subject to investigations with the aforementioned issue as an objective, and the collective evidence that we present unequivocally implicates the interplay between enzymatic catalysis and protein conformational fluctuations. Consequently, we will examine the nature of this linkage in the framework of three questions. First, does the motion of the protein generate a series of equilibrium conformational states that lie along the reaction coordinate and facilitate the conversion of substrate to product by decreasing the free energy barrier, or is there a requirement for the motion of the protein to be directly coupled to the substrate in order to cross over the reaction barrier? Second, how are the various timescales for motions of differing protein elements integrated into the catalytic cycle of the enzyme? Third, how inclusive with respect to the overall protein structure are the fluctuations that are linked to the catalytic event, i.e., is there a coupled network of residues extending throughout the protein that prepares and modulates the active site for the chemical transformation?

DIHYDROFOLATE REDUCTASE

DHFR catalyzes the reduction of 7,8-dihydrofolate (H_2F) to 5,6,7,8-tetrahydrofolate (H_4F) through a stereospecific

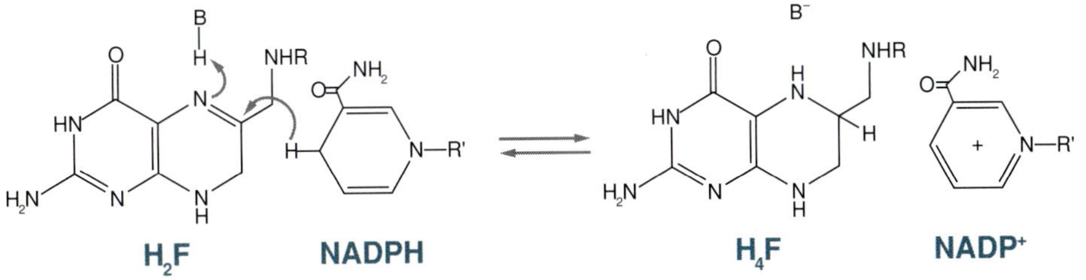

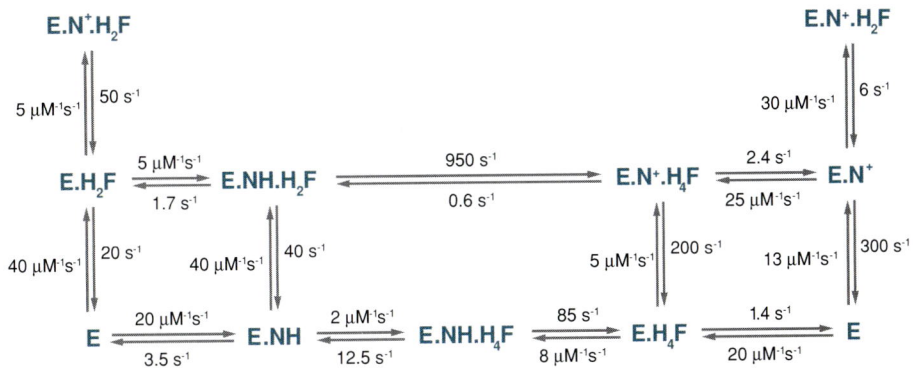

Figure 1

The reaction catalyzed by DHFR. The reactant substrate is H_2F (dihydrofolate), the product substrate is THF (tetrahydrofolate), and the cofactor is NADPH/NADP$^+$. Reproduced with permission from Reference 9.

transfer of the pro-R hydrogen from the cofactor NADPH to C6 of the pterin nucleus with concurrent protonation at the N5 position. This reaction is depicted in **Figure 1**. The pK_a of N5 in the *Escherichia coli* E·NADP$^+$·H_2F complex is 6.5, elevated by specific interactions at the active site from its solution pK_a of 2.6 (23). The enzyme from *E. coli* is the subject of this discussion and has been the subject of recent reviews (24, 25).

The data sets required for this inquiry were furnished by extensive presteady-state and steady-state kinetic studies of the reaction cycle catalyzed by the enzyme and by numerous crystallographic and more recent NMR determinations of the structural features of various ligand complexes of the enzyme (26–28). The kinetic scheme shows the enzyme cycling between five kinetically observable species: E·NH·H_2F, E·N$^+$·H_4F, E·H_4F, E·NH·H_4F, and E·NH with hydride transfer occurring at a rate of 950 s^{-1} (the pH-independent rate) and loss of H_4F from E·NH·H_4F at a rate of 12.5 s^{-1} (the rate-determining step in the steady state). The kinetic scheme and the pH-independent rate constants are given in **Figure 2**.

The earliest indication of structural flexibility derives from the crystallographic investigations (26). The *E. coli* enzyme is a polypeptide of 160 amino acids that folds into a structure containing an eight-stranded β-sheet and four α-helices connected by loops that shape a catalytic subdomain and adenosine-binding subdomain. X-ray crystal structures of various DHFR complexes revealed that three loops designated Met-20 (residues 9–24), βF-βG (residues 116–132), and βG-βH (residues 142–150) attached to the catalytic subdomain can assume different conformations, open, closed, or occluded, depending on the identity of the bound ligand(s). These conformations are illustrated in **Figure 3**. The structural transitions for

Figure 2

The pH-independent kinetic scheme for DHFR catalysis at 25°C. Abbreviations used: E, DHFR; NH, NADPH; N$^+$, NADP$^+$; H_2F, dihydrofolate; and H_4F, tetrahydrofolate. Reproduced with permission from Reference 9.

Figure 3

Depictions of the closed (*blue*), occluded (*green*), and open (*red*) forms of the loops in DHFR, as obtained from X-ray crystallography. The specific interactions of the βF-βG (*top left*) and βG-βH (*bottom left*) loops with the Met-20 loop are shown. Reproduced with permission from Reference 33.

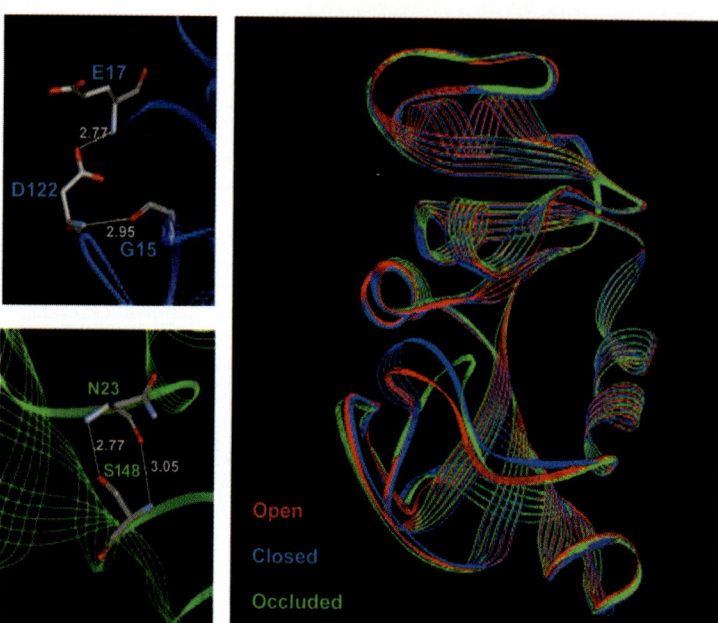

the principle species in the catalytic cycle have been determined from both X-ray and NMR studies and are depicted in **Figure 4** (29, 30).

The key ternary substrate and product complexes, E·NH·H$_2$F and E·N$^+$·H$_4$F, show a transition from a closed to an occluded state, as depicted in **Figures 3** and **4**. In the former, the Met-20 loop closes like a lid over the active site; in the latter, residues Met-16 and Glu-17 protrude into and sterically occlude the binding site for the nicotinamide-ribose moiety of the cofactor. Correspondingly smaller conformational changes occur in the βF-βG and βG-βH loops. The cycling of these loops between the two conformational states demonstrates their importance in facilitating the hydride transfer step. This conjecture is strongly supported by several additional lines of evidence, including extensive kinetic analysis of DHFR mutants (31–33) and molecular dynamics simulations discussed below.

Mutagenesis of DHFR

This section briefly summarizes the conclusions drawn from single-site mutagenesis of amino acids proximal and distal to the active site and from higher-order mutagenesis in conformationally active regions implicated

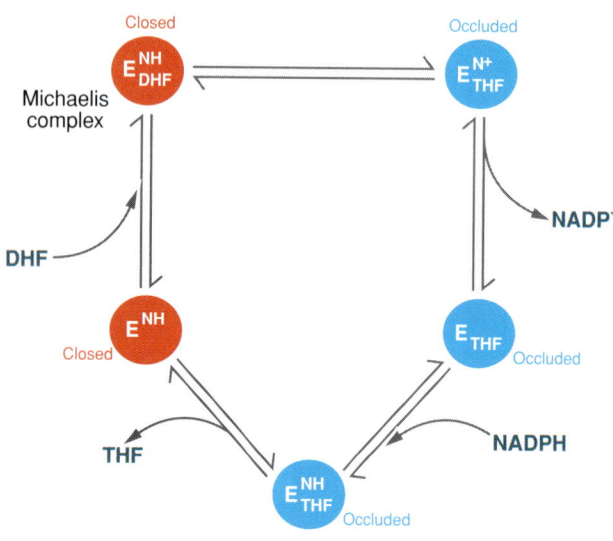

Figure 4

Schematic representation of the catalytic cycle of *E. coli* DHFR, showing the conformations of the principal species. DHF corresponds to H$_2$F, and THF corresponds to H$_4$F. Reproduced with permission from Reference 30.

by structural or computational studies. The effects of single mutations, which are generally conservative substitutions, are depicted in **Figure 5** for the binding of the cofactor and substrate and the rate of hydride transfer. Inspection reveals that the effect of a single mutation on ligand binding is not limited to the active site or a specific structural element but rather is felt throughout the protein. As expected, the larger unfavorable changes in

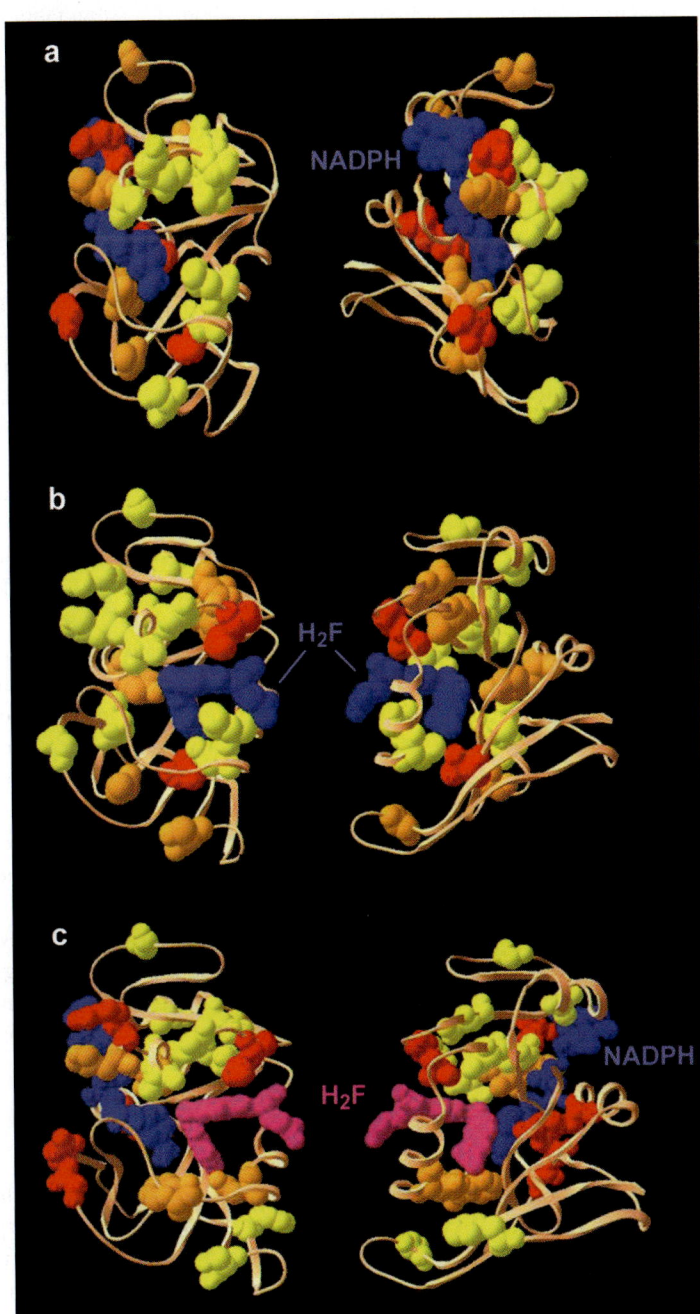

Figure 5

The effect of single mutation on (*a*) the dissociation constant for the binding of the cofactor to DHFR; (*b*) the dissociation constant for the binding of the substrate to DHFR; (*c*) the rate for hydride transfer. These properties are represented in terms of free energies; the red regions are >1 kcal/mol, the orange regions are 0.5–1 kcal/mol, and the yellow regions are <0.5 kcal/mol. This figure was obtained from the work of T. Selzer and S.J. Benkovic.

free energy for ligand affinity are observed for amino acids more proximal to the active site. Perturbations of the hydride transfer rate that are greater than 200-fold are caused by amino acid substitutions in the Met-20 and βF-βG loop elements. These observations are consistent with the need for conformational flexibility to retain ligand binding and facilitate hydride transfer and implicate the involvement of the entire protein structure.

The next question that arose was whether such mutational effects on the free energy of binding and the hydride transfer rate are coupled, suggesting that these structurally separated amino acids act as a unit, or whether these effects accumulate independently and, hence, are additive. Answers were sought by carrying out the kinetic analysis on a series of related single and double mutants and calculating via the cycle illustrated in **Figure 6** whether the interaction term, ΔG_I, has a nonzero value. This analysis shows that Gly-121 in the βF-βG loop couples to Ser-148 in the βG-βH loop, and Met-42 in the βB-strand couples to Thr-113 in the βF-strand (34). Conversely, Met-42 is not coupled to Ser-148, nor does Met-42 couple to Gly-121 (35). Nonadditivity of mutational effects is also found within the active site (37). This is only a partial listing of the findings from these types of studies. The most important point emerging from these studies is that specific regions of the protein are coupled.

One may, therefore, inquire as to whether such regions and specific amino acids within these regions are evolutionarily conserved. A genomic analysis for sequence conservation across 36 diverse species of DHFR from *E. coli* to human identified conserved or highly conserved residues that might implicate their preservation for a coupled network (38). The result is shown in **Figure 7**.

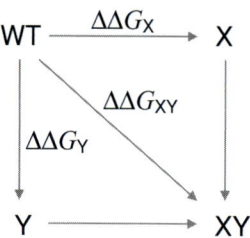

$$\Delta \Delta G_X = -RT \ln \left(k_X / k_{WT} \right)$$
$$\Delta \Delta G_Y = -RT \ln \left(k_Y / k_{WT} \right)$$
$$\Delta \Delta G_{XY} = -RT \ln \left(k_{XY} / k_{WT} \right)$$

$$\Delta G_I = \Delta \Delta G_{XY} - \left(\Delta \Delta G_X + \Delta \Delta G_Y \right)$$

Figure 6

Double mutational cycle involving general residues X and Y. Mutational effects are calculated as free energy changes relative to the wild-type enzyme. For example, the effect of mutating residue X is calculated as the free energy change $\Delta \Delta G_X$ relative to the wild-type enzyme. The degree of nonadditivity in the double mutational cycle is indicated by ΔG_I.

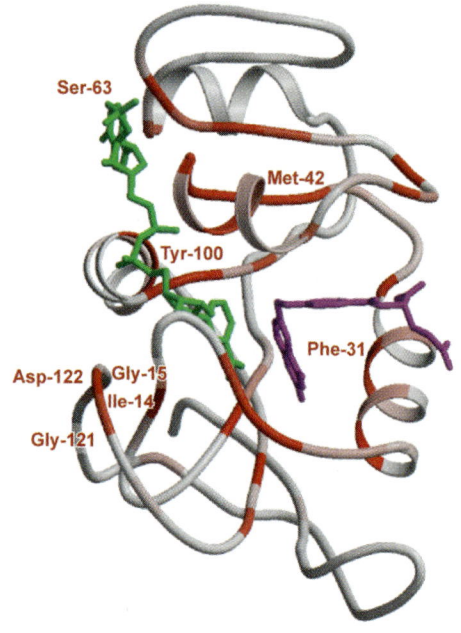

Figure 7

Sequence conservation in DHFR. Regions of conservation are mapped onto the structure of *E. coli* DHFR using a gradient color scheme (*gray* to *red*, with *red* as the most conserved). Reproduced with permission from Reference 38.

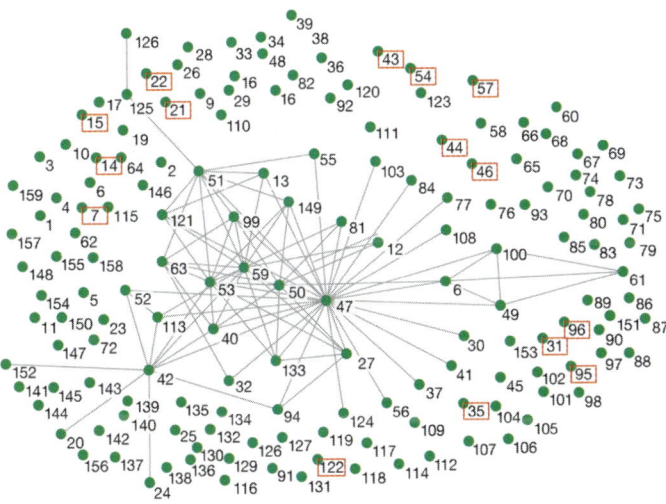
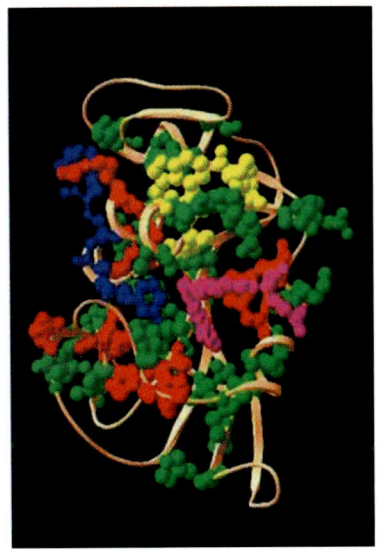

Figure 8

(*left*) The evolutionarily conserved network of amino acid residues in DHFR, obtained by application of a sequence-based statistical method by T. Selzer, R. Ranganathan, and S.J. Benkovic. (*right*) The highly braced residues (*yellow*) (i.e., those representing foci in the network), the other coevolving residues (*green*), the highly conserved residues (*red*), the cofactor (*blue*), and the substrate (*magenta*) are shown.

Many of the residues are in the active site and, hence, impact the binding of the substrate and coenzyme. Several conserved residues (i.e., residues 41–43, 60–63, and 121–123) are in the distal regions that are implicated by kinetic and structural studies to play an important role. Residues 122 and 15, which are hydrogen bonded in the closed conformation of the Met-20 loop, are absolutely conserved. These sequence conservation patterns were also noted in a more comprehensive alignment of 121 DHFR sequences. In all cases, the overall fold is retained. This analysis adds further supporting evidence for a coupled network.

The above analysis, of course, overlooks the importance of nonconserved residues in the creation of the complex interactions between the enzyme and the ligands. To address this issue, a sequence-based statistical method (39) for quantitatively mapping the global network of amino acid residues in a protein has been applied to DHFR by T. Selzer, R. Ranganathan, and S.J. Benkovic. In brief,

each amino acid position in DHFR is compared to that same position in a hypothetical protein derived from the mean distribution of amino acids found naturally in a large protein database. If a site makes a contribution to a structure/function parameter, the amino acid distribution at the site or sites should deviate from the mean. Functional coupling of two amino acid sites should exert an evolutionary bias that is reflected in the distribution of amino acids at these two sites, i.e., the identities of the residues depend on one another. The strength of the coupling can be quantitatively scored. The output from such an analysis is shown in **Figure 8**.

Unexpectedly, rather than a homogeneous pattern over many residues, one observes a fully interconnected unit with a highly heterogeneous coevolutionary pattern. Many residues that coevolve are within van der Waals contact. Some residues of the network, such as Gly-51–Gly-121 and Val-40–Ala-81, are robust, braced by many other interactions. The more highly braced residues lie somewhat outside the active site (yellow). Many are long-range, such as Met-42–Thr-113 and Met-42–Gly-121 (green). In total, nearly all regions of the protein are involved. This heterogeneous, highly braced network can be interpreted as a general design feature of enzymes: As binding and catalysis begin to develop in a given protein sequence, this information is protected from loss by reinforcing key residues.

Conformational Changes in DHFR

Investigations of the conformations of various surrogate DHFR complexes by ^{15}N spin relaxation generally were complementary to the crystallography studies and established that (*a*) the binary complex with the substrate/product analog folate (E·F) and the ternary complex with folate and dihydronicotinamide (E·F·DHNH) form occluded loop complexes and thus mimic the product complexes and that (*b*) the ternary complex with folate and NADP$^+$ (E·F·N$^+$) adopts a closed loop conformation in which substrate and cofactor are correctly positioned for hydride transfer within the active site, and this consequently serves as a model for the reactive ternary Michaelis complex (25, 28, 41). Thus, as noted earlier, the active-site loops Met-20, βF-βG, and βG-βH must cycle between the closed and occluded conformations during turnover. These conformational transitions require large-scale reorganization of various regions of the protein with backbone atoms moving as much as 10 Å. Loop mutations or deletions that destabilize the closed conformation impede hydride transfer (32, 33, 42).

Germane to this discussion is the timescale of these motions, which is admirably suited to NMR measurement. The ^{15}N spin relaxation studies showed large-amplitude picosecond to nanosecond timescale motions of the amides in the backbone of the Met-20 loop that are dampened upon loop closure, as depicted in **Figure 9** (28, 41). The extent of these motions is not limited to the active-site region but is visible throughout the protein and exhibits similar dampening during the closed

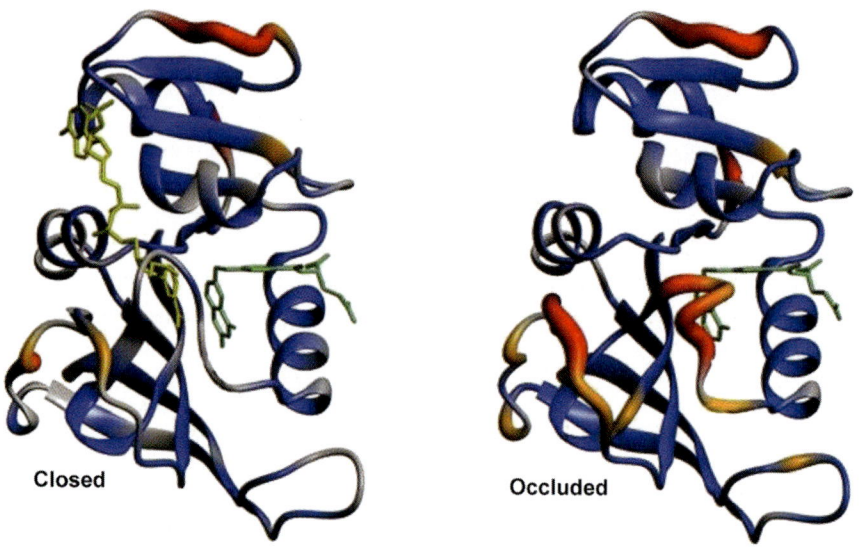

Figure 9

Depiction of the amide backbone motions determined by ^{15}N spin relaxation studies for the closed and occluded complexes of DHFR. The orange and red regions correspond to relaxation rates on the nanosecond to picosecond timescales, respectively. Reproduced with permission from Reference 41.

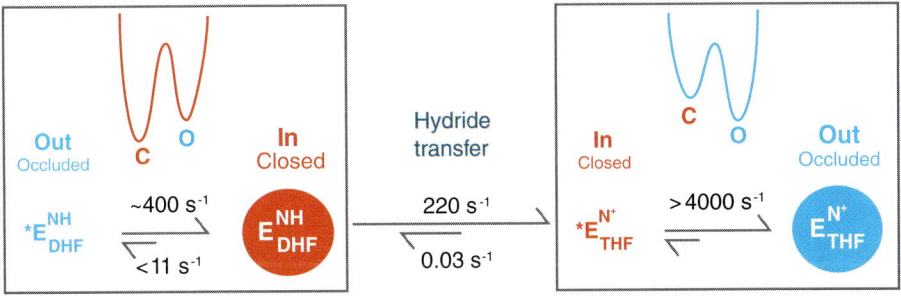

Figure 10

Schematic diagram showing the additional steps (in boxes) implicated in the transitions from the closed Michaelis complex to the occluded product ternary complex. The experimentally observed kinetic intermediates are circled, and closed (*red*) and occluded (*blue*) conformations are indicated. Schematic free energy diagrams, indicating the relative energies of the closed and occluded states, deduced from the present relaxation dispersion studies of the DHFR·folate·NADP$^+$ model of the Michaelis complex, are shown. Excited states associated with the closed-occluded equilibria are marked with an asterisk. "In" and "out" indicate whether the nicotinamide ring of the cofactor occupies the active site or is displaced by the Met-20 loop. The estimated rates, extrapolated to 25°C, for the conformational transitions are shown. The rates for hydride transfer are those at pH 7. Figure and caption reproduced with permission from Reference 30.

to occluded loop transition. The microsecond to millisecond timescale domain (i.e., the same regime as the catalytic cycle) expected for exchanging conformers is accessed by the application of relaxation dispersion protocols (30). Experiments on the closed E·F·NADP$^+$ complex, a surrogate for the Michaelis complex, revealed it to be favored in an equilibrium with an occluded form that would expel the nicotinamide ring from the active site, fluctuating at rates in the range of those seen in the catalytic cycle (30). The product complex (E·H$_4$F·N$^+$), which features the occluded conformation resulting from an increased pterin ring pucker that sterically repels the NADP$^+$, must arise from the decay of the initially closed ternary E·H$_4$F·N$^+$ product complex to the favored occluded form. This mechanistic scheme is depicted in **Figure 10** (30). From a combination of heteronuclear single-quantum coherence spectra and dispersion measurements, its rate of decay can also be estimated (29). Measurements of side-chain coupling constants for Ile-14, which packs against the nicotinamide ring of the cofactor, suggest that the methyl group movement is also on this timescale (43).

A second approach aimed at linking the conformational changes to the catalytic step is to monitor them directly as the enzyme loops through the catalytic cycle. This can be achieved by either ensemble or single-molecule fluorescence measurements that use DHFR labeled with donor/acceptor pairs at amino acid positions subject to conformational change (44). Specifically, the active-site Met-20 loop was labeled with a fluorescence quencher, QSY35, at amino acid position 17 and with a fluorescent probe, Alexa555, at amino acid 37 after both positions had been converted to cysteines. The double-labeled enzyme, which is depicted in **Figure 11**, retained full catalytic activity.

At pH 7.0–7.5, the average observed pH-independent rate of fluorescence change is approximately 220 s^{-1} and is found when both NADPH and H$_2$F are added, as well as when only H$_2$F is added. The change results from the two residues moving apart by <2.0 Å. Because the change occurs with H$_2$F alone and no change in rate is observed when NADPD

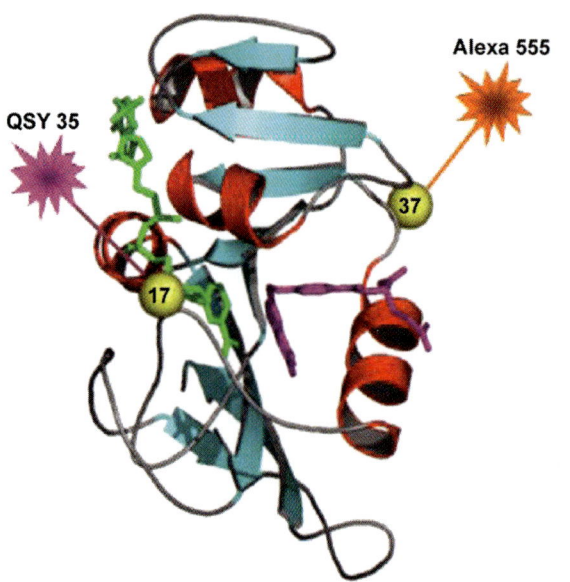

Figure 11

DHFR structure showing amino acid residues 17 and 37 (*yellow spheres*) and the probes (*purple* and *orange stars*) used in the fluorescence experiments. Figure reproduced with permission from Reference 44.

is substituted for NADPH, this conformational change must precede hydride transfer. Single-molecule measurements of a statistically large number of lifetimes for the interconversion of $E \cdot H_2F \cdot NH$ and $E \cdot H_4F \cdot N^+$ provide similar ensemble-averaged rates with no indication of an isotope effect. These results suggest that the observed conformational change does not include the actual chemical step of hydride transfer. However, introduction of a Gly121Val mutation, which is associated with a 160-fold reduction in the rate of hydride transfer, reduces the observed rate of conformational change to approximately 100 s^{-1}. Thus, the conformational change being observed is clearly associated with the catalytic process.

In summary, experimental evidence has been provided for two types of conformational fluctuations: those on the nano- to picosecond timescale and those on the milli- to microsecond timescale. The latter motions are clearly on the same timescale as the events making up the catalytic cycle. These rates are also sensitive to mutagenesis, yet to date, there is no experimental evidence for a protein motion concurrent with the hydride transfer. Theoretical studies have provided further insight into these issues.

Theoretical Studies of Motion in DHFR

The role of motion in DHFR has been studied with a variety of theoretical approaches. Mixed quantum mechanical/molecular mechanical calculations and classical molecular dynamics simulations have been used to investigate the domain motions and conformational changes that are relevant to hydride transfer (45–51). In one of these studies (47), classical molecular dynamics simulations were performed on the three ternary complexes in the preferred pathway. The starting points for the simulations were crystal structures that represented the $DHFR \cdot NADPH \cdot H_2F$, $DHFR \cdot NADP^+ \cdot H_4F$, and $DHFR \cdot NADPH \cdot H_4F$ complexes. Residue-residue-based maps of correlated motions representating fluctuations about an average structure were generated for all three complexes. These fluctuations are local to the average structure and typically occur on the femtosecond to picosecond timescale. In these maps, regions of correlated and anticorrelated motions were defined in which the two residues moved in concert either in the same or in opposite directions, respectively. Strong correlated and anticorrelated motions involving spatially distinct regions in the protein structure were observed for the Michaelis complex with dihydrofolate but were absent in the product complexes with tetrahydrofolate. These correlated motions appeared in many of the same spatial regions of the protein as implicated by the dynamic NMR measurements. Mutant DHFR enzymes with reduced activity exhibited a reduction in these correlated motions compared to the wild-type system (48, 52).

Hybrid quantum/classical molecular dynamics simulations have provided evidence of a network of coupled motions extending throughout the protein and ligands (38, 53, 54). These coupled motions represent equilibrium, thermally averaged conformational changes along the collective reaction coordinate, leading to configurations conducive

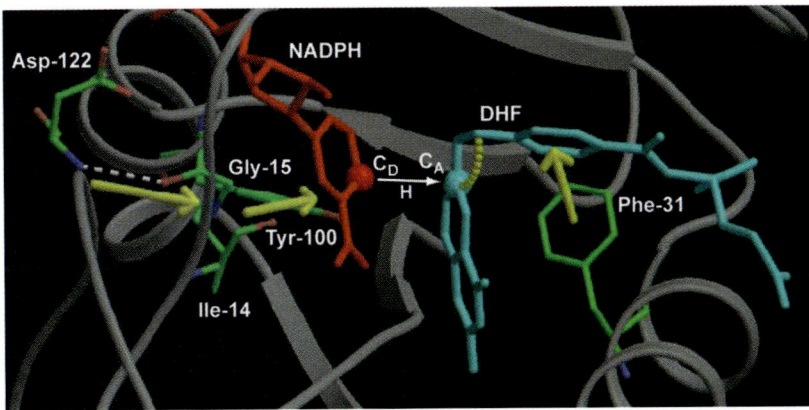

Figure 12

Schematic diagram of a portion of a network of coupled motions in DHFR. The yellow arrows and arc indicate the coupled motions. This picture does not represent a complete or unique network but illustrates the general concept of reorganization of the enzymatic environment to provide configurations conducive to the hydride transfer reaction. Reproduced with permission from Reference 38.

to reaction. The equilibrium motions in this network are not dynamically coupled to the chemical reaction but give rise to configurations that facilitate hydride transfer through short transfer distances, suitable orientation of the substrate and cofactor, and an appropriate electrostatic environment for charge transfer. A portion of this network of coupled motions is illustrated in **Figure 12**. Recent simulations based on alternative approaches identified similar variations in some of the geometrical properties in this network (52, 55, 56). Moreover, subsequent to the identification of Ile-14 as a key player in this network of coupled motions, NMR experiments confirmed the catalytic importance of this residue, as described above.

Applications of the hybrid quantum/classical approach to mutant DHFR enzymes have provided further insights into this network of coupled motions. Hybrid simulations of the Gly121Val mutant resulted in a rate reduction that is consistent with the experimental rate measurements and suggested that the mutation may modify the network of coupled motions through nonlocal structural perturbations, thereby increasing the free energy barrier and decreasing the reaction rate (54). A more comprehensive analysis of coupled motions correlated to hydride transfer was applied to the triple mutant M42F-G121S-S148A and the associated single and double mutants (35). The results illustrated that each mutant samples a unique distribution of motions. The observation of nonadditive changes to the network of coupled motions provided an explanation for the experimentally observed nonadditivity of the rates. Furthermore, these calculations indicated that site-specific mutations distal to the active site can introduce subtle structural perturbations that impact the catalytic rate by altering the conformational sampling of the entire enzyme. Because distal regions of the enzyme are coupled to each other through long-range electrostatics and extended hydrogen-bonding networks (57), the introduction of a site-specific mutation alters the thermal motions of the entire enzyme. Altering the thermal motions of the enzyme affects the probability of sampling conformations conducive to the catalyzed chemical reaction, thereby impacting the free energy barrier and the rate.

Similar conclusions were drawn from other theoretical studies. High-temperature classical molecular dynamics simulations of wild-type DHFR and the Gly121Val mutant indicated differences in the stability and unfolding for the wild-type and mutant enzymes (49). These observations were interpreted to suggest that the mutation causes nonlocal structural effects that may lead to perturbation of the network of coupled motions (49). In another theoretical investigation of wild-type and mutant DHFR enzymes, the mutations were found to influence the rate of hydride transfer by altering the equilibrium conformational distributions (52). These interpretations are consistent with those drawn from the hybrid quantum/classical molecular dynamics simulations.

In an effort to further understand the role of distal protein motions, hybrid quantum/classical molecular dynamics simulations in which a single constraint was applied to the distance between the alpha carbons of residues 17 and 37 were performed (58). The constrained residues are on the exterior of the enzyme and are separated by ∼28 Å. The choice of this constraint was motivated by the ensemble and single-molecule fluorescence experiments in which residues 17 and 37 were labeled with fluorescent probes, as illustrated in **Figure 11**. As discussed above, the experimental data indicate motion between residues 17 and 37 on a similar timescale as that of the hydride transfer reaction. Simulations of DHFR without the constraint predicted a change of ∼0.5 Å in the thermally averaged equilibrium distance between the alpha carbons of residues 17 and 37 during hydride transfer (57). Constraining the distance between these distal residues was found to increase the free energy barrier for hydride transfer by ∼3 kcal/mol (58). These simulations suggested that freezing a single motion between distal residues alters the network of coupled motions and the conformational sampling of the entire system enough to significantly increase the free energy barrier and decrease the rate of hydride transfer. Despite the changes in conformational sampling introduced by the constraint, the constrained system was found to assume a transition state conformation similar to that of the unconstrained system. Modified thermal sampling was found to result in a substantial increase in the average donor-acceptor distance for the reactant state, however, thereby decreasing the probability of sampling the transition state conformations with the shorter distances required for hydride transfer. Again, these studies illustrate the importance of conformational sampling of configurations favorable to the chemical reaction.

LIVER ALCOHOL DEHYDROGENASE

Horse LADH catalyzes the oxidation of a variety of primary, secondary, branched, and cyclic alcohols to the respective aldehydes with the concomitant reduction of NAD^+ to NADH. The active enzyme has a molecular weight of 80,000 and is a dimer of two identical subunits, each of which has a coenzyme and a substrate-binding catalytic domain containing a tetracoordinated zinc (59, 60). Numerous X-ray crystallographic structures have been determined for LADH with the coenzyme and different substrates (61, 62). On the basis of the structure in Reference 62, the LADH dimer is depicted in **Figure 13**, and a

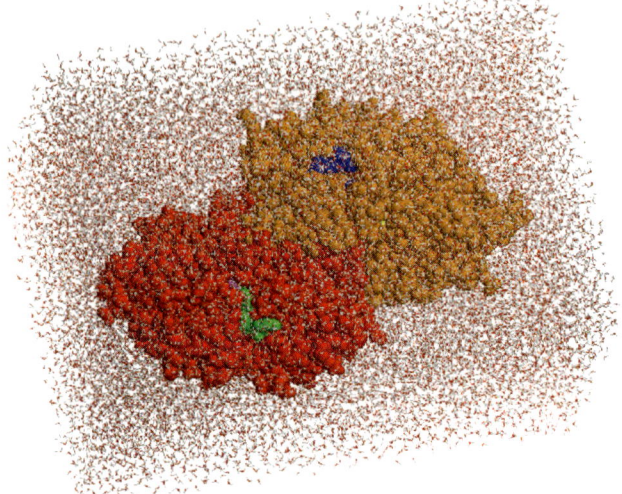

Figure 13

LADH dimer surrounded by water molecules. The protein structure was obtained from the X-ray crystallographic structure of Ramaswamy et al. (62).

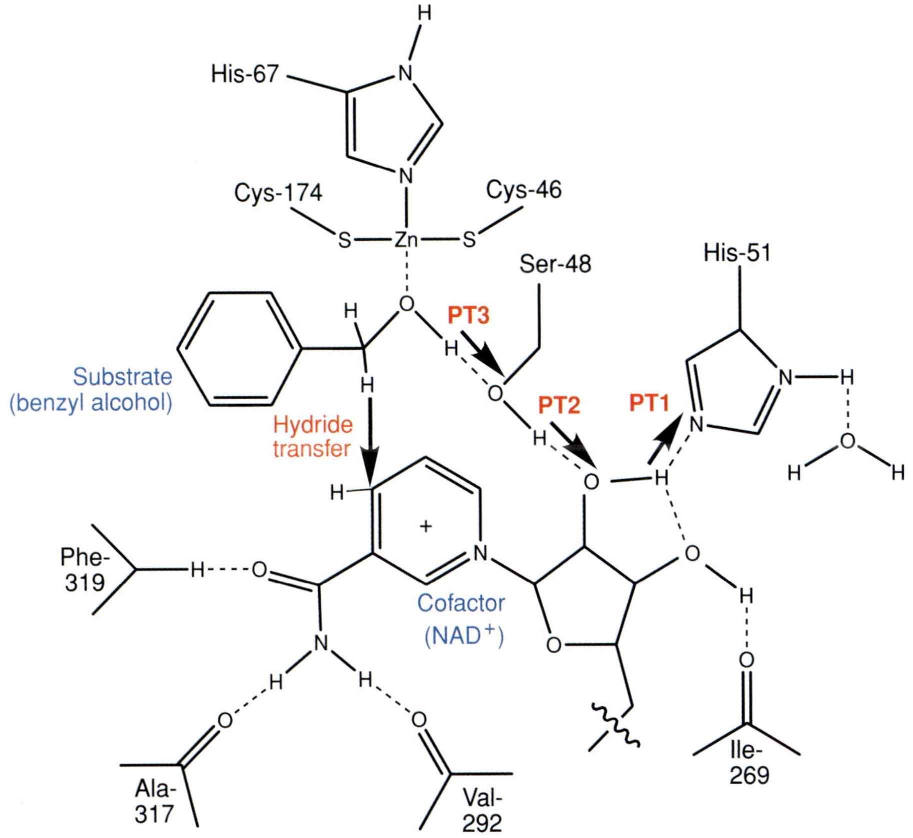

Figure 14

Schematic illustration of the active site of LADH with a benzyl alcohol substrate and an NAD$^+$ cofactor. The black arrows indicate the hydride transfer reaction, and three steps of the proton relay are labeled PT1, PT2, and PT3. Reproduced with permission from Reference 63.

schematic picture of the active site is provided in **Figure 14**. This crystal structure indicates that the substrate is coordinated to the catalytic zinc ion, which is also coordinated to residues Cys-46, Cys-174, and His-67.

The mechanism for alcohol oxidation by LADH has been studied with a variety of experimental methods. The hydride transfer is known to occur directly between the substrate and the NAD$^+$ cofactor, but the mechanism of proton transfer is not as well characterized. The crystal structures suggest a proton relay pathway in which the proton is transferred to His-51 through a hydrogen-bonded network containing the hydroxyl groups of Ser-48 and the nicotinamide ribose, as depicted in **Figure 14** (62, 63). Studies of the pH dependence of the LADH reaction were interpreted to implicate a mechanism in which the deprotonation of alcohol occurs prior to the hydride transfer (64, 65). Measurements of an inverse solvent deuterium isotope effect indicated that hydride transfer is accompanied by proton movement and was interpreted in the context of a low-barrier hydrogen bond between the zinc-bound alkoxide ion and the hydroxyl group of Ser-48 in the reactant state (66). The majority of kinetic studies support the following mechanism for alcohol oxidation by LADH (59, 60):

1. Binding of the coenzyme NAD$^+$;
2. Binding of the alcohol substrate by coordination to zinc;
3. Deprotonation of the alcohol, leading to a zinc-bound alkoxide ion;
4. Hydride transfer from the alkoxide ion to NAD$^+$, leading to NADH and a zinc-bound aldehyde or ketone;
5. Release of the product aldehyde; and
6. Dissociation of NADH.

Estimates of the relevant rate constants have been obtained from kinetic studies (67).

Conformational Changes and Mutagenesis in LADH

Similar to DHFR, LADH exhibits significant conformational changes during the catalytic cycle (61, 62, 68). The catalytic domain rotates about 10 degrees upon formation of the E·NAD$^+$·alcohol complex and undergoes a transition from an open to a closed conformation (66). A flexible loop containing residues 293–298 in the coenzyme-binding domain rearranges to accommodate the change from the open to the closed form (66). A combination of kinetic, crystallographic, and NMR studies of the Phe93Ala mutant indicated that the substitution of Phe-93 with alanine greatly increases the mobility of the benzyl alcohol substrate and concomitantly decreases the rate of hydride transfer (68). These results were interpreted to suggest a relationship between substrate mobility and catalytic efficiency. This partial summary of the extensive data is consistent with the concept of conformational change in the preorganization of the active site that facilitates catalysis.

The chemical step of hydride transfer can be "unmasked" from the complexity of the kinetic sequence for the wild-type enzyme by mutagenesis in the alcohol-binding site to enable the precise measurement of the hydride transfer rate (69). In particular, substitution of residues 57 and 93, which are in van der Waals contact with the bound alcohol substrate, with bulkier residues reduces the size of the substrate-binding pocket and therefore increases the rate of product release. The mutation of the hydrophobic residue Val-203, which is positioned against the backside of the bound nicotinamide cofactor, was also found to significantly impact the hydride transfer rate (70). The positions of these key residues are depicted in **Figure 15**.

To investigate the role of hydrogen tunneling in LADH, the primary and secondary k_H/k_T and k_D/k_T kinetic isotope effects for the oxidation of benzyl alcohol were measured for wild-type and mutant LADH enzymes. The relationship between these two kinetic isotope effects can be used to probe deviations from classical behavior and to detect hydrogen tunneling. The Swain-Schaad exponent (71), defined as the ratio of $\ln(k_H/k_T)$ to $\ln(k_D/k_T)$, is 3.3 for classical behavior but can be significantly larger than 3.3 in the presence of hydrogen tunneling (72, 73). Typically the effects of tunneling are more readily detected in the Swain-Schaad exponent for secondary than for primary kinetic isotope effects (70, 73). The unmasked enzymes Leu57Phe and Phe93Trp were found to have Swain-Schaad exponents of 8.5 and 6.1, respectively, for the secondary kinetic isotope effects. These relatively large values are indicative of tunneling. A decrease in the size of the 203 residue, i.e., changing Val to Leu, Ala, or Gly, leads to a decline in both catalytic efficiency and tunneling as measured by the Swain-Schaad exponent. This nearly linear relationship between catalytic efficiency and the Swain-Schaad exponent for the secondary kinetic isotope effect is depicted in **Figure 16**.

X-ray crystal structures of the Phe93Trp and Val203Ala mutants complexed with NAD$^+$ and the substrate analog trifluoroethanol revealed significant differences in the active-site geometries, as depicted in **Figure 17**. For the Phe93Trp mutant, the relative orientation of the cofactor and substrate is similar to that in the wild-type enzyme. For the Val203Ala mutant, however, the nicotinamide ring is rotated away from the substrate toward the gap left by the loss of the methyl group of the valine upon its replacement by alanine. This rotation leads to an increase in the distance between the hydride donor carbon of the substrate and the acceptor carbon of the cofactor, leading to a decrease in the extent of tunneling. These results provide an explanation for the decreased catalytic efficiency in the Val203Ala mutant.

A similar set of experiments using isotopic labeling was carried out on *E. coli* DHFR (74). In contrast to the results for LADH, the

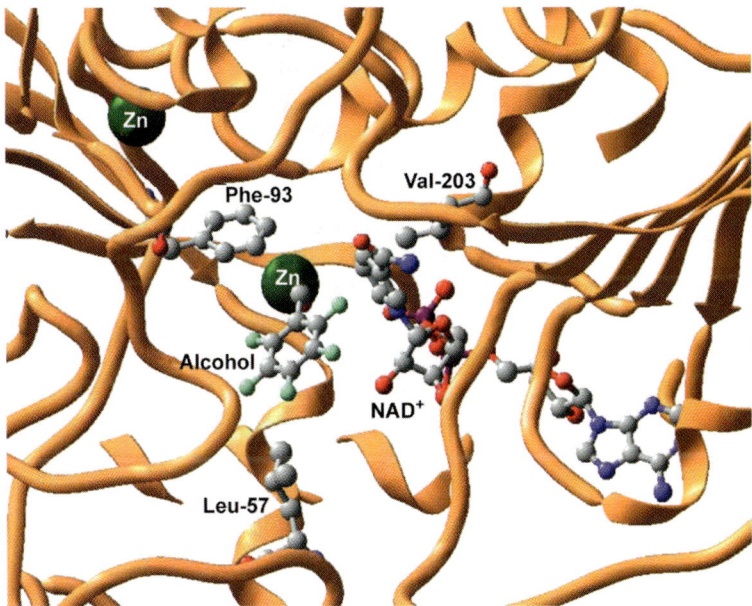

Figure 15
Portion of the X-ray crystallographic structure of LADH (62) depicting the bound pentafluorobenzyl alcohol substrate, the cofactor NAD^+, the structural and catalytic Zn atoms, and the residues Leu-57, Phe-93, and Val-203. The size of the alcohol-binding pocket is reduced when Leu-57 is mutated to Phe or Phe-93 is mutated to Trp, thereby increasing the rate of product release and "unmasking" the tunneling effects (69). Mutation of Val-203 to smaller residues leads to the rotation of NAD^+ away from the alcohol substrate and decreases the rate of hydride transfer (70).

Swain-Schaad exponent from the secondary kinetic isotope effects is only 3.5 for DHFR. This value does not provide an indication of tunneling or coupled motion between the primary and secondary hydrogens. The temperature dependence of kinetic isotope effects has also been used as a probe for tunneling. The intrinsic primary kinetic isotope effects for hydride transfer catalyzed by DHFR were found to be nearly temperature independent (74). The isotope effects on the Arrhenius preexponential factors were larger than the semiclassical limits and were interpreted to suggest extensive tunneling. The temperature dependence of the kinetic isotope effect for this reaction has also been studied theoretically in the framework of ensemble-averaged variational transition state theory with multidimensional tunneling (75). In general, the analysis of the temperature dependence of enzyme reactions is challenging because of the narrow range of temperatures available and the complexity of the reactions.

Theoretical Studies of Motion in LADH

Although early theoretical studies focused on mechanistic issues (76–80), a variety of more recent studies have probed the role of tunneling and protein motion in the hydride transfer reaction catalyzed by LADH. Variational transition state theory with semiclassical tunneling calculations confirmed the importance of tunneling inferred from the experimentally determined kinetic isotope effects (81–83). Hybrid quantum/classical molecular dynamics calculations also provided evidence of hydrogen tunneling (84). Centroid calculations indicated that these tunneling effects are similar in water and the protein and therefore do not represent the basis of catalysis

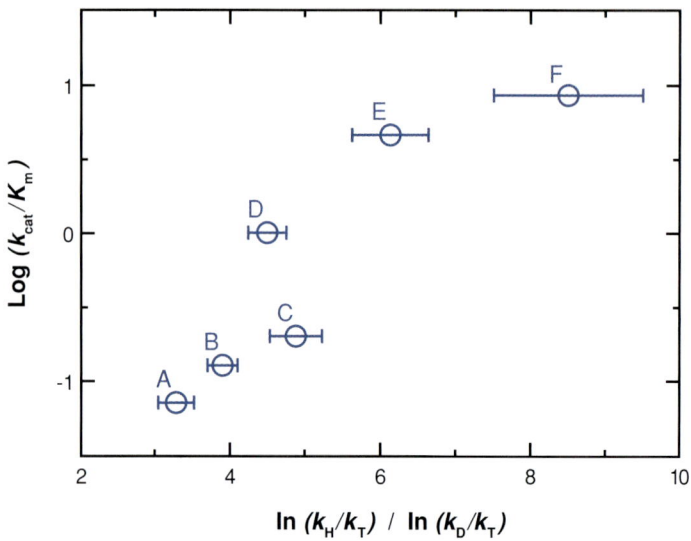

Figure 16

Correlation of the log(k_{cat}/K_m) and the Swain-Schaad exponent for a series of site-directed mutants of LADH. The points are labeled as follows: A, Val203Gly; B, Val203Ala:Phe93Trp; C, Val203Ala; D, Val203Leu; E, Phe93Trp; and F, Leu57Phe. Tunneling is thought to be indicated when the Swain-Schaad exponent is greater than 3.3. Reproduced with permission from Reference 70.

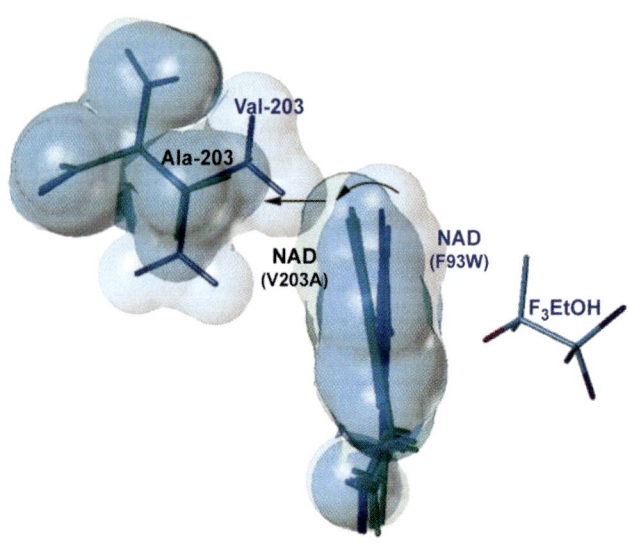

Figure 17

Comparison of the nicotinamide rings in Val203Ala and Phe93Trp. The surfaces are drawn at ≈90% of the van der Waals radii. In Phe93Trp, the nicotinamide ring is in van der Waals contact with one of the methyl groups of Val-203. In Val203Ala, the nicotinamide ring rotates (*curved arrow*) to fill the gap left by replacement to alanine (*straight arrow*). Reproduced with permission from Reference 70.

(85). These centroid calculations also illustrated that nonequilibrium dynamical effects do not contribute significantly to catalysis for this system and that the most important contribution to catalysis arises from the reduction of the activation free energy by electrostatic effects.

The relation between specific enzyme motions and enzyme activity in LADH has been analyzed by examining thermally averaged properties along the collective reaction coordinate (84). Similar to DHFR, the thermally averaged donor-acceptor distance decreases as the reaction evolves from the reactant to the transition state. Conversely, the thermally averaged distance between Val-203 and the acceptor carbon was found to increase as the reaction evolves from the reactant to the transition state, suggesting that the relative motion of Val-203 and the acceptor carbon contributes to the collective reaction coordinate and participates in the network of coupled motions that facilitates hydride transfer (84). The conformation of Val-203 relative to the substrate and cofactor in the X-ray crystal structure is illustrated in **Figure 18**. This analysis provided insight into the experimental observation that the enzyme activity decreases when Val-203 is substituted by the smaller residue alanine (70). Specifically, these results indicated that the impact of the mutation on the enzyme activity is due to alteration of the equilibrium conformational sampling and, hence, the free energy barrier rather than nonequilibrium dynamical factors.

Several other computational studies addressed the issues of protein motion and the role of residue Val-203 in LADH. Classical molecular dynamics of the reactive complex LADH·NAD$^+$·PhCH$_2$O$^-$ (86, 87) allowed the identification of the correlated and anticorrelated motions about the average structure. These correlated and anticorrelated motions were found to involve the entire protein, as observed for the hydride transfer reaction catalyzed by DHFR (47). Moreover, this cross-correlation analysis indicated a pushing motion of Val-203 that assists in moving the acceptor carbon of NAD$^+$ toward the substrate (87). The resulting reactive ground-state conformations, which are termed near-attack conformations, were shown to be associated with the transition states of lowest energy (87). Although these pushing motions occur on a much faster timescale than the turnover rate of the enzyme, these fast motions could influence the sampling of conformations favorable for hydride transfer.

In another theoretical study, a spectral density analysis of classical molecular dynamics simulations was interpreted to suggest that the substrate-NAD$^+$ motion plays the role of a promoting vibration symmetrically coupled to the reaction coordinate (88). This analysis indicated that the relative motions between the alcohol and NAD$^+$ and between NAD$^+$ and Val-203 are in resonance, leading the authors to conclude that the promoting vibration is induced by Val-203. Because this promoting vibration occurs on the femtosecond timescale, the authors claim that the enzyme goes through many cycles of the promoting vibration before an effective combination of motions occurs to produce catalysis. These fast donor-acceptor vibrations will not lead to catalysis, however, unless the average donor acceptor-distance is decreased substantially from the equilibrium reactant distance. Thus, the rate-limiting step is expected to be the decrease in the thermally averaged donor-acceptor distance and the associated conformational changes in the network of equilibrium-coupled motions characterized in Reference 84.

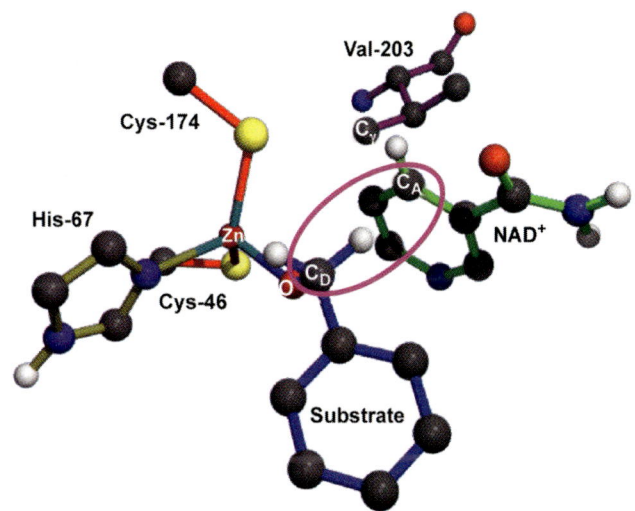

Figure 18

A portion of the crystal structure (62) of the active site of horse liver alcohol dehydrogenase with NAD$^+$ and benzyl alkoxide, wherein the positions of the heavy atoms of the substrate correspond to those of pentafluorobenzyl alcohol. The oval ring identifies the reaction core, consisting of the donor carbon atom C$_D$, the transferring hydride H, and the acceptor carbon atom C$_A$. Cγ is defined to be the Cγ of Val-203 closest to C$_A$. Reproduced with permission from Reference 84.

CONCLUSIONS

The above discussion clearly illustrates that the inherent mobility of a protein fold is manifest in the various steps constituting the catalytic cycle (89–91). In fact, one or more of those steps probably reflects an intrinsic motion of part of the enzyme's structure. We can with certainty state that this behavior is universal for enzymatic catalysis and believe that most would agree (92–97).

The nature of this linkage, posed as three questions in the introduction, is most surely complex. There are multiple examples of conformational changes within an enzyme-substrate complex following association of the substrate with the free enzyme, and many are associated with significant changes of structural elements both proximal and distal to the active site (41, 98–104). Perhaps all would agree that the steps to the reactive Michaelis complex represent the preorganization of substrate and active-site residues for initiating the actual chemical transformation. The nature of events along the climb to the transition state, however, is less clear. The picture emerging from experimental and theoretical studies of DHFR is that thermal motions of the enzyme, substrate, and cofactor lead to conformational sampling of configurations that facilitate hydride transfer. These thermal motions occur within the confines of the enzyme structure and form a coupled network that brings the donor and acceptor closer together, orients the substrate and cofactor properly, and provides a favorable electrostatic environment. In other words, a continuous series of thermally averaged equilibrium configurations is sampled along the collective reaction coordinate, which connects the reactant to the transition state to the product. Such a series of thermally averaged equilibrium configurations for DHFR is illustrated in **Figure 19**. In this figure, the free energy profile is depicted along a collective reaction coordinate that includes motions of the enzyme, substrate, and cofactor. The transition state corresponds to the set of configurations with a reaction coordinate of zero. In general, the free energy corresponds to the probability of sampling configurations with a specified reaction coordinate. The free energy barrier arises from the lower probability of sampling transition state configurations than reactant configurations. This figure also indicates that the thermally averaged donor-acceptor distance decreases as the reaction progresses from the reactant to the transition state.

In addition, there is the issue of the multiple timescales that characterize various motional elements within the fold. Which of these significantly determine the rate of enzymatic turnover? As noted above, there is direct evidence for DHFR of conformational changes on the same timescale as observed for hydride transfer in the kinetic analysis of the catalytic cycle. The experimentally measured hydride transfer rate on the millisecond timescale corresponds to the rate of transformation from the equilibrium reactant configuration to the equilibrium product configuration. As illustrated in **Figure 19**, the rate-limiting process can be viewed in terms of the conformational changes along the collective reaction coordinate. Note that the actual transfer of the hydrogen is virtually instantaneous relative to these conformational changes. The hydrogen remains bonded to the donor until the system achieves a transition state conformation, at which point the hydrogen transfers and becomes bonded to the acceptor for the remainder of the process. When the hydrogen is viewed as a quantum mechanical wavefunction, it is delocalized between the donor and acceptor only for the relatively small number of transition state configurations. The experimentally observed kinetic isotope effects can be explained within this framework. Zero-point energy effects influence the relative energies of the configurations and, hence, the conformational sampling, thereby altering the free energy barrier. For those reactions with significant tunneling, the probability of tunneling at the favorable configurations depends on the mass of the transferring nucleus, as well as the donor-acceptor distance.

The conformations depicted in **Figure 19** can be attained by protein motions that occur on the pico- to nanosecond timescales and therefore do not directly appear as steps in the catalytic cycle. These fast motions, although Brownian in nature, have restraints imposed by the protein fold and create configurations of substrate and cofactor within the confines

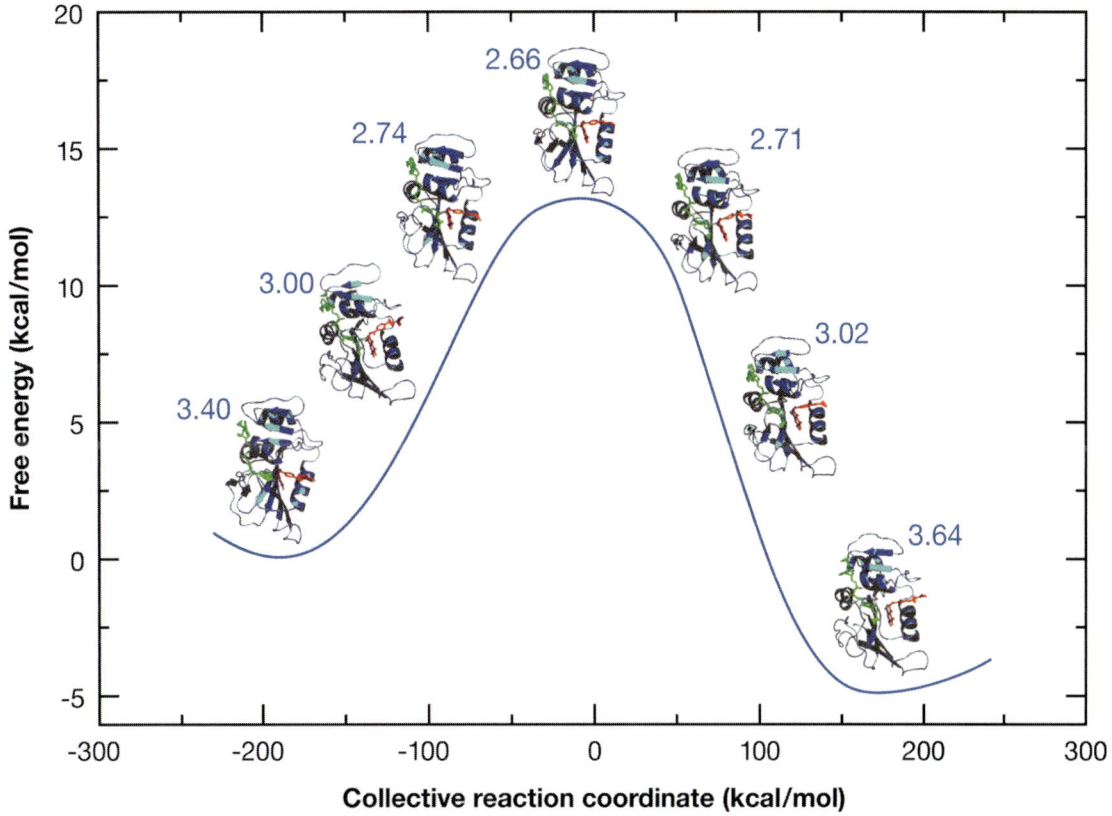

Figure 19
Free energy profile of the DHFR-catalyzed hydride transfer reaction as a function of a collective reaction coordinate that includes motions of the enzyme, substrate, and cofactor. The magnitude of the free energy barrier is determined by the relative probabilities of sampling the transition state and the reactant configurations. The thermally averaged equilibrium structures, as well as the average donor-acceptor distances in Angstroms, are provided for selected values of the reaction coordinate. Note that the donor-acceptor distance decreases as the reaction evolves from the reactant to the transition state. The conformational changes along the collective reaction coordinate are attained by equilibrium thermal motions occurring within the confines of the protein fold and reflect a network of coupled motions extending throughout the protein. This network facilitates the hydride transfer reaction by bringing the donor and acceptor closer together, orienting the substrate and cofactor properly, and providing a favorable electrostatic environment. Reproduced with permission from Reference 58.

of the active site that facilitate hydride transfer. We emphasize that these fast motions are in equilibrium as the reaction progresses along the reaction coordinate, and the overall average equilibrium conformational changes occur on the slower timescale measured experimentally. This view, also expounded by others, has the enzyme's rapid motions creating and sampling active-site populations. The question then is whether the motions on a milli- to microsecond timescale within the enzyme's structure that chaperone the chemical transformation are concurrent with the actual hydride transfer. So far, to our knowledge, there is no evidence for that type of coupling, but efforts to detect such phenomena are still in the early stages.

The extensive experimental studies on DHFR leave no doubt that all regions of the protein are involved and therefore contribute to the facilitation of substrate and cofactor binding as well as proton and hydride transfer. We expect this to be true of enzymes in general, and there are many examples of less conclusive studies of other enzymes that infer the same outcome. We also suspect that coupled networks of residues that act as units within the enzyme's framework, not necessarily confined to a region of β-sheet, α-helix, or loop, are also the norm. The observation of key amino acids that are conserved across species, as well as the evolutionary covariation of residues to construct a robust matrix that enables both folding and catalysis, supports our hypothesis. We can only marvel that evolution has constructed more than 200 DHFRs with nearly identical catalytic efficiency but only random amino acid sequence identity. Redundancy, therefore, occurs in the context of both the protein fold and catalysis (105). Much remains to be understood to allow de novo rather than combinatorial design of enzymes with high catalytic efficiency and specificity.

SUMMARY POINTS

1. The inherent mobility of a protein fold is manifest in the various steps constituting the catalytic cycle.

2. Thermal motions of the enzyme, substrate, and cofactor lead to conformational sampling of configurations that facilitate the chemical reaction. These thermal motions are Brownian in nature but have restraints imposed by the protein fold. They form a coupled network that brings the substrate and cofactor closer together, orients them properly, and provides a favorable electrostatic environment for the chemical reaction.

3. All regions of the protein are involved in this network of coupled motions and contribute to the facilitation of substrate and cofactor binding as well as the chemical reaction.

4. The picosecond to nanosecond thermal motions are in equilibrium as the reaction progresses along the reaction coordinate, and the overall average equilibrium conformational changes occur on the slower timescale measured experimentally.

5. There is direct experimental evidence in DHFR of conformational changes on the same timescale as observed experimentally for hydride transfer.

6. The conservation of key amino acids across species, as well as the evolutionary covariation of residues to construct a robust matrix that enables both folding and catalysis, provides evidence of coupled networks. The evolutionary analyses indicate that redundancy occurs in the context of both the protein fold and catalysis.

FUTURE ISSUES TO BE RESOLVED

1. An unresolved issue is whether there are motions on a milli- to microsecond timescale within the protein structure that are concurrent with and directly coupled to the chemical reaction.

2. The development of experimental and computational methods to enable de novo rather than combinatorial design of enzymes with high catalytic efficiency and specificity is needed.

LITERATURE CITED

1. Fischer E. 1894. *Ber. Dtsch. Chem. Ges.* 27:3189–232
2. Koshland DE Jr, Neet KE. 1968. *Annu. Rev. Biochem.* 37:359–410
3. Bennett WS, Huber R. 1984. *Crit. Rev. Biochem.* 15:291–384
4. Davis JH, Agard DA. 1998. *Biochemistry* 37:7696–707
5. Schnackerz KD, Mozzarelli A. 1998. *J. Biol. Chem.* 273:33247–53
6. Cannon WR, Singleton SF, Benkovic SJ. 1996. *Nat. Struct. Biol.* 3:821–33
7. Jencks WP. 1975. *Adv. Enzymes* 43:219–410
8. Bruice TC. 2002. *Acc. Chem. Res.* 35:139–48
9. Benkovic SJ, Hammes-Schiffer S. 2003. *Science* 301:1196–202
10. Perutz MF. 1967. *Proc. R. Soc. London Ser. B* 167:448
11. McCammon JA, Harvey SC. 1987. *Dynamics of Proteins and Nucleic Acids*. New York: Cambridge Univ. Press
12. Hammes GG. 2002. *Biochemistry* 41:8221–28
13. Wüthrich K, Wagner G. 1978. *Trends Biochem. Sci.* 9:152–54
14. Palmer AG. 2001. *Annu. Rev. Biophys. Biomol. Struct.* 30:129–55
15. Wand AJ. 2001. *Nat. Struct. Biol.* 8:926–31
16. Hubbell WL, Cafiso DS, Altenbach C. 2000. *Nat. Struct. Biol.* 7:735–39
17. Weiss S. 2000. *Nat. Struct. Biol.* 7:724–29
18. Moffat K. 2001. *Chem. Rev.* 101:1569–81
19. Daggett V. 2000. *Curr. Opin. Struct. Biol.* 10:160–64
20. Falke JJ, Koshland DE Jr. 1987. *Science* 237:1596–600
21. Careaga CL, Falke JJ. 1992. *J. Mol. Biol.* 226:1219–35
22. Huyghues-Despointes BM, Pace CN, Englander SW, Scholtz JM. 2001. *Methods Mol. Biol.* 168:69–92
23. Chen Y-Q, Kraut J, Blakley RL, Callendar R. 1994. *Biochemistry* 33:7021–26
24. Rajagopalan PTR, Benkovic SJ. 2002. *Chem. Record* 2:24–36
25. Schnell JR, Dyson HJ, Wright PE. 2004. *Annu. Rev. Biophys. Biomol. Struct.* 33:119–40
26. Sawaya MR, Kraut J. 1997. *Biochemistry* 36:586–603
27. Fierke CA, Johnson KA, Benkovic SJ. 1987. *Biochemistry* 26:4085–92
28. Epstein DM, Benkovic SJ, Wright PE. 1995. *Biochemistry* 34:11037–48
29. Venkitakrishnan RP, Zaborowski E, McElheny D, Benkovic SJ, Dyson HJ, Wright PE. 2004. *Biochemistry* 43:16046–55
30. McElheny D, Schnell JR, Lansing JC, Dyson HJ, Wright PE. 2005. *Proc. Natl. Acad. Sci. USA* 102:5032–37
31. Miller GP, Benkovic SJ. 1998. *Biochemistry* 37:6336–42
32. Miller GP, Benkovic SJ. 1998. *Biochemistry* 37:6327–35
33. Miller GP, Wahnon DC, Benkovic SJ. 2001. *Biochemistry* 40:867–75
34. Rajagopalan PTR, Lutz S, Benkovic SJ. 2002. *Biochemistry* 41:12618–28
35. Wong KF, Selzer T, Benkovic SJ, Hammes-Schiffer S. 2005. *Proc. Natl. Acad. Sci. USA* 102:6807–12
36. Deleted in proof
37. Huang Z, Wagner CR, Benkovic SJ. 1994. *Biochemistry* 33:11576–85
38. Agarwal PK, Billeter SR, Rajagopalan PTR, Benkovic SJ, Hammes-Schiffer S. 2002. *Proc. Natl. Acad. Sci. USA* 99:2794–99
39. Suel GM, Lockless SW, Wall MA, Ranganathan R. 2003. *Nat. Struct. Biol.* 10:59–69
40. Deleted in proof

41. Osborne MJ, Schnell J, Benkovic SJ, Dyson HJ, Wright PE. 2001. *Biochemistry* 40:9846–59
42. Li L, Falzone CJ, Wright PE, Benkovic SJ. 1992. *Biochemistry* 31:7826–33
43. Schnell JR, Dyson HJ, Wright PE. 2004. *Biochemistry* 43:374–83
44. Antikainen NM, Smiley RD, Benkovic SJ, Hammes GG. 2005. *Biochemistry* 44:16835–43
45. Shrimpton P, Allemann RK. 2002. *Protein Sci.* 11:1442–51
46. Verma CS, Caves LSD, Hubbard RE, Roberts GCK. 1997. *J. Mol. Biol.* 266:776–96
47. Radkiewicz JL, Brooks CL. 2000. *J. Am. Chem. Soc.* 122:225–31
48. Rod TH, Radkiewicz JL, Brooks CL III. 2003. *Proc. Natl. Acad. Sci. USA* 100:6980–85
49. Swanwick RS, Shrimpton PJ, Allemann RK. 2004. *Biochemistry* 43:4119–27
50. Cummins PL, Greatbanks SP, Rendell AP, Gready JE. 2002. *J. Phys. Chem. B* 106:9934–44
51. Castillo R, Andres J, Moliner V. 1999. *J. Am. Chem. Soc.* 121:12140–47
52. Thorpe IF, Brooks CL III. 2003. *J. Phys. Chem. B* 107:14042–51
53. Agarwal PK, Billeter SR, Hammes-Schiffer S. 2002. *J. Phys. Chem. B* 106:3283–93
54. Watney JB, Agarwal PK, Hammes-Schiffer S. 2003. *J. Am. Chem. Soc.* 125:3745–50
55. Thorpe IF, Brooks CL III. 2004. *Proteins: Struct. Funct. Bioinformatics* 57:444–57
56. Garcia-Viloca M, Truhlar DG, Gao J. 2003. *Biochemistry* 42:13558–75
57. Wong KF, Watney JB, Hammes-Schiffer S. 2004. *J. Phys. Chem. B* 108:12231–41
58. Sergi A, Watney JB, Wong KF, Hammes-Schiffer S. 2006. *J. Phys. Chem. B* 110:2435–41
59. Klinman JP. 1981. *Crit. Rev. Biochem.* 10:39–78
60. Pettersson G. 1987. *Crit. Rev. Biochem. Mol. Biol.* 21:349–88
61. Eklund H, Plapp BV, Samama JP, Branden CI. 1982. *J. Biol. Chem.* 257:14349–58
62. Ramaswamy S, Eklund H, Plapp BV. 1994. *Biochemistry* 33:5230–37
63. Agarwal PK, Webb SP, Hammes-Schiffer S. 2000. *J. Am. Chem. Soc.* 122:4803–12
64. Kvassman J, Pettersson G. 1980. *Eur. J. Biochem.* 103:557–64
65. Kvassman J, Pettersson G. 1980. *Eur. J. Biochem.* 103:565–75
66. Ramaswamy S, Park D-H, Plapp BV. 1999. *Biochemistry* 38:13951–59
67. Shearer GL, Kim K, Lee KM, Wang CK, Plapp BV. 1993. *Biochemistry* 32:11186–94
68. Rubach JK, Plapp BV. 2002. *Biochemistry* 41:15770–79
69. Bahnson BJ, Park DH, Kim K, Plapp BV, Klinman JP. 1993. *Biochemistry* 32:5503–7
70. Bahnson BJ, Colby TD, Chin JK, Goldstein BM, Klinman JP. 1997. *Proc. Natl. Acad. Sci. USA* 94:12797–802
71. Swain CG, Strivers EC, Reuwer JF, Schaad LJ. 1958. *J. Am. Chem. Soc.* 80:5885–93
72. Saunders WH Jr. 1985. *J. Am. Chem. Soc.* 107:164–69
73. Cha Y, Murray CJ, Klinman JP. 1989. *Science* 243:1325–30
74. Sikorski RS, Wang L, Markham KA, Rajagopalan PTR, Benkovic SJ, Kohen A. 2004. *J. Am. Chem. Soc.* 126:4778–79
75. Pu JZ, Ma SH, Gao JL, Truhlar DG. 2005. *J. Phys. Chem. B* 109:8551–56
76. von Onciul AR, Clark T. 1993. *J. Comput. Chem.* 14:392–400
77. Vanhommerig SAM, Meier RJ, Sluyterman LA, Meijer EM. 1996. *J. Mol. Struct. Theochem.* 364:33–43
78. Cardenas R, Andres J, Krechl J, Campillo M, Tapia O. 1996. *Int. J. Quantum Chem.* 57:245–57
79. Ryde U. 1996. *J. Comput. Aided Mol. Des.* 10:153–64
80. Olson LP, Luo J, Almarsson O, Bruice TC. 1996. *Biochemistry* 35:9782–91
81. Alhambra C, Corchado JC, Sanchez ML, Gao JL, Truhlar DG. 2000. *J. Am. Chem. Soc.* 122:8197–203

82. Cui Q, Elstner M, Karplus M. 2002. *J. Phys. Chem. B* 106:2721–40
83. Alhambra C, Corchado J, Sanchez ML, Garcia-Viloca M, Gao J, Truhlar DG. 2001. *J. Phys. Chem. B* 105:11326–40
84. Billeter SR, Webb SP, Agarwal PK, Iordanov T, Hammes-Schiffer S. 2001. *J. Am. Chem. Soc.* 123:11262–72
85. Villa J, Warshel A. 2001. *J. Phys. Chem. B* 105:7887–907
86. Luo J, Bruice TC. 2002. *Proc. Natl. Acad. Sci. USA* 99:16597–600
87. Luo J, Bruice TC. 2004. *Proc. Natl. Acad. Sci. USA* 101:13152–56
88. Caratzoulas S, Mincer JS, Schwartz SD. 2002. *J. Am. Chem. Soc.* 124:3270–76
89. Anfinsen CB. 1973. *Science* 181:223–30
90. Akasaka K. 2003. *Biochemistry* 42:10875–84
91. Parak FG. 2003. *Curr. Opin. Struct. Biol.* 13:552–57
92. Eisenmesser EZ, Bosco DA, Akke M, Kern D. 2002. *Science* 295:1520–23
93. Tousignant A, Pelletier JN. 2004. *Chem. Biol.* 11:1037–42
94. Knapp MJ, Rickert KW, Klinman JP. 2002. *J. Am. Chem. Soc.* 124:3865–74
95. Fersht A. 1999. *Structure and Mechanism in Protein Science. A Guide to Enzyme Catalysis and Protein Folding*. New York: Freeman
96. Karplus M. 2000. *J. Phys. Chem. B* 104:11–27
97. Young L, Post CB. 1996. *Biochemistry* 35:15129–33
98. Deneumax P, Schlick T. 1998. *Biophys. J.* 74:61–72
99. Desamero R, Rozovsky S, Zhadin U, McDermott A, Callendar R. 2003. *Biochemistry* 42:2941–51
100. Wang H, Borchardt RT, Schowen RL, Kuczera K. 2005. *Biochemistry* 44:7228–39
101. Guallar V, Jacobson M, McDermott A, Friesner RA. 2004. *J. Mol. Biol.* 337:227–39
102. Meroveh SO, Roblin P, Golemi D, Maveyraud L, Vakulenko SB, et al. 2002. *J. Am. Chem. Soc.* 124:9422–30
103. Joseph D, Petsko GA, Karplus M. 1990. *Science* 249:1425–28
104. Faber HR, Matthews BW. 1990. *Nature* 348:263–66
105. Todd AE, Orengo CA, Thornton JH. 2002. *Trends Biochem. Sci.* 27:419–26

Animal Cytokinesis: From Parts List to Mechanisms

Ulrike S. Eggert,[1] Timothy J. Mitchison,[2] and Christine M. Field[2]

[1]Dana-Farber Cancer Institute and Department of Biological Chemistry and Molecular Pharmacology, [2]Department of Systems Biology, Harvard Medical School, Boston, Massachusetts 02115; email: ulrike_eggert@hms.harvard.edu, timothy_mitchison@hms.harvard.edu, christine_field@hms.harvard.edu

Key Words

cleavage furrow, contractile ring, midzone microtubules, RNAi

Abstract

The mechanism underlying cytokinesis, the final step in cell division, remains one of the major unsolved questions in basic cell biology. Thanks to advances in functional genomics and proteomics, we are now able to assemble a "parts list" of proteins involved in cytokinesis. In this review, we discuss how to relate this parts list to biological mechanism. For easier analysis, we split cytokinesis into discrete steps: cleavage plane specification, rearrangement of microtubule structures, contractile ring assembly, ring ingression, and completion. We report on the advances that have been made to understand these steps and how they can be integrated into a global understanding of cytokinesis. We also discuss the extent to which classic questions have been answered and identify major outstanding questions.

Contents

INTRODUCTION AND
   RESEARCH APPROACHES 544
A PARTS LIST FOR
   CYTOKINESIS 546
SUBPROCESSES IN
   CYTOKINESIS 549
   Timing Cytokinesis.............. 549
   Cleavage Plane Specification:
      Evaluation of Classic Models... 550
   Cleavage Plane Specification:
      Molecular Mechanisms 552
   Organization of
      Cytokinesis-Associated
      Microtubules................. 553
   Contractile Ring Assembly 555
   Furrow Ingression............... 557
   Completion 558
CYTOKINESIS AND CANCER.... 560

RNAi: RNA interference

INTRODUCTION AND RESEARCH APPROACHES

Cytokinesis, the physical separation of one cell into two, is the last step in the cell cycle. It requires coordinated actions of the cytoskeleton, membrane systems, and the cell cycle engine, which are precisely controlled in space and time. Cytokinesis was described more than 100 years ago, and the roles of actin and myosin in cleavage, as well as that of the mitotic spindle in specifying the cleavage plane, were discovered more than 20 years ago (reviewed in Reference 1). The last few years have seen intensive focus on the identification of proteins involved in cytokinesis, using classical genetics and biochemistry. This effort, coupled with recent systematic RNA interference (RNAi) screens, has culminated in a more or less complete "parts lists." The next challenge, and the focus of this review, is to convert such lists into an interlinked system of molecular mechanisms.

In this review, we focus on cytokinesis in animal cells, especially the two model organisms whose parts lists are now available, nematode (*Caenorhabditis elegans*) embryos and cultured insect (*Drosophila*) cells. Other important systems include cultured mammalian cells, echinoderm and amphibian (*Xenopus*) eggs, and *Drosophila* embryos and spermatocytes. For recent reviews of cytokinesis in fungi and plants, see References 2 and 3. Each model system has unique features, including distinct regulatory and mechanical challenges faced by particular cell types to effect cytokinesis, as well as the more obvious technical pros and cons for different research methods. In cytokinesis research, as in other areas of basic cell biology, our molecular understanding has progressed to the point where the differences between cell types are becoming as interesting as their conserved features (4). For example, large eggs require longer microtubules to position spindles and furrows, and they may sacrifice accuracy-enhancing checkpoints in exchange for faster division. Also, the amount of new plasma membrane required to create two cells from one varies with the size and surface area of a given cell type, leading to differing levels of importance for vesicular trafficking pathways. These biological differences require that conserved mechanisms be implemented in distinct ways in different cells and may, perhaps, also require additional or different mechanisms in particular cell types. When controversies arise in the field, it is important to determine which differences arise from genuine differences between systems and which arise from personal interpretation of particular experiments in a fast-developing field.

Some differences between systems hinge primarily on the technical ease of implementing the three most important approaches in modern cell biology: genetics/genomics, microscopy, and biochemistry. For genetics/genomics, *C. elegans* embryos and *Drosophila* tissue culture cells are currently the most advanced, thanks to genome-wide RNAi screens. RNAi is far from a perfect genetic technology, typically providing an incomplete knockdown of protein levels that may take several days in cultured cells. It is also difficult to score a role in cytokinesis using RNAi for a protein that is important for earlier steps in the cell cycle, such as mitosis. Thus, parts lists determined solely by RNAi screens are incomplete. However, the power of this technology and its applicability to systems that lack traditional genetics have led to an explosion of interest and information. Given the current availability of technology and RNAi libraries, we expect publication of a list of proteins involved in cytokinesis in human cancer cells soon. These will be accomplished first for the commonly used HeLa cell line, originally derived from a cervical carcinoma, which is extensively adapted for life in culture and thus not truly representative of any normal human cell type. Generating parts lists for cytokinesis in untransformed human cells, or cancer cells in situ, will be more difficult.

Protein lists are not by themselves mechanistically informative, but the phenotypes observed using RNAi and traditional genetics provide functional information, especially in conjunction with high resolution microscopy. Microscopy has always been an important tool in cytokinesis research, and the past few years have seen dramatic improvement in optical methods. Genetically encoded fluorescent proteins of different colors are now in routine use in many systems, and the dynamic information they provide is being enhanced by methods for localized photobleaching and photoactivation. Instrumentation has also advanced, with highly sensitive digital cameras and new confocal technologies widely available. Together, these imaging methods are allowing the field to observe molecular events at play inside living cells. The confluence of microscopy with perturbations using genetics/genomics methods, or in some cases physical perturbation using microneedles or UV microbeams, is currently driving the field forward.

The biochemistry of cytokinesis, in contrast, has seen slower progress in the past few years. An extract system for all or part of cytokinesis has yet to be reported, and reconstitution with purified components has been achieved for only a few processes and proteins (5, 6). We believe this lack of biochemistry may become limiting in the quest for molecular mechanism, particularly as quantitative understanding is sought through modeling. For mathematical models to approach reality, it will be necessary to use real values for protein concentrations and association constants. Recently, a first attempt at protein quantitation has been made in fission yeast (6a).

Two less commonly used research approaches deserve comment. Modeling has a distinguished history in cytokinesis research (e.g., 7), and its importance is likely to grow in the future, driven in part by an influx into basic cell biology of young scientists with physical and mathematical training. Small molecule approaches have also been historically important, notably in the discovery of the role of actin in cytokinesis via the cytochalasins (reviewed in Reference 8). Because cytokinesis is a rapid and dynamic process that occurs during a small portion of the cell cycle, small molecules that can rapidly enter living cells and perturb specific processes are especially valuable tools. We also believe that interfering with specific aspects of cytokinesis may provide novel and useful anticancer therapeutics, with the recently introduced Aurora kinase inhibitors as examples (9–11). Despite their dual importance for research and potential therapeutics and the large number of potential protein targets, the number of well-characterized small molecules available for cytokinesis research is small (**Table 1**). We carried out a screen for small molecule inhibitors of cytokinesis and identified 50 small molecules that caused the formation of binucleate cells in *Drosophila* tissue culture cells (20). We are currently investigating the cellular targets and biochemical mechanisms of many of these compounds and hope to provide the field with more diverse and useful small molecule tools to study cytokinesis in the future.

Table 1 Small molecules that affect cytokinesis

Small molecule	Mechanism of action	Reference
Cytochalasins	Actin depolymerization: Bind barbed ends of actin	(12)
Latrunculin	Sequesters actin monomers	(13)
Swinholide	Severs actin filaments	(14)
Jasplakinolide	Stabilization of actin filaments	(15)
(−) Blebbistatin[a]	Inhibition of myosin II ATPase	(16)
Hesperadin,[b] ZM447439,[b] VX-680[b]	Inhibition of Aurora B kinase	(9–11)
Y27632[b]	Inhibition of Rho kinase	(17)
W7,[b] ML7[b]	Inhibition of myosin light chain kinase	(18, 19)

[a]Blebbistatin is sensitive to UV and blue light.
[b]Please note that these drugs target the active ATPase site of the kinase and can also inhibit other kinases, especially at higher concentrations.

A PARTS LIST FOR CYTOKINESIS

A prerequisite for a complete understanding of cytokinesis is a parts list, or inventory, of all the molecules involved. Effective RNAi technology and complete genome sequence information have allowed a preliminary listing of the full complement of genes involved in cytokinesis for two cell types and more fragmentary lists for several others (**Table 2**). Because of the limitations of RNAi mentioned above, we believe that the most significant omissions from current lists may be proteins involved both in mitosis and cytokinesis, for which the cytokinesis defect can be difficult to score. For example, Polo kinase has a central role in cytokinesis but does not score in the screens because it is required for mitosis.

Proteomics provides a complementary strategy for genome-wide understanding. It has the potential to supply more direct information on biochemical mechanisms by providing data on relative protein concentrations, posttranslational modifications, and the composition of protein complexes. The methodology for making such measurements on a global scale using mass spectrometry is only now emerging and has yet to be applied to cytokinesis. A major limitation in applying proteomic methods to cytokinesis has been the challenge of isolating relevant structures and complexes from defined cell cycle states.

Skop et al. (25) approached this by isolating midbodies, representing a very late stage of cytokinesis, from synchronized Chinese hamster ovary cells. Five hundred seventy-seven proteins enriched in midbodies were identified, and 160 candidates were tested for relevance by RNAi of orthologs in *C. elegans* (25). Some of the cytokinesis proteins identified in this study are included in **Table 2**.

Detailed examination of recently published screens (20–25, 29, 31, 34, 41, 49, 74) reveals that certain proteins appear in every screen, whereas many proteins were only found in one or a few screens. **Table 2** lists those that scored in two or more screens, as well as well-described cytokinesis proteins from the literature. Proteins that scored in only one screen are presumably a mixture of conserved proteins that escaped detection in other screens for technical reasons, and proteins required in only a subset of systems. [See Supplemental **Tables 1–3** for a summary of hits (follow the Supplemental Material link from the Annual Reviews home page at **http://www.annualreviews.org**)]. Only two screens, one in worms (23) and one in flies (20), covered most of a genome so it is currently difficult to distinguish these possibilities. Given the unprecedented rate of new data generation, **Table 2** will soon be out of date. The European MitoCheck consortium

Table 2 Parts list of proteins involved in cytokinesis[a]

Mammalian gene	*Drosophila* gene	*C. elegans* gene	Predicted protein function	Screen reference	Reference
Actin cytoskeleton					
Actin	Act5C	act-1	Actin	(20–23)	(12)
Myosin heavy chain	zip	nmy-2	Nonmuscle myosin II heavy chain	(20, 21, 24, 25)	(26–28)
MRLC	sqh	mlc-4	Myosin II regulatory light chain	(20–23, 25, 29)	(30)
Anillin	ani	ani-1	Anillin, actin binding	(20, 21, 24, 29, 31)	(32, 33)
Arp3	arp66B	arx-1	Actin nucleation	(20, 25)	
CAPZ	cpb	cap-2	Capping protein	(20, 25)	
Profilin	chic	pfn-1	Binding of actin monomers	(21, 23–25, 34)	(35)
Cofilin	tsr	unc-60	Actin severing	(20, 21, 24, 25, 29, 31)	(36)
mDia	dia	cyk-1	Formin-Rho effector, actin nucleator	(20–25, 31)	(37, 38)
RhoA	Rho1	rho-1	Rho GTPase	(20, 21, 24, 29, 31)	(39)
Ect2	pbl	let-21	RhoGEF	(20, 21, 23, 25, 29, 31)	(40)
MgcRacGAP	RacGAP50C	cyk-4	RhoGAP	(20–24, 29, 41)	(42, 43)
Citron kinase	CG10522	F59A6.5/ W02B8.2	Kinase-Rho effector	(20, 21, 24, 25, 31)	(44)
ROCK	rok	let-502	Rho kinase-Rho effector	(20, 21)	(45)
Microtubule associated					
Tubulin	tub84D	tba-2	Tubulin	(21, 25)	
γTubulin	γTub	tbg-1	Microtubule nucleation	(25, 41)	
PRC1	feo	spd-1	Microtubule bundling	(20, 21)	(46, 47)
CLASP1/2	orbit	cls-2	Microtubule-tip binding	(25)	(48)
MKLP1	pav	zen-4	Kinesin-6, microtubule motor	(20, 21, 23, 25, 29, 31, 49)	(50–52)
MKLP2/rab kinesin6			Kinesin-6, microtubule motor	(49)	(53)
Kif4A/B	klp3A	klp-12, klp-19	Kinesin-4, microtubule motor	(25, 49)	(54)
KIFC1	ncd	klp-16	Kinesin-14, microtubule motor	(20, 41)	
KIF18	klp67A		Kinesin-8, microtubule motor		(55)
Vesicle transport					
Clathrin heavy chain	chc	chc-1	Endocytosis	(20, 25)	(56)
Dynamin	shi	dyn-1	Endocytosis	(20, 25, 31)	(57)
Syntaxin 1A	syx1A	unc-64	Vesicle fusion	(21, 29)	
Syntaxin 5	syx5	syn-3	Vesicle fusion	(20, 21)	(58)
betaCOP	betaCOP	Y25C1A.5	COP1 coatomer	(20, 25)	
gammaCOP	gammaCOP		COP1 coatomer	(20, 21, 24, 31)	
NSF attachment protein	SNAP	phi-29	SNARE-mediated membrane fusion	(20, 21)	
Arfophilin/Fib3-Rab11			Recycling endosome		(59)

(*Continued*)

Table 2 (Continued)

Mammalian gene	Drosophila gene	C. elegans gene	Predicted protein function	Screen reference	Reference
Regulation					
Aurora B kinase	*ial*	*air-2*	Aurora B kinase complex	(20, 21, 31)	(60)
INCENP	INCENP	*icp-1*	Aurora B kinase complex		(61)
Survivin		*bir-1*	Aurora B kinase complex		(62)
Borealin	Borr	*csc-1*	Aurora B kinase complex	(20)	(63–65)
Cyclin B3	*cycB3*	*cyb-3*	Cyclin	(41)	(66)
CyclinB	*cycB*	*cyb-1, cyb2.1*	Cyclin	(41)	(66)
Polo	*polo*	*plk-1*	Polo kinase		(67)
Other					
Annexin11			Annexin		(68)
BRCA2			Oncogene		(69)
centriolin			Centrosome binding		(70)
Nir2	*rdgB*	*M01F1.7*	PI transferase		(71)
Orc6	*orc6*		Initiation of DNA replication		(72)
PI4 kinase	*fwd*		PI4 kinase	(20)	(73)
SEPT2[b] *SEPT9*[b]	*Sep2*[b], *pnut*[b]	*unc-59*[b], *unc-61*[b]	Septin	(21)	(73a–c)
SNW1	*Bx42*		Splicing factor	(20, 74)	
	CG7236		Kinase	(20, 21, 31)	
	Tra1		Transcription factor	(20, 21)	
		lin-5	Unknown	(41)	(75)
		spk-1	Kinase	(22, 23)	

[a]Gene products that scored in a screen and at least one other study as well as proteins identified in detailed studies are included. Putatvive human, Drosophila, and C. elegans orthologs are shown for each protein. If several copies of a gene are present in the genome (for example actin), only one is shown for clarity. For a full list of screening hits, see Supplemental **Tables 1–3**. Follow the Supplemental Material link from the Annual Reviews home page at http://www.annualreviews.org. Genes implicated in cytokinesis are shown in bold font; orthologs that have not yet been implicated are shown in regular font.

[b]Septins have been implicated in cytokinesis in a number of systems, but the ortholog relationships are unclear at this point.

Cleavage furrow or cytokinetic furrow or contractile ring: forms at cell equator during anaphase and ingresses during cytokinesis. Contains actin, myosin and other proteins

(http://www.mitocheck.org) is making a systematic effort to identify and annotate all proteins involved in mitosis and cytokinesis in human cells and has created a growing database. This resource will incorporate the results of genome-wide screening in human cells starting some time in 2006. **Table 2**, and related lists, are the first effort toward a complete parts list for cytokinesis.

Not all the molecules involved in cytokinesis will be proteins that can be identified by techniques such as RNAi; small molecule metabolites and nonprotein macromolecules will also have important roles. Because cytokinesis intimately involves the plasma membrane and organelles, we expect specific roles for lipids or their metabolites. Recent studies show that phosphatidylinositol-4,5-*bis* phosphate [PtdIns(4,5)P$_2$] accumulates at the cleavage furrow and is required for cytokinesis in HeLa cells and in *Drosophila* spermatocytes (76, 77), and **Table 2** includes two enzymes of lipid metabolism, the PtdIns transfer protein Nir2 and a PtdIns 4 kinase. Also, a glycosphingolipid, psychosine (1-β-D-galactosylsphingosine) is known to negatively regulate cytokinesis in certain cell types (78, 79). Further analysis of the role of specific lipids and other metabolites might provide interesting insights.

SUBPROCESSES IN CYTOKINESIS

In the rest of this review, we address new mechanistic information on various subprocesses in cytokinesis and relate these to the parts list in **Table 2**. **Figure 1** shows a breakdown of cytokinesis into a series of subprocesses based on timing and morphology. This is the traditional method of making cytokinesis more manageable from a reductionist perspective, but caution is required because biochemical subsystems may not fit neatly into categories defined by morphology and timing. For example, the biochemically defined Rho pathway is probably involved in several subprocesses, and some subprocesses that appear unitary, such as completion, may in fact be highly complex, requiring multiple biochemical pathways. The timing and physical requirements for cytokinesis were extensively explored in classic experiments, notably in echinoderm embryos by Rappaport (1). These constitute a platform for current research that has been extensively reviewed elsewhere (80). Determining the molecular mechanisms underlying the Rappaport phenomenology can be viewed as the central goal of modern cytokinesis research.

Timing Cytokinesis

An important Rappaport conclusion was that although the whole cell cortex can support assembly of a cleavage furrow, this ability is tightly restricted in time. In echinoderm embryos, the ability to form a cleavage furrow is only expressed in a short window starting after anaphase onset. Later, Margolis and coworkers (81) found that inhibition of cytokinesis in tissue culture cells with a drug that blocks actin depolymerization was reversible but only if the drug was washed out within a window of ~45 min following initiation of cytokinesis. Canman et al. (82) extended this work and coined the term "C phase" for the period during the cell cycle in which cytokinesis can occur. C phase is not part of the

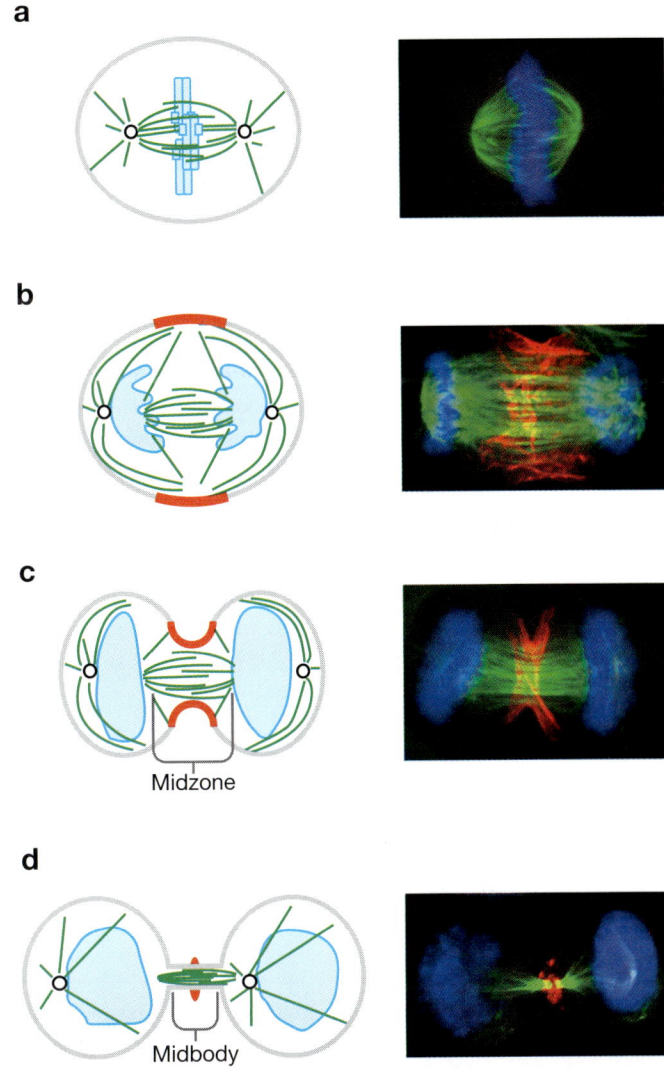

Figure 1

Schematic and immunofluorescence illustrations showing different stages of cytokinesis: DNA (*light blue*), microtubules (*green*), and the cleavage furrow protein Anillin (*red*) are shown. (*a*) Cell is in metaphase. Chromosomes are aligned at the metaphase plate by the microtubules of the mitotic spindle. (*b*) Cell is in late anaphase. Microtubules have elongated and contact the cortex. They have also rearranged to create a region of bundled microtubules between chromosomes termed the midzone. Cleavage furrow components (*red*) have assembled at the equator. (*c*) Cell is in early telophase. The cleavage furrow ingresses. (*d*) Cell is in late telophase. The cleavage furrow has fully ingressed compressing the midzone and creating an intercellular bridge containing a microtubule midbody. Completion occurs when the intercellular bridge is resolved creating two daughter cells.

Cell cortex: a meshwork attached to the plasma membrane that contains actin and other proteins

C phase: the time (~1 h) during which the cortex remains capable of contraction after anaphase onset

GEF: guanine nucleotide exchange factor

fundamental cell cycle oscillator like S or M phase but responds to that oscillator and appears to be a conserved aspect of animal cytokinesis. None of the proteins in **Table 2** has an obvious role in timing cytokinesis, suggesting C phase regulation may involve proteins required elsewhere in the cell cycle.

One might expect that onset of C phase is triggered by the reduction of Cdc2/Cdk1 kinase activity that accompanies anaphase onset because lowered Cdc2 activity is thought to trigger other cytoplasmic rearrangements at anaphase, such as nuclear envelope reformation (83). Consistent with this expectation, inhibiting Cdc2 in mammalian cells with drugs rapidly triggers cytokinesis-associated changes in cortical activity (83a). Studies in echinoderm embryos, however, suggest a different role for Cdc2. When anaphase spindles were pushed close to the cortex in echinoderm embryos in which Cdc2 kinase activity was kept artificially high, furrowing was induced (84). Thus, lowering Cdc2 kinase may not regulate the cortex directly in that system, rather it may function to increase the length of astral microtubules required for signaling (discussed below). The same manipulation did not induce furrowing when the anaphase-promoting complex (APC) was inhibited (85). APC is the E3 ligase that regulates exit from mitosis by ubiquitination of cyclin B and securin (86). Thus C phase entry may be induced by APC-dependent proteolysis. C phase exit may also involve ubiquitin-mediated proteolysis. A pharmacological study in cultured human cells found that, out of many drugs tested, the only one that altered the duration of C phase was a proteosome inhibitor, which approximately doubled its length when added after C phase was initiated (16).

Rape & Kirschner (87) recently proposed that the duration of G1 is regulated by ordered proteolysis of different APC substrates, with the ordering generated by different ubiquitination kinetics for varied substrates. Perhaps ordered proteolysis, mediated by APC and proteosomes, regulates both onset and exit from C phase in a similar manner. Consistent with this idea, Lindon & Pines (88) found that ordered proteolysis of mitotic regulators such as Polo-like kinase can contribute to the timing of mitotic exit and can influence the duration of C phase, and Echard & O'Farrell (66) proposed that sequential degradation of cyclin B and cyclin B3 controls the timing of C phase in *Drosophila*, possibly by regulating the Rho guanine nucleotide exchange factor (GEF) Ect2/pebble. In a screen for APC substrates in mammalian cells, two important cytokinesis proteins, the actin-myosin II-binding protein Anillin (89) and Aurora B kinase (89a), were found to be ubiquitinated and degraded late in the M to G1 transition. It seems likely that regulation of C phase involves both kinase-phosphatase and regulated protolysis systems. Dissecting their relative contributions might resolve apparent discrepancies in the literature.

Cleavage Plane Specification: Evaluation of Classic Models

Cleavage always occurs perpendicular to the axis of chromosome segregation, thus ensuring equal partition of the genome. Typically, it occurs in the middle of the cell, partitioning the cytoplasm equally. Unequal cleavage, following off-center positioning of the spindle, is common during embryonic development and in stem cell divisions during adult homeostasis. The special mechanisms involved in unequal division have recently been reviewed (90, 91), so we do not discuss them. It has long been known that cleavage plane specification involves communication between microtubules and the actin cortex (1), and the molecular mechanisms involved are now the subject of intense research efforts in many laboratories (92–95).

In both classical models and recent research, a large effort has been made to distinguish possible contributions of different types of signals from microtubules to the cortex (see **Figure 2**). One distinction focuses on the nature of the signal delivered. *Polar* *relaxation* refers to a negative signal from microtubules to the cortex at the poles (red arrows in **Figure 2**), preventing furrow assembly there, and specifying assembly at the equator by default. *Equatorial stimulation* refers to a positive signal from microtubules to the cortex at the equator (blue arrows in **Figure 2**), directly stimulating furrow assembly there. A second distinction focuses on the identity of the microtubules delivering the signal, and the primacy of asters vs midzones (**Figure 2**). *Asters* (pale green lines in **Figure 2**) are approximately radial microtubule arrays nucleated by centrosomes on the outsides of the spindle. *Midzones* (dark green lines in **Figure 2**) are antiparallel arrays of microtubules that assemble in between the separated chromosome masses during cytokinesis. Midzone organization mechanisms are discussed below. The two debates (relaxation vs stimulation, asters vs midzone) are related in the sense that any relaxing signal at the poles would presumably be delivered only by astral microtubules because only these come near the polar cortex. A stimulating signal at the equator, in contrast, could come from either type of microtubule, or both, because plus ends of astral and midzone microtubules intermix near the equatorial cortex in many cell types (see **Figure 2** [this review] and figure 5 in Reference 96). Although these hypotheses were classically seen as alternatives, its now seems likely that all of them may in fact operate in a single cell.

The asters vs midzone debate was recently resolved by an experiment proving that both arrays send signals to the cortex. These signals were distinguished by physically separating the two arrays, causing the induction of two distinct furrows (see **Figure 3**) (97). A UV microbeam was used to sever the connection between one aster and the midzone early in cytokinesis in *C. elegans* embryos (**Figure 3a**). The strong pulling forces that act on cortical microtubules in this system then caused the single aster and the midzone with the attached aster to move further apart than during

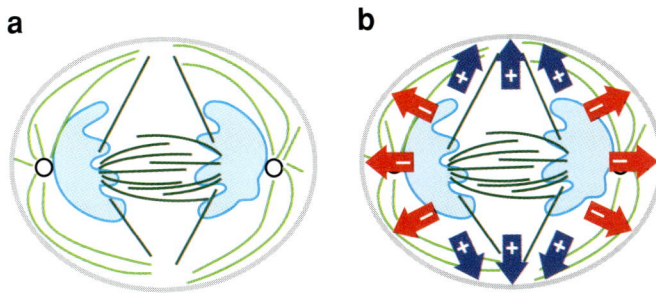

Figure 2
Organization of anaphase microtubule arrays and models for their roles in signaling to the cortex. Adapted from immunofluorescence images of PtK2 cells (99). (*a*) In anaphase, astral microtubules (*light green*) emanating from the centrosomes (*circles*) have elongated and many of their plus ends extend to the equatorial cortex. The midzone microtubule array (*dark green*) is made up of bundled, overlapping microtubules that extend between the chromatin masses (*light blue*). Note that some of the midzone microtubules extend to the equatorial cortex. The midzone microtubule array is typically more dense but has been reduced for clarity. (*b*) Models of how anaphase microtubules signal to the cortex. Polar relaxation signals (*red arrows*) and equatorial stimulation signals (*blue arrows*) are shown.

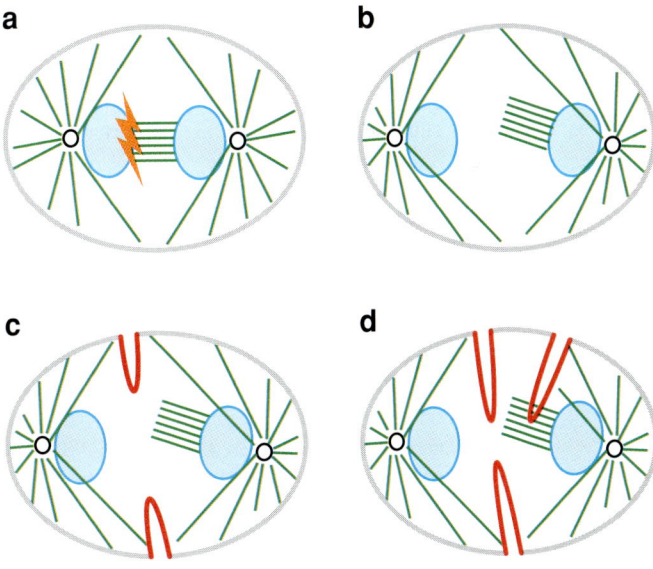

Figure 3
Cartoon of the experiments performed by Bringmann & Hyman (97). (*a*) Asymmetric spindle severing. (*b*) Severed aster is pulled away from the midzone and other aster. (*c*) First furrow ingresses at a point equidistant to both asters. (*d*) A second furrow ingresses at the middle of the midzone.

Midzone: bundled microtubule array between separating chromosomes, first formed during anaphase, sometimes called central spindle

Actomyosin: structures containing actin and nonmuscle myosin II

GAP: GTPase-activating protein

conventional cytokinesis (**Figure 3b**). The cell responded by first assembling a furrow equidistant between the two asters, ignoring the midzone (**Figure 3c**). This aster-promoted furrow ingressed deeply, but later it tended to regress and appeared incapable of completing cytokinesis alone. A few minutes after the first furrow formed, a second furrow assembled at the center of the midzone (**Figure 3d**). This midzone-promoted furrow ingressed rapidly, and remarkably, it often appeared to change direction to link up with the aster-promoted furrow, allowing cytokinesis to progress toward completion. Using RNAi, the authors began to assign some of the proteins from **Table 2** into roles in triggering each type of furrow. As expected, the proteins required for midzone assembly blocked only midzone-promoted furrows. Interestingly, some proteins, e.g., the kinesin Zen4, were involved in triggering both types of furrow, implying some overlap in signaling and/or delivery mechanisms.

The other debate, polar relaxation vs equatorial stimulation, has yet to be resolved decisively, but we favor the possibility that both hypotheses are again correct. There is abundant evidence that proteins enriched at the equator deliver positive signals (discussed below). There is currently no direct molecular evidence for a negative or relaxing signal delivered to the poles in cytokinesis. However, negative signals and/or the absence of proteins are harder to study. Nevertheless, strong phenomenological evidence has been produced for polar relaxation in cytokinesis. For example, a partially depolymerized microtubule array triggered flow of material away from the nearby cortex in *C. elegans* embryos (98), and abnormally strong contractions were noted when mammalian cells were forced to enter C phase in the absence of microtubules by overriding the mitotic checkpoint (99). Furthermore, microtubules have been shown to deliver a negative signal to cortical actomyosin in other systems (100), and one mechanism is thought to be sequestration of a GEF for the small GTPase RhoA (101).

If it is true that all the classically discussed mechanisms are in fact correct, and that they operate in parallel, why would cytokinesis be so complicated? An obvious idea is that overlapping mechanisms make site specification in cytokinesis robust to variation in the exact amounts of different proteins in the cell, the size and shape, or other variables that are difficult for a cell to control. Robustness—producing a well-defined outcome in the face of variation of input parameters—is a property of a molecular system that can be quantified if an appropriate model can be built, and it will be an interesting measure of future models of cytokinesis.

Cleavage Plane Specification: Molecular Mechanisms

The positive signal at the equator has been easiest to study; proteins that accumulate at the equator in a microtubule-dependent manner and are required for furrowing are candidates for participation in this signal. By these criteria, the Rho pathway, which regulates both myosin II activation and actin polymerization, is heavily implicated (4, 104). Rho (particularly RhoA in mammalian systems) and several Rho regulators, including the GEF Ect2/pebble and the GTPase-activating protein (GAP), MgcRacGAP, all accumulate at furrows and are required for cytokinesis (39, 40, 43, 102). The primacy of Rho in positive signaling is still controversial; enrichment of Rho at the furrow is clear in fixed cells by immunofluorescence (103), but reporters for active Rho in living cells have given somewhat contradictory results (104, 105). Furthermore, cytokinesis in the absence of RhoA function has been reported in strongly adherent mammalian cells, where it correlates with an absence of Rho activation using a fluorescence resonance energy transfer (FRET) reporter (106). Lack of a Rho pathway requirement in mammalian cells may correlate with the ability to cleave in the absence of myosin II function (discussed below).

Two kinases, Aurora B and Polo, also meet the criteria for involvement in the positive signal (reviewed in Reference 83). Their role in cytokinesis has been difficult to dissect because both are also involved in mitosis. The recent availability of small molecule Aurora B inhibitors (Table 1) may help elucidate its specific role in cytokinesis, and similar reagents for Polo are needed. To make progress in understanding the function of Aurora B and Polo in cytokinesis, we also need to identify substrates whose phosphorylation plays a specific role in regulating cytokinesis. There will probably be many such substrates. Two likely Aurora B substrates involved in cytokinesis are MgcRacGAP (107) and the kinesin Zen4/MKLP (108).

The mechanism by which microtubules spatially regulate the activity of the Rho pathway and the kinases is currently quite mysterious. Models under consideration include transport of signaling complexes along microtubules by motors, signaling by plus-end tracking complexes, and control of signaling simply by the local concentration of tubulin polymer. In support of the motor hypothesis, the plus-end-directed motor, kinesin-6 (Pavarotti/MKLP1/Zen4) has been implicated in cytokinesis in all systems examined (Table 2). This motor physically interacts with the Rho GAP, MgcRacGAP, (109) and participates in a defined molecular complex with the Rho GEF, Ect2/pebble, possibly through a Ect2/pebble-MgcRacGAP interaction (42, 110). Thus the motor protein could transport key Rho regulators to the cortex. In support of signaling by plus ends, a plus-end-tracking protein, CLASP/orbit, has been shown to be important for cytokinesis (48). However, taxol-stabilized asters can successfully signal (111, 112), implying dynamic plus ends are not always required. Local microtubule polymer concentration alone may also be important. An actin nucleator and essential cytokinesis protein, Diaphanous, binds to microtubules independent of Rho signaling and microtubule dynamics (113). Perhaps the presence of microtubules locally sequesters this protein (or other signaling factors) away from the plasma membrane. We still have much to learn about how microtubules direct the signaling pathways required for cytokinesis.

Organization of Cytokinesis-Associated Microtubules

Given their role in delivering spatially restricted signals to the cortex, it is important that microtubules are properly organized in space and time during cytokinesis. Cytokinesis-associated microtubules are dominated by asters and midzones (Figure 2). Astral microtubules elongate greatly at anaphase in many systems, so they often touch the cortex and elongate toward the equator. This elongation plays a central role in morphogenesis of the microtubule cytoskeleton during cytokinesis, and it is thought to be important for microtubule signaling to the cortex (84, 112). Elongation may be driven by a decrease in catastrophe rate after anaphase (114). This is correlated with decreased Cdc2 levels, although exactly how Cdc2 regulates microtubule dynamics is not understood. Rappaport (1) measured that the microtubule-derived signal propagates to the cortex in echinoderm eggs at a rate of 6 to 7 μm/min, and perhaps anaphase microtubule elongation sets this rate.

The midzone is initially formed between the separating chromosomes by bundling of elongating overlapping microtubules associated with the spindle (see the movies in Reference 115). Later, the midzone becomes self-organizing, and it can persist for many minutes, if furrow contraction is blocked (16). Although initially formed by the reorganization of existing microtubules, it is likely that new microtubules are nucleated within the midzone (115a). Some authors refer to midzones as "central spindles," which is confusing, as the same name has been applied to overlapping microtubules of the mitotic spindle during metaphase. Although midzones

Intercellular bridge: connects two daughter cells prior to abscission and is microtubule rich with the midbody at its center

Midbody: the center of the intercellular bridge. It contains microtubules and a high protein density area, the stembody

Figure 4

Thin section electron micrographs of a dividing HeLa cell. The upper panel shows an intercellular bridge between two daughter cells. The nucleus of the right-hand cell is visible. The lower panel is a higher resolution image of the same bridge. Note the bundled microtubules in the midbody and the electron-dense material (stembody) concentrated in a discrete zone at the center of the bridge (*lower panel*). Stembodies typically bulge outward at their center. The bar shown in the lower panel corresponds to 2 μm (*upper panel*) and 250 nm (*lower panel*). Images courtesy of Margaret Coughlin.

originate from spindle microtubules, distinct microtubule-associated proteins (MAPs) and motors organize the two arrays.

The midzone has several functions during cytokinesis. One is to help deliver the equatorial stimulation signal (discussed above), but this probably involves microtubules that elongate from the midzone to the cortex, rather than the midzone itself (48, 116). A second is to keep the separated genomes apart prior to completion; when microtubules were depolymerized before completion in mammalian cells, the nuclei collapsed back together (16). A third is to participate in completion and cell cycle regulation. These are functions of the midbody, a microtubule array within the intercellular bridge that connects two daugther cells. The midbody is a derivative of the midzone that forms by compression of the furrow during ingression, and its role is discussed below in the section on completion.

At the electron microscopy (EM) level, the midzone is dominated by electron-dense material that accumulates at the equator on overlapping microtubule bundles starting at anaphase (117). The electron-dense material coalesces as the midzone microtubule bundles are compressed by the furrow, eventually forming a small disc that was called a stembody, from the German Stemmkörper "pushing body" (see **Figure 4**). Despite its dominance in EM views, the molecular nature of the electron-dense material has not been determined, and its precise function is unknown.

The principal microtubule-interacting proteins implicated in morphogenesis of the midzone are bundling factors and kinesins. A conserved bundling factor, PRC1/Fascetto/spd-1 accumulates at the center of the midzone and ablating it blocks midzone assembly in all systems (**Table 2**). PRC1 binds and bundles microtubules in vitro and in cells (46). It is regulated by the cell cycle, possibly through phosphorylation by Cdc2-cyclin B. As mitosis progresses and Cdc2 activity decreases, PRC1 is dephosphorylated and becomes active (118). RNAi-mediated knockdown of PRC1 in mammalian cells prevented localization of other midzone markers and completely blocked midzone assembly (116). Signaling proteins such as Aurora B kinase still accumulated at the furrow, presumably because astral microtubules were unaffected. Ingression was normal, but completion failed. Given its biochemistry, localization, and genetics, PRC1 family members are probably the main bundling factor in midzones; an interesting unanswered question is whether they enforce antiparallel organization.

Two classes of plus-end-directed kinesins have been implicated in midzone assembly. In all systems, a kinesin-6 family member (MKLP1/Pavarotti/Zen4) accumulates at the center of the midzone and ablating it blocks midzone assembly (**Table 2**). MKLP1 is a plus-end-directed motor that can cross-link microtubules and slide one microtubule over another (52), making it an ideal candidate for organizing overlap interactions within the midzone. CHO1, a splice variant of MKLP1, may be the most relevant isoform for midzone morphogenesis in mammalian cells (119).

EM: electron microscopy

Stembody: the small electron-dense disk at the center of the midbody

A second plus-end-directed kinesin from the kinesin-4 family plays a less defined role in midzone assembly. Mammalian Kif4 is a chromokinesin during mitosis, with a poorly defined role in spindle assembly/function (120). At anaphase, Kif4 relocalizes to microtubule bundles and accumulates at the center of the midzone (120). Kif4 binds PRC1, with a preference for the dephosphorylated form, and may help to localize this microtubule-bundling protein correctly in the midzone (118).

Other proteins might also contribute to midzone formation. The plus-end-tracking protein CLASP/orbit may be involved (48). Annexin 11 is a new player in midzone organization in mammalian cells. Its biochemical function is not known, but it localizes strongly to midzones and removing it blocks their assembly (68).

Microtubule organization during cell division has typically been considered independent of the actin cytoskeleton, and signaling from microtubules to the cortex was classically considered unidirectional, but recent evidence has questioned this view. During mitosis, spindle organization depends on cortical actomyosin in some systems (121), and during cytokinesis, there may be feedback from the cortex to midzone organization. In *Drosophila* cells, damage to the actin cytoskeleton prevents assembly of a normal midzone (35); and, in mammalian cells, evidence from speckle imaging suggested that microtubule plus ends contacting the equatorial cortex are specifically stabilized against catastrophes (99). A feedback loop from the cortex to microtubule organization is appealing as a way to ensure robust self-organization of both cytoskeletal systems during cytokinesis.

Contractile Ring Assembly

The question of how the furrow assembles is closely connected to that of how microtubules signal to the cortex. The furrow consists of a contractile ring together with the plasma membrane to which it is connected.

We currently understand the biochemistry of the ring, which is dominated by actin and myosin II, much better than that of the membrane. Contractile rings are stable biochemical entities in the sense that they can be isolated and can be induced to contract in vitro (122), but these rings are highly dynamic in cells, with fast turnover of both actin and myosin (123, 124).

The organization of actin and myosin II in contractile rings is an unsolved problem. In **Figure 5a-c**, we depict three possible orthogonal alignments of actomyosin filaments to clarify their implications for force generation, noting that real cells presumably contain some mixture of these alignments. Schroeder proposed the "purse-string" model (**Figure 5a**) (125) on the basis of EM observations in echinoderm embryos. In this model, filament sliding shortens the ring, and ingression force comes from the component of the sliding force that is directed inward, which is given by contractile force in the ring multiplied by the reciprocal of the radius. In other words, this mechanism becomes more efficient as the ring gets smaller. Schroeder's purse string dominates textbooks, yet there is surprisingly little structural evidence for alignment of contractile fibers in this orientation. Fishkind & Wang (126) observed actin bundles connecting the dorsal part of the furrow to the ventral surface in adherent mammalian cells, leading them to propose the model in **Figure 5b**. In this model, all contractile force is directed inward. Paradoxically, the majority of fluorescently-labeled actin filaments observed in mammalian cells tend to show the organization in **Figure 5c** (33, 126, 127). Contraction of filaments with this orientation has no inwardly directed force component, if anything it would tend to oppose ingression. Perhaps these filaments do not participate in force generation and align passively in response to ingression. Finally, we note that anisotropic organization could potentially generate inward force. Cytoplasmic extracts containing actin and myosin II undergo "gelation-contraction" in which an anisotropic F-actin gel contracts

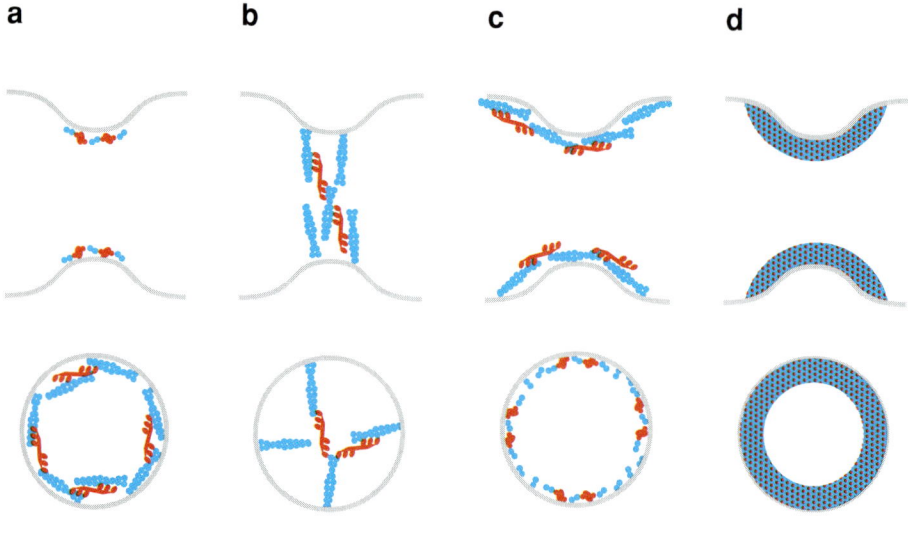

Figure 5

Models for different actomyosin alignments in the furrow. We cartoon three possible filament alignments to illustrate their implications for force production. Real cells presumably contain a mixture. The top row shows imaginary sections parallel to the axis of chromosome segregation, and the bottom row bisects that axis. Actin filaments are light blue and myosin motors red. (*a*) The purse-string model. Contraction of these filaments would promote ingression. (*b*) Filaments orthogonal to the ingressing membrane. Contraction of these filaments would also promote ingression. (*c*) Filaments parallel to the axis of chromosome segregation. Contraction of these filaments would, in theory, impede ingression, yet this is the major orientation observed in dividing cells. (*d*) Anisotropic orientation of filaments. Constriction could occur by gelation-contraction.

by myosin II activity (128). The structural basis by which an anisotropic gel contracts is unknown, and mixtures of pure F-actin and myosin II do not contract in the presence of ATP, rather they become more fluid (129). Nevertheless, observation of gelation-contraction implies that the contractile ring could contract even if its constituents are disorganized (**Figure 5d**), i.e., a mixture of the orientations in **Figure 5a-c**.

Recruitment of F-actin to the furrow might occur by nucleation in the furrow or by transport of filaments from elsewhere. Diaphanous, a formin, is a conserved, essential cytokinesis protein (**Table 2**), whose only known biochemical function is to nucleate actin filaments (130). The nucleation activity of Diaphanous is thought to be positively regulated by the Rho pathway (38). Thus it is a logical candidate to nucleate furrow actin. Whether it acts in furrows or nucleates else-where, followed by transport of filaments to the furrow, is unclear. White & Borisy (131) proposed that cleavage furrow actomyosin self-organizes by contraction-driven flow in the cortex, and subsequent microscopy studies found evidence for such flow in mammalian cells (123, 132–134) as well as in nematode embryos (98). In *Xenopus* eggs, Noguchi & Mabuchi (135) observed F-actin polymerization directly in the furrow as patch-like accumulations. In summary, there may be multiple mechanisms for accumulating furrow actin, and the precise function of Diaphanous is an important topic for further research in this area.

Recruitment of myosin II to furrows does not require its ATPase activity (16, 136, 137), and recent studies used live imaging and RNAi to show that recruitment of myosin to the cleavage furrow and its continued localization depend on different pathways

(96, 126a). Phosphorylated myosin regulatory light chain is observed at the furrow during early anaphase (138) and is needed to assemble myosin into the ring. DeBiasio et al. (127) carried out a comprehensive analysis of myosin II during cytokinesis in live 3T3 cells. They found that myosin fibers flow toward the equator during anaphase and form a meshwork, which contains fibers that are both parallel and perpendicular to the plane of cleavage. Cortical flow of myosin has also been observed in *Xenopus* eggs (135). Myosin is very dynamic at the furrow; phosphorylation of the heavy chain is required to maintain its dynamic behavior (124). Myosin transport along microtubules (139) is an appealing idea but has not been substantiated by biochemistry. In summary, cortical flow appears to contribute to myosin localization but is unlikely to be the only mechanism given the observations of recruitment in the absence of its own motor activity.

Furrow Ingression

Furrow ingression proceeds by some combination of force generation from the cytoskeleton (**Figure 5**), coupled to an increase of plasma membrane surface area. The textbook view of cytoskeletal force generation focuses on the purse-string contraction model proposed by Schroeder. Questions concerning this model were discussed above, highlighting the need for more information on the structural organization of contractile rings. Inhibition of myosin II immediately blocks furrow ingression in some mammalian cells (16), and myosin II subunits score as conserved, essential cytokinesis proteins (**Table 2**). It is thus very puzzling how Dictyostelium cells and highly adherent mammalian cells, under some conditions, are capable of executing a form of cytokinesis with myosin II absent or greatly reduced in activity (28, 124, 140–142). Wang (143) has proposed an alternative model for furrow ingression, named "equatorial relaxation," which posits that the whole cortex is under tension, and furrow ingression occurs at the equator because the cortex is softer there. Much of the evidence that supports this model comes from local drug perfusion experiments. Cytochalasin applied to the furrow tended to promote furrowing, whereas application to the poles inhibited it (144). Local application of the F-actin-stabilizing drug jasplakinolide, or the myosin II inhibitor blebbistatin, had essentially the opposite effects (136). These data support a model in which selective actin depolymerization at the equator is important for cytokinesis. However, direct measurements of cortical stiffness by atomic force measurements found increased stiffness at the equator (145). These observations might be reconciled if the equator was stiffer, yet depolymerized more rapidly, than the bulk cortex. In our view, the mechanics of cytoskeletal force production in cytokinesis is still an open question, and it is possible that more than one mechanism operates.

Independent of its effect on cortical stiffness, it is clear that actin depolymerization is an important aspect of furrow ingression. Actin and myosin both turn over rapidly in furrows (123, 124), implying that their local concentration depends on a dynamic balance between polymerization/recruitment and depolymerization/dissociation. Local concentration remains approximately constant during ingression, implying continuous decrease in contractile ring volume (125). The precise mechanism of actin depolymerization in cells is controversial, but proteins from the actin depolymerization factor (ADF)-cofilin family are key factors. Cofilin (twinstar in flies) is needed for cytokinesis (**Table 2**). In spermatocytes of *twinstar Drosophila* mutants, actin accumulates into abnormally large furrows, resulting in inhibition of cleavage (36). One model for keeping F-actin concentration constant while volume decreases would make the rate of recruiting new actin dependent on the surface area of the furrow, and the depolymerization rate dependent on the total amount of actin present. This could occur if recruitment depended on formin-driven nucleation within the furrow or, perhaps, if it also depended on cortical flow from neighboring regions.

It has been known for years that furrow ingression can be coupled to deposition of new plasma membrane (146). The source of this membrane is not well characterized. Embryonic systems contain stored membranes in the form of vesicles, but in general this problem is unsolved, and recent investigations have focused on vesicle trafficking pathways. Actin-independent addition of new membrane has been observed in *Xenopus* eggs, where the membrane is delivered by exocytic vesicles that seem to travel along microtubules (147). New membrane addition was also observed in sea urchin eggs, using the extracellular matrix protein hyalin as a membrane marker (148). Shuster & Burgess (148) show that new membrane is added specifically to the ingressing furrow after mitotic exit and that membrane addition is dependent on astral microtubules and calcium. Both of these studies were carried out in eggs, which are larger than most cells and have a greater surface area, thus requiring more membrane to be inserted during ingression. *Drosophila* embryo cellularization is a special form of cytokinesis that exhibits a particularly large requirement for new surface area, and the process by which new membrane is inserted just behind the ingressing furrows in this system has been analyzed by genetics and microscopy (149). It is possible that specialized mechanisms have evolved in embryonic systems, but a general requirement for membrane addition during ingression seems likely. RNAi screens have revealed a conserved requirement for the endocytosis proteins clathrin and dynamin (**Table 2**), suggesting that plasma membrane recycling may be important during ingression, in addition to the addition of membrane from vesicular stores.

Completion

Completion, also termed abscission or scission, is the final step of cytokinesis. It is in some sense an optional step in cytokinesis and the cell cycle. In embryos, blastomeres often remain connected by intercellular bridges for many cell cycles, perhaps explaining why some proteins involved in completion fail to score in cyokinesis screens performed in embryos (**Table 2**). Beyond its importance in cytokinesis, an emerging concept is that completion is a regulated process with a role in preventing accumulation of aneuploid cells (150). This exciting development may account for some of the biochemical complexity of completion. As the very last step in the old cell cycle or the first in the new one, it is logical to use completion as a sensor and regulator of cell cycle progression.

Completion occurs after the actomyosin ring has contracted and the cleavage furrow has ingressed to its fullest extent, creating an intercellular bridge. The bridge is packed with tightly bundled, antiparallel microtubules embedded in phase- and electron-dense material called the stembody (**Figure 4**). The morphology of completion in mammalian cells was described by live analysis and EM almost 30 years ago (151). When furrow ingression is complete, the intercellular bridge is approximately 1–1.5 μm in diameter. Separation of daughter cells is preceded by a reduction in the diameter of the bridge to approximately 0.2 μm. This occurs over the entire length of the bridge, except at the stembody, which retains its original diameter and projects out as a bulge in the center of the bridge (**Figure 4**). Microtubule bundles become further compacted and also begin to disappear across the entire length of the bridge, an observation duplicated recently by live imaging (21, 152).

The microtubule bundles and stembody in the intercellular bridge are required for completion. They form by compaction of the midzone during ingression of the furrow (16). Blocking midzone assembly prevents assembly of normal microtubule/stembody structures and causes completion to fail (68, 109, 116, 153). There have been recent suggestions that midbody microtubules are dynamic (153a), but the function of microtubules and the stembody in completion is unknown. Current ideas focus on a possible role of microtubules in directing vesicles to the stembody and a possible role of septins, polymerizing GTPases that accumulate in the bridge (**Table 2**), in directing vesicle fusion (153b).

To study completion, it is important to distinguish it from late stages of ingression. One way to do this is to test for dependence on F-actin. During ingression and before bridge maturation, low doses of the actin depolymerizing compound Latrunculin B cause the furrow to reopen. Once the bridge matures and completion begins, it becomes Latrunculin insensitive (21), implying that the plasma membrane is linked to the midbody by a connection that does not involve dynamic F-actin.

A conserved cleavage furrow component that may be involved in bridge stability is Anillin. This multidomain protein binds to F-actin, myosin II, septin complexes, and perhaps also membranes via its pleckstrin homology (PH) domain (5, 33, 96, 154, 155). Anillin plays a nonessential role during ingression (96, 154) but is required for completion in several systems (21, 29, 96). Anillin and its interaction partners, the septins, remain in the maturing intercellular bridge after myosin II and most F-actin have dissociated (21, 32, 33). They also remain permanently in the narrow intercellular bridges that persist after incomplete cytokinesis in *Drosophila* spermatocytes (156). Microtubules are absent in these long-lived spermatocyte bridges, implying the existence of a stable, membrane-bound collar that does not depend on a microtubule scaffold. Biochemically, septins polymerize into relatively stable filaments that associate with, and may regulate, membrane trafficking proteins (reviewed in References 157–159). We hypothesize that Anillin and septins together assemble into a stable, filamentous array, which shapes the plasma membrane of the stable intercellular bridge, and that may also regulate the vesicle fusion required for completion.

Completion requires remodeling of plasma membranes to create sister cells, and several recent reports focused on the role of vesicle trafficking components. Inhibition of the midbody-localized t-SNARE/v-SNARE pair syntaxin 2 and endobrevin/VAMP-8 by overexpression of dominant-negative constructs blocked completion but not ingression (160). Both the cellular localization and completion defects were specific for this SNARE pair. α-SNAP, part of a complex required for SNARE-mediated fusion, was found in a screen for completion defects in *Drosophila* cells (21). In SNAP-depleted cells, intercellular bridges formed normally but later disassembled without completion. Together, these studies imply that completion involves membrane fusion mediated by specific SNAREs, using mechanisms in common with other types of intracellular vesicle fusion. Whether these SNAREs promote direct fusion of plasma membranes, or work less directly by fusing transport vesicles to plasma membranes, remains to be determined, as does the possible role of midbody microtubules and septins in targeting fusion.

Centrosomes have long been implicated in cytokinesis biology as the nucleating sites of asters (**Figure 2**), but recently a new, and somewhat mysterious, role of centrosomes in completion has been discovered. Live imaging of GFP-tagged centrosomes revealed movement of one centrosome into the intercellular bridge late in cytokinesis. Completion correlated with this movement, suggesting a causal connection (152). The connection between centrosomes and completion was reinforced by the observation that RNAi depletion of the centrosome protein centriolin blocks completion (70). Centriolin localizes to centrosomes throughout the cell cycle, but during cytokinesis, it localizes in one or two spots adjacent to the stembody. Depletion of centriolin caused a defect in completion with persistent, elongated bridges and formation of multinucleated syncytia. Defects were also observed in cell cycle timing (70). Recently, centriolin has been shown to be required for the localization of exocyst components and SNARE proteins to a ring on the stembody, indicating a role in both vesicle targeting and fusion (160a). A role for intercellular bridge components in cell cycle timing was also found in a microsurgery study, where cells that failed to inherit a bridge experienced cell cycle delays (161). The biological logic behind the centrosome/completion/cell cycle connection is not yet clear, but clues may come from what is known about mitotic exit in yeast cells (162, 163). The mammalian centriolin implicated in completion shares homology with the MEN/SIN proteins Nud1p (budding yeast) and Cdc11p (fission yeast) (70). The relevance of the yeast MEN/SIN mitotic exit pathways for mammalian biology has been unclear; these data suggest that a related pathway may regulate completion.

A new direction in completion research was opened by a paper describing a connection between mistakes in chromosome segregation and completion failure in cultured human cells (150). Spontaneously arising binucleate cells, which result from failed completion, were found to exhibit a high frequency of chromosome missegregation, and drug treatments that promoted missegregation increased the frequency of cytokinesis failure. This study implies that failure in chromosome segregation that escapes detection by the mitotic checkpoint can nevertheless be detected by the cell, which responds by blocking completion. Shi & King (150) propose that this mechanism evolved to reduce carcinogenesis because, in their view, single chromosome aneuploidy is more dangerous than tetraploidy. The molecular basis of the connection between missegregation and failed completion is unknown. BRCA2, a protein involved in genome stability and protection from cancer, was recently implicated in cytokinesis, providing a possible molecular clue (69).

CYTOKINESIS AND CANCER

Failure in cytokinesis very likely contributes to cancer progression. Many cancers are aneuploid, facilitating genomic plasticity that allows rapid evolution of aggressive genotypes. Common solid tumors tend to exhibit polyploidy as well as single chromosome abnormalities, presumably resulting from failures in both mitosis and cytokinesis. The two may be closely connected both by chromosome missegregation triggering failure of completion (150) and by centrosome abnormalities, which cause defects in both spindle assembly and cytokinesis (reviewed in Reference 164). Pellman and coworkers (165) recently found that blocking cytokinesis causes primary cells lacking p53 to become much more carcinogenic in mice, directly demonstrating a causal connection between failed cytokinesis and carcinogenesis for the first time. Analysis of cytokinesis defects in human cancers is an important direction for future research.

Cytokinesis is also a point of possible therapeutic intervention in cancer. The first test of this idea will come from Aurora kinase inhibitors, currently in clinical trials (11). Cells treated with these drugs become polyploid before eventually dying, a mechanism of cell killing distinct from mitotic spindle poisons such as taxol. Given that blocking cytokinesis in $p53^-$ cells can cause cancer in mice (165), there is a risk that this treatment will cause cancer as well as treat it, a concept familiar for DNA-damaging agents. Cancer drugs that target generic cell division mechanisms kill normal stem cells and thus cause bone marrow and gut toxicity, which limits their therapeutic dose. It would be better to find drugs that selectively blocked cytokinesis (or mitosis) in cancer cells while sparing normal stem cells. Cytokinesis is highly conserved, and we cannot expect major mechanistic differences. However, different cell types probably vary in the extent to which overlapping cytokinesis pathways are used, so selective inhibition is not out of the question. In that light, it will be useful to extend research on cytokinesis mechanisms to comparative studies of stem and cancer cells, using both genome-wide and detailed mechanism approaches.

SUMMARY POINTS

1. A parts list of genes involved in cytokinesis has been assembled through a combination of RNAi and previous efforts.
2. We dissect cytokinesis into six subprocesses and discuss mechanistic progress in each subprocess:
   - Timing: Proteolysis is involved in regulating C phase, the time during which the cortex can contract.
   - Cleavage plane specification: Multiple mechanisms operate in parallel; the Rho pathway and kinases are involved.
   - Rearrangement of microtubule structures: Microtubules rearrange into different arrays that have varied functions.
   - Ring assembly: Cortical flow and local nucleation contribute to assembly of actomyosin filaments.
   - Ring ingression: Force generation mechanisms for different orientations relative to the furrow are discussed. The traditional purse-string model is most likely an oversimplification.
   - Completion: Midzone microtubules, vesicle transport, and centrosomes are important for completion.
3. The relevance of cytokinesis to cancer is discussed.

FUTURE ISSUES TO BE RESOLVED

1. We are beginning to generate a parts list of all proteins involved in cytokinesis but still know little about how they interact with each other to accomplish this process.
2. We have partial information for most of the fundamental mechanisms underlying cytokinesis, but our understanding of biochemical mechanisms is only just emerging.

ACKNOWLEDGMENTS

We thank Margaret Coughlin for the images in **Figure 4**. U.S.E was supported by a Merck-sponsored fellowship from the Helen Hay Whitney foundation. Cytokinesis research in the Mitchison lab is supported by National Institutes of Health grant R01 GM023928–25.

LITERATURE CITED

1. Rappaport R. 1996. *Cytokinesis in Animal Cells*. Cambridge, UK: Cambridge Univ. Press
2. Jurgens G. 2005. *Trends Cell Biol.* 15:277–83
3. Balasubramanian MK, Bi EF, Glotzer M. 2004. *Curr. Biol.* 14:R806–18
4. Uyeda TQ, Nagasaki A, Yumura S. 2004. *Int. Rev. Cytol.* 240:377–432
5. Kinoshita M, Field CM, Coughlin ML, Straight AF, Mitchison TJ. 2002. *Dev. Cell* 3:791–802
6. Mishima M, Pavicic V, Gruneberg U, Nigg EA, Glotzer M. 2004. *Nature* 430:908–13

6a. Wu JQ, Pollard TD. 2005. *Science* 310:310–14
7. Harris AK, Gewalt SL. 1989. *J. Cell Biol.* 109:2215–23
8. Peterson JR, Mitchison TJ. 2002. *Chem. Biol.* 9:1275–85
9. Hauf S, Cole RW, LaTerra S, Zimmer C, Schnapp G, et al. 2003. *J. Cell Biol.* 161:281–94
10. Ditchfield C, Johnson VL, Tighe A, Ellston R, Haworth C, et al. 2003. *J. Cell Biol.* 161:267–80
11. Harrington EA, Bebbington D, Moore J, Rasmussen RK, Ajose-Adeogun AO, et al. 2004. *Nat. Med.* 10:262–67
12. Schroeder TE. 1973. *Proc. Natl. Acad Sci. USA* 70:1688–92
13. Coue M, Brenner SL, Spector I, Korn ED. 1987. *FEBS Lett.* 213:316–18
14. Bubb MR, Spector I, Bershadsky AD, Korn ED. 1995. *J. Biol. Chem.* 270:3463–66
15. Bubb MR, Senderowicz AM, Sausville EA, Duncan KL, Korn ED. 1994. *J. Biol. Chem.* 269:14869–71
16. Straight AF, Cheung A, Limouze J, Chen I, Westwood NJ, et al. 2003. *Science* 299:1743–47
17. Ishizaki T, Uehata M, Tamechika I, Keel J, Nonomura K, et al. 2000. *Mol. Pharmacol.* 57:976–83
18. Nishikawa M, Tanaka T, Hidaka H. 1980. *Nature* 287:863–65
19. Silverman-Gavrila RV, Forer A. 2001. *Cell Motil. Cytoskelet.* 50:180–97
20. Eggert US, Kiger AA, Richter C, Perlman ZE, Perrimon N, et al. 2004. *PLoS Biol.* 2:e379
21. Echard A, Hickson GR, Foley E, O'Farrell PH. 2004. *Curr. Biol.* 14:1685–93
22. Gonczy P, Echeverri C, Oegema K, Coulson A, Jones SJ, et al. 2000. *Nature* 408:331–36
23. Sonnichsen B, Koski LB, Walsh A, Marschall P, Neumann B, et al. 2005. *Nature* 434:462–69
24. Rogers SL, Wiedemann U, Stuurman N, Vale RD. 2003. *J. Cell Biol.* 162:1079–88
25. Skop AR, Liu H, Yates J 3rd, Meyer BJ, Heald R. 2004. *Science* 305:61–66
26. Mabuchi I, Okuno M. 1977. *J. Cell Biol.* 74:251–63
27. Knecht DA, Loomis WF. 1987. *Science* 236:1081–86
28. De Lozanne A, Spudich JA. 1987. *Science* 236:1086–91
29. Somma MP, Fasulo B, Cenci G, Cundari E, Gatti M. 2002. *Mol. Biol. Cell* 13:2448–60
30. Karess RE, Chang XJ, Edwards KA, Kulkarni S, Aguilera I, Kiehart DP. 1991. *Cell* 65:1177–89
31. Kiger AA, Baum B, Jones S, Jones M, Coulson A, et al. 2003. *J. Biol.* 2:27
32. Field CM, Alberts BM. 1995. *J. Cell Biol.* 131:165–78
33. Oegema K, Savoian MS, Mitchison TJ, Field CM. 2000. *J. Cell Biol.* 150:539–52
34. Zipperlen P, Fraser AG, Kamath RS, Martinez-Campos M, Ahringer J. 2001. *EMBO J.* 20:3984–92
35. Giansanti MG, Bonaccorsi S, Williams B, Williams EV, Santolamazza C, et al. 1998. *Genes Dev.* 12:396–410
36. Gunsalus KC, Bonaccorsi S, Williams E, Verni F, Gatti M, Goldberg ML. 1995. *J. Cell Biol.* 131:1243–59
37. Castrillon DH, Wasserman SA. 1994. *Development* 120:3367–77
38. Watanabe N, Madaule P, Reid T, Ishizaki T, Watanabe G, et al. 1997. *EMBO J.* 16:3044–56
39. Mabuchi I, Hamaguchi Y, Fujimoto H, Morii N, Mishima M, Narumiya S. 1993. *Zygote* 1:325–31
40. Prokopenko SN, Brumby A, O'Keefe L, Prior L, He Y, et al. 1999. *Genes Dev.* 13:2301–14

41. Piano F, Schetter AJ, Morton DG, Gunsalus KC, Reinke V, et al. 2002. *Curr. Biol.* 12:1959–64
42. Somers WG, Saint R. 2003. *Dev. Cell* 4:29–39
43. Hirose K, Kawashima T, Iwamoto I, Nosaka T, Kitamura T. 2001. *J. Biol. Chem.* 276:5821–28
44. Madaule P, Eda M, Watanabe N, Fujisawa K, Matsuoka T, et al. 1998. *Nature* 394:491–94
45. Kosako H, Yoshida T, Matsumura F, Ishizaki T, Narumiya S, Inagaki M. 2000. *Oncogene* 19:6059–64
46. Mollinari C, Kleman JP, Jiang W, Schoehn G, Hunter T, Margolis RL. 2002. *J. Cell Biol.* 157:1175–86
47. Verni F, Somma MP, Gunsalus KC, Bonaccorsi S, Belloni G, et al. 2004. *Curr. Biol.* 14:1569–75
48. Inoue YH, Savoian MS, Suzuki T, Mathe E, Yamamoto MT, Glover DM. 2004. *J. Cell Biol.* 166:49–60
49. Zhu CJ, Zhao J, Bibikova M, Leverson JD, Bossy-Wetzel E, et al. 2005. *Mol. Biol. Cell* 16:3187–99
50. Adams RR, Tavares AA, Salzberg A, Bellen HJ, Glover DM. 1998. *Genes Dev.* 12:1483–94
51. Goshima G, Vale RD. 2003. *J. Cell Biol.* 162:1003–16
52. Nislow C, Lombillo VA, Kuriyama R, McIntosh JR. 1992. *Nature* 359:543–47
53. Fontijn RD, Goud B, Echard A, Jollivet F, van Marle J, et al. 2001. *Mol. Cell. Biol.* 21:2944–55
54. Williams BC, Riedy MF, Williams EV, Gatti M, Goldberg ML. 1995. *J. Cell Biol.* 129:709–23
55. Gatt MK, Savoian MS, Riparbelli MG, Massarelli C, Callaini G, Glover DM. 2005. *J. Cell Sci.* 118:2671–82
56. Niswonger ML, O'Halloran TJ. 1997. *Proc. Natl. Acad Sci. USA* 94:8575–78
57. Thompson HM, Skop AR, Euteneuer U, Meyer BJ, McNiven MA. 2002. *Curr. Biol.* 12:2111–17
58. Xu H, Brill JA, Hsien J, McBride R, Boulianne GL, Trimble WS. 2002. *Dev. Biol.* 251:294–306
59. Hickson GR, Matheson J, Riggs B, Maier VH, Fielding AB, et al. 2003. *Mol. Biol. Cell* 14:2908–20
60. Terada Y, Tatsuka M, Suzuki F, Yasuda Y, Fujita S, Otsu M. 1998. *EMBO J.* 17:667–76
61. Adams RR, Maiato H, Earnshaw WC, Carmena M. 2001. *J. Cell Biol.* 153:865–80
62. Fraser AG, James C, Evan GI, Hengartner MO. 1999. *Curr. Biol.* 9:292–301
63. Romano A, Guse A, Krascenicova I, Schnabel H, Schnabel R, Glotzer M. 2003. *J. Cell Biol.* 161:229–36
64. Gassmann R, Carvalho A, Henzing AJ, Ruchaud S, Hudson DF, et al. 2004. *J. Cell Biol.* 166:179–91
65. Sampath SC, Ohi R, Leismann O, Salic A, Pozniakovski A, Funabiki H. 2004. *Cell* 118:187–202
66. Echard A, O'Farrell PH. 2003. *Curr. Biol.* 13:373–83
67. Carmena M, Riparbelli MG, Minestrini G, Tavares AM, Adams R, et al. 1998. *J. Cell Biol.* 143:659–71
68. Tomas A, Futter C, Moss SE. 2004. *J. Cell Biol.* 165:813–22
69. Daniels MJ, Wang Y, Lee M, Venkitaraman AR. 2004. *Science* 306:876–79
70. Gromley A, Jurczyk A, Sillibourne J, Halilovic E, Mogensen M, et al. 2003. *J. Cell Biol.* 161:535–45

71. Litvak V, Tian D, Carmon S, Lev S. 2002. *Mol. Cell. Biol.* 22:5064–75
72. Prasanth SG, Prasanth KV, Stillman B. 2002. *Science* 297:1026–31
73. Brill JA, Hime GR, Scharer-Schuksz M, Fuller MT. 2000. *Development* 127:3855–64
73a. Neufeld TP, Rubin GM. 1994. *Cell* 77:371–79
73b. Surka MC, Tsang CW, Trimble WS. 2002. *Mol. Biol. Cell* 13:3532–45
73c. Kinoshita M, Kumar S, Mizoguchi A, Ide C, Kinoshita A, et al. 1997. *Genes Dev.* 11:1535–47
74. Kittler R, Putz G, Pelletier L, Poser I, Heninger AK, et al. 2004. *Nature* 432:1036–40
75. Lorson MA, Horvitz HR, van den Heuvel S. 2000. *J. Cell Biol.* 148:73–86
76. Field SJ, Madson N, Kerr ML, Galbraith KA, Kennedy CE, et al. 2005. *Curr. Biol.* 15:1407–12
77. Wong R, Hadjiyanni I, Wei HC, Polevoy G, McBride R, et al. 2005. *Curr. Biol.* 15:1401–6
78. Im DS, Heise CE, Nguyen T, O'Dowd BF, Lynch KR. 2001. *J. Cell Biol.* 153:429–34
79. Kanazawa T, Nakamura S, Momoi M, Yamaji T, Takematsu H, et al. 2000. *J. Cell Biol.* 149:943–50
80. Pollard TD. 2004. *J. Exp. Zool. A* 301:9–14
81. Martineau SN, Andreassen PR, Margolis RL. 1995. *J. Cell Biol.* 131:191–205
82. Canman JC, Hoffman DB, Salmon ED. 2000. *Curr. Biol.* 10:611–14
83. Nigg EA. 2001. *Nat. Rev. Mol. Cell Biol.* 2:21–32
83a. Niiya F, Xie X, Lee KS, Inoue H, Miki T. 2005. *J. Biol. Chem.* 280:36502–9
84. Shuster CB, Burgess DR. 1999. *J. Cell Biol.* 146:981–92
85. Shuster CB, Burgess DR. 2002. *Curr. Biol.* 12:854–58
86. King RW, Deshaies RJ, Peters JM, Kirschner MW. 1996. *Science* 274:1652–59
87. Rape M, Kirschner MW. 2004. *Nature* 432:588–95
88. Lindon C, Pines J. 2004. *J. Cell. Biol.* 164:233–41
89. Zhao WM, Fang G. 2005. *J. Biol. Chem.* 280:33516–24
89a. Stewart S, Fang G. 2005. *Cancer Res.* 65:8730–35
90. Betschinger J, Knoblich JA. 2004. *Curr. Biol.* 14:R674–85
91. Roegiers F, Jan YN. 2004. *Curr. Opin. Cell Biol.* 16:195–205
92. Maddox AS, Oegema K. 2003. *Nat. Cell Biol.* 5:773–76
93. Mandato CA, Benink HA, Bement WM. 2000. *Cell Motil. Cytoskelet.* 45:87–92
94. Glotzer M. 2004. *J. Cell Biol.* 164:347–51
95. D'Avino PP, Savoian MS, Glover DM. 2005. *J. Cell Sci.* 118:1549–58
96. Straight AF, Field CM, Mitchison TJ. 2005. *Mol. Biol. Cell* 16:193–201
97. Bringmann H, Hyman AA. 2005. *Nature* 436:731–34
98. Hird SN, White JG. 1993. *J. Cell Biol.* 121:1343–55
99. Canman JC, Cameron LA, Maddox PS, Straight A, Tirnauer JS, et al. 2003. *Nature* 424:1074–78
100. Canman JC, Bement WM. 1997. *J. Cell Sci.* 110(Pt.16):1907–17
101. Krendel M, Zenke FT, Bokoch GM. 2002. *Nat. Cell Biol.* 4:294–301
102. Tatsumoto T, Xie X, Blumenthal R, Okamoto I, Miki T. 1999. *J. Cell Biol.* 147:921–28
103. Yonemura S, Hirao-Minakuchi K, Nishimura Y. 2004. *Exp. Cell Res.* 295:300–14
104. Bement WM, Benink HA, von Dassow G. 2005. *J. Cell Biol.* 170:91–101
105. Yoshizaki H, Ohba Y, Kurokawa K, Itoh RE, Nakamura T, et al. 2003. *J. Cell Biol.* 162:223–32
106. Yoshizaki H, Ohba Y, Parrini MC, Dulyaninova NG, Bresnick AR, et al. 2004. *J. Biol. Chem.* 279:44756–62
107. Minoshima Y, Kawashima T, Hirose K, Tonozuka Y, Kawajiri A, et al. 2003. *Dev. Cell* 4:549–60

108. Guse A, Mishima M, Glotzer M. 2005. *Curr. Biol.* 15:778–86
109. Jantsch-Plunger V, Gonczy P, Romano A, Schnabel H, Hamill D, et al. 2000. *J. Cell Biol.* 149:1391–404
110. Yuce O, Piekny A, Glotzer M. 2005. *J. Cell Biol.* 170:571–82
111. Shannon KB, Canman JC, Moree CB, Tirnauer JS, Salmon ED. 2005. *Mol. Biol. Cell* 16:4423–36
112. Strickland LI, Donnelly EJ, Burgess DR. 2005. *Mol. Biol. Cell* 16:4485–94
113. Kato T, Watanabe N, Morishima Y, Fujita A, Ishizaki T, Narumiya S. 2001. *J. Cell Sci.* 114:775–84
114. Desai A, Mitchison TJ. 1997. *Annu. Rev. Cell Dev. Biol.* 13:83–117
115. Rusan NM, Wadsworth P. 2005. *J. Cell Biol.* 168:21–28
115a. Shu HB, Li Z, Palacios MJ, Li Q, Joshi HC. 1995. *J. Cell Sci.* 108:2955–62
116. Mollinari C, Kleman JP, Saoudi Y, Jablonski SA, Perard J, et al. 2005. *Mol. Biol. Cell* 16:1043–55
117. Buck RC, Tisdale JM. 1962. *J. Cell Biol.* 13:109–15
118. Zhu CJ, Jiang W. 2005. *Proc. Natl. Acad Sci. USA* 102:343–48
119. Kuriyama R, Gustus C, Terada Y, Uetake Y, Matuliene J. 2002. *J. Cell Biol.* 156:783–90
120. Vernos I, Raats J, Hirano T, Heasman J, Karsenti E, Wylie C. 1995. *Cell* 81:117–27
121. Rosenblatt J, Cramer LP, Baum B, McGee KM. 2004. *Cell* 117:361–72
122. Mabuchi I, Tsukita S, Tsukita S, Sawai T. 1988. *Proc. Natl. Acad Sci. USA* 85:5966–70
123. Murthy K, Wadsworth P. 2005. *Curr. Biol.* 15:724–31
124. Yumura S. 2001. *J. Cell Biol.* 154:137–46
125. Schroeder TE. 1972. *J. Cell Biol.* 53:419–34
126. Fishkind DJ, Wang YL. 1993. *J. Cell Biol.* 123:837–48
126a. Dean SO, Rogers SL, Stuurman N, Vale RD, Spudich JA. 2005. *Proc. Natl. Acad. Sci. USA* 102:13473–78
127. DeBiasio RL, LaRocca GM, Post PL, Taylor DL. 1996. *Mol. Biol. Cell* 7:1259–82
128. Pollard TD. 1976. *J. Cell Biol.* 68:579–601
129. Humphrey D, Duggan C, Saha D, Smith D, Kas J. 2002. *Nature* 416:413–16
130. Zigmond SH. 2004. *Curr. Opin. Cell Biol.* 16:99–105
131. White JG, Borisy GG. 1983. *J. Theor. Biol.* 101:289–316
132. Cao LG, Wang YL. 1990. *J. Cell Biol.* 111:1905–11
133. Cao LG, Wang YL. 1990. *J. Cell Biol.* 110:1089–95
134. Wang YL, Silverman JD, Cao LG. 1994. *J. Cell Biol.* 127:963–71
135. Noguchi T, Mabuchi I. 2001. *J. Cell Sci.* 114:401–12
136. Guha M, Zhou M, Wang YL. 2005. *Curr. Biol.* 15:732–36
137. Zang JH, Spudich JA. 1998. *Proc. Natl. Acad. Sci. USA* 95:13652–57
138. Matsumura F, Ono S, Yamakita Y, Totsukawa G, Yamashiro S. 1998. *J. Cell Biol.* 140:119–29
139. Foe VE, Field CM, Odell GM. 2000. *Development* 127:1767–87
140. Neujahr R, Heizer C, Gerisch G. 1997. *J. Cell Sci.* 110(Pt. 2):123–37
141. Kanada M, Nagasaki A, Uyeda TQ. 2005. *Mol. Biol. Cell* 16:3865–72
142. Gerisch G, Weber I. 2000. *Curr. Opin. Cell Biol.* 12:126–32
143. Wang YL. 2001. *Cell. Struct. Funct.* 26:633–38
144. O'Connell CB, Warner AK, Wang Y. 2001. *Curr. Biol.* 11:702–7
145. Matzke R, Jacobson K, Radmacher M. 2001. *Nat. Cell Biol.* 3:607–10
146. Bluemink JG, de Laat SW. 1973. *J. Cell Biol.* 59:89–108
147. Danilchik MV, Bedrick SD, Brown EE, Ray K. 2003. *J. Cell Sci.* 116:273–83

148. Shuster CB, Burgess DR. 2002. *Proc. Natl. Acad Sci. USA* 99:3633–38
149. Lecuit T, Wieschaus E. 2000. *J. Cell Biol.* 150:849–60
150. Shi Q, King R. 2005. *Nature* 437:1038–42
151. Mullins JM, Biesele JJ. 1977. *J. Cell Biol.* 73:672–84
152. Piel M, Nordberg J, Euteneuer U, Bornens M. 2001. *Science* 291:1550–53
153. Matuliene J, Kuriyama R. 2002. *Mol. Biol. Cell* 13:1832–45
153a. Rosa J, Canovas P, Islam A, Altieri DC, Doxsey SJ. 2006. *Mol. Biol. Cell* 17:1483–93
153b. Finger FP, White JG. 2002. *Cell* 108:127–30
154. Field CM, Coughlin M, Doberstein S, Marty T, Sullivan W. 2005. *Development* 132:2849–60
155. Maddox AS, Habermann B, Desai A, Oegema K. 2005. *Development* 132:2837–48
156. Robinson DN, Cooley L. 1996. *Trends Cell Biol.* 6:474–79
157. Versele M, Thorner J. 2005. *Trends Cell Biol.* 15:414–24
158. Joo E, Tsang CW, Trimble WS. 2005. *Traffic* 6:626–34
159. Beites CL, Campbell KA, Trimble WS. 2005. *Biochem. J.* 385:347–53
160. Low SH, Li X, Miura M, Kudo N, Quinones B, Weimbs T. 2003. *Dev. Cell* 4:753–59
160a. Gromley A, Yeaman C, Rosa J, Redick S, Chen CT, et al. 2005. *Cell* 123:75–87
161. Murata-Hori M, Sluder G, Wang YL. 2004. *BMC Cell Biol.* 5:49
162. McCollum D, Gould KL. 2001. *Trends Cell Biol.* 11:89–95
163. Pereira G, Schiebel E. 2001. *Curr. Opin. Cell Biol.* 13:762–69
164. Saunders W. 2005. *Semin. Cancer Biol.* 15:25–32
165. Fujiwara T, Bandi M, Nitta M, Ivanova EV, Bronson RT, Pellman D. 2005. *Nature* 437:1043–47

Mechanisms of Site-Specific Recombination

Nigel D.F. Grindley,[1] Katrine L. Whiteson,[2] and Phoebe A. Rice[2]

[1]Department of Molecular Biophysics and Biochemistry, Yale University, New Haven, Connecticut 06520-8114; email: nigel.grindley@yale.edu

[2]Department of Biochemistry and Molecular Biology, The University of Chicago, Chicago, Illinois 60637; email: katrine@uchicago.edu, price@uchicago.edu

We dedicate this review to the memory of Nick Cozzarelli, whose pioneering topological studies of site-specific recombination and rigorous approaches to the study of protein-DNA interactions were an inspiration to us all.

Key Words

DNA invertase, Holliday junction, integrase, resolvase, serine recombinase, tyrosine recombinase

Abstract

Integration, excision, and inversion of defined DNA segments commonly occur through site-specific recombination, a process of DNA breakage and reunion that requires no DNA synthesis or high-energy cofactor. Virtually all identified site-specific recombinases fall into one of just two families, the tyrosine recombinases and the serine recombinases, named after the amino acid residue that forms a covalent protein-DNA linkage in the reaction intermediate. Their recombination mechanisms are distinctly different. Tyrosine recombinases break and rejoin single strands in pairs to form a Holliday junction intermediate. By contrast, serine recombinases cut all strands in advance of strand exchange and religation. Many natural systems of site-specific recombination impose sophisticated regulatory mechanisms on the basic recombinational process to favor one particular outcome of recombination over another (for example, excision over inversion or deletion). Details of the site-specific recombination processes have been revealed by recent structural and biochemical studies of members of both families.

Contents

INTRODUCTION 568
THE MECHANISM OF
   SITE-SPECIFIC
   RECOMBINATION: AN
   OVERVIEW 569
TWO FAMILIES OF
   RECOMBINASES WITH
   DISTINCT MECHANISMS OF
   STRAND BREAKAGE,
   EXCHANGE, AND
   REUNION 570
TYROSINE RECOMBINASES 572
   The Process of Strand Exchange
     via a Holliday Junction
     Intermediate 572
   Structural Insights into Synapsis
     and Strand Exchange 574
   Controlling Catalytic Activity:
     Half-of-the-Sites Reactivity.... 577
   Controlling the Outcome of
     Recombination 578
SERINE RECOMBINASES 582
   Recombination by a Process of
     Double-Strand Break, Switch,
     and Rejoin 584
   Structural Insights into Synapsis,
     Cleavage, and Strand Exchange. 586
   Complex Systems of Serine
     Recombinases: Regulating the
     Outcome of Recombination.... 592
SUMMARY AND PERSPECTIVES. 600

INTRODUCTION

Site-specific recombination describes a variety of specialized recombination processes that involve reciprocal exchange between defined DNA sites. In its strictest definition (as used in this review), site-specific recombination involves (*a*) two DNA partners, (*b*) a specialized recombinase protein that is responsible for recognizing the sites and for breaking and rejoining the DNA, and (*c*) a mechanism that involves DNA breakage and reunion with conservation of the phosphodiester bond energy (i.e., lacking a requirement for either DNA synthesis or a high-energy nucleotide cofactor). The prototypes of site-specific recombination (thus defined) are the integration of bacteriophage λ into the *Escherichia coli* chromosome (1), the resolution of cointegrates derived from transposition of Tn*3*-related transposons (2), and the DNA inversions responsible for flagellar phase variation in *Salmonella* (3). [The strict definition excludes several other specialized recombination processes that have, on occasion, been described as site-specific; these include (*a*) VDJ recombination catalyzed by the RAG1/2 proteins during the development of the immune system; (*b*) most DNA transposition events (even when a specific target site is used), including integration of retroviral cDNAs; and (*c*) the "homing" of mobile introns.]

Depending on the initial arrangement of the parental recombination sites, site-specific recombination has one of three possible outcomes: integration, excision, or inversion (**Figure 1**). Integration results from recombination between sites on separate DNA molecules (provided that at least one of the parental chromosomes is circular) and occurs with a uniquely defined orientation. For sites located on the same chromosome, the outcome is determined by their relative orientation. Thus, excision results from recombination between sites in a head-to-tail orientation, whereas inversion results from exchange between inverted (head-to-head) sites. The three structural outcomes are used for a wide variety of purposes in biological systems. Most commonly, the use of site-specific recombination by an organism or a genetic element is driven by a primary need to physically join or separate DNA segments. In addition to phage integration and excision, and cointegrate resolution, examples include reduction of replicon dimers to monomers and DNA transposition (see **Table 1**). However, site-specific recombination is also used as a means of activating or switching gene expression as well as a means of generating genetic diversity through the acquisition of advantageous genes or gene segments.

THE MECHANISM OF SITE-SPECIFIC RECOMBINATION: AN OVERVIEW

The process of site-specific recombination can be divided into a series of conceptually simple steps. The recombinase binds to the two recombination sites. The two recombinase-bound sites pair, forming a synaptic complex with crossover sites juxtaposed. The recombinase then catalyzes cleavage, strand exchange, and the rejoining of the DNA within the synaptic complex. Finally, the synaptic complex breaks down, releasing the recombinant products.

From this description, it follows that the minimal components of a site-specific recombination system are a recombinase and a pair of recombination sites. The simplest sites are short duplex DNA segments, 20 to 30 bp in length, which contain an inverted pair of recognition sequences and bind one dimer (or two monomers) of the recombinase. Such sites contain at their center the point of DNA breakage and joining, and these are often referred to as the crossover sites. In nature, however, many recombination sites are more complicated, containing not only a crossover site,

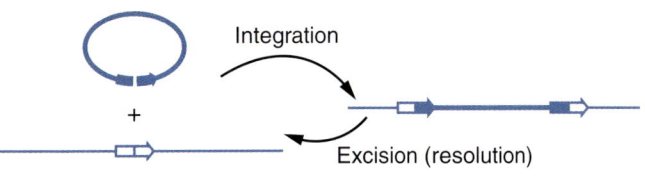

Figure 1
The three possible outcomes of site-specific recombination.

Table 1 Site-specific recombination: a sampling of enzymes and functions

Recombinase	Biological function
Tyrosine recombinase family	
λ Int and many other phage integrases	Integration and excision of phage genomes
Int of Tn916/Tn1545	Integration and excision: transposition of circular transposons
IntI	Integration and excision of gene cassettes in integrons
Cre	Excision: dimer reduction in phage P1 plasmids
XerC/D	Excision: dimer reduction in the *E. coli* chromosome as well as in many other bacterial chromosomes and some plasmids
TnpI of Tn4430	Excision: resolution of cointegrates resulting from transposition of Tn4430
FimB, FimE	Inversion: alternation of gene expression (fimbrial phase variation in *E. coli*)
Rci of R64	Inversion of shufflon segments in plasmid R64, producing various forms of pili
XisA, XisC	Excision: for developmentally regulated gene activation in *Anabaena*
Flp	Inversion: for amplification of yeast 2-μm plasmid
Serine recombinase family	
TnpR of Tn3/γδ and related transposons	Excision: resolution of cointegrates resulting from transposition
Sin of *Staphylococcus aureus*	Excision: dimer reduction in staphylococcal plasmids
ParA of RP4	Excision: dimer reduction in plasmid RP4
Hin	Inversion: alternation of gene expression (flagellar phase variation) in *Salmonella*
Gin, Cin	Inversion: alternation of gene expression (tail fiber proteins) in phages Mu and P1
OrfA of IS607/IS1535	Integration and excision: transposition of the *Helicobacter pylori* element IS607 (and others?)
Int of φC31/Bbv1/φRv1[a]	Integration and excision of *Streptomyces* and mycobacterial phages
TnpX of Tn4451[a]	Integration and excision: transposition of Tn4451 in *Clostridium*
SpoIVCA (CisA)[a]	Excision: for developmentally regulated gene activation in *Bacillus subtilis*
XisF[a]	Excision: for developmentally regulated gene activation in *Anabaena*

[a]Members of the large serine recombinase subfamily.

but also additional sequences spanning 100 or more base pairs. Such a complex site may operate in combination with a simple crossover site [as with λ integrase (λ Int) during integration] or with another complex partner (as with γδ resolvase). The extra DNA contains additional sites of protein recognition and may bind more copies of the recombinase or other protein factors encoded by the host cell or by the genetic element (e.g., phage or transposon) associated with the recombination system. The purpose of these additional DNA-bound proteins may be regulatory, structural, or both. They may initiate or stabilize the pairing of recombination sites or inhibit inappropriate pairings; they may deliver recombinase catalytic domains to the crossover site; they may activate the recombinase; and they may determine the directionality of recombination (for example, promoting deletion but preventing inversion, or vice versa).

As indicated above, breakage and rejoining of DNA in site-specific recombination occur with no loss or gain of nucleotides and with strict conservation of phosphodiester bond energy. To achieve this, a mechanism analogous to that of a topoisomerase is used; DNA strands are broken not by hydrolysis but rather by direct phosphoryl transfer to a side chain of the recombinase. This side chain, a tyrosine or a serine in all characterized cases, directly attacks the DNA sugar-phosphate backbone at the crossover site in a transesterification reaction, forming a covalent recombinase-DNA intermediate on one side of the break and a free hydroxyl group on the other. Rejoining the DNA strands is accomplished by reversing the process; the free hydroxyls from one recombination partner directly attack the phosphodiester linkage between recombinase and DNA of the other partner, releasing the recombinase and sealing the breaks to produce recombinant products. Intriguingly, the details of the process differ depending on whether the recombinase uses a tyrosine or a serine as the attacking nucleophile (see below).

TWO FAMILIES OF RECOMBINASES WITH DISTINCT MECHANISMS OF STRAND BREAKAGE, EXCHANGE, AND REUNION

Despite the many and distinct roles that site-specific recombination plays in biology and the large number of systems that have been identified, comparisons of the recombinase amino acid sequences indicate that nearly all fall into two families—the tyrosine recombinases (also known as the λ integrase family) (4) and the serine recombinases (also known as the resolvase family, named after the cointegrate-resolving recombinase encoded by transposons such as Tn3 and γδ)(5). Intriguingly, the tyrosine recombinases are also related to the eukaryotic type IB topoisomerases. The two recombinase families are unrelated in protein sequence or structure and employ different recombinational mechanisms; each family appears to have arisen and evolved separately.

There are interesting similarities and differences in the catalytic mechanims used by these recombinases. In both cases, DNA is cleaved by nucleophilic displacement of a DNA hydroxyl by a protein side chain (**Figure 2**). From studies of model compounds and other enzymes, the phosphotransfer reactions themselves are assumed to occur through the in-line nucleophilic displacement of one hydroxyl group by another with a pentacoordinate transition state. The degree of excess negative charge on the nonbridging oxygens in the transition state (relative to the ground state) depends on the degree of simultaneous bond formation to the nucleophile and the leaving group. For phosphodiesters, such as DNA, both bonds are probably only partially formed in the transition state (6). In the tyrosine recombinases, the DNA's 5′ bridging O is displaced to create a phosphotyrosyl bond to the 3′ end of the broken DNA strand, whereas in the serine case, the 3′ bridging O is displaced to form a 5′ phosphoserine linkage. Unlike many phosphotransferases,

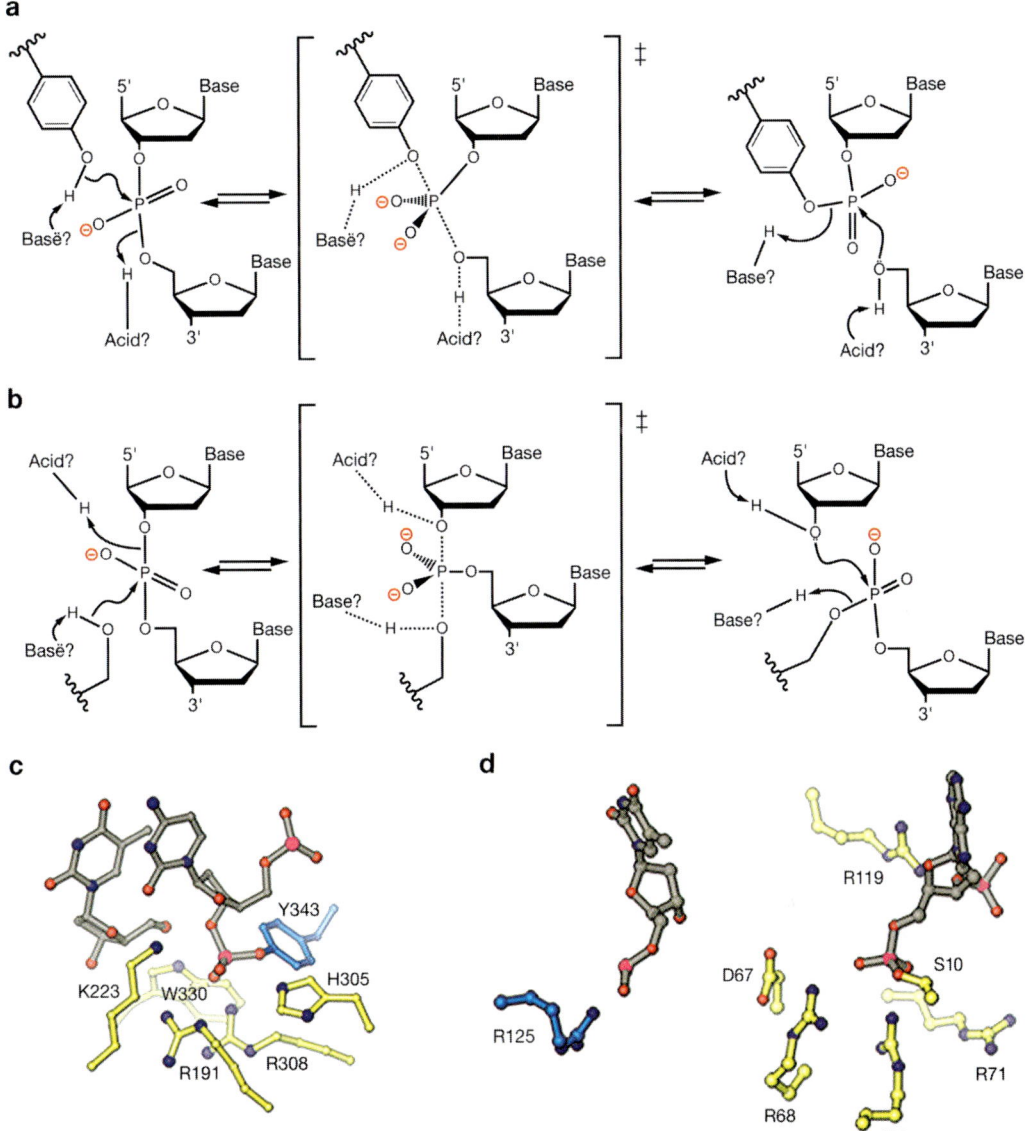

Figure 2

Phosphoryl transfer reactions catalyzed by (*a*) tyrosine and (*b*) serine recombinases. Cleavage is assumed to proceed through the in-line nucleophilic displacement of a DNA hydroxyl by the relevant protein side chain, and ligation by the reverse reaction. The cleavage direction may be assisted by a general base to accept a proton from the attacking side chain and by a general acid to protonate the leaving DNA oxygen (and vice versa for the ligation reaction). Their probable identities are discussed in the text for the tyrosine recombinases and are as yet unknown for the serine family. Both enzymes also position conserved arginine residues near the scissile phosphate, which may localize it and may stabilize the transition state geometrically and/or electrostatically. The nonbridging oxygens in the transition state are each given a formal negative charge for artistic simplicity; the degree and distribution of charge is not known in detail. (*c, d*) The constellation of conserved residues surrounding the scissile phosphate in (*c*) Flp and (*d*) γδ resolvase. **Figures 2c-d** as well as **Figures 4, 10, 12**, and **13** were made with Ribbons (156). Protein Data Bank (PDB) identification numbers (IDs) are 1M6X for Flp and 1ZR4 for γδ.

neither of these recombinase families exploit divalent metal ions for catalysis. Instead, they surround the scissile phosphate with several highly conserved positively charged amino acid side chains. Both recombinase types may also employ general/acid base catalysis, although experimental evidence for this is limited so far to the tyrosine family (discussed below).

The serine recombinases introduce double-strand breaks at both crossover sites; all strands are broken before any exchange is initiated (7). In contrast, the tyrosine recombinases only cleave one strand of each duplex at a time: After each crossover site is nicked, it must be joined to its partner before the second strand can be cut. This produces a cross-strand intermediate called a Holliday junction (8, 9).

TYROSINE RECOMBINASES

Tyrosine recombinases are most widespread among prokaryotes but are also found in archaea and even eukaryotes, where examples have been described in fungi, ciliates, and, most recently, certain families of retrotransposons (4, 10). The size of the family is illustrated by a recent iterative PSI-BLAST search (10a) that yielded ~1000 clearly related sequences.

The tyrosine recombinases share a catalytic domain with recognizable sequence motifs (4, 11). Structural studies have shown that the fold of the entire domain is well conserved even when the sequence identity outside of the active site region is insignificant. Although some family members, such as FimB and FimE, contain only this domain, in most the catalytic domain is preceded by a variable N-terminal domain that helps bind DNA. Some, such as λ Int, have a second N-terminal domain that binds different DNA sites, and sequencing projects are sure to reveal even more variety: For instance, database searches found two tyrosine recombinases of unknown function whose catalytic domains are followed by putative molybdate-binding domains (12).

The catalytic domain is shared with at least two other classes of enzyme. Members of the first, type IB topoisomerases, function as monomers to release supercoiling tension in DNA by cleaving and religating just one strand of DNA, but they do so through a similar 3′ phosphotyrosine intermediate with an almost identical active site. Enzymes of the second class, termed either telomere resolvases or protelomerases, maintain the covalently closed hairpin ends of the linear replicons found in certain prokaryotes and viruses (13, 14)

The Process of Strand Exchange via a Holliday Junction Intermediate

Each tyrosine recombinase has a specific DNA site, which is minimally comprised of a pair of inverted enzyme-binding sites separated by a 6–8-bp spacer, although many systems also include accessory sites where regulatory proteins can exert their influence. Cleavage and religation take place at the 5′ boundaries of the spacer. Although the sequences of the spacers can vary, there is generally a requirement for identity between recombination partners.

The requirement for sequence identity in the crossover region was originally interpreted to mean that each site would align with its partner, and the Holliday junction intermediate would branch migrate through this region, exchanging one base pair at a time (15). However, careful biochemical experiments led away from this toward a "strand-swapping isomerization" model, in which 2–3 bp adjacent to the cleavage site are melted, and the free end anneals to the complementary sequence in its recombination partner (16–18).

Current understanding of the reaction mechanism comes from many years of such biochemistry studies and just under a decade of structural knowledge. A generalized mechanism is cartooned in **Figure 3**. Recombination is initiated when one strand of each duplex is cleaved by a nucleophilic tyrosine,

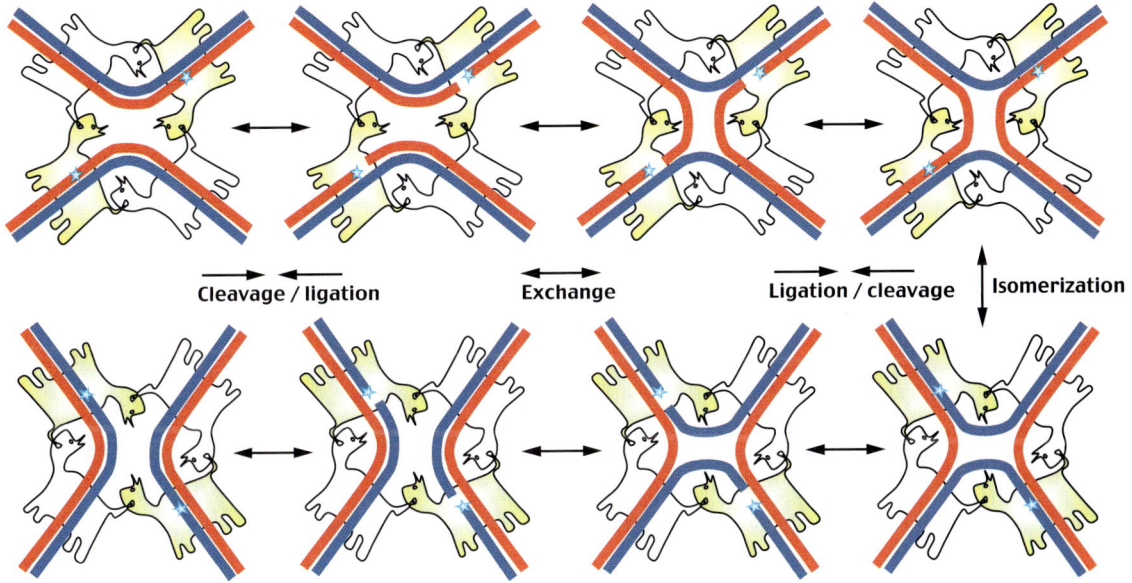

Figure 3

Cartoon of tyrosine recombinase–catalyzed strand exchange. The synaptic complex comprises two DNA duplexes bound by four recombinase protomers assembled in a head-to-tail fashion. Blue stars represent active catalytic centers in the active protomers (*pale yellow*). One strand from each duplex is cleaved, exchanged, and ligated to form a Holliday junction (*rightmost two panels*). Isomerization of this junction alternates the catalytic activity between the two pairs of protomers.

creating covalent DNA-protein phosphotyrosine linkages at the 3′ ends of the DNA and free hydroxyls at the 5′ ends. The energy from the phosphodiester bond in the DNA backbone is transferred to the phosphotyrosine. Although recombination requires synapsis of two sites, cleavage of a single strand, at least in some cases, requires only a dimer (19, 20). The next step involves an exchange where the free 5′ ends attack the 3′ phosphotyrosines of the opposing DNA substrates to form a Holliday junction. The complex can then isomerize so that the inactive monomers become active and vice versa. This enables a repeat of the whole process, i.e., the second, untouched strand is attacked, and the new 5′ ends migrate over and attack their partners' 3′ phosphotyrosine linkages, freeing the protein, resolving the Holliday junction, and completing the reaction. The approximately square planar conformation of the recombinase-bound DNA allows exchange of the DNA ends with remarkably little rearrangement of the protein component. The utility of this conformation was pointed out in 1989 (21) and firmly established by the Cre-DNA complex structure (22).

The phosphotransfer reaction catalyzed by tyrosine recombinases and type IB topoisomerases is diagrammed in **Figure 2**. Sequence comparisons and mutagenesis studies have identified five highly conserved active site residues that cluster (in three dimensions) near the critical tyrosine: RKHRH [see previous reviews (23–25)]. Studies of the topoisomerase from vaccinia virus implied that the invariant lysine acts as a general acid, protonating the leaving 5′ hydroxyl during the cleavage reaction (26). The first arginine is also important in this process, although its role is debated: It may form part of a proton shuttle, it may lower the pK_a of the nearby lysine, or it may act as the general acid itself with assistance from the lysine (27, 28). The first histidine may act as a general base, accepting a proton from the attacking tyrosine (29; Y. Chen, K.L. Whiteson, P.A. Rice, unpublished results). However, this residue is not as highly conserved as the others, and its mutagenesis is often not as deleterious to the reaction rate (1, 24). The need for a general base may not be as strong as the need for a general acid because the pK_a of tyrosine is lower than that of a 5′ hydroxyl. The positively charged catalytic residues may play multiple roles, such as localizing the scissile phosphate and stabilizing the pentacoordinate transition state both geometrically and electrostatically (28, 30). The second histidine also makes an important hydrogen bond to the scissile phosphate in the transition state of the vaccinia topoisomerase reaction (31). However, in Flp, the analogous tryptophan, W330, was found to have a largely architectural role, stabilizing the position of the helix containing the nucleophilic tyrosine (32). As more distantly related enzymes are studied in detail, new variations on these themes may emerge. For example, experiments with the hairpin telomere resolvase ResT did not find evidence that the lysine is acting as the general acid, although its mutation did nearly obliterate activity (33), and despite the presence of a serine in place of the normally crucial first arginine, wild-type CTnDOT integrase is fully functional (34).

Hydrolysis of the phosphotyrosine intermediate is normally much slower than religation. The Shuman group recently suggested that electrostatic repulsion of water by the phosphate itself plays an important role in preventing hydrolysis (30). When one of the nonbridging oxygens of the scissile phosphate was replaced with a methyl group, which has a similar size but lacks a charge, they found that vaccinia topoisomerase became a nuclease. Why is the incoming DNA's 5′ hydroxyl not similarly deterred by electrostatic repulsion? It may be that it is localized and oriented by other contacts within the complex (e.g., by base pairing with the opposite strand), whereas an incoming water molecule would have no such help. Many phosphotransferases that efficiently mediate the attack of water on a phosphodiester bond include divalent metal ions at the active site that can coordinate the attacking water.

Structural Insights into Synapsis and Strand Exchange

X-ray crystallography has provided a wealth of molecular detail on tyrosine recombinases. Structures are now available for the complete synaptic complexes of Cre, Flp, and λ Int, for monomeric DNA complexes of human topoisomerase I and λ Int, and for XerD, HP1 Int, λ Int, and vaccinia virus topoisomerase in the absence of DNA (16, 22, 35–41). Interestingly, in the absence of DNA, the active site tyrosine-containing helices are disordered or misoriented (with the exception of HP1 integrase, where a sulfate ion bound in the active site may mimic the scissile phosphate). In a recent crystallographic triumph, two structures were determined of full-length λ Int complexed with not only crossover but also accessory site DNAs (36).

Cre, Flp, and λ Int all form C-shaped clamps around the DNA substrate (**Figure 4**). The larger and mostly helical C-terminal domains are highly conserved and contain the catalytic residues, whereas the preceding domains are structurally varied. The latter interact with the major groove of the DNA near the substrate crossover region and can form significant protein-protein interfaces with the other protomers in an assembled tetrameric complex, as we have seen in the structures of Cre, Flp, and λ Int. The C-terminal domains interact with consecutive minor and major grooves on the opposite face of the DNA. The monomeric human topoisomerase I exhibits a similar architecture. It may be that the catalytic domain is the common ancestor of the tyrosine recombinases and that the N-terminal domains have been added independently to aid complex formation and regulation.

The overall architectures of the synaptic complexes of Cre, Flp, and λ Int are

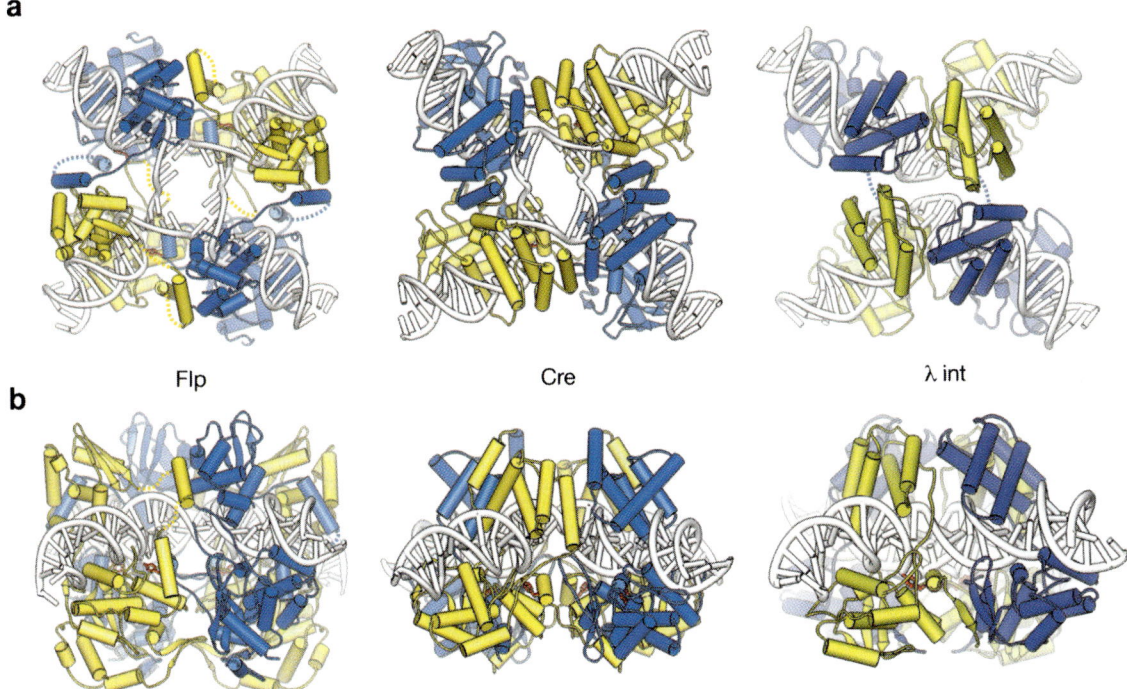

Figure 4

Comparison of Cre, Flp, and λ Int tetramers. (*a*) Tetramers viewed with their N-terminal domains in the foreground and catalytic domains in the background (Flp, *left*; Cre, *center*; λ Int, *right*). The catalytically active protomers are yellow, the inactive protomers are blue, and the nucleophilic tyrosines, where visible, are drawn in red. PDB IDs are Flp, 1M6X; Cre, 3Crx; lambda, 1Z19. (*b*) View of each rotated 90° such that their N-terminal domains are above the DNA and catalytic domains are beneath.

strikingly similar: The tetramers have twofold and pseudo fourfold symmetry, and hold the Holliday junction intermediate in a nearly square planar conformation (**Figure 4**). The structures also show that the catalytic domains interact by swapping part of the C terminus with a neighboring protomer. In all known cases, the segment immediately following the tyrosine-bearing helix (**Figure 5**) crosses from one protomer to the next. In Cre, the final helix nestles into a pocket on the neighboring subunit within the tetramer. In the HP1 apoprotein structure, a similar C-terminal helix swap occurs across a dimer. The C-terminal segment exchanged by λ Int forms a short beta strand when packed into the adjacent monomer in the synapsed complex. Flp displays a mechanistically important variation on this theme: The tyrosine-containing helix itself is swapped, so that each active site is assembled in *trans* (after this helix the chain returns to its original protomer) (42, 43). In addition to other Flp-like proteins from fungal plasmids, the thermophilic SSV1 integrase has also been reported to domain swap its tyrosine (44, 45). As discussed below, these *trans* segments are usually critical in enforcing half-of-the-sites activity.

The contributions of the N-terminal domains to the synaptic complexes vary. Cre's complex is the most rigid with close contacts between its globular N-terminal domains. Flp is more flexible: Its N-terminal domains interact through a second *trans* helix that packs into a hydrophobic groove in the neighbor's N-terminal domain but is connected to its own protomer by flexible turns. λ Int complexes, although still roughly square planar,

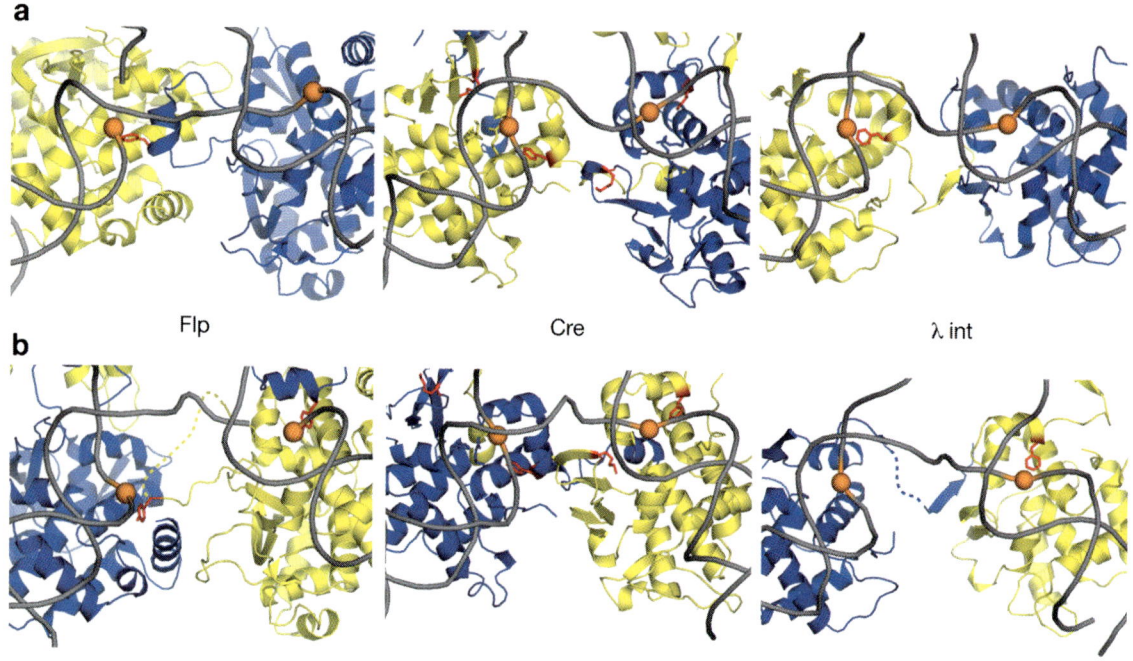

Figure 5

Comparison of interfaces within the Flp, Cre, and λ Int tetramers. (*a*) Two interacting protomers from each tetramer (viewed roughly as in **Figure 4a**). Only the catalytic domains and DNA are included. The catalytic tyrosines are shown in red, as is the lysine in Cre, and the scissile phosphates are marked by orange spheres. Even though the Cre and λ Int Holliday junctions shown here were trapped with inactive mutants, all the catalytic residues cluster near the scissile phosphate in the catalytically active protomers (*yellow*), whereas in the inactive protomers (*blue*) either the tyrosine (in Flp and λ Int) or the lysine (in Cre) is displaced. After Holliday junction resolution from this conformation, both protomers shown would be bound to the same product duplex. PDB IDs are Flp: 1M6X; Cre, 3Crx; and λ Int, 1Z1G. (*b*) The alternative interfaces from each tetramer, colored as in **Figure 4**. After Holliday junction resolution from this conformation, the protomers shown would be bound to different product duplexes. Made with PyMOL (157).

have much more skewed pseudo fourfold symmetry and little contact between the domains equivalent to Flp and Cre's N-terminal ones, which is consistent with λ's dependence on the accessory sites for synapsis. However, its additional N-terminal accessory site-binding domains (not shown in **Figure 4**) do make considerable contact with one another and with their neighbors' central domains.

The sequence specificity of these recombinases is of interest to those who exploit them as genetic tools as well as to basic scientists. In the Flp, Cre, and λ Int-DNA structures, each domain flanking the crossover site DNA inserts a helix into a major groove, but direct side-chain-base contacts are rather sparse and can rationalize only some of these proteins' sequence specificity (22, 35, 46, 48). λ Int binds this DNA rather weakly anyway, but in the Flp and Cre cases, water-mediated contacts and indirect readout of the DNA's sequence-dependent conformational parameters undoubtedly play important roles. Several clever approaches have recently been used to select Flp and Cre variants with relaxed and/or altered specificity (49–51). Structural studies from the Baldwin group (52) have highlighted the complexities of sequence recognition: The protein-DNA interface is a large hydrogen-bonded network involving many water molecules, and the connectivities of this network can shift in unexpected ways in response to mutation. Furthermore, specificity can be enforced at the catalytic step as well as at the binding step of the reaction (53).

Controlling Catalytic Activity: Half-of-the-Sites Reactivity

A recurring feature of the tyrosine recombinases is half-of-the-sites reactivity: Only alternating protomers within the synaptic tetramer are active at any given time. First noted by enzymologists in the 1960s (54), the practical consequence of this phenomenon for tyrosine recombinases is that double-strand breaks are avoided; one strand must be religated before its partner can be cleaved. This also avoids the formation of side products such as hairpins and three-way junctions that can result when adjacent protomers within a complex are simultaneously active (55–57).

In those cases that have been studied in detail, the key to mediating half-of-the-sites activity lies in the geometry of the synaptic tetramer. It has true twofold symmetry, such that protomers on opposite arms have similar conformations, and approximate fourfold symmetry, such that the isomerization between states is relatively straightforward. The alternating interfaces between protomers directly affect the active site conformations, but exactly how they do so varies in each case.

Flp's activity is primarily controlled by localization of the tyrosine nucleophile, which is donated *in trans* from an adjacent protomer. Within the synaptic tetramer, two interfaces (termed type I) allow proper docking of the tyrosine-bearing helix, while the alternate interfaces (type II) place the catalytic domains too far apart (**Figure 5**). Several lines of evidence imply that the remainder of the Flp active site is preassembled around the scissile phosphate.

1. Hydroxyl radical footprinting showed hypercleavage resulting from nucleophilic peroxide attack at the phosphate that lies in the active site, even in the context of Flp-binding sites that should not assemble into tetramers (58).
2. Structures of Flp revealed no significant differences among active site conformations other than the presence or absence of the tyrosine.
3. Kinetic studies imply that when the tyrosine is supplied on a synthetic peptide, only a monomer of Flp Y343F is needed for catalysis (K.L. Whiteson & P.A. Rice, unpublished results).
4. Flp can readily disassemble a synthetic three-armed DNA junction into a duplex and a hairpin product, which requires the simultaneous activity of adjacent Flp protomers (56, 59, 60).

Model building confirmed that, although adjacent type I interfaces could be formed within a trimer of Flp, addition of a fourth could only be accommodated by alternating type I interfaces with looser type II interactions (61). This implies that a Flp protomer is active whenever an exogenous tyrosine can reach into its active site.

λ Int is also regulated largely by positioning of the tyrosine. New structures of Int synaptic complexes show an even more skewed fourfold symmetry than Flp's (36). Here the tyrosine-bearing helix itself is *in cis*, but the polypeptide chain immediately following it crosses to the neighboring protomer, where it forms a short stretch of β strand. In two interfaces, the linker between these two secondary structure elements forms interactions that stabilize the tyrosine's position, whereas in the other two, the stabilizing interactions are disrupted, and the tyrosine-bearing helix is disordered (**Figure 5**).

Int's C-terminal tail also represses catalytic activity when the protein has not formed an appropriate oligomer. Mutations and deletions in this region greatly enhance topoisomerase-like DNA relaxation and cleavage by monomeric Int (62–64). In the crystal structure of the catalytic domain in the absence of DNA, this tail makes *cis* interactions similar to the *trans* ones seen in the tetramer, but in this context, they trigger a repacking of the previous segment, which removes the tyrosine from the active site (38, 48). NMR data found that the tail is flexible and is not altered by the addition of a single Int monomer-binding site (65). However, activity of the wild-type monomer can be restored by adding an excess of a peptide that mimics the C-terminal tail (66). Such negative regulation may be particularly important for Int, which binds crossover site DNA with relatively weak sequence specificity.

Structures of Cre-DNA complexes show more subtle differences between the two types of interfaces. Here, in the context of one type of interface, the catalytic lysine is displaced from the active site, whereas in the other type of interface, it lies near the scissile phosphate (16) (**Figure 5**). Cre's tyrosine-bearing helix is also *in cis* and is followed by a helix that packs *in trans* against the neighbor's catalytic domain. The linker between these two helices adopts different conformations as it crosses the two types of interface, which may affect the dynamics of the tyrosine-bearing helix but not its general placement. Cre will reluctantly recombine three-way junctions and has even been crystallized as a trimeric complex (57). In this structure, all three protomers are crystallographically identical, and although in some ways the active site most closely resembles that of the active protomers of tetramers, the catalytic lysine and the loop on which it lies are disordered.

In the XerCD system, the fourfold symmetry is broken not only by conformation but also by sequence: Here alternate monomers are actually different proteins (albeit closely related—36% identical) that bind different DNA sequences, one on each side of the central spacer. Sequence similarities suggest that both of these proteins contain a *trans*-packing C-terminal segment similar to Cre's, and biochemical studies have shown that the interactions made by these segments play a key role in regulating catalytic activity (67, 68).

Controlling the Outcome of Recombination

How do these systems specifically produce inversions, deletions, or insertions, and how do some of them drive the reaction in one direction despite the lack of obvious energy sources such as ATP?

Key factors include the conformation of the DNA in the initial synaptic complex and the stabilization of one conformation over another in the product complex. First, a crossover site's central spacer must adopt one of two possible bends to accommodate contacts between the two protomers that bind it (**Figure 6a**). The conformation that is adopted in the substrate complex determines which strand will be cleaved first. Second, the synaptic complex can form, and strand cleavages can be initiated with the asymmetric spacers in either relative orientation, but only the antiparallel orientation is productive (**Figure 6b**). After the second set of strand exchanges, the spacers in the resulting product complex will display the alternate bends to those in the substrate. In the simplest cases, the substrate and product are isoenergetic, and the reaction reaches equilibrium when there is 50% of each. However, as described below, many systems have evolved tricks (usually involving accessory factors) to tip the balance toward the desired product.

Simple systems: random synapsis and reliance on spacer complementarity.
Synaptic complexes formed with parallel spacers are unproductive: Ligation of the product strands would produce mismatches. These inhibit ligation, particularly near the cleavage sites, presumably by misorienting the attacking 5′ hydroxyl (69). The back reaction is thus favored, and the complex can dissociate. In simple systems, the sequence of the spacers is in fact the only feature determining the overall polarity of the sites (shown as arrows in **Figure 1**), which in turn determines the overall choice of inversion vs deletion/insertion reactions. This

is not utterly foolproof; however, if the products of incorrect (parallel) Cre synapses are selected for in vivo, they can readily be found (70).

Flp- and Cre-mediated recombination is rather simple and does not require accessory factors, nor is it strongly biased in one direction. Despite an asymmetric spacer sequence, Flp does not even display a significant preference for initial cleavage at one end of the spacer vs the other. Cre, however, does. In assays using 5′ bridging phosphorothioate suicide substrates, Cre preferentially cleaves at the GpC end of the spacer rather than the ApT end (19). Interestingly, cleavage of the preferred strand was more strongly stimulated by synapsis than cleavage of the other. When the reaction is started with Holliday junction substrates, Cre preferentially resolves them by cleavage at the ApT end. Combined, these preferences may improve the overall efficiency of recombination by minimizing the accumulation of junctions and biasing their resolution toward products (71). Cre's catalytic preferences probably reflect a balance of the sequence-dependent flexibility of the DNA itself and the interactions of the protein with its substrate DNA (72, 72a). Crystal structures of Cre-DNA complexes show strikingly asymmetric bends in the spacer regions that vary with sequence (16, 73, 74).

Using accessory proteins and other tricks to control outcome. Phage λ integration is a paradigm for the use of accessory proteins to help assemble the correct initial complex and to drive the overall reaction in the desired direction. Many other integrases appear to use similar overall schemes for regulating the directionality of recombination, although the details vary (75–79). The λ Int protein has an additional domain at its N terminus that binds tightly to "arm" DNA sites found on both sides of the crossover ("core") site within the phage attachment site, *attP* (**Figure 7**). DNA-bending proteins help bring the two types of site into close proximity. The bacterial attachment site, *attB*, lacks accessory sites, but the

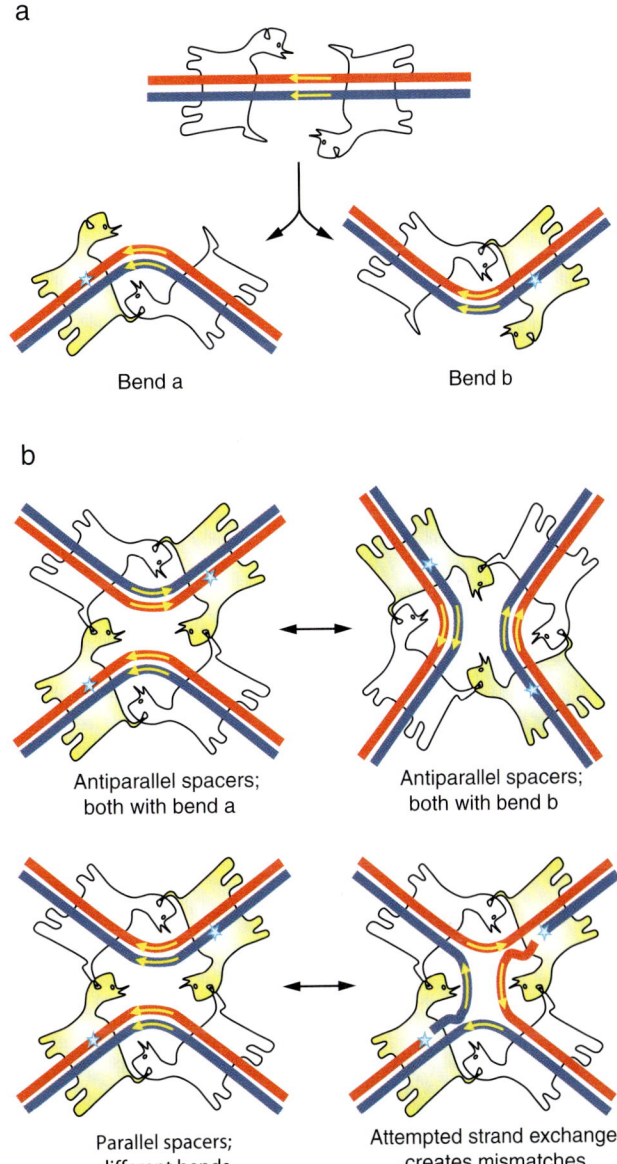

Figure 6

Controlling the outcome of recombination. (*a*) Determining which strand will be cleaved first. The initial recombinase-bound duplex bends in one of two different ways, and this determines which strand will be cleaved first (*blue stars*). In all cases studied, the activated protomer is the one whose C-terminal tail is bound by the other. (In Flp, this corresponds to the polypeptide chain returning after the *trans* tyrosine-bearing helix.) (*b*) These two differently bent duplexes can be combined into three different synaptic complexes: two productive ones with antiparallel spacers (*top*) and one unproductive one with parallel spacers (*bottom*). Note that if a productive synaptic complex initially has bend a in the spacers, its product will display bend b, and vice versa.

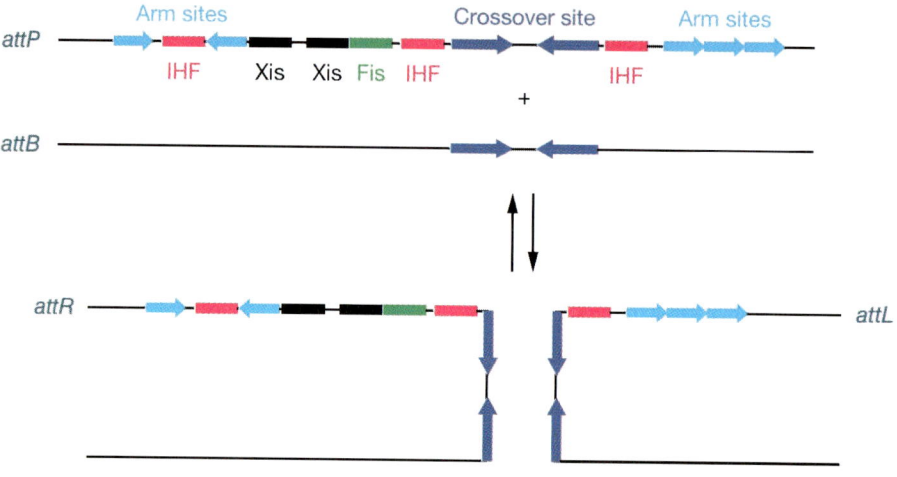

Figure 7

Cartoon of λ *att* sites. The crossover site within the phage attachment site, *attP* (*top*), is flanked by different arrays of binding sites for accessory proteins (IHF, Xis, and Fis) and for the N-terminal domain of Int (arm sites). Integrative recombination of *attP* with the simple bacterial *attB* requires IHF and creates *attR* and *attL* as products (*bottom*). Excisive recombination is not the direct reversal of this reaction and requires Xis and Fis as well as IHF.

Int protomers whose catalytic domains bind here also interact with the arm sites of *attP* via their N-terminal domains. Although the accessory proteins can only dictate the initial bend direction of *attP*, suicide substrates using bridging 5′ phosphorothioate linkages showed that synapsis with *attB* is not random (18). This implies that the initial bend in *attB* is determined by its DNA sequence. The strand that is preferentially cleaved first is also exchanged first (80, 81).

After integration, the recombinant sites flanking the phage are referred to as *attL* and *attR*. Excisive recombination, however, is not the simple reversal of integration: It is stimulated by different relative concentrations of DNA-bending proteins, leading to formation of an initial complex whose geometry differs from that of the integrative product complex. In both reactions, the same strand is cleaved and exchanged first, presumably because of similar bends in the core DNA (82).

The arm-binding domains of Int do more than just help deliver it to the crossover sites; they also regulate catalysis. If complexes are assembled with full-length Int on crossover-site Holliday junctions, addition of short duplexes bearing arm DNA sites not only stimulates resolution but also biases it toward appropriate duplex products rather than aberrant hairpins and three-armed junctions (55). The crystal structure of an entire Int tetramer bound to both core and arm DNA has recently been reported (36). The N-terminal arm-binding domains form an additional, intertwined layer of protein displaying twofold but not even pseudo fourfold symmetry. Each N-terminal domain interacts more closely with the core-binding domain of the neighboring protomer than its own. These interactions presumably bias the overall complex toward the product conformation, in a sense using product-binding energy to drive forward an otherwise isoenergetic reaction.

Accessory proteins also play crucial roles in XerCD-mediated recombination, but here the details vary greatly with context (67). However, one common feature is that two recombinases' preference for hetero- rather than homomeric interactions tidily avoids the formation of unproductive parallel synaptic complexes. Recombination at the chromosomal recombination site, *dif*, requires FtsK, an ATP-dependent DNA translocase that also plays other roles in chromosomal segregation. Without FtsK, XerC catalyzes cleavage and strand exchange, but the resulting Holliday junction can only be resolved backward to the substrate; XerD remains inactive. However, in the presence of FtsK, XerD is activated and actually catalyzes strand cleavage and exchange before XerC, implying a different initial synaptic complex is formed (83). FtsK's effects in vitro require only a short stretch of DNA extending past the XerD-binding site and involve direct contacts with XerD as well as ATP hydrolysis by FtsK (84).

XerCD-catalyzed resolution of dimers of the plasmids ColE1 (through its action at *cer* sites) and pSC101 (at *psi* sites) has also been well characterized (67). These systems do not require FtsK. However, their recombination sites are much larger than *dif* and include binding sites for accessory factors: the DNA-binding peptidase/transcription factor PepA in both cases, plus the arginine repressor ArgR for ColE1 and the phosphorylated form of the anaerobic growth regulator ArcA for pSC101 (**Figure 8**). These accessory proteins form a topologically defined synaptic complex that, much like the synaptic complexes formed by the canonical serine recombinases, dictates intra- rather than intermolecular recombination by a mechanism termed a topological filter. Depending on the spacer, the reaction is either completed by the sequential action of both XerC and XerD, or Holliday junctions formed by XerC action are resolved by other cellular enzymes. The conversion of negative supercoils, trapped within the synaptic complex, into intermolecular nodes, linking the catenated products, may also help drive the overall reaction forward. The X-ray structure of the hexameric PepA revealed a positively charged groove that is important for DNA binding and has greatly aided modeling of the synaptic complex (85, 86).

The synaptic complex formed by PepA not only channels but also stimulates recombination. It probably does so simply by aiding synapsis of the crossover sites by bringing them into close proximity; there is no evidence for direct recombinase-accessory protein contacts, and the XerC- and XerD-binding sites can be swapped without the deleterious effects that might be expected if such contacts were important (87). Furthermore, if the XerCD crossover site is replaced with the Cre

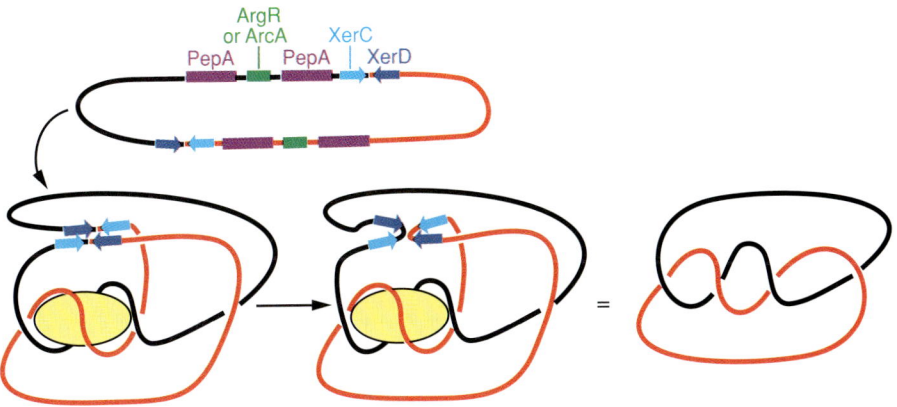

Figure 8

Cartoon of XerCD–mediated resolution of plasmid dimers. (*top*) Arrangement of protein-binding sites in a dimerized ColE1 or pSC101 plasmid (not to scale). (*bottom*) Synapsis as proposed by Reijns et al. (86) is mediated by a single PepA hexamer (*yellow*), which interacts with binding sites in both direct repeats. ArgR (at ColE1 *cer* sites) or ArcA (at pSC101 *psi* sites) assists synapsis. XerCD-mediated recombination of these substrates produces four-noded catenanes as products.

crossover site (*loxP*), Cre-mediated recombination also becomes dependent on PepA when Cre's protein-protein contacts are mutationally compromised (88). Surprisingly, it was recently reported that ArgR and PepA are required for stable maintenance of the plasmid state of phage P1 in vivo, although their mechanism of action on P1 has not yet been directly determined (89).

Vibrio cholerae phage CTXφ exploits its host's XerCD system for integration but requires none of the accessory proteins described above (90). The secondary structure in the single-stranded phage DNA creates a duplex recombination site with a bulged spacer (91). As shown previously for Flp-mediated recombination, such mismatches adjacent to the scissile phosphate will favor cleavage over religation of the substrate (69). However, after XerC-mediated cleavage, the new 5' end does match the chromosomal *dif* site spacer, so that a covalent Holliday junction can be formed. XerD's presence, but not its catalytic activity, is required; the junction is resolved by replication and/or repair. The resulting XerCD sites flanking the integrated prophage no longer have matching spacers, thus blocking excision (unlike λ, integration of this phage is a one-way street). Directionality in this case is thus dictated by the ability of the single-stranded phage to form a bulged recombination site.

Although the mechanistic details are less clear, the integrases that mobilize the gene cassettes of integrons may also exploit unusual DNA structures (92). The *att*C attachment site (also known as the 59-bp element) contains a set of repeats capable of forming a cruciform structure, and in vitro studies suggest that the enzyme preferentially binds such DNA hairpins (93, 94).

Formation of hairpins is the *raison d'etre* for hairpin telomere resolvases (also known as protelomerases), a novel type of tyrosine recombinase. These enzymes convert a single inverted repeat-containing site into two duplexes with hairpin ends. They contain most of the canonical conserved catalytic hexad, including the tyrosine that forms a covalent protein-DNA intermediate (33, 95). Recent work suggests that ResT from *Borrelia burgdorferi* combines this active site with a hairpin-binding module that may resemble one found in otherwise unrelated DNA transposases that form hairpin intermediates (96).

SERINE RECOMBINASES

The serine recombinases are a rather heterogeneous family of proteins, ranging in size from 180 to nearly 800 amino acid (aa) residues, and with unexpected variations in domain organization (**Figure 9a**) (5).

Most of our information regarding serine recombinase domain structure and function has come from the prototypical recombinase, γδ resolvase, which has been characterized extensively, both biochemically and structurally, by X-ray crystallography. This 183-residue protein has an N-terminal catalytic domain of ~100 residues, linked by a long (36-aa) α-helix (the E-helix) and an unstructured segment (10 aa) to a typical helix-turn-helix DNA-binding domain at the C terminus (**Figure 10**) (97). The serine nucleophile is close to the N terminus at position 10. γδ resolvase is a dimer in solution, with the N-terminal portion of the E-helix forming the bulk of the dimer interface. DNA binding (at least to the crossover site) involves not only the H-T-H domain but also the C-terminal portion of the E-helix and the intervening segment. The dimer's H-T-H domains bind symmetrically to the DNA, making sequence-specific major groove contacts ~10 bp from the central cleavage point; E-helix residues (particularly the conserved Arg-125) hold the DNA (via phosphate and minor groove contacts) close to the cleavage site and the 3' end of the DNA after cleavage (98); and the unstructured segment snakes along the minor groove between the two (see **Figure 10**) (97). The H-T-H domain appears to play no important roles outside of DNA binding because it could be replaced by a zinc finger DNA recognition domain in Tn3

a

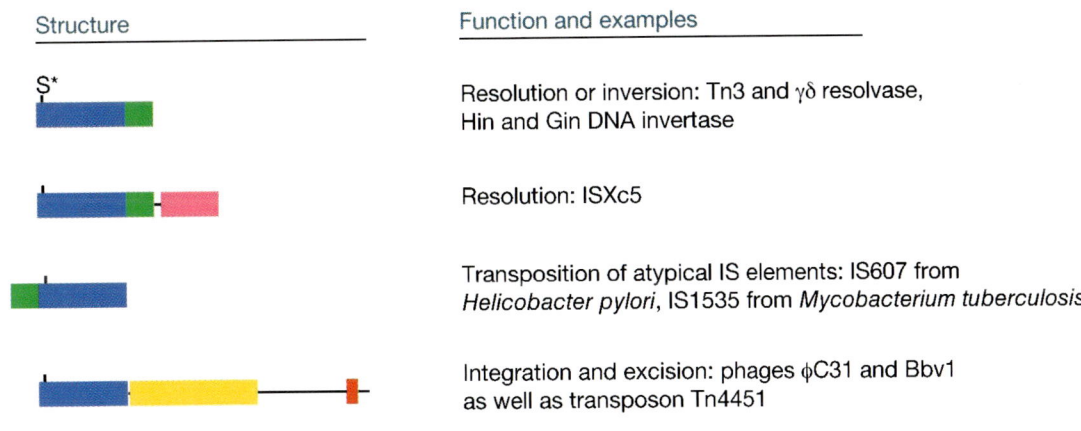

b

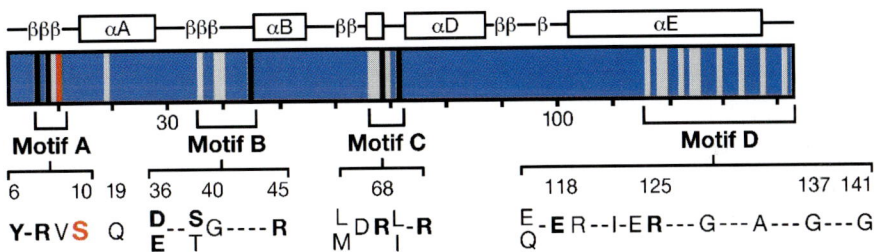

Figure 9

(*a*) Domain structure of serine recombinases. This figure shows the catalytic domain and E-helix (*blue*), with S* showing the position of the serine nucleophile; the DNA-binding domain (*green*) containing a recognizable helix-turn-helix (H-T-H) motif; and conserved domains of unknown function found in subsets of recombinases (*magenta*, *yellow*, and *red*) (5). (*b*) Conserved motifs within the catalytic domain and dimerization helix (αE) of serine recombinases. Motifs A and C contain the critical active site residues of the recombinase. Motif D, contained within the C-terminal portion of the E-helix plus a few residues beyond, is mostly involved with binding the DNA in the region abutting the cleavage site. Motif B forms a rather mobile loop whose function remains mysterious despite the remarkable conservation of the Ser-39, Gly-40, and Arg-45 residues.

resolvase without loss of recombination activity (99).

Although the well-studied DNA invertases, Hin and Gin, are similar in size and organization to γδ resolvase (and many others), some members of the serine recombinase family are considerably larger, and yet others have switched the DNA-binding domain to the N terminus of the protein (5). Despite these variations, all members

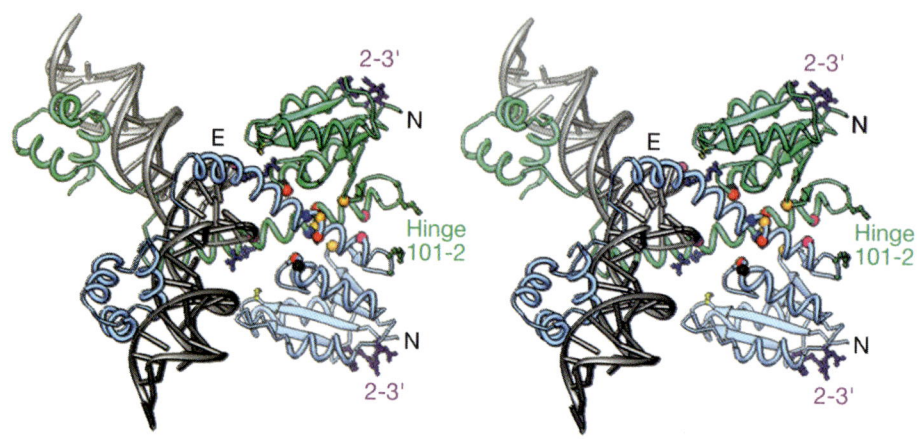

Figure 10

Structure of the γδ resolvase bound to its crossover site (97). Residues shown as ball-and-stick shapes are Ser-10 (*yellow*), Gly-101 and Glu-102 (*green*, labeled hinge), Arg-2, Arg-32, Glu-56, and lys54 (*purple*, labeled 2-3′), and Glu-124 and Arg-125 (*blue*). Cα atoms of positions of some informative cysteine substitutions (see text) are shown as colored balls: Met-106 (*pink*), Thr-73 and Ser-112 (*red*), Asp-95 and Ala-113 (*gold*), Ala-74 (*dark gray*), and Val-114 (*blue*). P atoms at the cleavage sites are shown as magenta balls. E marks the long E-helix at the dimer interface. N marks the N termini of resolvase. Figure made from PDB ID 1GDT.

contain a readily recognizable catalytic domain with two clusters of conserved residues, including the serine nucleophile, that form the recombinase active site (**Figure 9b**). The large recombinases also contain an obvious analogue of the E-helix immediately following the catalytic domain and in most cases have E-helix residues equivalent to the γδ resolvase residues Glu-118, Arg-119, Glu-124, Arg-125, and Gly-137 (those that are most conserved in the E-helices of the conventional resolvases and DNA invertases). Even the "domain-switched" recombinases appear to have an E-helix analog (with equivalents to Arg-119 and Arg-125), presumably allowing them to form dimeric assemblies and bind the 3′ end of the cleaved recombination site in a manner similar to resolvase.

Recombination by a Process of Double-Strand Break, Switch, and Rejoin

The salient features of the processes of DNA cleavage and strand exchange have been revealed by detailed biochemical and DNA topological analyses, primarily of the resolvases from Tn3 and γδ, and the DNA invertases, Hin and Gin (2, 3, 21, 100–103).

All catalytic processes usually occur within a synaptic complex with two crossover sites and four recombinase subunits (although the Sin recombinase appears to be at least one exception to this). It is now clear that, in synaptic complexes formed by the serine recombinases, the crossover sites are located on the outside, separated by the catalytic domains; this is in stark contrast to the synaptic complexes formed by the tyrosine recombinases. The idea that the recombinase was at the center of the synapse with DNA on the outside was an explicit feature of the earliest "precrystal structure" model of the resolvase synaptic complex (104, 105) and grew out of pioneering topological studies (21, 101, 102, 106), but lacked direct evidence. Subsequently, evidence for this arrangement was provided by three distinct and complementary experiments. First, recombination by an activated γδ resolvase between a pair of crossover sites separated by an IHF-induced DNA U-turn was shown to be sensitive to the helical phasing of the sites; the positions of maximum recombination efficiency indicated that the catalytic domains formed the core of the synapse with the DNA outside (107). A similar conclusion was obtained from low-angle X-ray and neutron scattering experiments performed with an activated Tn3 resolvase that formed a stable synaptic complex with two uncleaved crossover sites; moreover, these data indicated a substantial separation of the DNAs (108). Finally, an entirely different approach, chemical cross-linking at the sites of introduced cysteine residues in an activated Hin mutant, showed that the surface of the catalytic domain around the beginning of the E-helix (around residue 100) was at or very close to the synaptic interface (109). The recently solved crystal structure of a minimal synaptic complex formed by γδ resolvase has elegantly confirmed the "DNA-out" configuration of the crossover site synapse and has thrown new light on the processes of synapsis and strand exchange (see below) (98).

Once a synaptic complex is formed, the four recombinase subunits are activated to attack the two crossover sites, forming two double-strand breaks (**Figures 2** and **11**). This reaction covalently joins the four recombinase subunits by a phosphoserine linkage to the four 5′ ends of the broken strands, leaving free hydroxyls at the 3′ ends (7, 110). The spacing of the scissile phosphates is such that cleavage leaves a two-base single-strand extension at each 3′ end (7).

Cleavage is a coupled reaction and is performed by the recombinase "in *cis*" (111). Coordination is most pronounced between the two subunits bound to the same site; nicked sites are rarely seen with the wild-type recombinase and remain a minor species even when active and inactive subunits are simultaneously targeted to specific halves of a crossover site. Nevertheless, in such targeted experiments, nicking is seen to be highly strand specific, occurring at the scissile phosphate closest to the binding site of the active subunit.

Once both crossover sites are fully cleaved, the broken ends are rearranged to bring them into a recombinant configuration. Studies of the changes in DNA topology and linking number that accompany recombination indicate that strand exchange involves a motion equivalent to a single 180° rotation of one half of the complex relative to the other half (21, 101, 106, 112–115). The direction of rotation is generally right handed, serving to relax the natural negative superhelicity of the substrate. However, in reactions with relaxed DNA, it appears that rotations can occur in either direction (21, 116). Precisely how rotations of the broken ends occur is still a mystery, although recent structural studies of an active synaptic complex with cleaved crossover sites strongly favor a particular process of subunit exchange (see below) (98). Following the exchange, the free 3′ OH ends attack the 5′-phosphoseryl linkages to rejoin the crossover sites in recombinant configuration and release the resolvase subunits. For the initial cleavage step, the sequence of the central 2 bp is not very important (many 2-bp sequences can be cleaved, but not all). However, for the rejoining step, it is essential that the two-base single-strand extensions of the partner sites are able to form Watson:Crick base pairs. Thus, as in the tyrosine recombinases, asymmetry of the central 2-bp sequence can be used to dictate recombinational directionality, allowing recombination of two sites in one orientation but not in the other. Because crossover site heterologies are only detected when rejoining is attempted (that is, after cleavage and strand exchange), the presence of a mismatched site forces the recombinase to proceed through a second round of strand exchange. This restores the parental configuration and so is without genetic consequence; however, the double rotation leaves a topological footprint on the substrate DNA, removing supercoils and, in some cases, generating knotted products (113–115).

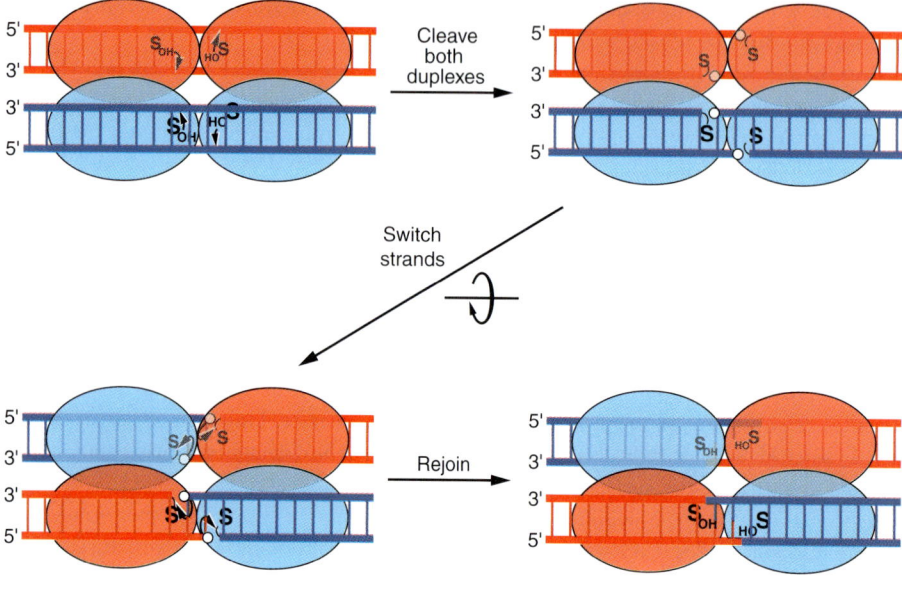

Figure 11

Mechanism of recombination by a serine recombinase. The cartoon shows a synaptic complex, formed from a pair of recombinase-bound crossover sites. Only the catalytic domains of the recombinase dimers are shown; these are responsible for the pairing and separate the two DNAs. The serine nucleophiles are represented by S_{OH} when free (*top left* and *bottom right*) or S- when attached to the DNA 5′ ends (*top right* and *bottom left*). The free 3′ OH groups at the cleavage sites are shown as o. During strand exchange, catalytic domains and DNAs move together because they are covalently joined.

Structural Insights into Synapsis, Cleavage, and Strand Exchange

For the serine recombinases, a molecular understanding of the processes of synapsis and strand exchange has long remained an elusive goal. A complicating factor for analysis of the serine recombinases is that the systems of recombination best characterized biochemically require large DNA sites in supercoiled plasmid vectors and either multiple copies of the recombinase bound at sites in addition to the crossover sites (as in the case of the resolvases) or an additional protein factor, FIS (as with the DNA invertases). To overcome these difficulties, it was first necessary to obtain mutants of the recombinase that were able to perform recombination on short, linear crossover sites in the absence of additional factors. Such activated mutants, first obtained with DNA invertases (117, 118) and subsequently with resolvases (119, 120), have played a crucial role in all experiments that have recently advanced our understanding of synapsis and strand exchange.

Synapsis. Using such an activated mutant, crystal structures of $\gamma\delta$ resolvase with cleaved crossover sites have recently been obtained and have provided clear confirmation of the DNA-out configuration of the synaptic complex (98). The activated mutant forms a tetramer in solution (with or without DNA) and binds, cleaves, and recombines pairs of isolated crossover sites (121).

Crystallization of the activated resolvase with a symmetrized crossover site yielded two crystal forms (98). In each, a tetramer of resolvase was bound to two cleaved crossover sites, with phosphoseryl covalent linkages joining the four 5′ phosphates at the cleavage sites to the Ser-10 residues of the recombinase subunits (**Figure 12**). The tetramers in the two crystal forms have pseudo 222 symmetry (that is, three orthogonal twofold symmetry axes) and essentially identical conformations. The core of the tetrameric complex is formed by the resolvase catalytic domains and the E-helices while the DNA and DNA-binding domains are on the outside.

The synaptic tetramer has a unique and unanticipated quaternary structure (98). It consists of two dimeric units with a quaternary structure related to, but distinct from, that of the dimer of resolvase bound to uncleaved site I (97). It is assumed that these represent the parental dimers that have just cleaved their crossover sites (or the recombinant dimers that are about to rejoin their crossover sites). These two site I-bound dimers are separated by an interface (the synaptic interface) that is extensive (1780 Å^2 per dimer) and unexpected—unexpected because the dimers are substantially interdigitated and the interface is not formed by docking contiguous pre-existing surfaces of the presynaptic dimer. Indeed, many side chains at the interface were not exposed on the surface of the presynaptic resolvase dimer but were buried at the interface between its subunits. A prominent feature of the tetramer (and one that is key to holding the complex together) is its central inner core, consisting of two pairs of antiparallel E-helices (associated over their N-terminal halves) that cross at an angle of 100° to form an extended X. Within this core, each subunit interacts with all three of its partners within the tetramer. In contrast, no contacts are formed between diagonally positioned protomers within the tyrosine recombinase complexes (22, 35). Remarkably, the other interface that separates the tetramer into left and right halves is almost totally flat; this may have functional significance for strand exchange.

The synaptic tetramer could not have been modeled readily, since its formation is accompanied by dramatic conformational transitions from the structure of the presynaptic resolvase dimer. These changes are seen both in the tertiary structure of individual subunits and in the interactions between them.

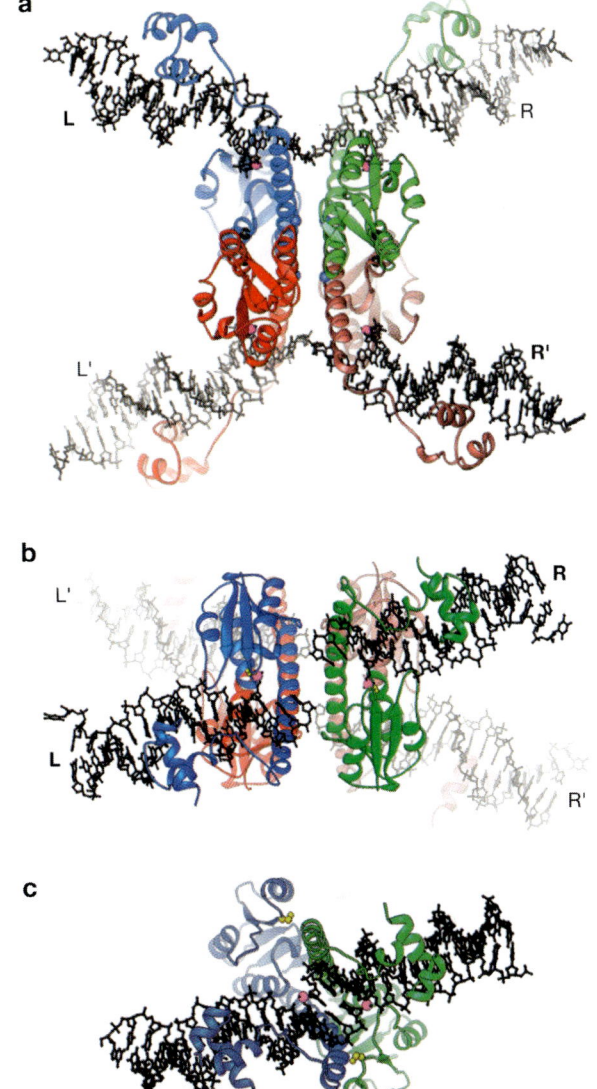

Figure 12

Structure of the γδ resolvase synaptic tetramer complexed with cleaved crossover sites (98). Magenta balls indicate the P atoms at the cleavage sites, and the yellow ball and stick side chains show the active site Ser-10 residues. (*a*) View through the complex showing the flat interface that separates the complex into two halves. Presumed parental sites are labeled L and R, and L' and R', respectively; it is not known, however, whether recombination will join L to R' or to L'. Colored balls indicate the Cα atoms of Ala-74 (*dark gray*) and Val-114 (*blue*) (see also **Figure 10** and text). (*b*) Structure in panel *a* rotated 90° about the horizontal axis. (*c*) Similar view of the presynaptic dimer, to emphasize the tertiary and quaternary changes that accompany the dimer-tetramer transition. Note the interlocked dimeric interface that exists prior to synapsis and DNA cleavage as well as the large distance between Ser-10 residues and the cleavage sites. Figures from PDB IDs 1ZR4 and 1GDT.

For individual subunits, if the DNA and DNA-binding portions (residues 120 to 183) of the presynaptic and synaptic states are superimposed, the catalytic Ser-10 residue of the activated resolvase is seen to have moved 11 Å directly toward the scissile phosphate (**Figure 13**) (98). This movement consists chiefly of two components: substantial rotation (40°–70°, depending on the subunit of the presynaptic dimer selected for comparison) of the catalytic domain relative to the E-helix using residues 101–102 as a hinge (see also **Figure 10**) and bending of the E-helix itself (which moves the helix N terminus by about 6 Å). This new tertiary structure of the resolvase subunit, with the Ser-10 hydroxyl well within bonding distance of the scissile phosphate, presumably represents the activated conformation necessary to catalyze cleavage of an initially unbroken crossover site. In the activated conformation, the D-helix of the catalytic domain swings toward the E-helix (the Thr-73-Ser-112 Cα-Cα distance changes from 9.8 Å to 5.9 Å). We find it intriguing that an engineered intrasubunit disulfide bond between T73C and S112C in an otherwise wild-type Tn3 resolvase is sufficient to activate its ability to cleave an isolated crossover site (122).

Substantial changes in the dimer interface accompany the subunit conformational changes (98). In the presynaptic state, the interface is formed largely by the E-helices, which cross at a 45° angle. In the tetramer, however, the E-helices open up like the blades of a pair of scissors, increasing the crossing angle to 100°. This scissoring of the E-helices, together with the subunit conformational changes, creates an altered and much reduced dimer interaction. Many interactions seen across the interface of the presynaptic dimer are lost (or exchanged for new, synaptic interactions), and only a few new ones are gained in the altered dimer.

The process by which a synaptic complex may be formed is illuminated by the structure of the γδ resolvase synaptic tetramer, even though the activated mutant has cir-

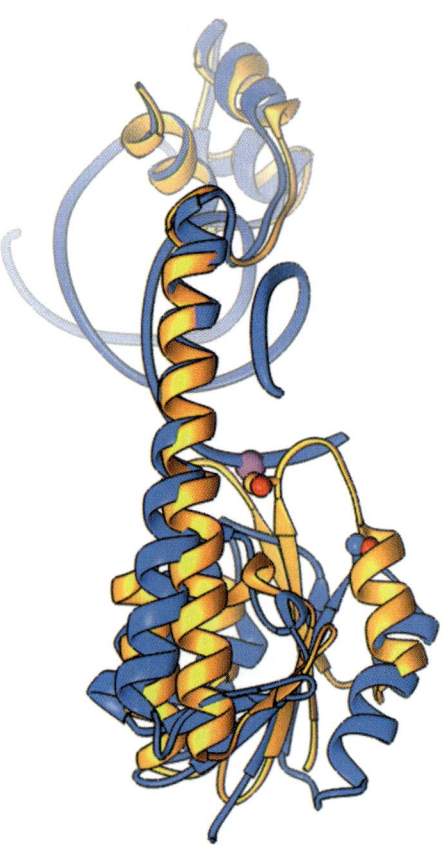

Figure 13

Motion of the resolvase catalytic domain associated with formation of the synaptic tetramer. A postsynaptic subunit (*yellow*) is superimposed on one half of the presynaptic γδ resolvase-DNA complex (*blue*) to illustrate the change in conformation of the catalytic domain and part of the E-helix relative to the DNA and DNA-binding domain (superimposed residues were 120–183). This motion brings Ser-10 (*red oxygen atom*) close to the scissile phosphate of the uncleaved DNA (*pink ball*).

cumvented the normal process of synapsis because it is already tetrameric. Presumably, in more natural circumstances, two crossover site-bound presynaptic dimers of a recombinase initially interact via the surface centered around the beginning of the E-helices, perhaps as suggested in the various models of Sarkis et al. (121), Nollmann et al. (108), or Dhar et al. (109). Li et al. (98) have modeled a plausible transition from this initial state to the interdigitated tetramer using the program MORPH (123). The transition shows that, as the N-terminal portions of the E-helices slide past each other to establish the antiparallel pairings at the synaptic interface, the E-helices across the dimer interface gradually scissor open.

Crossover site cleavage. The timing and mechanism of DNA cleavage during the transition are quite uncertain and are not addressed by the tetramer structure because cleavage has already occurred. Clearly the N-terminal catalytic domains must swing into the active configuration, allowing the Ser-10 residues to attack the scissile phosphates, and presumably, this conformational change is dependent on synapsis. However, although the activated tetramer can readily make double-strand breaks at the crossover site even when locked into the tetrameric conformation by a disulfide link between cysteines positioned in the antiparallel D helices (at position 74) (A. Sinha and N.D.F. Grindley, unpublished results), the precise conformation seen in the crystal structures appears inappropriate either for DNA cleavage or for ligation. Each Ser-10 residue is 30 Å from that of its dimer partner (compared with a distance of 16Å between the scissile phosphates in the presynaptic dimer site I complex) and is 14 Å from the 3′ OH that would be the attacking group in the joining step.

There are several possible solutions to this dilemma. (*a*) The overall structure of the synaptic complex protein core is retained, but motions of the DNA-binding region (residues from the middle of the E-helix to C terminus of resolvase) deliver the uncleaved DNA sequentially to the two Ser-10 nucleophiles (or the 3′ OHs sequentially to the phosphoseryl linkages). A potential problem with this scenario is that the covalent linkage of the DNA to the first Ser-10 is likely to impede its movement to the second Ser-10. (*b*) The quaternary structure of the crossed antiparallel pairs of E-helices is retained, but additional motions of the catalytic domains allow the two Ser-10 nucleophiles of each synapsed dimer to come closer together, facilitating attack on the scissile phosphates either simultaneously or sequentially but with minimal distortions (the DNA-binding domains would also need to move to accommodate binding to uncleaved DNA or deliver the 3′ ends to the phosphoseryl bonds). In the tetramer structure, the direct path between each phosphoserine and the appropriate 3′ OH is unimpeded, adding to the plausibility of this scenario. (*c*) Since one cannot simply model the uncleaved crossover site and its associated DNA-binding domains from the presynaptic structure onto the core of the synaptic tetramer, the possibility remains that a structure formed on the pathway of transition between the presynaptic dimer and the synaptic tetramer but with unknown conformation is responsible for cleavage.

Although the structures of $\gamma\delta$ resolvase with DNA provide two revealing views of the recombinase active site, some important details are missing, and the process of chemistry remains obscure. In particular, it is not at all clear how either the serine hydroxyl or the 3′ OH are activated for the cleavage and rejoining steps, respectively.

The most critical components of the active site, in addition to the Ser-10 nucleophile, are three arginine residues (residues 8, 68, and 71) and an aspartate (Asp-67) (**Figure 2d**). All these residues are very highly conserved, and mutation of any of them substantially reduces or abolishes crossover site cleavage and recombination. In the presynaptic structure, Ser-10, Arg-8, Arg-68, and Asp-67 of one subunit (along with the nonessential Glu-124′ residue provided *in trans* from the partner subunit) form a hydrogen-bonded network, but no residues appear to be appropriately positioned to act as a proton acceptor for Ser-10. The phosphate of the scissile bond is distant from the Ser-10 ($\sim$ 11 Å), and in one subunit, Arg-71 appears to coordinate it and the phosphate 5′ to it (97).

The synaptic tetramer is a product complex. The 3′ end (the leaving group of the cleavage reaction) has moved about 14 Å away from the 5′ phosphoserine and is held by an interaction between its phosphate and the well-conserved Arg-125′ (of the partner subunit) (**Figure 2d**) (98). Although the path between the 3′ end and the phosphoserine is open, it is not clear what conformational change would be necessary to bring them together. Arg-8 and Arg-68 interact with the nonbridging oxygens of the 5′ phosphoserine, suggesting that this was also their role immediately precleavage, and Asp-67 interacts with Arg-68. Arg-71 no longer appears to contact the DNA but instead packs with residues in the mobile loop (residues 40–45); however, if it adopted the configuration seen in the presynaptic structure, it would be well positioned to interact with the phosphate 3′ to the cleavage site.

Strand exchange. As the crystal structure of the resolvase synaptic tetramer indicates, strand exchange poses several substantial challenges (98). The free 3′ OH ends of the cleaved DNAs are far (about 50 Å) from the phosphoserine groups they must attack to produce recombinants (and they appear to be well bound by the C termini of each E-helix); the 5′ ends are covalently linked to the recombinase and are not free to move without accompanying protein motions; the space between the recombinant ends is filled with the resolvase catalytic domains, preventing diffusion of DNA ends across the gap; and both strands of each crossover site are broken, so that the complex is held together only by protein-protein interactions. In addition, topological analysis has shown that strand exchange involves a motion equivalent to a single 180° rotation of one half of the complex relative to the other half. How, then, is this movement of the ends achieved?

Potential mechanisms of strand exchange depend on which pairs of ends in the tetramer structure need to be joined to form recombinants. The crossover sites could be approximately parallel or approximately antiparallel. If the sites are parallel, then the half sites labeled L and L′ would be joined to R′ and R, respectively (**Figure 12a**). For antiparallel sites, however, L would be joined to L′ and R to R′. Each of these scenarios poses a very different strand exchange problem within the tetramer structure.

The structure of the tetramer provides a conceptually simple and elegant way to recombine parallel crossover sites (98). For this scenario, the left and right halves of the tetramer are separated by a remarkably flat interface with essentially no interlocking components (**Figure 12ab**). Indeed the only specific interactions that appear to hold the two halves together in the orientation observed are those between the positive patches formed in each subunit by Arg-121 and Arg-125 and the complementary negative patches formed in the dimeric partner subunit by Asp-95 and Asp-96. The rest of the interface is highly hydrophobic. Li et al. (98) propose that strand exchange is accomplished simply by allowing the two halves of the tetramer to rotate relative to each other—a process called subunit rotation (21)—using the flat, hydrophobic interface both as a bearing and to maintain stable contact regardless of the relative orientation of the two halves. Calculations and modeling suggested that each flat surface can readily adjust its precise conformation so as to avoid any steric clashes during the proposed rotation (98). The complementary positive and negative patches mentioned above may act to provide a gating mechanism that favors rotations in steps of 180° and stabilizes a state that favors rejoining the half sites. Not only does this proposed mechanism account for the topological changes observed in recombination by resolvases and DNA invertases, but it also readily accounts for the processive cycles of 360 rotations that occur when strand exchange of cleaved but nonidentical crossover sites creates mismatches between potentially recombinant ends (114).

If the appropriate alignment of crossover sites was antiparallel, not parallel, recombination by a simple process of subunit rotation would be sterically impossible. An alternative mechanism—domain swapping—has been proposed for facilitating strand exchange by serine recombinases (2, 120; M. Boocock, personal communication). In this process, the four E-helices forming the core of the synaptic complex are proposed to remain in place while a pair of catalytic domains, for example those attached to the two functional left half sites, break their interdomainal contacts and, rotating about the "hinge" that connects each to its E-helix, switch places. This would position the DNAs in a recombinant configuration. A conceptual advantage of the domain swap process is that the continued association of the E-helices holds the synaptic complex together; a disadvantage is that the DNA-binding domains of the pair of subunits that switch would need to release their hold on the half sites that move. If one looks at the tetramer as a synaptic complex with antiparallel crossover sites, the subunit arrangement does appear to be compatible with a domain swap. The putative hinge regions of the diagonally opposed subunits are close—a prerequisite of the domain swap process (the Cα positions of Ile-103 are only 11 Å apart). However, the antiparallel nature of the DNAs would mean that each moving half site would have to reverse its direction during the switch. Also, the two catalytic domains that would switch positions are not in contact and could not move as a single rigid body as originally proposed. Overall, the domain swap mechanism in the context of the tetramer structure appears rather implausible.

Since the mode of crossover site binding has a profound affect on possible strand exchange mechanisms, it remains crucial to determine whether recombination emanates from a parallel or antiparallel synapse. We are unaware of any definitive evidence for parallel rather than antiparallel crossover sites, although the parallel arrangement was favored by the phasing data of Leschziner & Grindley (107) and was also suggested to be the more likely alternative by Li et al. (98) on the basis of structural considerations of the complete resolvase synaptosome.

A variety of cross-linking experiments support the subunit rotation model. With the activated mutant of γδ resolvase, substitution of Val-114 with a cysteine enables efficient disulfide bond formation across the flat interface. The cross-linked species retains DNA cleavage activity but cannot perform recombination (unless the disulfide bond is reduced) (98). This cross-link would be expected to prevent subunit rotation using the flat interface as a bearing, but because it links the E-helices, it would not prevent domain swapping. By contrast, a disulfide between cysteines at position A74C across the synaptic interface allows both cleavage and recombination (A. Sinha and N.D.F. Grindley, unpublished data); this cross-link is not predicted to prevent recombination of parallel crossover sites by subunit rotation but should prevent recombination of antiparallel sites by domain swapping.

Further strong support for the approximately parallel alignment of the crossover sites and for rotation at the flat interface as the mode of strand transfer is provided by cross-linking data with an activated, Fis-independent mutant of Hin when these data are examined in the context of the structure of the γδ resolvase synaptic tetramer. Cysteines at Hin residues S94C (or S99C) (equivalent to γδ resolvase residues Gly-96 and Ser-101 in the tetramer structure) can be readily cross-linked in synaptic complexes with cleaved crossover sites, using a thiol reagent with an 8Å linker (109). The tetramer structure predicts that cross-links involving either of these residues would initially be between diagonally positioned resolvase subunits (e.g., L and R'). Following subunit rotation at the flat interface, L- and L'-linked subunits exchange places so predicted cross-links (still between diagonally positioned subunits) would join subunits at L and R. The Hin data indeed show a time-dependent switching of cross-linked pairs of subunits (identified by the DNA species to which each subunit is covalently linked) consistent with the predictions from the structure. Note that in the absence of the crystal structure, these Hin data are equally consistent with a domain swap model of strand exchange.

Overall, the available evidence supports the proposal that the flat interface separates the two recombining halves of the synaptic tetramer, adding credence to the model of subunit rotation using this interface as a rotational bearing. That the flat interface develops from interlocking dimeric interfaces during the transition from the presynaptic to the synaptic state is fully consistent with it playing a functionally important postsynaptic role.

Complex Systems of Serine Recombinases: Regulating the Outcome of Recombination

The simplest recombination system, consisting of just four subunits of a single recombinase operating on two identical minimal crossover sites, cannot distinguish between intermolecular and intramolecular reactions or (unless crossover sites with asymmetric central sequences are used) between inversions and excisions. Biological reactions that require specificity must impose regulatory processes on the simplest system to promote the desired reaction and suppress the undesirable ones. Examples of such reactions are the excision-specific recombination systems mediated by the resolvases and the inversion-specific recombination systems promoted by the DNA invertases. Regulatory processes may operate at the level of synapsis, for example, by promoting synapsis of appropriate recombination sites or recombination site orientations, or they may operate at the level of catalytic activation by promoting the cleavage and strand exchange steps only from appropriate synaptic complexes.

Resolvases: excision specificity and the "topological filter." Serine recombinases that specifically promote an excision prevent catalysis of inversions and intermolecular recombination by requiring the formation of a topologically elaborate synaptic complex for activation of the recombinase. This process, initially elucidated for Tn3 resolvase, is conceptually similar to that used to constrain XerCD activity on *dif* and *cer* (see above) but uses different protein components.

The Tn3 and γδ paradigm. The first resolvase site-specific recombination systems discovered were those encoded by the related bacterial transposons Tn3 and γδ, and these remain the most thoroughly studied and best understood. When transposons of the Tn3 family move from one replicon to another, they form an intermediate, called a cointegrate, in which the entire donor replicon is inserted into the target with a copy of the transposon at each of the donor-target junctions (see **Figure 14**). The transposon-encoded resolvase protein acting at the two recombination sites, termed *res* sites, within the duplicated transposons, excises the donor replicon (along with one of the transposon copies), leaving the other copy in the target DNA. Two features of the resolvase systems ensure their excision specificity: the complexity of their recombination sites and the requirement that the substrates be negatively supercoiled.

The *res* sites of γδ and Tn3 are nearly 120 bp in length and contain three binding sites

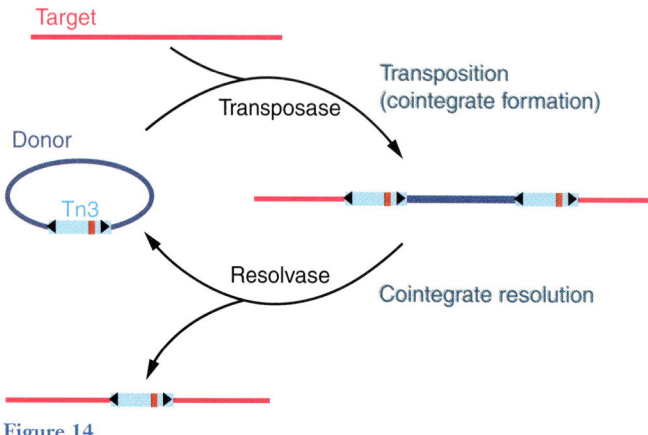

Figure 14

Two stages of Tn3 transposition: the formation of the donor-target cointegrate, and its subsequent resolution. Transposase, responsible (along with the host cell replication machinery) for the first step, acts at the transposon ends. The site-specific recombinase, resolvase, acts at a site, *res* (*red rectangle*), within the duplicated transposon (*light blue bars*). Adapted from Reference 2.

for dimers of resolvase (**Figure 15a**) (124). The point of DNA breakage and reunion, also called the crossover point, lies at the center of site I, but the other two sites (II and III, also called the accessory sites) are required for the recombination process. Spacings between the sites are critical for activity (125). A feature of the three binding sites, particularly unusual for DNA sites recognized by dimeric DNA-binding proteins, is that each has a different geometry; all consist of a head-to-head pair of 12-bp recognition sequences, but these sequences are separated by spacers of different lengths (4 bp, 10 bp, and 1 bp at sites I, II, and III, respectively). This variation indicates that the γδ resolvase dimer has an unusual flexibility (also seen in other serine recombinases) that enables the DNA-binding domains to reach over different distances along and around the DNA helix.

The accessory binding sites play a crucial regulatory role in resolvase-mediated recombination, preventing inversions or intermolecular recombination (i.e., integration) and, thus, ensuring that resolvase promotes only excisions. Synapsis is a prerequisite for site-specific recombination; nevertheless, single site I-bound dimers of γδ or Tn3 resolvase cannot form synaptic complexes. Rather, the minimal synaptically competent unit is a pair of dimers bound adjacently to sites II and III, with the subunits at II-L and III-L interacting specifically with each other using the crystallographically defined 2-3' interaction (126–129). Synapsis of two *res* sites is initiated when two such accessory site complexes interact (see **Figure 16**). Formation of this initial complex promotes the subsequent pairing of the two resolvase-bound site Is; this step also depends on the 2-3' interaction (this time, between subunits at site I-R and site III) as well as additional interactions between the site I-bound dimers (127). The role of the 2-3' interaction could be either to abut the subunits thus creating a larger (and thus a more stable and effective) interaction surface or to distort the dimer conformation into a quaternary structure that favors synapsis (such as the scissoring of the

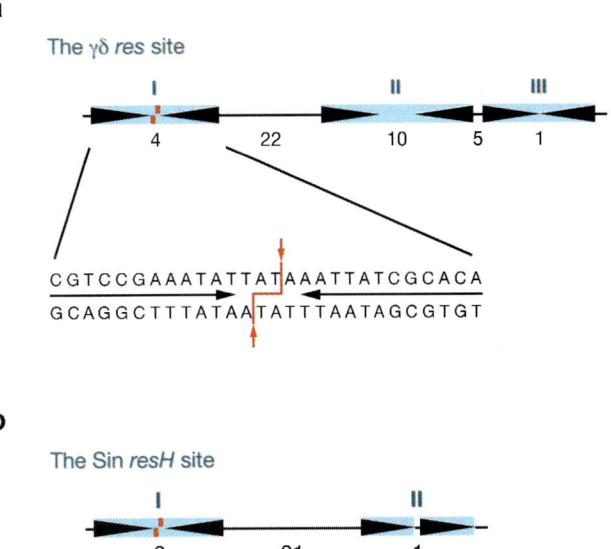

Figure 15

Two styles of *res* sites. (*a*) The γδ *res* site (a typical site) with three resolvase-binding sites; below it is shown the sequence of the crossover site. (*b*) The Sin *resH* site (an atypical site) with two resolvase-binding sites; note the unusual head-to-tail arrangement of the two halves of site II. The horizontal arrow heads represent the 12-bp binding sequences recognized by each recombinase subunit. The numbers indicate the lengths of DNA between these recognition sequences. The cleavage sites (*red*) are indicated.

E-helices to allow interdigitation of dimers as seen in the tetramer crystal structure).

The accessory site requirement inhibits inversion and intermolecular recombination by imposing what has been called a topological filter on the formation of a productive synapse (21, 102, 104, 105). When two *res* sites synapse, they wrap around each other, trapping three (−) superhelical turns. Remarkably, synapsis of just the accessory sites also traps three (−) nodes (126, 130). The *res* interwrapping is favored by negatively supercoiled substrate DNA but only under the specific circumstance of pairing two *res* sites in head-to-tail orientation on the same supercoiled DNA molecule (as shown in **Figure 16a**). In all other circumstances, for example when two *res* sites are inverted or are on separate molecules, negatively supercoiled DNA operates to inhibit productive synapsis because formation of the *res* site interwraps imposes a compensatory DNA tangle elsewhere in the substrate

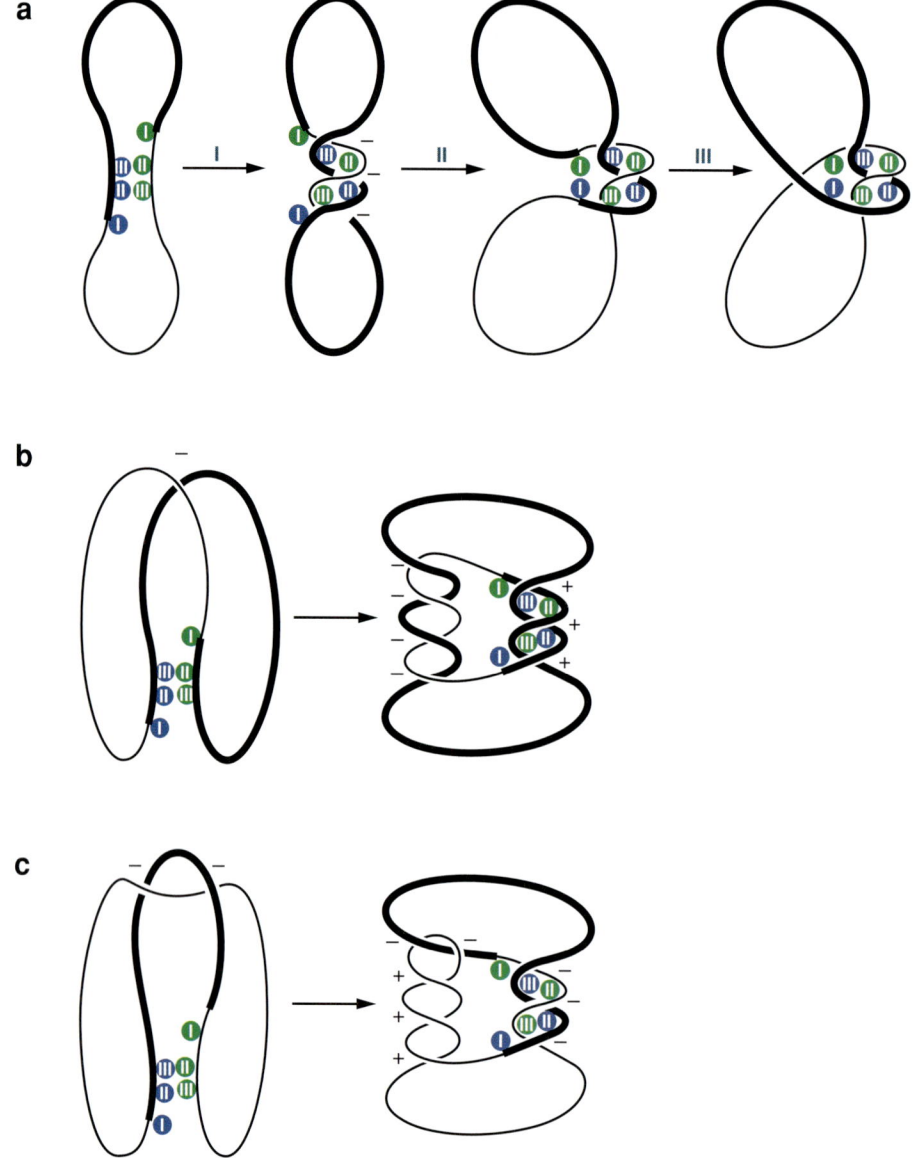

Figure 16

Two-step model for *res* site synapsis and the topological filter. (*a*) Two-step synapsis. Synapsis is initiated by antiparallel pairing of subsites II and III, trapping three (−) nodes. This facilitates the productive pairing of both site Is. (*b*) Consequence of pairing inverted *res* sites. To obtain antiparallel pairing of subsites II and III, at least one interdomain node must be formed. The subsequent interwrapping of the two *res* sites causes a further four interdomain nodes to form. (Note that nodes formed by the *res* interwrapping are (−) relative to the *res* site orientation but are (+) relative to the path of the DNA.) (*c*) Consequence of trapping two (−) interdomain nodes at the initiation stage. To compensate for the wrapping of subsites II and III around resolvase, three (+) intradomain nodes must be introduced into one of the two substrate domains. Adapted from Reference 2.

that would be energetically unfavorable (e.g., **Figure 16***b*). An interesting consequence of the topological filter is that it operates even when the *res* sites are in the correct orientation on the same molecule, but the initial synapsis passively traps two or more extra interdomain supercoils (as would very likely occur when the two sites synapse by random collision but are separated by many kilobases of DNA) (see **Figure 16***c*). This explains why resolution in vitro yields catenated product circles that are invariably singly linked (131). This property of resolvases contrasts dramatically with recombinases of the tyrosine family, such as λ Int, Flp, and Cre, which impose no topological filter and produce multiply catenated product circles when an excision reaction is performed on supercoiled substrates with well-separated recombination sites (126, 132, 133).

Molecular models for the synaptic complex formed by pairing accessory sites or complete *res* sites have been proposed on the basis of crystal structures of γδ resolvase in its presynaptic conformation (121, 134). Despite their structural plausibility and their compatibility with experimental data for the 2-3′ interaction between subunits at specific sites (127), none of the models provides a satisfactory fit with the latest crystal structure of the resolvase synaptic tetramer. Curiously, although data indicating a 2-3′ interaction between the site I-bound resolvase subunits and specific subunits within the accessory site synaptic complex appear to be very convincing, modeling additional subunits onto the synaptic tetramer via 2-3′ interactions places them in positions that seem to make any synaptic interactions impossible (98). Our conclusion is that the 2-3′ interactions between resolvase subunits at site I and at the accessory sites that are needed for assembly of the synaptic complex are likely to be broken during the formation of the synaptic tetramer.

The new structure of the synaptic tetramer raises the question of whether resolvase at synapsed accessory sites would also adopt the postsynaptic conformation. Existing data suggest that a transition to the conformation of the synaptic tetramer does not occur at the accessory sites. Resolvase dimers that are cross-linked by disulfide bonds, either at position 106 or between residues at 95 and 113, efficiently form accessory site synaptic complexes that readily support synapsis and recombination by wild-type resolvase at the two crossover sites (122, 135). Yet these cross-links should fix the dimers in a presynaptic state and prevent transition to the interdigitated conformation of the synaptic tetramer. This suggests that there is a stable synapsed state of resolvase that is distinct from that of the catalytically active tetramer. For a more complete molecular understanding of the role of the accessory site-bound resolvase subunits in the assembly of the crossover site synapse and the activation of catalysis, we await further structural information.

The Sin paradigm. The Sin recombinase is the prototype of a group of serine recombinases encoded by several large staphylococcal plasmids. The Sin recombination system differs from that of Tn3 and γδ resolvase in two significant ways (136, 137). First, its *res* site, although complex, is only 86 bp long and binds just two dimers of Sin (see **Figure 15**), and site II consists of direct (head-to-tail) repeats of the 12-bp binding sequence. Second, recombination requires an architectural, DNA-bending protein such as *E. coli* HU or *Bacillus subtilis* Hbsu. Nevertheless, like the transposon-encoded cointegrate resolvases, Sin is specific for an excision reaction (its biological role is likely to be reducing plasmid dimers to monomers to ensure their stability), and the product of recombination in vitro is a pair of singly linked, catenated circles. Furthermore, additional topological analysis indicated that the synaptic complex trapped three interdomain supercoils. Rowland et al. (136) have proposed that Sin together with Hbsu forms a synaptic complex topologically indistinguishable from that of γδ resolvase and that this complex creates a topological filter to inhibit inversions and

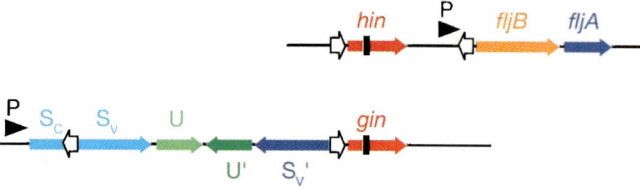

Figure 17

The *hin* and *gin* genes and inversion systems. The crossover sites (*open arrows*), bracketing the invertible DNA segments, and the enhancers (*black rectangles* within the *hin* and *gin* genes), bound by two dimers of Fis, are shown. P indicates the promoters of the inversion-regulated genes. Note that the left crossover site for Gin lies within the S gene so that the S gene product contains a constant and a variable portion: Sc-Sv in the orientation shown and Sc-Sv' after inversion.

integrations. In their model, they propose that Hbsu is used to facilitate a tight bend between sites I and II, allowing direct interactions between the Sin subunits at these sites. Consistent with this, protection assays have shown that when Sin is bound to sites I and II, Hbsu occupies the intervening DNA segment. In this model, Hbsu essentially replaces the DNA-bending role of γδ resolvase at site II of *res*, and Sin at site II of *resH* plays a similar role to γδ resolvase at site III.

One difference between Sin and Tn3/γδ resolvase is that Sin is catalytically active in the absence of synapsis (presumably as a dimer) and is able to cleave and rejoin isolated crossover sites (without site II or Hbsu) (136). Thus, for Sin, synapsis, which is essential for the resolution reaction, may simply be a way of bringing together a pair of recombination sites in a controlled (that is excision-specific) manner. By contrast, for Tn3/γδ resolvase, synapsis not only brings the crossover sites together but also activates the recombinase.

DNA invertases: inversion specificity and the Gin/Hin paradigm. A number of serine recombinases specifically promote inversion of DNA segments to provide a switch between two alternative and mutually exclusive genetic states (3). The best characterized of these are the very closely related proteins, Hin, which is responsible for the phenomenon of flagellar "phase variation" in *Salmonella*, and Gin,

which enables phage Mu to infect alternative bacterial host strains. As shown in **Figure 17**, inversion promoted by Hin switches the orientation of a promoter and, thus, turns on or off the expression of the adjacent genes; the *hin* gene lies within the 1-kb invertible segment. The action of Gin inverts an adjacent 3.0-kb DNA segment that contains alternative phage tail fiber genes. Remarkably, the Gin and Hin recombinases are interchangeable and are able to operate on each others recombination sites.

Complete in vitro recombination reactions with Hin or Gin generate inversions rapidly and efficiently, but intermolecular recombination is undetectable, and excisions (with substrates containing directly repeated recombination sites) are very rare (3, 117, 138). What specifies this pronounced directional bias? As with the resolvases, requirements for a superhelical substrate and for complex recombination sites are key determinants. However, the recombination site complexity contrasts with that of the resolvase systems. There are no requirements for additional recombinase subunits and binding sites; instead an additional protein, Fis (factor for inversion stimulation, a homodimer of 98 aa subunits), and a specific DNA sequence to which Fis binds, called the enhancer, are needed (3, 139–141). The enhancer, which contains two binding sites for the Fis dimer, separated by 48 bp (center-to-center), operates independent of orientation and can be placed virtually anywhere in a plasmid substrate but must be provided *in cis* (3, 142). The synaptic complex within which recombination takes place (also called the invertasome) is a three-looped structure that traps two interdomain negative supercoils and may form at the junction point of a branch in the supercoiled substrate (**Figure 18**) (112, 113, 143). The three DNA segments at the branch point consist of the two crossover sites and the enhancer, and the complex contains two dimers of the recombinase and two dimers of Fis. Direct interaction between Fis and the recombinase is needed to activate double-strand cleavage of the crossover sites. However,

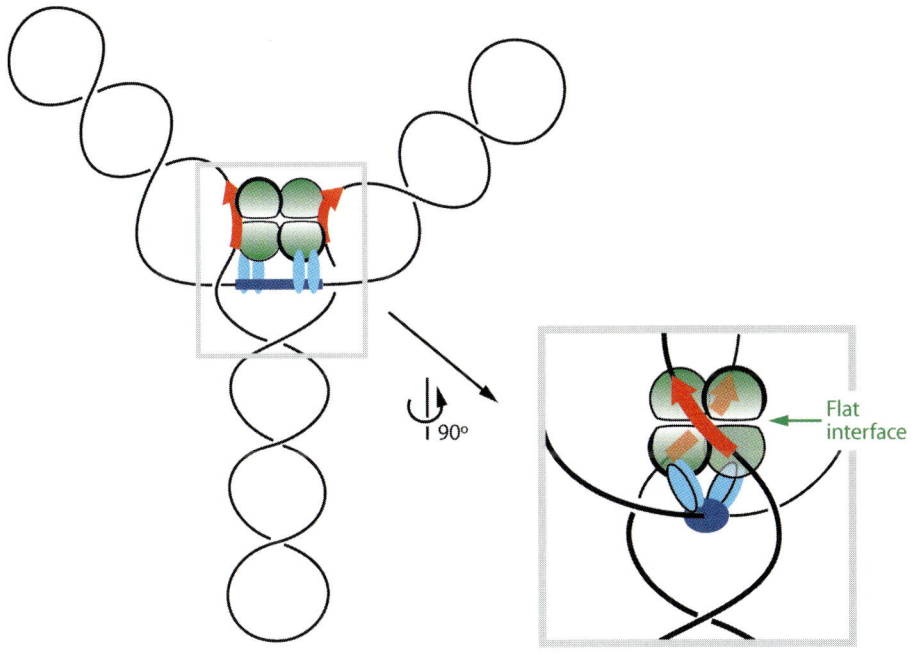

Figure 18

The invertasome, showing the −2 topology of the productive synapse. Crossover sites (*red arrows*) are bound by dimers of the recombinase (*green subunits*); the enhancer (*blue bar*) is bound by two dimers of Fis (*pale blue ellipses*). Two negative interdomain nodes are trapped by the threading of the enhancer. The recombinase tetramer has been drawn to show a flat interface (proposed to be the bearing for subunit exchange), separating the halves of the crossover sites to be exchanged. (*inset*) The invertasome, rotated 90°, showing how threading of the enhancer and the Fis-Hin interactions stabilize a proposed + node, which is necessary for appropriately orienting the Hin tetramer. Note that all other DNA crossings are negative.

following strand exchange, the rejoining step appears to be Fis independent (113, 142).

In contrast to the resolvase systems, the DNA invertases (along with Fis) do form active synaptic complexes with inappropriately oriented recombination sites. The DNA invertases synapse and cleave crossover sites that are directly repeated (with respect to their asymmetric central dinucleotides), but instead of producing the anticipated excision products, they efficiently convert the supercoiled DNA substrate into a knotted state (113, 138). Topological analysis indicated that the knotted circles result from two (or an even number of) 180° cycles of strand exchange. This showed that the crossover sites, despite their direct orientation in the substrate, were synapsed as if they had been inverted, form-

ing a synaptic complex identical in structure and topology to the normal invertasome. As a consequence, a single 180° cycle of strand exchange resulted in pairing of mismatched and unligatable crossover sequences (the 2-base 3′ single-strand extensions resulting from the double-strand cleavages), necessitating a second 180° rotation to occur before the sites could be rejoined. This second cycle restores the sites in their original (parental) configuration but because of the two negative supercoils, trapped by the synapse, converts the substrate into a trefoil knot (as this product is also a substrate, more complex knots are also produced by repetitions of the process). Not surprisingly, if crossover sites are modified to make the central dinucleotide symmetrical, then recombination by a DNA invertase yields inversion products regardless of whether the full recombination sites are inverted or directly repeated (138).

How Fis activates catalysis by the DNA invertases remains a mystery, although a number of important facts have been determined by experiments with Hin (3, 144, 145). Activation depends on a direct interaction between Fis and the recombinase. The Fis component appears to involve a surface at the tip of a flexibly connected β-hairpin arm formed by the N-terminal portion of Fis. Somewhat surprisingly, only one functional β-arm in each Fis dimer is needed for efficient activation (suggesting contact with only one subunit of each recombinase dimer). The region of Hin that contacts Fis remains to be determined, although the large separation of the two Fis-binding sites in the enhancer is most consistent with contacts to the outside of the Hin synaptic complex. It has been suggested that Fis induces a conformational change in the paired Hin dimers, which promotes cleavage and strand exchange.

Wild-type Hin, in contrast to Tn3/γδ resolvase, can efficiently pair isolated crossover sites, but these complexes are inactive in recombination (143). This raises the possibility that Fis activates Hin by contacting the preformed Hin synaptic complex. However, there are problems with this scenario. First, activation is absolutely dependent on topological linkage of the enhancer and crossover sites; even high concentrations of Fis and enhancer when added *in trans* fail to activate Hin synaptic complexes (3, 142). Second, the topology of the invertasome, with the requirement that two negative supercoils be trapped, is not easy to reconcile with action of Fis on a preformed Hin complex, since capturing these supercoils either depends on Hin synapsis randomly trapping a single enhancer-containing DNA segment within the supercoiled invertible domain or requires the Hin synapse to dissociate, allowing the enhancer-containing segment to pass between the crossover sites.

The structure of the resolvase synaptic tetramer and the strong hypothesis that the crossover sites form a local negative node and exchange by subunit rotation, using the flat interface as a swivel (98), offer a new way of thinking about the role of the enhancer. We suggest that the most important role of Fis at the enhancer may be to ensure that the inverted crossover sites cross with a local negative node and, thus, position the flat interface of the synaptic tetramer such that rotations will cause an inversion. This crossing creates a positive global node (**Figure 18**, inset), which would not be favored in a negatively supercoiled substrate in the absence of Fis [note that although negative with respect to the local (roughly parallel) orientations of the crossover sites, this node is globally positive because one site (and the arrow that represents it) is inverted with respect to the overall path of the DNA]. One way to facilitate assembly of this final structure would be for a Hin dimer bound to one crossover site to interact with a Fis dimer at the enhancer, forming the first ear of the invertasome. This half invertasome together with the second enhancer-bound Fis dimer would capture the second Hin-bound crossover site with the components of the initial half invertasome ensuring the correct final topology. Fis may activate the Hin recombination functions simply by stabilizing the Hin synaptic complex and providing a platform for the strand exchange steps, although a Fis-induced conformational change in the Hin tetramer remains a possibility (perhaps even mediated by mechanical/torsional effects imposed by twisting of the three DNA segments and Fis-invertase interactions).

Phage integrases: induced fit determines synapsis specificity. Phage integrases need to distinguish between intermolecular recombination, resulting in phage integration, and intramolecular recombination, resulting in prophage excision. The prototypical phage integrase (and tyrosine recombinase), λ Int, achieves this regulation by means of accessory sites and accessory proteins as described above.

A number of phage integrases that are serine recombinases and members of the large serine recombinase subgroup (5) appear to distinguish between integration and excision by a remarkably different (and still poorly understood) mechanism. The best studied of these integrases are those of the *Streptomyces* phage, φC31, and the mycobacteriophages, Bxb1 and φRv1. In each of these cases, the *attP* and *attB* sites are simple sites with central crossover points (*attP*s range from 40–52 bp, *attB*s from 34–40 bp) (146–149). Each integrase alone efficiently catalyzes integrative *attP* × *attB* recombination in vitro; however, despite binding with similar affinity to *attL* and *attR*, it cannot recombine them or any other pairs of sites (for example, *attP* × *attL* or *attP* × *attR*, even though these pairs have three of the four half sites arranged in the same way as the productive *attP* × *attB* pair). The block to recombining all other pairs of sites appears to be at the level of synapsis—the recombinase alone can only stably synapse *attP* with *attB* (149–151).

How is the *attP-attB* pairing specificity achieved in the absence of other factors when all four *att* sites are equally well bound? A clue comes from the sequences of the sites. *attP* and *attB* appear to be surprisingly dissimilar in sequence; moreover, although each binds just a dimer of the integrase, the difference in their minimal lengths suggests that they bind the recombinase in a different manner. It is proposed that each binding site induces an *att*-specific conformation on the bound integrase dimer and that only the *attP*- and *attB*-specific conformations have the necessary complementary interfaces to form a stable synaptic complex (149, 150). Following recombination, the conformations switch to the *attL* and *attR* specificities, the interface complementarity breaks down, and the complex dissociates into the separate integrase-bound *attL* and *attR* sites. Because these phages can also excise from their integrated state, the recombinases must be able to catalyze *attL* × *attR*

recombination. φRv1 encodes an Xis protein (and the other phages are expected to do so too), and Xis not only enables the φRv1 integrase to promote *attL* × *attR* recombination, but it inhibits *attP* × *attB* recombination (147, 152). These actions of Xis do not require any extra DNA sequences; minimal *attL* and *attR* sites are simply the recombinants of minimal *attP* and *attB* sites. Thus, it seems likely that Xis interacts directly with the *att*-bound integrase dimers to switch the conformation to a synapsis-competent state if they are bound to *attL* and *attR* but to a synapsis-incompetent state at *attP* and *attB*.

Serine recombinases as transposases: avoiding target-target recombination. A rather different regulatory issue occurs with the transposons such as Tn4451, Tn5397, or IS607. These elements move by sequential excision and insertion steps, with excision forming a circular transposon with abutted left and right ends. As with the phage integrases, the recombination sites are relatively small; by contrast, however, the recombinase is sufficient for both excision and integration (153, 154). Because the transposition process requires that the transposase recognizes the target sites in addition to the ends of the transposon, a conceptual problem is that simultaneous recognition and synapsis of two target sites could result in a substantial chromosomal deletion.

How is this avoided? TnpX, the recombinase from Tn4451 and a member of the large serine recombinase subfamily, has been shown to bind to target sites with much lower affinity (about 100- to 1000-fold) than to either transposon end (153). This helps reduce undesirable target-target interactions but (considering the large excess of potential target sites) is, alone, unlikely to eliminate them. An additional proposal is that binding of the recombinase dimer to an end-end junction induces a conformational change that enables it to capture and activate a second recombinase dimer, forming a tetrameric presynaptic complex poised for catalysis (153, 155). The second dimer could be loosely bound to target DNA (leading directly to cleavage of target and transposon crossover sites as well as to strand exchange), or it could be a free dimer that then captures (perhaps more tightly) a target sequence.

SUMMARY AND PERSPECTIVES

Despite performing identical biological processes, two distinct classes of recombinase have evolved, with each family using entirely different mechanisms of DNA synapsis, cleavage, and strand exchange. One of the most significant differences lies in the general nature of the synaptic complexes formed. The synapse of the tyrosine recombinases, with the DNA held within a protein scaffold, allows strand exchange to occur with only very minor adjustments of the quaternary structure. The serine recombinase synapse, however, with a solid protein core on which the DNA sites bind, necessitates dramatic movements of DNA-linked protein subunits to achieve strand exchange.

The relative rigidity of the tyrosine recombinase synaptic complexes has made it possible for structural studies to achieve an almost complete series of snapshots, greatly increasing our understanding of the entire recombination process (16, 24, 36). Thus far, perhaps hampered by the dynamics of the process, structural studies of serine recombinases have yielded two pictures (97, 98), and much needs to be revealed before a complete understanding of DNA cleavage and strand exchange is achieved. Although the latest crystal structure is highly suggestive of a plausible, if unprecedented, mechanism for strand exchange, namely subunit rotation, those with interests in either site-specific recombination or protein motion await a direct demonstration. The process of synapsis is also not well understood for either family of recombinases. What might a dimer of Cre or Flp bound to a single site look like and does synapsis involve conformational acrobatics? How does the resolvase dimer transition into the interdigitated structure of the synaptic tetramer?

As we have indicated, the natural examples of site-specific recombination are not only numerous but also highly varied in their regulation, yet only a relatively small number have been investigated biochemically or structurally. Mechanisms of recombinase activation are all mysterious and await a detailed examination. Do interactions of the recombinases with accessory proteins directly promote conformational changes? Might twisting or bending forces on the DNA sites set up by the topological interlinking of crossover sites and accessory protein-binding sites as well as protein-protein interactions play a significant role? Many serine recombinases display a remarkable and unusual capacity to bind to DNA sites with differing sequences and geometries; these differences affect the behavior of the recombinases in intriguing but poorly understood ways that deserve a thorough investigation.

Many questions remain regarding the catalytic mechanisms of these enzymes, particularly the serine family. What are the roles of the individual residues that surround the scissile phosphate? How plastic are these roles within a family?

Finally, both families have expanded to include noncanonical members with unusual domain arrangements and interesting variations on the standard recombination scheme. For example, what is the huge C-terminal extension of the large serine recombinases for? What enables some recombinases to show relaxed specificity for one of the partner sites? And does that play into the question of site-induced conformations? How do the telomere resolvases orchestrate hairpin formation rather than strand exchange? How do the integron integrases utilize their noncanonical recombination sites? And how many other uses are there for these versatile enzymes?

ACKNOWLEDGMENTS

We thank many in the community of site-specific recombination researchers, including our lab groups and especially Martin Boocock, Marshall Stark, Reid Johnson, Graham Hatfull, Satwik Kamtekar, Tom Steitz, Sean Colloms, Enoch Baldwin, and Kent Mouw for valuable discussions. Our research is supported by grants from the National Institutes of Health.

LITERATURE CITED

1. Azaro MA, Landy A. 2002. See Ref. 158, pp. 118–48
2. Grindley NDF. 2002. See Ref. 158, pp. 272–302
3. Johnson RC. 2002. See Ref. 158, pp. 230–71
4. Nunes-Duby SE, Kwon HJ, Tirumalai RS, Ellenberger T, Landy A. 1998. *Nucleic Acids Res.* 26:391–406
5. Smith MC, Thorpe HM. 2002. *Mol. Microbiol.* 44:299–307
6. Cassano AG, Anderson VE, Harris ME. 2004. *Biopolymers* 73:110–29
7. Reed RR, Grindley NDF. 1981. *Cell* 25:721–28
8. Nunes-Duby SE, Matsumoto L, Landy A. 1987. *Cell* 50:779–88
9. Kitts PA, Nash HA. 1987. *Nature* 329:346–48
10. Poulter RT, Goodwin TJ. 2005. *Cytogenet. Genome Res.* 110:575–88
10a. Altschul SF, Madden TL, Schaffer AA, Zhang J, Zhang Z, et al. 1997. *Nucleic Acids Res.* 25:3389–402
11. Esposito D, Scocca JJ. 1997. *Nucleic Acids Res.* 25:3605–14
12. Bateman A, Coin L, Durbin R, Finn RD, Hollich V, et al. 2004. *Nucleic Acids Res.* 32:D138–41
13. Chaconas G, Stewart PE, Tilly K, Bono JL, Rosa P. 2001. *EMBO J.* 20:3229–37
14. Deneke J, Ziegelin G, Lurz R, Lanka E. 2000. *Proc. Natl. Acad. Sci. USA* 97:7721–26
15. Ross W, Landy A. 1983. *Cell* 33:261–72
16. Van Duyne GD. 2001. *Annu. Rev. Biophys. Biomol. Struct.* 30:87–104
17. Nunes-Duby SE, Azaro MA, Landy A. 1995. *Curr. Biol.* 5:139–48
18. Burgin AB Jr, Nash HA. 1995. *Curr. Biol.* 5:1312–21
19. Ghosh K, Lau CK, Gupta K, Van Duyne GD. 2005. *Nat. Chem. Biol.* 1:275–82
20. Voziyanov Y, Lee Jehee, Whang I, Lee Joonsoo, Jayaram M. 1996. *J. Mol. Biol.* 256:720–35
21. Stark WM, Sherratt DJ, Boocock MR. 1989. *Cell* 58:779–90
22. Guo F, Gopaul DN, Van Duyne GD. 1997. *Nature* 389:40–46
23. Gopaul DN, Duyne GD. 1999. *Curr. Opin. Struct. Biol.* 9:14–20
24. Chen Y, Rice PA. 2003. *Annu. Rev. Biophys. Biomol. Struct.* 32:135–59
25. Champoux JJ. 2001. *Annu. Rev. Biochem.* 70:369–413
26. Krogh BO, Shuman S. 2000. *Mol. Cell* 5:1035–41
27. Krogh BO, Shuman S. 2002. *J. Biol. Chem.* 277:5711–14
28. Nagarajan R, Kwon K, Nawrot B, Stec WJ, Stivers JT. 2005. *Biochemistry* 44:11476–85
29. Van Duyne G. 2002. See Ref. 158, pp. 93–117
30. Tian LG, Claeboe CD, Hecht SM, Shuman S. 2005. *Structure* 13:513–20
31. Stivers JT, Jagadeesh GJ, Nawrot B, Stec WJ, Shuman S. 2000. *Biochemistry* 39:5561–72
32. Chen Y, Rice PA. 2003. *J. Biol. Chem.* 278:24800–7
33. Deneke J, Burgin AB, Wilson SL, Chaconas G. 2004. *J. Biol. Chem.* 279:53699–706
34. Cheng Q, Sutanto Y, Shoemaker NB, Gardner JF, Salyers AA. 2001. *Mol. Microbiol.* 41:625–32

35. Chen Y, Narendra U, Iype LE, Cox MM, Rice PA. 2000. *Mol. Cell* 6:885–97
36. Biswas T, Aihara H, Radman-Livaja M, Filman D, Landy A, Ellenberger T. 2005. *Nature* 435:1059–66
37. Redinbo MR, Stewart L, Kuhn P, Champoux JJ, Hol WG. 1998. *Science* 279:1504–13
38. Kwon HJ, Tirumalai R, Landy A, Ellenberger T. 1997. *Science* 276:126–31
39. Hickman AB, Waninger S, Scocca JJ, Dyda F. 1997. *Cell* 89:227–37
40. Subramanya HS, Arciszewska LK, Baker RA, Bird LE, Sherratt DJ, Wigley DB. 1997. *EMBO J.* 16:5178–87
41. Sharma A, Hanai R, Mondragon A. 1994. *Structure* 2:767–77
42. Chen JW, Lee Jehee, Jayaram M. 1992. *Cell* 69:647–58
43. Lee Jehee, Jayaram M, Grainge I. 1999. *EMBO J.* 18:784–91
44. Yang SH, Jayaram M. 1994. *J. Biol. Chem.* 269:12789–96
45. Letzelter C, Duguet M, Serre MC. 2004. *J. Biol. Chem.* 279:28936–44
46. Gopaul DN, Guo F, Van Duyne GD. 1998. *EMBO J.* 17:4175–87
47. Deleted in proof
48. Aihara H, Kwon HJ, Nunes-Duby SE, Landy A, Ellenberger T. 2003. *Mol. Cell* 12:187–98
49. Voziyanov Y, Konieczka JH, Stewart AF, Jayaram M. 2003. *J. Mol. Biol.* 326:65–76
50. Buchholz F, Stewart AF. 2001. *Nat. Biotechnol.* 19:1047–52
51. Santoro SW, Schultz PG. 2002. *Proc. Natl. Acad. Sci. USA* 99:4185–90
52. Baldwin EP, Martin SS, Abel J, Gelato KA, Kim H, et al. 2003. *Chem. Biol.* 10:1085–94
53. Martin SS, Chu VC, Baldwin E. 2003. *Biochemistry* 42:6814–26
54. Levitzki A, Stallcup WB, Koshland DE Jr. 1971. *Biochemistry* 10:3371–78
55. Radman-Livaja M, Shaw C, Azaro M, Biswas T, Ellenberger T, Landy A. 2003. *Mol. Cell* 11:783–94
56. Lee Jehee, Jayaram M. 1997. *Genes Dev.* 11:2438–47
57. Woods KC, Martin SS, Chu VC, Baldwin EP. 2001. *J. Mol. Biol.* 313:49–69
58. Kimball AS, Lee Jehee, Jayaram M, Tullius TD. 1993. *Biochemistry* 32:4698–701
59. Qian XH, Inman RB, Cox MM. 1990. *J. Biol. Chem.* 265:21779–88
60. Qian XH, Cox MM. 1995. *Genes Dev.* 9:2053–64
61. Conway AB, Chen Y, Rice PA. 2003. *J. Mol. Biol.* 326:425–34
62. Lee SY, Aihara H, Ellenberger T, Landy A. 2004. *Proc. Natl. Acad. Sci. USA* 101:2770–75
63. Kazmierczak RA, Swalla BM, Burgin AB, Gumport RI, Gardner JF. 2002. *Nucleic Acids Res.* 30:5193–204
64. Tekle M, Warren DJ, Biswas T, Ellenberger T, Landy A, Nunes-Duby SE. 2002. *J. Mol. Biol.* 324:649–65
65. Subramaniam S, Tewari AK, Nunes-Duby SE, Foster MP. 2003. *J. Mol. Biol.* 329:423–39
66. Hazelbaker D, Radman-Livaja M, Landy A. 2005. *J. Mol. Biol.* 351:948–55
67. Barre F-X, Sherratt DJ. 2002. See Ref. 158, pp. 149–61
68. Ferreira H, Butler-Cole B, Burgin A, Baker R, Sherratt DJ, Arciszewska LK. 2003. *J. Mol. Biol.* 330:15–27
69. Lee Joonsoo, Jayaram M. 1995. *J. Biol. Chem.* 270:4042–52
70. Aranda M, Kanellopoulou C, Christ N, Peitz M, Rajewsky K, Droge P. 2001. *J. Mol. Biol.* 311:453–59
71. Lee L, Sadowski PD. 2003. *J. Mol. Biol.* 326:397–412
72. Lee L, Sadowski PD. 2001. *J. Biol. Chem.* 276:31092–98
72a. Gelato KA, Martin SS, Baldwin EP. 2005. *J. Mol. Biol.* 354:233–45
73. Ennifar E, Meyer JE, Buchholz F, Stewart AF, Suck D. 2003. *Nucleic Acids Res.* 31:5449–60
74. Martin SS, Pulido E, Chu VC, Lechner TS, Baldwin EP. 2002. *J. Mol. Biol.* 319:107–27

75. Frumerie C, Sylwan L, Ahlgren-Berg A, Haggard-Ljungquist E. 2005. *Virology* 332:284–94
76. Sutanto Y, DiChiara JM, Shoemaker NB, Gardner JF, Salyers AA. 2004. *Plasmid* 52:119–30
77. Lewis JA, Hatfull GF. 2003. *J. Mol. Biol.* 326:805–21
78. Esposito D, Thrower JS, Scocca JJ. 2001. *Nucleic Acids Res.* 29:3955–64
79. Lewis JA, Hatfull GF. 2001. *Nucleic Acids Res.* 29:2205–16
80. Kitts PA, Nash HA. 1987. *Nature* 329:346–48
81. Nunes-Duby SE, Matsumoto L, Landy A. 1987. *Cell* 50:779–88
82. Kitts PA, Nash HA. 1988. *J. Mol. Biol.* 204:95–107
83. Aussel L, Barre FX, Aroyo M, Stasiak A, Stasiak AZ, Sherratt D. 2002. *Cell* 108:195–205
84. Massey TH, Aussel L, Barre FX, Sherratt DJ. 2004. *EMBO Rep.* 5:399–404
85. Strater N, Sherratt DJ, Colloms SD. 1999. *EMBO J.* 18:4513–22
86. Reijns M, Lu Y, Leach S, Colloms SD. 2005. *Mol. Microbiol.* 57:927–41
87. Bregu M, Sherratt DJ, Colloms SD. 2002. *EMBO J.* 21:3888–97
88. Gourlay SC, Colloms SD. 2004. *Mol. Microbiol.* 52:53–65
89. Paul S, Summers D. 2004. *Plasmid* 52:63–68
90. Huber KE, Waldor MK. 2002. *Nature* 417:656–59
91. Val ME, Bouvier M, Campos J, Sherratt D, Cornet F, et al. 2005. *Mol. Cell* 19:559–66
92. Recchia GD, Sherratt DJ. 2002. See Ref. 158, pp. 162–76
93. Johansson C, Kamali-Moghaddam M, Sundstrom L. 2004. *Nucleic Acids Res.* 32:4033–43
94. Segal H, Francia MV, Lobo JM, Elisha G. 1999. *Antimicrob. Agents Chemother.* 43:2538–41
95. Casjens SR, Gilcrease EB, Huang WM, Bunny KL, Pedulla ML, et al. 2004. *J. Bacteriol.* 186:1818–32
96. Kobryn K, Burgin AB, Chaconas G. 2005. *J. Biol. Chem.* 280:26788–95
97. Yang W, Steitz TA. 1995. *Cell* 82:193–207
98. Li WK, Kamtekar S, Xiong Y, Sarkis GJ, Grindley ND, Steitz TA. 2005. *Science* 309:1210–15
99. Akopian A, He J, Boocock MR, Stark WM. 2003. *Proc. Natl. Acad. Sci. USA* 100:8688–91
100. Grindley NDF. 1994. In *Nucleic Acids and Molecular Biology*, ed. F Eckstein, DMJ Lilley, pp. 236–67. Berlin: Springer-Verlag
101. Cozzarelli NR, Krasnow MA, Gerrard SP, White JH. 1984. *Cold Spring Harbor Symp. Quant. Biol.* 49:383–400
102. Stark WM, Boocock MR. 1995. In *Mobile Genetic Elements*, ed. DJ Sherratt, pp. 101–29. Oxford, UK: Oxford Univ. Press
103. Kanaar R, Cozzarelli NR. 1992. *Curr. Opin. Struct. Biol.* 2:369–79
104. Boocock MR, Brown JL, Sherratt DJ. 1986. *Biochem. Soc. Trans.* 14:214–16
105. Boocock MR, Brown JL, Sherratt DJ. 1987. In *DNA Replication and Recombination*, ed. TJ Kelly, R McMacken, pp. 703–8. New York: Liss
106. Wasserman SA, Dungan JM, Cozzarelli NR. 1985. *Science* 229:171–74
107. Leschziner AE, Grindley ND. 2003. *Mol. Cell* 12:775–81
108. Nollmann M, He J, Byron O, Stark WM. 2004. *Mol. Cell* 16:127–37
109. Dhar G, Sanders ER, Johnson RC. 2004. *Cell* 119:33–45
110. Reed RR, Moser CD. 1984. *Cold Spring Harbor Symp. Quant. Biol.* 49:245–49
111. Boocock MR, Zhu X, Grindley NDF. 1995. *EMBO J.* 14:5129–40
112. Kanaar R, van de Putte P, Cozzarelli NR. 1988. *Proc. Natl. Acad. Sci. USA* 85:752–56
113. Kanaar R, Klippel A, Shekhtman E, Dungan JM, Kahmann R, Cozzarelli NR. 1990. *Cell* 62:353–66

114. Stark WM, Grindley NDF, Hatfull GF, Boocock MR. 1991. *EMBO J.* 10:3541–48
115. Heichman KA, Moskowitz IP, Johnson RC. 1991. *Genes Dev.* 5:1622–34
116. Klippel A, Kanaar R, Kahmann R, Cozzarelli NR. 1993. *EMBO J.* 12:1047–57
117. Klippel A, Cloppenborg K, Kahmann R. 1988. *EMBO J.* 7:3983–89
118. Haffter P, Bickle TA. 1988. *EMBO J.* 7:3991–96
119. Arnold PH, Blake DG, Grindley NDF, Boocock MR, Stark WM. 1999. *EMBO J.* 18:1407–14
120. Burke ME, Arnold PH, He J, Wenwieser SVCT, Rowland S-J, et al. 2004. *Mol. Microbiol.* 51:937–48
121. Sarkis GJ, Murley LL, Leschziner AE, Boocock MR, Stark WM, Grindley ND. 2001. *Mol. Cell* 8:623–31
122. Wenwieser SVCT. 2001. *Subunit interactions in regulation and catalysis of site-specific recombination.* PhD thesis. Univ. Glasgow, Glasgow
123. Krebs WG, Gerstein M. 2000. *Nucleic Acids Res.* 28:1665–75
124. Grindley NDF, Lauth MR, Wells RG, Wityk RJ, Salvo JJ, Reed RR. 1982. *Cell* 30:19–27
125. Salvo JJ, Grindley NDF. 1988. *EMBO J.* 7:3609–16
126. Kilbride E, Boocock MR, Stark WM. 1999. *J. Mol. Biol.* 289:1219–30
127. Murley LL, Grindley NDF. 1998. *Cell* 95:553–62
128. Watson MA, Boocock MR, Stark WM. 1996. *J. Mol. Biol.* 257:317–29
129. Hughes RE, Hatfull GF, Rice PA, Steitz TA, Grindley NDF. 1990. *Cell* 63:1331–38
130. Benjamin HW, Cozzarelli NR. 1988. *EMBO J.* 7:1897–905
131. Krasnow MA, Cozzarelli NR. 1983. *Cell* 32:1313–24
132. Spengler SJ, Stasiak A, Cozzarelli NR. 1985. *Cell* 42:325–34
133. Abremski K, Hoess R. 1985. *J. Mol. Biol.* 184:211–20
134. Rice PA, Steitz TA. 1994. *EMBO J.* 13:1514–24
135. Hughes RE, Rice PA, Steitz TA, Grindley NDF. 1993. *EMBO J.* 12:1447–58
136. Rowland SJ, Stark WM, Boocock MR. 2002. *Mol. Microbiol.* 44:607–19
137. Rowland SJ, Boocock MR, Stark WM. 2005. *Mol. Microbiol.* 56:371–82
138. Moskowitz IP, Heichman KA, Johnson RC. 1991. *Genes Dev.* 5:1635–45
139. Huber HE, Iida S, Arber W, Bickle TA. 1985. *Proc. Natl. Acad. Sci. USA* 1985:3776–80
140. Johnson RC, Simon MI. 1985. *Cell* 41:781–91
141. Kahmann R, Rudt F, Koch C, Mertens G. 1985. *Cell* 41:771–80
142. Crisona NJ, Kanaar R, Gonzalez TN, Zechiedrich EL, Klippel A, Cozzarelli NR. 1994. *J. Mol. Biol.* 243:437–57
143. Heichman KA, Johnson RC. 1990. *Science* 249:511–17
144. Safo MK, Yang WZ, Corselli L, Cramton SE, Yuan HS, Johnson RC. 1997. *EMBO J.* 16:6860–73
145. Merickel SK, Haykinson MJ, Johnson RC. 1998. *Genes Dev.* 12:2803–16
146. Kim AI, Ghosh P, Aaron MA, Bibb LA, Jain S, Hatfull GF. 2003. *Mol. Microbiol.* 50:463–73
147. Bibb LA, Hancox MI, Hatfull GF. 2005. *Mol. Microbiol.* 55:1896–910
148. Groth AC, Olivares EC, Thyagarajan B, Calos MP. 2000. *Proc. Natl. Acad. Sci. USA* 97:5995–6000
149. Thorpe HM, Wilson SE, Smith MC. 2000. *Mol. Microbiol.* 38:232–41
150. Ghosh P, Pannunzio NR, Hatfull GF. 2005. *J. Mol. Biol.* 349:331–48
151. Smith MC, Till R, Brady K, Soultanas P, Thorpe H. 2004. *Nucleic Acids Res.* 32:2607–17
152. Bibb LA, Hatfull GF. 2002. *Mol. Microbiol.* 45:1515–26
153. Adams V, Lucet IS, Lyras D, Rood JI. 2004. *Mol. Microbiol.* 53:1195–207

154. Wang H, Mullany P. 2000. *J. Bacteriol.* 182:6577–83
155. Derbyshire KM, Grindley NDF. 2005. In *The Bacterial Chromosome*, ed. NP Higgins, pp. 467–97. Washington, DC: ASM Press
156. Carson M. 1997. *Methods Enzymol.* 277:493–505
157. DeLano WL. 2002. *The Pymol Molecular Graphics System*. San Carlos, CA: DeLano Sci.
158. Craig NL, Craigie R, Gellert M, Lambowitz AM, eds. 2002. *Mobile DNA II*. Washington, DC: ASM Press
159. Bouvier M, Demare G, Mazel D. 2005. *EMBO J.* 24:4356–67

NOTE ADDED IN PROOF

New data strongly support a mechanism for integration of integron cassettes involving a folded single-stranded substrate (159).

Axonal Transport and Alzheimer's Disease

Gorazd B. Stokin[1] and Lawrence S.B. Goldstein[2]

[1]Institute of Clinical Neurophysiology, Division of Neurology, University Medical Center, SI-1525 Ljubljana, Slovenia; email: gbstokin@alumni.ucsd.edu

[2]Howard Hughes Medical Institute, Department of Cellular and Molecular Medicine, School of Medicine, University of California San Diego, La Jolla, California 92093-0683; email: lgoldstein@ucsd.edu

Key Words

motor proteins, axons, axonal pathology, aging, neurodegeneration

Abstract

In contrast to most eukaryotic cells, neurons possess long, highly branched processes called axons and dendrites. In large mammals, such as humans, some axons reach lengths of over 1 m. These lengths pose a major challenge to the movement of proteins, vesicles, and organelles between presynaptic sites and cell bodies. To overcome this challenge axons and dendrites rely upon specialized transport machinery consisting of cytoskeletal motor proteins generating directed movements along cytoskeletal tracks. Not only are these transport systems crucial to maintain neuronal viability and differentiation, but considerable experimental evidence suggests that failure of axonal transport may play a role in the development or progression of neurological diseases such as Alzheimer's disease.

Contents

INTRODUCTION	608
AXONAL TRANSPORT	608
Overview of Axonal Transport	608
Proteins Required for Axonal Transport	609
Mechanisms of Axonal Transport	610
Regulation of Axonal Transport	611
AXONAL TRANSPORT, AGING, AND DISEASE	612
Axonal Transport and Aging	612
Impairments in Axonal Transport and the Pathogenesis of Neurodegenerative Diseases	613
Impaired Axonal Transport as a Result of Diseases	614
AXONAL TRANSPORT AND ALZHEIMER'S DISEASE	614
Overview of Alzheimer's Disease	615
Axonal Transport of Proteins Linked to the Pathogenesis of Alzheimer's Disease	617
Axonal Functions of Proteins Linked to the Pathogenesis of Alzheimer's Disease	617
Axonal Defects and the Pathology of Alzheimer's Disease	618
Impairments in Axonal Transport and the Pathogenesis of Alzheimer's Disease	619

Axonal cargo: any molecule or organelle transported within axons

INTRODUCTION

Effective communication between the cell bodies and presynaptic terminals of neurons requires reliable and timely movement of essential neuronal "cargoes" to their final destinations. Such cargoes include vesicles containing synaptic proteins, growth factors, signaling molecules, and ion channels. In addition, defined organelles such as endosomes and mitochondria are continuously transported within axons. Finally, protein complexes including proteins controlling cytoplasmic signaling, structure, and degradation are also actively transported within axons. Whereas some of these cargoes reside in axons where they play structural and other roles, other cargoes participate in a variety of neuronal activities such as synaptic plasticity and neurotransmission. Recent findings implicate many such axonal cargoes in disease processes and suggest the existence of interplay between axonal transport, damage signaling, synaptic plasticity, and diseases of the nervous system.

The past decade has seen an explosion in our knowledge of molecules involved in axonal transport. At the same time, there has been a dramatic expansion in the identification of molecules that may cause diseases of the nervous system. Not only are these two sets of molecules beginning to overlap, but accumulating evidence suggests that many such disorders affect axonal transport during the course of disease, perhaps at the earliest or causative stages. Although such disorders may vary in their ages of onset, the anatomical substrates affected, and the clinical symptoms and signs produced, they may share failures of axonal transport, which may provide opportunities for common therapeutic interventions. Dendrites may also be prone to similar defects, but for a variety of technical reasons this possibility has not been as well explored. This review focuses on work in axons, recognizing that similar issues may pertain to dendrites as well.

AXONAL TRANSPORT

Overview of Axonal Transport

Studies of axonal transport began with the observation that nerve cell bodies generate single, thin, nonbranching axis cylinders, which could be readily distinguished from multiple, highly branched protoplasmic processes. Shortly thereafter, axis cylinders became known as axons, and protoplasmic processes as dendrites, and their cytoplasmic continuity with the cell bodies was established. The realization that axons and dendrites were integral parts of neurons and neuronal circuits led to the suggestion that interruption between cell bodies and axons, and a resulting loss of trophic support, could account for the degeneration of fibers in the peripheral stump of a severed spinal nerve. This in turn led to the hypothesis that trophic material must be transported through axons to and from the cell bodies to maintain their structure and function.

Support for the proposal that active transport occurred in axons came from early studies of viral spread within neurons. Subsequent nerve constriction studies provided direct support for this hypothesis by demonstrating continuous transport of biological material along the axons (1). An explosion of work eventually culminated in the demonstration of anterograde axonal transport by radiolabeling cargo proteins in axons and of retrograde axonal transport by studying the uptake of horseradish peroxidase. This work revealed not only the bidirectional nature of axonal transport, but also established that axonal cargoes travel at different velocities with fast axonal transport occurring at speeds of circa (ca.) 100–400 mm/day (ca. 1.0–5.0 µm/s) and slow axonal transport at speeds of ca. 0.3–3 mm/day (ca. 0.004–0.04 µm/s). The finding that microtubules are the major tracks or "highways" for long-distance axonal transport emerged from experiments where microtubule assembly was chemically disrupted, which resulted in abrogation of axonal transport and provided direct evidence for the involvement of microtubules (2). Clues regarding the mechanism of such transport emerged from the determination of microtubule polarity, which in axons is highly organized so that microtubules have their plus ends oriented toward distal axons and presynaptic sites. These findings were soon followed by the identification of motor proteins that recognize microtubule polarity such as kinesin-1 and dynein, which transform chemical energy into mechanical movement toward microtubule plus and minus ends, respectively.

Proteins Required for Axonal Transport

Microtubules serve as tracks along which motor proteins such as kinesins and dyneins generate long-distance transport. Microtubules are also decorated with microtubule-associated proteins, which may modulate microtubule nucleation and elongation as well as control characteristics of motor protein transport. For example, there is compelling evidence that the tau protein, whose misbehavior is a prominent feature in many neurodegenerative disorders, can play a role in controlling motor protein–driven vesicle transport along microtubules (3).

Compelling evidence suggests that motor proteins play a pivotal role in both fast and slow microtubule-based axonal transports. Microtubule plus end–directed or anterograde axonal transport is thought to rely largely on the kinesin superfamily of motor proteins, which are currently divided into 14 families based on their sequence similarities (4). Strong evidence suggests members of the kinesin-1 (KIF5), -2 (KIF3), -3 (KIF1), -4 (KIF4), and -13 (KIF2) families participate in axonal transport; among them, kinesin-1 is the best studied. Structurally, kinesin-1 is a heterotetramer composed of two kinesin heavy chain (KHC) and two kinesin light chain (KLC) subunits (**Figure 1**). KHCs (KIF5A, KIF5B, and KIF5C) contain microtubule- and nucleotide-binding sites in their NH2-terminal motor heads, alpha-helical coiled-coil stalks in the center, and globular COOH termini that interact with KLC and perhaps with cargoes. KLCs (KLC1, KLC2, and KLC3) contain NH2-terminal heptad repeat regions, which probably form alpha-helical coiled coils and mediate interactions with KHC. KLCs also contain COOH-terminal tetratrico peptide repeats that appear to be involved in cargo binding and in the regulation of transport. KIF5B and KLC2 are ubiquitously expressed; KIF5A and KIF5C are restricted in expression to brain and other neuronal tissue; and KLC1

Neurodegenerative disorder: disorder characterized by progressive neuronal loss

KHC: kinesin heavy chain

KLC: kinesin light chain

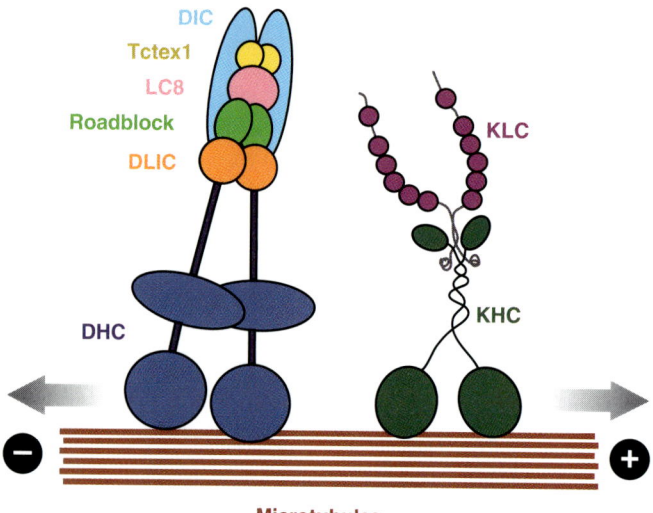

Figure 1

Schematic structure of anterograde and retrograde motor proteins. Abbreviations: DHC, dynein heavy chain; DIC, dynein intermediate chain; DLIC, dynein light intermediate chain; KHC, kinesin heavy chain; KLC, kinesin light chain.

expression is enriched, but not restricted to the nervous system. Genetic manipulation of kinesin-1 subunits in combination with numerous in vitro experiments revealed an important role of kinesin-1 in axonal transport (5–8), although knowledge of the specific transport functions carried out by the various kinesin-1 subunits is still needed.

In contrast to anterograde axonal transport, which takes advantage of several different kinesin motor proteins, a wealth of data suggests that dynein may be the major motor protein powering microtubule minus end–directed or retrograde axonal transport. Dynein is a multiprotein complex composed of two heavy chains and several intermediate, light intermediate, and light chains (**Figure 1**). Dynein heavy chains harbor microtubule- and nucleotide-binding sites within large COOH-terminal motor heads and NH2-terminal stemlike coiled coils. Dynein intermediate chains contain β-propeller group protein sequences and participate in the interaction with dynein heavy chains, dynein light chains, the dynactin complex, and perhaps cargoes. Dynein light intermediate chains and dynein light chains (Tctex1/rp3, roadblock, LC8) are thought to play a role in dynein-dynein and dynein-cargo interactions via ATP-binding loop motifs and amphiphilic alpha-helical segments, respectively. Dynein-mediated axonal transport is thought to be regulated by its interaction with the dynactin complex, which consists of several proteins including p150Glued, p62, p50-dynamitin, the actin-related protein Arp1, actin, actin capping protein α and β subunits, p27, p25, and p24. Although many details remain to be elucidated, in vivo and in vitro studies suggest that disruption of the dynein/dynactin complex results in a striking axonal phenotype (9, 10).

Mechanisms of Axonal Transport

Considerable effort has gone into elucidating the mechanisms by which motor proteins generate force and movement along microtubules (11, 12). There are, however, two principles of motor protein mechanism that may play major roles in axonal transport biology. The first principle relates to the processivity of individual molecular motor proteins. A large body of evidence suggests that for highly processive motor proteins such as kinesin-1, the number of motor proteins on a moving cargo does not change the velocity of movement but does change the probability of pausing or stalling during transport (13). In contrast, for poorly processive motor proteins such as dynein, motor number has a large influence on velocity as well as on the probability of pausing or stalling. Thus one expects the regulation of kinesin-1 or dynein motor number on moving cargoes to have different effects on the properties of flux and transport in axons. For example, decreasing kinesin-1 is predicted to change pause frequencies or perhaps the balance of anterograde and retrograde transport, but not velocity. This prediction has been experimentally observed (14). The second principle relates to the question of whether single-headed motor proteins can generate force and movement. For example, kinesin-1 consists of two KHCs and contains two identical motor domains. Because its motor heads are interconnected by a coiled-coil stalk, they cooperate in a coordinated manner to achieve efficient processive movement along microtubules. In contrast, KIF1A lacks an extended predicted coiled-coil stalk domain and appears to be monomeric in its native state. Although researchers have proposed that KIF1A is capable of processive movement in its monomeric state by exchanging its microtubule-binding loops in a cyclic interaction with microtubules during each ATP cycle (15), mounting evidence suggests that, to be efficiently processive, KIF1A monomers undergo a cargo-mediated transition to homodimers that move by a mechanism akin to that proposed for kinesin-1 (16, 17). Thus, cargo interaction may activate some motor proteins by stimulating multimerization needed for processive movement.

Most cargo transport in axons may be generated by motor proteins attaching to vesicles, organelles, or protein complexes and mediating movement along stationary microtubules. However, at least two additional mechanisms of motor protein–generated, microtubule-dependent axonal transport have been suggested to generate the array of movements observed. One mechanism may use stationary motor proteins attached to nonmotile axonal membranes to produce directed movement of microtubules and associated cargoes. In this mechanism, plus end–directed movement of microtubules and bound cargoes can be generated by minus end–directed motor proteins, and minus end–directed movements can be generated by plus end–directed motor proteins (18, 19). Another mechanism derives from the observation that some motor proteins can attach simultaneously to two independent microtubules. In this case, plus end–directed motor activity between parallel microtubules will generate plus end–directed movements; minus end–directed movements could be generated by a minus end–directed motor activity (20). In addition to the variety of data supporting these proposed mechanisms, considerable evidence suggests that in vivo several kinesin and dyneins work cooperatively to achieve cargo movement (21). How these cooperative interactions fit into the overall geometry of force generation in axons remains to be elucidated. Finally, several relatively recent observations suggest that "fast" motor proteins such as kinesin-1 may be responsible for slow axonal transport. Genetic manipulation of kinesin-1 subunits revealed that removal of a specific KHC subunit called KIF5A produced defective transport of neurofilaments, which are known to be transported by slow axonal transport (5). Indeed, in vivo imaging of neurofilament movement revealed these "slow" cargoes can travel intermittently at fast rates, but with slow average velocities owing to prolonged pauses and bidirectionality of movement (22).

Regulation of Axonal Transport

Regulation of kinesin and dynein activities and thus regulation of axonal transport remain poorly understood. In principle, regulation can occur at one of several steps including cargo recognition and binding by the motor protein, velocity and character of transport itself, and recognition of the correct destination by the motor-cargo complex. Indeed, accumulating evidence suggests that kinesin-1 may be regulated directly by cargo binding such that motor activation is coupled to the binding of the motor to vesicles and/or organelles (23–25). Similarly, as described above, at least one class of kinesin motor proteins may be activated by the clustering of phospholipids, which facilitate dimerization required for processive movement (16).

Among the many cargoes and binding partners identified for anterograde and retrograde motor proteins are several whose

Amyloid precursor protein (APP): type I glycoprotein that gives rise to amyloid-β peptides, when appropriately cleaved, and plays an important role in the development of Alzheimer's disease

JIP: JNK-interacting protein

GSK3β: glycogen-synthase kinase 3β

PS1: presenilin-1

Alzheimer's disease (AD): clinically the most frequent dementia, pathologically a neurodegenerative disorder characterized by senile plaques and neurofibrillary tangles

Microtubule tracks: a polymer of linearly arranged α/β-tubulin heterodimers

identities suggest possible mechanisms of regulation. Such candidates include amyloid precursor protein (APP) (26) and the group of c-Jun NH2-terminal kinase (JNK)-interacting proteins 1, 2, and 3 (JIP1, JIP2, and JIP3/Sunday driver) (27). These proteins are directly connected to kinase and protease systems, among others. Additional evidence suggests important roles of phosphorylation in motor protein regulation, including perhaps the action of glycogen-synthase kinase 3β(GSK3β), presenilin-1 (PS1), and cyclin-dependent kinase 5.

AXONAL TRANSPORT, AGING, AND DISEASE

Many lines of evidence raise the possibility that failures in axonal transport play a role in a variety of neurodegenerative diseases. This evidence comes from study of the transport machinery during aging, reports of a variety of axonal defects that could be caused by transport problems in a number of neurodegenerative disorders, the realization that many proteins implicated in disease are actively transported, and the demonstration that some neurodegenerative diseases can be caused by mutations in genes encoding likely components or regulators of the transport machinery. In the case of Alzheimer's disease (AD), new evidence is emerging that links proteins implicated in disease causation to axonal transport. In all of these cases, it remains unclear where in the timing of disease progression transport failures emerge and what role such failures play in progression versus causation. We review this evidence below beginning with aging and diverse neurological diseases and ending with a discussion of these issues in the context of AD.

Axonal Transport and Aging

Aging has become an an intensely studied issue as a result of the increased prevalence of people reaching advanced age and the resulting increase in the incidence of diseases related to advanced age such as atherosclerosis and AD. Surprisingly, however, relatively little is known about how axons and axonal transport perform during aging and whether such changes might predispose some people to the development of aging-related diseases such as AD. Most studies thus far have examined the effects of aging either on axonal compartment structure or on axonal transport rates. Several studies reported age-related reduction in axonal microtubules (28), shifts in the distribution of microtubule-associated proteins such as tau and neurofilaments (29, 30), and the appearance of axonal accumulations of proteins such as APP (31) and other materials such as glycogen and lipofuscins. Perhaps the most obvious and direct evidence for age-related changes within axons is provided by the observation of progressive increases in the number of focal axonal swellings with age (32). These are reminiscent of similar age-related accumulations of APP (31). Intriguingly, some of these changes resemble those found in the Klotho mice, which recapitulate several characteristics of premature aging (33). The substantial structural changes seen in axons with aging may hamper efficient axonal transport. This view is consistent with a number of radiolabeling experiments, which showed significant reductions in anterograde transport of axonal vesicles, various proteins, phospholipids, and steroid hormones. Reductions in the retrograde axonal transport of axonal vesicles, neurotransmitter-related proteins, growth factors, and steroid hormones and in the slow axonal transport of tubulin, neurofilaments, and synuclein have also been reported. The mechanisms underlying these age-related changes in axonal structure and transport remain obscure. In particular, it remains unknown whether these changes affect all transported proteins uniformly and thus might be a result of changes in the microtubule tracks or, alternatively, whether aging primarily affects only certain pathways (34) while sparing others.

Impairments in Axonal Transport and the Pathogenesis of Neurodegenerative Diseases

A number of neurodegenerative disorders appear to be caused by genetic defects in genes encoding proteins that play a direct role in axons and axonal transport. For example, Charcot-Marie-Tooth disease type II has been linked to mutations in genes encoding KIF1β(35) and the low molecular weight neurofilament protein (36). Similarly, mutations in the KIF5A gene are found in hereditary spastic paraplegia (37), and mutations in the $p150^{Glued}$ gene are found in amyotrophic lateral sclerosis (ALS) (38) and distal spinal bulbar muscular atrophy (39). These data corroborate results from studies with spastin (40), which support the hypothesis that impaired axonal transport could play a major role in the pathogenesis of hereditary spastic paraplegia. In addition, the identification of mutant $p150^{Glued}$ as a cause of ALS adds genetic support to the reported axonal defects and impairments in axonal transport in the pathogenesis of ALS. Intriguingly, dynein mutations can suppress ALS caused by mutant SOD in mice (41).

Genetic and cell biological evidence also implicate defective axonal transport in the pathogenesis of movement disorders such as Huntington's disease (HD). Although the expansion of CAG repeats in the gene encoding huntingtin is well established as the cause of HD, the mechanism(s) causing neurodegeneration and neurological defects remain(s) unknown. Some insights have emerged from reports that implicate the huntingtin protein in normal functions of the transport machinery. These findings are intriguing in light of reports suggesting that mutant huntingtin, and polyglutamine proteins in general, can cause impaired axonal transport and induce the formation of axonal protein aggregates in squid (42), fruit fly (43), and mammalian (44) HD. These observations together can account for the overt synaptic and axonal pathology in HD. Whether such a mechanism can account for the preferential death of some cell types in HD as well as defects seen in other polyglutamine diseases remains to be seen. At least some of the ataxin genes encode proteins whose normal functions are nuclear, suggesting nuclear pathology may be primary in at least some of these disorders. Compelling experimental evidence in mouse models of spinocerebellar ataxia type 1 (45) supports this view. Comparable phenomena may be part of the pathogenesis of at least one form of early onset dystonia. Intriguingly, in the case of early onset dystonia linked to mutations in the AAA^+ protein torsin A, a pathogenic mutation that encodes a mutant torsin A lacking a glutamic acid residue in the COOH termini has been reported to interrupt the binding of torsin A to KLC1 and result in its deficient transport (46).

With respect to the pathogenesis of some of the major dementias, a genetic association between abnormalities in the tau gene and some forms of fronto-temporal dementias, Pick's disease, corticobasal degeneration, and progressive supranuclear palsy is well established (47) and consistent with the proposal that defective tau function may cause defects in the assembly and stability of microtubules in turn leading to defective axonal transport. Similar changes in tau biology can cause abnormal axonal transport in cell culture including defective peroxisomal transport, perhaps leading to sensitivity to oxidative damage (3). Transport defects are also seen in tau animal models (48), although the precise mechanisms remain unknown. Intriguingly, abnormalities in tau protein are found diffusely in AD and represent one of its pathological hallmarks. Although the genetic associations between tau and AD are tentative, the biological roles of tau suggest that cytoskeletal defects and failed axonal transport can contribute to the pathogenesis of common dementias.

In summary, genetic and functional studies provide strong evidence for the involvement of impaired axonal transport in the pathogenesis of several neurodegenerative disorders.

Dementia: progressive decline in two or more cognitive functions that does not perturb consciousness, significantly affects independent everyday living, and is not caused by other known illnesses

DAI: diffuse axonal injury

TBI: traumatic brain injury

Aβ: amyloid-β peptide

Whether other disorders such as KIF21 defects in congenital fibrosis of the extraocular muscles type 1 and defects in KIF13 linked to schizophrenia have similar origins remains to be determined.

Impaired Axonal Transport as a Result of Diseases

Although axonal transport is not obviously causally related to the onset of many diseases of the nervous system, these diseases appear to affect, perhaps indirectly and significantly, axonal structure and transport. Understanding the mechanisms underlying axonal insult in these diseases may provide valuable clues for the development of new diagnostic assays and therapeutics. Examples of diseases where significant advances in understanding the mechanisms of axonal injury have been achieved are diffuse axonal injury (DAI) and demyelinating diseases.

DAI is a consequence of traumatic brain injury (TBI) and an epigenetic risk factor for AD. In DAI, a massive accumulation of APP and its potentially toxic proteolytic product, amyloid-β peptide (Aβ), takes place within swollen axons at injury sites. This finding is intriguing given the strong evidence that APP contributes to the pathogenesis of AD and that TBI is a risk factor for AD. Recent work suggests that impaired axonal transport as a result of TBI (49) can cause long-lasting accumulations of APP, its proteolytic machinery, and kinesin-1 within axonal swellings that contain intra-axonal Aβ. Release of axonally generated Aβ into the extracellular space may contribute to amyloid deposition in DAI.

Researchers have also proposed that axonal damage and loss may be responsible for the persistent neurological deficits observed in patients with demyelinating diseases such as multiple sclerosis and acute disseminated encephalomyelitis. APP accumulation within damaged axons also occurs in demyelinating diseases (50). Nevertheless, little is known about the mechanism by which axonal segments adjacent to compromised oligodendrocytes incur injury. Whereas an autoimmune destruction of the myelin sheaths surrounding axons and compromised conduction properties of the denuded axons are widely accepted, recent studies offer new insights into the mechanisms responsible for the accompanying axonal defects. Although the mechanisms remain to be clarified, these studies show that oligodendrocytes deficient in the myelin proteolipid protein (51) as well as activated microglia (52) can both cause impaired fast axonal transport.

That APP and Aβ accumulate within damaged axons in several diseases suggests that APP may represent a surrogate marker of axonal pathology. Thus APP and/or Aβ may be susceptible to reductions in axonal transport or play a general role in axonal repair or regeneration. Intriguingly, studies of Niemann-Pick type C disease provide another link between impaired axonal transport, axonal defects, and Aβ generation. In brief, Niemann-Pick disease type C is characterized by intracellular accumulation of unesterified cholesterol within the endocytic pathway, endosomal abnormalities, and well-described axonal defects. Coincidently, experimental models of Niemann-Pick disease type C exhibit impaired axonal transport of endogenously synthesized cholesterol, accumulations of APP and PS1, and aberrant generation of Aβ (53, 54). These studies support the view that deficient transport can play an important role in the generation of Aβ and are consistent with the findings from DAI studies.

AXONAL TRANSPORT AND ALZHEIMER'S DISEASE

That the observed pathological changes in AD are at least partly the result of abnormal axonal transport has been discussed for decades (55). However, until recently, too little was known about the axonal transport machinery to critically test these ideas. Here we summarize the major features of AD, the axonal transport and functions of AD-related molecules, and axonal defects seen in AD. We then discuss how a cascade of axonal blockages could cause the pathogenesis of AD.

Overview of Alzheimer's Disease

AD was first described in a report that associated bizarre psychiatric symptomatology with the postmortem observation of senile plaques and neurofibrillary tangles. Senile plaques were later described to consist of a network of dystrophic neurites embedded in extracellular, congophilic, and fibrillar amyloid (56). These neuritic changes and amyloid deposits are commonly accompanied by a prominent glial reaction. At about the same time neurofibrillary tangles were found to correspond to intracellular accumulations of paired helical filaments (57). AD is the most common dementia with senile plaques and neurofibrillary tangles as pathological hallmarks when found in diagnostically relevant brain regions (58, 59). In addition, AD brains generally exhibit severe perturbations of several neurotransmitters and widespread synaptic and neuronal loss in distinct anatomic areas such as the limbic system and the basal forebrain (60–63). These changes are accompanied by a severe disruption of the axonal as well as dendritic cytoskeleton, which suggests failed axonal transport at some point in the progression of the disease.

To date, mutations in three independent genes have been found to segregate with kindreds afflicted by rare familial AD (FAD). The first gene emerged from studies focused on the purification and sequencing of proteins found in amyloid deposits (64). These studies identified Aβ as the major constituent of amyloid, which led to the identification of the precursor protein, APP (65–67). A number of mutations in the human gene-encoding APP were found to cause some forms of FAD, perhaps as a result of aberrant Aβ generation (69, 70). In addition, the mapping of APP to chromosome 21, which is trisomic in Down's syndrome, provided a possible cause for the AD pathology consistently observed in the brains of Down's syndrome patients. Further genetic studies identified mutations in PS1 and presenilin-2 (PS2) as additional genetic causes of FAD (71). The finding that mutations in PS1 and PS2 also lead to aberrant Aβ generation was an important clue for deciphering the mechanism of Aβ formation and led to the discovery that presenilins are likely key components of the proteolytic processing machinery for APP and other proteins such as Notch (72). In fact, a combination of results from many studies revealed that Aβ formation can be either abrogated or stimulated by the proteolytic processing of APP. Cleavage of APP within the Aβ domain of APP (α-cleavage) abrogates Aβ formation, whereas sequential cleavage of APP at the N termini (β-cleavage) and C termini (γ-cleavage) of the Aβ domain of APP results in Aβ formation (**Figure 2**) (65, 73). Members of the disintegrin and metalloproteinase families were identified as participants in the α-cleavage, and the β-site APP cleaving enzyme (BACE) as a participant in the β-cleavage, whereas nicastrin, anterior-pharynx defective protein-1 and presenilin enhancer-2, along with PS1, orchestrate the γ-cleavage of APP. Finally, apolipoprotein ε4 alleles were found to be associated with earlier onset and more aggressive forms of AD. These findings, together with the discovery that abnormally hyperphosphorylated tau forms the neurofibrillary tangles (74), provide a framework for understanding the molecular basis of AD.

The genetic and biochemical data, combined with the observation that amyloid plaques are a characteristic feature of AD, led to the formulation of the amyloid-cascade hypothesis (75–77). In its simplest form this hypothesis suggests AD develops as a result of either increased production of Aβ throughout life owing to FAD mutations or owing to

Senile plaque: a network of dystrophic neurites embedded in extracellular, congophilic, and fibrillar amyloid

Neurofibrillary tangle: intracellular paired helical (twisted) filaments that form as a result of abnormal hyperphosphorylation of tau

BACE: β-site APP cleaving enzyme

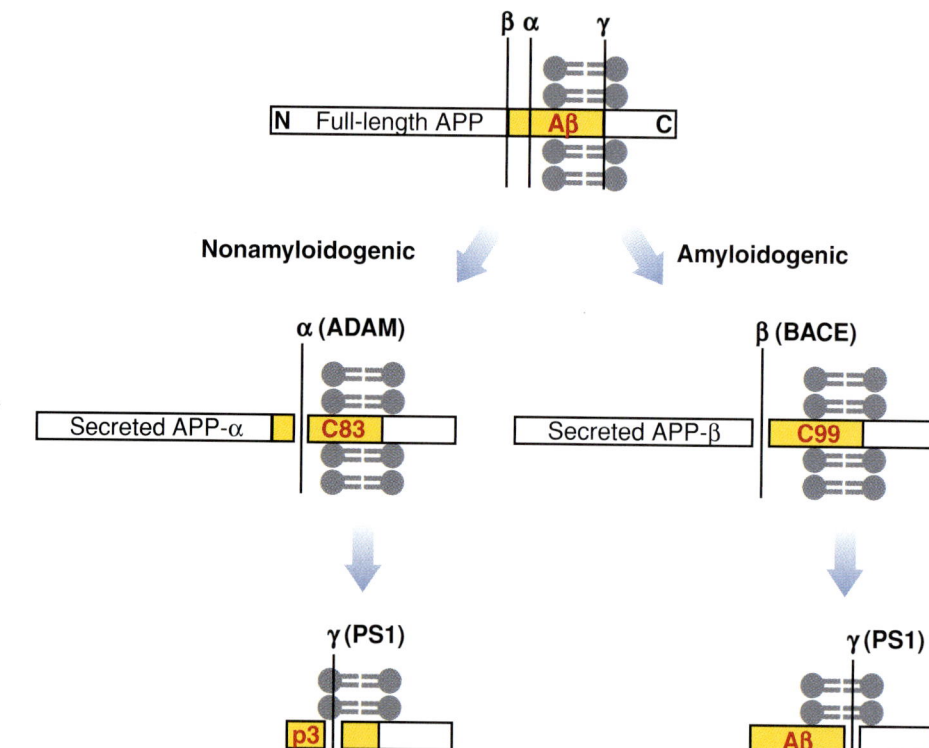

Figure 2
Major proteolytic processing pathways of amyloid precursor protein. Abbreviations: ADAM, a disintegrin and metalloproteinase; APP, amyloid precursor protein; BACE, β-site APP cleaving enzyme; PS1, presenilin-1.

a gradually increasing buildup of Aβ as a result of failed mechanisms of Aβ clearance thought to occur in sporadic AD cases. These events are proposed to result in accumulation, oligomerization, and deposition of Aβ. Aβ oligomers or deposits are thought to activate microglia, trigger an inflammatory response, and alter synaptic structure or functions. These events might in turn give rise to altered neuronal homeostasis, altered kinase and phosphatase activities, the formation of neurofibrillary tangles, and widespread synaptic and neuronal dysfunction and loss precipitating in dementia.

Although there are clear causative relationships between genes controlling Aβ formation and AD, several features of AD are not well accounted for by the amyloid-cascade hypothesis. A major issue is the apparently poor correlation between amyloid deposits and other aspects of the pathology observed in AD brains (78). For example, an obvious relationship between senile plaques and neurofibrillary tangles is lacking (79). Whereas neurofibrillary tangles start forming in the entorhinal region and then spread toward the cortices, the opposite appears true for senile plaques (80, 81). Similarly, synaptic numbers are generally not consistently affected by amyloid deposits in animal models of AD, and data establishing a role for Aβ in decimating synaptic and neuronal numbers in vivo are lacking. In addition, neuronal loss in AD often occurs independently and in anatomically distant regions from areas of amyloid deposition. Finally, amyloid burden is not well related to the clinical picture of AD. For example, significant amounts of amyloid can be found deposited in the brains of cognitively intact subjects (82). Amyloid burden in AD brains is variable and does not predict either the duration or severity of AD. Furthermore, many of the symptoms observed in AD result from damage to brain areas often devoid of amyloid deposits.

Axonal Transport of Proteins Linked to the Pathogenesis of Alzheimer's Disease

Many proteins associated with the pathogenesis of AD (including APP, BACE, PS1, nicastrin, Aph-1, presenilin enhancer-2, synuclein, and tau) have been observed in the axonal compartment of neurons, with many of them found at presynaptic terminals (83–86). Thus, transport is almost certainly necessary to deliver these proteins to their final destinations (87). For APP, there is strong evidence from cell culture, *Drosophila*, and mice that kinesin-1 is responsible for APP transport (26, 88–90). There are also reports that PS1, BACE, and synuclein undergo axonal transport at speeds consistent with fast anterograde axonal transport (91–95), and although there is some controversy (96), APP, BACE, and PS1 have been suggested to travel within the same membrane compartment in axons (92, 97–100). Some proposals for the mechanism of APP transport suggest the C terminus of APP interacts with the tetratrico repeats of the kinesin-1 KLCs (26), either directly, but possibly enhanced by JIP1 (101), or indirectly via complex formation with JIP1 (96, 102). Intriguingly, APP has also been reported to undergo fast anterograde axonal transport as a component of the herpes simplex virus-derived viral particles associated with kinesin-1 (103). Further work is required to better understand the mode of interaction between APP and KLCs; to evaluate the mechanisms of axonal transport of BACE, PS1, and other molecules involved in AD; and to identify additional partners, such as proteins interacting with APP tail 1 (104), that might take part in transporting axonal APP. Interestingly, recent observations indicated that unlike JIP1 and JIP2, which associate and phosphorylate APP, JIP3 does not associate with APP but may play a role in regulating its transport and phosphorylation (105).

Axonal Functions of Proteins Linked to the Pathogenesis of Alzheimer's Disease

Mounting evidence suggests that many proteins implicated in the pathogenesis of AD have functions in the axonal compartment. For example, early work on the biological functions of APP suggested a possible role in promoting axonal growth (106, 107). In fact, several cell-culture studies suggested that reductions (108, 109), overexpression (110), and other modifications of APP (111) could cause abnormalities in axonal growth. Similarly, deletions (112, 113) or overexpression of APP (114, 115) in mice both gave rise to reductions in white-matter brain structures and corroborated cell-culture studies by providing in vivo evidence for a role of APP in the maintenance of axonal structure and function. These data are consistent with the observed upregulation and axonal enrichment of APP during nervous system development and in axonal injury states when molecules involved in axonal growth are expected to be most active (106, 116). Of particular interest in this context are reports of axonal increases in APP in several disease states that exhibit axonal injury. These disease states include multiple sclerosis, TBI, brain infarctions and infections, and neurodegenerative diseases such as Creutzfeldt-Jakob's or AD. These observations therefore suggest APP may play an important role during axonal repair. This conjecture is supported by recent experiments in *Drosophila* indicating a role of the *Drosophila APP-like* (*APPL*) gene in recovery from brain injury (116). APP accumulation within axons in disease states could also be the result of abnormal or continued axonal transport in the presence of axonal blockages.

Recent work in mice suggests that APP may play a role within the axonal compartment by participating in axonal transport (92, 117). Although controversial (96), this proposal is based on the initial observation that reduction or overexpression of APP in *Drosophila* (90, 118) causes axonal transport deficits reminiscent of those observed in motor protein mutants (119). Similar defects have been observed in mice overexpressing APP (120) and in AD (14). Finally, axonal transport deficits caused by APP overexpression in *Drosophila* and mice can be enhanced by otherwise benign reductions of kinesin-1 (14, 90). Consistent with these data are the findings that synapses in flies and mice deficient in APP exhibit reduced numbers of synaptic boutons (121–123).

In addition to a role in APP processing, some evidence suggests that both PS1 and BACE possess additional functions within the axonal compartment. Similar to APP, manipulations of PS1 in cell-culture systems reveal a role in axonal growth and morphology (124–127), which has been shown to be important in development of the nervous system (126, 128) and in states of axonal injury (99). Examination of the means by which PS1 exerts its functions in the axonal compartment suggests several mechanisms ranging from direct (126, 129) to indirect actions (130) on the axonal cytoskeleton. Intimate involvement in the control of the axonal transport machinery via phosphorylation of KLC by GSK3β (131, 132) may also occur. Intriguingly, PS1 may also control the transport of APP and other proteins, possibly via GSK3β (133, 134). These ideas are consistent with a wealth of data that, although disputed (135), suggest that APP and PS1 interact (136–139). Although less is known about BACE, its overexpression produces overt axonal degeneration (140) and results in reduced axonal transport of APP (141). More specifically, BACE overexpression may shift β-cleavage of APP to the cell bodies, which results in inappropriate post-translational modifications of APP, in reduced targeting of APP into the axonal compartment, in reduced axonal transport of APP, and in diminished Aβ generation (141). Intriguingly, tau has also been proposed to regulate kinesin-1-mediated vesicle and organelle transport along microtubules as well as organizing axonal microtubules. Related roles in axonal transport may unite tau-related and APP-related pathologies in AD, perhaps via interactions including JIP1 and GSK3β (3, 52, 142). Such a view is consistent with the reported colocalization of APP, BACE, PS1, and Aβ in swollen axons induced by diffuse axonal injury (99).

Axonal Defects and the Pathology of Alzheimer's Disease

Years of pathological examination of AD brains have yielded many descriptions of abnormal axons. These axonal defects may reflect transport problems and can be divided into three classes: (*a*) those juxtaposed to amyloid within senile plaques; (*b*) those associated with neurofibrillary tangles, and (*c*) those spatially distinct from the hallmark lesions of AD.

First, dystrophic axons juxtaposed to amyloid are the best studied and are found associated with dense amyloid cores as well as with smaller amyloid bundles. Some workers have proposed that these dystrophic axons are more specific to AD than the amyloid itself (143). Some evidence suggests that amyloid-associated dystrophic neurites are best distinguished based on whether they harbor abnormally phosphorylated tau (144). In fact, most such axons are immunoreactive to axonal cargo proteins such as synaptophysin or tau (144, 145). In addition, a subpopulation of abnormal axons found adjacent to the amyloid was identified to accumulate actin, actin-depolymerizing factor, and cofilin (146). Overall, the tight relationship between amyloid and abnormal axons indicates that either amyloid underlies the formation of axonal defects or that axonal defects represent "hot spots" implicated in the amyloid deposition or both.

Second, abnormal axons associated with neurofibrillary tangles in AD are well described (147). Abnormal thin and tortuous axons immunoreactive to a battery of different phospho-tau epitopes represent an invariable feature of AD brains. These axons can be found in areas either spatially distinct from, or adjacent to, the amyloid deposits and neurofibrillary tangles. Although massive disorganization of the axonal compartment by the abnormally phosphorylated tau is associated with an abnormal microtubule network, it is unclear whether this can be accounted for by the action of abnormal tau (28). Interestingly, evaluation of abnormal axons associated with neurofibrillary tangles suggests that at least some of these are not related to amyloid (148). These tangle-associated neuritic clusters are enriched in tau and form dense aggregates and ghost tangles with the core made up of extracellular bundles of straight filaments. Although the significance of tangle-associated neuritic clusters in AD remains largely unexplored, the clusters provide important evidence that amyloid is not necessarily present in axonal defects at all stages and, alternatively, that amyloid is not required for the formation of some populations of abnormal axons encountered in AD brains.

Third, abnormal axons spatially distinct from the hallmark lesions of AD consist of shorter, more irregular, and tortuous processes (14, 149). These are most evident as focal axonal swellings that correspond to abnormal accumulations of axonal cargos and transport proteins (14). These swellings exhibit aberrant phosphorylation of neurofilaments but not of tau. They form early in AD and precede tau-immunoreactive axons spatially distinct from amyloid deposits and neurofibrillary tangles. Importantly, because abnormal axons form in brain areas that eventually develop hallmark lesions of AD, it is plausible they represent precursor lesions to those observed associated with amyloid and tangles.

Several studies reported axonal loss in brain regions typically afflicted by AD (150) and also in the olfactory and optic nerves, which are readily amenable to axonal quantification (151–153). Similarly, in vivo imaging studies of brains from AD patients showed selective white-matter changes (154, 155). These studies are consistent with the variety of axonal defects observed in AD brains and with the clinical disconnection syndrome demonstrated in subjects afflicted by AD (156).

A number of animal models expressing AD-related proteins exhibit a variety of axonal defects akin to those observed in AD. Intriguingly, these defects were observed not only in models based on molecules tightly linked to AD such as tau, APP, BACE, or PS1 (14, 140, 157, 158), but also in models based on molecules more distantly related to AD such as p25 and apolipoprotein ε4. In *Drosophila*, overexpression of APP alone or of APP and tau resulted in axonal defects reminiscent of those observed in *Drosophila* motor protein mutants (90, 118). Comparison of dystrophic neurites juxtaposed to amyloid in AD brains with those in brains of mouse models of AD also revealed striking similarities (159). Intriguingly, similar dystrophic axons could be found in areas devoid of amyloid or neurofibrillary tangles in AD (14). Therefore the formation of axonal defects may not require the presence of amyloid- or tau-related changes although these pathologies may modify or enhance axonal abnormalities.

Impairments in Axonal Transport and the Pathogenesis of Alzheimer's Disease

There is a considerable amount of data consistent with the hypothesis that impaired axonal transport plays a crucial role in the pathogenesis of AD (55). These data include the consistent observation of widespread axonal pathology in AD including abnormal axons that exhibit aberrant accumulations of APP (160) and its metabolites (161, 162); synapse- (163), endocytosis- (164), and neurotransmitter-related proteins (14, 165); resident axonal proteins such as neurofilaments and tau; glycogen (166); and organelles (167), in addition to the sequestration of tubulin (168); reduced number of microtubules tracks (28); and the emergence of abnormal filaments (169). These abnormal axons appear to comprise an important component of the dystrophic neurites that constitute senile plaques and, therefore, correspond to an essential pathological component of AD. The observation that development of axonal abnormalities in some brain regions precedes the rest of known AD pathology, whereas in others it occurs in spatially distinct areas from the rest of the known AD pathology, together with the finding that abnormal axons form at least one year prior to amyloid deposition in some mouse models (14), suggests the production of axonal defects could coincide with the earliest stages of AD pathogenesis. Intriguingly, reductions in kinesin-1 motor proteins promote the development of axonal defects in *Drosophila* (90, 118). In mouse similar kinesin-1 reduction enhances the development of axonal defects, increases aberrant Aβ generation, and enhances amyloid deposition (14, 90). These data suggest a causal relationship between failed axonal transport and the generation of axonal abnormalities and provide a direct link between changes in axonal transport, aberrant accumulation or production of Aβ, and the formation of senile plaques. Similarly, several studies showed that Aβ per se is sufficient to induce the formation of axonal abnormalities (170) and may directly contribute to the impairments in the axonal transport (171, 172). Whether aberrant axonal Aβ generation causes failed axonal transport and axonal defects or whether aberrant Aβ generation might be the result of impaired axonal transport that causes further deterioration of axonal transport is unclear.

Abnormal phosphorylation of tau and the corresponding axonal abnormalities are an invariable feature of AD that may directly impair axonal transport of APP and other molecules (173). This feature of AD is intriguing in light of reported improvements in axonal transport, increased numbers of microtubule tracks, and ameliorated motor impairments observed in tau transgenic mice upon treatment with microtubule-stabilizing drugs (174, 175). Unexpectedly, some mutations linked to kindreds afflicted by FAD alter axonal transport of axonal cargoes including APP (131, 133). This finding provides a link between abnormal processing of APP, aberrant Aβ generation, and impairments in axonal transport. Several risk factors for AD development including advanced age, apolipoprotein ε status, and repetitive trauma have all been linked to defects in axonal transport. For example, aging may be associated with reduced axonal transport in anatomically relevant areas to AD, such as in the basal cholinergic forebrain (176), whereas homozygosity for apolipoprotein ε4 produces white-matter changes in asymptomatic subjects (177).

We find it striking that impairments in axonal transport present an economical explanation for the early synaptic changes observed in AD along with the observations that cortical levels of nerve growth factor in AD are normal even though basal forebrain cholinergic neurons exhibit features diagnostic of nerve growth factor deprivation (178, 179). We also think it is relevant that cortical levels of cholinergic enzymes are reduced in AD even though basal forebrain cholinergic neurons retain normal expression of cholinergic markers for some time after disease onset (180). The finding of significant reductions in axonal transport in postmortem AD brains (181) corroborates these observations as do the obvious changes in tau and other cytoskeletal proteins in axons.

We suggest that small initial changes in axonal transport pathways or spontaneously occurring axonal blockages over time can trigger early abnormalities in synaptic and axonal structure and function and the development of sporadic AD. Such impairments might be a consequence of age-related reductions in microtubule number or transport, oxidative stress (182), or perhaps traumatic brain injury. Impairment of axonal transport could stimulate Aβ generation (14), which at some point cannot be efficiently cleared and starts accumulating. Abnormal accumulations in Aβ might induce further deterioration of axonal transport, more pronounced axonal pathology, and additional impairments in synaptic function. Importantly, impairments in axonal transport and aberrant Aβ generation could mutually potentiate each other in a vicious cycle, producing increasing damage over time including perhaps tau abnormalities. Such a vicious cycle, consisting of impaired axonal transport, aberrant Aβ generation, and tau abnormalities, might lead to the formation of paired helical filaments and clusters of axonal defects from multiple axons. This would mark the beginning of amyloid deposition and the appearance of neurofibrillary tangles. Notably, axonal blockages must also hamper retrograde axonal transport pathways (183) that provide cell bodies with the neurotrophic signals important for the maintenance of their differentiated state and survival (184), which could result in neuronal loss.

In FAD, blockages might be initiated by abnormal processing or transport of APP and in Down's syndrome by abnormal levels of APP similar to that reported in *Drosophila* and in mouse (14, 90). Finally, polymorphisms in the KLC1 subunit of kinesin-1 have been reported to increase the risk of AD (185), although large-scale studies are required to confirm or refute those data.

Although we favor a mechanism in which defective axonal transport or age-related defects in axonal transport precipitate AD, we recognize that this hypothesis does not account for all features of this disease. For example, similar to the amyloid-cascade hypothesis, an axonal blockage cascade does not explain the regional selectivity of the damage observed in AD. In addition, it does not explain the apparent lack of peripheral nervous system involvement. Furthermore, no hypothesis proposed thus far provides a clear account of what triggers the development of the prevalent sporadic form of AD. Further work is required to test rigorously the mechanisms involved in the pathogenesis of AD and to critically examine whether axonal transport plays a definitive role in the causation or the progression of AD.

SUMMARY POINTS

1. Many proteins linked to the pathogenesis of neurodegenerative disorders including AD undergo axonal transport and may be important to maintain not only synaptic but also axonal structure and function.

2. Recent data suggest that impaired axonal transport can promote aberrant Aβ generation and enhance amyloid deposition. Thus, an intimate link between axonal transport and protein deposition in the pathogenesis of AD may exist.

3. Axonal defects are observed early in AD, and APP accumulates in damaged axons in several disease and injury states.

FUTURE ISSUES TO BE RESOLVED

1. How is axonal transport regulated and what is the basis for age-related changes in axonal transport?

2. Are defects in axonal transport a cause or a consequence of the pathological changes in AD?

3. Do genetic variations in AD susceptibility identify genes involved in axonal transport functions?

DISCLOSURE STATEMENT

L.S.B. Goldstein is a consultant and shareholder for Cytokinetics, Inc. G.B. Stokin and L.S.B. Goldstein hold patents related to transport and AD.

LITERATURE CITED

1. Weiss P, Hiscoe HB. 1948. *J. Exp. Zool.* 107:315–95
2. Kreutzberg GW. 1969. *Proc. Natl. Acad. Sci. USA* 62:722–28
3. Ebneth A, Godemann R, Stamer K, Illenberger S, Trinczek B, Mandelkow E. 1998. *J. Cell. Biol.* 143:777–94
4. Lawrence CJ, Dawe RK, Christie KR, Cleveland DW, Dawson SC, et al. 2004. *J. Cell. Biol.* 167:19–22
5. Xia CH, Roberts EA, Her LS, Liu X, Williams DS, et al. 2003. *J. Cell. Biol.* 161:55–66
6. Tanaka Y, Kanai Y, Okada Y, Nonaka S, Takeda S, et al. 1998. *Cell* 93:1147–58
7. Gindhart JG Jr, Desai CJ, Beushausen S, Zinn K, Goldstein LS. 1998. *J. Cell. Biol.* 141:443–54
8. Rahman A, Kamal A, Roberts EA, Goldstein LS. 1999. *J. Cell. Biol.* 146:1277–88
9. Schroer TA. 2004. *Annu. Rev. Cell. Dev. Biol.* 20:759–79
10. LaMonte BH, Wallace KE, Holloway BA, Shelly SS, Ascano J, et al. 2002. *Neuron* 34:715–27
11. Yildiz A, Selvin PR. 2005. *Trends Cell Biol.* 15:112–20
12. Asbury CL. 2005. *Curr. Opin. Cell Biol.* 17:89–97
13. Spudich JA. 1990. *Nature* 348:284–85
14. Stokin GB, Lillo C, Falzone TL, Brusch RG, Rockenstein E, et al. 2005. *Science* 307:1282–88
15. Nitta R, Kikkawa M, Okada Y, Hirokawa N. 2004. *Science* 305:678–83
16. Tomishige M, Klopfenstein DR, Vale RD. 2002. *Science* 297:2263–67
17. Rashid DJ, Bononi J, Tripet BP, Hodges RS, Pierce DW. 2005. *J. Pept. Res.* 65:538–49
18. He Y, Francis F, Myers KA, Yu W, Black MM, Baas PW. 2005. *J. Cell. Biol.* 168:697–703
19. Orokos DD, Cole RW, Travis JL. 2000. *Cell Motil. Cytoskelet.* 47:296–306
20. Baas PW, Ahmad FJ. 2001. *Trends Cell Biol.* 11:244–29
21. Kural C, Kim H, Syed S, Goshima G, Gelfand VI, Selvin PR. 2005. *Science* 308:1469–72
22. Wang L, Ho CL, Sun DM, Liem RKH, Brown A. 2000. *Nat. Cell Biol.* 2:137–41
23. Friedman DS, Vale RD. 1999. *Nat. Cell Biol.* 1:293–97
24. Verhey KJ, Lizotte DL, Abramson T, Barenboim L, Schnapp BJ, Rapoport TA. 1998. *J. Cell. Biol.* 143:1053–66
25. Stock MF, Guerrero J, Cobb B, Eggers CT, Huang TG, et al. 1999. *J. Biol. Chem.* 274:14617–23
26. Kamal A, Stokin GB, Yang Z, Xia CH, Goldstein LS. 2000. *Neuron* 28:449–59
27. Cavalli V, Kujala P, Klumperman J, Goldstein LS. 2005. *J. Cell. Biol.* 168:775–87
28. Cash AD, Aliev G, Siedlak SL, Nunomura A, Fujioka H, et al. 2003. *Am. J. Pathol.* 162:1623–27
29. Niewiadomska G, Baksalerska-Pazera M. 2003. *Neuroreport* 14:1701–6
30. Uchida A, Tashiro T, Komiya Y, Yorifuji H, Kishimoto T, Hisanaga S. 2004. *J. Neurochem.* 88:735–45
31. Kawarabayashi T, Shoji M, Yamaguchi H, Tanaka M, Harigaya Y, et al. 1993. *Neurosci. Lett.* 153:73–76
32. Masuoka DT, Jonsson G, Finch CE. 1979. *Brain Res.* 169:335–41

33. Uchida A, Komiya Y, Tashiro T, Yorifuji H, Kishimoto T, et al. 2001. *J. Neurosci. Res.* 64:364–70
34. Goemaere-Vanneste J, Couraud JY, Hassig R, Di Giamberardino L, van den Bosch de Aguilar P. 1988. *J. Neurochem.* 51:1746–54
35. Zhao C, Takita J, Tanaka Y, Setou M, Nakagawa T, et al. 2001. *Cell* 105:587–97
36. Mersiyanova IV, Perepelov AV, Polyakov AV, Sitnikov VF, Dadali EL, et al. 2000. *Am. J. Hum. Genet.* 67:37–46
37. Reid E, Kloos M, Ashley-Koch A, Hughes L, Bevan S, et al. 2002. *Am. J. Hum. Genet.* 71:1189–94
38. Munch C, Sedlmeier R, Meyer T, Homberg V, Sperfeld AD, et al. 2004. *Neurology* 63:724–26
39. Puls I, Oh SJ, Sumner CJ, Wallace KE, Floeter MK, et al. 2005. *Ann. Neurol.* 57:687–94
40. McDermott CJ, Grierson AJ, Wood JD, Bingley M, Wharton SB, et al. 2003. *Ann. Neurol.* 54:748–59
41. Kieran D, Hafezparast M, Bohnert S, Dick JR, Martin J, et al. 2005. *J. Cell. Biol.* 169:561–67
42. Szebenyi G, Morfini GA, Babcock A, Gould M, Selkoe K, et al. 2003. *Neuron* 40:41–52
43. Gunawardena S, Her LS, Brusch RG, Laymon RA, Niesman IR, et al. 2003. *Neuron* 40:25–40
44. Trushina E, Dyer RB, Badger JD 2nd, Ure D, Eide L, et al. 2004. *Mol. Cell. Biol.* 24:8195–209
45. Orr HT, Zoghbi HY. 2001. *Hum. Mol. Genet.* 10:2307–11
46. Kamm C, Boston H, Hewett J, Wilbur J, Corey DP, et al. 2004. *J. Biol. Chem.* 279:19882–92
47. Goedert M, Spillantini MG. 2001. *Biochem. Soc. Symp.* (67):59–71
48. Ishihara T, Hong M, Zhang B, Nakagawa Y, Lee MK, et al. 1999. *Neuron* 24:751–62
49. Stone JR, Okonkwo DO, Dialo AO, Rubin DG, Mutlu LK, et al. 2004. *Exp. Neurol.* 190:59–69
50. Kuhlmann T, Lingfeld G, Bitsch A, Schuchardt J, Bruck W. 2002. *Brain* 125:2202–12
51. Edgar JM, McLaughlin M, Yool D, Zhang SC, Fowler JH, et al. 2004. *J. Cell. Biol.* 166:121–31
52. Stagi M, Dittrich PS, Frank N, Iliev AI, Schwille P, Neumann H. 2005. *J. Neurosci.* 25:352–62
53. Jin LW, Shie FS, Maezawa I, Vincent I, Bird T. 2004. *Am. J. Pathol.* 164:975–85
54. Burns M, Gaynor K, Olm V, Mercken M, LaFrancois J, et al. 2003. *J. Neurosci.* 23:5645–49
55. Terry RD. 1996. *J. Neuropathol. Exp. Neurol.* 55:1023–25
56. Terry RD, Gonatas NK, Weiss M. 1964. *Am. J. Pathol.* 44:269–87
57. Kidd M. 1963. *Nature* 197:192–93
58. Terry RD, Katzman R. 1983. *Ann. Neurol.* 14:497–506
59. Katzman R. 1986. *N. Engl. J. Med.* 314:964–73
60. Davies P, Maloney AJ. 1976. *Lancet* 2:1403
61. Hyman BT, Van Horsen GW, Damasio AR, Barnes CL. 1984. *Science* 225:1168–70
62. Whitehouse PJ, Price DL, Struble RG, Clark AW, Coyle JT, Delon MR. 1982. *Science* 215:1237–39
63. Terry RD, Masliah E, Salmon DP, Butters N, DeTeresa R, et al. 1991. *Ann. Neurol.* 30:572–80
64. Glenner GG, Wong CW. 1984. *Biochem. Biophys. Res. Commun.* 120:885–90

65. Haass C. 2004. *EMBO J.* 23:483–88
66. Goldgaber D, Lerman MI, McBride OW, Saffiotti U, Gajdusek DC. 1987. *Science* 235:877–80
67. Kang J, Lemaire HG, Unterbeck A, Salbaum JM, Masters CL, et al. 1987. *Nature* 325:733–36
68. Deleted in proof
69. Murrell J, Farlow M, Ghetti B, Benson MD. 1991. *Science* 254:97–99
70. Citron M, Oltersdorf T, Haass C, McConlogue L, Hung AY, et al. 1992. *Nature* 360:672–74
71. Sherrington R, Rogaev EI, Liang Y, Rogaeva EA, Levesque G, et al. 1995. *Nature* 375:754–60
72. Selkoe D, Kopan R. 2003. *Annu. Rev. Neurosci.* 26:565–97
73. Kojro E, Fahrenholz F. 2005. *Subcell. Biochem.* 38:105–27
74. Grundke-Iqbal I, Iqbal K, Tung YC, Quinlan M, Wisniewski HM, Binder LI. 1986. *Proc. Natl. Acad. Sci. USA* 83:4913–17
75. Hardy JA, Higgins GA. 1992. *Science* 256:184–85
76. Jarrett JT, Lansbury PT Jr. 1993. *Cell* 73:1055–58
77. Hardy J, Selkoe DJ. 2002. *Science* 297:353–56
78. Terry RD, Peck A, DeTeresa R, Schechter R, Horoupian DS. 1981. *Ann. Neurol.* 10:184–92
79. Armstrong RA, Myers D, Smith CU. 1993. *Dementia* 4:16–20
80. Thal DR, Rub U, Orantes M, Braak H. 2002. *Neurology* 58:1791–800
81. Braak H, Braak E. 1991. *Acta Neuropathol.* 82:239–59
82. Mackenzie IR, McLachlan RS, Kubu CS, Miller LA. 1996. *Neurology* 46:425–29
83. Siman R, Salidas S. 2004. *Neuroscience* 129:615–28
84. Capell A, Meyn L, Fluhrer R, Teplow DB, Walter J, Haass C. 2002. *J. Biol. Chem.* 277:5637–43
85. Beher D, Elle C, Underwood J, Davis JB, Ward R, et al. 1999. *J. Neurochem.* 72:1564–73
86. Iwai A, Masliah E, Yoshimoto M, Ge N, Flanagan L, et al. 1995. *Neuron* 14:467–75
87. Koo EH, Sisodia SS, Archer DR, Martin LJ, Weidemann A, et al. 1990. *Proc. Natl. Acad. Sci. USA* 87:1561–65
88. Kaether C, Skehel P, Dotti CG. 2000. *Mol. Biol. Cell* 11:1213–24
89. Amaratunga A, Leeman SE, Kosik KS, Fine RE. 1995. *J. Neurochem.* 64:2374–76
90. Gunawardena S, Goldstein LS. 2001. *Neuron* 32:389–401
91. Jensen PH, Li JY, Dahlstrom A, Dotti CG. 1999. *Eur. J. Neurosci.* 11:3369–76
92. Kamal A, Almenar-Queralt A, LeBlanc JF, Roberts EA, Goldstein LS. 2001. *Nature* 414:643–48
93. Papp H, Pakaski M, Kasa P. 2002. *Neurochem. Int.* 41:429–35
94. Kasa P, Papp H, Pakaski M. 2001. *Brain Res.* 909:159–69
95. Jensen PH, Nielsen MS, Jakes R, Dotti CG, Goedert M. 1998. *J. Biol. Chem.* 273:26292–94
96. Lazarov O, Morfini GA, Lee EB, Farah MH, Szodorai A, et al. 2005. *J. Neurosci.* 25:2386–95
97. Gu Y, Sanjo N, Chen F, Hasegawa H, Petit A, et al. 2004. *J. Biol. Chem.* 279:31329–36
98. Hook VY, Toneff T, Aaron W, Yasothornsrikul S, Bundey R, Reisine T. 2002. *J. Neurochem.* 81:237–56
99. Chen XH, Siman R, Iwata A, Meaney DF, Trojanowski JQ, Smith DH. 2004. *Am. J. Pathol.* 165:357–71

100. Yang J, Li T. 2005. *Exp. Cell Res.* 309:379–89
101. Inomata H, Nakamura Y, Hayakawa A, Takata H, Suzuki T, et al. 2003. *J. Biol. Chem.* 278:22946–55
102. Matsuda S, Matsuda Y, D'Adamio L. 2003. *J. Biol. Chem.* 278:38601–6
103. Satpute-Krishnan P, DeGiorgis JA, Bearer EL. 2003. *Aging Cell* 2:305–18
104. Zheng P, Eastman J, Vande Pol S, Pimplikar SW. 1998. *Proc. Natl. Acad. Sci. USA* 95:14745–50
105. Muresan Z, Muresan V. 2005. *J. Neurosci.* 25:3741–51
106. Masliah E, Mallory M, Ge N, Saitoh T. 1992. *Brain Res.* 593:323–28
107. Moya KL, Benowitz LI, Schneider GE, Allinquant B. 1994. *Dev. Biol.* 161:597–603
108. Perez RG, Zheng H, Van der Ploeg LH, Koo EH. 1997. *J. Neurosci.* 17:9407–14
109. Allinquant B, Hantraye P, Mailleux P, Moya K, Bouillot C, Prochiantz A. 1995. *J. Cell. Biol.* 128:919–27
110. Wang ZF, Wang JZ. 2004. *Chin. Med. J.* 117:775–78
111. Qiu WQ, Ferreira A, Miller C, Koo EH, Selkoe DJ. 1995. *J. Neurosci.* 15:2157–67
112. Muller U, Cristina N, Li ZW, Wolfer DP, Lipp HP, et al. 1994. *Cell* 79:755–65
113. Magara F, Muller U, Li ZW, Lipp HP, Weissmann C, et al. 1999. *Proc. Natl. Acad. Sci. USA* 96:4656–61
114. Gonzalez-Lima F, Berndt JD, Valla JE, Games D, Reiman EM. 2001. *Neuroreport* 12:2375–79
115. Redwine JM, Kosofsky B, Jacobs RE, Games D, Reilly JF, et al. 2003. *Proc. Natl. Acad. Sci. USA* 100:1381–86
116. Leyssen M, Ayaz D, Hebert SS, Reeve S, De Strooper B, Hassan BA. 2005. *EMBO J.* 24:2944–55
117. Cottrell BA, Galvan V, Banwait S, Gorostiza O, Lombardo CR, et al. 2005. *Ann. Neurol.* 58:277–89
118. Torroja L, Chu H, Kotovsky I, White K. 1999. *Curr. Biol.* 9:489–92
119. Hurd DD, Saxton WM. 1996. *Genetics* 144:1075–85
120. Cooper JD, Salehi A, Delcroix JD, Howe CL, Belichenko PV, et al. 2001. *Proc. Natl. Acad. Sci. USA* 98:10439–44
121. Yang G, Gong YD, Gong K, Jiang WL, Kwon E, et al. 2005. *Neurosci. Lett.* 384:66–71
122. Wang P, Yang G, Mosier DR, Chang P, Zaidi T, et al. 2005. *J. Neurosci.* 25:1219–25
123. Torroja L, Packard M, Gorczyca M, White K, Budnik V. 1999. *J. Neurosci.* 19:7793–803
124. Furukawa K, Guo Q, Schellenberg GD, Mattson MP. 1998. *J. Neurosci. Res.* 52:618–24
125. Dowjat WK, Wisniewski T, Efthimiopoulos S, Wisniewski HM. 1999. *Neurosci. Lett.* 267:141–44
126. Pigino G, Pelsman A, Mori H, Busciglio J. 2001. *J. Neurosci.* 21:834–42
127. Figueroa DJ, Morris JA, Ma L, Kandpal G, Chen E, et al. 2002. *Neurobiol. Dis.* 9:49–60
128. Levesque L, Annaert W, Craessaerts K, Mathews PM, Seeger M, et al. 1999. *Mol. Med.* 5:542–54
129. Dowjat WK, Wisniewski H, Wisniewski T. 2001. *Neuroscience* 103:1–8
130. Johnsingh AA, Johnston JM, Merz G, Xu J, Kotula L, et al. 2000. *FEBS Lett.* 465:53–58
131. Pigino G, Morfini G, Pelsman A, Mattson MP, Brady ST, Busciglio J. 2003. *J. Neurosci.* 23:4499–508
132. Morfini G, Szebenyi G, Elluru R, Ratner N, Brady ST. 2002. *EMBO J.* 21:281–93
133. Cai D, Leem JY, Greenfield JP, Wang P, Kim BS, et al. 2003. *J. Biol. Chem.* 278:3446–54
134. Saura CA, Chen G, Malkani S, Choi SY, Takahashi RH, et al. 2005. *J. Neurosci.* 25:6755–64

135. Thinakaran G, Regard JB, Bouton CM, Harris CL, Price DL, et al. 1998. *Neurobiol. Dis.* 4:438–53
136. Xia W, Zhang J, Perez R, Koo EH, Selkoe DJ. 1997. *Proc. Natl. Acad. Sci. USA* 94:8208–13
137. Waragai M, Imafuku I, Takeuchi S, Kanazawa I, Oyama F, et al. 1997. *Biochem. Biophys. Res. Commun.* 239:480–82
138. Verdile G, Martins RN, Duthie M, Holmes E, St George-Hyslop PH, Fraser PE. 2000. *J. Biol. Chem.* 275:20794–98
139. Pitsi D, Kienlen-Campard P, Octave JN. 2002. *J. Neurochem.* 83:390–99
140. Rockenstein E, Mante M, Alford M, Adame A, Crews L, et al. 2005. *J. Biol. Chem.* 280:32957–67
141. Lee EB, Zhang B, Liu K, Greenbaum EA, Doms RW, et al. 2005. *J. Cell. Biol.* 168:291–302
142. Caceres A, Potrebic S, Kosik KS. 1991. *J. Neurosci.* 11:1515–23
143. Dickson DW, Farlo J, Davies P, Crystal H, Fuld P, Yen SH. 1988. *Am. J. Pathol.* 132:86–101
144. Masliah E, Mallory M, Deerinck T, DeTeresa R, Lamont S, et al. 1993. *J. Neuropathol. Exp. Neurol.* 52:619–32
145. Yasuhara O, Kawamata T, Aimi Y, McGeer EG, McGeer PL. 1994. *Neurosci. Lett.* 171:73–76
146. Minamide LS, Striegl AM, Boyle JA, Meberg PJ, Bamburg JR. 2000. *Nat. Cell Biol.* 2:628–36
147. Kowall NW, Kosik KS. 1987. *Ann. Neurol.* 22:639–43
148. Munoz DG, Wang D. 1992. *Am. J. Pathol.* 140:1167–78
149. Booze RM, Mactutus CF, Gutman CR, Davis JN. 1993. *J. Neurol. Sci.* 119:110–18
150. Geula C, Mesulam MM. 1996. *Cereb. Cortex* 6:165–77
151. Davies DC, Brooks JW, Lewis DA. 1993. *Neurobiol. Aging* 14:353–57
152. Sadun AA, Bassi CJ. 1990. *Ophthalmology* 97:9–17
153. Syed AB, Armstrong RA, Smith CU. 2005. *Folia Neuropathol.* 43:1–6
154. Meyerhoff DJ, MacKay S, Constans JM, Norman D, Van Dyke C, et al. 1994. *Ann. Neurol.* 36:40–47
155. Stahl R, Dietrich O, Teipel S, Hampel H, Reiser MF, Schoenberg SO. 2003. *Radiologe* 43:566–75
156. Lakmache Y, Lassonde M, Gauthier S, Frigon JY, Lepore F. 1998. *Proc. Natl. Acad. Sci. USA* 95:9042–46
157. Boutajangout A, Authelet M, Blanchard V, Touchet N, Tremp G, et al. 2004. *Neurobiol. Dis.* 15:47–60
158. Spittaels K, Van den Haute C, Van Dorpe J, Bruynseels K, Vandezande K, et al. 1999. *Am. J. Pathol.* 155:2153–65
159. Masliah E, Sisk A, Mallory M, Mucke L, Schenk D, Games D. 1996. *J. Neurosci.* 16:5795–811
160. Cras P, Kawai M, Lowery D, Gonzalez-DeWhitt P, Greenberg B, Perry G. 1991. *Proc. Natl. Acad. Sci. USA* 88:7552–56
161. Sennvik K, Bogdanovic N, Volkmann I, Fastbom J, Benedikz E. 2004. *J. Cell Mol. Med.* 8:127–34
162. Takahashi RH, Almeida CG, Kearney PF, Yu F, Lin MT, et al. 2004. *J. Neurosci.* 24:3592–99
163. Dessi F, Colle MA, Hauw JJ, Duyckaerts C. 1997. *Neuroreport* 8:3685–89
164. Nakamura Y, Takeda M, Yoshimi K, Hattori H, Hariguchi S, et al. 1994. *Acta Neuropathol.* 87:23–31

165. Burke WJ, Park DH, Chung HD, Marshall GL, Haring JH, Joh TH. 1990. *Brain Res.* 537:83–87
166. Mann DM, Sumpter PQ, Davies CA, Yates PO. 1987. *Acta Neuropathol.* 73:181–84
167. Richard S, Brion JP, Couck AM, Flament-Durand J. 1989. *J. Submicrosc. Cytol. Pathol.* 21:461–67
168. Price DL, Altschuler RJ, Struble RG, Casanova MF, Cork LC, Murphy DB. 1986. *Brain Res.* 385:305–10
169. Praprotnik D, Smith MA, Richey PL, Vinters HV, Perry G. 1996. *Acta Neuropathol.* 91:226–35
170. Pike CJ, Cummings BJ, Cotman CW. 1992. *Neuroreport* 3:769–72
171. Hiruma H, Katakura T, Takahashi S, Ichikawa T, Kawakami T. 2003. *J. Neurosci.* 23:8967–77
172. Kasa P, Papp H, Kovacs I, Forgon M, Penke B, Yamaguchi H. 2000. *Neurosci. Lett.* 278:117–19
173. Stamer K, Vogel R, Thies E, Mandelkow E, Mandelkow EM. 2002. *J. Cell. Biol.* 156:1051–63
174. Trojanowski JQ, Smith AB, Huryn D, Lee VM. 2005. *Expert Opin. Pharmacother.* 6:683–86
175. Zhang B, Maiti A, Shively S, Lakhani F, McDonald-Jones G, et al. 2005. *Proc. Natl. Acad. Sci. USA* 102:227–31
176. Aston-Jones G, Rogers J, Shaver RD, Dinan TG, Moss DE. 1985. *Nature* 318:462–64
177. Nierenberg J, Pomara N, Hoptman MJ, Sidtis JJ, Ardekani BA, Lim KO. 2005. *Neuroreport* 16:1369–72
178. Mufson EJ, Kordower JH. 1992. *Proc. Natl. Acad. Sci. USA* 89:569–73
179. Mufson EJ, Conner JM, Kordower JH. 1995. *Neuroreport* 6:1063–66
180. Geula C, Mesulam MM. 1989. *Neuroscience* 33:469–81
181. Dai J, Buijs RM, Kamphorst W, Swaab DF. 2002. *Brain Res.* 948:138–44
182. Roediger B, Armati PJ. 2003. *Neurobiol. Dis.* 13:222–29
183. Carson C, Saleh M, Fung FW, Nicholson DW, Roskams AJ. 2005. *J. Neurosci.* 25:6092–104
184. Delcroix JD, Valletta J, Wu C, Howe CL, Lai CF, et al. 2004. *Prog. Brain Res.* 146:3–23
185. Dhaenens CM, Van Brussel E, Schraen-Maschke S, Pasquier F, Delacourte A, Sablonniere B. 2004. *Neurosci. Lett.* 368:290–92

Asparagine Synthetase Chemotherapy

Nigel G. J. Richards[1] and Michael S. Kilberg[2]

[1]Department of Chemistry and [2]Department of Biochemistry and Molecular Biology, University of Florida, Gainesville, Florida 32611; email: richards@qtp.ufl.edu, mkilberg@ufl.edu

Key Words

amino acids, asparaginase resistance, drug discovery, enzyme inhibitors, leukemia

Abstract

Modern clinical treatments of childhood acute lymphoblastic leukemia (ALL) employ enzyme-based methods for depletion of blood asparagine in combination with standard chemotherapeutic agents. Significant side effects can arise in these protocols and, in many cases, patients develop drug-resistant forms of the disease that may be correlated with up-regulation of the enzyme glutamine-dependent asparagine synthetase (ASNS). Though the precise molecular mechanisms that result in the appearance of drug resistance are the subject of active study, potent ASNS inhibitors may have clinical utility in treating asparaginase-resistant forms of childhood ALL. This review provides an overview of recent developments in our understanding of (*a*) the structure and catalytic mechanism of ASNS, and (*b*) the role that ASNS may play in the onset of drug-resistant childhood ALL. In addition, the first successful, mechanism-based efforts to prepare and characterize nanomolar ASNS inhibitors are discussed, together with the implications of these studies for future efforts to develop useful drugs.

Contents

- INTRODUCTION 630
- STRUCTURE AND MECHANISM OF ASPARAGINE SYNTHETASE 631
  - Structure of Asparagine Synthetase 631
  - Kinetic Mechanism of Asparagine Synthetase 633
- ASPARAGINE SYNTHETASE AND DRUG-RESISTANT LEUKEMIA 634
  - Asparagine Synthetase Expression and the Cell Cycle 634
  - Amino Acid Control of ASNS Gene Transcription 635
  - Asparagine Synthetase Expression and Asparaginase Therapy of Childhood ALL 637
- SYNTHESIS AND CHARACTERIZATION OF ASPARAGINE SYNTHETASE INHIBITORS 639
  - Evidence Supporting the Clinical Utility of Asparagine Synthetase Inhibitors 639
  - Using Structural Homology and Chemical Constraints to Model the Synthetase Active Site 640
  - Sulfonamide Derivatives as Inhibitors of Human Asparagine Synthetase 641
  - A Nanomolar Inhibitor of Human Asparagine Synthetase 643
- PERSPECTIVES 646

ASNS: asparagine synthetase

INTRODUCTION

The remarkable inverse correlation between the susceptibility of leukemia cells to drug therapy and their capacity for intracellular asparagine biosynthesis was first described almost forty years ago (1). This observation provides a rationale for the current widespread use of L-asparaginase (ASNase) in chemotherapeutic protocols for treating childhood acute lymphoblastic leukemia (ALL) and some forms of acute myeloblastic leukemia (AML) (2–7). The molecular basis for the therapeutic utility of ASNase remains ill defined (8), although it is believed that normal and malignant lymphocytes depend on the uptake of asparagine from circulating plasma for growth (9). ASNase therefore likely exerts its effects indirectly by depleting asparagine in the blood, which is followed by an efflux of cytoplasmic asparagine from leukemic blasts (1). The use of *Escherichia coli* ASNase as a single agent leads to nearly complete remission in 40% to 60% of cases of ALL (10, 11), and, in combination with vincristine and prednisone, increases the remission rate up to 95% in cases of childhood ALL. Unfortunately, three factors limit the clinical utility of ASNase in cancer therapy (8, 12). First, the treatment produces a wide variety of side effects, including immunosuppression and pancreatitis (13, 14). Second, 10% to 12% of patients who achieve remission suffer a relapse with tumors that are resistant to further ASNase therapy (5, 14–16). Finally, ASNase administration may enhance the growth of resistant tumors and increase their metastatic activity (10, 17). The molecular basis of ASNase resistance, which is a major clinical problem, remains poorly understood despite a significant amount of ongoing research (8, 18). Because ASNase sensitivity in tumors cannot yet be predicted reliably, the major use of this enzyme remains confined to the treatment of childhood ALL, despite estimates that 5–10% of all solid tumors may be sensitive to therapies based upon the depletion of blood asparagine (16).

Human asparagine synthetase (ASNS) catalyzes the biosynthesis of L-asparagine from L-aspartate in an ATP-dependent reaction for which L-glutamine is the nitrogen source under physiological conditions (**Scheme 1**) (19). Recent work has demonstrated the importance of ASNS overexpression in conferring ASNase resistance in cell lines (20), and several lines of evidence suggest that

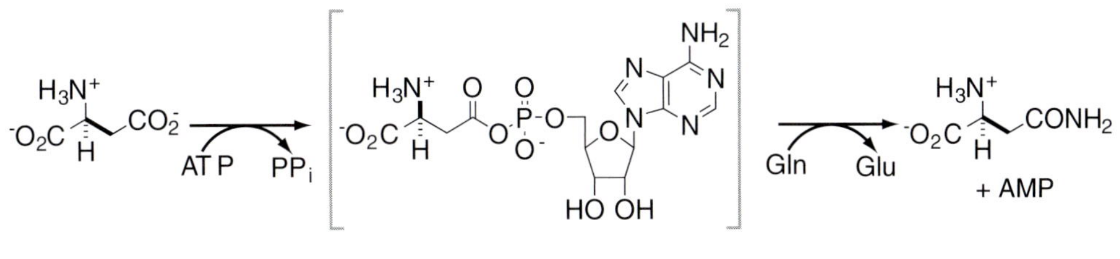

Scheme 1

Overall transformation catalyzed by glutamine-dependent ASNS, showing the β-aspartyl-AMP intermediate. Ammonia may replace glutamine as a nitrogen source in vitro.

inhibiting ASNS activity represents a viable strategy for treating ASNase-resistant leukemias in the clinic (1, 8, 21). Early large-scale screening studies employing a range of substrate and product analogs failed, however, to identify potent and selective ASNS inhibitors (22, 23). In part, the failure of these efforts reflected a lack of detailed knowledge concerning the structure of human ASNS and its functional role in cellular metabolism. Considerable progress has been made in all of these areas over the past few years, and several recent advances have set the stage for the identification and characterization of the first nanomolar inhibitors of human ASNS (24). This review provides an overview of recent developments in understanding (*a*) the structure and catalytic mechanism of ASNS, and (*b*) the role that ASNS plays in the onset of drug-resistant ALL. In addition, the molecular principles that have been employed to discover and characterize two potent classes of ASNS inhibitors are outlined, together with the implications of these studies for future efforts to develop clinically useful drugs against ASNase-resistant leukemia. Additional information, especially pertaining to the catalytic mechanism of ASNS, can be found in a previous review (19).

STRUCTURE AND MECHANISM OF ASPARAGINE SYNTHETASE

The use of rational, structure-based methods for obtaining small molecule enzyme inhibitors requires a detailed understanding of critical transition states and reaction intermediates that are bound tightly by the enzyme during catalysis. In addition, the identification of active site residues that have functional roles in the catalytic mechanism provides a set of "contact points" that can be exploited in inhibitor design because mutation of these residues will generally lead to loss of function in the protein. We therefore begin this review with an overview of ASNS structure and catalytic mechanism.

Structure of Asparagine Synthetase

The detailed kinetic and structural characterization of ASNS from mammalian sources historically proved to be difficult because of both reported low abundance and instability of the native enzyme during purification (25–29). In addition, only small amounts of recombinant, wild-type human ASNS could be obtained in a variety of early expression systems, the enzyme being purified to homogeneity using an affinity chromatography protocol (30–32). Access to recombinant human enzyme, albeit in small quantities, in these experiments did permit limited insights into properties such as substrate specificity (31, 32). As a result, until the very recent development of an efficient expression system for human ASNS (*vide infra*) (33), procedures for obtaining large amounts of the glutamine-dependent ASNS (AS-B) encoded by the asnB gene in *Escherichia coli* (34) were essential to detailed investigations of the structure and mechanism of the enzyme.

The C1A mutant of AS-B, in which the N-terminal cysteine residue is substituted by

AS-B: *Escherichia coli* glutamine-dependent ASNS

Prednisone: an orally active, synthetic corticosteroid used to suppress the immune system

L-asparaginase (ASNase): a serine-dependent hydrolase that catalyzes the hydrolysis of L-asparagine to L-aspartic acid

Acute lymphoblastic leukemia (ALL): a disease in which patients produce primitive lymphoid cells instead of cells that would normally develop along the lymphoid lineage into mature B- or T-lymphocytes

Acute myeloblastic leukemia (AML): a disease in which patients produce cancerous primitive cells instead of cells that would normally develop along the myeloid lineage into myeloid white blood cells

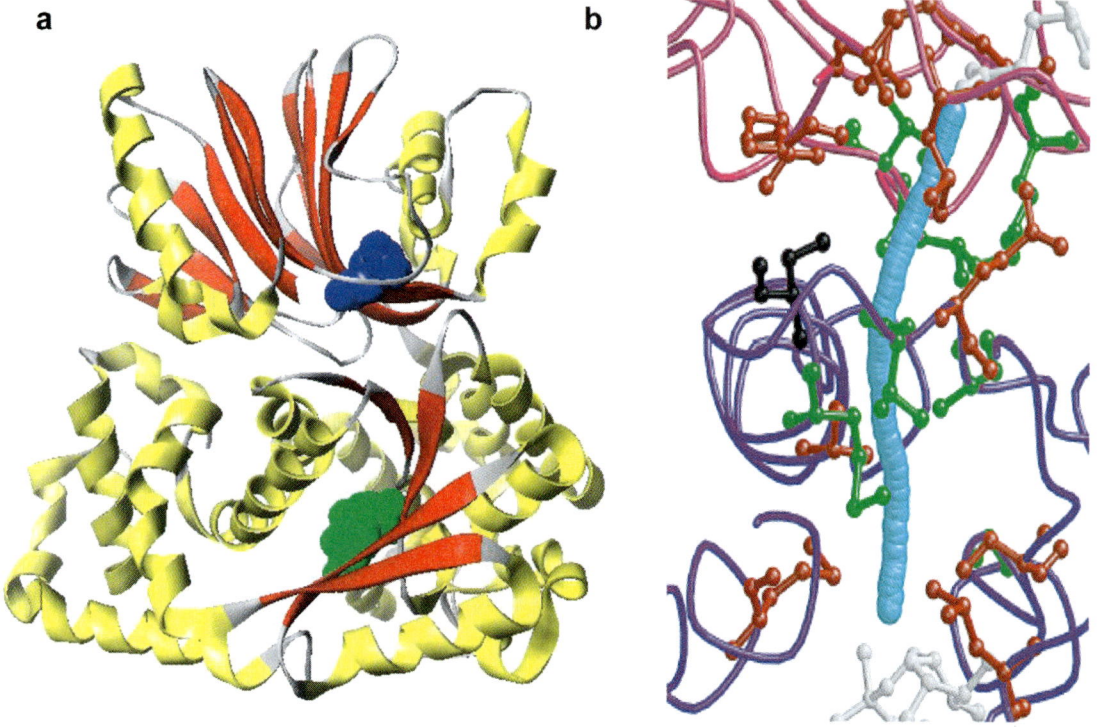

Figure 1

(*a*) Cartoon representation of the structure of the Cys-1-Ala mutant of *Escherichia coli* AS-B complexed with glutamine (*blue space-filling model*) and AMP (*green space-filling model*) showing the domain organization of the enzyme (35). Helices and β-strands are shown in yellow and red, respectively. The final 40 C-terminal residues are not observed in the crystal structure, presumably due to their disordered conformation in the absence of bound aspartate. (*b*) Cartoon showing the putative pathway by which ammonia (*light blue spheres*) travels between the glutaminase (*top*) and the synthetase (*bottom*) active sites in AS-B. The side chains of residues defining the ammonia tunnel that are variable and conserved throughout the family of known asparagine synthetases are colored green and red, respectively. Bound glutamine (*top*) and AMP (*bottom*) are rendered as gray-white "ball-and-stick" models. Reprinted from (63), Copyright 2003, with permission from Elsevier.

Vincristine: an alkaloid that binds to tubulin monomers thereby preventing the formation of spindle microtubules and stopping separation of duplicated chromosomes

Ntn: N-terminal amidohydrolase

alanine, exhibits no glutamine-dependent activity (19) but retains substantial affinity for L-glutamine (K_D of approximately 6 μM). Therefore this AS-B mutant could be crystallized as its ternary complex with glutamine and AMP, and the crystal structure of the complex determined to a resolution of 2.0 Å (**Figure 1***a*) (35). As expected on the basis of sequence alignment (36), and experimental studies of bovine ASNS using monoclonal antibodies (37), this structure showed that glutamine-dependent ASNS is composed of two distinct domains, each containing a separate catalytic site. The N-terminal domain of the enzyme has a tertiary structure that is observed in members of the N-terminal amidohydrolase (Ntn) enzyme superfamily (38, 39), which mediates a number of biologically important hydrolytic reactions (40–42). The N-terminal active site therefore catalyzes glutamine hydrolysis to yield glutamate and ammonia, and glutamine-dependent ASNS is classified as a Class II, or Ntn, amidotransferase (43). Other Ntn amidotransferases possessing structurally homologous glutamine-hydrolyzing domains include glutamine 5'-phosphoribosyl-1-pyrophosphate amidotransferase (GPATase) (43, 44) and glutamine fructose-6-phosphate amidotransferase (GFAT) (45, 46), which both play important functions in cellular metabolism. The mechanism of ASNS-catalyzed glutamine hydrolysis is discussed in detail elsewhere (19, 43). The complex three-dimensional fold of the C-terminal domain of AS-B is similar to that observed in ATP pyrophosphatases (47), an enzyme superfamily that includes guanosine-5′-monophosphate synthetase (GMPS) (48), argininosuccinate synthetase (49), ATP sulfurylase (50), carbapenam synthetase (51), β-lactam synthetase (BLS) (52, 53), and ThiI (54). Because all ATP pyrophosphatases convert ATP to AMP, the C-terminal active site of ASNS likely catalyzes activation of the side-chain carboxylate of aspartate to form an electrophilic intermediate, β-aspartyl-AMP (βAspAMP) 1, and inorganic pyrophosphate (PP$_i$) (**Scheme 2**) (28, 55). As observed in other glutamine-dependent amidotransferases (44, 46, 56–61), the two active sites of AS-B are linked by a solvent-inaccessible, intramolecular "tunnel" that is sufficiently wide to allow passage of an ammonia molecule (**Figure 1b**) (62). Glutamine-dependent asparagine production is therefore accomplished using ammonia as a common intermediate to couple the two "half-reactions" carried out in the independent active sites of the enzyme. Thus, after being released in, and channeled from, the glutaminase site, a molecule of ammonia attacks bound βAspAMP 1 to give asparagine and AMP via a tetrahedral intermediate (**Scheme 2**).

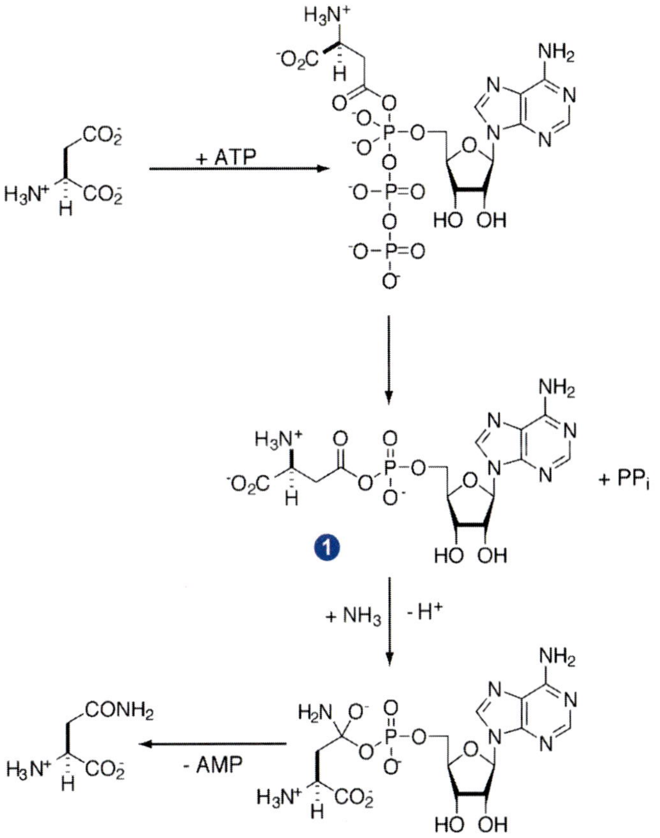

Kinetic Mechanism of Asparagine Synthetase

Given that all Ntn amidotransferases share a common N-terminal glutaminase domain, the development of selective ASNS inhibitors requires that they be targeted against the C-terminal, synthetase active site. As a result, recent work in this area has been concerned with obtaining a detailed understanding of (*a*) the steady-state kinetic mechanism

Scheme 2

Hypothetical mechanism for the ASNS-catalyzed formation of βAspAMP 1 and its subsequent reaction with ammonia to form asparagine and AMP.

for glutamine-dependent asparagine production (55, 63, 64), and (*b*) critical intermolecular interactions that play a role in mediating the synthesis of βAspAMP and its subsequent reaction with ammonia (**Scheme 2**). Although the general properties of ASNS from a variety of sources appear very similar, many different kinetic mechanisms have been proposed for their glutamine-dependent synthetase activity (65–67). As a result, it is only recently that a consensus has begun to emerge on the general details of substrate binding order and product release, particularly because kinetic analysis of ASNS-catalyzed asparagine formation is complicated by the high glutaminase activity of the enzyme. Steady-state experiments on AS-B

GMPS: guanosine-5′-monophosphate synthetase

BLS: β-lactam synthetase

βAspAMP: β-aspartyl-AMP

PP$_i$: inorganic pyrophosphate

BHK *ts11* cells: a hamster cell line that expresses a mutant form of ASNS and, consequently, is blocked at G1 in the cell cycle

(55), in combination with isotope partitioning techniques (68) and measurements of product stoichiometry, gave rise to an initial hypothesis that βAspAMP formation commits the enzyme to asparagine synthesis. Such a proposal is analogous to the kinetic mechanism of certain aminoacyl tRNA synthetases in which an acyl-AMP intermediate is synthesized prior to interaction of the enzyme with uncharged tRNA (69, 70). The question of when glutamine or ammonia binds to AS-B was experimentally unresolved in these studies, however (55), and it was proposed that glutamine (and therefore ammonia) binds to the E.βAspAMP complex after PP_i release. If correct, this order of catalytic events implies that βAspAMP must be stabilized within the synthetase site so as to prevent futile ATP hydrolysis prior to binding of the nitrogen source for asparagine production. On the other hand, numerical simulations (71) showed that the original kinetic model for AS-B (55) was inconsistent with the dependence of the glutamate/asparagine ratio upon glutamine concentration at saturating levels of ATP and aspartate. In addition, and as observed for other glutamine-dependent amidotransferases (36, 72), ASNS does not catalyze ATP/PP_i exchange (55). A new kinetic model has been recently developed that appears consistent with all known experimental data for the bacterial enzyme (63). Hence, in addition to suggesting that βAspAMP formation commits the enzyme to asparagine synthesis, implying that ammonia transfer from the N-terminal domain through the intramolecular tunnel is totally efficient, the model provides evidence for the hypothesis that glutamine can bind to the E.Asp.ATP ternary complex to yield a quaternary complex from which glutamate and ammonia can be released. Thus, coordination of catalytic activity in the two active sites of AS-B during glutamine-dependent asparagine synthesis appears to be remarkably small prior to βAspAMP formation, which is in sharp contrast to the coupling of glutaminase and synthetase activities seen for other Ntn glutamine-dependent amidotransferases (73–75). The lack of ATP/PP_i exchange has been rationalized by assuming that PP_i is released as the final product from the enzyme. Although this conclusion is supported by crystal structures of the BLS/CMA-AMP/PP_i and BLS/DGPC/PP_i complexes (53), and structural models for the AS-B/βAspAMP/PP_i complex (*vide infra*) in which PP_i is located deep in the active site cleft covered by the other ligands, PP_i release is reported to occur prior to glutamine binding for *Vibrio cholerae* ASNS (64). Whatever the timing of product release, the latest kinetic model (63) supports the hypothesis that ASNS must bind βAspAMP with high affinity, raising the possibility that stable analogs of this intermediate might be potent ASNS inhibitors. In addition, it seems likely that the enzyme also stabilizes the transition state for addition of ammonia to βAspAMP 1 (**Scheme 2**). As a consequence, compounds that mimic this transition state may also have significant potential as clinically useful drugs (76, 77).

ASPARAGINE SYNTHETASE AND DRUG-RESISTANT LEUKEMIA

Children with acute lymphoblastic leukemia (ALL) are treated with a multidrug regimen that includes the enzyme L-asparaginase (AS-Nase). Although modern therapeutic protocols lead to remission rates of greater than 80%, relapse and drug resistance remain a problem. Consequently, the relationship between the expression of the ASNS and development of ASNase resistance is of interest from the viewpoint of both metabolic regulatory mechanisms and development of new therapeutic strategies.

Asparagine Synthetase Expression and the Cell Cycle

Basilico and colleagues determined that ASNS could complement temperature-sensitive hamster BHK *ts11* cells, which are specifically blocked in progression through the G_1 phase of the cell cycle when grown at the nonpermissive temperature (78, 79). Those authors showed that because of a point mutation in the ASNS gene, at the nonpermissive temperature, the BHK *ts11* cells produce an inactive enzyme (79). This loss of ASNS activity leads to cell cycle arrest, as a consequence of a depletion of cellular asparagine, and a corresponding increase in ASNS mRNA due to regulatory mechanisms described below. Both of these effects could be reversed by the addition of exogenous asparagine to the cells maintained at the nonpermissive temperature. A link between asparagine content and the cell cycle is illustrated by several additional observations. For example, ASNS mRNA expression was substantially increased during the G_1 phase of the cell cycle by refeeding serum-deprived Balb/c 3T3 cells. After serum repletion, ASNS mRNA was induced, but the addition of asparagine to the culture medium could prevent the induction (80). Hongo et al. (81) showed that ASNS activity is induced during lymphocyte activation by phytohemagglutinin, and the increase in activity coincides with the rate of DNA synthesis. Likewise, Cirafici (82) showed that thyroid-stimulating hormone treatment of quiescent rat thyroid cells causes entry into the S phase and there was a concurrent increase in ASNS mRNA content. Collectively, these observations suggest that ASNS expression is linked to cell growth and ASNS mRNA content is controlled in accordance with changes in the cell cycle.

Amino Acid Control of ASNS Gene Transcription

Amino acids are known to modulate a number of fundamental processes in mammalian cells, especially with regard to the "central dogma" of DNA to RNA to protein (83). Although circulating amino acids and intracellular protein turnover both act to buffer cells from variations in dietary protein/amino acid intake, fluctuations in the intracellular levels of individual amino acids do occur in response to diet, disease, and metabolic status. In this context, amino acids serve as signal transduction messengers to transmit the nutritional status of the entire organism to individual cells. One of the mechanisms by which amino acids mediate this signaling is altered transcription for specific genes via a signal transduction process referred to as the amino acid response (AAR) pathway. Detection of a limiting amount of any single amino acid has been linked to a ribosome-associated kinase, GCN2, that binds and, therefore, monitors the level of uncharged tRNAs (84–86). Starvation-activated GCN2 phosphorylates the eukaryotic initiation factor eIF-2α and, as a consequence, global translation initiation is suppressed. However, certain mRNAs that contain short upstream open reading frames exhibit increased translation under amino acid limiting conditions. An example in yeast is the transcription factor GCN4 (84), which has been reported to alter the transcription rate of several hundred genes in response to amino acid limitation (87). The mammalian counterpart to yeast GCN4 appears to be the basic leucine zipper (bZIP) transcription factor ATF4. The translation of pre-existing ATF4 mRNA is rapidly increased following amino acid deprivation (88–90), and ATF4 protein has been shown to mediate the increased transcription of AAR pathway target genes, including ASNS (91, 92).

Gong et al. (93) were the first to determine that ASNS mRNA content increased in cells deprived of amino acids, and subsequently, Hutson & Kilberg (94) also demonstrated increased ASNS mRNA content following total amino acid deprivation or depletion of a single essential amino acid. In this circumstance, the definition of "essential amino acid" must be considered in the context of individual cell types rather than the entire organism. Guerrini et al. (95) analyzed the ASNS gene and determined that an amino acid response element (AARE) was present in the promoter. In vivo footprinting by Barbosa-Tessmann et al. (96) identified

Amino acid response (AAR) pathway: the pathway by which mammalian cells sense and respond to a deficiency of protein/amino acid

bZIP: basic leucine zipper transcription factor

NSRE: nutrient-sensing response element

UPR: unfolded protein response

EMSA: electrophoresis mobility shift analysis

ChIP: chromatin immunoprecipitation

five protein binding sites within the ASNS proximal promoter region that contribute to nutrient control of the human ASNS gene, three GC-rich sequences (GC-I, GC-II, and GC-III) and two sequences, originally labeled sites V and VI, and later renamed Nutrient Sensing Response Elements (NSRE-1, NRSE-2). The GC-rich sites are necessary to maintain the basal transcription rate and to permit maximal activation of the ASNS gene by the AAR pathway (97). Interestingly, expression of either Sp1 or Sp3 in *Drosophila* SL2 cells supported basal ASNS promoter activity, but only Sp3 expression permitted the starvation-induced ASNS-driven transcription. The location of the NSRE-1 footprint overlapped with the sequence identified by Guerrini et al. (95). Mutagenesis confirmed the presence of a second AARE, NSRE-2 (5'-GTTACA-3', nt –48 to –43), positioned eleven nucleotides downstream of NSRE-1. Mutagenesis studies have documented that these two elements must be aligned on the same side of the DNA helix and only one turn away from each other (98), presumably to permit protein-protein interactions that occur between the transcription factors that bind to these two sites. Barbosa-Tessmann et al. (96, 99) demonstrated that both NSRE-1 and NSRE-2 were required for activation of the human ASNS gene following activation by either the AAR pathway or an ER stress pathway referred to as the unfolded protein response (UPR). Experimentally, the UPR pathway is often activated by conditions that perturb ER calcium levels (thapsigargin treatment) or conditions that cause accumulation of misfolded glycoproteins (glucose starvation or tunicamycin treatment). Electrophoresis mobility shift analysis (EMSA) experiments revealed increased nuclear protein binding to the ASNS NSRE-1 binding site (5'-TGATGAAAC-3', nt –68 to –60) when extracts from either amino acid (AAR pathway) or glucose-deprived (UPR pathway) cells were tested (96). This broader nutrient detecting capability is the reason that the ASNS sites are referred to as nutrient-sensing response elements (NSRE), rather than solely as an AARE.

The binding proteins for the NSRE-2 site have not yet been identified, whereas progress has been made in identifying those for NSRE-1. In vitro binding analysis by EMSA revealed ATF4 binding to the NSRE-1 sequence that was increased when nuclear extracts from either histidine-deprived (AAR pathway) or glucose-deprived (UPR pathway) cells were tested (91). Consistent with translational control of ATF4 production following activation of the AAR pathway, the inhibition of protein synthesis blocked the starvation-dependent enhancement of ATF4/NSRE-1 complex formation. ATF4-dependent regulation of ASNS expression in vivo was suggested by the observation that ASNS promoter-driven transcription was induced by ATF4 overexpression (91, 92). Subsequently, chromatin immunoprecipitation (ChIP) analysis documented that there is increased ATF4 binding to the ASNS promoter in vivo following amino acid deprivation (92).

A survey of bZIP transcription factors by both EMSA experiments that ATF3 (91, 100) and C/EBPβ (101) also exhibited affinity for the ASNS NSRE-1 site and transient expression showed that both can modulate transcription driven by the ASNS promoter. Transient expression studies using combinations of ATF4, ATF3, and C/EBPβ then implied that ATF3 serves as the primary antagonist to ATF4 function (92, 102). Consistent with this hypothesis, Fawcett et al. (103) used nuclear extracts from arsenite-treated cells in EMSA studies to show a transient increase in ATF4 binding activity with affinity for a *cis*-element in the human C/EBP homology protein (CHOP) promoter referred to as a C/EBP-ATF composite site. ATF4 binding activity peaked at 2 h after arsenite exposure, but that activity was subsequently replaced at about 6 h by elevated ATF3 binding activity, a time at which the elevated transcription rate from the gene was declining back toward the basal rate. Those authors also used overexpression studies to document that elevated ATF4 activates the CHOP gene through the C/EBP-ATF composite site and that increased ATF3 production antagonizes the ATF4 function. The CHOP sequence (5'-TGATGCAAT-3') that Fawcett et al. (103) identified is only two nucleotides different than the ASNS NSRE-1 site, which is also a C/EBP-ATF composite site. Chen et al. (92) extended the observations of Fawcett et al. (103) by showing through ChIP analysis that a similar sequence of events, i.e., activation via ATF4 and subsequent antagonism by ATF3, can be observed in vivo at the ASNS NSRE-1 site.

Based on the in vivo ChIP analysis, Chen et al. (92) have proposed a working model for ASNS transcription that describes two distinct phases in response to amino acid limitation (**Figure 2**). Phase I encompasses the first 4 h after amino acid withdrawal and phase II covers the time from 4–24 h. Within 30 min of amino acid depletion, translational control of ATF4 mRNA results in increased de novo synthesis and subsequent binding of ATF4 to the NSRE-1 site (92). In parallel with ATF4 binding, acetylation of histones H3 and H4 is increased and the general transcription factors TBP and TFIIB, as well as RNA polymerase II, are recruited to the promoter. It is assumed that the ATF4 complex requires a coactivator and/or other bridging proteins to the general transcription machinery, which are shown as gray modules in the model (**Figure 2**). During phase I of amino acid deprivation, a low level of C/EBPβ is constitutively bound to the ASNS promoter, whereas in phase II, C/EBPβ de novo synthesis and subsequent binding increases at a time when the transcription rate has peaked. Likewise, the synthesis and action of ATF3 also increases during this period (92). It is proposed that ATF3 and C/EBPβ act in concert to suppress, but not to completely reverse, the increased ASNS transcription during phase II, such that even out to 24 h of amino acid deprivation, transcription remains elevated relative to the status of the gene in amino acid-complete medium (92).

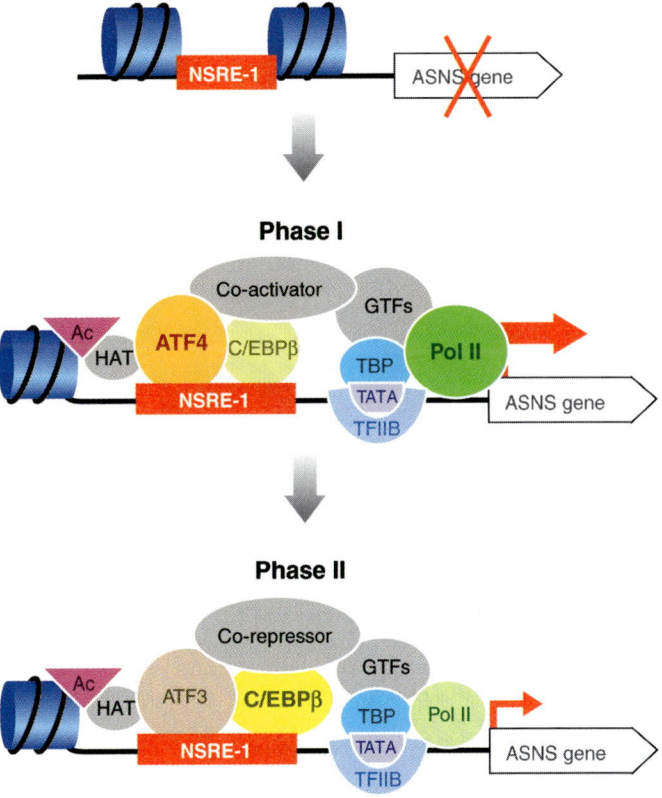

Figure 2

A working model for control of the asparagine synthetase (ASNS) gene by the AAR or UPR pathways (92). Transcription factors shown in color have been localized to the ASNS promoter by chromatin immunoprecipitation analysis. Unidentified or putative components are shown in gray. Transcription from the ASNS gene reaches its highest rate at 1–4 h (phase I) following nutrient stress. ASNS transcription is still elevated relative to the "fed" state between 4–24 h (phase II) following nutrient stress, but the rate is reduced.

Asparagine Synthetase Expression and Asparaginase Therapy of Childhood ALL

Cancer cells exhibit rapid growth and cell division, and therefore have an increased nutritional need. Consequently, therapies have been developed to take advantage of this dependency on circulating nutrients in those cases where critical enzymes are either not expressed at sufficient levels or can be selectively inhibited. As mentioned above, childhood ALL is treated using a combination of chemotherapeutic drugs that include the

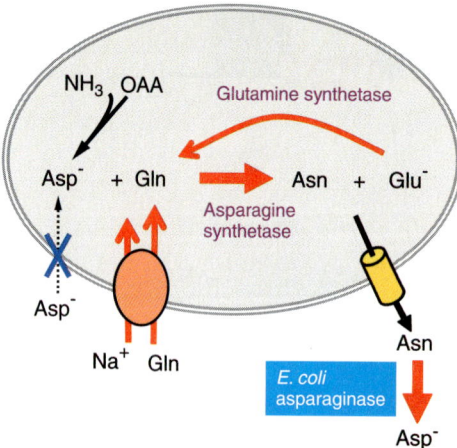

Figure 3

Selected metabolic changes that take place in ASNase-resistant MOLT-4 leukemia cells (123). Treatment with ASNase causes a rapid degradation of extracellular asparagine (Asn) and a subsequent depletion of intracellular Asn. Compensatory changes, shown with red arrows, include increases in: transcription from the asparagine synthetase gene, glutamine synthetase activity (post-transcriptional), and active glutamine transport. Conversely, there is a decrease, shown in yellow, in Asn efflux through Na^+-independent exchange. There is little or no aspartate uptake by these cells, so synthesis via transamination may play a role in supplying this substrate for the ASNS-catalyzed reaction.

MOLT-4 cells: a T cell–derived leukemic cell line isolated from a patient with ALL

enzyme ASNase (8, 13, 104, 105). The logic of ASNase therapy is that by delivering ASNase to the bloodstream plasma asparagine is quickly depleted causing a rapid efflux of cellular asparagine, which is also destroyed, and thus, the cells of the entire body are depleted of asparagine. Most cells express sufficient ASNS to counteract this asparagine starvation and survive, but in general, childhood ALL cells express ASNS at a low level, and therefore, treatment with ASNase is extremely effective in blocking growth of this particular form of leukemia. The bacterially derived ASNase enzymes currently used clinically also have an inherent glutaminase activity that is about 2% to 3% of the ASNase activity (19). As a result, because L-glutamine is the nitrogen donor for the ASNS-catalyzed reaction (see above), the depletion of this amino acid may also play a role in ASNase action (92).

Hutson et al. (108) showed that exposure to ASNase of a cultured human ALL-derived cell line, MOLT-4, can result in selection for a population of ALL cells that are drug resistant (107). Whether or not these resistant cells exist within the parental population or if they are induced by ASNase treatment is not yet known. Aslanian & Kilberg (109) documented that many adaptive changes occur in the ASNase-resistant cells to alter amino acid transport and metabolism (**Figure 3**). Collectively, these changes appear to support increased asparagine biosynthesis by increasing the intracellular levels of the ASNS substrates aspartate and glutamine, and by enhancing uptake of asparagine and glutamine via secondary active Na^+-dependent transporters (109). For example, the enzymatic activity of glutamine synthetase is increased, presumably by a post-transcriptional mechanism, as glutamine synthetase mRNA is not changed. The ASNase-resistant cells also exhibit reduced efflux of asparagine via Na^+-independent transporters. These drug-resistant MOLT-4 cells express elevated levels of ASNS mRNA and protein (108, 109), which Aslanian et al. (20) demonstrated remain elevated, even after removal of the ASNase from the culture medium for 6 weeks or more. In a result consistent with these observations it was also established that ASNase resistance, analyzed by cell growth or apoptosis following a drug challenge, was not completely reversible following long-term culture in the absence of ASNase (20). The relationship between ASNase resistance and increased ASNS expression had been noted before in non-ALL cell types (15, 110, 111), leading to the assumption that resistance to this particular drug was the result of a compensatory increase in ASNS expression. However, it was possible that elevated ASNS was a consequence of ASNase resistance rather than the cause. To establish which of these hypotheses was true, Aslanian et al. (20) showed that overexpression of ASNS in the ASNase-sensitive parental MOLT-4 cells caused these cells to acquire the ASNase-resistant phenotype without drug selection, thus proving that elevated ASNS levels alone are sufficient to generate drug resistance.

Despite these cell culture studies, microarray analyses and qPCR quantification of ASNS mRNA have illustrated that the relationship between AS expression and the manifestation of ASNase resistance in primary cells of childhood ALL patients is more complicated. For example, Holleman et al. (112) showed that in vitro resistance to ASNase was correlated with elevated ASNS expression, as assayed by array analysis of primary isolates of ALL cells. In contrast, Fine et al. (113) observed a similar correlation between high expression of ASNS and increased ASNase resistance in a collection of 16 different ALL-derived cultured cell lines, but that relationship did not hold true when primary ALL cells were tested in vitro for their ASNS mRNA content and sensitivity to ASNase. Among the ALL genetic subtypes, those cells with a t(12;21) chromosomal translocation, which leads to synthesis of a TEL/AML1 fusion protein (114), or those with hyperdiploidy, are more sensitive to ASNase (115, 116). Stams et al. (117) hypothesized that TEL/AML1(+) cells are more sensitive to ASNase because the fusion protein, which functions as a transcriptional repressor, may inhibit the expression of ASNS. Surprisingly, they observed that the TEL/AML1(+) cells contained a fivefold higher amount of ASNS mRNA compared to TEL/AML1(−) cells, a result consistent with the data of Krejci et al. (118). Furthermore, Stams et al. (117) discovered that in the TEL/AML1(+) cells there was no correlation between ASNS mRNA content and the in vitro sensitivity to ASNase. Interestingly, those authors went on to show that even though the TEL/AML1(−) ALL cells are less sensitive to ASNase, there is an inverse correlation between ASNS mRNA expression and drug sensitivity in vitro (119). Given that 78% of childhood ALL patients are TEL/AML1(−) (120), it is important to gain a further understanding of the relationship between ASNS expression and ASNase action. Collectively, the results obtained with patient samples show that we still do not fully understand the mechanisms of ASNase action and the development of drug resistance.

SYNTHESIS AND CHARACTERIZATION OF ASPARAGINE SYNTHETASE INHIBITORS

The need for cell-permeable, small molecules capable of inhibiting cellular ASNS in a potent and highly selective manner was recognized almost 40 years ago by the pioneers who developed L-ASNase-based protocols for the treatment of ALL (1). Even so, efforts to obtain such compounds were sporadic, perhaps because a lack of definitive evidence to support the hypothesis that ASNS overexpression really was a primary cause of drug-resistant ALL. As is evident from the previous section of this review, dissecting the cellular mechanisms that underpin the onset of refractory leukemia cells remains a complicated problem that has still not clarified this question. The increased knowledge of ASNS structure and mechanism has rekindled interest in the synthesis of transition state analogs, which might not only be clinically useful compounds but also tools for establishing the relevance of ASNS overexpression to drug resistance.

Evidence Supporting the Clinical Utility of Asparagine Synthetase Inhibitors

In the absence of small molecule inhibitors, preliminary evidence supporting the hypothesis that specific inhibitors of human ASNS have potential clinical utility in treating ASNase-resistant ALL and related leukemias was provided by electroporation studies involving delivery of anti-ASNS monoclonal antibodies to ASNase-resistant mouse L5178Y D10/R cells (21). Parental L5178 cells, derived from a murine leukemia cell line (121), have nearly undetectable ASNS activity and therefore their growth is dependent on the presence of asparagine in the external medium. In contrast, an ASNase-resistant subclone (D10/R) did not require extracellular asparagine because there was a high level of ASNS-catalyzed, intracellular asparagine production. Two mouse monoclonals (3F3 and 2B4) raised against native, bovine ASNS (122), were tested for their ability to inhibit the enzymatic activity of cellular ASNS using extracts from the L5178Y D10/R cells (21). Although both antibodies were able to recognize and bind to ASNS, only the 3F3 monoclonal could inhibit asparagine production in the cell extracts. This monoclonal antibody was therefore electroporated into the ASNase-resistant D10/R cells to evaluate whether suppressing ASNS activity could slow growth in the absence of extracellular asparagine. After establishing that there was no intracellular degradation of the heavy chains of the electroporated antibodies, cells that incorporated the antibody 3F3 were shown to require exogenous asparagine for growth. These experiments suggested that the effect of 3F3 was specifically caused by blocking the action of ASNS, its target enzyme (21). Subsequent controls demonstrated that both ASNase-sensitive and -resistant L5178Y cells contained immunoreactive material, and that the observations on the electroporated DR10/Y cells were not associated with genomic alterations in the form of translocations, gene amplification, or increased P-glycoprotein.

Using Structural Homology and Chemical Constraints to Model the Synthetase Active Site

Unfortunately for efforts to employ rational, structure-based strategies to discover potent, selective ASNS inhibitors that resemble βAspAMP or the transition state for ammonia addition, disorder in the C-terminal domain of the C1A/Gln/AMP complex does not permit observation of two loop regions (Ala-250 to Leu-267 and Cys-422 to Ala-426) and the final forty C-terminal residues of the enzyme by X-ray crystallography (35). Insight into the structure(s) of the ASNS synthetase site during catalytic turnover, and the potential role of residues involved in aspartate activation and ammonia addition, has been obtained, however, by recognizing the striking chemical similarity of the reactions employed to synthesize asparagine from aspartic acid and a functionalized β-lactam from the adenylated form of N^2-(carboxyethyl)-L-arginine (CEA) **2** (**Scheme 3**) (123, 124). Thus, the enzyme BLS, which is found in the biosynthetic pathway leading to the important secondary metabolite clavulanic acid (125), activates its substrate by forming an AMP derivative in which the carbonyl group is attacked intramolecularly by a nitrogen nucleophile. In agreement with modern models for enzyme evolution (126, 127), the common ancestry of ASNS and BLS is evident from the degree of similarity in their three-dimensional structures (**Figure 4**), and the catalytic machinery for substrate activation by adenylation is highly conserved in both enzymes. For example, the contiguous segment SGGLDSS in AS-B, which binds to PP_i during substrate adenylation, is also present in BLS. Using the crystal structure of a ternary complex between BLS, CEA and AMPCPP (an unreactive ATP analog), MgATP could be positioned within the synthetase site of AS-B. This initial model was then used to build a quaternary complex (C1A/Gln/Asp/MgATP) in which aspartate was placed in an analogous location within the protein to that observed for CEA in the BLS/CEA/AMPCPP crystal structure. This modeling exercise was facilitated by assuming two chemical constraints. First, by analogy to the chemistry of aminoacyl tRNA synthetases, attack of the side-chain carboxylate of bound aspartate on the α-phosphate of ATP likely proceeds via in-line attack with inversion at the phosphorus atom (**Scheme 2**) (128). Second, studies with non-natural, conformationally constrained aspartate analogs, prepared using the diastereoselective alkylation of L-aspartate diester derivatives (129, 130), demonstrated that the polar functional groups are located on one face in the bound conformation of

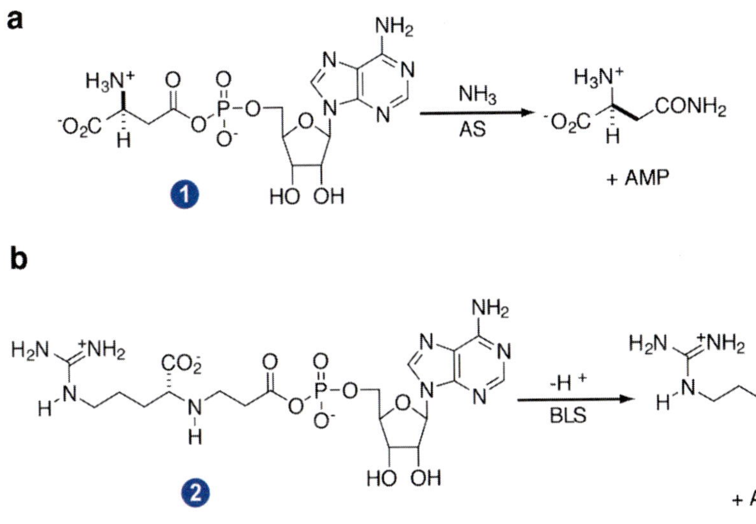

Scheme 3
A comparison of the chemical reactions catalyzed by BLS and ASNS. (*a*) Aspartate is activated by adenylation to yield βAspAMP, which undergoes intermolecular attack by ammonia to yield asparagine. (*b*) CEA, the substrate for BLS, is activated as an acyl-AMP derivative, which then undergoes intramolecular nitrogen attack to give the β-lactam product.

aspartate (131). After refinement using constrained molecular dynamics (MD) simulations (132) in combination with simulated annealing algorithms (133), and subsequent energy minimization, the resulting model was then modified by connecting the side-chain carboxylate of aspartate to the α-phosphate of the ATP moiety to form the βAspAMP intermediate. In this model, PP_i was positioned over the pyrophosphatase loop region of the enzyme so that (*a*) the hydrogen bonding interactions with the enzyme corresponded to those observed in the GMPS/AMP/PP_i crystal structure (48), and (*b*) noncovalent interactions with the active site Mg^{2+} ion were maintained. Further MD refinement and energy minimization then gave a computational model for the AS-B/βAspAMP/PP_i complex (**Figure 5**), which was checked for consistency with the results of mutagenesis and kinetic studies employing AS-B (19, 34, 131, 134–136). Importantly for inhibitor development, we were able to identify completely conserved residues having side chains positioned within 5 Å of either βAspAMP or PP_i. It is therefore likely that these residues have key functional roles in aspartate binding, acyl-AMP formation, and catalyzing the attack of ammonia on βAspAMP, although experiments testing this hypothesis have not yet been published.

Sulfonamide Derivatives as Inhibitors of Human Asparagine Synthetase

Early studies employing a variety of glutamine, ATP, and aspartic acid analogs failed to identify compounds exhibiting significant in vivo or in vitro activity against ASNS (22, 23). Aromatic sulfonylfluoride derivatives of glutamine and asparagine also failed to inhibit ASNS (137), and observations suggesting that cineoles, a group of monoterpenes isolated from the essential oils of aromatic plants (138), inhibit asparagine production in various plant tissues (139) have proven difficult to reproduce (140). Progress, however, (*a*) obtaining multi-milligram amounts of human ASNS from a baculovirus-based expression system (33), (*b*) determining the high-resolution, three-dimensional structure

Molecular dynamics (MD): a computational method in which Newton's equations of motion are solved to yield a trajectory showing the dynamical motions of a protein structure

Simulated annealing: a technique to optimize the structure of a protein by performing an MD simulation in which the temperature is systematically lowered to obtain a low-energy conformation

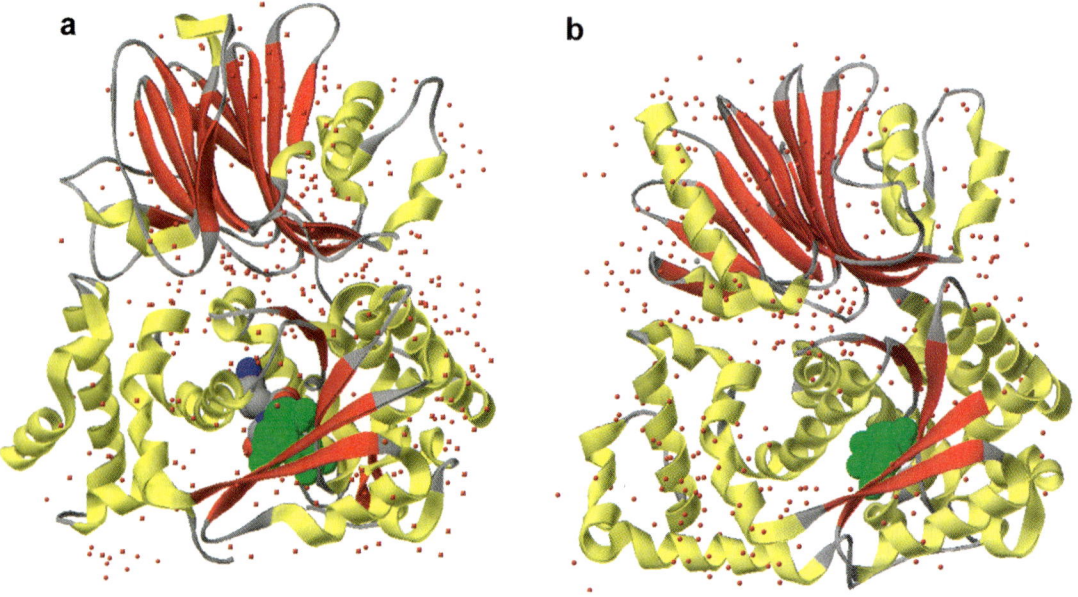

Figure 4

Cartoon representations of the X-ray crystal structures of (*a*) BLS complexed to its substrate (*CPK-colored space-filling model*) and AMPCPP (*green space-filling model*) in the C-terminal domain (52) and (*b*) *Escherichia coli* AS-B complexed with glutamine (not shown) and AMP (*green space-filling model*) (35). The striking structural conservation in both enzymes suggests either a common ancestor or recruitment of AS to provide BLS during the evolution of clavulanic acid biosynthesis. In both structures, helices and β-strands are shown in yellow and red, respectively, whereas water molecules are represented by red spheres.

of the bacterial enzyme AS-B (35), and (*c*) recognizing that aminoacyl tRNA synthetases are targets for new antibacterial and antifungal drugs (141–143) stimulated renewed efforts by Richards and coworkers to obtain potent, small molecule ASNS inhibitors (24, 144). For example, the antibiotic mupirocin **3** (**Figure 6**) exerts its action by selectively inhibiting prokaryotic isoleucyl-tRNA synthetases (145, 146), and phosmidosine **4** (**Figure 6**) (147, 148) likely inhibits prolyl-tRNA synthetase by effectively mimicking the acyladenylate intermediate formed by this enzyme. These findings therefore resulted in the development of efficient synthetic routes for preparing chemically stable analogs of acyladenylates (148–150) in the expectation that such compounds will exhibit antibacterial activity coupled with low mammalian toxicity (151, 152). In light of these studies (145–150),

β-asparaginyladenylate **5** (**Figure 6**) was synthesized (153) and proved to have micromolar affinity for the bacterial enzyme AS-B (S.K. Boehlein, J.-Q. Wang, Y. Ding, N.G.J. Richards & S.M. Schuster, unpublished results). The relatively lengthy synthetic route to obtain this compound and the presence of a charged phosphoamidate functional group argue against its likely clinical utility. In addition, the coupling reaction was insufficiently robust for use in constructing molecular libraries that could be assayed for ASNS inhibitory activity using modern high-throughput screening methods (154–157).

Given that *N*-acylsulfonamide derivatives are high-affinity, slow-binding inhibitors of isoleucyl-tRNA synthetases (158–160) and have "drug-like" physical properties (161, 162), a sulfonamide analog of the βAspAMP intermediate (**6**) (**Figure 6**) was also prepared and characterized for its ability to inhibit recombinant, wild-type human ASNS (24). These in vitro experiments clearly showed that sulfonamide derivative **6** is a slow-onset, tight-binding inhibitor that interacts with the free enzyme, and revealed the importance of carboxyl and amino groups in the recognition and binding of this class of βAspAMP analog. These observations are consistent with the computationally-derived model of the AS-B/βAspAMP/PP$_i$ complex (**Figure 3b**). Detailed steady-state kinetic analysis using standard models (163) revealed the K_I^* value for **6** to be approximately 700 nM, making this compound the first submicromolar ASNS inhibitor identified in the literature.

A Nanomolar Inhibitor of Human Asparagine Synthetase

Perhaps the most important progress toward the discovery of potent inhibitors of glutamine-dependent ASNS has been obtained from recent work employing the adenylated sulfoximine **7** (**Figure 6**). The conceptual impetus for synthesizing and characterizing this compound was the observation that methionine sulfoximine **8** (**Figure 6**) is a potent inhibitor of glutamine synthetase (164). Thus, this natural product binds to the enzyme in the glutamate site (165, 166) where it undergoes phosphorylation to yield the sulfoximine phosphate derivative **9** (167), which was proposed to be an analog for the transition state formed during the attack of ammonia on γ-glutamylphosphate **10** (**Figure 6**) (168). This interesting example of a mechanism-based inhibitor therefore stimulated the synthesis of the analogous sulfoximine **11** (**Figure 6**) from homocysteine with the goal of obtaining a potent inhibitor of ammonia-dependent asparagine synthetases (169), a class of enzymes that, to date, have only been found in prokaryotes (170, 171). On this point, we note that there is no evolutionary relationship of glutamine- and ammonia-dependent asparagine synthetases, the latter family being

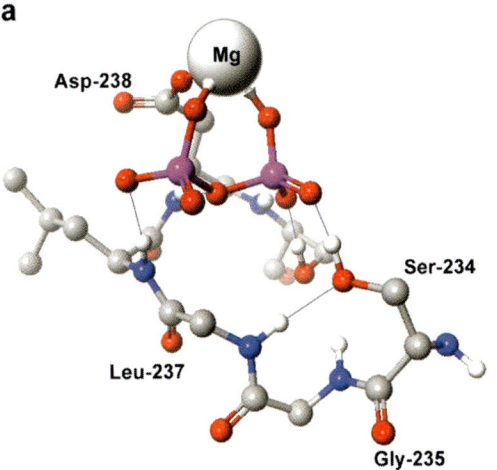

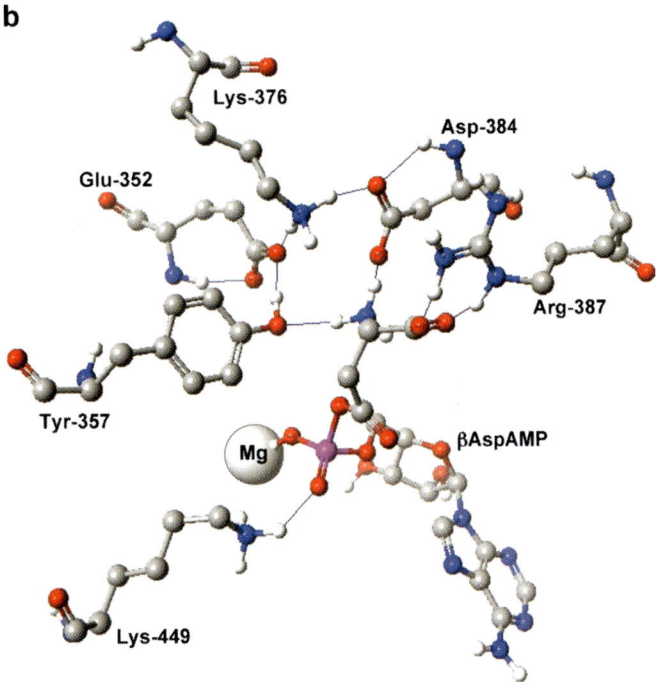

Figure 5

Views of the working molecular model for the AS-B/βAspAMP/PP$_i$ complex. (*a*) Interactions between PP$_i$ and residues defining the ATP pyrophosphatase loop motif. (*b*) Protein/βAspAMP interactions involving conserved residues Glu-352, Tyr-357, Lys-376, Asp-384, Arg-387, and Lys-449 that illustrate recognition of the α-amino and α-carboxylate groups present in the βAspAMP intermediate **1**. (Color coding: C – *gray*; H – *white*; O – *red*; N – *blue*; P – *purple*. *Light blue* lines show locations of putative hydrogen bonds.)

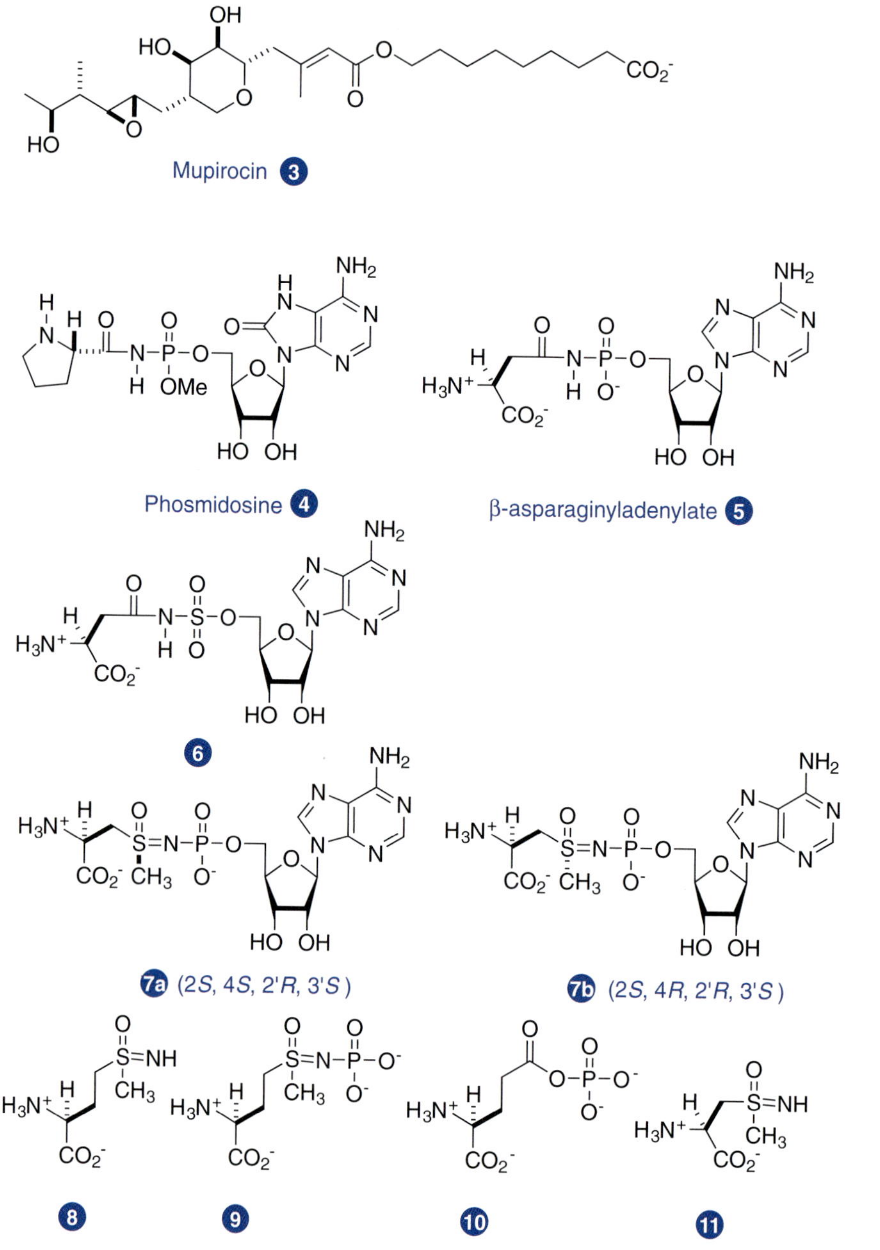

Figure 6

Structures of compounds 3–11 (see text for details).

related to aminoacyl tRNA synthetases (172, 173). Incubation of **11** with AS-A, the ammonia-dependent ASNS isolated from *Escherichia coli* (174, 175), however, showed that this compound was a poor (millimolar) inhibitor because it was not adenylated when bound within the synthetase active site. A synthetic route to the required adenylated sulfoximine was therefore developed by Hiratake and coworkers, yielding the target compound

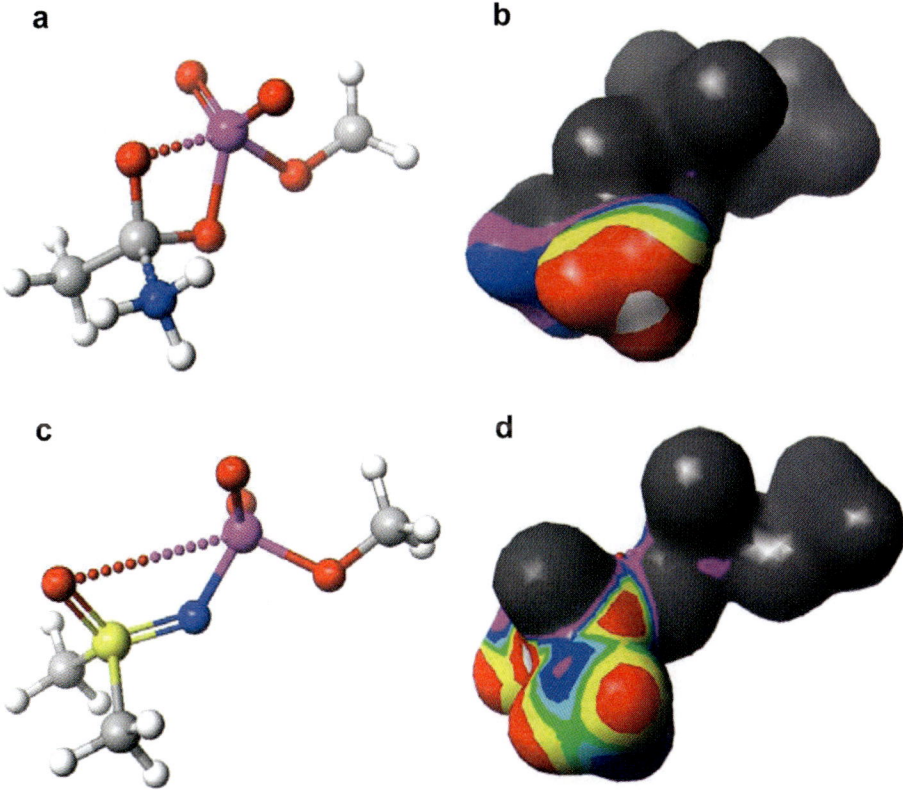

Figure 7

Computational visualization of transition state mimicry by the adenylated sulfoximine functional group. (*a*) Semiempirical (PM3) transition state for the attack of ammonia on a computational model of the acyladenylate intermediate formed in the ASNS synthetase site. (*b*) Graphical representation of the transition state structure showing the isodensity surface color-coded by the electrostatic potential. Note that the hydrogen atoms in ammonia gain substantial positive charge (*red*). (*c*) Optimized (PM3) structure for a model phosphorylated sulfoximine. (*d*) Graphical representation of the phosphorylated sulfoximine in (*c*) showing the isodensity surface color-coded by the electrostatic potential. Note that the hydrogen atoms of the methyl group have similar steric and electrostatic properties to those on ammonia in the transition state shown in (*b*). [Color coding in (*a*) and (*c*): C – *gray*; H – *white*; O – *red*; N – *blue*; P – *purple*; S – *yellow*. *Dotted lines* show noncovalent, electrostatic interactions.]

as a mixture of diastereoisomers (**7a** and **7b**) (**Figure 6**) at the sulfur center (169). Not only did this mixture prove to exhibit potent inhibition of AS-A (169) but it was also shown to be a slow-onset inhibitor of the glutamine-dependent AS-B when the bacterial enzyme was undergoing catalytic turnover (144). Although, at first sight, it may be difficult to see why this compound should resemble the transition state for ammonia addition to the βAspAMP intermediate, semiempirical quantum mechanical (QM) calculations (N.G.J. Richards, unpublished results), using the PM3 parameter set (176), clearly show that the methyl group in the sulfoximine moiety is electron deficient, giving a significant positive electrostatic potential about the three hydrogen atoms. The methyl group therefore resembles the steric and electrostatic properties of ammonia in the transition state for C-N bond formation (**Figure 7**). This analysis is supported by the observation that an aspartate side chain is positioned close to the methyl group of **7b** in the X-ray crystal structure of its complex with AS-A (169).

Incubation of **7**, as the diastereoisomeric mixture **7a** and **7b**, with recombinant, wild-type human ASNS at various ATP concentrations, showed this compound to be a potent, slow-onset, tight-binding inhibitor of the enzyme in vitro. Use of standard mathematical models (163) gave values for the constants K_I and K_I^* of 285 nM and 2.5 nM, respectively, making the adenylated sulfoximine the most potent inhibitor of the human enzyme reported to date (J.A. Gutierrez & N.G.J. Richards, unpublished results). In common with many other tight-binding inhibitors (177, 178), the adenylated sulfoximine **7** binds to the free form of the human enzyme, and reactivation and mass spectrometric experiments have demonstrated that it does not covalently modify ASNS. Efforts to cocrystallize the recombinant, human ASNS with **7** (albeit using the diastereoisomeric mixture) have failed, however, to yield crystals suitable for high-resolution X-ray structure determination. Because all kinetic experiments were performed using a 1:1 mixture of the sulfoximine diastereoisomers **7a** and **7b** (**Figure 6**), and given that residues defining the synthetase site are completely conserved throughout the glutamine-dependent ASNS family, computational docking methods (179, 180) were used to position each of diastereoisomeric adenylated sulfoximine derivatives into the working model of the AS-B/βAspAMP/PP$_i$ complex (N.G.J. Richards, unpublished results). After constrained MD-based refinement, as described above, only diastereoisomer **7b** could be positioned such that its methyl substituent was positioned correctly relative to the C-terminal end of the ammonia channel (**Figure 8**). In addition, the side chain of Glu-348 (AS-B numbering), which is conserved in the primary structures of known asparagine synthetases, was positioned adjacent to the electron-deficient methyl group of the docked inhibitor in the model, suggesting

that this residue might function as a general base to deprotonate ammonia in the transition state leading to the tetrahedral adduct (**Scheme 2**). In efforts to validate the computational model of the AS-B/**7b** active site complex, site-specific AS-B mutants were prepared in which this residue was replaced by aspartate (E348D) and alanine (E348A), and assayed for their ability to form asparagine and PP$_i$. In contrast to wild-type AS-B, which forms these products in a 1:1 ratio (63, 89), the E348D AS-B mutant formed two molecules of PP$_i$ per asparagine. In addition, the E348A AS-B mutant loses the ability to catalyze asparagine synthesis when glutamine or ammonia is the nitrogen source, although it retains significant levels of ATP pyrophosphatase activity (J.A. Gutierrez & N.G.J. Richards, unpublished results). Both of these observations are consistent with the role deduced for Glu-348 on the basis of the AS-B/inhibitor model, suggesting that this in silico structure may prove a useful asset in the discovery of simpler, and more cell permeable, structures with activity as ASNS inhibitors using computational (181) and other structure-based strategies (182, 183).

PERSPECTIVES

A number of valuable initiatives can be envisaged, some of which are ongoing, to further our knowledge of the relationship between ASNS expression and ASNase sensitivity or resistance in childhood ALL. First, to complement the microarray analysis of patient samples at the time of diagnosis, similar analyses performed on individual patients during the course of therapy should be of considerable interest. However, this proposed approach has some technical limitations because over the course of therapy in most patients there will be a significant decline in the number of lymphoblastic cells, and consequently the background of ASNS content in other cell types within the sample will complicate any quantitative analysis and interpretation of the data. It must also be recognized that ALL

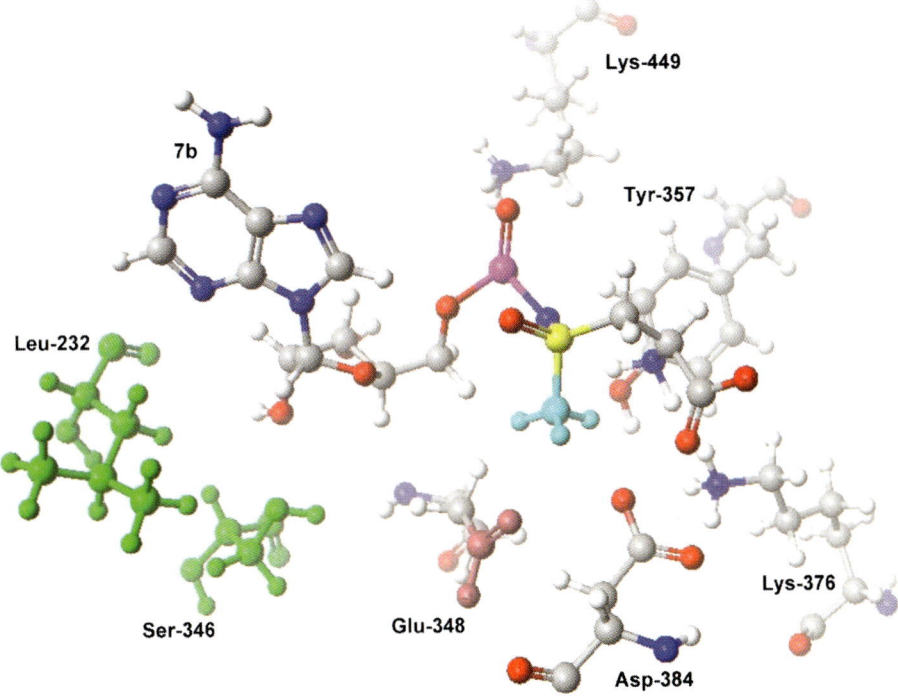

Figure 8
Computational model of the adenylated sulfoximine ASNS inhibitor **7b** docked into the synthetase site of AS-B. For ease of comprehension, only selected protein residues are shown, which are all conserved within glutamine-dependent ASNS. Note the distance between the Glu-348 side-chain carboxylate (*pink*) and the methyl group (*cyan*) of the ASNS inhibitor. The methyl substituent mimics the location of ammonia in the transition state formed during attack on the βAspAMP intermediate 1. Residues shown in green (Leu-232 and Ser-346) define the C-terminal end of the channel through which ammonia enters the synthetase active site after being released in the N-terminal glutaminase domain. (Color coding: C – *gray*; H – *white*; O – *red*; N – *blue*; P – *purple*; S – *yellow*.)

therapy involves a multidrug regimen, so that the effect of ASNase treatment is difficult to analyze in isolation. Parallel proteomic approaches to (*a*) document the expression of ASNS protein and enzymatic activity, thereby yielding a more definitive analysis of the actual cellular rates of asparagine synthesis, and (*b*) screen the entire proteome of the cell for the changes in specific protein expression that take place in ASNase-sensitive versus resistant cells, or during ASNase therapy, will also be important (184, 185). Such proteomic analyses will be critical because it is becoming increasingly clear that fluctuations in cellular amino acid content, similar to those in- duced by ASNase treatment, can cause significant effects on global mRNA translation rates, as well as on the translation of specific mRNA species. For example, data from cell culture studies indicate that in response to long-term ASNase treatment, cells are irreversibly altered by unknown mechanisms such that ASNase resistance and increased transcription from the ASNS gene is retained long after drug exposure has been terminated. It will be important to determine if such an irreversible elevation of ASNS expression occurs in subpopulations of patients undergoing therapy, and, if so, the identification of the molecular mechanism(s) by which that enhanced expression occurs might lead to new therapeutic targets. Finally, extending the current studies on the relationship between ASNS expression, ASNase effectiveness, hyperdiploidy, and the expression of the t(12;21) product TEL-AML1 is likely to be important in designing rational ASNase therapy in ALL genetic subtypes.

The demonstration that adenylated sulfoximine **7b** is a nanomolar ASNS inhibitor sets the stage for cell-based assays to determine the biological effects of inhibiting the enzyme in ASNase-resistant MOLT-4 cell lines. On the other hand, the ability of this compound to pass through cellular membranes is clearly limited by the presence of amine and carboxylate functional groups, which have been shown to be critical substituents for aspartate recognition and binding. Therefore obtaining functionalized sulfoximine derivatives that are "second-generation" ASNS inhibitors is dependent on identifying uncharged functional groups, such as hydroxyl groups or heteroaromatic rings, which can substitute for these ionizable groups. Accomplishing this goal will likely require the development of suitable, targeted molecular libraries (151, 186), for which access to (*a*) novel, and more efficient, synthetic methods for constructing the sulfoximine functionality, and (*b*) validated computational models of the AS-B/**7b** active site complex that can be employed in de novo design strategies (181) will be essential. Novel assays for detecting asparagine formation that are sufficiently sensitive for use in automated, high-throughput screening also remain to be developed. Finally, the use of profiling methods (187) should offer a rapid approach to delineating the specificity of adenylated sulfoximines when introduced into cells, particularly regarding their ability to bind to aspartyl tRNA synthetases.

SUMMARY POINTS

1. Crystal structures of the evolutionarily related enzymes AS-B and BLS have permitted the construction of a computational model for the complex between β-aspartyladenylate, a key reaction intermediate, and conserved residues in the ASNS synthetase active site.

2. Determination of a kinetic mechanism for ASNS has suggested that analogs of β-aspartyladenylate, and the transition state for nucleophilic attack of ammonia on this intermediate, will be potent inhibitors of the human enzyme.

3. ASNS transcription is enhanced by amino acid deprivation via a signal transduction pathway that senses amino acid availability, resulting in the enhanced synthesis and function of selected transcription factors belonging to the bZIP family.

4. When delivered to patients, ASNase degrades plasma asparagine and glutamine thereby depleting the cellular levels of these amino acids. ALL cells express a low level of ASNS activity and are therefore particularly sensitive to ASNase-based therapies.

5. Selection and characterization of ASNase-resistant ALL cell lines in culture suggests that ASNase resistance is correlated with the elevation of ASNS expression, although this relationship appears to be more complicated in primary ALL cells isolated from patients and probably depends on the genetic background of each individual.

6. The first slow-onset, tight-binding inhibitor for recombinant human ASNS, with nanomolar binding affinity, has been identified, and its interactions with the ASNS synthetase site characterized using molecular modeling methods.

FUTURE ISSUES TO BE RESOLVED

1. The effects of the *N*-acylsulfonamide and sulfoximine-derived ASNS inhibitors on ASNase-resistant cell lines need to be evaluated and assays developed to ensure their specificity of interaction with intracellular ASNS.

2. The validation, and refinement, of computational models describing the interaction of protein residues with the *N*-acylsulfonamide and sulfoximine-derived ASNS inhibitors must be undertaken prior to their subsequent use in designing molecular libraries fhat can be screened for compounds with "drug-like" properties and good cell permeability.

3. Further insight into the role that ASNS plays in the treatment of childhood ALL will require more intensive research into the regulation of its expression by amino acid deprivation and the consequences on cell growth of inhibiting its synthetase activity.

4. Also needed is the development of proteomics-based methods for (*a*) quantitating the expression of ASNS protein and enzymatic activity, thereby yielding a more definitive analysis of the actual cellular rates of asparagine synthesis, and (*b*) identifying changes in protein expression, which take place in ASNase-sensitive and -resistant cells.

ACKNOWLEDGMENTS

We thank the many coworkers and collaborators who have contributed to the studies discussed in this review, and apologize for omitting many details of their efforts because of space limitations. This work was supported by grants from the National Institutes of Health (CA09126, CA107437, and DK52064) and the Chiles Endowment Biomedical Research Program of the Florida Department of Health.

LITERATURE CITED

1. Cooney DA, Handschumacher RE. 1970. *Annu. Rev. Pharmacol.* 10:421–40
2. Ertel IJ, Nesbit MG, Hammon D, Weiner J, Sather H. 1979. *Cancer Res.* 39:3893–96
3. Amylon MD, Shuster J, Pullen J, Berard C, Link MP, et al. 1999. *Leukemia* 13:335–42
4. Sanz GF, Sanz MA, Rafecas FJ, Martinez JA, Martin-Aragonés G, et al. 1986. *Cancer Treat Rep.* 70:1321–24
5. Barr RD, DeVeber LL, Pai KM, Andrew M, Halton J, et al. 1992. *Am. J. Pediatr. Hematol. Oncol.* 14:136–39
6. Pui CH, Relling MV, Campana D, Evans WE. 2002. *Rev. Clin. Exp. Hematol.* 6:161–80

7. Ohnuma T, Holland JF, Meyer P. 1972. *Cancer* 30:376–81
8. Chakrabarti R, Schuster SM. 1997. *Int. J. Pediatr. Hemotol./Oncol.* 4:597–611
9. Graham ML. 2003. *Adv. Drug Deliv. Rev.* 55:1293–302
10. Tallal L, Tan C, Oettgen H, Wollner N, McCarthy M, et al. 1970. *Cancer* 25:306–20
11. Sutow WW, Garcia F, Starling KA, Williams TE, Lane DM, et al. 1971. *Cancer* 28:819–24
12. Muller HJ, Boos J. 1998. *Crit. Rev. Oncol. Hematol.* 28:97–113
13. Hersh EM. 1971. *Transplantation* 12:368–76
14. Terebello HR, Anderson K, Wiernik PH, Cuttner J, Cooper RM, et al. 1986. *Am. J. Clin. Oncol.* 9:411–15
15. Kiriyama Y, Kubota M, Takimoto T, Kitoh J, Tanizawa A, et al. 1989. *Leukemia* 3:294–97
16. Lobel JS, O'Brien RT, McIntosh S, Aspnes GT, Capizzi RL. 1979. *Cancer* 43:1089–94
17. Capizzi RL, Bertino JR, Skeel RT, Creasey WA, Zanes R, et al. 1971. *Ann. Intern. Med.* 74:893–901
18. Pieters R, Klumper E, Kaspers GJL, Veerman AJP. 1997. *Crit. Rev. Oncol. Hematol.* 25:11–26
19. Richards NGJ, Schuster SM. 1998. *Adv. Enzymol. Relat. Areas Mol. Biol.* 72:145–98
20. Aslanian AM, Fletcher BS, Kilberg MS. 2001. *Biochem. J.* 357:321–28
21. Chakrabarti R, Wylie DE, Schuster SM. 1989. *J. Biol. Chem.* 264:15494–500
22. Cooney DA, Driscoll JS, Milman HA, Jayaram HN, Davis RD. 1976. *Cancer Treat. Rep.* 60:1493–557
23. Cooney DA, Jones MT, Milman HA, Young DM, Jayaram HN. 1980. *Int. J. Biochem.* 11:519–39
24. Koroniak L, Ciustea M, Gutierrez JA, Richards NGJ. 2003. *Org. Lett.* 5:2033–36
25. Huang Y-Z, Knox EW. 1975. *Enzyme* 19:314–28
26. Hongo S, Motosugu F, Shioda S, Nakai Y, Takeda M, et al. 1992. *Arch. Biochem. Biophys.* 295:120–25
27. Mehlhaff P, Luehr CA, Schuster SM. 1985. *Biochemistry* 24:1104–10
28. Luehr CA, Schuster SM. 1985. *Arch. Biochem. Biophys.* 237:335–46
29. Patterson MK, Orr GR. 1968. *J. Biol. Chem.* 243:376–80
30. Van Heeke G, Schuster SM. 1990. *Protein Eng.* 3:739–44
31. Sheng S, Moraga-Amador DA, Van Heeke G, Schuster SM. 1992. *Prot. Exp. Purif.* 3:337–46
32. Sheng S, Moraga-Amador DA, Van Heeke G, Allison RD, Richards NGJ, et al. 1993. *J. Biol. Chem.* 268:16771–80
33. Ciustea M, Gutierrez JA, Abbatiello SE, Eyler JR, Richards NGJ. 2005. *Arch. Biochem. Biophys.* 440:18–27
34. Boehlein SK, Richards NGJ, Schuster SM. 1994. *J. Biol. Chem.* 269:7450–57
35. Larsen TM, Boehlein SK, Schuster SM, Richards NGJ, Thoden JB, et al. 1999. *Biochemistry* 38:16146–57
36. Zalkin H. 1993. *Adv. Enzymol. Relat. Areas Mol. Biol.* 66:203–309
37. Pfeiffer NE, Mehlhaff PM, Wylie DW, Schuster SM. 1986. *J. Biol. Chem.* 261:1914–19
38. Oinonen C, Rouvinen J. 2000. *Protein Sci.* 9:2329–37
39. Brannigan JA, Dodson G, Duggleby HJ, Moody PCE, Smith JL, et al. 1995. *Nature* 378:416–19
40. Voges D, Zwickl P, Baumeister W. 1999. *Annu. Rev. Biochem.* 68:1015–68
41. Duggleby HJ, Tolley SP, Hill CP, Dodson EJ, Dodson G, et al. 1995. *Nature* 373:264–68
42. Oinonen C, Tikkanen R, Rouvinen J, Peltonen L. 1995. *Nat. Struct. Biol.* 2:1102–8

43. Zalkin H, Smith JL. 1998. *Adv. Enzymol. Relat. Areas Mol. Biol.* 72:87–144
44. Krahn JM, Kim JH, Burns MR, Parry RJ, Zalkin H, et al. 1997. *Biochemistry* 36:11061–68
45. Massière F, Badet-Denisot M-A. 1998. *Cell. Mol. Life Sci.* 54:205–22
46. Teplyakov A, Obmolova G, Badet B, Badet-Denisot M-A. 2001. *J. Mol. Biol.* 313:1093–102
47. Bork P, Koonin EV. 1994. *Proteins: Struct. Funct. Genet.* 20:347–55
48. Tesmer JJG, Klem TJ, Deras ML, Davisson VJ, Smith JL. 1996. *Nat. Struct. Biol.* 3:74–86
49. Goto M, Omi R, Miyahara I, Sugahara M, Hirotsu K. 2003. *J. Biol. Chem.* 278:22964–71
50. MacRae IJ, Segel IH, Fisher AJ. 2001. *Biochemistry* 40:6795–804
51. Miller MT, Gerratana B, Stapon A, Townsend CA, Rosenzweig AC. 2003. *J. Biol. Chem.* 278:40996–1002
52. Miller MT, Bachmann BO, Townsend CA, Rosenzweig AC. 2001. *Nat. Struct. Biol.* 8:684–89
53. Miller MT, Bachmann BO, Townsend CA, Rosenzweig AC. 2002. *Proc. Natl. Acad. Sci. USA* 99:14752–57
54. Mueller EG, Palenchar PM. 1999. *Protein Sci.* 8:2424–27
55. Boehlein SK, Stewart JD, Walworth ES, Thirumoorthy R, Richards NGJ, Schuster SM. 1998. *Biochemistry* 37:13230–38
56. Anand R, Hoskins AA, Stubbe J, Ealick SE. 2004. *Biochemistry* 43:10328–42
57. Endrizzi JA, Kim HS, Anderson PM, Baldwin EP. 2004. *Biochemistry* 43:6447–63
58. van den Heuvel RHH, Curti B, Vanoni M, Mattevi A. 2004. *Cell. Mol. Life Sci.* 61:669–81
59. Myers RS, Jensen JR, Deras IL, Smith JL, Davisson VJ. 2003. *Biochemistry* 42:7013–22
60. Knöchel T, Ivens A, Hester G, Gonzalez A, Bauerle R, et al. 1999. *Proc. Natl. Acad. Sci. USA* 96:9479–84
61. Raushel F, Thoden JB, Holden HM. 2003. *Acc. Chem. Res.* 36:539–48
62. Huang XY, Holden HM, Raushel FM. 2001. *Annu. Rev. Biochem.* 70:149–80
63. Tesson AR, Soper TS, Ciustea M, Richards NGJ. 2003. *Arch. Biochem. Biophys.* 413:23–31
64. Fresquet V, Thoden JB, Holden HM, Raushel FM. 2004. *Bioorg. Chem.* 32:63–75
65. Hongo S, Sato T. 1985. *Arch. Biochem. Biophys.* 238:410–17
66. Markin RS, Luehr CA, Schuster SM. 1981. *Biochemistry* 20:7226–32
67. Rognes SE. 1975. *Phytochemistry* 14:1975–82
68. Rose IA. 1980. *Methods Enzymol.* 64:47–59
69. Ibba M, Söll D. 2000. *Annu. Rev. Biochem.* 69:617–50
70. Arnez JG, Moras D. 1997. *Trends Biochem. Sci.* 22:211–16
71. Barshop BA, Wrenn RF, Frieden C. 1983. *Anal. Biochem.* 130:134–45
72. Buchanan JM. 1973. *Adv. Enzymol. Relat. Areas Mol. Biol.* 39:91–183
73. Bera AK, Smith JL, Zalkin H. 2000. *J. Biol. Chem.* 275:7975–79
74. van den Heuvel RHH, Ferrari D, Bossi RT, Ravasio S, Curti B, et al. 2002. *J. Biol. Chem.* 277:24579–83
75. Bera AK, Chen SH, Smith JL, Zalkin H. 1999. *J. Biol. Chem.* 274:36498–504
76. Schramm VL. 1998. *Annu. Rev. Biochem.* 67:693–720
77. Radzicka A, Wolfenden R. 1995. *Methods Enzymol.* 249:284–312
78. Greco A, Gong SS, Ittmann M, Basilico C. 1989. *Mol. Cell. Biol.* 9:2350–59
79. Gong SS, Basilico C. 1990. *Nucleic Acids Res.* 18:3509–13
80. Greco A, Ittmann M, Basilico C. 1987. *Proc. Natl. Acad. Sci. USA* 84:1565–69
81. Hongo S, Takeda M, Sato T. 1989. *Biochem. Intern.* 18:661–66
82. Colletta G, Cirafici AM. 1992. *Biochem. Biophys. Res. Commun.* 183:265–72
83. Kilberg MS, Pan Y-X, Chen H, Leung-Pineda V. 2005. *Annu. Rev. Nutr.* 25:59–85

84. Hinnebusch AG. 1997. *J. Biol. Chem.* 272:21661–64
85. Sood R, Porter AC, Olsen DA, Cavener DR, Wek RC. 2000. *Genetics* 154:787–801
86. Zhang PC, McGrath BC, Reinert J, Olsen DS, Lei L, et al. 2002. *Mol. Cell. Biol.* 22:6681–88
87. Natarajan K, Meyer MR, Jackson BM, Slade D, Roberts C, et al. 2001. *Mol. Cell. Biol.* 21:4347–68
88. Harding HP, Novoa I, Zhang YH, Zeng HQ, Wek R, et al. 2000. *Mol. Cell* 6:1099–108
89. Vattem KM, Wek RC. 2004. *Proc. Natl. Acad. Sci. USA* 101:11269–74
90. Lu PD, Harding HP, Ron D. 2004. *J. Cell Biol.* 167:27–33
91. Siu F, Bain PJ, LeBlanc-Chaffin R, Chen H, Kilberg MS. 2002. *J. Biol. Chem.* 277:24120–27
92. Chen H, Pan Y-X, Dudenhausen EE, Kilberg MS. 2004. *J. Biol. Chem.* 279:50829–39
93. Gong SS, Guerrini L, Basilico C. 1991. *Mol. Cell. Biol.* 11:6059–66
94. Hutson RG, Kilberg MS. 1994. *Biochem. J.* 303:745–50
95. Guerrini L, Gong SS, Mangasarian K, Basilico C. 1993. *Mol. Cell. Biol.* 13:3202–12
96. Barbosa-Tessmann IP, Chen C, Zhong C, Siu F, Schuster SM, et al. 2000. *J. Biol. Chem.* 275:26976–85
97. Leung-Pineda V, Kilberg MS. 2002. *J. Biol. Chem.* 277:16585–91
98. Zhong C, Chen C, Kilberg MS. 2003. *Biochem. J.* 372:603–9
99. Barbosa-Tessmann IP, Pineda VL, Nick HS, Schuster SM, Kilberg MS. 1999. *Biochem. J.* 339:151–58
100. Waye MM, Stanners CP. 1981. *Cancer Res.* 41:3104–6
101. Siu FY, Chen C, Zhong C, Kilberg MS. 2001. *J. Biol. Chem.* 276:48100–7
102. Pan Y-X, Chen H, Siu F, Kilberg MS. 2003. *J. Biol. Chem.* 278:38402–12
103. Fawcett TW, Martindale JL, Guyton KZ, Hai T, Holbrook NJ. 1999. *Biochem. J.* 339:135–41
104. Capizzi RL, Holcenberg JS. 1993. In *Cancer Medicine*, ed. JF Holland, E Frei, RC Bast Jr, DW Kuffe, DL Morton, RR Weichselbaum, pp. 796–805. Philadelphia: Lea & Febiger
105. Chabner BA, Loo TL. 1996. In *Cancer Chemotherapy and Biotherapy*, ed. BA Chabner, DL Longo, pp. 485–92. Philadelphia: Lippincott-Raven
106. Deleted in proof
107. den Boer ML, Pieters R, Kazemier KM, Rottier MMA, Zwaan CM, et al. 1998. *Blood* 91:2092–98
108. Hutson RG, Kitoh T, Amador DAM, Cosic S, Schuster SM, Kilberg MS. 1997. *Am. J. Physiol. Cell Physiol.* 272:C1691–99
109. Aslanian AM, Kilberg MS. 2001. *Biochem. J.* 358:59–67
110. Prager MD, Bachynsky N. 1968. *Biochem. Biophys. Res. Commun.* 31:43–47
111. Andrulis IL, Argonza R, Cairney AEL. 1990. *Somat. Cell Mol. Genet.* 16:59–65
112. Holleman A, Cheok MH, den Boer ML, Yang W, Veerman AJ, et al. 2004. *N. Engl. J. Med.* 351:533–42
113. Fine BM, Kaspers GJ, Ho M, Loonen AH, Boxer LM. 2005. *Cancer Res.* 65:291–99
114. Fenrick R, Amann JM, Lutterbach B, Wang L, Westendorf JJ, et al. 1999. *Mol. Cell. Biol.* 19:6566–74
115. Ramakers-van Woerden NL, Pieters R, Loonen AH, Hubeek I, van Drunen E, et al. 2000. *Blood* 96:1094–99
116. Kaspers GJ, Smets LA, Pieters R, Van Zantwijk CH, Van Wering ER, Veerman AJ. 1995. *Blood* 85:751–56

117. Stams WA, den Boer ML, Beverloo HB, Meijerink JP, Stigter RL, et al. 2003. *Blood* 101:2743–47
118. Krejci O, Starkova J, Otova B, Madzo J, Kalinova M, et al. 2004. *Leukemia* 18:434–41
119. Stams WA, den Boer ML, Holleman A, Appel IM, Beverloo HB, et al. 2005. *Blood* 11:4223–25
120. Pui CH, Relling MV, Downing JR. 2004. *N. Engl. J. Med.* 350:1535–48
121. Horowitz B, Madras BK, Old LJ, Boyce EJ, Meister A. 1968. *Science* 160:533–35
122. Pfeiffer NE, Mehlhaff PM, Wylie DE, Schuster SM. 1987. *J. Biol. Chem.* 262:11565–70
123. Bachmann BO, Li RF, Townsend CA. 1998. *Proc. Natl. Acad. Sci. USA* 95:9082–86
124. McNaughton HJ, Thirkettle JE, Zhang ZH, Schofield CJ, Jensen SE, et al. 1998. *Chem. Commun.* 1998:2325–26
125. Baggaley KH, Brown AG, Schofield CJ. 1997. *Nat. Prod. Rep.* 14:309–33
126. Gerlt JA, Babbitt PC. 2001. *Annu. Rev. Biochem.* 70:209–46
127. O'Brien PJ, Herschlag D. 1999. *Chem. Biol.* 6:R91–105
128. Knowles JR. 1980. *Annu. Rev. Biochem.* 49:877–919
129. Parr IB, Dribben AB, Norris SR, Hinds MG, Richards NGJ. 1999. *J. Chem. Soc. Perkin Trans. 1* 8:1029–38
130. Baldwin JE, Moloney MG, North M. 1989. *Tetrahedron* 45:6319–25
131. Parr IB, Boehlein SK, Dribben AB, Schuster SM, Richards NGJ. 1996. *J. Med. Chem.* 39:2367–78
132. Brooks BR, Cheatham TE. 1998. *Theor. Chem. Acc.* 99:279–88
133. Brunger AT, Adams PD, Rice LM. 1998. *Curr. Opin. Struct. Biol.* 8:600–11
134. Boehlein SK, Richards NGJ, Schuster SM. 1994. *J. Biol. Chem.* 269:26789–95
135. Boehlein SK, Walworth ES, Richards NGJ, Schuster SM. 1997. *J. Biol. Chem.* 272:12384–92
136. Boehlein SK, Walworth ES, Schuster SM. 1997. *Biochemistry* 36:10168–77
137. Mokotoff M, Brynes S, Bagaglio JF. 1975. *J. Med. Chem.* 18:888–91
138. Ahmad A, Misra LN. 1994. *Phytochemistry* 37:183–86
139. Romagni JG, Duke SO, Dayan FE. 2000. *Plant Physiol.* 123:725–32
140. Romagni JG, Duke SO, Dayan FE. 2005. *Plant Physiol.* 137:1487
141. Winum JY, Scozzafava A, Montero JL, Supuran CT. 2005. *Med. Res. Rev.* 25:186–228
142. Kim S, Lee SW, Choi E-C, Choi SY. 2003. *Appl. Microbiol. Biotechnol.* 61:278–88
143. Tao JS, Schimmel P. 2000. *Exp. Opin. Investig. Drugs* 9:1767–75
144. Boehlein SK, Nakatsu T, Hiratake J, Thirumoorthy R, Stewart JD, et al. 2001. *Biochemistry* 40:11168–75
145. Nakama T, Nureki O, Yokoyama S. 2001. *J. Biol. Chem.* 276:48387–93
146. Yanagisawa T, Lee JT, Wu HC, Kawakami M. 1994. *J. Biol. Chem.* 269:24304–9
147. Phillips DR, Uramoto M, Isono K, McCloskey JA. 1993. *J. Org. Chem.* 58:854–59
148. Moriguchi T, Asai N, Okada K, Seio K, Sasaki T, et al. 2002. *J. Org. Chem.* 67:3290–300
149. Moriguchi T, Yanagi T, Kunimori M, Wada T, Sekine M. 2000. *J. Org. Chem.* 65:8229–38
150. Filippov D, Timmers CM, Roerdink AR, van der Marcel GA, van Boom JH. 1998. *Tetrahedron Lett.* 39:4891–94
151. von der Haar F, Gabius H-J, Cramer F. 1981. *Angew. Chem. Int. Ed. Engl.* 20:217–23
152. Pope AJ, Lapointe J, Mensah L, Brown MJB, Benson N, Moore KJ. 1998. 273:31691–701
153. Ding Y, Wang J, Schuster SM, Richards NGJ. 2002. *J. Org. Chem.* 67:4372–75
154. Sanchez-Martin RM, Mittoo S, Bradley M. 2004. *Curr. Top. Med. Chem.* 4:653–69
155. Ramström O, Lehn J-M. 2002. *Nat. Rev. Drug Discov.* 1:26–36
156. Bajorath J. 2002. *Nat. Rev. Drug Discov.* 1:882–94

157. Thompson LA, Ellman JA. 1996. *Chem. Rev.* 96:555–600
158. Brown MJB, Mensah LM, Doyle ML, Broom NJP, Osbourne N, et al. 2000. *Biochemistry* 39:6003–11
159. Brown P, Richardson CM, Mensah LM, O'Hanlon PJ, Osborne NF, et al. 1999. *Bioorg. Med. Chem.* 7:2473–85
160. Forrest AK, Jarvest RL, Mensah LM, O'Hanlon PJ, Pope AJ, et al. 2000. *Bioorg. Med. Chem. Lett.* 10:1871–74
161. Teague SJ, Davis AM, Leeson PD, Oprea T. 1999. *Angew. Chem. Int. Ed. Engl.* 38:3743–48
162. Lipinski CA, Lombardo F, Dominy BW, Feeney PJ. 1997. *Adv. Drug Deliv. Rev.* 23:3–25
163. Morrison JF, Walsh CT. 1988. *Adv. Enzymol. Relat. Areas Mol. Biol.* 61:201–301
164. Logusch EW, Walker DM, McDonald JF, Franz JE, Villafranca JJ, et al. 1990. *Biochemistry* 29:366–72
165. Abell LM, Villafranca JJ. 1991. *Biochemistry* 30:6135–41
166. Liaw SH, Eisenberg D. 1994. *Biochemistry* 33:675–81
167. Manning JM, Moore S, Rowe WB, Meister A. 1969. *Biochemistry* 8:2681–85
168. Weisbrod RE, Meister A. 1973. *J. Biol. Chem.* 248:3997–4002
169. Koizumi M, Hiratake J, Nakatsu T, Kato H, Oda J. 1999. *J. Am. Chem. Soc.* 121:5799–800
170. Kim S, Germond J-E, Pridmore D, Söll D. 1996. *J. Bacteriol.* 178:2459–61
171. Cedar H, Schwartz JH. 1969. *J. Biol. Chem.* 244:4112–21
172. Hinchman SK, Henikoff SA, Schuster SM. 1992. *J. Biol. Chem.* 267:144–49
173. Nakatsu T, Kato H, Oda J. 1998. *Nat. Struct. Biol.* 5:15–19
174. Hinchman SK, Schuster SM. 1992. *Protein Eng.* 5:279–83
175. Sugiyama A, Kato H, Nishioka T, Oda J. 1992. *Biosci. Biotechnol. Biochem.* 56:376–79
176. Stewart JJP. 1989. *J. Comput. Chem.* 10:209–20
177. Pang SS, Guddat LW, Duggleby RG. 2003. *J. Biol. Chem.* 278:7639–44
178. Singh V, Shi W, Evans GB, Tyler PC, Furneaux RH, et al. 2004. *Biochemistry* 43:9–18
179. Kitchen DB, Decornez H, Furr J, Bajorath J. 2004. *Nat. Rev. Drug Discov.* 3:935–49
180. Brooijmans N, Kuntz ID. 2003. *Annu. Rev. Biophys. Biomol. Struct.* 32:335–73
181. Schneider G, Fechner U. 2005. *Nat. Rev. Drug Discov.* 4:649–63
182. Gane PG, Dean PM. 2000. *Curr. Opin. Struct. Biol.* 10:401–4
183. Bohacek RS, McMartin C, Guida WC. 1996. *Med. Res. Dev.* 16:3–50
184. Bogdanov B, Smith RD. 2005. *Mass Spectrom. Rev.* 24:168–200
185. Saghatelian A, Jessani N, Joseph A, Humphrey M, Cravatt BF. 2004. *Proc. Natl. Acad. Sci. USA* 101:10000–5
186. Lipinski C, Hopkins A. 2004. *Nature* 432:855–61
187. Stoughton RB, Friend SH. 2005. *Nat. Rev. Drug Discov.* 4:345–50

RELATED RESOURCES

Schröder M, Kaufman RJ. 2005. *Annu. Rev. Biochem.* 74:739–89

Domains, Motifs, and Scaffolds: The Role of Modular Interactions in the Evolution and Wiring of Cell Signaling Circuits

Roby P. Bhattacharyya, Attila Reményi, Brian J. Yeh, and Wendell A. Lim

Department of Cellular and Molecular Pharmacology, University of California, San Francisco, California 94143; email: lim@cmp.ucsf.edu

Key Words

signal transduction, modularity, evolvability, docking, synthetic biology

Abstract

Living cells display complex signal processing behaviors, many of which are mediated by networks of proteins specialized for signal transduction. Here we focus on the question of how the remarkably diverse array of eukaryotic signaling circuits may have evolved. Many of the mechanisms that connect signaling proteins into networks are highly modular: The core catalytic activity of a signaling protein is physically and functionally separable from molecular domains or motifs that determine its linkage to both inputs and outputs. This high degree of modularity may make these systems more evolvable—in principle, novel circuits, and therefore highly innovative regulatory behaviors, can arise from relatively simple genetic events such as recombination, deletion, or insertion. In support of this hypothesis, recent studies show that such modular systems can be exploited to engineer nonnatural signaling proteins and pathways with novel behavior.

Contents

- INTRODUCTION 656
- EVOLVABILITY OF CELL CIRCUITRY: MAKING NEW CONNECTIONS 656
  - Connecting Transcriptional Nodes: Structural and Functional Modularity 657
  - Classical Regulatory Proteins 657
  - Modularity of Eukaryotic Signaling Proteins 659
- DOCKING INTERACTIONS: RECOGNITION BEYOND THE ACTIVE SITE 660
  - Use of Distributed Surfaces for Recognition 661
  - Versatility of Docking Interactions in Organizing Kinase Connectivity 662
  - Regulation via Docking Interactions 663
  - Evolvability of Kinase Circuits Using Docking Interactions 663
- MODULAR RECOGNITION DOMAINS: STRUCTURAL SEPARATION OF CONNECTIVITY AND CATALYSIS 664
  - Increased Recombinational Possibilities 665
  - Regulation by Modular Domains .. 665
  - The Problem of Domain Discrimination 665
- SCAFFOLDS AND ADAPTERS: GENETICALLY INDEPENDENT WIRING ELEMENTS 669
  - Organization of Signaling Complexes by Scaffold and Adapter Proteins 669
  - Regulation by Scaffolds and Adapters 669
- REWIRING SIGNALING PATHWAYS: SYNTHETIC BIOLOGY 671
  - Synthetic Scaffolds, Adapters, and Docking Interactions 672
  - Synthetic Signaling Switches 672
- CONCLUSIONS: MODULARITY AND EVOLVABILITY OF BIOLOGICAL REGULATORY SYSTEMS 674

INTRODUCTION

Living cells must constantly monitor and respond to their environment and internal conditions. In metazoans, individual cells must communicate with and respond to other cells in the organism. Thus, cells require a remarkable array of sophisticated signal processing behaviors that rivals or surpasses that of modern computers. Many of these responses involve processing by networks of cytoplasmic signaling proteins. Here we review recent advances in our understanding of the fundamental design principles underlying the structure and mechanism of eukaryotic signaling proteins, focusing particularly on how they are functionally linked to one another to form complex circuits capable of information processing. We discuss how the modular organization of the polypeptides that participate in signaling may help facilitate the evolution of innovative circuitry and corresponding phenotypes, providing increased fitness in a competitive and changing environment.

EVOLVABILITY OF CELL CIRCUITRY: MAKING NEW CONNECTIONS

How have the incredibly diverse and complex phenotypes observed in modern eukaryotic organisms evolved? A growing body of work suggests new phenotypes rarely arise through the evolution of radically new proteins (1). Rather, innovation is thought to occur through the establishment of novel connectivities between existing or duplicated proteins to generate new regulatory circuits and thereby new regulatory behaviors (**Figure 1a**). This model is consistent with the surprisingly small number of protein-coding genes in even very complex organisms and thus the limited number of protein or domain types observed (2–4). Phenotypic diversity and complexity appear to arise from new combinations of proteins and/or protein domains working as a network, not from the generation of completely new protein functions. This strategy is similar to that of electronic circuits—a huge variety of circuits can be built from a finite set of electronic components by wiring them together in different ways. Thus, a critical question is how new input-output connections can be established between biological components.

Connecting Transcriptional Nodes: Structural and Functional Modularity

Although this review focuses on protein-based signaling circuits, it is instructive to consider briefly how new connectivities are generated in transcriptional networks, a different class of biological regulatory networks (**Figure 1b**). Transcriptional control is mediated by promoters that respond to signals provided by upstream transcription factors and convert this input into gene expression. Transcriptional nodes are highly modular (1, 5–6a). First, they display structural modularity: The output region, the coding sequence to be transcribed, is physically separable from the input regions, the *cis*-acting elements that regulate expression. Second, and perhaps more importantly, they display functional modularity: Input and output components still function when separated and can be recombined to yield new input-output connectivities. For example, insertion of a new *cis*-acting element into a promoter can place a gene under the control of a new input pathway (6a, 7). Alternatively, insertion of a new gene behind a promoter can result in a radically new output in response to the same input signal. Even linking input and output elements that have had no previous physiological relationship will often work, in large part because gene expression is controlled by standardized general transcription machinery. Thus, the highly modular structure of promoters allows the input and output elements to be easily transferred to yield novel connectivities. Transcriptional nodes are therefore thought to provide a highly evolvable system (6a, 8). Recombination of transcriptional input and output components is thought to be a major source of phenotypic variation during evolution (1).

Classical Regulatory Proteins

Historically, the best-studied regulatory proteins are enzymes involved in metabolic pathways, which lack the modularity of transcriptional nodes and therefore present several fundamental problems with respect to generating new input-output connectivities. The output of an enzyme—the reaction it catalyzes and the products generated—is dependent on precise stereochemical requirements; thus, enzymes cannot easily undergo radical changes in output without compromising catalytic activity. Input control of enzymes can be mediated by allosteric effectors; binding of these effectors at allosteric sites is coupled to specific conformational changes at the active site (9, 10). The intimate and subtle coupling between allosteric sites and the catalytic center limits the possibility of radically modifying allosteric input control without concomitantly compromising function or stability of the active site. In summary, such metabolic enzymes rarely show structural or functional modularity; the elements that mediate input and output are often found within a single cooperatively folding unit and therefore cannot easily be independently modified. Such systems, which we refer to as being tightly functionally integrated, have less readily

Node: simplest element of a signaling network that can translate input into output

Structural modularity: the ability of a large molecule or system to be physically separated into multiple, structurally independent domains

Functional modularity: a domain's ability to function independent of context, allowing transfer of function between diverse molecular systems

Functional integration: the consolidation of multiple functions, e.g., input and output, into a single, nondecomposable structural module

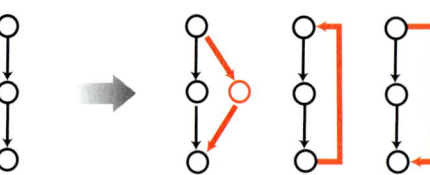

- New nodes
- New connectivities

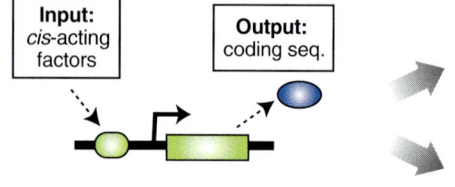

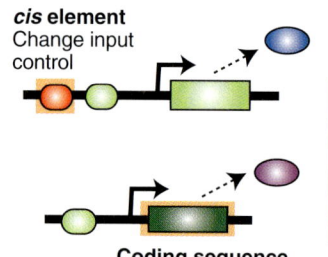

High modularity

High evolvability

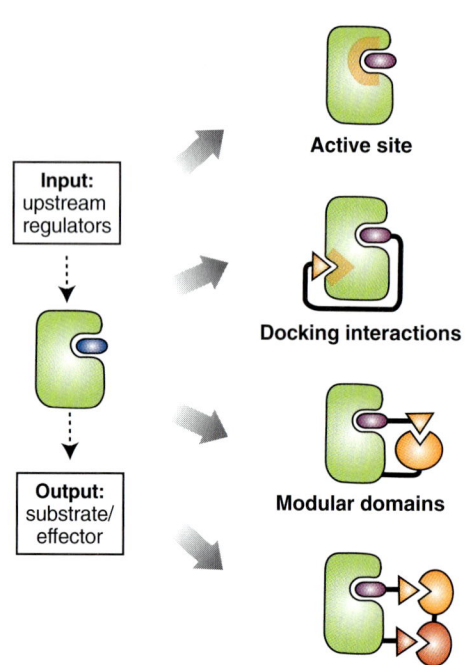

Highly integrated catalytic and regulatory functions

Separable, standardized components mediate connectivity and regulation of diverse catalytic units

transferable elements and thus are not as evolvable as functionally modular systems.

Modularity of Eukaryotic Signaling Proteins

Signaling pathways involve enzymes that catalyze reactions such as phosphorylation, dephosphorylation, and nucleotide exchange. The input control of such enzymes determines when, where, and by what they are activated. The output control determines what downstream partners these enzymes act upon once activated.

Signal transduction enzymes utilize far more modular mechanisms to determine their input-output connectivities than do classical metabolic enzymes (11). Over the last decade, our understanding of the design principles of signaling enzymes has increased dramatically as a result of mechanistic and structural studies as well as the sequencing of multiple eukaryotic genomes. Signaling enzymes often contain, in addition to their core catalytic function, multiple independently folding domains or motifs that mediate connectivity by interacting with other signaling elements. These modules are found in different combinations with diverse catalytic functions, suggesting insertion and recombination of modules may be a common mechanism of the evolution of new proteins and connections (2–4).

Eukaryotic signaling proteins appear to have developed a range of modular strategies for controlling their input and output connectivities, all of which involve increased functional separation between core catalytic elements and connectivity elements (**Figure 1c**). Here we review three basic mechanisms by which the catalytic activity of kinases and other signaling functions are directed and regulated in a modular manner: the use of peripheral docking sites, modular interaction domains, and scaffolding and adapter proteins. Each of these mechanisms can be used to select functional upstream and downstream partners as well as, in many cases, to allosterically regulate catalytic activity. These mechanisms represent a continuum of increasing structural modularity in which catalytic function is separated from the elements that determine its wiring (e.g., scaffolds or adapter proteins represent a separation of catalysis and input control into separate gene products).

We explore the hypothesis that the increasing modularity observed in signaling proteins correlates with higher evolvability: This framework may promote the formation of diverse linkages between catalytic functions via generic, standardized connecting elements. These modular connecting elements may facilitate the evolution of more complex phenotypes, much as standardized components facilitate the design of diverse and complex devices in engineering. We also review an

Module: an independently folding domain that can carry out a simple function

Docking: interaction between a catalytic domain and a partner protein that does not involve the active site

Scaffold: a protein that binds and colocalizes three or more members of a catalytic pathway

Adapter: protein that binds and colocalizes two functionally interacting members of a catalytic pathway

Evolvability: the ability of a system to generate new heritable traits or behaviors through genetic changes

Figure 1

Modularity and evolvability of cellular regulatory circuits and nodes. (*a*) Evolution of new regulatory pathways and responses. A simple linear pathway (*black*) can be converted to a more complex one through the addition of novel nodes that introduce branch points or by the generation of novel functional linkages between existing components, such as the feedback or feedforward circuits depicted. New components and connections are shown in red. (*b*) New connectivity with transcriptional nodes. Transcriptional circuits exemplify a highly modular network, as simple recombination events can alter input-output relationships. Introduction of new *cis*-acting elements such as promoters and enhancers can alter input control, and insertion of a new coding sequence downstream of an existing set of *cis*-acting elements can impose an existing mode of regulation upon expression of a different gene. (*c*) New connectivity with protein/enzyme nodes. Four means of mediating connections between protein nodes are depicted: active site recognition, docking interactions, recognition through modular domain/ligand pairs, and interactions mediated by organizing factors such as scaffolds or adapters. These connection strategies fall on a continuum of modularity versus integration; greater separation between catalytic functions and interactions that mediate connections lends itself to greater evolvability of the signaling network.

emerging body of work that demonstrates recombination of modular components can be used to rewire signaling pathways in nonnative ways, supporting this hypothesis. Because protein phosphorylation is an important currency of information in a cell and protein kinases are among the best studied of the signaling enzymes (12), much of our review focuses on the diversity of ways that protein kinases are integrated into signaling pathways.

DOCKING INTERACTIONS: RECOGNITION BEYOND THE ACTIVE SITE

Since Fischer (13) formulated his lock-and-key hypothesis for enzymes at the end of the nineteenth century, biochemists have generally assumed that the substrate specificity of an enzyme was determined primarily by stereochemical complementarity with its active site. Recently, however, a number of signaling enzymes have been characterized in which surfaces distinct from the active site play an equally important role in mediating substrate or partner recognition (**Figure 2a**). For example, many proteases have secondary substrate recognition sites referred to as exosites (14, 15). Similarly, many protein kinases have secondary partner recognition sites referred to as docking sites (16, 17). Here we focus on kinase docking sites, as these are well understood and most relevant to our focus on intracellular signaling.

As demonstrated over 30 years ago, many protein kinases display clear preferences for the amino acid sequence immediately surrounding the phosphorylated residue in the substrate (18). Such preferences can now be identified by peptide library–based phosphorylation studies (19, 19a). However, in many

Figure 2

Docking grooves can mediate connectivity and regulation of serine/threonine kinases. (*a*) Docking involves interactions between an enzyme and its substrate that take place away from the active site of the enzyme. Such interactions contribute to substrate selection and catalytic efficiency. (*b*) Docking grooves are found at various surfaces on an enzyme. The structure shown is of the budding yeast MAPK Fus3 but is meant to represent a generic kinase fold for the purposes of illustrating the different possible binding surfaces. Several examples are shown of kinase docking interactions. The mitogen-activated protein kinase (MAPK) docking groove for D-box ligands is on the back side of the kinase opposite the active site; the figure illustrates a Fus3/Ste7 D-box peptide complex crystal structure (26). The 3-phosphoinositide-dependent kinase (PDK)/AGC docking groove, also known as the PDK1 interaction fragment (PIF) pocket, mediates interactions at the N-terminal lobe of the kinase (39). The glycogen synthase kinase-3 (GSK3) docking groove for binding primed substrates is located on the N-terminal lobe adjacent to the active site (41), whereas the interacting surface for another ligand, axin/FRAT, is in the C-terminal lobe (140, 141). The MAPK DEF docking groove for FxFP ligands is adjacent to the active site on the C-terminal lobe (33). Thus, the highly conserved kinase structure has been exploited on several different surfaces for diverse types of docking interactions. (*c*) Docking interactions mediate many different types of connectivities within an MAP kinase cascade. MAPKs have docking grooves that interact with cognate docking motifs in activators [MAP kinase kinases (MAPKKs)], inactivators (MAPK phosphatases), substrates, and other pathway modulators such as scaffolds. In addition, MAP kinase kinase kinases (MAPKKKs) have docking grooves on their kinase domains that interact with DVD (domain for versatile docking) motifs on their MAPKK substrates. (*d*) Some docking interactions regulate enzyme activity in more complex ways than simple localization. Certain docking motifs alter the efficiency of an enzyme (k_{cat}) through classical allosteric effects, repositioning residues involved in catalysis. Others are involved in regulated interactions with substrates, in which covalent modifications such as phosphorylation can either promote or inhibit a docking interaction. Finally, some enzymes contain weak intramolecular docking motifs that autoinhibit their own activity; such an enzyme can then be activated by displacement of the intramolecular docking interaction by an external docking site on a substrate or other effector. (*e*) Docking motifs can direct the specificity of enzymes. Whereas a relatively large array of substrates may fit the stereochemical requirements for catalysis at the active site, those with appropriate docking motifs will be selectively used by kinases with cognate docking grooves.

cases, these substrate motif preferences are not sufficient to predict functional connectivity of kinases: Some ideal motifs do not appear to be endogenous substrates, and conversely, some known endogenous substrates do not match ideal profiles (16, 20, 20a). In addition, certain kinases appear to be quite promiscuous for minimal peptide substrates (16, 20).

Use of Distributed Surfaces for Recognition

The use of docking site interactions has emerged as a common mechanism used by certain serine (Ser)/threonine (Thr) kinases to achieve both selectivity and regulation (16, 17). Docking interactions involve a docking groove on the kinase that is distinct from the active site. The docking groove recognizes

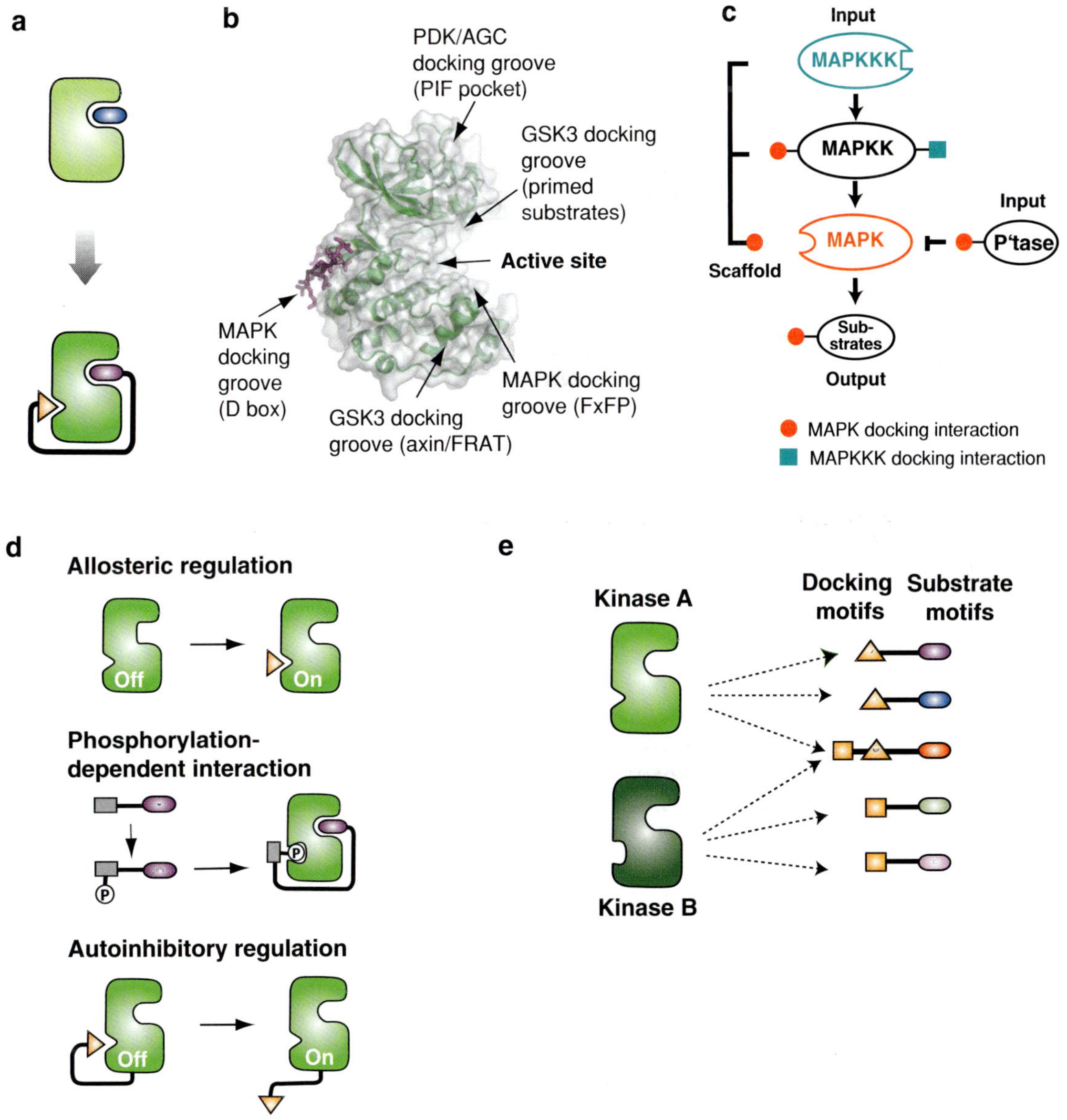

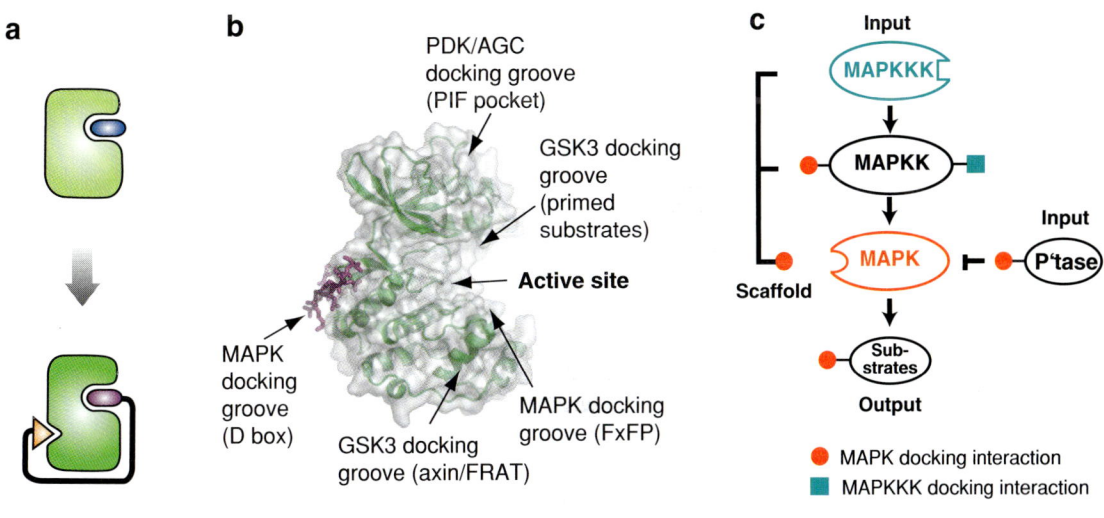

MAPK: mitogen-activated protein kinase

JNK: c-Jun N-terminal kinase

ERK: extracellular signal regulated kinase

a peptide docking motif, which is distinct from the actual phosphoacceptor substrate motif but on the same molecule. Docking interactions appear to function as extended recognition surfaces that can increase enzyme-substrate encounters (reduce K_m) and confer higher specificity than can be achieved by interactions between the active site and substrate motif alone. Moreover, such increases in efficiency and specificity can be achieved without alteration and compromise of active site function.

Docking grooves are found in several Ser/Thr kinase families; here we focus on the mitogen-activated protein kinases (MAPKs) (21, 22). The best-characterized MAPK docking motif is referred to as the D-box, which is recognized by a conserved groove on the MAPK (23). The structures of several D-box docking complexes have been solved (**Figure 2b**) (24–26), revealing the docking groove is on the opposite surface from the active site. Mutation of either the docking groove on MAPKs or of the docking motif on substrates disrupts proper signal transmission (26–28).

Many MAPKs have analogous D-box interacting sites, including the mammalian MAPKs p38, c-Jun N-terminal kinase (JNK), extracellular signal regulated kinase (ERK), and the yeast kinases Fus3 and Kss1 in *Saccharomyces cerevisiae* and Spc1 in *Schizosaccharomyces pombe* (22, 23, 28–31a). However, many of these kinases show distinct motif-sequence preferences (26, 32). Presumably the distinct docking and active site specificities work together to increase overall selectivity of kinase-substrate interactions.

Docking grooves can also be found at other locations on the surface of certain MAPKs, such as the groove that recognizes the consensus motif FxFP. This docking groove on the MAPK ERK, referred to as the DEF site (docking groove for ERK, FxFP), has been mapped by hydrogen-exchange studies to lie on the large domain of the kinase, below the active site (33). Interestingly, the positional relationship of the phosphoacceptor sites and the MAPK docking motifs within substrates can vary. Whereas D-box motifs are located variably with respect to the phosphoacceptor site, FxFP motifs are almost always 10 residues C-terminal to the phosphoacceptor site. Thus, such motifs can play a role in specifically directing which sites are effectively phosphorylated in a substrate bearing multiple potential phosphorylation sites (34).

Docking grooves have been identified in several families of Ser/Thr kinases, in addition to MAPKs (16). These docking grooves are distributed across the surface of the kinase domain (**Figure 2b**), illustrating how much of the kinase surface can potentially be tapped for this type of additional recognition function. The spatial relationship between the docking groove and active site on the kinase may set the distance constraints between the docking and phosphoacceptor sites in substrates.

Versatility of Docking Interactions in Organizing Kinase Connectivity

Studies of MAPK pathways reveal the importance and versatility of docking interactions in guiding many circuit connections (**Figure 2c**). Not only are docking motifs found in MAPK substrates, such as downstream transcription factors, but they are also found in upstream kinases [MAP kinase kinases (MAPKKs)], downregulatory phosphatases, and other regulatory partners, such as scaffold proteins (22, 26, 28, 29, 34a). More recently, docking interactions have been found to play an important role at a different level in MAPK cascades: Several MAP kinase kinase kinases (MAPKKKs) have been found to recognize peptide docking motifs found in their specific MAPKK downstream partners (35, 36). Such motifs have been found in yeast and mammalian systems. The motifs appear to bind directly to the kinase domain of the MAPKKK and to play a critical role in determining MAPKKK → MAPKK specificity.

Regulation via Docking Interactions

In most cases, docking interactions appear to play a relatively passive role as modular specificity control elements: They presumably increase the likelihood of enzyme-substrate encounter. However, in some cases, these interactions appear to regulate kinase function directly (**Figure 2d**). For example, there are now several reported cases in which peptide binding at the docking groove can allosterically activate kinase function. Certain D-box docking site peptides can stimulate MAPK catalytic activity or autophosphorylation (24), whereas others may inhibit activity (25). FxFP motif binding to ERK appears to be coupled to the positioning of the ERK activation loop (33). In addition, 3-phosphoinositide-dependent kinase-1 (PDK1) interacts with downstream substrate kinases that contain a conserved docking motif known as the PDK1 interaction fragment (PIF). Binding of PIF motifs to PDK1 increases kinase activity (37, 38).

Another way in which docking motifs can act as regulatory elements is when the docking interactions are themselves phosphorylation dependent. For example, PIF motifs must be phosphorylated before they bind effectively to the PIF pocket and activate PDK1 (PIF motif: Phe-X-X-Phe-pSer/pThr-Phe/Tyr). Thus, downstream substrates must be subjected to a priming phosphorylation prior to the interaction with and phosphorylation by PDK1 (38, 39). A similar priming event is required for phosphorylation of some substrates by glycogen synthase kinase-3 (GSK3), which is part of the insulin signaling pathway. GSK3 substrates must be phosphorylated on a residue that is C-terminal to the Ser/Thr site to be modified by GSK3 (40). This priming phosphorylation motif binds to a phospho-recognition docking groove adjacent to the active site (41) (**Figure 2b**). The priming phosphorylation scheme observed in GSK3 and PDK1 pathways provides a mechanism for making signal processing dependent on a sequence of catalytically distinct phosphorylation events, thereby increasing the specificity and complexity of control.

Finally, docking interactions, because they are critical for proper substrate recognition, can be used as targets for autoinhibition. For example, GSK3 can be inactivated by kinases that phosphorylate its N terminus. This phosphorylation event creates an intramolecular motif that mimics a docking site sequence, binding at the priming phosphate docking groove and occluding downstream substrate recognition (41, 42).

Evolvability of Kinase Circuits Using Docking Interactions

The development of substrate recognition sites distinct from the actual phosphoacceptor sequence dramatically increases the modularity of kinase interactions and connectivities. Related kinases can develop slightly different docking grooves, thus allowing them to have distinct specificities without evolutionarily taxing the structure and efficiency of the active site. For instance, the closely related yeast MAPKs, Fus3 and Kss1, which function in the mating and invasive growth pathways, respectively, retain docking grooves that equivalently recognize docking motifs on interacting partners shared by the two kinases, such as the MAPKK Ste7, which functions in both pathways (42a). However, they have evolved some degree of discrimination in binding to substrates specific to one pathway: Fus3 binds the docking motif from the mating pathway effector Far1 more tightly than does Kss1 (26), explaining its selectivity toward this substrate (43). These short docking peptides that mediate specific recognition can be spliced into potential substrates to mediate a new, specific connection (**Figure 2e**).

Nonetheless, docking motifs are limited in their degree of modularity and evolvability. The docking grooves are intimately tied to the core catalytic module, in this case the catalytic Ser/Thr kinase domain. Thus, although

PDK1: 3-phosphoinositide-dependent kinase-1

PIF: PDK1 interaction fragment

GSK3: glycogen synthase kinase-3

SH2: Src homology 2

SH3: Src homology 3

docking motifs can be easily transferred to new substrates, the docking grooves cannot be dramatically altered or transferred to unrelated catalytic activities. Docking grooves are a step toward the separation of recognition and catalysis, but they do not employ generic interactions that could be transferred to new functions. Thus, docking grooves may represent a more ancestral solution to achieving modular connectivities. Interestingly, although docking interactions are prevalent in many Ser/Thr kinases (the more ancient eukaryotic protein kinases), similar docking interactions have not been identified in the more recently evolved tyrosine kinases. Instead, as discussed below, many other catalytic functions utilize structurally independent recognition domains to mediate connectivity—a further step toward more standardized circuit connectivity.

MODULAR RECOGNITION DOMAINS: STRUCTURAL SEPARATION OF CONNECTIVITY AND CATALYSIS

The evolution of metazoans appears to have coincided with an explosion in the use of modular protein domains, including many recognition domains that play a major role in diverse cell signaling processes (2–4) (**Table 1**). These include, for example, domains that recognize peptides [e.g., Src homology 3 (SH3) domains], phosphopeptides (e.g., SH2 domains), and phospholipids [e.g., pleckstrin homology (PH) domains]. The detailed functions of these diverse domains are reviewed elsewhere (44–52). Compared with the more specialized Ser/Thr kinase docking sites, such domains represent an even more complete physical separation between elements that

Table 1 Abundance of selected modular domains (and proteins containing them) in commonly studied eukaryotes

	Homo sapiens	*Mus musculus*	*Drosophila melanogaster*	*Caenorhabditis elegans*	*Saccharomyces cerevisiae*
SH3[a]	223 (180)[b]	124 (92)	113 (76)	83 (68)	26 (22)
WW	91 (49)	27 (17)	21 (14)	40 (22)	9 (6)
PDZ	234 (126)	119 (78)	98 (71)	106 (79)	3 (2)
SH2	112 (98)	73 (67)	33 (30)	67 (66)	1 (1)
PTB	34 (30)	14 (12)	7 (7)	23 (20)	0 (0)
14-3-3	8 (8)	4 (4)	4 (4)	2 (2)	1 (1)
BRCT	39 (20)	23 (12)	28 (16)	44 (29)	9 (6)
FHA	16 (16)	9 (9)	17 (17)	12 (12)	13 (12)
C2	149 (99)	94 (63)	51 (36)	93 (64)	22 (11)
Total genes[c]	30,000	30,000	14,000	19,000	6,300

[a]Abbreviations and descriptions of domains in table: SH3 = Src homology 3 domain, binds PxxP peptide ligands (52); WW = PxxP binding domain named after two conserved Trp residues (52); PDZ = domain from PSD-95, Dlg, ZO-1, binds C-terminal peptide ligands (47); SH2 = Src homology 2 domain, binds phospho-Tyr peptide ligands (50); PTB = phospho-Tyr binding domain (50); 14-3-3 = phospho-Ser/Thr binding domain (44); BRCT = breast cancer susceptibility gene, C-terminal domain, binds phospho-Ser/Thr peptide ligands (46); FHA = forkhead-associated domain, binds phospho-Ser/Thr peptide ligands (45); C2 = domain from protein kinase C, binds phospholipids and occasionally phospho-Tyr peptide ligands (51).
[b]These data are gathered from the SMART (Simple Modular Architecture Research Tool) database (http://smart.embl-heidelberg.de) in Genomic Mode in May 2006 and reflect our current best estimates of the domain contents of the genomes of these organisms; however, since our knowledge of some of these genomes is less than total, some redundancies may exist, leading to artificially inflated domain counts in some cases (61, 62).
[c]Source: Human Genome Project Information, Functional and Comparative Genomics Fact Sheet (http://www.ornl.gov/sci/techresources/Human_Genome/faq/compgen.shtml).

mediate connectivity from those that mediate catalytic functions.

Increased Recombinational Possibilities

From a genetic perspective, modular interactions offer more flexibility than docking interactions: Both the peptide motifs and their cognate domains can be transferred through simple genetic exchanges such as recombination and insertion. Thus, both an enzyme and its substrate can make new connections by incorporating a relevant recognition domain or motif (**Figure 3a**). Circumstantial evidence for this higher degree of transferability can be found by comparing metazoan genomes. Increasing phenotypic complexity appears to correlate not with the development of new domains (only 7% of human protein families are vertebrate specific), but rather with an increase in the type and number of new domain combinations: Humans have 1.8-fold more distinct protein architectures (arrangements of domains in primary sequence) than do worms and flies (2). An example of domain mixing and matching is shown in **Figure 3b**, illustrating how specific regulatory and catalytic domains can be found in many combinations to yield proteins, and therefore pathways, with highly diverse input-output relationships.

Regulation by Modular Domains

Similar to docking sites, modular domains can be used not only to physically link partner proteins but also as regulatory elements (**Figure 3c**). Several classes of interaction domains display conditional recognition. These include phosphopeptide recognition domains such as SH2 domains, for which the linkage of a catalytic domain to its partners depends on a prior phosphorylation event (53). Similarly, regulated membrane localization can be achieved with lipid recognition modules that bind to rare phosphoinositide species such as phosphoinositol-(3,4,5)-trisphosphate that are only produced upon activation of phosphoinositide 3-kinase (48).

Modular recognition domains can also play more sophisticated roles in achieving allosteric regulation, most commonly through autoinhibitory mechanisms. Domains can interact in an intramolecular fashion with catalytic domains, either acting as pseudosubstrates or sterically occluding accessibility of the active site (54). The catalytic function can be specifically switched on by the binding of competitive ligands or by covalent modification events that disrupt the autoinhibitory interaction. In other cases, domains can interact with cognate motifs in a manner that conformationally disrupts catalytic function. In some cases, such as the Src family kinases or the actin regulator neuronal Wiskott-Aldrich syndrome protein (N-WASP), multiple domains function together to stabilize an inactive state of their respective catalytic output domains (55–60). In these cases, the proteins can act as sophisticated switches that are able to respond in complex ways to multiple inputs. For example, a protein might approximate an AND gate if two intramolecular interactions must both be disrupted to release the autoinhibited catalytic function. Interestingly, these modular allosteric switches show behavior similar to more conventional allosteric proteins: Switching involves preferential stabilization of a high-activity state by a ligand. However, in the case of modular switches, there is a clear physical and functional separation between the regions of the protein that mediate input regulation and those that mediate output catalytic activity. Not only does this architecture lend itself to increased transferability of function, but modularity may also allow the incremental construction of switch proteins with multiple layers of input control.

The Problem of Domain Discrimination

Although the use of modular domains may allow the rapid generation of new signaling input-output relationships, the expansion of

N-WASP: neuronal Wiskott-Aldrich syndrome protein

a
Docking motifs
Substrate motifs
New recognition module
Transfer of recogntion motif to new substrates/partners
Transfer of recognition module to new catalytic module
c
Phosphorylation-dependent interaction
P
Modular allostery
Steric or pseudosubstrate inhibition
Off
Conformational inhibition
Off
Competitive ligands
Covalent modification
ON
b
Regulatory domain
Modular protein
Catalytic domain
WHD
B
pro-rich
VCA
WAVE
EVH1
B
GBD
pro-rich
VCA
WASP
GBD
IS
Kinase
PAK
SH3
SH2
Kinase
Src
SH2
Phosphatase
pro-rich
SHIP
Voltage sensor
Phosphatase
Ci-VSP
Voltage sensor
Pore
Shaker
Bind activated Cdc42
GBD
Bind pY peptides
SH2
Sense membrane potential
Voltage sensor
VCA
Activation of actin polymerization
Kinase
Protein phosphorylation
Phosphatase
Phosphoinositide dephosphorylation

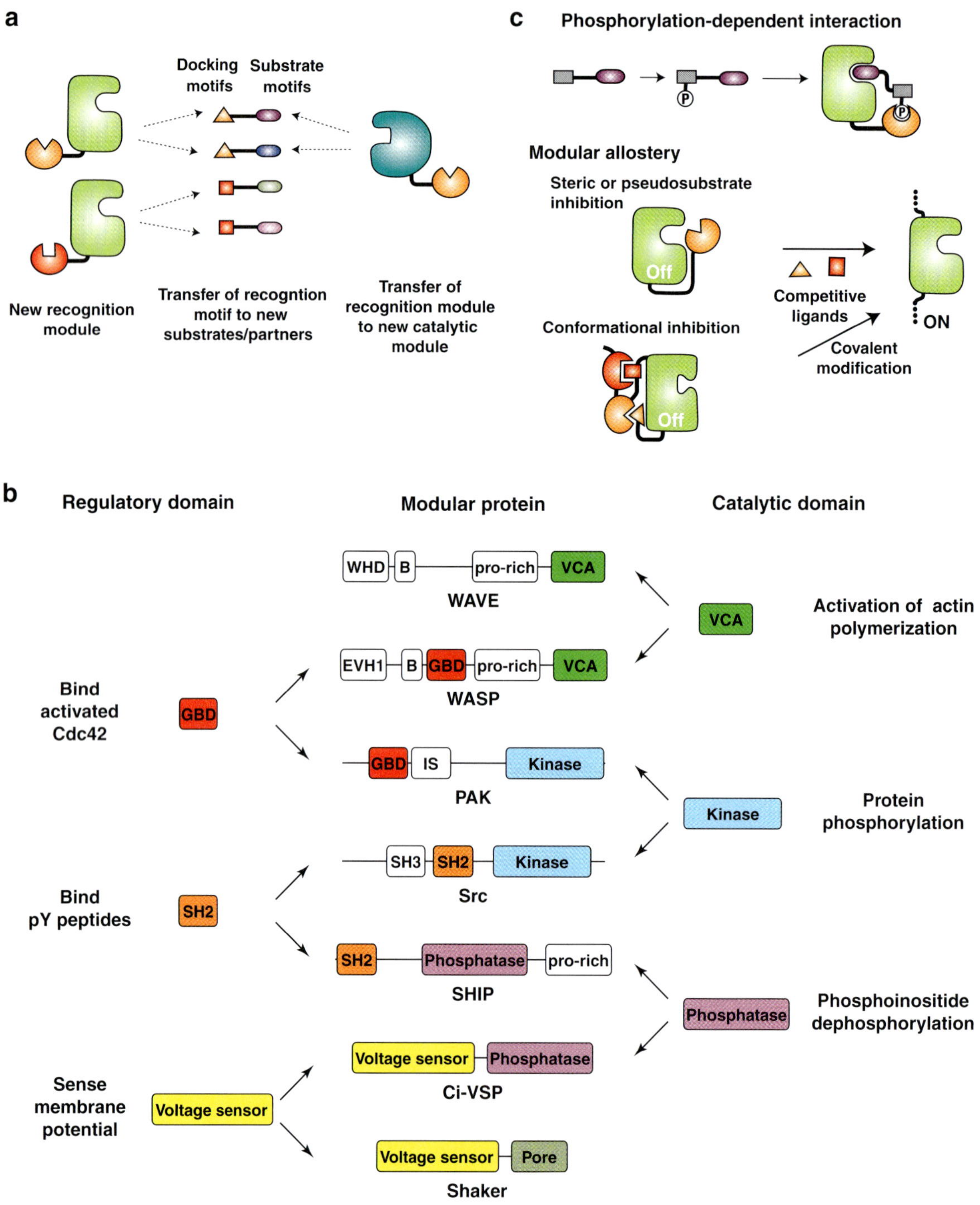

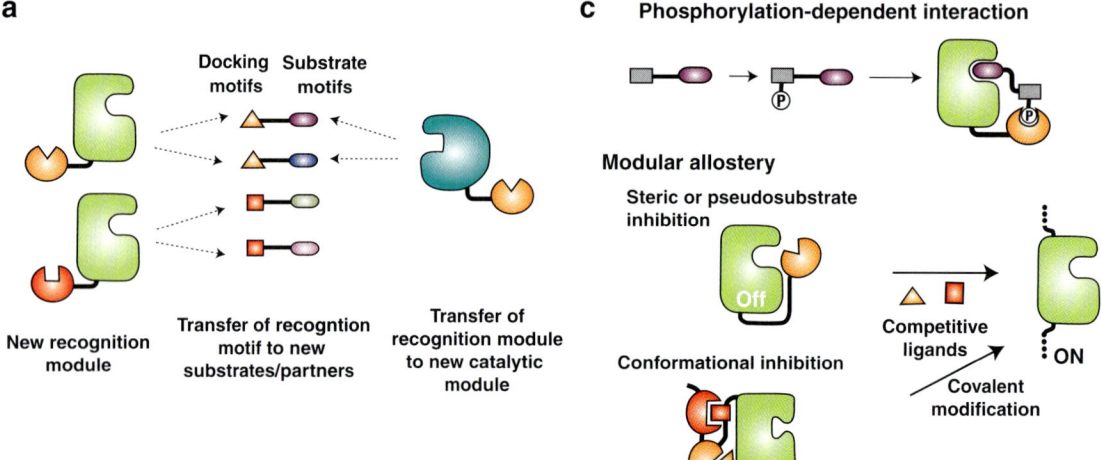

domain families presents a new problem: How can repeated domains in a proteome encode specific information in the context of many related family members (**Table 1**)? For example, let us consider the SH3 domain family, which in most cases binds to proline-rich peptides containing the core motif PxxP: The Simple Modular Architecture Research Tool (SMART) database predicts there are 31 SH3 domains in yeast, 132 in *Caenorhabditis elegans*, 273 in *Drosophila*, and 894 in humans (61, 62). How can an ordered array of component connectivities be maintained by such a large set of related domains?

Recent studies suggest several strategies have evolved for maintaining domain discrimination. First, domains can diverge so far from other family members that they display distinct, noncanonical recognition profiles (**Figure 4***a*). For example, some SH3 domains have diverged to no longer recognize PxxP motifs: The C-terminal SH3 domain of the T-cell adapter protein Gads (Grb2-related adapter downstream of Shc) instead recognizes RxxK motifs (63). This recognition event occurs on a surface distinct from the canonical proline binding pocket (64, 65). One Gads ligand, hematopoietic progenitor kinase-1, binds primarily through an RxxK motif, but its binding is augmented by a weak secondary PxxP motif, thus illustrating the versatility of this divergent domain (66). A pair of SH3 domains in p47phox has been found to act as a single unit, using the surface between the two domains to recognize a novel motif (67). Similarly, noncanonical domains have been found in many other domain families, including the SH2 domain from the protein SAP (also called SH2D1A) that binds unphosphorylated motifs (68) and the C2

Figure 3

Modular interaction domains can mediate new connectivity. (*a*) Transferability of modular recognition and catalytic functions. Modular domains facilitate the formation of new connections between proteins, as standardized recognition domains or their ligands can be swapped onto catalytic modules or substrates via recombination events, opening a new set of possible enzyme-substrate interactions. (*b*) Evidence of evolutionary input-output transfer. Naturally occurring examples are depicted in which domains are reused in various combinations to mediate distinct connections between catalytic activities and target molecules. The VCA (verprolin homology, cofilin homology, acidic) domain, which activates actin polymerization, is common to the actin regulatory proteins WAVE and WASP (142), but it is covalently linked to a different set of interaction domains in each case, contributing to distinct modes of deployment of this output activity. Of these interaction domains, the GTPase binding domain (GBD) of WASP is also found in p21-activated kinase (PAK) (143) and is used to direct its binding partner, activated Cdc42, to each of these two diverse proteins. The kinase domain from PAK is reused in many different contexts. The classical example of Src is depicted, in which the kinase domain is joined with several protein interaction domains, including the Src homology 2 (SH2) domain, which regulates the activation state of the kinase and mediates its interaction with phosphotyrosine-containing peptides (55, 58–60). This SH2 domain is, likewise, reused in many signaling components, such as the SHIP phosphatase (144). The phosphatase domain found in SHIP is reused in multitudes of signaling proteins as well, including the recently described voltage-sensing phosphatase from *Ciona intestinalis*, Ci-VSP. As a result of a fusion of the phosphatase domain with a voltage-sensing domain more traditionally found in voltage-gated channels such as Shaker, Ci-VSP exhibits regulation of its phosphatase activity by membrane potential (145, 146). Thus, many complex signaling proteins are built from a relatively small toolkit of standardized components that are combinatorially connected. (*c*) Enzyme regulation by modular domains. Some modular domains only recognize their ligands after covalent modification resulting from other cellular signaling processes, thus linking the connectivity of proteins containing these domains to the regulation of these other pathways. In addition, modular domains are often used to regulate enzyme activity more directly. These domains can participate in interactions that inhibit catalysis, either by sterically blocking access to the catalytic site or by preferentially stabilizing an inactive conformation of the catalytic domain. These inactive states can then be reversed upon exposure to competing ligands that bind to the domains or by covalent modification of the domains or ligands. Abbreviation: pro-rich, proline-rich peptides.

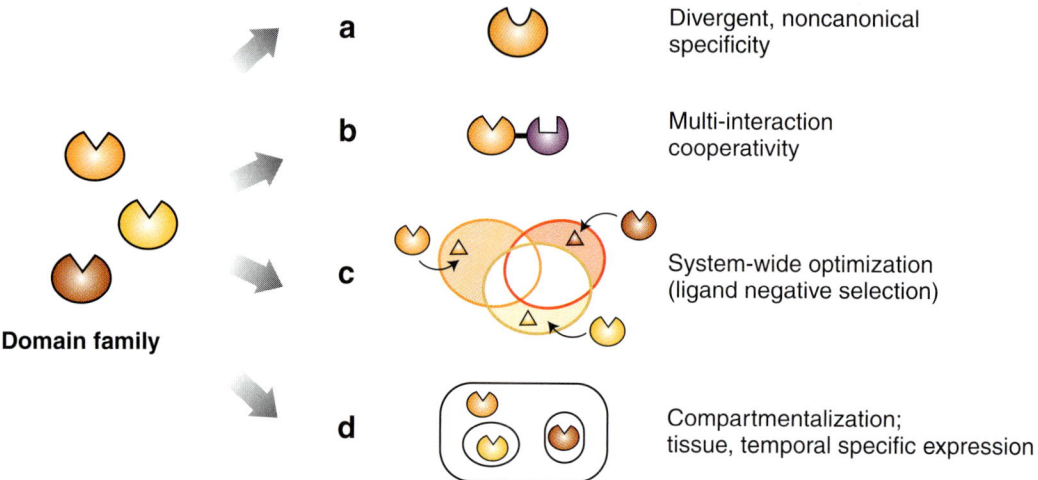

Figure 4

Mechanisms of domain discrimination. (*a*) Domains can evolve divergent ligand-binding pockets that recognize sequences that stray from the consensus for the domain family. (*b*) Multiple domains can be used in combination to generate a combinatorial increase in selectivity and/or affinity compared with the individual recognition events alone. (*c*) Domains and ligands within an organism can coevolve to occupy regions of recognition space with an acceptably low level of cross-recognition. (*d*) Domains and ligands can be segregated in space and time so they are more likely to be coexpressed with genuine interacting partners than with spurious cross-reactive partners.

domain from protein kinase C-δ, from a class of domains that normally binds phospholipids or unphosphorylated peptides, that recognizes phosphotyrosine motifs (51, 69).

A second mechanism for increasing domain-mediated specificity is to use multiple domains to recognize dual ligands in a cooperative manner (70–72) (**Figure 4***b*). A third mechanism is to use system-wide optimization of the domain interaction network (**Figure 4***c*). Recent studies in yeast have shown that although many of the ∼30 SH3 domains have overlapping specificity as determined by peptide libraries, there appears to be some level of negative selection against sequences that interact in a highly promiscuous manner (73). Many physiological partner peptides are optimized for specificity not only by positive selection for binding to the proper domain, but also by negative selection against interaction with competing domains in that genome. Thus, in many cases, individual SH3 domains are only observed to interact with a handful of more than 1500 potential PxxP partners within the genome (74). Finally, a fourth way to achieve specificity is to segregate domains either through subcellular compartmentalization, differential temporal expression, or tissue-specific expression (**Figure 4***d*) (73). Thus, domains with highly overlapping recognition properties might never have to compete for the same targets.

Nonetheless, even with these mechanisms, there likely comes a point at which the information-encoding capability of a domain family is saturated, and increasing signaling complexity may require the development of orthogonal domains. For example, SH2 domains are generally only found in metazoans, and the development of SH2 domains and tyrosine phosphorylation-based signaling in general may have been a prerequisite for the evolution of multicellularity, given its need for increased signaling bandwidth (cell-cell signaling in addition to cell-environment signaling). Interestingly, SH2 domains and receptor tyrosine kinases have recently been identified in choanoflagellates, the closest single-celled eukaryotes to the evolutionary branch point of multicellularity (75).

SCAFFOLDS AND ADAPTERS: GENETICALLY INDEPENDENT WIRING ELEMENTS

From an evolutionary perspective, the ultimate separation of catalytic and connectivity elements would be to segregate such functions into distinct proteins, each of which is genetically independent. Such a separation is achieved with scaffold and adapter proteins, which act as organizing platforms that recruit specific catalytic elements and their upstream and/or downstream partners to the same complex. In general, adapters are defined as organizing molecules that link together two partners, whereas scaffolds, in general, are defined as organizing molecules that link together more than two partners (**Figure 5a**).

Organization of Signaling Complexes by Scaffold and Adapter Proteins

Diverse scaffold- or adapter-mediated complexes are observed in eukaryotes (76). For example, cyclins can be thought of as adapters that change the substrate recruitment properties of associated cyclin-dependent kinases (77). Thus, changes in cyclin expression alter the set of target cyclin-dependent kinase substrates (78). Many MAPK cascades and protein kinase A (PKA) response pathways are coordinated by scaffold proteins that organize sequentially acting members of a pathway together (79–84). In some cases, multiple proteins work together to organize specific pathways (85). From an information-transfer perspective, the most important aspect of scaffolding is that it allows catalytic proteins such as kinases to play several distinct roles depending on the complex into which they are assembled (**Figure 5b**). For example, the yeast MAPKKK Ste11 is used in three distinct MAPK cascades, each of which responds to distinct inputs and yields distinct outputs: the mating, invasive growth, and high-osmolarity response pathways. Ste11 can be used for all of these functions because, in at least two of these pathways, scaffold proteins wire Ste11 such that it retains information about what input activated it and is directed to phosphorylate the appropriate substrates (85–89b). Thus, the subpopulations of Ste11 molecules that participate in different complexes have distinct functions.

Scaffolds and adapters can be built from various interaction components. For example, cyclins interact with both kinase and substrates using highly specialized binding surfaces (77). However, other organizing factors utilize more modular interaction components. The JNK scaffold JIP (JNK interacting protein) uses a canonical MAPK docking motif to bind to JNK (25). Some scaffold interactions are mediated by modular domains, with the scaffold bearing either a modular interaction domain or a motif recognized by such a domain (90, 91).

Regulation by Scaffolds and Adapters

Scaffolds provide many possibilities for complex pathway regulation (**Figure 5c**). For example, differential expression of a scaffold can determine if a pathway will function in a particular cell type (92). Moreover, in some cases, splice variants of scaffolds lacking specific interaction or localization modules have been identified; thus, temporal or tissue-specific control of splicing could alter pathway wiring (92–94). In addition, some scaffold-mediated interactions can be subject to independent regulation. For example, receptor tyrosine kinases and the immune signaling scaffolds LAT (linker for the activation of T-cells) and SLP-76 (SH2 domain-containing leukocyte phosphoprotein of 76 kDa) contain multiple protein recruitment sites that must first be tyrosine phosphorylated before they organize a complex of SH2 domain-containing proteins (71, 95, 96). In another case, phosphorylation of the mammalian Ras/MAPK scaffold kinase suppressor of Ras (KSR) by the Cdc25-associated kinase is used to

PKA: protein kinase A

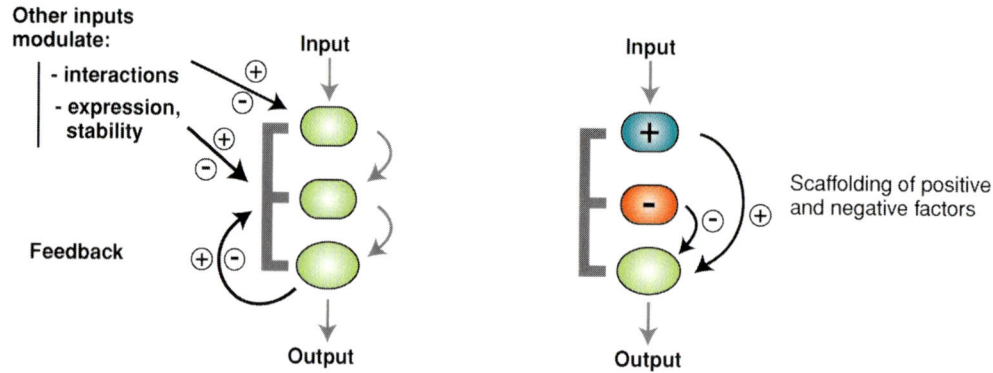

Figure 5

Scaffolds and adapters as mediators of new connectivity. (*a*) Adapters, such as the rounded molecule shown in orange on the left, link two components together. Scaffolds, such as the black and orange molecule on the right, link three or more components of a signaling pathway together. Both classes of molecules separate the catalytic functions of a signaling pathway from the recruitment functions. Either class of molecules may use standardized modular domains or more specific protein-interaction motifs to assemble their associated signaling complex. (*b*) Identical signaling molecules can signal through multiple distinct pathways, responding to different inputs and yielding different outputs despite being activated in chemically indistinguishable ways, by virtue of their recruitment to pathway-dedicated scaffold proteins. These scaffolds act to insulate the shared signaling component in the appropriate complex, encouraging the appropriate interactions for the pathway in question. Activation is often coupled to scaffold localization in such cases, ensuring the fidelity of signal transmission. (*c*) As elements that govern the assembly of components of a signaling complex, scaffolds or adapters can contribute in a number of ways to pathway regulation in addition to passive colocalization. First, they can themselves be the target of modifications, either from outside inputs or from elements of the pathway itself in instances of feedback regulation. Such modifications can alter the way in which interactions occur on these molecular platforms, or they can alter the expression or stability of these critical assembly factors. Second, they can contribute to pathway regulation by recruiting both positive and negative regulators, sometimes in temporally restricted ways, to alter pathway dynamics.

control its interactions and recruitment (97–99). KSR function can also be regulated by a novel protein called impedes mitogenic signal propagation (IMP), which negatively regulates KSR function by inducing its hyperphosphorylation and sequestration into detergent-insoluble domains, preventing it from assembling the normal Ras/MAPK signaling complex (100). The upstream input Ras apparently has two effects. First, activated Ras disrupts the IMP/KSR interaction and induces IMP degradation, thus freeing KSR. Second, Ras interacts with Raf and KSR to promote positive signaling in the pathway. In all of these cases, modulation of the scaffold or its interactions allows regulation of one specific pathway in a manner that does not necessarily affect other unrelated functions of the same catalytic proteins.

Scaffolds can also be used to precisely shape pathway behavior, in addition to simply determining pathway linkages. For example, pathway output, such as terminal kinase activation, might lead to phosphorylation of the scaffold or other components, leading to positive or negative feedback (101, 101a). In the well-studied mating pathway in budding yeast, the scaffold Ste5 allosterically activates one of its binding partners, the MAPK Fus3, initiating a negative feedback loop that regulates pathway output (101a). Such scaffold-mediated feedback loops could have strong effects on the temporal activation profile of the pathway as well as its quantitative dose-response behavior.

In several cases, scaffolds have been found to recruit not only positive acting factors, but also antagonistic negative factors, thus contributing in more complex ways to the shape of the overall signaling response. Some scaffolds recruit both an activating kinase and a deactivating phosphatase. For example, the scaffold JIP-1 (Jnk interacting protein-1) not only binds the MAPK JNK and upstream activators (MAPKKs and MAPKKKs) (102), it also binds the JNK phosphatase MKP7 (MAP kinase phosphatase-7), targeting it to dephosphorylate JNK (103). Another JNK scaffold, β-arrestin 2, binds MKP7 as well, but the phosphatase transiently dissociates from the scaffold upon pathway stimulation, rebinding after 30–60 min (104). Thus, β-arrestin 2 encodes a more sophisticated mechanism for time-dependent pathway activation and inactivation. In addition to MAPK scaffolds, several members of the A kinase anchoring protein family of PKA scaffolds recruit phosphatases as well as PKA (105–107). Some scaffolds use other recruitment strategies to achieve this sort of dynamic pattern of activation and inactivation. For example, a number of A kinase anchoring proteins recruit phosphodiesterases such as PDE4, which break down cyclic AMP, locally inactivating PKA in a spatially restricted negative feedback loop that alters pathway response and may contribute to cyclic AMP homeostasis (108, 109).

The multi-PDZ domain protein InaD acts as a scaffold that mediates *Drosophila* photoreceptor signaling. InaD assembles multiple members of the G-protein coupled phototransduction cascade, including the G-protein effector phospholipase C, the ion channel Trp (transient receptor potential), and an isoform of protein kinase C involved in downregulation of the response (110–112). Thus, InaD appears to play a role not only in accelerating and amplifying the phototransduction response, but also in limiting its timing.

REWIRING SIGNALING PATHWAYS: SYNTHETIC BIOLOGY

Many of the mechanisms described in this review are presented as elements that could facilitate the evolution of new pathway linkages and phenotypes. One approach to testing the evolvability of modular signaling components is to attempt to mimic evolution by using them to create new, synthetic pathways, an approach that is part of the new field of synthetic biology (54, 113, 114).

Synthetic biology: the discipline concerned with engineering biological systems to produce novel biological behaviors

Synthetic Scaffolds, Adapters, and Docking Interactions

Several studies have shown that chimeric or synthetic scaffolds and adapters can be used to generate novel, nonphysiological pathways. For example, a chimeric scaffold made from components of the yeast mating and high osmolarity response MAPK pathways can be used to generate a novel pathway that, in vivo, leads to the osmolarity response when cells are stimulated with mating pheromone (**Figure 6a**) (115). Even the simple step of transferring an osmoresponse-specific MAP-KKK docking motif from the osmolarity MAPKK (Pbs2) to the mating MAPKK (Ste7) is sufficient to induce cross talk to the mating response when cells are stimulated with high osmolarity (36). In mammalian cells, chimeric adapter proteins that link growth factor receptors to apoptotic signaling proteins can be used to convert proliferative signals into death signals (**Figure 6b**) (116).

Synthetic Signaling Switches

Modular recombination can be used to reprogram the input control of a signaling protein. The actin regulatory factor N-WASP is normally allosterically regulated; it stimulates actin-related protein (Arp)2/3 complex-mediated actin polymerization when bound to the GTPase Cdc42 and the phospholipid phosphoinositol-(4,5)-bisphosphate (56, 57). This regulation is achieved through modular autoinhibitory interactions. Dueber et al. (117) deleted the regulatory domains of N-WASP and replaced them with heterologous modular interaction domains and their cognate motifs. Many of the resulting synthetic proteins showed allosteric regulation: They were basally repressed but could be activated by the addition of competing ligands (**Figure 6c**). In many cases, when multiple modular domains were used, the proteins displayed complex signal integration, such as AND-gate behavior requiring the presence of two ligands for potent activation.

Interestingly, the relative ease of generating novel phenotypes from these modular connections can also contribute to the phenotypic manipulation of signaling systems by pathogens or even in stochastic mutations that lead to the development of cancer (49). For instance, the *Yersinia* virulence factor YopM acts as an adapter for the human kinases RSK1 and PRK2, directing a nonnative phosphorylation

---

Figure 6

Synthetic biology: rewiring modular signaling systems. (*a*) Park et al. (115) constructed a synthetic "diverter" scaffold by combining elements from the yeast mating and high osmolarity scaffolds such that the shared mitogen-activated protein kinase kinase kinase (MAPKKK) Ste11 could be activated by a pheromone on the diverter scaffold but could only transmit this signal to components of the high osmolarity pathway. In this manner, a pheromone input was transduced to the output of the high osmolarity response pathway. (*b*) Howard et al. (116) took advantage of the modular design of mammalian signaling components to construct a chimeric adapter protein that generated a novel input-output linkage. By fusing an Src homology 2 (SH2) domain that recognizes a phosphotyrosine on a growth factor-responsive receptor tyrosine kinase to a death-effector domain (DED) that recruits a caspase involved in apoptosis, the authors redirected a proliferative signal input into an output that favors cell death. (*c*) Dueber et al. (117) rebuilt the naturally occurring modular allosteric switch neuronal Wiskott-Aldrich syndrome protein (N-WASP) by replacing its normal regulatory domains with heterologous modular domains. In this manner, the authors generated variants of N-WASP whose activity was gated in different ways by nonnative inputs, the ligands for the appended modular domains. This simple strategy of recombining two modular domain-ligand pairs onto a catalytic domain yielded a set of proteins with surprisingly diverse gating behaviors depending on subtle variations in parameters such as linker length and binding affinity. Abbreviations: Arp, actin-related protein; EGF, epidermal growth factor; GBD, GTPase binding domain; PIP2, phosphoinositol-(4,5)-bisphosphate. Modified and reprinted from Reference 54, copyright 2004, with permission from Elsevier.

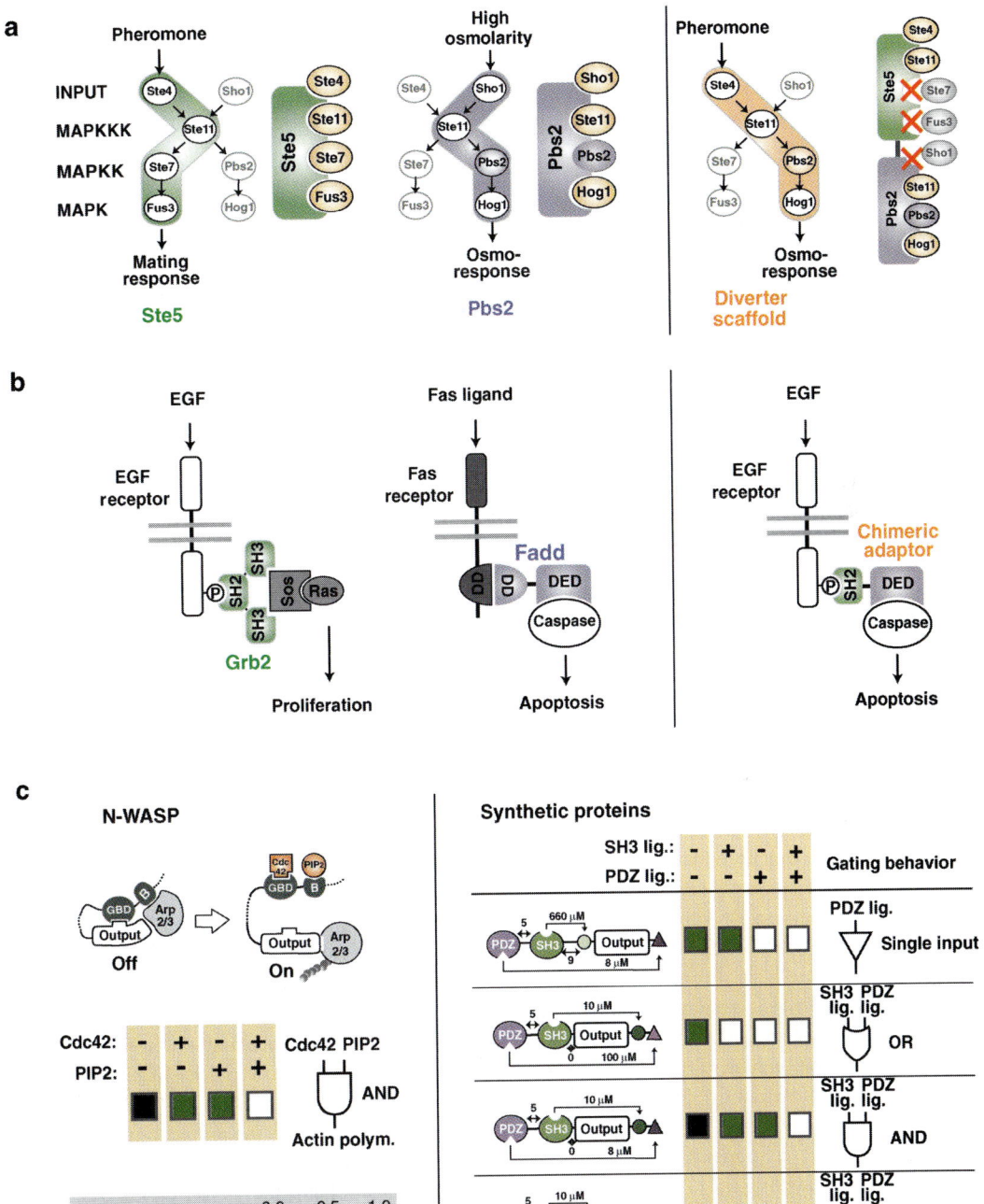

MODULAR DESIGN OF SYNTHETIC DNA MODIFYING ENZYMES

Recent studies have pushed the boundaries of modular design principles in the engineering of synthetic DNA binding proteins. Several groups (126–130a) have constructed libraries of synthetic or naturally occurring zinc finger domains and selected for those that bind to particular 3-bp DNA sequences. If these elements are modular, then multiple zinc fingers could be fused to form multidomain proteins capable of recognizing longer DNA sequences (131, 132). This strategy has succeeded in a number of cases, with as many as 6 zinc fingers fused to form 18-bp recognition elements (126, 132, 133). These extended DNA binding motifs can then be fused to output domains that activate or repress transcription or stimulate cleavage or recombination events, generating designer transcription factors, endonucleases, and recombinases with desired sequence specificity (133–139a). The principles of modular design are critical to the construction of these sophisticated proteins. The challenge of generating a novel, site-specific DNA modifying enzyme can be reduced by combining independently acting modules—multiple recognition domains as well as catalytic elements—into a larger protein with more complex behavior. The ability to generate targeted DNA modifying enzymes on demand has striking implications for functional genomics and gene therapy.

event between these two proteins (118). In addition, the modular organization of signaling proteins can contribute to oncogenesis. Mutational loss or recombination of these regulatory interactions can cause improper activation of important signaling molecules. In the classic case of the v-Src oncogene, a key tyrosine residue involved in an intramolecular interaction with an SH2 domain that autoinhibits the c-Src tyrosine kinase is lost, contributing to oncogenic transformation (119–121). Modularity can thus result in a trade-off between evolvability and fragility.

These studies are consistent with the hypothesis that the modular organization of signaling proteins allows for the facile reconnection of signaling components to yield new pathways and biological responses. In addition, these synthetic approaches present a potentially useful way to systematically perturb and alter complex signaling circuits in a way that may facilitate elucidation of basic systems properties controlling complex biological responses. Moreover, these approaches may allow the rational engineering of cells that could carry out new specific therapeutic functions.

CONCLUSIONS: MODULARITY AND EVOLVABILITY OF BIOLOGICAL REGULATORY SYSTEMS

Highly modular architectures are not only found in eukaryotic signaling systems but also in many other systems, including transcription, proteolysis, and cellular trafficking (1, 122, 123). These systems are characterized by the use of increasingly general, portable elements that can be genetically interchanged to mediate new regulatory connections. The exact domains and motifs that implement these connections vary to some extent, but there may be some pressure to maintain a degree of evolvability in such systems. The reuse of similar modular domains in different contexts represents standardization of the means of communication between protein nodes. Standardization is a central feature of highly complex and evolvable systems (124).

Why might there be selective pressure to maintain modularity and evolvability, given that cellular systems cannot actually foresee the need to change their response behaviors? Presumably, in a constantly changing and competitive environment, the lack of an ability to rapidly evolve novel responses might prove to be a disadvantage. In many cases, highly integrated, nonmodular systems perform in a more efficient, optimal manner, but such performance would only be optimal for a specific and unchanging environment. Hence, modular systems would be more robust to accommodating and buffering against change. During the course of evolution, as environmental pressures shift, there is likely a constant push and pull between the efficiency of integration on one hand and the flexibility and adaptability provided by modularity on the other. A study modeling network development by standard evolutionary algorithms found that modular network structures and motifs evolved spontaneously in response to shifting evolutionary goals (124a). Even in engineered systems such as electronic circuits, where modular components provide an advantage in circuit development, there is often pressure to minimize and integrate circuits once they are well developed. This optimization and integration can lead to a loss of the modularity that was critical during development (e.g., components in integrated circuits do not have transferable functions). Similarly, one might expect that modularity could easily be lost in fundamental housekeeping biological processes, which do not change significantly over evolution. In support of this model, recent bioinformatics studies indicate tissue-specific proteins, especially those associated with evolutionarily newer functions, tend to have a more modular composition than those proteins that are globally expressed and have a housekeeping function (125). Hopefully as more families of closely related genomes are sequenced, we will gain insight into the actual paths by which new signaling pathways have arisen over the course of evolution.

SUMMARY POINTS

1. Eukaryotic signaling proteins use modular strategies to achieve specific circuit connectivities. These strategies are often characterized by a physical and functional separation between elements of the protein that carry out catalytic functions and elements that determine upstream and downstream partner linkages.

2. Docking interactions between recognition grooves and peptide motifs allow for some degree of separation between recognition events and catalytic function, allowing for more adaptable interactions. In addition, the peptide motifs are readily transferable, creating new potential pathway linkages, whereas the docking grooves are less modular.

3. Protein interactions mediated by specialized modular domains allow for standardized, transferable interactions between catalytic elements and their targets or effectors. The interacting regions in this case are bidirectionally transferable, as either the domain or ligand can readily be exchanged onto a new protein, conferring new functional linkages.

4. Proteins specialized for protein interaction, termed scaffold or adapter proteins, further separate catalysis from molecular recognition. This separation allows the same catalytic molecule to be used in multiple distinct pathways with minimal cross-signaling.

5. The relative ease of transferability inherent to recognition events mediated by standardized modular domains may facilitate the evolution of new connections in signaling pathways, hence the development of complex signaling behaviors.

6. Common catalytic domains and protein interaction domains are recombined together in many different combinations in metazoans to yield complex targeting and regulation of catalytic activity. These multidomain proteins are enriched in cell signaling and other complex processes and are more likely to be expressed in a tissue-specific manner.

7. The principle that modular architecture contributes to the development of complexity can be exploited in the design of synthetic signaling systems, allowing the construction of proteins with novel regulation and behavior from a toolkit of common components.

FUTURE ISSUES TO BE RESOLVED

1. Comparative genomics studies will help illuminate the evolutionary origins of complex signaling proteins and networks.

2. Detailed analysis of gene expression in specific tissues will test the prediction that evolutionarily newer tissues and processes, such as the brain and immune system, will be enriched for the expression of highly modular proteins.

3. Given the separation between catalysis and molecular recognition in metazoan signaling systems, pharmacological disruption of protein interactions will continue to hold promise for more "surgical" interference with specific signaling events compared with the more general tactic of inhibiting catalytic activity.

4. Docking interactions, modular domains, and scaffold proteins will be combined in synthetic signaling pathways with increasingly sophisticated behaviors in vivo, with potential research and therapeutic applications.

ACKNOWLEDGMENTS

We thank C. Tang, C. Voigt, A. Arkin, J. Dueber, N. Helman, C. Bashor, and other members of the Lim lab for their helpful comments on the manuscript.

LITERATURE CITED

1. Carroll SB. 2005. *PLoS Biol.* 3:e245
2. Lander ES, Linton LM, Birren B, Nusbaum C, Zody MC, et al. 2001. *Nature* 409:860–921
3. Rubin GM. 2001. *Nature* 409:820–21
4. Venter JC, Adams MD, Myers EW, Li PW, Mural RJ, et al. 2001. *Science* 291:1304–51
5. Hartwell LH, Hopfield JJ, Leibler S, Murray AW. 1999. *Nature* 402:C47–52
6. Reményi A, Scholer HR, Wilmanns M. 2004. *Nat. Struct. Mol. Biol.* 11:812–15
6a. Ptashne M, Gann A. 2002. *Genes and Signals.* Cold Spring Harbor, NY: Cold Spring Harbor Lab.
7. Ihmels J, Bergmann S, Gerami-Nejad M, Yanai I, McClellan M, et al. 2005. *Science* 309:938–40
8. Kirschner M, Gerhart J. 1998. *Proc. Natl. Acad. Sci. USA* 95:8420–27
9. Monod J, Changeux JP, Jacob F. 1963. *J. Mol. Biol.* 6:306–29
10. Perutz MF. 1978. *Sci. Am.* 239:92–125
11. Pawson T, Nash P. 2003. *Science* 300:445–52
12. Johnson SA, Hunter T. 2005. *Nat. Methods* 2:17–25
13. Fischer E. 1894. *Ber. Dtsch. Chem. Ges.* 27:2985
14. Krishnaswamy S. 2005. *J. Thromb. Haemost.* 3:54–67

15. Overall CM. 2002. *Mol. Biotechnol.* 22:51–86
16. Biondi RM, Nebreda AR. 2003. *Biochem. J.* 372:1–13
17. Holland PM, Cooper JA. 1999. *Curr. Biol.* 9:R329–31
18. Kemp BE, Bylund DB, Huang TS, Krebs EG. 1975. *Proc. Natl. Acad. Sci. USA* 72:3448–52
19. Hutti JE, Jarrell ET, Chang JD, Abbott DW, Storz P, et al. 2004. *Nat. Methods* 1:27–29
19a. Songyang Z, Blechner S, Hoagland N, Hoekstra MF, Piwnica-Worms H, Cantley LC. 1994. *Curr. Biol.* 4:973–82
20. Miller WT. 2003. *Acc. Chem. Res.* 36:393–400
20a. Verma R, Annan RS, Huddleston MJ, Carr SA, Reynard G, Deshaies RJ. 1997. *Science* 278:455–60
21. Sharrocks AD, Yang SH, Galanis A. 2000. *Trends Biochem. Sci.* 25:448–53
22. Tanoue T, Adachi M, Moriguchi T, Nishida E. 2000. *Nat. Cell Biol.* 2:110–16
23. Jacobs D, Glossip D, Xing H, Muslin AJ, Kornfeld K. 1999. *Genes Dev.* 13:163–7524
24. Chang CI, Xu BE, Akella R, Cobb MH, Goldsmith EJ. 2002. *Mol. Cell* 9:1241–49
25. Heo YS, Kim SK, Seo CI, Kim YK, Sung BJ, et al. 2004. *EMBO J.* 23:2185–95
26. Reményi A, Good MC, Bhattacharyya RP, Lim WA. 2005. *Mol. Cell.* 20:951–62
27. Grewal S, Molina DM, Bardwell L. 2006. *Cell Signal.* 18:123–34
28. Kusari AB, Molina DM, Sabbagh W Jr, Lau CS, Bardwell L. 2004. *J. Cell. Biol.* 164:267–77
29. Ho DT, Bardwell AJ, Abdollahi M, Bardwell L. 2003. *J. Biol. Chem.* 278:32662–72
30. Nguyen AN, Ikner AD, Shiozaki M, Warren SM, Shiozaki K. 2002. *Mol. Biol. Cell* 13:2651–63
31. Tanoue T, Maeda R, Adachi M, Nishida E. 2001. *EMBO J.* 20:466–79
31a. Bardwell L, Thorner J. 1996. *Trends Biochem. Sci.* 21:373–74
32. Barsyte-Lovejoy D, Galanis A, Sharrocks AD. 2002. *J. Biol. Chem.* 277:9896–903
33. Lee T, Hoofnagle AN, Kabuyama Y, Stroud J, Min XS, et al. 2004. *Mol. Cell* 14:43–55
34. Fantz DA, Jacobs D, Glossip D, Kornfeld K. 2001. *J. Biol. Chem.* 276:27256–65
34a. Bardwell AJ, Flatauer LJ, Matsukuma K, Thorner J, Bardwell L. 2001. *J. Biol. Chem.* 276:10374–86
35. Takekawa M, Tatebayashi K, Saito H. 2005. *Mol. Cell* 18:295–306
36. Tatebayashi K, Takekawa M, Saito H. 2003. *EMBO J.* 22:3624–34
37. Biondi RM, Cheung PC, Casamayor A, Deak M, Currie RA, Alessi DR. 2000. *EMBO J.* 19:979–88
38. Frodin M, Jensen CJ, Merienne K, Gammeltoft S. 2000. *EMBO J.* 19:2924–34
39. Biondi RM, Komander D, Thomas CC, Lizcano JM, Deak M, et al. 2002. *EMBO J.* 21:4219–28
40. Fiol CJ, Haseman JH, Wang YH, Roach PJ, Roeske RW, et al. 1988. *Arch. Biochem. Biophys.* 267:797–802
41. Dajani R, Fraser E, Roe SM, Young N, Good V, et al. 2001. *Cell* 105:721–32
42. Frame S, Cohen P, Biondi RM. 2001. *Mol. Cell* 7:1321–27
42a. Bardwell L, Cook JG, Chang EC, Cairns BR, Thorner J. 1996. *Mol. Cell. Biol.* 16:3637–50
43. Breitkreutz A, Boucher L, Tyers M. 2001. *Curr. Biol.* 11:1266–71
44. Bridges D, Moorhead GB. 2004. *Science STKE* 2004:re10
45. Durocher D, Jackson SP. 2002. *FEBS Lett.* 513:58–66
46. Glover JN, Williams RS, Lee MS. 2004. *Trends Biochem. Sci.* 29:579–85
47. Harris BZ, Lim WA. 2001. *J. Cell Sci.* 114:3219–31

48. Lemmon MA, Ferguson KM. 2000. *Biochem. J.* 350(Pt. 1):1–18
49. Pawson T, Gish GD, Nash P. 2001. *Trends Cell Biol.* 11:504–11
50. Schlessinger J, Lemmon MA. 2003. *Science STKE* 2003:re12
51. Sondermann H, Kuriyan J. 2005. *Cell* 121:158–60
52. Zarrinpar A, Bhattacharyya RP, Lim WA. 2003. *Science STKE* 2003:re8
53. Marengere LE, Pawson T. 1994. *J. Cell Sci. Suppl.* 18:97–104
54. Dueber JE, Yeh BJ, Bhattacharyya RP, Lim WA. 2004. *Curr. Opin. Struct. Biol.* 14:690–99
55. Lim WA. 2002. *Curr. Opin. Struct. Biol.* 12:61–68
56. Prehoda KE, Scott JA, Mullins RD, Lim WA. 2000. *Science* 290:801–6
57. Rohatgi R, Ho HY, Kirschner MW. 2000. *J. Cell Biol.* 150:1299–310
58. Sicheri F, Moarefi I, Kuriyan J. 1997. *Nature* 385:602–9
59. Williams JC, Weijland A, Gonfloni S, Thompson A, Courtneidge SA, et al. 1997. *J. Mol. Biol.* 274:757–75
60. Xu W, Harrison SC, Eck MJ. 1997. *Nature* 385:595–602
61. Letunic I, Copley RR, Schmidt S, Ciccarelli FD, Doerks T, et al. 2004. *Nucleic Acids Res.* 32:D142–44
62. Schultz J, Milpetz F, Bork P, Ponting CP. 1998. *Proc. Natl. Acad. Sci. USA* 95:5857–64
63. Berry DM, Nash P, Liu SK, Pawson T, McGlade CJ. 2002. *Curr. Biol.* 12:1336–41
64. Harkiolaki M, Lewitzky M, Gilbert RJ, Jones EY, Bourette RP, et al. 2003. *EMBO J.* 22:2571–82
65. Liu Q, Berry D, Nash P, Pawson T, McGlade CJ, Li SC. 2003. *Mol. Cell* 11:471–81
66. Lewitzky M, Harkiolaki M, Domart MC, Jones EY, Feller SM. 2004. *J. Biol. Chem.* 279:28724–32
67. Groemping Y, Lapouge K, Smerdon SJ, Rittinger K. 2003. *Cell* 113:343–55
68. Li SC, Gish G, Yang D, Coffey AJ, Forman-Kay JD, et al. 1999. *Curr. Biol.* 9:1355–62
69. Benes CH, Wu N, Elia AE, Dharia T, Cantley LC, Soltoff SP. 2005. *Cell* 121:271–80
70. Bu JY, Shaw AS, Chan AC. 1995. *Proc. Natl. Acad. Sci. USA* 92:5106–10
71. Iwashima M, Irving BA, van Oers NS, Chan AC, Weiss A. 1994. *Science* 263:1136–39
72. Pluskey S, Wandless TJ, Walsh CT, Shoelson SE. 1995. *J. Biol. Chem.* 270:2897–900
73. Zarrinpar A, Park SH, Lim WA. 2003. *Nature* 426:676–80
74. Landgraf C, Panni S, Montecchi-Palazzi L, Castagnoli L, Schneider-Mergener J, et al. 2004. *PLoS Biol.* 2:e14
75. King N, Hittinger CT, Carroll SB. 2003. *Science* 301:361–63
76. Pawson T, Scott JD. 1997. *Science* 278:2075–80
77. Lowe ED, Tews I, Cheng KY, Brown NR, Gul S, et al. 2002. *Biochemistry* 41:15625–34
78. Loog M, Morgan DO. 2005. *Nature* 434:104–8
79. Burack WR, Shaw AS. 2000. *Curr. Opin. Cell. Biol.* 12:211–16
80. Elion EA. 2001. *J. Cell Sci.* 114:3967–78
81. Garrington TP, Johnson GL. 1999. *Curr. Opin. Cell. Biol.* 11:211–18
82. Morrison DK, Davis RJ. 2003. *Annu. Rev. Cell. Dev. Biol.* 19:91–118
83. Whitmarsh AJ, Davis RJ. 1998. *Trends Biochem. Sci.* 23:481–85
84. Wong W, Scott JD. 2004. *Nat. Rev. Mol. Cell Biol.* 5:959–70
85. Zarrinpar A, Bhattacharyya RP, Nittler MP, Lim WA. 2004. *Mol. Cell* 14:825–32
86. Choi KY, Satterberg B, Lyons DM, Elion EA. 1994. *Cell* 78:499–512
87. Marcus S, Polverino A, Barr M, Wigler M. 1994. *Proc. Natl. Acad. Sci. USA* 91:7762–66
88. Posas F, Saito H. 1997. *Science* 276:1702–5
89. Printen JA, Sprague GF Jr. 1994. *Genetics* 138:609–19
89a. Truckses DM, Bloomekatz JE, Thorner J. 2006. *Mol. Cell. Biol.* 26:912–28

89b. Harris K, Lamson RE, Nelson B, Hughes TR, Marton MJ, et al. 2001. *Curr. Biol.* 11:1815–24
90. Li W, Fan J, Woodley DT. 2001. *Oncogene* 20:6403–17
91. Zhang M, Wang W. 2003. *Acc. Chem. Res.* 36:530–38
92. Feliciello A, Gottesman ME, Avvedimento EV. 2001. *J. Mol. Biol.* 308:99–114
93. Muller J, Cacace AM, Lyons WE, McGill CB, Morrison DK. 2000. *Mol. Cell Biol.* 20:5529–39
94. Sierralta J, Mendoza C. 2004. *Brain Res. Brain Res. Rev.* 47:105–15
95. Raab M, da Silva AJ, Findell PR, Rudd CE. 1997. *Immunity* 6:155–64
96. Zhang W, Trible RP, Zhu M, Liu SK, McGlade CJ, Samelson LE. 2000. *J. Biol. Chem.* 275:23355–61
97. Muller J, Ory S, Copeland T, Piwnica-Worms H, Morrison DK. 2001. *Mol. Cell* 8:983–93
98. Nguyen A, Burack WR, Stock JL, Kortum R, Chaika OV, et al. 2002. *Mol. Cell Biol.* 22:3035–45
99. Ory S, Zhou M, Conrads TP, Veenstra TD, Morrison DK. 2003. *Curr. Biol.* 13:1356–64
100. Matheny SA, Chen C, Kortum RL, Razidlo GL, Lewis RE, White MA. 2004. *Nature* 427:256–60
101. Flotho A, Simpson DM, Qi M, Elion EA. 2004. *J. Biol. Chem.* 279:47391–401
101a. Bhattacharyya RP, Reményi A, Good MC, Bashor CJ, Falick AM, Lim WA. 2006. *Science* 311:822–26
102. Whitmarsh AJ, Cavanagh J, Tournier C, Yasuda J, Davis RJ. 1998. *Science* 281:1671–74
103. Willoughby EA, Perkins GR, Collins MK, Whitmarsh AJ. 2003. *J. Biol. Chem.* 278:10731–36
104. Willoughby EA, Collins MK. 2005. *J. Biol. Chem.* 280:25651–58
105. Coghlan VM, Perrino BA, Howard M, Langeberg LK, Hicks JB, et al. 1995. *Science* 267:108–11
106. Klauck TM, Faux MC, Labudda K, Langeberg LK, Jaken S, Scott JD. 1996. *Science* 271:1589–92
107. Schillace RV, Scott JD. 1999. *Curr. Biol.* 9:321–24
108. Dodge KL, Khouangsathiene S, Kapiloff MS, Mouton R, Hill EV, et al. 2001. *EMBO J.* 20:1921–30
109. McCahill A, McSorley T, Huston E, Hill EV, Lynch MJ, et al. 2005. *Cell Signal* 17:1158–73
110. Scott K, Zuker CS. 1998. *Nature* 395:805–8
111. Tsunoda S, Sierralta J, Sun Y, Bodner R, Suzuki E, et al. 1997. *Nature* 388:243–49
112. Tsunoda S, Zuker CS. 1999. *Cell Calcium* 26:165–71
113. Benner SA, Sismour AM. 2005. *Nat. Rev. Genet.* 6:533–43
114. Pawson T, Linding R. 2005. *FEBS Lett.* 579:1808–14
115. Park SH, Zarrinpar A, Lim WA. 2003. *Science* 299:1061–64
116. Howard PL, Chia MC, Del Rizzo S, Liu FF, Pawson T. 2003. *Proc. Natl. Acad. Sci. USA* 100:11267–72
117. Dueber JE, Yeh BJ, Chak K, Lim WA. 2003. *Science* 301:1904–8
118. McDonald C, Vacratsis PO, Bliska JB, Dixon JE. 2003. *J. Biol. Chem.* 278:18514–23
119. Cooper JA, Gould KL, Cartwright CA, Hunter T. 1986. *Science* 231:1431–34
120. Kmiecik TE, Shalloway D. 1987. *Cell* 49:65–73
121. Piwnica-Worms H, Saunders KB, Roberts TM, Smith AE, Cheng SH. 1987. *Cell* 49:75–82

122. Jackson PK, Eldridge AG, Freed E, Furstenthal L, Hsu JY, et al. 2000. *Trends Cell Biol.* 10:429–39
123. Vale RD, Milligan RA. 2000. *Science* 288:88–95
124. Csete M, Doyle J. 2004. *Trends Biotechnol.* 22:446–50
124a. Kashtan N, Alon U. 2005. *Proc. Natl. Acad. Sci. USA* 102:13773–78
125. Cohen-Gihon I, Lancet D, Yanai I. 2005. *Trends Genet.* 21:210–13
126. Bae KH, Kwon YD, Shin HC, Hwang MS, Ryu EH, et al. 2003. *Nat. Biotechnol.* 21:275–80
127. Dreier B, Beerli RR, Segal DJ, Flippin JD, Barbas CF 3rd. 2001. *J. Biol. Chem.* 276:29466–78
128. Dreier B, Fuller RP, Segal DJ, Lund C, Blancafort P, et al. 2005. *J. Biol. Chem.* 280:35588–97
129. Dreier B, Segal DJ, Barbas CF 3rd. 2000. *J. Mol. Biol.* 303:489–502
130. Segal DJ, Dreier B, Beerli RR, Barbas CF 3rd. 1999. *Proc. Natl. Acad. Sci. USA* 96:2758–63
130a. Rebar EJ, Pabo CO. 1994. *Science* 263:671–73
131. Klug A. 1999. *J. Mol. Biol.* 293:215–18
132. Segal DJ, Beerli RR, Blancafort P, Dreier B, Effertz K, et al. 2003. *Biochemistry* 42:2137–48
133. Beerli RR, Segal DJ, Dreier B, Barbas CF 3rd. 1998. *Proc. Natl. Acad. Sci. USA* 95:14628–33
134. Alwin S, Gere MB, Guhl E, Effertz K, Barbas CF 3rd, et al. 2005. *Mol. Ther.* 12:610–17
135. Beerli RR, Dreier B, Barbas CF 3rd. 2000. *Proc. Natl. Acad. Sci. USA* 97:1495–500
136. Blancafort P, Chen EI, Gonzalez B, Bergquist S, Zijlstra A, et al. 2005. *Proc. Natl. Acad. Sci. USA* 102:11716–21
137. Klug A. 2005. *FEBS Lett.* 579:892–94
138. Segal DJ, Goncalves J, Eberhardy S, Swan CH, Torbett BE, et al. 2004. *J. Biol. Chem.* 279:14509–19
139. Tan W, Zhu K, Segal DJ, Barbas CF 3rd, Chow SA. 2004. *J. Virol.* 78:1301–13
139a. Urnov FD, Miller JC, Lee YL, Beausejour CM, Rock JM, et al. 2005. *Nature* 435:646–51
140. Bax B, Carter PS, Lewis C, Guy AR, Bridges A, et al. 2001. *Structure* 9:1143–52
141. Dajani R, Fraser E, Roe SM, Yeo M, Good VM, et al. 2003. *EMBO J.* 22:494–501
142. Stradal TE, Rottner K, Disanza A, Confalonieri S, Innocenti M, Scita G. 2004. *Trends Cell Biol.* 14:303–11
143. Lei M, Lu W, Meng W, Parrini MC, Eck MJ, et al. 2000. *Cell* 102:387–97
144. Rohrschneider LR, Fuller JF, Wolf I, Liu Y, Lucas DM. 2000. *Genes Dev.* 14:505–20
145. Murata Y, Iwasaki H, Sasaki M, Inaba K, Okamura Y. 2005. *Nature* 435:1239–43
146. Sands Z, Grottesi A, Sansom MS. 2005. *Curr. Biol.* 15:R44–47

Ribonucleotide Reductases

Pär Nordlund[1] and Peter Reichard[2]

[1]Division of Biophysics and [2]Division of Biochemistry, Medical Nobel Institute, Department of Medical Biochemistry and Biophysics, Karolinska Institutet, S-17177 Stockholm, Sweden; email: par.nordlund@mbb.ki.se, peter.reichard@mbb.ki.se

Key Words

allosteric regulation, deoxyribonucleotide pools, evolution, protein radicals

Abstract

Ribonucleotide reductases (RNRs) transform RNA building blocks to DNA building blocks by catalyzing the substitution of the 2′OH-group of a ribonucleotide with a hydrogen by a mechanism involving protein radicals. Three classes of RNRs employ different mechanisms for the generation of the protein radical. Recent structural studies of members from each class have led to a deeper understanding of their catalytic mechanism and allosteric regulation by nucleoside triphosphates. The main emphasis of this review is on regulation of RNR at the molecular and cellular level. Conformational transitions induced by nucleotide binding determine the regulation of substrate specificity. An intricate interplay between gene activation, enzyme inhibition, and protein degradation regulates, together with the allosteric effects, enzyme activity and provides the appropriate amount of deoxynucleotides for DNA replication and repair. In spite of large differences in the amino acid sequences, basic structural features are remarkably similar and suggest a common evolutionary origin for the three classes.

Contents

INTRODUCTION 682
A SHORT CLASSIFICATION OF
   RNRs.......................... 682
      Class I RNRs..................... 682
      Class II RNRs.................... 683
      Class III RNRs................... 683
OCCURRENCE AND GENE
   ORGANIZATION.............. 683
SPECIAL ENZYMES.............. 684
      p53R2 684
      Yeast............................ 684
      E. gracilis...................... 684
      Chlamydia trachomatis............ 685
OVERALL STRUCTURE.......... 685
ACTIVE-SITE STRUCTURE AND
   CATALYTIC MECHANISM..... 687
RADICAL STORAGE AND
   TRANSFER..................... 690
GENE REGULATION............. 692
      Bacteria........................ 692
      Mammalian Cells................ 693
      Yeast............................ 693
ALLOSTERIC REGULATION 694
      The Specificiy Site 694
      The Activity Site (the ATP Cone) . 698
dNTP POOLS...................... 699
      Yeast............................ 699
      Mitochondria 699
EVOLUTION....................... 700

RNR:
ribonucleotide reductase

dNTP:
deoxyribonucleoside triphosphates

INTRODUCTION

In all cellular organisms, ribonucleotide reductases (RNRs) synthesize the four deoxyribonucleoside triphosphates (dNTPs) required for DNA replication and repair by substitution of the 2′-OH of a ribonucleoside di- or triphosphate by a hydrogen atom. A common feature of all reductases is their ability to provide an appropriate balance of the four DNA building blocks. A unique allosteric regulation of their substrate specificity makes this possible. All RNRs share a common basic catalytic mechanism involving the activation of the ribonucleotide by abstraction of the 3′-hydrogen atom of the ribose by a transient thiyl radical of the enzyme.

A general overview of the occurrence, catalytic function, regulation, and evolution of RNRs is found in a review from one of us in the *Annual Review of Biochemistry* of 1998 (1). Since then, many reviews have appeared, usually related to specific aspects, such as radical mechanisms in general (2–5), mechanism of class I (6), class II (7) or class III (8–11) enzymes, allosteric regulation (12), physiology (13), and evolution (14–17). Major advances in recent years concern the structure and regulation of RNRs. These have led to a better understanding of the structural basis for the allosteric transitions and the interplay between allosteric and genetic regulation. The emphasis of this review is on regulation.

A SHORT CLASSIFICATION OF RNRs

RNRs can be grouped into three different classes largely on the basis of their interaction with oxygen and the way in which they generate their thiyl radical. A more detailed description of the three classes is found in Reference 1.

Class I RNRs

Class I reductases contain two subgroups (Ia, NrdAB, and Ib, NrdEF) with slightly different primary structures. Both subgroups contain two nonidentical dimeric subunits (R1 and R2) and require oxygen for the generation of a stable tyrosyl radical by a Fe-O-Fe center in the smaller R2 subunit (NrdB or NrdF). During catalysis, the radical is continuously shuttled to a cysteine of the larger R1 subunit (NrdA or NrdE) and there generates a thiyl radical required for activation of the substrate. R1 contains both the catalytic site for the reduction of the ribotide and the allosteric sites for its regulation. The required electrons are provided from redox-active cysteines of small proteins, thioredoxin or glutaredoxin for class Ia and NrdH (18) for class Ib. Class Ib enzymes also depend on an additional protein (NrdI) of unknown function (19). Originally, it was thought that a major distinction between the two subgroups was the presence in Ia of an amino-terminal ATP cone (20), which is absent from Ib, codifying the allosteric activity site of the enzyme. However, some microbial Ia enzymes also lack an ATP cone. Class Ib is better identified from the presence of *nrdI* and *nrdH* in the genome together with the *nrdEF* genes.

Class II RNRs

Class II RNRs are indifferent to oxygen, neither requiring it nor inhibited by it. They contain a single subunit (NrdJ) and are isolated either as monomers or as dimers. They generate their thiyl radical with the aid of adenosylcobalamin, probably via formation of a deoxyadenosyl radical. Thus adenosylcobalamin fulfills the function of R2 as a radical generator. Class II enzymes use thioredoxin or glutaredoxin as electron donors. Their allosteric regulation is similar to that of class Ib, and many, but not all, lack an ATP cone.

Class III RNRs

Class III reductases are anaerobic enzymes that are inactivated by oxygen. Two dimeric proteins are required for activity, one large catalytic protein (NrdD) and one small (NrdG), catalyzing the activation of NrdD. The activase NrdG contains a redox-active [4Fe-4S] center, which together with *S*-adenosyl methionine and reduced flavodoxin generate a stable but oxygen-sensitive glycyl radical at the C terminus of NrdD. This radical is the counterpart of the stable tyrosyl radical of class I and continuously generates the thiyl radical required for the activation of the substrate. Once the glycyl radical is formed, NrdD alone catalyzes the reaction. This differs from class I enzymes that require R2 continuously for radical generation. Furthermore, class III enzymes use formate as electron donor. An earlier theory postulated a class IV enzyme, with a manganese center (21) instead of a dinuclear iron center, but it was shown to be an artifact (22).

OCCURRENCE AND GENE ORGANIZATION

As a result of the exponential increase in genome sequencing, databases are flooded with real, probable, and hypothetical sequences of genes for RNRs from different organisms (e.g., **http://www.rnrdb.molbio.su.se**). Even though the corresponding proteins have been identified only in a few instances, we can draw some tentative general rules from the data. With the exception of the unicellular *Euglena gracilis* (23), all eukaryotes from yeast to humans contain class Ia enzymes. This class is also found in eubacteria and in a few archea. Class Ib occurs in a large spectrum of eubacteria. Both classes require oxygen to function, and active enzymes are not found in strict anaerobes. Class II reductases are microbial enzymes that occur in both aerobic and anaerobic organisms. Class III reductases depend on anaerobiosis and are found in both strict anaerobic and facultative anaerobic organisms. Only class III enzymes from the latter category have been studied in some detail.

In many microorganisms, RNRs from different classes occur side by side. Facultative anaerobes rely on a class I enzyme during aerobiosis and a class III enzyme during anaerobiosis (18, 24). It is more difficult to understand why several bacteria contain both class I and II enzymes (25, 26), and some even have genes for all three classes (27). Unexplained is also the presence of *nrdEF* (class Ib) in *Escherichia coli* and some of its relatives (16), which code for a fully active RNR.

In eukaryotes the genes for R1 and R2 are located on separate chromosomes. In most bacteria, the *nrdA* and *nrdB* genes form a single transcriptional unit. Recently, a third gene (*nrdS*) was described in *Streptomyces coelicolor* (28) as part of an *nrdABS* operon. NrdS belongs to the large AraC family of transcription regulators found in diverse microorganisms, but its function in the *nrdABS* operon is unknown. Class Ib genes are organized as an *nrdHIEF* operon in *E. coli* (29) and *Lactococcus lactis* (18), but in many other bacteria one of the four genes may be missing, duplicated, or located on other parts of the chromosome. Class II reductases are generally coded for by a single *nrdJ* gene. In *S. coelicolor*, this gene forms a transcriptional unit with a preceding *nrdR* gene, reported to be involved in the regulation of the transcription of *nrdJ* (see below). *nrdR* is also present in other bacteria but not always linked to *nrdJ*. In class III, the two *nrdD* and *nrdG* reductase genes usually, but not always, reside in one operon.

SPECIAL ENZYMES

This section discusses recently found RNRs that do not fully conform with the general rules for ribonucleotide reductases and RNRs that have special functions.

p53R2

The discovery of p53R2 (30, 31), an analog of R2 in mammalian cells, provided an interesting link between RNR and cancer because its gene is a downstream target for the tumor supressor p53. The genes for p53R2 (*Rrm2b*) and R2 (*nrdB*) are located on separate chromosomes. p53R2 is 80% to 90% identical to R2 but lacks its 33 amino-terminal residues, including a KEN box (32) required for degradation during mitosis (see below). p53R2 can substitute for R2 and form a highly active RNR (33), believed to be active in DNA repair after DNA damage. Mutant $Rrm2b(-/-)$ mice lacking p53R2 grow apparently normally up to 6 weeks but then die from glomerular injury and kidney failure (34, 35).

Yeast

Budding yeast contains a class Ia reductase with unique features. Two genes (*RNR1* and *RNR3*) encode separate versions of R1 (Rrn1 and Rrn3) (36), and two genes encode two R2-like proteins, Rrn2 (37, 38) and Rrn4 (39, 40). *RNR3* is expressed poorly, and null mutants have no phenotype, not even after DNA damage when *RNR3* is strongly induced (41). Overexpression of Rnr3 can, however, rescue *RNR1* null mutants. Highly purified Rnr3 has a very low specific enzyme activity but shows a strong synergy with Rnr1. Rnr3 has an ATP cone but—unlike Rnr1—is not inhibited by dATP. A heterodimer between Rnr3 and a catalytically inactive Rnr1 mutant containing the ATP cone was, however, sensitive to dATP suggesting cross talk between the two subunits (41). An Rnr1/Rnr3 dimer with relative dATP insensitivity, permitting accumulation of larger dNTP pools, may be advantageous in DNA-damaged cells (41).

RNR2 and *RNR4* are both required to provide a functional reductase. Rnr2 has a typical eukaryotic R2 sequence, whereas Rnr4 lacks several conserved amino acids required for iron binding and cannot form the canonical tyrosyl radical (40). Rnr4 apparently has a structural function (40, 42, 43). Rnr4 and Rnr2 form a heterodimer ($\beta\beta'$) of the small subunit that, together with the large subunit, form the active $\alpha_2\beta\beta'$ heterotetramer of the yeast RNR. In an alternative hypothesis, Rrn4 was suggested to be a metallochaperone delivering iron to Rrn2 (44), but further work (43) makes this unlikely. A similar $\alpha_2\beta\beta'$ structure apparently provides the active RNR in *Plasmodium falciparum* (45).

E. gracilis

Unlike other eukaryotes, *E. gracilis* contains a class II RNR (23) with an amino acid sequence closely related to that of *Lactobacillus leichmannii* (45a), including the 130-residue insertion that provides the monomeric *L. leichmannii* enzyme with a mock dimeric interface, required for binding of allosteric effectors (see below). The allosteric regulation of the cloned enzyme is similar to that of the *L. leichmanii* reductase.

Chlamydia trachomatis

All sequenced chlamydial genomes contain an RNR with class I-type sequences but with a phenylalanine at the tyrosyl radical site of R2. The activity of the recombinant R2 is sensitive to hydroxyurea in vitro, indicating a radical reaction (46). This activity probably provides the dNTPs for DNA synthesis because the intracellular pathogen does not import deoxyribonucleosides from the host cell (47, 48). The hydroxyurea-sensitive radical is stored as an iron-coupled radical at the diiron site of the R2 subunit (49). Compared to other R2s, this iron site also shows characteristic amino acid changes in the iron coordination sphere. Tyrosine radicals are sensitive to NO, and it was suggested that the *Chlamydia* radical site adapts the enzyme to escape the NO-based intracellular defense system of mammalian cells (49). Similar characteristic sequence changes can also be found in R2s of other organisms, including some intracellular pathogens. However, an active RNR has as yet not been demonstrated in these cases.

OVERALL STRUCTURE

Crystallographic studies of RNRs have now provided high-resolution structures of at least one representative member of each major type of RNR subunit, except for the class III β-subunit (NrdG activase) (4).

The structures of the catalytic subunits of all three classes of RNR reveal a highly related α/β topology, and in particular, the structural similarity between class I and II is high (50, 51). This is consistent with a significant sequence identity between the most related members of the two classes (>25%). The structural conservation of residues in the active-site region of the two families is also high, supporting a similar catalytic mechanism. The sequence of the class III catalytic subunit instead reveals no significant sequence homology to the core α/β-barrel of the class I and II enzymes. Nevertheless, the folding topology and two critical active-site residues are conserved (52).

The first structure of a catalytic subunit of an RNR was that of the *E. coli* R1 protein (NrdA) (**Figure 1a**), which revealed the core domain of a 10-stranded α/β-barrel (53). Within the barrel, a central β-hairpin finger is found with the essential thiyl radical-forming cysteine residue of the active site at the tip. The two reducing cysteine residues are positioned on two neighboring β-strands. Structures of R1 in complex with effectors and substrate have revealed the position of the active site and the two effector-binding sites (54). The active-site residues are contributed from the β-strands of the core barrel and the thiyl radical-containing loop. The binding sites for effectors regulating substrate specificity are at the dimer interface, not far from the active site (**Figure 1a**). The specificity site is thus composed of residues from both subunits of the dimer. The *E. coli* R1 protein also contains an N-terminal ATP-cone domain carrying the binding site for effectors controlling the overall activity (**Figure 1a**). This domain is formed by a small α-helical bundle located relatively far from the active site. In the C-terminal region, R1 contains two cysteine residues, which can mediate the transfer of two electrons from thioredoxins or glutaredoxins to the active site by reversible disulphide formation. This C-terminal region is, however, not visible in the structure, which is consistent with the flexibility required for its function (53).

The structure of *Salmonella typhimurium* class Ib R1 has also recently been determined (55). The folded active-site structure is very similar to *E. coli* Ia R1 but lacks the N-terminal activity site (**Figure 1b**). Enzyme-effector complexes have revealed a specificity site very similar to that in class Ia.

The first structure of a class II RNR (NrdJ) was of the monomeric *L. leichmannii* enzyme (**Figure 1d**) (50). The overall fold and the key active-site residues are conserved with the class I R1 subunit. In addition, this monomeric class II enzyme has a small structural extension that substitutes for the part of

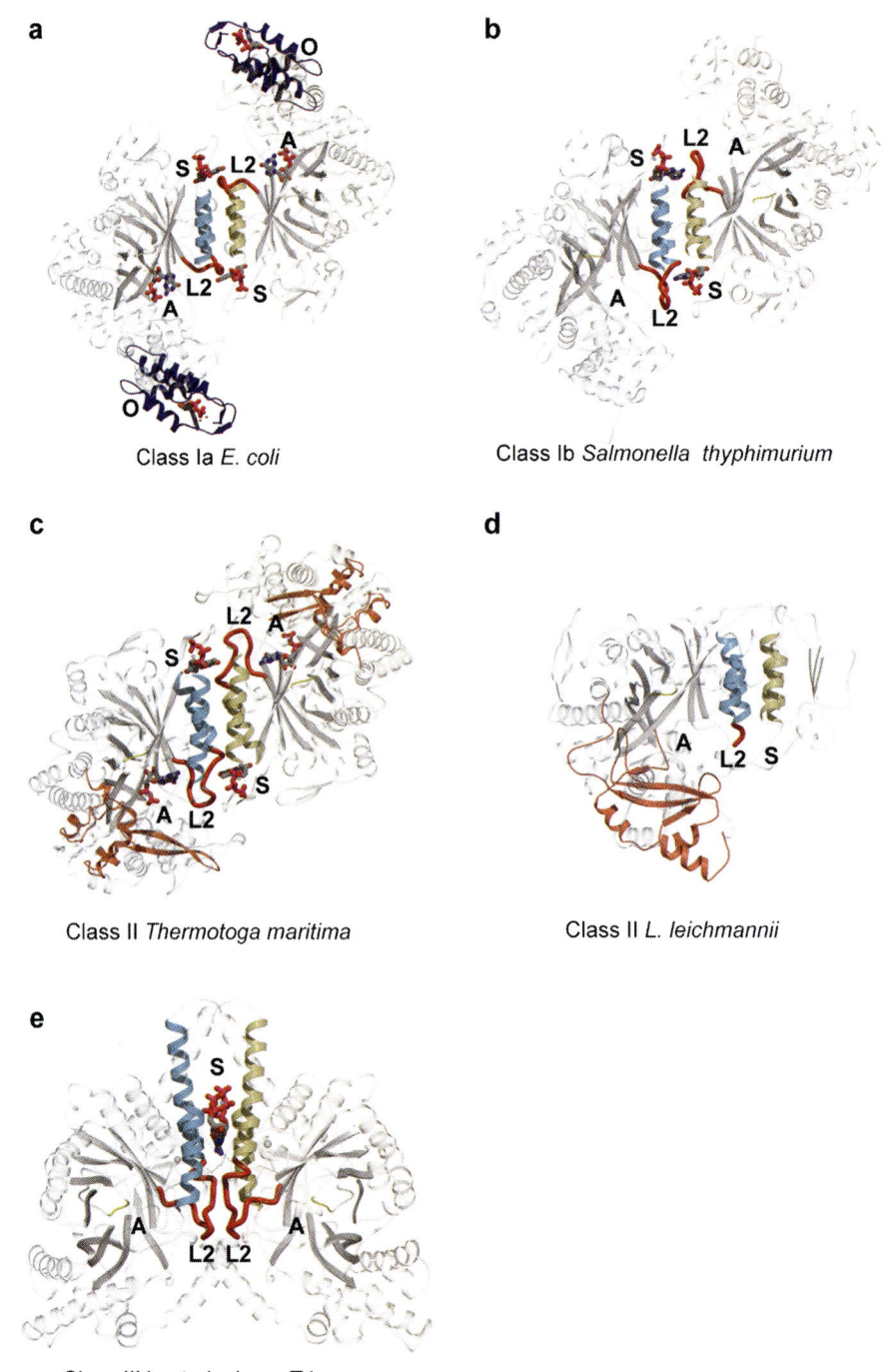

Figure 1

Overall organization of different RNRs. The ATP-cone domain of class I (*blue*) and B12-binding region of class II (*brown*) are shown. Loop 2 and dimer interactions essential for regulation are in bold. Abbreviations used: A, active site; L2, loop 2; O, overall activity site; and S, specificity site.

the specificity effector-binding site that is contributed by the neighboring subunit in the R1 dimer. The extension is a perfect mimic of the region of the missing subunit partner, which is essential for effector binding. This mock domain constitutes an interesting solution to retaining an allosteric binding site similar to the binding sites of the dimeric RNRs.

The structure of the complex of the *L. leichmannii* enzyme with a B12 analogue reveals a specific B12-binding region (**Figure 1*d***) (50), which is different from binding domains of other B12-dependent enzymes. A β-hairpin structure is the core component of this region, and intriguingly, this hairpin is also present in class I enzymes.

The structure of the dimeric class II RNR from *Thermotoga maritima* was recently determined in complex with several different specificity effectors and substrate nucleotides (51) as well as in complex with coenzyme B12 substructures (56). The overall structure of the enzyme resembles class I structures, with a similar dimer interaction site as well as similar effector- and substrate-binding sites (**Figure 1*c***). The coenzyme B12-binding region resembles that of the *L. leichmannii* enzyme.

The T4 class III catalytic α-subunit (NrdD) has a fold similar to the class I and II RNRs, with the active site occupying an identical position, although there is no significant sequence homology (52). The subunit interaction in the class III dimer is, however, very different, and the binding sites for specificity effectors are localized distant from the active site but still on the dimer interface (**Figure 1*e***). This is due to a 90° rotation of the two subunits as compared to the class I and class II dimers. However, the effector-binding sites on one of the subunits of the class III enzyme are in a similar position on the catalytic α/β-barrel as in the other classes.

The glycyl radical site of the class III enzyme is found in a β-hairpin loop in the region of the active site. This loop is found in a region similar to that of the adenosine of B12 in the class II enzyme.

ACTIVE-SITE STRUCTURE AND CATALYTIC MECHANISM

The *E. coli* R1 structure provided the first glimpse of substrate binding in a RNR (54). Although substrate binding of class I RNR was not completely defined in the structure, the positions of the phosphate and ribose moieties were sufficiently well localized to allow key active-site residues to be identified.

In the recent structures of the dimeric class II RNR from *T. maritima* complexed with four different substrates (ADP, GDP, CDP, UDP) and their cognate specificity effectors, the interactions with the substrates are well defined (51). The position and conformation of the ribose and the phosphate moieties are virtually identical in the four structures. The ribose ring of the substrate ribose has a C3′-endo pucker (**Figure 2*a***). C322 (C439 in *E. coli*), the residue carrying the initial thiyl radical in the reaction, is positioned in van der Waals contact with the 3′-carbon of the ribose. The two cysteine residues providing the reducing equivalents are found on the opposite side of the ribose, C134 (C225 in *E. coli*) proximal and C333 (C462 in *E. coli*) distal to the substrate. The catalytically important hydrogen-bonding network of the ribose OH groups is made by E324 (E441 in *E. coli*), N320 (N437 in *E. coli*), and C134, which is positioned close but not in hydrogen-bonding distance to the 2′OH group (**Figure 2*a***).

The conceptual foundation for the mechanism of ribonucleotide reduction is based on hydrogen isotope studies, which imply a hydrogen abstraction on the 3′carbon as the initial reaction step (6, 57). Potential catalytic residues have been probed by site-directed mutagenesis (3, 6, 58), and a number of different geometries for the reaction have been analyzed by density functional theory (DFT) (59–61). The detailed substrate geometry revealed by the high resolution structures of the *T. maritima* class II RNR shed new light on previous proposals for class I and II enzymes (51). In **Scheme 1**, a detailed reaction mechanism is outlined, which is based on features

DFT: density functional theory

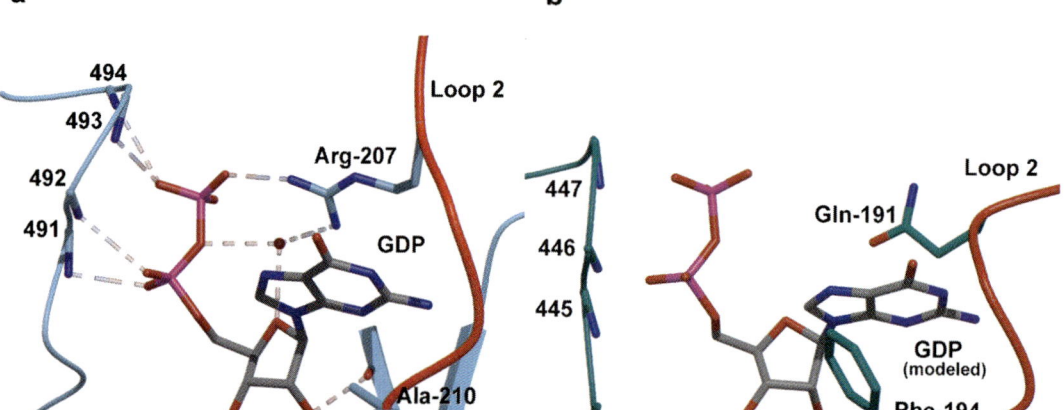

Figure 2

Active-site structure of (*a*) *T. maritima* class II RNR in complex with GDP (*E. coli* R1 residue numbers) and (*b*) T4 class III RNR with modeled GDP.

of previous mechanistic considerations as well as the new structural data.

The reaction is initiated by the thiyl radical in the active site, which abstracts a hydrogen atom from the 3′ carbon of the ribose and generates a substrate radical (S1, in **Scheme 1**). The cleavage of the 2′-C-O bond is coupled to the generation of a radical on the proximal reducing cysteine (C225 in *E. coli* R1) and the protonation of the water molecule (S2), which is leaving. The subsequent reduction of the ribose is promoted by the deprotonation of the 3′-OH group, concomitant with the formation of a disulphide radical and the protonation of the 2′-carbon (S3-S4). A deoxyribose radical is formed in the next step (S5), followed by the completion of the reaction through the regeneration of the initial thiyl radical (S6).

The detailed geometry defined by the *T. maritima* structure limits possible proton movements in the region. The conserved glutamate (E441 in *E. coli* R1) is positioned to serve exclusively as a base in the reaction. The proximal reducing cysteine is in the best position to protonate the leaving OH group. The protonation of the 3′ carbon is most likely driven by the distal cysteine (C461 in *E. coli*), which has the only remaining labile proton in the active site. The leaving water molecule is well positioned to play a role in mediating this proton transfer step. The dense structure of this region supports the idea of the product water fixed in the active site throughout the reaction (61). In Class II substrate structures, the proximal reducing cysteine is positioned relatively close to the 3′ carbon (∼5 Å), facilitating electron transfer (S4), which in previous geometries was a problematic reaction step (61).

In further support for the above-suggested reaction steps, a catalytically competent thiyl radical was observed directly for class II RNR (62), and substrate as well as thiol radicals

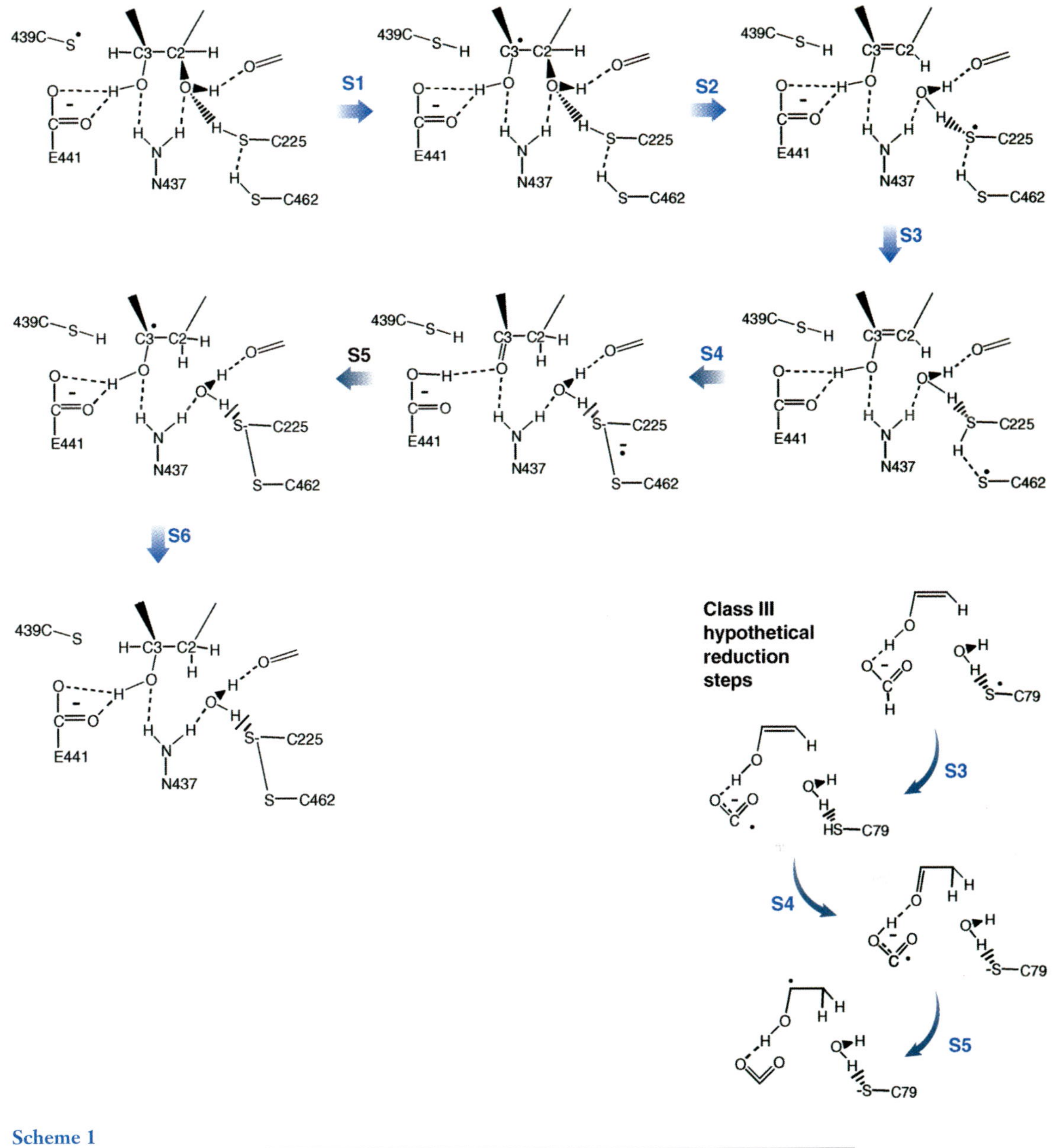

Scheme 1
Proposal for the catalytic mechanism of class I and II RNR. Residue numbering from *E. coli* class I RNR.

were found in mutated forms of RNRs (3, 58). DFT calculations demonstrated the feasibility of the 3′H abstraction step and of the initial reduction of the 2′C-O bond, using geometries similar to those in the outlined mechanism (59, 63).

In contrast to the class I and II systems, the structural basis for substrate interactions in class III is poorly understood because no structure of a class III enzyme with substrate or the reductant formate is known. However, the active-site position is defined by two catalytic cysteine residues of class I and II RNR, found in corresponding positions in class III (C290 and C79) (52). Modeling of substrate into the active site supports a substrate-binding geometry similar to that in class I and II but does not identify other candidate catalytic residues (**Figure 2b**). The differences in the catalytic machineries between class III and class I/II enzymes are likely because class III uses formate rather than redoxins as electron donor.

Despite the different active-site structure of class III RNR, it is likely that the initial steps of the reaction are conserved (S1 and S2 in **Scheme 1**) and that the structurally conserved C290 of class III plays the role of the thiyl radical cysteine and C79, the role of the proximal reducing cysteine, which protonates the leaving water molecule and harbors a radical intermediate. The reaction path of the subsequent reductive steps is, however, not clear, including the role of formate, which remains intriguing. DFT calculations were performed using modeled substrate and formate structures (60, 64). However, the quality of these models is probably limited owing to uncertainties of formate-binding modes. Because class III lacks a suitable base in the active site, formate could play the role of such a base. A formate radical preceding the formation of CO_2 has been proposed as a potential reaction intermediate (8). The source of the proton to be transferred to the 2′carbon is also unclear. The formate proton could potentially be driving this step, but as the formate proton is not incorporated into the ribose (65), proton transfer has to be mediated by the leaving water molecule. A hypothetical scheme explaining how formate could be involved in the reaction is shown in **Scheme 1**.

RADICAL STORAGE AND TRANSFER

In spite of the related overall and active-site structures, the three RNR classes use different solutions to store and transfer the catalytically essential radicals. Class I uses a tyrosyl radical of protein R2, class II, a cobalamin derived adenosyl radical, and class III, a glycyl radical. All are ultimately responsible for the generation of a thiyl radical on a structurally conserved cysteine residue at the active site. The structural requirements for harboring these radicals are very different. In class I, the first radical site is generated in protein R2 in a reaction involving molecular oxygen and a diiron site; in class II, cobalamin binds directly to the catalytic subunit; whereas in class III, the glycyl radical is generated in the catalytic subunit by an activase (NrdG) in a reaction involving S-adenosyl methionine and a iron sulfur center.

The structure of the *E. coli* R2 subunit revealed a compact dimer wherein each subunit is composed of an α-helix bundle (**Figure 3a**) (66). The structure is related to a family of diiron hydroxylases/oxidases (67). The diiron site of R2, as well as the radical site Y122, is buried within the α-barrel (**Figure 3a**). The diiron site is coordinated primarily by carboxylates and displays a large structural flexibility in the reaction with molecular oxygen, leading to the generation of the tyrosyl radical (67). Structures of R2 subunits from different organisms are very similar with high conservation of functionally important regions (43, 68–70).

The original structural finding that the tyrosyl radical was buried in the R2 subunit led to the proposal that a long-range electron-transfer event was required to mediate the radical-based reaction at the active site of the enzyme (66, 71). The conserved W48 on the R2 subunit was proposed to be a key residue in this electron-transfer event. Because the structure of the class I R1 + R2 holoenzyme was not available, a model was constructed of this complex (**Figure 3a**) in which the distance between Y122 and the substrate was estimated to be longer than 30 Å (53).

In addition to W48, Y376 on the R2 subunit and Y730 and Y731 on the R1 subunits were implied as key residues in a reversible

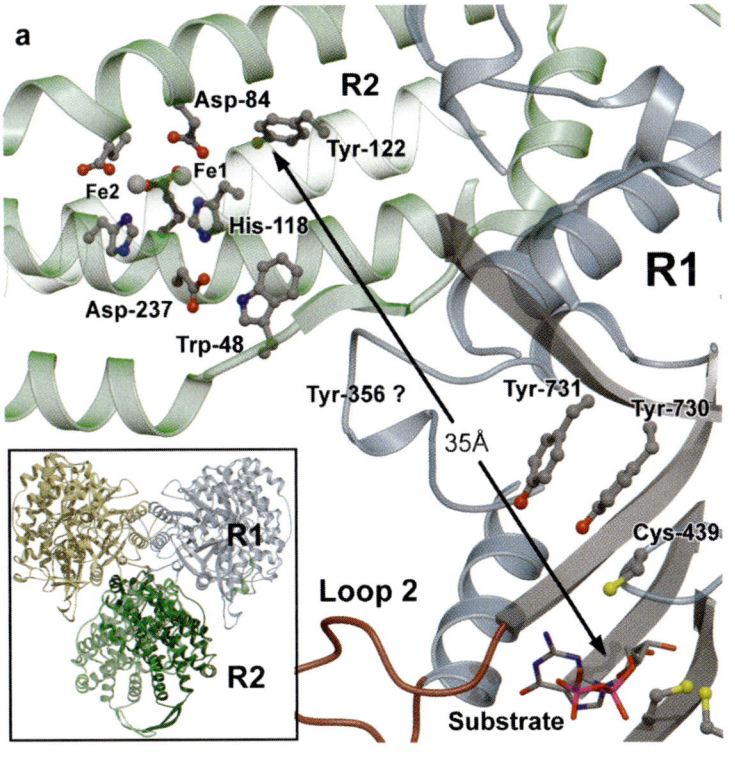

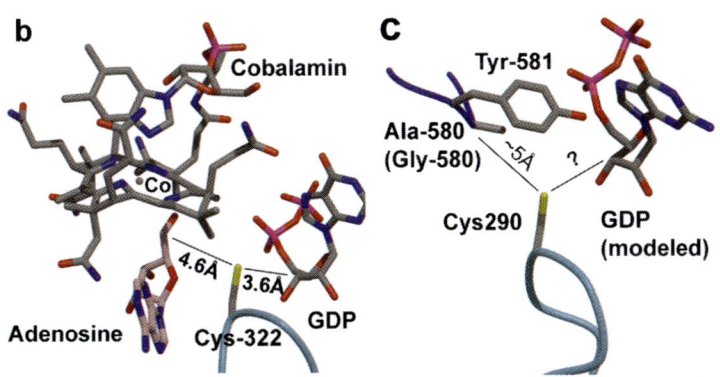

Figure 3

Radical transfer chain in the three classes of RNR. (*a*) Class I R1+R2 complex is modeled by Eklund & Uhlin (4). (*b*) Class II B12 binding is derived from adenosine and cobalamin structures of *T. maritima* RNR. (*c*) Class III structure shows a glycine to alanine mutant of the glycyl radical site.

radical/electron-transfer path between the tyrosyl radical and the active site (4, 53, 72, 73) (**Figure 3***a*). Such a radical transfer chain is unique among known enzyme systems. A concerted electron-proton transfer along a chain of hydrogen-bonded residues was proposed to occur (59, 72). However, a plausible proposal for a concerted long-range transfer of a proton and an electron has not yet been presented, and it is more likely that the observed hydrogen bonding plays a localized role related to the fine-tuning of the redox properties of the side-chain cofactors (66, 73, 74). Therefore, the key feature of the transfer path might be the establishment of a series of redox cofactors with similar redox potentials, which allows a useful equilibrium to be established between these sites, possibly regulated by substrate and effector binding. Effectors have been shown to strongly influence the interaction between the R1 and R2 subunits (75, 76), and it is possible that the binding of effectors directly affects the positions and redox properties of the proposed intermediate radical sites: on one end the tyrosyl radical site of R2, on the other the thiyl radical site of R1. The large distance between the electron-transfer sites on the two subunits in the current model (**Figure 3a**) is not consistent with an efficient electron transfer, implying that relatively large conformational rearrangements are required.

In the class II system, cobalamin binds close to the thiyl radical site (50, 51), and the generation of the radical is driven by homolysis of the C-C bond of the cofactor (**Figure 3b**) (6). Detailed studies of the *L. leichmannii* class II RNR imply that the rate of homolysis is, to some extent, controlled by effector and substrate binding (77). The recent structure of the *T. maritima* enzyme in complex with adenosine and cobalamine supports a direct interaction of the radical adenosine of the cofactor with the thiyl radical cysteine of the active site (**Figure 3b**). The close interaction between the adenosine 5′-carbon and the active-site thiyl radical suggests a direct and concerted electron-proton transfer step for the generation of the thiyl radical in class II RNRs.

In the Class III structure (**Figure 3c**), the loop containing the glycyl radical site (G580) is lining the active site with a neighboring Y581, which penetrates into the active site (52). In the absence of substrate, the glycyl radical site does not make direct van der Waals interactions with the thiyl radical cysteine but is located >5.0 Å away. Conversely, in the structure of the related glycyl radical enzyme pyruvate formate lyase, the glycyl radical site makes direct van der Waals interactions with the cysteine thiol site (78, 79). It is therefore not implausible that substrate or effector binding could induce the required readjustments of the active site in class III RNR and allow direct short-distance, concerted hydrogen-electron transfer also in the class III RNR.

GENE REGULATION

Our understanding of the expression of RNR genes is fragmentary. RNR activity is cell cycle related and highest early in the exponential phase in bacteria and during S phase in eukaryotes, when the requirement for dNTPs is largest.

Bacteria

E. coli produces dNTPs by expressing *nrdAB* during aerobic growth or *nrdDG* during anaerobiosis. A switch occurs from *nrdAB* to *nrdDG* when oxygen becomes sparse via two Fnr boxes upstream of the *nrdDG* operon (80, 81). During strict anaerobiosis, *nrdD*− mutants are not viable but during microanaerobiosis survive by overexpression of *nrdAB* (82). The complicated regulation of class Ia in *E. coli* (83) has not been further investigated since the previous review (1). The regulation of class Ib is incompletely understood (19). In *E. coli*, the transcription of the *nrdHIEF* operon as a single unit greatly depends on growth conditions and is largest during the early exponential phase (19, 84). DNA damage or inhibition of DNA replication in general induces only class Ia, and not class Ib, but hydroxyurea, an inhibitor of both subclasses, induces both. Only *nrdHIEF* is induced in cells lacking thioredoxin and glutaredoxin as well as in response to oxidative stress (84). A specific induction of *nrdEF* was also found in minimal medium (85). Furthermore, the presence of a FUR box immediately upstream of the *nrdH* gene (86) suggests that *nrdEF* may be used when iron is scarce.

The specific function of class Ib (87) in *Enterobacteria* containing both subclasses is not clearly resolved. *E. coli* mutants lacking an active Ib RNR are fully viable (19), and the chromosomal *nrdEF* does not complement *nrdAB* mutants. Complementation occurs if a second copy of *nrdEF* is introduced into the class Ia mutant (19).

In other bacteria, class Ia or Ib occur alone or together with class II and/or class III RNRs.

Corynebacterium ammoniagenes contains only a class Ib RNR, with nrdHIE and nrdF forming separate transcriptional units (88). Both are induced by hydroxyurea or H_2O_2 and also by manganese, explaining the previous belief of the existence of a Mn^{2+}-RNR (21). *Lactobacillus lactis* (24) and *Staphylococcus aureus* (25) have a class Ib together with a class III RNR. Aerobic growth requires class Ib; strict anaerobiosis requires class III. During microaerobiosis, Ib, and not class III, is overexpressed in *L. lactis* (24), reminiscent of the induction of nrdAB in the nrdDG$^-$ mutants of *E. coli* (80). *Pseudomonas aeruginosa* contains three different classes (Ia, II, and III). The class II RNR is unusual in that it is encoded by two genes separated by 16 bp (89). RNR Ia is maximally expressed during exponential growth, and class II, in the stationary phase (89). Examples of bacteria containing both class I and class II enzymes are *S. coelicolor* (Ia + II) (28) and *Mycobacterium tuberculosis* (Ib + II) (26). Either class supports aerobic vegetative growth in *S. coelicolor*, whereas class Ib is required in *M. tuberculosis*. An interesting switch between the two classes occurs in *S. coelicolor*. Adenosylcobalamin (the cofactor required by class II reductases) represses the class Ia enzyme by binding to a B_{12}-riboswitch element in the 5'-untranslated region of nrdAB (28, 90). Regulation of the class II gene (nrdRJ) appears to occur via an ATP cone (20) in NrdR (28). It was suggested that the protein senses the intracellular dATP concentration and that the NrdR-dATP complex binds to two 15-bp direct repeats upstream of the nrdRJ promoter to shut off enzyme activity. From comparative genomics, it appears that NrdR may be an important regulator also for other RNRs (91), leading to the possibility that dATP functions as corepressor for the transcription of RNR as well as an allosteric inhibitor of RNR activity.

Mammalian Cells

RNR activity increases greatly during S phase. R1 and R2 are regulated differently, with R2 being rate limiting for enzyme activity. R2 is regulated both transcriptionally and by enzyme degradation. Several promoter-active regions have been identified for R1 (92) and R2 (93). Binding of E2F4 represses R2 transcription during G_1 (94), which differs from the common activity of E2F as an activator of enzymes involved in DNA synthesis (94, 95). E2F apparently also regulates R2 transcription in plants (96). During mitosis, R2 is degraded (97). At positions 30–32, R2 contains a KEN box (33), which binds the Cdh1-anaphase-promoting complex formed during mitosis, leading to ubiquitination and proteolysis (98).

In resting cells, the R2 gene is not transcribed nor is it upregulated after DNA damage (97). A separate protein (p53R2) together with R1 provides dNTPs for DNA repair (30, 31, 99, 100). In contrast to R2, p53R2 is expressed at a low level throughout the whole cell cycle and is overexpressed in p53(+) cells after DNA damage when R1 is also overexpressed (99) and R2 is degraded (30, 99). In p53(-/-) cells, p53R2 is not induced after DNA damage (99, 101). Instead, R2 increases, and an R1/R2 complex may provide dNTPs (99). DNA repair is believed to take place within a few hours after damage (102), whereas, paradoxically, induction of p53R2 takes a longer time. p53 is controlled by ATM/CHK2, homologues of MEC1/RAD53 involved in the yeast DNA damage pathway (99), underscoring the general importance of a sufficient supply of dNTPs for cell survival. Some of the reported results are clouded by the methodology used to measure RNR activity in permeablized cells. Owing to incomplete hydrolysis of RNA, the used methods largely score incorporation of radioactive ribonucleotides into RNA and not incorporation of deoxyribonucleotides into DNA (103).

Yeast

In budding yeast, RNR is induced at the G_1/S boundary, after DNA damage and after DNA arrest. The main target for regulation is the large subunit and not the small subunit as in mammalian cells. Rnr1 is both transcriptionally regulated (36, 104) and inhibited by interaction with Sml1 (105–107), a protein under the surveillance of the Mec/Rad53 DNA damage checkpoint pathway (108, 109). Also, the transcription of the genes for the second large subunit *RNR3* (36) and for the two small subunits *RNR2* and *RNR4* (39, 40) is regulated similarly. These regulatory mechanisms affect not only recovery from DNA damage but also cellular and mitochondrial DNA replication during normal growth (110). An additional regulatory mechanism worthy of further exploration is the function of Cid13, a Poly(A) polymerase that in the cytosol may maintain the stability of the mRNA for *RNR2* (111). Also, redistribution of Rnr2 and Rnr4 from the nucleus to the cytoplasm appears to regulate RNR activity (112). It is a challenge to elucidate how these multilayered mechanisms interlock in the regulation of the yeast enzyme and to what extent they occur in mammalian cells.

Also in fission yeast, RNR activity is tightly controlled with the highest activity in S phase. The enzyme has the common $\alpha_2\beta_2$ structure of class Ia (113). Prior to DNA replication and in response to DNA damage, an inhibitory protein Spd1 is degraded under the control of the COP9/signalosome and Ddb1, resulting in activation of RNR (114–116). It has been suggested (114) that Spd1 sequesters the small subunit of the enzyme in the cell nucleus, whereas the large subunit is mainly localized in the cytosol. It was propsed that after degradation of Spd1 the small subunit moves to the cytosol to form an active RNR. Recent biochemical experiments (117) cast doubt on such a function for Spd1, as the protein inhibits RNR by specifically binding the large, and not the small, subunit, similar to Sml1 in budding yeast.

ALLOSTERIC REGULATION

Allosteric regulation of RNR makes it possible for a single protein to provide a balanced supply of all four deoxynucleotides and to adapt rapidly to changes in the requirements for dNTPs. For class Ia enzymes, the phenomenology was established by 1969 (118, 119), and since then, similar patterns were found for class II and III. Regulation involves the binding of effectors to two separate sites: the "specificity site," regulating substrate specificity, and the "activity site," regulating the general activity of RNRs. Nucleoside triphosphates are effectors, whereas either di- (class I and some class II enzymes) or triphosphates (class II and III) are substrates. Regulation becomes more sophisticated when substrates and effectors show different levels of phosphorylation as is the case for all class I enzymes. The effectors induce conformational changes of the protein structure providing the signal for the required adaptation at the catalytic site. The mechanism responsible for the regulation at the activity site is not well understood. The structural basis for the mechanism at the specificity site was elucidated recently and is described in some detail below.

The Specificiy Site

Structural studies of R1 from the *E. coli* class Ia RNR first identified the binding site for a specificity effector (dTTP) on the catalytic subunit and implicated Loop 2 as an important structural motif for specificity regulation (54) (**Figure 1a**). This loop forms a bridge between effector and substrate-binding sites (**Figure 4a–c**). Subsequently, the basic binding mode of three effectors was mapped in the R1 structure from the class Ib RNR from *S. typhimurium* (55). In this case, the effectors significantly modulate the conformation of Loop 2. The *S. typhimurium* structures were determined in the absence of substrate and the R2 protein, and the relevance of the observed conformational changes for specificity regulation remains to be determined.

Recently determined structures of complexes of the *T. maritima* class II RNR with effectors and substrates, as well as complexes

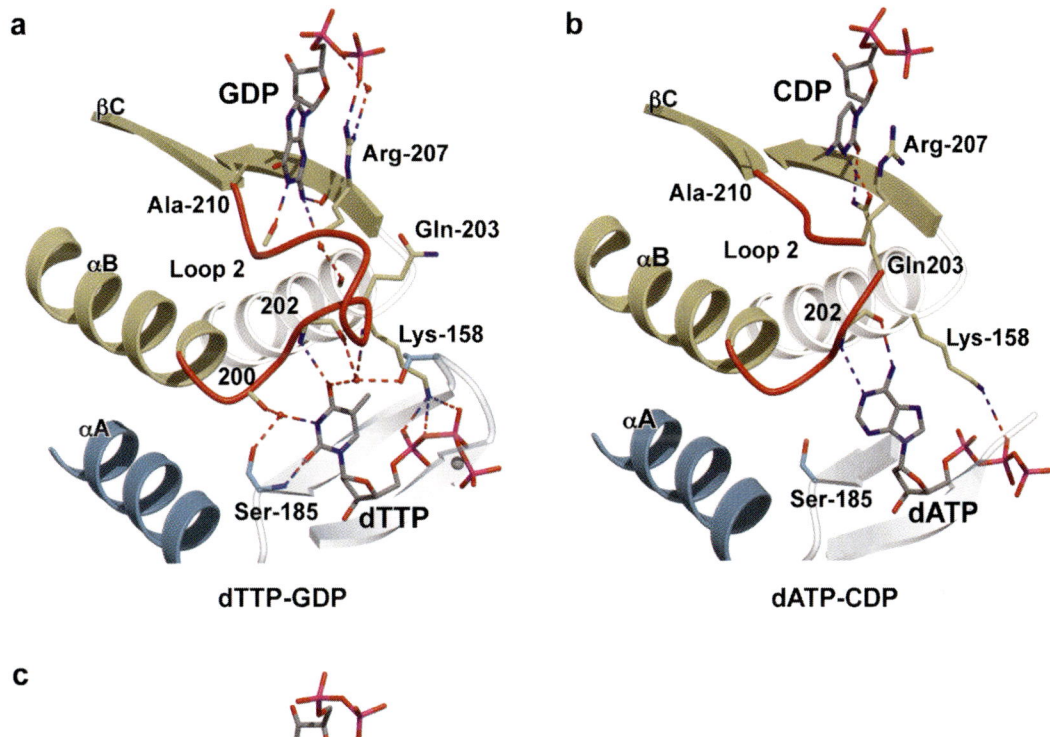
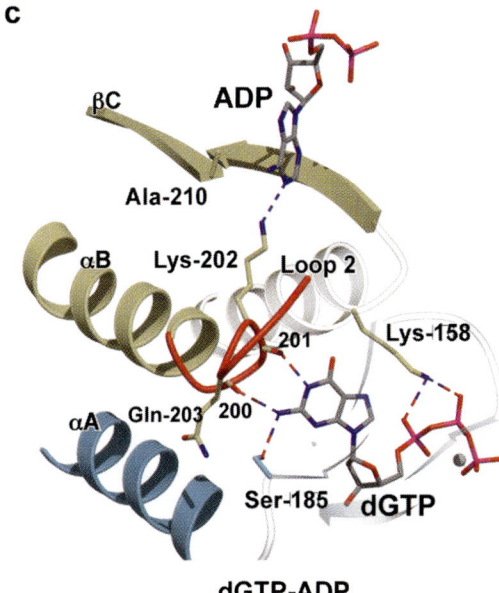

Figure 4

Allosteric control of Loop 2 rearrangements in class II RNR.

with only effectors, provided a detailed view of the structural basis for specificity regulation in a RNR (**Figure 4a–c**) (51). The general binding mode of effectors agrees well with the binding modes previously found in class I RNR structures. By comparing the effector only and effector plus substrate structures it became evident that the conformation of Loop 2 indeed was the major determinant for specificity regulation (51). The conformation of Loop 2 is modulated both by substrate and effector binding. The control of the active-site structure during catalysis is therefore made cooperatively by both nucleotides. Several key residues in Loop 2 are conserved between most class I and class II enzymes, suggesting that the mechanism of substrate specificity regulation is common for class I and class II RNRs.

In the class II complexes with cognate effector-substrate pairs (dTTP-GDP, dGTP-ADP, dATP-UDP, dATP-CDP), the substrate base fits into a tight pocket formed by Loop 2 and by a few residues from the core structural bundle (**Figure 2a, 4**). In all four complexes, the substrate base is clamped into the active site by direct interaction with the sidechain methyl group of A210, providing a small hydrophobic lid (**Figure 2a**). In two of the complexes (dTTP-GDP, dATP-CDP), R207 further clamps the substrate by stacking interactions with the base and direct charged interaction with the phosphate (**Figure 4**).

Each effector-substrate pair induces a distinct Loop 2 conformation. In the structures of the purine-pyrimidine pairs dTTP-GDP, dATP-CDP, and dATP-UDP, Loop 2 folds into the active site in a β-hairpin-like structure. In the purine-purine pair structure of dGTP-ADP, in which the base moieties require larger space, Loop 2 does not form a β-hairpin but forms instead an extra turn at the end of helix αB (residues 201–203).

The different Loop 2 conformations result in dramatically different positions of the two key side chains K202 and Q203 that form hydrogen bonds to the substrate base (**Figure 4a–c**). In the dGTP-ADP structural pair, the side chain of K202 extends into the active site to form an H-bond with the substrate base. In the dATP-CDP and dATP-UDP complexes, Loop 2 adopts very similar conformations and projects Q203 into the active site. However, the hydrogen bonding to the substrate base is slightly different for the two complexes (51). Finally, in the dTTP-GDP complex, side chains are not involved in substrate recognition. Instead, the substrate interacts with main-chain atoms of Loop 2 and a turn in the core barrel.

The three distinct conformations of Loop 2 have different degrees of ordering. In the dTTP-GDP complex, all residues of Loop 2 are ordered. In the dGTP-ADP complex, the residues 204–209 of Loop 2 are disordered; also disordered are residues 204–206 in the dATP-CDP complex and residues 204–207 in the dATP-UDP complex.

How is effector binding controlled by the conformations of Loop 2? All effectors show similar binding patterns, irrespective of the presence of their cognate substrates, with the bases forming hydrogen bonds to the main-chain atoms of residues 200–202, at the transition point between helix αB and Loop 2. The distinct interactions, made by the different effectors, induce different main-chain arrangements in this region. Relatively small modulations of the main-chain conformation at the transition point between helix αB and Loop 2 dramatically affect the conformation of Loop 2 for substrate binding (**Figure 4**). Apparently these changes provide the structural diversity required for allowing the recognition of different substrates in the active site. With dATP and dGTP, effector binding by itself provides a partially preformed active site for the correct substrate, with Q203 and K202 in the correct positions for hydrogen bonding to the cognate substrates. Binding of dTTP, however, does not preform a binding site for the cognate GDP. An important feature of the scheme of allosteric regulation is that, for all three effectors, the part of Loop 2 most proximal to the active site is disordered until the substrate is bound (51). When the substrate binds, the ordering of this part is induced. Most likely the clamping interactions made by A210 and R207 are important for the ordering of Loop 2 upon substrate binding.

Several of the key residues in the specificity regulation of *T. maritima* RNR are conserved in most class II RNRs as well as in many class I RNRs. These are A210 and R207 involved in clamping the substrate to the active site as well as Q203 involved in base recognition.

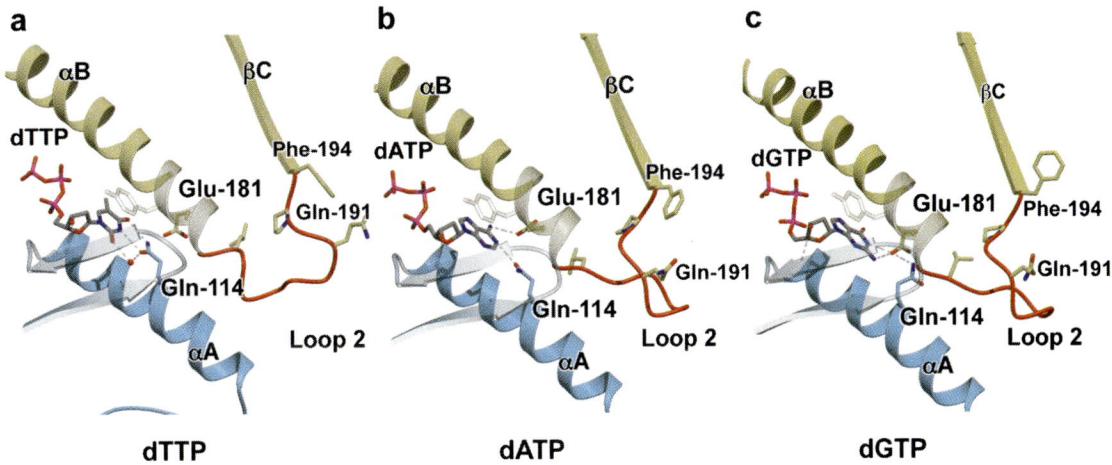

Figure 5
Allosteric control of Loop 2 rearrangements in class III RNR.

Surprisingly, K205 is not conserved in most class I/II enzymes, and in different enzymes, the specificity interaction in the dGTP-ADP complex has to be made differently. Nevertheless, the conservation of the key residues of Loop 2 supports the general validity of the allosteric specificity control of the *T. maritima* RNR for most class I and class II enzymes (51).

The structural basis for class III regulation was studied with phage T4 RNR complexed to four different effectors. At first glance, the results appear to provide a mechanism very different from class II (120) (**Figure 5**). Nevertheless, several intriguing similarities suggest a common basic mechanism for specificity regulation. A major difference comes from the very different dimer interactions in class III and class II. In both cases, effectors bind at the dimer interface, but in class III, the effector-binding site is at a larger distance from the closest active site (>25 Å) than in class II (<10 Å). In class III, the interaction between effector and substrate is mediated both by secondary structural elements of the α/β barrel (primarily helix αB) and by Loop 2, whereas in class II, primarily Loop 2 is involved.

In the T4 class III RNR, the key interactions for effector-base specificity are made by E181 and Q114, each coming from different subunits. These two residues are able to shift between a set of different hydrogen-bonding networks in an intricate manner. This allows Q114 to closely match its hydrogen-bonding pattern for the two pyrimidine bases, and both E181 and Q114 contribute to the binding of purines. A water molecule also mediates specificity recognition for dCTP. The effects of the different recognition of the bases by E181 and Q114 result in conformational transitions extending all the way to the transition point between helix αB and Loop 2. Subsequently, these transitions induce different conformations of Loop 2, as is the case in *Thermotoga* class II RNR (51) and the *S. typhimurium* class I (55).

Structures of complexes between the class III T4 reductase and substrates are not available, and we therefore do not know how the effector-induced conformational changes of Loop 2 control the specificity at the active site (121, 122). It is not clear to what extent the loop conformations represent preformed structures at the active site or if substrate binding will induce further changes. Modeling of class III substrate binding based on the recent class II structures (51) indicates that the conformation of the bound substrates in the two cases probably is similar but that some additional conformational changes of Loop 2 are required for recognition of the substrate (K.-M. Larsson and P. Nordlund, unpublished results) (**Figure 2b**). These modeling studies also implicate the highly conserved F194 as a possible residue for clamping the substrate base into the active site, similar to the action of A210 and R207 in the *Thermatoga* RNR. A conserved Q191 is positioned where it could either serve as a clamp or attach specific hydrogen bonds to the base.

The structural flexibility of Loop 2 is clearly at the heart of the mechanism for specificity regulation by both class I/II and class III RNRs. The effector nucleotides appear to modulate conformations around the transition point between helix αB and Loop 2, affecting a dynamic equilibrium between different Loop 2 conformations. The effectors thereby shift this equilibrium toward a state favoring the initial capture of the cognate substrate, with a subsequent additional ordering of Loop 2 that involves the substrate to various degrees. This apparent cooperativity in the ordering of Loop 2 in class I/II RNRs agrees with previous biochemical data, suggesting a two-way communication between active and specificity sites. Although both class I/II and class III use the same principle for regulation, the overall conformation of Loop 2 and the order/disorder transition in these structures differ between the two classes.

The Activity Site (the ATP Cone)

The ATP cones of RNRs are most readily recognized from the signature sequence VXKRDG a few amino acids downstream of the amino terminus. By this criterion, ATP cones are found in all nonviral eukaryotic R1 sequences. In prokaryotic RNRs, they occur in some, but not all, members of each class, except for class Ib. Also when present, the site is not always functional (dATP-binding is not inhibitory). Examples are the class Ia enzyme from *Trypanosoma brucei* (123) and the class II enzyme from *Thermoplasma acidophilum* (124). All nonviral class III RNRs apparently have an ATP cone, but its functionality has only been demonstrated in *E. coli* (125) and *L. lactis* (126). Originally (125, 126), the data were misinterpreted. They were obscured by the unrecognized presence of stoichiometric amounts of tightly bound dATP remaining with the protein after purification on a dATP-Sepharose column (R. Eliasson and P. Reichard, unpublished results). It is now clear that the two class III RNRs have one functional activity site and one functional specificity site per polypeptide chain.

What is the structural background for the ATP/dATP effect? Nucleotide binding is a prerequisite, but it does not suffice, as is evident from the above cited examples as well as from mouse (127) and *E. coli* (128) mutant enzymes that bind dATP without being inhibited. In these cases, the signal between the activity site and the catalytic site is not transmitted properly. The normal signal is specific in the sense that dATP binding inhibits, whereas ATP stimulates, enzyme activity. In early experiments with the R1/R2 complex of *E. coli* (118, 129), dATP increased the degree of oligomerization, and this effect was reversed by ATP. More detailed experiments with both the *E. coli* (75, 128, 130) and mouse reductases (76, 131) showed that many effectors, but in particular ATP and dATP, stabilized oligomers of R1 and complexes between R1 and R2 (76). It was proposed that the long-range electron transport between R2 and R1 may be disturbed in the tight complex (76). Experiments with *E. coli* R1 mutants identified a hydrogen bond between two histidines (H59 and H88) in wild-type R1, required for allosteric inhibition (130). Binding of ATP to the activity site—but not binding of dATP—was proposed to break this bond. All explanations that invoke an effect on R1/R2 interaction, however, do not fit class III RNRs, as subunit interaction in this case is not involved in enzyme activity. Structural studies of a class III enzyme binding ATP or dATP to its activity site are welcome.

A different explanation for the regulation of mouse reductase proposes the existence of a third allosteric site that exclusively binds ATP

(131). Binding of ATP induces the formation of an R1 hexamer that together with an R2 hexamer is claimed to constitute the active reductase. Although the formation of higher R1 oligomers at high ATP concentration was experimentally well documented, no evidence was presented for a third allosteric site

dNTP POOLS

RNR produces the correct amount of each of the four dNTPs because Nature invented diverse intricate mechanisms to regulate gene transcription, protein degradation, mRNA stability, and specific inhibitors, as well as a unique allosteric control of enzyme activity. The interplay of these mechanisms determines the size and turnover of the four dNTP pools. Measurements in intact cells of pool size and pool turnover are an important complement to measurements of RNR activity in cell extracts. An earlier review (132) discussed the large fluctuations of dNTP pools during the cell cycle of cultured mammalian cells and how pool imbalance may cause disease and increase mutation rates. Such experiments are not without problems. Complications arise from intra- and intercellular pool compartmentalization and cell heterogeneity (13). A striking example of intercellular compartmentalization, first interpreted to be intracellular, was recently described (133). Here we concentrate on pool studies in yeast and mitochondria in which these complications can be minimized.

Yeast

Yeast cells have well defined and controlled growth stages, and they can be genetically manipulated with greater ease than mammalian cells. Many of the upstream genes controlling RNR activity have their counterpart in higher cells (134), and specific human diseases have been mapped to some of these genes. Yeast cells cannot phosphorylate deoxynucleosides and depend exclusively on ribonucleotide reduction for the production of dNTPs, facilitating an understanding of the relation between the reductase and dNTP pools. As a consequence, yeast lacks substrate cycles, which in mammalian cells participate in the regulation of dNTP pools (135). On the negative side, the absence of kinases precludes studies of the turnover of pools with isotopic deoxynucleosides.

In budding yeast, analyses of dNTP pools provided the required fingerprints to distinguish between gene induction and allosteric effects for the regulation of the in situ activity of RNR (107, 136). During normal growth, the elimination of either the inhibitory protein Sml1 or the allosteric inhibition by dATP increased dNTP pools twofold. After DNA damage, which completely relieves inhibition by Sml1 and maximally upregulates RNR, dNTP pools increased fourfold in normal cells and another three- to fivefold in the dATP-mutant. It was suggested (13, 107) that the increased dNTP pools are required for optimal function of repair polymerases. Interestingly, the large pool expansions also increased mutation frequency (107). In fission yeast, mutation rates increased greatly when pool sizes decreased in the absence of the proper degradation of Spd1 (117). Thus in both budding and fission yeast, the RNR inhibitors Sml1 and Spd1, respectively, are important regulators of RNR activity that provides the critical size of dNTP pools.

Mitochondria

Mitochondrial DNA replication occurs within the limits of the mitochondrial double membrane and requires dNTP pools also outside S phase. Mammalian mitochondria do not contain a separate RNR and, for de novo production of dNTPs, therefore depend on the cytosolic enzyme. Mitochondria contain two deoxynucleoside kinases that are not cell-cycle regulated (137) as well as a dTMP-specific 5′-deoxynucleotidase (138). In quiescent cells, when RNR activity is very low or absent, the two kinases provide all four deoxynucleoside monophosphates that can be further phosphorylated to dNTPs. The deoxynucleosides required for this process originate from the extracellular fluid and reach the mitochondria via the cytosol. The 5′-deoxynucleotidase provides a substrate cycle against overproduction of dTTP. Isotope chase experiments demonstrated that the dTTP pools in cytosol and mitochondria had different kinetic properties but were in rapid interchange (139), probably via a specific deoxynucleotide transporter (140, 140a). The size of each mitochondrial pool amounted to approximately 2% to 4% of the corresponding cytosolic pools with similar large size fluctuations in cycling and resting cells (141). Deficiencies in various enzymes of nucleotide metabolism affect mitochondrial DNA replication and cause genetic diseases. The affected enzymes are the mitochondrial thymidine (142) or deoxyguanosine kinases (143), the cytosolic thymidine (144), and possibly purine nucleoside phosphorylases (145). A deficiency in the latter two enzymes results in the cytosolic accumulation of dTTP and dGTP, respectively. The pool imbalance is transmitted to mitochondria, resulting in abnormal mitochondrial DNA. The changes in mitochondrial DNA in thymidine phosphorylase-deficient cells were analyzed in great detail and shown to involve both deletions (146) and site-specific somatic mutations (147), demonstrating clearly the importance of balanced dNTP pools for mitochondrial DNA synthesis.

EVOLUTION

It is generally believed that RNA preceded DNA during the evolution of life and therefore that ribonucleotides existed before deoxyribonucleotides. With the appearance of ribonucleotide reduction, DNA could replace RNA as the repository of genetic information. Ribonucleotide reduction involves radical chemistry that is not likely to be catalyzed by RNA but should require the shielded environment of a protein. According to this line of thought, a RNR was a prerequisite for the appearance of DNA, and proteins preceded DNA during evolution.

One might expect that an enzyme performing such an essential function would have been conserved. Instead, there are three separate classes, with widely diverging amino acid sequences and different mechanisms for the generation of the free radical required for catalysis. The three classes show, however, related three-dimensional structures and striking similarities in their complicated allosteric regulation as well as in the catalytic mechanism. We have therefore suggested that the three classes arose by divergent evolution from a common ancestor that existed before the transition of the RNA to DNA world (52, 148). At that time, oxygen was absent or very low in the atmosphere, and life depended on anaerobic metabolism. Later, when oxygen appeared, new methods for radical generation were required to adapt ribonucleotide reduction to aerobic metabolism, and new enzyme classes evolved.

Which of the three classes existing today could be most closely related to the common ancestor? Class I can be ruled out because the enzymes require oxygen and could not function during the anaerobic conditions of early life. A distinction between class II and III enzymes is more difficult as both function during anaerobiosis. Both existed before the divergence of archea and eubacteria during evolution as today they are abundantly present in both domains of life, whereas class I is found only in eubacteria. Several arguments favor class III as the ancestral state. The arguments have been presented in some detail earlier (12, 16) but are summarized as follows:

- The connection between iron-sulfur minerals and primordial metabolism (149)
- The ubiquitous function of Fe-S clusters and S-adenosylmethionine in anaerobic metabolism (150)
- The paucity of cobalt on earth, compared to iron

- The simplicity of S-adenosyl methionine compared to adenosylcobalamin
- The close structural and functional relation of class III RNR to pyruvate formate lyase (an evolutionary very old anaerobic enzyme of intermediary metabolism) (78, 79)
- Strikingly, class III enzymes use formate, the product of the pyruvate formate lyase reaction, for the reduction of ribonucleotides.

A different scenario (17) was presented proposing the existence of class II RNRs before class III. We consider this less likely and propose that class II evolved later during evolution when the existence of an enzyme adapted to both aerobic and anaerobic life was required. Class I and II systems are highly similar in structural as well as catalytic and allosteric mechanisms, and class I likely diverged from class II at a later stage.

SUMMARY POINTS

1. RNRs are tightly controlled enzymes that provide cells with the appropriate amount of dNTPs required for DNA synthesis and repair. The enzymes were probably a prerequisite for the evolution of the DNA world from the RNA world.
2. Catalysis occurs by a free radical mechanism in which a cysteine of the enzyme becomes a free radical. Radical generation occurs by three different mechanisms, depending on the availability of oxygen (three classes of RNRs).
3. The structures of all three classes are closely related in spite of limited sequence homology.
4. Genetic regulation involves control of transcription of genes for RNR and inhibitor proteins, degradation of proteins, and mRNA stabilization.
5. The structural basis for allosteric mechanisms, regulating the substrate specificity of RNRs, contains common structural elements for the three classes.
6. Structural evidence and the common features of the allosteric regulation of members from the three classes suggest their divergent evolution from a common ancestor. Today's class III appears to be the closest relative of this ancestor.

FUTURE ISSUES TO BE RESOLVED

1. Structures of the complex between the two subunits of class I should be determined to establish the path of radical transfer and the mechanism for allosteric control.
2. An understanding of the catalytic mechanism of class III, in particular the participation of formate as a reductant, and of the structure of the glycyl radical activase are needed.
3. Knowledge of the structures of class III RNR in complex with substrates and formate is needed to establish the structural basis for specificity regulation and catalysis.
4. Increased information is needed about the multiplicity of bacterial RNRs, in particular the function of class Ib in the presence of class Ia.
5. The function of p53R2 and R2 during oncogenesis should be determined.

ACKNOWLEDGMENTS

We thank Vera Bianchi, Lars Thelander, and Eduard Torrents for criticism and for valuable suggestions that improved the original manuscript and thank Karl-Magnus Larsson for help with preparing the figures.

LITERATURE CITED

1. Jordan A, Reichard P. 1998. *Annu. Rev. Biochem.* 67:71–98
2. Fontecave M. 1998. *Cell. Mol. Life Sci.* 54:684–95
3. Sahlin M, Sjöberg B-M. 2000. *Subcell. Biochem.* 35:405–43
4. Eklund H, Uhlin U, Farnegardh M, Logan DT, Nordlund P. 2001. *Prog. Biophys. Mol. Biol.* 77:177–268
5. Kolberg M, Strand KR, Graff P, Andersson KK. 2004. *Biochim. Biophys. Acta* 1699:1–34
6. Stubbe J. 2003. *Curr. Opin. Chem. Biol.* 7:183–88
7. Lawrence CC, Stubbe J. 1998. *Curr. Opin. Chem. Biol.* 2:650–55
8. Eklund H, Fontecave M. 1999. *Structure* 7:R257–62
9. Frey PA, Booker SJ. 2001. *Adv. Protein Chem.* 58:1–45
10. Knappe J, Wagner AFV. 2001. *Adv. Protein Chem.* 58:277–315
11. Fontecave M, Mulliez E, Logan DT. 2002. *Prog. Nucleic Acid Res. Mol. Biol.* 72:95–127
12. Reichard P. 2002. *Arch. Biochem. Biophys.* 397:149–55
13. Chabes A, Thelander L. 2003. *Cell Cycle* 2:171–73
14. Stubbe J. 2000. *Curr. Opin. Struct. Biol.* 10:731–36
15. Stubbe J, Ge J, Yee CS. 2001. *Trends Biochem. Sci.* 26:93–99
16. Torrents E, Aloy P, Gibert I, Rodriguez-Trelles F. 2002. *J. Mol. Evol.* 55:138–52
17. Poole AM, Logan DT, Sjöberg BM. 2002. *J. Mol. Evol.* 55:180–96
18. Jordan A, Pontis E, Åslund F, Hellman U, Gibert I, Reichard P. 1996. *J. Biol. Chem.* 271:8779–85
19. Jordan A, Aragall E, Gibert I, Barbe J. 1996. *Mol. Microbiol.* 19:777–90
20. Aravind L, Wolf YI, Koonin EV. 2000. *J. Mol. Microbiol. Biotechnol.* 2:191–94
21. Willing A, Follmann H, Auling G. 1988. *Eur. J. Biochem.* 170:603–11
22. Huque Y, Fieschi F, Torrents E, Gibert I, Eliasson R, et al. 2000. *J. Biol. Chem.* 275:25365–71
23. Hamilton FD. 1974. *J. Biol. Chem.* 249:4428–34
24. Masalha M, Borovok I, Schreiber R, Aharonowitz Y, Cohen G. 2001. *J. Bacteriol.* 183:7260–72
25. Dawes SS, Warner DF, Tsenova L, Timm J, McKinney JD, et al. 2003. *Infect. Immun.* 71:6124–31
26. Borovok I, Kreisberg-Zakarin R, Yanko M, Schreiber R, Myslovati M, et al. 2002. *Microbiology* 1489(Part 2):391–404
27. Jordan A, Torrents E, Sala I, Hellman U, Gibert I, Reichard P. 1999. *J. Bacteriol.* 181:3974–80
28. Borovok I, Gorovitz B, Yanku M, Schreiber R, Gust B, et al. 2004. *Mol. Microbiol.* 54:1022–35
29. Jordan A, Åslund F, Pontis E, Reichard P, Holmgren A. 1997. *J. Biol. Chem.* 272:18044–50
30. Tanaka H, Arakawa H, Yamaguchi T, Shiraishi K, Fukuda S, et al. 2000. *Nature* 404:42–49
31. Nakano K, Balint E, Ashcroft M, Vousden KH. 2000. *Oncogene* 19:4283–89

32. Nasmyth K. 2001. *Annu. Rev. Genet.* 35:673–745
33. Guittet O, Håkansson P, Voevodskaya N, Fridd S, Gräslund A, et al. 2001. *J. Biol. Chem.* 276:40647–51
34. Kimura T, Takeda S, Sagiya Y, Gotoh M, Nakamura Y, Arakawa H. 2003. *Nat. Genet.* 34:440–45
35. Powell DR, Desai U, Sparks MJ, Hansen G, Gay J, et al. 2005. *Pediatr. Nephrol.* 20:432–40
36. Elledge SJ, Davis RW. 1990. *Genes Dev.* 4:740–51
37. Elledge SJ, Davis RW. 1987. *Mol. Cell. Biol.* 7:2783–93
38. Hurd HK, Roberts CW, Roberts JW. 1987. *Mol. Cell. Biol.* 7:3673–77
39. Huang MX, Elledge SJ. 1997. *Mol. Cell. Biol.* 17:6105–13
40. Wang PJ, Chabes A, Casagrande R, Tian XC, Thelander L, Huffaker TC. 1997. *Mol. Cell. Biol.* 17:6114–21
41. Domkin V, Thelander L, Chabes A. 2002. *J. Biol. Chem.* 277:18574–78
42. Chabes A, Domkin V, Larsson G, Liu AM, Gräslund A, et al. 2000. *Proc. Natl. Acad. Sci. USA* 97:2474–79
43. Voegtli WC, Ge J, Perlstein DL, Stubbe J, Rosenzweig AC. 2001. *Proc. Natl. Acad. Sci. USA* 98:10073–78
44. Ge H, Perlstein DL, Nguyen HH, Bar G, Griffin RG, Stubbe J. 2001. *Proc. Natl. Acad. Sci. USA* 98:10067–72
45a. Torrents E, Trevisiol C, Rotte C, Hellman U, Martin W, Reichard P. 2006. *J. Biol. Chem.* 281:5604–11 (doi:101074/jbc M512962200)
45. Bracchi-Ricard V, Moe D, Chakrabarti D. 2005. *J. Mol. Biol.* 347:749–58
46. Roshick C, Iliffe-Lee ER, McClarty G. 2000. *J. Biol. Chem.* 275:38111–19
47. McClarty G, Tipples G. 1991. *J. Bacteriol.* 173:4922–31
48. Tipples G, McClarty G. 1993. *Mol. Microbiol.* 8:1105–14
49. Högbom M, Stenmark P, Voevodskaya N, McClarty G, Gräslund A, Nordlund P. 2004. *Science* 305:245–48
50. Sintchak MD, Arjara G, Kellogg BA, Stubbe J, Drennan CL. 2002. *Nat. Struct. Biol.* 9:293–300
51. Larsson KM, Jordan A, Eliasson R, Reichard P, Logan DT, Nordlund P. 2004. *Nat. Struct. Mol. Biol.* 11:1142–49
52. Logan DT, Andersson J, Sjöberg BM, Nordlund P. 1999. *Science* 283:1499–504
53. Uhlin U, Eklund H. 1994. *Nature* 370:533–39
54. Eriksson M, Uhlin U, Ramaswamy S, Ekberg M, Regnström K, et al. 1997. *Structure* 5:1077–92
55. Uppsten M, Farnegardh M, Jordan A, Eliasson R, Eklund H, Uhlin U. 2003. *J. Mol. Biol.* 330:87–97
56. Larsson K-M. 2004. *Allosteric regulation and radical transfer in ribonucleotide reductase.* PhD thesis. Stockholm Univ. 56 pp.
57. Stubbe J, Ator M, Krenitsky T. 1983. *J. Biol. Chem.* 258:1625–31
58. Persson AL, Sahlin M, Sjöberg BM. 1998. *J. Biol. Chem.* 273:31016–20
59. Himo F, Siegbahn PEM. 2003. *Chem. Rev.* 103:2421–56
60. Cho KB, Pelmenschikov V, Gräslund A, Siegbahn PEM. 2004. *J. Phys. Chem. B* 108:2056–65
61. Pelmenschikov V, Cho KB, Siegbahn PEM. 2004. *J. Comp. Chem.* 25:311–21
62. Licht S, Gerfen GJ, Stubbe JA. 1996. *Science* 271:477–81
63. Siegbahn PEM. 1998. *J. Am. Chem. Soc.* 120:8417–29

64. Cho KB, Himo F, Gräslund A, Siegbahn PEM. 2001. *J. Phys. Chem. B* 105:6445–52
65. Mulliez E, Ollagnier S, Fontecave M, Eliasson R, Reichard P. 1995. *Proc. Natl. Acad. Sci. USA* 92:8759–62
66. Nordlund P, Eklund H. 1993. *J. Mol. Biol.* 232:123–64
67. Nordlund P. 2001. In *Handbook on Metalloproteins*, ed. I Bertini, A Sigel, H Sigel, pp. 320–445. New York: Marcel Dekker
68. Nielsen BB, Kauppi B, Thelander M, Thelander L, Larsen IK, Eklund H. 1995. *FEBS Lett.* 373:310–12
69. Eriksson M, Jordan A, Eklund H. 1998. *Biochemistry* 37:13359–69
70. Högbom M, Huque Y, Sjöberg BM, Nordlund P. 2002. *Biochemistry* 41:1381–89
71. Nordlund P, Sjöberg BM, Eklund H. 1990. *Nature* 345:593–98
72. Ekberg M, Potsch S, Sandin E, Thunissen M, Nordlund P, et al. 1998. *J. Biol. Chem.* 273:21003–8
73. Stubbe J, Nocera DG, Yee CS, Chang MC. 2003. *Chem. Rev.* 103:2167–201
74. Chang MC, Yee CS, Nocera DG, Stubbe J. 2004. *J. Am. Chem. Soc.* 126:16702–3
75. Kasrayan A, Birgander PL, Pappalardo L, Regnström K, Westman M, et al. 2004. *J. Biol. Chem.* 279:31050–57
76. Ingemarson R, Thelander L. 1996. *Biochemistry* 35:8603–9
77. Chen DW, Abend A, Stubbe J, Frey PA. 2003. *Biochemistry* 42:4578–84
78. Becker A, Fritz-Wolf K, Kabsch W, Knappe J, Schultz S, Wagner AFV. 1999. *Nat. Struct. Biol.* 6:969–75
79. Becker A, Kabsch W. 2002. *J. Biol. Chem.* 277:40036–42
80. Sun XY, Harder J, Krook M, Jörnvall H, Sjöberg BM, Reichard P. 1993. *Proc. Natl. Acad. Sci. USA* 90:577–81
81. Boston T, Atlung T. 2003. *J. Bacteriol.* 185:5310–13
82. Garriga X, Eliasson R, Torrents E, Jordan A, Barbe J, et al. 1996. *Biochem. Biophys. Res. Commun.* 229:189–92
83. Jacobson BA, Fuchs JA. 1998. *Mol. Microbiol.* 28:1315–22
84. Monje-Casas F, Jurado J, Prieto-Alamo MJ, Holmgren A, Pueyo C. 2001. *J. Biol. Chem.* 276:18031–37
85. Tao H, Bausch C, Richmond C, Blattner FR, Conway T. 1999. *J. Bacteriol.* 181:6425–40
86. Vassinova N, Kozyrev D. 2000. *Microbiology* 146(Part 12):3171–82
87. Jordan A, Gibert I, Barbe J. 1994. *J. Bacteriol.* 176:3420–27
88. Torrents E, Roca I, Gibert I. 2003. *Microbiology* 149(Part 4):1011–20
89. Torrents E, Poplawski A, Sjöberg BM. 2005. *J. Biol. Chem.* 280:16571–78
90. Rodionov DA, Vitreschak AG, Mironov AA, Gelfand MS. 2003. *J. Biol. Chem.* 278:41148–59
91. Rodionov DA, Gelfand MS. 2005. *Trends Genet.* 21:385–89
92. Johansson E, Hjortsberg K, Thelander L. 1998. *J. Biol. Chem.* 273:29816–21
93. Chabes AL, Björklund S, Thelander L. 2004. *J. Biol. Chem.* 279:10796–807
94. Degregori J, Kowalik T, Nevins JR. 1995. *Mol. Cell. Biol.* 15:4215–24
95. Trimarchi JM, Lees JA. 2002. *Nat. Rev. Mol. Cell Biol.* 3:11–20
96. Chaboute ME, Clement B, Sekine M, Philipps G, Chaubet-Gigot N. 2000. *Plant Cell* 12:1987–99
97. Chabes A, Thelander L. 2000. *J. Biol. Chem.* 275:17747–53
98. Chabes AL, Pfleger CM, Kirschner MW, Thelander L. 2003. *Proc. Natl. Acad. Sci. USA* 100:3925–29
99. Lin ZP, Belcourt MF, Cory JG, Sartorelli AC. 2004. *J. Biol. Chem.* 279:27030–38

100. Lozano G, Elledge SJ. 2000. *Nature* 404:24–25
101. Zhou BS, Liu XY, Mo XL, Xue LJ, Darwish D, et al. 2003. *Cancer Res.* 63:6583–94
102. Smith ML, Seo YR. 2002. *Mutagenesis* 17:149–56
103. Spyrou G, Reichard P. 1983. *Biochem. Biophys. Res. Commun.* 115:1022–26
104. Huang MX, Zhou Z, Elledge SJ. 1998. *Cell* 94:595–605
105. Zhao XL, Georgieva B, Chabes A, Domkin V, Ippel JH, et al. 2000. *Mol. Cell. Biol.* 20:9076–83
106. Chabes A, Domkin V, Thelander L. 1999. *J. Biol. Chem.* 274:36679–83
107. Chabes A, Georgieva B, Domkin V, Zhao XL, Rothstein R, Thelander L. 2003. *Cell* 112:391–401
108. Elledge SJ. 1996. *Science* 274:1664–72
109. Zhao XL, Chabes A, Domkin V, Thelander L, Rothstein R. 2001. *EMBO J.* 20:3544–53
110. Zhao XL, Rothstein R. 2002. *Proc. Natl. Acad. Sci. USA* 99:3746–51
111. Saitoh S, Chabes A, McDonald WH, Thelander L, Yates JR, Russell P. 2002. *Cell* 109:563–73
112. Yao RJ, Zhang Z, An XX, Bucci B, Perlstein DL, et al. 2003. *Proc. Natl. Acad. Sci. USA* 100:6628–33
113. Sarabia MJF, McInerny C, Harris P, Gordon C, Fantes P. 1993. *Mol. Gen. Genet.* 238:241–51
114. Liu C, Powell KA, Mundt K, Wu LJ, Carr AM, Caspari T. 2003. *Genes Dev.* 17:1130–40
115. Nielsen O. 2003. *Curr. Biol.* 13:R565–67
116. Holmberg C, Fleck O, Hansen HA, Liu C, Slaaby R, et al. 2005. *Genes Dev.* 19:853–62
117. Håkansson P, Dahlroth SL, Chilikova O, Domkin V, Thelander L. 2005. *J. Biol. Chem.* 281:1778–83 (doi: 101074/jbc M511716200)
118. Brown NC, Reichard P. 1969. *J. Mol. Biol.* 46:25–38
119. Brown NC, Reichard P. 1969. *J. Mol. Biol.* 46:39–55
120. Larsson KM, Andersson J, Sjöberg BM, Nordlund P, Logan DT. 2001. *Structure* 9:739–50
121. Eriksson S. 1983. *J. Biol. Chem.* 258:5674–78
122. Chimploy K, Mathews CK. 2001. *J. Biol. Chem.* 276:7093–100
123. Hofer A, Schmidt PP, Gräslund A, Thelander L. 1997. *Proc. Natl. Acad. Sci. USA* 94:6959–64
124. Eliasson R, Pontis E, Jordan A, Reichard P. 1999. *J. Biol. Chem.* 274:7182–89
125. Eliasson R, Pontis E, Sun XY, Reichard P. 1994. *J. Biol. Chem.* 269:26052–57
126. Torrents E, Buist G, Liu A, Eliasson R, Kok J, et al. 2000. *J. Biol. Chem.* 275:2463–71
127. Reichard P, Eliasson R, Ingemarson R, Thelander L. 2000. *J. Biol. Chem.* 275:33021–26
128. Birgander PL, Kasrayan A, Sjöberg BM. 2004. *J. Biol. Chem.* 279:14496–501
129. Thelander L. 1973. *J. Biol. Chem.* 248:4591–601
130. Birgander PL, Bug S, Kasrayan A, Dahlroth SL, Westman MA, et al. 2005. *J. Biol. Chem.* 280:14997–5003
131. Kashlan OB, Cooperman BS. 2003. *Biochemistry* 42:1696–706
132. Reichard P. 1988. *Annu. Rev. Biochem.* 57:349–74
133. Bianchi V, Borella S, Rampazzo C, Ferraro P, Calderazzo F, et al. 1997. *J. Biol. Chem.* 272:16118–24
134. Caspari T. 2000. *Curr. Biol.* 10:R315–17
135. Gazziola C, Ferraro P, Moras M, Reichard P, Bianchi V. 2001. *J. Biol. Chem.* 276:6185–90
136. Zhao XL, Muller EGD, Rothstein R. 1998. *Mol. Cell* 2:329–40

137. Arner ESJ, Eriksson S. 1995. *Pharmacol. Ther.* 67:155–86
138. Rampazzo C, Gallinaro L, Milanesi E, Frigimelica E, Reichard P, Bianchi V. 2000. *Proc. Natl. Acad. Sci. USA* 97:8239–44
139. Pontarin G, Gallinaro L, Ferraro P, Reichard P, Bianchi V. 2003. *Proc. Natl. Acad. Sci. USA* 100:12159–64
140. Bridges EG, Jiang ZL, Cheng YC. 1999. *J. Biol. Chem.* 274:4620–25
140a. Marobbio CMT, Di Noia MA, Palmieri F. 2006. *Biochem. J.* 393:441–46
141. Rampazzo C, Ferraro P, Pontarin G, Fabris S, Reichard P, Bianchi V. 2004. *J. Biol. Chem.* 279:17019–26
142. Saada A, Shaag A, Mandel H, Nevo Y, Eriksson S, Elpeleg O. 2001. *Nat. Genet.* 29:342–44
143. Mandel H, Szargel R, Labay V, Elpeleg O, Saada A, et al. 2001. *Nat. Genet.* 29:491
144. Nishino I, Spinazzola A, Hirano M. 1999. *Science* 283:689–92
145. Arpaia E, Benveniste P, Di Cristofano A, Gu YP, Dalal I, et al. 2000. *J. Exp. Med.* 191:2197–207
146. Nishigaki Y, Marti R, Hirano M. 2004. *Hum. Mol. Genet.* 13:91–101
147. Nishigaki Y, Marti R, Copeland WC, Hirano M. 2003. *J. Clin. Investig.* 111:1913–21
148. Reichard P. 1993. *Science* 260:1773–77
149. Huber C, Wächtershäuser G. 1997. *Science* 276:245–47
150. Sofia HJ, Chen G, Hetzler BG, Reyes-Spindola JF, Miller NE. 2001. *Nucleic Acids Res.* 29:1097–106

Introduction to the Membrane Protein Reviews: The Interplay of Structure, Dynamics, and Environment in Membrane Protein Function

Jonathan N. Sachs and Donald M. Engelman

Department of Molecular Biophysics and Biochemistry, Yale University, New Haven, Connecticut 06520; email: jsachs@csb.yale.edu, donald.engelman@yale.edu

Abstract

In our review, we introduce an organizational scheme for membrane protein function. It is the relationship between structure, dynamics, and environment that endows the membrane and its constituents with remarkable sensitivity and robustness. Our understanding begins with landmark advances like those presented in the following chapters. Membrane proteins are notoriously difficult to study, and so the work presented here on the ADP/ATP carrier [Nury et al. (2)], rhodopsin [Palczewski (24)], and the cytochrome b_6f complex [Cramer et al. (35)] represents incredible progress in this now blossoming field.

INTRODUCTION

Membrane protein structure determination is proceeding at an exciting pace, driven by the hope that structures can connect decades of biochemical and biophysical observations to protein function and mechanism. The reviews that follow demonstrate that structures are capable of resolving long-standing mysteries while suggesting new questions and opening new areas of scientific discovery. As science as a whole unravels increasingly complex and finely tuned functionality, can structure determination continue to keep pace, or will new methodologies and perspectives be required? The ultimate test will be to build robust molecular-level models capable of predicting the functional outcome of any arbitrary structural perturbation. The creation of such models will require adopting a broad perspective, one that takes into account dynamic deviations from static structures as well as the influence of the membrane environment.

It is helpful to establish a paradigm that categorizes and organizes the relevant contributors to function. Structures should, and will, act as the nucleation points for this organization. This is because understanding mechanisms is ultimately a matter of chemistry, and one cannot invoke chemical ideas of the mechanism without knowing where the atoms are. As we detail in the discussion that follows, evolution has exploited three categories of molecular-level organization in order to achieve efficient and diverse membrane protein functions: structure, molecular dynamics, and environmental constraints. As suggested schematically in **Figure 1**, they are inextricably linked, each influencing the other, collectively dictating membrane protein function. By recasting the structure-function relationship in this way, we suggest that a comprehensive view of membrane protein function may be more readily achieved.

STRUCTURE

Any modern model for predicting function must start with structure. As demonstrated in the following reviews, static structures can elegantly illuminate the structure-function relationship. Collectively, the articles detail a series of new features only now visible in the increasingly high-resolution structures. The emergence of these elements, which include protein oligomerization and complexation, lipidation, glycosylation, and the presence of structural waters, suggests that future models may need to account for a high level of complexity and structural variability.

Oligomerization is increasingly seen as a common motif in membrane proteins (1), and dimerization in particular is thought to be the rule for each of the three proteins discussed below. As the authors point out, distinguishing between dimers and higher order oligomers is far from trivial. Although this difficulty may be primarily due to experimental limitations such as dissociation by detergent, it may also reflect true biological variability: Perhaps there is a functionally

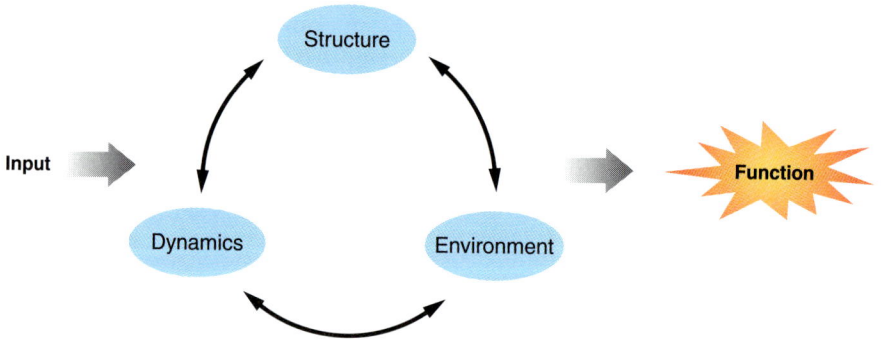

Figure 1
The three categories of molecular-level organization needed to achieve efficient and diverse membrane protein functions.

relevant equilibrium distribution of *n*-mers within cell membranes. This possibility suggests an investigation into the evolutionary origin of oligomerization. We suggest four distinct, though not exclusive, explanations. First, in most studied cases, oligomerization serves a functional role, and this function most likely drove its evolution. For example, the cytochrome b_6f dimer interface is thought to form an electron transfer bridge for "cross talk" between the two monomers. In the case of rhodopsin, dimerization is thought to regulate G-protein coupling. Second, stability of newly evolved proteins must have been critical, and oligomerization may provide an efficient way to select for stabilizing mutants. In the case of a homodimer, for example, a single mutation at the interface could be twice as efficient in stabilizing the protein compared with a single mutation in a monomer. Similarly, a third explanation is that formation of oligomeric structures can augment genetic efficiency. For example, in the several ion channel structures we now know, identical subunits surround the ion pathway, requiring the coding of only a single unit to form a larger structure. As a fourth possible explanation, we speculate that if dense packing of membrane proteins is evolutionarily important for optimizing functional output per unit area of membrane, then the evolution of oligomeric interfaces might have been a structural adaptation to support high packing density while minimizing energetically unfavorable protein-protein contacts. Evolution would then have built upon this through adaptations that functionalized the interface.

Like oligomerization, the presence of strongly bound lipids and specific sites for water molecules found in recent high-resolution structures suggests a potentially large degree of structural, and hence functional, variability. What is their role in structural stability and what is their functional significance? In the case of b_6f, structural lipids are suggested either to be stabilizers, acting as "structural struts," or functional, "imposing restraints on protein dynamics." In the case of the ADP/ATP carrier, removal of structural cardiolipin molecules, as Nury et al. (2) point out, leads to a 20% decrease in protein activity. How dynamic is the lipid binding? Should these lipids be considered as ligands or as parts of the structure? In some cases it seems that the lipid requirement is not specific, but in others it is. Imagine the impact if it is found that these interactions with lipids are a highly regulated and ubiquitous cellular phenomenon. Similar questions can and should be asked for the fascinating case of structural waters, as well as for all posttranslational modifications, including glycosylation. The potential combinatorial explosion because of variable glycosylation patterns on the surface of membrane proteins underscores our contention, as we now discuss, that the size of structure space may be staggeringly large if all variations have functional relevance.

WHY DYNAMICS?

In many cases, a protein must undergo a dynamic conformational transition between discrete structural states to carry out its function. Such transitions, for example from state A to state B, involve a change in the thermodynamic free energy of the system, $\Delta G_{A \to B}$. The ADP/ATP carrier protein located in the mitochondrial membrane, discussed by Nury et al. (2), is a good example. The authors point to the "induced transition fit" (ITF) mechanism to explain the dynamics of the protein. In this mechanism, described by Klingenberg (3), the membrane protein exists in multiple, discrete conformations, with the metabolite only binding perfectly to the highest energy state. The energy of this transition state is then utilized in triggering further conformational changes necessary for metabolite release. The total free energy change of such a process is the sum over all intermediate states:

$$\Delta G_{A \to B} = \sum_{i=1}^{N-1} \Delta G_{i \to i+1},$$

with $i = 1$ corresponding to state A, $i = N − 1$ corresponding to state B, and N commonly assumed to be a finite and reasonably small number of states.

Just how many relevant energy states exist for any given mechanism? Five such states are suggested in the case of the ITF mechanism. In the case of ion channels, models built from gating behavior suggest an even larger number (4). If one thinks in terms of a multidimensional free energy landscape for transitions in proteins, then these discrete states are found either in local energy minima (intermediates) or maxima (transition states). However, the energy wells corresponding to the minima may be quite broad, and transitions between them may follow multiple paths. Each of these paths may, in fact, be populated with an extremely large number of functionally pertinent conformational substates. Such an idea is supported by the case of calmodulin, wherein analysis of disorder in the crystal structure has suggested that the protein "may sample a quasi-continuous spectrum of conformations" (5). Similarly, a very large number of paths between any two states are theoretically possible, although the accessible number is likely diminished by environmental conditions (see below). This all suggests that the overall free energy change may also be expressed as

$$\Delta G_{A \to B} = \sum_{j=1}^{N'-1} \Delta G'_{j \to j+1}$$
$$= \sum_{k=1}^{N''-1} \Delta G''_{k \to k+1} = \ldots .$$

Which path is taken determines the work required in effecting a transition and thus has important consequences for protein efficiency and function. Could path choice be a variable parameter in cellular control, for example, through concerted variations in the proteins' surroundings (i.e., the membrane)?

Clearly, given the relevance of conformational states to function, single structures can only partially describe a mechanism. The range of available energy states along a transition path may be exploited by proteins for tuning function. How do we investigate the transition path? Is it possible to predict, solely from two static structures, the relevant path between the two states? The problem is especially complicated because structural dynamics span a range of frequencies, from low (large conformational changes or domain movements, more traditionally associated with function) to high (side-chain rotameric isomerizations, which may, for example, play an important role in electron transport proteins, which are otherwise relatively immobile). There is, however, significant progress being made in the area of computational biology that portends a numerical solution to this problem, although further improvements in the computational representation of chemistry are needed. For example, molecular dynamics simulations generally provide information about high-frequency fluctuations (6), and normal mode analysis can help predict the lower frequency paths (7). More sophisticated computational methods known as path sampling have been shown to yield highly detailed transition path information for simple systems (8–12), and more recently for biomolecules (13–16).

ENVIRONMENT

As we suggested above, a high degree of complexity exists owing to structural variability and dynamic transition paths. A third factor that contributes to complexity is the membrane environment. Structure determination generally requires isolation of the protein from the remainder of the biological system. However, any proper thermodynamic analysis must include all relevant components of the system and must pay particularly close attention to boundaries where energy is exchanged. The contribution of the membrane environment therefore deserves consideration in analysis of function. We have noted the multiplicity of available conformational transition paths, and that path choice determines the trade-off between energy spent and work done. The number of available paths, along with the paths themselves, must be altered by physical constraints placed on the protein by the membrane environment. Therefore, isolation of the protein from its native membrane provides only a partial story.

The membrane consists of, among other things, lipids and a multitude of proteins. In addition to severely biasing protein conformational states and the paths between them (17), the membrane itself is capable of storing energy through conformational flexibility of its own (18, 19). It has been clearly established that lipid composition varies profoundly in different membranes (20). Additionally, single membranes can be highly heterogeneous, as in the case of "lipid rafts," the functional manifestation of lipid domains known for decades to exist in synthetic mixtures (21–23). Sustaining such diversity, for example, through variable lipid composition, may cost the cell valuable resources, so specific functional rationales should be examined. Furthermore, biophysical measurements have established that physical properties of membranes, such as curvature elastic stress, profoundly affect the efficiency of membrane proteins. As one landmark example, a specific conformational transition in rhodopsin, the molecule discussed by Palczewki (24), is favored by the highly curved reverse hexagonal phase, rather than the standard lamellar phase (25). It seems likely that cells might take advantage of this type of specificity to tune function by genetically regulating their membrane-specific constituent lipid populations. Other significant membrane properties that are now appreciated as influencing structure and function of membrane proteins are, among others, lateral tension (26, 27), hydrophobic matching (28–32), and electrostatics (33, 34). By modifying these properties, cells have afforded themselves a highly tunable molecular environment and thus, we suggest, have gained incredibly fine control over protein function.

Molecular function underlies all of biology, from the processing of input in the simplest organisms to higher consciousness in humans. How intricate must the molecular machinery be to support such a diverse and elegant world? It is, of course, natural that we think in terms of discrete states. Primarily, it makes investigation of structure-function more tractable. Additionally, experimental techniques such as crystallization lend themselves to thinking in terms of single, or averaged, structures. The perspective we offer here poses a challenge to modern biologists given the immensity of combinatorial possibilities it suggests. It is hard to imagine how to begin to study this complexity; however, it is a challenge worth pursuing. If we seek a molecular-level explanation of phenomena as mystifying as consciousness, then our suggestion of near infinite combinatorial complexity seems less of a stretch. Clearly, the mere existence of complexity does not prove its evolutionary or functional significance. The important tasks are to digest the complexity and find the simplifications that remain true to the biology. We suggest that computational methods, along with increased experimental resolution, both spatially and temporally, should facilitate this effort. The progress reflected in the reviews that follow suggests that we are well on our way.

ACKNOWLEDGMENTS

We thank the members of the Engelman lab and Daniel M. Zuckerman for helpful discussion. J.N.S. is supported by an N.R.S.A. Grant (GM071134-02) and D.M.E. by an N.I.H. grant (GM070895).

LITERATURE CITED

1. Engelman DM. 2005. *Nature* 438:578–80
2. Nury AH, Dahout-Gonzalez C, Trézéguet V, Lauquin GJM, Brandolin G, Pebay-Peyroula E. 2006. *Annu. Rev. Biochem.* 75:713–41
3. Klingenberg M. 2005. *Biochemistry* 24:8563–70
4. Imredy JP, Yue DT. 1994. *Neuron* 12:1301–18
5. Wilson MA, Brunger AT. 2000. *J. Mol. Biol.* 301:1237–56
6. Crozier PS, Stevens MJ, Forrest LR, Woolf TB. 2005. *J. Mol. Biol.* 333:493–514
7. Valadie H, Lacapcre JJ, Sanejouand YH, Etchebest C. 2003. *J. Mol. Biol.* 332:657–74
8. Huber GA, Kim S. 1996. *Biophys. J.* 70:97–110
9. Pratt LR. 1986. *J. Chem. Phys.* 85:5045–48
10. Woolf TB. 1998. *Chem. Phys. Lett.* 289:433–41
11. Dellago C, Bolhuis PG, Csajka FS, Chandler D. 1998. *J. Chem. Phys.* 108:1964–77
12. Zuckerman DM, Woolf TB. 1999. *J. Chem. Phys.* 111:9475–84
13. Zuckerman DM. 2004. *J. Phys. Chem. B.* 108(16):5127–37
14. Bolhuis PG. 2005. *Biophys. J.* 88(1):50–61
15. Arora K, Schlick T. 2005 *J. Phys. Chem. B.* 109(11):5358–5367
16. Tobi D, Bahar I. 2005. *Proc. Nat. Acad. Sci.* 102(52):18908–13
17. Nanda H, Sachs JN, Petrache HI, Woolf TB. 2005. *J. Chem. Theory Comput.* 1:375–88
18. Mouritsen OG, Bloom M. 1993. *Annu. Rev. Biophys. Biomol. Struct.* 22:145–71
19. Hamill OP, Martinac B. 2001. *Physiol. Rev.* 81:685–740
20. Wenk MR, De Camilli P. 2004. *Proc. Natl. Acad. Sci. USA* 101:8262–69
21. Jain MK, White HB III. 1977. *Adv. Lipid Res.* 15:1–60
22. Klausner RD, Kleinfeld AM, Hoover RL, Karnovsky MJ. 1980. *J. Biol. Chem.* 255:1286–95
23. Simons K, van Meer G. 1988. *Biochemistry* 27:6197–202
24. Palczewki K. 2006. *Annu. Rev. Biochem.* 75:743–67
25. Brown MF. 1994. *Chem. Phys. Lipids* 73:159–80
26. Cantor RS. 1999. *Chem. Phys. Lipids* 101:45–56
27. Carrillo-Tripp M, Feller SE. 2005. *Biochemistry* 44:10164–69
28. Lee AG. 2003. *Biochim. Biophys. Acta* 1612:1–40
29. Williamson IM, Alvis SJ, East JM, Lee AG. 2002. *Biophys. J.* 83:2026–38
30. de Planque MRR, Killian JA. 2003. *Mol. Membr. Biol.* 20:271–84
31. Tieleman DP, Forrest LR, Sansom MSP, Berendsen HJC. 1998. *Biochemistry* 37:17554–61
32. Petrache HI, Zuckerman DM, Sachs JN, Killian JA, Koeppe RE, Woolf TB. 2002. *Langmuir* 18:1340–51
33. Aguilella VM, Bezrukov SM. 2001. *Eur. Biophys. J. Biophys. Lett.* 30:233–41
34. Jensen MO, Mouritsen OG. 2004. *Biochim. Biophys. Acta* 1666:205–26
35. Cramer WA, Zhang H, Yan J, Kurisu G, Smith JL. 2006. *Annu. Rev. Biochem.* 75:769–790

Relations Between Structure and Function of the Mitochondrial ADP/ATP Carrier

H. Nury,[1,*] C. Dahout-Gonzalez,[2,*] V. Trézéguet,[3] G.J.M. Lauquin,[3] G. Brandolin,[2] and E. Pebay-Peyroula[1]

[1]Institut de Biologie Structurale Jean-Pierre Ebel, UMR 5075 CEA-CNRS-Université Joseph Fourier, F-38027 Grenoble cedex 1, France; email: hugues.nury@ibs.fr, eva.pebay-peyroula@ibs.fr

[2]Laboratoire de Biochimie et Biophysique des Systèmes Intégrés, UMR 5092 CEA-CNRS-Université Joseph Fourier, Département de Réponse et Dynamique Cellulaires, F-38054 Grenoble cedex 9, France; email: cecile.gonzalez@ibs.fr, gbrandolin@cea.fr

[3]Laboratoire de Physiologie Moléculaire et Cellulaire, UMR 5095 CNRS-Université Bordeaux 2, Institut de Biochimie et Génétique Cellulaires, F-33077 Bordeaux cedex, France; email: guy.lauquin@ibgc.u-bordeaux2.fr, vero.trezeguet@ibgc.u-bordeaux2.fr

Key Words

membrane protein, nucleotide transport, oligomeric state

Abstract

Import and export of metabolites through mitochondrial membranes are vital processes that are highly controlled and regulated at the level of the inner membrane. Proteins of the mitochondrial carrier family (MCF) are embedded in this membrane, and each member of the family achieves the selective transport of a specific metabolite. Among these, the ADP/ATP carrier transports ADP into the mitochondrial matrix and exports ATP toward the cytosol after its synthesis. Because of its natural abundance, the ADP/ATP carrier is the best characterized within MCF, and a high-resolution structure of one conformation is known. The overall structure is basket shaped and formed by six transmembrane helices that are not only tilted with respect to the membrane, but three of them are also kinked at the level of prolines. The functional mechanisms, nucleotide recognition, and conformational changes for the transport, suggested from the structure, are discussed along with the large body of biochemical and functional results.

*These authors contributed equally to the work.

Contents

- INTRODUCTION 714
- BRIEF HISTORY 715
- STRUCTURE ANALYSIS 717
  - The Cavity 717
  - CATR Binding 722
  - The MCF Motif 722
  - Location of the ADP/ATP Carrier Signature 725
  - Lipids 726
  - Deviation to Pseudo-Threefold Symmetry 727
  - Conserved Residues: A Structural or Functional Role? 728
- NUCLEOTIDE ATTRACTION AND BINDING 728
  - Specificity for Binding and Transport 730
  - Biochemical Evidence for IMS-Binding Sites 730
  - Structural Features 731
  - Binding ATP from the Matrix 731
- TRANSPORT MECHANISM 732
  - Conformational Changes 732
  - What Triggers the Changes? Kinetic Aspects 733
  - Is the Functional Unit a Dimer? ... 734
- HYPOTHESES AND CONCLUSIONS 736
  - A Proposed Mechanism Involving CDLs 736
  - Open Questions 736

Mitochondrial carrier family (MCF): proteins of the inner mitochondrial membrane, which shuttle substrates in and out

IMS: intermembrane space

TM: transmembrane

INTRODUCTION

Cell compartmentalization into organelles limited by membranes, allowing segregation of various cell functions in different locations of the cellular space, implies that a complex communications network needs to be set up and regulated. In most cases, molecules involved in intracellular trafficking have to pass through membranes, and specific transport proteins generally catalyze these journeys. This is particularly true with charged species because the lipid bilayer of membranes is an effective barrier to anions and cations.

Solute transports through membranes are essential steps in many metabolic pathways. Mitochondria are particularly rich in solute carriers, and most of these constitute the so-called mitochondrial carrier family (MCF) (1), also referred to as SLC25. About 20 distinct transport functions are involved in the fluxes of the various metabolites through the inner mitochondrial membrane, which is the only mitochondrial permeability barrier, the outer membrane being freely permeable to molecules up to approximately 5000 Da. All of the characterized carriers are not always present in all tissues, according to the occurrence of tissue-specific metabolic pathways.

Among MCF members, the ADP/ATP carrier is the most abundant and, along with the phosphate carrier, appears indispensable to mitochondria. It can be considered the paradigm of mitochondrial metabolite carriers because several major findings of carrier structures and mechanisms have been first produced through studies of the ADP/ATP carrier. Albeit the function of all the members of the MCF have not yet been identified, they share main structural and functional characteristics that are summarized as follows: (*a*) all mitochondrial carriers are encoded by nuclear genes, (*b*) the primary structure of most carriers displays three repeated homologous regions of about 100 amino acids each, (*c*) the N and C termini face the intermembrane space (IMS) and six transmembrane (TM) segments can be delineated, (*d*) a common sequence, the MCF motif, can be found in each repeated region with slight deviations on one or two signature sequences for some carriers, and (*e*) comparison of primary structures indicates that mitochondrial carriers have no orthologues in prokaryotes, their emergence seems to be the evolutionary consequence of the capture of an ancient aerobic procaryotic cell by the primitive eucaryotic cell.

Although high-resolution structures are important tools to tackle functional mechanisms, structures of membrane proteins are still far behind expectations owing to difficulties in getting sufficient amounts of pure and well-folded proteins and in crystallizing such amphipatic molecules (2). Furthermore, transporters often undergo large conformational changes, and therefore solutions of solubilized proteins are highly inhomogeneous in terms of conformations. Obtaining well-ordered three-dimensional crystals is more likely to occur when a single conformation has been locked either by a mutation as for lactose permease (3) or by the presence of an inhibitor as discussed herein for the ADP/ATP carrier (4). Two-dimensional crystallization is an alternative approach to structure determination by X-ray crystallography, which might favor protein stability because the protein is embedded in a flat lipidic bilayer. Even though the resolution obtained by electron microscopy is usually lower, it is an interesting complementary approach that might help elucidate larger conformational changes. The ADP/ATP carrier is the only MCF carrier for which the three-dimensional structure of one conformation is known. The first projected structure of a yeast isoform complexed to atractyloside (ATR) was obtained by electron microscopy (5, 6). The high-resolution structure of the bovine carrier was solved to 2.2 Å in the presence of carboxyatractyloside (CATR) (4), and a second crystal form of the same protein gave additional information on cardiolipins (CDLs) (7). This review discusses mainly structure-function relationships, combining structural with biochemical and functional data.

BRIEF HISTORY

The concept of a mitochondrial carrier for adenine nucleotides arose in the 1960s as a result of the convergence of two independent series of observations described by researchers working at deciphering the mitochondrial phosphorylation process.

In 1955, a pioneering study by Siekevitz & Potter (8) established that two distinct pools of adenine nucleotides were operating cooperatively in the oxidative phosphorylation pathway, one located in the mitochondrial matrix and the other in the cytosol. Soon after, Pressman (9) reported that although the intramitochondrial concentrations of nucleotides were largely kept unaffected by the extramitochondrial concentration, "an extremely dynamic interchange of nucleotides" took place between the two pools. To rationalize these findings, the hypothesis put forward at that time was "the existence of a fixed quantity" of an intramitochondrial component with sufficient affinity for adenine nucleotides. Curiously enough, there was no mention of a putative adenine nucleotide transporter although Bartley & Davies (10) had already advanced the possibility of transporters for mitochondrial respiratory substrates.

The second series of findings came from studies aimed at elucidating the poisonous effect of a natural drug, ATR, extracted from the thistle *Atractylis gummifera*, which is widespread in Mediterranean countries. This plant had been already described two thousand sand years ago by Dioscorides (~40–90 AD) as a medicinal herb in his famous herbal

HUMAN DISEASE AND THE ADP/ATP CARRIER

Disorders in the mitochondrial energy-generating system are reflected by a variety of clinical symptoms ranging from myopathy, with lactic acidosis, to severe multisystem disease, involving the central nervous or cardiac system. In most cases, the origin of the disease is evidenced by a defect in one of the components of the respiratory chain and is due to mutations in either the mitochondrial or the nuclear genome. For some patients, however, impairment of the mitochondrial function cannot be ascribed to such defects. Almost 25% of the patients suffering from such a dysfunction do not show clear impairments of the respiratory chain. In some cases, metabolite carriers, including the ADP/ATP carrier, might be the cause of the pathology. Mitochondrial myopathies implicating the ADP/ATP carrier have been described in human and in mouse, and these myopathies may be involved in several cases of autosomal dominant progressive ophthalmoplegia.

High-resolution structure: three-dimensional protein structures with atomic details usually obtained by X-ray crystallography

ATR: atractyloside

CATR: carboxyatractyloside

Cardiolipin (CDL): an anionic phospholipid in which a glycerol moiety links two phosphatidyl groups

BA: bongkrekic acid

De Materia Medica. ATR is a diterpene heteroglucoside of which two sulfate residues make it strongly anionic at physiological pH and thereby nonpermeable through the mitochondrial membrane.

In 1962, Bruni et al. (11) and Vignais et al. (12) reported that ATR inhibited specifically the oxidative phosphorylation of extramitochondrial ADP. In the following years, these results were confirmed by several others groups (13–15), and it was clearly demonstrated that ATR did not inhibit the phosphorylation of intramitochondrial ADP. Then the obvious conclusion suddenly appeared that ATR was inhibiting the entry of ADP into mitochondria, and consequently, the existence of a transporter for ADP was postulated. Subsequently, detailed analyses of the kinetic properties of the adenine nucleotide transport (16, 17) allowed the characterization of an exchange-diffusion process between the intramitochondrial ADP and ATP with either ADP or ATP added to mitochondria; no other nucleotide could be a substrate for the carrier. The finding, not fully appreciated at that time, by Bruni et al. (11) that ATR inhibits the binding of adenine nucleotides to rat liver mitochondria, was instrumental later in the identification of the substrate carrier sites when radioactively labeled inhibitors became available.

In 1970, following preliminary studies by Welling et al. (18), Henderson & Lardy (19) introduced bongkrekic acid (BA), a complex fatty acid derivative, as an ADP/ATP carrier inhibitor. It is produced and secreted by the bacteria *Pseudomonas cocovenenans* and binds to the matrix side of the carrier after diffusion through the membrane. From that time to 1982, when the protein sequence of the ADP/ATP carrier was determined, a large body of results dealing with the properties of the carrier accumulated. These studies greatly benefited from the availability of the two families of inhibitors of exquisite specificity and affinity. The asymmetrical binding of the two classes of inhibitors allowed the characterization of two stable conformational states of the carrier, which possibly represent two extreme configurations close to that exhibited during the transport cycle by the carrier.

The other major findings are briefly summarized as follows: (*a*) When mitochondria are actively respiring in the presence of phosphate and ADP, the latter is exchanged against intramitochondrial ATP with a 1- to 1-stoichiometry; (*b*) the only physiological substrates are ADP and ATP, surprisingly, in their free forms, i.e., Mg-ADP and Mg-ATP are not recognized by the carrier; (*c*) the ADP/ATP exchange is electrogenic, which means one negative charge is extruded from the matrix to the cytosol for each cycle, and this process is driven by the membrane potential; (*d*) the kinetic parameters of the carrier are consistent with the mitochondrial ATP production and the cell nucleotide concentrations under physiological conditions; and (*e*) the carrier could be purified in detergent solutions, and transport activity could be reconstituted after reincorporation into liposomes.

To investigate the structure and characterize the conformational states of the ADP/ATP carrier, photoactivable derivatives of ADP/ATP or of the inhibitors were designed and tested either with the bovine carrier or with the yeast carrier (**Table 1**). Several authors found evidence that both conformational states expose quite different amino acid regions, and the differences were further highlighted with the use of specific proteases and antibodies directed against the ADP/ATP carrier. Site-directed mutagenesis was applied mainly to the yeast ADP/ATP carrier isoform 2 to identify residues potentially crucial for the binding or transport mechanism. Some positive and negative residues were thus proposed to be essential (**Table 1**). This approach, though promising, was limited to some residues in the absence of structural data. However, the search for second-site revertants starting from well-defined inactive mutants (**Table 1**) led to the proposed network of potentially interacting amino acids that can be now reexamined in the light of the three-dimensional bovine carrier structure. So far, only one charge pair suspected in the yeast carrier is confirmed in the bovine structure (D134-R234 of the bovine protein). However, the other pairs are in the vicinity of the ones evidenced in the bovine structure. All in all, this does not invalidate the general use of mutants because bovine and yeast amino acid sequences, although pretty similar, are different enough to suppose the involvement of different networks of structurally related amino acids.

It is interesting to note that in addition to the adenine nucleotide carrier, four other nucleotide transporters belonging to MCF have been identified: the mitochondrial carriers for GDP/GTP (20), dNDP/dNTP (21), and Mg-ATP/Pi (22) and, surprisingly, also the peroxisomal AMP/ADP/ATP carrier (23). All of these carriers are insensitive to ATR, and in contrast to the ADP/ATP carrier, the charges of their nucleotide substrates are partially or totally neutralized by Mg^{2+} or by H^+, with a possible exception of the dNDP/dNTP transporter. In addition, they also usually exhibit a lower specificity regarding either the base moiety or the polyphosphate chain length, and finally, their rate capacity is approximately 15- to 30-fold lower.

STRUCTURE ANALYSIS

The ADP/ATP carrier was crystallized in two different forms (7, 24). In both crystal forms, proteins surrounded by lipids are packed within layers as if they are sitting in a membrane. Three-dimensional crystals consist of stacks of these layers. The model of the protein structure was refined from the best diffracting crystal form (4) and deposited in the Protein Data Bank under accession code 1okc. From the N to the C termini, both located in the IMS, the overall architecture consists of six TM helices labeled H1 to H6, connected by three loops M1 to M3 on the matrix side and two loops C1 and C2 on the IMS (**Figures 1** and **2a**). Matrix loops are partially structured and contain short amphipatic helices (labeled h1-2, h3-4, and h5-6) spanning over 12 residues (**Figure 1c**). Odd-numbered TM helices are sharply kinked (20° to 35°), and all the six TM helices are also tilted with respect to the membrane. As a result, the backbone of the ADP/ATP carrier is shaped similar to a basket, closed toward the matrix and opened widely toward the IMS. The backbone also exhibits a pseudo-threefold symmetry consistent with the triplicated sequence of the gene. Most of the residues are well defined in the model, except a few residues at the N and C termini, which are probably disordered in the crystal. Residues located in IMS loops exhibit larger Debye-Waller factors than others, which is also indicative of partial disorder, preventing some of the side chains to be modeled.

The Cavity

The cavity, formed by the TM helices, is mainly hydrophilic and enters very deeply into the protein. It is shaped conically with an entrance diameter of about 20 Å, followed by a narrow funnel over 20 Å long with a diameter of 8 Å, and it is closed at 10 Å from the matrix side (**Figure 3**). Several patches of basic residues are observed. They are formed by residues K198/R104, R187/K91/K95, K22/R79/R279, and K32/R137/R234/R235 (labeled 1 to 4 in **Figure 4**) and located from the entrance to the bottom of the cavity. Facing the second patch at roughly 10 Å distance from it, Y194 is the first of three tyrosines, Y194/Y190/Y186; this arrangement forms a ladder along H4, entering the cavity. The ladder ends at the level of the third basic patch, where the conic cavity is constricted to 8 Å by four residues Y186/K22/R79/R279. The presence of many charged or polar residues within the cavity is balanced by well-ordered water molecules, forming extensive hydrogen-bond networks involving both water molecules and amino acid side chains. One of the largest networks connects all TM helices except H4 and involves the side chains of K22, E29, K32, Q36, N76, R79, D134, R137, T232,

Table 1 Compilation of chemically labeled and mutated amino acids of ADP/ATP carriers and their consequences on cell and kinetic properties

Location[a]	Bov isoform 1[b]	Hu isoform 1[c]	Sc isoform 2[d]	Sc isoform 1[e]	Nc isoform 1[f]	Organism	Mutation	Second site revertant of	Growth on glycerol	Protein content	ATR/CATR binding	BA binding	Nucleotide exchange	Other	References
Nt	M	M	M	M	M	Sc2								Maturation after translation	(44)
Nt	—	—	S1	S1	A1	Sc2								N-acetylation	(44)
Nt	D2	D2	P14	E4	—	Hu Hu	A, K E		No[g] Yes[g]	0[h]					(88)
H1	K9	K9	I24	V	A14	Hu Hu	A D, E		Yes[g] Yes[g]	148 pmoles/mg[h]	Yes		Yes[i,j]		(88)
H1	D10	D10	D25	D15	D15	Sc2 Sc2	E V	R253I R96A	Yes[k]					Isolated from the parent mutation	(89) (90)
H1	G14	G14	G29	G19	G19	Sc2	V C S[l]	R96T+D149G R253I	Yes Yes		Yes	246 nM[m]			(90) (89) (91)
H1	A17	A17	S32	S22	S22	Sc2 Sc2	N I	R253I R96L	Yes[k]						(89) (90)
H1	K22	K22	K37	K27	K27	Sc2 Nc[n] Bov	A A		No	57%[o]	57%P; 38%[q] 73%[v]	80% r; 89%[s] 97%[w]	0%; 49% t 22%[t]	1.2% u x	(33, 92) (93) (94)
H1	E29	E29	E44	E34	E34	Sc2 Sc2	Q G	R151A or R293A K37A or R293A	Yes Yes	49%[o]	46%P; 33%[q]	52% r; 86%[q]	83%[t]	37%[u]	(33) (92)
H1	K32	K32	K47	K37	K37	Sc2 Nc[n]	A A		No	43%[o]	6%P	5%[r]	9%[t] 1%[t]	No ATP synthesis	(92) (93)
H1	V37	V37	N52	N42	N42	Sc2	I	R252I	Yes[k]					Isolated from the parent mutation	(89)
M1	K42	K42	E55	E45	E45	Bov								qq	(94)
M1	K48	K48	T61	S51	R51	Bov								qq	(94)
M1	K51	K51	A67	K57	G57	Bov								Trimethylation	(63)
h12	C56	C56	C72	C62	C62	Sc2 Sc2 Sc1 Bov	S A S		Yes Yes		53%[v]		0.4%; 26%[oo]	+/−CL[z] aa bb,cc	(32) (95) (70) (51, 81, 96)
M1	W70	W70	W86	W76	W76	Sc2	Y		Yes	Yes	Yes	Yes		dd	(97, 98)
M1	R71	R71	R87	R77	R77	Sc2	A,D		No						(90)
H2	R79	R79	R95	R85	R85	Sc2 Sc2 Sc2 Nc[n]	A D, T, K, Q H[rr] L, P A		No No No No	0[o] 35% 0− WT[h]	0P; 28%P; 49%[q] Yes	0[r]; 36%[r]; 74%[w] Yes	0[v] 6%; 14% t,ee 13%[t]	0%−7%,i,u 3%−6%,i,u	(99) (99, G. Lauquin, unpublished data) (100, 101) (93)

718 Nury et al.

H2	Y80	Y80	Y96	Y86	Y86	Sc2	C		Yes[l]		Yes	223 nM[m]	Yes		(91)
H2	F88	F88	F104	F94	F94	Sc2	L, V	R253I							(93)
H2	A89	A89	A105	A95	A95	Sc2	T	R252I							(93)
H2	I97	L97	M113	L103	M103	Hu	P							Associated with adPEO	(102)
H2	D103	D103	M117	D107	K107	Hu	G							Associated with adPEO	(103)
C1	Y111	Y111	W125	W116	W116	Sc2	Y		Yes		Yes	Yes	Yes	ff	(97, 98)
H3	A113	A113	A127	A118	A118	Hu/Sc2	P/P		No[g]/Yes	0[h] WT[rh]	Yes		No WT gg	Associated with adPEO	(88, 104)
H3	L127	L127	L141	L132	L132	Sc2	S		Yes[l]		hh	215 nM[m]			(91)
H3	D134	D134	D148	D139	D139	Sc2/Sc2	S/N	R151A	No/Yes	46%[o]	5%[p]	6%[r]	0[t]	No ATP synthesis	(92)(30)
H3	R137	R137	R151	R142	R142	Sc2/Sc2/Nc[n]	A/K/A	R151A	No/Yes	18%[o]	21%P;22%q;19%ii,v	31%r;90%s;87%w	23%;92%t;34%t	No ATP synthesis	(92)(98)(93)
H3	R139	R139	R153	R144	R144	Nc[n]	A				84%[v]	95%[w]	5%[t]		(93)
H3	C159	C159	V175	V166	V166	Bov	jj							Very rapid labeling with EMA in inverted particles	(96, 105)
h34	K162	K162	K178	K169	K169	Bov/Sc2/Sc2	kk, ll/M/I		Yes/Yes	97%[o]	90%P;39%q	146%r;61%s	36%;149%r;32%s	60%[u] z	(42, 43, 94)(99, 101)(33)
h34	K165	K165	K181	K172	K172	Bov/Sc2	ll/mm		Yes	49%[o]	48%P;35%q	70%r;68%s	69%[t]	z	(42, 43)(92)
h34	S166	S166	S182	T173	S173	Sc2								2-AzNADP labeling of the Ser182-Arg190 peptide	(44)
h34	D167	D167	D183	D174	D174										
M2	G168	G168	G184	G175	G175										
M2	L169	L169	V185	L176	I176										
M2	R170	R170	A186	L177	A177										
M2	G171	G171	G187	G178	G178										
M2	L172	L172	L188	L179	L179										
M2	Y173	Y173	Y189	Y180	Y180										
M2	Q174	Q174	R190	R181	R181										
H4	V180	V180	V196	V187	V187	Hu	M, Fg		>		Yes	Yes	≫		(106)
H4	R187	R187	R203	R194	R194	Sc2/Nc[n]	L/A		No	0[h, o]	0P Yes	0[r] Yes	0%[t] 6%[t]	No ATP synthesis and no second site revertant	(33, 99, 100)(93)
H4	Y190	Y190	Y206	Y197	Y197	Sc2	H	R253I						No ATP synthesis	(89)

(Continued)

Table 1 (Continued)

Location[a]	Bov isoform 1[b]	Hu isoform 1[c]	Sc isoform 2[d]	Sc isoform 1[e]	Nc isoform 1[f]	Organism	Mutation	Second site revertant of	Growth on glycerol	Protein content	ATR/CATR binding	BA binding	Nucleotide exchange	Other	References
H4	G192	G192	G208	G199	G199	Sc2	S	R253I							(89)
H5	A216	A216	G233	G224	G224	Sc2	S	R253I							(89)
H5	Q217	Q217	W234	W225	W225	Sc2 / Sc2 / Sc2	Y / F / L	R253I	Yes / Yes / Yes		Yes	Yes	Yes	nn	(97, 98) / (89, 97) / (89, 97)
H5	V226	V226	C243	A234	A234	Sc2 / Sc2	S / A		Yes / Yes		93%,Y	Yes	95%,oo		(32)
H5	P229	P229	P246	P237	P237	Sc2	G		Yes					No ATP synthesis	(99)
H5	D231	D231	D248	D239	D239	Sc2	S		No	22%,o	0P	0r	0	No ATP synthesis	(92)
H5	R234	R234	R251	R242	R242	Sc2 / Nc[n]	I / A		No	32%,o	24%,P; 38%,v Yes	24%,r; 56%,w Yes	1%; 7%,t 23%,t	No ATP synthesis and no second site revertant	(33, 99, 100) / (93)
H5	R235	R235	R252	R243	R243	Sc2 / Nc[n]	I / A		No	23%,o	24%,P; 48%,v Yes	24%,r; 84%,w Yes	2%,t 15%,t	No ATP synthesis and 4 second site revertants	(33, 99, 100) / (93)
H5	R236	R236	R253	R244	R244	Sc2 / Nc[n]	I / A		No	29%,o	24%,P; 26%,v Yes	24%,r; 91%,w Yes	1%; 3%,t 127%,t	No ATP synthesis and 11 second site revertants	(33, 99, 100) / (93)
h56	C256	C256	C270	C261	A261	Sc2 / Sc2 / Sc1	S / A / A		Yes / Yes		46%,Y		55%,Y,oo		(32) / (95) / (70)
H6	A273	A273	C287	C278	A278	Sc2 / Sc2 / Sc1	Y / A / A	R96A	Yes		Yes	Yes			(32) / (95) / (70)
H6	R279	R279	R293	R284	R284	Sc2 / Nc[n]	A / A		No	74%,o	78%,P; 3%,v	94%,r; 43%,w	57%,Y,oo; 73% to 125%,t 21%,t	11%,u	(33, 99, 100) / (93)
H6	G283	G283	G297	A288	G288	Sc2	S		Yes		Yes	307 nM[m]	Yes	1	(91)
H6	A284	A284	A298	A289	A289	Sc2	S	R253I							(89)
H6	L287	L287	I301	I292	L292	Sc2	T	R253I							(89)
H6	V288	V288	S302	S293	S293	Hu / Sc2	M / M		No[g] / Yes,pp					Associated with adPEO	(88)
H6	Y290	Y290	Y304	Y295	Y295	Sc2	H	R96A or R252I							(89)
H6	I293	I293	L307	L298	L298	Sc2	P	R252I							(90)
H6	F296	Y296	I310	I301	L301	Sc2									(89)
H6														2-Az NADP labeling of the Ile310–Lys317 peptides prevented by CATR binding	(44)

[a]Location of the amino acids deduced from the three-dimensional structure of the bovine ADP/ATP carrier (isoform 1) and from the amino acid sequence alignment. Amino acids are indicated using the one-letter code and are followed with their numbering in the sequence. Those conserved in all known ADP/ATP carrier sequences are in red. Abbreviations: Nt: non-ordered N-terminal region; Ct: non-ordered C-terminal region; H1 to H6: TM helix 1 to 6; M1 and M3: matrix loop 1 and 3; h12: small helix in M1; h34: small helix in M2; h56: small helix in M3; C1: cytosolic loop 1, as defined in **Figure 2**.

[b]Bov: bovine.

[c]Hu: human.

[d]Sc: *Saccharomyces cerevisiae* with Sc isoform 2 or Sc2, which is isoform 2 of Sc ADP/ATP carrier.

[e]Sc isoform 1 or Sc1, which is isoform 1 of Sc ADP/ATP carrier.

[f]Nc: *Neurospora crassa*.

[g]Heterologous expression in yeast.

[h]Maximum number of [^{3}H]ATR-binding sites on isolated mitochondria in pmoles/mg mitochondrial proteins. This value is 450 pmoles/mg for the wild-type Sc isoform 2 and 174 pmoles/mg for the wild-type Hu isoform 1 produced in yeast.

[i]Exchange activity measured with isolated mitochondria.

[j]ADP/ATP exchange parameters with isolated mitochondria are $V_M^{ADP} = 32.6$ nmoles/mn/mg; $K_M = 3.7$ μM for the wild-type Hu isoform 1 and $V_M^{ADP} = 95.5$ nmoles/mn/mg; $K_M = 7.2$ μM for the K10A mutant.

[k]Does not support growth on glycerol at 37°C.

[l]Confers in vivo BA resistance.

[m]Corresponds to the inhibitor concentration yielding half the maximum value of the ADP/ATP exchange with isolated mitochondria (IC_{50}^{BA}). This value is 145 nM for the wild-type Sc isoform 2.

[n]ADP/ATP carrier expressed in *E. coli* inclusion body.

[o]ADP/ATP carrier content determined by competitive enzyme-linked immunosorbent assay.

[p][^{3}H]CATR binding to mitochondria expressed as a percentage of binding obtained with mitochondria isolated from yeast cells expressing the episomic wild-type Sc isoform 2 encoding gene.

[q]Percentage of inhibition of ADP/ATP exchange by CATR (10 μM) in reconstituted vesicles with isolated Sc isoform 2. Wild-type Sc isoform 2 ADP/ATP exchange inhibition is 7%.

[r][^{3}H]BA binding to mitochondria expressed as percentage of binding obtained with mitochondria isolated from yeast cells expressing the episomic wild-type Sc isoform 2 encoding gene.

[s]Percentage of inhibition of ADP/ATP exchange by BA (10 μM) in reconstituted vesicles with isolated Sc isoform 2. Wild-type Sc isoform 2 ADP/ATP exchange inhibition is 83%.

[t]Percentage of wild-type activity (ADP/ADP exchange) in a reconstituted system.

[u]Percentage of wild-type ATP synthesis activity with isolated mitochondria.

[v]Percentage of inhibition of ADP/ADP exchange by CATR (10 μM) in reconstituted vesicles with isolated Sc isoform 2. Wild-type Sc isoform 2 ADP/ADP exchange inhibition is 11%.

[w]Percentage of inhibition of ADP/ADP exchange BA (10 μM) in reconstituted vesicles with an isolated Nc ADP/ATP carrier. Wild-type carrier ADP/ADP exchange inhibition is 97%.

[x]Only lysine protected by CATR, not by ATR or BA, against Schiff base formation with pyridoxal phosphate incubated with intact mitochondria.

[y]Reconstitution/liposomes.

[z]Cardiolipin requirement during reconstitution experiments for ADP/ATP exchange activity.

[aa]Seventy-five percent decrease in dimer formation after import of Sc isoform 1 into the mitochondrial inner membrane.

[bb]Very rapid N-ethyl maleimide labeling on submitochondrial particles inhibited by CATR.

[cc]Intermolecular disulfide bridge catalyzed by copper-o-phenanthroline and bifunctional maleimides.

[dd]Very weak fluorescence change upon ATP binding: ΔF/F = 0.5% instead of 5% for wild-type Sc isoform 2.

[ee]The maximum rate of ADP/ATP exchange with isolated mitochondria is similar to that of wild-type Sc isoform 2 and the K_M constant is 500 times higher (V. Postis, personal communication).

[ff]No fluorescence change upon ATP and BA binding.

[gg]ADP/ATP exchange experiments were performed with isolated mitochondria. Kinetic parameters for the mutant are similar to that for the wild-type Sc isoform 2, which are $V_M^{ADP} = 69.6$ nmoles/mn/mg; $K_M = 0.67$ μM (free ADP).

[hh]$K_d^{ATR} = 3230$ nM as compared to 220 nM for the wild-type Sc isoform 2.

[ii]Nineteen percent inhibition of activity with 10 μM CATR instead of 55% for the wild-type protein.

[jj][^{3}H]7-azido-4-isopropylacridone labeling of Cys159 of Bov ADP/ATP carrier in mitochondria.

[kk]Schiff base formation with pyridoxal phosphate in mitochondria, not in submitochondrial particles, prevented by CATR and BA.

[ll]May be labeled with 2-azido ADP (Bov ADP/ATP carrier), or 2-azido and 8-azido ATP (yeast Sc isoform 2).

[mm]In combination with K178M mutation.

[nn]ΔF/F upon ATP or BA binding is similar to that of the wild-type protein.

[oo]Percentage of wild-type activity (ADP/ATP exchange) with added cardiolipins.

[pp]C. De Marcos Lousa, personal communication.

[qq]Schiff base formation with pyridoxal phosphate in SMP in the presence of BA but prevented by CATR.

[rr]The arginine to histidine change corresponds to the *op1* mutation first described by Slonimski and collaborators (107) and later identified by Kolarov and collaborators (108).

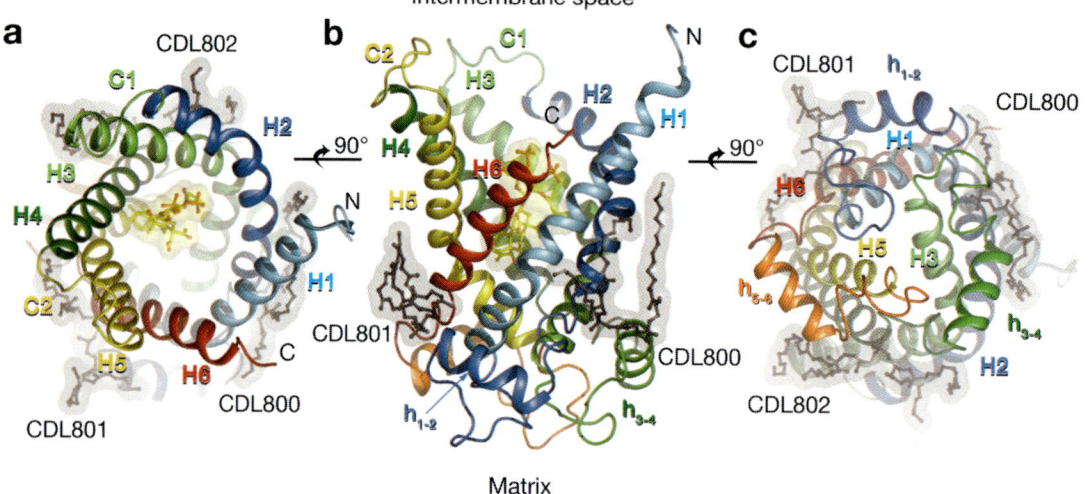

Figure 1
Overall structure of the bovine ADP/ATP carrier. The ribbon diagram, colored blue to red from the N terminus (N) to the C terminus (C), depicts transmembrane helices (H1–H6), loops facing the IMS (C1 and C2), and loops facing the matrix (M1–M3). Matrix loops are partially structured in short helices (h1–2, h3–4, and h5–6). Three cardiolipins (CDLs), CDL800, CDL801, and CDL802, are bound to the structure and represented as ball and sticks in gray. The inhibitor, CATR, complexed with the protein is depicted in yellow. Panels *a*, *b*, *c* are viewed from the IMS, the side, and the matrix, respectively. The color code for the ribbon diagram is the same as for **Figures 2, 3, 6, 7**, and **11**.

R234, R235, N276, R279 and about 20 water molecules (**Figure 5**). Almost all the residues described in this paragraph are conserved among ADP/ATP carriers (see the supplementary material). Follow the Supplemental Material link from the Annual Reviews home page at **http://www.annualreviews.org**.

CATR Binding

The inhibitor CATR, cocrystallized with the carrier, is located within the cavity with its diterpene moiety oriented toward the bottom and its sulfate groups toward the IMS (see figure 5 in Reference 4). The inhibitor binds to the carrier through numerous interactions that involve all the chemical groups except the primary alcohol located on the sugar ring. This structural observation is consistent with the inhibitory properties of CATR, which are reduced once the molecule is truncated or modified, except for the primary alcohol that can be modified without changing the inhibition capacity (25). The two carboxylates and the hydroxyl group on the diterpene moiety, as well as both sulfates on the glucose ring, are linked to the carrier through electrostatic interactions or hydrogen bonds, some of which also involve water molecules. The isovaleric chain and the diterpene ring interact via van der Waals contacts.

The MCF Motif

MCF members are characterized by triplication of the following motif: PxD/ExxK/RxK/R-(20 to 30 residues)-D/EGxxxxaK/RG, where the letter a represents an aromatic residue (**Figure 2b**). Several authors refined this motif by restricting some residues labeled as x (26). The PxD/ExxK/R sequence can be directly related to the basket shape of the carrier. Indeed, the prolines located in the second half of odd-numbered helices, close to the matrix, induce sharp kinks responsible for the closed form toward the matrix. It is

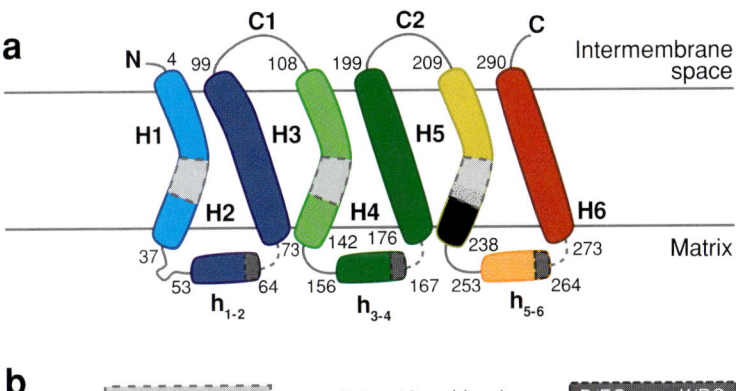

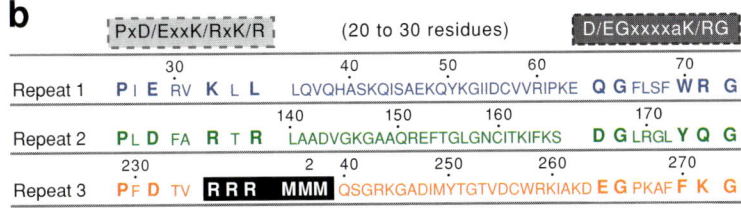

Figure 2

Overall topology and MCF motifs of the bovine ADP/ATP carrier. (*a*) Schematic diagram of the secondary structure. Regions containing MCF motif residues are colored in gray, and the RRRMMM motif is in back. Kinks in H1, H3, and H5 are induced by the prolines. (*b*) Alignment of the three MCF motifs. On the top, the consensus MCF sequence boxed in gray. The ADP/ATP carrier signature present in the third repeat is boxed in black.

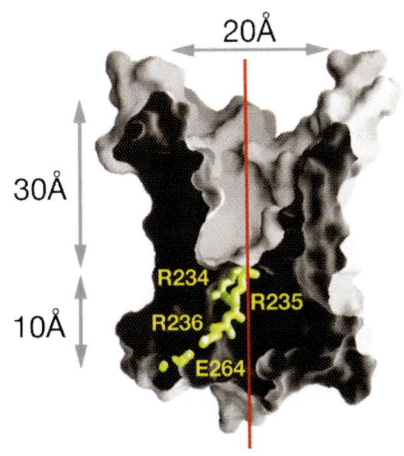

Figure 3

Surface representation of the cavity. The longitudinal section through the cavity shows the wide cavity present in the bovine ADP/ATP carrier and accessible from the IMS. R234, R235, and R236, the three arginines of the ADP/ATP carrier signature, located on the C-terminal end of H5 are shown, as well as E264, which forms a salt bridge with R236 (*yellow*). From Pebay-Peyroula et al. (4).

known that, in the absence of other specific interactions, proline residues can adopt a broad range of stable conformations (27). Therefore, when present in helices, surrounding interactions might induce a given kink angle. This is probably partly achieved by the acidic and basic residues that follow the prolines and form salt bridges that strengthen the closed conformation of the helix bundle (**Figure 6**). If the surrounding interactions are modified, the kink angles may be changed. Therefore, the prolines were proposed to act as hinges that could allow the opening toward the matrix, which could be triggered by the disruption of the salt bridges (28). However, it was also proposed that, during evolution, prolines could have mutated into serines and that subsequent packing defects were compensated by other mutations locking the kink in the structure (29). This could explain why the proline of the second motif is not conserved among ADP/ATP

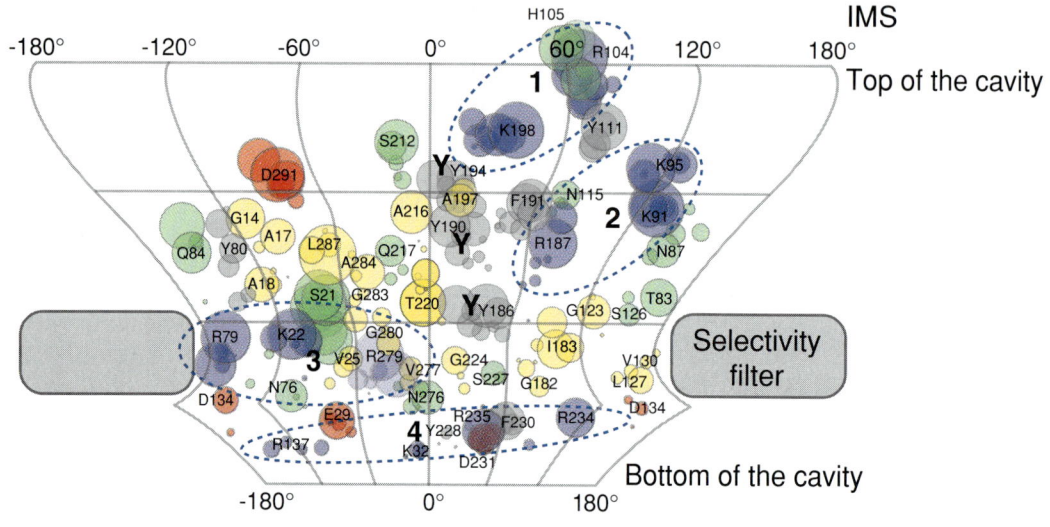

Figure 4

Distribution of residues within the cavity. The figure represents a two-dimensional projection of the residues present at the surface of the cavity. Each circle represents an atom of a residue located within the cavity, with a size proportional to its solvent accessibility. Residues are colored as follows: basic, K, R (*blue*); acidic, D, E (*red*); aromatic, F, Y, W (*gray*); hydrophobic, A, V, P, M, I, L, G (*yellow*); and polar, S, T, H, C, N, Q (*green*). The positive patches are labeled 1 to 4, and the tyrosine ladder is marked by Y.

carriers and is often found to be a serine. The second basic residue of the first part is replaced with a leucine, L34, in the first motif, although in the second and third, it is basic, R139 and R236. R139 participates in the interaction between M2 and M3 through E152, M238, and S241. The interaction of R236 with E264 is discussed below. The second part of the motif spans from the end of the short amphipatic helices to the N termini of the even-numbered helices. Each glycine therefore delineates a helix extremity and allows flexibility of the loop that links both helices. The acidic residues of the two first motifs, replaced with a glutamine Q64 in the first motif, participate with the interactions between M1/M2 and M2/M3, respectively. The residue of the third motif, E264, forms a salt bridge with R236, belonging to the ADP/ATP carrier signature, but also contacts K271, the basic residue of the third MCF motif (**Figure 7**). These interactions clamp together the C terminus of H5 with the C terminus of h5–6 and the N terminus of H6. The basic residue at the end of the first MCF motif, R71, interacts with a CDL through a water molecule and participates with the interaction of loop M1 with M2 (through D143 and G145), and in the second motif, the residue is replaced with a glutamine, Q174.

The sequence alignments of several MCF carriers highlight further conservations within the range of the second part of the MCF motif. Indeed, for ADP/ATP carriers, the short amphipatic helices and the second half of the three matrix loops are conserved in length and in sequences (alignments shown in supplementary material). Follow the Supplemental Material link from the Annual Reviews home page at **http://www.annualreviews.org**. On the contrary, the first half of the matrix loops, preceding the short amphipatic helices, are rather variable in length and in composition within MCF members or within the three motifs of a single carrier.

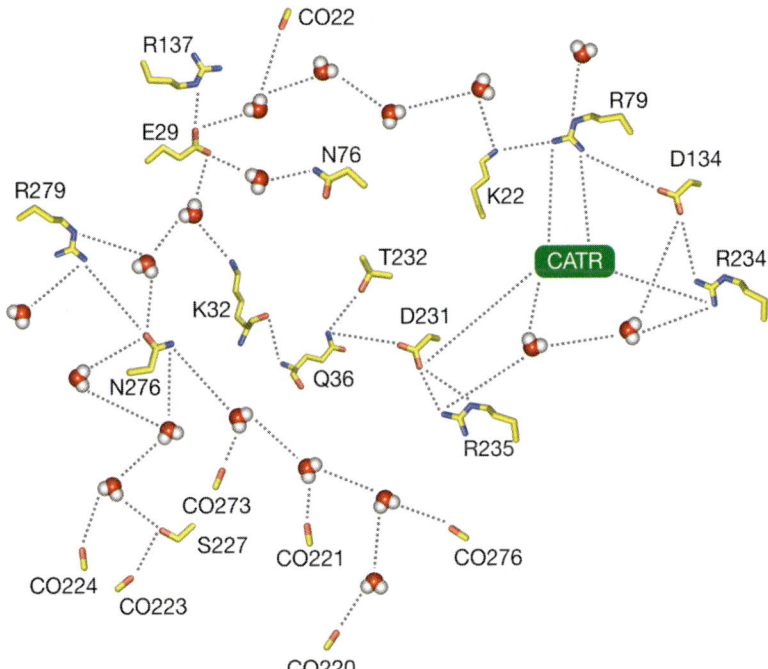

Figure 5

Schematic representation of a large hydrogen-bond network. The network connects all the TM helices, except H4. It implicates side chains of polar, acidic, and basic residues that are highly conserved within ADP/ATP carriers, as well as main-chain carbonyls (labeled CO) and water molecules. Hydrogen bonds are deduced from atomic distances and are represented as dotted lines.

Location of the ADP/ATP Carrier Signature

All ADP/ATP carriers that belong to the MCF are characterized by a unique signature, RRRMMM. These residues in the bovine carrier are located at the C terminus of H5. The first and the third arginines, R234 and R236, are part of the MCF motif (**Figure 2a**). The structure reveals that the three arginines span the thinnest part of the protein; the side chains of R234 and R235 are accessible from the cavity open toward the IMS, and R236 points toward the matrix but is shielded from the surface by a salt bridge involving E264. R234 and R235 are part of a large hydrogen-bond network involving water molecules at the bottom of the cavity (**Figure 5**). R234 is also involved in a salt bridge described above, with the MCF motif acidic residue D134, and R235 interacts with the MCF motif residue D231. Interestingly, the salt bridge between R234 and D134 was predicted from revertant studies (30). Both arginines, R234 and R235, interact with CATR. In addition to the salt bridge with E264, R236 is implicated in a network

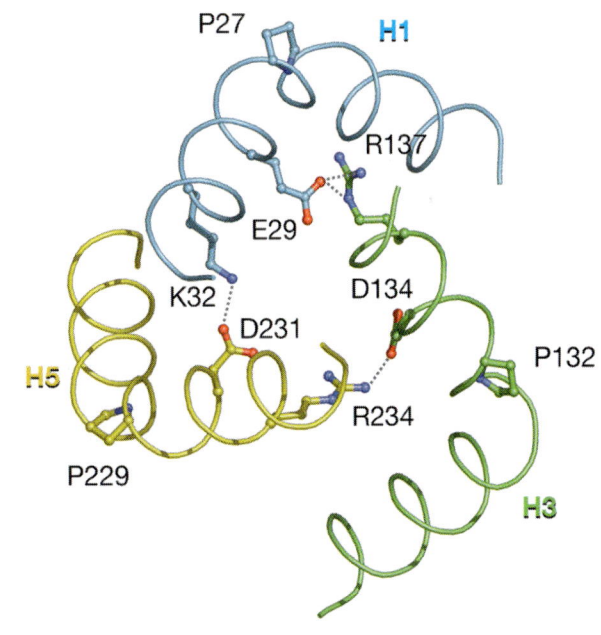

Figure 6

The kinked conformation of odd-numbered helices. H1, H3, and H5, represented as ribbons, are kinked after prolines P27, P132, and P229, which are the first residues in each MCF motif. Acidic and basic residues also belonging to the MCF motif form salt bridges (*dotted lines*) that tie the three helices together.

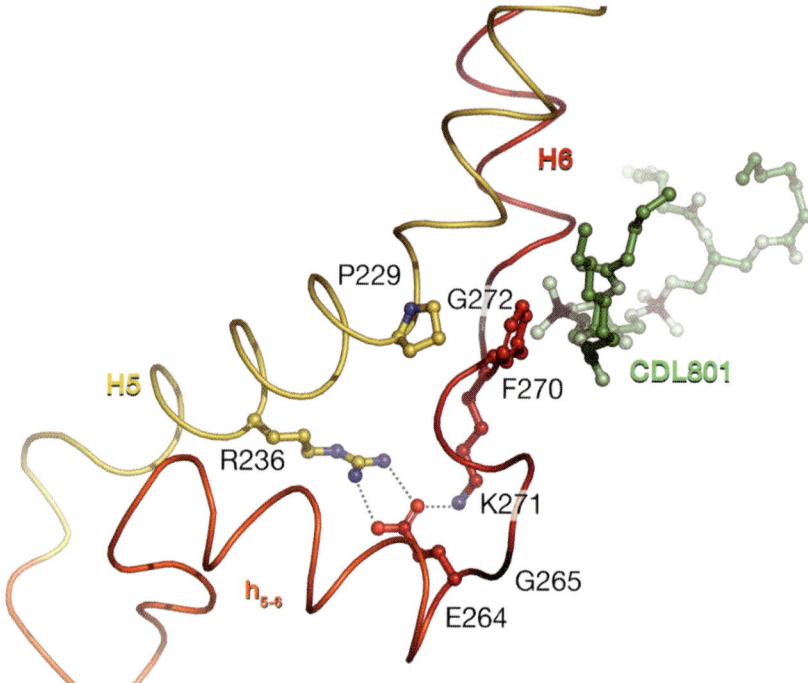

Figure 7
MCF motif of the third repeat. The protein between the C terminus of H5 and the N terminus of H6 is represented as ribbons. Side chains of MCF motif residues and CDL801 are shown in ball-and-stick form. A salt bridge between R236 (ADP/ATP carrier signature) and E264 (MCF motif) is highlighted. F270 is sandwiched between P229 and CDL801.

of hydrogen bonds including K271 and several water molecules, which link M3 to the first part of M1 (several interactions implicate residues L35, Q36, Q38, H39, I44, Q49, and Y50).

Lipids

The primitive crystal form that diffracted to 2.2-Å resolution showed the presence of two or three CDLs tightly bound to the carrier (**Figure 1b,c**). Though it could not be identified as a CDL from the experimental electron density maps, the third lipid was designated as a CDL because its position was related to the two others by the pseudo-threefold symmetry. The presence of three CDLs was confirmed by the second crystal form in which the three CDLs were nicely identified. CDLs are known to bind strongly to the carrier and to remain present after extraction and purification from the mitochondrial membrane (31). CDLs are located in the inner leaflet of the membrane and partially nested in small grooves formed by matrix loops. CDL801 clamps the N terminus of h1-2 to the C terminus of h5-6 and H6. CDL800 and CDL802 interact similarly. CDL acyl chains interact with the protein through hydrophobic or aromatic residues, whereas the phosphate groups and their glycerol linkers are involved in hydrogen bonds with main-chain nitrogens or carbonyls of symmetry-related residues located at the beginning of the short matrix helices and at the beginning of even-numbered helices. CDL801 interacts mainly with I53, I54, F270, G272, W274, and S275; CDL800 interacts with W70, G72, L74, L156, and G157; and CDL802 interacts with Y173, G175, V177, G252, T253, and V254 (7). Except for aromatic residues, most of the other residues have short side chains and interact with the lipids via main-chain atoms. Impaired activity of the carrier bearing a single mutation of either C56 or K162 could be retrieved by adding CDLs to the proteoliposomes (32, 33). Both mutations occur in the short helices, h1-2 and h3-4, respectively, and probably destabilize the structure. CDLs could compensate for this.

Deviation to Pseudo-Threefold Symmetry

The threefold symmetry results from evolution. The existence of a structural motif repeated three times in each MCF carrier is probably related to stability but also to a common transport mechanism that allows a small molecule to cross over the membrane. Divergence from a strict threefold repeat was induced by the necessity of selective transport. Therefore, internal symmetry and also deviations from it are interesting features that have to be analyzed. The three CDLs also follow the threefold symmetry. Maximum differences are at both extremities of each motif (close to IMS loops C1 and C2) and in the first part of the matrix loops (preceding the short helices). The three repeats have a very similar skeleton, and root mean square deviations between backbone atoms of the three motifs are about 2 Å [see figure 3 in (4)]. In particular, the first repeat has an additional turn in the first part of its matrix loop (one or two additional residues compared to the first parts of M3 and M2, respectively) (**Figure 8**), which allows the interaction between Q43 and the side chain of D143 and main-chain atoms of V144 and Q150, as well as between K42 and D247. As a result, the loop M1 protrudes toward the center of the protein surface on the matrix side (**Figure 1c**).

The backbone of H2 is bent and deviates therefore from a straight α-helix in contrast to H4 and H6. Negative charges are rather symmetrically distributed with the exception of E264, an extra negative charge, which is at

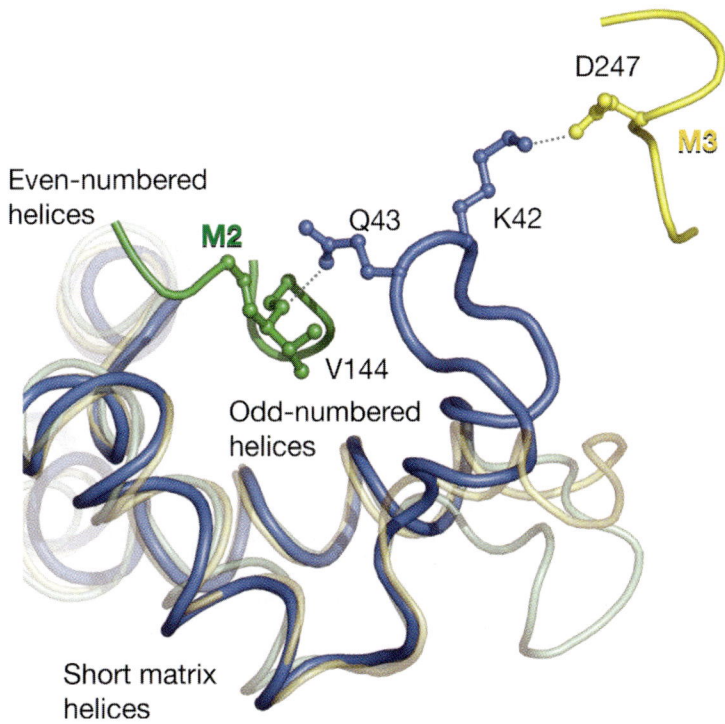

Figure 8

Deviation from the pseudo-threefold symmetry. The superposition of repeat 1 (*blue*), 2 (*pale green*), and 3 (*pale yellow*) highlights the similarity of helical parts. Repeats 2 and 3 were rotated to be superimposed on repeat 1. The figure shows also a significant difference at the beginning of matrix loops (between the odd-numbered and the short helices), which is slightly longer for M1. M1 folds back toward the center of the protein, as seen in **Figure 1c**, and thus allows interactions with M2 and M3. Small ribbon portions of M2 and M3 are depicted in green and yellow and represent these interactions.

the end of h5-6. Positively charged residues are numerous and asymmetrically distributed within the cavity, in particular at the entrance. Many charged residues are also present within the matrix loops. A global analysis of the distribution shows that all loops have an excess of positively charged residues with a ratio of positive to negative residues of 6/3, 5/3, and 7/4 for M1, M2, and M3, respectively. Aromatic residues are mainly located on the external surface of the protein, probably at the boundary between hydrophobic lipid chains and their hydrophilic head groups, as previously described for various proteins (34). However, the third MCF repeat contains less aromatic residues, especially in the IMS region (only three compared to eight for repeat 1 and seven for repeat 2). The tyrosine ladder of the second repeat, located on H4 and composed of Y186, Y190, Y194, and also F191, has side chains oriented toward the cavity, in contrast to most aromatic residues of this IMS region in repeat 1 or 3. Aromatic residues are not uniformly distributed within the cavity but are grouped along H4. Interestingly, H4 is also the only TM helix not to be implicated in the large hydrogen-bond network described above.

The two first MCF repeats also deviate from the strict consensus sequence, and the third MCF motif is the only one that strictly obeys the motif definition. Noticeable interactions are observed between K271 and E264 as well as between E264 and R236. In addition, the third motif contains many charged amino acids, one of which, K267, interacts with a neighboring molecule in the centered crystal form both directly and also through CDL801 (7).

Conserved Residues: A Structural or Functional Role?

In order to utilize the bovine ADP/ATP structure for a general analysis of mitochondrial transport, it is of interest to compare the sequences within the ADP/ATP carriers (see supplementary material; follow the Supplemental Material link from the Annual Reviews home page at http://www.annualreviews.org) and also within the MCF. In a first step, we relate the sequence similarities or differences within ADP/ATP carriers to structural elements observed for the bovine carrier. As for many membrane proteins, the parts protruding from the membrane are less conserved than the hydrophobic membrane-inserted parts. The most conserved region of the protein is the cavity, which has roughly twice as many strictly conserved residues (60%) (**Figure 9**) as the whole protein (33%) (**Figure 10**). The hydrophobic region facing CDL800 is also more conserved.

Analysis of conserved residues by type shows that aromatic residues are conserved to an unusual degree (92% of the phenylalanines and 86% of the tyrosines are conserved as aromatic residues within the ADP/ATP carrier subfamily) together with charged residues (80% of aspartic acids and 76% of arginines are conserved as acid or basic residues, respectively). The cavity of ADP/ATP carriers contains an unusually high amount of positively charged residues compared to corresponding residues in other MCF carriers.

Glycine, the smallest amino acid, is known to allow large conformational flexibility for its high entropic cost of insertion in secondary structures. It is often found at the extremity of α-helices for soluble protein. Glycines are also involved in dimerization motifs GxxG for monotopic membrane proteins and favor helix-helix interactions in polytopic membrane proteins (35). In the bovine ADP/ATP carrier structure, 70% of the glycines are conserved, and the glycine motifs present in TM helices probably play a role in intramolecular TM-TM interactions.

NUCLEOTIDE ATTRACTION AND BINDING

The specificity of nucleotide recognition and localization of their binding sites were deciphered using nucleotide derivatives and photolabeling approaches. The results obtained

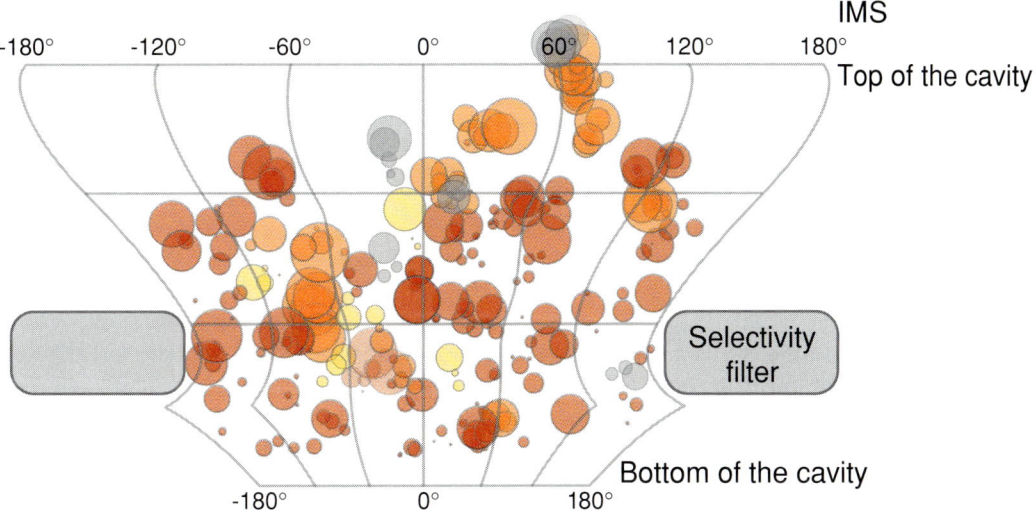

Figure 9
Conserved residues in the cavity. The representation is the same as in **Figure 4**, except that residues accessible within the cavity are colored according to their conservation among ADP/ATP carriers: no similarity (*gray*), medium or high similarities (*yellow* or *orange*), respectively, and identical (*red*).

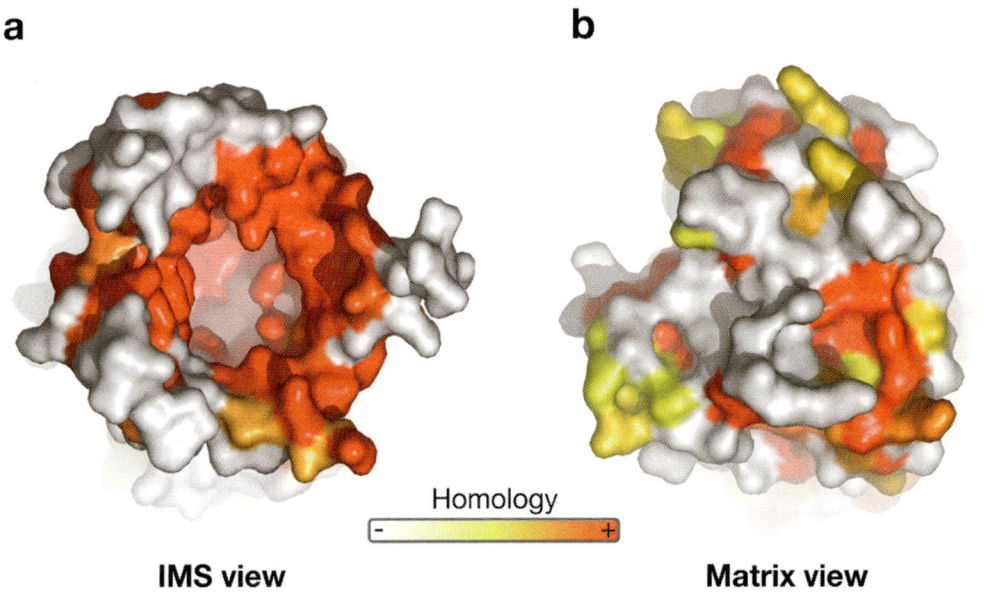

Figure 10
Conserved residues on external surfaces. Orientations in panels 10*a* and 10*b* are the same as in 1*a* and 1*c*, respectively. Residues are colored according to conservation among ADP/ATP carriers from white to red (0% to 100% of similarity).

can now be combined with the bovine structure, which revealed a cavity accessible to nucleotides from the IMS and highlighted the precise location of functionally important residues.

Specificity for Binding and Transport

The binding and transport properties of each MCF carrier are rather specific and were studied particularly for ADP/ATP carriers. None of the naturally occurring pyrimidine ribonucleotides 5′ diphosphate and triphosphate binds to the carrier. Using synthetic analogues of purine nucleotides, it became possible to characterize the two-step sequence of the transport process, namely the recognition of a nucleotide to a specific site, followed by the vectorial process of transport. Because several adenine nucleotide analogues can bind to the carrier with a high affinity without being transported, binding requires a lower specificity than transport (36). A central issue regarding the transport process involves the *anti* or *syn* conformation of nucleotides, which rely on the orientation of the planar purine base with respect to the ribose ring. To be transported by the ADP/ATP carrier, a nucleotide must have a nonfixed *anti* conformation, with an additional amino group on C6 and an unsubstituted C2 atom. For example, 8-Br ADP and the derived 8-azido ADP, blocked in the *syn* conformation, bind to the carrier but are not transported. 2-azido ADP adopts the *anti* conformation but substitution at position 2 prevents transport. In contrast, analogues, such as formycin or tubericidin diphosphate and triphosphate and 1-N oxide-ADP or -ATP, fulfill the required structural criteria and are transported. The presence of bulky substituents at positions 2′ or 3′ of the ribose moiety is tolerated for binding (37) or even for transport in mitochondria (38).

Fluorescein derivatives are structurally related to adenine nucleotides and therefore are recognized by the ADP/ATP carrier, although not transported. For example, eosin Y binds to the carrier from the matrix side and inhibits the ADP/ATP transport in bovine heart inside-out submitochondrial particles (39). These effects were interpreted on the basis of common structural features between the A/B and D rings of eosin Y, with the adenine ring and the ribose moiety of the *anti* form of ADP, respectively. In contrast, the absence of a negative potential at position N1 of guanine might explain why GDP is not recognized by the carrier.

The demonstration that the free forms of ADP and ATP are the actual substrates for the ADP/ATP carrier came from experiments carried out with the isolated carrier either in a detergent solution or after incorporation into the membrane of liposomes. For example, the fact that ADP- or ATP-induced conformational transitions of the isolated carrier (followed by tryptophanyl fluorescence changes) were abolished by Mg^{2+} ions afforded direct evidence that Mg nucleotides are not recognized by the carrier (40). Similarly, it was clearly demonstrated that Mg^{2+} ions inhibited the ADP/ATP exchange in reconstituted proteoliposomes (41). Finally, although AMP nucleotides are recognized by the carrier, only ADP or ATP is transported.

Biochemical Evidence for IMS-Binding Sites

Localization of two specific nucleotide-binding regions of the ADP/ATP carrier was achieved with photoaffinity radiolabeled nucleotides carrying a reactive azido group on the adenine ring. Two segments of the peptide chain of the bovine carrier, spanning residues F153 to M200 and Y250 to M281, were covalently labeled with 2-azido-ADP (42). Mapping of the yeast carrier with 2-azido-ADP led to the labeling of a segment delimited by residues G172 and M210 (43). A more precise assignment of the binding region was achieved with 2-azido-3′-O-naphthoyl-ADP and restricted to the S183-R191 segment, which is located in the second matrix loop between h3-4 and H4 (44). Although not directly accessible from the IMS in the CATR-ADP/ATP carrier structure, these residues are not far from the cavity, and it is possible that binding the ADP, even labeled, already induces a conformational change. An additional segment spanning residues I311-K318, corresponding to the C-terminal end of the carrier, was also labeled (44). It was shown from the characterization of deletion mutants that this nucleotide-binding region plays a critical role for ADP/ATP transport in yeast (G. Brandolin, unpublished data). Whether the two nucleotide-binding segments belong to the same carrier monomer or to adjacent monomers is still an open question. The other peptide segments implicated in nucleotide binding also remain to be elucidated.

Structural Features

From the IMS, the patch of positive charges located at the entrance of the cavity attracts ADP^{3-} in spite of the opposing membrane potential and possibly provides a first nucleotide-binding site. Furthermore, positive patches in the middle and at the bottom of the cavity attract nucleotides to the bottom. The cavity narrows at a bottleneck located at 20 Å from the entrance. The four residues surrounding the bottleneck are conserved only within the ADP/ATP carrier subfamily (except for the carrier involved in Grave's disease for which the transported molecule has not yet been clearly identified). These are three basic residues, K22, R79, and R279 (the two arginines related by the internal pseudosymmetry), and Y186, which belongs to the tyrosine ladder. Mutations of the residues, corresponding to K22 and R79 in yeast, severely impair the transport properties (**Table 1**). Coming from the IMS, a nucleotide could glide with its adenine ring along H4 using the tyrosine ladder, whereas the phosphates would follow the basic patches. At the level of the constriction, the four residues could restrict the entrance to the bottom of the cavity to adenine nucleotides. The cavity is limited to 8 Å. It is therefore interesting to note that the residues of the putative selectivity filter, K22, R79, Y186, and R279, correspond to R190, K241, Y338, and K434 in the human Mg-ATP/Pi carrier, which imports Mg-ATP and exports Pi. Although the lysines and arginines are switched, the same types of residues are found at the same locations. In addition, the Mg-ATP/Pi carrier also exhibits two out of the three tyrosines forming the ladder, and the positive patches are conserved, except for the second one (R187 and K91 do not correspond to positive residues). Regarding the cavity, the Mg-ATP/Pi carrier and the carrier implicated in Grave's disease are the most similar to the ADP/ATP carrier in terms of charged and aromatic residues.

The phosphate carrier is another MCF member in which several residues, including H32, T79, Y83, K90, Y94, and K98, were expected to be important in the transport pathway (45). Interestingly, H32 and T79, residues that are unique to phosphate carriers, correspond to K22 and N76, respectively, of the bovine ADP/ATP carrier. K22 is described herein as belonging to a putative selectivity filter, and N76 is located close to it.

Binding ATP from the Matrix

From the structure obtained in the presence of CATR, it is difficult to deduce ATP-binding sites. However, it highlights a salt bridge between R236 and E264 (**Figure 3**). R236 could be involved in ATP binding, thus modifying the salt bridge and inducing a conformational change. Binding of eosin Y, a non-thiol-reactive molecule, was shown to be displaced by ADP or ATP, and therefore eosin Y was supposed to bind in close vicinity to the nucleotide site. In addition, binding of eosin Y also prevented the thiol-reactive eosin-5-maleimide molecule to bind on C159 (39). The authors therefore suggested that ATP binds in the vicinity of C159. However, in the absence of further structural data, it is difficult to discuss the ATP-binding site from the matrix.

TRANSPORT MECHANISM

The binding of nucleotides to the carrier has to induce conformational changes that trigger a one-to-one ADP/ATP exchange. The structure in the presence of CATR and biochemical data offer a basis for the discussion on transport mechanism.

Conformational Changes

The very high selectivity of the ADP/ATP carrier for adenine nucleotides and the recognition of fully charged species instead of partially neutralized Mg^{2+}-nucleotide complexes considerably enhance the binding energy of nucleotide-carrier interaction, which in turn is used to drive the ADP and ATP translocation. Energy supply has to sustain the conformational rearrangements of the carrier required to transport large and charged nucleotides. The general concept of an induced transition fit mechanism that would provide energy for transport through the formation of a transient carrier-substrate complex has recently been discussed (46). Indeed, normal mode calculations based on the structure of the bovine ADP/ATP carrier show that low-energy movements are not sufficient to induce structural conformations that would allow the transport of the nucleotides (P. Amara, personal communication). However, conformation fluctuations in the unloaded carrier could be necessary to expose transiently the conformations to which ADP and ATP bind. The additional energy needed for the transport process could come from the relaxation of both ADP^{3-} and ATP^{4-} in their binding sites.

The conformation changes undergone by the carrier are a central issue with regard to the molecular mechanism of the ADP/ATP transport. This point was addressed by studying the modification of the carrier topography in the presence of CATR and BA. Conformers of the carrier were differentiated on the basis of their enzymatic, immunochemical, and chemical reactivities (for a review, see Reference 47). Experiments with the bovine heart ADP/ATP carrier have shown that the CATR-carrier complex is more iodinated than the BA-carrier complex or the denatured carrier (48). The N-terminal region of the membrane-bound bovine carrier, which is exposed to the IMS in intact mitochondria, is particularly sensitive to conformational changes; its reactivity to antibodies is much higher in the presence of CATR than of BA (49). In addition, systematic single-cysteine mutants on the yeast carrier followed by thiol accessibility exploration showed that the accessibility of residues 98 to 106 (equivalent to 81–89 in the bovine carrier) is drastically modified in the presence of BA compared to CATR (50). The modifications were interpreted by the authors as a 180° twist of H2. Residues 81 to 89 of the bovine carrier are located in the vicinity of the N-terminal end up to residue 12; therefore structural modifications of H2 could be correlated to that of the N terminus. The conformational changes undergone by the ADP/ATP carrier also affect peptide segments located on the matrix side. Thus, the extent of C56 alkylation by N-ethylmaleimide in the presence of ADP or ATP is enhanced by BA and counteracted by CATR (51). In addition, the K42-Q43, K146-G147, and K244-G245 bonds, which are not accessible in the native membrane of inverted mitochondrial particles, become unmasked and accessible for cleavage by specific proteases in the presence of BA (52). Additional conformation-dependent thiol-labeling and cross-linking experiments carried out on the bovine carrier with maleimide reagents were interpreted in terms of the participation of the matrix-exposed loops to the transport process. In particular, a critical gating role was proposed for loop M1, which is positively charged and could attract a nucleotide and convey it to a binding domain located on loop M2 (53). We have recently demonstrated by proteolytic and chemical labeling approaches that loop M2 of the yeast ADP/ATP carrier undergoes conformation-dependent swinging, probably related to the transport process (53a).

The bovine ADP/ATP carrier structure determined in the presence of CATR shows several interactions between matrix loops. Because of its difference in length and asymmetrical position, M1 plays a central role (**Figures 1c** and **8**). The first part of M1 (between H1 and h1-2) interacts with the C-terminal ends of H3 and H5 as well as with the first part of M3 (notably through a salt bridge between K42 and D247). The interactions between M1 and M2 involve the second part of M1 (between h1-2 and H2) and the first part of M2 (between H3 and h3-4). Similar interactions, obtained by circular permutation, are observed between M2 and M3. The interaction between M1 and M3 implicates the C-terminal ends of H1 and h5-6. Many polar or electrostatic interactions among the matrix loops are observed. It is therefore difficult to conceive that one single loop could move during transport without affecting the conformation of the other loops. Conversely, if dimerization of the carrier is necessary for its function, the dimerization interface breaks the threefold symmetry. Combining both aspects, it is conceivable that even if all three matrix loops move during transport, the amplitude of these movements might be different for each of them.

The structure also suggests that modifications of the kink angles in odd-numbered helices would induce large conformational changes, reorienting the short amphipatic loops (helped by the presence of conserved glycines of the MCF motif) and changing the accessibility of the cysteines as demonstrated by cross-linking experiments. Such modifications may be general for all MCF members.

What Triggers the Changes? Kinetic Aspects

The kinetic properties of the ADP/ATP carrier have been studied in detail in isolated mitochondria and after functional reconstitution in proteoliposomes. Most of the mitochondrial carriers were shown to catalyze strict solute exchange reactions (54, 55); this is also the case of the ADP/ATP carrier. In isolated mitochondria, stoichiometry of the exchange was assessed in experiments in which nonmetabolized transportable analogues were used instead of ADP. This avoided the difficult measurements linked to dephosphorylation and transphosphorylation of the external nucleotide once it has been transported into the matrix space. Using AOPCP, a methylene analogue of ADP, the stoichiometry of the exchange with intramitochondrial nucleotides is 1 (for a review, see Reference 56). A 1-to-1 stoichiometry of exchange was also determined for the ADP/ATP transport system reconstituted in proteoliposomes from the isolated bovine heart carrier (41). In contrast, a uniport function of the ADP/ATP carrier was deduced from investigation of the reconstituted ADP/ATP transport in black lipid membranes (57). In this approach, measurements of electrical currents associated with the functioning of the carrier upon photolysis of caged ADP/ATP were interpreted as reflecting net transport of nucleotides.

In isolated rat heart and rat liver mitochondria, the ADP/ATP transport was demonstrated to proceed according to a sequential mechanism in which both the external substrate and the internal substrate bind to the carrier before the translocation occurs (58, 59). This intermolecular mechanism implies the existence of positive interactions between distinct binding sites exposed to the outer and inner face of the membrane-embedded carrier. Consistent with these findings, the occurrence of distinct specific nucleotide-binding sites on the ADP/ATP carrier was deduced from the binding of ADP and ATP derivatives to the carrier either in the membrane-bound state or when isolated in detergent solution (60–62). For example, in the mitochondrial membrane, two specific nucleotide-binding sites are located on the same face of the membrane, either the IMS face of intact mitochondria or the matrix face of inside-out particles. Negative interactions between adjacent sites were evidenced for the binding of transported nucleotides but not for that of nontransportable nucleotides, such as naphthoyl-ADP, thus illustrating that the binding step triggers conformational changes responsible for negative cooperativity, which is set up prior to the transport process.

The structure shows that the negatively charged ADP attracted to the bottom of the cavity could interact with the basic residues of the MCF motifs and thus interfere with the salt bridges that strengthen the closed form. The binding of ADP itself could be one element that triggers the conformational changes, which result in nucleotide transport. An additional element could be the binding of ATP from the matrix. Combining the hypothesis of a selectivity filter with the conformational changes triggered by substrate binding would explain the selectivity of each MCF member, despite a common mechanism based on a similar structure induced by the MCF motifs for all the carriers of this family. The slow turnover of ADP/ATP carriers (about 1000 min^{-1}) (63) is compatible with the large structural changes hypothesized from the kinked helices and is necessary for the transport of nucleotides but not for ion transporters (64).

Is the Functional Unit a Dimer?

Experimental evidence for oligomerization. Ever since the discovery of the ADP/ATP carrier in the 1970s, several biochemical and biophysical experiments indicated that MCF members are dimeric. Mainly three types of experiments were performed. The first type led to a stoichiometry, the second gave a particle mass, and the third was based on distances between given residues of neighboring molecules. Klingenberg and coworkers first proposed the functional unit to be a dimer (65, 66) on the basis of inhibitor stoichiometry studies. Measuring CATR amounts by ^{35}S radioactivity and protein concentration with modified Biuret or Lowry protocols, a 0.55 molar ratio of CATR-to-protein was determined. Analytical ultracentrifugation was carried out with the ADP/ATP carrier (67) and the uncoupling protein (UCP) (68). Both proteins were solubilized in Triton X-100. The total CATR-ADP/ATP carrier micelle mass was ~180 kDa and dropped to 65 kDa, which corresponds to a dimer, after subtracting the contributions of detergent and lipids. The mass of UCP micelles was also compatible with a dimeric organization. However, similar experiments on the yeast ADP/ATP carrier solubilized in dodecyl maltoside (M. LeMaire, personal communication) or the bovine carrier (C. Ebel, personal communication) evidenced a monomer. Small-angle neutron scattering of the bovine ADP/ATP carrier solubilized in LAPAO led to a particle mass compatible with a dimer in the presence of both inhibitors CATR or BA (69). However, the bovine ADP/ATP carrier is usually solubilized with a large amount of detergent and lipids, and it is difficult to correct for these contributions to the scattering.

Native electrophoresis has been used for several mitochondrial carriers [ADP/ATP (70), 2-oxoglutarate (71), citrate (72), tricarboxylate (73)], and this led the authors to conclude that a dimeric organization exists. From in vivo and in vitro assembly of ADP/ATP and 2-oxoglutarate carriers, it was proposed that newly imported carriers assemble rapidly with the few preexisting monomers in the inner membrane, with the vast majority of the carrier pool being dimeric. Using double-tagged proteins, the phosphate carrier is the only one for which functional observations were related to the existence of a dimer (74).

Covalent dimers of the yeast ADP/ATP carrier, wherein the C terminus of the first monomer is fused to the N terminus of the second one, were found to be functional (75–77). The transport activity or inhibitor binding of chimeric proteins is similar to the native ones. From these experiments, the authors suggest a proximity of H1 from one monomer with H6 from the second one. However, the segment bridging H6 from one monomer to

H1 from the other could, if unstructured, span over more than 100 Å. A disulfide bridge between C28 of two yeast phosphate carriers inhibits transport (78), and thus, it was proposed that C28 is at the interface between monomers. An interface involving H3 and H4 also was proposed after homology modeling of the yeast citrate carrier structure (79).

The oligomeric state and the conformational changes of the ADP/ATP carrier were extensively investigated for years using thiol labeling or cross-linking with naturally occurring cysteines or single-cysteine mutants. Almost all the loops of the protein were explored in the presence of CATR or BA. Such experiments on the first IMS loop showed that the conformation of the first part of the loop is different with both inhibitors (80). The authors suggest that C1 could act as a swinging gate during ADP uptake from the IMS. The gate would be related to a partial unwinding of the C terminus of H2. These suggestions are in line with the proposed dimer in which C2 loops from both monomers interact on the IMS side, thus allowing flexibility for C1 loops. In addition, the secondary structure of H2 is bent toward its C-terminal end, possibly indicating a certain flexibility of the helix. On the basis of cysteine interactions (direct or induced through linkers), it was shown that C56 from both monomers located in loop M1 can be cross-linked in the presence of BA (53, 81).

Altogether, the experimental results do not converge clearly toward a single dimeric organization among all MCF carriers or even within a specific type of carriers. The discrepancies among the ADP/ATP carriers could result from a rather loose oligomeric organization in which protein-protein interactions could occur for the transport process but in which carriers are otherwise only in close vicinity. The oligomeric state could also depend on experimental conditions, such as native membrane versus detergent micelles or protein and detergent concentration. It could also be that other MCF carriers have evolved differently and that their dimeric interfaces are varied. Another possibility is the existence of higher oligomeric interactions.

Consequences for the mechanism. On the basis of inhibitor and substrate-binding studies, a single-binding-center gated pore (SBGP) mechanism has been advocated by Klingenberg (82). According to this model, a single binding site for substrates and inhibitors, located within the core of the protein, is assumed to be alternately accessible to each side of the membrane during the different steps of transport. The entrance and the exit of the substrate would be mediated by two gates, each gate facing one side of the membrane. A key feature of this model is that in order to prevent leakage, binding of the substrate controls the closing of the entrance gate and the opening of the exit gate on the opposite side of the membrane. Because of the possible dimeric organization of the carrier, it has been proposed that the SBGP mechanism operates with a single translocation pathway formed at the interface of two monomers (83) or formed by the merging of the 12 TM helices (84). Both arrangements are most unlikely because the bovine ADP/ATP carrier structure shows that a monomer presents a deep cavity, probably belonging to a translocation pathway. Therefore the transport function of a dimeric carrier would involve two transport pathways, which is consistent with a sequential type of mechanism because the carrier would be loaded simultaneously with an internal and an external nucleotide.

Gating of the channels is a compulsory function that can be generalized to most transport systems, including channels with either one or two gates opened simultaneously. It is conceivable that the switch of a carrier mode to a channel mode results from concerted opening of the gates, induced, for example, by site-directed mutagenesis or by chemical modifications of the carriers. Conversion of carrier function to channel-like function has been illustrated for a number of mitochondrial carriers, including the ADP/ATP carrier, as reviewed in Reference 85.

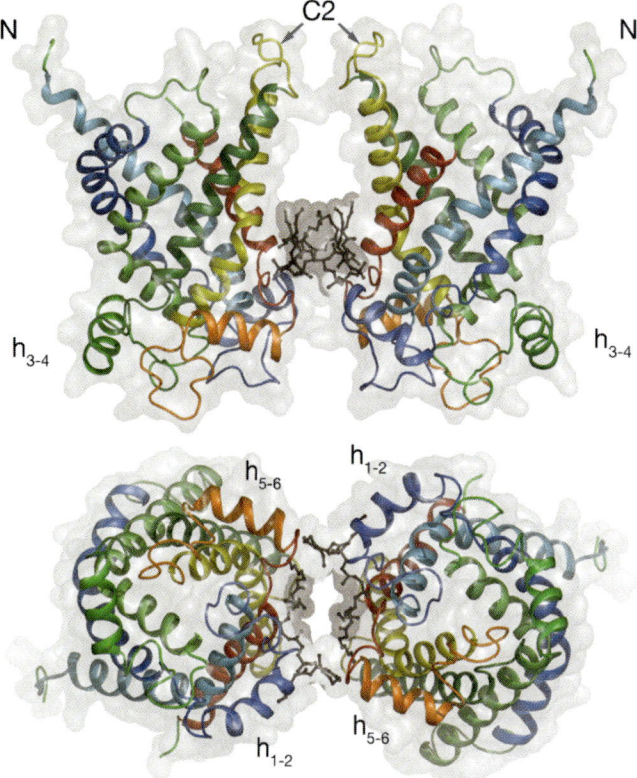

Figure 11
Protein-protein interaction mediated by CDLs. The two monomers seen in the crystal packing interact directly next to the matrix side and to the IMS. The interaction also involves cardiolipins (*gray*). Van der Waals surfaces of proteins and lipids are shown superposed on the ribbons for the protein and on the balls and sticks for the lipids.

HYPOTHESES AND CONCLUSIONS

A Proposed Mechanism Involving CDLs

A second crystal form of the bovine ADP/ATP carrier revealed possible monomer-monomer interactions favored by CDLs (**Figure 11**) (7). The importance of CDLs was already known for ADP/ATP carrier activity, and three CDLs were shown to be tightly associated to one monomer (31). Although CDLs are not essential for the growth of yeast cells on fermentable and nonfermentable carbon sources (86), the activity of ADP/ATP transport in the absence of CDL has been shown to be lowered to 20% of the level seen in mitochondria of wild-type cells (87). Retrieving the activity of different mutants in proteoliposomes in the presence of CDLs indicated that these lipids are important for the function and the stability of the carrier (32). On the matrix side, numerous interactions between the N-terminal end of h1-2 of one carrier with the C-terminal end of h5–6 from the second carrier involve K267 and K51 and also the phosphates of CDL801. On the IMS, the interaction is looser and implicates loops C2 from both monomers. Within the membrane, hydrophobic contacts are made through the lipid alkyl chains. Prolines of the MCF motif, P27, P132, and P229, have aromatic environments similar to W70, Y173, and F270, each of which stacks on one of the three CDLs. We postulate that the cross talk between monomers during the transport process could implicate the CDLs and that aromatic residues relay the signal from the proline kink modification to the lipid and act as a driving belt. Indeed, P229 (located in the third MCF repeat) interacts with CDL801 (**Figure 7**) close to the putative dimer interface (**Figure 11**). It is also interesting to note that residues Y228 and F230, which are close to P229, prolongate the tyrosine ladder present in the cavity. In this model, H2 is at the opposite side of the dimer interface. This location is consistent with the hypothesis that this helix twists during the transport. In addition, several residues belonging to the basic patches of the cavity belong to H2. From the entrance, the first patch comprises R104, belonging to C1, the second K91 and K95, and the third R79, known to be crucial for the binding and the transport of nucleotides. Therefore, binding of ADP to the cavity from the IMS could induce this twist (50).

Open Questions

Elucidating the transport mechanism of mitochondrial ADP/ATP carriers necessitates a combination of various approaches. High-resolution structural approaches such as X-ray crystallography are very powerful and highlight the quasi-atomic structure of the carrier locked into a precise conformational state. This structure provides a good starting point for exploring various hypotheses by guiding the choice of mutants and following their transport properties. **Table 1** summarizes the large body of mutations that have already been addressed. Most of the residues shown to have important functions are highlighted in the structural discussion. Nevertheless, many others still must be studied. However, membrane carriers are more difficult to handle than soluble enzymes for which enzymatic assays of modified proteins can be set up more easily. In addition, the oligomeric state of membrane proteins in their native lipidic environment is still difficult to approach because all the experiments in which the carrier is extracted from the membrane (solubilized in detergent or even reconstituted in liposomes) are potentially biased and not entirely satisfactory. Electron or atomic-force microscopies might offer alternative approaches to follow the oligomeric state under different conditions and environments.

Finally, the transport of charged nucleotides, which is triggered by in vivo conditions, nucleotide concentration, membrane potential, and possibly interactions with other proteins, needs to be explored along with their structure.

SUMMARY POINTS

1. The ADP/ATP carrier structure highlights a bundle of six tilted—with half of them kinked—helices forming a cavity that is wide open toward the IMS.

2. MCF members may share a common transport mechanism, which is based on a common scaffold and could rely on the kink and tilt modifications of the TM helices.

3. Substrate specificity may be related to the geometry and the chemical properties of the residues in the cavity, illustrated for instance by the distribution of patches of basic residues as well as by a ladder of aromatic residues.

4. The functional properties of many mutants were explored and are compiled in Table 1.

5 The sequential transport mechanism might be induced by the simultaneous binding of ADP and ATP on both sides of the membrane.

6 Many published results, such as cross-linking experiments, protein/inhibitor stoichiometries, chimeric dimers, analytical ultracentrifugation or neutron scattering, indicate that the ADP/ATP carrier is a dimer.

FUTURE ISSUES TO BE RESOLVED

1. The oligomeric state of the ADP/ATP carrier must be ascertained along with a description of the mechanism for cross talk between monomers.

2. Structural data on other conformations of the carrier should be obtained.

3. Descriptions are needed of the transport process at a molecular level, the common features for the whole MCF, and the differences that built specificity.

4. The putative partners of the ADP/ATP carrier, or supercomplexes to which it could belong, should be characterized.

LITERATURE CITED

1. Walker JE, Runswick MJ. 1993. *J. Bioenerg. Biomembr.* 25:435–46
2. White SH. 2004. *Protein Sci.* 13:1948–49
3. Abramson J, Smirnova I, Kasho V, Verner G, Kaback HR, Iwata S. 2003. *Science* 301:610–15
4. Pebay-Peyroula E, Dahout-Gonzalez C, Kahn R, Trézéguet V, Lauquin GJM, Brandolin G. 2003. *Nature* 426:39–44
5. Kunji ER, Harding M. 2003. *J. Biol. Chem.* 278:36985–88
6. Kunji ER. 2004. *FEBS Lett.* 564:239–44
7. Nury H, Dahout-Gonzalez C, Trézéguet V, Lauquin GJM, Brandolin G, Pebay-Peyroula E. 2005. *FEBS Lett.* 579:6031–36
8. Siekevitz P, Potter VR. 1955. *J. Biol. Chem.* 215:237–55
9. Pressman BC. 1958. *J. Biol. Chem.* 232:967–78
10. Bartley W, Davies RE. 1954. *Biochem. J.* 57:37–49
11. Bruni A, Contessa AR, Luciani S, 1962. *Biochim. Biophys. Acta* 60:301–11
12. Vignais PV, Vignais PM, Stanislas E. 1962. *Biochim. Biophys. Acta* 60:284–300
13. Kemp A Jr, Slater EC. 1964. *Biochim. Biophys. Acta* 92:178–80
14. Chappell JB, Crofts AR. 1965. *Biochem. J.* 95:707–16
15. Pfaff E, Klingenberg M, Heldt HW. 1965. *Biochim. Biophys. Acta* 104:312–15
16. Duee ED, Vignais PV. 1965. *Biochim. Biophys. Acta* 107:184–88
17. Heldt HW, Jacobs H, Klingenberg M. 1965. *Biochem. Biophys. Res. Commun.* 18:174–79
18. Welling W, Cohen JA, Berends W. 1960. *Biochem. Pharmacol.* 3:122–35
19. Henderson PJ, Lardy HA. 1970. *J. Biol. Chem.* 245:1319–26
20. Vozza A, Blanco E, Palmieri L, Palmieri F. 2004. *J. Biol. Chem.* 279:20850–57
21. Dolce V, Fiermonte G, Runswick MJ, Palmieri F, Walker JE. 2001. *Proc. Natl. Acad. Sci. USA* 98:2284–88
22. Fiermonte G, De Leonardis F, Todisco S, Palmieri L, Lasorsa FM, Palmieri F. 2004. *J. Biol. Chem.* 279:30722–30
23. Palmieri L, Rottensteiner H, Girzalsky W, Scarcia P, Palmieri F, Erdmann R. 2001. *EMBO J.* 20:5049–59
24. Dahout-Gonzalez C, Brandolin G, Pebay-Peyroula E. 2003. *Acta Crystallogr. D* 59:2353–55
25. Vignais PV. 1976. *Biochim. Biophys. Acta* 456:1–38
26. Jezek P, Jezek J. 2003. *FEBS Lett.* 534:15–25
27. Williams KA, Deber CM. 1991. *Biochemistry* 30:8919–23
28. Pebay-Peyroula E, Brandolin G. 2004. *Curr. Opin. Struct. Biol.* 14:420–25
29. Yohannan S, Faham S, Yang D, Whitelegge JP, Bowie JU. 2004. *Proc. Natl. Acad. Sci. USA* 101:959–63
30. Nelson DR, Felix CM, Swanson JM. 1998. *J. Mol. Biol.* 277:285–308
31. Beyer K, Klingenberg M. 1985. *Biochemistry* 24:3821–26
32. Hoffmann B, Stockl A, Schlame M, Beyer K, Klingenberg M. 1994. *J. Biol. Chem.* 269:1940–44
33. Klingenberg M, Nelson DR. 1994. *Biochim. Biophys. Acta* 1187:241–44
34. Pebay-Peyroula E, Garavito RM, Rosenbusch JP, Zulauf M, Timmins PA. 1995. *Structure* 3:1051–59
35. Curran AR, Engelman DM. 2003. *Curr. Opin. Struct. Biol.* 13:412–17

36. Vignais PV, Block MR, Boulay F, Brandolin G, Lauquin GJM. 1985. In *Structure and Properties of Cell Membranes*, ed. G Benga, 2:139–79. Boca Raton, FL: CRC Press
37. Lauquin GJ, Brandolin G, Lunardi J, Vignais PV. 1978. *Biochim. Biophys. Acta* 501:10–19
38. Klingenberg M, Mayer I, Dahms AS. 1984. *Biochemistry* 23:2442–49
39. Majima E, Yamaguchi N, Chuman H, Shinohara Y, Ishida M, et al. 1998. *Biochemistry* 37:424–32
40. Brandolin G, Dupont Y, Vignais PV. 1981. *Biochem. Biophys. Res. Commun.* 98:28–35
41. Brandolin G, Doussiere J, Gulik A, Gulik-Krzywicki T, Lauquin GJM, Vignais PV. 1980. *Biochim. Biophys. Acta* 592:592–614
42. Dalbon P, Brandolin G, Boulay F, Hoppe J, Vignais PV. 1988. *Biochemistry* 27:5141–49
43. Mayinger P, Winkler E, Klingenberg M. 1989. *FEBS Lett.* 244:421–26
44. Dianoux AC, Noel F, Fiore C, Trézéguet V, Kieffer S, et al. 2000. *Biochemistry* 39:11477–87
45. Wohlrab H. 2004. *Biochem. Biophys. Res. Commun.* 320:685–88
46. Klingenberg M. 2005. *Biochemistry* 44:8563–70
47. Brandolin G, Le Saux A, Trézéguet V, Lauquin GJM, Vignais PV. 1993. *J. Bioenerg. Biomembr.* 25:459–72
48. Brdiczka D, Schumacher D. 1976. *Biochem. Biophys. Res. Commun.* 73:823–32
49. Brandolin G, Boulay F, Dalbon P, Vignais PV. 1989. *Biochemistry* 28:1093–100
50. Kihira Y, Iwahashi A, Majima E, Terada H, Shinohara Y. 2004. *Biochemistry* 43:15204–9
51. Boulay F, Vignais PV. 1984. *Biochemistry* 23:4807–12
52. Marty I, Brandolin G, Gagnon J, Brasseur R, Vignais PV. 1992. *Biochemistry* 31:4058–65
53. Hashimoto M, Majima E, Goto S, Shinohara Y, Terada H. 1999. *Biochemistry* 38:1050–56
53a. Dahout-Gonzalez C, Ramus C, Dassa EP, Dianoux AC, Brandolin G. 2005. *Biochemistry* 44:16310–20
54. Kaplan RS. 2001. *J. Membr. Biol.* 179:165–83
55. Palmieri F. 2004. *Pflügers Arch.* 447:689–709
56. Vignais PV, Lunardi J. 1985. *Annu. Rev. Biochem.* 54:977–1014
57. Gropp T, Brustovetsky N, Klingenberg M, Muller V, Fendler K, Bamberg E. 1999. *Biophys. J.* 77:714–26
58. Duyckaerts C, Sluse-Goffart CM, Fux JP, Sluse FE, Liebecq C. 1980. *Eur. J. Biochem.* 106:1–6
59. Barbour RL, Chan SH. 1981. *J. Biol. Chem.* 256:1940–48
60. Brandolin G, Dupont Y, Vignais PV. 1982. *Biochemistry* 21:6348–53
61. Dupont Y, Brandolin G, Vignais PV. 1982. *Biochemistry* 21:6343–47
62. Block MR, Vignais PV. 1984. *Biochim. Biophys. Acta* 767:369–76
63. Klingenberg M. 1985. In *Enzymes of Biological Membranes*, ed. AM Martonosi, 4:511–53. New-York: Plenum
64. Hunte C, Screpanti E, Venturi M, Rimon A, Padan E, Michel H. 2005. *Nature* 435:1197–202
65. Riccio P, Aquila H, Klingenberg M. 1975. *FEBS Lett.* 56:133–38
66. Klingenberg M, Riccio P, Aquila H. 1978. *Biochim. Biophys. Acta* 503:193–210
67. Hackenberg H, Klingenberg M. 1980. *Biochemistry* 19:548–55
68. Lin CS, Hackenberg H, Klingenberg EM. 1980. *FEBS Lett.* 113:304–6
69. Block MR, Zaccai G, Lauquin GJ, Vignais PV. 1982. *Biochem. Biophys. Res. Commun.* 109:471–77
70. Dyall SD, Agius SC, De Marcos Lousa C, Trézéguet V, Tokatlidis K. 2003. *J. Biol. Chem.* 278:26757–64

71. Palmisano A, Zara V, Honlinger A, Vozza A, Dekker PJ, et al. 1998. *Biochem. J.* 333(Part 1):151–58
72. Kotaria R, Mayor JA, Walters DE, Kaplan RS. 1999. *J. Bioenerg. Biomembr.* 31:543–49
73. Capobianco L, Impagnatiello T, Ferramosca A, Zara V. 2004. *J. Biochem. Mol. Biol.* 37:515–21
74. Schroers A, Burkovski A, Wohlrab H, Kramer R. 1998. *J. Biol. Chem.* 273:14269–76
75. Hatanaka T, Hashimoto M, Majima E, Shinohara Y, Terada H. 1999. *Biochem. Biophys. Res. Commun.* 262:726–30
76. Trézéguet V, Le Saux A, David C, Gourdet C, Fiore C, et al. 2000. *Biochim. Biophys. Acta* 1457:81–93
77. Huang SG, Odoy S, Klingenberg M. 2001. *Arch. Biochem. Biophys.* 394:67–75
78. Phelps A, Wohlrab H. 2004. *Biochemistry* 43:6200–7
79. Ma C, Kotaria R, Mayor JA, Remani S, Walters DE, Kaplan RS. 2005. *J. Biol. Chem.* 280:2331–40
80. Kihira Y, Majima E, Shinohara Y, Terada H. 2005. *Biochemistry* 44:184–92
81. Majima E, Ikawa K, Takeda M, Hashimoto M, Shinohara Y, Terada H. 1995. *J. Biol. Chem.* 270:29548–54
82. Klingenberg M. 1991. In *A Study of Enzymes. Vol II : Mechanism of Enzyme Action*, ed. SA Kuby, pp. 367–90. Boca Raton, FL: CRC Press
83. Klingenberg M. 1992. *Biochem. Soc. Trans.* 20:547–50
84. Terada H, Majima E. 1997. *Prog. Colloid Polym. Sci.* 106:192–97
85. Ledesma A, de Lacoba MG, Arechaga I, Rial E. 2002. *J. Bioenerg. Biomembr.* 34:473–86
86. Jiang F, Rizavi HS, Greenberg ML. 1997. *Mol. Microbiol.* 26:481–91
87. Jiang F, Ryan MT, Schlame M, Zhao M, Gu Z, et al. 2000. *J. Biol. Chem.* 275:22387–94
88. De Marcos Lousa C, Trézéguet V, Dianoux AC, Brandolin G, Lauquin GJM. 2002. *Biochemistry* 41:14412–20
89. Nelson DR, Douglas MG. 1993. *J. Mol. Biol.* 230:1171–82
90. Nelson DR. 1996. *Biochim. Biophys. Acta* 1275:133–37
91. Zeman I, Schwimmer C, Postis V, Brandolin G, David C, et al. 2003. *J. Bioenerg. Biomembr.* 35:243–56
92. Muller V, Heidkamper D, Nelson DR, Klingenberg M. 1997. *Biochemistry* 36:16008–18
93. Heimpel S, Basset G, Odoy S, Klingenberg M. 2001. *J. Biol. Chem.* 276:11499–506
94. Bogner W, Aquila H, Klingenberg M. 1986. *Eur. J. Biochem.* 161:611–20
95. Hatanaka T, Kihira Y, Shinohara Y, Majima E, Terada H. 2001. *Biochem. Biophys. Res. Commun.* 286:936–42
96. Majima E, Koike H, Hong YM, Shinohara Y, Terada H. 1993. *J. Biol. Chem.* 268:22181–87
97. Le Saux A, Roux P, Trézéguet V, Fiore C, Schwimmer C, et al. 1996. *Biochemistry* 35:16116–24
98. Roux P, Le Saux A, Trézéguet V, Fiore C, Schwimmer C, et al. 1996. *Biochemistry* 35:16125–31
99. Nelson DR, Lawson JE, Klingenberg M, Douglas MG. 1993. *J. Mol. Biol.* 230:1159–70
100. Heidkamper D, Muller V, Nelson DR, Klingenberg M. 1996. *Biochemistry* 35:16144–52
101. Muller V, Basset G, Nelson DR, Klingenberg M. 1996. *Biochemistry* 35:16132–43
102. Napoli L, Bordoni A, Zeviani M, Hadjigeorgiou GM, Sciacco M, et al. 2001. *Neurology* 57:2295–98
103. Komaki H, Fukazawa T, Houzen H, Yoshida K, Nonaka I, Goto Y. 2002. *Ann. Neurol.* 51:645–48

104. Kaukonen J, Juselius JK, Tiranti V, Kyttala A, Zeviani M, et al. 2000. *Science* 289:782–85
105. Oettmeier W, Masson K, Kalinna S. 1995. *Eur. J. Biochem.* 227:730–33
106. De Marcos Lousa C, Trézéguet V, David C, Postis V, Arnou B, et al. 2005. *Biochemistry* 44:4342–48
107. Kovac L, Lachowicz TM, Slonimski PP. 1967. *Science* 158:1564–67
108. Kolarov J, Kolarova N, Nelson N. 1990. *J. Biol. Chem.* 265:12711–16

RELATED REVIEWS

Neupert W. 1997. *Annu. Rev. Biochem.* 66:863–917
Popot J-L, Engleman DM. 2000. *Annu. Rev. Biochem.* 69:881–922

G Protein–Coupled Receptor Rhodopsin

Krzysztof Palczewski

Department of Pharmacology, School of Medicine, Case Western Reserve University, Cleveland, Ohio 44106–4965; email: kxp65@case.edu

Key Words

visual pigments, crystal structure, phototransduction, signal transduction, all-*trans*-retinal

Abstract

The rhodopsin crystal structure provides a structural basis for understanding the function of this and other G protein–coupled receptors (GPCRs). The major structural motifs observed for rhodopsin are expected to carry over to other GPCRs, and the mechanism of transformation of the receptor from inactive to active forms is thus likely conserved. Moreover, the high expression level of rhodopsin in the retina, its specific localization in the internal disks of the photoreceptor structures [termed rod outer segments (ROS)], and the lack of other highly abundant membrane proteins allow rhodopsin to be examined in the native disk membranes by a number of methods. The results of these investigations provide evidence of the propensity of rhodopsin and, most likely, other GPCRs to dimerize, a property that may be pertinent to their function.

Contents

BACKGROUND AND SCOPE..... 744
RHODOPSIN...................... 745
OVERVIEW OF THE
   RHODOPSIN STRUCTURE... 748
CHROMOPHORE OF
   RHODOPSIN.................. 753
RECENT STRUCTURAL DATA
   ON RHODOPSIN............. 753
DIMERIZATION OF
   RHODOPSIN.................. 757
INTERACTION OF RHODOPSIN
   WITH G PROTEIN............ 758
CONCLUSIONS AND
   PERSPECTIVES............... 760

G protein–coupled receptors (GPCRs): cell surface receptors with a seven-transmembrane helical structure

G proteins: trimeric intracellular proteins so named because they bind to guanine nucleotides GDP and GTP

Rhodopsin: the light-sensitive receptor of rod photoreceptor cells and a well-known GPCR

Chromophore: an organic compound that absorbs light. In vision, the chromophore is 11-*cis*-retinal

BACKGROUND AND SCOPE

G protein–coupled receptors (GPCRs) constitute by far the largest family of cell surface proteins involved in signaling across biological membranes. All GPCRs share a common seven α-helical transmembrane architecture (1). For most GPCRs, the external signal is a small molecule that binds to the membrane-embedded receptor and causes it to undergo a conformational change. The conformational change on the intracellular surface of the receptor results in the binding and activation of several (2, 3) or hundreds (4, 5) of heterotrimeric guanylate nucleotide-binding protein (G protein) molecules by a universal mechanism. Although GPCRs couple to G proteins, these receptors are also referred to as seven-transmembrane receptors, reflecting their seven membrane-embedded helices and additional signaling independent of G proteins (6, 7).

The GPCR superfamily encompasses approximately 950 genes in the human genome, including ∼500 sensory GPCRs (8–10). GPCRs modulate an extremely wide range of physiological processes, and mutations in the genes encoding these receptors have been implicated in numerous diseases. It is, thus, not surprising that these receptors form the largest class of therapeutic targets (e.g., see 11–13). Mammalian GPCRs are usually grouped by amino acid sequence similarities into the three distinct families A, B, and C (e.g., see 6, 7). More recently, the International Union of Pharmacology Committee on Receptor Nomenclature and Drug Classification published reports on the nomenclature and pharmacology of GPCRs that consider their predicted structure, pharmacology, and roles in physiology and pathology [(14); see also **http://www.iuphar.org/nciuphar_arti.html** and **http://www.iuphar-db.org/iuphar-rd/**].

In vision, rhodopsin in rod photoreceptors and cone opsins in cone photoreceptors respond to light (15, 16). Their chromophore, 11-*cis*-retinal, is covalently bound via a protonated Schiff base to the polypeptide chains of each opsin, embedded within the transmembrane domain. Upon absorption of a photon, the chromophore undergoes photoisomerization to all-*trans*-retinylidene, inducing a correspondent change in the opsin from its inactive to its active conformation. The active form, known as Meta II, then recruits and binds intracellular G proteins, continuing the visual signal cascade that culminates in an electrical impulse to the visual cortex of the brain. Soon after, opsin and the chromophore recombine to regenerate fresh rhodopsin. Progress in understanding how rhodopsin works has been steady during the past 120 years; however, recognition of rhodopsin as a member of the GPCR family, roughly 20 years ago, greatly enhanced interest in this receptor (16a). Remarkable advancements, which have benefited the GPCR field in general, have been achieved from studying rhodopsin. More recently, structural, genetic, and biochemical studies of rhodopsin have revealed unanticipated properties of this receptor, and a number of summary publications are available (References 1 and 16–25 to cite a few). The activation mechanism of rhodopsin has been extensively discussed on the basis of existing data (16, 21, 24, 26). Here, I summarize recent work on rhodopsin in relation to the receptor structure, the ligand-binding site, dimerization as a widespread property of GPCRs, and the interaction with the cognate G protein.

RHODOPSIN

Retinal rod cells (also known as photoreceptor cells) are highly differentiated neurons responsible for detecting photons (**Figure 1**) (27). A specialized part of the rod cell, the rod outer segment (ROS) (**Figure 1***a,b*), contains rhodopsin and auxiliary proteins, which convert and amplify the light signal (28). The system is so exquisitely sensitive that a single photon can be detected [(29), and more recently (30)]. Each mammalian ROS consists of a pancake-like stack of 1000–2000 distinct disks enclosed by the plasma membrane (**Figure 1***c*). The main protein component (>90%) of the bilayered disk membranes is light-sensitive rhodopsin. Approximately 50% of the disk membrane area is occupied by rhodopsin, whereas the remaining space is filled with phospholipids and cholesterol (**Figure 1***d*). Rhodopsin is also present at a lower density in the plasma membrane (31). The ROS of wild-type mice have an average length of 23.6 ± 0.4 μm (32) or 23.8 ± 1.0 μm (33) and contain on average 810 ± 10 disks (33). Given the approximately 6.4×10^6 rods in the mouse retina (34), this translates to $\sim 5 \times 10^9$ disks per retina. The total amount of rhodopsin per eye is ~ 650 pmoles ($650 \times 10^{-12} \times 6.022 \times 10^{23} = 3.96 \times 10^{14}$ rhodopsin molecules); thus there are $\sim 8 \times 10^4$ rhodopsin molecules per disk. The size of ROS is reduced when animals are exposed for a prolonged period to light, a phenomenon described as photostasis (35) and for which the molecular mechanism has not yet been elucidated. Rhodopsin expression is essential for the formation of the ROS, which are absent in knockout rhodopsin-/- mice (36, 37). The ROS of mice heterozygous for the rhodopsin gene deletion (rhodopsin+/−) have a similar density of rhodopsin, but the ROS volume is reduced by $\sim 60\%$ compared with wild-type mice (33).

Synthesis of the seven-transmembrane apoprotein portion of rhodopsin, called opsin, begins in the inner segments of photoreceptors, where it undergoes maturation in the endoplasmic reticulum (ER) and Golgi membranes before it is transported vectorially to the ROS. The C-terminal region of the protein is essential for interaction with the transport machinery that delivers the cargo of transmembrane- and membrane-associated proteins on membrane vesicles to the ROS (38–40). Specifically, the C-terminal-sorting motif of rhodopsin binds to the small GTPase ARF4, a member of the ARF family of membrane-budding and protein-sorting regulators (41), whereas the sorting protein rab8 is implicated in docking of post-Golgi membranes containing rhodopsin in rods (42). The transport of a rhodopsin mutant lacking the C-terminal region to the ROS does not occur in vivo, and the mutant is piggybacked to the ROS only in the presence of wild-type rhodopsin (43, 44). The regeneration of rhodopsin from opsin and its chromophore 11-*cis*-retinal is not essential for vectorial transport to the ROS, as mice deficient in chromophore production still develop ROS (45, 46). However, because of the continuous coupling of opsin with G proteins, these rods slowly degenerate (47–51).

On the basis of the predicted structure, conservation of few amino acids in the region critical for G protein activation, and activation by small ligand, rhodopsin belongs to the largest subfamily of GPCRs (family A) (10). More than a century of extensive biochemical, biophysical, and structural information collected on rhodopsin has given rise to its status as a prototypical receptor of this family. Several efficient methods were developed to isolate rhodopsin, including selective extraction from the ROS in the presence of divalent metal ions (52, 53) and immunoaffinity chromatography (54) (summarized in Reference 6). Bovine retina is an extraordinary

Schiff base: a functional group containing a carbon-nitrogen double bond

Photoisomerization: a molecule's shift from one isomer to another upon excitation by light

Rod cells: the photoreceptor cells of the retina sensitive to low levels of light

Rod outer segment (ROS): the cylindrical portion of the rod cell containing several to 2000 membranous disks

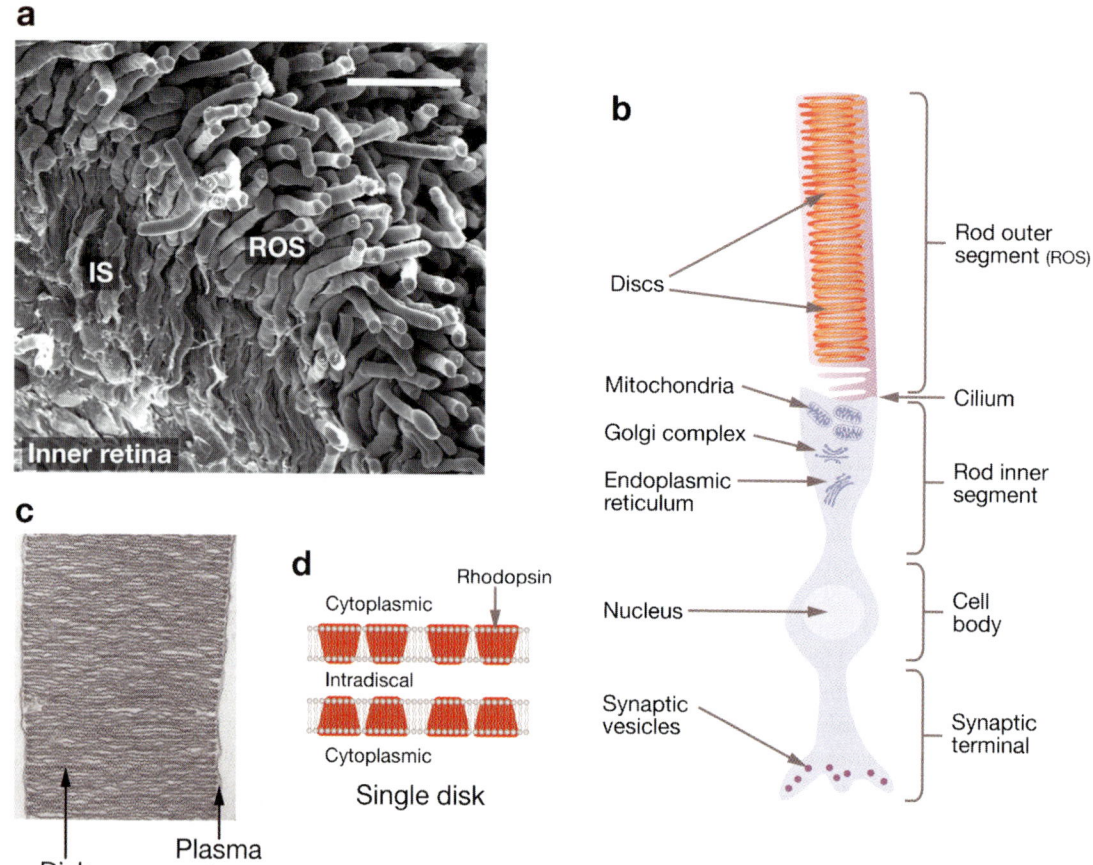

Figure 1

Vertebrate retina and rhodopsin. (*a*) Scanning electroretinogram of mouse retina [courtesy of Yan Liang (33)]. Rod cells comprise ~70% of all 6.4 million retinal cells, and cone cells represent <2%. Rods are postmitotic neurons with highly differentiated rod outer segments (ROS) connected to the inner segments (IS), which generate proteins and energy to sustain phototransduction events. (*b*) Diagram depicting the rod cell. The processes in ROS allow rapid transduction of the light signal to graded hyperpolarization of the plasma membrane, ensuing from the decrease of light-sensitive conductance in the ROS cGMP-gated cation channels. In ROS, hundreds of distinct, rhodopsin-loaded disk membranes (20) are enveloped by the plasma membrane. (*c*) Electron micrograph of isolated ROS from the mouse retina [courtesy of Yan Liang (33)]. The disk membranes consist of a phospholipid bilayer studded with rhodopsin. (*d*) Diagram of disk membranes. The main protein of ROS disk membranes is light-sensitive rhodopsin, which occupies 50% of the disk area. The molar ratio between rhodopsin and phospholipids is about 1:60 (for example 138 and 139; reviewed in 140). Multiple techniques suggest that rhodopsin forms oligomeric structures in the native membranes, with the rhodopsin dimer most likely being the signaling unit.

source of native protein that yields ~0.7 mg rhodopsin per retina (16). Bovine rhodopsin consists of a 348-amino acid apoprotein opsin and 11-*cis*-retinal, which is bound to the protein through a Schiff-base linkage to a Lys[296] side chain (**Figure 2a,b**). Protein posttranslational modifications include double palmitoylation, acetylation of the N terminus, glycosylation with two $(Man)_3(GlcNAc)_3$ groups via two asparagine residues, and a disulfide bond (**Figure 2a**). In addition, rhodopsin undergoes a light-dependent phosphorylation at one or a few of the six to seven Ser/Thr residues at the C-terminal region (reviewed in 55).

Rhodopsin changes in color upon exposure to light, as described by Kühne (see translations in References 56 and 57; also see the side bar Discovery of Rhodopsin). In ordinary conditions, absorption of a photon of light causes photoisomerization of 11-*cis*-retinylidene to all-*trans*-retinylidene, with an accompanying shift in the λ_{max} of absorption bovine rhodopsin from 498 nm to 380 nm (**Figure 3a–c**). Ultimately, the Schiff base is hydrolyzed, and all-*trans*-retinal is reduced by retinol dehydrogenase to all-*trans*-retinol (reviewed in 15 and 58). The change in the λ_{max} of absorption after the illumination of rhodopsin is a very sensitive parameter, which has been correlated with the receptor conformation in numerous studies (**Figure 3d**). A number of intermediates were trapped at low temperatures, and the equilibrium between slowly formed species of photoisomerized rhodopsin was shown to be affected by ionic strength, pH, glycerol, and temperature (**Figure 3d**). The main conclusions of these spectroscopic studies in conjunction with the current structural understanding of rhodopsin can be summarized as follows:

1. Prior to any change in protein conformation, the energy of the photon is stored by the chromophore in its highly distorted all-*trans*-retinylidene form in the same binding pocket where it resides in the dark state as 11-*cis*-retinylidene. Rohring et al. (60) suggested that the protein-binding pocket selects and accelerates the isomerization exclusively around the C_{11}-C_{12} bond, resulting in the formation of a twisted structure; "hence, the initial step of vision can be viewed as the compression of a molecular spring that can then release its strain by altering the protein environment in a highly specific manner" (60).

2. Photoisomerization is ultrafast and occurs within 200 femtoseconds (59).
3. Meta I is transiently formed and decays to Meta II (19).
4. Meta II is the heterogeneous form of several photoactivated conformations (61). This physiologically important intermediate of rhodopsin is responsible for interaction with peripheral membrane proteins, including the heterotrimeric G protein transducin.
5. Opsin spontaneously combines with 11-*cis*-retinal chromophore to regenerate rhodopsin. In contrast with opsin, rhodopsin has no basal activity toward the G protein transducin. The rapid transformation of opsin to rhodopsin terminates signaling activity, allowing the rod cells to maintain a low activation threshold. The extraordinary sensitivity of rods is illustrated by

DISCOVERY OF RHODOPSIN

The reddish-purple coloration of rod cells was noted in 1851 by Heinrich Müller, who attributed it to hemoglobin. In 1876, Franz Boll recognized that frog retina is photosensitive and when exposed to light, the pigment bleached to a yellowish color and then became colorless. Boll demonstrated that frogs exposed to sunlight and then kept in darkness regenerated the red pigment. After many observations, he concluded, "The basic color of the retina is constantly consumed *in vivo* by the light falling on the eye.... In the dark, *in vivo*, the color is regenerated." Willy Kühne, pursuing Boll's findings, determined the pigmented material to be a rod outer segment protein he named "visual purple" (rhodopsin). Kühne isolated frog retinas and retinal pigment epithelium (RPE) layers in experiments, proving that the RPE is necessary for the regeneration of rhodopsin. He extracted rhodopsin using bile salts, matched its spectral absorption profile to that of dissected retinas, proposed that the yellow and colorless products of bleaching must be chemically distinct substances, and correlated the electrical impulses emitted by isolated retinas to their illumination. Kühne's extensive investigation of the visual system began to elucidate the now-familiar story of a photochemical reaction whose products stimulate nerve impulses to the brain.

the photoactivation of 1 to 10,000 rhodopsin molecules out of the $\sim 8 \times 10^7$ per rod. A single photon is sufficient to consistently generate a measurable response. As mentioned above, in cases when opsin fails to reunite with the chromophore to regenerate rhodopsin, the persistent activation of G proteins by opsin destabilizes and eventually damages the rod cell (40–44). As a note of interest, rhodopsin in the lancelet *Branchiostoma* is regenerated by the agonist all-*trans*-retinal as well as by 11-*cis*-retinal, suggesting that the structures of opsin and rhodopsin are similar (62, 63).

OVERVIEW OF THE RHODOPSIN STRUCTURE

The structure of rhodopsin is only briefly described because the specific structural features of the receptor were described fully in the original research (20, 64–67) and previous review publications (16, 18, 19). The overall elliptic, cylindrical shape of the rhodopsin molecule is due to arrangement of its seven transmembrane helices, which vary in length from 20 to 33 residues (**Figure 2a,b**). The N-terminal region is located intradiscally (extracellularly) (**Figures 1d** and **2a,b**), and the C-terminal region is cytoplasmic. The dimensions of rhodopsin, described by a ellipsoid, are ~ 75 Å perpendicular to the membrane, ~ 48 Å wide in the standard view, and ~ 35 Å thick. The surface area of the portions projecting from the membrane is ~ 1200 Å^2, with the cytoplasmic projection larger in volume and surface area than the intradiscal face (**Figure 2b**). The distribution of mass for the intra- and extracellular regions is comparable, whereas the transmembrane region encompasses $\sim 65\%$ of the amino acids. The chromophore is located within this hydrophobic transmembrane core (**Figure 2b**; see also the Chromophore side bar).

The transmembrane helices are irregular, particularly with respect to the degree of

Figure 2

Modification of rhodopsin molecule and orientation in the membranes. (*a*) Two-dimensional model of rhodopsin. The polypeptide of rhodopsin crosses the membrane seven times. C-I, C-II, and C-III correspond to the cytoplasmic loops, and E-I, E-II, and E-III correspond to extracellular loops. The transmembrane segment is α-helical (*yellow cylinders*), although the helices are highly distorted and tilted. The stability of the helical segment is increased by the Cys^{110}-Cys^{187} bridge (141) (depicted in *dark yellow*) a highly conserved feature among many GPCRs. The chromophore, 11-*cis*-retinal, not depicted here, is attached to Lys^{296} (*dark red*) via a protonated Schiff base. The positive charge of the base is neutralized by counterion Glu^{113} (*blue*). During postisomerization changes in the receptor, it was proposed that the counterion migrates to Glu^{181} (*blue*) (76). Asn^2 and Asn^{15} (*red*) are sites of glycosylation within the conserved glycan composition, and Met^1 (*orange*) is acetylated. Cys^{322} and Cys^{323} (*light green*) are palmitoylated, whereas two other Cys, Cys^{140} and Cys^{316} (*brown*), are reactive to many chemical probes and are used to explore rhodopsin's structure. Rhodopsin, when exposed to light, is phosphorylated by rhodopsin kinase (or G protein–coupled receptor kinase 1). The predominant phosphorylation sites are Ser^{334}, Ser^{338}, and Ser^{343} (*green*) (55), and the whole C-terminal region is highly mobile (142). However, as shown using a model peptide, the C-terminal region may become structured when bound to arrestin (143). The highly conserved domains among GPCRs, D(E)RY in helix 3 and NPXXY in helix VII (*gray*), are important in transformation of the receptor from an inactive to a G protein–coupled conformation. Different versions of this figure were published previously (for example in 19 and 64), and all of them are refinements of the pioneering work on rhodopsin topology by Paul Hargrave (144, 145). (*b*) Location of the chromophore and charges on the cytoplasmic and intradiscal (extracellular) surface of rhodopsin in relation to the hypothetical membrane bilayer. The negative charges (*red*) and basic residues (*blue*) are shown. The proposed location of the membrane is shown in gray, and the location of the chromophore 11-*cis*-retinylidene is shown by deleting fragments of transmembrane helices. Two sides of rhodopsin are depicted.

bending around Gly–Pro residues, and they tilt at various angles with respect to the expected membrane surface, as described elsewhere (20, 64). The strongest distortion is imposed by Pro[267] in helix VI, one of the most conserved residues among the GPCRs.

The presence of Pro[291] and Pro[303] in the region around the retinal attachment site Lys[296] elongates helix VII. Pro[303] is a part of the NPXXY (AsnProXaaXaaTyr) motif located at the end of helix VII and the beginning of helix 8. A high-affinity Zn^{2+} coordination

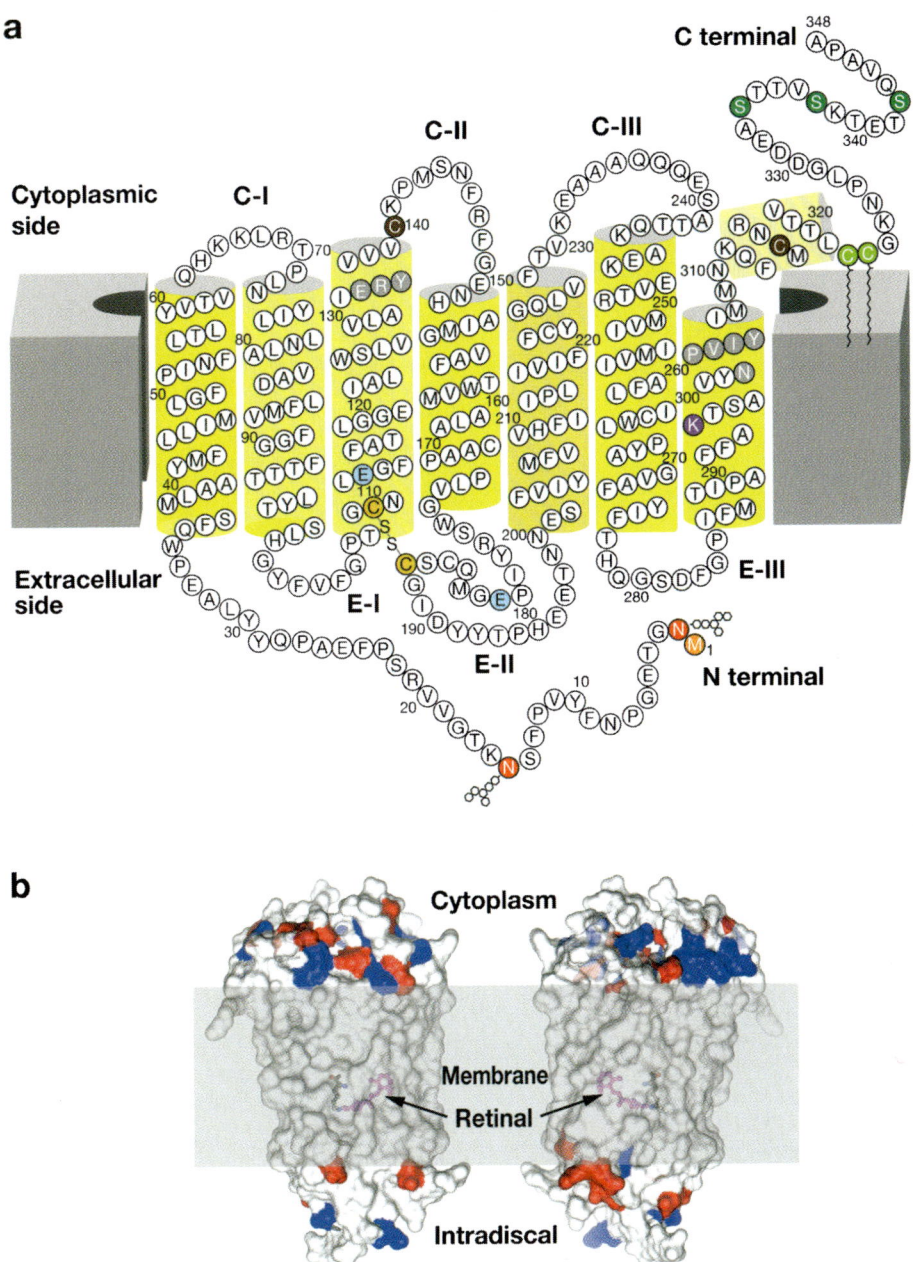

site has been identified within the transmembrane domain of rhodopsin, coordinated by the side chains of two highly conserved residues, Glu^{122} of helix III and His^{211} of helix V (68). It is not clear whether (or there is no evidence one way or the other to support the conclusion that) the ability of rhodosin to bind Zn^{2+} in vitro reflects a physiologically relevant role for this divalent cation in rhodopsin function in vivo. In detergent solutions, Zn^{2+} lowers the stability of rhodopsin and the extent of rhodopsin regeneration (69).

The extracellular (intradiscal) and intracellular regions of rhodopsin each consist of three interhelical loop and a terminal tail regions. Bourne & Meng (70) describe the extracellular N-terminal domain as a

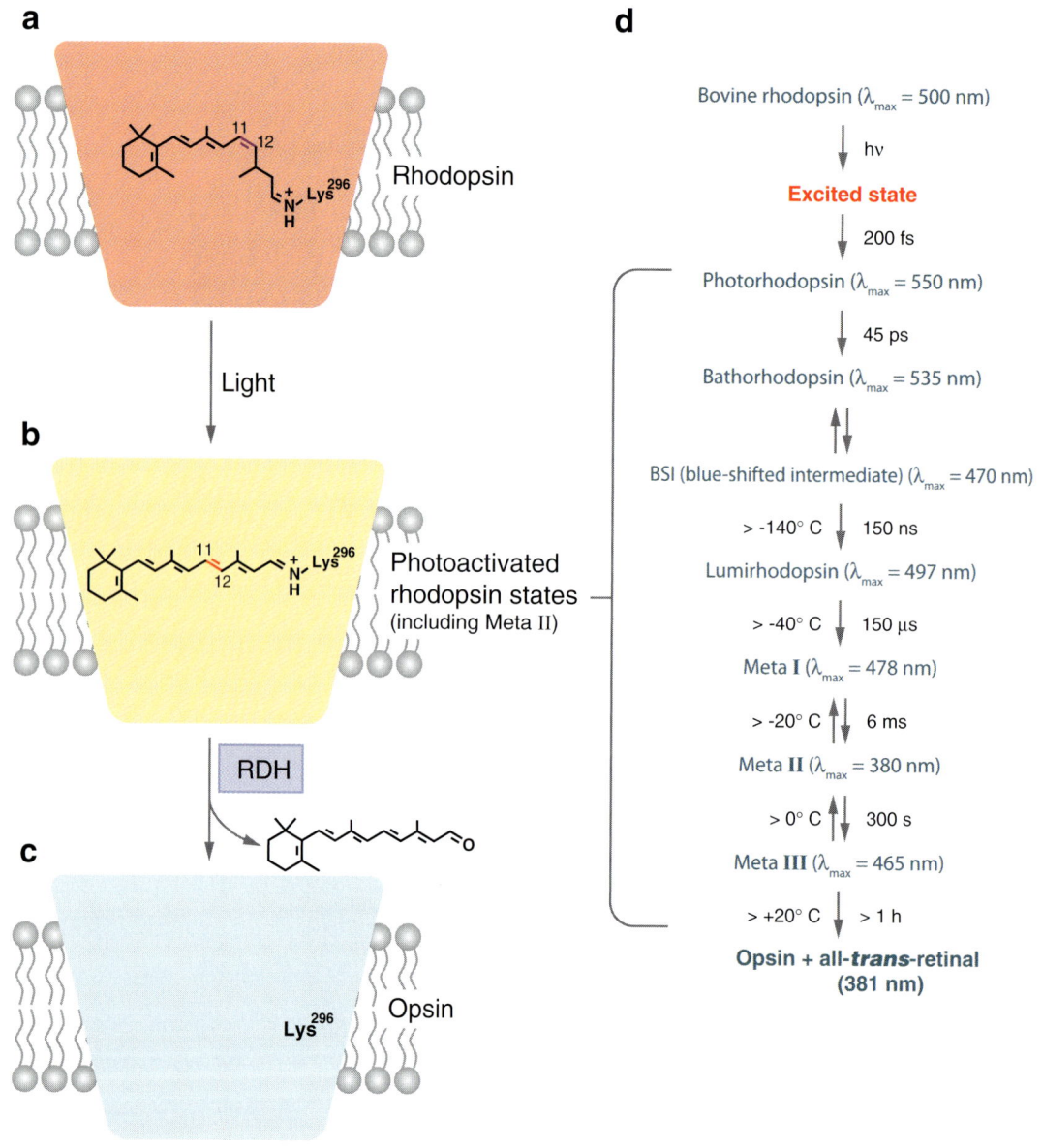

"plug" for the chromophore-binding pocket (**Figure 4b**). This globular domain is formed by residues 1 to 33, the short loop of residues 101–105, "plug" residues 173–198 between helices IV and V, and residues 277–285 between helices VI and VII (**Figure 2a**). The Asn^2 and Asn^{15} residues are the glycosylation acceptor sites for GlcNAc-(β1,4)-GlcNAc-(β1,4)-mannose (**Figures 2a** and **4a**). There are four extracellular structural elements associated significantly with each other (pairs of β1-β2 and β3-β4 hairpins) (**Figure 4b**). The extracellular loop II is connected to helix III via a disulfide bridge and fits tightly into a limited space inside the bundle of helices. A detailed description of the region can be found in References 16, 20, 64, and 67.

The cytoplasmic side includes the three loops, located between helices, encompassing residues Gln^{64}-Pro^{71}, Glu^{134}-His^{152}, and Gln^{225}-Arg^{252}; the peripheral helix 8; and the C-terminal tail. The residues of a highly conserved (D/E)R(Y/W) motif, found in GPCR subfamily A, are formed by the tripeptide Glu^{134}-Arg^{135}-Tyr^{136} located in this region (10) (**Figure 2a**). The carboxylate of Glu^{134} forms a salt bridge with Arg^{135}, whereas Arg^{135} also interacts with Glu^{247} and Thr^{251} in helix VI. The ionization state of Glu^{134} is sensitive to its environment, and when this residue is protonated, rhodopsin could become activated (71). Val^{137}, Val^{138}, and Val^{139} are also located close together and partially cover the cytoplasmic side of Glu^{134} and Arg^{135}. This region is likely to be a critical constraint, which keeps rhodopsin in the inactive conformation. Cytoplasmic helix 8 (residues 311–321), structurally similar in all crystal forms, is fastened to the membrane by the palmitoylation of residues Cys^{322} and Cys^{323}. The helix VII/helix 8 kink is stabilized by residues Glu^{249} to Met^{309}, Asn^{310} to Phe^{313}, and Arg^{314} to Ile^{307}, and helix 8 is further stabilized by hydrophobic side chain residues buried within the hydrophobic residues of the transmembrane domain located in helices I and VII (**Figure 4c**). Another region conserved among GPCRs and located close to helix 8 is

---

Figure 3

Light-cycle of rhodopsin. (*a*) Rhodopsin and 11-*cis*-retinal. Rhodopsin consists of a colorless protein moiety (the opsin) and the chromophore, 11-*cis*-retinylidene, which imparts a red color to rhodopsin. The chromophore, a geometric isomer of vitamin A in aldehyde form, is coupled to opsin via the protonated Schiff base at Lys^{296}, located in the transmembrane domain of the protein. Bovine rhodopsin absorbs at a λ_{max} = 498 nm. (*b*) Photoactivated rhodopsin. Absorption of light by rhodopsin leads with high probability (~65%) to photoisomerization of the *cis* C_{11}-C_{12} chromophore double bond to a *trans* configuration. The probability of isomerization depends only modestly on the wavelength of the light (146). This reaction, one of the fastest photochemical reactions known in biology, produces multiple intermediates that culminate in the formation of the G protein–activating state, termed metarhodopsin II, or Meta II. (*c*) Opsin without chromophore. Ultimately the photoisomerized chromophore, all-*trans*-retinylidene, is released from the opsin as all-*trans*-retinal and reduced to alcohol by short-chain alcohol dehydrogenases, such as prRDH, retSDR, and RDH12. The all-*trans* chromophore diffuses to the adjacent retinal pigment epithelium, where it undergoes enzymatic transformation back to 11-*cis*-retinal in a metabolic pathway known as the retinoid cycle. Opsin recombines with replenished 11-*cis*-retinal to form rhodopsin. (*d*) Reaction scheme of rhodopsin photoactivation. Upon absorption of a photon by rhodopsin and electronic excitation, fast isomerization of 11-*cis*-retinylidene to all-*trans*-retinylidene takes place. At body temperature, the Meta I and Meta II exist in equilibrium shifted toward Meta II. In vitro, further decay of rhodopsin to both opsin and free all-*trans*-retinal or to Meta III is possible. In vivo, Meta III is not formed at significant levels because it decomposes in the presence of G protein transducin (147). In vitro, prolonged incubation of Meta II involves a thermal isomerization of the chromophore double bond with Lys^{296} to an all-*trans*-15-syn configuration. This isomerization step is catalyzed by the opsin itself (148). On the left are maximal temperatures at which indicated intermediates can be trapped, and on the right is time required for that particular transformation. In the brackets are λ_{max} of absorption for different intermediates. The reaction scheme is based on Shichida & Imai [(149); see also the thermodynamic properties of these reactions (19)].

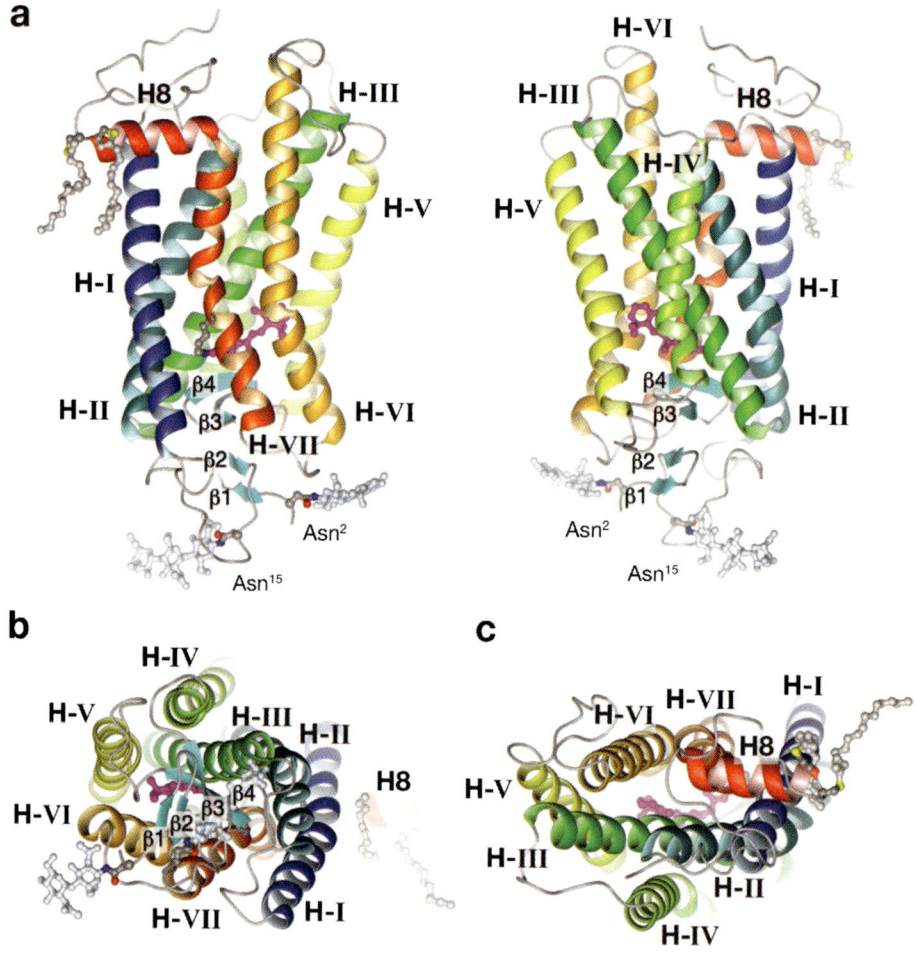

Figure 4

Three-dimensional model of rhodopsin. (*a*) Ribbon drawings of rhodopsin parallel to the plane of the membrane. (*b*) View into the membrane plane as seen from the intradiscal side of the membrane. The carbohydrate moieties are at Asn^2 and Asn^{15}. The pairs of β1-β2 and β3-β4 hairpins, the transmembrane helices (Hs) I–VII, and the cytoplasmic helix 8 (H8) are labeled. A palmitoyl group is attached to each of the two Cys residues at the end of helix 8. The removal of the palmitoyl groups has only a minor effect on phototransduction processes (e.g., 150 and 151). (*c*) A view into the membrane plane as seen from the cytoplasmic side. The cytoplasmic side has greater surface area than the intradiscal side. (The roman numeral convention is related to transmembrane helices, whereas an Arabic numeral indicates a solvent-exposed helix.)

the NPXXY sequence (NPVIY in rhodopsin) near the cytoplasmic end (10), which is likely to be involved in G protein coupling (72) (**Figure 2***a*). The side chains of the two polar residues in this region, Asn^{302} and Tyr^{306} in bovine rhodopsin, project toward the transmembrane core of the protein and Phe^{313}, respectively, in helix 8. The -OH group of Tyr^{306} is close to Asn^{73} and is engaged in the interhelical hydrogen-bonding constraints between helix VII and helix II. These interactions most likely occur through water molecules.

CHROMOPHORE OF RHODOPSIN

Rhodopsin of the rod cell and other visual pigments of the cone cells contain the 11-*cis*-retinal chromophore, bound via a protonated Schiff-base linkage to a Lys side chain (Lys^{296} in bovine rhodopsin) in the middle of helix VII (**Figures 2a, 4**). The interaction between the protein moiety and the chromophore produces a specific absorption shift for visual pigments compared with the retinal protonated Schiff-base model compounds formed between alkylamines and retinal, which absorb light maximally at ∼440 nm. In addition to the retinal cavity formed by helices of the transmembrane segment, an antiparallel β sheet of the plug, the part of extracellular loop II that includes Glu^{181}, penetrates deep into rhodopsin's interior, close to the chromophore. The retinylidene moiety is located closer to the extracellular side in the hypothetical lipid bilayer (**Figure 2b**). The counterion for the protonated Schiff base is provided by Glu^{113}, which is highly conserved among all known vertebrate visual pigments. Kim et al. (73) reported that this salt bridge, formed between a protonated Schiff base and Glu^{113}, is a key constraint in maintaining the resting state of the receptor and that disruption of the salt bridge is the cause, rather than a consequence, of the helix VI motion that occurs upon photoactivation. The counterion has two other important functions: (*a*) It stabilizes the protonated Schiff base by increasing the K_a for this group by as much as 10^7, thus preventing its spontaneous hydrolysis (reviewed in 74); and (*b*) it causes a bathochromic shift in the maximum absorption for visual pigments, which makes them more sensitive to longer wavelengths because UV light is filtered out by the front of the eye in most animals. The 11-*cis*-retinylidene group is surrounded by the 20 residues depicted in **Figure 5**.

The energy of a photon enables protein conformational changes that culminate in the formation of active Meta II, which can be considered analogous to the agonist-bound state

CHROMOPHORE

George Wald and his Harvard colleagues were the first to reveal that rhodopsin contains two distinct components, a colorless protein termed opsin and a yellow pigment, 11-*cis*-retinal, that serves as its chromophore. In 1933, following a hunch that rhodopsin might contain a carotenoid, Wald isolated vitamin A from retinal tissue, a finding consistent with literature linking night blindness with vitamin A deficiency. At the time, little was known about the biochemical roles played by vitamins. Using frog retinal extracts in fat solvents, Wald proceeded to show that rhodopsin and its orange bleached intermediate both released a yellow material he termed retinene, which was then replaced by vitamin A as the purplish retinal color faded. Retinene (retinal) proved to be vitamin A aldehyde and could be mixed in the dark with bleached rhodopsin or with opsin to regenerate fresh rhodopsin. Wald's team later showed that only the bent 11-*cis*-isomer combined with opsin to form rhodopsin. Wald received the 1967 Nobel Prize in Physiology and Medicine for characterizing the molecular components of vision and discovering the biochemical role for vitamin A.

of many ligand-binding GPCRs. Recently, Ernst and colleagues (75) demonstrated that *cis* acyclic retinals, lacking four carbon atoms of the β-ionone ring, can only partially activate rhodopsin when photoisomerized. Detailed analysis of rhodopsin regenerated with these acyclic retinals revealed that a lack of the ring structure destabilizes the active state. This study describes for the first time the molecular mechanism of activation by a partial agonist, a mechanism that possibly extends to other GPCRs. The partial agonism is due to instability of the only partially active state of the receptor.

RECENT STRUCTURAL DATA ON RHODOPSIN

Progress continues in further refining the structure of rhodopsin and obtaining new details of intermediate photobleaching states. Okada and colleagues (66) investigated the functional role of water molecules in the transmembrane regions of bovine rhodopsin.

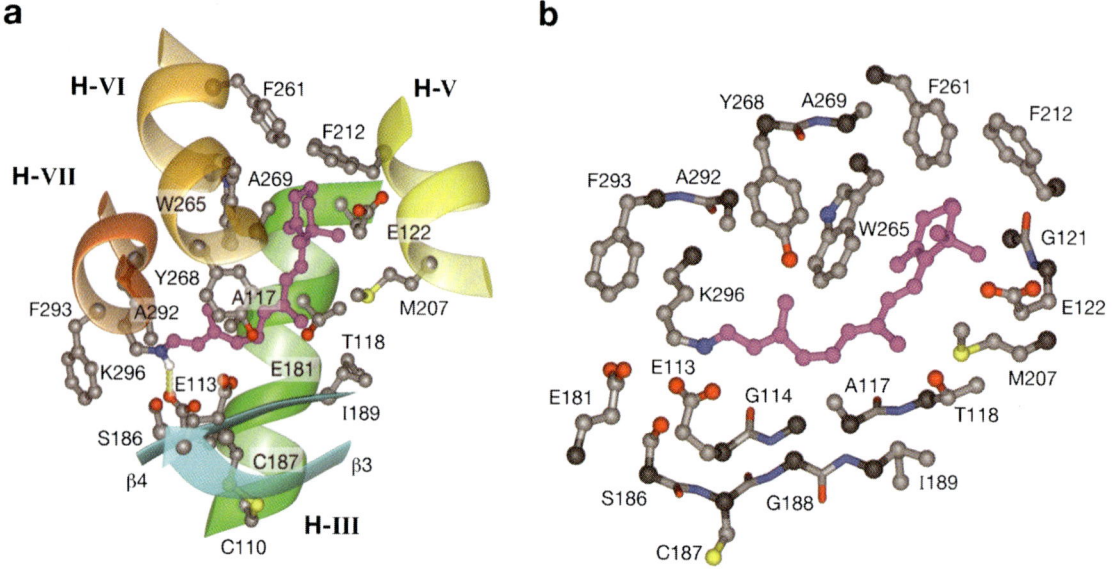

Figure 5

The amino acid residues in the vicinity of the chromophore. (*a*) Schematic showing the side chains surrounding the 11-*cis*-retinylidene group (*pink*); side view through helices III, V, and VI. (*b*) Schematic presenting the residues within 5 Å distance from the 11-*cis*-retinylidene group (*pink*). Note that the chromophore is coupled via the protonated Schiff base with Lys[296].

They successfully improved the original rhodopsin crystals, increasing their resolution to 2.6 Å. This improvement led to unambiguous differentiation between water molecules and Zn^{2+} used for rhodopsin purification. Seven water molecules were found in the transmembrane segment. The cluster 1 containing water molecules 1a, 1b, and 1c is linked to Asn[302] and Asp[83] in helix II and a residue of the NPXXY motif of the helix VII. The second cluster, containing water molecules 2a and 2b, is located in the vicinity of the retinal Schiff base. Water molecule 2a is located between the side chains of Glu[181] and Ser[186] in the vicinity of the counterion Glu[113]. Water molecule 2a may play a key role in transferring the counterion from Glu[113] in rhodopsin to Glu[181] in Meta I (76) (**Figure 5***a*,*b*). A similar switch of counterions was proposed for the UV visual pigment (77). Water molecule 3 is surrounded by the peptide main chains of helices VI and VII, and water molecule 4 facilitates interaction between helices I and II at the cytoplasmic surface. It is expected that water molecules play a key role in visual pigment spectra sensitivity (response to a specific wavelength of light), conformational changes during activation, and transmission of the signal from the photoisomerized chromophore to the D(E)RY and NPXXY regions on the cytoplasmic surface of rhodopsin during photoactivation (for example 19 and 20). In invertebrate rhodopsin, the counterion of the Schiff base is a residue, which corresponds to Glu[181] (78), the proposed counterion of Meta I in vertebrate rhodopsin (76).

A major contribution to our understanding of how rhodopsin works at the molecular level was made by the Schertler laboratory. Krebs et al. (79), using electron cryomicroscopy of two-dimensional crystals with $p22_12_1$ symmetry, produced a rhodopsin map with a resolution of 5.5 Å in the membrane plane and 13 Å perpendicular to the membrane, obtaining information about the orientation of the molecule relative to the bilayer.

Li et al. (67) generated new three-dimensional crystals of highly purified rhodopsin, using C_8E_4 (*n*-octyltetraoxyethylene) and LDAO (*N,N*-dimethyldodecylamine-*N*-oxide) detergents with Li_2SO_4 and PEG 800 as precipitants. The crystals belong to the trigonal space group $P3_1$ and diffract to 2.55 Å. The rhodopsin structure has been determined using data obtained from these crystals (67). As in the previous studies, ordered water molecules were found, for example, linking Trp^{265} in the retinal-binding pocket to the NPXXY motif and stabilizing the Glu^{113} counterion with the protonated Schiff base at the extracellular surface. The cytoplasmic ends of helix V and helix VI are extended by one turn, distinguishing this structure from ones previously determined (**Figure 6a–c**). However, the cytoplasmic loops have the highest temperature factor (B factor), a measure of certainty of a location of an atom within the crystal structure, in all crystal structures (**Figure 6d,e**), suggesting flexibility in this region in addition to the possibility of a crystal-packing artifact. Ruprecht and coworkers (80) used electron crystallography to determine a density map of Meta I to a resolution of 5.5 Å in the membrane plane, and this suggested that Meta I formation does not involve large helical movements. They provided some evidence that the changes in Meta I involve a rearrangement close to the bend of helix VI in the vicinity of the chromophore-binding pocket. The spectra of these crystals studied with Fourier transform infrared difference spectroscopy revealed that the formation of the active state, Meta II, is blocked in the crystalline environment, as indicated by a lack of spectral features in Meta II and a lack of activation of the G protein transducin (81).

Buss and colleagues (65) used a theoretical study of the chromophore geometry in combination with rhodopsin crystals that diffracted to 2.2 Å to focus on the conformation of the chromophore, providing new insight into the twist of the 6-s-*cis*-bond and the C_{11}-C_{12} double bond (65). The comparison of these structures is depicted in **Figure 6a–c**. The panels in this figure reveal differences in the structure only at the flexible cytoplasmic region of rhodopsin (**Figure 6b,c**), characterized by large B factors (**Figure 6d,e**).

Creemers and colleagues (82) determined the complete ^1H and ^{13}C assignments of the 11-*cis*-retinylidene chromophore in its ligand-binding site using ultra-high field magic-angle-spinning NMR. A gallant synthesis of 99% enriched uniformly ^{13}C-labeled 11-*cis*-retinal by Lugtenburg's laboratory made this work possible (83). Authors found interactions between the chromophore's H^{16}/H^{17} and Phe^{208}, Phe^{212}, and H^{18} to be in close contact with Trp^{265}. This NMR study revealed that binding of the chromophore involves a chiral selection of the ring conformation, resulting in equatorial and axial positions for CH_3-16 and CH_3-17 (82).

High-resolution solid-state NMR studies of the Meta I photointermediate are in agreement with the electron microscopy data (84). The β-ionone ring retains strong contacts in its binding pocket prior to activation of the receptor without any major protein rearrangements around the chromophore-binding site. Further studies reveal an increase in steric clashes and an adjustment of the protein structure in Meta II, without substantial changes in the location of the all-*trans*-retinylidene chromophore (85). These results are at odds with another NMR study and biochemical studies that predict changes in the location of the chromophore during transition to Meta II (86, 87).

Patel et al. (88), using solid-state magic-angle-spinning NMR spectroscopy, found that Trp^{126} and Trp^{265} become more weakly hydrogen bonded during the transformation of rhodopsin to Meta II and that both the side chain of Glu^{122} and the backbone carbonyl of His^{211} are disrupted in Meta II. Clearly, the full picture of rhodopsin transition from an inactive to active state will require a high-resolution structure of Meta II obtained by X-ray crystallography.

The isomerization of 11-*cis*-retinylidene is followed by hydrolysis of the photobleached product all-*trans*-retinylidene and the subsequent release of all-*trans*-retinal from the binding pocket (**Figure 3a–c**). The rapid regeneration and recombination of 11-*cis*-retinal with opsin restores dark conditions to permit subsequent photon absorption, allowing our vision to work in an uninterrupted manner. Perhaps the same key residues involved in isomerization are involved in the hydrolysis process, such as Glu[113] and Glu[181] via the carbinol ammonium ion (**Figure 7a**), and in a coupling reaction between 11-*cis*-retinal and opsin (**Figure 7b**). To form the Schiff base, Lys[296] must be deprotonated, the

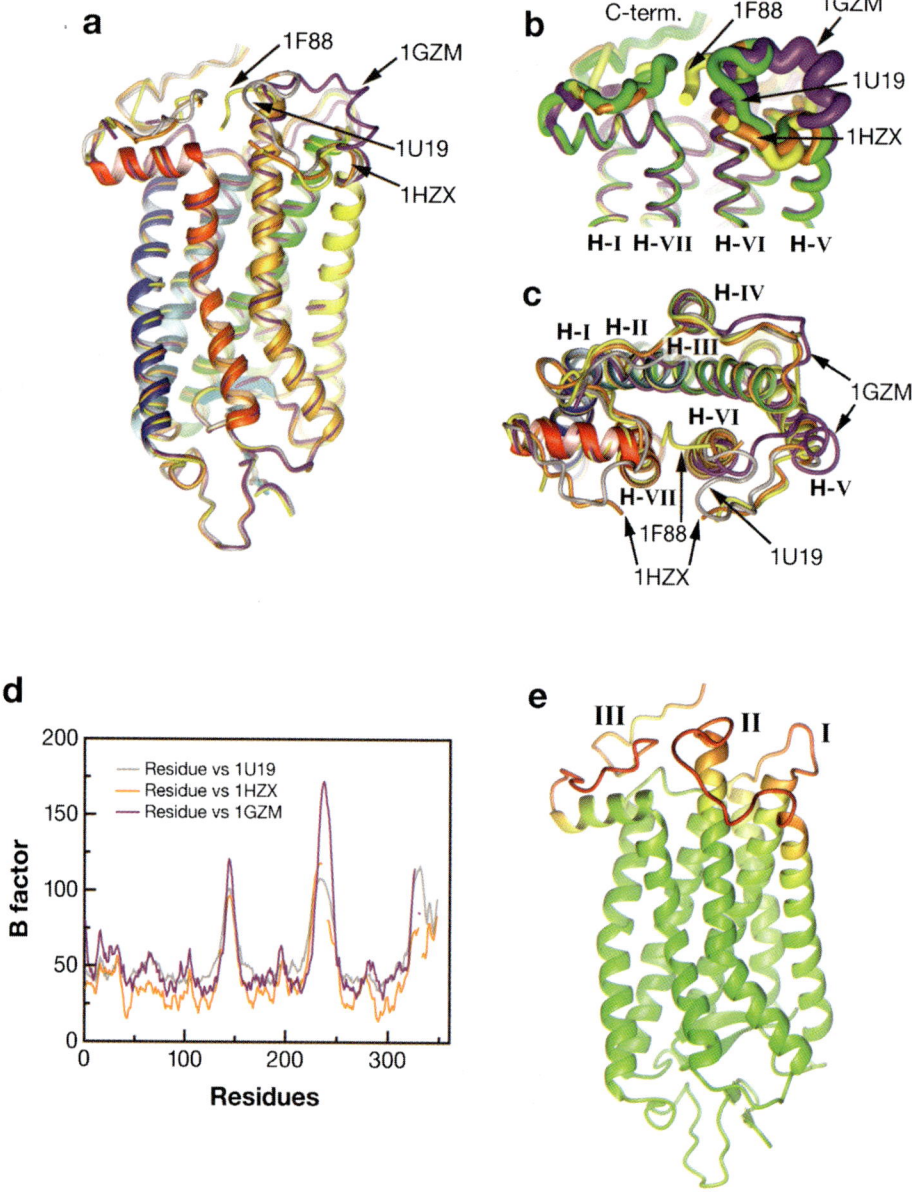

carbonyl group must be polarized, and water excluded or organized in the chromophore-binding site. This mechanism or alternative proposals require rigorous experimental analysis to reveal the correct mechanism.

Another challenging question in studies of the rhodopsin cycle is how the chromophore inserts in and out of the binding pocket. Hofmann and colleagues (89) proposed that, in addition to the retinylidene pocket (site I), there are two other retinoid-binding sites within opsin. Site II is an entrance site to the binding site involved in the uptake signal, and the exit site (site III) is occupied when retinal remains bound after its release from site I. This chromophore-channeling mechanism, movement of the chromophore from one to another site in a sequential way, is supported by the rhodopsin crystal structure, which unveiled two putative hydrophobic-binding sites. Importantly, this proposed mechanism enables a unidirectional process for the release of a photoisomerized chromophore and the uptake of newly synthesized 11-*cis*-retinal for the regeneration of rhodopsin. Arrestin, the capping protein that binds to activated phosphorylated rhodopsin and blocks G protein (transducin) activation, may play an important role in chromophore release. Farrens and colleagues (90) showed that arrestin and all-*trans*-retinal release are linked and require similar activation energies.

DIMERIZATION OF RHODOPSIN

Rhodopsin has been visualized in the native disk membranes by atomic force microscopy and transmission electron microscopy under various temperatures and other conditions (91–93). Rhodopsin was found to form rows of dimers containing densely organized higher-order structures. Moreover, native and denaturing sodium dodecyl sulfate polyacrylamide gel electrophoresis, chemical cross-linking, and proteolysis experiments corroborated that rhodopsin consists mainly of dimers and higher oligomers in disk membranes (94, 95). Rhodopsin dimerization was also observed by luminescence and fluorescence resonance energy transfer approaches, using fluorescently labeled rhodopsin samples in an asolectin liposome-reconstituted system (S.E. Mansoor, K. Palczewski, and D.L. Farrens, unpublished findings). Medina and colleagues (91) reported that rhodopsin and photoactivated rhodopsin retained a dimeric quaternary structure in n-dodecyl-β-maltoside. Jastrzebska et al. (95) found also that the dimeric structure is preserved in low concentrations of this detergent. Understandably, static techniques such as atomic force microscopy and transmission electron microscopy may not reflect the kinetic aspects of formation and disassembly of these higher-order structures.

Figure 6

Comparison of the current rhodopsin structures. There are currently five crystallographic entries for rhodopsin in the Protein Data Bank (PDB). The structures deposited under the accession number 1F88, 1HZX, 1GZM, and 1U19 are superimposed. Accession number 1F88 (*yellow thread*), 1HZX (*orange*), 1GZM (*purple*), and 1U19 (*gray*) are represented in the cartoon. Entries 1F88, 1HZX, 1L9H, and 1U19 are for a tetragonal crystal obtained by very similar methods. Entry 1U19 is at the highest resolution reported, 2.2 Å. Entry 1GZM is for a trigonal crystal form obtained in a different condition than the other listed crystals. (*a*) Side view. (*b*) Side view, with a close-up of the cytoplasmic region. (*c*) View from the cytoplasmic side. (*d,e*) A plot (*d*) and three-dimensional representation (*e*) of the B factor for rhodopsin structures from three data sets. The B factor is also known as the temperature factor or Debye-Waller factor and describes the degree to which the electron density is spread out, indicating the static or dynamic mobility of an atom or incorrectly built models. In panel *d*, the orange line represents 1HZX, the purple line represents 1GZM, and the gray line represents 1U19. In panel *e*, spectral grading of the B factor, green represents a low B factor, and red indicates the highest B factor. Note that the loop II is incomplete in 1HZX because of ambiguity in the electron density, and this region in the 1GZM set has the highest B factor.

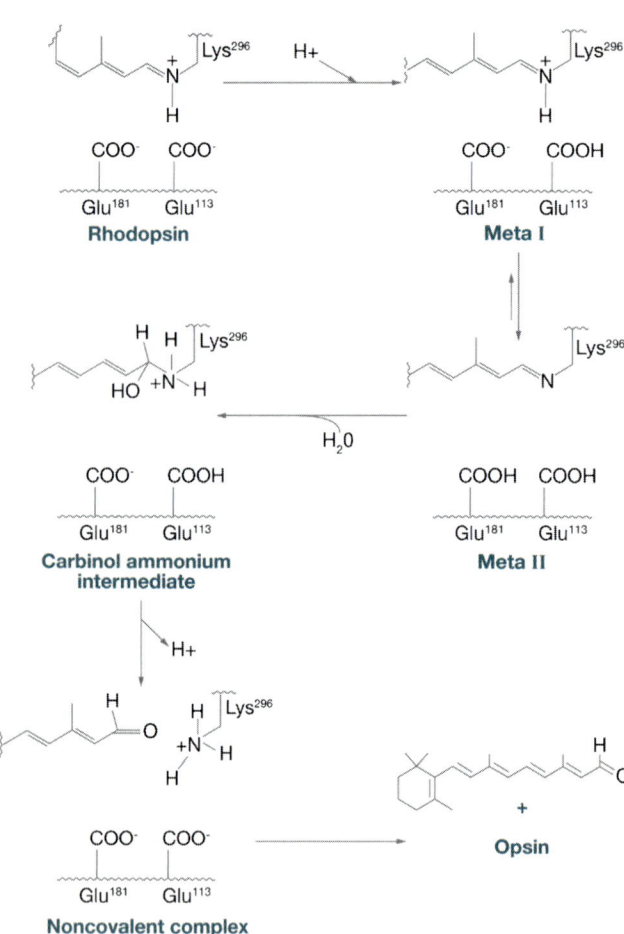

Figure 7

Hydrolysis of the all-*trans*-retinylidene chromophore and regeneration of rhodopsin with newly synthesized 11-*cis*-retinal. (*a*) The scheme of the retinylidene group hydrolysis. The role of Glu^{181} and Glu^{113} is hypothetical. Note that Glu^{181} is protonated in rhodopsin (71). (*b*) Formation of rhodopsin. Polarization of the carbonyl group of 11-*cis*-retinal and deprotonation of the Schiff-base group is required before the Schiff base can be formed.

Reconstitution of rhodopsin into two-dimensional crystals produces dimers where both rhodopsin molecules are correctly oriented, but the dimers contact other dimers that are rotated 180° along their long axis in their orientation (for example, see 79–81). The dimerization of rhodopsin can explain the autosomal dominant character of rhodopsin mutants, e.g., P23H rhodopsin. A cell line expressing P23H mutant rhodopsin retains both the mutant and wild-type proteins in intracellular inclusion bodies (109).

In a first approximation, these results are in conflict with a model of rapidly diffusing and rotating rhodopsin in homogenous fluid disk membranes and lack of any symmetry within ROS as determined by low-resolution neutron diffraction (96–100). In addition, the concept of rapidly diffusing monomeric molecules is also at variance with the dimeric forms of other GPCRs (101, 102), although diffusible dimers could be compatible with these recent structural and biochemical results and earlier biophysical measurements. Oligomerization of GPCRs has been a very active research area recently (102–107), resulting in strong evidence that the higher-order structures play a central role in GPCR signal transduction and desensitization (108).

INTERACTION OF RHODOPSIN WITH G PROTEIN

A model was generated for the interaction of photoactivated rhodopsin with G protein (110). This model, the so-called IV-V arrangement of rhodopsin in the membranes (**Figure 8***a*), is based on a structure of rhodopsin determined from X-ray crystallography and on the dimensions of rhodopsin found by atomic force microscopy in the native ROS membrane (111). The model proposes that rhodopsin molecules in the dimer contact each other via transmembrane helices IV-V. Activation of the dimer is accompanied by changes within helix VI and the NPXXY region of only one rhodopsin molecule within the dimer (72). Particularly interesting is the interaction model of photoactivated rhodopsin with transducin. This model of G protein activation considers the size of partner proteins, structural constraints, and organization of GPCRs in the membranes. The N- and C-terminal regions of transducin's α-subunit are engaged in the interaction with photoactivated rhodopsin (112–114). Only a narrow region of transducin containing hydrophobic posttranslational modification of α- (myristoylation) and γ-subunits (farnesylation) anchors this protein to the membrane, and the remaining protein surface is available to interact with rhodopsin molecules (115). The C-terminal tail binds to the inner face of helix VI in an activation-dependent manner (114) in a configuration (116), consistent with our model. This model, for which a short movie is available (106), reveals structural details about the critically important interface between a GPCR and a G protein. The IV-V model suggests that the view of GPCR signaling of oligomers of G protein and the receptors put forward by Rodbell (117, 118) are consistent with the structural evidence. Although not every detail may withstand experimental scrutiny, the general concept is probably correct. Our model is in substantial agreement with the transactivation of family A GPCRs, employing fusion proteins between active and inactive receptors and G proteins (119) and the identified pentameric complex between dimeric leukotriene B4 receptor BLT1 and heterotrimeric Gi (120). The studies on the chimeric rhodopsin/β2-adrenergic receptor clearly confirm that specificity for G proteins is confined to the cytoplasmic surface. In the chimera, the surface of rhodopsin or β2-adrenergic receptor undergoes changes as a result of chromophore isomerization and recruits the G protein (121). These studies are reminiscent of the innovative and insightful work of Kobilka et al. on chimeras of adrenergic receptors (122), which led to the

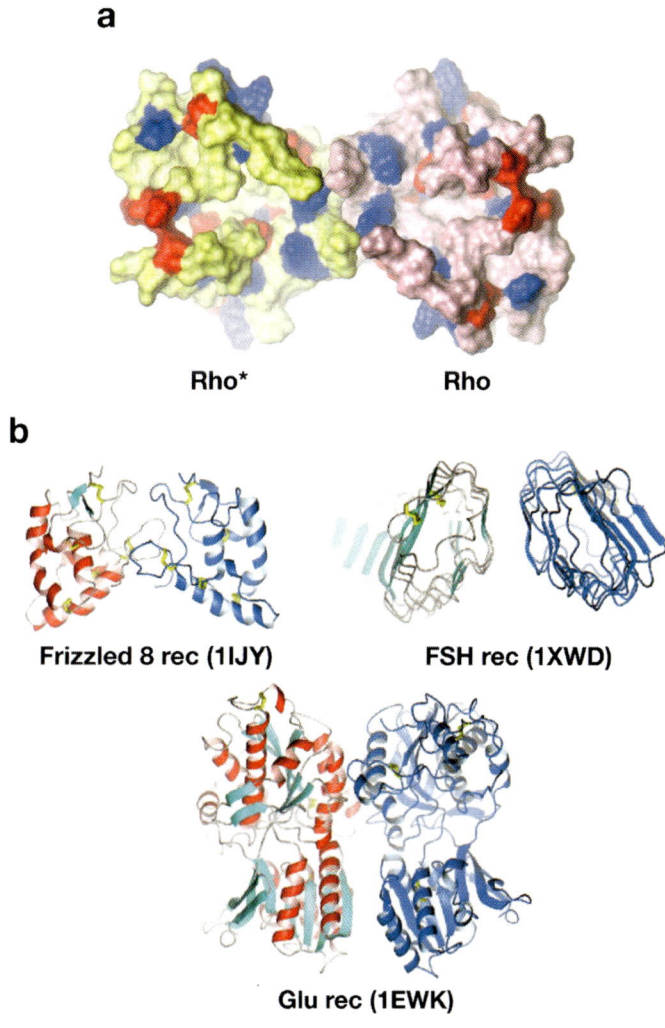

Figure 8

Models of GPCR dimerization. (*a*) Top view from the cytoplasmic side of the rhodopsin dimer. The model was generated by Dr. S. Filipek using structural constraints of rhodopsin and experimental data obtained by atomic force microscopy on the organization of rhodopsin in native membranes (33, 92, 93, 110, 111). This model is in agreement with cross-linking experiments (94). Photoactivated rhodopsin is depicted in yellow (Rho*), and rhodopsin is shown in pink. The acidic residues are shown in red and basic residues in blue. This cytoplasmic surface is involved in the interaction with G protein transducin. (*b*) The crystal structures of extracellular domains of different GPCRs. Crystal structures of the extracellular domains of the frizzled 8 receptor, the follicle-stimulating hormone (FSH) receptor, and the Glu receptor revealed that the extracellular domains formed a dimer. These structures may represent a physiological dimer that would stabilize the transmembrane domain and result in a dimeric platform for interaction with G proteins and other partner proteins. Protein Data Bank accession numbers are shown in parentheses. Panels *a* and *b* are not drawn to the same scale.

identification of major determinants of ligand and G protein specificities.

CONCLUSIONS AND PERSPECTIVES

Studies on rhodopsin continue at a rapid pace, as is evident in this account of recent progress. The key questions that require further investigation include the following:

1. What are the changes in structure that transform rhodopsin from its inactive to active conformation capable of interacting with partner proteins?
2. What are the structures of complexes between photoactivated rhodopsin and G protein transducin, and between a photoactivated phosphorylated monomer or dimer of rhodopsin and arrestin or its spliced form, p^{44}?
3. What is the role of GPCR dimerization in signaling and desensitization?
4. What is the mechanism of all-*trans*-retinal's release from opsin and regeneration of rhodopsin with 11-*cis*-retinal?
5. How does the cell transport and degradation of rhodopsin?

This is an ambitious set of goals that will keep rhodopsin studies at the cutting edge of GPCR research.

An image of the photoactivated structure of rhodopsin appears to be achievable by X-ray crystallography. The question that will inevitably need to be answered is whether this crystal structure will accurately reflect the conformation of photoactivated rhodopsin in membranes. Possibly, activated rhodopsin forms a constellation of conformations (106). Obtaining a high-resolution structure of the complexes between rhodopsin and its partner proteins transducin and arrestin is a paramount challenge, but it will yield major insight into how rhodopsin and other GPCRs work. What is the role of rhodopsin dimerization during signaling, biosynthesis in ER and ROS formation, and degradation? Dimerization of rhodopsin, like other GPCRs, may positively or negatively modulate G protein coupling (see for example 106 and 123). Because of the transient nature of the photoactivated rhodopsin-rhodopsin kinase complex, an image of this complex will be achieved when breakthroughs in biochemical techniques allow stabilization of the complex.

Great progress is anticipated in understanding the mechanism of rhodopsin regeneration and photoisomerized chromophore release. The methodology is well developed (e.g., see 89, 124, and 125), and fast kinetic recording devices are available. Activation of a single rhodopsin molecule reliably triggers the enzymatic events of the phototransduction pathway. A precise structure and kinetic base model of a single molecular event is needed. Moreover, at such low bleaches, a specific tunneling of the chromophore from one binding site to another must take place in order for it to be reduced by dehydrogenase, or for 11-*cis*-retinal to locate and bind to bleached opsin if the regeneration takes place under such conditions. The dehydrogenase responsible for reduction of all-*trans*-retinal is under investigation (e.g., see 126 for recent discussion).

The way in which the transmembrane core and extracellular plug structure of rhodopsin is held together will provide clues as to why so many mutations in this region cause rod degeneration (127). Some of these mutations may cause problems with membrane insertion or with maturation of rhodopsin during biosynthesis, and some may lead to opsin instability or to novel conformations. For example, the Thr4Arg mutation destabilizes opsin (128), causing light-dependent degeneration of the retina (129), and the Thr94Ile mutation is linked to thermal instability (130). Interestingly, a pharmacological chaperone may stabilize mutant proteins (131–133), a phenomenon which must be explored further in animal models of retinitis pigmentosa. Opsin unfolding can be studied by atomic force spectroscopy, as was exemplified for bacteriorhodopsin (134).

The biological lifetime of rhodopsin was previously impossible to study because rod cells, which are postmitotic neurons, cannot be cultured while maintaining the proper morphology of a tight link between photoreceptors and the adjacent retinal pigment epithelium, essential for their functioning (12). These problems have now been overcome by two major technical innovations, in vivo two-photon fluorescent microscopy, which can noninvasively penetrate the sclera of the eye (135, 136), and genetically engineered mice (137). Green fluorescent protein-tagged rhodopsin can be now traced in truly in vivo conditions.

Progress in structure determination of the extracellular domain of several GPCRs (**Figure 8b**) illustrates that, although the general topology of GPCRs is conserved, the extracellular domain has evolved to provide the most specific platform for activation by agonists, and the cytoplasmic domains may contain common structural features because of their interaction with structurally conserved G proteins, arrestins, and GPCR kinases. The cytoplasmic surfaces allow individual receptors to be differentially and selectively regulated. Even though rhodopsin is a prototypical model system for the other GPCRs, progress on the structure of other receptors, including transmembrane domains, will add a new molecular understanding of receptor function. Thus, the study of rhodopsin and other GPCRs is poised for an exciting expansion in which many concepts will be challenged and new ones will emerge, ultimately leading to a fundamental understanding of the process of cellular signaling.

Two-photon fluorescent microscopy: imaging with the near-simultaneous absorption of two photons by a molecule

SUMMARY POINTS

1. The structure of rhodopsin provides the fundamental basis for understanding how this G protein works. The importance of rhodopsin arises from its primary role in vision and also from being part of a large family of cell surface receptors termed G protein–coupled receptors or TM7 receptors.

2. New X-ray and NMR data provide additional detailed information on the conformation of the chromophore, alternative loop conformation, and the first view of the Meta I rhodopsin intermediate, as well as additional molecular particulars of rhodopsin structure and function.

3. Like most or all other G protein–coupled receptors, rhodopsin displays a propensity to oligomerize. This property appears to be fundamental in the function and interaction of these receptors with their partner proteins.

4. New details are emerging on how rhodopsin interacts with G protein transducin. However, we are still far from having a comprehensive molecular picture of the coupling of these two proteins.

5. An extensive list of additional challenges to understanding how rhodopsin works is presented. Our understanding of many basic features of rhodopsin in the context of the rod cell awaits further intellectual and technical developments.

FUTURE ISSUES TO BE RESOLVED

1. The changes in structure that transform rhodopsin from its inactive to active conformation capable of interacting with partner proteins should be determined.

2. High-resolution structures of the complex formed by the photoactivated monomer or dimer of rhodopsin and G protein transducin and the complex formed by the photoactivated phosphorylated monomer or dimer of rhodopsin and arrestin (or its splice form) should be identified.

3. The role of GPCR dimerization and its functions during signaling and desensitization should be explained.

4. The mechanism of all-*trans*-retinal's release and regeneration as 11-*cis*-retinal should be identified.

5. The cell biological cycle of rhodopsin, including synthesis, intracellular transport, phagocytosis, and degradation, should be determined.

ACKNOWLEDGMENTS

I thank Dr. Slawomir Filipek (Warsaw, Poland) for help in the preparation of many figures, of which only a few are used in this publication; Dr. Yan Liang for electron microscope images of the retina; Dr. David Lodowski, Dr. Kevin Ridge, and Dr. Jack S. Saari for comments on the manuscript; Rebecca Birdsong for contributions to side bars and to preparation of the manuscript; and the members of my laboratory for valuable comments. K.P. was supported by National Institutes of Health grant EY09339.

LITERATURE CITED

1. Filipek S, Teller DC, Palczewski K, Stenkamp R. 2003. *Annu. Rev. Biophys. Biomol. Struct.* 32:375–97
2. Bhandawat V, Reisert J, Yau KW. 2005. *Science* 308:1931–34
3. Minke B, Cook B. 2002. *Physiol. Rev.* 82:429–72
4. Heck M, Hofmann KP. 2001. *J. Biol. Chem.* 276:10000–9
5. Leskov IB, Klenchin VA, Handy JW, Whitlock GG, Govardovskii VI, et al. 2000. *Neuron* 27:525–37
6. Pierce KL, Premont RT, Lefkowitz RJ. 2002. *Nat. Rev. Mol. Cell Biol.* 3:639–50
7. Lefkowitz RJ. 2004. *Trends Pharmacol. Sci.* 25:413–22
8. Fredriksson R, Schioth HB. 2005. *Mol. Pharmacol.* 67:1414–25
9. Takeda S, Kadowaki S, Haga T, Takaesu H, Mitaku S. 2002. *FEBS Lett.* 520:97–101
10. Mirzadegan T, Benko G, Filipek S, Palczewski K. 2003. *Biochemistry* 42:2759–67
11. Dahl SG, Sylte I. 2005. *Basic Clin. Pharmacol. Toxicol.* 96:151–55
12. Doggrell SA. 2004. *Drug News Perspect.* 17:615–32
13. Bjenning C, Al-Shamma H, Thomsen W, Leonard J, Behan D. 2004. *Curr. Opin. Investig. Drugs* 5:1051–62
14. Foord SM, Bonner TI, Neubig RR, Rosser EM, Pin JP, et al. 2005. *Pharmacol. Rev.* 57:279–88

15. McBee JK, Palczewski K, Baehr W, Pepperberg DR. 2001. *Prog. Retin. Eye Res.* 20:469–529
16. Filipek S, Stenkamp RE, Teller DC, Palczewski K. 2003. *Annu. Rev. Physiol.* 65:851–79
16a. Dixon RA, Kobilka BK, Strader DJ, Benovic JL, Dohlman HG, et al. 1986. *Nature* 321:75–79
17. Ridge KD, Abdulaev NG, Sousa M, Palczewski K. 2003. *Trends Biochem. Sci.* 28:479–87
18. Okada T, Palczewski K. 2001. *Curr. Opin. Struct. Biol.* 11:420–26
19. Okada T, Ernst OP, Palczewski K, Hofmann KP. 2001. *Trends Biochem. Sci.* 26:318–24
20. Teller DC, Okada T, Behnke CA, Palczewski K, Stenkamp RE. 2001. *Biochemistry* 40:7761–72
21. Hubbell WL, Altenbach C, Hubbell CM, Khorana HG. 2003. *Adv. Protein Chem.* 63:243–90
22. Sakmar TP. 1998. *Prog. Nucleic Acid Res. Mol. Biol.* 59:1–34
23. Sakmar TP. 2002. *Curr. Opin. Cell Biol.* 14:189–95
24. Sakmar TP, Menon ST, Marin EP, Awad ES. 2002. *Annu. Rev. Biophys. Biomol. Struct.* 31:443–84
25. Abdulaev NG. 2003. *Trends Biochem. Sci.* 28:399–402
26. Teller DC, Stenkamp RE, Palczewski K. 2003. *FEBS Lett.* 555:151–59
27. Molday RS. 1998. *Investig. Ophthalmol. Vis. Sci.* 39:2491–13
28. Polans A, Baehr W, Palczewski K. 1996. *Trends Neurosci.* 19:547–54
29. Baylor DA, Lamb TD, Yau KW. 1979. *J. Physiol.* 288:613–34
30. Sampath AP, Rieke F. 2004. *Neuron* 41:431–43
31. Molday RS, Molday LL. 1987. *J. Cell Biol.* 105:2589–601
32. Carter-Dawson LD, LaVail MM. 1979. *J. Comp. Neurol.* 188:245–62
33. Liang Y, Fotiadis D, Maeda T, Maeda A, Modzelewska A, et al. 2004. *J. Biol. Chem.* 279:48189–96
34. Jeon CJ, Strettoi E, Masland RH. 1998. *J. Neurosci.* 18:8936–46
35. Penn JS, Williams TP. 1986. *Exp. Eye Res.* 43:915–28
36. Lem J, Krasnoperova NV, Calvert PD, Kosaras B, Cameron DA, et al. 1999. *Proc. Natl. Acad. Sci. USA* 96:736–41
37. Humphries MM, Rancourt D, Farrar GJ, Kenna P, Hazel M, et al. 1997. *Nat. Genet.* 15:216–19
38. Sung CH, Makino C, Baylor D, Nathans J. 1994. *J. Neurosci.* 14:5818–33
39. Tam BM, Moritz OL, Hurd LB, Papermaster DS. 2000. *J. Cell Biol.* 151:1369–80
40. Deretic D, Schmerl S, Hargrave PA, Arendt A, McDowell JH. 1998. *Proc. Natl. Acad. Sci. USA* 95:10620–25
41. Deretic D, Williams AH, Ransom N, Morel V, Hargrave PA, Arendt A. 2005. *Proc. Natl. Acad. Sci. USA* 102:3301–6
42. Moritz OL, Tam BM, Hurd LL, Peranen J, Deretic D, Papermaster DS. 2001. *Mol. Biol. Cell* 12:2341–51
43. Frederick JM, Krasnoperova NV, Hoffmann K, Church-Kopish J, Ruther K, et al. 2001. *Investig. Ophthalmol. Vis. Sci.* 42:826–33
44. Deretic D, Traverso V, Parkins N, Jackson F, de Turco EBR, Ransom N. 2004. *Mol. Biol. Cell* 15:359–70
45. Redmond TM, Yu S, Lee E, Bok D, Hamasaki D, et al. 1998. *Nat. Genet.* 20:344–51
46. Batten ML, Imanishi Y, Maeda T, Tu DC, Moise AR, et al. 2004. *J. Biol. Chem.* 279:10422–32
47. Jin S, Cornwall MC, Oprian DD. 2003. *Nat. Neurosci.* 6:731–35

48. Jager S, Palczewski K, Hofmann KP. 1996. *Biochemistry* 35:2901–8
49. Melia TJ Jr, Cowan CW, Angleson JK, Wensel TG. 1997. *Biophys. J.* 73:3182–91
50. Woodruff ML, Wang Z, Chung HY, Redmond TM, Fain GL, Lem J. 2003. *Nat. Genet.* 35:158–64
51. Lem J, Fain GL. 2004. *Trends Mol. Med.* 10:150–57
52. Okada T, Takeda K, Kouyama T. 1998. *Photochem. Photobiol.* 67:495–99
53. Okada T, Le Trong I, Fox BA, Behnke CA, Stenkamp RE, Palczewski K. 2000. *J. Struct. Biol.* 130:73–80
54. Oprian DD, Molday RS, Kaufman RJ, Khorana HG. 1987. *Proc. Natl. Acad. Sci. USA* 84:8874–78
55. Maeda T, Imanishi Y, Palczewski K. 2003. *Prog. Retin. Eye Res.* 22:417–34
56. Crescitelli F. 1977. *Arch. Ophthalmol.* 95:1766
57. Marmor MF, Martin LJ. 1978. *Surv. Ophthalmol* 22:279–85
58. Kuksa V, Imanishi Y, Batten M, Palczewski K, Moise AR. 2003. *Vis. Res.* 43:2959–81
59. Peteanu LA, Schoenlein RW, Wang Q, Mathies RA, Shank CV. 1993. *Proc. Natl. Acad. Sci. USA* 90:11762–66
60. Rohrig UF, Guidoni L, Laio A, Frank I, Rothlisberger U. 2004. *J. Am. Chem. Soc.* 126:15328–29
61. Arnis S, Hofmann KP. 1993. *Proc. Natl. Acad. Sci. USA* 90:7849–53
62. Tsukamoto H, Terakita A, Shichida Y. 2005. *Proc. Natl. Acad. Sci. USA* 102:6303–8
63. Koyanagi M, Terakita A, Kubokawa K, Shichida Y. 2002. *FEBS Lett.* 531:525–28
64. Palczewski K, Kumasaka T, Hori T, Behnke CA, Motoshima H, et al. 2000. *Science* 289:739–45
65. Okada T, Sugihara M, Bondar AN, Elstner M, Entel P, Buss V. 2004. *J. Mol. Biol.* 342:571–83
66. Okada T, Fujiyoshi Y, Silow M, Navarro J, Landau EM, Shichida Y. 2002. *Proc. Natl. Acad. Sci. USA* 99:5982–87
67. Li J, Edwards PC, Burghammer M, Villa C, Schertler GF. 2004. *J. Mol. Biol.* 343:1409–38
68. Stojanovic A, Stitham J, Hwa J. 2004. *J. Biol. Chem.* 279:35932–41
69. del Valle LJ, Ramon E, Canavate X, Dias P, Garriga P. 2003. *J. Biol. Chem.* 278:4719–24
70. Bourne HR, Meng EC. 2000. *Science* 289:733–34
71. Periole X, Ceruso MA, Mehler EL. 2004. *Biochemistry* 43:6858–64
72. Fritze O, Filipek S, Kuksa V, Palczewski K, Hofmann KP, Ernst OP. 2003. *Proc. Natl. Acad. Sci. USA* 100:2290–95
73. Kim JM, Altenbach C, Kono M, Oprian DD, Hubbell WL, Khorana HG. 2004. *Proc. Natl. Acad. Sci. USA* 101:12508–13
74. Ebrey T, Koutalos Y. 2001. *Prog. Retin. Eye Res.* 20:49–94
75. Bartl FJ, Fritze O, Ritter E, Herrmann R, Kuksa V, et al. 2005. *J. Biol. Chem.* 280:34259–67
76. Yan EC, Kazmi MA, Ganim Z, Hou JM, Pan D, et al. 2003. *Proc. Natl. Acad. Sci. USA* 100:9262–67
77. Kusnetzow AK, Dukkipati A, Babu KR, Ramos L, Knox BE, Birge RR. 2004. *Proc. Natl. Acad. Sci. USA* 101:941–46
78. Terakita A, Koyanagi M, Tsukamoto H, Yamashita T, Miyata T, Shichida Y. 2004. *Nat. Struct. Mol. Biol.* 11:284–89
79. Krebs A, Edwards PC, Villa C, Li J, Schertler GF. 2003. *J. Biol. Chem.* 278:50217–25
80. Ruprecht JJ, Mielke T, Vogel R, Villa C, Schertler GF. 2004. *EMBO J.* 23:3609–20

81. Vogel R, Ruprecht J, Villa C, Mielke T, Schertler GF, Siebert F. 2004. *J. Mol. Biol.* 338:597–609
82. Creemers AF, Kiihne S, Bovee-Geurts PH, DeGrip WJ, Lugtenburg J, de Groot HJ. 2002. *Proc. Natl. Acad. Sci. USA* 99:9101–6
83. Lugtenburg J. 1996. *Eur. J. Clin. Nutr.* 50(Suppl. 3):S17–20
84. Spooner PJ, Sharples JM, Goodall SC, Seedorf H, Verhoeven MA, et al. 2003. *Biochemistry* 42:13371–78
85. Spooner PJ, Sharples JM, Goodall SC, Bovee-Geurts PH, Verhoeven MA, et al. 2004. *J. Mol. Biol.* 343:719–30
86. Patel AB, Crocker E, Eilers M, Hirshfeld A, Sheves M, Smith SO. 2004. *Proc. Natl. Acad. Sci. USA* 101:10048–53
87. Borhan B, Souto ML, Imai H, Shichida Y, Nakanishi K. 2000. *Science* 288:2209–12
88. Patel AB, Crocker E, Reeves PJ, Getmanova EV, Eilers M, et al. 2005. *J. Mol. Biol.* 347:803–12
89. Schadel SA, Heck M, Maretzki D, Filipek S, Teller DC, et al. 2003. *J. Biol. Chem.* 278:24896–903
90. Sommer ME, Smith WC, Farrens DL. 2005. *J. Biol. Chem.* 280:6861–71
91. Medina R, Perdomo D, Bubis J. 2004. *J. Biol. Chem.* 279:39565–73
92. Fotiadis D, Liang Y, Filipek S, Saperstein DA, Engel A, Palczewski K. 2003. *Nature* 421:127–28
93. Fotiadis D, Liang Y, Filipek S, Saperstein DA, Engel A, Palczewski K. 2004. *FEBS Lett.* 564:281–88
94. Suda K, Filipek S, Palczewski K, Engel A, Fotiadis D. 2004. *Mol. Membr. Biol.* 21:435–46
95. Jastrzebska B, Maeda T, Zhu L, Fotiadis D, Filipek S, et al. 2004. *J. Biol. Chem.* 279:54663–75
96. Liebman PA, Parker KR, Dratz EA. 1987. *Annu. Rev. Physiol.* 49:765–91
97. Liebman PA, Entine G. 1974. *Science* 185:457–59
98. Poo M, Cone RA. 1974. *Nature* 247:438–41
99. Cone RA. 1972. *Nat. New Biol.* 236:39–43
100. Saibil H, Chabre M, Worcester D. 1976. *Nature* 262:266–70
101. Angers S, Salahpour A, Bouvier M. 2002. *Annu. Rev. Pharmacol. Toxicol.* 42:409–35
102. Terrillon S, Bouvier M. 2004. *EMBO Rep.* 5:30–34
103. Milligan G, Ramsay D, Pascal G, Carrillo JJ. 2003. *Life Sci.* 74:181–88
104. Angers S, Salahpour A, Bouvier M. 2001. *Life Sci.* 68:2243–50
105. Dean MK, Higgs C, Smith RE, Bywater RP, Snell CR, et al. 2001. *J. Med. Chem.* 44:4595–614
106. Park PS, Filipek S, Wells JW, Palczewski K. 2004. *Biochemistry* 43:15643–56
107. Javitch JA. 2004. *Mol. Pharmacol.* 66:1077–82
108. Park PS, Palczewski K. 2005. *Proc. Natl. Acad. Sci. USA* 102:8793–94
109. Rajan RS, Kopito RR. 2005. *J. Biol. Chem.* 280:1284–91
110. Filipek S, Krzysko KA, Fotiadis D, Liang Y, Saperstein DA, et al. 2004. *Photochem. Photobiol. Sci.* 3:628–38
111. Liang Y, Fotiadis D, Filipek S, Saperstein DA, Palczewski K, Engel A. 2003. *J. Biol. Chem.* 278:21655–62
112. Natochin M, Gasimov KG, Moussaif M, Artemyev NO. 2003. *J. Biol. Chem.* 278:37574–81
113. Wang X, Kim SH, Ablonczy Z, Crouch RK, Knapp DR. 2004. *Biochemistry* 43:11153–62
114. Janz JM, Farrens DL. 2004. *J. Biol. Chem.* 279:29767–73

115. Zhang Z, Melia TJ, He F, Yuan C, McGough A, et al. 2004. *J. Biol. Chem.* 279:33937–45
116. Brabazon DM, Abdulaev NG, Marino JP, Ridge KD. 2003. *Biochemistry* 42:302–11
117. Rodbell M. 1995. *Biosci. Rep.* 15:117–33
118. Schlegel W, Kempner ES, Rodbell M. 1979. *J. Biol. Chem.* 254:5168–76
119. Carrillo JJ, Pediani J, Milligan G. 2003. *J. Biol. Chem.* 278:42578–87
120. Baneres JL, Parello J. 2003. *J. Mol. Biol.* 329:815–29
121. Kim JM, Hwa J, Garriga P, Reeves PJ, RajBhandary UL, Khorana HG. 2005. *Biochemistry* 44:2284–92
122. Kobilka BK, Kobilka TS, Daniel K, Regan JW, Caron MG, Lefkowitz RJ. 1988. *Science* 240:1310–16
123. Urizar E, Montanelli L, Loy T, Bonomi M, Swillens S, et al. 2005. *EMBO J.* 24:1954–64
124. Janz JM, Farrens DL. 2004. *J. Biol. Chem.* 279:55886–94
125. Heck M, Schadel SA, Maretzki D, Bartl FJ, Ritter E, et al. 2003. *J. Biol. Chem.* 278:3162–69
126. Maeda A, Maeda T, Imanishi Y, Kuksa V, Alekseev A, et al. 2005. *J. Biol. Chem.* 280:18822–32
127. Rader AJ, Anderson G, Isin B, Khorana HG, Bahar I, Klein-Seetharaman J. 2004. *Proc. Natl. Acad. Sci. USA* 101:7246–51
128. Zhu L, Jang GF, Jastrzebska B, Filipek S, Pearce-Kelling SE, et al. 2004. *J. Biol. Chem.* 279:53828–39
129. Cideciyan AV, Jacobson SG, Aleman TS, Gu D, Pearce-Kelling SE, et al. 2005. *Proc. Natl. Acad. Sci. USA* 102:5233–38
130. Ramon E, del Valle LJ, Garriga P. 2003. *J. Biol. Chem.* 278:6427–32
131. Noorwez SM, Kuksa V, Imanishi Y, Zhu L, Filipek S, et al. 2003. *J. Biol. Chem.* 278:14442–50
132. Saliba RS, Munro PM, Luthert PJ, Cheetham ME. 2002. *J. Cell Sci.* 115:2907–18
133. Li T, Sandberg MA, Pawlyk BS, Rosner B, Hayes KC, et al. 1998. *Proc. Natl. Acad. Sci. USA* 95:11933–38
134. Oesterhelt F, Oesterhelt D, Pfeiffer M, Engel A, Gaub HE, Muller DJ. 2000. *Science* 288:143–46
135. Imanishi Y, Batten ML, Piston DW, Baehr W, Palczewski K. 2004. *J. Cell Biol.* 164:373–83
136. Imanishi Y, Gerke V, Palczewski K. 2004. *J. Cell Biol.* 166:447–53
137. Chan F, Bradley A, Wensel TG, Wilson JH. 2004. *Proc. Natl. Acad. Sci. USA* 101:9109–14
138. Stone WL, Farnsworth CC, Dratz EA. 1979. *Exp. Eye Res.* 28:387–97
139. Calvert PD, Govardovskii VI, Krasnoperova N, Anderson RE, Lem J, Makino CL. 2001. *Nature* 411:90–94
140. Aveldano MI. 1995. *Arch. Biochem. Biophys.* 324:331–43
141. Davidson FF, Loewen PC, Khorana HG. 1994. *Proc. Natl. Acad. Sci. USA* 91:4029–33
142. Getmanova E, Patel AB, Klein-Seetharaman J, Loewen MC, Reeves PJ, et al. 2004. *Biochemistry* 43:1126–33
143. Kisselev OG, Downs MA, McDowell JH, Hargrave PA. 2004. *J. Biol. Chem.* 279:51203–7
144. Argos P, Rao JK, Hargrave PA. 1982. *Eur. J. Biochem.* 128:565–75
145. Hargrave PA, McDowell JH, Curtis DR, Wang JK, Juszczak E, et al. 1983. *Biophys. Struct. Mech.* 9:235–44
146. Kim JE, Tauber MJ, Mathies RA. 2001. *Biochemistry* 40:13774–78
147. Zimmermann K, Ritter E, Bartl FJ, Hofmann KP, Heck M. 2004. *J. Biol. Chem.* 279:48112–19

148. Vogel R, Siebert F, Mathias G, Tavan P, Fan G, Sheves M. 2003. *Biochemistry* 42:9863–74
149. Shichida Y, Imai H. 1998. *Cell Mol. Life Sci.* 54:1299–15
150. Ohguro H, Rudnicka-Nawrot M, Buczylko J, Zhao X, Taylor JA, et al. 1996. *J. Biol. Chem.* 271:5215–24
151. Wang ZY, Wen XH, Ablonczy Z, Crouch RK, Makino CL, Lem J. 2005. *J. Biol. Chem.* 280:24293–300

RELATED REVIEWS

1. Arshavsky VY, Lamb TD, Pugh EN Jr. 2002. *Annu. Rev. Physiol.* 64:153–87
2. Rattner A, Sun H, Nathans J. 1999. *Annu. Rev. Genet.* 33:89–131
3. Shi L, Javitch JA. 2002. *Annu. Rev. Pharmacol. Toxicol.* 42:437–67

Transmembrane Traffic in the Cytochrome $b_6 f$ Complex

William A. Cramer,[1] Huamin Zhang,[1] Jiusheng Yan,[2] Genji Kurisu,[3] and Janet L. Smith[4]

[1] Department of Biological Sciences, Purdue University, West Lafayette, Indiana 47907-2054; email: waclab@purdue.edu, hzhang@bilbo.bio.purdue.edu

[2] Department of Pharmacology, University of California Davis, California 95616-8635; email: jshyan@ucdavis.edu

[3] Department of Life Sciences, Graduate School of Arts and Sciences, University of Tokyo, Tokyo 153-8902, Japan; email: gkurisu@bio.c.u-tokyo.ac.jp

[4] Life Sciences Institute, University of Michigan, Ann Arbor, Michigan 48109-2216; email: janetsmith@umich.edu

Key Words

lipophilic quinone, electron, proton transfer, photosynthesis

Abstract

Crystal structures and their implications for function are described for the energy transducing hetero-oligomeric dimeric cytochrome $b_6 f$ complex of oxygenic photosynthesis from the thermophilic cyanobacterium, *Mastigocladus laminosus*, and the green alga, *Chlamydomonas reinhardtii*. The complex has a cytochrome b core and a central quinone exchange cavity, defined by the two monomers that are very similar to those in the respiratory cytochrome bc_1 complex. The pathway of quinol/quinone (Q/QH$_2$) transfer emphasizes the labyrinthine internal structure of the complex, including an 11 × 12 Å portal through which Q/QH$_2$, containing a 45-carbon isoprenoid chain, must pass. Three prosthetic groups are present in the $b_6 f$ complex that are not found in the related bc_1 complex: a chlorophyll (Chl) a, a β-carotene, and a structurally unique covalently bound heme that does not possess amino acid side chains as axial ligands. It is hypothesized that this heme, exposed to the cavity and a neighboring plastoquinone and close to the positive surface potential of the complex, can function in cyclic electron transport via anionic ferredoxin.

Contents

1. INTRODUCTION 770
2. OVERVIEW OF THE STRUCTURE 771
   - 2.1. Subunit Composition and Unique Prosthetic Groups 771
   - 2.2. Weakly Bound Protein Components of the Cytochrome b_6f Complex 772
   - 2.3. Dimer Organization 772
   - 2.4. Lipid Requirement for Crystallization of the *M. laminosus* Complex 774
   - 2.5. Asymmetry of Surface Potential 775
3. QUINOL OXIDATION SITE ... 775
   - 3.1. Oxidant-Induced Reduction .. 775
   - 3.2. Q_p Pocket 778
   - 3.3. The Meaning of the TDS Ring-Out Conformation 779
   - 3.4. Intermonomer Cross Talk 780
   - 3.5. Extrinsic Domains on the *p* Side: Cytochrome *f* and the Rieske ISP 780
4. QUINONE REDUCTION SITE (Q_n SITE) 781
   - 4.1. *n* Side of the Cycle for Quinone Turnover 781
   - 4.2. The Unique Heme *x* 783
   - 4.3. Cyclic Electron Transport 784
   - 4.4. Other Pathways for Heme *x* .. 785
5. BOUND PIGMENT MOLECULES 786

E_{m7}: midpoint oxidation-reduction potential at pH 7

bc_1-resp: cytochrome bc_1 complex of mitochondrial respiratory chain

DBMIB: 2,5-dimethyl-6-isopropyl-benzoquinone

1. INTRODUCTION

In oxygenic photosynthesis, the hetero-oligomeric cytochrome b_6f complex mediates electron transfer (ET) between the photosystem II (PSII) and photosystem I (PSI) reaction center complexes by oxidizing the relatively low-potential ($E_{m7} \approx +0.1$ V) lipophilic plastoquinol and reducing soluble plastocyanin ($E_{m7} = +0.35$ V) or cytochrome c_6 (1). These ET events are linked to proton translocation across the low-dielectric membrane. The net transfer of electrical charge contributes to a proton electrochemical potential gradient of approximately 250 mV, which is positive on the side of the membrane to which the proton flow is directed. Recent mechanistic studies of the cytochrome bc_1 complex from the mitochondrial respiratory (bc_1-resp) chain and photosynthetic bacteria have utilized structure information derived from X-ray diffraction analysis (2–8). Three-dimensional structures of the b_6f complex with the quinone (Q)-analogue inhibitor, tridecylstigmatellin (TDS), have been obtained at a resolution of 3.0 Å and 3.1 Å, respectively, from the thermophilic filamentous cyanobacterium, *Mastigocladus laminosus* (9), and the green alga, *Chlamydomonas reinhardtii* (10). Structures of the native complex and complex cocrystallized with the quinone analogue 2,5-dimethyl-6-isopropyl-benzoquinone (DBMIB) have also been obtained from *M. laminosus* at a resolution of 3.4 Å and 3.8 Å. Together with structures of the PSI (11) and PSII (12–16) reaction center complexes, obtained from cyanobacterial sources, a structure description has been completed of the three integral membrane protein complexes that sustain linear electron transport in oxygenic photosyntheisis (**Figure 1**). This structural detail has provided deeper insight into the pathways and mechanisms that govern electron and proton traffic across the photosynthetic membrane. The present discussion is focused on unique structure-function aspects associated with the b_6f complex, emphasizing the presence of three novel prosthetic groups discovered in the structure, and the internal binding sites of lipophilic quinones and Q-analogue inhibitors. The latter may provide markers to the traffic pattern of lipophilic quinone across the complex and the associated transmembrane charge transfer. Binding sites of Q-analogue inhibitors and quinone (quinol) in the bc_1 complex have been described and reviewed recently (17, 18). Some aspects of the b_6f structures have previously been reviewed (19–22).

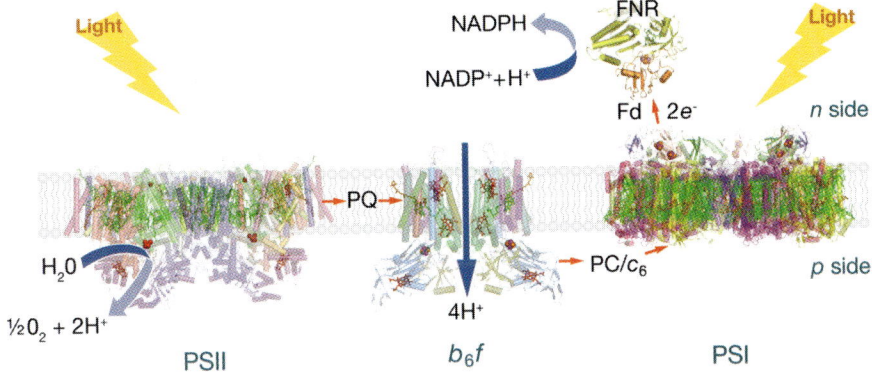

Figure 1

Cross section of the three hetero-oligomeric integral membrane protein complexes, whose structures have been solved by X-ray diffraction, that sustain linear ET from H_2O as the electron donor to $NADP^+$ as the terminal acceptor. All three integral membrane protein structures in this figure were derived from thermophilic cyanobacteria [PSI, Protein Data Bank identification number (PDB id): 1JB0; PSII, PDB id: 1FE1; b_6f, PDB id: 1VF5]. The b_6f complex was also solved from the green alga, *C. reinhardtii* (PDB id: 1Q90). Figure is modified from that in Reference 21.

2. OVERVIEW OF THE STRUCTURE

2.1. Subunit Composition and Unique Prosthetic Groups

The complexes from *M. laminosus* and *C. reinhardtii* contain the same number of polypeptide subunits, and their structures are very similar. This might not have been anticipated because the emergence of cyanobacteria and green algae in evolution is separated by a time interval of $\sim 10^9$ years (23). However, the sequence identity of the photosynthetic electron transport (Pet) protein A-D subunits between *C. reinhardtii* and *M. laminosus* is 58%, 80%, 80%, and 58%. In contrast, large changes in the quaternary structure of the bc_1 complex have occurred in evolution, so that the purple photosynthetic bacterium, *Rhodobacter capsulatus* (8) and the respiratory chain in yeast (6) contain 3 and 9 subunits, respectively. The complex from bovine (2–5) and avian mitochondria (3) is the largest bc_1 heterooligomer and contains 11 types of subunits.

The b_6f-TDS complex from the cyanobacterial and algal sources is considered to be a functional homodimer. Some structural asymmetry is shown by a difference of 12° in rotation and a 2-Å translation of the iron-sulfur protein (ISP) in the two monomers of the *M. laminosus* complex. The issue of symmetry is of interest because an asymmetry in the yeast bc_1 complex is indicated by a structure showing cytochrome c bound to only one monomer (24). Each monomer of the b_6f complex consists of eight polypeptide subunits: One subset of four large subunits, as originally defined by Hurt & Hauska (25), consists of the membrane bound c-type cytochrome f (PetA), cytochrome b_6 (PetB), the ISP (PetC), and subunit IV (PetD, suIV) that corresponds to the C-terminal half of bc_1 cytochrome b (26); the second small subset consists of four low-molecular-weight (3–4 kDa) subunits, PetG, PetL, PetM, and PetN, which form a "picket fence" type structure around the core of the large subunits. Each of these small hydrophobic subunits contains 1–2 positively charged residues on the n (stromal) side of the complex. Thus, the direction of their insertion into the membrane obeys the *cis*-positive rule for assembly of membrane proteins (27). The total molecular weight of the protein and covalently bound heme in the *M. laminosus* monomer is 108,500, as determined by electrospray mass spectrometry (28). The eight

Q/QH$_2$:
quinone/quinol

FNR: ferredoxin: NADP⁺ reductase

ISP(s): soluble domain of ISP

subunits of each monomer form a bundle of 13 transmembrane helices inside the membrane bilayer, four helices from cytochrome b_6, three from suIV, and one each from cytochrome f, the ISP, and the four small subunits (**Figure 2a, b**) (**Table 1**). Each monomer contains an electron and proton transfer chain consisting of six redox prosthetic groups (**Figure 2**), (a, b) the lumen- (p-) side [2Fe-2S] cluster of the ISP and covalently bound heme of cytochrome f; (c, d) the membrane-spanning noncovalently bound b-type hemes, b_p and b_n; (e) the n-side novel covalently bound heme x or heme c_i, and (f) plastoquinone near heme x in the *M. laminosus* structure.[1] The distances between redox prosthetic groups in the *M. laminosus* structure, which are similar in *C. reinhardtii*, are summarized in **Table 2**. The structures of the extrinsic domains of cytochrome f and the Rieske ISP in the *M. laminosus* structure were the same as those established previously for the soluble domain of cytochrome f in a higher plant (turnip), green alga (*C. reinhardtii*), and cyanobacterium, *Phormidium laminosus* (29–31) and in ISP from spinach thylakoid membranes (32).

Although the core (cytochrome b) structure of the b_6f complex and several aspects of function are similar to those of the bc_1-resp complex (22, 33) and purple photosynthetic bacteria (8, 33, 34), the presence of three additional and novel prosthetic groups in the b_6f complex is unique. These three groups, chlorophyll (Chl) a, β-carotene, and the covalently bound unique high-spin heme x, present at one copy per monomer, imply the existence of one or more functions that, in comparison with the bc_1 complex, are unique to oxygenic photosynthesis.

---

[1] "Heme x" is unique among protein-bound heme prosthetic groups because it does not have any axial ligands. An alternative notation, heme c_i, expresses the covalent linkage of the heme to the protein (i.e., "c-type" through a single thioether linkage). The subscripts i and o are ambiguous. They can be interpreted as the sides of proton input or as the "inside" and "outside" of the membrane. Because the unique heme in the b_6f complex is located on the outside of the oxygenic photosynthetic membrane, the i notation is confusing. A better notation would be heme c_n.

2.2. Weakly Bound Protein Components of the Cytochrome b_6f Complex

Other soluble n- or p-side proteins dock to the b_6f complex. Ferredoxin:NADP⁺ reductase (FNR) in the complex isolated from spinach thylakoids (35, 36) and the PetO subunit from *C. reinhardtii* (37), which may mediate state transitions, are examples of proteins that can be bound peripherally to the complex. Incomplete binding of such peripheral proteins to the purified complex would result in heterogeneity of the complex, which could explain why it has not yet been possible to crystallize the complex from higher plant chloroplasts.

2.3. Dimer Organization

The dimeric cytochrome b_6f complex is stabilized by contacts of the transmembrane helices of the two copies of cytochrome b_6 and by the TM helix of the ISP, whose soluble domain of ISP [ISP(s)] and transmembrane domains reside in different monomers. The interface between the two monomers contains facing cytochrome b aromatic residues Phe52, Phe56, and Phe189, which stabilize the dimeric state through van der Waals or π-π interactions and also form a possible ET bridge between the two hemes b_p that would allow ET "cross talk" between the monomers (38–40). One clear function of the dimer is to provide an intermonomer cavity where the lipophilic quinone(ol) [Q/QH$_2$] from the membrane bilayer can be concentrated. The two monomers form a protein-free central cavity (30 Å high, 15 Å deep, and 25 Å wide at its n-side base), which occupies a volume of ∼5500 Å³ (**Figure 3**). One cavity is on each side of the transmembrane intermonomer interface, as previously described for bc_1-resp (2–4, 6, 8). Electron density near heme x on the n side of the cavity was interpreted to be a bound plastoquinone molecule, thus further implicating the open space around it as a quinone exchange cavity,

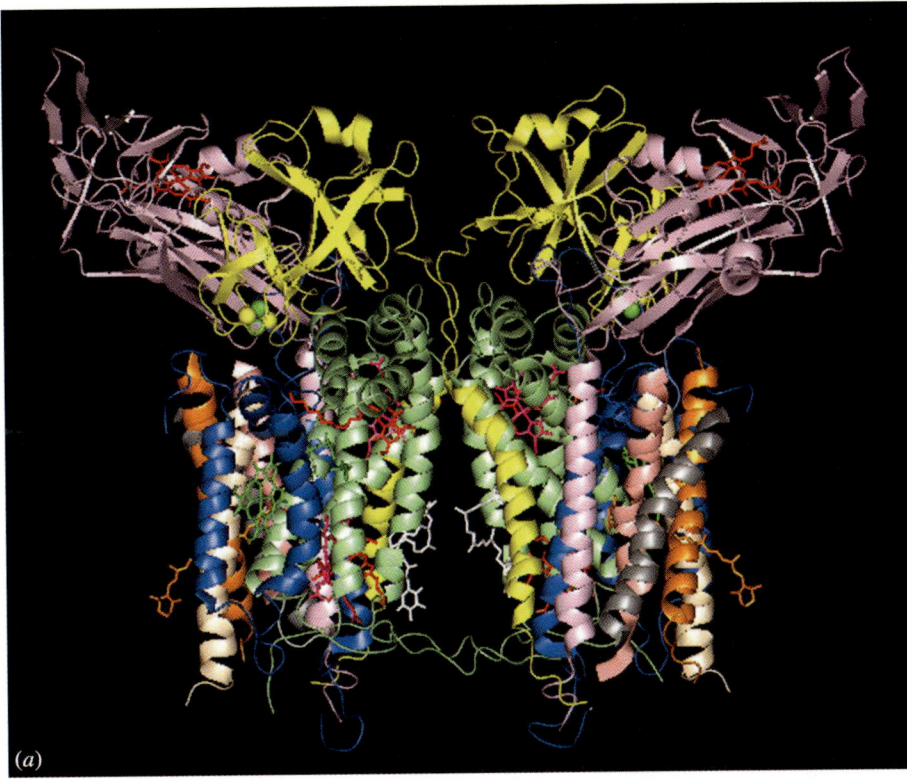

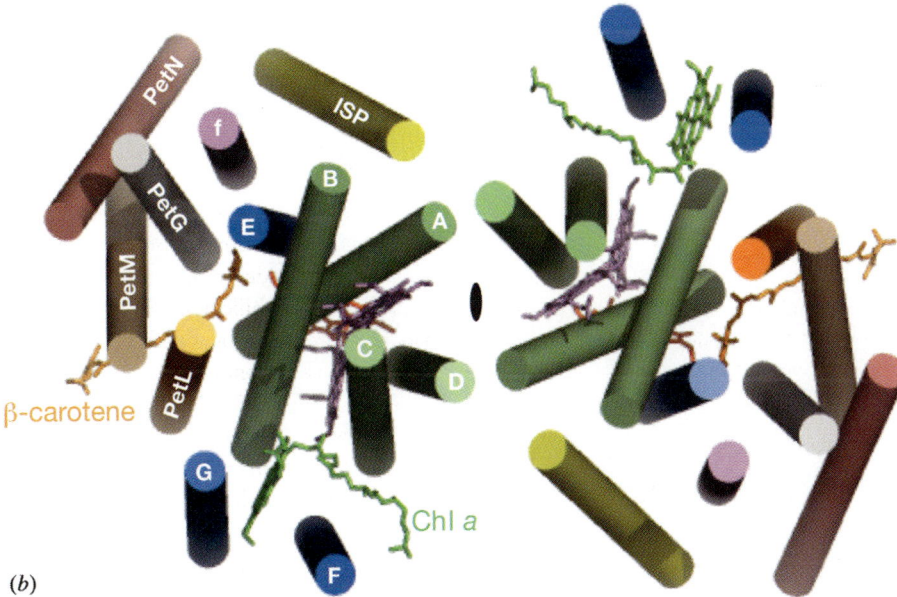

Figure 2

(*a*) Side view of the b_6f complex showing subunit composition and bound cofactors. The color code is as follows: cytochrome b_6 (*green*); subunit IV (*blue*); cytochrome *f* (*pink*); ISP (*yellow*); the four small subunits Pet G, L, M and N are colored in salmon, wheat, orange and gray, respectively. Bound cofactors are shown as sticks: heme *f* and *b* (*pink*); heme *x* and TDS (*red*); Chl *a* (*green*) and β-carotene (*orange*). (*b*) View from the *p* side of the 26 transmembrane helices along an axis normal to membrane plane, showing only transmembrane helices and prosthetic groups inside the complex. Color codes are as in (*a*), except: two *b* hemes (*pink*), heme *x* (*red*), Chl (*green*), and β-carotene (*orange*). Transmembrane helices of cytochrome b_6 and subunit IV are labeled A-D and E-G, respectively.

Table 1 Subunit composition and properties of *M. laminosus* cytochrome b_6f complex

Subunit	Measured mass (Da)[a]	pI	E_m (mV)
Cytochrome f	32,273	6.7	350–380
Cytochrome b_6	24,712	9.0	-50 (b_n), $-50 - -150$, (b_p), $+100$ (heme x)
ISP	19,295	6.8	300–320
suIV	17,528	8.1	NA[b]
PetG	4,058	4.5	NA
PetM	3,841	4.3	NA
PetL	3,530	10.2	NA
PetN	3,304	5.7	NA

[a]Reference 28.
[b]NA, not applicable.

DOPC: dioleolylphosphatidylcholine

which transfers and exchanges quinone(ol) to and from the Q/QH$_2$ pool consisting of ~6 Q/QH$_2$ per cytochrome f in the lipid bilayer membrane (41). In each of the two cavities, the quinone exchange cavity provides a conduit for transmembrane Q/QH$_2$ transfer between the p-side quinol oxidation site located in one monomer and the n-side reduction site in the other, as depicted in **Figure 4**, and as described below in reaction formulae 1–7.

The walls of the quinone exchange cavity are formed by the C, D (cytochrome b), and F (suIV) helices of one monomer and the A, E, and ISP transmembrane helices of the other. The n-side "floor" of the cavity is formed by the N-terminal 25 residues of cytochrome b_6, and the p-side roof by helices cd1 and cd2 (between transmembrane helices C and D) of cytochrome b_6. Two molecules of the lipid dioleolylphosphatidylcholine (DOPC), whose addition was necessary for crystallization of the *M. laminosus* complex (42), may act as structural struts in the cavity, or they may impose restraints on protein dynamics in the portal near the roof of the cavity (9) (**Figure 3a**) (also discussed in section 3.2, below).

2.4. Lipid Requirement for Crystallization of the *M. laminosus* Complex

It has thus far not been possible to obtain a structure of the b_6f complex from higher plant chloroplasts nor from any other

Table 2 Distances between prosthetic groups of b_6f complex from *M. laminosus*[a]

Prosthetic group	Distance (Å)	
	Fe to Fe	Edge to edge
[2Fe-2S] - cytochrome f	29	25
[2Fe-2S] - heme b_p	28	24
Heme b_p - heme b_n	20	7
Heme b_p - heme b_p	22	11
Heme b_n - heme b_n	35	29
Heme b_n - heme x	9.6	4(b_n Fe to x propionate)
Chl a - heme b_p	NA[b]	12
Chl a - hem b_n	NA	5.5
Chl a - β-carotene	NA	14

[a]Distances are similar in the b_6f complex from *C. reinhardtii*.
[b]NA, not applicable.

cyanobacterium, e.g., *Synechocystis* PCC sp. 6803 and *Synechococcus* PCC sp. 7002. In the latter case, detergent extraction of the complex from the membranes results in its monomerization, which may result from a protease activity as yet undefined. Proteolysis of the purified active complex from *M. laminosus* was most noticeable in the ISP subunit and occurred over several days. This degradative cleavage occurred at a site that is similar to the one used for systematic proteolysis of the b_1 (43) and b_6f (44) complexes that were used for structure analysis (32, 45). The complex was also degraded at the N and C termini of the cytochrome b_6 polypeptide, which had earlier been shown to be protease accessible (46). The unfavorable proteolysis necessitated a rapid (1–2 day) crystallization. This was achieved by augmenting the purified complex with a stoichiometric amount of synthetic lipid (lipid:protein, 10:1, mol:mol) immediately after the last purification step. Two molecules of the lipid are seen in the quinone exchange cavity (**Figure 3**). The stabilizing role of lipid in the cavity may be to reduce the extent or amplitude of the dynamics of proteins that border the cavity. Another possible consequence of lipid depletion is that it might result in an increased level of protein dynamics and susceptibility to protease (42, 47). The head group of the lipid initially employed, DOPC, is not native to cyanobacteria (48) and does not satisfy a specific intraprotein structure requirement (49, 50) apart from that of the quinone exchange cavity for lipophilicity. These added lipids do not interfere with ET activity or the efficiency of inhibitors (H. Zhang & W.A. Cramer, unpublished data). Regarding natural lipids, an endogenous sulfoquinovosyldiacylglycerol anionic lipid is seen near the *n* side of the cavity near heme *x* in *C. reinhardtii*. This *n*-side position is consistent with the inference, based on analysis of internal lipid-binding sites of several energy transducing membrane protein structures, that anionic lipids tend to be found near the side with the more positive surface potential (50).

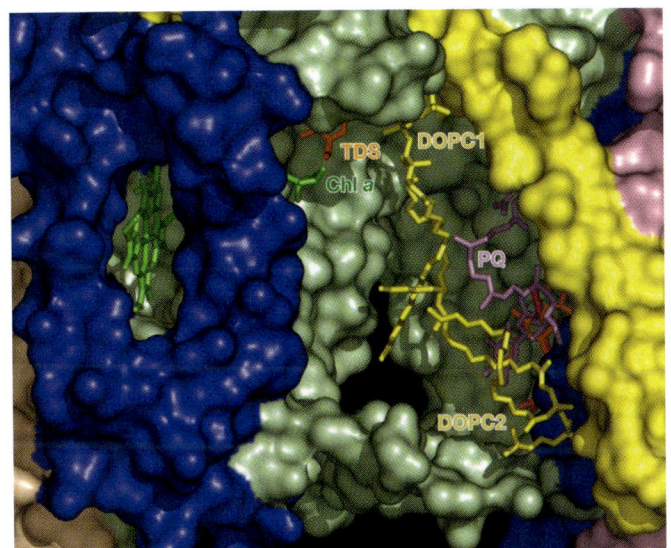

2.5. Asymmetry of Surface Potential

As a result of an asymmetric distribution of charged residues, which includes positive charges at the C termini of the subset of small subunits, and the acidic soluble domains of cytochrome *f* and the ISP, the surface potential (ψ°) of the *n* side of the complex is positive [av ψ° = +4.6 kT (*M. laminosus*) and +3.3 kT (*C. reinhardtii*)], calculated according to 51. It is negative on the *p* side, −5.4 kT in *M. laminosus* (**Figure 6**).

3. QUINOL OXIDATION SITE

3.1. Oxidant-Induced Reduction

The process of oxidant-induced reduction (52, 53) in the b_6f complex, involving the consecutive transfer of two electrons of the quinol [E_{m7} (Q/QH$_2$) = +0.1 V] in a "bifurcated reaction" to high- and low-potential ET chains, is believed to occur by mechanisms and pathways discussed extensively for the bc_1 complex (33, 34, 54–62). In summary, the first electron to be transferred to the high-potential chain is supplied by the donor isoprenoid quinol [UQH$_2$ in bc_1; PQH$_2$ in b_6f; CH$_3$, CH$_3$O, and CH$_3$O in UQ are replaced by H,

Figure 3

View of the central quinone exchange cavity with two molecules of embedded dioleoylphosphatidylcholine (DOPC) lipid (*yellow*, DOPC1 and DOPC2) and plastoquinone (*purple*), and heme *x* (*orange*), showing an aperture (*black*, in the middle of the figure) of 120–150 Å^2 that connects the two central cavities. The color code for TM helices is as in **Figure 2a**.

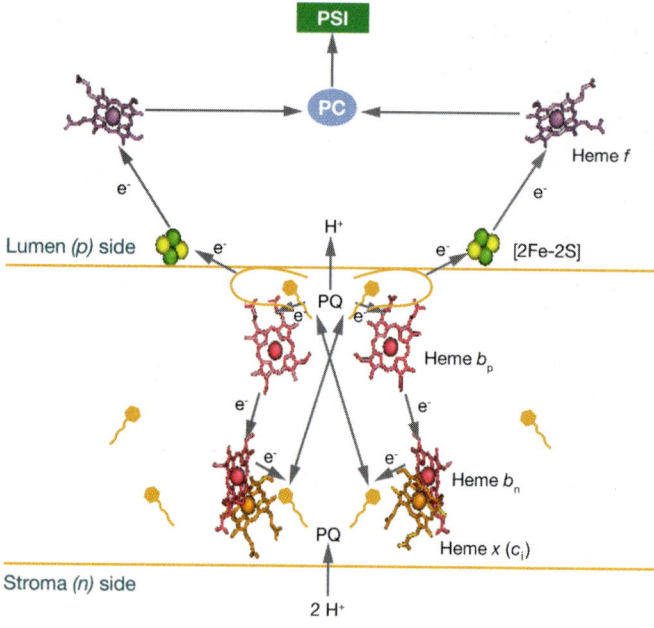

Figure 4

Schematic of electron, proton, and plastoquinone(ol) transfer pathway in the dimeric b_6f complex. After entry from the pool in the lipid bilayer membrane into the quinone exchange cavity, the Q/QH$_2$ are transferred across the between the reducing site near heme b_n-heme x on the stromal (n) side of the complex to the oxidation site near the [2Fe-2S] cluster of the Rieske ISP on the lumen (p) side. The plastoquinone (PQ)/PQH$_2$ transfer pathway includes passage through the narrow portal (**Figures 5a** and **7**) on the p side of the exchange cavity into the "Q$_p$ pocket" (**Figures 5b** and **7**) to the [2Fe-2S] cluster. The high-potential chain that accepts the first electron from PQH$_2$ and extends from the [2Fe-2S] cluster via cytochrome f and plastocyanin (PC) to P700 in the photosystem I (PSI) reaction center is shown. The transfer of 4H$^+$ to the p-side aqueous phase coupled to the uptake of 2H$^+$ from the n side corresponds to p-side oxidation of 2 PQH$_2$. Detailed pathways of H$^+$ transfer at the p- and n-side aqueous interfaces are not shown.

CH$_3$, and CH$_3$ at quinone side-chain positions 1, 4, and 5] to the high-potential (E$_{m7}$ = 0.3–0.45 V) ET chain in oxygenic photosynthesis, thus generating a lower potential semiquinone species, PQ$^{\bullet-}$. The concept of the more reducing lower potential quinone species generated by oxidation is unusual. The formalism of this Q/QH$_2$ redox chemistry is described in Reference 63.

There are details in the transfer of the first electron that are not shown in reaction 1. The quinol must be deprotonated (QH$_2$ → QH$^-$) to create a sufficiently reducing redox potential (64) before the first electron can be transferred or the electron and proton transferred in concert to the His ligand of the [2Fe-2S] cluster (65). Similarly the protonated semiquinone QH$^{\bullet}$ must be deprotonated (QH$^{\bullet}$ → Q$^{\bullet-}$ + H$^+$). The transfer pathway of the second proton to the p side has been proposed to occur via the glutamate residue in the conserved Pro-Glu-Trp-Tyr (PEWY) sequence. Mutagenesis of Glu78 in this sequence inhibited turnover of the complex (66).

The reducing potential of PQ$^{\bullet-}$ is sufficient to reduce O$_2$ to the reactive oxygen species, superoxide (O$_2^{\bullet-}$) at the Q$_p$ site (reaction 3a below) at low rates, a few percent of the control ET rate. The highest rate occurs when the natural rate of oxidation of the PQ$^{\bullet-}$ is decreased, e.g., by inhibitors (67–69). It is not known whether the b_6f complex requires a mechanism to protect itself against this reactive oxygen species.

Use of the semiquinone to bridge the 24–29 Å distance from the [2Fe-2S] cluster to heme b_p results in transfer of the second electron through a low-potential chain consisting of cytochrome hemes b_p and b_n from the p to the n side of the membrane (reaction 3 below).

ET through the high-potential chain of the b_6f complex proceeds via the [2Fe-2S] cluster (E$_{m7}$ = +0.3 V) through cytochrome f-plastocyanin (E$_{m7}$ ≈ +0.35 V) to the P700 reaction center (E$_{m7}$ = +0.45 V), as shown in **Figures 1** and **4**, and also in reaction formulae 1 and 2, below. The 29–32 Å distance from the [2Fe-2S] cluster to the cytochrome f, as in the bc_1 complex (2, 3, 6) is too large to sustain a competent [2Fe-2S] → cytochrome f ET rate (70, 71). Although a structure with the [2Fe-2S] cluster close to the cytochrome f heme has not yet been obtained, it is necessary that the ISP(s) in the b_6f complex can move to enable ET to cytochrome f at a competent rate. This inference follows from the existence of bc_1 structures showing the ISP(s) in a position proximal to the cytochrome c_1 heme (3–5) as well as to the UQ/UQH$_2$ site, the mobility of the ISP(s) in the b_6f complex (see section 2.1 above), and the glycine-enriched flexible hinge region of the ISP (72). The existence of a residence site for the [2Fe-2S] cluster within 10–15 Å of the cytochrome f heme would allow ET from the cluster to the heme to occur on a timescale of 10^{-3} s instead of 10^3 s. However, the difference in position of cytochrome f compared to cytochrome c_1 in bc_1-resp implies that the constrained diffusional motion of the ISP that closes the distance gap must be different in the b_6f complex. The kinetically competent electron-transfer pathway between the [Fe-2S] center and the cytochrome f heme over a distance of 10–15 Å likely includes intraprotein ET through cytochrome f (9). The movement of the ISP(s), which can be described as a "tethered diffusion," would seem a legitimate candidate for the rate-limiting step through the complex. Alternately, a detailed argument for proton-coupled ET from quinol to the ISP as the rate-limiting step has been presented (65). A first approximation to the bifurcated mechanism of p-side quinol oxidation would occur according to the following reaction sequence:

$$PQH_2 + [2Fe-2S(ox)] \rightarrow$$
$$PQ^{\bullet} + [2Fe-2S(red)] + 2H^+$$
(p-side aqueous phase) 1.

$$FeS(red) + Cyt\,f(ox) \rightarrow$$
$$FeS(ox) + Cyt\,f$$
(red) [high-potential chain] 2.

$$PQ^{\bullet} + heme\,b_p\,(ox) \rightarrow$$
$$PQ + heme\,b_p\,(red)$$
[low-potential chain] 3.

$$PQ^{\bullet} + O_2 \rightarrow PQ + O_2^{\bullet}$$ 3a.

$$heme\,b_p\,(red) + heme\,b_n\,(ox) \rightarrow$$
$$b_p\,(ox) + b_n\,(red)$$
(transmembrane reaction), 4.

where (red) and (ox) are reduced and oxidized species, respectively.

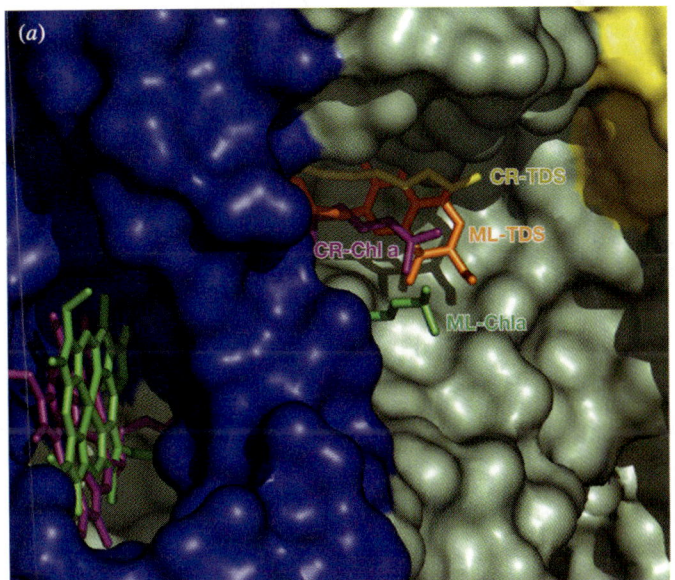

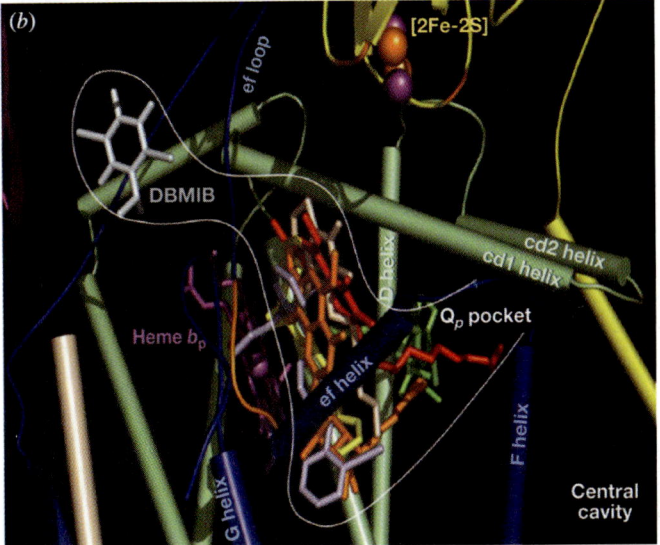

Figure 5

(a) The Q_p portal showing inserted TDS and Chl phytyl chains in "bent up" and "bent down" configurations. The color code is as follows: *C. reinhardtii* TDS (*yellow*) and Chl *a* (*purple*); *M. laminosus* TDS (*orange*) and Chl *a* (*green*). (b) Comparison of the Q_p inhibitor-binding sites in the bc_1 and b_6f complexes. All structures are aligned with the *M. laminosus* structure. The bound inhibitors are shown as sticks. The structures used in the alignment are the following: *M. laminosus* b_6f complex with TDS (*dark blue*, PDB id: 1VF5) and DBMIB (*gray*, PDB id: 2D2C); *C. reinhardtii* b_6f complex with TDS (PDB id: 1Q90); yeast bc_1 complex with UHDBT (*cyan*, PDB id: 1PP9); bovine bc_1 complex with stigmatellin (*orange*, PDB id: 1PP9), famoxadone (*yellow*, PDB id:1LOL) and azoystrobin (*light purple*, PDB id: 1SQB). The color code for subunits in both panels is as in **Figure 2**.

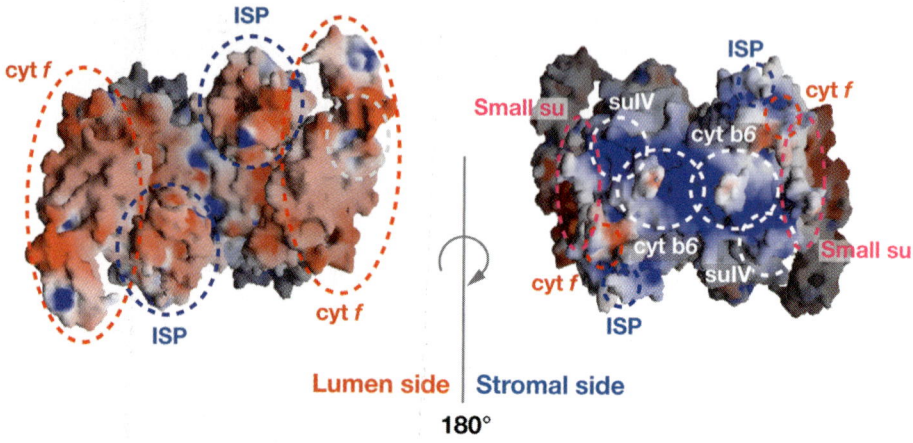

Figure 6

Asymmetric distribution of the surface potential, which is negative (*red*) and positive (*blue*) on the lumen and stromal sides of the complex, respectively. The average potential on the n side is +4.6 kT (*M. laminosus*) and +3.3 kT (*C. reinhardtii*); on the p side of *M. laminosus*, it is −5.4 kT. Abbreviation: cyt, cytochrome.

3.2. Q_p Pocket

A small portal (∼11 Å × 12 Å near the p side of each quinone exchange cavity, with an approximate cross-sectional area of 130 Å^2) leads to a p-side antechamber, or "Q_p pocket," which contains the [2Fe-2S] cluster of the Rieske ISP at the p-side aqueous interface that is the high-potential (E_{m7} = +0.3 V) electron and H$^+$ acceptor for the plastoquinol. The aperture of the entry portal is defined by a distance of separation in each monomer of 13 Å between Val98 (F helix-suIV) and Val126 (C helix cytochrome b_6) and 11 Å between Val126 and Thr134 (cytochrome b_6) (**Figure 5a**). A similar portal is seen in the bovine bc_1 complex, where the cross section at its entrance is 9 Å (Ileu146-Met124) × 11 Å (Ala125-Leu294) (D. Xia, personal communication). The b_6f Q_p pocket is bounded by the [2Fe-2S] cluster, heme b_p, and the "ef loop" connecting helices E and F of subunit IV. In bc_1-resp, a large number of Q-analogue inhibitors have been shown to bind in the Q_p pocket (**Figure 5b**), although native ubiquinone has not been detected (58) either because of antiferromagnetic coupling or because its residence time in the pocket is very short (∼1 ms). A set of Q_p-analogue inhibitors, defined by structure analysis, is distributed over the Q_p pocket. Stigmatellin is an example of an inhibitor that is closer to, and interacts more strongly with, the [2Fe-2S] cluster (**Figure 5b**) (stigmatellin shown in orange and TDS in red and green for *C. reinhardtii* and *M. laminosus*, respectively); other inhibitors are farther by a few angstroms from the cluster and closer to heme b_p, with which they interact more strongly (e.g., azoxystrobin, light purple). UHDBT (wheat) and famoxadone (yellow) have intermediate locations and interaction properties. The defined interactions of the inhibitors on the two spatially separated redox groups do not necessarily imply that two molecules of ubi- or plastoquinone can simultaneously fit into the Q_p pocket. However, the distribution of inhibitor positions in the Q_p pocket can reasonably be interpreted to define boundaries for quinone occupancy that allow sufficient space and degrees of freedom for diffusion of physiological quinones between the Q_p pocket and the exchange cavity.

In the *C. reinhardtii* structure (10), as in bc_1-resp (6), the TDS chromone ring protrudes through the portal into the Q_p pocket, placing its O-4 atom within H-bonding distance of the His155 ligand to the [2Fe-2S] cluster ("ring-in" mode). In the yeast bc_1 structure, the other side of the stigmatellin chromone ring (O-8) forms an H-bond with the glutamate residue (Oε1) of the PEWY sequence that is conserved in the cytochrome bc_1 complex (26). However, in *M. laminosus*, the TDS 13-C hydrocarbon tail is plugged into the portal, and its chromone ring remains in the quinone exchange cavity ("ring out"), approximately 20 Å from the histidine ligand of the [2Fe-2S] cluster. Plugging of the portal will accomplish partial inhibition of *p*-side ET (9). However, total loss of inhibitor sensitivity by mutagenesis of the ISP-proximal residue Leu111 in *Synechococcus* sp. PCC 7002 showed that the primary site of inhibitor action is at the [2Fe-2S] cluster. This is the site for the ring-in conformation of TDS in *C. reinhardtii*. The sequence of residues in the portal determines the orientation of the TDS inserted into it. Changing suIV residue Leu81 in the cyanobacterial portal to the Phe in *C. reinhardtii* resulted in a 10–100-fold increase of TDS sensitivity. Residue 81 is one of a residues within 5 Å of TDS with a different side chain in cyanobacteria and *C. reinhardtii*. This implies that the distribution between ring in and ring out is determined by the steric properties of residue 81, which may influence local protein dynamics in the portal (73).

3.3. The Meaning of the TDS Ring-Out Conformation

This conformation might be indicative of out-going flow of quinone in two-way traffic through the portal. However, it seems unlikely that quinone export through the portal would be associated with tight binding in the portal. It is more likely that the TDS ring is excluded from the portal containing Leu81 in cyanobacteria because of steric considerations or insufficient portal plasticity. Under these circumstances, the ring-out conformation with the ring 20 Å from the [2Fe-2S] cluster is the thermodynamically favorable, but inactive, conformation (73).

The two different TDS ring orientations in the *M. laminosus* and *C. reinhardtii* structures are associated with (*a*) different depths of penetration of the [2Fe-2S] cluster into the Q_p pocket and (*b*) a different orientation of the Chl phytyl chain that is inserted into the portal from the Q_p side (**Figure 5***a*). The [2Fe-2S] cluster in the algal complex is inserted 4 Å further into the hydrophobic phase, and the Q_p pocket is extended toward the *p* side by 2 Å, presumably to accommodate formation of an H-bond between the TDS chromone ring and the cluster His155 ligand.

It is reasonable to assume that the binding sites of the Q-analogue inhibitors (**Figure 5***b*) define the accessible space and binding sites that can accompany Q/QH$_2$ movement in the complex. Although some such sites could be artifactual, at least 10 inhibitors have been shown to occupy the Q_p pocket, and it therefore seems likely that the ensemble of these binding sites defines an intracomplex space that is accessible to the physiological Q/QH$_2$. In addition, for the *P* side inhibitor, DBMIB (74,75), a niche more peripheral than the Q_p pocket and 20 Å from the [2Fe-2S] cluster has been found for the high-affinity-binding site in the structure of the *M. laminosus* complex (**Figure 5***b*). Occupancy of this site, which is high affinity but distant from the [2Fe-2S] cluster, is consistent with the finding that the first DBMIB molecule that binds to the b_6f complex does not perturb the [2Fe-2S] electron paramagnetic resonance (EPR) signal (76). After activation by a light flash, initially oxidized DBMIB, bound at the high-affinity site, can inhibit cytochrome *f* reduction, implying that there is an internal passage from the peripheral site to the [2Fe-2S] cluster (77).

The existence of the internal spaces and passages for a large set of quinone analogue inhibitors implies that Q/QH$_2$, in spite of its long isoprenoid chain, can negotiate transmembrane passages through the internal labyrinth of the cytochrome *bc* complexes. The existence of the narrow portal between the exchange cavity and the Q_p pocket,

EPR: electron paramagnetic resonance

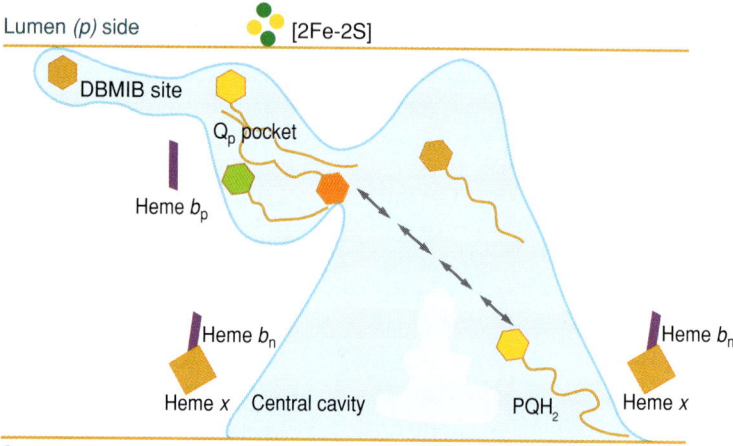

Figure 7
Schematic diagram showing steric complexity of quinone movement through the labyrinthine interior of the cytochrome b_6f complex.

through which Q/QH$_2$ must pass, by itself defines a labyrinthine route of Q/QH$_2$ across the complex (**Figure 7**). Furthermore, taking the rate-limiting step (∼ a few ms) of the complex into account, the diffusion or transfer pathway defined by the exchange cavity-portal-Q$_p$ pocket must allow Q/QH$_2$ traffic through the portal on this timescale. The necessary passage of the quinone through the cavity and portal shows that the concept for movement of the long-chain quinone across the complex is much more complicated than originally proposed (78) before structures of energy-transducing membrane protein complexes were known.

3.4. Intermonomer Cross Talk

Discussion of the function of these cavities has been focused on the movement of the lipophilic quinone across the cavity between the n and p sides of the complex and the binding sites on each side of the cavity. It should be noted, however, that the back wall of the cavity, as viewed in the representation of **Figure 3** (the central black patch), has a ∼150-Å^2 passage that connects the two cavities. The passage is bordered on each side by half of transmembrane helix D and the helix-ending Arg207, contributed by each monomer. The central area of this passage is as large as the portal linking the Q-exchange cavity and the Q$_p$ pocket, described above, and could accommodate insertion of a lipophilic quinone, perhaps as part of the semiquinone cycle (79, 80). The effective size of this passage would depend on the extent to which it is shielded by lipids bound in the cavity. Quinone communication through this intercavity passage could be involved in electron equilibration between the heme b chains in the two cavities (38–40).

3.5. Extrinsic Domains on the p Side: Cytochrome f and the Rieske ISP

As in the bc_1 complex, the soluble domain of the ∼180-residue ISP of the b_6f complex is connected to its transmembrane (TM) helix by a flexible hinge. This flexible hinge is essential for the ISP rotation/translational motion that is necessary to allow ET on an energetically competent timescale to cytochrome f. The hinge region is readily identified by the presence of an eight-residue polyglycine domain that separates the N-terminal ∼45-residue domain containing the 25-residue transmembrane helix from the ISP(s) that consists of ∼140 amino acids. Although the function of the b_6f complex must require the ISP flexible hinge, it was found to be less sensitive than bc_1 to changes in hinge length and flexibility (72). The ISP TM helix is domain swapped. It crosses over from the 140-residue ISP(s) domain containing the [2Fe-2S] cluster in one monomer and spans, in an oblique orientation, the TM domain in the other monomer. The 25-residue TM "crossover helix" of the ISP in the b_6f complex (see section 2.3) makes van der Waals contacts with Phe261 and Met271 of the TM of cytochrome f and with Tyr71, Val76, and Phe78 of helix B of cytochrome b_6 as observed in the bc_1 complex (81). The dependence of the rate of cytochrome f reduction on ambient viscosity (82) suggests that the movement of the ISP is the rate-limiting step in ET through the high-potential chain. Alternatively, in studies on the bc_1 complex in the photosynthetic bacterium, *Rhodobacter sphaeroides*, a cogent argument has been made for proton-coupled ET of the first electron from the quinol to the [2Fe-2S] cluster being the rate-limiting step (65).

Cytochrome f is the other p-side subunit with a well-defined redox function. Like cytochrome c_1 in the bc_1 complex, it is anchored in the complex by one TM helix (**Figure 2a,b**). In each complex, this cytochrome has a similar function, oxidation of the ISP in the high-potential electron transport chain. Convergent evolution, selecting for necessary function, resulted in a membrane-anchored, covalently bound, high-potential heme in the respiratory and photosynthetic electron transport chains. However, these proteins are otherwise completely different in sequence and structure. Cytochrome c_1 is an α-helical protein. Cytochrome f is not only different from cytochrome c_1, but it is distinct from all other c-type cytochromes in its predominantly β-strand structure, heme ligation by the α-amino group of the N-terminal Tyr residue (29), and its internal extended five-water chain that is also conserved in evolution (30, 83, 84). No homolog of cytochrome f has been identified, and its evolutionary origin is therefore unknown.

Function of the cytochrome f internal water chain. It has been proposed that the internal extended (11-Å) water chain could function as a H^+ channel in the pathway of H^+ transfer to the p-side bulk aqueous phase following oxidation of PQH_2 by the [2Fe-2S] cluster of the ISP (83–86). Alternatively, the water chain could provide a dipole relaxation mechanism associated with the electron-transfer reactions of the cytochrome. It is now clear from the structure of the TDS complex that H^+ transfer to the cytochrome f H_2O chain is very unlikely when the ISP [2Fe-2S] cluster is in the Q-proximal position. In this state, Leu27 and Asn153, the residues H-bonded to the first water in the water chain, are 16–17 Å and 20 Å, respectively, from Pro-Leu 72–73 of the suIV ef loop. This does not preclude the possibility of H^+ transfer from the cluster when it is in the cytochrome f-proximal position. However, any proton transfer to the water chain is most likely to occur concomitantly with ET.

4. QUINONE REDUCTION SITE (Q_n SITE)

4.1. n Side of the Cycle for Quinone Turnover

Studies on the turnover of the bc_1 complex from mitochondria and photosynthetic bacteria have led to the following summary of reactions that can complete a cycle on the n side of the membrane through reduction of quinone by two consecutive one-electron transfers (see below, n-side quinone reduction, reactions 5–6) or one concerted transfer of two electrons stored in hemes b_p and b_n (reaction 7 below). The evidence for these n-side reactions has been strengthened greatly by data for the structure of the bc_1 complex showing both the bound n-side quinone (6, 7) and a structure with the inhibitor, antimycin A (18, 87). Antimycin A has long been known to block the oxidation of cytochrome b and is inferred to displace the natural ubiquinone from its binding site. The history of the use of antimycin A in oxygenic photosynthesis, using chloroplasts or thylakoid membranes, has been confusing. Initial studies implied that the ferredoxin-dependent cyclic electron and phosphorylation pathway

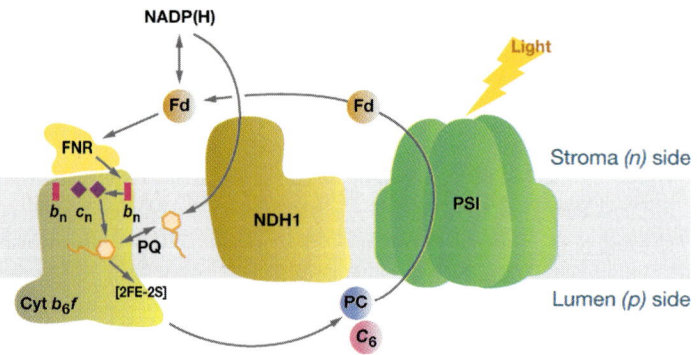

Figure 8
Potential pathways for cyclic electron transport. In addition to the well-known role of ferredoxin in the pathway of cyclic electron transport and phosphorylation (88), a role of FNR in series with the b_6f complex was suggested by studies in which FNR copurified with the chloroplast b_6f complex (25, 35, 36). It has been proposed that the b_6f complex in the stromal nonappressed membranes contains FNR, and this can sustain cyclic electron transport (101). See References 102–105 for a recent discussion of the physiological role of the cyclic pathway. A role for the NADH dehydrogenase (NDH1) in the cyclic pathway has been proposed (132, 133). The binding site of the pgr5 protein, implicated from mutation and fluorescence-quenching studies in *Arabidopsis* to be involved in cyclic electron transport around PSI, is not included in the diagram (113). pgr5 is a soluble and extremely basic (pI = 11.4) protein. The conserved 64-residue C-terminal domain of pgr5 contains 12 basic (Arg, Lys) and four acidic (Glu, Asp) residues. Thus, it must bind to an acidic surface-exposed site of a protein or protein complex that functions in the cyclic pathway. The cytochrome b_6f complex is included in this list of pgr5 partners. A recent inference that the b_6f complex does not participate in the cyclic pathway, on the basis of a pgr1 mutant that impairs the function of the Rieske protein but not of the cyclic pathway (114), has the consequence that it eliminates the plastoquinol oxidase function that is necessary for cyclic electron transport in any scheme.

NQNO: nonyl-4-hydroxyquinoline N-oxide

around the PSI reaction center (**Figure 8**) is sensitive to antimycin A and, by analogy with antimycin inhibition of electron transport in mitochondria, proposed to involve a cytochrome b (88). Antimycin-sensitive flash-induced turnover of cytochrome b_6 appeared consistent with this view (89). However, other studies subsequently contradicted an effect of antimycin on the amplitude of flash-induced cytochrome b_6 reduction and thereby its involvement in the cyclic pathway. These studies implied the presence of another protein complex, a ferredoxin:plastoquinone reductase, as the site of inhibition of cyclic electron transport by antimycin A (90, 91). To reconcile these contradictory results, it could be that effects of antimycin depended on the precise state of the membranes. However, the structure of the b_6f complex has now established that antimycin A cannot act on the b_6f complex as it does on bc_1 because a new redox component, heme x, occupies the n-side site in b_6f that is occupied by ubiquinone or antimycin A in the bc_1 complex (**Figure 9a**). The action of the quinone analogue, N(H)QNO, on the n side of the b_6f complex, through which it causes an increase in the flash-induced amplitude of heme b reduction (92, 93), qualitatively resembles that of antimycin A on the bc_1 complex. However, the increase in reduction occurred at nonyl-4-hydroxyquinoline N-oxide (NQNO) concentrations that had no significant effect on the rate of linear electron transport (94), implying weak coupling at best between the linear pathway and ET through the low-potential chain. The absence of a well-defined n-site inhibitor for the b_6f complex, analogous to antimycin A action on the bc_1 complex, has been a major obstacle in unequivocal definition of low-potential ET pathways through the b_6f complex. Continuing the numbering for the reactions in the cycle from those used for the p-side reactions (section 3.1), n-side reduction of quinone would occur according to the following reaction sequences:

$$\text{heme } b_n \text{ (red)} + PQ$$
$$\to \text{heme } b_n \text{ (ox)} + PQ^{\bullet} \qquad 5.$$

and, after a second turnover,

$$\text{heme } b_n \text{ (red)} + PQ^{\bullet} + 2H^+$$
$$\to b_n \text{ (ox)} + PQH_2. \qquad 6.$$

Reactions 1–6 above define a Q cycle.

4.2. The Unique Heme x

Heme x is covalently linked to the protein by a single thioether bond to an invariant Cys35 on the n side of helix A of cytochrome b_6. The covalent binding and uniqueness of a heme without any amino acid side chains as axial ligands have led to the notation heme x (9) or heme c_i (10), respectively, in the *M. laminosus* and *C. reinhardtii* structures. The two hemes are close (edge-edge separation, ~4.0 Å) and oriented orthogonally to each other (**Figure 9a**). The heme edges are connected through a propionate of heme b_n via a water or hydroxyl molecule, implying that interactions between the two hemes should be strong and interheme ET facile.

Given the many measurements of actinic light-induced difference spectra in situ and in vivo of the b_6f complex made by photosynthesis experimentalists, why was the presence of heme x not well known long ago? In retrospect, the only experiments that detected heme x, then called G (95, 96), were those that utilized instrumentation with the highest signal:noise capability, a photosynthetic system (mutant cells of the alga, *Chlorella*) in which one of the reaction centers was missing so as to minimize background redox absorbance changes, and measurement in the blue region (400–450 nm) of the cytochrome's visible spectrum. The latter spectral region is generally avoided because of its large Chl absorbance, but it turned out that absorbance changes of heme x, on the short-wavelength

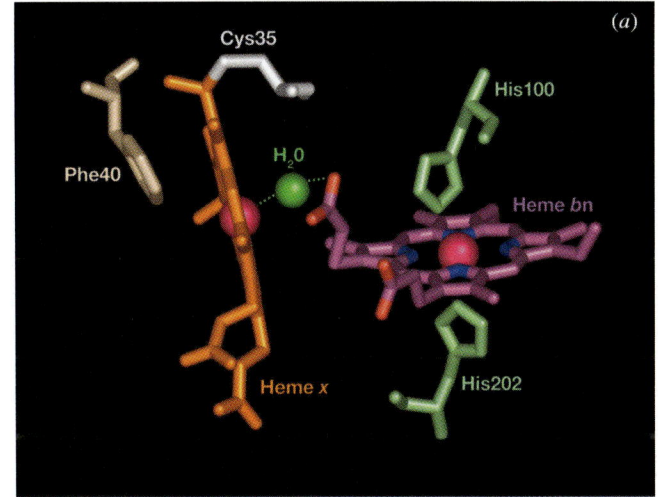

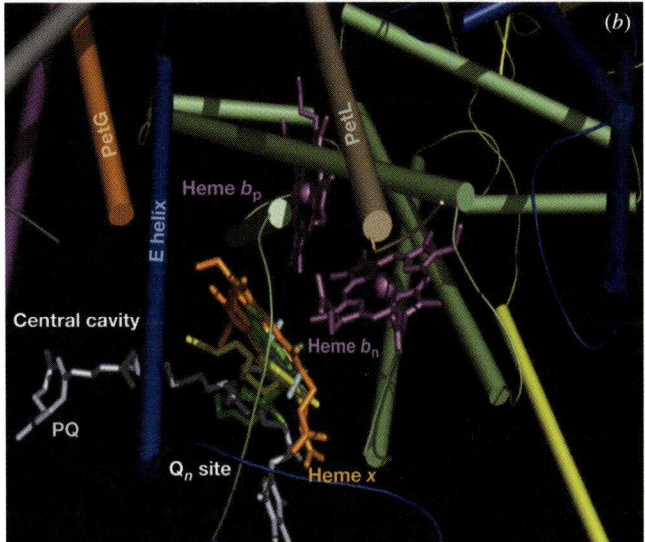

Figure 9

(*a*) Ligation of hemes b_n and x. The covalent bond to Cys35 on the n side of helix A is shown, as are the histidine ligands to heme b_n and neighboring residue Phe40. (*b*) Q_n inhibitor-binding sites in the bc_1 and b_6f complexes. Structures used in the alignment are *M. laminosus* b_6f complex with bound PQ (*light blue*) (PDB id:1VF5), ubiquinone (*cyan*) (PDB id:1NTZ), NQNO (*yellow*) (PDB id:1NU1), and antimycin (*green*) (PDB id:1NTK), as positioned in bovine bc_1 complex.

side of cytochrome b_6, are better separated than in the α-band region, which most workers believed to be "the green window." However, the cytochrome b polypeptide on sodium dodecyl sulfate-polyacrylamide electrophoresis could be heme stained (25, 97). In addition, pyridine hemachromagen spectra of the denatured cytochrome b polypeptide showed a defined spectral peak, implying the existence of a covalently bound heme (98). In the *Chlorella* studies, equilibrium of ET between the two hemes implied that heme x or c_i is located near the n-side interface of the complex, where it can exchange electrons with heme b_n, has a high-spin character, and binds the strong heme ligand CO, properties that have been confirmed by EPR analysis of the purified crystallizable complex. The latter studies showed heme x in the untreated complex from *M. laminosus* and spinach chloroplasts to be ferric high spin in a rhombic environment with characteristic g values of 7.4, 6.7, and 4.7 (98). Heme x has a somewhat more positive midpoint redox potential than heme b_n [ΔE_m = 20–30 mV (95–96) or 150 mV (98a)].

The presence of heme x implies that the traditional Q-cycle reaction may have to be modified. One obvious change is to describe the possibility that heme b_n can reduce heme x, instead of the oxidized quinone (reaction 7 below). In fact, heme x in the $b_6 f$ complex occupies most of the site that is occupied by bound ubiquinone in the bc_1 complex, as well as the binding site in the bc_1 complex of the classical n-site inhibitor, antimycin A (**Figure 9b**). Electron density near heme x in the *M. laminosus* structure has been interpreted as bound plastoquinone in which a quinone ring O is 3.9 Å from the heme x propionate O. Other residues that could participate in the binding of this PQ$_n$ are cytochrome b residues Tyr25 (within 5 Å of the quinone head group), Leu45, and Phe203, which are near the quinone isoprenoid tail. An electron density corresponding to this plastoquinone was not resolved in the *C. reinhardtii* complex, where a sulfoquinovosyl-diacylglycerol molecule is seen that is farther (>10 Å) from heme x. The structure and redox potential data imply a pathway, $b_n \to x \to$ PQ, possibly involving ferredoxin as a second electron donor to heme x as an alternative pathway for n-side reduction of quinone (reactions 8 and 9 below). A role for heme x in the pathway of ferredoxin-linked cyclic electron-transfer pathway is an obvious possibility for both *M. laminosus* and *C. reinhardtii* because heme x is uniquely associated with the electron transport pathways of light-dependent oxygenic photosynthesis. Reactions 8 and 9 below describe PQ reduction by ferredoxin via heme x in a putative n-side pathway for cyclic electron transport.

Proposed utilization of heme x for cyclic electron transport is as follows:

$$b_n \text{ (red)} + \text{heme } x \text{ (ox)} \to \text{heme } b_n \text{ (ox)}$$
$$+ \text{heme } x \text{ (red), or} \qquad 7.$$

$$\text{ferredoxin (red)} + \text{heme } x \text{ (ox)}$$
$$\to \text{ferredoxin (ox)}$$
$$+ \text{heme } x \text{ (red), and} \qquad 8.$$

$$\text{heme } x \text{ (red)} + \text{ferredoxin (red)} + \text{PQ}$$
$$+ 2\text{H}^+ \to \text{heme } x \text{ (ox)}$$
$$+ \text{ferredoxin (ox)} + \text{PQH}_2. \qquad 9.$$

4.3. Cyclic Electron Transport

Ferredoxin-linked cyclic electron transport (19, 88, 91), in which there is feedback of electrons from the reducing side of the PSI reaction center via ferredoxin to reduce plastoquinone, is considered a metabolic requirement to preserve a proper balance of NADPH and ATP for CO_2 fixation (99). Neither the absolute magnitude of the contribution of the cyclic pathway to the net synthesis of ATP under different conditions of illumination and plant stress nor the mechanism of regulation between the linear and cyclic pathways (100) is understood. The contribution of this pathway in unstressed C3 plants (101) to the electrochemical proton gradient has been the subject of debate (102–105). There are many aspects to the subject of cyclic electron transport, ranging from the physiological to the mechanism at the electron transport level. The present discussion focuses on the latter.

The ferredoxin-dependent pathway requires that ferredoxin and FNR reduce PQ to PQH_2 via the n-side membrane interface. Alternatively, and particularly in cyanobacteria, NADH can reduce the quinone pool through a dehydrogenase, ndh-1 (106, 107) (**Figure 8**). FNR was found to be bound to a significant fraction (≥ 0.5) of the population of isolated spinach b_6f complex (35, 36). This implies that the pathway of ferredoxin-dependent cyclic ET could reduce plastoquinone (108, 109) but in a pathway that does not necessarily include the cytochrome b_6f complex (90). The FNR-b_6f population may be concentrated in the nonappressed stromal membranes (101), possibly in an efficient "supercomplex" with the PSI reaction center (36, 101). In this case, the fraction of FNR-b_6f could readily be regulated by state transitions that alter the fraction of nonappressed membranes (100, 110).

The presence of heme x on the n side of the oxygenic photosynthetic membrane and its exposure to the lipophilic cavity suggest the possibility that heme x provides a connection to ferredoxin and cyclic electron transport. Furthermore, given its free ligand position exposed to the cavity and the fact that cyanobacterial ferredoxin has a short C-terminal helix (approx. 10 Å) extending from the vicinity of its [2Fe-2S] cluster (111), it might be proposed that the ferredoxin carboxy-terminal carboxylate can insert into the complex and provide a ligand to heme x. It could thereby establish a connection for a cyclic ET pathway from ferredoxin to the PQ near heme x. However, such an insertion would require a large rearrangement of the n-side segments of cytochrome b_6 and subunit IV. This is not inconceivable but presently is unsupported by data. An alternative proposal is that the acidic (pI = 3.5 for ferredoxin from *Synechocystis* sp. PCC 6803) ferredoxin could dock to the positively charged n-side surface of the b_6f complex (**Figures 6** and **8**) and reside within 15 Å of heme x, with the capability to reduce it via an intraprotein ET pathway (70, 71). It is noted that this surface is calculated to have a positive surface potential for both *M. laminosus* and *C. reinhardtii*, as noted above

(**Figure 6**) (section 2.5). The 120 disordered residues missing from the refined structure of the *M. laminosus* are charge neutral and have no effect on the net surface charge. The constraints on the electron-transfer rate are sufficiently minimal, and the redox potential of the ferredoxin ($E_m = -0.41$ V) negative enough that a specific docking site might not be necessary. However, a specific site is suggested, as follows. It has been noted for the *C. reinhardtii* structure that heme x is within H-bonding distance of the positively charged n-side residue Arg207 (10). Changes in this residue affect the ET rate of heme b_n (112). The neighboring Lys208, which is also conserved, creates a patch of very positive surface charge density on the n-side surface of the b_6f complex near heme x. This is a proposed specific interface between the b_6f complex and ferredoxin that would enable reaction 8 above. Transfer of two electrons to bound PQ through ferredoxin-heme x, coupled to the uptake of two H^+, could provide a pathway for reduction of PQ at an n site (reaction 9 above). This pathway would utilize only the heme x component of the b_6f complex. Heme x could, in fact, be the putative ferredoxin:quinone reductase component proposed by Moss & Bendall (90). The pathway to heme x could include a structural role for the highly charged pgr5 protein that has been implicated in PSI-cyclic electron transport (113, 114). The inhibitors NQNO, and antimycin A under some conditions, could act by displacing the PQ bound near heme x in the cavity near heme x (**Figure 3**).

4.4. Other Pathways for Heme x

Other possibilities for a function of heme x relate to its high-spin character (98), which suggests that it could act as an oxidase in a light-dependent or dark pathway. A role as an oxidase in chlororespiration (115, 116) suggests itself. However, the midpoint potential, $E_m \cong 0.0$ V, of heme x argues against this possibility. Other possibilities are suggested by genomics considerations.

Implications of the existence of heme x in gram-positive bacteria; conservation of the Cys residue in helix A of cytochrome b_6.

It has been noted that a modified cytochrome $b_6 f$ complex is present in nonphotosynthetic gram-positive bacteria (10, 117). Such a complex, characterized by a small four-TM helix cytochrome b, has been purified from the thermophilic *bacillus*, PS3 (118). The likelihood of "split" cytochrome b can be seen in the genes of gram-positive bacteria, *Bacillus stearothermophilus* (119) and *Bacillus subtilis*, and such a complex has been isolated from the two sources (120, 121). Furthermore, the Cys residue near the n side of the first transmembrane helix (A) of the cytochrome b_6 subunit, which is diagnostic for the covalent-binding site of heme x, is found not only in the cyanobacteria, but also in all bacteria of primitive origin that possess the four-helix cytochrome b_6 and subunit IV. It is found in the firmicutes, which include gram-positive bacteria such as *B. subtilis*. It was shown in the latter that a heme is covalently attached to the cytochrome b polypeptide subunit (122), as observed previously in *C. reinhardtii* (123). It was inferred in these studies that one of the b hemes was covalently bound to the protein. Alternatively, one would presently infer that these nonphotosynthetic bacteria also contain heme x, presumably as part of a metabolic pathway.

5. BOUND PIGMENT MOLECULES

The crystal structures of the $b_6 f$ complex each reveal two noncovalently bound pigment molecules, a Chl a and a β-carotene. One chlorophyll a per monomer was found in higher plants (124) and *C. reinhardtii* (125) from biochemical and spectroscopic analysis. After the presence of this Chl a was established, it made sense to search for a carotenoid in the complex that presumably would be needed to protect the protein around the Chl from singlet O_2 damage. Indeed, one β-carotene per Chl was found in the $b_6 f$ complex from spinach thylakoids, *C. reinhardtii*, and *M. laminosus* (126). Both crystal structures of the complex show the Chl ring to be inserted between helices F and G of suIV (**Figures 2b, 3,** and **5a**). The Chl ring plane is approximately perpendicular to the membrane plane and parallel to and separated by 5.5 Å, edge-edge, from heme b_n (**Table 2**). Spectroscopic evidence for coupling between the Chl and the heme(s) has been presented (127). Ligands from the protein or solvent to the Chl Mg could not be identified. Thus, the presence of the Chl in the complex does not provide evidence for a proposed evolutionary connection between bc complexes and type 2 photosynthetic reaction centers (128). The Chl ring is close (6–8 Å) to three aromatic residues of subunit 4, at least one of which participates in electron exchange that shortens the Chl fluorescence lifetime by a factor of 20–25, thereby reducing the yield of the Chl triplet state and singlet oxygen (129). The 20-carbon Chl phytyl tail is wrapped around part of the suIV G helix and inserted from the p side through the Q-transfer portal into the Q-exchange central cavity (**Figures 2b** and **5a**). This "wrapping" feature of the Chl a phytyl tail and the interhelix position of the Chl ring resemble the proposed structure-stabilizing role of lipids and the role in assembly, as described for the yeast bc_1 complex (50). A role of bound quinone in transmembrane signaling and activation of the light-harvesting chlorophyll protein complex kinase was proposed by Vener et al. (130). Regarding this proposal, it was also suggested that the intercalated Chl might have a sensor role between p-side quinone and kinase activation in *C. reinhardtii* (10). In this context, the two different conformations of the phytyl tail inserted through the portal, "bent in/bent out," seen in the *M. laminosus* and *C. reinhardtii* (**Figure 5a**) may be relevant.

The 9-*cis* β-carotene, inserted between the PetL and PetM helices, is 14 Å from the Chl a. This distance was believed to be much too long for efficient transfer of triplet state energy from the Chl, which requires overlap of wave functions and therefore should not occur over distances >4 Å. The 14-Å distance was therefore a surprise because the sole known function of β-carotene is to quench the Chl-excited triplet state. However, transfer of triplet-excited state energy from the Chl to the β-carotene has been observed. The long-distance transfer mechanism was proposed to occur through O_2 and an O_2 channel (131). The question arises as to why the spatial separation of β-carotene and Chl is so great. Presumably, the β-carotene has a function in addition to that of protecting against O_2 damage.

ACKNOWLEDGMENTS

For helpful discussions during the preparation of this article, the authors thank our colleagues at Purdue, E. Yamashita, X. Shi, A. Aronson, S. Savikhin, N. Dashdorj, H. Kim, and elsewhere, J. Whitelegge, E-M. Aro, E.A. Berry, F. Daldal, and D. Xia. We also thank the staff of synchrotron beam lines SBC-19 at the Advanced Photon Source (Argonne National Laboratory, Illinois) and at Spring-8 BL44XU (Hyogo, Japan), the Japanese Ministry of Science and Education for a Fellowship (GK), and grant support for the studies described herein from the National Institutes of Health, GM-38323, and the Henry Koffler Professorship (WAC).

LITERATURE CITED

1. Kallas T. 1994. In *The Molecular Biology of Cyanobacteria*, ed. DA Bryant, pp. 259–317. Dordrecht: Kluwer Acad.
2. Xia D, Yu C-A, Kim H, Xia J-Z, Kachurin AM, et al. 1997. *Science* 277:60–66
3. Zhang Z, Huang L, Shulmeister VM, Chi YI, Kim KK, et al. 1998. *Nature* 392:677–84
4. Iwata S, Lee JW, Okada K, Lee JK, Iwata M, et al. 1998. *Science* 281:64–71
5. Kim H, Xia D, Yu CA, Xia JZ, Kachurin AM, et al. 1998. *Proc. Natl. Acad. Sci. USA* 95:8026–33
6. Hunte C, Koepke J, Lange C, Rossmanith T, Michel H. 2000. *Struct. Fold. Des.* 8:669–84
7. Gao XG, Wen XL, Esser L, Quinn B, Yu L, et al. 2003. *Biochemistry* 42:9067–80
8. Berry EA, Huang L-S, Saechao L-K, Pon NG, Valkova-Valchanova M, Daldal F. 2004. *Photosynth. Res.* 81:251–75
9. Kurisu G, Zhang HM, Smith JL, Cramer WA. 2003. *Science* 302:1009–14
10. Stroebel D, Choquet Y, Popot J-L, Picot D. 2003. *Nature* 426:413–18
11. Jordan P, Fromme P, Witt HT, Klukas O, Saenger W, Krauss N. 2001. *Nature* 411:909–17
12. Zouni A, Witt HT, Kern J, Fromme P, Krauss N, et al. 2001. *Nature* 409:739–43
13. Kamiya N, Shen J-R. 2003. *Proc. Natl. Acad. Sci. USA* 100:98–103
14. Ferreira KN, Iverson TM, Maghlaoui K, Barber J, Iwata S. 2004. *Science* 303:1831–38
15. Kern J, Loll B, Zouni A, Saenger W, Irrgang KD, Biesiadka J. 2005. *Photosynth. Res.* 84:153–59
16. Yano J, Kern J, Irrgang K-D, Latimer MJ, Bergmann U, et al. 2005. *Proc. Natl. Acad. Sci. USA* 102:12047–52
17. Esser L, Quinn B, Li Y-F, Zhang MQ, Elberry M, et al. 2004. *J. Mol. Biol.* 341:281–302
18. Huang LS, Cobessi D, Tung EY, Berry EA. 2005. *J. Mol. Biol.* 351:573–97
19. Allen JF. 2004. *Trends Plant Sci.* 9:130–37
20. Cramer WA, Zhang HM, Yan JS, Kurisu G, Smith JL. 2004. *Biochemistry* 43:5921–29
21. Cramer WA, Yan JS, Zhang HM, Kurisu G, Smith JL. 2005. *Photosynth. Res.* 85:133–43

22. Smith JL, Zhang HM, Yan JS, Kurisu G, Cramer WA. 2004. *Curr. Opin. Struct. Biol.* 14:432–39
23. Knoll AH. 2003. *Life on a New Planet: The First Three Billion Years of Evolution on Earth.* Princeton, NJ: Princeton Univ. Press
24. Lange C, Hunte C. 2002. *Proc. Natl. Acad. Sci. USA* 99:2800–5
25. Hurt EC, Hauska G. 1981. *Eur. J. Biochem.* 117:591–99
26. Widger WR, Cramer WA, Herrmann RG, Trebst A. 1984. *Proc. Natl. Acad. Sci. USA* 81:674–78
27. von Heijne G. 1994. *Annu. Rev. Biophys. Biomol. Struct.* 23:167–92
28. Whitelegge JP, Zhang HM, Aguilera R, Taylor RM, Cramer WA. 2002. *Mol. Cell. Proteomics* 1:816–27
29. Martinez SE, Huang D, Szczepaniak A, Cramer WA, Smith JL. 1994. *Struct. Fold. Des.* 2:95–105
30. Carrell CJ, Schlarb BG, Bendall DS, Howe CJ, Cramer WA, Smith JL. 1999. *Biochemistry* 38:9590–99
31. Chi Y-I, Huang L-S, Zhang Z, Fernandez-Velasco JG, Berry EA. 2000. *Biochemistry* 39:7689–701
32. Carrell CJ, Zhang H, Cramer WA, Smith JL. 1997. *Structure* 5:1613–25
33. Berry EA, Guergova-Kuras M, Huang L-S, Crofts AR. 2000. *Annu. Rev. Biochem.* 69:1005–75
34. Crofts AR. 2004. *Annu. Rev. Psychol.* 66:689–733
35. Clark RD, Hawkesford MJ, Coughlan SJ, Hind G. 1984. *FEBS Lett.* 174:137–42
36. Zhang HM, Whitelegge JP, Cramer WA. 2001. *J. Biol. Chem.* 276:38159–65
37. Hamel P, Olive J, Pierre Y, Wollman FA, de Vitry C. 2000. *J. Biol. Chem.* 275:17072–79
38. Soriano GM, Ponamarev MV, Carrell CJ, Xia D, Smith JL, Cramer WA. 1999. *J. Bioenerg. Biomembr.* 31:201–13
39. Covian R, Trumpower BL. 2005. *J. Biol. Chem.* 280:22732–40
40. Gong X, Yu L, Xia D, Yu CA. 2005. *J. Biol. Chem.* 280:9251–57
41. Stiehl HH, Witt HT. 1969. *Z. Naturforsch. Teil B* 24:1588–98
42. Zhang HM, Kurisu G, Smith JL, Cramer WA. 2003. *Proc. Natl. Acad. Sci. USA* 100:5160–63
43. Link TA, Hagen WR, Pierik AJ, Assmann C, von Jagow G. 1992. *Eur. J. Biochem.* 208:685–91
44. Zhang H, Carrell CJ, Huang D, Sled V, Ohnishi T, et al. 1996. *J. Biol. Chem.* 271:31360–66
45. Iwata S, Saynovits M, Link TA, Michel H. 1996. *Structure* 4:567–79
46. Szczepaniak A, Black MT, Cramer WA. 1989. *Z. Naturforsch. Teil C* 44:453–61
47. Zhang H, Cramer WA. 2004. In *Photosynthesis Research Protocols*, ed. R Carpentier, pp. 67–78. Totowa, NJ: Humana Press
48. Murata N, Wada H, Gombos Z. 1992. *Plant Cell Physiol.* 33:933–41
49. Garavito RM, Ferguson-Miller S. 2001. *J. Biol. Chem.* 276:32403–6
50. Palsdottir H, Hunte C. 2004. *Biochim. Biophys. Acta* 1666:2–18
51. Nicholls A, Sharp K, Honig B. 1991. *Proteins: Struct. Funct. Genet.* 11:281–96
52. Wikström MKF, Berden JA. 1972. *Biochim. Biophys. Acta* 283:403–20
53. Mitchell P. 1976. *J. Theor. Biol.* 62:327–67
54. Crofts AR, Meinhardt SW, Jones KR, Snozzi M. 1983. *Biochim. Biophys. Acta* 723:202–18
55. Girvin ME, Cramer WA. 1984. *Biochim. Biophys. Acta* 767:29–38
56. Joliot P, Joliot A. 1986. *Photosynth. Res.* 9:113–24

57. Trumpower BL. 1990. *J. Biol. Chem.* 265:11409–12
58. Jünemann S, Heathcote P, Rich PR. 1998. *J. Biol. Chem.* 273:21603–7
59. Sacksteder CA, Kanazawa A, Jacoby ME, Kramer DM. 2000. *Proc. Natl. Acad. Sci. USA* 97:14283–88
60. Osyczka A, Moser CC, Dutton PL. 2004. *Nature* 427:607–12
61. Rich PR. 2004. *Biochim. Biophys. Acta* 1658:165–71
62. Mulkidjanian AY. 2005. *Biochim. Biophys. Acta* 1709:5–34
63. Cramer WA, Knaff DB. 1991. In *Energy Transduction in Biological Membranes. A Textbook of Bioenergetics.* New York: Springer-Verlag
64. Rich PR. 1985. *Photosynth. Res.* 6:335–48
65. Crofts AR. 2004. *Biochim. Biophys. Acta* 1655:77–92
66. Zito F, Finazzi G, Joliot P, Wollman FA. 1998. *Biochemistry* 37:10395–403
67. Muller FL, Roberts AG, Bowman MK, Kramer DM. 2003. *Biochemistry* 42:6493–99
68. Sun J, Trumpower BL. 2003. *Arch. Biochem. Biophys.* 419:198–206
69. Ouyang Y, Horn D, Grebe R, Guo L, Nelson ME, et al. 2004. *Photosynthesis: fundamental aspects to global perspectives.* Presented at 13th Int. Congr. Photosynth., Montreal, Can.
70. Gray HB, Winker JR. 2003. *Q. Rev. Biophys.* 36:341–72
71. Page CC, Moser CC, Chen X, Dutton PL. 1999. *Nature* 402:47–52
72. Yan J, Cramer WA. 2003. *J. Biol. Chem.* 278:20925–33
73. Yan J, Cramer WA. 2004. *J. Mol. Biol.* 344:481–93
74. Böhme H, Trebst A. 1969. *Biochim. Biophys. Acta* 180:137–48
75. Roberts AG, Kramer DM. 2001. *Biochemistry* 40:13407–12
76. Roberts AG, Bowman MK, Kramer DM. 2004. *Biochemistry* 43:7707–16
77. Yan J, Kurisu G, Cramer WA. 2006. *Proc. Natl. Acad. Sci. USA* 103:69–74
78. Mitchell P. 1966. *Biol. Rev.* 41:445–502
79. Wikström M, Krab K. 1986. *J. Bioenerg. Biomembr.* 18:181–93
80. Joliot P, Joliot A. 1994. *Proc. Natl. Acad. Sci. USA* 91:1034–38
81. Xiao KH, Yu L, Yu CA. 2000. *J. Biol. Chem.* 275:38597–604
82. Heimann S, Ponamarev MV, Cramer WA. 2000. *Biochemistry* 39:2692–99
83. Martinez SE, Huang D, Szczepaniak A, Cramer WA, Smith JL. 1994. *Structure* 2:95–105
84. Sainz G, Carrell CJ, Ponamarev MV, Soriano GM, Cramer WA, Smith JL. 2000. *Biochemistry* 39:9164–73
85. Ponamarev MV, Cramer WA. 1998. *Biochemistry* 37:17199–208
86. Soriano GM, Smith JL, Cramer WA. 2001. In *Handbook of Metalloproteins*, ed. A Messerschmidt, R Huber, K Wieghardt, T Poulos, pp. 172–81. London: Wiley
87. Xia D, Yu CA, Deisenhofer J, Xia J-Z, Yu L. 1996. *Biophys. J.* 70:A253
88. Tagawa K, Tsujimoto HY, Arnon DL. 1963. *Biochemistry* 49:567–72
89. Slovacek R, Hind G. 1980. *Plant Physiol.* 65:526–32
90. Moss DA, Bendall DS. 1984. *Biochim. Biophys. Acta* 767:389–95
91. Bendall DS, Manasse RS. 1995. *Biochim. Biophys. Acta* 1229:23–38
92. Selak MA, Whitmarsh J. 1982. *FEBS Lett.* 150:286–92
93. Rich PR, Madgwick SA, Moss DA. 1991. *Biochim. Biophys. Acta* 1058:312–28
94. Jones RW, Whitmarsh J. 1988. *Biochim. Biophys. Acta* 933:258–68
95. Lavergne J. 1983. *Biochim. Biophys. Acta* 725:25–33
96. Joliot P, Joliot A. 1988. *Biochim. Biophys. Acta* 933:319–33
97. de Vitry C, Desbois A, Redeker V, Zito F, Wollman F-A. 2004. *Biochemistry* 43:3956–68
98. Zhang H, Primak A, Bowman MK, Kramer DM, Cramer WA. 2004. *Biochemistry* 43:16329–36

98a. Alric J, Pierre Y, Picot D, Lavergne J, Rappaport F. 2005. *Proc. Natl. Acad. Sci. USA* 102:15860–65
99. Blankenship RE. 2002. *Molecular Mechanism of Photosynthesis*. Williston, VT: Blackwell Sci.
100. Finazzi G, Forti G. 2004. *Photosynth. Res.* 82:327–38
101. Joliot P, Joliot A. 2002. *Proc. Natl. Acad. Sci. USA* 99:10209–14
102. Kramer DM, Avenson TJ, Edwards GE. 2004. *Trends Plant Sci.* 9:349–57
103. Johnson GN. 2004. *Trends Plant Sci.* 9:570–71
104. Kramer DM, Avenson TJ, Edwards GE. 2004. *Trends Plant Sci.* 9:571–72
105. Johnson GN. 2005. *J. Exp. Bot.* 56:407–16
106. Zhang P, Battchikova N, Jansen T, Appel J, Ogawa T, Aro EM. 2004. *Plant Cell* 16:3326–40
107. Battchikova N, Zhang PP, Rudd S, Ogawa T, Aro EM. 2005. *J. Biol. Chem.* 280:2587–96
108. Lam E, Malkin R. 1982. *FEBS Lett.* 141:98–101
109. Furbacher PN, Girvin ME, Cramer WA. 1989. *Biochemistry* 28:8990–98
110. Finazzi G, Rappaport F, Furia A, Fleischmann M, Rochaix J-D, et al. 2002. *EMBO J.* 3:280–85
111. Fukuyama K, Hase T, Matsumoto S, Tsukihara T, Katsube Y, et al. 1980. *Nature* 286:522–24
112. Nelson ME, Finazzi G, Wang QJ, Middleton-Zarka KA, Whitmarsh J, Kallas T. 2005. *J. Biol. Chem.* 280:10395–402
113. Munekage Y, Hashimoto M, Miyake C, Tomizawa K, Endo T, et al. 2004. *Nature* 429:579–82
114. Okegawa Y, Tsuyama M, Kobayashi Y, Shikanai T. 2005. *J. Biol. Chem.* 280:28332–36
115. Nixon PJ. 2000. *Philos. Trans. R. Soc. London Ser. B* 355:1541–47
116. Peltier G, Cournac L. 2002. *Annu. Rev. Plant Biol.* 53:523–50
117. Schütz M, Brugna M, Lebrun E, Baymann F, Huber R, et al. 2000. *J. Mol. Biol.* 300:663–75
118. Kutoh E, Sone N. 1988. *J. Biol. Chem.* 263:9020–26
119. Sone N, Sawa G, Sone T, Noguchi S. 1995. *J. Biol. Chem.* 270:10612–17
120. Yu J, Hederstedt L, Piggot PJ. 1995. *J. Bacteriol.* 177:6751–60
121. Sone N, Tsuchiya N, Inoue M, Noguchi S. 1996. *J. Biol. Chem.* 271:12457–62
122. Yu J, Le Brun NE. 1998. *J. Biol. Chem.* 273:8860–66
123. Kuras R, de Vitry C, Choquet Y, Girardbascou J, Culler D, et al. 1997. *J. Biol. Chem.* 272:32427–35
124. Huang D, Everly RM, Cheng RH, Heymann JB, Schägger H, et al. 1994. *Biochemistry* 33:4401–9
125. Pierre Y, Breyton C, Lemoine Y, Robert B, Vernotte C, Popot J-L. 1997. *J. Biol. Chem.* 272:21901–8
126. Zhang H, Huang D, Cramer WA. 1999. *J. Biol. Chem.* 274:1581–87
127. Wenk S-O, Schneider D, Boronowsky U, Jager C, Klughammer C, et al. 2005. *FEBS Lett.* 272:582–92
128. Xiong J, Bauer CE. 2002. *J. Mol. Biol.* 322:1025–37
129. Dashdorj N, Zhang HM, Kim HY, Yan JS, Cramer WA, Savikhin S. 2005. *Biophys. J.* 88:4178–87
130. Vener A, van Kan PJ, Rich PR, Ohad II, Andersson B. 1997. *Proc. Natl. Acad. Sci. USA* 94:1585–90
131. Kim HY, Dashdorj N, Zhang HM, Yan JS, Cramer WA, Savikhin S. 2005. *Biophys. J.* 89:28–30
132. Burrows PA, Sazanov LA, Svab Z, Maliga P, Nixon PJ. 1998. *EMBO J.* 17:868–76
133. Joët T, Cournac L, Horvath EM, Medgyesy P, Peltier G. 2001. *Plant Physiol.* 125:1919–29

Subject Index

A

ABCD1 gene
    mammalian peroxisomes and, 317, 319, 321, 324
ABC transporter
    capsular polysaccharides in E. coli and, 54, 57–59
    mammalian peroxisomes and, 295, 297, 322–24
Abd-B gene
    chromosome and plasmid segregation, 224
Acatalasemia
    mammalian peroxisomes and, 316, 318, 321
Acceptors
    endogenous
        capsular polysaccharides in E. coli and, 56–57
Accessory proteins
    site-specific recombination and, 579–82
2′-O-Acetyl-ADP-ribose
    sirtuins and, 435–36, 441–42
Acetylation
    gene expression regulation and chromatin modifications, 243
    sirtuins and, 455, 457
ACOX1 gene
    mammalian peroxisomes and, 318, 320
Actins
    animal cytokinesis and, 547, 552, 555–56, 558
    axonal transport and Alzheimer's disease, 610, 618
    death-associated protein kinases and, 194, 202, 206
    dynamic filaments of bacterial cytoskeleton and, 467, 476–83
Activation
    cold-adapted enzymes and, 405, 408
    Hsp90 molecular chaperone and, 289
    LKB1-dependent signaling pathways and, 146–49
    sirtuins and, 454, 460–61
    site-specific recombination and, 569
Active site
    asparagine synthetase chemotherapy and, 640–41
    cold-adapted enzymes and, 403
    dynamic filaments of bacterial cytoskeleton and, 471–72
    eukaryotic signaling circuits and, 658, 660–64
    proton transfer through respiratory complexes and, 171–74
    ribonucleotide reductases and, 687–90, 698–99
Activity-stability relationship
    cold-adapted enzymes and, 415–17
Actomyosin
    animal cytokinesis and, 552, 555–56
Acute lymphoblastic leukemia (ALL)
    asparagine synthetase chemotherapy and, 629–49
N-Acylsulfonamide derivatives
    asparagine synthetase chemotherapy and, 642, 649
ADAM protein
    axonal transport and Alzheimer's disease, 616
Adaphostin
    tyrphostins and other tyrosine kinase inhibitors, 97–98
Adaptor proteins
    eukaryotic signaling circuits and, 658–59, 669–70, 672
"Addiction module"

chromosome and plasmid segregation, 226
Adenylate kinase
  cold-adapted enzymes and, 420–22
Adipose tissue
  obesity-related derangements in metabolic regulation and, 367, 370–74
ADP/ATP carrier
  mitochondrial, 713–37
ADP-ribosyl transferase
  sirtuins and, 445–46
ADPR-peptidyl-imidate
  sirtuins and, 450–52
Adrenomyeloneuropathy
  mammalian peroxisomes and, 321
AG 957
  tyrphostins and other tyrosine kinase inhibitors, 97–98
AG 1112
  tyrphostins and other tyrosine kinase inhibitors, 97, 99
AG 1318
  tyrphostins and other tyrosine kinase inhibitors, 97
AG 1478
  tyrphostins and other tyrosine kinase inhibitors, 99
Aging
  axonal transport and Alzheimer's disease, 607–21
  sirtuins and, 444–45
AGPS gene
  mammalian peroxisomes and, 303, 316, 318
Agrobacterium spp.
  capsular polysaccharides in *E. coli* and, 52
AGXT gene
  mammalian peroxisomes and, 316
AHA α-amylase
  cold-adapted enzymes and, 405–7, 409–16, 418–20, 422, 424–27
Aha1 protein
  Hsp90 molecular chaperone and, 285–86
Akt protein
  signaling pathways in skeletal muscle remodeling and, 27–28
Alcaligenes eutrophus
  NADH:quinone oxidoreductase (complex I) and, 76–77
Alcohol dehydrogenase
  protein motion and catalysis, 519–20, 530–35
Alkaline phosphatase
  cold-adapted enzymes and, 406, 411–12, 415–16, 422–23, 426

Allelic determinants
  capsular polysaccharides in *E. coli* and, 47
Allosteric regulation
  ribonucleotide reductases and, 681, 694–99
All-*trans*-retinal
  rhodopsin and, 743, 748–49
ALT pathway
  break-induced replication and recombinational telomere elongation in yeast, 131–32
Alzheimer's disease
  amyloid structures and protein misfolding, 333, 336–37, 356
  axonal transport and, 607–21
  LKB1-dependent signaling pathways and, 155
AMACR gene
  mammalian peroxisomes and, 316, 318–19, 321
amber gene
  history of research, 9, 11–12
Amidohydrolases
  N-terminal
    asparagine synthetase chemotherapy and, 632
Amino acid response pathway
  asparagine synthetase chemotherapy and, 635–37
5-Aminoimidazole-4-carboxamide riboside (AICAR)
  LKB1-dependent signaling pathways and, 147–48
AMN107
  tyrphostins and other tyrosine kinase inhibitors, 98–99
AMP-activated protein kinase (AMPK)
  LKB1-dependent signaling pathways and, 137, 144, 146–51, 156–57
α-Amylase
  cold-adapted enzymes and, 405–7, 409–16, 418–20, 422, 424–27
  obesity-related derangements in metabolic regulation and, 369
Amyloid-β peptide
  axonal transport and Alzheimer's disease, 614–16, 618–21
Amyloid structures
  protein misfolding and human disease
    aggregation rate, 353–54
    charge, 353
    common determinants, 356–57
    conformational states, 350–52
    disease, 335–41
    exploitation, 339
    fibril formation, 346–52

future research, 360
globular proteins, 349–50
hydrophobicity, 353
introduction, 334–35
nonchromosomal genetic elements, 339–41
normal biology, 335–41
nucleated growth, 347
oligomers, 347–49
partial unfolding, 349
pathogenesis of protein deposition diseases, 357–59
perspectives, 359
polymorphism, 345–46
polypeptide chains, 350–52
protein aggregation, 336–38
protofibrils, 347–49
secondary structure, 353
sequence, 352–57
solid-state NMR, 341–43
structure of amyloid fibrils, 341–46
summary points, 359–60
toxicity of prefibrillar aggregates, 358–59
unstructured aggregates, 348–49
various systems, 344–45
X-ray crystallography, 343
Anabaena spp.
site-specific recombination and, 569
Anabolic steroids
signaling pathways in skeletal muscle remodeling and, 32
Anillin
animal cytokinesis and, 547
Annexins
animal cytokinesis and, 548
Anoikis
death-associated protein kinases and, 203
Anterograde transport
axonal transport and Alzheimer's disease, 610–11, 617
ant genes
gene expression regulation and chromatin modifications, 248
mammalian peroxisomes and, 324
Antimycin A
transmembrane traffic in cytochrome $b_6 f$ complex and, 781–82
Aph-1 protein
axonal transport and Alzheimer's disease, 617
Aplysia californica
amyloid structures and protein misfolding, 340
Apoptosis
death-associated protein kinases and, 189–207
sirtuins and, 436, 439–40
tyrphostins and other tyrosine kinase inhibitors, 93, 97
APPL gene
axonal transport and Alzheimer's disease, 617
APP protein
axonal transport and Alzheimer's disease, 612, 615–19, 621
Aquaspirillum arcticum
cold-adapted enzymes and, 420–21
Aquifex aeolicus
NADH:quinone oxidoreductase (complex I) and, 72
Arabidopsis thaliana
NADH:quinone oxidoreductase (complex I) and, 73, 81
transmembrane traffic in cytochrome $b_6 f$ complex and, 782
Archaea
cold-adapted enzymes and, 404, 409, 419, 423–24, 427
Archaeoglobus fulgidus
NADH:quinone oxidoreductase (complex I) and, 76
sirtuins and, 438
Arginine-mediated interactions
cold-adapted enzymes and, 424
Arginine methyltransferases
histone
gene expression regulation and chromatin modifications, 258–59
Aromatic interactions
cold-adapted enzymes and, 424–25
Arp1 protein
axonal transport and Alzheimer's disease, 610
Arthrobacter spp.
cold-adapted enzymes and, 406, 412, 415–16, 420–21
AS-B enzyme
asparagine synthetase chemotherapy and, 642, 646
Asparagine synthetase (ASNS) chemotherapy
active site, 640–41
amino acids, 635–37
cell cycle, 634–35
chemical constraints, 640–41
childhood ALL, 637–39
drug-resistant leukemia, 634–39
enzyme inhibitors, 639–46
expression, 637–39
future research, 649
gene transcription, 635–37

introduction, 630–31
kinetic mechanism, 633–34
nanomolecular inhibitor, 643–46
perspectives, 646–48
related resources, 649
structure, 631–33, 640–41
sulfonamide derivatives, 641–43
summary points, 648–49
Aspartate aminotransferase
cold-adapted enzymes and, 409
β-Aspartyl-AMP
asparagine synthetase chemotherapy and, 633–34
Atomic force microscopy (AFM)
amyloid structures and protein misfolding, 336, 347
ATPases
chromosome and plasmid segregation, 226–27
dynamic filaments of bacterial cytoskeleton and, 483–86
Hsp90 molecular chaperone and, 271–90
ATP cone
ribonucleotide reductases and, 698–99
ATP mimics
tyrphostins and other tyrosine kinase inhibitors, 99
Atractyloside
mitochondrial ADP/ATP carrier and, 715
Atrophy
signaling pathways in skeletal muscle remodeling and, 28, 31–32
att genes
site-specific recombination and, 579–80, 582, 599
Autophagy
death-associated protein kinases and, 189, 195, 198, 200, 206
Axonal transport
Alzheimer's disease and
aging, 612–13
future research, 621
introduction, 608
mechanisms of axonal transport, 610–11
neurodegenerative diseases, 613–14
overview of Alzheimer's disease, 614–16
overview of axonal transport, 608–9
pathogenesis of Alzheimer's disease, 617–21
pathology of Alzheimer's disease, 618–19
proteins, 617, 609–10
regulation of axonal transport, 611–12
summary points, 621
Azotobacter spp.

cold-adapted enzymes and, 406
history of research, 5

B

BACE enzyme
axonal transport and Alzheimer's disease, 615–19
Bacillus spp.
cold-adapted enzymes and, 406, 411, 413, 416, 420–21
Bacillus subtilis
chromosome and plasmid segregation, 213–14, 223–24, 229
cold-adapted enzymes and, 406, 417
dynamic filaments of bacterial cytoskeleton and, 469, 479–82, 485
NADH:quinone oxidoreductase (complex I) and, 77, 79
site-specific recombination and, 569, 595
Base exchange
sirtuins and, 447–48
Basic leucine zipper transcription factor (bZIP)
asparagine synthetase chemotherapy and, 635–36
Bcr-Abl kinase
tyrphostins and other tyrosine kinase inhibitors, 97–99
β-Lactamase
cold-adapted enzymes and, 409
β-Lactam synthetase (BLS)
asparagine synthetase chemotherapy and, 633, 640, 642, 648
BHK *ts11* cells
asparagine synthetase chemotherapy and, 634–35
Bithorax gene
gene expression regulation and chromatin modifications, 248
Blebbistatin
animal cytokinesis and, 557
BMS-354825
tyrphostins and other tyrosine kinase inhibitors, 98–99
Bombyx mori
amyloid structures and protein misfolding, 340
Bongkrekic acid
mitochondrial ADP/ATP carrier and, 716, 732
Borealin
animal cytokinesis and, 548
Borrelia burgdorferi
site-specific recombination and, 582

Bos taurus
    NADH:quinone oxidoreductase (complex I) and, 73
Brain
    axonal transport and Alzheimer's disease, 607–21
Branchiostoma spp.
    rhodopsin and, 748
BRCA1/2 genes
    animal cytokinesis and, 560
    history of research, 14
Breakage
    strand
        site-specific recombination and, 567, 570–72, 584–86
Break-induced replication
    recombinational telomere elongation in yeast and
        ALT pathway, 131–32
        break-induced replication, 112–16
        chromosomal ends, 129–30
        growth senescence, 116–19
        introduction, 112
        mitochondrial DNAs, 130
        molecular mechanisms, 115–16
        recombinational telomere elongation, 119–30
        repair of nontelomeric broken ends, 112–15
        "roll and spread" model, 123–28
        survivor formation by recombination-based mechanisms, 119–23
        telomerase, 128–32
        terminal palindromes, 130–31
BRSK kinases
    LKB1-dependent signaling pathways and, 154–56
BtubA/B proteins
    dynamic filaments of bacterial cytoskeleton and, 475–76

C

Caenorhabditis elegans
    animal cytokinesis and, 544, 547–48, 551–52
    chromosome and plasmid segregation, 222–23
    eukaryotic signaling circuits and, 667
    LKB1-dependent signaling pathways and, 143–44, 147, 154–55
    mammalian peroxisomes and, 303, 311
    NADH:quinone oxidoreductase (complex I) and, 73
    sirtuins and, 440, 444
    telomerase reverse transcriptase and, 496–98
Calcineurin
    signaling pathways in skeletal muscle remodeling and, 19–20, 24–26
Calcium/calmodulin-dependent protein kinase (CaMK)
    signaling pathways in skeletal muscle remodeling and, 19, 26–27
Calmodulin
    death-associated protein kinases and, 190–91, 193
Calmodulin-regulated kinase kinase (CaMKK)
    death-associated protein kinases and, 203, 205
Caloric restriction
    sirtuins and, 444–45
Campylobacter jejuni
    NADH:quinone oxidoreductase (complex I) and, 76
Cancer
    animal cytokinesis and, 560
    asparagine synthetase chemotherapy and, 629–49
    LKB1-dependent signaling pathways and, 139, 149–52
    tyrphostins and other tyrosine kinase inhibitors, 93–105
Candida spp.
    amyloid structures and protein misfolding, 346
    break-induced replication and recombinational telomere elongation in yeast, 130
    cold-adapted enzymes and, 416
    mammalian peroxisomes and, 324
    sirtuins and, 436
Capsular polysaccharides in *E. coli*
    ABC transporter, 57–59
    additional proteins, 59–61
    allelic determinants, 47
    assembly, 45–62
    biosynthesis, 45–62
    chain elongation, 55–56
    colanic acid, 44–47, 50–51
    conclusions, 62
    cps loci, 45–46
    endogenous acceptor, 56–57
    export, 57–59
    future research, 63–64
    genetics, 45–46, 54–55
    group 1 capsules, 42–53
    group 2 capsules, 44–62
    group 3 capsules, 44–62
    group 4 capsules, 42–44, 47–49, 52–53

initiation reactions, 56–57
introduction, 40–41
K antigens, 47
K_{LPS}, 49
kps loci, 54–55, 61
LPS, 42–44, 49
O antigens, 47
OMA proteins, 52–53
polymerization, 50–51
structures, 41–45
summary points, 62–63
surface association, 41–45
translocation, 52–53
type 4 secretion systems, 53
wza, 52–53
wzb, 52–53
wzc, 50–53
wzi, 53
wzy, 47–49
wzz, 49

Carboxyatractyloside
mitochondrial ADP/ATP carrier and, 715, 722

Cardiolipins
mitochondrial ADP/ATP carrier and, 715, 736

Cardiovascular disease
obesity-related derangements in metabolic regulation and, 367–68

Cargo proteins
axonal transport and Alzheimer's disease, 608–9, 611, 618, 620

CARM1 gene
gene expression regulation and chromatin modifications, 258

Carnitine palmitoyltransferase 1
obesity-related derangements in metabolic regulation and, 369

β-Carotene
transmembrane traffic in cytochrome b_6f complex and, 769, 787

Casein
cold-adapted enzymes and, 415

Caspases
death-associated protein kinases and, 206

Catabolism
mammalian peroxisomes and, 313

Catalysis
death-associated protein kinases and, 192–96
eukaryotic signaling circuits and, 664–69
Hsp90 molecular chaperone and, 274–75
NADH:quinone oxidoreductase (complex I) and, 83–84
protein motion and, 519–38

ribonucleotide reductases and, 687–90
sirtuins and, 454–56
site-specific recombination and, 577–78, 583
telomerase reverse transcriptase and, 499–500

CAT genes
mammalian peroxisomes and, 318, 321

Caulobacter crescentus
chromosome and plasmid segregation, 215, 223, 230, 233
dynamic filaments of bacterial cytoskeleton and, 480–81, 483

Cdc37 protein
Hsp90 molecular chaperone and, 284–85

Celecoxib
LKB1-dependent signaling pathways and, 152

Cell cycle
animal cytokinesis and, 549–51, 557
asparagine synthetase chemotherapy and, 634–35
chromosome and plasmid segregation, 211–37
dynamic filaments of bacterial cytoskeleton and, 467

Cell polarity
LKB1-dependent signaling pathways and, 154–56

Cell-surface biogenesis
capsular polysaccharides in *E. coli* and, 39–64

Cellulase
cold-adapted enzymes and, 406, 420–22

Centriolin
animal cytokinesis and, 548, 559–60

Centromeres
chromosome and plasmid segregation, 229–30

Ceramide
death-associated protein kinases and, 202, 206

CGP 57148
tyrphostins and other tyrosine kinase inhibitors, 97

CGP 76030
tyrphostins and other tyrosine kinase inhibitors, 98

Chaenocephalus aceratus
cold-adapted enzymes and, 406, 411

Chain elongation
capsular polysaccharides in *E. coli* and, 55–56

Chaperones
Hsp90 molecular chaperone and, 271–90

Chaplins
amyloid structures and protein misfolding, 340

Chemotherapy
asparagine synthetase chemotherapy and, 629–49

tyrphostins and other tyrosine kinase
inhibitors, 93
Children
asparagine synthetase chemotherapy and,
629–49
Chitinases
cold-adapted enzymes and, 406–7
Chitobiase
cold-adapted enzymes and, 406, 409, 415, 424
Chlamydia trachomatis
ribonucleotide reductases and, 685–87
Chlamydomonas reinhardtii
NADH:quinone oxidoreductase (complex I)
and, 73, 81
transmembrane traffic in cytochrome $b_6 f$
complex and, 769–72, 775, 777–79, 783–86
Chlorella spp.
transmembrane traffic in cytochrome $b_6 f$
complex and, 783–84
Chlorophyll *a*
transmembrane traffic in cytochrome $b_6 f$
complex and, 769
Cholesterol
mammalian peroxisomes and, 314–15
Chromatin
sirtuins and, 438
Chromatin immunoprecipitation (ChIP)
asparagine synthetase chemotherapy and,
636–37
Chromophores
rhodopsin and, 753–54
Chromosome segregation
dynamic filaments of bacterial cytoskeleton
and, 467, 480
plasmid segregation and
addendum, 235–36
bacteria, 213–16, 223–30, 233
cell division, 233
centromeres, 229–30
chromosome segregation, 213–25, 229–30,
233–35
cohesin, 224–25, 231
cohesion, 217–20
coils, 233
condensation, 223–24
condensins, 222–25
counting chromosomes, 217
cytoskeleton, 233
DNA compaction, 224
DNA segregation, 223–24
DNA translocases, 233–35
dynamic protein rings, 233

eukaryotes, 216–22
faithful chromosome segregation, 233–35
future research, 237
genome maintenance, 233–35
introduction, 212–13
meiosis, 220–22
mitotic spindle, 231–32
nascent duplexes, 215–16
nucleoid, 213
oscillators, 233
pairing, 220–22
par, 229–30
partitioning, 226–29, 231
plasmid segregation, 225–33
proteases, 220
replication, 213–15, 217
separase, 220
site-specific recombinases, 233–35
summary points, 236–37
topoisomerases, 233–35
unpairing, 220–22
yeast plasmid, 230–33
Chronic myelogenous leukemia (CML)
tyrphostins and other tyrosine kinase
inhibitors, 97–99
Ciona intestinalis
eukaryotic signaling circuits and, 667
Circular dichroism (CD)
amyloid structures and protein misfolding,
339, 348
cold-adapted enzymes and, 410, 417–18
cis-acting factors
eukaryotic signaling circuits and, 658
Citrate synthase
cold-adapted enzymes and, 416, 419–22, 424
c-Jun N-terminal kinase
eukaryotic signaling circuits and, 662
Cleavage furrow
animal cytokinesis and, 543–61
Client proteins
Hsp90 molecular chaperone and, 274–75, 289
Clostridium spp.
cold-adapted enzymes and, 411, 413
NADH:quinone oxidoreductase (complex I)
and, 76
site-specific recombination and, 569
Coactivators
obesity-related derangements in metabolic
regulation and, 375
signaling pathways in skeletal muscle
remodeling and, 19, 27
Cochaperones

Hsp90 molecular chaperone and, 271–90
Coevolution
   protein motion and catalysis, 525
Cofilin
   animal cytokinesis and, 547, 557
   axonal transport and Alzheimer's disease, 618
Cohesin
   chromosome and plasmid segregation, 216–19, 224–25, 231–32
Cohesion
   chromosome and plasmid segregation, 211, 215, 217–20
Coiled-coils
   chromosome and plasmid segregation, 218, 225, 233
Colanic acid
   capsular polysaccharides in E. coli and, 44–47, 50–51
Cold-adapted enzymes
   activation energy, 408
   activity, 405–10
   activity-stability relationship, 415–17
   arginine-mediated interactions, 424
   aromatic interactions, 424–25
   conformational stability, 412–14
   core hydrophobicity, 421–22
   denaturation, 414–15
   electrostatic interactions, 423–25
   flexibility, 417–27
   folding funnel model, 419
   future research, 427–28
   global flexibility, 418–19
   H-bonds, 423–24
   introduction, 404–5
   k_{cat}, 407–8
   kinetic stability, 411–12
   k_m, 408–10
   local flexibility, 418–19
   miscellaneous factors, 426–27
   salt bridges, 425
   secondary structure elements, 425–26
   stability, 410–17
   structural adaptation, 419–27
   summary points, 427
   surface hydrophilicity, 423
   surface hydrophobicity, 422–23
   surface loops, 426
   unfolding, 410
   viscosity, 408
Colwellia maris
   cold-adapted enzymes and, 406
COMPASS proteins
   gene expression regulation and chromatin modifications, 251–53
Completion
   animal cytokinesis and, 558–60
Condensation
   chromosome and plasmid segregation, 211, 223–24
Condensins
   chromosome and plasmid segregation, 217, 219, 222–25
Conformational change
   cold-adapted enzymes and, 412–14
   dynamic filaments of bacterial cytoskeleton and, 474–75
   Hsp90 molecular chaperone and, 271–90
   mitochondrial ADP/ATP carrier and, 732–34
   protein motion and catalysis, 526–28, 532–33
Congo Red
   amyloid structures and protein misfolding, 336, 339, 348, 350, 353, 358
Connectivity
   eukaryotic signaling circuits and, 664–69
Conserved residues
   mitochondrial ADP/ATP carrier and, 728
Contractile ring
   animal cytokinesis and, 543–61
Cooperative assembly
   dynamic filaments of bacterial cytoskeleton and, 472–73
Corepressors
   signaling pathways in skeletal muscle remodeling and, 19
Coulombic interactions
   cold-adapted enzymes and, 408
Coupling
   proton transfer through respiratory complexes and, 171
Cowden's syndrome
   LKB1-dependent signaling pathways and, 149–50
Cox-2 gene
   LKB1-dependent signaling pathways and, 152
cps genes
   capsular polysaccharides in E. coli and, 45–46
CPT-malonyl-CoA system
   obesity-related derangements in metabolic regulation and, 375–76, 379
Crescentin
   dynamic filaments of bacterial cytoskeleton and, 483
Cross talk
   death-associated protein kinases and, 203

transmembrane traffic in cytochrome $b_6 f$
   complex and, 780
Crossover site cleavage
   site-specific recombination and, 586, 589–90
C-terminal domains
   Hsp90 molecular chaperone and, 275–76
Cu^{2+}
   proton transfer through respiratory complexes
      and, 168–69, 171
Curlin
   amyloid structures and protein misfolding, 340
"Cut and separate" mechanism
   chromosome and plasmid segregation, 218
Cyanidioschyzon merolae
   chromosome and plasmid segregation, 222
Cyclins
   animal cytokinesis and, 548
Cytochalasin
   animal cytokinesis and, 557
Cytochrome $b_6 f$ complex
   transmembrane traffic in
      bound pigment molecules, 786–87
      cyclic electron transport, 784–85
      cytochrome f, 780–81
      dimer organization, 772, 774
      extrinsic domains on p side, 780
      heme x, 783–86
      intermonomer cross talk, 780
      introduction, 770
      lipids, 774–75
      Mastigocladus laminosus, 774–75
      oxidant-induced reduction, 775–77
      prosthetic groups, 771–73
      Q_p pocket, 778–79
      quinol oxidation state, 775–81
      quinone reduction site, 781–86
      Rieske ISP, 780–81
      structure, 771–75
      subunit composition, 771–73
      TDS ring-out conformation, 779–80
      weakly bound protein components, 772
Cytochrome c oxidase
   energy transduction and proton transfer
      through respiratory complexes, 165–80
Cytochrome c reductase
   history of research, 4
Cytokinesis
   animal
      cancer, 560
      classic models, 550–52
      cleavage plane specification, 550–53
      completion, 558–60
      contractile ring assembly, 555–57
      furrow ingression, 557–58
      future research, 561
      introduction, 544–46
      microtubule organization, 553–55
      molecular mechanisms, 552–53
      molecules involved, 546–48
      research approaches, 544–46
      subprocesses, 549–60
      summary points, 561
      timing, 549–50
Cytoskeleton
   chromosome and plasmid segregation, 233
   death-associated protein kinases and, 201–3,
      206
   dynamic filaments of bacterial cytoskeleton
      and, 467–88

D

Deacetylation
   signaling pathways in skeletal muscle
      remodeling and, 23–25
   sirtuins and, 435–61
Death-associated protein kinases (DAPKs)
   catalytic domain, 192–96
   cross talk, 203
   cytoskeleton, 201–3
   extracatalytic domains, 193–96
   function, 197–207
   future research, 207
   introduction, 190–91
   molecular dissection, 199–201
   nomenclature, 198–99
   nonapoptotic functions, 203–4
   phosphorylation, 191
   regulation, 192–93, 196–97
   related proteins, 191–97
   signaling network, 204–7
   structure, 192–96
   substrates, 191
   summary points, 207
n-Decylubiquinone
   NADH:quinone oxidoreductase (complex I)
      and, 83
Degeneration
   axonal transport and Alzheimer's disease,
      607–21
Deletion
   eukaryotic signaling circuits and, 655
   site-specific recombination and, 567
Dementia

axonal transport and Alzheimer's disease, 607–21
Demethylation
    gene expression regulation and chromatin modifications, 261–63
Denaturation
    cold-adapted enzymes and, 411, 414–15
Density-function theory
    ribonucleotide reductases and, 687
Desulfovibrio fructosovorans
    NADH:quinone oxidoreductase (complex I) and, 77
Diabetes
    LKB1-dependent signaling pathways and, 151
    obesity-related derangements in metabolic regulation and, 367–68, 382
    signaling pathways in skeletal muscle remodeling and, 29–31
Differential scanning calorimetry (DSC)
    cold-adapted enzymes and, 410, 417–18
Diffuse axonal injury
    axonal transport and Alzheimer's disease, 614
dif genes
    chromosome and plasmid segregation, 234
Dihydrofolate reductase
    cold-adapted enzymes and, 410, 413–14
    protein motion and catalysis, 519–30
Dimerization
    Hsp90 molecular chaperone and, 275–76
    mitochondrial ADP/ATP carrier and, 734–35
    rhodopsin and, 743, 757–59
    telomerase reverse transcriptase and, 505–7
    transmembrane traffic in cytochrome $b_6 f$ complex and, 772, 774
Disintegrins
    axonal transport and Alzheimer's disease, 616
Distributed surfaces
    eukaryotic signaling circuits and, 661–62
Diversity
    signaling pathways in skeletal muscle remodeling and, 21–22
DNA compaction
    chromosome and plasmid segregation, 224
DNA invertases
    site-specific recombination and, 567, 583–84, 596–98
DNA ligases
    cold-adapted enzymes and, 409–13, 422, 426
    history of research, 1, 10–11
DNA polymerase
    history of research, 1, 4–12
DNA replication
    history of research, 1, 12–15
DNA segregation
    chromosome and plasmid segregation, 223–24
DNA translocases
    chromosome and plasmid segregation, 233–35
dNTP pools
    ribonucleotide reductases and, 681, 699–700
Docking
    eukaryotic signaling circuits and, 658–64, 672
Domain discrimination
    eukaryotic signaling circuits and, 665, 667–69
Dot1 gene
    gene expression regulation and chromatin modifications, 256–57
Double-strand break repair
    break-induced replication and recombinational telomere elongation in yeast, 111–32
D pathway
    proton transfer through respiratory complexes and, 168–70, 175–77, 179
Drosophila melanogaster
    animal cytokinesis and, 544–45, 547–48, 550, 557–59
    asparagine synthetase chemotherapy and, 636
    axonal transport and Alzheimer's disease, 617–21
    chromosome and plasmid segregation, 218, 224
    death-associated protein kinases and, 194
    eukaryotic signaling circuits and, 667
    gene expression regulation and chromatin modifications, 244–45, 248, 254–55, 262
    history of research, 12–14
    LKB1-dependent signaling pathways and, 143–44, 147, 154–56
    NADH:quinone oxidoreductase (complex I) and, 73
    sirtuins and, 444
    telomerase reverse transcriptase and, 498
DRP-1 protein
    death-associated protein kinases and, 191
Drug binding
    Hsp90 molecular chaperone and, 273–74
Drug resistance
    tyrphostins and other tyrosine kinase inhibitors, 98–100
Duchenne muscle dystrophy
    signaling pathways in skeletal muscle remodeling and, 28–29
Duplexes
    chromosome and plasmid segregation, 215–16
Dynactin

axonal transport and Alzheimer's disease, 610
Dynamic protein rings
    chromosome and plasmid segregation, 233
Dynamics
    cold-adapted enzymes and, 418
    dynamic filaments of bacterial cytoskeleton and, 472
    membrane proteins and, 709–10
Dynamin
    animal cytokinesis and, 547
Dynein
    axonal transport and Alzheimer's disease, 609–11

E

EcoB
    history of research, 9
EF-2 protein
    cold-adapted enzymes and, 410, 412
Effectors
    sirtuins and, 441–42
EGFR kinase inhibitors
    tyrphostins and other tyrosine kinase inhibitors, 99–101
Ehhadh gene
    mammalian peroxisomes and, 321
Elastase
    cold-adapted enzymes and, 420–21
Electron microscopy (EM)
    animal cytokinesis and, 554, 558
Electron paramagnetic resonance (EPR)
    transmembrane traffic in cytochrome $b_6 f$ complex and, 779
Electron transport
    history of research, 4
    NADH:quinone oxidoreductase (complex I) and, 69–87
    proton transfer through respiratory complexes and, 165, 168–80
    transmembrane traffic in cytochrome $b_6 f$ complex and, 769, 781–82, 784–85
Electrophoresis mobility shift analysis (EMSA)
    asparagine synthetase chemotherapy and, 636
Electrostatic interactions
    amyloid structures and protein misfolding, 353
    cold-adapted enzymes and, 408–9, 423–25
    protein motion and catalysis, 529
    proton transfer through respiratory complexes and, 180
    site-specific recombination and, 571, 574
Elongation
    break-induced replication and recombinational telomere elongation in yeast, 111–32
    gene expression regulation and chromatin modifications, 243
    telomerase reverse transcriptase and, 493–512
"Embrace" mechanism
    chromosome and plasmid segregation, 225
Encephalitozoon cuniculi
    telomerase reverse transcriptase and, 498
Endocellulase
    cold-adapted enzymes and, 406
Energy transduction
    proton transfer through respiratory complexes and, 165–82
Enterobacteria
    ribonucleotide reductases and, 692
Enthalpy
    cold-adapted enzymes and, 403, 412–14
Entropy
    cold-adapted enzymes and, 407, 412, 414, 422
Enzyme inhibitors
    asparagine synthetase chemotherapy and, 629, 639–46
Epigenetic regulation
    gene expression regulation and chromatin modifications, 263, 243
Equatorial relaxation
    animal cytokinesis and, 557
Erbstatin
    tyrphostins and other tyrosine kinase inhibitors, 96–97
Erlotinib
    tyrphostins and other tyrosine kinase inhibitors, 99–100
Erwinia chrysanthemi
    cold-adapted enzymes and, 406
Escherichia coli
    amyloid structures and protein misfolding, 339–40, 354, 358
    asparagine synthetase chemotherapy and, 631–32, 642
    capsular polysaccharides and, 39–64
    chromosome and plasmid segregation, 213–14, 216, 223, 227–28, 230, 233–34, 236
    cold-adapted enzymes and, 411, 413
    dynamic filaments of bacterial cytoskeleton and, 469–70, 477, 480–81, 484–85
    history of research, 3–14
    Hsp90 molecular chaperone and, 274–75, 279
    NADH:quinone oxidoreductase (complex I) and, 72, 76–78, 84
    protein motion and catalysis, 521–22, 524, 532

proton transfer through respiratory complexes and, 167
ribonucleotide reductases and, 683–90, 692–93, 698
site-specific recombination and, 568–69, 595

Est2 protein
    telomerase reverse transcriptase and, 501, 509
Esterase
    cold-adapted enzymes and, 417
Ether phospholipids
    mammalian peroxisomes and, 302–5
Etoposide
    death-associated protein kinases and, 202
Eucarya
    cold-adapted enzymes and, 419, 427
Euglena gracilis
    ribonucleotide reductases and, 683–84
Euplotes crassus
    telomerase reverse transcriptase and, 498, 500, 505, 507
"Eversporting displacement"
    gene expression regulation and chromatin modifications, 248
Evolution
    dynamic filaments of bacterial cytoskeleton and, 467
    NADH:quinone oxidoreductase (complex I) and, 69–87
    protein motion and catalysis, 525
    ribonucleotide reductases and, 681, 700–1
    telomerase reverse transcriptase and, 496–98
Evolvability
    eukaryotic signaling circuits and, 655–60, 674–75
Exchange
    strand
        site-specific recombination and, 567, 570–77, 584–86, 590–92
Excision
    site-specific recombination and, 567, 569, 583, 592–96
Exercise adaptation
    signaling pathways in skeletal muscle remodeling and, 19
Exit simulations
    proton transfer through respiratory complexes and, 179–80
Exit/backflow pathway
    proton transfer through respiratory complexes and, 178
Exonuclease I
    history of research, 8, 9

Export
    capsular polysaccharides in *E. coli* and, 39, 57–59
Extracatalytic domains
    death-associated protein kinases and, 193–96
Extracellular signal-regulated kinase (ERK)
    death-associated protein kinases and, 195, 197, 203, 205–6
    eukaryotic signaling circuits and, 662–63
Extremophiles
    cold-adapted enzymes and, 403

F

Faithful chromosome segregation
    chromosome and plasmid segregation, 233–35
Fast-twitch myofibers
    signaling pathways in skeletal muscle remodeling and, 21, 25
Fatty acid oxidation
    mammalian peroxisomes and, 295–325
FEAR complex
    chromosome and plasmid segregation, 219
Ferredoxin
    transmembrane traffic in cytochrome $b_6 f$ complex and, 769, 784–85
Fibrillar aggregates
    amyloid structures and protein misfolding, 333–60
Fibronectin
    death-associated protein kinases and, 202–3
Filaments
    bacterial cytoskeleton
        actin, 476–77
        actin homologues, 476–83
        active site, 471–72
        conformational change, 474–75
        cooperative assembly, 472–73
        crescentin, 483
        FtsA, 482–83
        FtsZ, 468–75
        future research, 487–88
        intermediate filament homologue, 483
        introduction, 468
        isodesmic assembly, 472–73
        molecular function, 481–82
        MreB, 479, 480–82
        nucleotide exchange, 473–74
        nucleotide hydrolysis, 476
        ParM, 477–79
        self-assembly in vitro, 469–70, 475–76

summary points, 486–87
tubulin homologues, 468–76
Walker A cytoskeletal ATPase, 483–86
Flavin mononucleotide
NADH:quinone oxidoreductase (complex I) and, 69–87
Flexibility
cold-adapted enzymes and, 403, 405, 409–10, 414–15, 417–27
Fluorescence recovery after photobleaching (FRAP)
dynamic filaments of bacterial cytoskeleton and, 469
Fluorescence resonance energy transfer (FRET)
animal cytokinesis and, 552
Folding funnel model
cold-adapted enzymes and, 419
Formin
animal cytokinesis and, 557
Fourier transform infrared (FTIR) spectroscopy
amyloid structures and protein misfolding, 345
proton transfer through respiratory complexes and, 170, 172–73, 180
FOXO factors
sirtuins and, 440
FRT genes
chromosome and plasmid segregation, 230, 234–35
Fts genes
dynamic filaments of bacterial cytoskeleton and, 467–75, 482–83
Functional modularity
eukaryotic signaling circuits and, 657–58
Fungi
amyloid structures and protein misfolding, 340
NADH:quinone oxidoreductase (complex I) and, 75, 86
sirtuins and, 437–45
Furrow ingression
animal cytokinesis and, 557–58

G

Gadus morhua
cold-adapted enzymes and, 406, 420–21
β-Galactosidase
cold-adapted enzymes and, 410, 424
gal genes
capsular polysaccharides in *E. coli* and, 46–47
gene expression regulation and chromatin modifications, 253, 260
Gating
proton transfer through respiratory complexes and, 170
Gefitinib
tyrphostins and other tyrosine kinase inhibitors, 99–100
Gelation-contraction
animal cytokinesis and, 555–56
Gene expression regulation
chromatin modifications by methylation and ubiquitination in
COMPASS, 251–53
enzymes, 246–48
future research, 263–64
histone arginine methyltransferases, 258–59
histone demethylation, 261–63
histone methylation, 245–63
histone methyltransferases, 256–58
introduction, 244
lysine, 248–58
MLL, 251–53
polycomb-group silencing, 254–56
posttranslational modification, 245–46
Set2, 253–54
summary points, 263–64
transcriptional memory, 260–61
ubiquitination, 254–56
X-chromosome inactivation, 254–56
Gene silencing
animal cytokinesis and, 543–46, 548, 552, 554, 556
gene expression regulation and chromatin modifications, 248–51, 254–56
sirtuins and, 435, 438, 440
Genomics
animal cytokinesis and, 543, 560
chromosome and plasmid segregation, 211–12, 233–35
NADH:quinone oxidoreductase (complex I) and, 69–87
Gentobiose
history of research, 9
Gin DNA invertase
site-specific recombination and, 583–84
Gleevec
tyrphostins and other tyrosine kinase inhibitors, 99
Global flexibility
cold-adapted enzymes and, 418–19
Global proteomic analysis (GPS)
gene expression regulation and chromatin modifications, 259

Globular proteins
   amyloid structures and protein misfolding, 349–50
Glucanases
   cold-adapted enzymes and, 409
Glucose-stimulated insulin secretion (GSIS)
   obesity-related derangements in metabolic regulation and, 386–91, 393
β-Glucosidase
   cold-adapted enzymes and, 416–17
GLUT4 transporter
   signaling pathways in skeletal muscle remodeling and, 29
Glutamate dehydrogenase
   cold-adapted enzymes and, 406, 411–12
"Glutamate trap"
   proton transfer through respiratory complexes and, 172
Glutamine
   asparagine synthetase chemotherapy and, 629–49
Glyceraldehyde-3-phosphate dehydrogenase
   cold-adapted enzymes and, 422
Glycogen synthase kinase-3 (GSK3)
   eukaryotic signaling circuits and, 663
Glycosyl hydrolase
   cold-adapted enzymes and, 411–13
Glycosyltransferases
   capsular polysaccharides in *E. coli* and, 39, 45, 49, 56, 60–61
Glyoxylates
   mammalian peroxisomes and, 312–13
gnd genes
   capsular polysaccharides in *E. coli* and, 46
GNPAT gene
   mammalian peroxisomes and, 303, 316, 318, 320
G protein-coupled receptors
   rhodopsin and, 743–62
Green fluorescent protein (GFP)
   animal cytokinesis and, 559
   death-associated protein kinases and, 194
Growth senescence
   break-induced replication and recombinational telomere elongation in yeast, 116–19
GSK3β protein
   axonal transport and Alzheimer's disease, 612, 618
GTPase-activating protein (GAP)
   animal cytokinesis and, 552–53
GTPases
   cold-adapted enzymes and, 406–7, 409, 411

Guanine nucleotide exchange factor (GEF)
   animal cytokinesis and, 550, 552–53
Guanosine-5′-monophosphate synthetase
   asparagine synthetase chemotherapy and, 633

H

Haemophilus influenzae
   capsular polysaccharides in *E. coli* and, 44
   chromosome and plasmid segregation, 229
Half-of-the-sites reactivity
   site-specific recombination and, 577–78
"Hand-in-hand" mechanism
   chromosome and plasmid segregation, 225
Heat lability
   cold-adapted enzymes and, 403–28
Helicobacter pylori
   NADH:quinone oxidoreductase (complex I) and, 76
   site-specific recombination and, 569, 583
Heme
   proton transfer through respiratory complexes and, 168, 170–72, 174, 180–81
   transmembrane traffic in cytochrome $b_6 f$ complex and, 769, 783–86
Herpes simplex virus type 1 (HSV-1)
   history of research, 14–15
Heterochromatic silencing
   gene expression regulation and chromatin modifications, 248–51
HET-s protein
   amyloid structures and protein misfolding, 340
HIEF operon
   ribonucleotide reductases and, 692
Hin DNA invertase
   site-specific recombination and, 583–84
Histone deacetylases (HDAC)
   signaling pathways in skeletal muscle remodeling and, 23–25
Histones
   gene expression regulation and chromatin modifications, 243, 245–56, 259–63
   sirtuins and, 438–39
HM genes
   chromosome and plasmid segregation, 224
   sirtuins and, 437, 441, 454
Holliday junctions
   site-specific recombination and, 567, 572–74, 582
Homeostasis
   obesity-related derangements in metabolic regulation and, 375

Homo sapiens
  amyloid structures and protein misfolding, 340
  NADH:quinone oxidoreductase (complex I) and, 73
  sirtuins and, 439
  telomerase reverse transcriptase and, 496–97, 504
hox genes
  gene expression regulation and chromatin modifications, 253–54, 257
  NADH:quinone oxidoreductase (complex I) and, 76, 78
HSD17B4 gene
  mammalian peroxisomes and, 318, 321
Hsp90 molecular chaperone
  activation, 289
  Aha1, 285–86
  ATP, 279–81
  ATPases, 276–79, 283–86
  catalytic loop, 274–75
  Cdc37, 284–85
  client proteins, 274–75, 289
  cochaperones, 281–89
  complexes, 281–89
  conformation, 273–81
  C-terminal domain, 275–76
  dimerization, 275–76
  drug binding, 273–74
  future research, 290
  introduction, 272–73
  middle segment, 274–75
  N-terminal domain, 273–74
  nucleotide binding, 273–74
  p23/Sba1, 286–89
  perspectives, 272–73
  structure, 273–81
  summary points, 290
  TPR-domain cochaperones, 281–83
Hydrogen bonds
  cold-adapted enzymes and, 409, 423–24
  energy transduction and proton transfer through respiratory complexes, 165, 169, 178–80
  protein motion and catalysis, 529
Hydrophilicity
  cold-adapted enzymes and, 423
Hydrophobicity
  amyloid structures and protein misfolding, 353
  cold-adapted enzymes and, 408–9, 412, 419, 421–23, 425
Hydrophobin
  amyloid structures and protein misfolding, 340

Hyperoxaluria type 1
  mammalian peroxisomes and, 316, 318
Hypertension
  obesity-related derangements in metabolic regulation and, 367–68
Hypertrophy
  signaling pathways in skeletal muscle remodeling and, 19, 26, 28

I

IDH1 gene
  mammalian peroxisomes and, 301
Immune disorders
  tyrphostins and other tyrosine kinase inhibitors, 93, 104–5
IMS-binding sites
  mitochondrial ADP/ATP carrier and, 730–31
Induced fit
  site-specific recombination and, 598–99
Inhibitors
  NADH:quinone oxidoreductase (complex I) and, 82–83
  sirtuins and, 461
  tyrphostins and other tyrosine kinase inhibitors, 93–105
Initiation
  capsular polysaccharides in *E. coli* and, 56–57
Insertion
  eukaryotic signaling circuits and, 655
Insulin-like growth factor (IGF)
  signaling pathways in skeletal muscle remodeling and, 27–28
Insulin resistance
  obesity-related derangements in metabolic regulation and, 376–78, 386–91, 393
Integrases
  site-specific recombination and, 567, 570, 598–99
Integration
  eukaryotic signaling circuits and, 657
  site-specific recombination and, 567, 569, 583
Integrins
  death-associated protein kinases and, 203, 206
Intercellular bridge
  animal cytokinesis and, 553
Interferon (IFN)
  death-associated protein kinases and, 196
Intermediate filaments
  dynamic filaments of bacterial cytoskeleton and, 483
Intermonomer cross talk

Inversion
    transmembrane traffic in cytochrome b_6f complex and, 780
Inversion
    site-specific recombination and, 567, 569, 583
Invertases
    site-specific recombination and, 596–98
Iron-sulfur clusters
    NADH:quinone oxidoreductase (complex I) and, 69–87
Irreversible inhibitors
    tyrphostins and other tyrosine kinase inhibitors, 100–1
Isocitrate dehydrogenase
    cold-adapted enzymes and, 406–7, 409
    obesity-related derangements in metabolic regulation and, 389
Isodesmic assembly
    dynamic filaments of bacterial cytoskeleton and, 472–73
Isoprenoids
    mammalian peroxisomes and, 314–15
3-Isopropylmalate dehydrogenase
    cold-adapted enzymes, 417, 426
Isoquinolines
    tyrphostins and other tyrosine kinase inhibitors, 97

J

Jak inhibitors
    tyrphostins and other tyrosine kinase inhibitors, 104–5
Jasplakinolide
    animal cytokinesis and, 557
JIP proteins
    axonal transport and Alzheimer's disease, 612, 617–18
"Jumping"
    dynamic filaments of bacterial cytoskeleton and, 485

K

K antigen
    capsular polysaccharides in *E. coli* and, 40–41, 47
Kinesins
    animal cytokinesis and, 547, 555
    axonal transport and Alzheimer's disease, 609–11, 614, 617–18, 620–21
Kinetics
    asparagine synthetase chemotherapy and, 633–34
    cold-adapted enzymes and, 406–7, 411–12
    mitochondrial ADP/ATP carrier and, 733–34
Klebsiella spp.
    capsular polysaccharides in *E. coli* and, 45, 47
    NADH:quinone oxidoreductase (complex I) and, 84
Klenow fragment
    history of research, 12
K_{LPS}
    capsular polysaccharides in *E. coli* and, 49
Kluyveromyces lactis
    break-induced replication and recombinational telomere elongation in yeast, 112, 116, 119, 121–23, 125, 127–30, 132
K pathway
    proton transfer through respiratory complexes and, 168–70, 176–79
kps genes
    capsular polysaccharides in *E. coli* and, 53–61

L

Lability
    cold-adapted enzymes and, 403–28
Lactobacillus leischmanii
    ribonucleotide reductases and, 684–87, 692
Lactococcus lactis
    ribonucleotide reductases and, 684, 693, 698
lacZ gene
    signaling pathways in skeletal muscle remodeling and, 24
Laminin
    death-associated protein kinases and, 203
Lateral interaction
    dynamic filaments of bacterial cytoskeleton and, 470
Lavendustin
    tyrphostins and other tyrosine kinase inhibitors, 96
Lehman IR, 1–16
Leukemia
    asparagine synthetase chemotherapy and, 629–49
    tyrphostins and other tyrosine kinase inhibitors, 97–99
lipAB genes
    capsular polysaccharides in *E. coli* and, 60
Lipase B
    cold-adapted enzymes and, 415–16
Lipid metabolism
    obesity-related derangements in metabolic regulation and, 367–94

Lipid rafts
  membrane proteins and, 711
Lipopolysaccharide (LPS)
  capsular polysaccharides in *E. coli* and, 42–44, 49, 56
Liver
  obesity-related derangements in metabolic regulation and, 367–68, 370–74
Liver alcohol dehydrogenase
  protein motion and catalysis, 519–20, 530–35
LKB1-dependent signaling pathways
  activation, 146–49
  AMP-activated protein kinases, 146–49
  BRSK kinases, 154–56
  cancer, 151–52
  cell polarity, 154–56
  conclusions, 157
  future research, 158
  introduction, 138–39
  LKB1:STRAD:MO25 complex, 144–46
  MARK kinases, 154–56
  master upstream kinase, 152–54
  perspectives, 157
  Peutz-Jeghers syndrome, 151–52
  posttranslational modifications, 143–44
  pseudokinases, 146
  regulation, 154–56
  summary points, 158
  tumor suppressor, 139–42
Local flexibility
  cold-adapted enzymes and, 418–19
Longitudinal interaction
  dynamic filaments of bacterial cytoskeleton and, 472
loxP gene
  site-specific recombination and, 582
Lysine
  gene expression regulation and chromatin modifications, 248–58

M

Malate dehydrogenase
  cold-adapted enzymes and, 409, 420–21, 424, 426
Mammalian target of rapamycin (mTOR)
  LKB1-dependent signaling pathways and, 149–52
  signaling pathways in skeletal muscle remodeling and, 27–28
mar1-1 gene
  sirtuins and, 437
MARK kinases
  LKB1-dependent signaling pathways and, 154–56
Mastigocladus laminosus
  transmembrane traffic in cytochrome $b_6 f$ complex and, 769–72, 774–75, 777–79, 783–86
MAT genes
  sirtuins and, 436–37
MCF motif
  mitochondrial ADP/ATP carrier and, 722–24
MEF2 genes
  signaling pathways in skeletal muscle remodeling and, 19, 23–25
Meiosis
  chromosome and plasmid segregation, 211, 220–22
Membrane arm
  NADH:quinone oxidoreductase (complex I) and, 81
Membrane proteins
  dynamics, 709–10
  environment, 710–11
  introduction, 708
  mitochondrial ADP/ATP carrier and, 713–37
  structure, 708–9
Membrane trafficking
  capsular polysaccharides in *E. coli* and, 39
MEN/SIN proteins
  animal cytokinesis and, 560
MEN1 gene
  gene expression regulation and chromatin modifications, 253
Metabolism
  history of research, 3
  mammalian peroxisomes and, 295–325
  obesity-related derangements in metabolic regulation and, 367–94
  signaling pathways in skeletal muscle remodeling and, 19, 21
  sirtuins and, 435, 438, 440, 442–44
Metal inhibition
  proton transfer through respiratory complexes and, 176–77
Metalloproteases
  cold-adapted enzymes and, 420–21, 424
Metalloproteinases
  axonal transport and Alzheimer's disease, 616
Metazoans
  sirtuins and, 437–45
Metformin
  LKB1-dependent signaling pathways and, 151–52

Methanococcoides burtonii
   cold-adapted enzymes and, 409, 411
Methanococcus jannaschii
   dynamic filaments of bacterial cytoskeleton and, 470–71
Methanosarcina spp.
   cold-adapted enzymes and, 406, 409, 411
   NADH:quinone oxidoreductase (complex I) and, 77–78
Methylation
   gene expression regulation and chromatin modifications, 243, 245–56, 259–63
Methyltransferases
   histone
      gene expression regulation and chromatin modifications, 256–58
MFP proteins
   capsular polysaccharides in *E. coli* and, 61
Mg^{2+}
   history of research, 5–7
Micrococcus luteus
   capsular polysaccharides in *E. coli* and, 57
Microgravity
   signaling pathways in skeletal muscle remodeling and, 19
Microtubule affinity-regulating kinase
   LKB1-dependent signaling pathways and, 152
Microtubules
   animal cytokinesis and, 543–61
   axonal transport and Alzheimer's disease, 609–12, 619–20
Midpoint potential
   NADH:quinone oxidoreductase (complex I) and, 82, 85–86
Mimicry
   tyrphostins and other tyrosine kinase inhibitors, 99
Mitochondria
   break-induced replication and recombinational telomere elongation in yeast, 130
   death-associated protein kinases and, 205
   NADH:quinone oxidoreductase (complex I) and, 69–87
   obesity-related derangements in metabolic regulation and, 367, 382–85
   proton transfer through respiratory complexes and, 166
   ribonucleotide reductases and, 699–700
Mitochondrial ADP/ATP carrier
   ATP binding from matrix, 731
   brief history, 715–17
   carboxyatractyloside, 722
   cardiolipins, 736
   cavity, 717–22
   conclusions, 736–37
   conformational changes, 732–34
   conserved residues, 728
   deviation, 727–28
   dimerization, 734–35
   functional unit, 734–35
   future research, 737
   human disease, 715
   hypotheses, 736–37
   IMS-binding sites, 730–31
   introduction, 714–15
   kinetics, 733–34
   lipids, 726
   MCF motif, 722–24
   nucleotide attraction and binding, 728–31
   oligomerization, 734–35
   open questions, 736–37
   pseudo-threefold symmetry, 727–28
   signature, 725–26
   structure analysis, 717–22, 731
   summary points, 737
   transport mechanism, 732–34
Mitochondrial carrier family (MCF)
   mitochondrial ADP/ATP carrier and, 713–37
Mitogen-activated protein kinases (MAPKs)
   death-associated protein kinases and, 197
   eukaryotic signaling circuits and, 662–63
Mitosis
   chromosome and plasmid segregation, 211, 218, 231–32
MLL genes
   gene expression regulation and chromatin modifications, 251–53
MO25 gene
   LKB1-dependent signaling pathways and, 144–46
Modularity
   eukaryotic signaling circuits and, 655, 657–59, 664–69, 674–75
   NADH:quinone oxidoreductase (complex I) and, 75–79
Molecular dynamics (MD) simulations
   asparagine synthetase chemotherapy and, 641
   proton transfer through respiratory complexes and, 167, 172, 178–79
MOLT-4 cells
   asparagine synthetase chemotherapy and, 638, 648
Monoubiquitination

gene expression regulation and chromatin modifications, 259–60
Moritella profunda
  cold-adapted enzymes and, 413
Motor proteins
  axonal transport and Alzheimer's disease, 607–21
MPA proteins
  capsular polysaccharides in *E. coli* and, 51, 61
MreB protein
  dynamic filaments of bacterial cytoskeleton and, 467, 479–82
mrp genes
  NADH:quinone oxidoreductase (complex I) and, 79
mThB gene
  mammalian peroxisomes and, 321
mTOR gene
  LKB1-dependent signaling pathways and, 149–52
  signaling pathways in skeletal muscle remodeling and, 27–28
mukB gene
  chromosome and plasmid segregation, 223
Multimerization
  sirtuins and, 459–60
  telomerase reverse transcriptase and, 505–7
Mupirocin
  asparagine synthetase chemotherapy and, 642, 644
MuRF gene
  signaling pathways in skeletal muscle remodeling and, 31
Muscle
  obesity-related derangements in metabolic regulation and, 367, 374–85
  signaling pathways in skeletal muscle remodeling and, 19–33
Muscular dystrophy
  signaling pathways in skeletal muscle remodeling and, 28–29
Mus musculus
  NADH:quinone oxidoreductase (complex I) and, 73
  sirtuins and, 439
Mycobacterium tuberculosis
  ribonucleotide reductases and, 693
  site-specific recombination and, 583
Myocyte enhancer factor-2 (MEF2)
  signaling pathways in skeletal muscle remodeling and, 19, 23–25
Myofibers
  signaling pathways in skeletal muscle remodeling and, 19–33
Myoglobin
  signaling pathways in skeletal muscle remodeling and, 24
Myosin
  animal cytokinesis and, 547, 552, 555–56, 559
  death-associated protein kinases and, 191, 195, 202–5
  signaling pathways in skeletal muscle remodeling and, 21, 24–25

N

Na^+/H^+ antiporters
  NADH:quinone oxidoreductase (complex I) and, 69, 78–79
NAD^+
  sirtuins and, 435, 441–55, 457–61
NADH:quinone oxidoreductase (complex I)
  accessory subunits, 73–75
  central subunits, 71–73
  function, 81–86
  future research, 87
  inhibitors, 82–83
  introduction, 70
  medical aspects, 86
  membrane arm, 81
  modular design, 75–79
  N module, 76–77
  peripheral arm, 79–81
  P module, 78–79
  proton-pumping mechanism, 84–86
  Q module, 78
  quinine-reducing catalytic core, 83–84
  redox centers, 82–83
  structure, 80–81
  substrates, 82–83
  subunits, 70–71
  summary points, 86–87
NADP-dependent isocitrate dehydrogenase
  obesity-related derangements in metabolic regulation and, 389
Nanomolecular inhibitor
  asparagine synthetase chemotherapy and, 643–46, 648
Neisseria meningitidis
  capsular polysaccharides in *E. coli* and, 44, 59
Nephila edulis
  amyloid structures and protein misfolding, 340
neu genes

capsular polysaccharides in *E. coli* and, 55–57, 59–61
Neurofibrillary tangles
   axonal transport and Alzheimer's disease, 615–16, 618–19, 621
Neurons
   axonal transport and Alzheimer's disease, 607–21
   signaling pathways in skeletal muscle remodeling and, 22
Neurospora crassa
   amyloid structures and protein misfolding, 340
   history of research, 4
   NADH:quinone oxidoreductase (complex I) and, 75, 86
NFB42 neuron-specific protein
   history of research, 15
Nicastrin
   axonal transport and Alzheimer's disease, 615, 617
Nicotinamide
   sirtuins and, 452–54
Nitrogen metabolism
   mammalian peroxisomes and, 301–2
N module
   NADH:quinone oxidoreductase (complex I) and, 76–77
Nodes
   transcriptional
      eukaryotic signaling circuits and, 657–58
Nonchromosomal genetic elements
   amyloid structures and protein misfolding, 339–41
Nonyl-4-hydroxyquinolone *N*-oxide (NQNO)
   transmembrane traffic in cytochrome $b_6 f$ complex and, 782–83
NQR protein
   NADH:quinone oxidoreductase (complex I) and, 70
nrd genes
   ribonucleotide reductases and, 683–84, 692–93
N-terminal domains
   Hsp90 molecular chaperone and, 273–74
Nuclear factor of activated T cells (NFAT)
   signaling pathways in skeletal muscle remodeling and, 25–26
Nuclear magnetic resonance (NMR)
   amyloid structures and protein misfolding, 341–43, 346
   sirtuins and, 448
Nucleases
   history of research, 8–9

telomerase reverse transcriptase and, 504–5
Nucleated growth mechanism
   amyloid structures and protein misfolding, 347
Nucleoids
   chromosome and plasmid segregation, 213, 216
Nucleophilic attack
   sirtuins and, 449–50
Nucleosomes
   gene expression regulation and chromatin modifications, 243, 245
Nucleotide addition processivity
   telomerase reverse transcriptase and, 501
Nucleotide binding
   Hsp90 molecular chaperone and, 273–74
   mitochondrial ADP/ATP carrier and, 728–31
Nucleotide exchange
   dynamic filaments of bacterial cytoskeleton and, 473–74
Nucleotide hydrolysis
   dynamic filaments of bacterial cytoskeleton and, 476
Nucleotide transport
   mitochondrial ADP/ATP carrier and, 713–37
Nutrient-sensing reponse element (NSRE)
   asparagine synthetase chemotherapy and, 636–37
N-WASP protein
   eukaryotic signaling circuits and, 665

O

O antigens
   capsular polysaccharides in *E. coli* and, 47–49, 56
Obesity
   signaling pathways in skeletal muscle remodeling and, 29–31
Obesity-related derangements
   in metabolic regulation
      adipose tissue, 370–74
      conclusions, 393–94
      CPT-malonyl-CoA system, 375–76
      fuel homeostasis, 375
      future research, 394
      insulin resistance, 376–78
      introduction, 368–70
      liver, 370–74
      mitochondria, 382–85
      pancreatic islets, 385–93
      PPARγ coactivator 1, 375
      regulation, 375–76

responder peripheral tissues, 374–93
skeletal muscle, 374–85
substrate utilization, 378–82
summary points, 394
transcriptional reprogramming, 374–75
Okazaki fragments
history of research, 12
Oligomerization
amyloid structures and protein misfolding, 333, 347–51
capsular polysaccharides in *E. coli* and, 52
mitochondrial ADP/ATP carrier and, 713, 734–35
OMA proteins
capsular polysaccharides in *E. coli* and, 52–53, 58, 61
ops genes
capsular polysaccharides in *E. coli* and, 46
oriC replication irigin
chromosome and plasmid segregation, 213, 216, 230, 233
Ornithine carbamoyltransferase
cold-adapted enzymes and, 410, 417
Oryza sativa
NADH:quinone oxidoreductase (complex I) and, 73
Oscillators
chromosome and plasmid segregation, 233
OSI-774
tyrphostins and other tyrosine kinase inhibitors, 99
Oxidation
mammalian peroxisomes and, 295–325
transmembrane traffic in cytochrome $b_6 f$ complex and, 775–81
Oxidative phosphorylation
proton transfer through respiratory complexes and, 181–82
Oxygen labeling
sirtuins and, 448
Oxygen metabolism
mammalian peroxisomes and, 301–2

P

p23/Sba1 protein
Hsp90 molecular chaperone and, 286–89
p53R2 gene
ribonucleotide reductases and, 684
p150^{Glued} gene
axonal transport and Alzheimer's disease, 613
Paenibacillus polymyxa

cold-adapted enzymes and, 416
PAHX/PHYH gene
mammalian peroxisomes and, 318
Pairing
chromosome and plasmid segregation, 220–22
Pancreas
obesity-related derangements in metabolic regulation and, 367–68, 385–93
Pandalus borealis
cold-adapted enzymes and, 406, 411
par genes
chromosome and plasmid segregation, 211, 226–30, 233, 235
dynamic filaments of bacterial cytoskeleton and, 467, 477–79
LKB1-dependent signaling pathways and, 154
Paracoccus denitrificans
NADH:quinone oxidoreductase (complex I) and, 72, 76, 80
proton transfer through respiratory complexes and, 168, 180
Parkinson's disease
amyloid structures and protein misfolding, 333, 336–37, 356
Partial unfolding
amyloid structures and protein misfolding, 349
Partitioning
chromosome and plasmid segregation, 211–12, 226–29, 231
sirtuins and, 453–54
Pasteurella multocida
capsular polysaccharides in *E. coli* and, 56
Pausing
axonal transport and Alzheimer's disease, 610
PCG1 protein
obesity-related derangements in metabolic regulation and, 375
PCP proteins
capsular polysaccharides in *E. coli* and, 49, 51
PDK1 interaction fragment (PIF)
eukaryotic signaling circuits and, 663
pds genes
chromosome and plasmid segregation, 218
Pentose phosphate pathway
mammalian peroxisomes and, 313
Pepsin
cold-adapted enzymes and, 420–21
Peripheral arm
NADH:quinone oxidoreductase (complex I) and, 79–81
Permeability
mammalian peroxisomes and, 319–20

Peroxisome proliferators-activated receptors (PPARs)
    obesity-related derangements in metabolic regulation and, 374–75
    signaling pathways in skeletal muscle remodeling and, 27
Peroxisomes
    mammalian
        ABC transporter, 322–24
        amino acid catabolism, 313
        cholesterol, 314–15
        conclusions, 325
        enzymes, 308–11, 314
        ether phospholipids, 302–5
        fatty acid α-oxidation, 311–12
        fatty acid β-oxidation, 305–11
        genetic diseases, 315–21
        glyoxylates, 312–13
        human disorders, 315–18
        intraperoxisomal pH, 322
        introduction, 296–97
        isoprenoids, 314–15
        membrane transport, 306–8
        metabolic pathways, 302–15
        metabolite transport, 319, 322–25
        mouse models, 316, 319–21
        oxygen metabolism, 301–2
        pentose phosphate pathway, 313
        permeability, 319–20
        peroxisomal proteins, 297–300
        peroxisomal transporters, 325
        polyamine oxidation, 313–14
        reactive nitrogen species, 301–2
        reactive oxygen species, 301–2
        substrate specificity, 306
Peutz-Jeghers syndrome
    LKB1-dependent signaling pathways and, 137–40, 145, 151–55, 157
PEX genes
    mammalian peroxisomes and, 316–17, 320
pH
    mammalian peroxisomes and, 322
Phage integrases
    site-specific recombination and, 598–99
Phenformin
    LKB1-dependent signaling pathways and, 147
Phosmidosine
    asparagine synthetase chemotherapy and, 644
Phosphatases
    death-associated protein kinases and, 206
Phosphate
    asparagine synthetase chemotherapy and, 643

history of research, 3
Phosphoglycerate kinase
    cold-adapted enzymes and, 410, 415, 424
3-Phosphoinositide-dependent kinase-1 (PDK1)
    eukaryotic signaling circuits and, 663
Phosphorylation
    death-associated protein kinases and, 191–93, 195–97, 201–6
    gene expression regulation and chromatin modifications, 243
    proton transfer through respiratory complexes and, 181–82
    tyrphostins and other tyrosine kinase inhibitors, 93–105
Photosynthesis
    transmembrane traffic in cytochrome $b_6 f$ complex and, 769
Phototransduction
    rhodopsin and, 743–62
Pichia philodendra
    break-induced replication and recombinational telomere elongation in yeast, 130
Piericidin A
    NADH:quinone oxidoreductase (complex I) and, 83
Pigment molecules
    transmembrane traffic in cytochrome $b_6 f$ complex and, 786–87
Plasmalogens
    mammalian peroxisomes and, 295
Plasma membrane
    NADH:quinone oxidoreductase (complex I) and, 69–87
Plasmid segregation
    chromosome segregation and, 211
Plasmodium falciparum
    ribonucleotide reductases and, 684
    telomerase reverse transcriptase and, 496–98
Plastoquinones
    transmembrane traffic in cytochrome $b_6 f$ complex and, 769, 784
Platelet-derived growth factor receptor (PDGFR)
    tyrphostins and other tyrosine kinase inhibitors, 103–4
P module
    NADH:quinone oxidoreductase (complex I) and, 78–79
Podospera anserina
    amyloid structures and protein misfolding, 340
pol A gene
    history of research, 11–12

Polyamine oxidation
  mammalian peroxisomes and, 313–14
polycomb genes
  gene expression regulation and chromatin modifications, 248, 254
Polymerases
  chromosome and plasmid segregation, 218, 493
Polymerization
  capsular polysaccharides in *E. coli* and, 50–51
Polymorphism
  amyloid structures and protein misfolding, 345–46
Polyphenols
  sirtuins and, 460–61
Position-effect variegation (PEV)
  gene expression regulation and chromatin modifications, 248
Posttranslational modification
  gene expression regulation and chromatin modifications, 245–46
  LKB1-dependent signaling pathways and, 143–44
PP1
  tyrphostins and other tyrosine kinase inhibitors, 98
PPA α-amylase
  cold-adapted enzymes and, 406, 410–11, 413, 425
PRC1 gene
  gene expression regulation and chromatin modifications, 254
Prednisone
  asparagine synthetase chemotherapy and, 631
Presenilins
  axonal transport and Alzheimer's disease, 612, 615–19
Prions
  amyloid structures and protein misfolding, 333, 339–40, 354
PRMT1 gene
  gene expression regulation and chromatin modifications, 258
Processivity
  telomerase reverse transcriptase and, 493–512
Profilin
  animal cytokinesis and, 547
Programmed cell death
  death-associated protein kinases and, 189–207
Prosthecobacter dejongeii
  dynamic filaments of bacterial cytoskeleton and, 471, 475
Prosthetic groups
  transmembrane traffic in cytochrome $b_6 f$ complex and, 771–73
Proteases
  chromosome and plasmid segregation, 220
Protein aggregation
  amyloid structures and protein misfolding, 333–60
Protein conformational motions
  enzyme catalysis and
    conclusions, 535–38
    conformational changes, 526–28, 532–33
    dihydrofolate reductase, 520–30
    future research, 538
    introduction, 520
    liver alcohol dehydrogenase, 530–35
    mutagenesis, 522–33
    summary points, 538
    theoretical studies of motion, 528–35
Protein kinases
  death-associated protein kinases and, 189–207
  eukaryotic signaling circuits and, 662–64, 669
  LKB1-dependent signaling pathways and, 137, 139, 144, 146–52, 156–57
  signaling pathways in skeletal muscle remodeling and, 19, 26–27
Protein radicals
  ribonucleotide reductases and, 681, 690–92
Protein tyrosine kinases (PTKs)
  tyrphostins and other tyrosine kinase inhibitors, 93–105
Protein unfolding
  cold-adapted enzymes and, 410–11
Protelomerases
  site-specific recombination and, 582
Proteomics
  animal cytokinesis and, 543
  asparagine synthetase chemotherapy and, 647, 649
  gene expression regulation and chromatin modifications, 259
Protofibrils
  amyloid structures and protein misfolding, 347–49
Proton pumping
  NADH:quinone oxidoreductase (complex I) and, 69–87
  proton transfer through respiratory complexes and, 170–74
Proton transfer
  through respiratory complexes
    active site, 171–74

computational insights, 178–80
coupling, 171
CuA, 171
cytochrome *c* oxidase, 168–80
D pathway, 177, 179
electron transfer, 171–74
electron transfer-driven pump, 168–80
electrostatic calculations, 180
exit/backflow pathway, 178
experimental insights, 174–78
future research, 183
heme *a*, 171
introduction, 166–67
K pathway, 177–79
long-range proton movement through proteins, 167–68
metal inhibition of proton uptake, 176–77
mutants affecting proton movement, 177–78
oxidative phosphorylation, 181–82
proton exit simulations, 179–80
proton motive complexes, 181–82
proton pathways, 168–70
proton pumping, 170–74
subunit III, 174–76
summary points, 182–83
water and proton dynamics, 178–79

Pseudoalteromonas haloplanktis
cold-adapted enzymes and, 405–7, 411, 413–14, 420–21, 423

Pseudokinases
LKB1-dependent signaling pathways and, 144–49

Pseudomonas spp.
chromosome and plasmid segregation, 230
ribonucleotide reductases and, 693

Pseudo-threefold symmetry
mitochondrial ADP/ATP carrier and, 727–28

Psychrophiles
cold-adapted enzymes and, 403

Purse-string model
animal cytokinesis and, 555

Pyrococcus furiosus
cold-adapted enzymes and, 417
dynamic filaments of bacterial cytoskeleton and, 484

Pyrophosphate
inorganic
asparagine synthetase chemotherapy and, 633–34

Pyruvate cycling
obesity-related derangements in metabolic regulation and, 367, 386–88, 392–93

Q

Q module
NADH:quinone oxidoreductase (complex I) and, 78

QN 012380
tyrphostins and other tyrosine kinase inhibitors, 98

Q_p pocket
transmembrane traffic in cytochrome $b_6 f$ complex and, 778–79

Quercetin
tyrphostins and other tyrosine kinase inhibitors, 96

Quinazolines
tyrphostins and other tyrosine kinase inhibitors, 97

Quinolines
tyrphostins and other tyrosine kinase inhibitors, 97

Quinols
transmembrane traffic in cytochrome $b_6 f$ complex and, 769, 775–81

Quinones
transmembrane traffic in cytochrome $b_6 f$ complex and, 769, 781–86

Quinoxalines
tyrphostins and other tyrosine kinase inhibitors, 97

R

rad51 gene
break-induced replication and recombinational telomere elongation in yeast, 111, 114–15, 120

Random synapsis
site-specific recombination and, 578–79

Rapamycin
LKB1-dependent signaling pathways and, 149–52
signaling pathways in skeletal muscle remodeling and, 27–28

Ras/mitogen-activated protein kinase
signaling pathways in skeletal muscle remodeling and, 27

Reactive nitrogen species (RNS)
mammalian peroxisomes and, 301–2

Reactive oxygen species (ROS)
mammalian peroxisomes and, 295, 301–2

obesity-related derangements in metabolic regulation and, 384–85
proton transfer through respiratory complexes and, 166
recA gene
history of research, 13–14
Recombinases
break-induced replication and recombinational telomere elongation in yeast, 111, 114–15, 120
chromosome and plasmid segregation, 233–35
Recombination
break-induced replication and recombinational telomere elongation in yeast, 111–32
eukaryotic signaling circuits and, 655, 665
site-specific recombination and, 567–600
Redox mechanism
NADH:quinone oxidoreductase (complex I) and, 69–87
Refsum disease
mammalian peroxisomes and, 316, 318–19
Regulation
axonal transport and Alzheimer's disease, 611–12
death-associated protein kinases and, 192–93, 196–97
eukaryotic signaling circuits and, 657, 659, 663, 665, 669–71, 674–75
gene expression regulation and chromatin modifications, 243
LKB1-dependent signaling pathways and, 154–56
obesity-related derangements in metabolic regulation and, 367–94
ribonucleotide reductases and, 681–99
site-specific recombination and, 567, 569, 578, 592–600
Rejoining
strand
site-specific recombination and, 567, 570–72, 584–86
Religation
site-specific recombination and, 567
Remodeling
signaling pathways in skeletal muscle remodeling and, 19–33
Repeat addition processivity
telomerase reverse transcriptase and, 502
Replication
break-induced replication and recombinational telomere elongation in yeast, 111–32

chromosome and plasmid segregation, 213–15, 217
Rep proteins
chromosome and plasmid segregation, 211, 230–31
Resolution
site-specific recombination and, 583
Resolvases
site-specific recombination and, 567, 570, 582–85, 592–96
Respiratory complexes
proton transfer through, 165–82
Restenosis
tyrphostins and other tyrosine kinase inhibitors, 93, 104
Restriction endonucleases
history of research, 9
Resveratrol
sirtuins and, 460–61
Retina
rhodopsin and, 743–62
Retrograde transport
axonal transport and Alzheimer's disease, 610–11
Reverse transcriptase
telomerase reverse transcriptase and, 493–512
Reversible inhibitors
tyrphostins and other tyrosine kinase inhibitors, 99–100
Rho-activated protein kinase (ROCK)
death-associated protein kinases and, 201–2, 206
Rhodobacter sphaeroides
dynamic filaments of bacterial cytoskeleton and, 481
NADH:quinone oxidoreductase (complex I) and, 76
proton transfer through respiratory complexes and, 168–69, 173, 176, 178–79
transmembrane traffic in cytochrome b_6f complex and, 781
Rhodomonas salina
NADH:quinone oxidoreductase (complex I) and, 73
Rhodopsin
background, 744–45
chromophore, 753–54
conclusions, 760–61
dimerization, 757–59
discovery, 747
future research, 762
G protein interaction, 758–60

overview, 745–48
perspectives, 760–61
scope of review, 744–45
structure, 748–57
summary points, 761
Rhodospirillum rubrum
    NADH:quinone oxidoreductase (complex I) and, 78
Rhodothermus marinus
    proton transfer through respiratory complexes and, 180
Ribonucleoproteins (RNPs)
    telomerase reverse transcriptase and, 493, 510
Ribonucleotide reductases (RNRs)
    active site, 687–90, 698–99
    allosteric regulation, 694–99
    ATP cone, 698–99
    bacteria, 692–93
    catalytic mechanism, 687–90
    Chlamydia trachomatis, 685–87
    class I, 682–83
    class II, 683
    class III, 683
    classification, 682–83
    dNTP pools, 699–700
    Euglena gracilis, 684
    evolution, 700–1
    future research, 701
    gene organization, 683–84
    gene regulation, 692–94
    introduction, 682
    mammalian cells, 693
    mitochondria, 699–700
    occurence, 683–84
    p53R2, 684
    radical storage and transfer, 690–92
    special enzymes, 684–85
    specificity site, 694–98
    structure, 685–90
    summary points, 701
    yeast, 684, 693–94, 699
Ribosomal DNA (rDNA) genes
    sirtuins and, 454
Ribosylation
    sirtuins and, 439
Rieske ISP
    transmembrane traffic in cytochrome $b_6 f$ complex and, 780–81
RNA binding
    telomerase reverse transcriptase and, 502–3
RNA interference (RNAi)
    animal cytokinesis and, 543–46, 548, 552, 554, 556
    sirtuins and, 438, 440
Rnq1 protein
    amyloid structures and protein misfolding, 340
RNR genes
    ribonucleotide reductases and, 684, 694
Rod outer segments
    rhodopsin and, 743, 745–46, 758
"Roll and spread" mechanism
    break-induced replication and recombinational telomere elongation in yeast, 111, 123–28
Rolling circles
    break-induced replication and recombinational telomere elongation in yeast, 111, 127, 132
Rolliniastain-1
    NADH:quinone oxidoreductase (complex I) and, 83
Rotenone
    NADH:quinone oxidoreductase (complex I) and, 83
Rrm2b gene
    ribonucleotide reductases and, 684

S

Saccharomyces cerevisiae
    amyloid structures and protein misfolding, 339–40, 346
    break-induced replication and recombinational telomere elongation in yeast, 112, 116–23, 125, 127–28, 130
    chromosome and plasmid segregation, 217–18, 230
    eukaryotic signaling circuits and, 662
    gene expression regulation and chromatin modifications, 251, 253, 259
    mammalian peroxisomes and, 323–24
    NADH:quinone oxidoreductase (complex I) and, 70, 73
    sirtuins and, 436, 438–40
    telomerase reverse transcriptase and, 496–97, 502, 504
Salmonella spp.
    capsular polysaccharides in *E. coli* and, 47–49
    ribonucleotide reductases and, 685–86, 694, 697
    sirtuins and, 439, 445
    site-specific recombination and, 568–69, 596
Salmo salar
    cold-adapted enzymes and, 420–21
Salt bridges

cold-adapted enzymes and, 410, 412, 423, 425
Salt-inducible kinase
  LKB1-dependent signaling pathways and, 152
Sba1 protein
  Hsp90 molecular chaperone and, 286–89
Scaffolds
  eukaryotic signaling circuits and, 658–59, 669–70, 672
Scanning transmission electron microscopy (STEM)
  amyloid structures and protein misfolding, 341, 343
Schizosaccharomyces pombe
  break-induced replication and recombinational telomere elongation in yeast, 116, 129–30
  chromosome and plasmid segregation, 217
  eukaryotic signaling circuits and, 662
  gene expression regulation and chromatin modifications, 248–49
  sirtuins and, 438–40
SCL25 proteins
  mitochondrial ADP/ATP carrier and, 713–37
Scopulariopsis spp.
  cold-adapted enzymes and, 407
scp genes
  chromosome and plasmid segregation, 224
  mammalian peroxisomes and, 321
Secondary structure
  amyloid structures and protein misfolding, 353
  cold-adapted enzymes and, 425–26
Secretins
  capsular polysaccharides in *E. coli* and, 52
Self-assembly
  in vitro
    dynamic filaments of bacterial cytoskeleton and, 469–70, 475–76
Semiquinones
  NADH:quinone oxidoreductase (complex I) and, 69–87
  proton transfer through respiratory complexes and, 166
Senile plaques
  axonal transport and Alzheimer's disease, 615–16, 618–20
Separase
  chromosome and plasmid segregation, 220
Septins
  animal cytokinesis and, 559
serA gene
  capsular polysaccharides in *E. coli* and, 53
Serine recombinases
  site-specific recombination and, 567, 570–72, 582–600
Serine/threonine kinases
  death-associated protein kinases and, 189–207
Serratia marcescens
  cold-adapted enzymes and, 406–7
Set2 gene
  gene expression regulation and chromatin modifications, 253–54
SH2/SH3 homology
  eukaryotic signaling circuits and, 664
Signaling
  animal cytokinesis and, 551
  death-associated protein kinases and, 189, 204–7
  rhodopsin and, 743–62
  sirtuins and, 435
  skeletal muscle remodeling and, 19–33
  tyrphostins and other tyrosine kinase inhibitors, 93–105
Signaling circuits
  eukaryotic
    adaptors, 669–70, 672
    catalysis, 664–69
    classical regulatory proteins, 657, 659
    conclusions, 674–75
    connecting transcriptional nodes, 657–58
    connectivity, 664–69
    distributed surfaces, 661–62
    docking interactions, 660–64, 672
    domain discrimination, 665, 667–69
    evolvability of new circuitry, 656–60, 674–75
    functional modularity, 657–58
    future research, 676
    genetically-independent wiring elements, 669–70, 672
    introduction, 656
    kinases, 662–64
    modularity, 674–75
    modular recognition domains, 664–69
    recognition beyond active site, 660–64
    recombination, 665
    regulation, 663, 665, 669–71, 674–75
    scaffolds, 669–70, 672
    signaling pathways, 671–74
    signaling switches, 672, 674
    structural modularity, 657–59
    summary points, 675–76
    synthetic biology, 671–74
Signal transduction therapy

tyrphostins and other tyrosine kinase
inhibitors, 95–96
Simulated annealing
asparagine synthetase chemotherapy and, 641
Single-particle analysis
NADH:quinone oxidoreductase (complex I)
and, 79
Sin paradigm
site-specific recombination and, 595–96
Sir2 genes
sirtuins and, 437–45, 448–49
Sirtinol
sirtuins and, 461
Sirtuins
acetylation, 455, 457
activators, 454, 460–61
ADP-ribosyl transferase, 445–46
ADPR-peptidyl-imidate mechanism, 450–52
aging, 444–45
bacteria, 437–45
base exchange, 447–48
binding, 455, 457–58
caloric restriction, 444–45
chemical mechanism, 445–52, 454
conclusions, 461
conserved catalytic core, 454–56
deacetylase, 445
disclosure statement, 462
effectors, 441–42
enzymes, 450, 458–59
fungi, 437–45
history, 436–37
inhibitors, 461
introduction, 436–37
metabolism, 442–44
metazoans, 437–45
multimerization, 459–60
NAD^+, 442–44, 452–53, 457–58
nicotinamide, 452–54
NMR studies, 448
nucleophilic attack, 449–50
O-acetyl-ADP-ribose, 441–42
oxygen labeling, 448
partition model of imidate reactivity, 453–54
plant polyphenols, 460–61
products, 446–48
resveratrol, 460–61
Sir2 gene families, 437–49
sirtinol, 461
small molecule modulators, 460–61
solvent labeling patterns, 446–48
splitomicins, 461

structural properties, 454–60
Site-directed spin labeling coupled to electron
paramagnetic resonance (SDSL-EPR)
amyloid structures and protein misfolding,
341, 344
Site-specific recombinases
chromosome and plasmid segregation, 233–35
Site-specific recombination
accessory proteins, 579–82
catalysis, 577–78
complex systems, 592–600
controlling outcome, 578–82, 592–600
crossover site cleavage, 586, 589–90
DNA invertases, 596–98
excision specificity, 592–96
Gin/Hin paradigm, 596–98
half-of-the-sites reactivity, 577–78
Holliday junctions, 572–74
induced fit, 598–99
integrases, 598–99
introduction, 568–69
inversion specificity, 596–98
invertases, 596–98
mechanism overview, 569–70
perspectives, 600
phage integrases, 598–99
random synapsis, 578–79
resolvases, 592–96
serine recombinases, 570–72, 582–600
simple systems, 578–79
Sin paradigm, 595–96
spacer complementarity, 578–79
specificity, 598–99
strand breakage, 570–72, 584–86
strand exchange, 570–86, 590–92
strand reunion, 570–72, 584–86
summary, 600
synapsis, 574–79, 586–99
target-target recombination, 599–600
Tn3/$\gamma\delta$ paradigm, 592–95
"topological filter," 592–96
transposases, 599–600
tyrosine recombinases, 570–82
Sk27a2 gene
mammalian peroxisomes and, 321
Skeletal muscle
obesity-related derangements in metabolic
regulation and, 367, 374–85
signaling pathways in remodeling
Akt, 27–28
anabolic steroids, 32
calcineurin, 25–26

calcium/calmodulin-dependent protein
kinase, 26–27
clinical significance, 29–32
conclusions, 32–33
diabetes mellitus type 2, 29–31
discussion, 22–29
future research, 33
histone deacetylases, 23–25
insulin-like growth factor, 27–28
introduction, 20
mammalian target of rapamycin, 27–28
muscular dystrophy, 28–29
muscle atrophy, 31–32
myocyte enhancer factor-2, 23–25
myofiber adaptability, 22
myofiber diversity, 21–22
NFAT, 25–26
obesity, 29–31
PPAR δ, 27
PPAR coactivator-1α, 27
protein kinase C, 26–27
protein kinase D, 26–27
Ras/mitogen-activated protein kinase, 27
summary points, 33
Slow-twitch myofibers
signaling pathways in skeletal muscle
remodeling and, 21, 25
Small molecule modulators
sirtuins and, 460–61
smc genes
chromosome and plasmid segregation,
223–25
SNARE proteins
animal cytokinesis and, 559–60
death-associated protein kinases and, 204
Solid-state nuclear magnetic resonance (SSNMR)
amyloid structures and protein misfolding,
341–44, 346
Solvent labeling patterns
sirtuins and, 446–48
Spacer complementarity
site-specific recombination and, 578–79
Specificity
mammalian peroxisomes and, 306
ribonucleotide reductases and, 681, 694–98
site-specific recombination and, 592–94,
596–99
Splitomicins
sirtuins and, 461
spoOJ gene
chromosome and plasmid segregation, 230
Srpk3 gene

signaling pathways in skeletal muscle
remodeling and, 24–25
Stability
cold-adapted enzymes and, 403, 410–17
Stalling
axonal transport and Alzheimer's disease, 610
Static flexibility
cold-adapted enzymes and, 417
STB genes
chromosome and plasmid segregation, 211,
227, 230–32, 236
Stembody
animal cytokinesis and, 554
Steroids
signaling pathways in skeletal muscle
remodeling and, 32
STI-571
tyrphostins and other tyrosine kinase
inhibitors, 97–99
STRAD genes
LKB1-dependent signaling pathways and,
144–46
Streptococcus pneumoniae
capsular polysaccharides in *E. coli* and, 56
dynamic filaments of bacterial cytoskeleton
and, 482
Streptomyces coelicolor
amyloid structures and protein misfolding,
339–40
chromosome and plasmid segregation, 230
ribonucleotide reductases and, 683–84, 693
Streptomyces spp.
cold-adapted enzymes and, 407
site-specific recombination and, 569
Streptomycin
history of research, 5
Structural adaptation
cold-adapted enzymes and, 415, 419–27
Structural modularity
eukaryotic signaling circuits and, 657–58
ribonucleotide reductases and, 681
Subtilisin
cold-adapted enzymes and, 406, 415–16,
420–21, 424, 426–27
"Suicide inactivation" process
proton transfer through respiratory complexes
and, 175
Sulfobetaine
cold-adapted enzymes and, 413
Sulfolobus solfataricus
amyloid structures and protein misfolding, 348
sirtuins and, 439

Sulfonamide derivatives
    asparagine synthetase chemotherapy and, 641–43
Sulfoximine
    asparagine synthetase chemotherapy and, 643, 645–47, 649
Sumoylation
    gene expression regulation and chromatin modifications, 243
Sup35 protein
    amyloid structures and protein misfolding, 340
Surface loops
    cold-adapted enzymes and, 426
Survivin
    animal cytokinesis and, 548
SUV genes
    gene expression regulation and chromatin modifications, 248–49
Switches
    synthetic
        eukaryotic signaling circuits and, 672, 674
Synapsis
    site-specific recombination and, 574–79, 586–89, 598–99
Synechococcus spp.
    transmembrane traffic in cytochrome $b_6 f$ complex and, 775, 779
Synechocystis spp.
    NADH:quinone oxidoreductase (complex I) and, 73
    transmembrane traffic in cytochrome $b_6 f$ complex and, 775
Syntaxins
    animal cytokinesis and, 547, 559
    death-associated protein kinases and, 206
Synthetic biology
    eukaryotic signaling circuits and, 671–74
Synuclein
    axonal transport and Alzheimer's disease, 617

T

T2 phage
    history of research, 3–4
Target-target recombination
    death-associated protein kinases and, 189
    site-specific recombination and, 599–600
tau protein
    axonal transport and Alzheimer's disease, 617–20
T cells
    signaling pathways in skeletal muscle remodeling and, 25–26
TDS-ring-out conformation
    transmembrane traffic in cytochrome $b_6 f$ complex and, 779–80
TEL genes
    sirtuins and, 454
Telomerase reverse transcriptase (TERT)
    catalysis, 499–500
    conclusions, 511–12
    dimerization, 505–7
    elongation, 500–1
    evolution, 496–98
    future research, 512
    introduction, 494–95
    multimerization, 505–7
    nuclease, 504–5
    nucleotide addition processivity, 501
    processivity, 500–2
    repeat addition processivity, 502
    RNA binding, 502–3
    structure, 496–500
    summary points, 512
    telomere maintenance, 509–11
    template utilization, 503–4
    TERT-associated proteins, 507–9
Telomerases
    break-induced replication and recombinational telomere elongation in yeast, 111, 128–29, 131–32
Template utilization
    telomerase reverse transcriptase and, 503–4
terC gene
    chromosome and plasmid segregation, 213, 216, 234
Terminal palindromes
    break-induced replication and recombinational telomere elongation in yeast, 130–31
5,6,7,8-Tetrahydrofolate (H$_4$F)
    protein motion and catalysis, 520–21
Tetrahymena thermophila
    telomerase reverse transcriptase and, 495–97, 500–6
Thermoactinomycetes vulgaris
    cold-adapted enzymes and, 427
Thermodynamics
    cold-adapted enzymes and, 403–28
Thermophiles
    transmembrane traffic in cytochrome $b_6 f$ complex and, 769–72, 774–75, 777–79, 783–86
Thermoplasma acidophilum

ribonucleotide reductases and, 698
Thermotoga maritima
    cold-adapted enzymes and, 413
    dynamic filaments of bacterial cytoskeleton and, 477, 480
    ribonucleotide reductases and, 686, 688, 691–92, 694, 696–97
Thermus thermophilus
    amyloid structures and protein misfolding, 349
    cold-adapted enzymes and, 417
    NADH:quinone oxidoreductase (complex I) and, 72, 79–82
    proton transfer through respiratory complexes and, 180
Thioflavin T
    amyloid structures and protein misfolding, 336, 339, 348–50, 353
Tissue rejection
    tyrphostins and other tyrosine kinase inhibitors, 93
Tn3/γδ paradigm
    site-specific recombination and, 592–95
Topoisomerases
    chromosome and plasmid segregation, 224–25, 233–35
"Topological filter"
    site-specific recombination and, 592–96
TPR-domain cochaperones
    Hsp90 molecular chaperone and, 281–83
Trafficking
    capsular polysaccharides in *E. coli* and, 39
    transmembrane traffic in cytochrome $b_6 f$ complex and, 769–87
Transcription
    asparagine synthetase chemotherapy and, 635–37
    eukaryotic signaling circuits and, 657–58
    gene expression regulation and chromatin modifications, 243
    obesity-related derangements in metabolic regulation and, 374–75
Transcriptional memory
    gene expression regulation and chromatin modifications, 260–61
Transcription factors
    signaling pathways in skeletal muscle remodeling and, 19, 23–25
trans-envelope assembly complex
    capsular polysaccharides in *E. coli* and, 39, 52
Transforming growth factor (TGF)
    death-associated protein kinases and, 194, 197, 206
Transgenes
    signaling pathways in skeletal muscle remodeling and, 19, 24–25
Transition state mimicry
    asparagine synthetase chemotherapy and, 645
Translocases
    chromosome and plasmid segregation, 233–35
Translocation
    capsular polysaccharides in *E. coli* and, 52–53
Transmembrane traffic
    cytochrome $b_6 f$ complex and, 769–87
Transmission electron microscopy (TEM)
    amyloid structures and protein misfolding, 336, 343, 346–47
Transport
    axonal transport and Alzheimer's disease, 607–21
    mammalian peroxisomes and, 306–8, 319, 322–25
    mitochondrial ADP/ATP carrier and, 713, 732–34
Transposases
    site-specific recombination and, 599–600
Transposition
    site-specific recombination and, 569, 583
Transverse urea-gradient gel electrophoresis (TUG-GE)
    cold-adapted enzymes and, 410, 418
Traumatic brain injury
    axonal transport and Alzheimer's disease, 614
Triacylglycerol
    obesity-related derangements in metabolic regulation and, 376–77, 390–91
Tricarboxylic acid (TCA) cycle
    obesity-related derangements in metabolic regulation and, 380–82, 386–89, 394
Triosephosphate isomerase
    cold-adapted enzymes and, 416, 420–21, 426
Troponin
    signaling pathways in skeletal muscle remodeling and, 24
trx genes
    gene expression regulation and chromatin modifications, 248
Trypanosoma brucei
    mammalian peroxisomes and, 322
    NADH:quinone oxidoreductase (complex I) and, 72
    ribonucleotide reductases and, 698
    sirtuins and, 439
Trypsin

cold-adapted enzymes and, 406, 411, 422, 424, 426
TSC2 gene
   LKB1-dependent signaling pathways and, 149–51
Tuberous sclerosis
   LKB1-dependent signaling pathways and, 149–51
Tubulin
   animal cytokinesis and, 547
   axonal transport and Alzheimer's disease, 619
   dynamic filaments of bacterial cytoskeleton and, 467–76
   sirtuins and, 438–39
Tumor necrosis factor (TNF)
   death-associated protein kinases and, 194–95
Tumor suppressor gene
   LKB1-dependent signaling pathways and, 139–42
Tunicamycin
   death-associated protein kinases and, 206
twinstar gene
   animal cytokinesis and, 557
Type 4 secretion systems
   capsular polysaccharides in *E. coli* and, 53
Tyrosine kinase inhibitors
   tyrphostins and other tyrosine kinase inhibitors, 93–105
Tyrosine recombinases
   site-specific recombination and, 567, 570–82
Tyrphostins
   other tyrosine kinase inhibitors and
      ATP mimics, 99
      Bcr-Abl kinase, 97–99
      chronic myelogenous leukemia, 97–99
      conclusions, 105
      EGFR-directed tyrphostins in combination with other agents, 101
      EGFR kinase inhibitors, 99–101
      erlotinib, 100
      gefitinib, 100
      introduction, 94–95
      irreversible inhibitors, 100–1
      Jak-2 inhibitors, 104
      Jak-3 inhibitors, 104–5
      PDGFR, 103–4
      PDGFR kinase inhibitors, 104
      protein tyrosine kinases, 95–96
      resistance, 98–100
      restenosis, 104
      reversible inhibitors, 99–100
      signal transduction therapy, 95–96
      STI-571, 98–99
      structural considerations, 101–2
      tyrphostins, 96–97
      VEGF receptor tyrosine kinase inhibitors, 102–4

U

Ubiquinones
   NADH:quinone oxidoreductase (complex I) and, 69–87
   proton transfer through respiratory complexes and, 181
Ubiquitination
   gene expression regulation and chromatin modifications, 243, 247, 254–56
UCP proteins
   obesity-related derangements in metabolic regulation and, 375, 385, 391–92
UL9 protein
   history of research, 15
Undecaprenylphosphate (und-P)
   capsular polysaccharides in *E. coli* and, 47–48
Unfolded protein response (UPR)
   asparagine synthetase chemotherapy and, 636–37
Unpairing
   chromosome and plasmid segregation, 220–22
Unstructured aggregates
   amyloid structures and protein misfolding, 348–49
Uracil-DNA glycosylase
   cold-adapted enzymes and, 420–22, 424
Ure2 protein
   amyloid structures and protein misfolding, 340

V

VAMP-8 protein
   animal cytokinesis and, 559
van der Waals contacts
   cold-adapted enzymes and, 408, 421
   protein motion and catalysis, 525, 532, 534
VEGF receptor tyrosine kinase inhibitors
   tyrphostins and other tyrosine kinase inhibitors, 102–3
Velocity
   axonal transport and Alzheimer's disease, 610–11
Vibrio spp.
   asparagine synthetase chemotherapy and, 634
   cold-adapted enzymes and, 416, 420–21, 427
   site-specific recombination and, 582

Vincristine
asparagine synthetase chemotherapy and, 632
Viral latency
history of research, 1, 14–15
Viscosity
cold-adapted enzymes and, 408
Visual pigments
rhodopsin and, 743–62
Vlcs gene
mammalian peroxisomes and, 323

W

Walker A cytoskeletal ATPase (WACA) proteins
dynamic filaments of bacterial cytoskeleton and, 482–86
wecA gene
capsular polysaccharides in *E. coli* and, 47
"Worse and random" segregation
chromosome and plasmid segregation, 229
wz genes
capsular polysaccharides in *E. coli* and, 45–54, 61

X

X-chromosome inactivation
gene expression regulation and chromatin modifications, 254–56
Xenopus laevis
animal cytokinesis and, 556–58
LKB1-dependent signaling pathways and, 143–44, 154
XerCD gene
chromosome and plasmid segregation, 234–35
site-specific recombination and, 578, 581–82
X-linked adrenoleukodystrophy
mammalian peroxisomes and, 316–17
X-ray crystallography
amyloid structures and protein misfolding, 343
asparagine synthetase chemotherapy and, 642, 646
cold-adapted enzymes and, 420–21, 427
death-associated protein kinases and, 192
Hsp90 molecular chaperone and, 271–90
mitochondrial ADP/ATP carrier and, 715
NADH:quinone oxidoreductase (complex I) and, 79
protein motion and catalysis, 530–33

proton transfer through respiratory complexes and, 176, 178
rhodopsin and, 754–58
ribonucleotide reductases and, 685–86
transmembrane traffic in cytochrome $b_6 f$ complex and, 769–71, 773, 786
Xylanase
cold-adapted enzymes and, 407, 410, 420–22
Xylose isomerase
cold-adapted enzymes and, 417

Y

Yarrowia lipolytica
NADH:quinone oxidoreductase (complex I) and, 73–74, 76, 79–81, 84, 86
Yeasts
amyloid structures and protein misfolding, 339–40, 346
animal cytokinesis and, 560
asparagine synthetase chemotherapy and, 635
break-induced replication and recombinational telomere elongation in yeast, 111–32
chromosome and plasmid segregation, 211, 217–18, 222, 227, 230–33
eukaryotic signaling circuits and, 662
gene expression regulation and chromatin modifications, 248–49, 251, 253, 259
history of research, 13
LKB1-dependent signaling pathways and, 144
mammalian peroxisomes and, 295–325
NADH:quinone oxidoreductase (complex I) and, 70, 73–74, 76, 79–81, 84, 86
ribonucleotide reductases and, 684, 693–94, 699
sirtuins and, 436, 438–40
telomerase reverse transcriptase and, 495–97, 500, 502–5, 508–9
yjbHGFE operon
capsular polysaccharides in *E. coli* and, 53
ymcABCD operon
capsular polysaccharides in *E. coli* and, 52–53

Z

Zellweger syndrome
mammalian peroxisomes and, 315–18, 320
ZIP kinase (ZIPk)
death-associated protein kinases and, 189, 191, 193, 195–97, 202–4

Author Index

A

Aagaard A, 169, 173–76
Aaron MA, 599
Aaron W, 617
Aaronson W, 54, 61
Abbatiello SE, 631, 641
Abbott DW, 660
Abdollahi M, 662
Abdrakhmanova A, 70, 73, 74, 76
Abdulaev NG, 744, 759
Abel ED, 372, 373
Abel J, 576
Abell LM, 643
Abelmann A, 75, 181
Abend A, 692
Ablonczy Z, 752, 759
Abo-Hashema KAH, 379
Abourbeh G, 101
Aboushadi N, 315
Abraham M, 503, 504
Abramson J, 168, 169, 172, 173, 180, 715
Abramson T, 611
Abremski K, 595
Abresch EC, 176
Abud HE, 496
Abu-Elheiga L, 376, 379
Abumrad NA, 369, 383
Acin-Perez R, 70
Adachi M, 662
Adalsteinsson T, 348
Adame A, 618, 619
Adams MD, 657, 659, 664

Adams PD, 641
Adams R, 548
Adams RR, 547, 548
Adams V, 599, 600
Adamski J, 309
Addinall SG, 469, 474, 482
Adelroth P, 167, 169, 170, 173, 174, 176, 177, 179, 180
Adelmant G, 28
Adhya S, 353
Adler J, 7, 245
Adler N, 7
Aebi U, 336, 342–45
Aebischer P, 357
Affar EB, 262
Affourtit C, 86
Afzal AJ, 407, 408
Agard DA, 273, 275, 276, 279, 520
Agarwal PK, 524, 528, 529, 531, 533–35
Aggeler R, 75
Aghajari N, 418, 421–25
Agius SC, 718, 720, 734
Aguilella VM, 711
Aguilera I, 547
Aguilera R, 771, 774
Aharonowitz Y, 683, 693
Ahlers P, 78
Ahlgren-Berg A, 579
Ahmad A, 641
Ahmad FJ, 611
Ahmad K, 244, 245, 261, 262
Ahn SH, 259

Ahringer J, 546, 547
Aigner S, 500
Aihara H, 574, 576–78, 580, 600
Aihara K, 324
Aimi Y, 618
Aisner DL, 272, 507
Aittaleb M, 407, 409, 415, 418, 419, 424
Aizawa T, 386, 388
Aizono Y, 408, 419, 422, 424
Ajose-Adeogun AO, 545, 546, 560
Akamine P, 144
Akasaka K, 535
Akimoto T, 27, 28, 375, 380, 382, 384
Akira S, 191, 195–99, 201
Akiyama T, 96
Akke M, 535
Akopian A, 583
Al-Amoudi A, 50
Alani E, 121
Alaranta S, 55
Alarcon C, 389, 393
Albar JP, 496
Albers C, 310
Alberts BM, 115, 547, 559
Albracht SPJ, 71, 75, 78, 81, 82
Albrecht B, 27, 377
Alcalay M, 197, 199
Aldemir H, 385
Alekseev A, 760
Aleman TS, 760

ALESSI DR, 137–63; 140–46, 148, 149, 151, 152, 156, 663
Alexandrescu AT, 336
Alexandru G, 219
Alexson SE, 308
Alfonso C, 473
Alford M, 618, 619
Al-Hakim AK, 156, 157
Alhambra C, 533
Ali MM, 284
Ali S, 407
Aliev G, 612, 619
Alitalo K, 140, 141
Al-Khalili L, 23
Allemand JF, 214, 234
Allemann RK, 528, 530
Allen DG, 22
Allen JF, 770, 784
Allfrey VG, 245
Allinquant B, 617
Allis CD, 245, 249, 251, 255, 259, 260, 438
Allison RD, 631
Allshire R, 249
Allshire RC, 438, 510
Allsop J, 313
Allsopp RC, 496
Almarsson O, 533
Almeida CG, 619
Almenar-Queralt A, 617
Almgren P, 382
Alon U, 675
Aloy P, 682, 683, 700
Alric J, 784
Al-Shamma H, 744
Alsina MM, 96, 101
Altenbach C, 520, 744, 753
Altieri DC, 558
Altman B, 500
Altschul SF, 572
Altschuler RJ, 619
Alvarez M, 416, 421, 424, 426
Alvarez S, 510
Alvis SJ, 711
Alwin S, 674
Amacker M, 505, 506
Amador DAM, 638
Amann JM, 639
Amaral MD, 335
Amaratunga A, 617
Ambler A, 78
Amery L, 297, 310

Ames BN, 383
Amini AA, 31
Amini AR, 31
Amon A, 220, 437
Amor P, 49
Amor PA, 44, 47–50
Amoresano A, 356
Amos CI, 140–42
Amos LA, 469–71, 475, 477, 478, 480, 481
Amylon MD, 630
An J, 372, 373, 375, 378, 380, 382, 384, 389
An XX, 694
Anafi M, 97
Anand R, 633
Andalis AA, 436, 444
Andaluz E, 130
Anderson C, 61
Anderson DE, 218, 222, 469
Anderson G, 760
Anderson I, 475
Anderson K, 630
Anderson KA, 148
Anderson MJ, 377
Anderson PM, 633
Anderson RC, 383
Anderson RE, 746
Anderson RM, 436, 442, 444, 452, 454
Anderson S, 253
Anderson VE, 570
Andersson B, 786
Andersson J, 685, 687, 690, 692, 697, 700
Andersson KK, 682
Andersson U, 28
Andewelt A, 372
Andras LM, 378
Andre J, 383
Andreassen PR, 549
Andreelli F, 379
Andrei-Selmer C, 224
Andres J, 528, 533
Andreu JM, 469, 470, 473–76
Andrew M, 630
Andrews D, 404
Andrews JL, 386
Andrews SC, 78
Andrews WH, 496
Andreyev AY, 384, 385
Andrianopoulos K, 47

Andronesi OC, 344
Andrulis IL, 638
Anfinsen CB, 535
Angell JE, 75
Angelletti RH, 447, 450
Angermuller S, 302
Angers S, 758
Angleson JK, 745
Angus LM, 29
Anis M, 147, 149
Anjum R, 197
Annaert W, 618
Annan RS, 661
Annilo T, 322
Anraku N, 11
Antikainen NM, 527, 528
Antinozzi PA, 389, 393
Antonarakis SE, 157
Antonenkov VD, 310, 311, 313, 314, 322
Antos CL, 27
Antunes JL, 393
Antzutkin ON, 341
Anusuya DM, 383
Aoyama H, 171
Aoyama T, 312, 324
Aparicio OM, 437
Apazidis A, 375
App H, 102
Appanah R, 255
Appel IM, 639
Appel J, 785
Appel S, 196, 201
Appelkvist EL, 315
Aquila H, 718, 719, 734
Aragall E, 683, 692
Arai K, 505, 506
Arakawa H, 684, 693
Araki T, 436, 442, 460
Arand M, 302
Aranda M, 579
Aratani S, 439
Aravind L, 498, 683, 693
Arbel-Eden A, 117
Arber W, 596
Arboit P, 385
Archer DR, 617
Arciszewska LK, 574, 578
Ardekani BA, 620
Arechaga I, 735
Arendt A, 745, 748
Argonza R, 638

Argos P, 454, 461, 748
Arisaka F, 417
Ariyan C, 382
Arjara G, 685, 687, 692
Arkan MC, 374, 377
Arlow D, 375
Armati PJ, 620
Armbruster BN, 496, 500, 505, 508
Armitage JP, 481
Armstrong CM, 436, 438, 439, 446, 447
Armstrong EA, 101
Armstrong RA, 616, 619
Arnaud D, 255
Arndt KT, 289
Arner ESJ, 699
Arneric M, 127, 510
Arnez JG, 634
Arnio E, 305
Arnis S, 747
Arnold FH, 415–17
Arnold M, 24
Arnold PH, 586, 591
Arnon DL, 782, 784
Arnorsdottir J, 421, 423, 427
Arnou B, 719
Aro EM, 785
Aronson D, 29
Arora K, 710
Aroyo M, 581
Arpaia E, 104, 700
Arpigny JL, 407, 408, 420
Arrecubieta C, 61, 62
Arriaga EA, 419
Artaza J, 32
Artemyev NO, 759
Arthur JS, 143, 144
Artsimovitch I, 46
Artzatbanov VY, 169
Asai N, 642
Asakawa S, 357
Asayama K, 301, 302
Asbury CL, 610
Ascano J, 610
Asfari M, 386
Asgeirsson B, 406, 427
Ashcroft M, 684, 693
Ashley-Koch A, 613
Ashmarina LI, 297
Ashworth A, 145, 155
Asilmaz E, 373

Askree SH, 116
Aslanian AM, 630, 638
Åslund F, 683, 684
Aspnes GT, 630
Asselberghs S, 309, 312
Assimacopoulos-Jeannet F, 386, 390, 391, 393
Assmann C, 775
Assmann G, 310
Assselberghs S, 308, 313
Assuncao J, 70, 73
Aston-Jones G, 620
Astrof NS, 341, 343
Atkins E, 341, 343
Atlung T, 692
Atmakuri K, 52
Ator M, 687
Attardi G, 81
Au HC, 75
Aubourg P, 323
Aufiero DJ, 253
Auling G, 683, 693
Auluck PK, 357
Aung S, 229
Aurora R, 426
Ausmees N, 483
Aussel L, 234, 581
Austin S, 216, 223, 227–29, 235, 484
Austin SJ, 226
AUTEXIER C, 493–517; 496, 500–8
Authelet M, 619
Avalos J, 447, 448, 450, 452, 458
Avalos JL, 448, 450, 454, 455, 457–59
Aveldano MI, 746
Avenson TJ, 782, 784
Avigan J, 311
Aviles FX, 349, 353
Avilion AA, 500
Aviram M, 315
Avizienyte E, 138–40, 142, 157
Avner P, 255
Avraham S, 301
Avram D, 439, 440, 448
Avvedimento EV, 669
Awad ES, 744
Awad WM Jr, 424
Axelman K, 357
Axelrod A, 438
Axelrod HL, 176

Ayaz D, 617
Azaro M, 577, 580
Azaro MA, 568, 572, 574
Azevedo JE, 73
Azevedo JL, 30, 31
Azhar S, 27
Azzalin CM, 509

B

Baar K, 28
Baas AF, 143–46, 154
Baas PW, 611
Babbitt PC, 640
Babcock A, 613
Babcock E, 390
Babcock GT, 168, 170, 171
Babitzke P, 41
Babu KR, 754
Baccarini Melandri A, 70
Bachand F, 500, 503, 504
Bachmann BO, 633, 634, 638, 640, 642
Bachynsky N, 638
Backgren C, 169, 173
Bada JL, 405
Bader R, 348, 349
Badet B, 633
Badet-Denisot M-A, 633
Badger JD 2nd, 613
Badrinath A, 221
Bae E, 421, 422, 425
Bae KH, 674
Baehr W, 744, 745, 747, 761
Baekelandt V, 357
Baerlocher G, 509
Baes M, 300, 309, 315, 320, 321, 324
Baffy G, 391
Bagaglio JF, 641
Baggaley KH, 640
Baggenstoss BA, 282
Bahar I, 710, 760
Bahnson BJ, 532–35
Bai X, 391
Bai Y, 114, 156
Bai YD, 81
Bailey D, 244, 248
Bailey MJ, 46
Bailey SM, 123, 132
Bailey ST, 371, 372
Bailly A, 117

Bailly D, 251
Bain J, 148
Bain PJ, 635, 636
Baise E, 406, 407, 416
Bajaj VS, 343
Bajard L, 20
Bajorath J, 642, 646
Baker CS, 41
Baker MB, 370, 372
Baker ME, 75
Baker R, 578
Baker RA, 574
Baker SJ, 98, 99
Baksalerska-Pazera M, 612
Balasubramanian B, 116, 496
Balasubramanian MK, 544
Balbach JJ, 341
Balbirnie M, 341–43
Baldus M, 344
Baldwin AS Jr, 31
Baldwin EP, 576–79, 633
Baldwin J, 405
Baldwin JE, 640
Bales KR, 357
Balguerie A, 355
Balijja A, 121
Balint E, 684, 693
Ball DJ, 105
Ballif BA, 197
Ballinger CA, 282, 283
Balu M, 383
Bambara R, 12
Bamberg E, 169, 170, 173, 733
Bamburg JR, 618
Bamford R, 348, 349
Ban C, 278
Banai S, 104
Bandi M, 560
Baneres JL, 759
Banerjee A, 474
Banerjee M, 437
Banerjee RR, 371, 372
Banik SSR, 496, 500, 505, 508, 509
Banks GR, 13
Bannister AJ, 260, 262
Bansal NP, 508
Banwait S, 617
Baptista AM, 180
Bar G, 684
Bar J, 378
Barakat HA, 31

Baran N, 495, 502
Barbas CF 3rd, 146, 674
Barbe J, 683, 692
Barbeau A, 389, 393
Barber CM, 251
Barber D, 146
Barber J, 770
Barbier V, 272, 282
Barbosa-Tessmann IP, 635, 636
Barbour RL, 733
Bardeesy N, 139, 147–49, 153, 154, 156
Bardelli A, 155
Bardwell AJ, 662
Bardwell L, 662–64
Barenboim L, 611
Barford D, 282
Bargmann CI, 155
Barilla D, 485
Barnes CL, 615
Barnes R, 407
Barnhart BC, 156
Barone M, 370, 372
Baroni F, 353, 355, 358
Barquera B, 86, 181, 182
Barr K, 49
Barr M, 669
Barr RD, 630
Barrachina I, 83
Barre F-X, 213, 214, 234, 578, 580, 581
Barrett B, 54, 55, 61, 62
Barrientos T, 26
Barshop BA, 634
Barsyte-Lovejoy D, 662
Bartek J, 217
Bartel DP, 249
Bartha B, 276
Bartl FJ, 751, 753, 760
Bartley SM, 256
Bartley W, 715
Bartoov-Shifman R, 387, 393
Bartoschek S, 78
Bartsch U, 319, 321
Barucci N, 27, 376, 379
Bashmakov Y, 372
Bashor CJ, 671
Basilico C, 635, 636
Basit M, 385
Baskakov IV, 336
BASSEL-DUBY R, 19–37; 30
Bassel-Duby RS, 21

Basset G, 718–20
Bassi CJ, 619
Bassilian S, 308
Basso AD, 273, 289
Bateman A, 572
Bates AD, 276
Bates AL, 377
Bates D, 216, 480
Bath J, 215
Batlle E, 154
Battchikova N, 785
Batten M, 747
Batten ML, 745, 761
Baudry M, 156
Bauer CE, 786
Bauer HH, 336, 345
Bauer MF, 70
Bauer UM, 258, 440
Bauerle R, 633
Baulieu EE, 282
Baum B, 546–48, 555
Baumeister W, 632
Baumgart E, 300, 308, 313, 315, 320
Baumhueter S, 31, 32
Baur J, 272, 507
Bausch C, 692
Bavenholm PN, 379
Bax B, 660
Baxa U, 346, 355
Baxter J, 255
Baxter M, 101
BAYASCAS JR, 137–63
Bayer ME, 52
Baylor D, 745
Baylor DA, 745
Baymann F, 786
Bayona-Bafaluy MP, 70
Beall B, 468
Bearer EL, 617
Beattie TL, 500, 502–9
Beausejour CM, 674
Beaver A, 385
Bebbington D, 545, 546, 560
Bechter OE, 132
Becker A, 692, 701
Becker E, 223
Becker S, 344
Becker TC, 389
Beckers A, 151
Beckman K, 303
Bedalov A, 461

Bedford MT, 245, 258
Bednenko J, 498, 500, 505
Bedrick SD, 558
Beech PL, 468, 472
Beerli RR, 146, 674
Beets AL, 369, 383
Befroy D, 382
Begg KJ, 482
Begueret J, 340
Behan D, 744
Beher D, 617
Behnke CA, 744–46, 748, 749, 751
Behr J, 170, 180
Beijersbergen RL, 508
Beis K, 50, 51, 53
Beites CL, 559
Bejarano ER, 286
Bejma J, 385
Belcourt MF, 693
Belevich I, 170, 173
Belfort R, 390
Belichenko PV, 618
Bell CE, 459
Bell DW, 100, 101
Bell GI, 373
Bell GS, 412, 415, 419, 422
Bell K, 27, 376, 379
Bell SD, 439, 452
Bellamy HD, 176, 181, 182
Bellefroid E, 197, 199
Bellen HJ, 547
Belloni G, 547
Bellotti V, 359
Belogrudov G, 79, 85
Bement WM, 550, 552
Bemporad F, 350
Benarous R, 323
Benavente S, 101
Bendall DS, 772, 781, 782, 784, 785
Benedek GB, 347, 348
Benedikz E, 619
Benes CH, 668
Beness AM, 49–51, 61
Benhattar J, 510
Benink HA, 550, 552
Bénit P, 86
Benjamin HW, 593
Benko G, 744, 745, 751, 752
Benkovic PA, 8
Benkovic S, 8

BENKOVIC SJ, 519–41; 520–22, 524, 526–29, 532, 533, 536
Ben-Neriah Y, 97
Benner SA, 671
Bennett MJ, 305, 507
Bennett WS, 520
Benovic JL, 744
Benowitz LI, 617
Ben-Porath I, 509
Benson MD, 336, 341, 615
Benson N, 642
Benson WG, 386
Bentahir M, 408, 409, 415, 419, 424
Bentley P, 302
Benton R, 154
Benveniste P, 700
Ben-Yaakov S, 102
Ben-Yehuda S, 214, 469
Bera AK, 634
Berant M, 305, 315
Berard C, 630
Berchtold MW, 22
Berden JA, 775
Berends W, 716
Berendsen HJC, 711
Berg JM, 301
Berger J, 305, 321, 324
Berger MB, 146
Berger S, 258
Bergerat A, 273
Bergeron C, 357
Berggren JR, 376, 385
Berglund GI, 421
Bergmann S, 657
Bergmann U, 770
Bergquist S, 674
Bergseth S, 308
Berissi H, 190, 194, 197, 198
Berks BC, 78
Berlin I, 340
Berman J, 117
Bermejo A, 83
Bermudes D, 468
Bernal-Mizrachi C, 379, 383
Bernal-Mizrachi E, 391, 436
Berndt JD, 617
Bernstein BE, 260
Bernstein E, 438
Berria R, 382, 390
Berriman J, 344, 345

Berry D, 667
Berry DM, 667
Berry EA, 176, 181, 182, 770–72, 775, 781
Berry R, 138
Bershadsky AD, 546
Berson A, 383
Berson JF, 339, 340
Berteaux-Lecellier V, 323
Bertino JR, 630
Bertoncini CW, 356
Bertos NR, 24
Bertrand G, 388
Bertrand P, 122
Bertsch L, 7, 8
Bertsova YV, 84
Bertuch AA, 118, 122
Bessman MJ, 6, 7
Best JL, 156
Best RB, 350
Bestman J, 147, 154
Bestor TH, 261
Betarbet R, 86
Betschinger J, 550
Bettencourt-Dias M, 156
Beuria TK, 474
Beushausen S, 610
Bevan S, 613
Bever KM, 454, 457, 459
Beveridge TJ, 41, 50
Beverloo HB, 639
Beyer K, 718, 720, 726, 736
Bezman L, 316
Bezrukov SM, 711
Bhandawat V, 744
Bhat S, 371, 372
Bhattacharya S, 423
Bhattacharyya A, 505
BHATTACHARYYA RP, 655–80; 660, 662–65, 669, 671, 672
Bi E, 469, 482
Bi EF, 468, 469, 544
BIALIK S, 189–210; 190, 192–201, 203, 205
Bian F, 308
Bianchi A, 127, 508, 511
Bianchi V, 699, 700
Bianco MR, 358
Bibb LA, 599
Bibikova M, 546, 547
Bickle TA, 586, 596

Biden TJ, 378
Bidnenko V, 113
Bieganowski P, 442
Biek DP, 226
Bieman M, 440
Bierbaum H, 195–98, 201, 204
Bieri F, 302
Biermann J, 304
Biernat J, 155
Biesele JJ, 558
Biesiadka J, 770
Biessmann H, 120
Bigard X, 25
Bignell CR, 230
Bigot S, 214, 234
Billaud M, 146
Billeter SR, 524, 528, 529, 533–35
Billiar TR, 302
Billington BL, 437
Binas B, 377
Binder LI, 615
Bingley M, 613
Binkley SB, 43
Biondi RM, 660–64
Birck MR, 59, 60
Bird LE, 574
Bird T, 614
Birdsey GM, 300, 313
Birgander PL, 691, 692, 698
Birge RR, 754
Birnbaum MJ, 30
Birolo L, 409
Birren B, 657, 659, 664, 665
Bischof T, 78
Bishop JE, 303
Biswas T, 574, 577, 580, 600
Bitan G, 348
Bitsch A, 614
Bitter-Suermann D, 58–60
Bitterman KJ, 436, 439, 442, 444, 452, 454, 460
Bitton R, 273
Bjarnason JB, 406
Bjenning C, 744
Bjorbaek C, 372
Bjork A, 415
Björklund S, 693
Bjorkman J, 320
Bjornholm M, 30
Bjornson K, 473

Blaak EE, 379
Black BL, 23, 25
Black MM, 611
Black MT, 775
Black PN, 305
Blackburn EH, 116, 117, 119–23, 128, 129, 494, 495, 500–6, 509
Blackwell HE, 461
Blackwell TK, 222
Blagden C, 32
Blagosklonny MV, 273
Blair DF, 171
Blaise V, 404, 405, 407, 410–13, 415, 417–19, 422
Blake C, 336, 341
Blake DG, 586
Blakley RL, 521
Blanaco L, 8
Blancafort P, 674
Blanchard V, 619
Blanco E, 717
Blanke CD, 98
Blankenship RE, 784
Blasco MA, 496, 504
Blattner FR, 692
Blau HM, 30
Blechner S, 660
Bledsoe B, 341
Blenis J, 197
Bliska JB, 674
Bliss JM, 57–59
Bloch D, 170, 173
Bloch Y, 52
Block MR, 730, 733, 734
Bloom M, 711
Bloomekatz JE, 669
Blue EK, 194–96, 199–202
Bluemink JG, 558
Blum G, 99
Blumenthal R, 552
Boardman LA, 138
Bobbioni-Harsch E, 385
Bochar DA, 262
Böck A, 75, 78
Böcker C, 76
Bodine SC, 26, 28, 31, 32
Bodnar AG, 510
Bodner R, 671
Boehlein SK, 631–34, 641, 642, 645
Boehmer PE, 14

BOEKE JD, 435–65; 436–39, 442, 444, 447, 448, 450, 452, 454, 455, 457, 458, 499
Boekema EJ, 73, 81
Boerger AL, 122
Bogachev A, 172
Bogachev AV, 84
Bogdanov B, 647
Bogdanovic N, 619
Boggs BA, 255
Bogner W, 718, 719
Bohacek RS, 646
Böhm R, 75, 78
Böhme H, 779
Bohmer A, 96, 97, 103, 104
Bohmer FD, 99, 103, 104
Bohnert S, 613
Bok D, 745
Bokhari SA, 408
Bokoch GM, 552
Bokranz W, 41
Boldin M, 194
Bolhuis PG, 710
Bollag G, 103
Bolotin-Fukuhara M, 130
Bolton EC, 499
Bolwien C, 170, 172
Bolzi JA, 13
Bonaccorsi S, 547, 555, 557
Bonasera TA, 101
Bondar AN, 748, 755
Bone JR, 253
Bonen A, 377
Bonini NM, 357
Bonner J, 245
Bonner TI, 744
Bonner WM, 117
Bonner-Weir S, 390
Bono JL, 572
Bonomi M, 760
Bononi J, 611
Boocock MR, 573, 583–86, 588, 590, 593, 595, 596
Booker SJ, 682
Boor PP, 138
Boos J, 630
Boosen M, 196, 201
Booze RM, 619
Borch J, 229, 479
Borchardt RT, 536
Bordo D, 74
Bordoni A, 719

Borella S, 699
Borhan B, 755
Borhani DW, 474
Borisy GG, 556
Bork P, 74, 226, 476, 480, 482, 633, 664, 667
Bork-Jensen J, 482
Borkovich KA, 272
Borman MA, 196, 202
Bornens M, 558, 559
Bornmann W, 98
Bornmann WG, 102
Boronowsky U, 786
Borovok I, 683, 693
Borra MT, 438, 439, 441, 444, 452, 453, 457, 461
Borst P, 322
Borts RH, 121
Boschelli F, 272, 289
Bosco DA, 535
Bosco G, 113, 114, 129
Bose S, 274
Bosoy D, 498, 500, 501, 503, 505, 509
Boss O, 372, 373
Bossa F, 420, 422, 423, 425
Bossi RT, 634
Bossy-Wetzel E, 546, 547
Bostina M, 70, 73, 79–81, 83
Boston H, 613
Boston T, 692
Böttcher B, 79, 80
Botting CH, 439, 452
Bottomley SP, 349
Bouchard M, 343, 345, 349
Boucher A, 386, 390, 392, 393
Boucher L, 663, 664
Boudeau J, 140–48, 154
Bouet JY, 226, 229
Bouffard GG, 75
Bouhouche I, 276, 277, 287
Bouillot C, 617
Boulay F, 718, 719, 730, 732
Boulianne GL, 547
Boulnois G, 58–61
Boulnois GJ, 59–61
Bouman P, 260
Boumedienne M, 48
Bouras T, 439, 440
Bourenkov G, 282
Bourette RP, 667
Bourey RE, 30

Bouriotis V, 406, 407, 409, 410, 415, 419, 424
Bourne HR, 750
Bousché O, 173
Bousset L, 336, 346, 350
Boutajangout A, 619
Bouton CM, 618
Bouvier M, 582, 605, 758
Bouwmeester T, 144
Bove KE, 316
Bovee-Geurts PH, 755
Bovina C, 86
Bowie JU, 723
Bowman MK, 776, 779, 784, 785
Bowtell DD, 496
Boxer LM, 639
Boyce EJ, 639
Boyle JA, 618
Braak E, 616
Braak H, 616
Brabazon DM, 759
Bracchi-Ricard V, 684
Brachmann CB, 436, 438, 446, 447
Brachova L, 357
Bradford WW, 104
Bradley A, 761
Bradley M, 274, 642
Bradwin G, 374
Brady K, 599
Brady ST, 618, 620
Braiman MS, 173
Braiterman LT, 323, 324
Brajenovic M, 144
Brame CJ, 258
Bramhill D, 469, 474
Brand MD, 86, 166, 391
Branden CI, 530, 532
Brändén G, 180
Brändén M, 169, 180
Brandenburg K, 61
BRANDOLIN G, 713–41; 709, 715, 717–20, 722, 723, 726–28, 730, 732, 733, 736
Brandsdal BO, 418, 419
BRANDT U, 69–92; 70, 74, 76, 78–81, 83–86, 181
Brannigan JA, 632
Brasseur R, 732
Bratton M, 174, 175
Braun HP, 73, 81

Braun S, 96, 99
Brdiczka D, 732
Brees C, 313, 324
Bregu M, 581
Breier AM, 213, 215
Breitenstein W, 98, 102
Breitkreutz A, 663, 664
Bremer J, 308
Brenchley JE, 404, 419
Brenneman MA, 132
Brenner C, 442
Brenner SL, 546
Brensinger JD, 138
Bresnick AR, 194, 201, 203, 552
Brewsaugh T, 316
Breyton C, 786
Briaud I, 386, 390
Brickner H, 272, 282
Bridges A, 660
Bridges AJ, 101
Bridges D, 664
Bridges EG, 700
Brieger A, 142
Briere JJ, 86
Briggs SD, 245, 251–53, 258, 259
Briki F, 346, 350
Brill JA, 547, 548
Bringmann H, 551
Brinker A, 282
Brinkmeier H, 22
Brion JP, 619
Briscoe CP, 386
Brites P, 305, 316, 320
Brito RMM, 348
Britton RA, 223
Brocard MP, 24
Broccoli D, 508, 509
Brock HW, 252
Brockmann C, 74
Brockway MJ, 283
Brody JP, 419
Brody S, 75
Bron P, 417
Bronner D, 47, 56, 57, 59–61
Bronson RT, 560
Brooijmans N, 646
Brooks BR, 641
Brooks CL, 528, 535
Brooks CL III, 528–30
Brooks JW, 619

Broom NJP, 642
Broome BM, 353
Brors B, 75, 78, 82, 83, 181
Brown A, 611
Brown AG, 640
Brown EE, 558
Brown EJ, 371, 372
Brown JL, 584, 593
Brown KK, 379
Brown LJ, 386, 389
Brown LS, 167, 168, 173
Brown MF, 711
Brown MJB, 642
Brown MS, 314, 372
Brown NC, 694, 698
Brown NR, 669
Brown P, 642
Brown RH Jr, 349
Brozinick JT Jr, 30
Bruce CR, 377
Bruck W, 614
Bruder G, 302
Bruford E, 8
Bruford MW, 300, 313
Brugge JS, 284
Brüggemann H, 76
Bruggen J, 98, 102
Brugger B, 320
Brugna M, 786
Bruice TC, 520, 533, 535
Brumby A, 547, 552
Brun YV, 482, 483
Brunet A, 436
Brunger AT, 641, 710
Bruni A, 716
Brunzelle J, 192, 193
Brusch RG, 611, 613, 618–21
Bruschi CV, 113
Brusselmans K, 151
Brustovetsky N, 733
Brutlag D, 7, 12
Bruynseels K, 619
Bryan TM, 131, 496, 500, 501, 503, 505–7
Bryk M, 245, 251, 437
Brynes S, 641
Brysch P, 310
Brzezinski P, 167–70, 172–77, 179, 180, 182
Bu JY, 668, 669
Bubb MR, 546
Bubis J, 757

Bucan M, 30
Bucci B, 694
Bucciantini M, 349, 358
Buchanan JM, 634
Buchdunger E, 97, 103
Buchenau P, 196, 198, 204
Bucher P, 509
Buchet-Poyau K, 157
Buchholz F, 576, 579
Buchner J, 273, 274, 276, 283, 287
Buck RC, 554
Buckingham JA, 166
Buckingham M, 20
Buckley K, 128, 508, 511
Buczylko J, 752
Budas GR, 148, 149
Buder K, 356
Budnik V, 618
Bueno OF, 25, 26
Bug S, 698
Buijs RM, 620
Buist G, 698
Bukau B, 283
Buller AJ, 20, 22
Bundey R, 617
Bundle DR, 43
Bungert S, 84
Bunick GJ, 258
Bunny KL, 582
Bunya M, 310
Buonanno A, 26
Buonomo SB, 221
Burack WR, 669, 671
Burbaev DS, 78, 83
Burdelya L, 104
Burden SJ, 32
Burdett K, 304
Burgart LJ, 138
Burgers PMJ, 8
Burgess AW, 101
Burgess DR, 550, 553, 558
Burgess R, 262
Burgess SC, 386, 388–90, 392, 393
Burghammer M, 748, 751, 755
Burgin A, 578
Burgin AB, 574, 577, 582
Burgin AB Jr, 572, 580
Burk DH, 482
Burke ME, 586, 591
Burke WD, 498, 499

Burke WJ, 619
Burkovski A, 734
Burlina AB, 388
Burn P, 368
Burns M, 614
Burns MR, 632, 633
Burrows PA, 782
Burtis KC, 8
Busche G, 98
Busciglio J, 618, 620
Buser AS, 104
Bush AI, 356
Buss JD, 386, 389
Buss V, 748, 755
Butler AE, 390
Butler D, 311
Butler PC, 390
Butler PJ, 484, 485
Butler-Cole B, 578
Butte A, 382
Butte AJ, 382
Butters N, 615
Buxbaum JN, 336, 358
Buzzai M, 151
Bylund DB, 660
Byron O, 585, 588
Byvoet P, 261
Bywater RP, 758

C

Cabrita LD, 349
Cacace AM, 669
Caceres A, 618
Cacicedo JM, 379
Caddle SD, 508
Cafiso DS, 520
Caflisch A, 354
Cagney G, 253
Cai D, 31, 32, 618, 620
Cai DS, 372, 373
Cairney AEL, 638
Cairns BR, 663, 664
Cairns J, 10, 11
Calabria E, 25, 26, 29, 31
Calamai M, 350, 353
Calcutt MJ, 230
Calderazzo F, 699
Caldwell JA, 258, 259
Calheiros-Lourenco F, 195, 196, 198
Calhoun MW, 169

Callahan PM, 171
Callaini G, 547
Callendar R, 521, 536
Callizot N, 319, 321
Calloni G, 355, 356
Calos MP, 599
Calvert PD, 745, 746
Calvo S, 26
Camardella L, 406, 411, 412
Cameron DA, 745
Cameron LA, 551, 552, 555
Camilleri S, 380
Campana D, 630
Campbell CS, 478, 479
Campbell DG, 156, 157
Campbell ID, 349
Campbell KA, 559
Campillo M, 533
Campos J, 582
Camps C, 324
Canale C, 349
Canavate X, 750
Cande C, 86
Canet D, 349
Canettieri G, 156
Canman JC, 549, 551–53, 555
Cannon WR, 520
Canovas P, 558
Cantley LC, 660, 668
Cantor RS, 711
Cao C, 482
Cao LG, 556
Cao P, 273
Cao Q, 374
Cao R, 254, 255
Cao WH, 391
Capaldi RA, 75
Capanni C, 353, 355
Capell A, 617
Capizzi RL, 630, 638
Caplan AJ, 272, 286, 289
Caplan MR, 473
Capobianco L, 734
Capowski EE, 223
Caratzoulas S, 535
Carballido-Lopez R, 468, 479, 481, 485
Carbonetto S, 30
Carchon H, 309
Cardenas M, 282
Cardenas R, 533
Cardol P, 70, 73, 74, 76, 81

Cardoso I, 348, 358, 359
Careaga CL, 520
Carey AL, 377
Carey JO, 30, 31
Carling D, 30, 147
Carlo-Stella C, 97
Carlow DC, 409
Carlson M, 147
Carlson RW, 60
Carlsson E, 382
Carmeliet P, 309, 315, 321
Carmen AA, 438, 441
Carmena M, 548
Carmon S, 548
Carneheim CM, 300
Caron MG, 759
Carpenter FH, 447
Carper MJ, 385
Carr AM, 694
Carr DB, 390
Carr SA, 661
Carratore V, 406, 411, 412
Carrell CJ, 772, 775, 780, 781
Carrell RW, 335
Carrello A, 282
Carretero J, 139, 140, 142
Carrillo JJ, 758, 759
Carrillo-Tripp M, 711
Carroll J, 70, 71, 73, 75, 79, 81, 85, 86
Carroll ML, 101
Carroll SB, 657, 669, 674
Carson C, 621
Carson DR, 219
Carson M, 117, 128, 571
Cartailler J-P, 167
Cartee RT, 56, 57
Carter C, 103
Carter PS, 660
Carter-Dawson LD, 745
Cartier N, 323
Cartwright CA, 674
Carucci DJ, 498
Caruso U, 303
Carvalho A, 548
Casagrande R, 684, 694
Casamayor A, 663
Casanova MF, 619
Cascales E, 52
Cases S, 391
Cash AD, 612, 619
Cashikar AG, 347, 348

Casjens SR, 582
Caspari T, 694, 699
Cassano AG, 570
Cassell GH, 468
Cassia R, 101
Castagnoli L, 668
Casteels M, 312, 315
Castelluccio C, 86
Castillo R, 528
Castrillon DH, 547
Catlett-Falcone R, 104
Caucutt G, 276
Cavalli A, 354
Cavalli V, 612
Cavanagh J, 671
Cavenee WK, 101
Cavener DR, 635
Caves LSD, 528
CAVICCHIOLI R, 403–33; 404–7, 409–13, 415, 416, 418–20, 426
Cech TR, 116, 117, 129, 494, 496, 498–508
Cedar H, 643
Cedergren R, 72, 73
Celic I, 436, 439, 442, 446–50, 452, 454, 455, 457, 458
Cenci G, 546, 547, 559
Ceruso MA, 751, 758
Cesare A, 125, 126
Cesare AJ, 132
Cetkovic-Cvrlje M, 104
Cha Y, 532
Chabes A, 682, 684, 693, 694, 699
Chabes AL, 693
Chabner BA, 638
Chaboute ME, 693
Chabre M, 758
Chaconas G, 572, 574, 582
Chadha AD, 129
Chadli A, 276, 277, 287
Chahwan C, 225
Chai WH, 508
Chai XM, 442, 454, 457–60
Chaignepain S, 355
Chaika OV, 671
Chak K, 672
Chakhparonian M, 494
Chakkalakal JV, 29, 30
Chakrabarti D, 684
Chakrabarti R, 630, 631, 638–40

Chakraborti S, 385
Chakraborti T, 385
Chakraborty AK, 43
Chalberg MD, 7
Chambers JK, 386
Chambraud B, 282
Champoux JJ, 573, 574
Chan AC, 668, 669
Chan CB, 391
Chan CS, 113, 230, 231
Chan F, 761
Chan HY, 357
Chan OH, 101
Chan RC, 223
Chan SH, 733
Chan SI, 171
Chan SR, 500
Chance B, 82
Chandler D, 710
Chandra J, 97
Chaney MO, 348
Chang C, 176
Chang EC, 663, 664
Chang G, 59
Chang H-CJ, 272, 282
Chang HH, 128
Chang J, 120, 122, 127
Chang JD, 27, 660
Chang JH, 442
Chang KT, 101
Chang LF, 377
Chang MC, 691
Chang MP, 259
Chang P, 618
Chang PY, 29
Chang S, 27
Chang SH, 121
Chang T, 20
Chang XJ, 547
Changelian PS, 105
Changeux JP, 657
Chao ED, 461
Chapman KB, 496
Chapman MR, 339, 340
Chappell AS, 503
Chappell JB, 716
Charbonneau M, 117, 120, 121, 128, 129
Chareonsudjai S, 61, 62
Charge SB, 20
Charnock SJ, 56
Charon MH, 84

Chater KF, 230
Chattoraj DK, 229, 230
Chaubet-Gigot N, 693
Chavany C, 273
Chavez JA, 378
Chayen NE, 347
Chazin WJ, 282
Cheatham TE, 641
Checover S, 174
Cheeseman IM, 222
Cheetham ME, 760
Chen C, 121, 507, 635, 636, 671
Chen CC, 190, 194, 197–201
Chen CH, 190, 194, 195, 197–203
Chen CT, 560
Chen D, 245, 258, 436, 439, 440
Chen DG, 258
Chen DJ, 123
Chen DL, 439, 440
Chen DW, 692
Chen E, 618
Chen EI, 674
Chen EY, 468
Chen FS, 617
Chen G, 386, 618, 700
Chen GQ, 273
Chen H, 635–38
Chen I, 546, 550, 553, 554, 556–58
Chen J, 14, 58, 344, 354
Chen JL, 496, 501–4
Chen JR, 12
Chen JW, 575
Chen JY, 190, 194, 197–201
Chen KS, 388
Chen L, 374
Chen ML, 348
Chen M-S, 282, 283
Chen P, 98
Chen QJ, 119, 120, 126
Chen RH, 190, 194, 197–204
Chen RM, 75
Chen S, 76, 144, 273
Chen SF, 500
Chen SH, 634
Chen SY, 275, 282, 283
Chen W, 122, 297, 314
Chen X, 776, 785
Chen XH, 617, 618

Chen Y, 27, 180, 373, 376–79, 469, 473, 474, 573, 574, 576, 577, 587, 600
Chen YH, 508
Chen YJ, 255
Chen YL, 147
Chen Y-Q, 521
Cheng CC, 101
Cheng F, 104
Cheng HL, 440
Cheng J, 26
Cheng JB, 305
Cheng KY, 669
Cheng NQ, 346, 355
Cheng NS, 154
Cheng Q, 574
Cheng RH, 786
Cheng SH, 674
Cheng XD, 258
Cheng YC, 700
Chenoweth J, 262
Cheok MH, 639
Cherny D, 356
Cherny RA, 357
Cherny V, 176
Chertkova R, 348
Chessa J, 409, 424, 426
Chessa JP, 407, 418, 421, 424
Chester MW, 385, 386
Chester R, 102
Cheung A, 546, 550, 553, 554, 556–58
Cheung P, 255
Cheung PC, 663
Cheung WL, 245, 251
Chevalier B, 260
Chevallet M, 70, 82
Chevillard G, 319, 321
Chi Y-I, 770–72, 776
Chia MC, 672
Chiang AS, 357
Chiba T, 282
Chibalin AV, 23, 30
Chien D, 375, 376
Chien KR, 297
Chien P, 339, 340, 346, 347
Chien W, 311
Chilikova O, 694, 699
Chimploy K, 697
Chin E, 30
Chin ER, 22, 24, 25, 27
Chin JK, 532–35

Chin L, 496
Chinault AC, 255
Chinkers M, 282–84
Chinnaiyan P, 101
Chinwalla V, 244, 248
Chiodi P, 383
Chiosis G, 273, 289
Chisholm DJ, 380, 384
CHITI F, 333–66; 348–50, 353–56, 358
Chiu C-P, 510
Chiu TK, 346
Cho CH, 509
Cho HK, 28, 31, 375
Cho JW, 55, 56
Cho KB, 687, 688, 690
Cho KR, 140
Cho Y, 409, 417, 421, 423–25
Chohan S, 272, 276–78, 281, 287
Choi E-C, 642
Choi H, 369, 383
Choi KY, 669
Choi SY, 618, 642
Chopra R, 501
Choquet Y, 770, 778, 783, 785, 786
Chorev M, 96, 99
Chorny M, 104
Chow MK, 349
Chow SA, 674
Chow Y-H, 289
Choy Y-M, 43
Chretien D, 474
Christ N, 579
Christensen G, 347, 349
Christian S, 191, 195–97, 201
Christiansen EN, 308
Christiansen G, 336, 345, 346
Christie KR, 609
Christie PJ, 52
Christman MF, 219
Christodoulou J, 356
Chu AM, 340
Chu C, 311
Chu H, 617, 619, 620
Chu R, 320
Chu VC, 577–79
Chua KF, 436, 440
Chuman H, 730, 731
Chung HD, 619
Chung HY, 745

Chung IK, 509
Chung N, 436, 441, 460
Church-Kopish J, 745, 748
Chwalla B, 221
Ciardiello MA, 406, 411, 412
Ciccarelli FD, 664, 667
Ciciliot S, 26
Cideciyan AV, 760
Cieslewicz M, 55, 59, 60
Cimpean L, 104
Ciosk R, 219
Cirafici AM, 635
Cirino G, 272
Citron M, 615
Ciudad T, 130
Ciustea M, 631–34, 641–43, 646
Claborn KA, 338
Claeboe CD, 574
Claessen D, 339, 340
Claeyssens M, 407, 410
Clark AW, 615
Clark J, 276
Clark RD, 772, 782, 785
Clark T, 533
Clarke BA, 28
Clarke BR, 50, 55, 56
Clarke J, 347, 353, 418
Clarke S, 245
Clarkson B, 98
Claverie P, 404, 407–9, 414, 415, 417–19
Clayton PT, 316
Cleary ML, 251
Clemencet MC, 319, 321
Clement B, 693
Clements A, 454, 458
Cleveland DW, 609
Cliby WA, 140
Clifton IJ, 311
Cline GW, 27, 372, 373, 376–80, 384
Clippe A, 302
Cloppenborg K, 586, 596
Clyne RK, 221
Cobb B, 611
Cobessi D, 770, 781
Coburn CT, 369, 383
Cocco T, 75
Cochran BH, 272, 284
Coffey AJ, 667
Coghlan VM, 671
Cohen AS, 336

Cohen FE, 336, 359
Cohen G, 683, 693
Cohen H, 121
Cohen HY, 436, 439, 442, 444, 452, 454, 460
Cohen J, 389
Cohen JA, 716
Cohen JS, 43
Cohen O, 190–96, 198, 199, 201
Cohen P, 373, 663, 664
Cohen PTW, 282
Cohen SS, 3, 4
Cohen-Fix O, 220
Cohen-Gihon I, 675
Cohn M, 500, 504
Coin L, 572
Coindre JM, 510
Colaneri A, 73
Colbeau A, 75
Colby TD, 532–35
Colditz GA, 368
Cole R, 49
Cole RN, 436, 439
Cole RW, 545, 546, 611
Coleman RA, 377
Coleman T, 379, 383
Colla S, 97
Colle MA, 619
Collen D, 309, 321
Colletta G, 635
Collins K, 495, 496, 500–5, 508
Collins MK, 671
Collins RF, 50, 51
Collins SP, 143, 144
Collins T, 404, 405, 407–15, 417–19, 421, 422
Colloms SD, 581, 582
Colombo C, 372
Condron MM, 347, 348, 357
Cone RA, 758
Confalonieri S, 667
Cong YS, 507
Coniglio P, 25
Conkright MD, 156
Connaris H, 412, 415, 419, 422
Connell P, 283
Connelly C, 113, 115, 340
Conner JM, 620
Connerton I, 407
Connolly DC, 140
Conrads TP, 671

Conry D, 309
Constans JM, 619
Contessa AR, 716
Conti R, 383
Converse AK, 101
Conway AB, 577
Conway KA, 348
Conway T, 692
Conzelmann E, 310
Cook B, 744
Cook DL, 386
Cook JG, 663, 664
Cook KD, 341, 353, 355
Cook N, 56, 61
Cook NJ, 56
Cook PN, 99
Cooley L, 559
Coombs JM, 404, 419
Cooney DA, 630, 631, 639, 641
Cooney GJ, 377, 380
Cooper GJ, 343, 345
Cooper JA, 660, 661, 674
Cooper JD, 618
Cooper JP, 116, 129, 494
Cooper RM, 630
Cooperman BS, 698, 699
Copeland NG, 191, 196, 198, 199, 201
Copeland T, 671
Copeland TD, 253
Copeland V, 26
Copeland WC, 700
Copley RR, 664, 667
Corchado J, 533
Corchado JC, 533
Corda D, 446
Cordell SC, 218, 470, 471, 473–75, 484, 486
Cordero M, 480, 481
Corey DP, 613
Cork LC, 619
Corkey BE, 388
Corkey P, 386, 390
Corless CL, 98
Cornet F, 582
Cornforth MN, 123
Cornwall MC, 745
Corradetti MN, 149, 151
Corselli L, 598
Corsetti JP, 390
Cortes D, 83
Corton JM, 147

Cortright RN, 376, 377, 385
Cory JG, 693
Cosenza SC, 98, 99
Cosgrove MS, 448, 450, 454, 455, 457, 458
Cosic S, 638
Cost GJ, 214, 234
Costa PJ, 259
Costerton JW, 41
Cotman CW, 357, 620
Cottrell BA, 617
Couch FJ, 138
Couck AM, 619
Coue M, 546
Coughlan SJ, 772, 782, 785
Coughlin M, 559
Coughlin ML, 545, 559
Coulary-Salin B, 355
Coulson A, 546–48
Counter CM, 496, 500, 505, 508, 509
Couraud JY, 612
Cournac L, 782, 785
Courtneidge SA, 665, 667
Coustou V, 340
Covian R, 772, 780
Cowan CW, 745
Cowan-Jacob SW, 98, 102
Cox BS, 340
Cox MB, 286
Cox MM, 574, 576, 577, 587
Cox S, 484
Coyle JT, 615
Cozzarelli NR, 213, 215, 217, 223, 584, 585, 593, 595–98
Cozzone AJ, 50
Crabbe L, 123
Crabtree GR, 25
Craessaerts K, 618
Craig C, 248
Craig DB, 419
Craig DL, 378
Craig NL, 13
Cramer F, 642, 648
Cramer LP, 555
Cramer P, 505
CRAMER WA, 769–90; 770–72, 774–77, 779–87
Crameri A, 418
Cramton SE, 598
Cras P, 619
Cravatt BF, 647

Crawford DL, 405
Creamer TP, 426
Creasey WA, 630
Creemers AF, 755
Crescenzo R, 382
Crescitelli F, 747
Crews L, 618, 619
Crick FHC, 7
Crisona NJ, 596–98
Crispino M, 156
Cristina N, 617
Cristobal S, 300
Cristofari G, 496
Croce CM, 260
Crocker E, 755
Croes K, 311, 312
Crofts AR, 181, 182, 716, 772, 775–77, 781
Crosato M, 24
Cross JC, 255
Crossley R, 468, 472
Crossley RE, 485
Crouch RK, 752, 759
Crouse GF, 122
Crowther DC, 357
Crozier PS, 710
Cruickshank J, 509
Crump JG, 155
Crunkhorn S, 382
Crystal H, 618
Csajka FS, 710
Csermely P, 274, 276, 358
Csete M, 674
Cuddihy AR, 224
Cuebas DA, 309
Cui Q, 533
Cukier RI, 169, 170, 172, 178–80, 182
Cullen MJ, 370, 372
Culler D, 786
Cummings BJ, 620
Cummings LM, 498
Cummins JM, 155
Cummins PL, 528
Cundari E, 546, 547, 559
Cunningham AA, 300, 313
Cuperus G, 438
Cupo P, 424
Curd H, 47
Curmi PA, 474
Curmi PM, 410–13, 418–20
Curmi PMG, 404, 409, 422–24

Curran AR, 728
Currie RA, 663
Currier JL, 509
Curry JA, 357
Curtet S, 24
Curti B, 633, 634
Curtis DR, 748
Cusi K, 382
Cuthbert GL, 262
Cuthbertson L, 56
Cutson JJ, 372
Cuttner J, 630
Cyr DM, 282
Czar MJ, 282, 283
Czjzek M, 421, 423
Czubryt MP, 26

D

Dabrazhynetskaya A, 229
Dacremont G, 297, 309, 310, 312, 314
Dadali EL, 613
D'Adamio L, 617
Dadi H, 104
Dafforn TR, 474
Daga RR, 286
Daggett V, 520
Dahia PL, 151
Dahl HA, 22
Dahl SG, 744
Dahlroth SL, 694, 698, 699
Dahlstrom A, 617
Dahms AS, 730
DAHOUT-GONZALEZ C, 713–41; 709, 715, 717, 722, 723, 726–28, 736
Dai J, 620
Dai K, 468, 472, 482
Dai Q, 282
Daitoku H, 439
Dajani R, 660, 663
Dalal I, 700
Dalbon P, 719, 730, 732
Daldal F, 82, 181, 770–72
Dallner G, 315
Dam M, 227, 229, 479
Damaschun G, 348, 349
Damasio AR, 615
Damay P, 127
d'Amico D, 408, 410, 412, 414, 418

D'Amico S, 404–7, 409–19, 424, 425, 427
Damien B, 410–13, 418, 422
Damm K, 505, 506, 512
Damon C, 128
D'Amours D, 220
Dams T, 413
Dandjinou AT, 503, 504
Daniel JA, 260
Daniel K, 347, 759
Daniel RA, 468, 480, 481
Daniel RB, 101
Daniels C, 48–50
Daniels MB, 385, 386
Daniels MJ, 548, 560
Danilchik MV, 558
Danilevskaya ON, 498
Danpure CJ, 300, 313
Dansen TB, 322
Danson MJ, 409, 412, 415, 416, 418, 419, 421–26
Darden T, 504
Darinskas A, 358
Darrouzet E, 70, 82, 84, 181
Darwis D, 305
Darwish D, 693
Das A, 274
Das AK, 282, 304, 314
Das S, 196, 204
Daser A, 251
Dasgupta S, 224
Dashdorj N, 786, 787
DasSarma S, 423
Datta A, 59, 60, 121
Datta AK, 56, 60
Datta SC, 305
Daubas P, 20
Daujat S, 258, 262
Davail S, 406, 408, 420, 423–25, 427
David C, 718–20, 734
David F, 308
Davidson AL, 58
Davidson FF, 748
Davie JK, 245, 251
Davies CA, 619
Davies DC, 619
Davies DD, 245
Davies DH, 99
Davies DR, 346
Davies GJ, 56

Davies P, 615, 618
Davies RE, 715
D'Avino PP, 550
Davis AM, 642
Davis AP, 115
Davis B, 502
Davis BJ, 147
Davis C, 122, 127, 129
Davis JB, 617
Davis JH, 520
Davis JN, 619
Davis M, 484
Davis MA, 226
Davis RD, 631, 641
Davis RJ, 273, 669, 671
Davis RW, 684, 694
Davisson VJ, 633, 641
Dawe RK, 609
Dawes SS, 683, 693
Dawson SC, 609
Dayan FE, 641
Deak M, 143–46, 148, 152, 154–57, 660, 663
Dean D, 375, 376
Dean K, 259
Dean M, 322
Dean MK, 758
Dean PM, 646
Dean SO, 557
Dean SR, 505–7
DeAngelis PL, 56
de Baar E, 138
de Backer M, 422, 423
DeBaryshe PG, 498
Deber CM, 723
DeBiasio RL, 555, 557
Debilly S, 155
de Boer P, 468, 472
de Boer PA, 484, 485
De Camilli P, 711
Decanniere K, 349
Decatur SM, 348
Declercq PE, 309, 315, 321
de Coo R, 70
Decornez H, 646
De Costa B, 273
Decottignies P, 336, 346, 350
DeCoursey T, 176
Dedmon MM, 356
Deechongkit S, 358
Deems RO, 383

Deeney JT, 388
Deerinck T, 618
Defossez PA, 444
de Fourmestraux V, 380
De Francisci D, 424
DeGiorgis JA, 617
Degli Esposti M, 83
Degl'Innocenti D, 353
Deglon N, 357
Degott C, 383
DeGrado WF, 426
Degregori J, 693
DeGrip WJ, 755
de Groot H, 378
de Groot HJ, 755
Deisenhofer J, 781
DeJesus OT, 101
de Jong AMP, 82
De Jong M, 322
de Jong W, 339, 340
Dejonghe S, 389
Dekker PJ, 734
de Klerk JB, 316
De Koninck P, 27
de Kromme I, 313
de Laat SW, 558
de Lacoba MG, 735
Delacourte A, 621
DeLange RJ, 245
de Lange T, 123, 128, 132, 494, 496, 507, 509–11
DeLano WL, 576
Delcroix JD, 618, 621
De Leonardis F, 717
Deleu C, 340
Delille D, 407, 414, 415, 417–19
Dellago C, 167, 710
Delling U, 25
Del Mar C, 344
Delon MR, 615
De Lozanne A, 547, 557
Del Rizzo S, 672
De Lucia P, 10, 11
del Valle LJ, 750, 760
De Marcos Lousa C, 718–20, 734
Demare G, 605
de Massy B, 273
DeMattos RB, 357
Demetri GD, 103
Demin P, 104
Deming JW, 423

Dempke W, 152
Dempster M, 510
De Mulder RHM, 103
de Napoles M, 255, 256
den Blaauwen T, 471, 474
den Boer ML, 638, 639
Dendle MA, 356
Deneke J, 572, 574, 582
Deneumax P, 536
Deng HT, 447, 448, 450, 452, 458
Dengler RE, 223
Dengler T, 43
Denhardt GH, 10
Denis S, 297, 308–10, 316
Denny WA, 101
Denu JM, 438, 439, 441, 444, 446–48, 450, 452, 453, 457, 459, 461
De Nys K, 297, 310
DePace AH, 339, 340
Depiereux E, 410–13, 418, 422
DePinho RA, 149, 154, 190, 197–99, 204, 496
de Planque MRR, 711
Deppenmeier U, 76
Depristo MA, 350
Deras IL, 633
Deras ML, 633, 641
Derbyshire KM, 600
Deretic D, 745, 748
Derijck AA, 255
DeRisi JL, 219
Derman AI, 235
Deroanne C, 407
Der-Sarkissian A, 344, 354
Desai A, 222, 472, 553, 559
Desai CJ, 610
Desai J, 103
Desai U, 684
Desamero R, 536
Desautel SL, 373
Desbois A, 783
Deschamps D, 383
Deschenes-Furry J, 29
Descoteaux A, 482
Deshaies RJ, 284, 437, 550, 661
Desmyter A, 349
Dessain SK, 436, 439, 440
Dessi F, 619
De Strooper B, 617
Deszo EL, 55, 56, 60

DeTeresa R, 615, 616, 618
Detimary P, 386
de Turco EBR, 745, 748
DeVeber LL, 630
de Vet EC, 304
de Vet S, 485
de Vitry C, 772, 783, 786
Devos N, 405
de Vos WM, 417
Devreese B, 70, 73, 74, 76
de Vreugd P, 339, 340
de Vries-Smits LM, 439
DeWaard A, 10
Dewachter I, 357
Dewar SJ, 482
de Weijer KJM, 103
De Windt LJ, 25
de Wit JG, 471
Dey B, 272
De Zoysa PA, 301
Dhaenens CM, 621
Dhar G, 585, 588, 591
Dhar S, 507
Dharia T, 668
Dhaunsi GS, 301, 302
Dialo AO, 614
Dianoux AC, 718–20, 730, 731
Dias P, 750
Dias S, 510
Diaz JF, 475
di Bard BLG, 356
Di Bernardo S, 70, 81, 84, 86
DiChiara JM, 579
Dick DW, 101
Dick JR, 613
Dickerson K, 147, 148
Dickson DW, 357, 618
Di Cosimo S, 101
Di Cristofano A, 700
Dieckman LJ, 349
Diegelman P, 314
Diehl A, 74
Diehl T, 357
Diekmann S, 356
Dietrich O, 619
Dietrich V, 248
Dieuaide M, 309
Dieuaide-Noubhani M, 309
Diffley JF, 114
Di Giacomo M, 157
Di Giamberardino L, 612
Di Girolamo M, 446

Dikov D, 336, 345, 346
Dill KA, 417
DiMauro S, 70, 86
Dimroth P, 84, 85
Din N, 482
Din TM, 76
Dinan TG, 620
Ding Y, 642
Di Noia MA, 700
Dinouel N, 130
DiPietro L, 376
di Prisco G, 406, 411, 412
DiRusso CC, 305
Disanza A, 667
Discafani CM, 101
Dissoki S, 101
Distel B, 322–24
Ditchfield C, 545, 546
D'Itri E, 169
Dittmar KD, 289
Dittrich PS, 614, 618
Divakaruni AV, 480, 481
Divecha N, 436, 439, 440
Dixon JE, 674
Dixon RA, 744
Dixon S, 194–96, 199–202
Djafarzadeh R, 74, 76, 79
Dobbins RL, 386
Doberstein S, 559
Dobrowsky RT, 378
Dobrynin K, 74
Dobrzyn A, 373, 378
Dobrzyn P, 373
DOBSON CM, 333–66; 338, 339, 341, 343, 345, 347–50, 353–56, 358
Dobson M, 230, 231
Dobson MJ, 231
Dodart JC, 357
Dodge KL, 671
Dodgson C, 43, 47, 49
Dodson EJ, 632
Dodson G, 632
Dodt G, 313
Doerks T, 664, 667
Doggrell SA, 98, 744, 761
Doherty A, 101
Dohlman HG, 744
Dohm GL, 31, 377, 385
Dohner H, 103
Doi J, 156
Doi M, 479, 480

Dolce V, 717
Dole S, 259
Dolinski K, 282
Dollins DE, 278, 280
Dolmetsch RE, 25
Domart MC, 667
Domingues P, 323
Dominguez O, 149, 151
Dominy BW, 642
Domkin V, 684, 694, 699
Dommes V, 300
Doms RW, 618
Donachie WD, 468, 482
Donald AM, 338
Donato NJ, 98
Dong CK, 494
Dong GX, 320
Dong JX, 149
Dong Y, 262
Doniach S, 347
Donigian JR, 454
Donnelly EJ, 553
Donnelly LA, 151
Donnelly R, 27
Doolittle RF, 355, 470
Doolittle WF, 472
Dopazo A, 149, 151, 482
Dorn R, 244, 248, 249
Dorow D, 43
Dos Reis S, 355
Dotti CG, 617
Doty P, 9
Dou Y, 253
Doublet P, 50
Doubrovinski K, 485
Doucet J, 346, 350
Douglas MG, 718–20
Douma AC, 322
Doumont G, 197, 199
Doussiere J, 730, 733
Douziech M, 146
Dover J, 245, 251, 252, 259–61
Dovichi NJ, 419
Dowd GJ, 437, 445, 457
Dowds CA, 324
Dowhan W, 469
Dowjat WK, 618
Downie DL, 508
Downing JR, 639
Downing KH, 470, 471
Downs JA, 508
Downs MA, 748

Doxsey SJ, 558
Doyle J, 674
Doyle ML, 642
Draborg H, 418
Draghici O, 312, 314
Drapeau GR, 474, 482
Draper GC, 213, 214
Dratz EA, 746, 758
Dreier B, 674
Drennan CL, 685, 687, 692
Dresner A, 376
Drew DA, 485
Drewes G, 144, 152, 154, 155
Dribben AB, 640, 641
Driemel C, 140–42
Driessen AJ, 471, 474
Driscoll JS, 631, 641
Driscoll M, 356
Drissen R, 324
Droge P, 579
Droguett G, 190, 197–99, 204
Dröse S, 83, 85, 181
Druker BJ, 97
Drummelsmith J, 45, 47, 48, 50–52, 61, 62
Duarte M, 70, 73, 78, 86
Dubail F, 409, 410, 424
DuBay KF, 354, 355
Dubey AK, 41
Dubochet J, 50
DuBois ML, 118
Dubois P, 409, 410, 424
Dubois S, 336, 346, 350
Duborjal H, 70, 82
Dubuisson M, 302
Duclos B, 50
Dudek RW, 376, 385
Dudenhausen EE, 635–38
Dudkina NV, 73, 81
Dueber JE, 665, 671, 672
Duee ED, 716
Dueker J, 246, 252
Duerre JA, 261
Dufour S, 382
Dufresne SD, 29
Duggan C, 556
Duggan LJ, 260
Duggleby HJ, 632
Duggleby RG, 646
Duguet M, 575
Duina AA, 272, 282
Dujon B, 442

Duke SO, 641
Dukes M, 102
Dukkipati A, 754
Dulyaninova NG, 552
Dumas J, 482
Dumoulin M, 349
Duncan KL, 546
Dungan JM, 584, 585, 596, 597
Dunham MA, 132
Dunham WR, 83, 85
Dunker AK, 417
Dunlop LE, 338
Dunlop MG, 510
Dunn B, 113
Dunn TA, 308
Dupont Y, 730, 733
Dupont-Versteegden EE, 26
Dupuis A, 70, 82, 84
Dupuis S, 496, 503, 505, 507
Durbin R, 572
Durham B, 178–80
Durkin AS, 436
Durocher D, 664
Durr K, 178
Duthie M, 618
Dutta A, 148, 149
Dutta R, 274
Dutton PL, 82, 181, 775, 776, 785
Duyckaerts C, 619, 733
Duyne GD, 573
Dvorak HF, 102
Dworkin J, 214
Dy M, 272, 507
Dyall SD, 718, 720, 734
Dyda F, 574
Dye NA, 233, 480, 481
Dyer RB, 613
Dykes DJ, 101
Dyson HJ, 521, 522, 526, 527, 536
Dzekov C, 32
Dziura J, 382

E

Eaglestone SS, 340
Ealick SE, 633
Earnshaw WC, 548
East JM, 711
Easter J Jr, 230
Eastman J, 617

Easton Esposito R, 437
Eaton EN, 496
Eaton JW, 316
Ebah L, 54, 55, 61
Eberhardy S, 674
Ebersbach G, 229, 233, 235, 484, 485
Ebihara K, 372
Ebneth A, 155, 609, 613, 618
Ebrey T, 753
Eccles JC, 20, 22
Eccles RM, 20, 22
Echard A, 546–48, 550, 558, 559
Echeverri C, 546–48
Eck MJ, 665, 667
Eckman CB, 357
Eda M, 547
Edgar JM, 614
Edgar R, 229
Edgar RS, 10
Edlund H, 387, 393
Edman L, 419
Edmonds B, 348
Edskes HK, 340
Edwards GE, 782, 784
Edwards KA, 547
Edwards PC, 748, 751, 754, 755, 758
Efendic S, 379
Effertz K, 674
Efthimiopoulos S, 618
Egan W, 43
Egeberg K, 391
Egelman EH, 480
Eggers CT, 611
EGGERT US, 543–66; 545–48
Eggertsen G, 300
Eggertsson G, 421
Egli M, 192, 193, 202, 207
Ehli E, 226
Ehrhardt DW, 469, 482
Ehrlich SD, 224
Eickbush TH, 498, 499
Eide L, 613
Eigner EA, 408
Eijsink VG, 415
Eilers M, 755
Einarsdottir OO, 174
Eisen JA, 436
Eisenberg D, 459, 507, 643
Eisenhaber F, 221, 222, 224, 245, 248, 249, 251

Eisenmesser EZ, 535
Eisenstein M, 193–96
Eissenberg JC, 248
Ek MS, 170, 177
Ekberg M, 685, 691, 694
Eklund H, 530–33, 535, 682, 685, 690, 691, 694, 697
El-Agnaf OMA, 354
Elberry M, 770
El-Deiry W, 424
Eldridge AG, 674
Elela SA, 503, 504
Elgersma Y, 325
Elgin SC, 244, 245, 248
Elia AE, 668
Eliasson R, 683, 685, 687, 690, 692, 694–98
Elion EA, 669, 671
Elisha G, 582
Elkan CP, 75
El Khissiin A, 272, 282
Elle C, 617
Elledge SJ, 117, 684, 693, 694
Ellenberger T, 570, 572, 574, 576–78, 580, 600
Ellinghaus P, 310, 321
Elliott R, 155
Ellisdon AM, 349
Ellisen LW, 496
Ellman JA, 642
Ellston R, 545, 546
Elluru R, 618
Elmore LW, 286, 507
Elmquist JK, 372
Elofsson A, 300
Elpeleg O, 700
Elstner M, 533, 748, 755
Emanuelsson O, 300
Emerson SG, 98
Emre NC, 260
Emslie-Smith AM, 151
Endo A, 196, 202
Endo T, 782, 785
Endoh M, 255, 256
Endrizzi JA, 633
Enenkel B, 505, 506
Enerback S, 391
Engel A, 757–60
Engel M, 96
Engel T, 309, 310
Engelberg-Kulka H, 226

ENGELMAN DM, 707–12; 708, 728
Engemann H, 195, 198
Engfelt WH, 315
Englander SW, 344, 520
Engman DM, 274
Ennifar E, 579
Enomoto S, 117
Entel P, 748, 755
Entine G, 758
Entius MM, 138
Eom C-Y, 15
Epstein DM, 521, 526
Epstein RH, 10
Erdjument-Bromage H, 245, 251, 254–56, 258, 262, 439, 440, 461
Erdmann N, 229, 509
Erdmann NJ, 509
Erdmann R, 322, 324, 717
Erickson HP, 218, 222, 468, 469, 471, 473–76
Erikson E, 96, 138
Erikson RL, 96
Eriksson A, 32
Eriksson KF, 382
Eriksson M, 685, 690, 694
Eriksson S, 697, 699, 700
Erlich RL, 260
Ermler U, 168
Ernst M, 341, 342
Ernst OP, 744, 747, 748, 751, 752, 759
Erol E, 377
Errington J, 214, 215, 230, 468, 469, 479–82, 484, 485, 488
Ertan H, 404
Ertel IJ, 630
Escalante-Semerena JC, 436, 439, 445, 446
Espe MP, 170
Espeel M, 300, 301, 315
Espeli O, 233, 234
Espinosa R III, 251
Esposito D, 572, 579
Esposito MS, 113
Esposito RE, 436, 444
Esser L, 770, 781
Esser V, 386, 389, 390, 393
Esteller M, 139
Esteves TC, 86, 166, 391
Estornell E, 83

Esue O, 474, 480, 481
Esumeh F, 61
Esumi H, 151, 156
Eswaran J, 51, 61
Etchebest C, 710
Etheridge KT, 508
Etienne F, 302
Etienne-Manneville S, 155
Eto K, 391
Eubel H, 73, 81
Euteneuer U, 547, 558, 559
Evan GI, 548
Evans GB, 646
Evans JM, 151
Evans SK, 116, 129, 508, 511
Evans WE, 630
Everly RM, 786
Eydmann T, 224
Eyler JR, 631, 641
Eyring H, 405
Eyssen HJ, 308, 313

F

Fabbro D, 98
Faber HR, 536
Fabian M, 169
Fabre F, 118
Fabris S, 700
Fadden P, 276
Faguy DM, 472
Faham S, 723
Fahien LA, 386, 388, 389
Fahimi HD, 301, 302, 313
Fahlman R, 30
Fahrenholz F, 615
Fahy E, 70, 73
Fain GL, 745
Falick AM, 671
Falinski F, 76
Falk PK, 376
Falke JJ, 520
Falquet L, 380
Falzone CJ, 526
Falzone TL, 611, 618–21
Fambrough DM, 245
Fan CY, 320
Fan F, 102
Fan G, 751
Fan J, 468, 470, 669
Fan JH, 32
Fan R, 272

Fan YX, 408
Fandrich M, 343, 345, 353, 356
Fang G, 550
Fang J, 255
Fang Y, 272, 286
Fanning AS, 145
Fantes P, 694
Fantz DA, 662
Farah MH, 617
Farese RV Jr, 391
Faretta M, 440
Farlo J, 618
Farlow M, 615
Farmer P, 380
Farmer S, 224
Farnegardh M, 682, 685, 691, 694, 697
Farnsworth CC, 746
Farrar GJ, 745
Farrell A, 284
Farrelly FW, 272
Farrens DL, 757, 759, 760
Fasching CL, 132
Fastbom J, 619
Fasulo B, 546, 547, 559
Fato R, 86
Faust PL, 320
Faux MC, 671
Fawcett TW, 636, 637
Fearnley IM, 70, 71, 73, 75, 76, 78, 79, 81, 86
Febbraio M, 369, 383
Fechner U, 646, 648
Fecke W, 75, 76
Fee JA, 180
Feeback DL, 22
Feeney PJ, 642
Feher G, 176, 177
Feher MD, 316
Fehmel F, 43
Feinstein E, 190, 193, 194, 198, 199, 201
Fekete RA, 230
Feldman JD, 156
Feldman MF, 49
Feldser DM, 117, 118
Feliciello A, 669
Felix CM, 719, 725
Feller G, 404–27
Feller SE, 711
Feller SM, 667
Felton HM, 26

Felts SJ, 286, 287
Fendler K, 169, 170, 173, 733
Feng Q, 251, 256–58
Fenrick R, 639
Ferdinandusse S, 297, 306, 308–10, 316
Ferguson KM, 664, 665
FERGUSON-MILLER S, 165–87; 166, 168–70, 172, 173, 175–80, 775
Ferland-McCollough D, 146
Fernandes N, 144
Fernandes R, 358, 359
Fernandez CO, 353, 356
Fernandez P, 139, 140, 142, 149, 151
Fernandez-Escamilla AM, 355
Fernandez-Silva P, 70
Fernandez-Velasco JG, 772
Ferramosca A, 734
Ferrandiz MJ, 473, 475
Ferrao-Gonzales AD, 349
Ferrari D, 634
Ferraro P, 699, 700
Ferreira A, 617
Ferreira H, 578
Ferreira KN, 770
Ferreira MG, 494
Ferrell RE, 382
Ferrer I, 324
Ferrer-Martinez A, 297
Fersht A, 407, 409, 535
Fersht AR, 273, 418
Fetter JR, 169, 170, 177
Feucht A, 469, 482
Fezoui Y, 347, 348
Ficner R, 421, 423, 427
Fiebig R, 385
Fiedler W, 103
FIELD CM, 543–66; 545, 547, 551, 555, 557, 559
Field SJ, 548
Fielding AB, 547
Fields PA, 409, 415, 417, 418, 420, 423, 426
Fields S, 444
Fierke CA, 436, 447, 521
Fiermonte G, 717
Fiers W, 9
Fieschi F, 683
Figadere B, 83
Figge RM, 480, 481

Figueiredo LM, 498
Figueroa DJ, 618
Filimonov VV, 349, 353
Filipe SR, 213
Filipek S, 744–46, 748, 751, 752, 757–60
Filippov D, 642
Fillmore JJ, 27, 373, 377
Filman D, 574, 577, 580, 600
Filosi N, 122
Filppula SA, 311
Finazzi G, 776, 784, 785
Finch CE, 612
Finch JT, 344, 345
Finck BN, 379, 383
Findell PR, 669
Fine BM, 639
Fine RE, 617
Finel M, 71, 75, 76, 78, 81–83, 86, 171, 172
Finger FP, 559
Fingerhut R, 310
Fink AL, 273, 335, 336, 339, 349, 353
Fink GR, 436
Finke A, 56, 57
Finkelstein DB, 272
Finley J, 389
Finn CW, 54
Finn RD, 572
Finne J, 56
Finnie J, 320
Finnin MS, 454
Fiol CJ, 663
Fiorani S, 121
Fiore C, 718–20, 730, 731, 734
Fisch C, 383
Fischer E, 520, 660
Fischer R, 144
Fischer W, 43
Fisher AJ, 633
Fisher N, 83
Fisher SJ, 374
Fisher TS, 504, 508
Fishkind DJ, 555
FitzPatrick DR, 311
Fjelstad M, 56
Flaks JG, 4
Flament-Durand J, 619
Flamez D, 379
Flanagan L, 617
Flanagan ME, 105

Flardh K, 469
Flatauer LJ, 662
Flatt JP, 375, 376
Flaxen K, 170
Fleck O, 694
Fleckenstein B, 282
Fleischmann M, 785
Flemming D, 78, 84
Fletcher BS, 630, 638
Fletcher CDM, 103
Fletcher JA, 103
Flier JS, 372
Flink JL, 336, 345, 346
Flippin JD, 674
Fliss AE, 272, 286
Floering LM, 391
Floeter MK, 613
Florens L, 170
Flores LG, 101
Flores MJ, 113
Flotho A, 671
Floyd MB, 101
Floyd MB Jr, 101
Fluhrer R, 617
Fobker M, 310, 321
Foe VE, 557
Fogel S, 437
Foguel D, 349
Foiani M, 118
Foldes-Papp Z, 419
Foley E, 546–48, 558, 559
Foley JE, 383
Follmann H, 683, 693
Fong TA, 102
Fontanella B, 409
Fontecave M, 682, 690
Fontecilla-Camps JC, 84
Fontijn RD, 547
Foord SM, 744
Foran JM, 103
Forcet C, 155
Ford E, 391, 439, 445, 452
Ford LP, 508
Ford RC, 50, 51
Forer A, 546
Forge V, 356
Forgon M, 620
Forleo G, 157
Forman MS, 353
Forman-Kay JD, 667
Formiggini G, 86
Formigli L, 358

Formosa T, 115
Formstone A, 479
Forrest AK, 642
Forrest LR, 710, 711
Forrester C, 50, 52
Forsburg SL, 438
Forsee WT, 56, 57
Forss-Petter S, 321, 324
Forstemann K, 494, 498, 504, 506
Forsythe HL, 286, 507
Forte E, 169
Forterre P, 273
Forti G, 784, 785
Foster DW, 378
Foster JW, 442
Foster MP, 578
Fotiadis D, 745, 746, 757–59
Foulon V, 311, 312
Fouquet F, 323
Fournier L, 146, 155
Fowler JH, 614
Fox BA, 745
Fox JW, 406
Frame S, 663, 664
Franch T, 226
Francia MV, 582
Francis F, 611
Francis NJ, 254
Franco AV, 49
Franco S, 440, 510
Frangione B, 336
Frank D, 26
Frank I, 747
Frank N, 43, 614, 618
Frankel G, 52
Fransen M, 297, 300, 308, 310, 311, 324
Frantz DF, 372, 373
Frantz JD, 31, 32, 372
Franz JE, 643
Franzone JM, 148
Fraser AG, 546–48
Fraser BA, 44, 60
Fraser E, 660, 663
Fraser FW, 357
Fraser PE, 336, 341, 618
Frayn KN, 378, 386
Frazier AE, 75
Fredenrich A, 31
Frederick JM, 745, 748
Frederiks WM, 302

Fredriksson R, 744
Freed E, 674
Freedberg DI, 55, 56
Freedman NM, 101
Freeman BC, 286
Freeman-Cook LL, 438
Frémont L, 460
Frerman FE, 47
Freskgard PO, 417
Fresquet V, 633, 634
Freund SMV, 273
Frey M, 84
Frey N, 26
Frey PA, 682, 692
Friday BB, 24, 26
Fridd S, 684, 693
Fridkin M, 190, 194, 197, 198
Friebolin H, 43
Friedel W, 138–40, 142
Frieden C, 634
Friedman DS, 611
Friedman JM, 370, 372
Friedman KL, 496, 500, 503, 507, 508
Friedman SA, 226
Friedrich B, 76
Friedrich T, 70, 73, 75, 76, 78–80, 82–86, 181
Friend SH, 648
Friesner RA, 536
Frigimelica E, 699
Frigon JY, 619
Fritze CE, 437
Fritze O, 752, 753, 759
Fritzsch G, 74, 79
Fritz-Wolf K, 692, 701
Frodin M, 663
Frolkis M, 510
Fromenty B, 383
Fromme P, 770
Frosch M, 60
Frumerie C, 579
Fry DW, 101
Frye RA, 436, 438–40, 448
Fryer LG, 147, 148, 371, 387
Fryer P, 313
Fu C, 145
Fu MF, 439, 440
Fuchs J, 221
Fuchs JA, 692
Fujii R, 386
Fujiki Y, 324

Fujimoto H, 547, 552
Fujimoto Y, 372, 373, 378, 380, 384
Fujioka H, 612, 619
Fujisawa K, 547
Fujita A, 553
Fujita M, 214
Fujita N, 274, 289
Fujita S, 548
Fujiwara T, 560
Fujiyoshi Y, 748, 753
Fukazawa T, 719
Fukuda A, 253
Fukuda S, 684, 693
Fukuhara H, 130
Fukumori Y, 76
Fukuoka M, 101
Fukushima T, 251, 254
Fukuyama K, 785
Fulco M, 439, 448
Fuld P, 618
Fuller JF, 667
Fuller MT, 548
Fuller RP, 674
Fuller S, 474
Fuller SJ, 357
Fulton TB, 123, 500, 503, 504
Funabiki H, 548
Fung E, 229
Fung FW, 621
Funnell BE, 226, 227, 229, 235
Furbacher PN, 785
Furia A, 785
Furneaux RH, 646
Furr J, 646
Furstenthal L, 674
Furukawa K, 618
Futcher AB, 235
Futter C, 548, 555, 558
Fux JP, 733

G

Gabaldon T, 73, 74
Gaber RF, 272, 282
Gabius H-J, 642, 648
Gadelle D, 273
Gadzicki D, 98
Gafken PR, 251, 256, 261
Gafvels M, 300
Gagnon J, 732
Gaida GM, 122

Gajdusek DC, 615
Galanis A, 662
Galante YM, 76
Galbraith KA, 548
Galgoczy DJ, 117
Galigniana MD, 282, 283, 289
Galkin AS, 84
Galkin VE, 480
Gallagher PJ, 194–96, 199–202
Gallick G, 98
Gallinaro L, 699, 700
Gallo CM, 442, 444, 454
Galloo B, 24
Galova M, 219, 221
Galuska D, 30
Galvan V, 617
Games D, 357, 617, 619
Gamm DM, 143, 144
Gammeltoft S, 663
Gandhi L, 496, 505
Gane PG, 646
Ganim Z, 748, 754
Gann A, 657
Gannon KS, 357
Ganz P, 276
Gao J, 529, 533
Gao JL, 533
Gao XG, 770, 781
Gao XS, 149
Gao ZY, 386
Gaposchkin DP, 305
Garau D, 97
Garavito RM, 728, 775
Garcia F, 630
Garcia R, 382
Garcia-Cardena G, 272
García-Horsman JA, 169, 170, 177
Garcia-Salcedo JA, 439, 445, 454
Garcia-Viloca M, 529, 533
Garczarek F, 173
Gardner JF, 574, 577, 579
Gardner MJ, 498
Gardonyi M, 417
Garegg PJ, 43
Garfinkel DJ, 437
Garin-Chesa P, 512
Garland PB, 378, 384
Garner EC, 478, 479
Garofano A, 78, 86
Garon CF, 58, 59

Garriga P, 750, 759, 760
Garriga X, 692
Garrington TP, 669
Garry DJ, 21
Garsoux G, 406, 419, 422, 423
Garvik B, 117, 128
Gaseidnes S, 415
Gasimov KG, 759
Gasmi L, 297, 314
Gasser SM, 438
Gassmann R, 548
Gast K, 348, 349
Gastaldelli A, 390
Gastaldi G, 385
Gatbonton T, 461
Gatt MK, 547
Gatti M, 546, 547, 557, 559
Gatto GJ Jr, 301
Gaub HE, 760
Gaude H, 146, 155
Gauthier S, 619
Gavish L, 104
Gavrilova O, 372
Gay J, 684
Gaynor K, 614
Gazit A, 96, 97, 99, 101–4
Gazziola C, 699
Ge H, 684
Ge J, 682, 684, 690
Ge N, 617
Gebhard W, 476, 477, 481
Geday MA, 338
Gee R, 314
Geider K, 52
Geier S, 70
Geisbrecht BV, 311
Geiselmann J, 48
Gejyo F, 347
Gelato KA, 576, 579
Gelbart WM, 244
Gelfand MS, 693
Gelfand VI, 611
Gelles J, 171
Gembal M, 386
Gemperli AC, 84, 85
Genicot S, 408–10, 420, 424
Genier S, 221
Gennis R, 168, 169, 178
Gennis RB, 169, 170, 172, 177, 179–82
George JP, 437
Georgieva B, 694, 699

Georgievskii Y, 175
Georlette D, 404, 405, 407–13, 415, 417–19, 422, 424, 426
Gephart J, 26
Gerami-Nejad M, 657
Gerber M, 245, 251, 254, 259–61
Gerber MR, 284
Gerber SA, 220
Gerday C, 404–25, 427
Gerdes K, 226–29, 233, 235, 477–82, 484, 485
Gere MB, 674
Geremia RA, 48
Geren L, 170, 178–80
Gerfen GJ, 688
Gerhart J, 657
Gerike U, 409, 415, 416, 418, 421–26
Gerisch G, 557
Gerke V, 761
Gerlich D, 220
Gerlt JA, 640
Germond J-E, 643
Gerrard SP, 584, 585
Gerratana B, 633
Gershenson A, 417
Gershey EL, 245
Gerstein M, 589
Gerton JL, 219
Gertz HJ, 357
Gertz SD, 104
Gerwert K, 169, 172, 173
Getmanova E, 748
Getmanova EV, 755
Getsis M, 315
Geula C, 357, 619, 620
Geva M, 348
Gewalt SL, 545
Gewirth DT, 274, 278, 280
Gewirtz AM, 98
Ghaderi N, 459
Ghetti B, 615
Ghosh G, 193
Ghosh K, 573, 579
Ghosh MK, 305
Ghosh P, 599
GHOSH SK, 211–41
Ghosh SS, 75
Giaever G, 340
Gianese G, 420, 422, 423, 425
Giannakou ME, 440

Giannessi F, 383
Giannoni E, 358
Giansanti MG, 547, 555
Giaquinto L, 410, 415, 418
Giardiello FM, 138
Giasson BI, 353
Gibberd FB, 316
Gibbs D, 252
Gibert I, 682–84, 692, 693, 700
Gibson Q, 170
Gierasch LM, 355
Giet R, 156
Gigant B, 474
Gijon P, 439, 445, 454
Gilbert A, 31
Gilbert DJ, 191, 196, 198, 199, 201
Gilbert RJ, 667
Gilcrease EB, 582
Gilde AJ, 374, 375
Gilderson G, 169, 170, 173, 174, 180
Gileadi O, 193, 195, 196
Gill HJ, 251
Gill RM, 509
Gill SJ, 421, 422
Gillespie JG, 147
Gillespie JR, 348
Gilon C, 96, 97, 99
Gilon P, 386
Gilson PR, 472
Gimenez-Abian JF, 220
Gimeno RE, 305, 369, 383
Gindhart JG Jr, 610
Giorgetti S, 344
Girardbascou J, 786
Giresi PG, 31
Giri B, 273, 289
Girvin ME, 775, 785
Girzalsky W, 322, 324, 717
Gish G, 667
Gish GD, 664, 672
Gitai Z, 233, 480, 481
Giver L, 417
Glabe CG, 357
Glansdorff N, 410, 413–15, 417
Glaser G, 226
Glass JI, 468
Glass JS, 468
Glatz JFC, 305
Glenner GG, 615
Glennon MC, 388

Glossip D, 662
Glotzer M, 224, 544, 545, 548, 550, 553
Glover DM, 547, 550, 553–55
Glover JN, 664
Glowczewski L, 117
Glumoff T, 309
Gobelet C, 385
Gober JW, 213, 214, 230, 480, 481
Godemann R, 609, 613, 618
Godfrin-Estevenon AM, 230
Godin-Heymann N, 100, 101
Godinot C, 70
Goecke T, 140–42
Goeddel DV, 273
Goedert M, 470, 613, 617
Goemaere-Vanneste J, 612
Goers J, 353
Gogol EP, 485
Golbe LI, 353
Goldberg AL, 31
Goldberg ML, 547, 557
Goldfarb T, 121
Goldgaber D, 615
Goldie KN, 343, 345, 470
Golding GB, 72, 73
Goldman H, 509
Goldman RC, 43, 44
Goldsbury CS, 341, 343, 345
Goldstein BM, 532–35
Goldstein JL, 314, 372
GOLDSTEIN LSB, 607–27; 610, 612, 617–21
Golecki JR, 58–61
Golemi D, 536
Goll MG, 261
Golomb G, 104
Golshani A, 253
Gombos Z, 775
Gomes CM, 173
Gomes MD, 31
Gomez EB, 438
Gomez-Puertas P, 474
Gonatas NK, 615
Goncalves J, 674
Gonczy P, 546–48, 553, 558
Goneska E, 501
Gonfloni S, 665, 667
Gong K, 618
Gong SS, 635, 636
Gong X, 772, 780

Gong YD, 618
Gonzalez A, 633
Gonzalez B, 416, 674
Gonzalez JM, 469, 473, 475
Gonzalez M, 26, 28
Gonzalez TN, 596–98
Gonzalez-Blasco G, 416
Gonzalez-Cadavid N, 32
Gonzalez-DeWhitt P, 619
Gonzalez-Lima F, 617
Good MC, 660, 662, 663, 671
Good V, 660, 663
Good VM, 660
Goodall SC, 755
Goodarzi AA, 144
Goodchild A, 404
Goodman JM, 324
Goodman MF, 8
Goodman SN, 138
Goodnow CC, 25
Goodpaster B, 379, 382
Goodpaster BH, 376, 377, 382
Goodrich KJ, 500, 501, 503, 505–7
Goodwin EH, 123, 132
Goodwin TJ, 572
Goodyear LJ, 29
Gopaul DN, 573, 574, 576, 587
Goransson O, 148, 149, 152, 154–57
Gorbatyuk B, 214
Gorczyca M, 618
Gordon C, 694
Gordon DJ, 341
Gordon GS, 213, 227
Gordon I, 407
Gordon R, 405
Gordon S, 236
Gore J, 214, 234
Gore M, 385
Gorgas K, 303, 304
Gorgun CZ, 377
Gorostiza O, 617
Gorovitz B, 683, 693
Gorski J, 378
Gosal WS, 348, 349
Goshima G, 547, 611
Goto M, 142, 633
Goto S, 732, 735
Goto Y, 341, 344, 719
Gotoh M, 684
Gotschlich EC, 44, 60

Gottesman ME, 669
Gottesman S, 44
Gottfridson E, 56, 61
Gottlieb S, 436, 444
Gottschalk WK, 379
Gottschling DE, 116-18, 251, 256, 436, 437, 461, 508
Goud B, 547
Gould KL, 560, 674
Gould M, 613
Gould SJ, 300, 301, 311, 315, 316, 320
Gourdet C, 734
Gourlay SC, 582
Govardovskii VI, 744, 746
Gow NA, 130
Goy MF, 245
Gozuacik D, 195, 196–98, 200, 205
Graff P, 682
Graham A, 316
Graham D, 372
Graham LL, 41
Graham ML, 630
Graham TE, 372, 373
Graige M, 177
Grainge I, 575
Grammatikakis A, 272
Grammatikakis N, 272, 284
Grandin N, 117, 120, 121, 128, 129
Grangeasse C, 50
Grant PA, 253
Grantcharova N, 469
Gräslund A, 684, 685, 687, 690, 693, 698
Gratia E, 404, 405, 407, 409, 415, 417–19
Graumann PL, 213, 224, 230, 479–81
Graves CL, 353
Graves PR, 193, 195, 197, 198
Graves R, 28, 375
Graves RA, 372
Gray HB, 171, 776, 785
Gray J, 174
Gray MW, 72, 73
Grayling RA, 415, 416
Grayson J, 21
Graziani Y, 96
Gready JE, 528
Greatbanks SP, 528

Grebe R, 776
Greco A, 635
Green DK, 510
Green JB, 154
Green K, 86, 166
Green KA, 139, 147–49, 156, 157
Green LR, 99
Green S, 374
Greenbaum EA, 344, 618
Greenberg B, 619
Greenberg CR, 388
Greenberg ML, 736
Greenberg RA, 496
Greenblatt JF, 245, 251, 252, 259
Greene EC, 500, 504, 505
Greene SJ, 357
Greenfield JP, 618, 620
Greenfield TJ, 226
Greenleaf AL, 254
Greenwood C, 170
Greer PL, 436
Gregan J, 221, 222
Gregson NA, 62
Greider CW, 116–20, 126, 495, 496, 500–5
Greider K, 7
Greis KD, 101
Grenert JP, 276
Gressens P, 315, 320
Greten FR, 374, 377
Greulich H, 100
Grewal S, 662
Grewal SI, 244, 248, 249, 251, 439, 440
Grgic L, 78, 86
Grierson AJ, 613
Griesenbeck J, 251
Griesinger C, 356
Griffin ME, 27, 376, 379
Griffin RG, 341, 343, 684
Griffith DW, 43
Griffith JD, 125, 126, 130, 132
Griffiths DE, 70, 82
Griffiths G, 55, 56, 61
Griffiths J, 31
Griffiths S, 313
Grigorieff N, 73, 79
Grill V, 391
Grill VE, 386, 390, 392
Grimaldi PA, 28

Grimm AA, 391, 436, 442, 444
GRINDLEY NDF, 567–605; 568, 572, 582, 584–88, 590–95, 598, 600
Grivennikova VG, 78, 82–85
Groemping Y, 667
Groff-Vindman C, 122, 125, 126
Gromley A, 548, 559, 560
Grompone G, 113
Grone HJ, 369
Gropp T, 733
Gross H, 470
Grossi S, 127
Grossman AD, 214, 223, 229, 230, 484, 485, 487
Groth AC, 599
Grothey A, 152
Grottesi A, 667
Grounds MD, 30
Grozinger CM, 461
Gruber BC, 101
Gruber S, 218
Grummt I, 436, 439
Grunberger T, 104
Grunberg-Manago M, 5
Grundke-Iqbal I, 615
Grundy WN, 75
Gruneberg U, 545
Grunstein M, 438, 441
Gryaznov S, 502
Gu D, 760
Gu L, 104
Gu YJ, 617
Gu YP, 700
Gu YS, 440
Gu Z, 736
Guacci V, 219
Guallar V, 536
Guan KL, 149, 151
Guanti G, 140
Guarente L, 121, 436, 438–40, 444–47, 452
Gubin AN, 75
Guddat LW, 646
Gueiros-Filho FJ, 469
Guenebaut V, 73, 79, 181
Guenther MG, 260
Guergova-Kuras M, 182, 772, 775
Guerre-Millo M, 383
Guerrero J, 611

Guerrini L, 635, 636
Guglielmi B, 509
Guha M, 556, 557
Guhl E, 674
Guida WC, 646
Guidoni L, 747
Guijarro JI, 343, 345, 349
Guilhaus M, 404, 410, 424, 427
Guillard S, 118
Guillory RJ, 76
Guillou L, 510
Guimaraes A, 358, 359
Guimaraes CP, 323
Guimard NK, 353
Guittet O, 684, 693
Guix C, 195, 196, 198
Gul S, 669
Gulati S, 302
Gulick T, 376
Gulik A, 730, 733
Gulik-Krzywicki T, 730, 733
Gultyaev AP, 226
Gumbs C, 28
Gumireddy K, 98, 99
Gumport RI, 577
Gunawardena S, 613, 617–21
Gunner MR, 180
Gunsalus KC, 546–48, 557
Gunther NW 4th, 230
Guo CH, 496, 500, 505, 508
Guo F, 573, 574, 576, 587
Guo JT, 341, 353, 355
Guo L, 776
Guo Q, 618
Guo S, 154
Guo Z, 346
Gupta K, 573, 579
Gupta MP, 301
Gurevich R, 116
Gurley CM, 26
Gurrath M, 78
Gurubhagavatula S, 100
Guse A, 548, 553
Gust B, 683, 693
Gustafsson M, 353
Gustus C, 554
Gutierrez JA, 631, 641–43
Gutman CR, 619
Gutman M, 174
Guttridge DC, 31
Guven KL, 32
Guy AR, 660

Guyton KZ, 636, 637
Gygi SP, 197, 220, 253, 439, 441

H

Haapalainen AM, 309
Haas W, 439, 441
Haas-Kogan DA, 100
Haass C, 615, 617
Haavik AG, 70, 82
HABER JE, 111–35; 113–15, 117, 118, 120, 121, 127, 437
Habermann B, 559
Hackenberg H, 734
Hacker B, 181
Hackett JA, 117, 118
Hadchouel J, 20
Hadjigeorgiou GM, 719
Hadjiyanni I, 548
Haering CH, 218
Hafezparast M, 613
Haffter P, 586
Haga T, 744
Hagen WR, 775
Hagenfeldt L, 315
Hagerhall C, 81
Haggard-Ljungquist E, 579
Haggblom CI, 123
Hagge-Greenberg A, 373
Hagihara B, 70
Hagihara Y, 341, 344
Hagiwara T, 262
Hagman DK, 391
Hagstrom KA, 217, 224, 225
Hahn WC, 121, 494, 508
Hahn-Hagerdal B, 417
Hahnel R, 283
Hai T, 636, 637
Haikawa Y, 468
Haiker H, 79
Haimberger ZW, 118, 508
Hainzl O, 283
Hajdu I, 417, 418
Hajra AK, 303–5, 314
HAJRA S, 211–41
Hajri T, 369
Håkansson P, 684, 693, 694, 699
Hakimi MA, 262
Halban PA, 386
Halder SK, 156
Hale PJ, 384
Hales CN, 378, 384, 386

Halilovic E, 548, 559, 560
Hall CK, 349
Hall H, 253, 254
Hall IM, 249
Hall MC, 253
Hall ZW, 10
Halle B, 417
Hallén S, 170
Hallewell RA, 302
Hallgren CG, 97
Haltia T, 175
Halton J, 630
Ham J, 28
Hamad NM, 509
Hamada H, 24
Hamada M, 96
Hamaguchi Y, 547, 552
Hamamoto T, 301
Hamasaki D, 745
Hamel P, 772
Hamer KM, 255
Hamill D, 553, 558
Hamill OP, 711
Hamilton FD, 683, 684
Hammarstrom P, 349
Hammarton TC, 55, 61, 62
Hammer RE, 372
Hammerton T, 61
Hammes GG, 520, 527, 528
HAMMES-SCHIFFER S, 519–41; 520, 521, 524, 528–31, 533–35, 537
Hammon D, 630
Hammond PW, 502
Hamodrakas SJ, 340
Hampel H, 619
Hampsey M, 254, 259
Han DH, 379, 383
Han HY, 354
Han X, 391
Han YS, 409, 417, 421, 423–25
Hanai R, 574
Hancox MI, 599
Hande MP, 509
Handler P, 442
Handschin C, 375
Handschumacher RE, 630, 631, 639
Handy JW, 744
Haner M, 336, 345
Hanks SK, 96
Hannett NM, 259

Hanratty WP, 146
Hansen G, 684
Hansen HA, 694
Hansen L, 372, 373, 377
Hansen LK, 421
Hansen MT, 419
Hantraye P, 617
Hao AJ, 75
Haorah J, 151
Harada M, 386
Harari PM, 101
Harauz G, 51
Hardeman D, 303
Harder J, 692, 693
Harder W, 322
Hardie DG, 30, 139, 144,
  147–49, 152, 156, 157
Hardin JM, 261
Harding HP, 635
Harding M, 715
Hardwick KG, 217
Hardy CD, 500, 501
Hardy J, 615
Hardy JA, 615
Hargrave PA, 745, 748
Hargreaves M, 27
Harigaya Y, 612
Hariguchi S, 619
Haring JH, 619
Harkiolaki M, 667
Harmon JS, 386, 390
Harp JM, 258
Harper DC, 339, 340
Harper JD, 341, 347, 348
Harper JW, 284
Harrenga A, 168, 177
Harrington EA, 545, 546, 560
Harrington L, 494, 496, 500,
  502–7, 509
Harrington LA, 495, 502
Harrington WW, 379
Harris AK, 545
Harris BZ, 664
Harris CL, 618
Harris F, 61
Harris GW, 407
Harris K, 669
Harris ME, 570
Harris P, 694
Harris R, 41
Harris SF, 273, 275, 276, 279
Harrison BC, 27

Harrison MA, 29, 30
Harrison SC, 501, 665, 667
Harshaw R, 454, 457–59
Harst A, 285
Hart CM, 254
Harte PJ, 244, 248
Hartl FU, 272–75, 278, 282, 286
Hartley BS, 418
Hartley DM, 347, 348, 357
Hartshorne DJ, 196, 202
Hartson SD, 284
Hartwell L, 117, 128
Hartwell LH, 117, 657
Harvey SC, 520
Harvey SH, 224
Hasan NM, 386, 389
Hase T, 785
Hasegawa H, 617
Hasegawa K, 341, 344
Hasegawa M, 344
Haseman JH, 663
Hasenkampf CA, 223
Haser R, 410, 421, 422, 424,
  425, 427
Hashimoto M, 357, 718, 732,
  734, 735, 782, 785
Hashimoto N, 347
Hashimoto T, 309, 311, 442
Hashizume K, 388
Haslett GW, 245
Hassan BA, 617
Hassig R, 612
Hassinen IE, 83
Hassing H, 233
Hastie ND, 510
Hastings ML, 140
Hatanaka T, 718, 720, 734
Hatcher ME, 459
Hatchikian EC, 84
Hatefi Y, 70, 76, 79, 82, 85
Hatfull GF, 579, 585, 590, 593,
  599
Hathaway NA, 139
Hatta M, 439
Hatten ME, 320
Hattori H, 619
Hattori N, 357
Hauel N, 512
Hauf S, 217, 220, 545, 546
Haupt V, 75
Hauska G, 771, 782, 783
Hauw JJ, 619

Haverkamp RG, 340
Haviv Y, 495, 502
Hawk J, 116
Hawkesford MJ, 772, 782, 785
Hawkins PN, 336
Hawley JA, 377
Hawley SA, 144, 147–49
Haworth C, 545, 546
Hayaishi O, 445, 448
Hayakawa A, 617
Hayashi M, 70
Hayatsu N, 120
Hayes F, 226, 485
Hayes KC, 760
Hayes MJ, 231
Hayes SF, 58
Haykinson MJ, 598
Häyrinen J, 56
Hays AP, 30
Haystead TAJ, 193, 195–98,
  202, 276
Hazel M, 745
Hazelbaker D, 578
He F, 759
He J, 382, 583, 585, 586, 588,
  591
He XY, 311
He Y, 547, 552, 611
Head E, 348
Heald R, 546, 547
Healy JI, 25
Heard E, 255
Hearn SA, 357
Heasman J, 555
Heath EC, 47
Heath R, 148
Heathcote P, 181, 775, 778
Heazlewood JL, 70, 73
Heberle J, 170, 172
Hebert SS, 617
Hebron KJ, 324
Hecht A, 438
Hecht MH, 353
Hecht R, 370, 372
Hecht SM, 574
Heck M, 744, 751, 757, 760
Hedderich R, 78
Hederstedt L, 786
Hedlund BP, 475
Hedrick SM, 192
Heichman KA, 585, 596–98
Heid H, 303, 304

Heidelberg JF, 436
Heidkamper D, 718–20
Heidt J, 259–61
Heimann S, 781
Heimpel S, 718–20
Heimstad ES, 418, 421–23
Heinemeyer J, 73
Heiner CR, 468
Heinrich H, 73
Heinzel V, 195, 198
Heinzer AK, 321, 323
Heise CE, 548
Heise H, 344
Heit JJ, 507, 508
Heitbrink D, 170, 172
Heitman J, 282
Heizer C, 557
Hekking B, 436, 439
Helander HM, 297, 309, 310
Heldin CH, 103
Heldt HW, 716
Helfand SL, 444
Helge JW, 378
Heller RC, 436, 446, 447
Hellingwerf KJ, 325
Hellman U, 683, 684
Hellwig P, 83, 84, 170, 180
Helms V, 178
Heltweg B, 444, 452, 461
Hemmann U, 512
Hemminki A, 138–40, 142, 157
Henderson PJ, 716
Hendrix M, 122
Henes ST, 377
Hengartner MO, 548
Henikoff S, 244, 245, 261, 262
Henikoff SA, 644
Heninger AK, 546, 548
Henis-Korenblit S, 193, 195, 196
Henkel-Rieger R, 386
Hennequin LF, 102
Hennig R, 22
Henquin JC, 386, 388
Henriksen EJ, 30
Henriksson J, 27
Henrissat B, 56
Henry KW, 260
Henson JD, 494
Henzing AJ, 548
Heo YS, 662, 663, 669
Her LS, 610, 611, 613

Herbert S, 468
Hermann R, 336, 345
Hermans J, 167, 178
Hermoso JA, 416
Hernandez MP, 283
Hernandez-Verdun D, 498
Hernick M, 436, 447
Herrera E, 496
Herriott RM, 3
Herrmann C, 282
Herrmann R, 753
Herrmann RG, 771, 779
Herrmann T, 369
Herschlag D, 640
Herschman HR, 156
Hersh EM, 630, 638
Herskowitz I, 284, 437
Hesman T, 253
Hess JL, 245, 251, 254, 257
Hesselink MK, 385
Hester G, 633
Hesterberg M, 79, 80
Hettema EH, 303, 311, 322–25
Hetzler BG, 700
Heubi JE, 316
Heuser J, 339, 340
Hevener AL, 374, 377
Hewett J, 613
Hewitt EW, 348, 349
Heymann JB, 786
Hezel AF, 139
Hibner B, 103
Hickey MS, 30, 31
Hickman AB, 574
Hickner RC, 385
Hicks JB, 130, 437, 671
Hicks JL, 308
Hickson GR, 546–48, 558, 559
Hickson ID, 121
Hidaka H, 546
Hieter P, 113, 115
Higa HH, 56
Higa M, 393
Higashimori T, 27, 369, 377, 383
Higgins GA, 615
Higgins MK, 61
Higgins NP, 215
Higginson J, 25
Higgs C, 758
Higuchi M, 353
Higuchi T, 220

Hill CM, 51
Hill CP, 632
Hill EV, 671
Hill J, 169
Hill JA, 27
Hillard CJ, 196, 198
Hillier W, 170
Hiltunen JK, 309, 310, 322
Hilz H, 445, 457
Hime GR, 496, 548
Himo F, 687, 689–91
Himri T, 409, 415, 419, 424
Hinchliffe P, 70, 79, 80, 82, 85
Hinchman SK, 644
Hind G, 772, 782, 785
Hindelang C, 319, 321, 324
Hinds MG, 640
Hinkle G, 468
Hinkle PC, 85
Hinnebusch AG, 635
Hino Y, 468
Hioki T, 190, 196, 198
Hippler M, 73
Hiraga S, 213, 215, 223, 226, 227, 229, 484
Hirano M, 224, 226, 229, 700
Hirano T, 217, 218, 222, 224, 555
Hirao M, 461
Hirao-Minakuchi K, 552
Hiratake J, 642, 643, 645, 646
Hird SN, 552, 556
Hirokawa N, 611
Hirose H, 386, 389, 390
Hirose K, 547, 552, 553
Hirosumi J, 377
Hirota M, 139
Hirota T, 220
Hirotsu K, 633
Hirsch D, 369
Hirschfield GM, 336
Hirshfeld A, 755
Hirst J, 70, 71, 73, 75, 81, 86
Hirth P, 102
Hiruma H, 620
Hisanaga S, 612
Hiscoe HB, 609
Hiser C, 166, 176, 178–80
Hittinger CT, 669
Hixon J, 444, 452, 461
Hjortsberg K, 693
Hjuler HA, 336, 345, 346

Hlaing J, 283
Ho CL, 611
Ho DS, 319, 321
Ho DT, 662
Ho HY, 665, 672
Ho M, 639
Ho S, 284
Ho TQ, 229
Ho YS, 319, 321
Hoagland N, 660
Hobbs M, 47
Hoberg JE, 439, 444
Hobson A, 248
Hochachka PW, 405
Hochwagen A, 218
Hocking J, 50, 52
Hodge VJ, 315
Hodges RS, 611
Hodson N, 55, 56, 61, 62
Hoe W, 156
Hoefler G, 315
Hoekstra MF, 660
Hoess R, 595
Hofacker I, 172, 178, 182
Hofer A, 698
Hoffman DB, 549
Hoffman P, 43
Hoffmann B, 718, 720, 726, 736
Hoffmann K, 282, 283, 745, 748
Hoffmeier K, 177
Hofhaus G, 73
Hofmann A, 244, 248, 249
Hofmann ER, 75
Hofmann KP, 744, 745, 747, 748, 751, 752, 759
Hofmann S, 70
Hogan E, 223
Hogan PG, 25
Hoganson CW, 177
Högbom M, 685, 690
Hogenboom S, 300, 301, 315
Hogenhout EM, 303
Hogness D, 12
Hoh JF, 30
Hohmeier HE, 386
Hojrup P, 427
Hol WG, 574
Holbro T, 146
Holbrook NJ, 636, 637
Holcenberg JS, 638
Holden HM, 633, 634
Holdgate GA, 99

Holland AJ, 217
Holland JF, 630
Holland PM, 660, 661
Holland WL, 378
Hollander IJ, 101
Hollander J, 385
Holleman A, 639
Hollich V, 572
Hollosi M, 274, 276
Holloszy JO, 30
Holloway BA, 610
Hollunger G, 82
Holly S, 274, 276
Holmback UC, 379
Holmberg C, 694
Holme T, 43
Holmes A, 117
Holmes E, 618
Holmes KC, 476, 477, 481
Holmes VF, 217, 223
Holmgren A, 684, 692
Holst D, 31
Holt P, 79, 167
Holt PJ, 79, 81, 85
Holt SE, 272, 286, 507, 510
Holzinger A, 324
Homberg V, 613
Honda S, 336, 343
Honda Y, 321
Hong H, 245, 258
Hong M, 613
Hong SP, 147, 148
Hong YM, 718, 719
Hongo S, 631, 633–35
Honig B, 775
Honlinger A, 734
Honsho M, 324
Hood L, 423
Hoofnagle AN, 660, 662, 663
Hoogenraad NJ, 282
Hook VY, 617
Hoover RL, 711
Hopfield JJ, 657
Hopfner KP, 224
Hopkins A, 648
Hoppe J, 719, 730
Hoptman MJ, 620
Hordvik A, 421–23
Hori T, 748, 749, 751
Horikawa H, 468
Horike N, 156
Horikoshi M, 441

Horn D, 776
Horoupian DS, 616
Horowitz B, 639
Horowitz H, 127
Horsley V, 26
Horvath EM, 782
Horvitz HR, 548
Horwitz MS, 190, 197–99, 204
Hoshino K, 191, 196, 198, 199, 201
Hoshino M, 341, 344
Hoskins AA, 633
Hosler J, 174, 175, 177
HOSLER JP, 165–87; 169, 170, 174–78, 182
Hosoya H, 194, 201, 202, 204
Hoss M, 509
Hossain S, 501, 507
Hotamisligil GS, 371–73
Hothorn M, 442
Hou JM, 748, 754
Hou L, 195, 196, 199–202
Houchens T, 75
Hough DW, 409, 412, 415, 416, 418, 419, 421–26
Hough E, 421
Houle JD, 26
Houlston RS, 138
Houmard JA, 30, 31, 375, 380, 383
Houseman DE, 409
Houzen H, 719
Howard AE, 311
Howard JB, 484
Howard M, 485, 671
Howard PJ, 231
Howard PL, 672
Howe CJ, 772, 781
Howe CL, 618, 621
Howe L, 253
Howell AW, 385
Howell KA, 70, 73
Howitz K, 444
Howitz KT, 436, 444, 452, 460
Hoyer W, 344, 353, 356
Hoyoux A, 409, 410, 424
Hruby H, 461
Hsieh B, 314
Hsien J, 547
Hsu BY, 388
Hsu JM, 231
Hsu JY, 674

Hu B, 272, 274, 277, 280, 285, 287
Hu Z, 283, 484, 485
Huai Q, 274, 279, 281
Huang D, 772, 775, 781, 786
Huang GC, 75
Huang H, 501, 508
Huang HJ, 101
Huang J, 231, 445, 457, 482
Huang L, 176, 181, 182, 770–72, 776
Huang L-S, 770–72, 775, 781
Huang MX, 684, 694
Huang P, 121
Huang S, 101
Huang SG, 734
Huang SM, 245, 258
Huang TG, 611
Huang TS, 660
Huang WJ, 284, 289
Huang WM, 582
Huang XY, 633
Huang Y-Z, 631
Huang Z, 524
Huang Z-Q, 258
Huard S, 496, 500–5, 507
Hubbard RE, 528
Hubbell CM, 744
Hubbell WL, 520, 744, 753
Hubeek I, 639
Huber C, 700
Huber GA, 710
Huber HE, 596
Huber KE, 582
Huber R, 520, 786
Hubner R, 418
Huddleston MJ, 661
Hudson DF, 548
Hudson ER, 147
Huecas S, 469, 473, 474
Huff ME, 356
Huffaker TC, 684, 694
Huffman JL, 225
Hughes C, 46, 51, 61
Hughes CM, 253
Hughes L, 613
Hughes RE, 593, 595
Hughes TR, 116, 117, 496, 499, 669
Hughson AG, 358
Hull RL, 390
Hulley M, 50

Huls PG, 487
Hulver MW, 375, 376, 382, 385
Hummer G, 167, 169, 170, 172, 173, 180
Humphrey D, 556
Humphrey EL, 260
Humphrey M, 647
Humphries C, 24, 25
Humphries MM, 745
Hung AY, 615
Hungerer D, 43
Hunsicker-Wang L, 180
Hunt MC, 300, 308
Hunte C, 79–81, 83, 167, 181, 182, 734, 770–72, 775–78, 781, 786
Hunter RB, 26
Hunter T, 95, 96, 139, 146, 152, 153, 155, 547, 554, 660, 674
Hunziker A, 303, 304
Huque Y, 683, 690
Hurd DD, 618
Hurd HK, 684
Hurd LB, 745
Hurd LL, 745, 748
Hurley LA, 498
Hurley LH, 500
Hurley RL, 147, 148, 153, 154
Hurshman AR, 349
Hurt EC, 771, 782, 783
Huryn D, 620
Husain Z, 101
Huss JM, 28
Hussain A, 503, 504
Huston E, 671
Hutchison KA, 282, 283, 289
Hutson RG, 635, 638
Hutter MC, 178
Hutti JE, 660
Huyghe S, 309, 321
Huyghues-Despointes BM, 520
Huynen MA, 73, 74
Huysmans E, 312
Hwa J, 750, 759
Hwang ES, 437
Hwang KY, 409, 417, 421, 423–25, 442
Hwang MS, 674
Hyman AA, 551
Hyman BT, 615
Hynes NE, 146

I

Iannuzzi C, 358
Ibba M, 634
Ichikawa T, 620
Ichinose C, 226
Iconomidou VA, 340
Ide C, 548
Iezzi S, 439, 448
Ignatova Z, 355
Ihmels J, 657
Iida S, 596
Iijima K, 357
Iinuma H, 96
Iizuka K, 392
Ijlst L, 297, 304, 310, 313, 316, 322, 324, 325
Ijpma A, 119, 120, 126
Ijpma AS, 116, 117
Ikawa K, 718, 735
Ikebe M, 196, 201, 202, 204
Ikemoto S, 372
Ikner AD, 662
Ilan O, 44, 47, 52, 53
Iliev AI, 614, 618
Iliffe-Lee ER, 685
Illenberger S, 196, 201, 609, 613, 618
Im DS, 548
Imafuku I, 618
Imai H, 751, 755
Imai S, 436, 438–40, 442, 444, 446, 447
Imai SI, 436, 439, 440
Imamura K, 151
Imamura M, 139, 140
Imamura R, 223
Imanaka T, 324
Imanishi Y, 745, 747, 748, 760, 761
Immormino RM, 278, 280
Imoto M, 96
Impagnatiello T, 734
Imredy JP, 709, 710
Inaba K, 667
Inagaki M, 547
Inbal B, 190–95, 198–201
Ingemarson R, 692, 698
Ingledew WJ, 83, 84
Ingley E, 282
Ingouff M, 405
Ingui C, 273
Ingvarsdottir K, 260

Inman RB, 577
Innocenti M, 667
Inoki K, 149, 151
Inomata H, 617
Inoue H, 550
Inoue M, 786
Inoue YH, 547, 553–55
Inouye M, 274
Ionescu RM, 413
Ionescu-Zanetti C, 348
Iordanov T, 533–35
Iossa S, 382
Ippel JH, 694
Iqbal K, 615
Ira G, 113–15, 117, 121
Ireton K, 230
Irrgang K-D, 770
Irvine GB, 354
Irvine WP, 461
Irving BA, 668
Irwin A, 284
Ishida J, 96
Ishida M, 730, 731
Ishihara T, 613
Ishii T, 190, 198
Ishii Y, 341
Ishikawa F, 120, 509
Ishikawa TO, 139, 140
Ishino F, 479, 480
Ishiyama N, 388
Ishizaki T, 546, 547, 553, 556
Isin B, 760
Islam A, 558
Isono K, 642
Issartel JP, 70, 82, 84
Issemann I, 374
Isshiki K, 96
Istrail S, 353
Itani S, 379
Ito M, 139, 479, 480
Itoh K, 421
Itoh RE, 552
Itoh Y, 386
Ittmann M, 635
Ivanov D, 236
Ivanov EL, 114, 118
Ivanova EV, 560
Ivarsson R, 389
Ivens A, 633
Iverson TM, 770
Iwahashi A, 732, 736

Iwai A, 617
Iwai N, 480
Iwaki M, 170
Iwakoshi NN, 374
Iwamoto I, 547, 552
Iwasaki H, 667
Iwashima M, 668
Iwata A, 617, 618
Iwata J, 83
Iwata M, 181, 770–72, 776
Iwata S, 168, 169, 172, 173, 180, 181, 715, 770–72, 775, 776
Iwata T, 139, 140
Iwatsubo T, 344
Iyer S, 118, 129
Iype LE, 574, 576, 587
Izumi M, 56

J

Jablonski SA, 554, 558
Jackson BM, 635
Jackson F, 745, 748
Jackson JF, 7, 8
Jackson MD, 441, 446–48, 450, 452, 453, 459
Jackson PK, 674
Jackson SE, 277, 283, 287
Jackson SP, 439, 442, 452, 508, 664
Jacob F, 657
Jacob J, 282
Jacobs D, 662
Jacobs H, 716
Jacobs RE, 617
Jacobs SA, 249, 507
Jacobson BA, 692
Jacobson K, 557
Jacobson M, 536
Jacobson O, 101
Jacobson SG, 760
Jacobs-Wagner C, 483
Jacoby ME, 775
Jacqueminet S, 390
Jaehning JA, 259
Jaenicke R, 276, 413
Jaffe A, 223
Jagadeesh GJ, 574
Jagdale GB, 405
Jager C, 786
Jager S, 745
Jagoe RT, 31

Jahnke U, 75
Jain MK, 711
Jain S, 599
Jaken S, 671
Jakes R, 617
Jakob U, 273, 274
Jakobs BS, 308
Jakobs C, 308, 312, 316
Jakobsen SN, 156
Jakubczak JL, 499
Jakubowski H, 408
Jakubowski S, 52
Jaleel M, 152, 154, 156
Jalsovszky G, 274, 276
James C, 548
James MN, 409
James TC, 248
Jan YN, 194, 550
Jang CW, 190, 194, 197–201
Jang GF, 760
Janjic D, 386
Jann B, 40, 41, 43–45, 47, 56, 57, 59
Jann K, 40, 41, 43–45, 47, 56, 57, 59, 61
Janne PA, 100
Jansen GA, 306, 311, 312, 314, 316
Jansen T, 785
Janson J, 390
Janssen RJRJ, 75
Jantsch-Plunger V, 553, 558
Janvier K, 323
Janz JM, 759, 760
Jao CC, 344, 354
Jardine K, 440
Jares-Erijman EA, 353, 356
Jarmuz A, 224
Jaroniec CP, 341, 343
Jarrell ET, 660
Jarrett JT, 615
Jarvest RL, 642
Jarvinen H, 138
Jarvis JL, 286, 507
Jasaitis A, 169, 170, 173
Jascur T, 272, 282
Jasmin BJ, 30
Jass J, 348
Jastrzebska B, 757, 760
Javerzat JP, 221, 222
Javitch JA, 758
Jayaram HN, 631, 641

JAYARAM M, 211–41; 230, 231, 573, 575–78, 582
Jayaratne P, 47
Jayasinghe SA, 344
Jeghers H, 138
Jencks WP, 520
Jenkins AB, 384
Jenkins C, 475
Jenkins JA, 407
Jenne DE, 138–40, 142
Jenner RG, 260
Jennes I, 409, 410, 424
Jennings PR, 301
Jensen CJ, 663
Jensen JR, 633
Jensen MO, 711
Jensen MV, 386, 388–90, 392, 393
Jensen PH, 617
Jensen RB, 223, 226–29, 477–79
Jensen SE, 640
Jenuwein T, 244, 245, 248, 249, 262
Jeon CJ, 745
Jessani N, 647
Jessberger R, 217
Jetton TL, 392
Jezek J, 722
Jezek P, 722
Jia T, 98
Jiang F, 736
Jiang J, 282, 283
Jiang LL, 309
Jiang Q, 256, 257
Jiang S, 301
Jiang W, 547, 554, 555
Jiang WL, 618
Jiang X, 349, 358
Jiang XJ, 30
Jiang ZL, 700
Jiao J, 155
Jimenez AI, 139, 140, 142, 149, 151
Jimenez J, 286
Jimenez JL, 343, 345
Jimenez M, 473
Jimenez-Sanchez G, 320, 324
Jin LW, 338, 614
Jin S, 745
Jin Y, 155, 195, 196, 199–202
Jin YJ, 194
Jinks-Robertson S, 122

Jishage KI, 139, 140
Jivan A, 274
Jobbins KA, 308
Joberty G, 144
Joët T, 782
Joh TH, 619
Johansson C, 582
Johansson E, 693
Johansson J, 341, 343, 353
John P, 98, 99
Johns A, 459
Johnsingh AA, 618
Johnson AD, 436, 438
Johnson BD, 101
Johnson FB, 121
Johnson GL, 669
Johnson GN, 782, 784
Johnson JL, 286
Johnson KA, 8, 521
Johnson LM, 390
Johnson LN, 148
Johnson RC, 568, 584, 585, 588, 591, 596–98
Johnson RD, 386
Johnson SA, 660
Johnson VL, 545, 546
Johnston JM, 618
Johnston M, 259
Johnstone C, 102
Johnstone L, 75
Johnstone M, 260
Johnstone SR, 147, 148
Joliot A, 775, 780, 782–85
Joliot P, 775, 776, 780, 782–85
Jollivet F, 547
Jona G, 193, 195, 196
Jones AC, 230
Jones CM, 145
Jones DR, 439, 444
Jones EM, 346
Jones EY, 667
Jones JA, 349
Jones KR, 775
Jones LJ, 479, 481, 485
Jones M, 546–48
Jones MT, 631, 641
Jones RG, 151
Jones RS, 244
Jones RW, 782
Jones S, 546–48
Jones SJ, 546–48
Jones TA, 419

Jones TE, 28
Jonietz EL, 482
Jonsson G, 612
Jonsson ZO, 409, 424, 426
Joo E, 559
Jordan A, 682–85, 687, 690, 692, 694–98
Jordan P, 770
Jorgensen IH, 391
Jorgensen SB, 379
Jörnvall H, 692, 693
Joseph A, 647
Joseph D, 536
Joseph J, 196, 198
Joseph JW, 391
Joshi HC, 553
Jourdain I, 474
Jove R, 289
Jovin TM, 356
Joya JE, 32
Joza N, 86
Juers DH, 419
Juguilon H, 24
Jünemann S, 172, 177, 181, 775, 778
Jung M, 461
Jung YS, 356
Junop M, 278
Jurado J, 692
Jurczyk A, 548, 559, 560
Jurgens G, 544
Juselius JK, 719
Just M, 98
Just WW, 303, 304
Juszczak E, 748

K

Kaant A, 178
Kaback HR, 70, 715
Kabsch W, 476, 477, 481, 692, 701
Kabuyama Y, 660, 662, 663
Kachurin AM, 181, 770–72, 776
Kadenbach B, 166
Kadi F, 32
Kadowaki S, 744
Kadowaki T, 372
Kaeberlein M, 436, 438, 439, 444, 446, 447, 452, 461
Kaehlcke K, 439, 461
Kaercher T, 320

Kaether C, 617
Kaguni LS, 13
Kahana A, 256, 260
Kahana JA, 213, 230
Kahmann R, 585, 586, 596, 597
Kahn CR, 375, 379
Kahn R, 715, 717, 722, 723, 727
Kahn SE, 390
Kahne D, 62
Kaiser JT, 484
Kaitna S, 224
Kaito A, 336, 343
Kajava AV, 342–44, 346
Kajimura J, 49
Kajtar J, 274, 276
Kakuma T, 390
Kalhovde JM, 29
Kalinna S, 719
Kalinova M, 639
Kalivoda KA, 60
Kallas T, 770, 785
Kallberg Y, 353
Kalman S, 468
Kalota A, 98
Kalvakolanu DV, 75
Kalyanaraman B, 196, 198
Kamal A, 610, 612, 617
Kamali-Moghaddam M, 582
Kamath RS, 546, 547
Kambhampati S, 104
Kamel C, 436, 439, 440
Kamijo K, 301, 312, 324
Kaminishi M, 151
Kamiryo T, 310
Kamis AB, 50, 51
Kamiya N, 770
Kamm C, 613
Kamon J, 371
Kamphorst W, 620
Kampranis SC, 276
Kamtekar S, 582, 585–88, 590,
    591, 595, 598, 600
Kanaar R, 584, 585, 596–98
Kanada M, 557
Kanai Y, 610
Kanatous S, 24, 25
Kanazawa A, 775
Kanazawa I, 618
Kanazawa T, 548
Kandarian SC, 31
Kandel ER, 340
Kandpal G, 618

Kaneko S, 505, 506
Kanellopoulou C, 579
Kang AD, 98, 99
Kang J, 615
Kang MR, 508
Kannenberg F, 310, 321
Kannt A, 169, 170, 173, 177, 180
Kansagra NV, 26
Kanter EM, 384
Kao CF, 259
Kao HY, 24
Kao MC, 80, 81, 86
Kaper T, 417
Kapiloff MS, 671
Kaplan N, 446
Kaplan RS, 733–35
Kapp GT, 425
Kar S, 470
Karamysheva Z, 498
Kardos J, 417, 418
Karess RE, 547
Karlseder J, 123
Karlsen S, 421
Karlsson C, 24
Karlyshev AV, 62
Karni R, 102
Karnovsky MJ, 711
Karos M, 144
Karpefors M, 169, 170, 180
Karplus M, 533, 535, 536
Karsenti E, 474, 555
Karshikoff AD, 423
Kas J, 556
Kasa P, 617, 620
Kashani-Poor N, 78, 84
Kashefi K, 404
Kashlan OB, 698, 699
Kasho V, 715
Kashtan N, 675
Kashyap S, 390
Kasiwayama Y, 323
Kaspers GJ, 639
Kaspers GJL, 630
Kasrayan A, 691, 692, 698
Kasumov T, 308
Katabuchi H, 140
Katajisto P, 152
Katakura T, 620
Katayama Y, 168, 171
Katis VL, 220, 221
Kato H, 643–46
Kato T, 553

Katoh A, 442
Katoh Y, 156
Katou H, 341, 344
Katou Y, 219
Katsonouri A, 175
Katsube Y, 785
Katz KH, 138
Katzman R, 615
Kauffmann I, 512
Kaufman RJ, 745
Kaukonen J, 719
Kaulen A, 170, 177
Kauppi B, 690
Kaushik N, 501
Kawaguchi K, 70
Kawai M, 619
Kawai T, 191, 195–99, 201
Kawajiri A, 553
Kawakami M, 642
Kawakami T, 620
Kawamata T, 618
Kawamata Y, 386
Kawano Y, 30
Kawarabayashi T, 612
Kawarabayasi Y, 468
Kawasaki H, 275
Kawasaki K, 336, 343
Kawase Y, 139, 140
Kawashima SA, 221, 222
Kawashima T, 547, 552, 553
Kayano T, 373
Kayed R, 341, 348
Kazemier KM, 638
Kazmi MA, 748, 754
Kazmierczak RA, 577
Ke HM, 274, 279, 281
Keane EM, 378
Kearney PF, 619
Kearsey SE, 114
Keegstra W, 73, 81
Keel J, 546
Keenleyside WJ, 43, 47
Kegley KM, 24, 26
Keitheri-Cheteri MB, 140–42
Kelleher C, 494, 498, 505, 506
Keller GA, 302
Keller MD, 439, 444
Kelley DE, 376, 377, 379, 382
Kellogg BA, 685, 687, 692
Kelly DP, 28, 374, 375, 383
Kelly GP, 219
Kelly JW, 349, 358, 359

Kelly TJ Jr, 7
Kelpe CL, 386, 390, 391
Kemp A Jr, 716
Kemp BE, 27, 148, 660
Kemp S, 321, 323, 324
Kemphues KJ, 147, 154
Kempner ES, 759
Kendrew J, 102
Kendrick MA, 386, 389
Kendziorski CM, 373
Kenna MA, 436, 446, 447
Kenna P, 745
Kenne L, 43
Kennedy BK, 444
Kennedy CE, 548
Kennedy CJ, 384
Kennedy SP, 423
Kenney JM, 340
Kent CR, 105
Kenyon C, 440
Kern D, 535
Kern J, 770
Kerr ML, 548
Kerscher S, 70, 74, 76, 78, 79, 84–86, 181
Kershaw NJ, 311
Kervinen M, 83
Ketting R, 303, 311
Khan SA, 130
Kharbanda S, 507
Kheterpal I, 341, 348, 353, 355
Khochbin S, 24
Khorana HG, 7, 8, 744, 745, 748, 753, 759, 760
Khorasanizadeh S, 249
Khorasanizadeh SA, 249
Khorrami S, 245, 251, 252
Khouangsathiene S, 671
Khurana R, 336, 348
Khurts S, 505, 506, 509
Kickhoefer VA, 509
Kidd M, 615
Kidner C, 249
Kieffer S, 718–20, 730, 731
Kiehart DP, 547
Kieloch A, 143–46, 154
Kienlen-Campard P, 618
Kieran D, 613
Kiger AA, 545–48
Kihira Y, 718, 720, 732, 735, 736
Kiihne S, 755
Kikkawa M, 611

KILBERG MS, 629–54; 630, 635–38
Kilbride E, 593, 595
Kilian A, 496
Killian JA, 711
Kilpelainen SH, 311
Kim AI, 599
Kim BS, 618, 620
Kim DG, 311–13, 324
Kim H, 181, 576, 611, 770–72, 776
Kim HC, 442
Kim HJ, 230, 369, 383
Kim HS, 378, 633
Kim HY, 274, 279, 281, 786, 787
Kim JE, 751
Kim JH, 508, 632, 633
Kim JK, 27, 369, 372, 373, 377, 383
Kim JL, 102
Kim JM, 753, 759
Kim JY, 376
Kim K, 532, 533
Kim KK, 770–72, 776
Kim M, 253, 259
Kim S, 62, 258, 642, 643, 710
Kim SH, 409, 417, 421, 423–25, 759
Kim SK, 662, 663, 669
Kim SY, 409, 417, 421, 423–25
Kim Y, 177
Kim YB, 371–73, 387
Kim YJ, 373, 377
Kim YK, 662, 663, 669
Kimata K, 56
Kimball AS, 577
KIMCHI A, 189–210; 190–95, 197–201, 203, 204
Kimmins S, 272
Kimura A, 441
Kimura H, 347
Kimura T, 684
Kimura Y, 275
King GF, 484, 485
King J, 353
King MT, 439, 448
King N, 669
King R, 558, 560
King RW, 550
Kinghorn KJ, 357
Kingston RE, 245
Kinoshita A, 548

Kinoshita M, 545, 548, 559
Kintscher L, 78, 82, 83, 181
Kirchmaier AL, 245
Kiriyama Y, 630, 638
Kirkitadze MD, 348
Kirkland KT, 444
Kirkpatrick SS, 147
Kirmizis A, 256
Kirschner M, 487, 657
Kirschner MW, 220, 550, 665, 672, 693
Kirshenmann T, 226
Kishi M, 155
Kishimoto A, 151, 156
Kishimoto T, 612
Kisselev OG, 748
Kissil JL, 190–95, 198, 199, 201
Kistler A, 438
Kitada T, 357
Kitajima TS, 221, 222
Kitamura T, 547, 552
Kitchen DB, 646
Kitmitto A, 418
Kitoh J, 630, 638
Kitoh T, 638
Kittler M, 356
Kittler R, 546, 548
Kitts PA, 572, 580
Kizer KO, 253, 254
Klar AJ, 437
Klauck TM, 671
Klausner RD, 711
Kleckner N, 216, 480
Klein H, 113–15, 121
Klein S, 96
Kleinfeld AM, 711
Klein-Seetharaman J, 748, 760
Klem TJ, 633, 641
Kleman JP, 547, 554, 558
Klenchin VA, 744
Kline WO, 26, 28
Klingenberg EM, 734
Klingenberg M, 82, 709, 716, 718–20, 726, 730, 732–36
Klinman JP, 530–35
Klippel A, 585, 586, 596–98
Kloos M, 613
Klopfenstein DR, 611
Kluckman KD, 320
Klug A, 674
Klughammer C, 786
Klukas O, 770

Klumper E, 630
Klumperman J, 612
Kmiecik TE, 674
Knaff DB, 776
Knapp DR, 759
Knapp FF Jr, 369, 383
Knapp MJ, 535
Knappe J, 682, 692, 701
Knecht DA, 547
Knehr M, 302
Knight D, 340
Knoblich JA, 550
Knöchel T, 633
Knoll AH, 771
Knotts TA, 378
Knowles JR, 640
Knox BE, 754
Knox EW, 631
Knox M, 26
Knudsen KE, 437
Kobayashi H, 321
Kobayashi K, 224, 321, 505, 506
Kobayashi M, 386
Kobayashi Y, 782, 785
Kobilka BK, 744, 759
Kobilka TS, 759
Kobor MS, 259
Kobryn K, 582
Koch C, 596
Koch J, 78
Kodama K, 390
Koeller KM, 56
Koepke J, 182, 770–72, 776, 778, 781
Koeppe RE, 711
Koerten HK, 154
Kogel D, 191, 195–98, 201
Kogoma T, 113
Koh SS, 245, 258
Kohane I, 382
Kohen A, 532, 533
Koike H, 718, 719
Koizumi M, 643, 645, 646
Kojro E, 615
Kok J, 698
Kok RM, 308, 312, 316
Kolarov J, 721
Kolarova N, 721
Kolberg M, 682
Kolli VS, 59, 60
Kolodner R, 117
Kolodner RD, 118, 121, 217

Komada S, 415
Komaki H, 719
Komander D, 148, 149, 152, 660, 663
Komatsu M, 386, 388
Komatsu S, 196, 201, 202, 204
Komiya Y, 612
Koncarevic A, 31
Kondo H, 192
Kondo N, 312, 324
Kong LB, 509
Konieczka JH, 576
Konno T, 353
Kono M, 753
Konrad EB, 11, 12
Konsolaki M, 357
Konstantinov AA, 169, 170, 177, 178
Konstantopoulos N, 373, 377
Kontogiannis L, 144
Koo BK, 194
Koo EH, 348, 617, 618
Koo SH, 148, 156
Koole R, 138
Koonin EV, 8, 226, 483, 633, 683, 693
Koopman WJH, 70, 86
Kopan R, 615
Kopito RR, 758
Köplin R, 49, 50
Kopolovic J, 190
Koranyi L, 30
Kordari P, 376
Kordower JH, 620
Korge G, 244, 248, 249
Korman MH, 499
Korn ED, 546
Kornberg A, 6–9, 12, 13
Kornberg RD, 244, 245, 251
Kornberg SR, 9
Kornfeld K, 662
Korolev VG, 118
Korolik V, 76
Koronakis E, 61
Koronakis V, 46, 51, 61
Koroniak L, 631, 642, 643
Korsisaari N, 139–41, 152
Kortum R, 671
Kortum RL, 671
Kosako H, 547
Kosaras B, 745
Koshkin V, 391

Koshland D, 219, 223
Koshland DE, 223
Koshland DE Jr, 520, 577
Kosik KS, 617, 618
Koski LB, 546–48
Kosmatka M, 147–49, 153, 154
Kosofsky B, 617
Kostakioti M, 51
Koster J, 300, 301, 315
Kostner GM, 315
Kotaria R, 734, 735
Kotovic KM, 253
Kotovsky I, 617, 619, 620
Kotti TJ, 297, 310, 319, 321
Kotula L, 618
Koutalos Y, 753
Kouyama T, 745
Kouzarides T, 24, 245, 258, 262
Kovac L, 130, 721
Kovacs I, 620
Kovacs WJ, 315
Kovalenko M, 96, 97, 103, 104
Koves TR, 375–77, 380, 382, 384
Kowal AS, 347, 348
Kowalik T, 693
Kowall NW, 618
Koyama K, 386
Koyama N, 316
Koyanagi M, 748, 754
Koza RA, 375, 382
Koziczak M, 146
Kozyrev D, 692
Krab K, 170, 780
Kraegen EW, 372, 376, 379, 380, 384
Krahn JM, 632, 633
Krainer AR, 140
Krais R, 101
Krajewski W, 252
Kralicek A, 475
Kralovics R, 104
Kramer DL, 314
Kramer DM, 775, 776, 779, 782, 784, 785
Kramer G, 283
Kramer R, 734
Krascenicova I, 548
Krasnoperova N, 746
Krasnoperova NV, 745, 748
Krasnow MA, 584, 585, 595
Kraus RM, 385

Krauskopf A, 117
Krauss N, 770
Krauss S, 391
Krauss V, 244, 248, 249
Kraut J, 521
Krebs A, 754, 758
Krebs EG, 660
Krebs MRH, 338, 345, 347
Krebs W, 84, 85
Krebs WG, 589
Krechl J, 533
Kreisberg-Zakarin R, 683, 693
Krejci O, 639
Krendel M, 552
Krenitsky T, 687
Krentz AJ, 384
Kreutzberg GW, 609
Kriegstein H, 12
Kriketos AD, 377
Krisans S, 315
Krisans SK, 315
Krishna P, 282
Krishnamoorthy G, 85
Krishnamoorthy V, 52
Krishnan R, 342–44, 348
Krishnaswamy S, 660
Kristensen O, 353
Kristjansson MM, 421, 423, 427
Krogan NJ, 245, 251–53, 259–61
Krogh A, 72
Krogh BO, 573
Kröncke K-D, 58–61
Krook A, 30
Krook M, 692, 693
Krssak M, 376
Kruger RG, 438, 442, 445
Kruhlak MJ, 24
Krüll C, 75
Kruse K, 485
Kruse T, 229, 233, 480–82
Krylov D, 372
Krzysko KA, 758, 759
Kubek S, 151
Kubicek J, 196, 201
Kubicek S, 255, 262
Kubokawa K, 748
Kubota M, 630, 638
Kubota N, 371
Kubu CS, 616
Kuchinke J, 357
Kuchler WR, 102

Kuchta RD, 8
Kuczek E, 404, 409, 419, 422–25
Kuczera K, 536
Kudo N, 559
Küffner R, 75
Kuhl J, 379
Kuhlmann T, 614
Kuhn JR, 483
Kuhn P, 574
Kuijpers MA, 509
Kuipers J, 154
Kujala P, 612
Kuksa V, 747, 752, 753, 759, 760
Kulkarni S, 547
Kumar M, 468
Kumar N, 409, 410
Kumar S, 408, 415, 419, 422, 423, 425, 548
Kumar V, 507
Kumaran S, 383
Kumasaka T, 748, 749, 751
Kumer KE, 259
Kunau WH, 300
Kunimori M, 642
Kuninaga S, 421
Kunji ER, 324, 715
Künkel A, 78
Kuntz ID, 646
Kuo JC, 190, 194, 195, 197–204
Kuo YM, 348, 357
Kural C, 611
Kuras R, 182, 786
Kurimasa A, 123
Kurimoto M, 338
KURISU G, 769–90; 770–72, 774, 775, 777, 779, 783
Kurisu M, 323
Kuriyama R, 547, 554, 558
Kuriyan J, 98, 102, 664, 665, 667, 668
Kurki S, 83
Kurokawa K, 552
Kuroki T, 27
Kurosawa T, 309, 310
Kurowski TG, 375
Kurtev M, 436, 439, 441, 445, 452, 460
Kurth EJ, 30
Kurtz RB, 282
Kurtzman CP, 340
Kusakai G, 156
Kusari AB, 662

Kusche-Gullberg M, 55
Kushnareva YE, 384, 385
Kushnirov VV, 355
Kusnetzow AK, 754
Kustatscher G, 442
Kuster B, 144
Kusumoto Y, 347, 348
Kutoh E, 786
Kuzmichev A, 256
Kuzminov A, 112
Kuznicki J, 282
Kvassman J, 531
Kvist AP, 309
Kwak EL, 100, 101
Kwon E, 618
Kwon HJ, 570, 572, 574, 576, 578
Kwon K, 573, 574
Kwon YD, 674
Kyte J, 355
Kyttala A, 719

L

Labay V, 700
Laberge G, 146
Labib K, 114
Labudda K, 671
Lacapcre JJ, 710
Lachkar S, 474
Lachner M, 249
Lachowicz TM, 721
Ladbury JE, 272–74, 278
Ladenstein R, 423
Ladurner AG, 442
LaFrancois J, 614
Lahesmaa AM, 152
Lahue R, 122
Lai CF, 621
Lai CK, 500, 501, 503, 504
Lai EC, 194
Lai KM, 28
Lai VK, 31, 32
Laible G, 244, 248
Lain S, 143, 144
Laio A, 747
Laird AD, 102
Lake AC, 305
Lakhani F, 620
Lakmache Y, 619
Laky D, 101
Laloraya S, 219

Lam E, 785
Lamb TD, 745
Lambert AJ, 86, 166
Lameloise N, 386, 390, 391
Lamina KA, 148, 156
Lammel C, 468
Lammens A, 224
Lamming DW, 436, 444, 445, 452, 460
Lamont S, 618
LaMonte BH, 610
Lamotte J, 406, 419, 422, 423
Lamotte-Brasseur J, 409, 415, 418, 419, 423, 424
Lamprecht AK, 49
Lamson RE, 669
Lan F, 262
Lan H, 373
Lancaster CRD, 180
Lancet D, 675
Landaker EJ, 382
Landau EM, 748, 753
Lander ES, 657, 659, 664, 665
Landgraf C, 668
Landick R, 46
Landry J, 253, 436, 446–48, 450, 452, 454, 457
Landsberg G, 191, 195–98, 201
Landy A, 568, 570, 572, 574, 576–78, 580, 600
Lane D, 226, 229, 230, 236
Lane DM, 630
Lane KE, 386
Lane WS, 262
Lanes O, 419, 421, 422, 424
Lang BF, 72, 73
Lang DB, 375, 380, 383
Langbein L, 369
Lange C, 181, 182, 770–72, 776, 778, 781
Langeberg LK, 671
Langen R, 344, 354
Langer MR, 452, 453, 457
Langley E, 24, 440
Langmann T, 322
Lanka E, 572
Lansbury PT Jr, 341, 347–49, 354, 615
Lansdorp PM, 494
Lansing JC, 522, 527
Lanyi JK, 167, 168, 173
Lanzilotta WN, 76

Lapointe J, 642
Lapouge K, 667
Lara B, 482
Lardy HA, 716
Larkins LK, 304
LaRocca GM, 555, 557
Laroche T, 438
Laroche Y, 424
Larose S, 503, 504
Larsen IK, 690
Larsen TM, 632, 642
Larson J, 357
Larsson B, 72
Larsson G, 168, 169, 172, 173, 180, 182, 684
Larsson KM, 685, 687, 692, 695–97
Lasorsa FM, 322, 717
Lassmann H, 321
Lassonde M, 619
Last AM, 349
Lastovica AJ, 62
LaTerra S, 545, 546
Latif F, 407
Latimer MJ, 770
Latorre-Esteves M, 442, 445, 452, 454
Latres E, 31, 32
Latruffe N, 319, 321
Lau CK, 573, 579
Lau CS, 662
Lau IF, 213
Lauquin GJ, 730, 734
LAUQUIN GJM, 713–41; 709, 715, 717–20, 722, 723, 726–28, 730, 732, 733, 736
Laurent BC, 231
Lauth MR, 593
LaVail MM, 745
Lavergne J, 783, 784
Lavoie BD, 223
Lavu S, 436, 444, 452, 460
Law BK, 151
Lawler AM, 321
Lawlor MA, 146
Lawrence CC, 408, 682
Lawrence CJ, 609
Lawson JE, 718–20
Laybutt DR, 384
Laymon RA, 613
Lazarevic D, 408
Lazarov O, 617

Lazarow PB, 301
Le S, 120
Le T, 481, 485
Leach S, 581
Leahy JL, 392
Leapman RD, 341, 346
Leaver M, 482, 488
Lebbink JH, 417
LeBlanc JF, 617
LeBlanc-Chaffin R, 635, 636
Lebrun E, 786
Le Brun NE, 786
Lechner TS, 579
Lecker SH, 31
Lecomte JT, 423
Lecuit T, 558
Ledesma A, 735
Lee A, 120
Lee AG, 711
Lee AH, 374
Lee AI, 419
Lee C, 234
Lee CH, 28
Lee CT, 261
Lee D, 320, 439, 440, 461
Lee DY, 258
Lee E, 745
Lee EB, 617, 618
Lee FY, 98
Lee GE, 509
Lee HC, 446
Lee J, 509
Lee JC, 100
Lee JK, 181, 770–72, 776
Lee Jehee, 573, 575, 577
Lee Joonsoo, 573, 578, 582
Lee JS, 246, 252, 373, 377
Lee JT, 642
Lee JW, 181, 442, 770–72, 776
Lee KM, 532
Lee KS, 550
Lee L, 579
Lee M, 495, 502, 548, 560
Lee MK, 613
Lee ML, 32
Lee MS, 664
Lee S, 146
Lee SE, 117, 118
Lee SR, 502, 508
Lee SW, 642
Lee SY, 577
Lee T, 660, 662, 663

Lee TH, 508
Lee VM, 357, 620
Lee WNP, 308
Lee Y, 386, 389, 390, 393
Lee YH, 258, 262, 386, 390
Lee YL, 674
Lee YT, 282
Leech A, 147
Leem JY, 618, 620
Leeman SE, 617
Leenders F, 309
Lees JA, 693
Lees-Miller SP, 508
Leeson PD, 642
Lefebvre DL, 156
Lefkowitz EJ, 468
Lefkowitz RJ, 744, 745, 759
Legembre P, 156
Legname G, 336
Legner MA, 305
Le Goff X, 152
Legrain C, 417
Lehman AR, 224
LEHMAN IR, 1–17; 3, 6–15
Lehn J-M, 642
Lei L, 635
Lei M, 500, 667
Lei Z, 297, 314
Leibler S, 657
Leif H, 70, 76, 78
Leighton F, 308
Leiper FC, 147, 148
Leiros HK, 404, 409, 418–26
Leiros I, 419, 421, 422, 424
Leismann O, 548
Lem J, 745, 746, 752
Lemaire HG, 615
Le Marechal P, 336, 346, 350
Lemercier C, 24
Lemieux C, 72, 73
Lemieux M, 436, 439, 440
Lemmon MA, 146, 664, 665
Lemoine Y, 786
Lemon BJ, 76
Lemon KP, 214, 487
Lenaz G, 83, 86
Lendvay TS, 116, 496
Lenertz L, 508
Leng SY, 500
Leng XH, 284
Lengronne A, 219
Lennon AM, 73

Leonard J, 744
Leonard K, 73, 79
Leonard TA, 484, 485
Leone TC, 374, 375, 383
Leopold L, 370, 372
Leopold WR, 101
Lepisto A, 152
Lepore F, 619
Lerin C, 439, 441
Lerman MI, 615
Leroux P, 320
LeRoy G, 249
Le Saux A, 718–20, 732, 734
Leschziner AE, 585, 586, 588, 591, 595
Lesk A, 344, 345
Leskov IB, 744
Lester D, 121
Le Trong I, 745
Letunic I, 664, 667
Letzelter C, 575
Leube B, 140–42
Leung-Pineda V, 635, 636
Lev S, 548
Leverson JD, 546, 547
Levesque G, 615
Levesque L, 618
Levesque N, 503, 504
Levi M, 391
Levin PA, 470, 479
Levine C, 233
Levine SS, 253
LEVITZKI A, 93–109; 95–97, 99, 101–4, 577
Levy O, 214, 234
Lewin J, 300, 313
Lewis C, 660
Lewis DA, 619
Lewis JA, 579
Lewis RA, 230
Lewis RE, 671
Lewis RJ, 278
Lewis RS, 25
Lewis SE, 391
Lewitzky M, 667
Leyssen M, 617
Li B, 253
Li G, 378, 386
Li H, 305, 471
Li J, 151, 253, 347, 748, 751, 754, 755, 758
Li JP, 55

Li JY, 617
Li L, 526
Li LX, 391
Li P, 28, 375, 380, 382, 384
Li PM, 171
Li PW, 657, 659, 664
Li Q, 553
Li RF, 638, 640
Li S, 375, 380, 383
Li SC, 667
Li SSC, 667
Li T, 617, 760
Li W, 669
Li WK, 582, 585–88, 590, 591, 595, 598, 600
Li X, 320, 559
Li XY, 155
Li Y, 147, 216, 227–29, 235
Li Y-F, 770
Li YQ, 256, 257
Li Z, 553
Li ZW, 373, 374, 377, 617
Liang C, 102
Liang P, 314
Liang Y, 615, 745, 746, 757–59
Liao CY, 285
Liaw D, 151
Liaw SH, 643
Libante V, 229
Liberi G, 117
Libioulle C, 407
Licht JD, 503
Licht S, 688
Licht SS, 408
Lidholt K, 56
Lieb JD, 223
Liebecq C, 733
Lieber CM, 341, 347, 348
Lieberthal W, 305
Liebman PA, 758
Liedtke W, 373
Liem RKH, 611
Lienhard GE, 408
Lieu YK, 388
Lightbody JJ, 272
Lilie H, 276
Lill MA, 178
Lillo C, 611, 618–21
Lim GE, 235
Lim HS, 194
Lim HW, 25
Lim KO, 620

Lim S, 308
LIM WA, 655–80; 660, 662–65, 667–69, 671, 672
Limouze J, 546, 550, 553, 554, 556–58
Lin CC, 121, 128
Lin CS, 734
Lin CY, 128
Lin DC, 223, 229, 484, 485
Lin H, 98, 285
Lin J, 28, 32, 375, 503, 504, 509
Lin J-H, 272
Lin JR, 194, 195, 197, 201–4
Lin MT, 619
Lin SF, 102
Lin SJ, 444
Lin YC, 496, 502
Lin YH, 256, 257
Lin ZP, 693
Linardic CM, 508
Lind T, 56
Lindahl M, 74, 76, 79
Lindahl U, 55, 56
Lindberg B, 43
Lindgren CM, 382
Linding R, 671
Lindley P, 422, 423
Lindner DJ, 75
Lindon C, 550
Lindorff-Larsen K, 350, 356
Lindquist S, 272, 281, 282, 285, 340
Lindquist SL, 340, 342–44, 348
Lindroos HB, 224
Lindstrom DL, 259
Ling C, 382
Ling ZC, 390, 392
Lingfeld G, 614
Lingner J, 117, 127, 494, 496, 498, 499, 504–6, 509, 510
Link MP, 630
Link T, 176, 182
Link TA, 775
Linn SC, 378
Linton D, 62
Linton LM, 657, 659, 664, 665
Lionetti L, 382
Liou GG, 438, 442, 445
Lipinski C, 648
Lipinski CA, 642
Lipp HP, 617
Liscum L, 305

Lisowsky T, 75
Listenberger LL, 391
Liszt G, 439, 445, 452
Lithgow T, 468
Liti G, 118, 120
Litman P, 102
Litvak V, 548
Liu A, 698
Liu AM, 684
Liu C, 694
Liu D, 48, 49, 305
Liu FF, 672
Liu H, 546, 547
Liu HP, 357
Liu J, 383
Liu JK, 496, 501, 503–5
Liu K, 618
Liu L, 459
Liu LX, 323
Liu Q, 667
Liu SK, 667, 669
Liu T-Y, 43, 44, 60
Liu W, 102
Liu X, 610, 611
Liu XS, 214
Liu XY, 693
Liu Y, 24, 27, 507, 509, 667
Liu YD, 274, 279, 281
Liu YQ, 392
Liutkevicius E, 358
Lively TN, 502
Liverini G, 382
Livingstone D, 221
Livnah O, 99, 102
Lizcano JM, 143, 144, 148, 152, 154–57, 660, 663
Lizotte DL, 611
Llambi F, 195, 196, 198
Llorente B, 442
Lloyd MD, 311, 316
Lo WS, 260
Lobbezoo B, 277
Lobel JS, 630
Lo Bianco C, 357
Lobner-Olesen A, 480, 481
Lobo JM, 582
Lobritz MA, 60
Lock WG, 156
Lockless SW, 524, 525
Lodige I, 195, 198
Lodish HF, 305, 369
Loes DJ, 316

Loewen MC, 748
Loewen PC, 748
Loff S, 138–40, 142
Loftfield RB, 408
Logan DT, 682, 685, 687, 690–92, 695–97, 700, 701
Logusch EW, 643
Loidl J, 221
Loll B, 770
Lolli G, 148
Lomakin A, 347, 348
Lomas DA, 335
Lombard DB, 440
Lombardo CR, 617
Lombardo F, 642
Lombet A, 320
Lombillo VA, 547, 554
Lomo T, 22, 29
London R, 301
Long D, 507, 508
Long SR, 472
Lonhienne T, 405–10, 415, 419, 424
Lonn A, 417
Loo JA, 101
Loo TL, 638
Loog M, 669
Loomis WF, 547
Loonen AH, 639
Lopes M, 118
Lopez A, 473, 475
Lopez L, 149
Lopez-Guajardo CC, 500
Lopez-Soriano J, 31
Loppes R, 405
Lorch Y, 244, 245
Lord L, 61
Lorent K, 357
Lorenzo A, 357
Lorson MA, 548
Losada A, 218, 222
Losick R, 213, 214, 230, 469, 479
Lotz GP, 285
Loubtchenkov M, 103
Loughran PA, 302
Louis EJ, 118, 120, 121
Louvion JF, 276
Lovett ST, 425, 426
Lovgren TN, 408
Lovley DR, 404
Low PS, 405, 407

Low SH, 559
Lowe ED, 669
LÖWE J, 467–92; 218, 228, 229, 233, 468–71, 473–82, 484–86
Lowell BB, 382, 383, 391
Lowenhaupt K, 498
Lowery D, 619
Loy T, 760
Loyola A, 249
Lozano G, 693
Lu B, 155
Lu C, 469, 474
Lu D, 386, 388, 392
Lu DH, 386, 389, 390, 392, 393
Lu H, 75, 419
Lu HP, 419
Lu J, 24, 26, 27, 156
Lu JF, 321, 323, 324
Lu KP, 509
Lu PD, 635
Lu R, 156
Lu W, 667
Lu Y, 581
Luan DD, 499
Lubben M, 169, 172
Lucas DM, 667
Lucas RL, 177
Lucca C, 118
Lucet I, 469, 482
Lucet IS, 599, 600
Luciani S, 716
Lucier JF, 503, 504
Ludwig B, 168, 169, 178
Lue LF, 357
LUE NF, 493–517; 116, 130, 495, 496, 498, 500–5, 507–9
Luecke H, 167
Luehr CA, 631, 633
Luengo JM, 57
Lufei CC, 75
Luger K, 245, 256
Lugtenburg J, 755
Luhrs T, 341, 342
Luisi BF, 348, 349
Lukas C, 217
Lukas J, 217
Lumb MJ, 301
Lümmen P, 83
Luna E, 180
Lunardi J, 70, 82, 84, 730, 733
Lund C, 674

Lundblad V, 116–22, 128, 496, 499, 503, 508, 511
Lundmark K, 336
Lunyak VV, 262
Luo C, 25
Luo H, 146
Luo J, 533, 535
Luo JY, 439, 440
Luo K, 441
Luquet S, 31
Lurz R, 229, 572
Lustig AJ, 126
Lustig U, 469
Luthert PJ, 760
Lutkenhaus J, 233, 468, 469, 472, 474, 482–85
Lutkenhaus JF, 468, 482
Lutsch G, 348, 349
Lutterbach B, 639
Lutz S, 524
Luukko K, 139–41, 152
Ly H, 501, 503–5
Lyall RM, 96, 101
Lydall D, 118, 122, 130, 131
Lynch G, 357
Lynch KR, 548
Lynch MJ, 671
Lynch TJ, 100
Lyons DM, 669
Lyons JF, 103
Lyons WE, 669
Lyras D, 599, 600

M

Ma B, 408, 419, 423, 425
Ma C, 735
Ma H, 245, 258
Ma J, 75
Ma L, 618
Ma LY, 484, 499
Ma SH, 533
Ma W, 319, 321
Ma X, 469, 482
Mabuchi I, 547, 552, 555–57
MacDonald KG, 31
MacDonald LA, 196, 202
MacDonald MJ, 386, 388, 389
Machin F, 224
Mackay JP, 340
Mackay LG, 340
MacKay S, 619

Mackenzie IR, 616
MacKenzie KL, 510
MacLachlan PR, 43, 47
MacLean PS, 375, 380, 383
Macleod K, 437
MacPhee CE, 338, 341, 343
MacRae IJ, 633
MacRae TH, 272
Mactutus CF, 619
Madaule P, 547, 556
Maddelein ML, 341, 342, 355
Madden B, 282
Madden TL, 572
Maddison RL, 121
Maddox AS, 550, 559
Maddox P, 469, 474
Maddox PS, 222, 551, 552, 555
Mader AW, 245, 256
Madgwick SA, 782
Madhusudan, 144
Madison V, 415, 416
Madras BK, 639
Madrid LV, 31
Madsen AO, 341–43
Madson N, 548
Madzo J, 639
Maeba R, 305
Maebuchi M, 310
Maechler P, 388
Maeda A, 745, 746, 759, 760
Maeda N, 309, 319, 321
Maeda R, 662
Maeda S, 374, 377
Maeda T, 745–48, 757, 759, 760
Maezawa I, 338, 614
Maffei M, 370, 372
Magara F, 617
Mager J, 255
Maghlaoui K, 770
Magliano L, 32
Magnuson KS, 105
Magnusson OT, 421
Magyar C, 425, 426
Mahieu V, 312
Mahmoudi T, 244, 248
Maiato H, 548
Maier VH, 547
Mailleux P, 617
Maine IP, 500
Mainfroid V, 416, 421, 424, 426
Mainolfi K, 409
Maison C, 251

Maiti A, 620
Maizel JV Jr, 422
Majander AS, 82
Majdalani N, 44
Majima E, 718–20, 730–32, 734–36
Mak W, 255
Makela TP, 139, 144, 145, 151
Makhatadze GI, 415, 421
Makhov AM, 130
Makin OS, 341, 343
Makino C, 745
Makino CL, 746, 752
Male R, 421–23
Maliga P, 782
Malik HS, 498, 499
Malinverni J, 62
Malisauskas M, 348, 358
Malkani S, 618
Malkin R, 785
Malkova A, 113–15, 121
Malle E, 315
Maller JL, 138
Mallory JC, 121, 128
Mallory M, 357, 617–19
Mallory SB, 138
Malm C, 32
Malmo C, 358
Malmstrom BG, 170, 171
Maloney A, 272, 283, 285, 287
Maloney AJ, 615
Mamat B, 169, 172
Mamedova AA, 79, 85
Mamuya N, 101
Manasse RS, 782, 784
Mancio-Silva L, 498
Mandarino LJ, 379
Mandato CA, 550
Mande SC, 426
Mandel G, 262
Mandel H, 305, 315, 700
Mandel JL, 319, 321
Mandelkow E, 155, 609, 613, 618, 620
Mandelkow EM, 155, 620
Mandrup S, 391
Mangasarian K, 635, 636
Mangione P, 344
Manis JP, 440
Manley PW, 98, 102
Mann DM, 619
Mann M, 496, 499

Mannaerts GP, 297, 306, 308–13, 324, 378
Manning BD, 149
Manning G, 95, 139, 146, 152, 153, 155
Manning JM, 643
Manor H, 495, 502
Manske DD, 386
Mante M, 618, 619
Mäntele W, 84, 170, 180
Mar V, 509
Marantz Y, 174
Marash L, 196–98, 200, 205
Marathe R, 468
Marciniak RA, 121
Marcon G, 349
Marcucci MJ, 27, 376, 379
Marcus S, 669
Marcus-Samuels B, 372
Marengere LE, 665
Maretzki D, 757, 760
Marfatia SM, 145
Margesin R, 404
Margittai M, 344
Margolin W, 233, 468, 469, 472, 474, 482
Margolis PS, 479
Margolis RL, 547, 549, 554
Margolis SS, 500, 508
Margueron R, 256
Margulis L, 468
Marians KJ, 233, 234
Marie-Egyptienne DT, 501–7
Marin EP, 744
Marine JC, 197, 199
Maringele L, 118, 122, 130, 131
Marino JP, 759
Marino MW, 371, 372
Marion TN, 282
Marison M, 28
Markert C, 73
Markham KA, 532, 533
Markie D, 138–40, 142, 157
Markin RS, 633
Marks MS, 339, 340
Markus M, 309
Marlett J, 438
Marmor MF, 747
Marmorstein R, 436, 442, 454, 457–60
Marmur J, 9
Maro B, 221

Marobbio CMT, 700
Marolda CL, 49
Marques I, 70, 73
Marrington R, 474
Marschall P, 546–48
Marsh DJ, 151
Marsh JA, 272, 282
Marsh JW, 485
Marshall B, 441
Marshall BL, 438, 439, 461
Marshall CJ, 415
Marshall GL, 619
Marston AL, 484, 485
Marti R, 700
Martial JA, 416, 421, 424, 426
Martin FJ, 31
Martin J, 613
Martin LJ, 617, 747
Martin ME, 252
Martin SG, 144, 147, 154
Martin SS, 576–79
Martin W, 684
Martinac B, 711
Martin-Aragonés G, 630
Martin-Benito J, 469
Martindale JL, 636, 637
Martineau SN, 549
Martin-Rivera L, 496
Martinez JA, 630
Martinez R, 95, 139, 146, 152, 153, 155
Martinez SE, 772, 781
Martinez-Campos M, 546, 547
Martins-Salas E, 482
Martins RN, 618
Marton MJ, 669
Martoriati A, 197, 199
Marttila MS, 309
Marty I, 732
Marty T, 559
Maruya M, 276, 277
Marx JC, 406, 410–16, 418, 419
Marynen P, 308
Masalha M, 683, 693
Mascarenhas J, 224
Masendycz P, 284
Masison DC, 355
Masland RH, 745
Masliah E, 357, 615, 617–19
Mason DX, 501
Mason JM, 120
Mason MM, 372

Massarelli C, 547
Massey TH, 234, 581
Massière F, 633
Masson K, 719
Massow M, 75
Masters CL, 615
Masuoka DT, 612
Masutomi K, 494, 505, 506, 509
Masuzaki H, 372
Matagne RF, 70, 73, 74, 76, 81, 405
Matar P, 101
Mateo P, 25
Matern D, 305
Mathe E, 547, 553–55
Matheny SA, 671
Mathes R, 98
Matheson J, 547
Mathews CK, 697
Mathews PM, 618
Mathias G, 751
Mathies RA, 747, 751
Matias VR, 50
Matos J, 221
Matschinsky FM, 386, 388
Matson C, 262
Matsuda N, 96
Matsuda S, 617
Matsuda Y, 617
Matsukuma K, 662
Matsumine H, 357
Matsumoto K, 118
Matsumoto L, 572, 580
Matsumoto M, 191, 195–99, 201
Matsumoto S, 785
Matsumura F, 547, 557
Matsuno Y, 100
Matsuno-Yagi A, 70, 75, 80, 81, 86
Matsuoka T, 547
Matsushiro A, 144
Matsushita K, 70
Matsuura A, 120, 509
Matsuzaki H, 151, 439
Mattern KA, 509
Matteson S, 26
Mattevi A, 633
Matthews BW, 419, 536
Matthews CR, 413
Matthews JM, 340
Mattick JS, 404, 409, 419, 422–25

Matts RL, 284, 289
Mattson MP, 346, 618, 620
Matuliene J, 554, 558
Matzke R, 557
Matzuk MM, 379
Mauerer R, 322
Maurer F, 146
Maurer-Stroh S, 224
Maveyraud L, 536
Maxwell A, 276, 278
Maxwell M, 320
May C, 510
Mayer I, 730
Mayer MP, 283
Mayinger P, 719, 730
Maynard EL, 301
Mayne L, 344
Mayo MW, 31
Mayor JA, 734, 735
Mazel D, 605
Mazumdar A, 156
McArdle F, 385
McBee JK, 744, 747
McBride A, 152, 154, 156
McBride AA, 232
McBride OW, 615
McBride R, 547, 548
McBurney MW, 440
McCabe K, 313
McCabe NR, 251
McCahill A, 671
McCammon JA, 520
McCammon MG, 345
McCammon MT, 324
McCarthy A, 139, 147, 148, 156, 157
McCarthy M, 630
McCartney RR, 147
McCarty MF, 102
McClarty G, 685
McClay J, 78
McClellan M, 657
McCloskey JA, 642
McCollum D, 560
McConlogue L, 615
McCormick F, 96
McCowan B, 120, 122, 127
McCracken J, 170
McCullagh KJ, 26
McDermott A, 536
McDermott CJ, 613
McDonagh T, 444, 452, 461

McDonald C, 674
McDonald JF, 643
McDonald WH, 694
McDonald-Jones G, 620
McDowell JH, 745, 748
McEACHERN MJ, 111–35; 116–23, 125–30, 132
McElheny D, 522, 527
McEntee K, 13, 14
McGargill MA, 192
McGarrity TJ, 140–42
McGarry JD, 378, 379, 386, 388–90, 393
McGee KM, 555
McGeer EG, 618
McGeer PL, 618
McGill CB, 669
McGlade CJ, 667, 669
McGough A, 759
McGowen MM, 55–57
McGrath BC, 49, 635
McGrath PT, 215
McGroarty E, 314
McGroarty EJ, 43, 44
McGuinness M, 313
McGuinness MC, 324
McGuire EJ, 43
McInerny C, 694
McIntire T, 341
McIntire TM, 348
McIntire WS, 314
McIntosh JR, 547, 554
McIntosh MW, 118
McIntosh S, 630
McKinney JD, 683, 693
McKinsey TA, 24–27
McKittrick E, 261
McKusick VA, 138
McLachlan RS, 616
McLarty G, 685
McLaughlin M, 614
McLaughlin SH, 277, 283, 287
McLean CA, 357
McLennan AG, 297, 314
McMahon G, 102
McMahon MT, 343
McMahon N, 276, 277, 287
McMartin C, 646
McMurray MA, 436
McNabola A, 103
McNamara DJ, 101
McNaughton HJ, 640
McNew JA, 324

McNiven MA, 547
McNulty C, 61
McPhail T, 496, 509
McPhie P, 408
McPhillips MG, 232
McSorley T, 671
McSweeney S, 422, 423
McTigue MA, 102
McVey M, 121, 436
Meadows E, 27
Meaney DF, 617, 618
Means AR, 148
Meberg PJ, 618
Mechtler K, 219, 249, 251, 255
Meda P, 386
Medema R, 144, 145
Medema RH, 439
Medgyesy P, 782
Medina PP, 139, 140, 142
Medina R, 757
Medvedev AV, 391
Medvedev DM, 178
Medvedev ES, 175
Medvedik O, 436, 442, 444, 445
Meewan M, 215
Megee PC, 219
Mehenni H, 157
Mehler EL, 751, 758
Mehlhaff P, 631
Mehlhaff PM, 632, 640
Mehta S, 230, 231
Meier RJ, 533
Meijer EM, 533
Meijerink JP, 639
Meinhardt H, 485
Meinhardt SW, 76, 775
Meinke A, 220
Meisinger LL, 256, 260
Meister A, 639, 643
Melamud E, 423
Melandri BA, 70
Melek M, 500, 502, 504, 505
Melendez PA, 31, 32, 372, 373
Melhem ER, 316
Melia TJ, 759
Melia TJ Jr, 745
Melki J, 31
Melki R, 336, 346, 350
Mellinghoff IK, 100
Mendelsohn ME, 196, 202

Mendelson NH, 481
Mendoza C, 669
Mendrola JM, 146
Mendz GL, 76
Meneely P, 223
Meng EC, 750
Meng W, 667
Menon ST, 744
Mensah L, 642
Mensah LM, 642
Menshikova EV, 382
Mensink M, 385
Mentele R, 74, 79
Mercer B, 24, 25
Mercken M, 614
Meredith TC, 59
Merickel SK, 598
Merienne K, 663
Merkel JS, 349
Merkle HP, 336, 345
Merlini G, 344, 359
Mermoud JE, 255, 256
Meroveh SO, 536
Merrill AH Jr, 378
Merrill GF, 30
Mersiyanova IV, 613
Mertens G, 596
Merz A, 418
Merz G, 618
Meselson M, 112
Mesulam MM, 619, 620
Methe BA, 423
Mett H, 103
Metzger E, 324
Meuer J, 78
Meunier B, 169, 172, 177
Meuwis MA, 407, 410–13, 421, 422
Meydan N, 104
Meyer BJ, 217, 223–25, 546, 547
Meyer JE, 579
Meyer MR, 635
Meyer P, 272, 274, 277, 280, 283–85, 287, 630
Meyer SG, 378
Meyer T, 103, 613
Meyerhoff DJ, 619
Meyerson M, 496
Meyn L, 617
Mezzogiorno A, 358
Miake H, 344

Mian IS, 130, 496, 498, 500–3, 505–9
Michel B, 113
Michel H, 168–70, 173, 177, 178, 180, 182, 734, 770–72, 775, 776, 778, 781
Michel RN, 30
Michell BJ, 27
MICHIE KA, 467–92; 484
Michowski W, 282
Middleton-Zarka KA, 785
Mielke T, 755, 758
Mihalik SJ, 311–13, 324
Mikami B, 419, 422, 424
Miki T, 227, 550, 552
Mikkelsen ND, 226
Mikolajczyk S, 75
Milanesi E, 699
Milbrandt J, 436, 442, 460
Milburn CC, 145
Milburn JL Jr, 386, 390
Mildvan AS, 499
Miles EW, 408
Mileykovskaya E, 469
Millar AH, 70, 73
Millay DP, 25, 26
Miller AF, 338
Miller BG, 409
Miller C, 436, 439, 617
Miller CA 3rd, 286
Miller ES, 436
Miller GP, 522, 526
Miller JC, 674
Miller JL, 75
Miller KM, 494
Miller LA, 616
Miller MC, 496, 501, 503–5
Miller MT, 633, 634, 642
Miller NE, 700
Miller P, 273
Miller T, 245, 251
Miller VA, 100
Miller WT, 98, 661
Millett F, 170
Millett IS, 347
Milligan G, 758, 759
Milligan RA, 470, 674
Mills D, 170, 177
MILLS DA, 165–87; 166, 169, 172, 173, 175–80
Millson SH, 277
Milman HA, 631, 641

Milne L, 438
Milne TA, 252, 253
Milpetz F, 664, 667
Milton S, 341
Milton SC, 348
Mimnaugh EG, 273
Min J, 253
Min JR, 446–48, 450, 454, 457
Min L, 156
Min XS, 660, 662, 663
Minami M, 282
Minami Y, 275, 282
Minamide LS, 618
Mincer JS, 535
Minchin RF, 282
Minestrini G, 548
Mingorance J, 469, 473–75
Mingrone G, 383
Minke B, 744
Minokoshi Y, 371, 387
Minoshima Y, 553
Miranda E, 357
Mironov AA, 693
Mirzadegan T, 744, 745, 751, 752
MISHANI E, 93–109; 101
Mishima K, 101
Mishima M, 545, 547, 552, 553
Miska EA, 24
Misra LN, 641
Mitaku S, 744
Mitchell C, 284
Mitchell D, 169, 170, 177, 178
Mitchell DM, 170, 177
Mitchell GA, 297
Mitchell JR, 500, 503, 504
Mitchell P, 171, 775, 780
Mitchell PO, 24
Mitchell RM, 169, 171
MITCHISON TJ, 543–66; 472, 474, 487, 545, 547, 551, 553, 555, 557, 559
Mitina O, 98
Mitsuya T, 142
Mittoo S, 642
Miura M, 559
Miyagawa J, 392
Miyahara I, 633
Miyake C, 782, 785
Miyamoto H, 144
Miyanaga F, 372
Miyata T, 754

Miyata Y, 274
Miyazaki J, 392
Miyazaki K, 415, 416
Miyazaki M, 373, 385
Miyazaki Y, 221
Miyazawa S, 311
Miyoshi H, 83, 139, 140
Miziorko HM, 297
Mizoguchi A, 548
Mizrachi S, 102
Mizrahi V, 8
Mizuno K, 27
Mizusawa H, 344
Mlynarczyk-Evans SK, 255
Mo XL, 693
Moarefi I, 665, 667
Moazed D, 244, 248, 249, 253, 438–40, 442, 445, 447–50, 457, 461
Mochizuki N, 196, 202
Mochizuki S, 196, 202
Modak MJ, 501
Modler AJ, 348, 349
Modrich P, 11, 122
Mody N, 372, 373
Modzelewska A, 745, 746, 759
Moe D, 684
Moe E, 419, 421, 422, 424
Moechars D, 357
Moehle C, 322
Moffat K, 520
Mogensen M, 548, 559, 560
Mogk A, 272
Mohl DA, 230
Moir R, 442, 452–54
Moise AR, 745, 747
Moitra J, 372
Mokotoff M, 641
Molday LL, 745
Molday RS, 745
Molina DM, 662
Moliner V, 528
Molkentin JD, 25
Mollapour M, 277
Moller DE, 29, 371
Møller-Jensen J, 226, 228, 229, 233, 468, 477–81, 484, 485
Mollica MP, 382
Mollinari C, 547, 554, 558
Moloney MG, 640
Molzer B, 321, 324
Momany FA, 47

Momcilovic M, 148
Momen B, 423
Momoi M, 548
Momose H, 415
Monasterio O, 470
Mondragon A, 574
Monje-Casas F, 692
Monod J, 657
Montanelli L, 760
Montecchi-Palazzi L, 668
Monteiro MA, 49
Montero JL, 642
Monti M, 344, 356
Montuenga LM, 139, 140
Moody AJ, 169
Moody PCE, 632
Moomaw CR, 324
Moore GD, 474
Moore I, 373, 377
Moore J, 545, 546, 560
Moore JK, 117, 120
Moore K, 374
Moore KJ, 642
Moore MAS, 510
Moore P, 389, 393
Moore PC, 391
Moore S, 643
Moorhead GB, 664
Mootha VK, 375, 382
Mooyer PA, 320
Mora A, 148, 149, 152
Mora LB, 104
Moraga-Amador DA, 631
Morand OH, 305
Moras D, 289, 634
Moras M, 699
Moree CB, 553
Morel F, 302
Morel V, 745, 748
Moreno-Bueno G, 101
Moreno-Loshuertos R, 70
Morfini G, 618, 620
Morfini GA, 613, 617
Morgam DO, 284
Morgan DJ, 81, 167
Morgan DO, 669
Morgan J, 170, 177
Morgan JA, 103
Morgan JE, 82, 170–73
Mori H, 226, 229, 618
Mori S, 219, 221
Mori T, 252

Mori Y, 226
Moriarty TJ, 496, 500–8
Moriguchi T, 642, 662
Morii N, 547, 552
Morin GB, 495, 496, 500, 502
Morio-Liondore B, 379
Morishima Y, 289, 553
Morita M, 323
Moritz OL, 745, 748
Moroder L, 282
Morona R, 48–50
Moroz IA, 78, 83
Morozova-Roche LA, 347, 348
Morrell JC, 320
Morrice N, 143, 144, 156
Morrice NA, 143–46, 148, 152, 154–57
Morris AD, 151
Morris DK, 116, 496
Morris JA, 618
Morrison DK, 669, 671
Morrison JF, 643, 646
Morrow DM, 113, 115
Morten IJ, 348, 349
Morton DG, 147, 154, 546–48
Moser A, 320
Moser AB, 321, 323
Moser CC, 82, 181, 775, 776, 785
Moser CD, 585
Moser HW, 315, 316, 320
Mosier DR, 618
Moskowitz IP, 585, 596–98
Moslehi JJ, 347, 348
Moss DA, 782, 785
Moss DE, 620
Moss SE, 548, 555, 558
Mostoslavsky R, 440
Motamedi MR, 112, 116, 439, 440
Motley A, 301
Motley AM, 303, 311, 320
Motoshima H, 748, 749, 751
Motosugu F, 631
Motta MC, 436, 439, 440
Mouchou MA, 316
Mount LA, 393
Mouritsen OG, 711
Moussaif M, 759
Mouton R, 671
Mow BM, 97
Moya K, 617

Moya KL, 617
Moyens S, 426
Moynihan KA, 391, 436
Mozzarelli A, 520
Mrabet NT, 424
Mu J, 30, 151
Mucke L, 619
Mueller CL, 259
Mueller EG, 633
Mueller LJ, 459
Mufson EJ, 620
Muhammad S, 436, 446–48, 450, 454, 455, 457, 458
Muijsers AO, 303
Muir S, 282
Mukherjee A, 468, 469, 472, 474, 482, 485
Mukherji M, 311
Mulder E, 487
Mulder H, 386, 388, 389, 392
Mulders PFA, 103
Mulkidjanian AY, 775
Mullany P, 599
Müller A, 60
Muller C, 371, 387
Muller DJ, 760
Muller EGD, 699
Muller FL, 776
Muller HJ, 248, 630
Muller J, 254, 669, 671
Muller M, 103, 336, 345
Muller MT, 509
Muller SA, 341, 343, 345
Muller U, 617
Muller V, 718–20, 733
Mulliez E, 682, 690
Mulligan P, 262
Mullins JM, 558
Mullins RD, 478, 479, 665, 672
Munakata R, 301
Munch C, 613
Mundt K, 694
Munekage Y, 782, 785
Munn NJ, 305
Munoz DG, 619
Munoz MJ, 286
Munro PM, 760
Muntener M, 22
MUOIO DM, 367–401; 372, 373, 375, 377, 378, 380, 383, 384
Murakami S, 505, 506

Mural RJ, 657, 659, 664
Murali D, 101
Muramatsu Y, 508
Muramoto K, 168, 171
Muranyi A, 196, 202
Murata N, 775
Murata S, 282
Murata Y, 667
Murata-Hori M, 560
Murayama T, 22
Muresan V, 617
Muresan Z, 617
Murgia M, 25, 29
Murley LL, 586, 588, 593, 595
Murphy C, 341
Murphy CL, 336
Murphy DB, 619
Murphy M, 437
Murphy PJ, 289
Murray AW, 657
Murray CJ, 532
Murray J, 70, 73, 75
Murray JA, 231
Murray K, 245
Murrell J, 615
Murrey HE, 371
Murthy K, 555–57
Musacchio A, 217
Muschler P, 283
Muslin AJ, 662
Mustafi D, 484
Mustard KJ, 147, 148
Muth V, 436, 439
Mutlu LK, 614
Muzzin P, 391
Myers CE, 273
Myers D, 616
Myers EW, 657, 659, 664
Myers KA, 611
Myers R, 147
Myers RS, 633
Myint AT, 75
Mynatt R, 375, 382
Myslovati M, 683, 693
Myung K, 121, 217

N

Na E, 28
Nabholz M, 509
Nachliel E, 174

Nadal-Ginard B, 23
Nadaud S, 436, 439
Nadeau K, 274
Nadler-Yona C, 44, 47, 52, 53
Nagai K, 480
Nagan N, 303, 305
Nagane M, 101
Nagar B, 102
Nagarajan R, 573, 574
Nagasaki A, 544, 552, 557
Nagasawa Y, 386, 390
Nagase T, 316
Nagayama S, 196
Nagel-Steger L, 79, 80
Nagle JF, 167
Nagy PL, 251
Nahmias AJ, 15
Naiki H, 341, 344, 347
Nair SC, 273
Naismith JH, 50, 51, 53, 380
Nakada D, 118
Nakagawa M, 24
Nakagawa O, 24
Nakagawa S, 96
Nakagawa T, 613
Nakagawa Y, 613
Nakai Y, 631
Nakajima H, 392
Nakajima-Iijima S, 190, 198
Nakakuki K, 347
Nakama T, 642
Nakamaru-Ogiso E, 76, 86
Nakamura H, 509
Nakamura S, 548
Nakamura T, 196, 252, 260, 552
Nakamura TM, 116, 129, 496, 498
Nakamura Y, 617, 619, 684
Nakanishi K, 755
Nakanishi M, 156
Nakano K, 684, 693
Nakata Y, 98
Nakatsu T, 642–46
Nakau M, 139, 140
Nakayama J, 249, 251
Nakayama J-I, 509
Nakayama Y, 70
Nakayashiki T, 340
Namba M, 392
Namslauer A, 169, 170, 175–77
Nanda H, 711
Naor MM, 167

Napoli JL, 297, 314
Napoli L, 719
Narendra U, 574, 576, 587
Narinx E, 406, 408, 416, 420, 423–25, 427
Narita Y, 101
Narlikar JV, 404
Narumiya S, 547, 552, 553
Nash HA, 572, 580
Nash P, 659, 664, 667, 672
Nasmyth K, 217–19, 221, 222, 224, 236, 684
Natarajan K, 635
Natarajan S, 122, 123, 125, 126, 128, 132
Nath N, 147
Nathan DF, 272, 281, 285
Nathans J, 745
Nations L, 140
Natochin M, 759
Nattrass M, 384
Naumov G, 120
Naumova ES, 120
Nauser T, 302
Navarro J, 748, 753
Navon A, 31
Nawrot B, 573, 574
Naya FJ, 24, 25
Naylor ML, 113–15, 121
Nebreda AR, 660–62
Neckers L, 273
Neckers LM, 273
Neet KE, 520
Negrini S, 508, 511
Nelson B, 669
Nelson C, 262
Nelson DH, 378
Nelson DR, 718–20, 725, 726
Nelson JS, 147
Nelson KE, 423
Nelson M, 461
Nelson MC, 28, 31, 375
Nelson ME, 776, 785
Nelson N, 721
Nelson R, 341–43
Nelson S, 244, 248
Nelson WC, 436
Nemeth A, 417, 418
Nemoto T, 276, 277
Nenquin M, 388
Nesbit MG, 630
Nesper J, 51

Nesterova TB, 255
Netik A, 324
Nettleton EJ, 343, 345, 349
Neubauer H, 380
Neubig RR, 744
Neufeld TP, 548
Neufer PD, 21, 28
Neuhard J, 474
Neujahr R, 557
Neumann AA, 132, 494
Neumann B, 546–48
Neumann H, 614, 618
Nevins JR, 693
Nevo Y, 700
Newall AE, 255
Newatia A, 384
Newcomb JR, 102
NEWGARD CB, 367–401; 386, 389, 393
Newman CL, 51
Newman DG, 377
Newsholme EA, 378, 384
Nezu J, 138–40, 142, 144
Nezu JI, 139, 140
Ng DCH, 75
Ng HH, 251, 256, 259–61
Ng KK, 421
Ng WV, 423
Ng-Eaton E, 436, 439, 440
Nguyen A, 671
Nguyen AN, 662
Nguyen HD, 349
Nguyen HH, 684
Nguyen HO, 336
Nguyen P, 273
Nguyen T, 320, 548
Nheu T, 468
Ni L, 340
Nibbeling HA, 509
Nicchitta CV, 274
Nicholls A, 775
Nicholls P, 170, 177
Nicholson A, 122
Nicholson DW, 621
Nick HS, 636
Nickles K, 122, 127, 129
Nicolas A, 273
Nicolas-Frances V, 319, 321
Nicolay K, 322
Niederlander C, 145
Nielsen BB, 690
Nielsen BN, 427

Nielsen L, 336
Nielsen MS, 617
Nielsen O, 694
Nielsen SJ, 24
Nierenberg J, 620
Niesman IR, 613
Nieva J, 356
Niewiadomska G, 612
Nigg EA, 545, 550, 553
Niida H, 156
Niiya F, 550
Nijtmans LGJ, 70, 75, 86
Niki H, 213, 223, 226, 227, 229, 230
Nikolaev AV, 56, 57
Nikolaev AY, 439, 440
Nikolay R, 283
Nilsberth C, 357
Nilsson MR, 336, 356
Nilsson T, 170
Nimtz M, 41
Ninomiya T, 56
Nishida E, 481, 662
Nishigaki Y, 700
Nishikawa M, 546
Nishimura I, 155
Nishimura O, 44, 60
Nishimura Y, 552
Nishino I, 700
Nishioka T, 644
Nislow C, 547, 554
Niswonger ML, 547
Nitta M, 560
Nitta R, 611
Nittler MP, 669
Niu HW, 505
Nixon PJ, 73, 782, 785
Njoroge JM, 75
Nobbmann U, 485
Nocera DG, 691
Noda M, 386
Noel F, 718–20, 730, 731
Nogales E, 470, 471, 474
Noguchi S, 786
Noguchi T, 314, 556, 557
Noirot-Gros MF, 224
Nolan C, 389, 393
Nolan DP, 439, 445, 454
Noland BJ, 261
Noland RC, 377
Nolen B, 193
Nollmann M, 585, 588

Nolte LA, 28
Noma K, 249
Nomoto S, 139
Nomura F, 191, 196, 198, 199, 201
Nonaka I, 719
Nonaka S, 610
Nonaka Y, 156
Nonomura K, 546
Nony P, 146
Noorwez SM, 760
Noppe W, 348
Nordberg J, 558, 559
NORDLUND P, 681–706; 682, 685, 687, 690–92, 695–97, 700
Nordstrom K, 224, 229, 233
Norman D, 619
Norman DG, 144
Norris AW, 374
Norris D, 98
Norris SR, 640
North BJ, 438, 439, 461
North M, 640
Nosaka T, 547, 552
Nosek J, 130
Nov S, 44, 47, 52, 53
Novick RP, 130
Novikov D, 311
Novikov DK, 297, 309, 310
Novoa I, 635
Novotny-Diermayr V, 75
Nowak RJ, 349
Nowak SJ, 32
Nowotny M, 282
Nozaki M, 144, 251
Nsahlai C, 58, 62
Ntambi JM, 373, 385
Nugent CI, 116, 129
Nunes-Duby SE, 570, 572, 576–78, 580
Nunez L, 28, 31, 32
Nunomura A, 612, 619
Nureki O, 642
Nurse P, 233
NURY H, 713–41; 709, 715, 717, 726, 728, 736
Nusbaum C, 657, 659, 664, 665
Nussinov R, 408, 415, 419, 422, 423, 425
Nygaard P, 474

Nylen C, 23
Nyman PO, 12
Nyquist RM, 170, 172
Nystrom S, 336

O

Oakes ND, 380, 384
Oas TG, 425
Oatey PB, 301
Obadia B, 50
Obermann WMJ, 272, 274, 275, 282, 285
Obmolova G, 633
Obradovic Z, 417
O'Brien PJ, 640
O'Brien R, 272–74, 276–78, 281, 283, 287
O'Brien RT, 630
O'Carroll D, 245, 248, 249, 251, 255
Ochoa S, 5
O'Connell CB, 557
O'Connell MJ, 224
O'Connell NC, 316
O'Connor CM, 503, 504
Octave JN, 618
Oda J, 643–46
Oda T, 301
Odell GM, 557
O'Dowd BF, 548
Odoy S, 718–20, 734
Oegema K, 222, 546–48, 550, 555, 559
Oesch F, 302
Oestergaard PR, 418
Oesterhelt D, 224, 760
Oesterhelt F, 760
Oettgen H, 630
Oettl K, 315
Oettmeier W, 719
O'Farrell AM, 103
O'Farrell PH, 546–48, 550, 558, 559
Ofman R, 303, 312, 314
Ogawa A, 386, 390
Ogawa H, 117, 121
Ogawa T, 785
Ogawa Y, 22, 372
Ogawara H, 96
Ogilvie DJ, 102
Ogilvie RW, 22

Oglesbee D, 70, 73
Ogura T, 151, 156, 223, 226
Oh BC, 31, 32
Oh M, 26
Oh SJ, 613
Oh SY, 508
Oh W, 376
Ohad II, 786
O'Halloran TJ, 547
O'Hanlon PJ, 642
O'Hare T, 98
Ohba Y, 552
Ohe Y, 100
Ohguro H, 752
Ohi R, 548
Ohko K, 508
Ohman A, 348
Ohne O, 102
Ohnishi T, 70, 76, 78, 82–85, 775
Ohno S, 27, 97
Ohnuma T, 630
Ohshima M, 83
Ohshima T, 439
Ohsumi M, 301
Ohta T, 251
Oinonen C, 632
Okada K, 181, 642, 770–72, 776
Okada T, 744–49, 751, 753, 755
Okada Y, 256, 257, 610, 611
Okamoto I, 552
Okamoto M, 156
Okamura MY, 176, 177
Okamura Y, 667
Okazaki Y, 391
O'Keefe L, 547, 552
Okegawa Y, 782, 785
Okimoto RA, 100, 101
Okonkwo DO, 614
Okstad OA, 213
Oku A, 144
Okun JG, 83
Okun P, 78, 86
Okuno M, 547
Old LJ, 639
Oliva MA, 469–71, 473–76, 486
Olivares EC, 599
Olive J, 772
Olive M, 372
Oliveberg M, 170, 353
Oliveira JG, 232
Olivera BM, 10

Olivier LM, 315
Olkhova E, 178
Ollagnier S, 690
Olm V, 614
Olovnikov AM, 494
Olsen A, 336
Olsen DA, 635
Olsen DS, 635
Olsen RA, 459
Olsen RL, 421
OLSON EN, 19–37; 23–27, 30
Olson LP, 533
Olsson MH, 179, 180
Oltersdorf T, 615
Olufsen M, 419
O'Mahoney JV, 32
Omi R, 633
O'Neil KT, 426
O'Neill FJ, 441
O'Neill JC Jr, 413
Onishi Y, 144
Onn T, 43
Ono S, 557
Onoda T, 96
Onogi T, 215, 226, 227, 229
Ookawara T, 385
Oostheim W, 304, 313
Oppenheimer NJ, 446, 450, 452, 453, 459
Opperdoes FR, 322
Opperman KK, 504
Oprea T, 642
Oprian DD, 745, 753
Orak JK, 301, 302
Oral EA, 372
Orantes M, 616
O'Reilly M, 496
Orengo CA, 538
Organe S, 151
Orlando V, 244, 248
Orlova A, 480
Orlova E, 343, 345
Orokos DD, 611
Orr GR, 631
Orr HT, 613
Ørskov F, 40, 41, 43, 44
Ørskov I, 40, 41, 43, 44
Orth K, 324
Ortiz AI, 57
Ortu G, 101
Ory S, 671
Os V, 404, 409, 418–21, 423–26

Osada S, 27
Osborn AJ, 117
Osborn MJ, 49, 485
Osborne MJ, 526, 536
Osborne NF, 642
Osbourne N, 642
Osherov N, 96, 97, 99
Oshima M, 139, 140
Oshima T, 417
Ossipova O, 154
Ostaszewski BL, 348
Ostermeier C, 168
Osteryoung KW, 468
Ostman A, 103
Ostman J, 358
O'Sullivan RJ, 255
Osyczka A, 181, 775
Otero RC, 417
Otova B, 639
Otsu M, 548
Otsuki Y, 142
Otte AP, 255
Ottmann OG, 103
Otzen DE, 336, 345–47, 349, 353
Ouellette M, 499, 510
Oulton R, 496, 504, 505, 509
Ouyang Y, 776
Overall CM, 660
Overbeek E, 101
Overbeek P, 24
Owen BA, 287
Owens-Grillo JK, 282, 283, 289
Oyama F, 618
Oyler NA, 341
Ozcan U, 374

P

Pabo CO, 674
Pace CN, 520
Packard M, 618
Pacoma RL, 180
Padan E, 734
Paddock ML, 176, 177
Paddon-Jones D, 379
PAEK A, 211–41
Paez JG, 100
Pagacova B, 130
Pagani M, 255
Pagans S, 439, 461
Page CC, 776, 785

Page G, 195, 196, 198, 201
Page R, 357
Pahan K, 302
Pai KM, 630
Paik WK, 258
Paiment A, 47, 50–52
Pakaski M, 617
Pakkiri LS, 57
Pal C, 358
Palacios JM, 469, 475
Palacios MJ, 553
PALCZEWSKI K, 743–67; 711, 744–49, 751, 752, 757–61
Palenchar PM, 633
Pallafacchina G, 25, 26, 29
Palmer AG, 520
Palmer G, 169
Palmieri F, 322, 324, 700, 717, 733
Palmieri L, 322, 324, 717
Palmisano A, 734
Palsdottir H, 167, 181, 182, 775, 786
Pan D, 748, 754
Pan DA, 147, 148
Pan DJ, 149
Pan J, 309, 319–21
Pan Q, 100
Pan YA, 155
Pan Y-X, 635–38
Panaretou B, 272, 274, 276–78, 281, 283–85, 287
Panasik N, 404, 419
Panchenko LF, 314
Panda D, 470, 474
Pandey P, 507
Pandey VN, 501
Pang SS, 646
Pang WW, 440
Panico M, 418
Panneerselvam C, 383
Panni S, 668
Pannunzio NR, 599
Pao W, 100
Pap EH, 322
Papa S, 75
Papermaster DS, 745, 748
Papp B, 358
Papp H, 617, 620
Pappalardo L, 691, 692, 698
Pappenberger G, 418
Papst PJ, 27

Paquette RL, 103
Parak FG, 535
Parast CV, 102
Parazzoli SD, 391
Pardue ML, 113, 498
Parello J, 759
Parenteau J, 123
Park CW, 28, 375
Park DH, 531–33, 619
Park EC, 441
Park JT, 468, 472
Park PS, 758, 760
Park SH, 668, 672
Park SJ, 272
Park SM, 508
Parker KR, 758
Parker SB, 284
Parkins N, 745, 748
Parmar S, 104
Parmryd I, 315
Parodi AJ, 49
Parr IB, 640, 641
Parrella P, 139
Parrini C, 353
Parrini MC, 552, 667
Parry RJ, 632, 633
Parslow TG, 501, 505
Parsons DW, 155
Parsons SA, 25, 26
Partridge L, 440
Parul D, 169
Pascal G, 758
Pascarella S, 420, 422, 423, 425
Pasierbek P, 221
Pasokony J, 461
Pasquier F, 621
Passamonti F, 104
Pasta F, 230
Pastorakova A, 130
Pastuszyn A, 408
Patel AB, 748, 755
Patel Y, 251
Patkar S, 419
Paton VG, 315
Patterson BD, 245
Patterson MK, 631
Patti ME, 375, 379, 382
Paul AV, 10
Paul B, 495, 502
Paul P, 144
Paul S, 582
Paulsen IT, 49–51, 61

Paushkin SV, 355
Pavelka MS Jr, 58
Pavicic V, 545
Pavlath GK, 24, 26
Pavletich NP, 272–74, 278, 454
Pawar AP, 354, 355
Pawate AS, 170, 177
Pawlyk BS, 760
Pawson T, 659, 664, 665, 667, 669, 671, 672
Payan F, 410, 424, 427
Payne DJ, 482
Pays E, 439, 445, 454
Pays L, 195, 196, 198
Pazzani C, 59–61
Peak-Chew SY, 71, 75, 81
Pearce KH, 482
Pearce R, 55, 61
Pearce-Kelling SE, 760
PEARL LH, 271–94; 272, 273, 277, 278, 284, 287
Pearl S, 503, 504
Pearle A, 104
Pearson M, 440
Pease PJ, 214, 234
PEBAY-PEYROULA E, 713–41; 709, 715, 717, 722, 723, 726–28, 736
Peck A, 616
Pecoraro C, 169, 178
Pedal A, 439, 461
Pedersen JS, 336, 345–47, 349
Pedersen K, 226
Pediani J, 759
Pedulla ML, 582
Peggie M, 283
Pegoraro S, 282
Peitz M, 579
Peleg A, 44, 47, 52, 53
Pelicci PG, 197, 199
Pelkonen S, 56
Pellarin R, 354
Pelled D, 190, 194, 197, 198
Pelletier JN, 535
Pelletier L, 546, 548
Pelletier N, 24
Pelleymounter MA, 370, 372
Pellicena P, 98, 102
Pellicioli A, 117, 118, 121
Pellman D, 560
Pelmenschikov V, 687, 688, 690

Pelosi L, 48
Pelsman A, 618, 620
Peltier G, 782, 785
Peltonen L, 632
Penfound T, 442
Peng G, 74, 79
Peng Y, 495, 496, 498, 500–5, 507–9
Penke B, 620
Penn JS, 745
Pennaneach V, 118
Pennock E, 128, 508, 511
Pepperberg DR, 744, 747
Pepys MB, 336, 341, 359
Perales M, 73
Peranen J, 745, 748
Perard J, 554, 558
Perdew GH, 282, 283, 284, 289
Perdomo D, 757
Perego M, 86
Pereira G, 560
Pereira MM, 180
Perepelov AV, 613
Perez LS, 104
Perez R, 618
Perez RG, 617
Perez-Martin J, 436
Perez-Martos A, 70
Periole X, 751, 758
Perkins GR, 671
Perlin JR, 262
Perlman ZE, 545–48
Perlstein DL, 684, 690, 694
Permutt MA, 30
Peroni O, 372, 373, 391
Peroni OD, 371–73, 387
Perraud AL, 441
Perret P, 391
Perrimon N, 545–48
Perrin C, 379
Perrini S, 27
Perrino BA, 671
Perry G, 619
Perry MB, 49
Perseghin G, 377
Persons PE, 96, 101
Persson AL, 687, 689
Persson B, 353
Perutz MF, 344, 345, 520, 657
Pessayre D, 383
Pessin JE, 373, 377
Pessotto P, 383

Peteanu LA, 747
Peter ME, 156
Peters AHFM, 251, 255
Peters JM, 220, 221, 550
Peters JW, 76
Petersen GM, 138
Petersen K, 377
Petersen KF, 382
Peterson AA, 43, 44
Peterson CA, 26
Peterson JR, 545
Peterson RG, 390
Peterson SE, 256, 260
Peterson VJ, 439, 440, 448
Petes TD, 113, 121, 128
Petit A, 617
Petit E, 302
Petkova AT, 341, 346
Petrache HI, 711
Petrick JS, 348
Petrini JH, 117, 121
Petroff T, 229
Petronczki M, 221
Petroni G, 475
Petrunka C, 357
Petruzzelli S, 482
Petsko GA, 536
Pette D, 21
Pettersson G, 530, 531
Pettijohn DE, 215
Petty SA, 348
Peutz JLA, 138, 157
Pfaff E, 716
Pfanner N, 75
Pfeiffer K, 70
Pfeiffer M, 174, 760
Pfeiffer NE, 632, 640
Pfitzner U, 177
Pfleger CM, 693
Phatnani HP, 254
Phelps A, 735
Philipps G, 693
Phillips DR, 642
Phillips GN Jr, 421, 422, 425
Phoenix DA, 61
Phoenix P, 474
Piano F, 546–48
Picard D, 276
Picard F, 436, 441, 444, 460
Pichoff S, 482, 483, 485
Pickersgill RW, 407
Pico A, 102

Picot D, 770, 778, 783–86
Piekny A, 553
Piel M, 558, 559
Pierce DW, 611
Pierce KL, 744, 745
Pierik AJ, 775
Pierre Y, 772, 784, 786
Pieters R, 630, 638, 639
Piggot PJ, 786
Pigino G, 618, 620
Pigon J, 379
Pijnappel WW, 251
Pike CJ, 357, 620
Pilch D, 117
Pilegaard H, 28
Pimplikar SW, 617
Pin JP, 744
Pineda VL, 636
Pines J, 24, 550
Pinkner JS, 339, 340
Pinko C, 102
Pio R, 139, 140
Piper PW, 273, 277, 278, 287
Piras C, 84
Pirozhkov SV, 314
Pirrotta V, 244, 248
Piston DW, 761
Piton A, 302
Pitsi D, 618
Piwnica-Worms H, 660, 671, 674
Pla J, 482
Plakoutsi G, 348–50
Plapp BV, 530–33, 535
Plas DR, 151
Plasterk R, 303, 311
Platanias LC, 104
Plath K, 255
Plé PA, 102
Ploegaert I, 320
Plottner O, 191, 195–97, 201
Plueger MM, 391, 436
Plug A, 14
Pluskey S, 668
Podell ER, 500, 507
Podlisny MB, 348
Pogliano J, 223, 229, 235
Pogliano K, 223
Pohnert SC, 28
Pohorelic B, 508
Poitout V, 383, 386, 390, 391, 393

Pokholok DK, 259
Pokkuluri PR, 349
Polaina J, 416
Polak-Charcon S, 190
Polancic JE, 219
Polans A, 745
Polevoy G, 548
Politi KA, 100
Poljak A, 406, 410–12, 415, 416, 424, 427
Pollard TD, 545, 549, 556
Polverino A, 669
Polyakov AV, 613
Pomara N, 620
Pomes R, 170, 172, 180
Pon NG, 770–72
Ponamarev MV, 772, 780, 781
Pongracz K, 502
Ponniah S, 75
Pontarin G, 700
Ponting CP, 664, 667
Pontis E, 683, 684, 698
Poo M, 758
Poole AM, 682, 701
Poon R, 156
Pope AJ, 642
Poplawski A, 693
Popot J-L, 770, 778, 783, 785, 786
Popovic DM, 170, 180
Popp D, 476, 477, 481
Populo H, 78
Poremba C, 131
Pories WJ, 30, 31
Portelius E, 341, 353, 355
Porter AC, 635
Porter CW, 314
Porter SE, 259
Posakony J, 461
Posas F, 669
Poser I, 546, 548
Posner I, 96, 97
Posokony J, 461
Possoz C, 236
Post CB, 535
Post PL, 555, 557
Postis V, 718–20
Potluri P, 75
Potrebic S, 618
Potsch S, 691
Potter VR, 715
Poueymirou WT, 28

Poulain P, 383
Pouleur AC, 148, 149
Poulos A, 303
Poulsen P, 382
Poulter RT, 572
Poupart MA, 219
Pourhossein M, 56, 61
Poussin C, 380
Poutanen MH, 309
Powell DR, 684
Powell KA, 694
Powers DA, 405
Powers ET, 349, 356
Powers JM, 321
Pozniakovski A, 548
Prager MD, 638
Prapapanich V, 273
Praprotnik D, 619
Prasad VR, 504
Prasanth KV, 548
Prasanth SG, 548
Pratipanawatr T, 390
Pratt EA, 9
Pratt LR, 710
Pratt WB, 272, 282–84, 289
Prefontaine GG, 262
Prehoda KE, 665, 672
Preiss J, 442
Preitner F, 372, 373
Premkumar A, 372
Premont RT, 744, 745
Prentki M, 386, 388, 389, 391, 393
Prescott AG, 311
Prescott J, 502, 504–6
Pressler M, 174, 175, 177
Pressman BC, 715
Preuss U, 155, 195–98, 201, 204
Prevost C, 498
Price DL, 615, 618, 619
Price NP, 47
Pridmore D, 643
Priess JR, 154
Prieto-Alamo MJ, 692
Prigeon RL, 390
Primak A, 784, 785
Prince T, 289
Principe S, 344
Printen JA, 669
Prior K, 58, 62
Prior L, 547, 552
Privalov PL, 410, 415, 421, 422

Probst M, 322
Prochaska LJ, 175
Prochiantz A, 617
PRODROMOU C, 271–94; 272–74, 276–78, 280, 281, 283, 285, 287
Proenca R, 370, 372
Prokopenko SN, 547, 552
Prommeenate P, 73
Proteau G, 177
Prowse KR, 500
Prutsch A, 169, 172
Pruzan R, 499, 502
Pryde FE, 121
Ptacin JL, 214, 234
Ptashne M, 657
Ptasznik A, 98
Ptock A, 70
Pu JZ, 533
Pueyo C, 692
Pugieux C, 442
Pui CH, 630, 639
Puigserver P, 28, 375, 441
Pujol A, 319, 321, 323, 324
Pulido E, 579
Pullen J, 630
Puls I, 613
Purdue PE, 301, 313
Putnam CD, 217
Putz G, 546, 548
Puustinen A, 82, 169–73
Puype M, 308

Q

Qi C, 309, 319, 321
Qi M, 671
Qian J, 169, 170, 177
Qian M, 410, 424, 427
Qian XH, 577
Qin L, 168, 169, 173, 176, 178–80
Qin XL, 194
Qin YM, 309
Qiu WQ, 617
Qu W, 47
Quada J, 500
Quail MA, 78
Quardokus EM, 482, 483
Quinlan M, 615
Quinn AM, 96
Quinn B, 770, 781

Quinones B, 559
Quintas A, 348
Quintens R, 389
Quisel JD, 484, 485
Quivy JP, 251
Quong AA, 439, 440

R

Raab M, 669
Raabe M, 310, 321
Raats J, 555
Rabbitts TH, 251
Rabindran SK, 101
Rabitsch KP, 221, 222
Race RE, 358
Rackwitz HR, 303, 304
Radanyi C, 282
Radding CM, 7
Rader AJ, 760
Radermacher M, 70, 73, 74, 76, 79–81, 83
Radford SE, 346, 348–50
Radhakrishna U, 157
Radivojac P, 417
Radkiewicz JL, 528, 535
Radmacher M, 557
Radman M, 122
Radman-Livaja M, 574, 577, 578, 580, 600
Radnedge L, 226, 484
Radzicka A, 634
Radzio-Andzelm E, 144
Raedle J, 142
Raetz CR, 305, 315
Raetz CRH, 46–49, 60
Rafecas FJ, 630
Raffen R, 349
Rafii S, 510
Raftery M, 404
Rafty LA, 441
Raha S, 75
Rahman A, 610
Rahn A, 45, 46, 53
Rainey D, 73, 74
Rajagopalan PTR, 521, 524, 528, 529, 532, 533
Rajagukguk S, 170
Rajan RS, 758
Rajaratnam P, 404
RajBhandary UL, 759
Rajewsky K, 579

Rajoka MI, 407, 411
Raleigh DP, 356
Ramakers-van Woerden NL, 639
Ramaswamy S, 530–33, 535, 685, 694
Ramazzotti M, 353
Ramirez RM, 113
Ramirez-Alvarado M, 349
Ramon E, 750, 760
Ramos L, 754
Rampazzo C, 699, 700
Ramponi G, 349, 353, 354
Ramsay D, 758
Ramsay RR, 380
Ramsey CS, 439, 444
Ramström O, 642
Rancourt D, 745
Randall WR, 24, 27
Randle PJ, 378, 384
Ranganathan R, 524, 525
Rangarajan M, 418
Rangnekar V, 196, 198, 201
Ransom N, 745, 748
Rao A, 25
Rao J, 272, 286
Rao JK, 748
Rape M, 550
Rapoport TA, 611
Raposo G, 339, 340
Rappaport F, 784, 785
Rappaport R, 544, 549, 550, 553
Rashid A, 140
Rashid DJ, 611
Rashid MH, 411
Raskin DM, 484
Rasmussen BB, 379
Rasmussen HB, 422, 423
Rasmussen RK, 545, 546, 560
Rasmussen T, 78, 82, 83
Ratajczak T, 282, 283
Ratkevicius A, 25
Ratner N, 618
Rauch C, 302
Raushel F, 633
Raushel FM, 633, 634
Ravasio S, 634
Raveh T, 190, 194, 197–99, 204
Ravelli RBG, 474
Ravier MA, 388
Ravnskjaer K, 391
Ray K, 558

Ray SS, 349
RayChaudhuri D, 468, 472, 485
Raychowdhury MK, 27
Raymond GJ, 358
Raymond GV, 316
Rayssiguier C, 122
Rayter S, 155
Rayter SI, 145
Razidlo GL, 671
Rea S, 245, 248, 249, 251
Reach G, 390
Reaven GM, 27
Rebar EJ, 674
Recchia GD, 582
Rech J, 236
Reddel RR, 131, 132, 494
Redeker V, 336, 346, 350, 783
Redick S, 560
Redick SD, 473
Redinbo MR, 574
Redmond TM, 745
Redwine JM, 617
Reed J, 341
Reed JC, 196
Reed MJ, 27
Reed RR, 572, 585, 593
Reed SI, 128
Reedy M, 469, 474
Rees DC, 484
Rees JE, 316
Reeve S, 617
Reeves PJ, 748, 755, 759
Reeves PR, 47–49
Regan JW, 759
Regan L, 283, 349
Regard JB, 618
Regazzi E, 97
Regazzi R, 388
Rege N, 501
Reggiani C, 21
Reglero A, 57
Regnström K, 685, 691, 692, 694, 698
Rehbein K, 74
Reich MF, 101
REICHARD P, 681–706; 682–85, 687, 690, 692–700
Reichenbach P, 509
Reid AN, 41, 44, 50, 51
Reid E, 613
Reid JL, 147, 148
Reid T, 547, 556

Reidl J, 49
Reijns M, 581
Reilly JF, 617
Reiman EM, 617
Reimann H, 138–40, 142
Reimann SA, 313
Reinberg D, 245, 249, 254, 256, 259
Reinert J, 635
Reinhart BJ, 249
Reinke V, 546–48
Reinstein J, 274, 283
Reisenauer A, 233, 480, 481
Reiser MF, 619
Reisert J, 744
Reisine T, 617
Reiss-Sklan E, 102
Reitman ML, 372
Reixach N, 358
Reizer A, 58
Reizer J, 58
Relini A, 349
Relling MV, 630, 639
Remacle C, 70, 73, 74, 76, 81
Remani S, 735
REMÉNYI A, 655–80; 657, 660, 662, 663, 671
Renaud JP, 289
Rendell AP, 528
Renstrom E, 389
Rentier-Delrue F, 416, 421, 424, 426
Reoma JL, 143, 144
Resta N, 140, 145, 146, 154, 157
Reuber BE, 313
Reuter G, 244, 248, 249
Reuwer JF, 532
Reverse D, 357
Revollo JR, 442, 444
Rewcastle GW, 101
Reyes CL, 59
Reyes-Spindola JF, 700
Reynard G, 661
Reynolds RC, 474
Rhodes CJ, 391
Rhodes D, 496
Rial E, 735
Ribacka C, 170, 173
Ribar TJ, 27
Riberty M, 50
Riccio P, 734
Rice A, 102

Rice JC, 249, 251
Rice LM, 641
RICE PA, 567–605; 573, 574, 576, 577, 587, 593, 595, 600
Rich BE, 104
Rich PR, 83, 169–72, 177, 181, 775, 776, 778, 782, 786
Richard S, 245, 258, 619
Richards EJ, 244, 245, 248
RICHARDS NGJ, 629–54; 630–34, 638, 640–43, 646
Richardson CM, 642
Richardson DA, 500, 508
Richardson DC, 426
Richardson JA, 21, 24–26, 372
Richardson JR, 86
Richardson JS, 425, 426
Richey PL, 619
Richmond C, 692
Richmond RK, 245, 256
Richmond TJ, 245, 256
Richter C, 545–48
Richter H-T, 167
Richter K, 283, 287
Richter O-MH, 169, 178
Rick PD, 40, 41, 43, 44, 49
Rickert KW, 535
Rico AI, 482
Ridderstrale M, 382
Ridet JL, 357
Ridge KD, 744, 759
Rie C, 152
Riebeling C, 190, 194, 197, 198
Riedel D, 344
Riedy MF, 547
Rieke F, 745
Riekel C, 341–43
Riely GJ, 100
Riesner D, 336
Rigg GP, 61
Riggs B, 547
Riggs CD, 223
Rigler R, 419
Riise BW, 422, 423
Riise EK, 419
Riles L, 340
Rill RL, 244
Rimerman RA, 273, 282
Rimon A, 734
Rinaldo P, 305
Rincon J, 30
Rine J, 245, 437, 438

Rink R, 339, 340
Riparbelli MG, 547, 548
Rist W, 283
Ristow M, 374
Ritchie KB, 121, 128
Ritov VB, 382
Ritter C, 341, 342, 355
Ritter E, 751, 753, 760
Rittinger K, 667
Ritzel R, 390
Rivas G, 469, 473, 475
Rivera MA, 501, 505, 506
Rizavi HS, 736
Rizki A, 121, 122
Rizza RA, 390
Rizzo NW, 341
Roach PJ, 663
Robbins JB, 44, 60
Robbins JD, 43
Robert B, 786
Robert F, 259–61
Roberts AG, 776, 779
Roberts C, 635
Roberts CR, 27
Roberts CW, 13, 684
Roberts EA, 610, 611, 617
Roberts GCK, 528
Roberts I, 58–60
Roberts IS, 41, 54–56, 59–61
Roberts JW, 13, 684
Roberts TM, 674
Robertson DA, 384
Robertson RP, 386, 390, 393
Robidoux J, 391
Robinson BH, 75
Robinson CS, 32
Robinson CV, 343, 345, 347, 349
Robinson DN, 559
Robinson DR, 102
Robinson EJ, 470
Robinson LS, 339, 340
Robinson MO, 500, 502–7, 509
Roblin P, 536
Roca I, 693
Rocha EPC, 498
Rochaix J-D, 785
Roche D, 251
Roche E, 386, 390, 393
Rock JM, 674
Rockenstein E, 357, 611, 618–21

Rod TH, 528
Rodbell M, 759
Rodemer C, 320
Rodger A, 474
Rodgers JT, 439, 441
Rodionov DA, 693
Rodnick KJ, 30
Rodriguez ML, 43
Rodríguez-Aparicio LB, 57
Rodriguez-Perales S, 139, 140, 142
Rodriguez-Trelles F, 682, 683, 700
Roduit R, 389, 393
Roe C, 308, 312, 316
Roe DS, 308, 312, 316
Roe SM, 272–74, 277, 278, 280, 283–85, 660, 663
Roeder GS, 113
Roediger B, 620
Roegiers F, 194, 550
Roehlen D, 75
Roepstorff P, 229, 479
Roerdink AR, 642
Roeske RW, 663
Rogaev EI, 615
Rogaeva EA, 615
Rogers J, 620
Rogers SL, 546, 547, 557
Rogina B, 444
Rognes SE, 633
Roguev A, 251, 253
Rohatgi R, 665, 672
Rohde M, 41
Roher AE, 348, 357
Rohr A, 75
Rohr TE, 57
Rohrig UF, 747
Rohrschneider LR, 667
Roiniotis J, 284
Roizman B, 15
Rojo F, 101
Romagni JG, 641
Romano A, 548, 553, 558
Romberg L, 470, 473, 474
Rome LH, 509
Romeo T, 41
Romero DP, 504
Romeyn GJ, 313
Romling U, 41
Rommel C, 28
Ron D, 635

Ronnebaum SM, 389
Ronnstrand L, 96, 97, 103, 104
Rontani JF, 314
Rood JI, 599, 600
Roos JM, 154
Roovers M, 417
Rorsman C, 96, 97, 103, 104
Rortveit T, 308
Rosa J, 558, 560
Rosa P, 572
Rose AJ, 27
Rose GD, 426
Rose IA, 634
Rose P, 146
Rose PE, 102
Rosen ED, 374
Rosen N, 273, 278, 289
Rosenberg MF, 485
Rosenberg P, 24, 25
Rosenberg SM, 112, 116
Rosenblat M, 315
Rosenblatt J, 555
Rosenbusch JP, 728
Rosenow C, 59, 61
Rosenshine I, 52
Rosenthal N, 30
Rosenzweig AC, 633, 634, 642, 684, 690
Rosfjord EC, 101
Roshick C, 685
Roskams AJ, 621
Rosner B, 760
Ross KE, 220
Ross L, 148
Ross LJ, 474
Ross LO, 129
Ross SR, 372
Ross W, 572
Rossel M, 146
Rosser EM, 744
Rossi DJ, 139–41, 152
Rossignol JM, 13
Rossman R, 28
Rossmanith T, 182, 770–72, 776, 778, 781
Rossmann MG, 454, 461
Roth R, 81, 339, 340
Roth S, 138–40, 142, 157
Roth WK, 142
Rothermel B, 24, 25
Rothfield L, 468, 472, 481, 484, 485

Rothfield LI, 485
Rothlisberger U, 747
Rothschild KJ, 173
Rothstein R, 694, 699
Rotte C, 684
Rottensteiner H, 322, 324, 717
Rottier MMA, 638
Rottner K, 667
Rotwein P, 25
Rouault C, 383, 390
Rouault JP, 146
Rougeulle C, 255
Rounova O, 104
Rousseau F, 355
Rousseau V, 25
Rouvinen J, 632
Roux P, 718–20
Roux PP, 197
Row RH, 75
Rowe S, 55
Rowe WB, 643
Rowland S-J, 586, 591, 595, 596
Rowland SL, 485
Rowley JD, 245, 251, 257
Roy F, 146
Roy J, 123, 503, 504
Rozen Y, 101
Rozenblatt-Rosen O, 253
Rozovskaia T, 252
Rozovsky S, 536
Rub U, 616
Rubach JK, 532
Rubin DG, 614
Rubin GM, 194, 548, 657, 659, 664
Rubingh DN, 415, 416
Rubins N, 387, 393
Ruchaud S, 548
Rudd CE, 669
Rudd S, 785
Ruderman NB, 147, 372, 375, 376, 379
Rudiger S, 273
Rudnicka-Nawrot M, 752
Rudnicki MA, 20
Rudt F, 596
Rueda S, 469, 474
Rufibach M, 84, 85
Ruitenberg M, 169, 170, 173
Ruiz E, 372
Ruiz N, 62
Ruiz T, 79–81, 83

Ruiz-Lozano P, 297
Runswick MJ, 75, 86, 714, 717
Rupp H, 383
Ruprecht J, 755, 758
Ruprecht JJ, 755, 758
Rusan NM, 553
Rusche LN, 245
Rush JS, 49
Rusnak N, 297
Russell AP, 385
Russell DW, 305
Russell NJ, 426
Russell P, 694
Russell RJ, 409, 412, 415, 418, 419, 421–26
Russo AA, 272–74, 278
Rutenberg AD, 485
Ruther K, 745, 748
Rutten H, 26
Ryan MT, 736
Rybkin II, 26
Rycovska A, 130
Ryde U, 533
Rydel RE, 348
Rydell RE, 348
Ryder JW, 30
Ryu EH, 674

S

Saada A, 700
Saafi EL, 343, 345
Sabanay I, 190, 195, 198–201
Sabbagh W Jr, 662
Sablonniere B, 621
Sabripour M, 140–42
Sacchettini JC, 349
Saccone R, 375, 379
SACHS JN, 707–12; 711
Sackett MJ, 482
Sackmann U, 75, 76
Sacksteder CA, 775
Sacksteder KA, 300
Sadelain M, 510
Sadoski RC, 170
Sadowski PD, 579
Sadun AA, 619
Saechao L-K, 770–72
Saenger W, 770
Saez C, 484
Saffiotti U, 615
Safina B, 104

Safo MK, 598
Saghatelian A, 647
Sagiya Y, 684
Sah J, 116, 496
Saha AK, 147, 371, 372, 375, 376, 379
Saha D, 556
Saharinen P, 146
Sahlin M, 682, 687, 689
Saibil H, 758
Saibil HR, 343, 345
Saido TC, 27
Saier MH Jr, 58
Saier MHJ, 49–51, 61
Saiki M, 336, 343
Saint R, 547, 553
St George-Hyslop PH, 618
St Johnston D, 144, 147, 154
St Pierre J, 375
Sainz G, 781
Saito H, 662, 669, 672
Saito M, 509
Saito S, 440
Saitoh S, 694
Saitoh T, 617
Sakagami H, 192
Sakai H, 481
Sakamoto H, 100
SAKAMOTO K, 137–63; 83, 139, 147–49, 152, 156, 157
Sakmar TP, 744
Saksouk N, 48
Sakuraba H, 314
Sala I, 683
Salahpour A, 758
Salbaum JM, 615
Saleh M, 621
Saleh MC, 391
Saleh OA, 214, 234
Salehi A, 618
Salerno JC, 85
Saliba RS, 760
Salic A, 548
Salidas S, 617
Salinger AP, 129
Salitra Y, 99, 102
Salles-Passador I, 272, 282
Salmivirta M, 55
Salmon DP, 615
Salmon ED, 469, 474, 549, 553
Salmond GP, 482
Salmons S, 22

Salomonsson L, 174
Salovaara R, 139, 152
Saltiel AR, 377
Saltin B, 28, 378
Salvo JJ, 593
Salyers AA, 574, 579
Salzberg A, 547
Salzberg SL, 423
Samama JP, 530, 532
Sambandam N, 379, 383
Samelson LE, 669
Sameshima M, 276, 277
Sammarelli G, 97
Sampath AP, 745
Sampath SC, 548
Samudrala R, 475
Samuels Y, 155
Sanchez H, 25
Sanchez ML, 533
Sanchez R, 417
Sanchez-Cespedes M, 139, 140, 149, 151
Sanchez-Martin RM, 642
Sandberg MA, 760
Sandell LL, 117
Sander C, 226, 476, 480, 482
Sanders ER, 585, 588, 591
Sandhoff K, 378
Sandin E, 691
Sandmeier JJ, 442, 454
Sandri C, 31
Sandri M, 31
Sands Z, 667
Sanejouand YH, 710
Sanes JR, 155
Sanjo H, 191, 195–99, 201
Sanjo N, 617
Sankaran K, 49
Sansom MSP, 667, 711
Santarella RA, 470
Santolamazza C, 547, 555
Santona A, 482
Santoro SW, 576
Santos-Rosa H, 260
Santra MK, 470, 474
Sanyal M, 253
Sanz GF, 630
Sanz MA, 630
Sanz-Aparicio J, 416
Saoudi Y, 554, 558
Sapag A, 322, 323
Saperstein DA, 757–59

Sapkota G, 140–42, 145
Sapkota GP, 143–45
Saqib AA, 411
Sarabia MJF, 694
Sarafianos SG, 501
Saraiva MJ, 358, 359
Saraiva MJM, 348
Saraste M, 170, 175
Sardanelli A, 75
Sardanelli AM, 75
Sargent DF, 245, 256
Sarkis GJ, 582, 585–88, 590, 591, 595, 598, 600
Saroff D, 357
Sarov M, 253
Sarraf P, 374
Sarti P, 169
Sartorelli AC, 693
Sasaki M, 667
Sasaki T, 642
Sasaki Y, 96, 436, 442, 460
Sather H, 630
Sato M, 309, 310
Sato R, 324
Sato S, 274, 289
Sato T, 633–35
Sato Y, 388
Satpute-Krishnan P, 617
Satterberg B, 669
Sattler W, 315
Sauer S, 251
Saunders KB, 674
Saunders NF, 404, 409, 419, 422–25
Saunders W, 560
Saunders WH Jr, 532
Saupe S, 340
Saupe SJ, 340
Saura CA, 618
Sausville EA, 546
Sauter M, 75, 78
SAUVE AA, 435–65; 439, 440, 442, 446–48, 450, 452–54, 458, 459
Savikhin S, 786, 787
Savitha S, 383
Savoian MS, 547, 550, 553–55, 559
Savolainen K, 297, 310, 319, 321
Savolainen TI, 319, 321
Sawa G, 786
Sawa T, 96

Sawada H, 508
Sawada J, 262
Sawai T, 555
Sawaya MR, 341–43, 521
Sawicki GJ, 347, 348
Sawitzke JA, 223
Sawyers CL, 98
Saxton WM, 618
Sayegh J, 262
Saynovits M, 775
Sazanov LA, 70, 71, 79–82, 85, 167, 782
Scacco SC, 75
Scandurra F, 169, 178
Scarcia P, 322, 324, 717
Scarpulla RC, 75
Schaad LJ, 532
Schachman HK, 7
Schachner M, 319, 321
Schadel SA, 757, 760
Schaefer CB, 114
Schaffer AA, 572
Schaft D, 251, 253
Schägger H, 70, 73, 74, 79, 83, 786
Scharenberg AM, 441
Scharer-Shuksz M, 548
Schechter R, 616
Scheel C, 131
Scheffers DJ, 468, 471, 474
Scheffler IE, 75
Scheffzek K, 442
Scheibel T, 274, 276
Scheide D, 78–80, 82, 84
Scheidtmann KH, 191, 195–98, 201, 204
Schekman R, 7, 13
Schele A, 224
Schellenberg GD, 618
Schenk D, 619
Scher M, 439, 440, 461
Scherf A, 498
Schermerhorn T, 386
Scherthan H, 251
Schertler GF, 748, 751, 754, 755, 758
Schetter AJ, 546–48
Scheufler C, 282
Scheuring J, 450
Schiaffino S, 21, 29
Schickel R, 156
Schiebel E, 560

Schieltz D, 260
Schild S, 49
Schildkraut CL, 9
Schillace RV, 671
Schilling B, 75
Schiltz RL, 439, 448
Schimmel P, 642
Schindler T, 98, 102
Schinner F, 404
Schioth HB, 744
Schjerling P, 25
Schlame M, 718, 720, 726, 736
Schlarb BG, 772, 781
Schlegel W, 759
Schleiffer A, 219, 221, 222, 224
Schlessinger J, 664
Schlick T, 536, 710
Schlieper D, 475, 476
Schlitt A, 73, 78, 79
Schmerl S, 745, 748
Schmid C, 84, 85
Schmidt B, 166, 170, 176, 178–80
Schmidt FJ, 230
Schmidt MA, 43, 44
Schmidt MC, 147
Schmidt MT, 441, 446, 450, 452, 453, 459
Schmidt PP, 698
Schmidt S, 664, 667
Schmiede A, 75
Schmieder P, 74
Schmittschmitt JP, 353
Schmitz W, 310, 319, 321
Schmitz-Peiffer C, 378
Schmoll HJ, 152
Schnabel H, 548, 553, 558
Schnabel R, 548
Schnackerz KD, 520
Schnaider T, 274
Schnapp BJ, 611
Schnapp G, 545, 546
Schneerson R, 54
Schneider BL, 357
Schneider C, 273, 278
Schneider D, 786
Schneider E, 255
Schneider G, 646, 648
Schneider GE, 617
Schneider J, 245, 246, 251, 252, 259–61
Schneider KF, 43

Schneider MB, 151
Schneider MF, 24, 27
Schneider R, 75, 260, 262
Schneider-Mergener J, 668
Schnell J, 526, 536
Schnell JR, 521, 522, 526, 527
Schobert B, 167, 168, 173
Schoehn G, 547, 554
Schoenberg SO, 619
Schoenlein RW, 747
Schofield CJ, 311, 316, 640
Scholer HR, 657
Scholtz JM, 353, 520
Scholz GM, 284
Schowen RL, 536
Schrader M, 301
Schraen-Maschke S, 621
Schrage K, 310
SCHRAMM VL, 435–65; 442, 446–48, 450, 452–54, 458, 459, 634
Schrauwen P, 385
Schreck R, 102
Schreiber R, 683, 693
Schreiber SL, 461
Schroeder TE, 546, 547, 555, 557
Schroer TA, 610
Schroers A, 734
Schuchardt J, 614
Schuck P, 473
Schuit FC, 389
Schulenberg B, 75
Schuler F, 84
Schulman H, 27
Schulte TW, 273
Schulte U, 75, 78, 86
Schulten K, 172, 178, 182
Schultz CS, 500, 501
Schultz J, 664, 667
Schultz PG, 576
Schultz S, 692, 701
Schultz T, 468
Schulz H, 300, 311
Schulze R, 484
Schumacher AM, 193, 202, 207
Schumacher D, 732
Schumacher MA, 235
Schumacher V, 140–42
Schumann W, 272
Schurter BT, 258
Schuster R, 246, 252

Schuster SM, 630–36, 638–42, 644
Schutgens RBH, 296, 313, 316
Schütz M, 786
Schwartz BE, 262
Schwartz D, 138
Schwartz JH, 643
Schwartz R, 353
Schwartz SD, 535
Schweizer D, 255
Schwille P, 614, 618
Schwimmer C, 718–20
Schwittay M, 103
Schymkowitz J, 355
Sciacco M, 719
Scita G, 667
Scocca JJ, 572, 574, 579
Scott JA, 665, 672
Scott JD, 669, 671
Scott JW, 144–47, 149, 154
Scott K, 671
Scott MI, 217
Scott-Drew S, 231
Scott-Drew SR, 231
Scozzafava A, 642
Screaton RA, 156
Screpanti E, 734
Scroggins BT, 284, 289
Scully R, 14
Searls T, 311
Sedlmeier R, 613
Seedorf H, 755
Seedorf U, 309, 310, 321
Seefeldt LC, 76
Seeger M, 618
Sega M, 348
Segal DJ, 674
Segal GM, 97
Segal H, 582
Segall L, 386, 389, 390, 393
Segel IH, 633
Segers J, 151
Segu VB, 386, 390
Seibold SA, 172, 178–80
Seimiya H, 508
Seio K, 642
Seitz LE, 474
Sekine M, 642, 693
Sekizaki H, 421
Selak MA, 782
Seldin MF, 139, 140
Selkoe D, 615

Selkoe DJ, 348, 615–18
Selkoe K, 613
Selvin PR, 610, 611
Selzer T, 524, 529
Senawong T, 436, 439–41, 448, 460
Senderowicz AM, 546
Senderovitz A, 99
Sengupta A, 254
Sennvik K, 619
Seo BB, 75, 86
Seo CI, 662, 663, 669
Seo YR, 693
Sergi A, 530, 537
Sergueev K, 216
Serio TR, 347, 348
Serpell LC, 336, 341, 343
Serrano AL, 25, 26, 29
Serrano L, 349, 353, 355
Serre MC, 575
Serve H, 103
Setchell KD, 316
Setlow P, 12, 479
Seto AG, 503, 504
Setou M, 613
Seubert P, 357
Severson AF, 223
Seydel U, 61
Shaag A, 700
Shabanowitz J, 251, 259
Shafaatian R, 438
Shaffer PL, 278, 280
Shah NP, 98
Shah V, 258, 272
Shahar M, 104
Shahmolky N, 156
Shakeley RM, 81
Shalloway D, 674
Shang T, 196, 198
Shani G, 190–93, 195–201, 205
Shani N, 322, 323
Shank CV, 747
Shankaranarayana GD, 439, 440
Shannon KB, 553
Shannon RJ, 70, 71, 73, 75, 81, 86
Shao J, 284, 289
Shao ZL, 194
Shapiro L, 223, 233, 480, 481
Shapiro PS, 75
Shapleigh J, 169, 170, 177
Shapleigh JP, 177

Sharf R, 305
Sharfe N, 104
Sharma A, 574
Sharma M, 156
Sharma PK, 179, 180
Sharp GW, 386
Sharp K, 775
Sharp P, 320
Sharpe M, 170, 177
Sharpe MA, 169, 173, 179, 180
Sharpe ME, 214
Sharples JM, 755
Sharrocks AD, 662
Shaul M, 101
Shaver RD, 620
Shavlakadze T, 30
Shaw AS, 668, 669
Shaw BR, 244
Shaw C, 577, 580
Shaw RJ, 144, 147–49, 153, 154, 156
Shawver LK, 102
Shay JW, 132, 494, 507, 508, 512
Shearer GL, 532
Shechter Y, 96, 99
Sheedy DM, 224
Shekhtman E, 585, 596, 597
Shelly SS, 610
Shelton J, 25
Shelton JM, 24–26
Shen GJ, 56
Shen J-R, 770
Shen XL, 147
Shen Y, 357
Sheng S, 631
Sheng ZH, 196, 204
Shepherd DM, 439, 440, 448
Shepherd GR, 261
Sherer TB, 86
Sheridan PP, 404, 419
Sherman JM, 438
Sherratt D, 581, 582
Sherratt DJ, 213, 233, 234, 236, 573, 574, 578, 580–82, 584, 585, 590, 593
Sherriff J, 260
Sherrington R, 615
Sherry AD, 389
Sherwood DJ, 29
Shevchenko A, 219, 251, 253, 496, 499

Sheves M, 751, 755
Shi Q, 558, 560
Shi W, 177, 646
Shi YG, 368
Shi YJ, 262
Shiau AK, 273, 275, 276, 279
Shibata H, 139
Shibata Y, 253, 254
Shichida Y, 748, 751, 753–55
Shie FS, 614
Shiekhattar R, 262
Shifrin Y, 44, 47, 52, 53
Shih YL, 481, 485
Shikanai T, 782, 785
SHILATIFARD A, 243–69; 245, 251, 254, 257, 259–61
Shimabukuro M, 386, 391, 393
Shimada H, 168, 171
Shimane M, 144
Shimizu M, 508
Shimizu S, 142
Shimokata K, 168, 171
Shimomura I, 372
Shimozawa N, 316
Shin DS, 225
Shin HC, 674
Shinkai Y, 251, 254
Shinmura K, 142
Shinohara Y, 718–20, 730–32, 734–36
Shintani M, 372
Shioda S, 631
Shiota M, 372, 373, 378, 380, 384
Shioya H, 139
Shiozaki K, 662
Shiozaki M, 662
Shippen DE, 498, 500, 502, 504–7
Shirahige K, 221, 224
Shiraishi K, 684, 693
Shirayama M, 219
Shively S, 620
Shoelson SE, 372, 377, 668
Shoemaker GK, 419
Shoemaker NB, 574, 579
Shohat G, 193, 196
Shoji M, 612
Shore D, 127, 438, 508, 511
Short SA, 409
Shou W, 437
Shoukry K, 311

Showalter HD, 101
Showalter RE, 102
Shrimpton P, 528
Shrimpton PJ, 528, 530
Shrode T, 498
Shroff R, 117
Shu HB, 553
Shulman GI, 372, 376, 377, 382, 383
Shulman JM, 154
Shulmeister VM, 770–72, 776
Shuman S, 573, 574
Shuster CB, 550, 553, 558
Shuster J, 630
Si K, 340
Sicheri F, 102, 665, 667
SIDDIQUI KS, 403–33; 404–8, 410–13, 415, 416, 418, 420, 424, 427
Siddiqui NU, 223
Sidtis JJ, 620
Siebarth V, 43
Siebert F, 751, 755, 758
Sieberth V, 59–61
Siebring J, 339, 340
Siedlak SL, 612, 619
Siegbahn PEM, 687–91
Siegers A, 357
Siegmund HI, 276
Siekevitz P, 715
Siemer AB, 341, 342
Sierralta J, 669, 671
Signon L, 114, 115, 121
Signoretti S, 139
Sigurdson H, 169
Sihag S, 382
Siivari KM, 309
Sikorski P, 341, 343
Sikorski RS, 532, 533
Silberstein L, 30
Siletsky S, 170, 177
Siletsky SA, 170, 177
Silhavy TJ, 62
Siligardi G, 272, 276–78, 281, 283–85, 287
Sillibourne J, 548, 559, 560
Silow M, 748, 753
Silva J, 255
Silva JL, 349
Silva-Zolezzi I, 324
Silveira JR, 358
Silvennoinen O, 146

Silver PA, 213, 230
Silver RP, 40, 41, 54, 57–59, 61, 62
Silverman JD, 556
Silverman-Gavrila RV, 546
Silverstein AM, 282–84
Silverstein RL, 369, 383
Sim VL, 358
Siman R, 617, 618
Simha V, 372
Simic R, 259
Simms ES, 6, 7
Simon C, 24
Simon I, 72
Simon JA, 461
Simon M, 473, 474
Simon MI, 596
Simon MN, 346, 355
Simon N, 98
Simoneau JA, 379, 382
Simons K, 711
Simpson DA, 55
Simpson DM, 671
Sinclair DA, 121, 436, 442, 444, 445, 452, 454
Singer MS, 116, 117, 256, 260
Singh AB, 32
Singh AK, 301, 302
Singh BM, 384
Singh G, 482
Singh H, 303
Singh I, 301, 302
Singh N, 325
Singh OM, 278
Singh S, 283, 284, 408, 501, 507
Singh SM, 130
Singh SS, 284
Singh V, 646
Singleton SF, 520
Sinha M, 139
Sinha-Hikim I, 32
Sinka R, 156
Sinsheimer RL, 9
Sintchak MD, 685, 687, 692
Siomos MF, 221
Sirangelo I, 358
Sirevag R, 415
Sisk A, 619
Sismour AM, 671
Sisodia SS, 617
Sistonen P, 138
Sitnikov D, 227

Sitnikov VF, 613
Siu F, 635, 636
Siu FY, 636
Sivitz WI, 373, 391
Sjöberg BM, 682, 685, 687, 689, 690, 692, 693, 697, 698, 700, 701
Sjoblom T, 103, 104
Sjodin B, 23
Sjogren C, 224
Skalka A, 112
Skarzynski T, 278
Skeel RT, 630
Skehel JM, 71, 75, 81
Skehel P, 617
Skiniotis G, 470
Skjeldal OH, 316
Sklar J, 76
Skoko D, 214
Skop AR, 546, 547
Skorpen F, 391
Skulachev VP, 169
Skurk C, 31
Slaaby R, 694
Slade D, 635
Slama JT, 446–48, 450, 452, 453, 457
Slater AM, 99
Slater EC, 716
Slater R, 96
Slattery ML, 152
Slaughter CA, 324
Slavcev RA, 226, 227
Sled V, 775
Sled VD, 76, 78, 82, 83
Slenczka W, 82
Slentz D, 375, 380, 382, 384
Slonimski PP, 721
Slovacek R, 782
Slovak PM, 481
Sluder G, 560
Sluse FE, 733
Sluse-Goffart CM, 733
Sluyterman LA, 533
Smaill JB, 101
Smalas AO, 404, 409, 418–26
Small E, 474
Smaradottir RB, 421
Smeitink JAM, 70, 75, 86
Smerdon SJ, 667
Smets LA, 639
Smiley RD, 527, 528

Smirnov VN, 355
Smirnova I, 715
Smit L, 144, 145
Smith A, 61
Smith AB, 620
Smith AC, 496, 500, 505, 508
Smith AE, 674
Smith BC, 444, 452, 461
Smith BT, 302
Smith CM, 144
Smith CU, 616, 619
Smith CV, 278
Smith D, 139, 147, 148, 155–57, 556
Smith DA, 348, 349
Smith DF, 273, 275, 282, 283
Smith DH, 617, 618
Smith DK, 417
Smith DL Jr, 442, 444, 454
Smith DP, 145
Smith EL, 245
Smith ER, 253
Smith HW, 283, 287
SMITH JL, 769–90; 632–34, 641, 770–72, 774, 775, 777, 779–81, 783
Smith JS, 436, 437, 442, 444, 446, 447, 454
Smith KD, 323
Smith MA, 76, 619
Smith MC, 570, 582, 583, 599
Smith MJ, 357, 470
Smith ML, 693
Smith RD, 647
Smith RE, 758
Smith RL, 390
Smith S, 446
Smith SO, 755
Smith VF, 413
Smith WC, 757
Smogorzewska A, 123, 128, 132, 494, 496, 507, 510, 511
Smolikov S, 116, 117
Smonou I, 409
Snell CR, 758
Snider MJ, 407–9
Sniekers M, 312
Snow BE, 509
Snowden AW, 262
Snozzi M, 775
Soares CM, 180
Soballe B, 213

Socci ND, 373
Soderling TR, 193
Söderstrom E, 43
Sofia HJ, 700
Sogo JM, 118
Sokolov Y, 348
Soldano KL, 274, 278, 280
Solit DB, 273, 289
Söll D, 634, 643
Solomon A, 336
Soltoff SP, 668
Somero GN, 404, 405, 407, 409, 415, 417–19
Somers WG, 547, 553
Somlyo AV, 196, 202
Somma MP, 546, 547, 559
Sommer ME, 757
Somwar R, 100
Sonan GK, 421, 423
Sondermann H, 272, 274, 664, 668
Sone N, 786
Sone T, 786
Song R, 194
Song XM, 30
Songyang Z, 145, 660
Sonnhammer EL, 72
Sonnichsen B, 546–48
Sontag CA, 475, 476
Sood R, 635
Soper TS, 632–34, 646
Soppa J, 224
Sordella R, 100, 101
Soriano GM, 772, 780, 781
Sormunen RT, 319, 321, 322
Sorrentino R, 272
Soti C, 358
Soufo HJD, 479–81
Souillac P, 347
Soultanas P, 599
Sousa M, 744
Sousa MM, 358, 359
Souto ML, 755
Souto SO, 349
Sowers KR, 404
Spada AP, 96, 101
Sparks CE, 390
Sparks JD, 390
Sparks LM, 375, 382
Sparks MJ, 684
Spassov VZ, 423
Spector A, 319, 321

Spector DL, 255
Spector I, 546
Spedale EJ, 438
Spehr V, 78
Speicher DW, 262
Spencer A, 349
Spengler SJ, 595
Speranza F, 75
Sperfeld AD, 613
Sperger JM, 496
Sperisen P, 127, 510
Spicer J, 145, 155
Spiegelman BM, 28, 372, 375, 439, 441
Spiers S, 385
Spillantini MG, 613
Spinazzola A, 700
Spittaels K, 619
Spivak T, 194
Spivak-Kroizman T, 193, 196
Spooner PJ, 755
Sprague GF Jr, 669
Sprecher H, 311
Springer MS, 245
Spudich JA, 547, 556, 557, 610
Spyrou G, 693
Squazzo SL, 259, 348
Srinivasan R, 426
Staddon JM, 194, 201, 202, 204
Stagi M, 614, 618
Stahl A, 305, 369
Stahl FW, 112
Stahl R, 619
Staley JT, 475, 476
Stallcup MR, 258
Stallcup WB, 577
Stals I, 407, 410
Stamer K, 609, 613, 618, 620
Stams WA, 639
Stancato LF, 282, 289
Standop J, 151
Stanislas E, 716
Stanley CA, 388
Stanners CP, 636
Stanssens P, 424
Stapley J, 98
Stapon A, 633
Starai VJ, 436, 439
Stark MJ, 147
Stark WM, 573, 583–86, 588, 590, 593, 595, 596

Starkov AA, 384, 385
Starkova J, 639
Starling KA, 630
Staron RS, 21
Stasiak A, 581, 595
Stasiak AZ, 581
Stassen MJ, 244, 248
Stathopoulos C, 51
Staubli W, 302
Stearn AE, 405
Stebbins CE, 273, 278
Stebbins J, 436, 446, 447
Stec WJ, 573, 574
Steenbergen SM, 55–57, 60, 61
Stefani M, 339, 348–50, 353–55
Stegmeier F, 220
Steimer KS, 302
Stein CJ, 368
Stein DT, 385, 386
Steinberg SJ, 311, 312, 324
Steinberg-Neifach O, 130
Steinemann D, 98
Steiner P, 496
Steinmetz AC, 289
Steitz TA, 499, 582, 584–91, 593, 595, 598, 600
Stella A, 140, 157
Stellwagen AE, 508
Stemmann O, 220
Steneberg P, 387, 393
Stenkamp RE, 744–46, 748, 749, 751
Stenmark P, 685
Stensgard B, 276, 277, 287
Stepanova L, 284
Stephen AG, 509
Stephens DS, 60
Stephens RS, 468
Steppan CM, 371, 372
Sternglanz R, 446–48, 450, 452, 454, 457
Steuber J, 84–86
Steven AC, 342–44, 346
Stevens MJ, 710
Stevens MP, 55
Stevens S, 461
Stevenson BE, 385, 386
Stevenson EJ, 26, 31
Stevenson G, 47
Steward A, 353
Stewart AF, 369, 576, 579
Stewart G, 147, 149

Stewart GC, 479
Stewart JD, 633, 634, 642, 645
Stewart JJP, 645
Stewart L, 574
Stewart PE, 572
Stewart PL, 509
Stewart S, 550
Stewart SA, 121
Stich A, 313
Stiehl HH, 774
Stigter RL, 639
Stillman B, 548
Stine WB, 348
Stirm S, 43
Stitham J, 750
Stitt TN, 26, 28
Stivers JT, 573, 574
Stock JL, 671
Stock MF, 611
Stockdale FE, 21
Stockl A, 718, 720, 726, 736
Stocksley MA, 29
Stoehr JP, 373
Stoeltzing O, 102
Stoffregen EP, 98
Stojanovic A, 750
Stokes ES, 102
STOKIN GB, 607–27; 611, 612, 617–21
Stolpe S, 84, 86
Stolz DB, 302
Stone JR, 614
Stone WL, 746
Storch J, 305
Storer TW, 32
Storz P, 660
Stough DB 4th, 138
Stoughton RB, 648
Stover DR, 102
Stover GL, 26, 28
Stradal TE, 667
Strader DJ, 744
Strahl BD, 245, 248, 249, 251, 253, 258
Strahl-Bolsinger S, 438
Straight A, 227, 551, 552, 555
Straight AF, 437, 545, 546, 550, 551, 553, 554, 556–59
Strand KR, 682
Strater N, 581
Strathern JN, 437
Straub SG, 386

Strauss H, 74
Strawn LM, 102
Strettoi E, 745
Strich R, 437
Stricker J, 469, 474
Strickland LI, 553
Striegl AM, 618
Strijland A, 316
Strivers EC, 532
Stroebel D, 770, 778, 783, 785, 786
Strole C, 59, 60
Strole CA, 60
Strom L, 224
Stronghill PE, 223
Stroud J, 660, 662, 663
Struble RG, 615, 619
Struhl K, 259–61
Strunnikov A, 223
Strunnikov AV, 224
Strynadka NC, 409
Stubbe J, 408, 633, 682, 684, 685, 687, 690–92
Stubbe JA, 688
Stuchebrukhov AA, 169, 170, 175, 178, 180
Studamire B, 121
Sturgill JF, 436
Sturtz LA, 444
Stuurman N, 546, 547, 557
Su F, 439, 440
Su HM, 320
Su JY, 138
Su YH, 190, 194, 197–201
Subathra M, 383
Subramani S, 301, 311, 315
Subramaniam S, 578
Subramaniam V, 353, 356
Subramanian A, 382
Subramanian SV, 23
Subramanya HS, 574
Suck D, 579
Suda K, 757, 759
Sudarsanam S, 95, 139, 146, 152, 153, 155
Suefuji K, 485
Suel GM, 524, 525
Suen WC, 415, 416
Suetsugu S, 272
Sugahara M, 633
Sugawara N, 121
Sugihara M, 748, 755

Sugimoto K, 56, 118, 251, 254
Sugiura N, 56
Sugiyama A, 644
Sui GC, 262
Suka N, 441
Suka Y, 441
Sullivan JK, 272, 283, 285, 287
Sullivan M, 220
Sullivan MA, 274
Sullivan W, 276, 277, 287, 559
Sullivan WP, 275, 276, 282, 283, 287
Sumara I, 220
Summers D, 582
Summers SA, 378
Sumner CJ, 613
Sumpter PQ, 619
Sun D, 479, 500
Sun DM, 611
Sun J, 776
Sun L, 102
Sun Q, 469, 482
Sun XY, 692, 693, 698
Sun Y, 144, 671
Sun ZW, 245, 248, 249, 251, 253, 259, 260
Sunako Y, 215
Sundaram PV, 411
Sunde M, 336, 341, 349
Sunderji S, 341
Sundseth SS, 379
Sundstrom L, 582
Sung BJ, 662, 663, 669
Sung CH, 745
Sung HC, 409, 417, 421, 423–25
Sunshine MJ, 27, 377
Supuran CT, 642
Surewicz WK, 346
Surka MC, 548
Surks HK, 196, 202
Susca FC, 157
Sutanto Y, 574, 579
Sutherland CM, 147
Sutow WW, 630
Sutton A, 253, 436, 446, 447
Suyama Y, 130
Suzuki A, 156
Suzuki E, 671
Suzuki F, 548
Suzuki K, 27, 41, 275
Suzuki S, 347

Suzuki T, 142, 417, 547, 553–55, 617
Suzuki Y, 309
Svab Z, 782
Svendsen A, 418
Svensson H, 391
Svensson-Ek M, 168, 169, 172, 173, 180
Svingen PA, 97
Svingor A, 417, 418, 425, 426
Swaab DF, 620
Swain CG, 532
Swalla BM, 577
Swan CH, 674
Swanson J, 178
Swanson JM, 719, 725
Swanwick RS, 528, 530
Swartzman EE, 322
Sweeney LB, 436
Swenberger JR, 380
Swiggers SJ, 509
Swillens S, 760
Swinkels BW, 301
Swinnen JV, 151
Swoap SJ, 26
Syed AB, 619
Syed S, 611
Sylte I, 418, 744
Sylwan L, 579
Symington LS, 114, 115, 121
Synstad B, 415
Szadkowski H, 418
Szanto I, 374
Szargel R, 700
Szauter P, 113
Szczepaniak A, 772, 775, 781
Sze SH, 262
Szebenyi G, 613, 618
Szeto TH, 485
Szigety SK, 112, 116
Szilagyi A, 425, 426
Szodorai A, 617
Szostak JW, 113, 116, 117, 119, 441
Szpunar M, 349

T

Tabak HF, 303, 311, 325
Tabata I, 28
Taboski MAS, 502, 508
Tachibana M, 251, 254
Tackett AJ, 253
Tada S, 252
Tadayyon M, 386
Taddei N, 348–50, 353–55
Tadros M, 469, 473, 475
Taegtmeyer H, 377
Tafrov ST, 436, 446, 447
Tagawa K, 782, 784
Tager JM, 296, 313
Taggart AKP, 116, 118, 120, 508, 509
Taguchi S, 415
Tainer JA, 225
Tajiri R, 301
Takada Y, 406, 409, 410, 418
Takaesu H, 744
Takahashi H, 196
Takahashi RH, 618, 619
Takahashi S, 620
Takano T, 100, 324
Takata H, 617
Takeda K, 191, 195–99, 201, 745
Takeda M, 619, 631, 635, 718, 735
Takeda S, 610, 684, 744
Takegami K, 83
Takekawa M, 662, 672
Takematsu H, 548
Takemori H, 156
Takemoto Y, 316
Takenawa T, 272
Taketo MM, 139, 140
Takeuchi S, 618
Takimoto T, 630, 638
Takita J, 613
Talbot D, 255
Tallal L, 630
Tam BM, 745, 748
Tamechika I, 546
Tamura K, 101
Tamura S, 97
Tan C, 630
Tan W, 674
Tan Z, 175, 177
Tanabe K, 323
Tanaka AR, 323
Tanaka H, 684, 693
Tanaka I, 406, 409, 410, 418
Tanaka K, 282
Tanaka M, 346, 347, 612
Tanaka T, 219, 546
Tanaka TU, 217
Tanaka Y, 610, 613
Tang C, 102
Tang F, 102
Tang KE, 417
Tang L, 103
Tang YA, 255, 256
Taniuchi I, 32
Tanizawa A, 630, 638
Tanner CJ, 31
Tanner KG, 446–48, 450
Tanny JC, 438, 442, 445, 447–50, 457
Tanoue T, 662
Tao H, 142, 692
Tao JS, 642
Tapia O, 533
Tapscott EB, 377
Tarr PT, 28, 375
Tartaglia GG, 354
Tartaglia LA, 305
Tashiro M, 169
Tashiro T, 612
Tassan JP, 152
Tasso R, 382
Tassoni E, 383
Tate G, 142
Tatebayashi K, 662, 672
Tatsuka M, 548
Tatsumoto T, 552
Tauber MJ, 751
Taulien J, 272
Tavan P, 751
Tavares AA, 547
Tavares AM, 548
Tawa NE Jr, 31, 32
Tawada A, 56
Tawiah-Boateng MA, 259
Taylor DL, 555, 557
Taylor DW, 469, 474
Taylor GL, 409, 412, 415, 418, 419, 421–26
Taylor JA, 752
Taylor KA, 469, 474
Taylor KL, 346, 355
Taylor RK, 485
Taylor RM, 771, 774
Taylor S, 193
Taylor SS, 144, 217
Taylor SW, 70, 73, 75
Teague SJ, 642
Tebabi P, 439, 445, 454
Technikova-Dobrova Z, 75

Teichmann SA, 347, 496
Teipel S, 619
Teitelbaum L, 196–98, 200, 205
Teixeira M, 173, 180
Teixeira MT, 127, 494, 498, 506, 510
Tekle M, 577
Teleman A, 227
Teleman AA, 213, 230
Telemaque-Potts S, 386, 390, 392, 393
Teller DC, 744, 746, 748, 749, 751, 757, 760
Tempst P, 245, 251, 255, 256, 439, 440, 461
Teng G, 249
Teng SC, 116, 117, 119–22, 127, 509
Tennent GA, 336, 341
Tenney K, 245, 251, 254, 257, 260
Tenza D, 339, 340
Teo SS, 104
Teplova M, 192, 193
Teplow DB, 347, 348, 617
Teplyakov A, 633
Terada H, 718–20, 732, 734–36
Terada S, 28
Terada Y, 548, 554
Terakita A, 748, 754
Terauchi Y, 371
Ter-Avanesyan MD, 355
Terebello HR, 630
Tereshko V, 192, 193
Terns RM, 508
Terpstra P, 426
Terret ME, 221
Terrillon S, 758
Terry RD, 614, 615, 616, 619
Tersmette AC, 138
Tertoolen LG, 439
Tesche N, 57
Tesdorpf JG, 309
Tesmer JJG, 633, 641
Tesmer VM, 272, 499, 507
Tesson AR, 632–34, 646
Testa CM, 86
Tettelin H, 498
Tewari AK, 578
Tews I, 669
Teyssier C, 258
Tezcan FA, 484

Thai TP, 303, 304, 320
Thal DR, 616
Thaler DS, 122
Thanassi DG, 51
Thanbichler M, 215
Thanedar S, 469
Theis JF, 114
Thelander L, 682, 684, 690, 692–94, 698, 699
Thelander M, 690
Theos AC, 339, 340
Theret N, 302
Therrien M, 146
Theys F, 410, 424, 427
Thieringer R, 315
Thies E, 620
Thinakaran G, 618
Thion L, 229
Thirkettle JE, 640
Thirumoorthy R, 633, 634, 642, 645
Th'ng J, 440
Thoden JB, 632–34, 642
Thomas AP, 102
Thomas CC, 660, 663
Thomas CM, 230
Thomas H, 302
Thomas JO, 244, 245
Thomas JW, 169, 170, 177
Thomas KR, 308
Thomas T, 404–7, 409–13, 418–20, 422–26
Thompson A, 665, 667
Thompson AM, 510
Thompson CM, 469, 474
Thompson HM, 547
Thompson J, 61
Thompson JL, 348
Thompson LA, 642
Thompson LJ, 283
Thompson RP, 376, 385
Thomsen W, 744
Thomson NH, 346, 348–50
Thorbjarnardottir SH, 421
Thornell LE, 32
Thorner J, 322, 559, 662–64, 669
Thornton JH, 538
Thorpe H, 599
Thorpe HM, 570, 582, 583, 599
Thorpe IF, 528–30
Thrower JS, 579

Thumelin S, 393
Thunissen M, 691
Thurnher A, 102
Thurow H, 52
Thyagarajan B, 599
Thyberg J, 353
Thyfault JP, 385
Tiainen M, 139, 144, 145, 151
Tian D, 548
Tian JH, 196, 204
Tian LG, 574
Tian XC, 684, 694
Tiede DM, 180
Tiedt R, 104
Tieleman DP, 711
Tiep S, 28
Tighe A, 545, 546
Tikkanen R, 632
Till R, 599
Tilly K, 572
Timm J, 683, 693
Timm T, 155
Timmers CM, 642
Timmins PA, 728
Tindbaek N, 418
Ting NSY, 508
Tipples G, 685
Tiranti V, 719
Tirnauer JS, 551–53, 555
Tirumalai R, 574, 578
Tirumalai RS, 570, 572
Tisdale JM, 554
Tishkoff DX, 122
Tissenbaum HA, 436, 440, 444
Tito P, 349
Tittmann P, 470
Tobi D, 710
Toczyski DP, 117
Todd AE, 538
Todisco S, 717
Toft D, 274, 279, 281
Toft DO, 272, 275, 282, 283, 286, 287
Togashi S, 144
Toime L, 79
Tokatlidis K, 718, 720, 734
Tolbert NE, 314
Tolley SP, 632
Tomas A, 548, 555, 558
Tomas E, 371
Tomaska L, 130
Tomioka S, 479, 480

Tomishige M, 611
Tomizaki T, 171
Tomizawa K, 782, 785
Tomlinson I, 138–40, 142, 157
Tomlinson IP, 138
Tomson FL, 169, 170, 180
Toneff T, 617
Tong A, 253
Tonozuka Y, 553
Topark-Ngarm A, 436, 441, 460
Topcu Z, 122, 127, 129
Torbett BE, 674
Tornheim K, 392
Tornqvist HE, 156
Tornroth S, 168, 169, 172, 173, 180
Torok M, 341
Torok MS, 260
Torrents E, 682–84, 692, 693, 698, 700
Torres-Rosell J, 224
Torroja L, 617–20
Toth A, 219
Toth R, 148, 152, 154–57
Totsukawa G, 557
Touchet N, 619
Tournier C, 671
Tousignant A, 535
Townsend CA, 633, 634, 638, 640, 642
Toyama K, 139
Toyota E, 421
Trabachino LC, 348
Tracy S, 100
Tran C, 98
Tran N, 102
Tran PO, 390
Tran-Betcke A, 76
Traum D, 140
Traverse KL, 498
Traverso V, 745, 748
Travis JL, 611
Trebst A, 771, 779
Tremp G, 619
Trevisiol C, 684
TRÉZÉGUET V, 713–41; 709, 715, 717–20, 722, 723, 726–28, 730–32, 734, 736
Trible RP, 669
Trimarchi JM, 693
Trimble WS, 547, 548, 559

Trinczek B, 609, 613, 618
Trink B, 139
Tripet BP, 611
Tristram-Nagle S, 167
Trojan J, 142
Trojanowski JQ, 357, 617, 618, 620
Troy AE, 374
Troy FA, 41, 47, 57
Troy FA 2nd, 49, 55–57
Truckses DM, 669
True HL, 340
Truhlar DG, 529, 533
Trumana A, 277
Trumpower BL, 167, 181, 182, 772, 775, 776, 780
Trushina E, 613
Tsai CC, 24
Tsai CJ, 408, 415, 419, 422, 423, 425
Tsai CM, 101
Tsai HC, 195, 197, 202, 203
Tsai S, 282
Tsai YL, 121
Tsang AW, 445, 446
Tsang CW, 548, 559
Tsang FA, 445
Tsao TS, 371
Tschiersh B, 244, 248, 249
Tseng BP, 348
Tseng SF, 121, 128
Tseng Y, 474, 480, 481
Tsenova L, 683, 693
Tsichlis P, 273, 289
Tsichlis PN, 272
Tsigos I, 409
Tsika GL, 25, 26
Tsou HR, 101
Tsuchiya N, 786
Tsujimoto HY, 782, 784
Tsukamoto H, 748, 754
Tsukamoto KV, 389
Tsukamoto T, 316
Tsukamoto Y, 118, 120
Tsukihara T, 168, 171, 785
Tsukita S, 555
Tsunoda S, 671
Tsuruo T, 274, 289
Tsuruta H, 408, 419, 422, 424
Tsuta K, 100
Tsuyama M, 782, 785
Tu DC, 745

Tuite MF, 340
Tullius TD, 577
Tuncman G, 377
Tung EY, 770, 781
Tung YC, 615
Turck CW, 437
Turcotte LP, 380
Tureckova J, 25
Turner B, 258
Turner JW, 286, 507
Tusnady GE, 72
Tutino ML, 409
Tuyp JJM, 300, 301, 315
Tycko R, 341, 342, 346
Tye BK, 12, 113
Tyers M, 663, 664
Tyler PC, 646
Tzeng YL, 59, 60
Tzfati Y, 503, 504

U

Uchida A, 612
Uchida Y, 312, 324
Uchino S, 190, 198
Uchiyama A, 312, 324
Uckun FM, 104
Udd L, 147, 148, 152
Ueberheide B, 251
Uecker A, 103, 104
Ueda J, 251
Ueda K, 445, 448
Ueda M, 321
Uehata M, 546
Ueta N, 305
Uetake Y, 554
Ugalde C, 70, 75, 86
Ugas MA, 391
Uhler MD, 143, 144, 314
Uhlin U, 682, 685, 690, 691, 694, 697
Uhlmann F, 217, 219–21
Ullrich H, 52
Ulrich A, 151
Ulrich HD, 224
Uma S, 284, 289
Umansky K, 503, 504
Umehara T, 441
Umezawa H, 96
Umezu K, 117
Underwood DH, 123, 125, 129

Underwood J, 617
Unemoto T, 70
Unger RH, 386, 389–91
Unterbeck A, 615
Unterrainer G, 324
Uppenberg J, 419
Uppsten M, 685, 694, 697
Uramoto M, 642
Ure D, 613
Uria JA, 436
Urizar E, 760
Urnov FD, 674
Usada N, 309, 319
Ushakova AV, 70, 73, 86
Usuda N, 321
Usui T, 117, 121
Utzschneider KM, 390
Uversky VN, 335, 336, 339, 347, 349, 353, 410–13, 418, 422
Uyeda TQ, 544, 552, 557
Uyemura D, 12
Uyemura DG, 8, 11
Uysal KT, 371, 372, 377
Uzri D, 230, 231

V

Vaahtomeri K, 144, 145, 151
Vacratsis PO, 674
Vadai E, 190
Vaganay E, 50
Vahsen N, 86
Vajkoczy P, 102
Vakulenko SB, 536
Val ME, 582
Valach M, 130
Valadie H, 710
Vale RD, 546, 547, 557, 611, 674
Valencia A, 226, 474, 476, 480, 482
Valkova-Valchanova M, 770–72
Valla JE, 617
Valladares O, 30
Valle D, 300, 316, 320, 322, 323
Valletta J, 621
Valluzzi R, 485
Valpuesta JM, 469
Valvano MA, 47–49
van Aalten DM, 145, 148, 149, 152
van Beckhoven JR, 315

Van Beeumen J, 70, 73, 74, 76, 407, 409, 424, 426
van Berkel E, 316
Van Bilsen M, 374, 375
van Boom JH, 642
Van Brussel E, 621
van den Berg M, 322–25
van den Bosch H, 296, 303, 304, 316
Van den Bosch L, 49
van den Bosch de Aguilar P, 612
Van den Brande I, 424
van den Brink DM, 312, 314
Van den Broeck A, 424
van den Ent F, 218, 470, 477, 478, 480–82
Van den Haute C, 619
Vanden Heuvel JP, 284
van den Heuvel L, 70, 86
van den Heuvel LWPJ, 70, 86
van den Heuvel RHH, 633, 634
van den Heuvel S, 548
Vande Pol S, 617
van de Putte P, 585, 596
Vander SD, 302
van der Ende MA, 316
van der Hoeven F, 369
van der Horst A, 439
van der Marcel GA, 642
van der Oost J, 417
Van der Ploeg LH, 617
van der Wel NN, 154
Vandezande K, 619
Van Dorpe J, 619
van Drunen E, 639
Van Duyne GD, 572–74, 576, 578, 579, 587, 600
Van Dyke C, 619
van Grunsven EG, 316
Van Heeke G, 631
van Helvoort JM, 213
Van Heusden GP, 315
van Holde KE, 244
Vanhommerig SAM, 533
Vanhooren JC, 300, 308
Vanhorebeek I, 315
Van Horsen GW, 615
Vanhove G, 308, 313
Vanhove GF, 308
van Kan PJ, 786
van Leeuwen F, 251, 256
VanLoock MS, 480

van Marle J, 322, 547
van Meer G, 711
van Miert JNI, 314
Vann WF, 43, 54–57, 61
van Oers NS, 668
Vanoni M, 633
Van Petegem F, 409, 418, 421, 422, 424, 426
Vanrobaeys F, 70, 73, 74, 76
van Roermund CW, 325
van Roermund CWT, 309, 316, 322–25
van Spronsen DJ, 103
Van Veldhoven P, 320
Van Veldhoven PP, 297, 301, 306, 308–13, 324
Van Wering ER, 639
Van Zantwijk CH, 639
van Zyl W, 417
Vaquero A, 439, 440, 461
Varfolomeev E, 194
Vargas ML, 254
Varki A, 56
Varley AW, 479
Varoutas PC, 273
Vasella D, 96
Vashishta A-C, 180
Vasilaki A, 385
Vasquez D, 148, 156
Vasquez S, 357
Vassinova N, 692
Vattem KM, 635, 646
Vaughan C, 272, 274, 277, 280, 285
Vaughan CK, 284
Vavvas D, 375
Vaz DC, 348
Vaze MB, 117
Vaziri H, 436, 439, 440
Veach DR, 102
Veatch JR, 508
Veenhuis M, 322
Veenstra TD, 671
Veerman AJP, 630, 639
Vega RB, 27, 28
Veinbergs I, 357
Veldman T, 508
Velentza AV, 193, 202, 207
Velez M, 469, 473, 475
Velmurugan S, 230, 231
Velonia K, 409
Vendruscolo M, 350, 354–56

Venepally P, 26
Vener A, 786
Venizelos N, 315
Venkatesh R, 411
Venkitakrishnan RP, 522, 527
Venkitaraman AR, 548, 560
Venter DJ, 496
Venter JC, 657, 659, 664
Ventouras LA, 277
Venturi M, 734
Veprintsev DB, 273
Verdel A, 24
Verdile G, 618
Verdin E, 438, 439, 441
Verdine GL, 501
Verdun RE, 123
Vergari R, 75
Verhey KJ, 611
Verhoeven MA, 755
Verhoeven NM, 308, 312, 316
Verkhovskaya ML, 170, 173
Verkhovsky MI, 82, 169–71, 173, 180
Verlhac MH, 221
Verma A, 104
Verma CS, 528
Verma R, 661
Vermaak D, 245
Verner G, 715
Verni F, 547, 557
Vernos I, 555
Vernotte C, 786
Verrijzer CP, 244, 248
Verschueren K, 437
Versele M, 559
Versteeg S, 272
Vetterkind S, 196, 201
Vezmar M, 24
Vicarioli J, 49
Vicente M, 469, 473–75, 482
Vician L, 156
Vidal M, 255, 256
Vidali G, 245
Videira A, 70, 73, 74, 78, 86
Vieille C, 423, 425
Vierling E, 468
Vignais PM, 75, 716
Vignais PV, 716, 718, 719, 722, 730, 732–34
Vihinen M, 146
Vijay IK, 57
Vilela C, 322–24

Villa C, 748, 751, 754, 755, 758
Villa J, 534
Villafranca JJ, 643
Villegas V, 349, 353
Villeret V, 418, 421
Villiger W, 41
Vimr E, 55, 59, 60
Vimr ER, 41, 55–57, 60, 61
Vincent C, 50
Vincent I, 614
Vincent PW, 101
Vincente M, 482
Vindurampulle C, 48
Vinogradov AD, 78, 82–85
Vinson C, 372
Vinters HV, 619
Viollet B, 379
Viollier PH, 215
Violot S, 421, 423
Vionnet J, 55–57
Virkki M, 171
Vischer NOE, 487
Vischer S, 388
Visintin R, 220, 437
Visser N, 322
Visser WF, 324, 325
Viswanathan MN, 322
Vitari AC, 148
Vitreschak AG, 693
Vivanco I, 100
Vodovotz Y, 302
Voegtli WC, 684, 690
Voelkel-Meiman K, 113
Voevodskaya N, 684, 685, 693
Vogel H, 407
Vogel R, 620, 751, 755, 758
Vogel SM, 376
Vogel T, 140–42
Voges D, 632
Voit R, 436, 439
Volbeda A, 84
Volkl A, 302, 313
Volkmann I, 619
Volkov A, 224
Vollers SS, 348
Vollrath F, 340
Volpe TA, 249
Volpi E, 379
von Dassow G, 552
von der Haar F, 642, 648
von der Mulbe F, 282

von Heijne G, 72, 300, 771
Von Hoff DD, 500
von Jagow G, 176, 182, 775
von Neuhoff N, 98
von Onciul AR, 533
Vorgias CE, 406, 409, 410, 415, 419, 424
Vos MH, 285
Voth GA, 167, 170, 179
Vousden KH, 684, 693
Voziyanov Y, 230, 573, 576
Vozza A, 717, 734
Vrbova G, 22
Vreeling-Sindelarova H, 302
Vreken P, 305, 306, 316
Vriend G, 340
Vujcic S, 314
Vygodina TV, 169, 178

W

Wachi M, 233, 479–81
Wächtershäuser G, 700
Wackerhage H, 25
Wada H, 775
Wada T, 642
Wadhams GH, 481
Wadsworth P, 553, 555–57
Waechter CJ, 49, 57
Waechter F, 302
Waelkens E, 310, 311
Wagner AFV, 682, 692, 701
Wagner CR, 524
Wagner G, 520
Wahnon DC, 522, 526
Wainwright M, 404
Waizenegger I, 221
Waizenegger IC, 220
Wakao R, 255, 256
Wakeling AE, 99
Waki H, 371
Wakil SJ, 376, 379
Wakimoto BT, 248
Waldor MK, 582
Walencewicz AJ, 357
Waletko A, 76
Walker AK, 222
Walker DM, 643
Walker JE, 70, 71, 73, 75, 76, 78, 79, 81, 86, 714, 717
Walker MA, 278, 280
Walker MD, 387, 393

Walker W, 132
Wall JS, 346, 355
Wall MA, 524, 525
Wall NR, 436, 439
Wallace KE, 610, 613
Wallach D, 194
Wallon G, 425, 426
Wallweber G, 502
Walmsley RW, 113
Walsh A, 546–48
Walsh CM, 192
Walsh CT, 274, 643, 646, 668
Walsh DM, 347, 348, 357
Waltenberger J, 104
Walter J, 617
Walter S, 287
Walters DE, 734, 735
Walters DK, 98
Walton MY, 484
Walworth ES, 633, 634, 641
Walz T, 438, 442, 445
Wand AJ, 520
WANDERS RJA, 295–332; 296, 297, 300, 301, 303–6, 308–16, 322, 324, 325
Wandless TJ, 668
Wang AH, 24
Wang CK, 532
Wang CY, 31
Wang D, 619
Wang D-C, 76
Wang E, 378
Wang F, 439, 440
Wang H, 251, 255, 256, 501, 502, 505, 536, 599
Wang HB, 254, 255, 256, 258
Wang HC, 274, 279, 281
Wang HW, 474
Wang J, 357, 391, 642
Wang JB, 255
Wang JC, 215, 217
Wang JK, 748
Wang JZ, 617
Wang L, 47–49, 532, 533, 611, 639
Wang LB, 498, 505–7
Wang LJ, 254–56
Wang LP, 378
Wang MY, 100, 386, 390, 393
Wang P, 618, 620
Wang PJ, 684, 694
Wang Q, 747
Wang QJ, 785
Wang R, 468, 482
Wang RC, 123, 128, 132
Wang SI, 151
Wang SJ, 311, 312, 324
Wang TL, 155
Wang TY, 100
Wang W, 669
Wang WJ, 190, 195, 197–203
Wang X, 41, 121, 236, 482, 759
Wang XX, 32
Wang Y, 262, 548, 557, 560
Wang YH, 249, 663
Wang YL, 555–57, 560
Wang YX, 28, 31, 375
Wang Z, 156, 253, 745
Wang ZF, 617
Wang ZW, 390
Wang ZY, 752
Waninger S, 574
Waragai M, 618
Ward R, 617
Ward RJ, 502, 508
Ward WH, 99
Wardleworth BN, 439, 452
Ware JA, 27
Ware TL, 501, 505
Warmuth M, 98
Warnecke U, 76
Warner AK, 557
Warner DF, 683, 693
Warner TG, 302
Warren DJ, 577
Warren GL, 26
Warren SM, 662
Warshel A, 179, 180, 534
Warth R, 276
Wartmann M, 273
Washburn TM, 259
Wasserman SA, 547, 584, 585
Wassmann K, 221
Watanabe G, 547, 556
Watanabe H, 56
Watanabe M, 139, 140
Watanabe N, 547, 553, 556
Watanabe S, 96, 406, 409, 410, 418
Watanabe Y, 217, 221, 222
WATERHAM HR, 295–332; 297, 300, 301, 305, 306, 310, 315, 316, 324, 325
Watkins P, 324
Watkins PA, 305, 311–13, 321–24
Watney JB, 528–30, 537
Watson JD, 7, 494
Watson MA, 593
Watt PM, 121
Watterson DM, 192, 193, 202, 207
Watts JL, 147, 154
Way JM, 379
Waye MM, 636
Weaver AJ, 287
Weaver D, 14
Weaver KE, 226
Webb CD, 213, 227, 230
Webb SP, 531, 533–35
Weber I, 557
Weber SA, 219
Webster C, 30
Webster SD, 348
Wedge SR, 102
Wei C, 140–42
Wei HC, 548
Wei HM, 323, 324
Wei W, 508
Weidemann A, 617
Weidner U, 70
Weigle J, 112
Weijland A, 665, 667
Weiler KS, 248
Weimbs T, 559
Weinberg RA, 496
Weiner A, 13
Weiner J, 630
Weinheimer CJ, 374, 375, 383
Weinreb PH, 354
Weinrich SL, 496, 499
Weinstock GM, 13, 14
Weisberg E, 97, 98, 102
Weisbrod RE, 643
Weisgerber C, 57
Weiss A, 668
Weiss C, 193, 202, 207
Weiss DS, 234
Weiss H, 70, 73, 75, 76, 78, 79, 82
Weiss K, 170, 177
Weiss M, 615
Weiss P, 609
Weiss S, 520
Weissman JS, 339, 340, 346, 347
Weissman SM, 508

Weissmann C, 617
Weitao T, 224
Wek R, 635
Wek RC, 635, 646
Wellen KE, 373
Weller S, 316
Welling W, 716
Wellinger RJ, 123, 494, 503, 504
Wells JW, 758, 760
Wells RG, 593
Wen BG, 192
Wen XH, 752
Wen XL, 770, 781
Wende AR, 28
Wenk MR, 711
Wenk S-O, 786
Wennmalm S, 419
Wensel TG, 745, 761
Wenwieser SVCT, 586, 588, 591, 595
Wenz C, 505, 506
Werner H, 320, 321
Werner M, 509
Werner S, 73
Wesch H, 302
West L, 215
Westendorf JJ, 639
Westerblad H, 22
Westergaard O, 7
Westerman AM, 138
Westermark GT, 336
Westermark P, 336, 359
Westgaard RH, 22
Westley E, 166, 176
Westlind-Danielsson A, 357
Westman EA, 444, 452, 461
Westman M, 691, 692, 698
Westman MA, 698
Westra R, 316
Westwood NJ, 546, 550, 553, 554, 556–58
Wetzel R, 341
Wey J, 102
Weyer L, 357
Whang I, 573
Wharton SB, 613
Wheeler MB, 391
Whetstine JR, 262
White CL, 56
White D, 43, 44
White EL, 474
White HB III, 711

White J, 30, 221
White JG, 552, 556, 559
White JH, 584, 585
White K, 617–20
White MA, 671
White MF, 439, 452
White P, 349
White SH, 715
Whitehouse PJ, 615
Whitelegge JP, 723, 771, 772, 774, 782, 785
Whitesell L, 273
Whiteside JH, 386
WHITESON KL, 567–605
WHITFIELD C, 39–68; 41, 43–54, 56, 60–62
Whitlock GG, 744
Whitmarsh AJ, 669, 671
Whitmarsh J, 782, 785
Whitney PL, 424
Whittaker M, 470
Whyte DB, 95, 139, 146, 152, 153, 155
Wickersham JA, 102
Wicki P, 341
Wickner RB, 340, 346, 355
Wickner W, 7
Wickramasinghe NC, 404
Wicksteed B, 391
Widegren U, 27
Widger WR, 771, 779
Wiech H, 273
Wied DM, 314
Wiedemann N, 75
Wiedemann U, 546, 547
Wiederkehr T, 283
Wiegand H, 284
Wiernik PH, 630
Wierzbicki AS, 311, 316
Wiesbrock SM, 371, 372
Wieschaus E, 558
Wigler M, 669
Wigley DB, 574
Wikström M, 169–73, 175, 180
Wikström MK, 170
Wikström MKF, 82, 84, 86, 775
Wilbur J, 613
Wiles WG 4th, 147
Wilhelm S, 103
Wilhelm SM, 103
Wilkie D, 103
Wilkins BJ, 25, 26

Willassen NP, 404, 409, 418–26
Willems PHGM, 70, 86
Willer A, 152
Williams A, 341
Williams AD, 341, 348, 353, 355
Williams AH, 745, 748
Williams B, 547, 555
Williams BC, 547
Williams DS, 610, 611
Williams E, 547, 557
Williams EV, 547, 555
Williams JC, 665, 667
Williams KA, 723
Williams RS, 21, 25, 27, 664
Williams TE, 630
Williams TP, 745
Williamson IM, 711
Willing A, 683, 693
Willis I, 442, 452–54
Willoughby EA, 671
Willy PJ, 324
Wilm M, 251
Wilmanns M, 657
Wilson JH, 761
Wilson KS, 175
Wilson MA, 710
Wilson SE, 599
Wilson SL, 574, 582
Winder WW, 30, 147
Windheim M, 283
Windle B, 500
Windsor W, 415, 416
Winefield RD, 340
Wing RR, 379, 382
Winker JR, 776, 785
Winkfield KM, 193, 195, 197, 198
Winkler DC, 346
Winkler E, 719, 730
Winkler JR, 171
Winters D, 370, 372
Wintrode PL, 415, 416
Winum JY, 642
Winzell MS, 391
Wirtz D, 474, 480, 481
Wirtz KW, 310, 315, 322
Wirtz S, 341
Wise MJ, 340
Wisniewski H, 618
Wisniewski HM, 615, 618
Wisniewski T, 618
Wissner A, 101

Witt HT, 770, 774
Witters L, 375
Witters LA, 147, 148, 153, 154
Wityk RJ, 593
Wohlrab H, 731, 734, 735
WOLBERGER C, 435–65; 448, 450, 454, 455, 457–59
Woldringh CL, 213, 487
Wolf AJ, 256, 260
Wolf I, 667
Wolf SG, 470
Wolf Y, 104
Wolf YI, 683, 693
Wolf-Watz H, 468
Wolfe RR, 379, 385
Wolfenden R, 407–9, 634
Wolfer DP, 617
Wollheim CB, 386, 388
Wollman F-A, 772, 776, 783
Wollner N, 630
Wong CH, 56
Wong CM, 231
Wong CW, 615
Wong DA, 308
Wong JCY, 419
Wong JMY, 502, 508
Wong KF, 524, 529, 530, 537
Wong MC, 231
Wong R, 548
Wong SN, 445
Wong SS, 347
Wong W, 669
Wood A, 246, 252, 259–61
Wood AN, 260
Wood JD, 613
Wood JG, 436, 442, 444, 445
Woodard RW, 59
Woodhouse L, 32
Woodley DT, 669
Woodruff ML, 745
Woods A, 147, 148
Woods KC, 577, 578
Woods N, 25
Woods VL Jr, 344
Woolf TB, 710, 711
Woolfson DN, 283, 284
Worcester D, 758
Workman JL, 245, 253
Worringer KA, 255
Wortinger MA, 483
Wouters F, 310
Wren BW, 62

Wrenn RF, 634
Wrigglesworth JM, 177
Wright A, 213, 236
Wright CF, 347
Wright LF, 58, 62
Wright M, 28, 375
Wright PE, 521, 522, 526, 527, 536
Wright WE, 132, 494, 507, 508, 512
Wrona TJ, 55–57, 61
Wu C, 24, 621
Wu CK, 357
Wu H, 24, 25, 28, 375
Wu HC, 642
Wu J, 24, 441
Wu JQ, 545
Wu JY, 98
Wu KJ, 128
Wu LH, 101
Wu LJ, 214, 215, 230, 694
Wu N, 668
Wu P, 341
Wu T, 62, 314
Wu Y, 282, 283
Wu YJ, 167
Wu YM, 102
Wu Z, 28
Wu ZD, 375
Wugeditsch T, 50, 52
Wunschl C, 349
Wurth C, 353
Wüthrich K, 520
Wuttke DS, 507
Wyatt GR, 3
Wyce A, 260
Wylie C, 555
Wylie DE, 631, 639, 640
Wylie DW, 632
Wylin T, 324
Wysocka J, 253, 262
Wysolmerski RB, 195, 196, 199–202

X

Xia CH, 610–12, 617
Xia D, 181, 770–72, 776, 780, 781
Xia JQ, 496, 502, 505, 507, 508
Xia J-Z, 181, 770–72, 776, 781
Xia L, 254, 255, 258

Xia W, 618
Xiao KH, 781
Xiao L, 415, 416
Xiao TJ, 253, 259
Xiao XY, 104
Xiao Y, 14
Xie H, 375, 382
Xie X, 375, 550, 552
Xie XS, 419
Xing H, 662
Xiong J, 786
Xiong Y, 319, 321, 498, 582, 585–88, 590, 591, 595, 598, 600
Xu H, 547
Xu J, 145, 618
Xu JC, 170, 179
Xu K, 25
Xu LF, 505
Xu RM, 253, 259, 446–48, 450, 454, 457
Xu W, 665, 667
Xu Y, 27, 410, 413–15
Xue LJ, 693
Xue Q, 419
Xun L, 419

Y

Yadav PNS, 501
Yadava N, 75
Yagi A, 297, 310
Yagi AI, 311
Yagi T, 70, 75, 76, 80, 81, 84, 86
Yagihashi N, 379
Yahara I, 274, 275–77
Yaish P, 96, 99
Yamada KA, 384
Yamada M, 120
Yamada S, 388
Yamagishi A, 417
Yamaguchi H, 171, 612, 620
Yamaguchi K, 344
Yamaguchi M, 79, 85, 113–15
Yamaguchi N, 730, 731
Yamaguchi T, 684, 693
Yamaichi Y, 229, 230
Yamaji T, 548
Yamakita Y, 557
Yamamoto KR, 286
Yamamoto M, 190, 196, 198, 221

Yasuhara O, 618
Yasugi M, 417
Yasuda Y, 548
Yasuda J, 671
Yasothornsrikul S, 617
Yarmolinsky M, 229
Yao RJ, 94
Yao CC, 190, 198–203
Yano T, 70, 76, 81, 83–85
Yano J, 770
Yanku M, 683, 693
Yankovskaya V, 84, 314
Yanko M, 683, 693
Yankner BA, 357
Yang Z, 612, 617
Yang YZ, 300
Yang YH, 508
Yang Y, 155
Yang XM, 230, 231
Yang XF, 440
Yang WZ, 598
Yang WJ, 639
Yang W, 278, 582, 584, 587, 589, 600
Yang TP, 255
Yang SH, 575, 662
Yang Q, 24, 25, 372, 373
Yang J, 30, 311, 617
Yang G, 618
Yang D, 667, 723
Yanai I, 657, 675
Yanagisawa T, 642
Yanagi T, 642
Yan Z, 27
Yan P, 510
Yan K, 482
Yan JS, 770–72, 777, 779, 780, 786, 787
YAN J, 769–90
Yan EC, 748, 754
Yamazoe M, 226, 229
Yamazaki S, 468
Yamauchi T, 371, 372, 391
Yamauchi K, 388
Yamashita T, 391, 754
Yamashita S, 391
Yamashita H, 301
Yamashita E, 171
Yamashita A, 324
Yamashiro S, 557
Yamamura Y, 357
Yamamoto MT, 547, 553–55

Yasusake Y, 406, 409, 410, 418
Yates J 3rd, 546, 547
Yates JR, 694
Yates JR 3rd, 253
Yates R, 226
Yates PO, 619
Yeaman C, 560
Yeates TO, 459
Yechezkel T, 99, 102
Yechoor VK, 375, 379
Yeadon JE, 32
Yeager TR, 494
Ye ZH, 482
Ye CP, 357
Yau WM, 346
Yau KW, 744, 745
Yates JR 3rd, 253
Yates JR, 694
Yates R, 226
Yates PO, 619
Yasusake Y, 406, 409, 410, 418
Yao CC, 190, 198–203
YEH BJ, 655–80, 665, 671, 672
Yee MC, 418
Yee K, 103, 104
Yee CS, 682, 691
Yee AJ, 380
Yehuda T, 116
Yen SH, 618
Yeo M, 660
Yeung D, 509
Yeung DS, 496
Yeung ES, 419
Yeung F, 439, 444
Yildiz A, 610
Yilmaz E, 374
Ylikoskala A, 139–41, 144, 145, 151, 152
Yohanan S, 723
Yokobayashi S, 219
Yokota S, 301, 311
Yokoyama A, 253
Yokoyama S, 642
Yoneda T, 96, 101
Yonekura K, 346, 347
Yonemura S, 552
Yoo KW, 194
Yool D, 614
Yoon KJ, 194
Yoon MJ, 194
York AL, 470
Yoshida K, 719
Yoshida T, 547
Yoshihara T, 301
Yoshimi K, 619
Yoshimoto M, 617
Yoshizaki H, 552

Zandomeneghi G, 345
Zamotin V, 348, 358
Zamora R, 302
Zalkin H, 632–34
Zaks A, 415, 416, 509
Zakian VA, 116–22, 127, 508
Zakharova NV, 82
Zaitsu K, 282
Zaidi T, 618
Zahn A, 322
Zafra-Polo MC, 83
Zachariae W, 221
Zaccai G, 734
Zaborowski E, 522, 527
Zaborosch C, 76
Zabaleta E, 73

Z

Yun CH, 181
Yumura S, 544, 552, 555, 557
Yue DT, 709, 710
Yudkin MD, 469, 482
Yuce O, 553
Yuan MS, 357, 372, 373, 377
Yuan HS, 598
Yuan C, 759
Yu YP, 508
Yu XC, 469, 474
Yu W, 611
Yu S, 745
Yu RT, 28, 31, 375
Yu M, 23
Yu L, 770, 772, 780, 781
Yu J, 786
Yu HG, 223
Yu GS, 376
Yu F, 619
Yu EY, 509
Yu CL, 376–79
Yu C-A, 181, 770–72, 776, 780, 781
Yu C, 27
Youngren B, 229, 484
Young RA, 259–61
Young N, 155, 660, 663
Young L, 535
Young JW, 509
Young JC, 275, 282, 286
Young DM, 631, 641
Yocher J, 56, 57

Zanes R, 630
Zang JH, 556
Zannoni D, 70
Zappulla DC, 503, 504
Zara V, 734
Zarain-Herzberg A, 383
Zarinpar A, 664, 668, 669, 672
Zarrinpashneh E, 148, 149
Zaslavsky D, 170
Zaug AJ, 500, 503, 504
Zavitz TM, 380
Zavodszky P, 417, 418
Zecchion L, 407, 414, 415, 417–19
Zechiedrich EL, 596–98
Zeelen JP, 416, 421, 424, 426
Zehavi A, 97
Zeikus GJ, 423, 425
Zeilemaker A, 509
Zekhnini Z, 409, 423, 424
Zeller HD, 302
Zeman I, 718–20
Zeng HQ, 635
Zeng W, 230
Zenke FT, 552
Zensen R, 75
Zeuzem S, 142
Zevaiani M, 719
Zha S, 308
Zhadin U, 536
Zhang B, 70, 73, 75, 613, 618, 620
Zhang C, 28
Zhang CC, 371
Zhang CL, 24–28, 31, 375
Zhang CY, 28, 375, 391, 508
Zhang D, 27, 311
Zhang DY, 376–79
Zhang H, 769–90
Zhang HM, 770–72, 774, 775, 777, 779, 782–87
Zhang J, 452, 572, 618
Zhang L, 167, 178
Zhang M, 28, 283, 297, 314, 669
Zhang MQ, 770
Zhang NY, 415, 416
Zhang PC, 635
Zhang PP, 785
Zhang Q, 356
Zhang SC, 614
Zhang T, 75
Zhang W, 669
Zhang X, 258
Zhang Y, 149, 245, 259, 315, 370, 372
Zhang YH, 635
Zhang Z, 176, 181, 182, 572, 694, 759, 770–72, 776
Zhang ZH, 640
Zhao C, 613
Zhao J, 546, 547
Zhao KH, 442, 454, 457–60
Zhao M, 736
Zhao P, 439, 448
Zhao PY, 386, 388, 392
Zhao Q, 289
Zhao WM, 550
Zhao X, 752
Zhao XL, 694, 699
Zharova TV, 82
Zhen M, 155
Zhen Y, 170, 177
Zheng H, 617
Zheng P, 617
Zheng XH, 178
Zheng C, 635, 636
Zhong G, 259
Zhong X, 357
Zhong Z, 229
Zhou BS, 693
Zhou D, 336, 343
Zhou GC, 147
Zhou GP, 49
Zhou H, 101, 484, 485
Zhou M, 556, 557, 671
Zhou S, 341
Zhou W, 496, 500, 502–7, 509
Zhou XZ, 509
Zhou YP, 386, 390, 392
Zhou YT, 386, 389–91, 393
Zhou Z, 694
Zhu CJ, 546, 547, 554, 555
Zhu G, 417
Zhu K, 674
Zhu L, 757, 760
Zhu M, 669
Zhu S, 100
Zhu T, 149
Zhu X, 32, 102, 585
Zhu Y, 309, 319, 321
Zhu YS, 508
Zhuang H, 104
Zickermann V, 70, 73, 74, 76, 79–81, 83–86, 181
Ziegelin G, 572
Ziemin-van der Poel S, 251
Zierath JR, 27, 30
Zigmond SH, 556
Zijlmans JM, 509
Zijlstra A, 674
Zimmer C, 545, 546
Zimmermann SB, 7, 9
Zimmermann J, 103
Zimmermann K, 751
Zimmermann R, 273
Zinn K, 610
Zinzen RP, 125, 129
Zipperlen P, 546, 547
Zito F, 776, 783
Zody MC, 657, 659, 664, 665
Zoeller RA, 303, 305
Zoffoli S, 355
Zogaj X, 41
Zoghbi HY, 613
Zoidakis J, 406, 409, 410, 415, 419, 424
Zoll J, 25
Zong H, 27
Zong HH, 376–79
Zorrilla S, 473, 475
Zou H, 220
Zou L, 117
Zou MH, 147
Zou Y, 132
Zouni A, 770
Zubko MK, 118
Zuckerman DM, 710, 711
Zuker CS, 671
Zulauf M, 728
Zurdo J, 343, 345, 348, 349, 353–55
Zvetkova I, 255
Zwaan CM, 638
Zweckstetter M, 353, 356
Zwicker K, 74, 76, 78, 79, 83–86, 181
Zwickl P, 632

ANNUAL REVIEWS
Intelligent Synthesis of the Scientific Literature

Annual Reviews — Your Starting Point for Research Online
http://arjournals.annualreviews.org

- Over 900 Annual Reviews volumes—more than 25,000 critical, authoritative review articles in 32 disciplines spanning the Biomedical, Physical, and Social sciences—available online, including all Annual Reviews back volumes, dating to 1932.
- Current individual subscriptions include seamless online access to full-text articles, PDFs, Reviews in Advance (as much as 6 months ahead of print publication), bibliographies, and other supplementary material in the current volume and the prior 4 years' volumes.
- All articles are fully supplemented, searchable, and downloadable — see http://biochem.annualreviews.org
- Access links to the reviewed references (when available online)
- Site features include customized alerting services, citation tracking, and saved searches

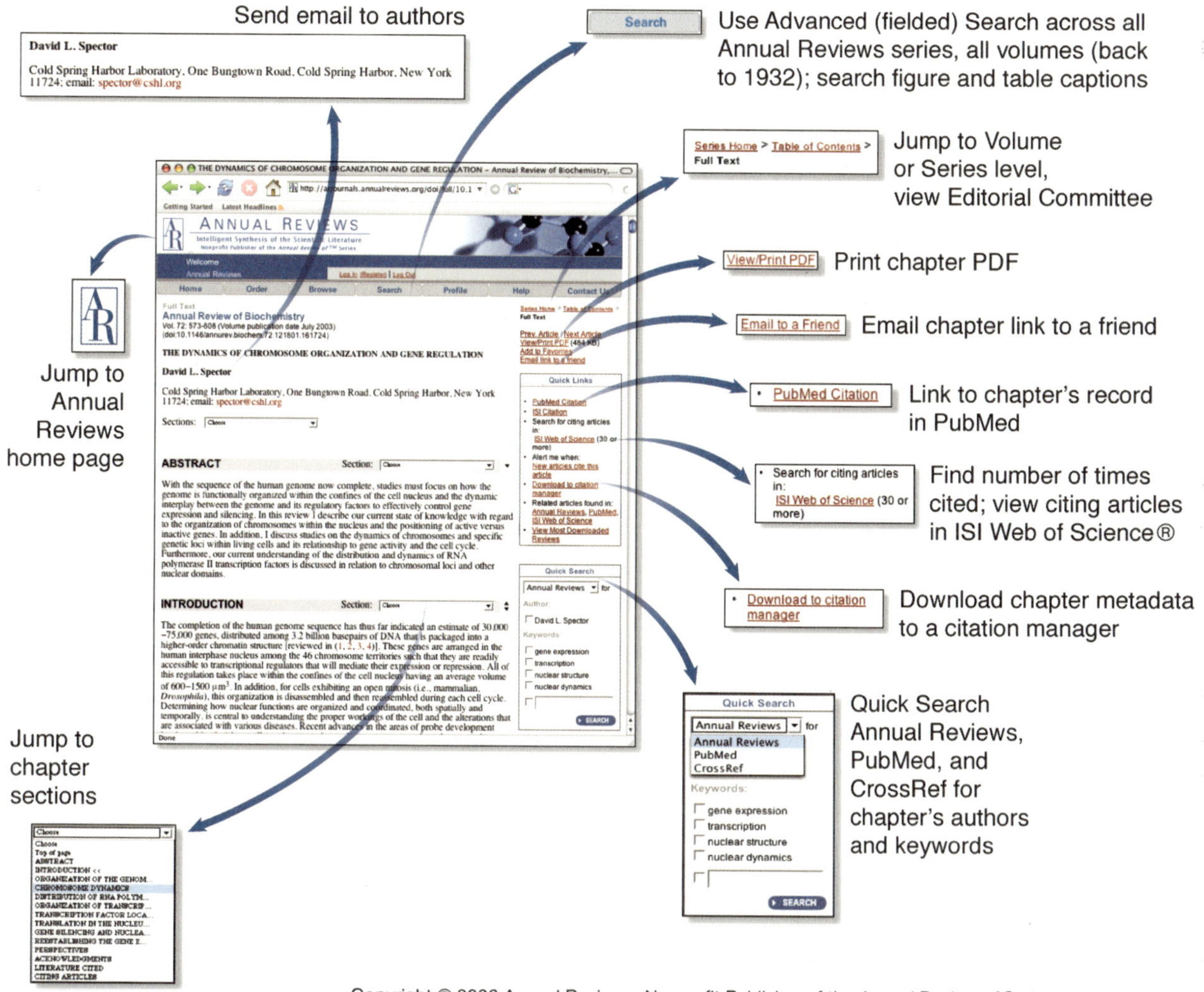

- Jump to Annual Reviews home page
- Send email to authors
- Jump to Volume or Series level, Full Text
- Jump to chapter sections
- Use Advanced (fielded) Search across all Annual Reviews series, all volumes (back to 1932); search figure and table captions
- View Editorial Committee
- Print chapter PDF — View/Print PDF
- Email chapter link to a friend — Email to a Friend
- Link to chapter's record in PubMed — PubMed Citation
- Find number of times cited; view citing articles in ISI Web of Science®
- Download chapter metadata to a citation manager — Download to citation manager
- Quick Search Annual Reviews, PubMed, and CrossRef for chapter's authors and keywords

Copyright © 2006 Annual Reviews, Nonprofit Publisher of the Annual Review of Series